Human Factors of a Global Society

A SYSTEM OF SYSTEMS PERSPECTIVE

Ergonomics Design and Management: Theory and Applications

Series Editor

Waldemar Karwowski

Industrial Engineering and Management Systems
University of Central Florida (UCF) – Orlando, Florida

Published Titles

Ergonomics: Foundational Principles, Applications, and Technologies
Pamela McCauley Bush

Aircraft Interior Comfort and Design
Peter Vink and Klaus Brauer

Ergonomics and Psychology: Developments in Theory and Practice
Olexiy Ya Chebykin, Gregory Z. Bedny, and Waldemar Karwowski

Ergonomics in Developing Regions: Needs and Applications
Patricia A. Scott

Handbook of Human Factors in Consumer Product Design, 2 vol. set
Waldemar Karwowski, Marcelo M. Soares, and Neville A. Stanton

Volume I: Methods and Techniques
Volume II: Uses and Applications

Human–Computer Interaction and Operators' Performance: Optimizing Work Design with Activity Theory
Gregory Z. Bedny and Waldemar Karwowski

Human Factors of a Global Society: A System of Systems Perspective
Tadeusz Marek, Waldemar Karwowski, Marek Frankowicz, Jussi I. Kantola, and Pavel Zgaga

Knowledge Service Engineering Handbook
Jussi Kantola and Waldemar Karwowski

Trust Management in Virtual Organizations: A Human Factors Perspective
Wiesław M. Grudzewski, Irena K. Hejduk, Anna Sankowska, and Monika Wańtuchowicz

Manual Lifting: A Guide to the Study of Simple and Complex Lifting Tasks
Daniela Colombiani, Enrico Ochipinti, Enrique Alvarez-Casado, and Thomas R. Waters

Neuroadaptive Systems: Theory and Applications
Magdalena Fafrowicz, Tadeusz Marek, Waldemar Karwowski, and Dylan Schmorrow

Safety Management in a Competitive Business Environment
Juraj Sinay

Forthcoming Titles

Organizational Resource Management: Theories, Methodologies, and Applications
Jussi Kantola

Self-Regulation in Activity Theory: Applied Work Design for Human-Computer Systems
Gregory Bedny, Waldemar Karwowski, and Inna Bedny

Human Factors of a Global Society

A SYSTEM OF SYSTEMS PERSPECTIVE

EDITED BY

TADEUSZ MAREK • WALDEMAR KARWOWSKI

MAREK FRANKOWICZ • JUSSI KANTOLA • PAVEL ZGAGA

CRC Press is an imprint of the
Taylor & Francis Group, an **informa** business

CRC Press
Taylor & Francis Group
6000 Broken Sound Parkway NW, Suite 300
Boca Raton, FL 33487-2742

First issued in paperback 2017

ISBN-13: 978-1-4665-7286-7 (hbk)
ISBN-13: 978-1-138-07168-1 (pbk)

Library of Congress Cataloging-in-Publication Data

Human factors of a global society : a system of systems perspective / edited by Tadeusz Marek, Waldemar Karwowski, Marek Frankowicz, Jussi Kantola, and Pavel Zgaga.
pages cm. -- (Ergonomics design and management theory and applications)
Includes bibliographical references and index.
ISBN 978-1-4665-7286-7 (hardcover : alk. paper)
1. Technology--Social aspects. 2. Human engineering. I. Marek, Tadeusz, 1951-

T14.5.H855 2014
920.8'2--dc23 2013041051

Visit the Taylor & Francis Web site at
http://www.taylorandfrancis.com

and the CRC Press Web site at
http://www.crcpress.com

Printed and bound by CPI Group (UK) Ltd, Croydon, CR0 4YY

Contents

SECTION II Psychology

SECTION IV Higher Education

SECTION V Education in Modern Society

Preface

Over the last more than 60 years, the modern discipline of human factors and ergonomics (HF/E) has been evolving in parallel with rapid and often disruptive advances in science, engineering, and technology, accompanied by the increasing trends of massive urbanization and a dramatic shift toward a wired, service-oriented, and knowledge-based global society. The contemporary focus of HF/E is clearly shifting toward understanding progressively larger, interconnected, and dynamic socioeconomic systems worldwide. The design of such systems requires adopting a new perspective that considers unexplored HF/E issues related to human roles as creative, purposeful, and dynamic elements of much larger systems that often exhibit emergent properties, that is, unpredictable behaviors, due to unparalleled technological, social, economic, and political complexity that challenges human progress worldwide.

This book offers a unique system-of-systems perspective on the human factors and ergonomics of a modern society. The future of human, social, and economic developments is explored as seen through the eyes of the HF/E discipline and is considered in the context of unprecedented advancements in science, engineering, technology, and education that will shape the future of a growing global society in the 21st century.

The book is composed of a total of 97 chapters organized into five main sections. Section I: Human Factors and Technology contains 21 chapters and focuses on the advances in science and engineering that lead to developments of novel technologies that significantly impact the nature and quality of human condition. Section II is devoted to psychology and contains 16 chapters that discuss a variety of topics related to individual and collective human experiences in contemporary workplaces. These include occupational stress and burnout, marketing, as well as public policy making, leadership, and organizational issues. Section III: Management contains 18 chapters that mainly focus on the human factors of entrepreneurship, business management of enterprises, and knowledge-based economy. Section IV: Higher Education contains 29 chapters devoted to new paradigms and policies for higher education, as well as their delivery, administration, and funding in the near future and long-term perspectives. Finally, Section V: Education in Modern Society is composed of 13 chapters that explore the issues related to modern pedagogy, models of the teaching profession, education leadership, and lifelong learning.

We hope that this book will help cultivate new ideas and induce creative polemics about the direction of the future development of our global society. We also hope that this book will encourage the HF/E community and others, including the science, engineering, and technology professionals, policy makers, education leaders, industry, and governments worldwide, to take a broad, system-of-systems perspective when considering the critical needs of the citizens at large and incorporating relevant human–system interaction considerations in the design of sustainable socioeconomic environments that will serve the progress of our human race well into the future.

Tadeusz Marek
Jagiellonian University in Krakow, Poland
Waldemar Karwowski
University of Central Florida, Orlando, USA
Marek Frankowicz
Jagiellonian University in Krakow, Poland
Jussi Kantola
University of Vaasa, Finland
Pavel Zgaga
University of Ljubljana, Slovenia

MATLAB® is a registered trademark of The MathWorks, Inc. For product information, please contact:

The MathWorks, Inc.
3 Apple Hill Drive
Natick, MA 01760-2098, USA
Tel: 508-647-7000
Fax: 508-647-7001
E-mail: info@mathworks.com
Web: www.mathworks.com

Editors

Tadeusz Marek is a chair professor and the head of the Department of Neuroergonomics and the Neurobiology Department of the Malopolska Center of Biotechnology at Jagiellonian University in Krakow and a professor of psychology at Warsaw School of Social Science and Humanities, Warsaw, Poland. He is the author or a co-author of over 300 scientific publications. His research deals with mental stress and workload, fatigue, professional burnout, cognitive neuroscience, chronopsychology, neuroadaptive technologies, and neuroergonomics. His current scientific interests include the neuronal mechanisms of mental fatigue and stress, chronic sleep restriction influence on activity of neuronal networks related to the processes of visual attention, and executive functions, and diurnal variability of brain activity. Professor Marek is a past president of the Committee on Ergonomics of Polish Academy of Sciences and past vice-rector of Jagiellonian University. He was the editor of the international journal *Ergonomics*. He is a member of many Polish and international learned societies.

Waldemar Karwowski, PhD, DSc, PE, is a professor and the chairman of the Department of Industrial Engineering and Management Systems at the University of Central Florida, Orlando, Florida. He is also the executive director of the Institute for Advanced Systems Engineering at the University of Central Florida. He earned a master's (1978) in production engineering and management from the Technical University of Wroclaw, Poland, and his PhD (1982) in industrial engineering at Texas Tech University. He earned a DSc (dr habil.) in management science at the State Institute for Organization and Management in Industry, Poland (2004). He also received honorary doctoral degrees from three European universities. He is a past president of the Human Factors and Ergonomics Society (2007) and the International Ergonomics Association (2000–2003). Dr. Karwowski served on the Committee on Human Systems Integration, National Research Council, the National Academies, USA (2007–2011). He is a co-editor of the *Human Factors and Ergonomics in Manufacturing* journal and the editor-in-chief of *Theoretical Issues in Ergonomics Science* journal. He is an author or editor of over 400 scientific publications in the areas of human systems integration, cognitive engineering, activity theory, systems engineering, human–computer interaction, fuzzy logic and neuro-fuzzy modeling, applications of nonlinear dynamics to human performance, and neuroergonomics.

Marek Frankowicz is an associate professor at the Theoretical Chemistry Department of Jagiellonian University in Krakow. He is also the vice-chair of the Jagiellonian University Centre for Research on Higher Education. He was a post-doctoral fellow at the Free University of Brussels (1981–1982) and Tokyo University (1982–1983) and a senior lecturer at University Paris 6 (1988–1990). He is a member of the advisory board of the International Journal *Interdisciplinary Description of Complex Systems* (INDECS). He is also a member of EURASHE Council and Bologna Expert for Poland. His current research interests include stochastic dynamics of nonlinear chemical systems, foundations of thermodynamics, applications of complexity theory (complex adaptive systems, active walks in adaptive landscapes) to natural and social systems, and modeling of dynamics of higher education reforms in Europe.

Jussi Kantola is a professor in the Industrial Management Department at the University of Vaasa, Finland. Before that, he was an associate professor in the Knowledge Service Engineering Department at the Korea Advanced Institute of Science and Technology. From 2003 to 2008, he worked at the Tampere University of Technology, Finland, and the University of Turku, Finland, in various research roles, including research director in the IE and IT departments. He earned a PhD in industrial engineering at the University of Louisville, Kentucky, USA, in 1998 and a second PhD at the Industrial Management and Engineering Department at the Tampere University of Technology, Finland, in 2006. From 1999 to 2002, he worked as an IT and business and process consultant in the United States and in Finland. His current research interests include service and new product development as well as various soft-computing applications.

Pavel Zgaga is a professor in the philosophy of education and education policy at the University of Ljubljana, Slovenia. During the period of social and political transition, he was the State Secretary for Higher Education (1992–1999) and Minister of Education (1999–2000). In 2001, after his return to academe, he cofounded the Centre for Educational Policy Studies at the university. He has held several research grants, directed a number of national and international projects on education, and published extensively in the area of his expertise. He has also led projects concerned with policy issues and the development of higher education (particularly the Bologna Process) and teacher education as a specific segment of higher education. In this area, he has been cooperating as an expert or invited speaker with the European Commission (DG EAC), the Council of Europe, UNESCO, OECD, the World Bank, and other organizations. In the Bologna process, he was engaged as a general rapporteur (2001–2003), as a member of the Board of the Bologna Follow-Up Group (2004–2005), and as the rapporteur of the Working Group on External Dimension of the Bologna Process (2005–2007). He is a cofounder of the South East European Educational Cooperation Network (SEE ECN, 2001) and Teacher Education Policy in Europe (TEPE, 2006) network. He has published broadly in domestic and international journals and written ten monographs, including *Looking Out: The Bologna Process in a Global Setting* (Oslo, 2006) and *Higher Education in Transition: Reconsiderations on Higher Education in Europe at the Turn of Millennium* (Umeå University, 2007).

Contributors

Andrzej Adamski
Institute of Automatics and Bioengineering
AGH University of Science and Technology
Krakow, Poland

Jolanta Babiak
Institute of Organisation and Management
Wroclaw University of Technology
Wroclaw, Poland

Beata Bajcar
Institute of Organisation and Management
Wroclaw University of Technology
Wroclaw, Poland

Adela Barabasz
Department of Management System Design
University of Economics
Wroclaw, Poland

Tibor Baráth
Hungarian-Netherlands School of Educational Management
University of Szeged
Szeged, Hungary

David F. Barbe
Maryland Technology Enterprise Institute (Mtech)
A. James Clark School of Engineering
University of Maryland
College Park, Maryland

Ronald Barnett
Institute of Education
University of London
London, United Kingdom

Danny Barrantes Acuña
University Teaching Department (DEDUN)
University of Costa Rica
San José, Costa Rica

Joanna Bartnicka
Faculty of Organization and Management
Silesian University of Technology
Zabrze, Poland

Roman Batko
Faculty of Management and Social Communication
Jagiellonian University
Krakow, Poland

Gregory Z. Bedny
Evolute
Louisville, Kentucky

Inna S. Bedny
Evolute
Louisville, Kentucky

Barbara Behrnd-Wenzel
Higher Vocational School
Merseburg, Germany

Christophe Bevilacqua
CITYTAK
Hellemmes-Lille, France

Dmitriy Borodin
Production Information Systems Lab
University College Ghent
Ghent, Belgium

Justyna Bugaj
Faculty of Management and Social Communication
Institute of Economy and Management
Jagiellonian University
Krakow, Poland

Stefania Camplone
Ergonomics and Design for Sustainability Research Unit
Department of Architecture
University of Chieti-Pescara
Pescara, Italy

Diogo Casanova
Research Centre for Didactics and Technology in Teacher Education
University of Aveiro
Aveiro, Portugal

Elżbieta Chwalibog
Department of Management System Design
University of Economics
Wroclaw, Poland

Małgorzata Cieciora
Faculty of Information Management
Polish-Japanese Institute of Information Technology
Warsaw, Poland

Nilza Costa
Research Centre for Didactics and Technology in Teacher Education
University of Aveiro
Aveiro, Portugal

Wojciech Cwalina
University of Social Sciences and Humanities
Warsaw, Poland

Agnieszka Cybal-Michalska
Department of Pedagogical Problems of Youth
Faculty of Educational Studies
Adam Mickiewicz University
Poznan, Poland

Estela Daukšienė
Department of Education
Vytautas Magnus University
Kaunas, Lithuania

Trevor Davies
The Institute of Education
University of Reading
Reading, United Kingdom

Wim De Bruyn
Production Information Systems Lab
University College Ghent
Ghent, Belgium

Giuseppe Di Bucchianico
Ergonomics and Design for Sustainability Research Unit
Department of Architecture
University of Chieti-Pescara
Pescara, Italy

Tymoteusz Doligalski
Centre of Customer Value
Institute of Value Management
Warsaw School of Economics
Warsaw, Poland

Roman Dorczak
Department of Educational Management
Institute of Public Affairs
Jagiellonian University
Krakow, Poland

Golde Dudell
Department of Neonatology
Children's Hospital and Research Institute of Northern California
Oakland, California

Anna Dyląg
Faculty of Management and Social Communication
Jagiellonian University
Krakow, Poland

Juris Dzelme
Higher Education Quality Evaluation Centre (AIKNC)
Riga, Latvia

Maria L. Ekiel-Jeżewska
Institute of Fundamental Technological Research
Polish Academy of Sciences
Warsaw, Poland

Andrzej Falkowski
University of Social Sciences and Humanities
Warsaw, Poland

Aleksandra Fedaczynska
Faculty of Management and Social Communication
Jagiellonian University
Krakow, Poland

Magdalena Gasiorowska
University of Social Sciences and Humanities
Warsaw, Poland

Marcin Geryk
Gdansk Management College
Gdansk, Poland

and

College of Management and Infrastructure
Warsaw, Poland

Zofia Godzwon
Centre for Research on Higher Education
Jagiellonian University
Krakow, Poland

Krystyna Golonka
Faculty of Management and Social Communication
Jagiellonian University
Krakow, Poland

Marek Goral
Research and Development Laboratory for Aerospace Materials
Rzeszow University of Technology
Rzeszow, Poland

Victor Gorelik
Simulation Modeling and Operations Research Department
Computing Center of the Russian Academy of Sciences
Moscow, Russia

Piotr Górski
Faculty of Management
University of Science and Technology
Krakow, Poland

Damian Grabowski
Faculty of Pedagogy and Psychology
Institute of Psychology
University of Silesia
Katowice, Poland

Alicja Grochowska
Department of Marketing Psychology
University of Social Sciences and Humanities
Warsaw, Poland

Krystian Gurba
Institute of Economics and Management
Jagiellonian University
Krakow, Poland

Krzysztof Hanusz
Institute of Psychology
Jagiellonian University
Krakow, Poland

David Brian Hay
King's Learning Institute
King's College London
London, United Kingdom

Jürgen Heene
DeuZert Deutsche Zertifizierung in Bildung und Wirtschaft GmbH
Wildau, Germany

Grzegorz Heldak
ITS-ILS Laboratories: Management and Control in Transport and Logistics Section
Civil Engineering Faculty
Krakow University of Technology
Krakow, Poland

Marek Hetmański
Department of Philosophy and Sociology
Marie Curie-Sklodowska University
Lublin, Poland

Tomasz Ingram
Entrepreneurship and Innovation Management
University of Economics
Katowice, Poland

Toshiaki Isozaki
Academic Foundation Programs
Kanazawa Institute of Technology
Nonoichi, Japan

Magdalena Anna Jaworek
Faculty of Management and Social Communication
Jagiellonian University
Krakow, Poland

Lilianna Jodkowska
Hochschule für Technik und Wirtschaft
Berlin, Germany

Verka Jovanović
Department of Tourism and Hospitality Management
Singidunum University
Belgrade, Serbia

Joanna Kalkowska
Faculty of Engineering Management
Poznan University of Technology
Poznan, Poland

Jussi Kantola
Industrial Management
University of Vaasa
Vaasa, Finland

Ludmyla M. Karamushka
Institute of Psychology
Kiev, Ukraine

Victor I. Karamushka
University of Educational Management
Kiev, Ukraine

Waldemar Karwowski
Department of Industrial Engineering and Management Systems
University of Central Florida
Orlando, Florida

Jacek Klich
Department of Public Economy and Administration
Cracow University of Economics
Krakow, Poland

Leonid Kompanets
Department of Theoretic Physics and Informatics
University of Lodz
Lodz, Poland

Paweł Koniak
Department of Marketing Psychology
University of Social Sciences and Humanities
Warsaw, Poland

Małgorzata Kossowska
Institute of Psychology
Jagiellonian University
Krakow, Poland

Janina Kostkiewicz
Institute of Educational Sciences
Jagiellonian University
Krakow, Poland

Zoltán Kovács
Department of Management
Faculty of Economics and Business
University of Pannonia
Veszprém, Hungary

Agnieszka Kowalska-Styczeń
Faculty of Organization and Management
Silesian University of Technology
Zabrze, Poland

Barbara Kożusznik
Institute of Psychology Faculty of Pedagogy and Psychology
University of Silesia
Katowice, Poland

Małgorzata W. Kożusznik
University of València, IDOCAL
València, Spain

Małgorzata Krzeczkowska
Department of Chemical Education
Jagiellonian University
Krakow, Poland

Daniel Kubek
ITS-ILS Laboratories: Management and Control in Transport and Logistics Section
Civil Engineering Faculty
Krakow University of Technology
Krakow, Poland

Krzysztof Kubiak
Research and Development Laboratory for Aerospace Materials
Rzeszow University of Technology
Rzeszow, Poland

Damian Kurach
Institute of Computational Intelligence
Czestochowa University of Technology
Czestochowa, Poland

Zofia Lacala
Institute of Economy and Management
Jagiellonian University
Krakow, Poland

Dong Yun Lee
Department of Knowledge Service Engineering
Korea Advanced Institute of Science and Technology (KAIST)
Daejeon, South Korea

Ho Lom Lee
University of California, San Francisco Medical Center
San Francisco, California

Geoffrey Lepoutre
CITYTAK
Hellemmes-Lille, France

Koryna Lewandowska
University of Information Technology and Management
Rzeszow, Poland

and

Faculty of Management and Social Communication
Jagiellonian University
Krakow, Poland

Ivars Linde
Riga International School of Economics and Business Administration (RISEBA)
Riga, Latvia

Anna Lubecka
Institute of Public Affairs
Jagiellonian University
Krakow, Poland

Iwona Maciejowska
Department of Chemical Education
Jagiellonian University
Krakow, Poland

Jozef Maciuszek
Institute of Applied Psychology
Jagiellonian University
Krakow, Poland

Robert Mackiewicz
Department of Marketing Psychology
University of Social Science and Humanities
Warsaw, Poland

Valery D. Magazannik
N. Bauman Moscow State Technical University
Moscow, Russia

Magdalena Majowska
Human Resource Management
University of Economics
Katowice, Poland

Liliana Mammino
Department of Chemistry
University of Venda
Thohoyandou, South Africa

Tadeusz Marek
Faculty of Management and Social Communication
Jagiellonian University
Krakow, Poland

Dariusz Masły
Faculty of Architecture
Silesian University of Technology
Gliwice, Poland

Masakatsu Matsuishi
Project Education Center
Kanazawa Institute of Technology
Nonoichi, Japan

Shigeo Matsumoto
Academic Foundation Programs
Kanazawa Institute of Technology
Nonoichi, Japan

Grzegorz Mazurkiewicz
Department of Educational Management
Institute of Public Affairs
Jagiellonian University
Krakow, Poland

Bartlomiej Melges
Faculty of Management and Social Communication
Krakow University
Krakow, Poland

James Moir
Abertay University
Dundee, United Kingdom

Justyna Mojsa-Kaja
Faculty of Management and Social Communication
Jagiellonian University
Krakow, Poland

António Moreira
Research Centre for Didactics and Technology in Teacher Education
University of Aveiro
Aveiro, Portugal

Joyce Nacario
University of California, San Francisco Medical Center
San Francisco, California

Bruce I. Newman
DePaul University
Chicago, Illinois

Elżbieta Niezabitowska
Department of Strategy of Design and New Technologies in Architecture
Silesian University of Technology
Gliwice, Poland

Czeslaw S. Nosal
University of Social Sciences and Humanities Faculty
Wroclaw, Poland

Aleksander Noworol
Institute of Public Affairs
Jagiellonian University
Krakow, Poland

Andrzej Olak
Institute of Organization and Management
Bronisław Markiewicz State School of Higher Vocational Education
Jarosław, Poland

Tadeusz Oleksyn
Collegium of Business Administration
Warsaw School of Economics
Warsaw, Poland

Arkadiusz Onyszko*
Research and Development Laboratory for Aerospace Materials
Rzeszow University of Technology
Rzeszow, Poland

Edmund Pawlowski
Faculty of Engineering Management
Poznan University of Technology
Poznan, Poland

Jarosław Polak
Institute of Psychology Faculty of Pedagogy and Psychology
University of Silesia
Katowice, Poland

Grażyna Prawelska-Skrzypek
Institute of Public Affairs
Jagiellonian University
Krakow, Poland

Robin Precey
Centre for Education Leadership and School Improvement (CELSI)
Canterbury Christ Church University (CCCU)
Canterbury, United Kingdom

Anna Ptak
Faculty of Organization and Management
Silesian University of Technology
Gliwice, Poland

* Deceased.

Alexander Rodyukov
Department of Operations Research
Borisoglebsk State Pedagogical University
Borisoglebsk, Russia

Jerzy Rosinski
Institute of Economics and Management
Jagiellonian University
Krakow, Poland

Radosław Rybkowski
Institute of American Studies and Polish Diaspora
Jagiellonian University
Krakow, Poland

Janusz Sasak
Institute of Public Affairs
Jagiellonian University
Krakow, Poland

Luiza Seklecka
Institute of Applied Psychology
Jagiellonian University
Krakow, Poland

N. C. Shivaprakash
Department of Instrumentation and Applied Physics
Indian Institute of Science
Bangalora University
Bangalore, India

Jan Sieniawski
Research and Development Laboratory for Aerospace Materials
Rzeszow University of Technology
Rzeszow, Poland

Sylwia Słupik
Department of Social and Economic Policy
Faculty of Economics
University of Economics
Katowice, Poland

Kaz Sobczak
University of California, San Francisco Medical Center
San Francisco, California

Joanna Sokolowska
University of Social Sciences and Humanities
Warsaw, Poland

Agata Stachowicz-Stanusch
Department of Management and Marketing
Faculty of Organization and Management
Silesian University of Technology
Gliwice, Poland

Izabela Stańczyk
Faculty of Management and Social Communication
Institute of Economy and Management
Jagiellonian University
Krakow, Poland

Nemanja M. Stanišić
Department of Business Economy
Singidunum University
Belgrade, Serbia

Svetlana M. Stanišić
Department of Tourism and Hospitality Management
Singidunum University
Belgrade, Serbia

Janusz Strużyna
Human Resource Management
University of Economics
Katowice, Poland

Kateryna Synytsya
International Research and Training Center for Information Technologies and Systems
Kiev, Ukraine

Dorota Szablisty
University of Social Sciences and Humanities
Warsaw, Poland

István Szalkai
Department of Management
Faculty of Economics and Business
University of Pannonia
Veszprém, Hungary

Elżbieta Izabela Szczepankiewicz
Department of Accounting
Poznan University of Economics
Poznan, Poland

Pawel Szewczyk
Faculty of Organization and Management
Institute of Production Engineering
Silesian University of Technology
Zabrze, Poland

Kazuya Takemata
Academic Foundation Programs
Kanazawa Institute of Technology
Nonoichi, Japan

Dóra Tasner
Department of Management
Faculty of Economics and Business
University of Pannonia
Veszprém, Hungary

Margarita Teresevičienė
Department of Education
Vytautas Magnus University
Kaunas, Lithuania

Mariusz Trejtowicz
Institute of Psychology
Jagiellonian University
Krakow, Poland

Stefan Trzcielinski
Faculty of Engineering Management
Poznan University of Technology
Poznan, Poland

Dariusz Turek
Institute of Enterprise
Collegium of Business Administration
Warsaw School of Economics
Warsaw, Poland

Roksana Ulatowska
Faculty of Management and Social Communication
Jagiellonian University
Krakow, Poland

Barbara Urbanowicz
Department of Strategy of Design and New Technologies in Architecture
Silesian University of Technology
Gliwice, Poland

Anna G. Usowicz
Department of Neonatology
Children's Hospital and Research Institute of Northern California
Oakland, California

Thaddeus W. Usowicz
Department of Information Systems
College of Business
San Francisco State University
San Francisco, California

Gerard Uzan
THIM Laboratory
University of Paris 8
Paris, France

Bert Van Vreckem
Production Information Systems Lab
University College Ghent
Ghent, Belgium

Vilmos Vass
Department of the Theory of Pedagogy
Faculty of Education and Psychology
Institute of Education Science
Eötövös Loránd University
Budapest, Hungary

Péter Volf
Department of Management
Faculty of Economics and Business
University of Pannonia
Veszprém, Hungary

Airina Volungevičienė
Department of Education
Vytautas Magnus University
Kaunas, Lithuania

Hansjörg von Brevern
Independent Consultant
Zurich, Switzerland

Fred Voskoboynikov
Baltic Academy of Education
St. Petersburg, Russia

Barbara Wachowicz
Neurobiology Department
Malopolska Centre of Biotechnology
and
Faculty of Management and Social Communication
Jagiellonian University
Krakow, Poland

Peter Wagstaff
CEREMH
Velizy, France

Iwona Ewa Waldzińska
Higher Vocational School
Oświęcim, Poland

Hartmut Wenzel
Institute for Pedagogy and Dydactics
University of Halle-Wittenberg
Halle (Saale), Germany

Teodor Winkler
Faculty of Organization and Management
Silesian University of Technology
Zabrze, Poland

Hanna Wlodarkiewicz-Klimek
Faculty of Engineering Management
Poznan University of Technology
Poznan, Poland

Jacek Wojcik
Centre of Customer Value
Institute of Value Management
Warsaw School of Economics
Warsaw, Poland

Agnieszka Wojtczuk-Turek
Department of Human Capital Development
University of Economics
Warsaw, Poland

Agnieszka Woźnica-Kowalewska
Department of Psychology of Marketing
Warsaw School of Social Science and Humanities
Warsaw, Poland

Taketo Yamakawa
Academic Foundation Programs
Kanazawa Institute of Technology
Nonoichi, Japan

Georgy M. Zarakovsky
All-Russian Research Institute of Ergo-Design
Moscow, Russia

Sergey Zhdanov
Department of Web-Based Learning Solutions
State Educational Institution Academy of Postgraduate Pedagogical Education
Moscow, Russia

Vaiva Zuzevičiūtė
Department of Humanities
Faculty of Public Security
Mykolas Romeris University
Kaunas, Lithuania

Section I

Human Factors and Technology

1 Birth of Asymmetrical Face Biometrics and Its Transdisciplinarian Near Future

Leonid Kompanets and Damian Kurach

CONTENTS

PROBLEM STATEMENT AND SHORT INTRODUCTION TO KNOWLEDGE GLOBALIZATION AND OBJECTIFICATION

The main purpose of this chapter is to obtain a modern methodology that allows us to compose in automatic mode an *appropriate* psychological description of any person based on ophthalmic geometry pattern (OGP) (Muldashev 2002) and objective (x, y)-frontal facial projection (Anuashvili 2008); in such a manner, her/his vertical, facial photo scientific converts in a rational, quantitative, biometrics making-decision. The methodology analysis must indicate that the correlation would be a significant one, that is, the correlation coefficient must be more than ± 0.11, which is the level of invisibility (sensitivity) of a modern human being (HB) perception. In the end, we would like to solve the original problem: choosing and switching during (e-)learning path, in automatic mode, for talented personalities (such as M. Tatcher and W. Churchill) to do their education not as dull as the ditchwater. Our visual system or system advisor, which may be elaborated based on phase portraits, observes attentively a concrete face personality and defines whether the person is a left-sided, neutral, or right-sided thinker. Based on this idea, the system finds an appropriate (e-)learning path to personality demand in an automatic manner.

Identification of psychological traits is a task widely used in theoretical and practical biometrics, psychology, education, coaching, career guidance process, business and political affairs, psychotherapeutics and diagnostics, self-exploration, awareness, etc. But according to Boeree (2010), people are different in fundamental ways. Psychology is, first and foremost, about people, *real people in real lives*, and not about computer models, statistical analysis, rat behavior, or test scores. Personality is *not yet a science*, at least not in the sense that biology and chemistry are sciences.

Worldwide outstanding scientist, Professor M. Kleiber, president of the Polish Academy of Sciences, states in "Canon scientific books that you must know" (Kleiber 2008) that a modern situation is characterized as follows: "...without the knowledge of elementary particles, artificial intelligence (AI), chaos theory, non-equilibrium processes, and the network society fully understand the modern world is virtually impossible." He proposed 16 main books for reading. The authors have a consensus with Kleiber and use more updated references for professionals (Drucker 2000; Penrose 2000). A good review is provided in Newth (1999).

People have studied HBs—the most complex system we know. For these studies, people elaborate a scientific method based on measurement. The conventionalities of the scientific method are as follows: (1) state of the art of system approximation; (2) measurable signal; (3) the system has quantitative effect(s); (4) signal with "smallness" of system output effect(s) is not considered; (5) signal has a form *exp(st)* only; (6) linearity, small dimension, stationarity, and other simplifications of technical system use; (7) effect of smoothing signal (energetic but does not form a signal is a determined trait); (8) usually, a system researches in constant mode; (9) available situation prognosis (also statistics); etc. On limited capability of the constant mode, N. Wiener joked: "For incomprehensible cases, mathematicians invent notion *an infinity* but people—*the God*." For this, a technical system performs in a more abstract way than any biological or social one, that is, it is a more approximative system (Siebert 1988).

Modern system functions under the following conditions: (1) spectrum system spreads from computer (programmable) to a HB (learning process used in natural language [NL]); (2) researches exploit mainly the AI strong version; (3) precise and descriptive sciences exist; (4) a HB system exists under Maslow's spectrum of needs; (5) an external system environment is a dynamic one; (6) a HB creativity is described in terms of Figure 1.5 components; (7) object description moves toward objectification, that is, a scientific one; and (8) in AI, spectrum system spreads from productive theory of the calculus predicates to decision-making theory based on incorrect information (Zadeh 2011; Martin and Mendel 1995; Mendel et al. 2011; Luger 2005). In Moiseev (1987), the origin of life on Earth in a dynamic environment is described as a necessary phenomenon; life is a process of aspiration of symmetry to asymmetry: total symmetry of living entities will result to death. This fact scientifically connects stagnant, biological, and social matter and asymmetry of life (Anuashvili 2008). The shortage of external conditions run to system symmetrization, but in biology—to entity death. So life is a process of aspiration of symmetry to asymmetry. It means that total symmetry implies death of the biological and sociological system. In Maslow (1986), deductive system discerns as system, a HB left/right hemisphere functioning during the mind rebuilding on adaptive mode; therefore, we will discern the system (Figures 1.1 through 1.4).

The phenomenon was discovered only about 2000 years ago: the OGP is "invisible" for HB senses and objective personality classification in minimum 2D dimensions. Based on those and other facts, direction of asymmetrical facial biometrics was created. Modern man-made machine (computer) system moves toward an alive/social a HB one (Figure 1.5 and Table 1.1).

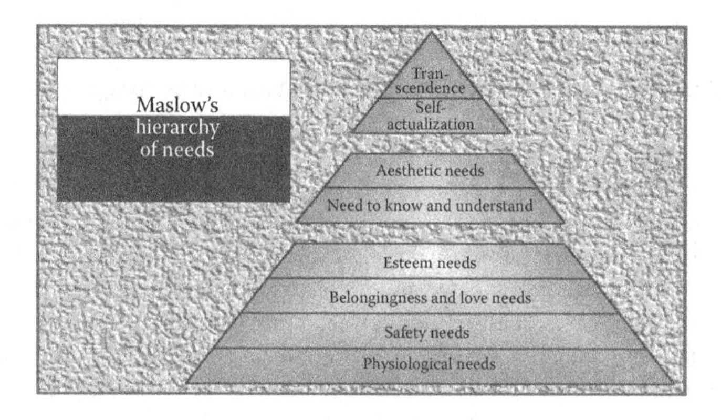

FIGURE 1.1 Maslow's pyramid of deficit needs that act on a HB concurrently (about 1954). (From Maslow, Abraham, 1997. Available at http://www.edpsycinteractive.org/topics/maslow.html. Accessed on February 2011.)

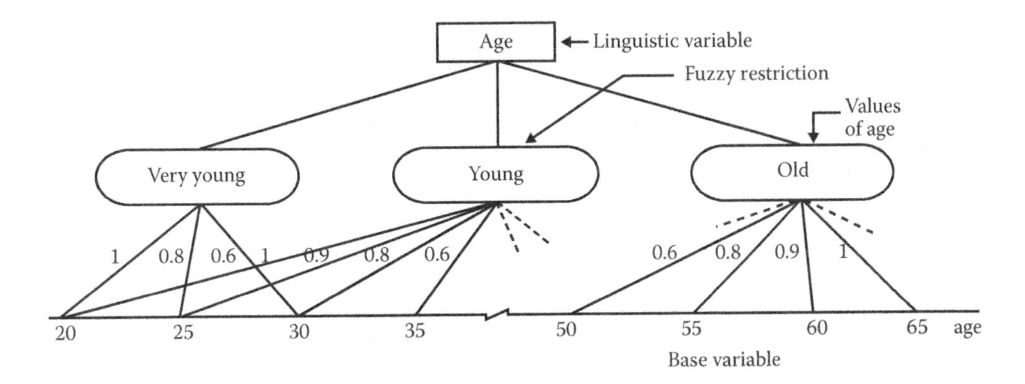

FIGURE 1.2 Illustration of notion of "linguistic variable" or *Zadeh's variable* and its normalization for a case of three values of notion "age": very young, young, and old. (From Zadeh, Lotfi A., From computing with number to computing with words—From manipulation of measurement to manipulation of perception. Available at http://www.cs.berkeley.edu/~zadeh/papers/. Accessed on February 2011.)

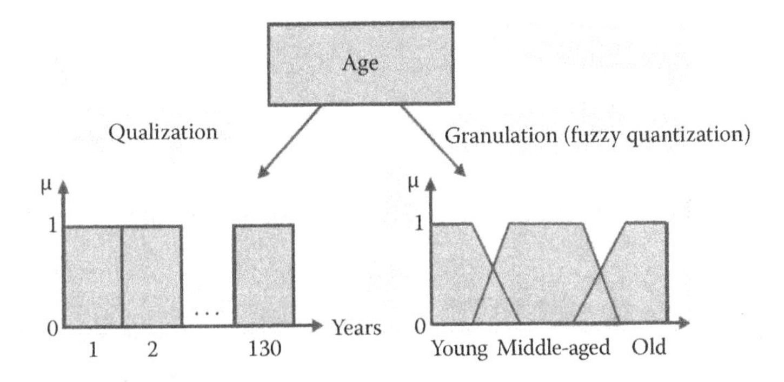

FIGURE 1.3 The fuzzy notion "granulation" illustrated by L. Zadeh. (From Zadeh, Lotfi A., From computing with number to computing with words—From manipulation of measurement to manipulation of perception. Available at http://www.cs.berkeley.edu/~zadeh/papers/. Accessed on February 2011.)

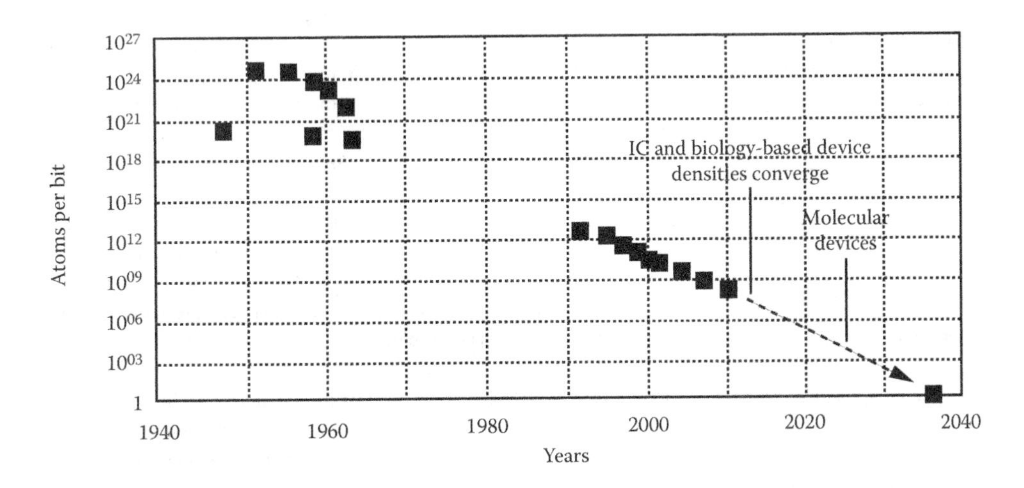

FIGURE 1.4 Modern state of bulk technologies moving toward atom-by-atom ones (enthusiastic prognosis). (From Zhirnov, Victor; Herr, David, *IEEE Computer*, 34–45, © 2001 IEEE.)

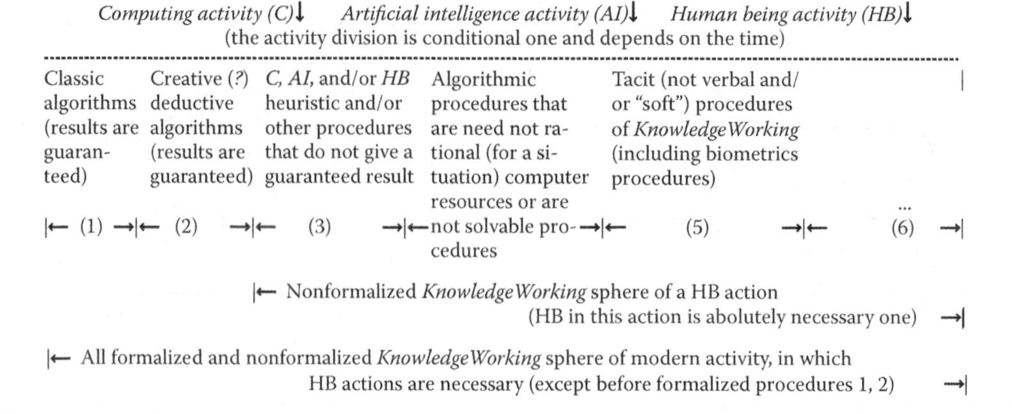

FIGURE 1.5 Structure of modern (state-of-the-art) *Knowledge Working* sphere of modern activity oriented on the science integration (globalization) precise and/or descriptive processes.

TABLE 1.1
Essential Distinctions between Traits of Modern Machine/(Computer) versus Alive/Social System

#	Relationship	Modern Man-Made Machine/(Computer) System	Alive/Social HB System
1	Input signal	Data	Knowledge
2	AI version	Weak version of AI	Strong version of AI
3	Method of construction	Modern construction of any technical machine	Self-organization of matter with having intelligence
4	Relationship to a life cycle/death	It passes a life cycle of technical system	From birth to death passes through reproduction stage (before and past stages are auxiliary ones)
5	Where the energy for functioning system is going from	Energy resources to system give a projecting HB	Open system that during its functioning/rest in any HB cells has energy transmitters that execute the so-called Maslow's physiologic needs
6	Design/fabrication principle	Fabrication on the level man-made bulk technology designed by man functional parties	Self-organization and/or self-assembly based on precise Atom-by-Atom (A-b-A) technology and/or modern micro-miniaturized bulk technology (MMBT) that reaches A-b-A one
7	Design/fabrication technology	Man-made bulk technology without modern MMBT that does not reach A-b-A one	MMBT that might reach A-b-A one
8	Functioning principle	Modern science/engineering (technical) functioning	Birth–death functioning based on Darwin's natural selection (with the Eastern meditation methods)
9	Type of the outer world	Known	Not known, dynamic changing one
10	Having consciousness	No consciousness	Full consciousness

The OGP and objective classification are phenomena that shift an accepted Newtonian–Cartesian scientific paradigm with Kuhn's normal science (knowledge that students learn at any university and which is like "a straitjacket" for new ideas).

Drucker (2000), a worldwide visionary and an outstanding scientist in the field of management, proposes to any HB of about 20 years old to know his or her personality type for a good selection of specialty on the near future and for creation of theory of modern mental worker productivity. As known, during the last 100 years, the productivity theory of the manual worker, based on principles of Henry Ford and some brilliant programmers, had been created. One analogy, the creation and usage by the state of the productivity theory of mental workers, which in the United States are about 30% now, will make the state economically competitive heading to the twenty-first century.

OGP E

The author uses the OGP, which is a radical, novel knowledge (Muldashev 2002). The essence of invisibility/visibility has been interpreted before. Most of the works of L. Kompanets will require incursions in Google. For lack of place, we describe the essence of only some works. For own research, eminent Russian ophthalmologist Professor E. Muldashev discovers (Muldashev 2002) that the visible part of retina of any HB after 4–5 years of age is 10 ± 0.56 mm if the person lives a normal life; a person's OGP does not change (or changes very weakly) during the entire life, but the photo changes significantly. The OGP contains about 30 parameters in a HB and may be an indicator for diagnosing appropriate somatic and/or mental states (including Meyers-Briggs [2007] or Keirsey [2007] cognitive mental traits). This fact may be interpreted as follows: it discovers visible measure of face/body of a HB during natural selection for postnatal condition. In biometrics sense, this is a procedure of Muld normalization—using a person's photo for measuring a HB's face/body in absolute unit *Muld*^{2}*/Muld* without any intrusion; a HB increases to 16–18 years old in *z* dimension only. The number of pixels of the right eye is accepted as a measure *Muld*^{2}*/Muld* (area of circle and its diameter). In Figures 1.6 and 1.7, two stages of the "invisible" OGP extraction (Valchuk and Kompanets 2002) are illustrated.

Relations between parameters and algorithmic nuances are given in Figures 1.6 and 1.7. To form stability of the quadrangle 1234, the tangents define points; we do agree about leading suitable 15°

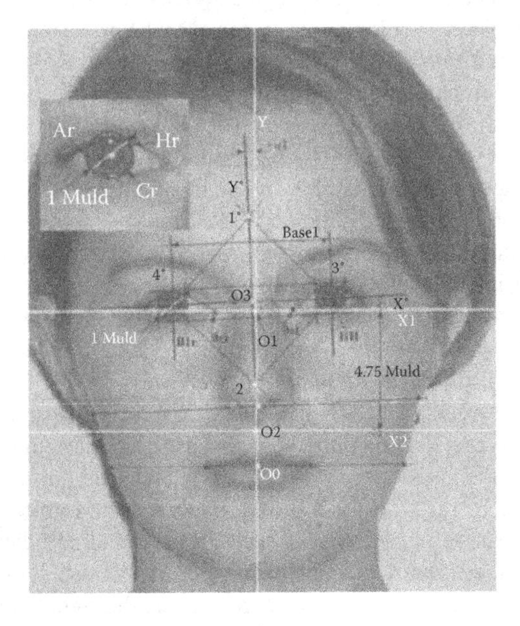

FIGURE 1.6 Illustration of the first stages of personality's OGP extraction.

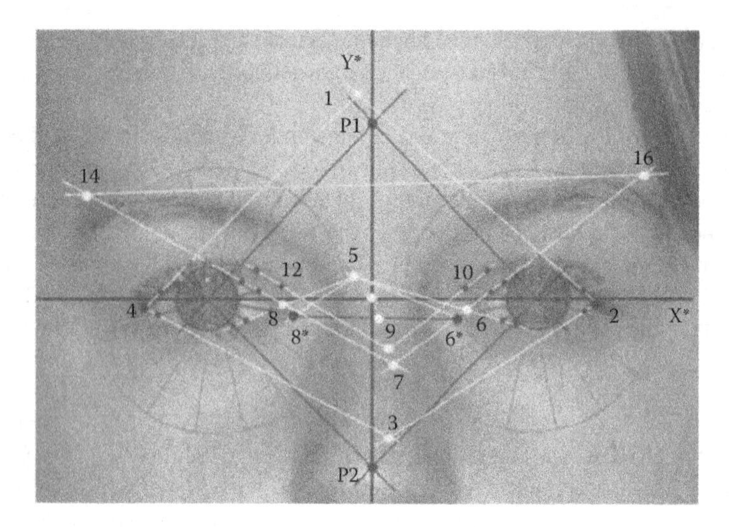

FIGURE 1.7 Illustration of the second stages of the OGP extraction.

point lines from eye centrums. Muldashev (2002) shows that if the area of quadrangle 5678 is out of the area of quadrangle 1234, then the person has a mental or somatic disease. Kompanets (2004, 2007) shows that point 1 corresponds to a person's mental ability.

THE FIRST ANUASHVILI'S OBJECTIVE PERSONALITY TYPOLOGY AND HEMISPHERE SPECIALIZATION

After scientific analysis of personality types (including C. Jung classification), outstanding scientist and a professor of technical sciences (physician) and psychology A. Anuashvili metaphorically said that "in modern psychology, there cannot be ten or more, different, Archimedes laws that all are simultaneously right. In science, it is indisputable fact, but in modern psychology such coincidence may be. But the fact is, it means that method regularity has not been established yet" (Anuashvili 2008). He shows that all existent typologies are linear ones; he proves that usable for determination of living object must be minimum 2D typologies. So he elaborates the first objective (scientific, repeatable) typology with comparison given in Figure 1.8.

The authors use radical, novel knowledge on a HB (Anuashvili 2008). The essence of elaborated typology and other things is interpreted below. Anuashvili was the first who created and introduced in a wide international practice the measurable and automatic 2D typology for a not separate group of HBs but for the entire population. As a matter of fact, the psychological type of a HB is defined to be a relative (not absolute) one. At the moment, the population of the Caucasian races is about 18%–20% only, and their number fell down in time. Besides, according to Szczekin (1995), about 50% people are now the accentuated ones. Anuashvili's typology and their partial comparison with other typologies are given in Figure 1.8. Some established functions of left/right hemisphere for modern phenomenon of a HB brain asymmetry are given in Figure 1.9. As can be seen from Figure 1.8, the new types LISD, LIS, LS, IS, LSD, ISD, LID, LD, and ID are established as basic types, but the rest (49 psychological types) depend on the preciseness of modern measured equipment.

Anuashvili's psyche-mathematical mind/psyche model is based on formulation of the *information background principle* of recognition of invisible objects; examples of such may be aircraft stealth, which is "invisible" to radar. It cannot see an object and sufficiently researches its background model only. A HB face is a background of functioning hemispheres on which functioning of the left hemisphere reveals itself on the left part of a face and the right hemisphere on the right part of a face. Anuashvili establishes that a HB psychological state changes nonlinearly and minimum in

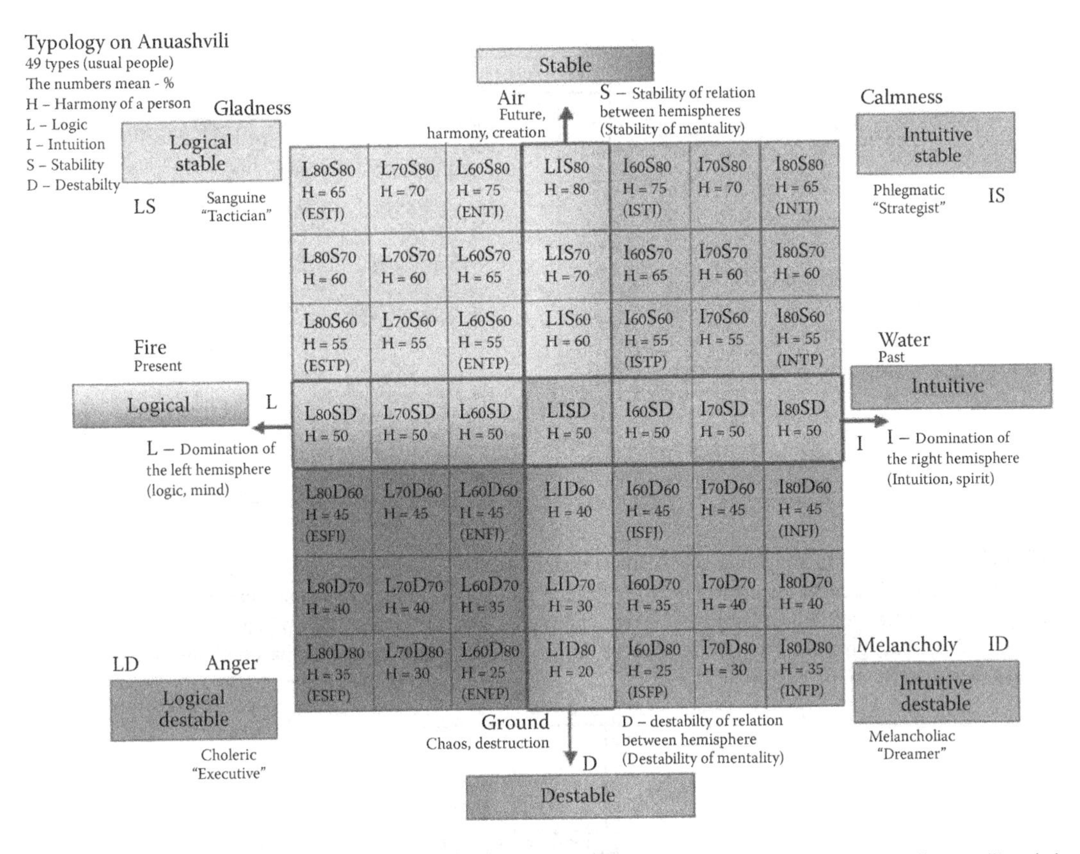

FIGURE 1.8 Anuashvili's 2D objective psychological typology for usual people in coordinates "Logic/Intuitive—Stable/Destable." (From Anuashvili, Avtandil, *Objective psychology based on wave brain model.* 6th ed. Corrected and refilled. Moscow, Published by Econ-Inform, 2008 (in Russian). Available at http://anuashvili.ru/images/knigal.doc. Accessed on April 2011.)

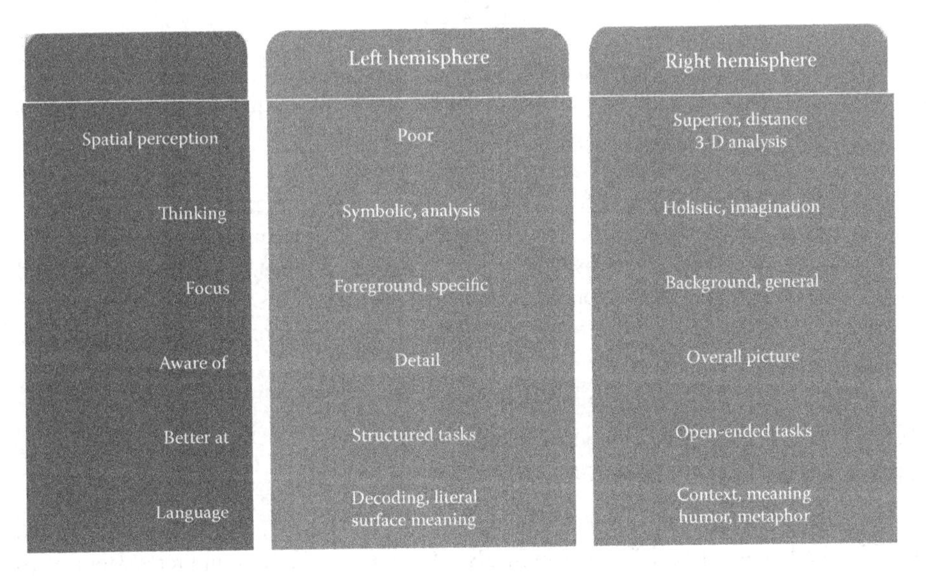

	Left hemisphere	Right hemisphere
Spatial perception	Poor	Superior, distance 3-D analysis
Thinking	Symbolic, analysis	Holistic, imagination
Focus	Foreground, specific	Background, general
Aware of	Detail	Overall picture
Better at	Structured tasks	Open-ended tasks
Language	Decoding, literal surface meaning	Context, meaning humor, metaphor

FIGURE 1.9 Function sorting out to left/right hemisphere specialization for modern phenomenon of a HB mind asymmetry (the Internet).

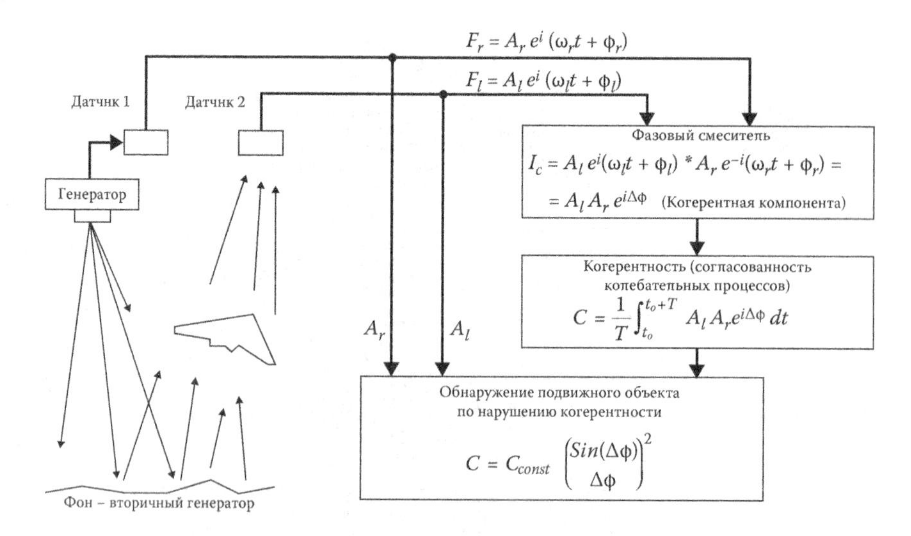

FIGURE 1.10 Fragment of patent no. 2201131 from March 27, 2003 (Anuashvili 2002). Definitions of Russian notions: detector, датчик; generator, генератор; phase mixer, фазовый смеситель; coherency, когерентность; recognition of (invisible) moving object based on calculation of C, обнаружение подвижного объекта по C; background, фон. (From Anuashvili, Avtandil, Video-computer psychodiagnostics and correction, 2002 (in Russian). Available at http://www.anuashvili.ru/. Accessed on April 2011.)

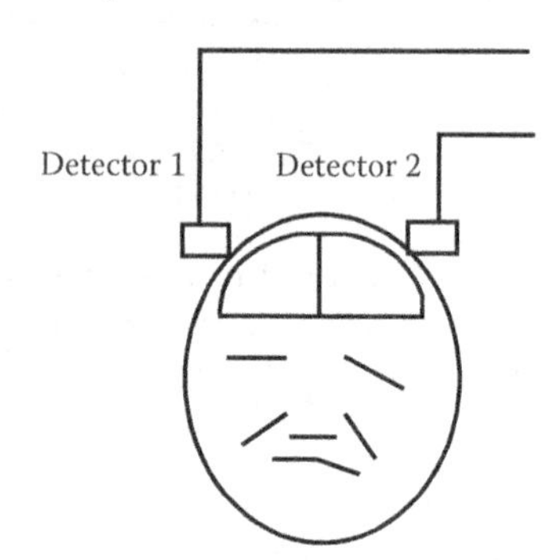

FIGURE 1.11 Scheme of measuring hemisphere waves. (From Anuashvili, Avtandil, Video-computer psychodiagnostics and correction, 2002 (in Russian). Available at http://www.anuashvili.ru/. Accessed on April 2011.)

2D *L–I* and *S–D* coordination (Figure 1.8), where *L* means *logic*, *I* means *intuition*, *S* means *stability*, and *D* means *destability*. Values *L* and *I* are obtained by calculating ΔA, which is the difference between the signal amplitude of hemisphere waves A_l and A_r. Values *S* and *D* are calculated using the equation in Figure 1.10. The personality coefficient harmony *H* is also calculated (Figure 1.11). Values ΔA and I_c can be measured directly and without any invasion; the others can be calculated only.

STRUCTURES OF INVISIBLE ANUASHVILI'S PHASE PORTRAIT AND VISIBLE KOMPANETS' ONE

According to Russian Federation patent no. 2201131 (27.03.2003), "Determination manner of Anuashvili's psychological type," Anuashvili also designed a video-computer system of psychodiagnostic and biologic feedback correction (VCP). VCP works on the basis of Anuashvili's phase portrait (Figure 1.12). VCP occupies 1 min and psycho-correction occupies 10 min.

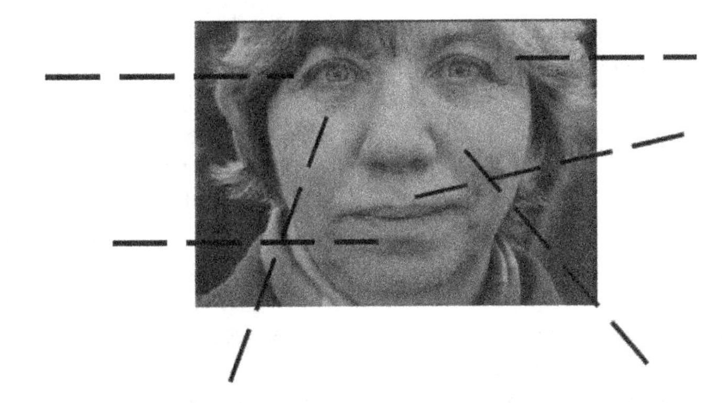

FIGURE 1.12 Elements of Anuashvili's phase portrait: *lines* of corner eye angles, of mouth–nasal wrinkles, and of corners of mouth and middle mouth point. (From Anuashvili, Avtandil, Video-computer psychodiagnostics and correction, 2002 (in Russian). Available at http://www.anuashvili.ru/. Accessed on April 2011.)

There are about 400 experts scattered worldwide (United States, United Kingdom, Germany, Spain, Poland, Hungary, Ukraine, etc.). The VCP system is used in cosmos on station Sojuz TM32 and on MKS. One of the composites consists of two right halves of the face and figuratively displays a condition of the right hemisphere—"spiritual." Computer builds the left/left composite, which is referred to as "vital."

The main difference between nonmeasurable and scientific methods is a possibility of using a HB trait that estimates modern micro miniaturization process, bulk-IT (i.e., large mass or shape), and its limited method—an individual atom one (i.e., *atom by atom*). To understand the problem, HB memory was divided into genetic and learning parts: in the former, in any herd/flock, the leader rule "You must act as I will do" exists; in the latter, scientific knowledge is used by the social community.

Then, on the basis of a visual biological feedback, the mechanism of psychophysical self-control (harmonization) starts. As a result of psychocorrection, stability of mental processes, and degree of harmony of the personality, a corrected person becomes more adequate and capable to operate processes.

Hypothesis about N types of the "truthful" axis projection and experimental result are given in Kompanets (2004) and references therein. Defining the "truthful" (x, y)-projection is not a trivial technique of photo image processing. Besides, it accents on building precise facial composites, for example, for Christ the Redeemer (a well-known statue in Rio de Janeiro), for creating the test "alive/virtual" facial image on the well-known Ramona avatar (R. Kurzweil's site), etc. (Kompanets 2007).

Using lines (see Figure 1.12) during building portrait/composites, we use a notion HB in general.

ALGORITHM OF FINDING "TRUTHFUL" AXIS PROJECTION AND INTRODUCED EXTRA AREA MEASURES

Results of finding the area facial asymmetry measure values and calculating results concerning brain hemisphere dominance are given in Tables 1.2 through 1.4 and are shown in Figures 1.14 through 1.16, respectively. For this, it uses the normalized RGB color space representation to detect a face on image and gradient intensity to localize eye position. The idea of finding the possible iris edge by using information of regional average pixel intensity is the original one. After an estimation of edge candidate points, the iris diameter *Dr* and the center *S* of the right eye are obtained. The diameter becomes a unit (Muld, area = Muld^2) in *Muld* normalization procedure. After that, uniqueness of eyelid contours is achieved by means of approximation with three-degree polynomials. The point coordinates of upper and lower eyelids are detected based on a pixel's illumination variation and unique mapping of inner and outer eye corners. Afterward, the algorithm of the facial asymmetry measures is applied. The algorithm includes automatic selection of facial asymmetry regions using the "truthful" projection and the absolute facial size is defined.

Furthermore, introducing two new measures, AAs^+ and AAs^-, of appropriate facial regions presents significant improvement of the method. The values AAs^+ and AAs^- are calculated as the difference in pixels of the left and right parts (defined by the vertical axis) of the facial binary images (BIs) (Tables 1.2 through 1.4 and Figures 1.14 through 1.16). The values represent the right- and left-sided facial asymmetry.

The *AAs-measure* algorithm is applied to calculate the values of the measures $\pm AAs$, AAs^+, and AAs^- in an adaptive way, as presented in the following:

Input data: Oce, anthropological point coordinates; φ (°), rotation angle; BI, silhouette's BI; *Dr*, right eye iris diameter. *Output data*: *AAs*, AAs^+, and AAs^- values.

Step 1. Define the vertical axis projection, *FA*(φ), on a binary facial silhouette's image BI and the value of angle φ (°) (see BIs in Figure 1.13).

Step 2. Create a vertically mirrored BI (MBI) based on the information on the BI image and the axis projection *FA*(φ) (Figure 1.13).

Step 3. Calculate the AAs^+ and AAs^- values as the difference between the right and left sides of BIs and MBIs.

Step 4. Use the AAs^+ and AAs^- values to determine the person's asymmetry type and other problem aspects.

BIs represent facial silhouette and are obtained from the frontal input image using Otsu's threshold algorithm. The minimal value of $\pm AAs$-measure defines the "truthful" axis.

There are many educational situations, but "we are attempting to educate 21st-century engineers with a 20th-century curriculum taught in 19th-century institutions" (Duderstadt 2008).

We choose automatic selection procedure of determining the (e-)learning path for talented personalities (such as M. Tatcher and W. Churchill) to do her/his education not as dull as ditchwater. A visual system, which may be elaborated, observes attentively a face personality and defines whether the person is a left-sided, neutral, or right-sided thinker. And then, the system finds an appropriate (e-)learning path in an automatic manner. To examine objectively the author's method, we compare the obtained results with those of Anuashvili (2002) – which is only one existing method.

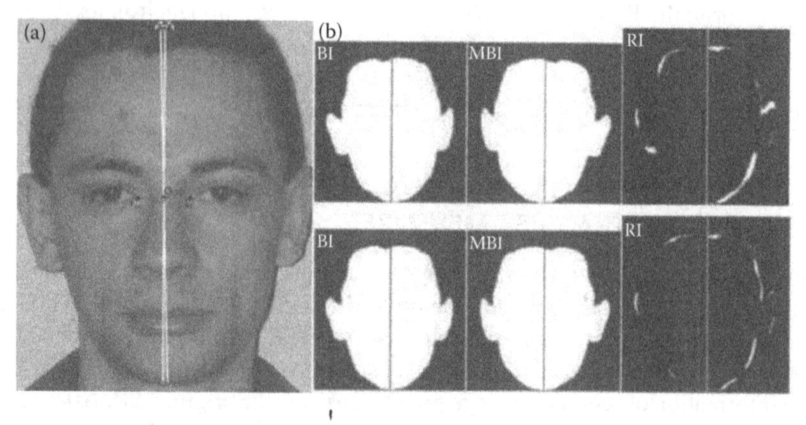

FIGURE 1.13 Graphic illustration of measuring the facial asymmetry traits: (a) input image with two candidate axes (0° and 1.5°); (b) BIs, MBIs, and result images. For 0° axis: upper images; for 1.5° axis: lower images.

INITIAL RESULT OF ELABORATION OF THE METHOD OF (E-)LEARNING TALENTED PERSONALITY

It takes a color image of a person's vertical face, for example, produced by a web camera, as an input signal, before a program can determine the facial asymmetry type. We believe that the computed asymmetry types correspond to the brain thinker type (Milczarski et al. 2010). Examples of the psychological mind thinker types may be the following: a person is a bilateral (neutral), right-sided, or left-sided thinker. If our assumption is true, the dominance of the right asymmetry means that the person is a right-sided thinker and the left asymmetry dominance indicates a left-sided one; when the asymmetry is balanced, it means that the person is a neutral one.

Results of finding the area facial asymmetry measure values and calculating results concerning brain hemisphere dominance are given in Tables 1.2 through 1.4 and shown in Figures 1.14 through 1.16. The beginning is the same as in the algorithm of finding the "truthful axis." It uses the normalized RGB color space representation to detect a facial image and gradient intensity information to localize eye position. Then we find the possible iris edge by using information of regional average pixel intensity. After that, eyelid contours are achieved by means of approximation with three-degree polynomials. The point coordinates of upper and lower eyelids are used to point the inner and outer eye corners. *Muld* normalization process is used to define a unique unit and real face proportion. After that, the algorithm of the facial asymmetry measure is applied. The algorithm includes automatic selection of facial asymmetry regions and defines absolutely the facial size.

TABLE 1.2
Results of "Truthful" Procedure Axis Projection Defining and Asymmetry Measuring for the Left-Sided Thinker

	AAs^-		AAs^+		$\pm AAs$	
φ (°)	Pixel	$Muld^2$	Pixel	$Muld^2$	Pixel	$Muld^2$
−2	205	0.31	741	1.14	946	1.45
*−1	**261**	**0.40**	69	0.10	330	0.50
0	676	1.04	0	0.00	676	1.04
1	1682	2.59	0	0.00	1682	2.59
2	4787	7.37	80	0.12	4867	7.49

TABLE 1.3
Results of "Truthful" Procedure Axis Projection Defining and Asymmetry Measuring for the Right-Sided Thinker

	AAs^-		AAs^+		$\pm AAs$	
φ (°)	Pixel	$Muld^2$	Pixel	$Muld^2$	Pixel	$Muld^2$
−2	381	0.59	4558	7.02	4939	7.61
−1	416	0.64	3090	4.76	3506	5.40
0	554	0.85	1908	2.94	2462	3.79
*1	757	1.17	860	1.32	1617	2.49
2	1498	2.30	232	0.36	1730	2.66

TABLE 1.4
Results of "Truthful" Procedure Axis Projection Defining and Asymmetry Measuring for the Neutral Thinker

	AAs^-		AAs^+		$\pm AAs$	
φ (°)	Pixel	*Muld*²	Pixel	*Muld*²	Pixel	*Muld*²
−2	202	0.31	1835	2.83	2037	3.14
−1	158	0.24	1197	1.84	1355	2.08
0	579	0.89	170	0.26	749	1.15
*1	**348**	**0.53**	219	0.34	567	0.87
2	690	1.06	82	0.12	772	1.18

Furthermore, like before, two extra measures of appropriate facial regions, AAs^+ and AAs^-, are used. The values are calculated as the difference of the left and right parts of the facial BI (given in Tables 1.2 through 1.4 and Figures 1.14 through 1.16).

To objectively examine the method, comparison of the obtained results of the personality asymmetry types with the psychological types described by Anuashvili (2002, 2008) is performed. In the first step of the comparison, the algorithm chooses a base axis calculated from Anuashvili's

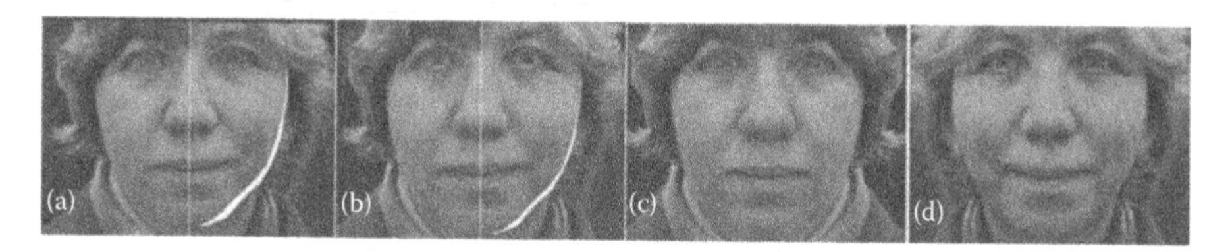

FIGURE 1.14 Person with dominance of the left hemisphere according to Anuashvili: (a) image with 0° axis (Anuashvili's axis projection); (b) image with −1° axis (Kompanets's axis projection); (c) right-sided composite (Kompanets's axis projection); and (d) left-sided composite (Kompanets's axis projection).

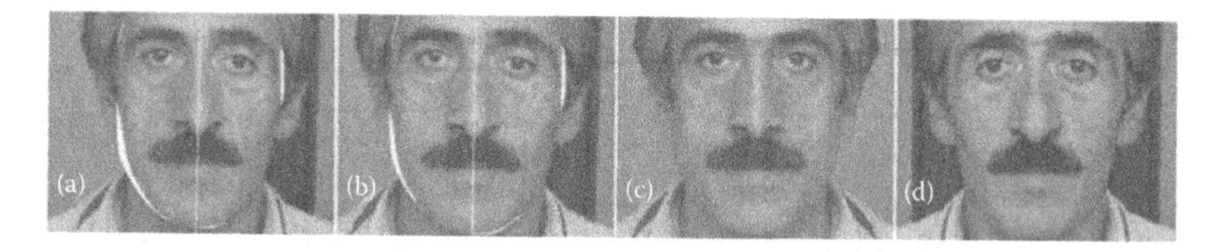

FIGURE 1.15 Person with dominance of the right hemisphere: (a) image with 0° axis; (b) image with +1° axis; (c) right-sided composite; and (d) left-sided composite.

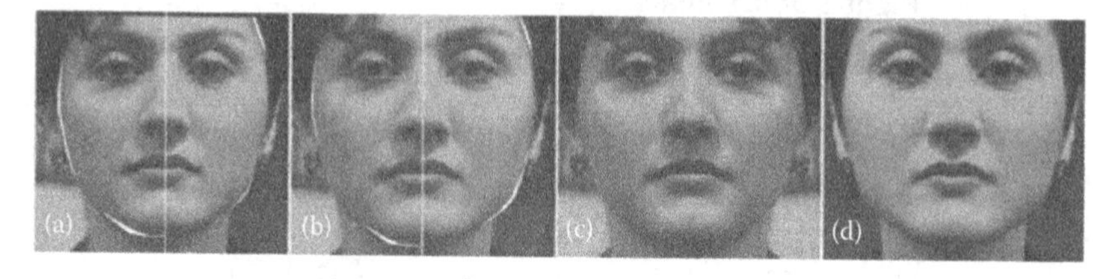

FIGURE 1.16 Person with no dominance (bilateral): (a) image with 0° axis; (b) image with +1° axis; (c) right-sided composite; and (d) left-sided composite.

composites. The composites are synthesized images from left–left and right–right parts of the face. In our opinion, Anuashvili's method of finding the axis is based on an intuitive technique. Usually we suppose the method is not objective, that is, the axis found has an error. In Kompanet's method, the "truthful" vertical axis is determined by the automatic adaptation procedure, and it is more precise. There is a possibility for further classification, especially when we define the interrelation of "personality type—values of measures" (see white areas in Figures 1.14 through 1.16).

Tables 1.2 through 1.4 correspond to Figures 1.14 through 1.16, respectively. The tables show the results of the area asymmetry measures AAs (the appropriate axis is selected from five axis candidates). The base axis rotates left and right according to angles $-2°$, $-1°$, $1°$, and $2°$ (the sign "+" means clockwise axis rotation) around the anthropological point Oec. The values of AAs, AAs^+, and AAs^- are given in pixels and $Muld^2$ units. It shows the results for finding the vertical axis with the accuracy 1°, which is enough to illustrate the idea of the proposed method. The proper vertical axis is indicated by the lowest value of the asymmetry measure AAs; the symbol * in Tables 1.2 through 1.4 denotes the case of minimal $\pm AAs$. The gray color shows the results for the 0° axis, that is, Anuashvili's imprecise phase portraits), while bold emphasis indicates Kompanets's precise phase portraits.

The adaptive method of finding the "truthful" axis balances the difference between the AAs^+ and AAs^- values of asymmetry (see rows indicated by * in Tables 1.2 through 1.4). The minimal value of the asymmetry AAs for the left-sided thinker (see Figure 1.14 and Table 1.2) equals 0.5 $Muld^2$. It is a small value that means that the person is close to a bilateral one. The value AAs^- equals 0.4 $Muld^2$ and is bigger in comparison with AAs^+. The bigger left-sided asymmetry indicates that this person is a left-sided thinker type.

For a person with right hemisphere dominance (see Figure 1.15 and Table 1.3), the minimal value of facial asymmetry measure equals 2.49 $Muld^2$. The difference between AAs^- and AAs^+ is rather small but with majority of AA^+ value. This indicates that the person is classified as a right-sided thinker.

A person with no dominance of a hemisphere (see Figure 1.16 and Table 1.4) shows that AAs equals 0.87 $Muld^2$ while AAs^+ equals 0.34 $Muld^2$ and AAs^- equals 0.53 $Muld^2$. The small difference between the sided asymmetry measures and the asymmetry region locations classifies this person as a neutral thinker.

Modified or precise method has much additional information about the type of thinker. In the author's opinion, therefore, it gives possibility to find an appropriate relationship between personality's psychological type and personality's (x, y)-projection traits.

CONCLUSION AND PROPOSITIONS

The paper's principal goal is to promote the up-to-date idea—*the facial asymmetric biometrics*—which may begin shifting the Newtonian–Cartesian paradigm to Eastern methodology.

The authors propose to create a suitable international transdisciplinary unit for the multimedia productivity theory of mental worker creation and other significant problems to be solved. It may be the International Association of Trans-disciplinary Research in Informatics, Biology, Psychology, Sociology, Business, and Medicine (IATR IBPSBM) that will conciliate a scientific knowledge of traditional and nontraditional directions.

REFERENCES

Anuashvili, Avtandil: *Objective Psychology Based on Wave Brain Model.* 6th ed. Corrected and refilled. Econ Inform, Moscow, Russia, 2008 (in Russian). Available at: http://anuashvili.ru/images/kniga1.doc (accessed on April 2011).

Anuashvili, Avtandil: Video-computer psychodiagnostics and correction. 2002 (in Russian). Available at: http://www.anuashvili.ru/ (accessed on April 2011).

Boeree, Gorge: 2010. *Personality Theories*. Available at: http://webspace.ship.edu/cgboer/personalityintroduction.html (accessed on January 2011).

Drucker, Peter F.: *Management Challenges for the 21st Century*. Translated from English. Williams: Moscow, Russia 2000 (in Russian).

Duderstadt, J. J.: Engineering for a Changing World. A Roadmap to the Future of Engineering Practice, Research, and Education. The Millennium Project of the Univ. of Michigan, 2008. Available at: http://milproj.ummu.umichdu/publications/EngFlex_report/download/Eng Flex%20Report.pdf (accessed on March 2011).

Keirsey, David: 2007. Keirsey Temperament Web. Available at: http://www.kiersey.com/ (accessed on December 2010).

Kleiber, Michal: Canon scientific books that you must know. 2008. Available at: http://www.polskatimes.pl/forumpolska/krakow/55415,michal-kleiber-kanon-ksiazek-naukowych-ktore-nalezy-znac.id,t.html (accessed on February 2011).

Kompanets, Leonid: Biometrics of Asymmetrical Face. Proc. of 1st Intern. Conf. on *Biometric Authentication – ICBA 2004*. Hong Kong, China, July. In: *LNCS* # 3072, pp. 67–73.

Kompanets, Leonid: Facial Composites, Ophthalmic Geometry Pattern, and Based on Stated Phenomena the Test of Person/Personality Aliveness. Proc. 7th Intern. Conf. on *Intelligent Systems Design and Applications,* October 22–24, 2007. Rio de Janeiro, pp. 831–836.

Luger, George F.: *Artificial Intelligence. Structures and Strategies for Complex Problem Solving*. 4th ed. Williams: Moscow, Russia, 2005 (in Russian).

Maslow, Abraham: 1997. Available at: http://www.edpsycinteractive.org/topics/maslow.html (accessed on February 2011).

Maslow, Sergej: *Theory of Deductive Systems and its Use*. Moscow, Radio and Communications (in Russian), 1986.

Martin, Matthew; Mendel, Jerry: Flirtation, A Very Fuzzy Prospect, 1995. Available at: http://sipi.usc.edu/~mendel/ (accessed on February 2011).

Mendel, Jerry et al.: What Computing with Words Means to Me. Available at: http://www.cs.berkeley.edu/~zadeh/papers/ (accessed on February 2011).

Meyers-Briggs: MBTI @ Personality Type Applications. Available at: http://www.personalitypathways.com/typeinventory.html. (accessed on December 2010), 2007.

Milczarski, Piotr; Kompanets, Leonid; Kurach, Damian: An Approach to Brain Thinker Recognition Based on Facial Asymmetry. Proc. 10th Intern. Conf. on *Artificial Intelligence and Soft Computing*. Zakopane, Poland, June 13–17, 2010. In: *LNAI, # 6113*, 643-650. Available at: http://www.springerlink.com/content/62071865q741803w/fulltext.pdf (accessed on February 2011).

Moiseev, Nikita: *Algorithms of Progress*. Moscow, Science (in Russian), 1987.

Muldashev, Ernst: *Whom Did We Descend From?* OLMA Press, Moscow, Russia (in Russian), 2002.

Newth, Eirik: *In Search of Truth. Stories about a Science*. 3rd ed. Translated from Norwegian. Warsaw, WNT (in Polish), 1999.

Penrose, Roger: *The Imperor's New Mind. Concerning Computers, Minds, and the Laws of Physics*. 3rd ed. Translated from English. Warsaw, PWN (in Polish), 2000.

Siebert, W. McC: *Circuits, Signals, and Systems*. Vol. 2. Translated from English. Moscow, Mir (in Russian), 1988.

Szczekin, Georgy: *Visual Psychodiagnostics: To Get to Know People on Appearance and Behavior*. Intern. Academy of Personnel Control Press, Kiev, (in Russian), 1995.

Valchuk, Tatiana; Kompanets, Leonid: Identification/Authentication of Person Cognitive Characteristics. The IEEE Proc. of 3rd Workshop on *Automatic Identification Advanced Technologies: AutoID'02*, 14–15 March, Tarrytown, New York, 2002, pp. 12–16.

Zadeh, Lotfi A.: From Computing with Number to Computing with Words – from Manipulation of Measurement to Manipulation of Perception. Available at: http://www.cs.berkeley.edu/~zadeh/papers/ (accessed on February 2011).

Zhirnov, Victor; Herr, David: New frontiers: Self-assembly and nanoelectronics. *IEEE Comput.*, January 2001, 34–45.

2 HITS
Advanced City Logistics Systems

Andrzej Adamski and Daniel Kubek

CONTENTS

INTRODUCTION

The solution of integrated vehicle routing problems (VRPs) in intelligent logistics system (ILS)–integrated transportation system (ITS) multilayer intelligent, integrated systems is a challenging research area (Adamski 2002a,b, 2003, 2006a,b, 2007a,b,c,d, 2011). The main challenges concentrate on application-oriented practical use of existing functionalities of hierarchical ITS-ILS systems in which context-oriented intelligent exploration of available system resources is of paramount importance. This system-wide approach guarantees many practically important advantages: several solution options for a wide range of network logistics problems (flexible logistics, reverse logistics, stochastic problems, and different customers level of service [LoS] specifications) with adequate vehicle routing options, dynamic and multicriteria solutions, cooperative solutions in terms of SupNet system operational environment (Adamski 2003), and capabilities of generation of dedicated ILS system options for individual problems. Most VRPs, such as capacitated (CVRP)/noncapacitated (NCVRP), arc common address redundancy protocol (CARP)/node, common name resolution protocol (CNRP) based with location clustered-loss retransmission protocol (CLRP), virtual router redundancy protocol (VRPP), and several others, concentrate mainly on solution methods and metaheuristics. In this context, we observe the growing importance of evolutionary/population-based methods with the use of local search procedures and development of matheuristics with strictly optimal solutions of subproblems. The transportation and logistics systems belong to the family of artificial systems developed by humans in order to satisfy their demands. These systems are in a natural way embedded in an interconnected complex of physical or virtual networks called SupNet (Adamski 2003). The SupNet, through a wide spectrum of complex interactions, creates the key operational dynamic and behavioral environment for transportation and logistics processes (Adamski 1998, 1999, 2002a, 2003). The real-time identification, estimation, and prediction of these interactions for ITS-ILS management, surveillance, and control purposes are crucial problems conditioning the efficiency and productivity of these systems. In this context, the necessary condition for management and control of such systems requires both deep understanding of the features and behavior of these systems as well as the general understanding of operational environment (SupNet) that essentially influences the operation of these systems. Fulfillment of these conditions is necessary to rationally influence the operation of these systems both in terms of system primary goals (e.g., minimization of negative primary goal impacts) and secondary goals of system operational general influences (e.g., minimization of environmental impacts). The new approaches for professional ITS and ILS system development were presented in Adamski (1999, 2002a,b, 2003, 2006b, 2007c,d). These approaches are unified and expanded by methodology (Adamski 2007a). In this modern ITS and ILS system development approach, the following general principles formulated in Adamski (2003) are obligatory as well as very well motivated:

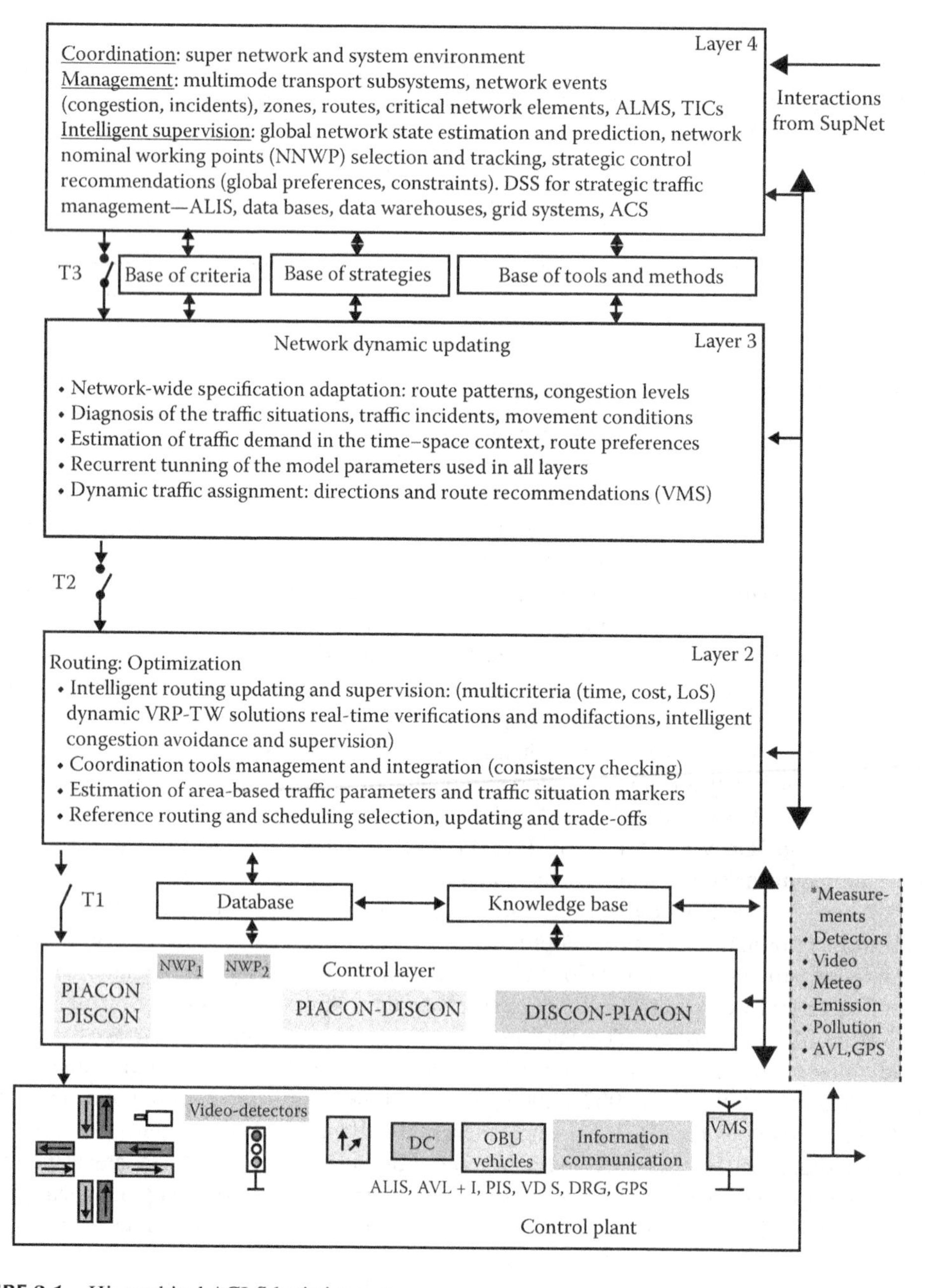

FIGURE 2.1 Hierarchical ACLS logistics system.

- TRIAD I: integration–intelligence–information
- TRIAD E: efficiency–emergency–ecology
- TRIAD U: user-oriented–up-to-date–unified

The basic logistic system functional hardware-in-the-loop simulation (HILS) platform with management, surveillance, and control system structure was presented in Adamski (2006a,b, 2011). Five layers of this system in a natural way integrate and vertically order (in time, frequencies of interventions, aggregation levels) a wide spectrum of decision-making and optimal control functions that additionally are supported by integrated data, knowledge, and tools basis equipped with dedicated decision support system (DSS) and computer-aided scheduling and dispatch (CASD). In this paper, the advanced city logistics system (ACLS) HITS option was proposed for the wide spectrum of the city logistics problems (see Figure 2.1).

HITS: Hierarchical Integrated Intelligent Transportation System Platform

The cooperative HITS multilayer operation may be presented by the following cycle:

1. Management and coordination layer offers advanced transport information service (ALIS; dedicated to logistics) concerning SupNet interactions, essential traffic events (e.g., traffic incidents, critical network elements), network actual state specifications, global preferences, behavioral patterns, and constraint and coordination premises.
2. Adaptation layer realizes dynamic network updating process: network structure (available nodes, links), routes (patterns, nodes/links, route specifications), and levels of congestion (traffic incidents, available throughput) with short-term prediction estimates. Very modern solutions and advanced real-time traffic situational analysis tools dedicated to transportation and logistics systems were presented in Leiths and Adamski (2011).
3. Optimization layer solves vehicle routing and scheduling problem with time windows (VRP-TW) after robust estimation of travel times T_i^* from depot to customers. The following specifications (depots, customers, vehicles) are used in this optimization problem:
   (S1): $n_0 \geq 2$, the number of depots in routes.
   (S2): $j \in J = \{1,..,n\}$, set of customers, $\forall j$: demand of customer is given d_j and time windows $TW_j = \left[t_j^s, t_j^e\right]$ for customer visit and service is specified.
   (S3): $i \in I = \{1,..,m\}$, fleet of vehicles, $\forall i$: load i at j is lower than vehicle capacity $L_{ij} \leq Cap_i$; J_i – represents the total number of customers visited by the ith vehicle; $S_i = \{n(j): j = 1,...,J_i\}$ assignment of customer numbers $n(j)$ and sequence $j = 1,...,J_i$ of visiting customers for the ith vehicle.
   (S4): The assignment matrix for all vehicles $S = \{S_i. | i \in I\} = \{S_{i.n(j)}: i \in I, n(j) \in S_i, j = 1,..., J_i\}$.

Assumptions

(A1) All customers must be assigned to exactly one route of the vehicle:

$$\sum_{i \in I} J_i = J \tag{2.1}$$

(A2) Each customer is visited only once, that is, all demand/supply in reverse logistics must be loaded on the vehicle simultaneously:

$$\sum_{j^* \in Si} D_{j^*} = L_i(S_i) \tag{2.2}$$

$$L_i\,(S_i) \leq \mathrm{Cap}_i \tag{2.3}$$

(A3) Departure t_{i0}/arrival t_{0i} times of vehicles from/to depot must belong to time window $\mathrm{TW}_0 = \left[t_0^{\mathrm{s}}, t_0^{\mathrm{e}}\right]$; however, due to $t_{i0} < t_{0i}$ inequality, the nonredundant constraints may be formulated as follows:

$$t_{i0} \geq t_0^{\mathrm{s}} \quad t_{0i} \leq t_0^{\mathrm{e}} \quad t_{0i} = t_{i0} + \sum\nolimits_{j=0,Ji} T_{in(j)} + \tau_{in(j)}^{\mathrm{s}} \tag{2.4}$$

(A4) For the customer sequence unit $(n(j), n(j+1))$, the following recurrence equation for arrival times is obligatory:

$$t_{in(j+1)}^{\mathrm{a}} = t_{in(j)}^{\mathrm{a}} + \tau_{in(j)} + T_{in(j)} \tag{2.5}$$

where $t_{in(j)}^{\mathrm{a}}$ is the arrival time at customer $n(j)$; $\tau_{in(j)}$ is the waiting time, $w_{in(j)} = \max\left[0, t_{n(j)}^{\mathrm{s}} - t_{in(j)}^{\mathrm{a}}\right]$ (if the window $\mathrm{TW}_{n(j)}$ is not open) plus service time $\tau_{in(j)}^{\mathrm{s}}$; and $T_{in(j)}$ is the travel time between $n(j)$ and $n(j+1)$ customers.

Remark: The solution of Equation 2.5 is a function of the initial condition (i.e., $t_{i0}^{\mathrm{a}} = t_{i0}^{\mathrm{d}} = t_{i0}$) and S_i (i.e., assignment and sequence of visiting customers for vehicle "i").

To illustrate VRP-TW problem optimization, the single criterion model of total costs TC minimization with one depot and fleet of identical vehicles is formulated. The cost specifications cs ∈{fixed (f), operational (o), penalty: early (e)/delayed (d) visits at customers}, that is, cs ∈ {f, o, e, d} represent unit costs for vehicles and customers. In the first updating iteration, the optimization problem is formulated as follows:

$$\mathrm{TC}(t_0, S) = \sum\nolimits_i \sum\nolimits_{\mathrm{cs}} C_i^{\mathrm{cs}}(t_{i0}, S_i) = \sum\nolimits_i \Big\{ c_{fi}\delta(S_i) + c_{oi}(t_{oi}(S_i) - t_{io}) + \sum\nolimits_{j=0,Ji} \left[c_{en(j)} w_{in(j)} + c_{dn(j)} d_{in(j)} \right] \tag{2.6}$$

where $w_{in(j)} = \max\left[0, t_{n(j)}^{\mathrm{s}} - t_{in(j)}^{\mathrm{a}}\right]$; $d_{in(j)} = \max\left[0, t_{in(j)}^{\mathrm{a}} - t_{n(j)}^{\mathrm{e}}\right]$; and $\delta(S_i)$ is the Kronecker delta.

$$\mathrm{PO}_{\min} t_0, S\ TC(t_0, S) = \sum\nolimits_i \sum\nolimits_{\mathrm{cs}} C_i^{\mathrm{cs}}(t_{i0}, S_i) \quad \|\ (1)\text{-}(6)$$

In the second updating iteration, all customers that have been serviced are removed with corresponding arcs from the route, and all remaining arcs have their travel times functions updated. Formally, several short path problems (SPPs; node/arc routing problems [NRPs/ARPs]) for the remaining customers with time windows may be formulated and solved.

In a simple way, we try to improve (in terms of travel times, costs, reliability measures) route traveled by the ith vehicle by inserting beneficial arcs (e.g., ARP problem with required arcs represented beneficial service functions solved by VND heuristics).

We can easily minimize a weighted sum of travel time and lateness at customer locations and at the depot ($c_{fi} = 0$, $c_{en(j)} = 0$, $c_{oi} = c_{dn(j)} = 1$) and

$$\mathrm{TT}(S) = \sum\nolimits_i \Big\{ w_1 \sum\nolimits_{j=0,Ji} T_{in(j)} + w_2 \sum\nolimits_{j=0,Ji} \max\left[0, t_{in(j)}^{\mathrm{a}} - t_{n(j)}^{e}\right] + w_3 \max\left[0, t_{oi} - t_0^{e}\right] \Big\} \tag{2.7}$$

Real-time routing may be realized by dynamic travel times with added dynamic perturbations.

4. Direct control layer solves dispatching control problem by the DISCON method. For DISCON dispatching control method purposes, the control variables $u_{in(j)}$ are added to Equation 2.5, and additionally, corresponding scheduling Equation 2.5′ is formulated:

$$t^{\mathrm{as}}_{in(j+1)} = t^{\mathrm{as}}_{in(j)} + \tau^{s}_{in(j)} + T^{s}_{in(j)} \tag{2.5'}$$

where $t^{\mathrm{as}}_{in(j)}$ is the scheduled arrival time at customer $n(j)$, for example, $t^{\mathrm{as}}_{in(j)} = t^{\mathrm{as}}_{n(j)} + \alpha_{n(j)}\left(t^{\mathrm{e}}_{n(j)} - t^{\mathrm{s}}_{n(j)}\right)$ with $\alpha_{n(j)} \in (0,1)$ reliability coefficient selected for $n(j)$ customer (e.g., $\alpha_{n(j)} = 1/3$).

Denoting deviations from the schedule by $x_{in(j)} = t^{\mathrm{a}}_{in(j)} - t^{\mathrm{as}}_{in(j)}$ and $z_{in(j)} = T_{in(j)} - T^{\mathrm{s}}_{in(j)} + w_{in(j)}$, according to Equations 2.5 and 2.5′, the following punctuality control model for the DISCON method (Equation 2.8) with state and control constraints 2.9 and 2.10 can be written in the vector notation form:

$$x_{n(j+1)} = x_{n(j)} + u_{n(j)} + z_{n(j)} \text{ with } x_0 = t_0 - t^{\mathrm{s}}_0 \tag{2.8}$$

$$x_{\,n(j)} \in [x\mathrm{LB}_{n(j)}, x\mathrm{UB}_{n(j)}] \tag{2.9}$$

$$u_{n(j)} \in [u\mathrm{LB}_{n(j)}, u\mathrm{UB}_{n(j)}] \tag{2.10}$$

where $x\mathrm{LB}_{n(j)} = -\,\alpha_{n(j)}\,\left|\mathrm{TW}_{n(j)}\right|$, $x\mathrm{UB}_{n(j)} = (1 + \alpha_{n(j)})\,\left|\mathrm{TW}_{n(j)}\right|$ and $\left|\mathrm{TW}_{n(j)}\right| = t^{e}_{n(j)} - t^{s}_{n(j)}$.

At this point, all DISCON dynamic dispatching control options (deterministic, stochastic, single/multicriteria, robust, anticipative, priority control) are available (see Adamski 1993, 1998, 2003, 2007, 2011) for HIILS application. For example, LQ/LOG control problems may be formulated as follows:

$$\mathrm{PO}_{\min} u_{n(j)} \quad J_{T-j} = \left\|x_T\right\|^2_{Q_T} + \sum_{k=j}^{T-1} \left\|x_k\right\|^2_{Q_k} + \left\|u_k\right\|^2_{R_k} \tag{2.11}$$

where Q_k and R_k are symmetric nonnegative definite weighting matrices, and the first and second terms present the off-reference trajectory deviation penalties at terminal (T) and all customer points. The last term penalizes the weighted sum of squares of control actions.

ANALYZED OPTIONS OF VRPS

VRP, VRP-SPDTW

Vehicular Routing Problem (VRP) with simultaneous pickup and deliveries and time windows (VRP-SPDTW) is a good example of integration of forward and reverse logistics tasks in HITS platform. Formally, it is some mixed integer-programming problem to be some generalization of the classical routing problems of the types VRP, Vehicular Routing Problem-Time Windows (VRP-TW), VRP-with simultaneous pickup and deliveries, Vehicular Routing Problem with Backhauling, and Transit Signal Priority (TSP). The formulation of the VRP-SPDTW problem is as follows. We have transportation network $G = (V,L)$ with customers located in nodes: v_0 is a depot (customer 0). The problem specifications are as follows: $\forall i \in V$, $\left\{d_i, p_i, \mathrm{TW}_i = [ts_i, te_i], \tau_i, t^a_i \; i = 1,..,n\right\}$ represents, respectively, delivery, pickup of some amount of goods, time window, service time, and arrival time specifications with LoS represented by requirements of exactly one visit by one vehicle for both operations. The identical fleet of vehicles $k = 1,..,K$ specifications $\forall k \in K$: {limited capacity $\mathrm{cap}_k = \mathrm{cap}$}. The link specifications $\forall(i, j) \in L$: ($T = \{T_{ij}\}$, $D = (D_{ij})$ represent travel time and distance. Decision variables $x_{ijk} \in \{0,1\}$—assignment of arc (i,j) to the route of the kth vehicle; $y_{ij} \geq 0/z_{ij} \geq 0$—demand pickup from/delivered to customers routed up to/after node i. Problem VRP-SPDTW is formulated as follows:

$$\mathrm{PO}_{\min} X, Y, Z \quad Q = \sum_{k=1}^{K} \sum_{i=0}^{n} \sum_{j=0}^{n} c_{ij} \cdot x_{ijk} \tag{2.12}$$

$$\sum_{i=0}^{n} \sum_{k=1}^{K} x_{ijk} = 1 \quad \forall j = 1 \tag{2.13}$$

$$\sum_{i=0}^{n} x_{ijk} - \sum_{i=0}^{n} x_{jik} = 0 \tag{2.14}$$

$$\sum_{i=1}^{n} x_{ojk} \leq 1 \tag{2.15}$$

$$\sum_{i=0}^{n} y_{ji} - \sum_{i=1}^{n} y_{ij} = p_j \quad \forall j \neq 0 \tag{2.16}$$

$$\sum_{i=0}^{n} z_{ij} - \sum_{i=1}^{n} z_{ji} = d_j \quad \forall j \neq 0 \tag{2.17}$$

$$y_{ij} + z_{ij} \leq cap \sum_{k=1}^{K} x_{ijk} \tag{2.18}$$

$$ts_i \leq t_{ik}^{a} \leq te_i \tag{2.19}$$

$$t_{ik}^{a} + \tau_i + T_{ij} - M(1 - x_{ijk}) \leq t_{jk}^{a} \tag{2.20}$$

$$\sum_{i=0}^{n} \sum_{j=0}^{n} c_{ij} x_{ijk} \leq L \tag{2.21}$$

$$x_{ijk} \in \{0,1\};\ y_{ij} \geq 0;\ z_{ij} \geq 0 \tag{2.22}$$

where constraints 2.13 and 2.14 ensure customer visit specifications, Equation 2.15 gives the limited number of vehicles, Equations 2.16 and 2.17 are the flow conservation equations for pickup and delivery demands, Equation 2.18 shows that demands will be transported using arcs included in the solution, Equations 2.19 and 2.20 are the time-window constraints, Equation 2.21 is the maximum distance constraint with upper limit L on the total load transported, and Equation 2.22 specifies the nature of decision variables. Remark: taking into account some premises to use MATLAB® tools, this problem was reformulated to continuous problem with $x_{ijk} \in [0,1]$, and additional penalty components are added in the criteria function. The testing example was selected from Mingyong and Erbato (2010). The calculation time was strictly related to quality of the starting point (it must be admissible what creates some difficulty requiring some intelligent support to the selection of this point).

Example: K = 3; N = 8; cap = 8 tons

Vehicle 1: k = 1;{depot-3-5-1-depot}; C = [0 75 200 40;75 0 50 40;200 50 0 200;40 40 50 0]
p = [0 4.5 1.5 2]'; d = [0 2 3 3]'; NI = 10^8; L = 400; tau = [1 1 1]'; ts = [1 3 6]'; te = [3 5 7]'; Xopt = [0 1 0 0; 0 0 1 0;0 0 0 1; 1 0 0 0]
Vehicle 2: k = 2 {depot-8-7-2-depot}; C = [0 80 160 60;80 0 100 75;160 100 0 75;60 75 75 0];
p = [0 3 1.5 1]'; d = [0 3 2.5 1.5]'; NI = 10^8; L = 400;
tau = [0.8 1 0.5]'; ts = [1.5 4 5]'; te = [4 6 7]';
Xopt = [0 1 0 0; 0 0 1 0;0 0 0 1; 1 0 0 0]
Vehicle 3: k = 3 {depot-6-4-depot}; C = [0 100 90;100 0 75; 90 75 0]; tau = [1.5 1]';
ts = [2 4]'; te = [5 7]'; p = [0 4 2]'; d = [0 4 3]'; NI = 10^8;L = 400;
Xopt = [0 1 0; 0 0 1; 1 0 0]

Optimization terminated: first-order optimality measure less than options. TolFun and maximum constraint violation is less than optionsTolCon.

VRP with Time Windows (VRP-TW)

The VRP with time windows (VRP-TW) is an extension of well-known classical VRPs, and it is often encountered in decisions dedicated to distributions of goods. Let us assume that there is a directed graph without self-loops $G = (V, L)$, where V is a set of nodes including depot node V_1, and L is a set of links with traveling costs c_{ij} but K is the number of vehicles with identical limited capacity Cap. Each node has defined a service time τ_i, and a specified time window $TW_i = [e_i, l_i]$ for $i = 2, 3,\ldots, n + 1$ within each customer must be serviced by vehicle. Lower bound e_i and upper bound l_i of TW_i determine, respectively, earliest arrival time and latest arrival time. Each customer i required a delivery operation of a certain amount of goods d_i, and service at customer must begin within time window TW_i. A vehicle is not allowed to start service at i customer after l_i. If a vehicle arrives before the lower limit e_i of the time window, it must wait w_i and increase a cost of route. Every customer i must be visited exactly once by vehicle k, and all vehicles should start and end their route at the depot node. Each vehicle serviced only one route in the network. The global customer demand serviced vehicle in one route must be smaller or equal to vehicle capacity Cap. In this problem, there are two types of decision variables. First, principle decision variable $x_{ijk} = \{0,1\}$ with 1 if vehicle k is traveling through link (i, j) and 0 otherwise. Second is the arrival time t_i at customer i, which denotes the beginning of service by vehicle. The objective is to minimize the sum of cost (mostly time) and the number of vehicles needed to supply all customers in their required time windows:

$$PO_{\min} \quad X,t \quad PI = \sum_{i=1}^{n+1} \sum_{j=1}^{n+1} \sum_{k=1}^{K} c_{ij} \cdot x_{ijk}$$

$$\sum_{k=1}^{K} \sum_{j=1}^{n+1} x_{ijk} = K \text{ for } i = 1 \tag{2.23}$$

$$\sum_{i=1}^{n+1} \sum_{k=1}^{K} x_{ijk} = 1 \text{ for } j = 2,\ldots,n+1 \tag{2.24}$$

$$\sum_{i=1}^{n+1} x_{ijk} - \sum_{i=1}^{n+1} x_{ijk} = 0 \text{ for } j = 1,\ldots,n+1;\ k = 1,\ldots,K \tag{2.25}$$

$$\sum_{i=1}^{n+1} d_i \cdot \sum_{j=1, j \neq i}^{n+1} x_{ijk} \leq Cap \text{ for } \forall k \in [1, \ldots K] \tag{2.26}$$

$$e_i \leq t_i \leq l_i \text{ for } i,j = 1,\ldots, n+1; k = 1,\ldots, K \tag{2.27}$$

$$t_i + x_{ijk}(t_{ij} + \tau_i + w_i) \leq t_j \text{ for } i,j = 1,\ldots, n+1; i \neq j; k = 1,\ldots, K \tag{2.28}$$

where constraints 2.23 ensure the starting point for routes at depot; Equations 2.24 and 2.25 are the customer visit specifications, Equation 2.26 are the vehicle capacity constraints, and Equations 2.27 and 2.28 are the time-window constraints. This problem was solved using Matlog Toolbox coded in MATLAB language. The sample is shown in Figure 2.2—the street network of Cracow city center. There are 158 edges and 62 nodes, but the number of nodes with demand is less: 46 (Figure 2.2).

By default, the depot node is set as node $i = 1$ (sets in square in Figure 2.2). The vehicle capacity is Cap = 6000 kg; unloading time in each node is τ = 15 minutes. Time windows were randomly created within time interval from 7 a.m. to 6 p.m. And at least, every arc has defined cost C_{ij} assumed as the traveling time from one node to another. The traveling cost is simply calculated by division of the distances matrix and the speed matrix. The obtained results are as follows: 4 numbers of routes, and total costs (time) is 12,597,493 hours. It is well known that real-life sample is composed of many dynamic processes, for example, constantly speed changing at an urban network. Dispatching control for routing solution, presented in the previous section, would be most suitable in these cases, but the control actions should start when there are any changes in the network, and these changes cause the deviation in time schedule in the route. Hence, it is needed to know which speed modification causes the deviation in time schedule, and which changes cause changes in the sequence of nodes in the route. For this, some computational experiments were

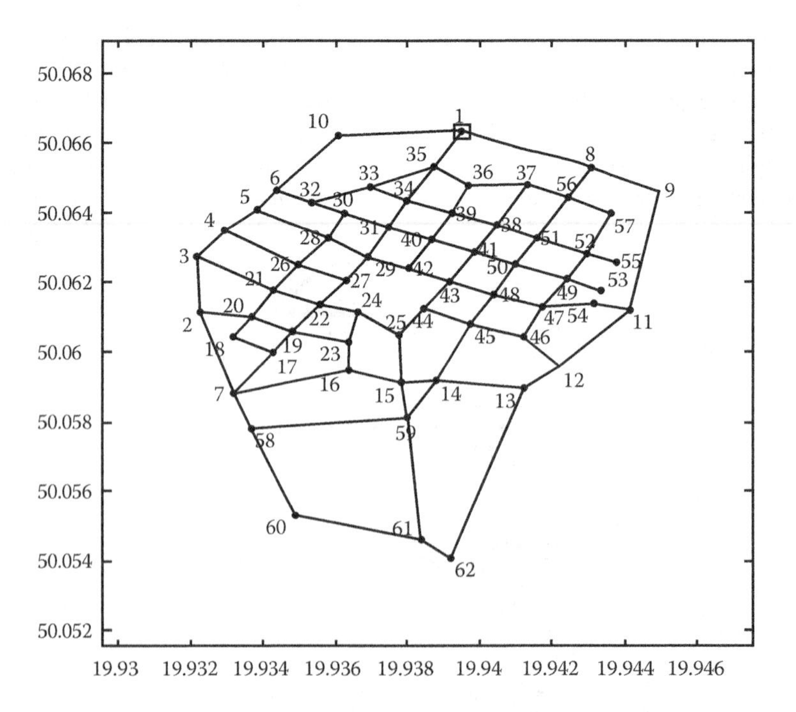

FIGURE 2.2 Network used to solve example of VRPTW.

generated to get information about solution sensitivity for speed changing in the network. Table 2.1 presents the results of the experiment—computed 30 instances of original solution obtained previously. The original solution was modified in three cases (see Figures 2.3 through 2.5): first, changing speed at one randomly selected edge; second, changing speed value at two randomly selected edges; and third, the change was in five randomly selected edges. In every case, the value of change was ± 10% with 2% step.

TABLE 2.1
Computational Results for Sensitivity of VRPTW Solution

No.	Value of Change	No. of Routes	Total Costs	New Solution? ('1' for 'Yes', '0' for 'No')	% Change in Total Cost
			Speed Change in 1 Edge		
1	−10%	4	12,596,347	0	−0.104%
2	−8%	4	12,596,576	0	−0.084%
3	−6%	4	12,596,805	0	−0.063%
4	−4%	4	12,597,034	0	−0.042%
5	−2%	4	12,597,264	0	−0.021%
6	0%	4	12,597,493	0	0.000%
7	2%	4	12,597,722	0	0.021%
8	4%	4	12,597,896	0	0.037%
9	6%	4	12,559,855	1	−3.429%
10	8%	4	12,560,008	1	−3.416%
11	10%	4	12,560,161	1	−3.402%
			Speed Change in 2 Edges		
1	−10%	5	12,599,069	1	0.144%
2	−8%	4	12,596,316	0	−0.107%
3	−6%	5	12,599,657	1	0.197%
4	−4%	4	12,596,904	0	−0.054%
5	−2%	5	12,600,245	1	0.251%
6	0%	4	12,597,493	0	0.000%
7	2%	5	12,600,834	1	0.304%
8	4%	4	12,598,026	0	0.049%
9	6%	5	12,601,422	1	0.358%
10	8%	4	12,560,268	1	−3.392%
11	10%	5	12,602,011	1	0.412%
			Speed Change in 5 Edges		
1	−10%	4	12,639,661	1	0.335%
2	−8%	4	12,555,906	1	−0.330%
3	−6%	4	12,645,362	1	0.380%
4	−4%	4	12,561,269	1	−0.288%
5	−2%	4	12,597,083	1	−0.003%
6	0%	4	12,597,493	0	0.000%
7	2%	5	12,594,766	1	−0.022%
8	4%	5	12,543,951	1	−0.425%
9	6%	5	12,649,094	1	0.410%
10	8%	5	12,525,320	1	−0.573%
11	10%	4	12,596,356	1	−0.009%

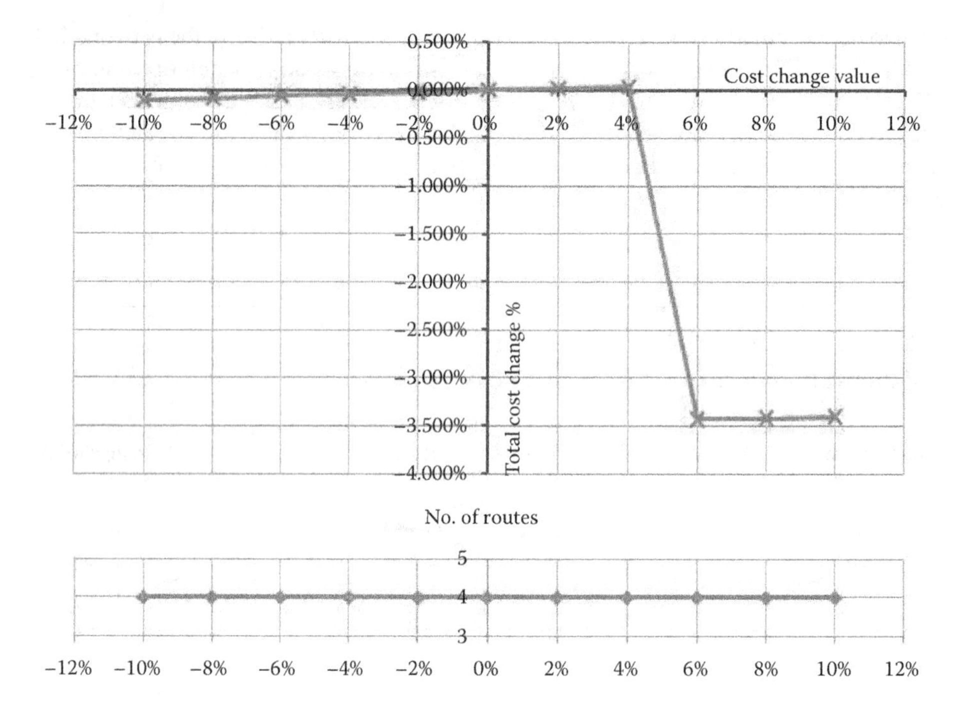

FIGURE 2.3 Results of cost changes in 1 edge.

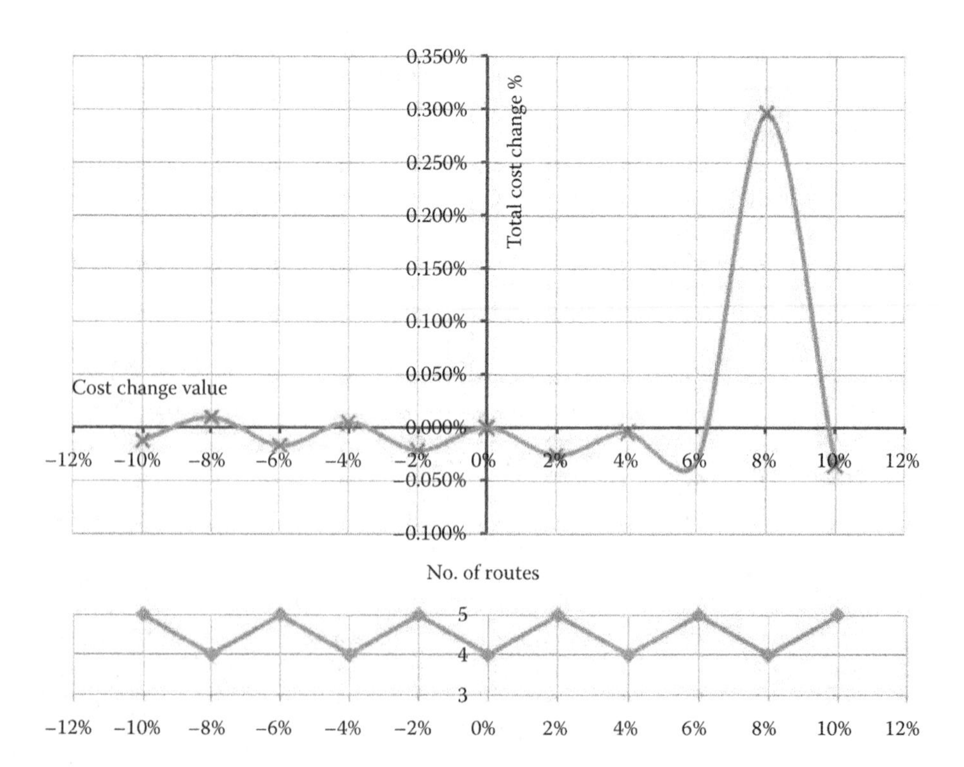

FIGURE 2.4 Results of cost changes in 2 edges.

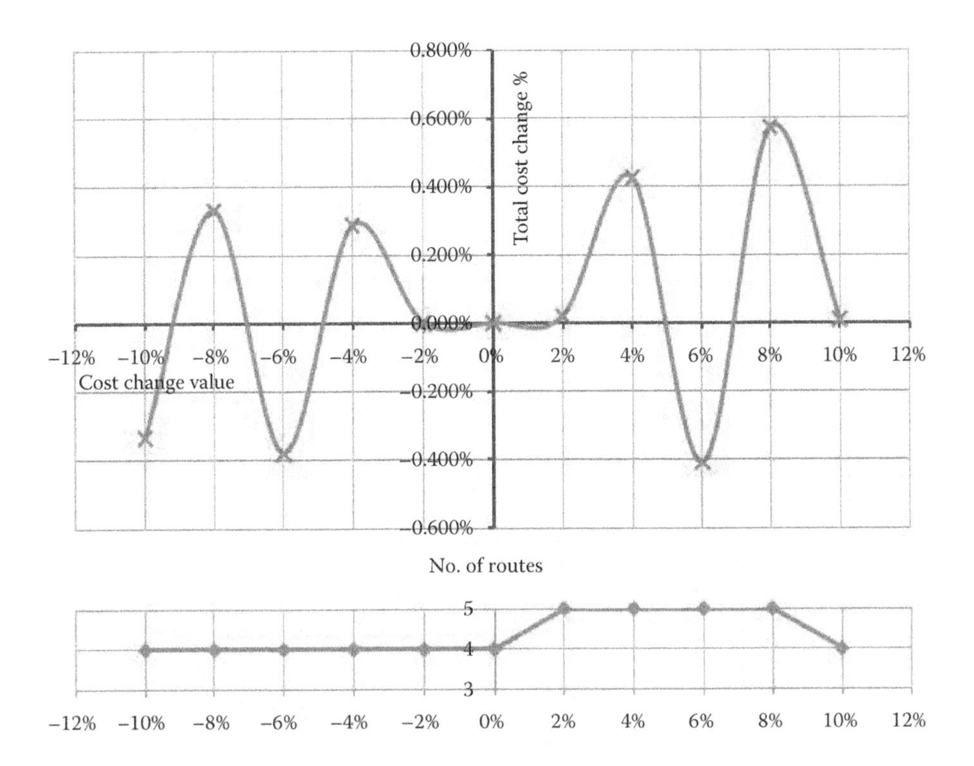

FIGURE 2.5 Results of cost changes in 5 edges.

In the first case, the original solution was kept to a certain value of change, and after exceeding it, the solution was modified into a new one. The solution changes almost in every step when speed was changed at least in the two edges in the network. Results show that the VRP-TW is very sensitive for cost changing in the network. In some instances, during decreasing/increasing speed in edges, the time schedule of routes changed by creating a deviation in arrival time t_i at customers, even when the computed solution kept the same sequence of node in routes. In this case, to compensate those deviations, we can use the DISCON method. The example route is presented in Table 2.2.

ACLS Logistics System Control Layer

A very important new functional element at the bottom direct control layer concerns the full integration of the tasks of intelligent supervision with the intelligent adaptive control actions (see Figure 2.6) realized by the DISCON control method (Adamski 2007, 2011). DISCON optimal dispatching control actions for VRP-TW solutions (t_0^*, S^*) compensate off-reference trajectory deviations increasing the robustness of the actual obligatory trajectory.

Dispatching control actions dynamically evolving in time and space are integrated in the DISCON method (Adamski 1993, 1998, 2007a,b, 2011) in an optimal dynamic control strategy resulting from the minimization of some selected measures of service standards, for example, off-reference routing trajectory deviations. A wide spectrum of DISCON control tasks (punctuality, regularity, synchronizing priority control) calls for a multicriteria integrated approach. In Adamski (1993, 1998, 1999, 2002, 2007), the 1-D and 2-D (primal and dual) dynamic control plant representations have been developed and illustrated by a family of single criteria optimal control DISCON solutions of dead-beat, LQ, LQG type. In Figure 2.7, DISCON LQG stabilizing control was implemented for three vehicles in the first example.

TABLE 2.2
Route Example with Time Deviation Caused by Speed Changes

						Original Route						
Loc	1	33	51	42	27	28	30	29	39	38	50	1
Arrive	0	14.223	14.4838	14.7442	15	15.2553	15.5073	15.7938	16.0579	16.3099	16.5659	16.8654
Wait	0	0	0	0	0	0	0	0	0	0	0	0
Start	14.1858	14.223	14.4838	14.7442	15	15.2553	15.5073	15.7938	16.0579	16.3099	16.5659	16.8654
Depart	14.1858	14.473	14.7338	14.9942	15.25	15.5053	15.7573	16.0438	16.3079	16.5599	16.8159	16.8654
						Changed Route						
Loc	1	33	51	42	27	28	30	29	39	38	50	1
Arrive	0	14.223	14.4838	14.7442	15.1841	15.4394	15.6914	15.9779	16.242	16.494	16.75	17.0495
Wait	0	0	0	0	0	0	0	0	0	0	0	0
Start	14.1858	14.223	14.4838	14.7442	15.1841	15.4394	15.6914	15.9779	16.242	16.494	16.75	17.0495
Depart	14.1858	14.473	14.7338	14.9942	15.4341	15.6894	15.9414	16.2279	16.492	16.744	17	17.0495

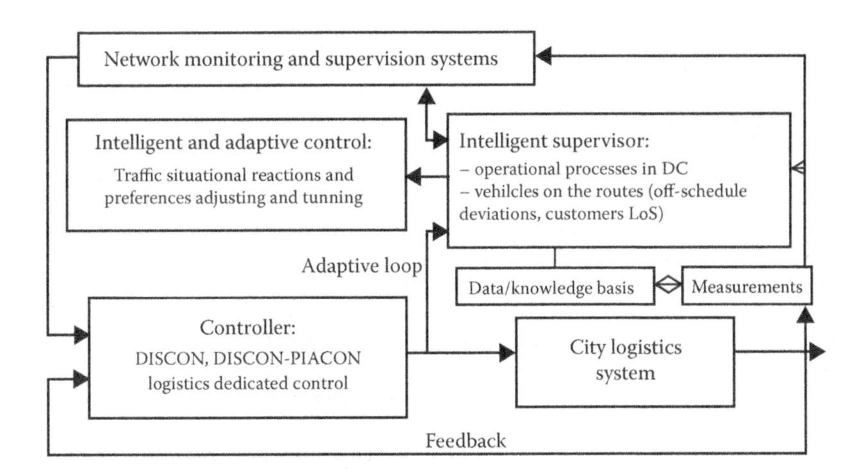

FIGURE 2.6 Intelligent supervision and DISCON control realized in bottom direct control layer.

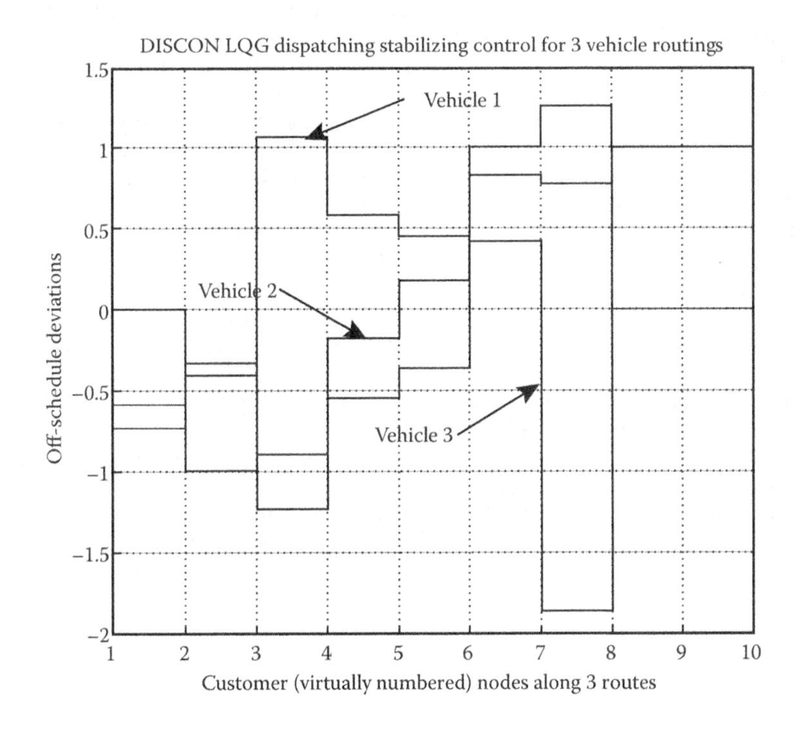

FIGURE 2.7 DISCON stabilizing control for vehicle routings in the first example.

CONCLUSIONS

In this paper, the ACLS embedded city logistics option of the HITS system is presented and illustrated by implementation of advanced enabling technologies and intelligent operational computer tools like the DISCON control method. The new ACLS dedicated functionalities of HITS system management and coordination, adaptation, scheduling, monitoring, and surveillance and direct control layers are presented and illustrated by representative practical examples. The special emphasis in the functional HITS system structure is placed on the real-time vehicle routing and scheduling problems supported by dispatching control methods.

REFERENCES

Adamski A. Hierarchical Integrated Intelligent Logistics System Platform 26th Mini EURO Conference, Intelligent Decision making in Transport and Logistics, Procedia Social and Behavioral Sciences, vol. 20, pp. 1004–1016, 2011, Elsevier, New York.

Adamski A. IILS: Integrated intelligent logistics systems. Krakow University of Technology Logistics Conference, pp. 25–38, 2006b.

Adamski A. Integrated traffic management, surveillance and control systems in urban areas. *TiBT*$_{06}$ *Transport Telematics and Safety*, 104–112, 2006a.

Adamski A. Integrated transportation and logistics systems. ITS ILS'07: Transportation and Logistics Integrated Systems: International Conference: Cracow, October 11–12, 2007. AGH University of Science and Technology. Faculty of Electric, Automatics, Informatics and Electronics. Institute of Automatics, pp. 15–28, 2007c.

Adamski A. Integrated transportation systems. *Modeling and Management in Transportation EURO WG on Transportation*, 1, 21–34, 1999.

Adamski A. Intelligent anticipative dispatching system of public transport. In *Transactions on Transport Systems Telematics: Monograph*, ed. Jan Piecha. Gliwice, Poland: Wydawnictwo Politechniki Śląskiej, pp. 70–75, 2007d.

Adamski A. Intelligent integrated transportation systems. In 9th Meeting of the EURO WG Intermodality, Sustainability and ITS, Bari, pp. 565–570, 2002b.

Adamski A. ITS: Integrated transportation systems. *Archives of Transport Quarterly, Polish Academy of Science*, 14(2), 5–22, 2002a.

Adamski A. ITS: Management, surveillance and control. Monograph: AGH University Scientific Publications, 2003, Krakow.

Adamski A. Optimal adaptive dispatching control in an integrated public transport management system. In 2nd EURO WG Conf. INRETS ACTES, pp. 917–938, 1993.

Adamski A. Robust dispatching control in public transport—DISCON-R. In *Transactions on Transport Systems Telematics: Monograph*, ed. Jan Piecha. Gliwice, Poland: Wydawnictwo Politechniki Śląskiej, pp. 56–69, 2007a.

Adamski A. and S. Habdank-Wojewodzki. The integrated intelligent logistic systems—ILS. In *Transactions on Transport Systems Telematics: Monograph*, ed. Jan Piecha. Gliwice, Poland: Wydawnictwo Politechniki Śląskiej, pp. 39–45, 2007b.

Adamski A. and A. Turnau. Simulation support tool for real-time dispatching control in public transport. *Transportation Research A*, 32(2), 73–87, 1998.

Leiths D. and A. Adamski. Situational analysis in real-time traffic systems. *Procedia Social and Behavioral Sciences*, vol. 20, 506–513, 2011, Elsevier.

Mingyong L. and C. Erbato. An improved differential evolution algorithm for vehicle routing problem with simultaneous pickups and deliveries and time windows. *Engineering Applications of AI*, vol. 23, 188–195, 2010.

3 HITS
Hierarchical Intelligent Transportation Systems

Andrzej Adamski

CONTENTS

INTRODUCTION

In this paper, the hierarchical intelligent transportation system (HITS) multinetwork, multilayer, multicriteria original system platform is proposed. This platform is a crucial step enabling the development of practically efficient ITS proposals dedicated to very complex dynamic, stochastic, and behavioral interactions existing in transportation processes (Adamski 1999, 2002a,b, 2003, 2006a,b,c, 2011). The real-time system-wide identification, intelligent diagnosis, estimation, and prediction of these interactions condition the efficiency and productivity of the crucial ITS integrated functionalities (e.g., management, scheduling, surveillance, monitoring, and direct control actions). In this context, the deep understanding of the system processes governing mechanisms and behavioral patterns as well as interactions from system operational environment (SupNet: Super-Network) creates the necessary conditions for exploration of these ITS integrated functionalities. These functionalities are realized in an integrated way in the hierarchical multilayer functional structure and are characterized by different task specifications (e.g., decision time horizons, types of process representations and optimization problems, reaction times, etc.). These layer/task specifications condition both efficiency and productivity of individual system layers as well as integrated operation of the whole system. The proposed system platform and professional ITS development methodology enable first of all the practical realization of above system-wide specifications. This platform from one side is embedded in a nowadays available advanced sensing, information, computer, and communication enabling technologies supported by capabilities of vehicle platforms (e.g., vehicle navigation and location, v-v, v-i communication, vehicle probe, Advanced Driver Assistance Systems [ADAS], etc.). Additionally, this platform is supported by professional exploration of advanced integration (cooperative complex system approach with multinetwork/multilayer/multilevel/multiuser/multioption/multimode/multiservice/multiobjective specifications) and intelligence (recognizing, diagnosing, and understanding complex interactions and behavioral patterns, detection of abnormal traffic events, and opportunities for very efficient system actions) tools. In the paper, the fundamental HITS multilayer, multilevel functional system structure is described. The functionalities of HITS management and coordination, adaptation, scheduling, monitoring, and surveillance and direct control layers are presented and illustrated by representative practical examples.

HITS PLATFORM FUNCTIONAL PREMISES AND PRINCIPLES

The development of urbanization processes in our cities and the increase in socioeconomic activity of urban population lead to the rapid rise in transportation demand. Such a situation has caused intolerable increase in traffic volumes in urban networks and consequently has led to heavy congestion. Well-known symptoms of congestion that became visible first of all as a decrease in *travel safety*, *travel efficiency* (delays, stops, queues, longer trip times), and *travel economy* (excessive fuel and energy consumption) are accompanied by essential increase in negative environment impacts from transport (*noise, air pollution, vibrations*) and transport fatigue. It is important that all these congestion effects occur mainly in central parts of cities, where many trip destinations are concentrated, and due to compact building, there are difficult natural cleaning conditions for toxic air pollution ingredients. The observed energy and ecological economized requirement has inserted the *social and energy context* into the transportation problems. It seems that presently a rational and economic solution to these problems can be found first of all in *the most effective and intelligent use of the existing transportation systems*, for example, *in the form of advanced HITSs*. The problem-oriented effective integrated traffic management, monitoring, and surveillance and control system activities are the cornerstone of successful minimization of enormous transportation costs expressed roughly in terms of perceptible transport service standards, energy consumption, and environmental impacts. To achieve essential benefits, however, it is necessary to integrate the potential of existing *new AMI intelligent and ITS technologies with new on-board vehicle capabilities* in hierarchical multilayer system platform HITS with advanced intelligent layers offering completely new traffic control philosophy, which is based on truly adaptive, intelligent, and multicriteria traffic control and surveillance methods.

HITS Premises and Development Principles

The main integration, technological, and intelligence premises for this HITS platform proposal dedicated to nowadays transport systems are as follows:

- *Integration premises*: Transportation system features: large-scale, complex spatiotemporal phenomena multicommodity flows, O-D random fields, network-wide interactions, behavioral feedback reactions, multigoal decisions
- *Technological HITS enabling premises*: Automation, wireless communication, progress in computer networks, navigation systems, real-time optimization and decision-making tools, multisensor technologies, on-board vehicle platforms
- *Intelligence premises*:
  - Transportation system general features: high uncertainty, behavioral anisotropy aspects, traffic complex interactions, fast nonlinear dynamics, stochastic phenomena and disturbances, wide range of unpredictable phenomena (illegal parking, road works, incidents, weather conditions); the challenging operational system requirements for real-time monitoring, intelligent supervision, management, and control system-wide activities
  - Integration of different technologies (high level of automation of transportation systems)

- Traffic control and management problem challenges: ill-structured, real-time data and knowledge dependent, traffic situational dedicated robust multicriteria network problems
- Large hardware and software real-time systems (operation, monitoring, maintenance)
- Fast progress in advanced intelligent technological tools and methods

The transportation systems belong to the family of artificial systems developed by humans in order to satisfy their demands. These systems are in a natural way embedded in operational environment, that is, an interconnected complex of physical or virtual networks called SupNet (Adamski 2002a,b, 2003, 2006a,b,c). The SupNet through a wide spectrum of complex interactions creates the key operational dynamic and behavioral environment for transportation processes. The real-time identification, estimation, and prediction of these interactions for ITS management, surveillance, and control purposes are crucial problems determining the efficiency and productivity of these systems. In this context, the necessary condition for management and control of such systems requires both deep understanding of the features and behavior of these systems as well as the general understanding of operational environment (SupNet) that essentially influences the operation of these systems. Only when this condition is fulfilled will we be able to rationally influence the operation of these systems both in terms of system primary goals (e.g., minimization of negative primary goal impacts) and secondary goals of system operational general influences (e.g., minimization of environmental impacts). The proposals of approaches and development methodologies for ITS (Intelligent Integrated Transportation Systems) were presented in Adamski (2002a,b, 2003, 2006a,b,c, 2011). In a professional approach to HITS platform, the following general principles formulated in Adamski (2003) are obligatory:

TRIAD I: integration–information–intelligence
TRIAD E: efficiency–economy–environment
TRIAD U: up-to-date–unified–user-oriented

Integration Premises

Integrated operation offers essential benefits, that is, system-wide synchronized response to a full range of mobility needs with opportunity to bridge multimodal demand management and real-time supervision and control over different time horizons. However, there are several natural and formal integration premises.

1. System features: large-scale systems, complex phenomena (uncertainty, randomness, complex interactions, structural instabilities, very fast dynamics, essential nonlinearities, human behavioral anisotropy).
2. Complex decision-making environment with a broad spectrum of multigoal decision-making tasks, existent beneficial synergic effects, for example, "control by opportunity modes," high requirements in terms of decision robust features, and multitime horizon compatibility.
3. Nowadays technological premises: information technologies with communication revolution and computer technology advances create the formal base for integration.

In Figure 3.1, the integration premises are used in the solution of the complex system efficiency problem. Most of decision problems in ITSs can be in a natural way formulated in terms of elements of multilayer multilevel space of integration (MMSI) (Adamski 2003, 2006a,b,c). In this space, the system-wide integrated tasks are formulated by knowledge-based decision support system (DSS) polyoptimal integrated adaptive control (PIACON) in terms of available resources, control, supervision, scheduling, and management actions (tools) distributed among different hierarchical system levels and layers.

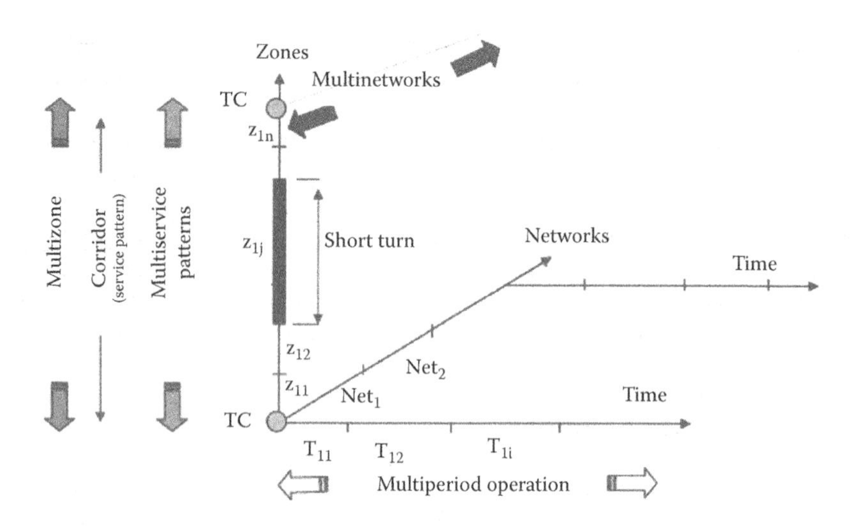

FIGURE 3.1 Illustration of integration premises.

Intelligence Premises

The existence of nonpredictable operational incidents demands intelligent detection, identification, and system-wide reaction in a very short response time. The existence of very beneficial synergic effects requires the intelligent recognition of symptoms and localization of actions enabling the exploration of synergic effects and necessitates integration of many technologies, platforms, subsystems, and users. Ambient intelligence system environment provides intelligent link between travellers, vehicles, and infrastructure. The application of intelligent management, supervision, and control systems with knowledge-intensive expert systems supporting tools enables us to understand the process evolution in 2-D {time × space} scene. The traffic control problems are usually ill-structured and are not amenable to be formulated and solved by purely algorithmic techniques. Typical applications of intelligence concern supervisory/monitoring pattern detection, recognition, and prediction in incomplete knowledge situations, for example, user behavioral patterns and realization of adaptive tuning procedures based on transition state analysis and heuristic rules.

Up-to-Date and Unified Premises

Technological premises stimulate the development of multilayer and multilevel systems with distributed automatic actuators, for example, cyber-cars. The technical architecture of the system should be to a high degree unified with provision of open standards and mechanisms to share data between systems and users. According to the integration principle, the interactions generated by **SNet** environment require the integrated representation in ITSs. This integration may concern several levels, for example, integration of structures of several interactive component networks $\mathbf{Net}_i$ (transport, logistic communication, financial, information, business, distribution) in super net $\mathbf{SNet} = \{\mathbf{Net}_i\}_1^n$. The integration of **SNet** functional structure and direct operational and decisional environment offers the greatest capabilities for transport policy objective realization. The 3-D (networks × time × space) space of integration makes it possible to represent most importantly from the ITSs perspective internetwork interactions generated by heterogeneous SupNet components. These interactions are frequently primary sources of social choices, traffic incidents, and nonrecurrent congestion, that is, nonpredictable uncertain and disturbance events. The internetwork integration may concern many functional and structural aspects of ITSs, for example, multiperiod, multizone, multiservice pattern operation, and behavioral aspects directly influencing demand patterns. The HITS platform–related ITS development methodology is presented in Table 3.1.

TABLE 3.1
HITS Platform Methodology for ITS Development

Complex System →	**Principles Triad: I–E–U** →	**System Efficiency Problem** ↓
Complex System-wide efficiency: Measures of efficiency problem specifications, system interactions importance	**Complex Decision-Making Tasks:** Many interconnected DM heterogeneous tasks, uncertainty and behavioral aspects	**Decomposition:** **Layers:** control, monitoring surveillance, adaptive optimization (scheduling), management, and coordination **Levels:** infrastructure, subsystems, networks, SupNet
System-Wide Integration	**Problems:** ←	**Representation** ↓ ←
Problems and interactions: multilayer architecture; integration options: subsystems, info, tools, functions; times of layer/task activation	• criteria • constraints • structure • properties	**MMSI:** multilayer multilevel space of integration **DSS-PIACON:** task identifications, synchronization, coordination, harmonization

HITS PLATFORM OPTIONS

The main premises of the traffic management, surveillance, and control (TMSC) system proposal are given in Figure 3.2.

The control plant features, that is, the wide area of operation, multiform of possible actions, and essential interactions with other systems, suggest the integrated approach; the broad spectrum of conflicting goal suggests the multicriteria approach; whereas the high level of uncertainty and ambiguity suggests the intelligent approach; and finally, the fast, dynamic, essential plant instabilities and incomplete and inaccurate data suggest the adaptive robust approach. Recently, available technologies and wide-area intelligent network analysis and management tools with high-quality real-time information supply from integrated data and knowledge bases updated by traffic data sources (like video-detectors, AID, AVL/GPS systems supplemented by meteo, pollutant concentration data) created a new platform for TMSC systems development. In this context, the planning processes have to be integrated with real-time operation of the transportation system that gives the continuous real-time information feedback about the system operation and enables searching through practically a continuum of potential alternatives to match options against policies and objectives. The management and control actions have to be integrated with data-rich real-time environment and advanced knowledge, methods, and tool bases. Computer DSS should be equipped with high proportions of automated monitoring, surveillance, and intelligent management functions to ensure real-time beneficial actions in response to online detected traffic, safety, and environmental situations. Very high potential of integrated actions results both from demand reduction in time–space context (more rational choices, reduced wasted times) and from more efficient multicriteria use of the existing resources (domination of system-level NWP optimal solutions, real-time professional reaction time on sudden/unforeseen traffic events, anticipatory and control by opportunity synergic actions) on the supply side.

Functionality of HITS Platform Layers

The general problem of the HITS upper layers is to cover at the least service-oriented or system-oriented generalized costs a set of multitasks by choosing in a network a family of flows satisfying the macro-user level of service (LoS) preferences and system operating constraints. The first general problem concerns the recognition and identification of network 2-D traffic macrogenerators

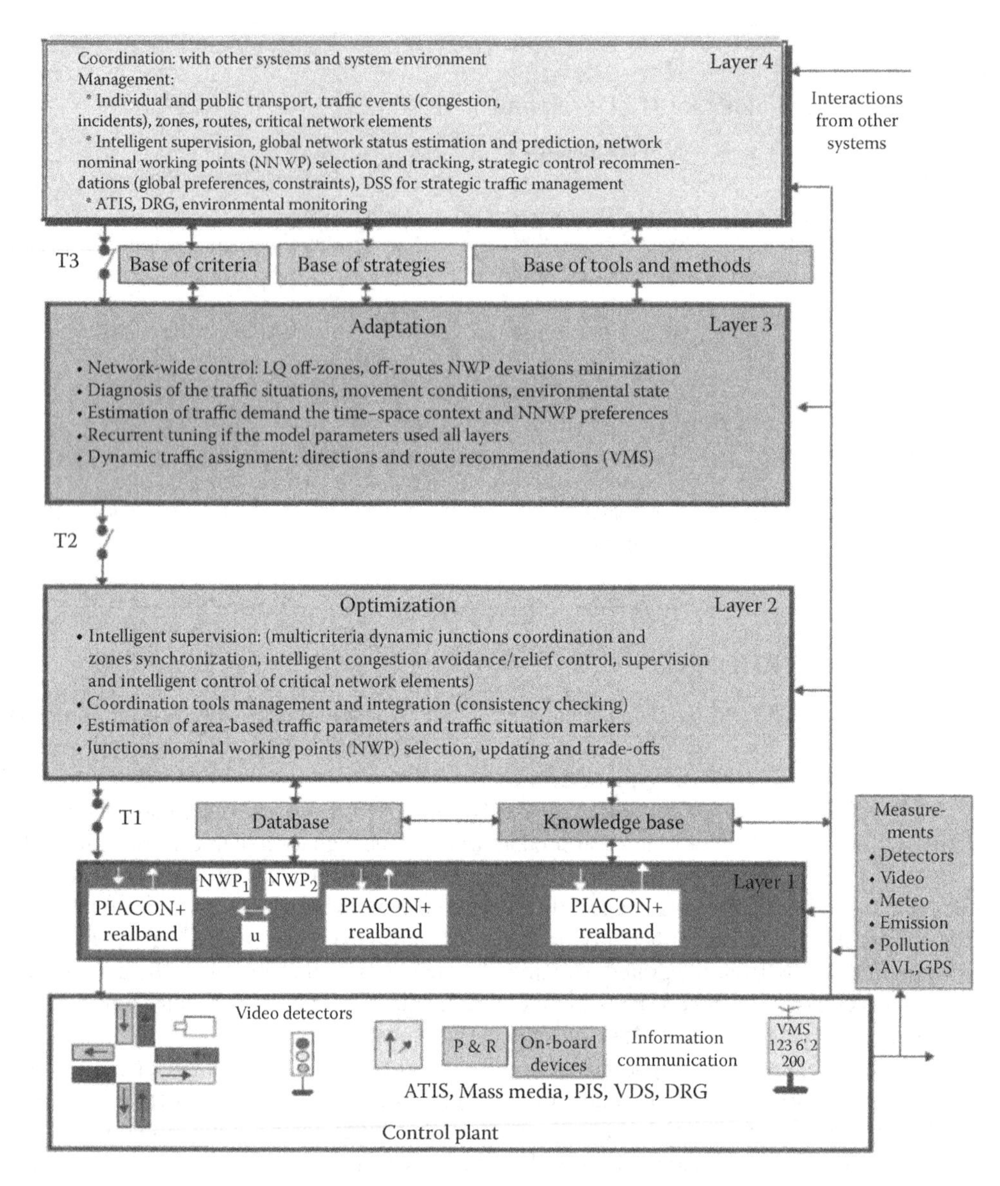

FIGURE 3.2 Multilayer multilevel hierarchical traffic management, surveillance, and control system.

and attractors that conditioned the existence of macrotraffic demand. In general, this is a complex dynamic and uncertain problem in transport area (the results of complex spatial interactions of human activities at different locations, SupNet interactions, urban area operational characteristics, behavioral patterns, unpredictable events, etc.). The second general problem appears in intergenerator/attractor flow integration by user-behavioral/system decision-making integrators. Let O_i and D_j represent the zones generators and attractors, respectively, and $f(c_{ij})$ the decreasing function of c_{ij}, which is the unit cost of travel between these zones (e.g., $f(c_{ij}) = \phi(\beta)^{-cij}$ with β a parameter). Assuming the probability of trip from zone i to zone j in the form $p_{ij} = O_i D_j f(c_{ij})$, we can formulate the maximum likelihood estimation (MLE) problem of the trip distribution parameters O_i, D_j, and those associated with $f(c_{ij})$ (i.e., β, c_{ij} parameters) based on a sample of observed trips $X = \{x_{ij}\}$. The maximum likelihood estimators of p_{ij} parameters are the solutions of the likelihood function $L = \prod_i \prod_j p_{ij}^{xij}$

maximization problem under $\sum_{i,j} p_{ij} = 1$ constraints. In the analyzed case, the adequate maximization problem may be formulated as follows: $\mathrm{PO}_{\max} T_{ij} = p_{ij} T Q(T) = \prod_i \prod_j (1/T_{ij})^{T_{ij}} T^T$ with the following constraints:

$$\sum_j T_{ij} = \sum_j x_{ij} = O_i; \sum_i T_{ij} = \sum_i x_{ij} = D_j; \sum_i \sum_j c_{ij} T_{ij} = \sum_i \sum_j c_{ij} x_{ij} = C;$$

$$T = \sum_i \sum_j T_{ij} \tag{3.1}$$

where the constraints represent the sample compatible level of generators/attractors and the total traveling cost. It can be shown that equivalently this problem is an entropy maximization problem with $Q_2(T) = -\sum_i \sum_j T_{ij} \ln(T_{ij})$ under constraints 3.1 and $f(c_{ij}) = \exp(-\beta c_{ij})$. In the case of fixed parameter β, the associated objective function equals $Q_2^*(T) = Q_2(T) - \beta \sum_i \sum_j c_{ij} T_{ij}$. The entropy maximizing the spatial interaction model is particularly suitable for the trip distribution problem where we want to find trip matrix $\{T_{ij}\}$ with the greatest number of assignments leading to this matrix under constraint 3.1. In practice, the additional objectives in transportation processes can supplement the entropy objective (e.g., total cost). The most existing, for example, BMW elastic demand, gravity model formulations using, in general, the costs, in the form of $Q_2 = \sum_i \sum_j \int_0^{Tij} g_{ij}(y)\,dy$ with $g_{ij}(y) = y - 1 - \ln(y)$ lead to the same solutions as entropy under the following constraints:

$$\sum_j T_{ij} = O_i; \sum_i T_{ij} = D_j; T_{ij} \ge 0 \tag{3.2}$$

The next generic problem appears in recognition and identification of the modal/mixed-modal split mesogenerators/attractors and their integration through adequate flows by using behavioral decision-making integrators. In consequence, the general network problem may be decomposed on different modes T_{ij}^m, $m \in M$ (individual traffic, public transport, logistics modes, etc., with possible interactions in the case of mixed modes). Formally this decomposition introduces in Equation 3.2, instead of $T_{ij} \ge 0$, the constraint $T_{ij} = \sum_m T_{ij}^m$ and corresponding modal entropy $Q_3(T) = -\sum_m \sum_i \sum_j T_{ij}^m \ln\left(T_{ij}^m\right)$. The equivalent multicriteria optimization problem with Equation 3.2 may be formulated as follows (Cea et al. 2008):

$$\mathrm{PO}_{\min} T_{ij}, T_{ij}^m Q(T) = \left\{ Q_1 = \sum_m \sum_i \sum_j T_{ij}^m c_{ij}^m, Q_2 = \sum_i \sum_j T_{ij} \ln(T_{ij} - 1), Q_3 = Q_3(T) - Q_2 \right\}$$

The decomposition process may be extended to nested and mixed modes. In the combined distribution-assignment problem, the additional criterion represents the total arc network costs $Q_4 = \sum_a C(f_a)$, but constraints are extended on the arc $a \in A$ and routes $r \in R_{ij}$ flow conservation laws; $f_a = \sum_r \delta_{ar} T_{ij}^r$ and $T_{ij} = \sum_r T_{ij}^r$. The adaptation layer in this context offers feedback mechanism to verify the adequacy of identified generators/attractors and behavioral integrators. This is a crucial feedback system functionality enabling the generation in the management and coordination layer of the decisions that are first of all dedicated to guarantee system stability (i.e., fulfill the assumptions of the Banach fixed-point theorem). In the context modern enabling technologies, the adaptation layer functionality evolution will concern not only on updating tasks but also on multimodal network-wide flow estimation and prediction and behavioral pattern recognition tasks. The generic problems in the 2-D scheduling and optimization layer concern the optimal space scheduling, that is, routing problems (e.g., optimal traffic assignments to routes, optimal bus route geometry) and time scheduling problems for task sequences resulting from route (i.e., 2-D scheduling in individual traffic with arc flow assignment, buses, and crew scheduling in public transport). The operational constraints are the rules of labor agreements (e.g., maximal/minimal worked time), obligatory LoS, operational specifications related to routes, and resulting sequence of tasks (e.g., flow conservation laws, available resources, LoS: time windows, number of node/arc visits, maximal lengths of routes, cooperative activities like transfers). The simple illustrative example of 2-D (space–time) network scheduling problem may be the shortest path problem with time windows, that is, the last costly path between two given vertices respecting the time intervals associated with the visited vertices. It may be a simple logistics routing problem with the visited customers but also an offline artery synchronizing problem with bandwidth-related time windows (Adamski 2003). The bottom layer generic problems concern the 2-D reference schedule stabilizing multicriteria control actions by real-time feedback mechanisms (e.g., minimization off-schedule bus/logistics vehicle deviations by real-time DISCON control method; Adamski 1989, 1993, 1995, 1996a,b, 1998a,b, 2005, 2007a,b) or multicriteria traffic situational reference (traffic situation markers) based traffic control by real-time PIACON multicriteria control method (Adamski 1998a,b, 2003, 2005, 2006a,b,c). The fundamental requirements concern the 2-D reference schedule validity; therefore, the control-scheduling layer feedback is supported by adaptation–scheduling–control layer feedback mechanisms for updating/recognition of valid behavioral selection patterns for the 2-D scheduling problem and guarantees their validity for control purposes. Similarly as before, the higher priority for these feedback mechanisms is first of all the guarantee of the whole system stability (i.e., fulfill the premises of implementation of the Banach fixed-point theorem).

APTS: Advanced Public Transport System Option

The provision of high-quality on-time and reliable service in public transport is a key operational problem that affects both travelers and city community. The additional APTS system features concern the existence of positive feedback mechanisms amplifying off-schedule deviations (e.g., bunching phenomenon) and possible negative influences on the service level of individual transport vehicles. The proposal of APTS option of ITS hierarchical multilayer management, surveillance, and control system was proposed in Adamski (2003). The embedded APTS control is split into the direct dispatching control layer realizing regulatory (schedule follow-up) control providing a stable steady-state operation of the lines and optimization layer realizing optimizing (transit schedule set point) control of the steady-state operated processes. In Adamski (1989, 1992, 1993, 1995, 1996a, 2003, 2005, 2006), the flexible 1-D and 2-D dynamic network control DISCON method is presented, which offers a wide spectrum of control capabilities, starting from control of interacting network elements with different levels of aggregation, in space (individual stops, route zones, common segments of different routes, overall routes), in vehicle population (one vehicle, groups of vehicles), or in time periods (rush hours, transient service periods during a day), and on multicriteria and robustness features of the dispatching control actions ending. The wide spectrum of 1-D and 2-D, aggregated/disaggregated bus line representations has been developed in the context of dispatching control (Adamski 1998b, 2003). The 2-D vehicle trajectory-based model (Adamski 2002a,b, 2005) consists of state vector $x_{i,j}$ with coordinates $i \in I$ vehicle index i and $j \in J$ index of possible service-oriented points along the route (e.g., traffic intersections). The input $u_{i,j}$ and output $y_{i,j}$ vectors are connected with states through real matrices and create state-space and output equations in the form

$$x_{i+1,j+1} = \sum_{l=0}^{2} [A_l x_{i+l(l-1)/2,j+l(2-l)} + B_l u_{i+l(l-1)/2,j+l(2-l)}];\ y_{i,j} = Cx_{i,j} \tag{3.3}$$

The state vector represents the measure of deviation between planned and actual vehicle trajectories supplemented by appropriate deviations in busload (Adamski 2003). In these models, dynamic and stochastic interactions between the traffic demand and service supply are adequately represented. The 1-D models can be treated as special cases of Equation 3.3:

- PCM: 1-D punctuality control models (Adamski 1993, 1995, 1996, 1998). $x_{j+1} = A_j^1 x_j + A_j^n x_{j-n} + B_j u_j + A_j^1 z_j$ with $x_k = \text{traj}_k - \text{traj}_k^s$ state vector (off-schedule deviation actual trajectories from scheduled).
- u_k are control variables both constrained by $x_k \in [xL_k, xU_k]$; $u_k \in [L_k, U_k]$.
- z_k are disturbances in travel and service times.
- $A_j^1 \in R^{m \times m}$.
- $B_j \in R^{m \times r}$ are the lower triangular diagonally dominated matrices with nonzero *elements:* $a_{ik} = \lambda_k \prod_{1=k+1,..,j} (1-\lambda_1)$.
- $b_{ik} = a_{ik}/\lambda_k$
- λ_k are the eigenvalues of A^1.
- $A_j^n \in R^{m \times m}$ is a matrix with the last nonzero column equal to $c_{im} = (-1)^{i+1} \prod_{1=1,..,i} (1-\lambda_1)$.

Dispatching control actions dynamically evolving in time and space are integrated in the DISCON method in an optimal dispatching control strategy resulting from the minimization of performance indexes to be service standard measures (e.g., off-schedule/off-regular headway deviations). The 1-D and 2-D dynamic control plant representations have been used for a family of single criteria optimal control DISCON solutions of dead-beat, LQ, LQG type with state and control constraints (Adamski 1989, 1992, 1993, 2003). Based on state–space representations, the properties (stability, controllability, observability) have been investigated (Adamski 2007). The dispatching control problem may be formulated as follows:

$$\text{PO}_{\min} J_{T-j} = x_T Q_T x_T' + \sum_k \left[x_k' Q_k x_k + u_k' R_k u_k + (u_k - u_{k-1})' S_k (u_k - u_{k-1}) \right]$$

with the state equations and state and control variable constraints given above. To illustrate the solution of this problem, the LQG control option solutions are presented in Figure 3.3.

In particular, DISCON optimal priority control solutions realized at traffic signalized intersections are developed. In Figure 3.4, the selected DISCON control options (punctuality, regularity control, LQG control, reference trajectory based, intelligent priority control with cooperation with intelligent supervisor layer) are presented (also see Figure 3.5).

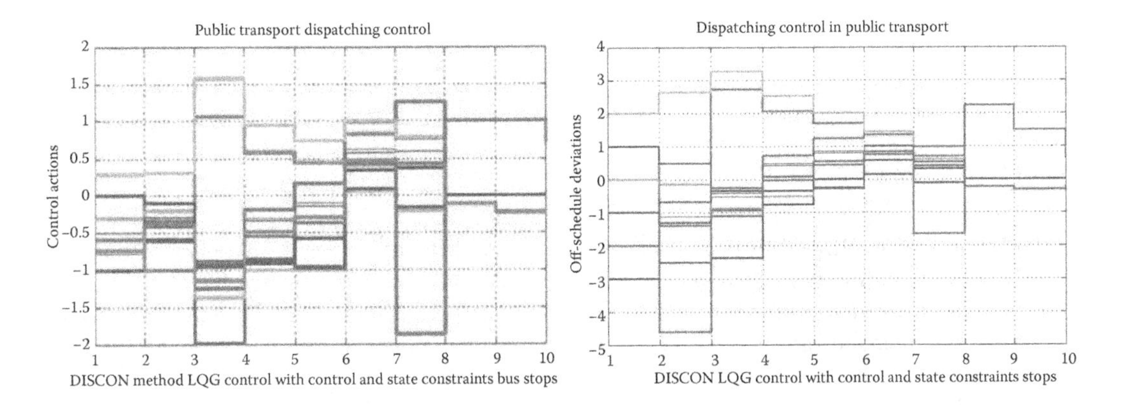

FIGURE 3.3 LQG-DISCON punctuality control of the bus route.

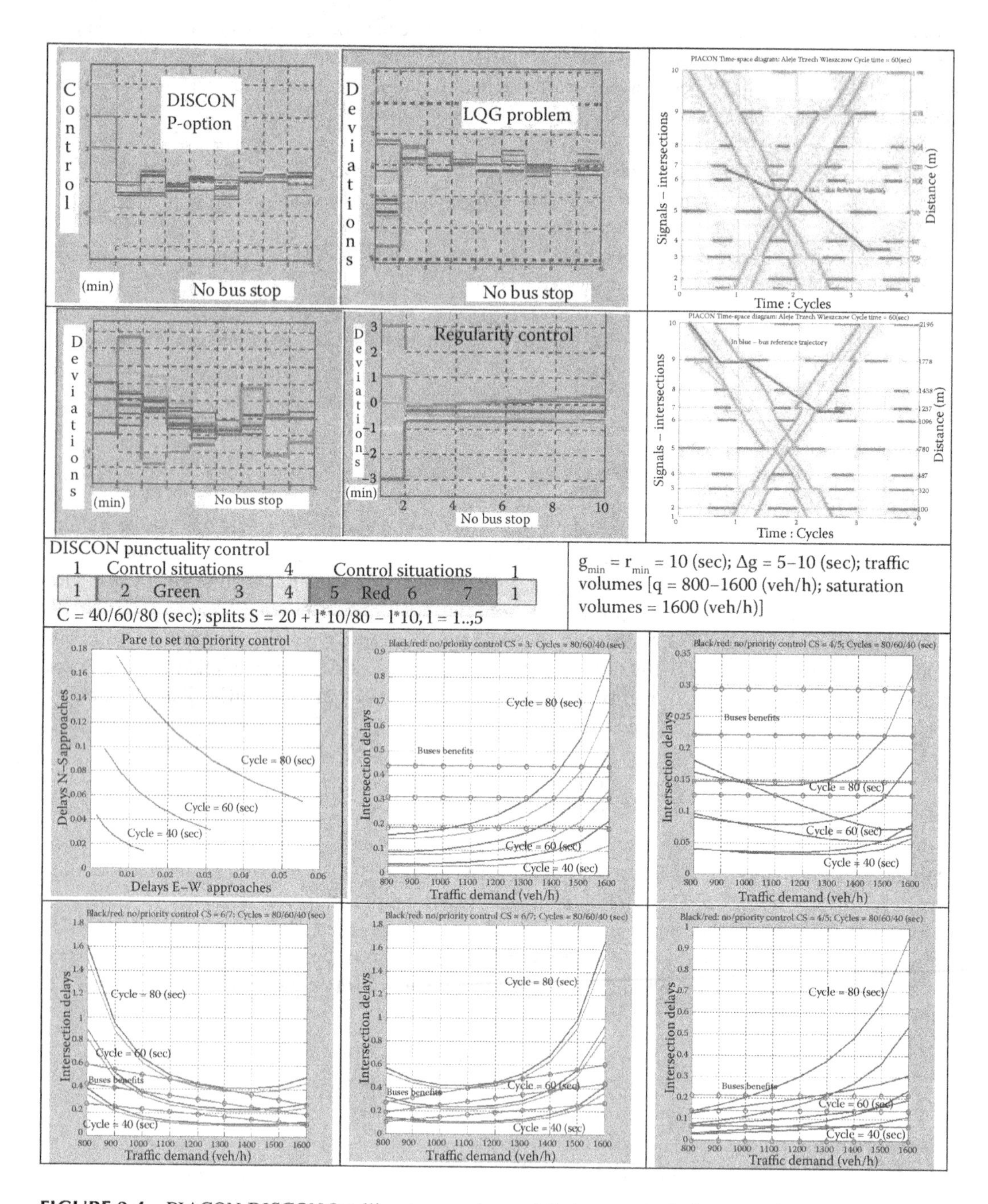

FIGURE 3.4 PIACON-DISCON-Intelligent supervisor public transport priority control.

The results of intelligent priority control illustrate the impacts of a simple priority control action realized on a single signalized intersection within a coordinated artery. This priority control impacts are analyzed in the context of the PIACON control method Delay Mode realized usually in this demand range (see PARETO SET for different cycle times in Figure 3.4). The Intelligent Supervisor (IS) suggests intelligent modifications of priority control actions in order to marginalize individual traffic disbenefits. The recognition by IS in a given traffic situation of

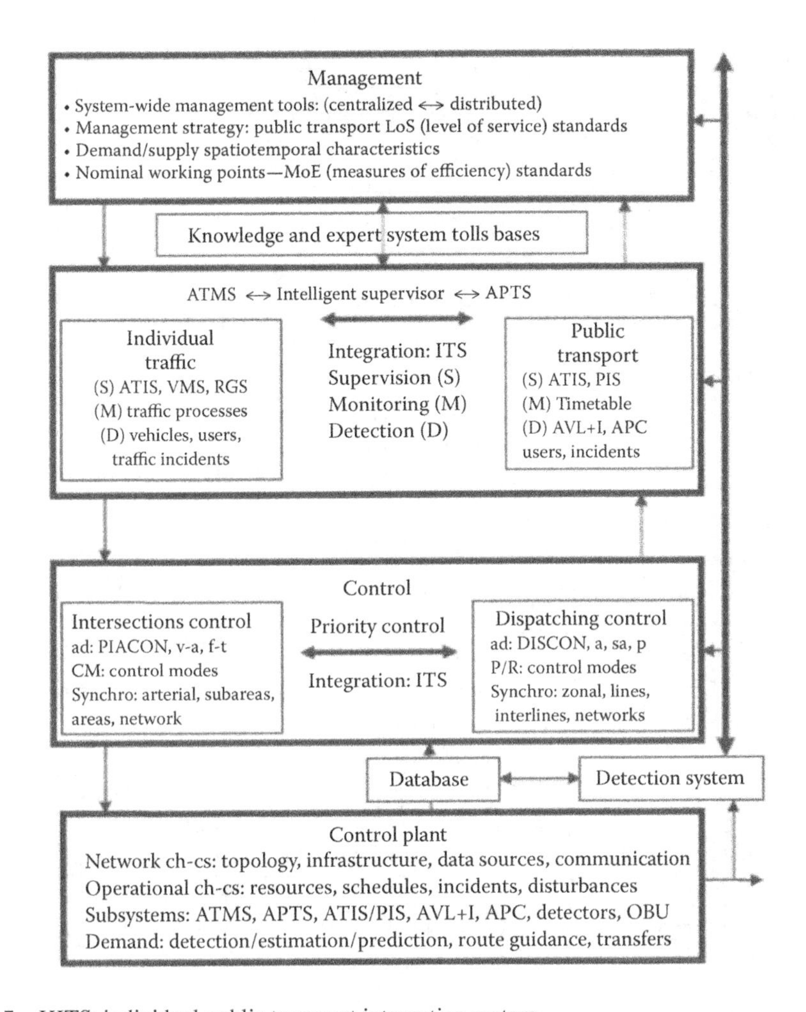

FIGURE 3.5 HITS: individual-public transport integration system.

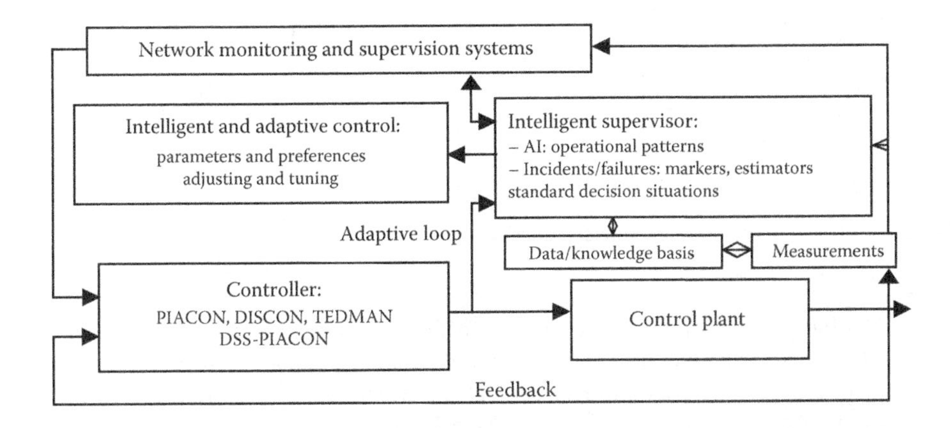

FIGURE 3.6 HITS control layer.

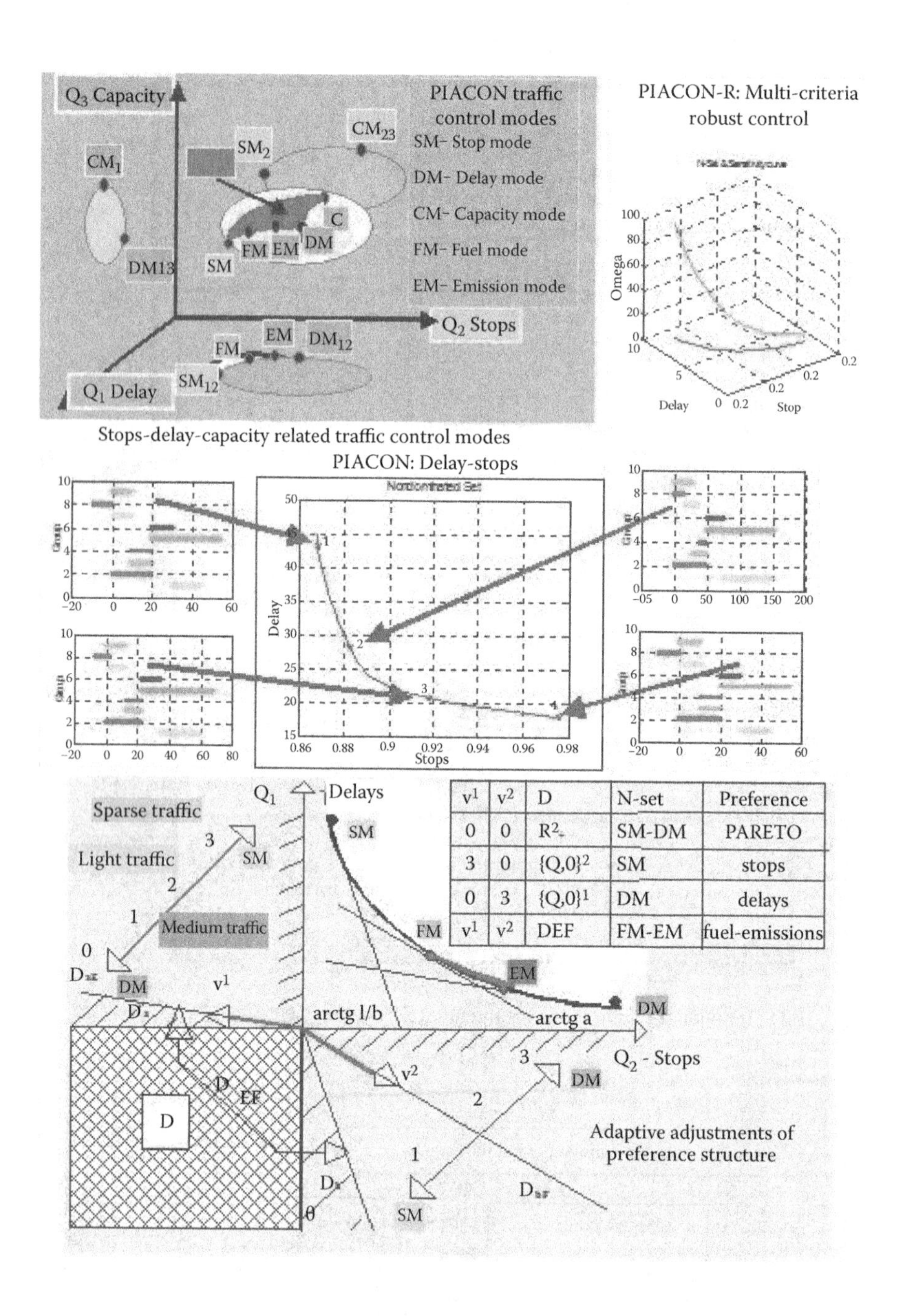

v^1	v^2	D	N-set	Preference
0	0	R^2_+	SM-DM	PARETO
3	0	$\{Q,0\}^2$	SM	stops
0	3	$\{Q,0\}^1$	DM	delays
v^1	v^2	DEF	FM-EM	fuel-emissions

FIGURE 3.7 PIACON Multicriteria Robust Adaptive.

symptoms of synergic effects of priority actions offered also benefits to individual traffic (see Figure 3.4) The other examples illustrate the increase in the robust features by IS actions of the PIACON reference trajectory priority control on artery Al. Trzech Wieszczów in Cracow (Adamski 2003). The robust margins increase by 6%–11% by intelligent suggestions of IS (see Figure 3.3).

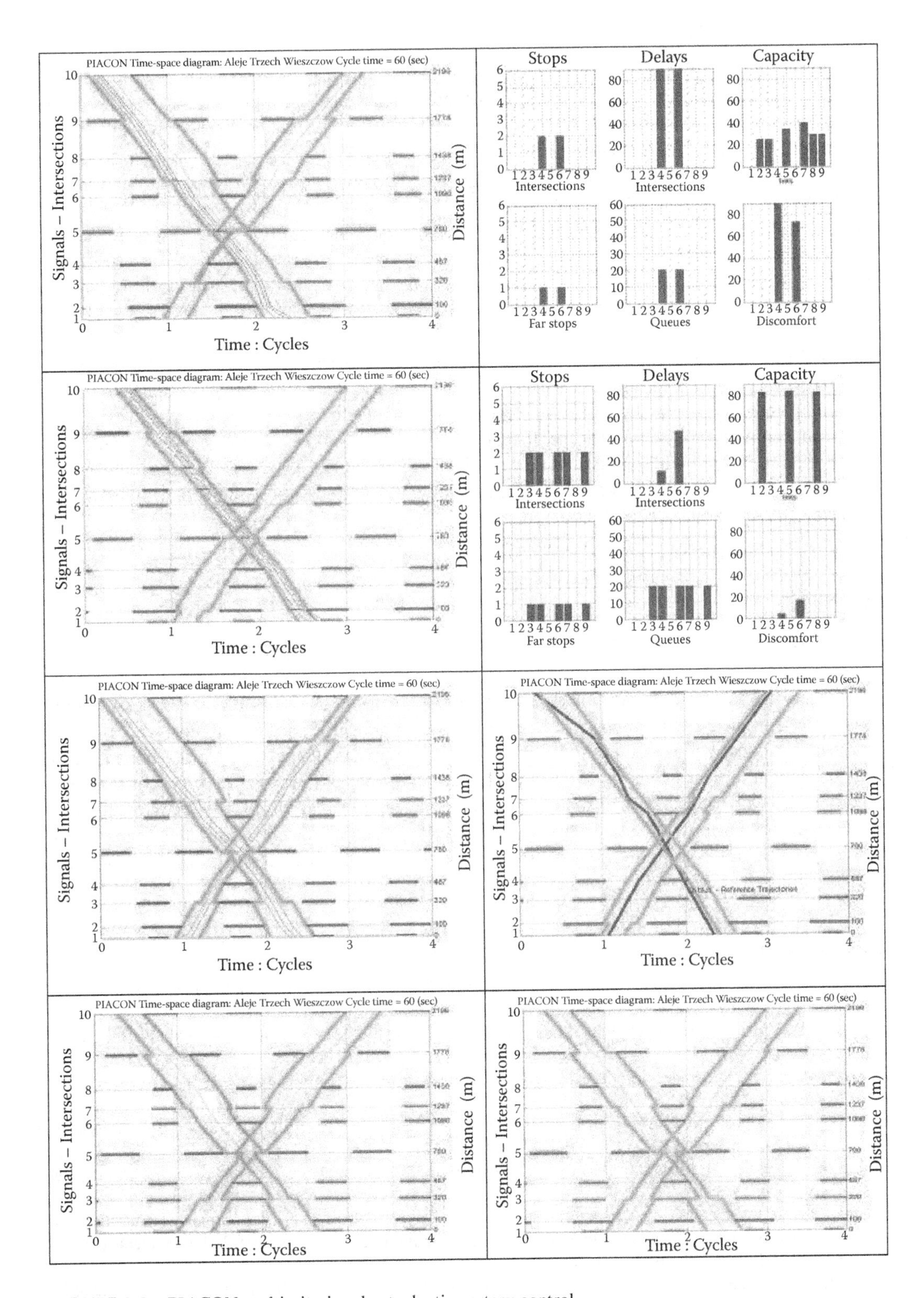

FIGURE 3.8 PIACON multicriteria robust adaptive artery control.

HITS Control Layer

A very important new functional element at the bottom direct control layer concerns the full integration of the tasks of intelligent supervision with the intelligent adaptive control actions (see Figure 3.6 realized by the PIACON control method; Adamski 2003, 2005). The practical proposals of PIACON traffic multicriteria control capabilities realized in hierarchical multilayer adaptive, optimization, and direct control structure were presented in Adamski (1998a,b, 2003, 2006a,b,c). In this context, very important new adaptive layer tasks are concerned with RG and AIDM of vehicle route guidance in the network and automatic incident detection and management functions, respectively.

The main multicriteria mechanisms of the PIACON traffic control method are illustrated in Figure 3.7 (for traffic intersections) and in Figure 3.8 (for arteries). The main features of this sophisticated control method are as follows:

- Multicriteria robust control with traffic adaptive related structure of preferences (the Pareto Cone is only one possibility for such structures; see Figure 3.4, bottom figure). The control robust measure is one of the criteria used in the PIACON method.
- The completely original are the PIACON multicriteria robust adaptive artery control options realized with 2-D random field traffic model (Adamski 2002a) and online real-time estimated (by data from Vehicle Probe and V2V/V2I wireless communication) reference trajectories (Adamski 2003, 2006b,c).

CONCLUSIONS

Professional development of HITS platform for ITS and ILS that support transport and logistics very complex tasks is of fundamental importance. For urban transportation systems, it offers the maintenance in the long-term horizon of the guaranteed mobility standards (efficiency, costs, safety, and comfort) in the city and their surroundings for different traffic participants (drivers, bicycles, pedestrians, goods) while essentially reducing negative impacts of the road traffic. For logistics, additional client service standards will be guaranteed. In the paper, the system approach is proposed in conformity with triads I–E–U: I (integration, intelligence, information), E (efficiency, emergency, ecology), and U (up-to-date, user-oriented, unified). The presented results constitute a fundamental step to the generation of highly efficient robust adaptive multicriteria solutions for the whole network problems.

REFERENCES

Adamski A. Flexible Dispatching Control Tools in Public Transport. In *Advanced Methods in Transportation Analysis*, eds. L. Bianco and P. Toth. Springer, Berlin, pp. 481–506, 1996a.

Adamski A. Hierarchical integrated intelligent logistics system platform. *Procedia Social and Behavioral Sciences*, 20, 1004–1016, 2011.

Adamski A. IILS: Integrated intelligent logistics systems. Cracow University of Technology Logistics Conference, pp. 25–38, 2006c.

Adamski A. Integrated traffic management, surveillance and control systems in urban areas. *TiBT$_{06}$ Transport Telematics and Safety*, 104–112, 2006b.

Adamski A. Integrated transportation systems. *Modeling and Management in Transportation EURO WG on Transportation*, 1, 21–34, 1999.

Adamski A. Intelligent anticipative dispatching control system in public transport. Monograph: Transactions on Transport Systems Telematics: Modeling, Management and Image Processing, Silesian University Press, Katowice, Poland, pp. 70–75, 2007b.

Adamski A. Intelligent integrated transportation systems. In 9th Meeting of the EURO WG Intermodality, Sustainability and ITS, Bari, pp. 565–570, 2002b.

Adamski A. ITS: Integrated transportation systems. *Archives of Transport Quarterly, Polish Academy of Science*, 14(2), 5–22, 2002a.

Adamski A. ITS: Management, surveillance and control. Monograph: AGH University Scientific Publications, Krakow, 2003.

Adamski A. Multi-criteria traffic control with video feedback. In *Applications of Advanced Technologies in Transportation Engineering*, eds. Y.J. Stephanedes and F. Filippi. ASCE Publications, New York, pp. 620–627, 1996b.

Adamski A. Optimal adaptive dispatching control in an integrated public transport management system. In 2nd EURO WG Conf. INRETS ACTES, pp. 917–938, 1993.

Adamski A. Optimal dispatching control in public transport. Habilitation. Thesis. Scientific Bulletins of AGH, AUTOMATICS no. 50, 1989.

Adamski A. PIACON: Robust vehicle trajectory based arterial multi-criteria traffic signal control. Extra EURO Conf. on Handling Uncertainty in Transport Bari, pp. 608–615, 2006a.

Adamski A. PIACON: Traffic control method with video feedback. AATT Conference, eds. C.T. Chendrickson and S. Ritchie. ASCE Publ. 1801, pp. 217–224, 1998a.

Adamski A. PIACON-DISCON integrated approach to public transport priority control at traffic signals. Advanced OR and AI Methods in Transportation. Poznan University of Technology, pp. 417–422, 2005.

Adamski A. Probabilistic models of passengers service processes at bus stops. *Transportation Research B*, 26, 253–259, 1992.

Adamski A. Real-time computer-aided control in public transport from the point of view of schedule reliability. *Lecture Notes in Economics and Mathematics*, 430, 23–38, 1995.

Adamski A. Robust dispatching control in public transport- DISCON-R. Monograph: Transactions on Transport Systems Telematics: Modeling, Management and Image Processing, Silesian University Press, Katowice, Poland, pp. 56–69, 2007a.

Adamski A. and A. Turnau. Simulation support tool for real-time dispatching control in public transport. *Transportation Research A*, 32(2), 73–87, 1998b.

Cea J., J. Fernandez and L. Grange. Combined models with hierarchical demand choices: A multi-objective entropy optimisation approach. *Transport Reviews*, 28(4), 415–438, 2008.

4 Hierarchical Intelligent Traffic System

Application of Vehicular Telematics over Wireless Networks for Intelligent Traffic Incident Detection and Diagnosis

Andrzej Adamski and Grzegorz Heldak

CONTENTS

INTRODUCTION

The adequately working monitoring and intelligent surveillance layer in the hierarchical intelligent traffic system (HITS) system is potentially able directly or indirectly, through other layers, essentially improve wide-area operational efficiency and productivity of the whole transportation system. This chapter presents a proposal that collaborative implementations of vehicular telematics working in heterogeneous wireless networks environment are essential enabling technology for the HITS system. Advanced heterogeneous vehicular network (AHVN) architectures and intervehicle (vehicle-to-vehicle [V2V]) communication (IVC) and vehicle-to-infrastructure (V2R) communication tools use multiple access technologies and multiple radios in a collaborative manner. The modern HITS multilayer integrated traffic management, surveillance, and control activities can be essentially improved by real-time intelligent reactions on properly detected traffic incidents. This is a challenging task in the context of the very high complexity of traffic phenomena. The collaborative AHVN–IVC support activities for traffic incident detection, recognition, and diagnosis should establish the challenges and rules in designing the essential functional components of AHVN and the corresponding protocols. The illustrative practical examples of real-time control preference recognition based on AHVN–IVC support creates very attractive practical proposals for HITS system operation. Vehicular platforms with telematics services provide key support for development of integrated intelligent transportation system (ITS) options to achieve safety and productivity in transportation. Wireless communications and networking technologies such as IEEE 802.11 (Wi-Fi), IEEE 802.16 (Worldwide Interoperability for Microwave Access [WiMAX]), 3G cellular, and satellite technologies to support data communications for vehicular telematics. ITS applications can be supported by vehicle-to-roadside (V2R) and V2V communications. V2R involves vehicular nodes and roadside base stations. IEEE 802.11 (Wi-Fi), IEEE 802.16 (WiMAX), and dedicated short-range communications (DSRC) technologies can be used in this model of communication. In particular, with the DSRC standard, onboard units (OBUs) placed at each vehicle can send/receive data to/from roadside units (RSUs). However, if a vehicle cannot directly send its data to an RSU, it can relay its data to other vehicles until the data reach the RSU using a multihop transmission strategy. It is also possible that OBUs form a group and elect the group leader. In this case, all group member OBUs will send their reports to the leader OBU, which will aggregate them and forward the resulting message(s) to the RSU. There are several applications for this communication model, such as electronic toll collection, infotainment services, safety message dissemination, and Web browsing. The OBUs are typically equipped with an onboard computer, communication interface, and global positioning system (GPS), which provides information on vehicle position in real time, and an event data recorder, which stores relevant data that, in case of an accident, can be used in forensic analysis. RSUs act as base stations or access points (APs) and are connected to application servers. V2V communications with vehicular nodes on a road form a vehicular ad hoc network (VANET), which is mainly used in safety warning systems, traffic information systems, and multimedia services. Collision avoidance, road obstacle warning, intersection collision warning, and lane change assistance are example applications of V2V communications. Most V2V safety applications require low transfer latency since these applications are used in a dynamic and unpredictable traffic environment. Most research tends to find improvements in such medium access control (MAC) protocols, transmission strategies, and wireless technologies in order to reduce the latency. By definition, emerging vehicular telematic applications will be wireless. In the past, the wireless services for individuals were dedicated to alternative route selection in a more or less independent manner. This chapter provides a HITS dedicated survey on the research issues, challenges, and possible approaches to tackle these challenges for vehicular telematics over heterogeneous wireless networks.

ITS HIERARCHICAL MULTILAYER SYSTEM

The professionally developed ITS systems create a universal platform and operational environment for current transportation systems. Figure 4.1 presents the intelligent supervisor layer embedded in an ITS system and functioning in the range of integrated, coordinated hierarchy.

Integrated hierarchical management, surveillance, and control structures of ITS multilayer systems create the base for completely new capabilities (Adamski 2003):

- Integrated stimulating very beneficial synergic effects system-wide solutions
- Intelligent real-time opportunity-based efficient management strategy selection
- Real-time fully adaptive "control by opportunity" and robust control modes

New features include high efficiency (due to system components' interoperability) and productivity (multicriteria dynamic approach); flexibility (modularity, expandability); and transparency (sophisticated expert systems and visualization tools). The implementation of effective management and control strategies requires continuous monitoring and surveillance, that is, real-time assessment of network components' (like intersection/artery/subarea modules [IN/AM/SM]) operating conditions based on real-time traffic data (Adamski 1979, 2003, 2005). The assessment of operating conditions requires robust recognition and identification of traffic modes and their dynamic transitions in time (Adamski 2000, 2003, 2005). Traffic modes are represented by traffic markers to be an appropriate transformation of representative traffic variables into network component performance

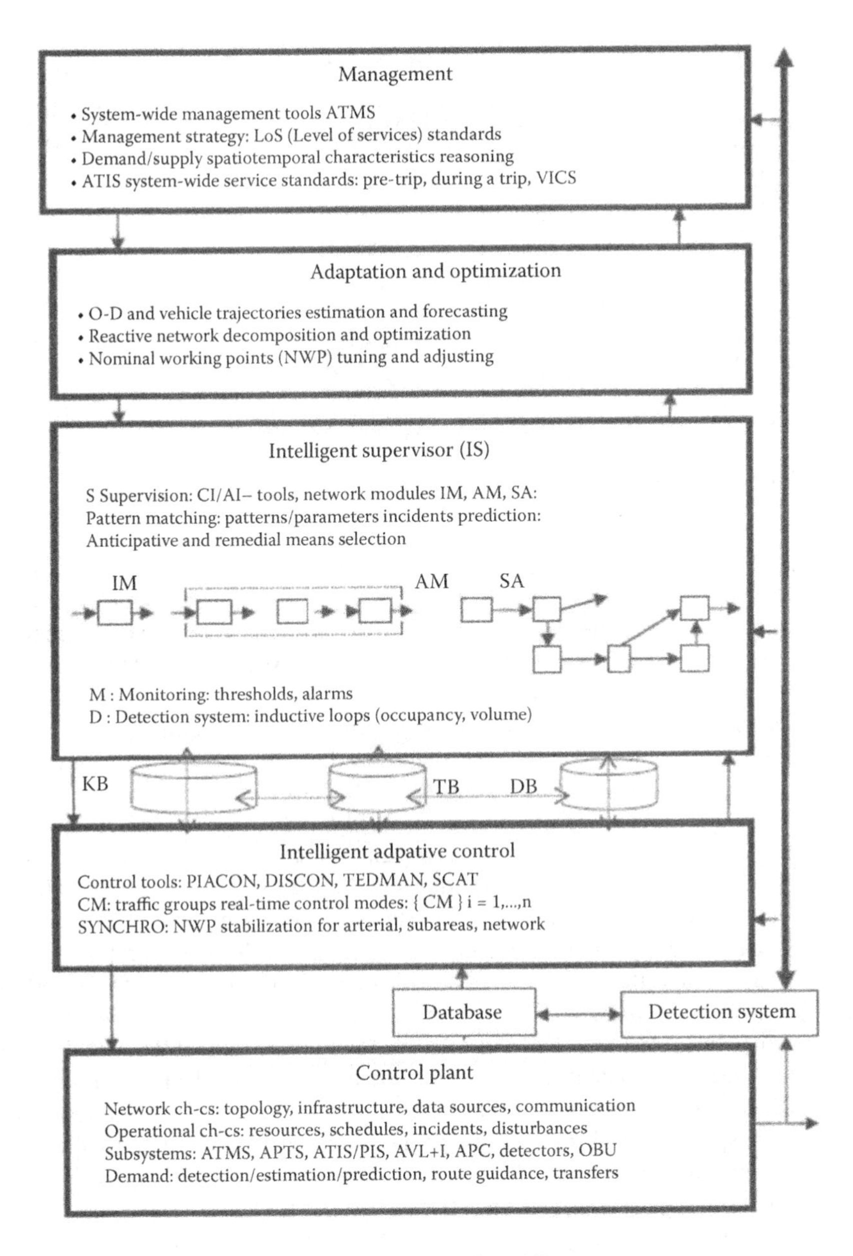

FIGURE 4.1 Intelligent supervisor layer embedded in ITS multilayer system.

measures, for example, for arterials delays, stops, queue lengths, and travel times (Adamski 2003, 2006). The recent observed developments in wireless mobile communications technologies motivate their implementation in the HITS platform and ITS systems in particular. This enabling technology at present may be viewed as a key element of the HITS cooperative systems with interuser interactions allowing the generation of the new, more professional cooperative HITS system services, for example, a new generation of the advanced driver assistance system (ADAS) with new cooperative capabilities offered by new generic sensors (V2V, vehicle-to-infrastructure [V2I], vehicle to pedestrians). The new quality of these services is related to advantages of new communication-based "generic sensors" over existing sensors (video camera, radar, laser scanner, infrared) in terms of more applications dedicated to range, better accuracy, and better transparency with respect to

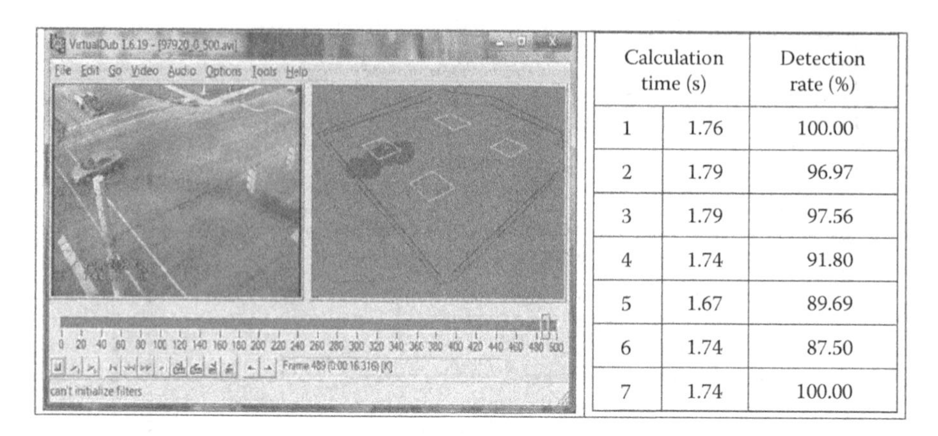

Calculation time (s)		Detection rate (%)
1	1.76	100.00
2	1.79	96.97
3	1.79	97.56
4	1.74	91.80
5	1.67	89.69
6	1.74	87.50
7	1.74	100.00

FIGURE 4.2 Safety monitoring system.

type of shared information. In consequence, due to V2V, the completely new types of information, for example, driver awareness, the braking capacity of the previous vehicles, and kinematical and dynamic vehicle parameters (weight, dimensions, steering capacities, adherence), can be utilized in risk estimation and prediction and exchanged between ITS system components and vehicle platforms. This essentially influences the efficiency of operation of all the HITS system layers (management and coordination, adaptation and optimization, monitoring and supervision, and direct control) with the bottom system processes layer. In this context, there are many new existing sophisticated ITS system dedicated control and management methods and tools: several versions of polyoptimal intelligent integrated adaptive control (PIACON), dispatching control (DISCON), PIACON–DISCON, traffic environmental management and control (TEDMAN-C), and decision support system for ITS-ILS (DSS-PIACON) (Adamski 2000, 2007) can be supported by this information, to mitigate the sophisticated and time-consuming estimation and prediction processes. In the field of ITS systems safety, the implementation of such technology was proposed by Adamski (2000, 2007). It was based on the USA VII system with a specifically designated DSRC (300–1000 m) of 5.9 GHz frequency. The basic VII architecture contained five elements: onboard equipment (OBE), a human machine interface (HMI), an OBU to communicate by DSRC with the entities outside the vehicle, roadside equipment (RSE) to collect and disseminate data, and a traffic operation center (TOC). The main safety services include warning drivers of unsafe and precrash conditions, road runoffs or overspeeding, real-time advantageous congestion information, weather conditions, lane change warnings, and more. The OBUs transmit messages every 100 ms (a safety application on the intersection required regular communication every 100 ms). The video sensor–based intersection safety analysis project was realized by AGH Krakow with the cooperation of Purdue University, USA (Adamski and Mikrut 2004). The VirtualDub program supplemented by a set of filters (programmed in the Visual C++ Studio 6.0) was developed. To save time, only extracted object data and dedicated accuracy of the optical flow field calculations were used. A traffic analysis system is presented in Figure 4.2. The tests were carried out on a PC with a Pentium IV 2.4 GHz processor in batch mode. The comparison video sensors with wireless communication parameters are visible.

APPLICATION OF VEHICLE TELEMATICS

In Table 4.1, three types of vehicular networks are presented. The main concern here is that ITS elements need communication—that is, first of all, the vehicles, the operators of transportation systems, and the service providers. A large number of data dissemination mechanisms were proposed in the literature. While many of these works rely on repeaters and (or) APs for disseminating the

TABLE 4.1
Types of Application of Vehicular Telematics

Applications	Activity	Requirements
Safety applications	Improve safety of passengers on roads by notifying vehicles about any dangerous situation in their neighborhood (warnings about collision, road conditions, approaching emergency vehicle)	Finding low-latency, reliable, and efficient methods for disseminating safety data among neighboring vehicles
Traffic applications	Deployment of traffic information systems (TISs), which carry out traffic management and provide drivers with traffic situation and road information	Minimum delays at far distances, confirmation of TIS receipt, TIS data aggregation algorithm
Nonsafety applications (comfort or entertainment)	Varied: from real-time, non–real-time multimedia streaming and interactive communications (videoconferencing and interactive games), roadside service application (location and price lists of restaurants or gas stations, weather information, or Internet access)	Varied, depend on type of service, but lower priority than for safety and traffic application

Source: Hossain, E. et al., Vehicular telematics over heterogeneous wireless networks: A survey. *Computer Communications*, *33*, 7, 775–793, 2010 (May).

safety data, some other works suggest that infrastructure-independent fully ad hoc communications suffice (Hossain et al. 2010).

The safety applications have a lot of different sources of information. In contrary to the first-generation vehicular telematic applications, the emerging generation of wireless applications capitalizes on the extraction of data from fleets of multiple vehicles. Data mining on the extracted data can aid better decision support of difficult-to-predict emergent behavior. The conceptual objective of these second-generation applications is to translate or convert multisource data and then to estimate or infer extant conditions back to the users as well as interested third parties. Examples of this include highway safety systems where wireless intravehicle relays can be used to impede or halt oncoming traffic in a freeway accident scenario (forward collision warning). These are somewhat obvious applications that are built upon layers of sophisticated technology with vehicle safety systems leading the development.

In addition, there is a class of applications that rely on inference from probes or floating cars, with the intent to capture a statistically significant portion of vehicle behavior such that meaningful inferences can be made and potentially generalized to the entire population of vehicles (Kerner et al. 2005). Applications include infrastructure monitoring as well as sampling empirical data input for traffic flow control (Hossain et al. 2010). Schoch et al. (2008) collected and categorized envisioned applications from various sources and classified the unique network characteristics of vehicular networks (Table 4.2). Based on this analysis, they proposed five distinct communication patterns that form the basis of almost all VANET applications.

Schoch et al. (2008) classified a variety of applications into logical groups to get a more concise picture of the applications. In addition, node and network characteristics clarify influences on the design of mechanisms. Based on this analysis, they propose five "communication patterns," which satisfy the communication needs of virtually all VANET applications that are currently under discussion. These patterns can serve as a basis for future development. The classification also reflects that the close coupling between applications and communication in VANETs shifts the focus to a more integrated system architecture, which ultimately also includes information aggregation (Table 4.2).

TABLE 4.2
Overview of Application on VANET

	Situation/Purpose	Application Examples
I. Active Safety	1. Dangerous road features	1. Curve speed warning 2. Low bridge warning 3. Warning about violated traffic lights or stop signals
	2. Abnormal traffic and road conditions	1. Vehicle-based road condition warning 2. Infrastructure-based road condition warning 3. Visibility enhancer 4. Work zone warning
	3. Danger of collision	1. Blind spot warning 2. Lane change warning 3. Intersection collision warning 4. Forward/rear collision warning 5. Emergency electronic brake lights 6. Rail collision warning 7. Warning about pedestrians crossing
	4. Crash imminent	1. Precrash sensing
	5. Incident occurred	1. Postcrash warning 2. Breakdown warning 3. SOS service
II. Public Service	1. Emergency response	1. Approaching emergency vehicle warning 2. Emergency vehicle signal preemption 3. Emergency vehicle at scene warning
	2. Support for authorities	1. Electronic license plate 2. Electronic driver's license 3. Vehicle safety inspection 4. Stolen vehicle tracking
III. Improved Driving	1. Enhanced driving	1. Highway merge assistant 2. Left turn assistant 3. Cooperative adaptive cruise control 4. Cooperative glare reduction 5. In-vehicle signage 6. Adaptive drive train management
	2. Traffic efficiency	1. Notification of crash or road surface conditions to a traffic operation center 2. Intelligent traffic flow control 3. Enhanced route guidance and navigation 4. Map download/update 5. Parking spot locator service
IV. Business/Entertainment	1. Vehicle maintenance	1. Wireless diagnostics 2. Software update/flashing 3. Safety recall notice 4. Just-in-time repair notification
	2. Mobile services	1. Internet service provisioning 2. Instant messaging 3. Point-of-interest notification
	3. Enterprise solutions	1. Fleet management 2. Rental car processing 3. Area access control 4. Hazardous material cargo tracking
	4. E-payment	1. Toll collection 2. Parking payment 3. Gas payment

Source: Capkun, S. et al., Technical Report EPFL-IC, EPFL, 2004.

WIRELESS NETWORKS

Most of the studies in the field of vehicular communications and vehicular networks only deal with a single type of applications in these networks; that is, none of them addresses all types of safety, traffic, and comfort applications, and neither purely infrastructure-based or purely ad hoc vehicular networks nor even traditional single-radio wireless mesh networks (WMNs) meet the requirements of all the applications at the same time. An AHVN that uses multiple radios and multiple access technologies in a collaborative manner could be the best candidate for a vehicular network (Hossain et al. 2010). A key motivation in considering the AHVN is that DRSC will only be effective when it is ubiquitously deployed, but this will not happen until the needed infrastructure is in place, governments legislate for DSRC deployment in passenger vehicles, and older noncompliant vehicles have retired. This is not very likely to happen within the next 10–20 years. Therefore, in the meantime, a heterogeneous platform is the best way forward. In fact, on the issue of infrastructure deployment, the ITS branch of the US Department of Transportation (DOT) is still exploring different business models, as it is not certain that the public sector can or wishes to cover all the costs (Hossain et al. 2010). In the AHVN, the vehicles request services with different requirements in terms of latency, bandwidth, error rate, area of coverage, and so forth at any time and any place. Existing access technologies such as wireless LANs (WLAN IEEE 802.11a/b/g/n/p standards), WiMAX (IEEE 802.16 a/e standards), ultra wideband (UWB IEEE 802.15.3a standard), third- and fourth-generation cellular wireless (3G and 4G), satellite communications, and so forth are designed for specific service requirements (Hossain et al. 2010) (Table 4.3).

Designing any heterogeneous network should, as much as possible, be on the basis of intelligent integration of readily available technologies in order to minimize the deployment cost and to make the deployment fast. However, the design should also be left open to any better substituting or complementing alternatives (Hossain et al. 2010). Some studies proposed the integration of infrastructure into ad hoc wireless networks, commonly referred to as WMNs, for addressing the vehicular applications we mentioned earlier. The deployment of WMNs enables vehicles to get access to the APs outside their transmission range via multihop communications, which improves the network coverage and also enhances the capacity of the network by allowing more parallel transmissions (Hossain et al. 2010). When the AHVN architecture is being designed, the important element is the network selection. In order to guarantee the scalability of the AHVN, a hierarchical structure should be considered for network components. The next important element is the operating system, which should have the possibility to automatically upload and should have services-oriented architectures (SOAs); in addition, there is a need for an open architecture standard to be developed to relieve the application developers of the details of supporting layers in the protocol stack (Hossain et al. 2010). Some wireless access strategies include the following.

Transmission strategies determine how data packets are delivered from a vehicular node to an RSU (and vice versa) or from a vehicular node to another vehicular node. A direct transmission or single-hop transmission strategy can be used where an RSU or a vehicular node can be reached directly from a vehicular node. If an RSU or a vehicular node is located far away from a source node, a multihop transmission strategy can be employed. In this scenario, data packets from a vehicular node are relayed by other vehicular nodes until they reach the destination. Finally, a cluster-based transmission strategy forms groups (i.e., clusters) of vehicles, selects a representative (i.e., a cluster head or a gateway) for each group, and transmits data through the selected representative. The cluster head receives data packets from its cluster members and then relays the packets to the RSU (and vice versa). This strategy is efficient to decrease request/data congestion at the RSUs. For V2V communications, a multihop broadcast scheme is used when a target vehicle is out of the transmission range of a broadcasting vehicle. In vehicular safety applications, this transmission strategy is used to transmit warning messages to other vehicles. However, due to the unpredictable

TABLE 4.3
Access Technology

Access Technology	Description
IEEE 802.11–based WLAN	It has achieved great acceptance in the market and supports short-range relatively high-speed data transmission. The maximum achievable data rate in its latest version 802.11n is about 100 Mbps. As for its performance reduction due to interference, on the other hand, the short transmission range leads to frequent transmission interruption, particularly when vehicle speed is high, and consequently, many access points have to be deployed along the road, which incurs high deployment cost.
IEEE 802.11p	The new communication standard in the IEEE 802.11 family, which is based on the IEEE 802.11a. IEEE 802.11p, which is also referred to as the DSRC standard, is designed for wireless access in the vehicular environment (WAVE) to support ITS applications. For DSRC, 75 MHz of licensed spectrum at 5.9 GHz has been allocated, which consists of seven channels (10 MHz each) for supporting safety and nonsafety applications.
DSRC	Supports a very high data rate (6–27 Mbps) with a maximum communication range of 1000 m. Presently, DSRC is mainly used in electronic toll collection. Potential applications of DSRC are emergency warning systems for vehicles, adaptive cruise control, forward collision warning, electronic parking payments, approaching emergency vehicle warning, transit or emergency vehicle signal priority, and in-vehicle signing. Vehicles equipped with DSRC can communicate directly with each other, making it possible to send warning messages to neighboring vehicles. DSRC can also be used to provide in-vehicle entertainment for drivers and passengers.
IEEE 802.16–based WiMAX (Worldwide Interoperability for Microwave Access)	Systems are able to cover a large geographical area, up to 50 km, and to deliver significant bandwidth to end users up to 72 Mbps theoretically. While the IEEE 802.16 standard only supports fixed broadband wireless communications, the IEEE 802.16e/mobile WiMAX standard supports speeds up to 160 km/h and different classes of quality of service, even for non–line-of-sight transmissions. In WLAN, a contention-based channel access mechanism is used, which can cause subscriber stations distant from the AP to be repeatedly interrupted by closer stations. The key advantage of WiMAX compared to WLAN is that the channel access method in WiMAX uses a scheduling algorithm for which the subscriber station needs to compete only once for initial entry into the network. After that, it is allocated an access slot by the broadcasting system (BS). Table 4.1 shows a list of some applications envisioned by mobile WiMAX and their quality of service (QoS) requirements.
3G/4G cellular	For vehicular telematic services, 3G cellular wireless technology can provide a very broad coverage and support high-mobility vehicles. Current 3G networks deliver a data rate of 384 kbps to moving vehicles, which goes up to 2 Mbps for fixed nodes. 3G systems deliver smoother handoffs compared to WLAN and WiMAX systems; however, due to centralized switching at the mobile switching center (MSC) or serving general packet radio service (GPRS) support node (SGSN), their latency may become an issue for many applications. Although the data rates of cellular networks are expected to go higher in 3.5G high-speed downlink packet access (HSDPA) and 4G, the technology may not be available in the very near future.
Satellite communications	It provides ubiquitous coverage at any location but incurs high cost and large propagation delays. Some suggested key conceptual models and services in a satellite-based vehicular broadband network as well as the design considerations at different levels are discussed.

Source: Hossain, E. et al., Vehicular telematics over heterogeneous wireless networks: A survey. *Computer Communications, 33*, 7, 775–793, 2010 (May).

network topology, interference, packet collisions, and hidden nodes, multihop transmissions over V2V networks can be very challenging. Some groups consider other problems like data dissemination protocols (e.g., delivery traffic information to every vehicle, which need these), data aggregation protocols, routing protocols, congestion control protocols, security protocols, and privacy protocols (Hossain et al. 2010).

Security and Privacy Aspects

Ma et al. (2009) examined a location privacy problem. They presented a trip-based location privacy metric for measuring user location privacy in the V2V/V2X systems. The applications of vehicle systems include intersection collision warning, traffic monitoring through probe vehicle data, location-based services, and so forth. The sending and dissemination of personal location information has the potential to infringe on a user's privacy, especially location privacy. The basic consideration behind this is that location privacy of users is not only determined by vehicle tracking but also by linking vehicle trips to the individuals generating them. Based on snapshots of the V2X communication systems, we capture the information on location privacy in terms of individuals in the system and their trips, which are defined by the origins and destinations of the trips. Assuming that an adversary has information on the linking between vehicles and trips expressed in probabilities, the location privacy of an individual is measured by the uncertainty of such information and quantified as entropy. Then, the location privacy of a specific user can be determined by the ratio of its current entropy and the maximum possible entropy within the given system. The feasibility of the approach is supported by means of different case studies. Project PRECIOSA (2008–2010; http://www.preciosa-project.org/) dealt with the protection of privacy in vehicular communication, including security aspects: SeVeCom (2006–2009; http://www.sevecom.org/) dealt with the protection of external vehicular communication; EVITA (2008–2011) dealt with the protection of onboard networks. As has been mentioned before, drivers have the incentive to "try to bend the rules" and avoid liability for violating traffic rules, by modifying the information about their speed, position, acceleration/deceleration, and other relevant variables in the safety messages, in order to avoid liability. On the other hand, as noted in (Ma et al. 2009), the majority of drivers are honest and would not feel comfortable tinkering with their OBUs. Such drivers will gladly contribute safety messages, but under the condition that these messages will not reveal their ID, name, location, and driving patterns to an unauthorized third party. However, drivers' privacy in a VANET is conditional, since in case of a dispute, information can be revealed by legal authorities. This issue has launched significant research activity, and basically, all aforementioned VANET security architectures differ in their view and implementation of drivers' privacy. Drivers' privacy is generally achieved using pseudonyms, which can be time or geographically based. Time-based pseudonyms were defined by Capkun et al. (2004) as

$$P_X(t) = \mathrm{HMAC}_{K_{\bar{X}}}(\mathrm{ID}_X, t)$$

where the real identity of the OBU (and, by extension, the vehicle) is the vehicle pseudonym calculated at time, hash message authentication code (HMAC) is the keyed hash function, and is the secret key. Mapping between pseudonyms and real identities is performed by the trusted authority, which stores the real identities and can, thus, easily perform the translation in either direction (Hossain et al. 2010).

PRACTICAL EXAMPLES OF COOPERATIVE IMPLEMENTATION IN ITS SYSTEMS

The adaptive PIACON control method (Adamski 2003) mechanism consists of several tasks realized recurrently:

1. New set of traffic data handling (number and type of vehicles, queue lengths, vehicle headways, etc.).
2. Adaptive estimation of the traffic parameters (traffic stream volumes, saturation volumes, queue lengths, traffic situation markers, vehicle speeds, and concentration). The real-time

traffic control requires predictions of what would be the values of traffic parameters during the next control period. In the PIACON method, these parameters are used for updating of the compromise set and traffic control model. In such a situation, it is necessary to use a smoothing or filtering model in order to extract the trends in traffic data or make short-term predictions of the control variables. Basically, the smoothing algorithm processes the incoming detector data as they are received to produce estimates of the underlying traffic parameters.
3. Adaptive adjustment of the preference structure according to the currently identified traffic situation.
4. Determining the corner points and/or compromise sets in the criteria and control spaces for a given preferential structure.
5. Defining the potential nominal intersection working (NIW) point with its robustness a registered memory module (RDIM) indicator and proximity measure (in the Euclidean or "street metric" sense) with respect to the last NIW point.
6. Checking the robustness and proximity tests for a given potential NIW point.
7. Transfer to the new NIW point and returning to step 1.

To illustrate the above procedure, we present the example of a real intersection in Krakow.

Example 1

The volumes on the approaches to this intersection were the variable in time traffic parameters. The saturation volumes were fixed, $s_1 = s_2 = 1640$ [vehicles per hour]. The set of traffic volume data is represented by a second-order Aw-Rascle (AR) model described by the following second-order real-valued difference equation: $\boldsymbol{q(n) + a_1q(n-1) + a_2q(n-2) = z(n)}$, where constants a_1 and a_2 are AR parameters and $z(n)$ is the white Gaussian noise process with zero-mean and constant variance $V(z)$, that is, $E[z(n)z'(n)] = V(z)\delta(k-n)$, where $\delta(k)$ is the Kronecker delta. The AR parameters may be effectively determined by the solution of the set of linear Yule-Walker equations: $\boldsymbol{a = -R^{-1}r}$, where the matrix $\boldsymbol{R}$ and vector $\boldsymbol{r}$ elements are equal to the values of the correlation function $r(.)$ in appropriate time instants. Similarly, the white-noise variance $V(z)$ is expressed by a linear combination of the AR model parameters a_1 and a_2 and correlation function values $r(.)$. The eigenvalue location analysis supplemented by the requirements of asymptotic stationarity of the AR process determine the admissible triangular region for AR parameters a_1 and a_2 defined by $|a_2| \le 1$, $a_2 \ge \pm a_1 - 1$. The correlation function $r(m)$ of an asymptotically stationary AR process for lag m satisfies the following difference equation: $\boldsymbol{r(m) + a_1r(m-1) + a_2r(m-2) = 0}$; $m > 0$ with the initial values $r(0) = V(q)$; $r(1) = -a_1V(q)/(1 + a_2)$, where $V(q)$ is the variance of the traffic volume. The solution of this equation has the simple form $r(m) = V(q)\left[w_1^{m+1}(w_2^2-1) - w_2^{m+1}(w_1^2-1)\right]/(w_2-w_1)(w_1w_2+1)$ where w_1 and w_2 are eigenvalues $w_{1,2} = \left(-a_1 \pm \sqrt{a_1^2 - 4a_2}\right)/2$. Solving the Yule-Walker equation and denoting the normalized correlation coefficients by $\rho_1 = \boldsymbol{r}(1)/\boldsymbol{r}(0)$, $\rho_2 = \boldsymbol{r}(2)/\boldsymbol{r}(0)$, we get the following formulas for AR parameters: $a_1 = \rho_1[\rho_2 - 1]/(1-\rho_1^2)$; $a_2 = [\rho_1^2 - \rho_2]/(1-\rho_1^2)$. Whence the admissible region for a_1 and a_2 parameters considered previously may be described in terms of ρ_1 and ρ_2 correlation coefficients as follows: $|\rho_i| \le 1$; $\rho_1^2 < (1+\rho_2)/2$. We may express also the variance of the white-noise process as $V(z) = V(q)\left[(1+a_2)^2 - a_1^2\right](1-a_2)/(1+a_2)$. For a new traffic volume data set, the corresponding Pareto set may be easily determined and analyzed if the new potential NIW point fulfills the robustness and proximity tests.

Example 2

Let us consider for illustration the problem of adaptive adjustment of the domination structure (Figure 4.3 and Table 4.4). Real-time traffic control requires the real-time identification of the current traffic situations (traffic modes) on the basis of extracted rough features from detector

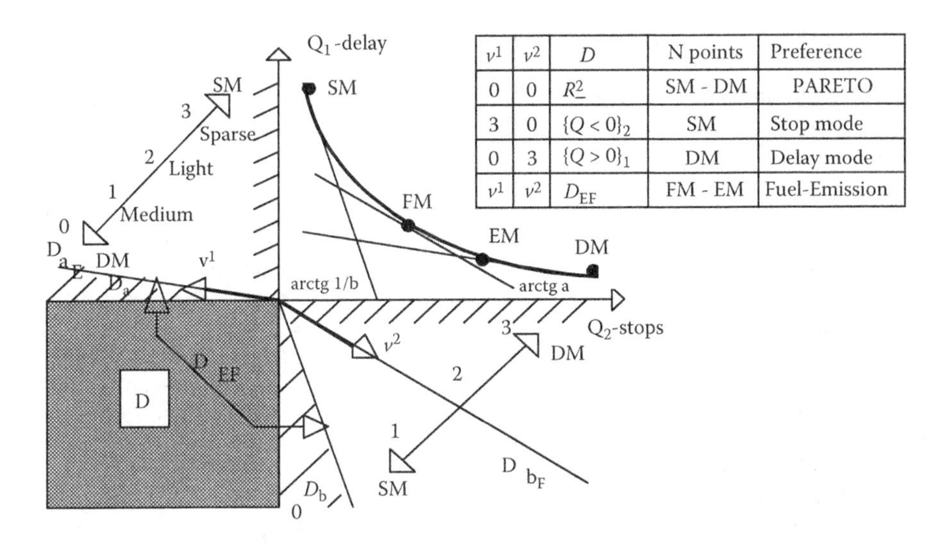

FIGURE 4.3 Adaptive adjustment of the domination structure.

TABLE 4.4
Preference Structure

v_1	v_2	Comments	U	a	b	c	v_1	v_2
0	0	PARETO	1	0	0	0	3	0
3	0	SM Mode	2	0	1	1	1	0
0	3	DM Mode	3	0	1	1	1	1
1	0	SM–EM	4	1	1	1	1	2
2	0	SM–FM	5	1	2	2	1	2
1	1	FM–EM	6	1	1	1	1	2
0	1	FM–DM	7	2	1	1	0	3
0	2	EM–DM	8	2	2	1	0	2
1	2	EM	9	2	1	1	0	1
2	1	FM	10	2	2	1	1	2

data and calls for a self-adjusting mechanism for the selection of adequate control modes. In PIACON, this task is realized by real-time adjustment of the domination structure, which is represented by polyhedral convex cone $D = \{a \in R^2 : Av \geq 0, \text{Ker}A = (0)\}$ generated in real-time by preference directions v^i of the currently most representative traffic and environmental situation modes. In Figure 4.3, we can see that for the D_{EF} domination cone, the corresponding $N[Y|D_{EF}]$ set contains the N points between fuel mode (FM) and emission mode (EM). This is why the polyhedral D_{EF} cone can be treated as the cone generated by preference directions $\{v^1,v^2\}$ connected with the FM and EM modes, that is, $D_{EF} = \left\{\sum_{i=1}^{2} a_i v^i \middle| a \geq 0\right\}$. Similarly, the traffic modes between dense mode (DM) and Sparse Mode (SM) modes can be generated. In the example, the control situations are represented by traffic (***a***), environmental (pollution level ***b*: EM**), and service quality (fuel consumption ***c*: FM**) modes. To increase clarity, we assume for condition attributes {a,*b*,*c*} the domains {0,1,2} (i.e., *a*: 0—sparse traffic, 1—light traffic, 2—medium traffic; *b*: 0—normal EM, 1—high EM, 2—pollution alarm; *c*: 0—low FM,

1—medium FM, 2—high FM), while for decision attributes $\{v_1,v_2\}$ the domain {0,1,2,3} with marking values as in Table 4.4 and Figure 4.3. The knowledge is represented in Table 4.4. The universe U consists of 10 elements $U = \{1,.,10\}$, the set of attributes $A = \{a,b,c,v_1,v_2\} = \{C,D\}$. We analyze the sets of condition $C = \{a,b,c\}$ and decision $D = \{v_1,v_2\}$ attributes: $U|I(C) = \{\{1\},\{2,3\}, \{4,6\},\{5\},\{7,9\}, \{8,10\}\} = \{Y_i{:}i = 1,.,6\}$; $U|I(D) = \{\{1\},\{2\},\{3\},\{4,5,6,10\},\{7\},\{8\},\{9\}\} = \{X_i{:} i = 1,.,7\}$.

From analysis of C and D, we get the following: the $C = \{a,b,c\}$ is dependent, the attribute c is superfluous. There is only one reduct (core) of $C{:}\{a,b\}$; thus, Table 4.1 shows dependency of $\{a,b\} \Rightarrow \{c\}$. The degree of dependency of attributes $D = \{v_1,v_2\}$ on the attributes $C = \{a,b,c\}$ is 0.4. The set of attributes C is independent of D, and $\{a\}$ attribute is D-core of C. This means that there are, in Table 4.4, the following dependencies: $\{a,b\} \Rightarrow \{v_1,v_2\}$; $\{a,c\} \Rightarrow \{v_1,v_2\}$.

The simple sorting rules generated from the reduced decision table (i.e., reduct $\{a,b\}$) have the following form: (1) If the traffic mode is "light" or traffic and environmental modes are highest, then the pollutant emission EM control mode $\{v_1,v_2\} = \{1,2\}$ is preferred. (2) If the traffic mode is "medium," then DM-related control modes are preferred, that is, $v_1 = 0$ with DM–EM control mode $\{v_1,v_2\} = \{0,2\}$ in the case of EM alarm. (3) If the traffic mode is "sparse," then SM-related control modes are preferred, that is, $v_2 = 0$ with FM = EM control mode $\{v_1,v_2\} = \{1,1\}$ when the EM is "high." In a similar way, we can analyze and support the other classification, detection, allocation, and selection tasks that have to be realized in the traffic control system.

CONCLUSIONS

The ITS system is potentially able to offer many important advantages, but it needs to communicate with its nodes: mobile (vehicles) and RSUs. There are a lot of used wireless protocols, and there are some projects for creating better solutions, but there are some tasks needed to improve mechanisms and adjust cooperation between units of ITS hierarchical systems. An important aim is to secure data transmission and making ITS a high priority. Moreover, all applications should be classified into logical groups, and node and network characteristics must clarify influences on the design of mechanisms.

REFERENCES

Adamski A.: Hierarchical traffic control system for the City of Cracow. *System Science VI* 21, 57, 20–37, 1979.

Adamski A.: ITS: Management, surveillance and control. Monograph: AGH University Scientific Publications, Krakow, 2003.

Adamski A.: Multi-criteria robust real-time dispatching control method. In *Proc. 8th Meeting of the EURO WG on Transportation*, M. Bielli and P. Carotenuto (Eds.) pp. 313–318, 2000.

Adamski A.: PIACON: Robust vehicle trajectory based arterial multi-criteria traffic signal control. In *Extra EURO Conf. on Handling Uncertainty in Transport Bari*, pp. 608–615, 2006.

Adamski A. and S. Habdank-Wojewódzki: Traffic congestion and incident detector realized by fuzzy discrete dynamic system. *The Archives of Transport, Polish Academy of Science*, 1, 2, 5–13, 2005.

Adamski A. and A. Kwaśniak: ITS: Hierarchical multi-layer system traffic safety option, ITS-ILS EURO WG on Transportation Conf., Krakow 2007.

Adamski A. and Z. Mikrut: The Cracovian prototype of video detector's feedback in transportation systems. In *Monograph: Transactions on Transport Systems Telematics*, pp. 140–151, 2004.

Capkun S., J.-P. Hubaux, and M. Jakobsson: Secure and privacy preserving communications in hybrid ad hoc networks. Technical Report EPFL-IC, EPFL, 2004.

Hossain E., G. Chow, V.C.M. Leung, R.D. McLeod, J. Mišic, V.W.S. Wong, and O. Yang: Vehicular telematics over heterogeneous wireless networks: A survey. *Computer Communications*, 33, 7, 775–793, 2010 (May).

Kerner B.S., C. Demir, R.G. Herrtwich, S.L. Klenov, H. Rehborn, M. Aleksic, and A. Haug: Traffic state detection with floating car data in road networks. In *Proceedings of the 8th International IEEE Conference on ITS*, September, pp. 44–49, 2005.

Ma Z., F. Kargl, and M. Weber: Measuring location privacy in V2X communication systems with accumulated information, 2009.

Schoch E., F. Kargl, T. Leinmuller, and M. Weber: Communication patterns in VANETs. *IEEE Communications Magazine*, 46, 11, 2–8, 2008.

5 New Design Philosophies in Architecture as a Way of Achieving Substantial Improvements of Office Buildings' Quality in Consideration of Sustainable Development

Dariusz Masły

CONTENTS

INTRODUCTION

The world of the 21st century is characterized by an increasing need for making building projects more ecologically viable. In the current situation where buildings in developed countries are responsible for around 50% of greenhouse gas emissions, being the most significant cause of environmental damage, it is critical that we create buildings with thorough knowledge of their impact on the planet's ecosystems over the long term.

There is a common knowledge about the building industry that it is slow to embrace change and that buildings are commonly created for short-term profits. Insufficient information and inadequate communication among building professionals are the most important cause of the building problems identified after occupancy. There is a lack of information about how buildings respond to the needs, requirements, expectations, and lifestyles of the people who use them. Davis and Szigeti (1996) emphasize that the building industry is "a fragmented industry, with a poor institutional memory, little research about the building product as a whole, and almost no feedback loop."

However, the global concern about sustainability requires the building industry to act differently from the conventional approaches, and a number of drivers for change in the building sector have been already developed by researchers involved in the process of enhancing building performance. Building quality evaluations are such drivers of change that will shape and influence our thinking about future sustainable high-performance buildings. It is increasingly evidenced in building delivery processes that professionals take a more systematic approach to controlling and shaping building performance. The primary goals are to achieve better building performance and better building quality as perceived by its users.

Rather than waiting for the way the building industry works to cause irreversible damage to our planet, acquiring and sharing knowledge about building performance helps us make the right decisions about our constructed and future buildings. Building evaluation methods have the ability to cause the performance of the buildings of the 21st century to be better than the performance of those of the 20th century. In this case, our future buildings will enable users to feel comfortable, as well as to make sustainable energy choices. These are the reasons why the building industry should make more use of this approach and why the new design philosophies in architecture will have to influence our design thinking.

SUSTAINABLE DEVELOPMENT CONCEPT AND ITS IMPLEMENTATION IN OFFICE BUILDINGS

The concept of sustainability involves a wide range of issues and concerns, so it is evident that it has been open to wide interpretations in different contexts. For example, Preiser and Vischer (2005) suggest that the fundamental concepts of sustainable development are as follows:

- "Equity"—the resource consumption should be equitable across the world.
- "Futurity"—current generations must not transfer the costs associated with their activities to future generations and should let them enjoy equal resources and development opportunities.
- "Climate stabilization"—some scientific hypotheses say that the dramatic increase of greenhouse gases emissions caused by human activities contributes to ozone depletion, global warming, and, as a consequence, irreversible changes in climate. Accordingly, there is a need to reduce the emissions to the extent that the climate will not be altered.

The concept of a buildings' sustainability can involve different components, but the following objectives are usually considered: ecological sustainability, economic sustainability, social sustainability, cultural sustainability, and levels of sustainability (Langford et al. 1999). Among the objectives, the key issue of sustainable development central to the area of professional designers' concern is ecological sustainability. Buildings should be environmentally friendly (they are expected to use less energy and water, generate less greenhouse gas, use materials efficiently, and reduce waste, as well as air and water pollution), as far as possible naturally ventilated, and adequately exposed to daylight for the health of their occupants; the quality of life of buildings' users should be optimized. A new model of sustainable development is based on reducing energy demand, rather than increasing energy supply, and on improving performance and process (ARUP et al. 2009). The energy efficiency can be achieved through both passive and active design strategies, such as optimal building form, skin, and orientation for daylight; utilization of direct solar exposure; narrow-plan buildings to maximize daylight and natural ventilation; thermal mass; efficient lighting; and ventilation systems with heat recovery. The design strategies offer enormous potential for radical improvement in building environmental performance, reducing unnecessary mechanical systems.

The process of making our office buildings more environmentally responsible starts at the beginning of a building life cycle, at stages of planning, architectural programming, and conceptual design, and the role of an investor or developer in the sustainable building shaping process is crucial. Preiser and Vischer (2005) suggest, "If the client wishes to make all of its activities fundamentally environmentally sustainable, the buildings it commissions and operates will reflect its core policies; and if not, then they probably won't." However, the main goal is to consistently apply strategies to achieve sustainability at each subsequent stage of the building life cycle, that is, from design to construction, occupation, and adaptive reuse.

This article recognizes the significant changes that office buildings have gone through. Offices have become more than just "the amount of square meters of workspace"; they have evolved into workplaces that fully meet employees' needs, requirements, and expectations. The results of analyses conducted at the turn of the 20th century on numerous office buildings indicate that the three key issues central to all sustainable office buildings, which will be further explored in this article, are as follows (Masły 2009):

1. Flexibility—spatial and functional solutions that respond easily to the changes in the size and structure of organizations and in the nature of office work
2. Provision of high quality of indoor environment, which results in occupant comfort, health, and well-being
3. Efficiency of the use of floor area

Flexibility

Baird et al. (1996) suggest, "Among the qualities that distinguish a good building from a bad one is the good building's capability to provide for different demands by owner and occupant groups and to respond quickly to the changing demands of its occupants and owners." The DEGW's Office Research, Buildings and Information Technology (ORBIT) studies, which investigated the impact of information technology on organizational structure and office building design, emphasized "the absolute imperative for offices of the future to incorporate provision for rapid change" (Duffy et al. 1998). Finally, Duffy (1996) emphasizes that "the real design problem for architects involved in office design is (...) to design the best way to use a changing stock of space to meet changing organisational needs over time." Therefore, the capability of an office building to respond easily and promptly to rapid changes in the size and structure of organizations and work practices—the office building's flexibility—is one of the three key issues of sustainable office buildings design (Cole and Brown 2009). Flexible offices can accommodate a wealth of programs from different types of workspaces to shops for daily needs, restaurants for a business lunch or dinner, banks, post offices, and recreational facilities during lunchtime.

Brand (1995) emphasizes that the leading theorist of change rate in buildings is Frank Duffy, cofounder of the Europe-wide architectural practice DEGW. In the early 1970s, Duffy said that there was no such thing as a building. He distinguished four layers of the modern office, each of them having a its own life cycle: shell (the building shell lasts usually the lifetime of the building—50 years); services (electricity and telecommunications, plumbing, HVAC [heating, ventilation, and air-conditioning], lifts—they are replaced every 7 to 15 years); scenery (the interior layout: partitions, dropped ceilings, raised floors—it changes every 5 to 7 years); and setting (mainly office furniture and lighting—often, they last weeks or months) (Duffy et al. 1998). Brand (1995) has expanded Duffy's "4 *S*s" into "6 *S*s," including site, structure, skin (20-year life cycle), services, space plan, and stuff. Brand says, "Duffy's time-layered perspective is fundamental to understanding how buildings actually behave." Crucially, each layer of the modern office has to be designed with a high degree of independence from the others. Many offices were demolished too early because their outdated layers of shorter life cycle were too expensive to change.

The following performance issues of office buildings' flexibility are of fundamental importance to the ongoing operation of a building:

- Type of main bearing structure:
  - Lack of structures impossible or difficult to remove, for example, intermediate columns or load-bearing walls, within an office workspace
  - Floor-to-ceiling height
  - Structural potential for layouts in accordance with defined modular arrangements
- Possibility to divide an office floor area into independent units
- Flexible under-floor services distribution (electricity and telecommunications, ventilation)

Provision of High Quality of Indoor Environment

Today, we are surrounded by office buildings (glass, high-rise, air-conditioned offices) embedded with technologies designed to provide optimum comfort conditions more reliably and effectively than users. Cole and Brown (2009) suggest that during the 20th century, "technological innovation led to a shifting of design responsibility in comfort provision from architects to mechanical engineering consultants, and control responsibility from occupants to technology." However, outcomes of the latest research confirm that while a universally applicable set of optimum comfort conditions is convenient to engineers, they may not reflect the actual needs of users. Paramount to occupant comfort, satisfaction, and optimal performance is some level of direct control over the environmental conditions in the workplace; users should be enabled to easily intervene to adapt the indoor environment and to relieve discomfort. The high quality of an indoor environment or healthier interiors can be achieved through the use of a narrow plan and an optimal building form and orientation. The sustainable high-performance office buildings should be capable of natural ventilation for extended periods and have to be properly day-lit whenever weather conditions allow. Nevertheless, the fundamental step to creating healthy indoor environments, which are linked to gains in productivity, decreased absenteeism, improved employee, morale and increased occupant satisfaction, is enabling buildings' users to make appropriate adjustments in the environmental conditions in their workplace.

What is interesting, in light of this discussion, is that there is a need for a systematic approach toward how to reduce the building industry impact on the natural environment. This systematic innovative approach to the planning, programming, design, construction, and occupancy of buildings is based on holistic, multidisciplinary building quality evaluations. Moreover, this approach is being increasingly incorporated into best practice management of building delivery processes.

MAIN DIRECTIONS AND FUNDAMENTAL PRINCIPLES OF QUALITY ANALYSES OF BUILDINGS

Baird et al. (1996) state that "building evaluation is the systematic assessment of building performance relative to defined objectives and requirements," and he emphasizes that it is well established as a concept. Building evaluations are a mechanism through which the actual performance of buildings can be compared to explicitly documented required performance. The evaluations' main objectives are to assess the physical environments in terms of how people are using them and to develop knowledge about it. Quality assessments enable us to "produce new knowledge about a specific building or new knowledge about some general aspects of all or most buildings of a certain type" (Baird et al. 1996).

We can identify many methods of building evaluation, most of which have reached their current level of sophistication over the last three decades. However, at the broadest level of classification, we distinguish two types of building quality assessments (BQAs):

- Flexible methods focusing on users' experience of the performance of buildings
- Methods based on a systematic set of building performance categories

Flexible Methods Focusing on Users' Experience of the Performance of Buildings

Methods focusing on users' experience of the performance of buildings reflect an inclusive and participatory people-centered design approach. This means that needs, expectations, behavior or lifestyles of the people using a facility are taken into account as the prime issue and that the users are participating in a research process as the evaluators and experts. Flexible methods are based on the postoccupancy evaluation (POE) process model. The theoretical foundation of POE is the building performance concept with its four principles: measurement, comparison, evaluation, and feedback (Preiser 1989). Clements-Croome (2004) emphasizes that the POE method is one of the key instruments to appropriately assess building performance. Besides, the evaluation method is widely used around the world. Flexible methods can be applied to any type of building or environment. Using them is the best or most appropriate way to evaluate a building when we want to, for example, better match the demand and supply in the building, increase user satisfaction, or improve productivity within the workplace. While there is not a single formula for evaluation focusing on users' experience of the performance of buildings, because there are many ways to plan and conduct such a research process, well-established measurement and evaluation techniques are used to collect the relevant data. Examples of these techniques include direct observation, still photography, focused interviews, focus groups, occupancy surveys, questionnaires, and walkthroughs. Methods focusing on users' experience of the performance of buildings represent an open approach. This means that they can be adapted easily to reflect the particular requirements of an analyzed organization.

Methods Based on a Systematic Set of Building Performance Categories

Among the most widely used methods of this type are BQA (Bruhns and Isaacs 1996), Serviceability Tools and Methods (STM) (Davis and Szigeti 1996), the Real Estate Norm (REN) (Real Estate Norm Netherlands Foundation 1992), and Vacancy Risk Meter (VRM) (van der Voordt and Geraedts 2003). Methods based on a systematic set of building performance categories provide very rigorous and systematized means of measuring the performance of buildings. The data on an evaluated quality are investigated and recorded in a consistent, coherent structure. At the heart of these methods is a set of performance categories describing a particular building type, such as an office. Examples of these categories include "access to building," "vertical movement of goods," "main staircase capacity," and "main structure flexibility," as well as "identity outside building." Every category has its own scale for rating using precise technical and performance terms (i.e., "access of daylight into the building—floor area from the facade with windows [6 m deep] as percentage of total gross floor area": 1, less than 70%; 2, between 70% and 80%; 3, between 80% and 90%; 4, between 90% and 90%; 5, between 95% and 100%) (Real Estate Norm Netherlands Foundation 1992). The methods enable evaluators to rate an analyzed building in a short time without being oversimplifying; the results of an evaluation are usually available as a profile of performance categories' scores. Davis and Szigeti (1996) emphasize that the profile is "far more informative than a single score and far less misleading," because it enables comparison between outcomes and the project goals or comparison of category scores for two analyzed buildings, while a "bottom line," single score only indicates that an evaluated building is better or worse (Davis and Szigeti 1996). Evaluated performance categories are not of equal importance; a system of weighting enables their establishment within a clear framework. Weightings can be based on the experience of evaluators or a user opinion survey. Methods based on a systematic set of building performance categories include various kinds of tools (i.e., multiple-choice questionnaires or a booklet containing a specification or description for each performance category), along with procedures, and even computer software for using them. Some common uses of the assessment methods are as follows:

- Defining users' requirements in the form of goals, recommendations, and performance criteria documented in the functional program or brief

- Rating the capabilities of buildings and facilities to meet users' requirements and needs, while selecting properties to rent or buy—comparing competing buildings
- Matching of occupant requirements (demand) and buildings' capabilities to meet those requirements (supply)
- Evaluating opportunities for improving the performance and attractiveness of a building, identifying the benefits and drawbacks of an analyzed building
- Reviewing architectural designs

Methods based on a systematic set of building performance categories provide clear and explicit performance criteria against which the quality of buildings can be measured easily and regularly; therefore, they can be used for benchmarking and worldwide standardization. Examples of worldwide sustainable building assessment methods include the following (Niezabitowska and Masły 2007): Building Research Establishment Environmental Assessment Method (BREEAM), Building Environmental Performance Assessment Criteria (BEPAC), Green Building Rating System, Leadership in Energy and Environmental Design (LEED), Green Building Challenge (GBC)—GBTool, Comprehensive Assessment System for Building Environmental Efficiency (CASBEE). These evaluation methods focus on analyzing buildings from the point of view of their influence on the natural environment, energy consumption, and preservation of natural resources.

Building Performance Evaluation

Currently, it is Building Performance Evaluation (BPE) that offers the most complex, advanced, and holistic approach to the improvement of the quality of buildings (Preiser and Vischer 2005; Preiser and Wang 2008). The method originated from POE, and the experience acquired from quality evaluations of buildings occupied for some period of time helped in developing this unique approach. The BPE method is a way of systematizing the research activities needed to acquire feedback from users at every stage of a building's life cycle. Preiser and Vischer (2005) emphasize that each of the six major phases of building delivery and life cycle, that is, planning, programming, design, construction, occupancy and facility management, and adaptive reuse of buildings, is crucial as the time of collecting data about our buildings to improve the quality of decisions made throughout the building delivery process. The BPE approach is a most appropriate way to assess the impact of design and construction decisions over the long term, and it is applicable to all building types. The BPE framework is depicted by its authors as "an ever-expanding helix of knowledge on building performance." Each of BPE's subphases (planning, programming, etc.) has internal review and feedback loops; moreover, the whole process is the main enormous loop. The knowledge gained throughout one building delivery and life cycle is fed forward into the next building cycle. The main objective of this approach is to gradually develop a knowledge base that will be translated into building performance criteria and accumulated in building type–specific databases and published guidelines. Preiser and Schramm (2005) emphasize, "BPE constitutes an important step in validating performance standards that may already exist, or that have to be developed for a given building type" (Preiser and Vischer 2005).

CONCLUSIONS

Increasingly, research confirms the importance of conducting building quality evaluations as being paramount to making the best of existing buildings and creating new sustainable, high-performance buildings of the 21st century. Duffy et al. (1998) emphasize that design proposals should always be tested by research and that consultancy needs to be constantly challenged by built reality. The benefits, to both providers of buildings and users, of using building quality evaluations are evident and are as significant and diverse as the following (Preiser et al. 1988; Baird et al. 1996):

- Better communication between supply side (providers of buildings) and demand side (buildings' users)
- Better matching of occupant requirements (demand) and buildings' capabilities to meet those requirements (supply), resulting in increased occupant comfort, satisfaction, health, and well-being
- Informed decision making and better understanding of design's consequences
- Minimization of costs throughout the building life cycle
- Update of building performance standards, guidelines, norms, design databases, architectural manuals, and regulations
- Development and dissemination of the latest knowledge on sustainable buildings and direct contribution to the improvement of the quality of design services and the quality of the construction industry as a whole

Finally, it should be noted that building evaluations can be used in many different ways to help plan, cost, design, use, maintain, operate, and manage buildings. The Faculty of Architecture at the Silesian University of Technology, Gliwice, and especially the Department of Strategy of Design and New Technologies in Architecture, has been involved in building quality evaluations since the late 1990s. The results of the studies are published and widely disseminated (Niezabitowska 2004, 2005, 2006; Niezabitowska and Masły 2007; Masły 2009). Further research is being conducted, including several habilitation projects devoted to sustainable intelligent high-performance offices, facades and their impact on the indoor environment quality, and residential homes for seniors, as well as university buildings.

REFERENCES

ARUP, Sauerbruch Hutton, Experientia, Galley Eco Capital (2009), *C_life. City as living factory of ecology*, Manual, Low2No Design Competition, Jatkasaari, Helsinki, Finland (www.low2no.org).

Baird G., Gray J., Isaacs N., Kernohan D., McIndoe G. (1996), *Building Evaluation Techniques*, McGraw-Hill, New York.

Brand S. (1995), *How Buildings Learn. What Happens After They're Built*, Penguin Books, New York.

Bruhns H., Isaacs N. (1996), Building Quality Assessment, in Baird G., Gray J., Isaacs N., Kernohan D., McIndoe G., *Building Evaluation Techniques*, McGraw-Hill, Inc., New York.

Clements-Croome D. (2004), *Intelligent Buildings: Design, Management and Operation*, American Society of Civil Engineers, Thomas Telford Ltd.

Cole R. J., Brown Z. (2009), *Reconciling Human and Automated Intelligence in the Provision of Occupant Comfort*, Intelligent Buildings International, Number 1, Earthscan.

Davis G., Szigeti F. (1996), Serviceability Tools and Methods (STM): Matching Occupant Requirements and Facilities, in Baird G., Gray J., Isaacs N., Kernohan D., McIndoe G., *Building Evaluation Techniques*, McGraw-Hill, Inc., New York.

Duffy F. (1996), Building Appraisal: Measuring the Capacity of Office Buildings to Accommodate Change, in Baird G., Gray J., Isaacs N., Kernohan D., McIndoe G., *Building Evaluation Techniques*, McGraw-Hill, Inc., New York.

Duffy F., Greenberg S., Myerson J., Powell K., Thomson T., Worthington J. (1998), *Design for Change. The Architecture of DEGW*, Watermark Publications, Birkhauser Verlag, Boston.

Langford D. A., Zhang X. Q., MacLeod I., Dimitrijevic B. (1999), Design and Managing for Sustainable Buildings in the UK, in Ogunlana S.O., *Profitable Partnering in Construction Procurement*, CIB W92 (Procurement Systems) and CIB TG 23 (Culture in Construction) Joint Symposium, E. & F.N. Spon, London.

Masły D. (2009), *Jakość budynków biurowych w świetle najnowszych metod oceny jakości środowiska zbudowanego*, Wydawnictwo Politechniki Śląskiej, Gliwice, Poland (in Polish).

Niezabitowska E. (ed.) (2004), Baron A., Bielak M., Fross K., Komar B., Masły D., Niezabitowska E., Niezabitowski A., Sitek M., Staniszewski Z., Tymkiewicz J., Winnicka-Jasłowska D., *Wybrane elementy Facility Management w architekturze*, Wydawnictwo Politechniki Śląskiej, Gliwice, Poland (in Polish).

Niezabitowska E. (ed.) (2005), *Budynek Inteligentny Tom I. Potrzeby użytkownika a standard budynku inteligentnego*, Wydawnictwo Politechniki Śląskiej, Gliwice, Poland (in Polish).
Niezabitowska E. (ed.) (2006), Badania jakościowe środowiska zbudowanego, Seria Wydawnicza pod redakcją A. Bańki Stowarzyszenia *Psychologia i Architektura: Zachowanie, Środowisko, Architektura*, Poznań, Poland (in Polish).
Niezabitowska E., Masły D. (ed.) (2007), Komar B., Kucharczyk-Brus B., Masły D., Niezabitowska E., Niezabitowski A., Niezabitowski M., *Oceny jakości środowiska zbudowanego i ich znaczenie dla rozwoju koncepcji budynku zrównoważonego*, Wydawnictwo Politechniki Śląskiej, Gliwice, Poland (in Polish).
Preiser W. F. E. (1989), *Building Evaluation*, Plenum Press, New York.
Preiser W. F. E., Rabinowitz H. Z., White E. T. (1988), *Post-Occupancy Evaluation*, Van Nostrand Reinhold, New York.
Preiser W. F. E., Schramm U. (2005), A Conceptual Framework for Building Performance Evaluation, in Preiser W. F. E., Vischer J. C., *Assessing Building Performance*, Elsevier Butterworth-Heinemann, Oxford.
Preiser W. F. E., Vischer J. C. (ed.) (2005), *Assessing Building Performance*, Elsevier Butterworth-Heinemann, Oxford.
Preiser W. F. E., Wang X. (2008), Quantitative (GIS) and Qualitative (BPE) Assessments of Library Performance, Archnet–IJAR, *International Journal of Architectural Research*, 2, 1, March.
Real Estate Norm Netherlands Foundation (1992), *Real Estate Norm. Method for Advising On and Evaluating Office Locations and Office Buildings. Second Completely Revised Edition*, REN Netherlands Foundation, Nieuwegein, The Netherlands.
van der Voordt D. J. M., Geraedts R. P. (2003), Good Buildings Drive Out Bad Buildings. An Instrument for Defining the Lower End of the Office Premises Market, Proceedings of the CIB Conference W104: Open Building Implementation, Hong Kong.

6 The Role of Emotions in Cognitive Processes

Theoretical and Empirical Basis for Complex Brain–Computer Interfaces—Electroencephalography and Functional Magnetic Resonance Imaging

Krystyna Golonka, Justyna Mojsa-Kaja, and Tadeusz Marek

CONTENTS

INTRODUCTION

Technological development and a growing amount of neuroimaging research results in more and more advanced solutions, which have become applicable in many contexts, for example, rehabilitation, neuroscientific experiments, the military, and entertainment. The progress of systems that enable direct communication between the brain and external devices, such as brain–computer interfaces (BCIs), results in modifying such systems into more complex schemes. For example, modifications of existing systems into, as yet, less available and more difficult in applying systems exploit the advanced technology of electroencephalography (EEG) and functional magnetic resonance imaging (fMRI). Understanding the limited possibilities for extensive applications of such systems, for example, fMRI-BCI and dense EEG (dEEG)–BCI, the presented chapter concentrates on the theoretical and empirical premises for complex neuroadaptive systems as a significant and promising field (Fafrowicz et al. 2012).

The aim of the chapter is to analyze the processes that influence human behavior and to describe the methods and techniques that allow us to characterize them. Especially, the influence of emotional processes on cognition and executive functions are explored. This may lead further to the prediction and control of specific actions and complex behavior. The chapter is written from the perspective of neuroergonomics (Fafrowicz and Marek 2007).

EMOTIONS AND THEIR INFLUENCE ON EXECUTIVE PROCESSING

Emotions and their impact on executive processing may be viewed in two ways—emotions considered as the situational factors and as the individual factors. In the first case, we think about emotion evoked in a specific context as the response to a current stimulus. In the case of individual factors, we consider emotions as the individual disposition to experience particular emotions that define an individual's personality and one's tendency to react and behave in a peculiar manner.

Emotions play a significant role in regulating an individual's behavior, and they are inherent with attentional, cognitive, and motivational processes. Attention predetermines further information processing—what aspects of current situation will be analyzed, how it will be incorporated into the present state of knowledge, which information will be evoked, and so forth. Emotions and cognition are inherent aspects of the motivational processes that lead to behavioral tendencies and actions. Although there were attempts to contrast the state of *feeling* with the state of *knowing*, more and more evidence reveals the intercorrelated character of emotional, cognitive, and motivational processes that define complex cognitive and behavioral manifestations.

The attempt to divide cognitive and emotional processes into separate processes was the most emphasized in philosophical dualism and metapsychological theories, where cognition is opposed to emotion and affect, and vice versa. In psychological theories, the fact that this evaluation of the stimulus is automatic and uncontrolled, and thus an unconscious process, shows the conceptualization of emotion as being in opposition to the cognitive process, which is usually described as voluntary and conscious.

In psychological cognitive approach, emotions are evoked by the automatic detection that a stimulus has an important implication considering the possibility of realizing the main goals of the individual. This *evaluation* of the stimulus, regarding its novelty, importance, imminence, and unexpectedness, has direct influence on what kind of emotion will be caused and regulates its strength. Scherer (2001) describes other multiple complementary criteria including agreeableness, goal conduciveness, coping potential, and norm compatibility.

Regarding many theories, it may be concluded that this initial evaluation in the emotional process plays a significant role in cognition (LeDoux 1989; Williams et al. 1997), surviving (Oatley and Johnson-Laird 1987; Frijda 1986), and informing and regulating mechanisms (Frijda 1986), as well as in appraisal mechanism (Scherer 2003; Ellsworth and Scherer 2003). These emotion's functionalities are not excluded, but depending on the theory, different goals of emotions are underlined.

A fundamental characteristic of emotion is its functionality regarding the ability to regulate behavior so as to help the individual to achieve his/her goals. The main goal is to *survive*—in the physical as well as in psychological and social meanings (Lewandowska et al. 2012). For some researchers (Schwarz and Clore 1983), the most fundamental role of emotions is to *inform*. Emotions inform an individual about his/her state and the meaning of surrounding stimuli and the present situation; they are very important with regard to the ability to realize one's aims and adapt to current circumstances. Informing also has value for others and plays a significant role in regulating social relations. Emotions express an individual's intentions and behavioral tendencies.

In the appraisal model of emotion, it has been emphasized that elicitation of a specific emotion depends on the complex process of evaluation of the significance of an event, which depends on individual needs, values, goals, and the current context. Emotions would be conceptualized as a continuum of the appraisal mechanisms (Sander 2008).

Emotions affect cognitive processing—attention, perception, and complex information processing—they activate or inhibit particular perceptual and cognitive schemata and influence the specific categorization in the current situation. Emotions affect memory encoding and recall—influencing its speed and capacity—as well as reasoning, judgment, situation assessment and decision making, and planning and learning (Williams et al. 1997; LeDoux 1989).

Emotions have significant influence on information processing—they can amplify or attenuate this process, depending on their intensity and time duration. Schwarz and Clore (1983) highlight that emotions that have an interfering effect are connected with strong intensity and/or long-lasting character. This is associated mainly, but not only, with negative emotions (mania). According to some researchers (Gasper and Clore 2002), the affective states of weaker intensity, like moods and propensity for experiencing certain affective emotions, have an influence on the manner of information processing—negative affective disposition is related to more analytical and detailed processing, in which more precise analyses are revealed. In positive affective states, more holistic, stereotypical processing is revealed, which is related to more general analysis defined by a more superficial approach but which has a broad scope. It is difficult to point to the predominance of any affective disposition—the character of the task and its complexity is of more importance. In tasks that require greater accuracy, a moderate negative affect may be more conducive. Conversely, in tasks that require creative thinking aimed at an originative solution, moderate positive affect seems to be more effective.

There is more and more evidence that emotional and cognitive processes are strongly related with each other. Especially, the results of neuroimaging studies call into question the dualistic approach. The isolated modules of emotions (e.g., the amygdala) or emotion circuits (e.g., limbic system) have been replaced with the view of a pervasive neuromodulatory role of emotion across the brain circuitry (Fellous 2004).

For example, Phelps (2006) suggests that fear can enhance perception. In general, emotional situations can enhance attention, which was proved in the experiments in which a reduced amount of attentional blinks was observed—emotions improved perception. Although there are still considerable controversies, it seems that emotions change the range of the visual area.

"Investigations into the neural systems underlying human behavior demonstrate that the mechanisms of emotion and cognition are intertwined from early perception to reasoning. These findings suggest that the classic division between the study of emotion and cognition may be unrealistic and that an understanding of human cognition requires the consideration of emotion" (Phelps 2006).

The other piece of evidence of the intercorrelation between emotions and cognition comes from the literature concerning different emotional disorders. For example, according to Etkin et al. (2006), people who suffer from posttraumatic stress disorder (PTSD) or depression are unable to control emotional intrusion into their thoughts. In their study, they analyzed the interconnection between the anterior cingulate cortex (ACC) and that amygdala as structures that are recognized as the brain centers of cognitive and emotional processing, respectively. The aim of their experiment was to distinguish if the ACC was responsible for "monitoring" conflict between cognitive and emotional processing or for actively "resolving" that conflict. Coping with distracting emotions due to "executive processing" restrains activity of "emotion processing"—executive processing helps to inhibit emotions that could interfere with mental functioning.

NEURAL BASIS OF EMOTIONS AND ITS RELATIONS TO COGNITIVE PROCESSES

Many theorists and researchers are interested in relationships between emotions and cognition (Damasio 1994; LeDoux 1996). One of the possibilities to explore this problem is to analyze the neural basis of those processes.

The neuroimaging technologies allow us to conclude that emotions and cognition are not separate processes. Theoretical questions and theories that led to philosophical dualism and psychological dualistic approaches (Breuer and Freud 1895) have given the answer from empirical studies.

The *peripheralist* theory of William James was the earliest and most influential approach to the neural basis of emotions (James 1884). We do not escape because we feel fear—we feel fear because we escape; experiencing of emotion is due to peripheral responses, and the conscious aspect of emotion arises later as the mental consequence of observing one's response/behavior. James emphasized the automatic involuntary character of emotional responses. Still, researchers underline a central role for afferent information from the body that influences emotional expression and subjective feelings:

"This dynamic modulation of visceral state is neurally mediated by sympathetic and parasympathetic divisions of the autonomic nervous system. Moreover, neural afferents convey representations of the internal state of the body back to the brain to further influence emotion and cognition" (Critchley 2009).

According to Critchley (2005): "Feedback representations of changing states of bodily arousal influence learning and facilitate concurrent and prospective decision-making. (...) In particular, anterior cingulate cortex is implicated in generating autonomic changes, while insula and orbitofrontal cortices may be specialized in mapping visceral responses."

Although the *peripheralist* theory has been important in describing the neural basis of emotions, the criticism of the following researchers led to the accenting of the central nervous system's role in generating and regulating emotions. The *centralist* theory proposed by Cannon (1929) was a subsequent theory based on Bard's (1929) experimental findings in animals. According to Cannon's theory of emotion, the thalamus mediates emotional perception, and the hypothalamus mediates emotional expression (Frackowiak et al. 2004). The neural basis of emotions was described later by Papez (1937), who expanded Cannon's circuit with additional structures: the hippocampus and mammillary bodies. Further, Maclean (1952) incorporated his own clinical observation and Freudian metapsychological structuralism into a *limbic* theory of emotion, in which the hippocampus plays a central role, being described as an "emotional keyboard" (Frackowiak et al. 2004). *Centralist* theories located emotions in the visceral brain, more primitive and unconscious, while cognition resided in the more recently evolved cerebral neocortex (Frackowiak et al. 2004). Future research, using more advanced technologies, opened up new possibilities in analyzing brain activity, and it gave some evidence that was not consistent with the limbic approach (LeDoux 1996).

New technologies and methodology allow us to indicate a more detailed description of emotional processing and its connectivity with cognitive processes. In trying to analyze particular structures involved in emotional processing, researchers describe its functional segregation; while in delineating structures' connectivity, mutual influences, and relationships, they focus on functional integration of emotional processing.

In functional segregation, Frackowiak et al. (2004) specify the following brain structures involved in emotional processing: amygdala, orbitofrontal cortex (OFC), insula, and ACC. Davidson et al. (2000) describe the circuitry underlying emotion regulation, which includes prefrontal areas—orbital prefrontal cortex, ventromedial prefrontal cortex, dorsolateral prefrontal cortex, and ACC and amygdala. But with regard to differentiating aspects of emotions, affective style, and regulatory processes, other interconnected regions are included in this circuit, like the hippocampus, hypothalamus, insular cortex, ventral striatum, and others. Any disturbances in functioning or structure in one or more of these interconnected regions cause failures of emotion regulation (Davidson et al. 2000). The key structures of this circuit are analyzed below.

OFC

The cortex of the ventral (orbital) surface of the frontal lobe has been divided into rostral, lateral, caudal, and medial areas. It has the main thalamic input from the medial part of the mediodorsal nucleus, which receives inputs from temporal lobe structures, including the amygdala (Krettek and Price 1974); the caudal OFC receives a direct projection from the amygdala and olfactory, gustatory, somatosensory, and visual inputs (Carmichael and Price 1995b); medial and lateral areas receive direct inputs from the ACC (Carmichael and Price 1995a). The OFC has outputs to the basal ganglia, hypothalamus, brainstem, inferior temporal cortex, amygdala, and ACC (Cavada et al. 2000; Frackowiak et al. 2004).

"The human orbitofrontal cortex is an important brain region for the processing of rewards and punishments, which is a prerequisite for the complex and flexible emotional and social behaviour which contributes to the evolutionary success of humans" (Kringelbach and Rolls 2004). Animals with lesions to the OFC demonstrate changed behaviors towards aversive and appetitive stimuli and show disruptive reward learning. According to Rolls (1999), the OFC is engaged in reassessment and relearning processes of emotional associations. The role of the OFC in the regulation of behavior is connected with the modification of emotional responses in a dynamic and flexible way (Frackowiak et al. 2004). This function of the OFC seems to be crucial in adaptation to changing context and, thus, the ability to moderate individual responses and being responsive to change.

ACC

The ACC is a region around the rostrum of the corpus callosum that is described as the "anterior executive region" (Devinsky et al. 1995). Initially it was assumed that the cingulate cortex was associated with the limbic system and was usually linked to emotional processes.

Some experiments revealed, however, that the anterior part of the cingulate gyrus is related to emotion and motor functions, while the posterior region is involved in visual, spatial, and memory functions and has little or no involvement in emotional processes (Devinsky et al. 1995).

Focusing on the aspect of emotion, there are many functional neuroimaging data that support the thesis of the ACC's role in emotional processing. Frackowiak et al. (2004) quote studies that indicate the ACC's responses to facial expressions of emotion, processing emotional pictures and words, fear-conditioned stimuli, cardiovascular arousal, and pain: intensity, affect, and expectation. Further data led to the conclusion that ACC may be divided into dorsal and ventral subdivisions—the dorsal region in nonemotional, cognitive tasks was linked to stimulus selection and response competition, while the ventral area was specifically activated in emotional tasks, and its function was described as selecting suitable responses with respect to current priorities and goals (Frackowiak et al. 2004).

From meta-analysis of emotion activation studies, Phana et al. (2002) concluded that induction by emotional recall/imagery and emotional tasks with cognitive demand involves both the ACC and insula.

Insula

The "insula cortex is a multimodal sensory region with visceral, gustatory, somatosensory, visual, and auditory afferents and reciprocal connections to amygdala, hypothalamus, cingulate gyrus, and orbitofrontal cortex" (Frackowiak et al. 2004).

In describing the function of the insula, emotional and homeostatic regulations are usually emphasized. Critchley et al. (2004) pointed out that the greater right anterior insular gray matter volume, the more accurate the subjective sense of the inner body, proving that the right anterior insula mediates interoceptive awareness. In the study by Critchley et al., activity in right anterior insula correlated with interoceptive accuracy and with measures of subjective negative emotional experience. Researchers underlined the role of this region in explicit awareness of bodily processes and conscious awareness as subjective feeling states.

Williamson et al. (2001) conducted an experiment whose results indicate that the level of insular activation differs depending on the intensity of exercise and perceived exertion.

Amygdala

The amygdala, a small nucleus in the medial temporal lobe, is involved in integration of biologically salient stimuli—it is a critical component of the brain's fear circuitry. In human and animal research, this structure is an essential center in inducing the automatic, reflexive response in relation to threats and dangers in the environment (Frackowiak et al. 2004).

The amygdala has a complex internal structure and extensive external anatomical connections. It consists of at least 13 anatomically and functionally distinct subnuclei (Amaral et al. 1992), and its extrinsic connectivity gives the amygdala the potential to integrate sensory information from all modalities and to influence autonomic and motor output systems (Frackowiak et al. 2004). Its efferents project to many regions of the brain, including the hypothalamus, OFC, basal forebrain, entorhinal cortex, ventral striatum, cingulate gyrus, and brainstem autonomic centers, as well as back to early sensory processing regions (e.g., occipitotemporal and insula cortices) from which the amygdala receives inputs (Amaral et al. 1992). The amygdala's feedback connections give potential evidence for its role in modulating early sensory processing (Frackowiak et al. 2004).

In numerous animal studies, it has been reported that the amygdala plays an important role in generating fear response—electrical stimulation evokes fear responses, while bilateral ablation or lesion of the temporal lobes (including amygdala) eliminate or inhibit fear responses (Kluver and Bucy 1939; Weiskrantz 1956; Greenberg et al. 1984; LeDoux 1987; Amalar 2002). Researchers emphasize also the role of the amygdala in learning about threats and dangers; for example, LeDoux (1996) reports that destruction of the amygdala prevents the learning of temporal associations between neutral and aversive stimuli.

In human studies, the analogous observation was reported in patients with restricted bilateral amygdala damage—they demonstrated a specific deficit in the recognition of fearful facial expressions, whereas they performed correctly other psychological tests (Adolphs et al. 1994; Calder et al. 1996).

Additionally, neuropsychological studies on healthy subjects revealed the lateralization of the amygdala—in some research on basic emotional facial expressions, fearful faces evoked left, but not right, amygdala activity (Morris et al. 1996), or they evoked higher/stronger left amygdala activation than the right one (Breiter et al. 1996).

In some modifications of studies on emotional facial expressions, researchers used kind of incongruent tests where fearful eyes were combined with neutral mouths and vice versa (Morris et al. 2002). For recognizing such emotions, eyes were sufficient to evoke the brain's activation, which was observed in the right amygdala, superior colliculus, and posterior thalamus. Mouths expressing fear were not enough to elicit increased responses, which suggests a more dominant role of the eyes in expressing emotions. Concerning higher activation in the right than in the left amygdala, it may be hypothesized that in case of incomplete information, more intuitive strategies are used, and they activate the right hemisphere, which is more connected with this kind of information processing.

Concerning other basic emotions that elicit amygdala responses, sadness, anger, and disgust were reported to evoke increased amygdala activity (Blair et al. 1999; Hariri et al. 2000; Gorno-Tempini et al.

2001). It seems that the amygdala is activated in the context of salient stimuli that evoke the negative individual response.

Analyzing the amygdala's connections, there is also evidence of a subcortical pathway for visual signals, which reach the amygdala independently of the visual neocortex, which supports the proposal of a colliculo-thalamo-amygdala visual pathway—the superior colliculus receives direct retinal input and projects to the pulvinar and extrageniculate visual nuclei in the posterior thalamus, which sends direct projections to the lateral nucleus of the amygdala (Linke et al. 1999, 2000). The reflexive nature of the fear response is better understood in the light of evidence of subcortical pathways, which limit the time of the individual reaction in response to salient stimuli. Comparing this with automatic responses, more complex analysis of the same stimuli leading to semantic or verbal categorization requires the sensory neocortex (LeDoux 1996).

Amalar (2002) underlines the regulating role of the amygdala in individuals' behavior, depicting the amygdala's function as "a protective *brake* on engagement of objects or organisms while an evaluation of potential threat is carried out," suggesting that social anxiety may come from a dysregulation or hyperactivity of the amygdala's evaluative process.

Although the amygdala is recognized to be an emotion center in the brain, it is specifically activated in basic emotions with negative valence, such as fear. A meta-analysis of emotion activation studies with positron emission tomography (PET) and fMRI led to the conclusion that fear specifically engaged the amygdala, and this region as well as the occipital cortex are also activated by emotional induction by visual stimuli (Phana et al. 2002).

In LeDoux's (2000) research, the amygdala has been pinpointed as "an important component of the system involved in the acquisition, storage, and expression of fear memory."

Etkin et al. (2006) designed an experiment in which they presented emotional stimuli (faces and words) in congruent and incongruent modifications. In the case of incongruent face–word combinations (e.g., happy face with the word "fear"), a response conflict between emotional and cognitive stimuli was evoked. It was observed that if incongruent stimuli were presented one after another, it helped to resolve the conflict more readily in the second trial.

Using fMRI techniques, Etkin and his colleagues (2006) found that emotional stimuli activated the amygdala, but this activity was inhibited by specific activation of the rostral ACC (rACC). They suggest that rACC has an inhibitory control over the amygdala.

The amygdala's interconnections with frontal areas were investigated in many studies where affective disturbances were the side effect or the component of an unbalanced state of individual homeostasis. For example, Yoo et al. (2007) research on sleep deficit revealed differences in intensity and volumetric extent of amygdala activation between the two groups, sleep control and sleep deprivation, observing higher activity in the latter one. Moreover, level differences in amygdala functional connectivity were observed—significantly greater connectivity in the medial prefrontal cortex for the sleep control group, but significantly stronger connectivity with autonomic brainstem regions in the sleep deprivation group.

AFFECTIVE DISORDERS AND EXECUTIVE FUNCTIONS

Emotions and their regulating functions are unbalanced in all neuropsychiatric disorders, such as anxiety disorders, attention deficit/hyperactivity disorders, depression, anorexia and bulimia nervosa, schizophrenia, and so forth. What is typical in all emotional disorders is the disparity between the situational demands and the amount of attention and strength of reaction as the response to current stimuli. An emotion's intensity is disproportional to the requirements of specific circumstances.

MacLeod et al. (1986) reported that anxious subjects have a tendency to shift attention toward emotionally threatening stimuli in their visual field, while normal control subjects tend to shift their attention away from such stimuli. A similar tendency was reported by Neisser (1976), who, in anxious subjects, observed facilitation of the encoding of threat-related stimuli, whereas nonanxious subjects tended to actively inhibit such encoding.

According to MacLeod et al. (1986), in nonanxious individuals, the danger must exceed some threshold to direct their attention and facilitate the encoding process of apparently danger-related information. Anxious individuals, on the other hand, seem to automatically react to any threatening cues of current stimuli. In the experiment with threatening stimuli, researchers reported also the negative affective state evoked as a response to given stimuli in the group of anxious subjects (MacLeod et al. 1986).

The individual who has a higher level of anxiety is more sensitive to threatening stimuli—this tendency is observed on all levels of information processing and executive control. Mathews and MacLeod (1985) emphasize the existence of an attentional bias in anxious patients and suggest that anxious individuals show evidence of overactive danger schemata.

MacLeod argues that in the case of depression, there is little evidence to suggest an attentional bias favoring the encoding of depression-related material (e.g., Gerrig and Bower 1982; Gotlib and McCann 1984). Opposite to anxiety that is "characterized by attentional biases operating at a perceptual level to facilitate the pickup of mood congruent, emotionally threatening information," depression is "characterized by biases operating at the retrieval stage that facilitate the recall of mood-congruent, emotionally negative information" (MacLeod et al. 1986). In both disorders, the typical negative affect is attributed to distinct sources—to internal sources in case of depression and to external ones in case of anxiety, which may suggest that the cognitive bias in depression and anxiety primarily influences retrieval and encoding, respectively (MacLeod et al. 1986).

Depression is associated with dysfunction of the medial prefrontal cortex involved in cognitive control and emotional response. In the study of Lemogne et al. (2009), it has been proven that the dysfunction of the medial prefrontal region may help to explain specific complaints of depressed patients such as self-blame, rumination, and feeling of guilt.

Etkin et al. (2006) report that in subjects who suffer from PTSD and depression resistant to treatment, lower rostral cingulate activity is observed during emotional processing. In many studies dedicated to PTSD, inhibited activation in the rostral (pregenual) ACC was observed. For example, Whalen et al. (2006) designed an experiment with an emotional variant of the counting Stroop where combat veterans with and without PTSD were compared. In the presented task, subjects were exposed to neutral, general-negative, and disorder-specific words. The researchers revealed that on the performance level, accuracy in PTSD subjects was significantly lower for disorder-specific words, while no difference in neutral and general-negative words was observed. From fMRI analysis, they report greater activation of the rACC in the control group, which was linked to "balancing" task performance, avoiding the interfering influence of emotional stimuli.

In the Shin et al. (2001) study, the functional integrity of ACC in PTSD was tested. In a similar procedure as the previous study, subjects divided into two groups, non-PTSD and PTSD group (combat veterans), performed the emotional counting Stroop. As mentioned before, the non-PTSD group exhibited a significant fMRI blood oxygenation level–dependent signal increase in the rACC. The findings of this study suggest a diminished response in rACC in the presence of emotionally relevant stimuli in PTSD.

Kasai et al. (2008) report regional gray matter density reduction in the hippocampus, pregenual ACC, and insula in the group of subject suffering from PTSD. "The pattern of results obtained for pregenual ACC suggests that gray matter reduction in this region represents an acquired sign of PTSD consistent with stress-induced loss."

Describing the regulatory influence of rACC on amygdala activation, Etkin et al. (2006) suggest that in understanding a variety of psychiatric disorders, "the elevated amygdalar activity and exaggerated behavioral interference may be due to deficient amygdalar inhibition by the rostral cingulate, which leads to an inability to deal with emotional conflict." According to the authors, the level of the activity of the rostral cingulated may have a regulatory role in coping with the interference of negative emotional stimuli and may determine how well an individual can solve an emotional conflict.

Davidson et al. (2000) analyzed the neural base of impulsive, affective aggression, which they associated with a lower threshold for activating negative affect, such as anger, distress, and agitation. Describing other studies, using fMRI, EEG, as well as PET methods, Davidson et al. (2000) point to two particular brain structures whose role in regulating negative emotions seems to be crucial—they prove that decreased activation in the OFC and ACC may be linked with aggression and violence. The increased activation in these regions is considered as an automatic regulatory process that helps to control the intensity of experienced anger. In studies with humans who had selective damage to the OFC or ventromedial prefrontal cortex (PFC), the role of these prefrontal areas in reversal learning was underlined (Bechara et al. 1997). Lesion studies indicate that damage to the OFC and PFC produces syndromes characterized by impulsivity and aggression (Davidson et al. 2000). Davidson et al. (2000) describe the neural mechanism for suppression of negative emotions pointing to the role of prefrontal areas in inhibiting the amygdala and their strong reciprocal connections. In their experiment, subjects were presented unpleasant stimuli (pictures) and were instructed to maintain a negative emotion or just to passively view given stimuli. Using the fMRI method, they observed greater activation of the amygdala in the group "maintaining" evoked emotion. Evidence of serotonergic and glucose abnormalities in subjects with impulsive aggression in the prefrontal cortex (Davidson et al. 2000) allows us to conclude that the role of prefrontal areas in regulating negative affect seems to be evident. On the other hand, increased metabolic rate was observed in subcortical regions in the right hemisphere (including the hippocampus, amygdala, thalamus, and midbrain).

Raine et al. (2000) revealed that patients with antisocial personality disorder who were prone to impulsive aggression exhibited an 11% reduction of overall prefrontal gray matter volume. On the other hand, Davidson et al. (2000) emphasize that increased activation of the amygdala may result in excessive negative affect, while reduced activation of this area may result in diminished sensitivity to social cues that regulate emotion.

The complex interconnections between the PFC and amygdala and the negative correlations between them (Etkin et al. 2006) are responsible for the abnormalities in regulating negative emotions in case of any structural or/and functional changes in these brain areas. In some cases, though, the problem is revealed in the changed relations between the PFC and amygdala—Dougherty et al. (2004) indicate that in unmedicated patients with major depressive disorder with anger attacks, there is a positive correlation between the left ventromedial prefrontal cortex and the left amygdala during anger induction. There was no significant relationship between these brain regions during anger induction in patients with major depressive disorder who did not manifest impulsive anger.

EMOTIONS AND NEUROADAPTIVE SYSTEMS

Knowledge from research on emotions may be used in designing a neuroadaptive system—new techniques and methodologies in neuroimaging give the possibility to develop current knowledge on the connectivity between cortical and subcortical areas. This leads to the comprehension of how data from brain activity and peripheral signals may be used in anticipating individual behavior—impulsive, spontaneous responses or planned and controlled actions (Trzopek et al. 2012). Feedback from the BCI system based on registered brain activity and registration of autonomic reactions will allow us to more accurately appraise the risk of losing or limiting consciousness and control over performed tasks. This may occur in the states of strong emotions, as well as in the states of tiredness, sleep deficit, and other activities in which affective disturbances are important and inherent components.

dEEG is one of the neuroimaging techniques that provide us with an opportunity to understand cognitive functions from spatial and temporal patterns of fluctuating voltages measured on the scalp. In recent years, scalp electrophysiological studies in humans have provided insight into a variety of event-related potential (ERP) components that are considered to reflect mechanisms/processes engaged in executive control. Previously conducted research established several of these components of the ERP associated with error commission (the response error-related negativity [rERN]) and error feedback (feedback error-related negativity [fERN]) and error positivity (PE).

Each of these ERP components is characterized by their specific features such as different latencies, polarities, and scalp distribution. Thus, the development of BCI becomes possible due to the discovery of the aforementioned ERP features and their relationship with particular cognitive functions and specific brain regions.

Error-Related Negativity

rERN

The rERN occurs when a user makes an error and recognizes it immediately. Falkenstein et al. (1991) identified a negative deflection in the ERP that is associated with an error response and named it error negativity (Ne). Other researchers independently observed this effect (Gehring et al. 1993) and termed it error-related negativity (ERN). This negativity can be observed in both stimulus-locked and response-locked ERPs, has a frontocentral scalp distribution, and peaks within 80–110 ms after the response in response-locked averages (Luu et al. 2000b). This rERN is related to error processing during response execution and reflects a difference in neural processing between error responses and correct responses in speeded response time (RT) tasks (Holroyd et al. 2009).

fERN

The second ERP component, the feedback-related negativity, called also "feedback ERN" (fERN), reflects a difference in neural processing between error feedback and correct feedback in guessing and trial-and-error learning tasks (Holroyd et al. 2009). It can be observed about 200–400 ms after the onset of a negative feedback and shows a scalp distribution similar to ERN. The fERN occurs when a user makes an error but is unaware of it until a subject is informed by feedback. The brain response in this case is similar to that of rERN, but is a reaction to the feedback rather than the incorrect action.

PE

The ERN is typically followed by the PE, which is a slow positive wave with a diffuse scalp distribution and maximum amplitude between 200 and 400 ms. It manifests aspects of error processing that seem to be independent of those reflected by ERN (Overbeek et al. 2005).

The ERNs can be elicited by stimuli presented in different modalities (somatosensory, auditory, and visual) (Falkenstein et al. 1991; Holroyd and Coles 2002). Additionally, ERNs and PE are elicited by incorrect responses in a wide variety of cognitive tasks such as go/no-go (Elton et al. 2004), oddball (Brázdil et al. 2002), Stroop (Alain et al. 2002; Hajcak et al. 2004), antisaccade (Nieuwenhuis et al. 2001), and flanker task (Ullsperger et al. 2002; Davies et al. 2004).

ACC as the Neural Source of Error-Related ERP Components

Electrophysiological source analyses indicated the ACC to be the neural source of both rERN and fERN (Luu et al. 2003). Moreover, Miltner et al. (2003) demonstrated the existence of a magnetic equivalent of the ERN (the "mERN") generated within ACC that provides additional arguments for this source localization.

Previous neuroimaging evidence suggests that the neuroanatomical areas activated during error-related processing include the pre–supplementary motor area (pre-SMA), left lateral prefrontal cortex, inferior parietal lobule and bilateral insula cortex, and, finally the ACC (Hester et al. 2004). Van Veen and Carter (2002) underline the role of ACC in detecting the presence of conflict during information processing and in altering the system involved in top–down control that leads to conflict resolution. Van Veen and Carter point out that conflict and control are often confounded. The

ACC activity is related to the specific type of control when strong cognitive engagement is involved to solve the conflict situation. Two streams of inconsistent information generate the specific situation that may cause a state of uncertainty and hesitation. Competing information leads to more complex processing. Botvinick et al. (1999) proved that ACC was activated more strongly in high-conflict situations—the ACC's activation was observed in the incongruent trial, but when the trial was preceded by a congruent trial, the activation was more notable compared with the incongruent trial preceded by an incongruent trial. (A version of the Eriksen flanker interference paradigm was used). In the first type of task, the control was lower and the conflict was higher compared to the second type, in which control was lower and conflict was highest. Botvinick and his colleagues (1999) state that these results support conflict theory rather than the selection-for-action hypothesis of ACC functioning. Holroyd et al. (2004), in their event-related fMRI experiment, showed that error responses and error feedback activated the same region of the dorsal area of the ACC. On the basis of these outcomes, they proved that the dorsal ACC is sensitive to internal and external sources of error information. Similarly, van Schie et al. (2004), describing the ACC's engagement in error processing, report that similar neural mechanisms are involved in monitoring both one's own actions and the actions of others.

These fMRI results are consistent with electrophysiological studies that implicate an engagement of ACC in error processing. Nevertheless, Jessup et al. (2010) state that activity in the medial prefrontal cortex, especially ACC, may be not involved particularly in error-related activity, proving that, in fact, it is more the process of a comparison between actual and expected outcomes. Their study supports the ACC's critical role in performance monitoring and cognitive control (e.g., Carter et al. 1998; Botvinick et al. 1999; Scheffers and Coles 2000), but researchers provide a new view on the ACC's role in error detection—the typical ERN may appear if no error is committed but when the correct result of one's action is highly improbable. Inversely, committing the error that was highly probable may not evoke ERN. Thus, the "error" should be clearly defined as the discrepancy between current and expected results.

ERN and Its Relation to Post-Error Slowing

An increase in cognitive control has been observed in responses produced immediately after the commission of an error (Marco-Pallarés et al. 2008). Such responses show a delay in reaction time called "post-error slowing," which is considered as a tendency to compensate an error in the form of being slower and more careful. Thus, after recognizing an error, people are able to adjust their behavior to reduce/prevent future errors. The degree of post-error slowing is related to the ERN (Debener et al. 2005); that is, the greater ERN's amplitude, the slower people tend to be on trials following the error (Van Veen and Carter 2006). Interestingly, post-error slowing is automatically triggered following the commission of an error, and it appears to occur independently of participant awareness. In a study (Rabbitt 2002), subjects were asked to signal whether they had committed an error or not. The post-error slowing effects were visible even in those trials that were not consciously registered, thus providing the arguments for the automatic and involuntary character of post-error slowing.

Previous neuroimaging studies evidently showed that the activation of the ACC (Kerns et al. 2004) and the pre-SMA (Klein et al. 2007) in erroneous trials was positively related to post-error slowing. Kerns et al. (2004), in the fMRI study, observed a relationship between the amount of post-error slowing and control-related activation of the right dorsolateral prefrontal cortex (DLPFC). Later, in an EEG study, Marco-Pallarés et al. (2008) investigated the neuropsychological mechanisms involved in post-error slowing using ERP analysis. The results suggest that the adaptive actions after commission of an error may be evoked by a neural circuit involving the right DLPFC and the inferior parietal cortex. This activation is related to an increase in EEG beta activity 600–800 ms after the error, which is correlated to post-error slowing and theta activity linked with ERN.

Therefore, implementing the knowledge of post-error slowing effects into the area of neuroadaptive interfaces seems to be particularly promising.

Individual Differences in the Error-Related Neural Response

Increasing interests in the nature of human error processing provided a wide array of aspects of ERN characteristics and arguments for high system flexibility. It was found that the system that produces the ERN is sensitive to the degree of error and the importance of error commission to the participant (for review, see Holroyd et al. 2002; Van Veen and Carter 2006).

There is also a growing body of research that describes the influence of motivation, affective style, and personality factors on ERN amplitude. Boksem et al. (2006) investigated the relationship between error-related ERP components and individual differences in reward and punishment sensitivity. The authors based their research on Gray's (1987) biopsychological theory of personality, which describes a reward-seeking system (Behavioral Activation System [BAS]) and punishment sensitivity system (Behavioral Inhibition System [BIS]). They found that subjects scoring high on the BIS scale displayed larger ERN amplitudes, while subjects scoring high on the BAS scale displayed larger PE amplitudes. Boksem et al. (2006) suggest that the ERN may reflect a motivation toward punishment avoidance (as measured by the BIS), while the PE may reflect a post-error process of engaging in proactive control to prevent future errors and maximize future rewards (reward seeking, as measured by the BAS). Therefore, the ERP studies support the theory that the ERN reflects responses to punishment or nonreward. These results are consistent with a number of previous studies demonstrating a relationship between negative affectivity/punishment sensitivity and ERN amplitude, such as research conducted by Tucker et al. (1999). These authors observed a centromedial frontal negativity (stimulus-locked, at 480 ms) in response to task feedback, with a more negative deflection for negative than for positive feedback. This effect was larger for subjects who reported either pleasant or unpleasant arousal during the task, suggesting that it may have reflected the subject's level of motivation. Long-lasting emotional and motivational factors have also been shown previously to modulate ERN amplitude: It is increased in subjects suffering from increased anxiety and negative affect (Hajcak et al. 2004). In an early study, Luu et al. (2000a) similarly found that college students who had high negative affect and negative emotionality displayed larger ERN amplitudes compared to participants who displayed low negative affect and emotionality. Luu et al. (2000a) found that in the initial stages of the experiment, high-negative-affect subjects (characterized as subjects who experience high levels of subjective distress) initially exhibited larger ERN amplitudes compared with the low-negative-affect subjects. Thus, subjects who are characterized as being high or low on the dimension of subjective distress and who engage in excessive self-monitoring show exaggerated ERNs. Compton et al. (2008) demonstrated that individuals who better distinguished between errors and correct responses, reflected by the ERN measure, were less reactive to stressors.

The enhanced brain activity, located in ACC, was also related to a pathophysiology of multiple anxiety disorders such as panic disorder (PD) (Bystritsky et al. 2001), PTSD (Shin et al. 2001), and obsessive–compulsive disorder (OCD) (Hajcak et al. 2003). Therefore, these results suggest that the ERN is larger in groups with emotional disorders and plays the role of an affective signal related to response monitoring (Luu et al. 2000b).

Error-Related ERP Components Implemented in the Field of BCIs

The development of BCIs becomes possible due to the discovery of the spatial location of specific brain wave phenomena and their relationship with specific cognitive processes.

The ERN and PE are generally thought to reflect the activity of the neural system responsible for error monitoring. Thus, despite their complexity, they can be favorably used in the study of BCI based on the neural substrates of the error-processing system involved in error detection and the improvement of human performance.

A BCI consists of components (sensors) for the acquisition of signals from the brain's activity, for the analysis and classification of these signals, and for driving a computer or other external devices based on the classifier output (Maye et al. 2011). A functioning BCI system is a closed-loop, real-time system. In the case of BCI communication systems, the external device serves as a channel for the user to interact and communicate with the environment. In a wider sense, any system that translates brain activity into control signals for a device can be considered a BCI.

Nowadays, there are two different types of BCIs. The first one, called "invasive" BCI, includes the recording of field potentials and multiunit neuronal activity from implanted electrodes (Parasuman 2008). The subjects do not require intensive practice to control the outputs, although they may learn how to improve their accuracy. Secondly, the number of tasks that can be performed is constrained only by the number of electrodes implanted in different brain areas. Such invasive techniques have a superior signal-to-noise ratio, but are obviously limited in use to animals or patients with severe motor diseases in whom electrode implantation is clinically recommended.

An alternative approach to EEG-based BCI is "noninvasive" BCI, which is a computer-actuated electronic recording system using electrode contacts on the scalp to record EEG signals from the subject's cerebral cortex. Over 80% of BCIs record information from noninvasive sensors (Allison et al. 2008).

Typically, the subjects must use biofeedback methods to learn how to control their own "brain waves" with sufficient accuracy to move a cursor on a computer screen. After much practice, the cursor control becomes sufficiently accurate to spell words by using the BCI to control the computer cursor to choose letters from the alphabet. This BCI approach has been highly successful and seems to provide great solutions for providing severe spinal cord injury (SCI) patients with some ways of controlling external devices.

Noninvasive EEG recordings provide an alternative to invasive methods that may provide useful BCI communication devices for individuals with disabilities. Moreover, a variety of studies have shown that the scalp-recorded EEG can be also used as the basis for a BCI (McFarland 2008).

Although the creation of a BCI system was originally driven by the motivation for the development of a technique as a rehabilitation method for patients suffering from motor impairments, recently, there are applications of BCI technology for feedback and assistance systems, person authentication and identification, and finally, games, which are used mainly by healthy users (Maye et al. 2011).

From the perspective of expanding research in the area of BCI on executive function, enormously interesting is developing BCIs that would combine the knowledge of neuronal structures responsible for error processing, emotional processing, and their neural features measured with EEG.

Research in the area of BCI went into the direction of identifying a new type of error potential that is generated in response to errors made by the BCI rather than the user. Buttfield et al. (2006) discriminated EEG error potentials generated in response to decoding errors made by the interface, not by the users. These signals can be fed back to the BCI to correct mistakes and improve system performance. These findings could also be used in online learning that is focused on facilitating adaptation between the user and the BCI.

CONCLUSIONS

In the future, BCIs will be able to not only analyze cognitive parameters but also discern different emotional states and motivation. This information could be essential for error correction and therefore use information provided by neuronal systems to design neuroadaptive interfaces adjusted to cognitive states and moods of users.

Schroeder and Cowie (2007) started developing systems that can register, model, and influence human emotional states and processes. Combining the ability to decode emotions from EEG recordings with such emotion-focused computing should enable BCI users to express their emotions and enable the BCI to respond adequately to the users' emotional state.

Finally, it is worth noticing that although dEEG systems work satisfactorily to train certain brain signals, EEG-BCIs have certain drawbacks; mainly, they provide only a low spatial resolution and ambiguous localization of neuronal activity, and they are not well suited to localize other specific brain areas linked with emotional processing, such as insula or amygdala. Real-time fMRI neurofeedback is more appropriate to self-regulate those brain activities; however, major disadvantages of fMRI-BCI are its high cost and complexity of development and usage. A BCI based on real-time fMRI allows for noninvasive recordings of neuronal activity across the entire brain with relatively high spatial resolution and moderate temporal resolution. Unlike EEG-BCI, fMRI-BCI allows brain activity in very specific parts of the cortical and subcortical regions of the brain. Studies that have been reported so far (for review, see Sitaram et al. 2007) have demonstrated that human subjects using real-time fMRI can learn voluntary self-regulation of localized brain regions, including those related with emotions, such as the ACC, insula, and amygdala.

In a study described by Caria et al. (2007), participants were able to successfully regulate blood oxygen level-dependent (BOLD) signal magnitude in the right anterior insular cortex, which demonstrates that volitional control of an emotional area can be learned by training with an fMRI-BCI. In another study Posse et al. (2003), subjects were trained to self-regulate their amygdala activation by a strategy of self-induced sadness. Therefore, these researchers have shown that the emotional system can be self-regulated. It is a promising domain for the application of fMRI-BCI in the area of emotional regulation training for patients suffering from depression, anxiety, and posttraumatic disorder. If the neurobiological basis of the emotional disturbances is known in terms of abnormal activity in a certain region of the brain, fMRI-BCI can be targeted to those regions with greater specificity for treatment.

On the one hand, neuroadaptive systems based on fMRI (fMRI-BCI) enable the volitional control of anatomically specific regions of the brain (Sitaram et al. 2007); on the other hand, the activation of these brain areas, treated as an independent variable, make it possible to identify its effect on behavior and may help to predict concrete actions. This is especially important in complex, high-risk tasks, where neuroadaptive systems may limit or eliminate the possibility of error and danger.

The application of knowledge on emotion in designing neuroadaptive systems seems to be unquestionable. Understanding the influence of emotions on executive processing enables us to further predict decision making and individual behavior. Numerous data on human behavior reveal that brain control on the basis of biofeedback and neurofeedback is a promising domain.

REFERENCES

Adolphs, R., D. Tranel, H. Damasio, and A. Damasio. 1994. Impaired recognition of emotion in facial expressions following bilateral damage to the human amygdala. *Nature* 372, 669–672.

Alain, C., H.E. McNeely, Y. He, B.K. Christensen, and R. West. 2002. Neurophysiological evidence of error-monitoring deficits in patients with schizophrenia. *Cerebral Cortex* 12, 840–846.

Allison, B.Z., D.J. McFarland, G. Schalk, S.D. Zheng, M.M. Jackson, and J.R. Wolpaw. 2008. Toward an independent brain-computer interface using state visual evoked potentials. *Clinical Neurophysiology* 119, 399–408.

Amalar, D.G. 2002. The primate amygdala and the neurobiology of social behavior: Implications for understanding social anxiety. *Biological Psychiatry* 51 (1), 11–17.

Amaral, D.G., J.L. Price, A. Pitkanen, and S.T. Carmichael. 1992. Anatomical organization of the primate amygdaloid complex. In *The Amygdala: Neurobiological Aspects of Emotion, Memory, and Mental Dysfunction*, ed. J.P. Aggleton, pp. 1–66, New York: Wiley-Liss.

Bard, P. 1929. The central representation of the sympathetic system: As indicated by certain physiological observations. *Archives of Neurology and Psychiatry* 22, 230–246.

Bechara, A., H. Damasio, D. Tranel, and A.R. Damasio. 1997. Deciding advantageously before knowing the advantageous strategy. *Science* 275, 1293–1294.

Blair, R.J.R., J.S. Morris, C.D. Frith, D.I. Perret, and R.J. Dolan. 1999. Dissociable neural responses to facial expressions of sadness and anger. *Brain* 122, 883–893.

Boksem, M.A.S., M. Tops, A.E. Wester, M.M. Lorist, and T.F. Meijman. 2006. Error related ERP components and individual differences in punishment and reward sensitivity. *Brain Research* 1101, 92–101.

Botvinick, M., L.E. Nystrom, K. Fissel, C.S. Carter, and J.D. Cohen. 1999. Conflict monitoring versus selection-for-action in anterior cingulate cortex. *Nature* 402, 179–181.

Breiter, H.C., N.L. Etcoff, P.J. Whalen, W.A. Kennedy, S.L. Rauch, R.L. Buckner, M.M. Strauss, S.E. Hyman, and B.R. Rose. 1996. Response and habituation of the human amygdale during visual processing of facial expression. *Neuron* 17, 875–887.

Breuer, J. and S. Freud. 1895. *Studies on Hysteria.* Standard Edition, London: Hogarth Press.

Brázdil, M., R. Roman, M. Falkenstein, P. Daniel, P. Jurák, and I. Rektor. 2002. Error processing—Evidence from intracerebral ERP recordings. *Experimental Brain Research* 146, 460–466.

Buttfield, A., P. Ferrez, and J. Millan. 2006. Towards a robust BCI error potentials and online learning. *IEE Transactions on Neural Systems and Rehabilitation Engineering* 14, 2, 164–168.

Bystritsky, A., D. Pontillo, M. Powers, F.W. Sabb, M.G. Craske, and S.Y. Bookheimer. 2001. Functional MRI changes during panic anticipation and imagery exposure. *Neuroreport* 12, 3953–3957.

Calder, A.J., A.W. Young, D. Rowland, D.I. Perrett, J.R. Hodges, and N.L. Etcoff. 1996. Facial emotion recognition after bilateral amygdala damage: Differentially severe impairment of fear. *Cognitive Neuropsychology* 13, 699–745.

Cannon, W.B. 1929. *Bodily Changes in Pain, Hunger, Fear and Rage*. New York: Appleton Publishers.

Caria, A., R. Veit, R. Sitaram, M. Lotze, N. Weiskopf, W. Grodd, and N. Birbaumer. 2007. Regulation of anterior insular cortex activity using real-time fMRI. *Neuroimage* 35, 1238–1246.

Carmichael, S.T. and J.L. Price. 1995a. Limbic connections of the orbital and medial prefrontal cortex in macaque monkeys. *Journal of Comparative Neurology* 363(4), 615–641.

Carmichael, S.T. and J.L. Price. 1995b. Sensory and premotor connections of the orbital and medial prefrontal cortex of macaque monkeys. *Journal of Comparative Neurology* 363 (4), 642–664.

Carter, C.S., T.S. Braver, D.M. Barch, M.M. Botvinick, D. Noll, and J.D. Cohen. 1998. Anterior cingulate cortex, error detection, and the online monitoring of performance. *Science* 280, 747–749.

Cavada, C., T. Company, J. Tejedor, R.J. Cruz-Rizzolo, and F. Reinoso-Suarez. 2000. The anatomical connections of the macaque monkey orbitofrontal cortex. A review. *Cerebral Cortex* 10, 220–242.

Compton, R., M. Robinson, S. Ode, L. Quandt, S. Fineman, and J. Carp. 2008. Error-monitoring ability predicts daily stress regulation. *Psychological Science* 19, 702–708.

Critchley, H.D. 2005. Neural mechanisms of autonomic, affective, and cognitive integration. *Journal of Comparative Neurology* 493 (1), 154–166.

Critchley, H.D. 2009. Psychophysiology of neural, cognitive and affective integration: fMRI and *autonomic* indicants. *International Journal of Psychophysiology* 73 (2), 88–94.

Critchley, H.D., S. Wiens, P. Rotshtein, A. Ohman, and R.J. Dolan. 2004. Neural systems supporting interoceptive awareness. *Nature Neuroscience* 7 (2), 189–195.

Damasio, A. 1994. *Descartes' Error: Emotion, Reason, and the Human Brain*. New York: Putnam.

Davidson, R.J., K.M. Putnam, and C.L. Larson. 2000. Dysfunction in the neural circuitry of emotion regulation—A possible prelude to violence. *Science* 289, 591–594.

Davies, P.L., S.J. Segalowitz, and W.J. Gavin. 2004. Development of response-monitoring ERPs in 7–25 year olds. *Developmental Neuropsychology* 25, 355–376.

Debener, S., M. Ullsperger, M. Siegel, K. Fiehler, D.Y. von Cramon, and A.K. Engel. 2005. Trial-by-trial coupling of concurrent electroencephalogram and functional magnetic resonance imaging identifies the dynamics of performance monitoring. *Journal of Neuroscience* 25, 11730–11737.

Devinsky, O., M.J. Morrell, and B.A. Vogt. 1995. Contributions of anterior cingulate cortex to behaviour. *Brain* 118, 279–306.

Dougherty, D.D., S.L. Rauch, T. Deckersbach, C. Marci, R. Loh, L.M. Shin, N.M. Alpert, A.J. Fischman, and M. Fava. 2004. Ventromedial prefrontal cortex and amygdale dysfunction during an anger induction positron emission tomography study in patients with major depressive disorder with anger attacks. *Archives of General Psychiatry* 61, 795–804.

Ellsworth, P.C. and K.R. Scherer. 2003. Appraisal processes in emotion. In *Handbook of Affective Sciences*, eds. R.J. Davidson, K.R. Scherer, H.H. Goldsmith. New York: Oxford.

Elton, M., M. Spaan and K.R. Ridderinkhof. 2004. Why do we produce errors of commission? An ERP study of stimulus deviance detection and error monitoring in a choice go/no-go task. *European Journal of Neuroscience* 20 (7), 1960–1968.

Etkin, A., T. Egner, D.M. Peraza, E.R. Kandel, and J. Hirsch. 2006. Resolving emotional conflict: A role for the rostral anterior cingulate cortex in modulating activity in the amygdala. *Neuron* 51, 871–882.
Fafrowicz, M. and T. Marek. 2007. Quo vadis, neuroergonomics? *Ergonomics* 50 (11), 1941–1949.
Fafrowicz, M., T. Marek, W. Karwowski, and D. Schmorrow. (eds). 2012. *Neuroadaptive Systems: Theory and Applications*, pp. 43–67. Boca Raton, FL: CRC Press.
Falkenstein, M., J. Hohnsbein, J. Hoormann, and L. Blanke. 1991. Effects of crossmodal divided attention on late ERP components. II: Error processing in choice reaction tasks. *Electroencephalography and Clinical Neurophysiology* 78 (6), 447–455.
Fellous, J.-M. 2004. From human emotions to robot emotions. *Proceedings of the AAAI Spring Symposium-Architectures for Modeling Emotion.* TR SS-04-02. Menlo Park, CA: AAAI Press.
Frackowiak, R.S.J., K.J. Friston, C.D. Frith, R.J. Dolan, C.J. Price, S. Zeki, J.T. Ashburner, and W.D. Penny (ed). 2004. *Human Brain Function*, 2nd edition.
Frijda, N. 1986. *The Emotions.* New York: Cambridge University Press.
Gasper, K. and G.L. Clore. 2002. Attending to the big picture: Mood and global versus local processing of visual information. *Psychological Science* 13 (1), 34–40.
Gehring, W.J., B. Goss, M.G.H. Coles, D.E. Meyer, and E. Donchin. 1993. A neural system for error detection and compensation. *Psychological Science* 4, 385–390.
Gerrig, R.J. and G.H. Bower. 1982. Emotional influences on word recognition. *Bulletin of the Psychonomic Society* 19, 197–200.
Gorno-Tempini, M.L., S. Pradelli, M. Serafini, G. Pagnoni, P. Baraldi, C. Porro, R. Nicoletti, C. Umita, and P. Nichelli. 2001. Explicit and incidental facial expression processing: An fMRI study. *Neuroimage* 14, 465–473.
Gotlib, I.H. and C.D. McCann. 1984. Construct accessibility and depression: An examination of cognitive and affective factors. *Journal of Personality and Social Psychology* 47, 427–439.
Gray, J.A. 1987. The neuropsychology of emotion and personality. In *Cognitive Neurochemistry*, eds. S.M. Stahl, S.D. Iverson, E.C. Goodman, pp. 171–190.
Greenberg, N., M. Scott, and D. Crews. 1984. The role of the amygdala in the reproductive and aggressive behavior of the lizard, Anolis carolinensis. *Physiology Behaviour* 32, 147–151.
Hajcak, G., N. McDonald, and R.F. Simons. 2003. Anxiety and error-related brain activity. *Biological Psychology* 64, 77–90.
Hajcak, G., N. McDonald, and R.F. Simons. 2004. Error-related psychophysiology and negative affect. *Brain and Cognition* 56, 189–197.
Hariri, A.R., S.Y. Bookheimer, and J.C. Mazziotta. 2000. Modulating emotional responses: Effects of a neocortical network on the limbic system. *Neuroreport* 11, 43–48.
Hester, R., C. Fassbender, and H. Garavan. 2004. Individual differences in error processing: A review and reanalysis of three event-related fMRI studies using the GO/NOGO task. *Cerebral Cortex* 14, 986–994.
Holroyd, C.B. and M.G.H. Coles. 2002. The neural basis of human error processing: Reinforcement learning, dopamine, and the error-related negativity. *Psychological Review* 109, 679–709.
Holroyd, C.B., O.E. Krigolson, R. Baker, S. Lee, and J. Gibson. 2009. When is an error not a prediction error? An electrophysiological investigation. *Cognitive, Affective and Behavioral Neuroscience* 9, 9–70.
Holroyd, C.B., S. Nieuwenhuis, N. Yeung, L. Nystrom, R. Mars, M.G.H. Coles, and J.D. Cohen. 2004. Dorsal anterior cingulate cortex shows fMRI response to internal and external error signals. *Nature Neuroscience* 7, 497–498.
James, W. 1884. What is an emotion? *Mind* 9, 188–205.
Jessup, R.K., J.R. Busemeyer, and J.W. Brown. 2010. Error effects in anterior cingulate cortex reverse when error likelihood is high. *Journal of Neuroscience* 30, 3467–3472.
Kasai, K., H. Yamasue, M.W. Gilbertson, M.E. Shenton, S.L. Rauch, and R.K. Pitman. 2008. Evidence for acquired pregenual anterior cingulate gray matter loss from a twin study of combat-related post-traumatic stress disorder. *Biological Psychiatry* 63 (6), 550–556.
Kerns, J., J. Cohen, A. MacDonald, R. Cho, A. Stenger, and C. Carter. 2004. Anterior cingulate conflict monitoring and adjustments in control. *Science* 303, 1023–1026.
Klein, T., T. Endrass, N. Kathmann, J. Neumann, D. von Cramon, and M. Ullsperger. 2007. Neural correlates of error awareness. *Neuroimage* 34, 1774–1781.
Kluver, H. and P.C. Bucy. 1937. "Psychic blindness" and other symptoms following bilateral temporal lobectomy in rhesus monkeys. *American Journal of Physiology* 119, 352–353.
Krettek, J.E. and J.L. Price. 1974. A direct input from amygdala to the thalamus and the cerebral cortex. *Brain Research* 67, 169–174.
Kringelbach, M.L. and E.T. Rolls. 2004. The functional neuroanatomy of the human orbitofrontal cortex: Evidence from neuroimaging and neuropsychology. *Progress in Neurobiology* 72 (5), 341–372.

LeDoux, J.E. 1987. Emotion. In *Handbook of Physiology, The Nervous System, Higher Functions of the Brain*, ed. F. Plum, pp. 419–460. Bethesda, MD: American Physiological Society.
LeDoux, J.E. 1989. Cognitive-emotional interactions in the brain. *Cognition and Emotion* 1, 3–28.
LeDoux, J.E. 1996. *The Emotional Brain*. New York: Simon and Schuster.
LeDoux, J.E. 2000. Emotion circuits in the brain. *Annual Review of Neuroscience* 23, 155–184.
Lemogne, C., G. le Bastard, H. Mayberg, H.E. Volle, L. Bergouignan, S. Lehéricy, J.F. Allilaire, and P. Fossati. 2009. In search of the depressive self: Extended medial prefrontal network during self-referential processing in major depression. *Social Cognitive and Affective Neuroscience* 4, 305–312.
Lewandowska, K., B. Wachowicz, E. Beldzik, A. Domagalik, M. Fafrowicz, J. Mojsa-Kaja, H. Oginska, and T. Marek. 2012. A new neural framework for adaptive and maladaptive behaviors in changeable and demanding environments. In *Neuroadaptive Systems: Theory and Applications*, eds. M. Fafrowicz, T. Marek, W. Karwowski, and D. Schmorrow.
Linke, R., A.D. De Lima, H. Schwegler, and H.C. Pape. 1999. Direct synaptic connections of axons from superior colliculus with identified thalamo-amygdaloid projection neurons in the rat: Possible substrates of a subcortical visual pathway to the amygdala. *Journal of Computer Neurology* 403 (2), 158–170.
Linke, R., G. Braune, and H. Schwegler. 2000. Differential projection of the posterior paralaminar thalamic nuclei to the amygdaloid complex in the rat. *Experimental Brain Research* 134, 520–532.
Luu, P., P. Collins, and D.M. Tucker. 2000a. Mood, personality, and self-monitoring: Negative affect and emotionality in relation to frontal lobe mechanisms of error monitoring. *Journal of Experimental Psychology: General* 129, 43–60.
Luu, P., T. Flaisch and D.M. Tucker. 2000b. Medial frontal cortex in action monitoring. *Journal of Neuroscience* 20, 464–469.
Luu, P., D.M. Tucker, D. Derryberry, M. Reed, and C. Poulsen. 2003. Electrophysiological responses to errors and feedback in the process of action regulation. *Psychological Science* 14, 47–54.
Maclean, P.D. 1952. Some psychiatric implications of physiological studies on frontotemporal portion of limbic system (visceral brain). *Electroencephalography and Clinical Neurophysiology* 4, 407–418.
MacLeod, C., A. Mathews, and P. Tata. 1986. Attentional bias in emotional disorders. *Journal of Abnormal Psychology* 95, 15–20.
Marco-Pallarés, J., E. Camara, T. Münte, and A. Rodríguez-Fornells. 2008. Neural mechanisms underlying adaptive actions after slips. *Journal of Cognitive Neuroscience* 20, 1595–1610.
Mathews, A. and C. MacLeod. 1985. Selective processing of threat cues in anxiety states. *Behaviour Research and Therapy* 23, 563–569.
Maye, A., D. Zhang, Y. Wang, S. Gao, and A. Engel. 2011. Multimodal brain-computer interfaces. *Tsinghua Science and Technology* 16, 133–139.
McFarland, D.J. 2008. Noninvasive communication Systems. In *Brain-Computer Interfaces. International Assessment of Research and Development Trends*, eds. W.T. Berger, J.K. Chapin, G.A. Gerhardt, D.J. McFarland, J.C. Principe, W.V. Soussou, D.M. Taylor, and P.A. Tresco, pp. 95–108.
Miltner, W.H.R., U. Lemke, T. Weiss, C. Holroyd, M.K. Scheffersand, and M.G.H. Coles. 2003. Implementation of error-processing in the human anterior cingulate cortex: A source analysis of the magnetic equivalent of the error-related negativity. *Biological Psychology* 64, 157–166.
Morris, J.S., M. deBonis, and R.J. Dolan. 2002. Human amygdale responses to fearful eyes. *NeuroImage* 17, 214–222.
Morris, J.S., C.D. Frith, D.I. Perrett, D. Rowland, A.W. Young, A.J. Calder, and R.J. Dolan. 1996. A differential neural response in the human amygdala to fearful and happy facial expressions. *Nature* 383, 812–815.
Neisser, U. 1976. *Cognition and Reality: Principles and Implications of Cognitive Psychology*. San Francisco: Freeman.
Nieuwenhuis, S., K.R. Ridderinkhof, J. Blom, G.P.H. Band, and A. Kok. 2001. Error-related brain potentials are differentially related to awareness of response errors: Evidence from an antisaccade task. *Psychophysiology* 38, 752–760.
Oatley, K. and P. Johnson-Laird. 1987. Towards a cognitive theory of emotion. *Cognition and Emotion* 1, 51–58.
Overbeek, T.J.M., S. Nieuwenhuis, and K.R. Ridderinkhof. 2005. Dissociable components of error processing. On the functional significance of the Pe vis-a-vis the ERN/Ne. *Journal of Psychophysiology* 19, 319–329.
Papez, J.W. 1937. A proposed mechanism for emotion. *Archives of Neurology and Psychiatry* 38, 725–743.
Parasuman, R. 2008. Putting the brain to work: Neuroergonomics past, present and future. *Human Factors* 50, 468–474.
Phana, K.L., T. Wagerb, S.F. Taylora, and I. Liberzo. 2002. Functional neuroanatomy of emotion: A meta-analysis of emotion activation studies in PET and fMRI. *NeuroImage* 16, 2, 331–348.

Phelps, E.A. 2006. Emotion and cognition: Insights from studies of the human amygdala. *Annual Review of Psychology* 57, 27–53.

Posse, S., D. Fitzgerald, K. Gao, U. Habel, D. Rosenberg, G.J. Moore, and F. Schneider. 2003. Real-time fMRI of temporolimbic regions detects amygdala activation during single-trial self-induced sadness. *NeuroImage* 18, 760–768.

Rabbitt, P. 2002. Consciousness is slower than you think. *Quarterly Journal of Experimental Psychology* 55, 1081–1092.

Raine, A., T. Lencz, S. Bihrle, L. LaCasse, and P. Colletti. 2000. Reduced prefrontal gray matter volume and reduced autonomic activity in antisocial personality disorder. *Archives of General Psychiatry* 57, 119–127.

Rolls, E.T. 1992. *The Brain and Emotion*. London: Oxford University Press.

Rolls, E.T. 1999. *The Brain and Emotion*. Oxford: Oxford University Press.

Sander, D. 2008. Basic tastes and basic emotions: Basic problems and perspectives for a nonbasic solution. *Behavioral And Brain Science* 31 (1), 88.

Scheffers, M.K. and M.G. Coles. 2000. Performance monitoring in a confusing world: Error-related brain activity, judgments of response accuracy, and types of errors. *Journal of Experimental Psychology: Human Perception and Performance* 26, 141–151.

Scherer, K.R. 2001. Appraisal considered as a process of multi-level sequential checking. In *Appraisal Processes in Emotion: Theory, Methods, Research*, eds. K.R. Scherer, A. Schorr, and T. Johnstone, pp. 92–120. Oxford: Oxford University Press.

Scherer, K.R. 2003. Cognitive components of emotion. In *Handbook of Affective Sciences*, eds. R.J. Davidson, K.R. Scherer, and H.H. Goldsmith. New York: Oxford.

Schroeder, M. and R. Cowie. 2007. HUMAINE Emotion Research Website at http://emotion-research.net.

Schwarz, N. and G.L. Clore. 1983. Mood, misattribution, and judgments of well-being: Informative and directive functions of affective states. *Journal of Personality and Social Psychology* 45, 513–523.

Shin, L.M., P.J. Whalen, R.K. Pitman, G. Bush, M.L. Macklin, N.B. Lasko, S.P. Orr, S.C. McInerney, and S.L. Rauch. 2001. An fMRI study of anterior cingulate function in posttraumatic stress disorder. *Biological Psychiatry* 50 (12), 932–942.

Sitaram, R., A. Caria, R. Veit, T. Gaber, G. Rota, A. Kuebler, and N. Birbaumer. 2007. fMRI Brain-Computer Interface: A tool for neuroscientific research and treatment. *Computational Intelligence Neuroscience* 2007, 1–10.

Trzopek, J., M. Fafrowicz, T. Marek, and W. Karwowski. 2012. Psychological Constructs versus neural mechanisms: Different perspectives for advanced research of cognitive processes and development of neuroadaptive technologies. In *Neuroadaptive Systems: Theory and Applications*, eds. M. Fafrowicz, T. Marek, W. Karwowski, and D. Schmorrow.

Tucker, D.M., A. Hartry-Speiser, L. McDougal, P. Luu, and D. de Grandpre. 1999. Mood and spatial memory: Emotion and right hemisphere contribution to spatial cognition. *Biological Psychology* 50, 103–125.

Ullsperger, M., D.Y. von Cramon, and N.G. Müller. 2002. Interactions of focal cortical lesions with error processing: Evidence from event-related brain potentials. *Neuropsychology* 16, 548–561.

Van Schie, H.T., R.B. Mars, M.G. Coles, and H. Bekkering. 2004. Modulation of activity in medial frontal and motor cortices during error observation. *Nature Neuroscience* 7 (5): 549–554.

Van Veen, V. and C.S. Carter. 2002. The anterior cingulate as a conflict monitor: fMRI and ERP studies. *Physiology and Behavior* 77, 477–482.

Van Veen, V. and C.S. Carter. 2006. Error detection, correction, and prevention in the brain: A brief review of data and theories. *Clinical EEG and Neuroscience* 37, 330–335.

Weiskrantz, L. 1956. Behavioral changes associated with ablation of the amygdaloid complex in monkeys. *Journal of Comparative and Physiological Psychology* 49, 381–391.

Whalen, P.J., G. Bush, L.M. Shin, and S.L. Rauch. 2006. The emotional counting Stroop: A task for assessing emotional interference during brain imaging. *Nature Protocols* 1 (1), 293–296.

Williams, J.M.G., F.N. Watts, C. MacLeod, and A. Mathews. 1997. *Cognitive Psychology and Emotional Disorders*. New York: John Wiley.

Williamson, J.W., R. McColl, D. Mathews, J.H. Mitchell, P.B. Raven, and W.P. Morgan. 2001. Hypnotic manipulation of effort sense during dynamic exercise: Cardiovascular responses and brain activation. *Journal of Applied Physiology* 90 (4), 1392–1399.

Yoo, S.S., N. Gujar, P. Hu, F.A. Jolesz, and M.P. Walker. 2007. The human emotional brain without sleep—A prefrontal amygdala disconnect. *Current Biology* 17, 877–878.

7 Evidence-Based Practice

Evaluation of the Effectiveness of Three Forced-Air Warming Blankets in Preventing Perioperative Hypothermia

Kaz Sobczak

CONTENTS

PROBLEM IDENTIFICATION

Hypothermia related to surgery is an issue. Researchers have shown that a high percentage (70%–77%) of surgical patients become hypothermic, with a significant percentage (22%) having a body core temperature less than 35°C during surgery (Akca et al. 2000; Beattie et al. 1997; Kumar et al. 2005; Kurz 2007; Sessler 1997). Clear evidence has linked hypothermia to adverse effects on various organ systems, including serious complications to the heart, lungs and liver, as well as blood coagulation and wound-healing.

Since hypothermia (a body core temperature of less than 36°C) is recognized as a common occurrence during surgery and anesthesia (Kurz 2007), managing such hypothermia poses a challenge not only because of its frequency but also due to the risk of deleterious effects, such as heat loss, that often lead to intraoperative and postoperative complications and threaten patient recovery (Chan et al. 2003; Kurz et al. 1996). This issue affects half the patients scheduled for surgeries in the United States and is associated with serious medical risks (Kumar et al. 2005). Many of these adverse consequences are preventable through the use of various warming techniques; however, there is no standard except for the current, but less effective, practice of applying passive or reflective blankets (surgical drapes, plastic sheeting, and one or two cotton blankets placed on surgical patients). Despite the fact that other preventive strategies, in particular, active warming blankets using forced-air warming techniques, have been found much more effective in preventing surgical-related hypothermia (Chan et al. 2003), no managing protocol for surgical hypothermia has emerged that takes these more effective strategies into account.

Nevertheless, much research exists about the seriousness of hypothermia's clinical manifestations as well as ways to prevent, manage, and treat it. During the last two decades, one method—the forced-air warming modality—has emerged as having unique preventive and therapeutic potential for hypothermia prevention. This active warming system uses heat absorbed from air passing over the skin to warm the patient (Lenhardt 2003). It includes a thermostatically controlled fan heater (blower) and a warming blanket. Warm air entering the blanket fills channels in the blanket, causing it to inflate and flex concavely around the patient, thus preventing perioperative hypothermia. Forced-air warming systems not only eliminate metabolic heat loss but even transfer some heat across the skin surface (Mahoney and Odom 1999). Unlike the passive warming technique, this system has the ability to maintain normothermia (>36°C) even during an extensive operation (Kumar et al. 2005). Yet, it is unclear why this way of managing operative hypothermia is not yet part of the clinical surgery protocol.

SIGNIFICANCE OF THE PROBLEM

Specifically, hypothermia becomes an issue for adult surgical patients because it may cause short- or long-term postoperative complications. For the short term, these could include impaired wound healing, coagulopathy, decreased metabolism of most drugs, and prolonged postoperative recovery periods (Kurz 2007; Sessler 1997). For the long term, these could include bradycardia, atrial fibrillation, premature ventricular contractions, and ventricular fibrillation (Alfonsi et al. 2003; Brauer et al. 2003; Chan et al. 2003; Kumar et al. 2005). The consequences are significant since both short- and long-term consequences require longer hospitalization and much higher treatment costs and typically result in adverse health and economic consequences for the patient (Ali et al. 2001; Mahoney and Odom 1999).

Thus, establishing strategies to prevent perioperative hypothermia and its severe complications has become a major surgical health concern for health care organizations, surgeons, anesthesiologists, and nurses, as well as for the patient. This becomes important work for nurse leaders since role competence is a predictor of performance, fostering practice excellence and rewarding staff for excellence. In addition, making sure nurses at all levels of the organization have a clear and complete understanding of the professional role and related obligations is essential (Wolf et al. 2004). According to Adams and O'Neil (2008), to lead a change in a positive direction, nurse leaders and advance practice nurses (APNs) must inspire others with a goal or vision. Hence, it is important that the nurse leader can project a value of nursing practice, in terms of authority, responsibility, and accountability, that demonstrates the positive outcomes for patients, ensuring that patients receive the best possible care in a safe, high-quality, effective, and efficient manner.

Cost-effectiveness is also important to consider. Patients experiencing perioperative hypothermia often require a longer hospital stay and extended treatment; thus, minimizing perioperative hypothermia risk is critical in addressing this administrative component in patient care. In today's competitive health care environment, the number of these surgeries in the general population has increased, with an estimated 50% of surgical patients developing hypothermia with complications (Akca et al. 2000; Beattie et al. 1997; Kumar et al. 2005). Lack of effectively implemented strategies to decrease hypothermia occurrence (or to prevent it) means that complications and increased costs will continue to rise (Mahoney and Odom 1999).

Reasons for this hypothermia upsurge in surgical patients are numerous: cold surgical waiting rooms, patients without adequate warming blankets applied prior to and during surgery, cold operating rooms (ORs), cold intravenous (IV) fluids, length of surgery, antimicrobial skin preparations, and various forms of anesthesia (Brauer et al. 2004; Chan et al. 2003; Leslie and Sessler 2003). However, the primary cause of inadvertent hypothermia is the cold temperatures of OR suites, which, on average, are about 19.4°C and thereby expose the large body surface area to the typical low temperature and humidity in that extrinsic environment (Lenhardt 2003). Age and comorbidity (intrinsic) also increase the risk of hypothermia. Older surgical patients usually have less subcutaneous tissue and are not as efficient in their thermoregulation. Comorbidities, such as chronic heart failure (CHF), diabetes, and hypothyroidism, further accentuate the potential for and effects of hypothermia (Beattie et al. 1997). The quality of a patient's recovery may also suffer because of shivering and thermal discomfort (Cooper 2006; Sessler 1997). High-risk surgical patients with a body core temperature of less than 35°C also have a twofold to threefold increased incidence of early postoperative myocardial ischemia (MI), independent of age and anesthetic factors (Beattie et al. 1997). These factors clearly affect patient outcomes.

The quality of a patient's recovery may also suffer because of shivering and thermal discomfort (Cooper 2006; Sessler 1997). Shivering is the most recognized effect of hypothermia, and the majority of perioperative shivering is believed to be caused by thermoregulatory imbalances. Shivering stresses the cardiac system by increasing metabolic demand, which increases oxygen consumption by up to 400% to 500% (Beattie et al. 1997; Mahoney and Odom 2003; Kumar et al. 2005). This metabolic and oxygen stressor can cause hypoxemia and MI (Paulikas 2008). Shivering also causes patient discomfort. Therefore, perioperative temperature drops should be monitored very closely with surgical patients, and measures need to be taken to mitigate hypothermia. The question remains: Which warming method works best to patients' advantage and aligns well with organizational needs?

RESEARCH METHODS

Purpose: The purpose of this small test of change was to examine the efficacy and efficiency of using a forced-air system to prevent hypothermia in surgical patients in clinical practice.

Background: This test project investigated three forced-air warming systems: upper body, lower body, and combination of upper and lower forced-air body warming blankets for existing differences in heat transfer.

Setting: The evidence-based project trial was conducted in the OR unit of an academic medical center at University of California San Francisco (UCSF). It took place from December 2009 to January 2010.

Methods: Convenient samples of 45 ($N = 45$) patients undergoing elective surgical procedures were selected. Three different warming blankets (Bair Huggers) were investigated in the project trial (Device A, $N = 14$, Device B, $N = 19$; and Device C, $N = 12$). After anesthesia induction and after the patient had been positioned on the OR table, one of the warming blankets, such as device A (upper and lower blanket), device B (lower warming blanket), or device C (upper-body blanket), were applied to patients. Data collection included (1) which device was used (A, B, or C); (2) gender; (3) room temperature; (4) type of surgery; (5) preoperative patient body core temperature; (6) patient body core temperature at surgical incision; (7) patient body core temperature each hour after incision; and (8) body core temperature at the end of surgery.

An esophageal temperature catheter measured body core temperature. The forced-air warming unit temperature output was set at 38°C for lower-body, upper-body, and combination upper- and lower-body units. The OR temperature was adjusted thermostatically to between 16°C and 20°C. Measurement was done before transfer to the OR (baseline), at the induction of anesthesia (induction), at the start of the operation (start-op), at 1-hour intervals, and at the end of the operation (end-op). The validity or reliability of the measurements, body core temperature, room temperature, and forced-air warming units routinely are tested to detect discrepancies and were standardized to this particular setting.

RESULTS

Mean values were calculated for the following variables: for preoperative patient body core temperature, patient body core temperature at surgical incision, at each hour after incision, and at the end of surgery for devices A, B, and C. The values are presented as standard error of the mean, median, mode, standard deviation, percentiles, and skewness (Tables 7.1 and 7.2; Figure 7.1). Repeated measures analysis of variance (ANOVA) group by time was used in this study. A P value of less than .05 was considered to indicate a statistically significant difference. Statistically significant results based on outcome were identified between subject effects in groups 1, 2, and 3, with $P = .008$.

For device A (combination of lower and upper blanket), mean body core temperature at the end of surgery was 36.7786°C, standard error of mean 0.10802, and standard deviation 0.40417 (Figures 7.2 through 7.5).

For device B (lower-body blanket), mean body core temperature at the end of surgery was 35.6895°C, standard error of mean 0.19070, and standard deviation 0.83126 (Figures 7.6 through 7.9).

For device C (upper-body blanket), mean body core temperature at the end of surgery was 35.3833°C, standard error of mean 0.21028, and standard deviation 0.72843 (Figures 7.10 through 7.13).

Initial preoperative and incision temperatures were similar, but temperatures in forced-air device A were higher at the start until the end of surgery than device C or device B by 1°C (device A, 36.7786°C; device B, 35.6895°C; and device C, 35.3833°C).

TABLE 7.1
Repeated Measures ANOVA

Descriptive Statistics

	Group	Mean	Standard Deviation	*N*
Bctmp_ (body core temperature in pre-op)	1: Upper and lower	36.5786	0.44407	14
	2: Lower	36.6211	0.30107	19
	3: Upper	36.4917	0.41001	12
	Total	36.5733	0.37441	45
Bctmp_inc_ (body core temperature at surgical incision)	1: Upper and lower	35.2214	0.32387	14
	2: Lower	35.1421	0.88651	19
	3: Upper	34.9250	0.73500	12
	Total	35.1089	0.70801	45
Bctmp_end_ (body core temperature at end of surgery)	1: Upper and lower	36.7786	0.40417	14
	2: Lower	35.6895	0.83126	19
	3: Upper	35.3833	0.72843	12
	Total	35.9467	0.89382	45

Source: Sobczak, K., *AORN Journal, 93*(6), 707–708, 2011.

TABLE 7.2
Statistically Significant Results Based on Outcome Identified between Subject Effects in Groups 1, 2, and 3

Tests of Between-Subject Effects

Measure: MEASURE_1
Transformed Variable: Average

Source	Type III Sum of Squares	Degrees of Freedom	Mean Square	F	Significance	Partial Ratio of Variance (Eta) Squared
Intercept	167,507.651	1	167,507.651	254,439.118	.000	1.000
Group	7.154	2	3.577	5.433	.008	.206
Error	27.650	42	0.658			

Note: $P = .008$. Eta 2 and partial Eta 2 are measure of effect size.

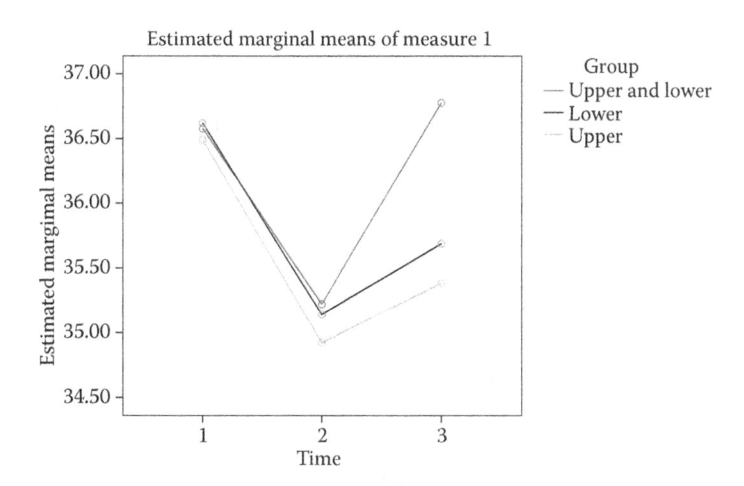

FIGURE 7.1 Difference in heat transfer in three warming blankets: upper and lower, lower, and upper. Vertical line (temperature in Celsius), Horizontal Line (Time in hours). (From Sobczak, K., *AORN Journal, 93*(6), 707–708, 2011.)

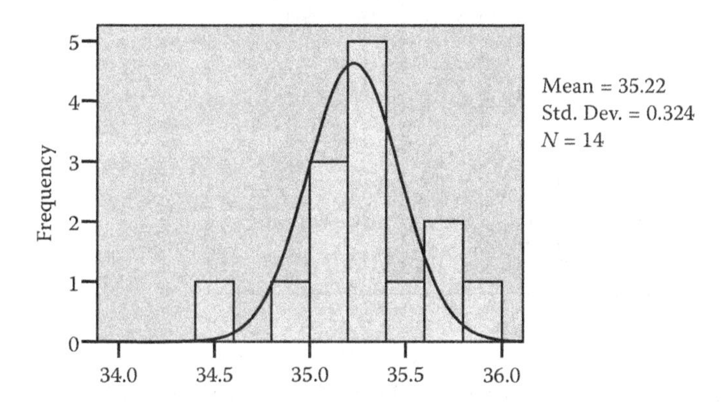

FIGURE 7.2 Device A: body core temperature at surgical incision. (From Sobczak, K., *AORN Journal, 93*(6), 707–708, 2011.)

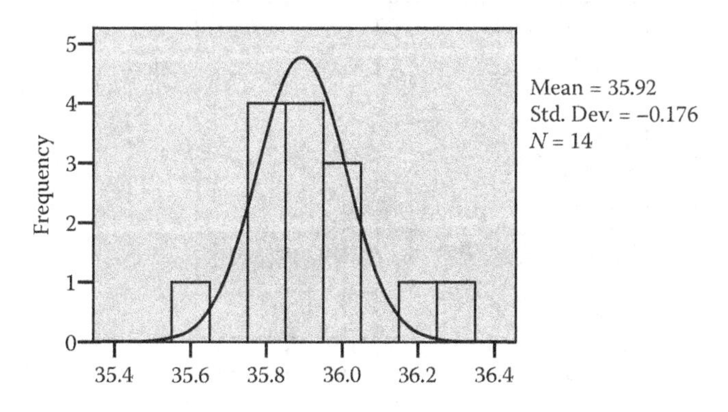

FIGURE 7.3 Device A: body core temperature at 1 h after incision. (From Sobczak, K., *AORN Journal, 93*(6), 707–708, 2011.)

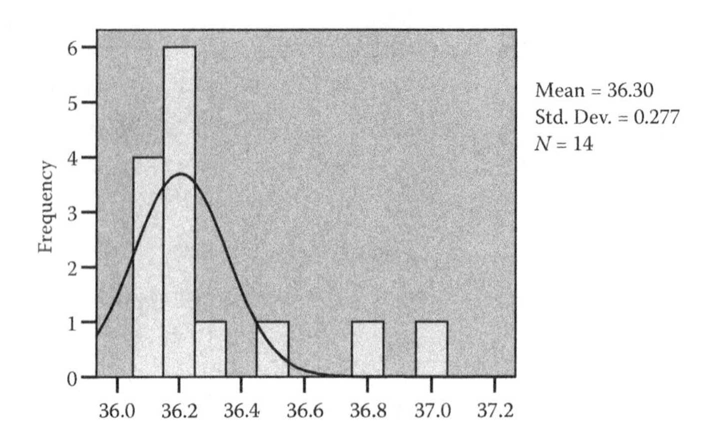

FIGURE 7.4 Device A: body core temperature at 1 h after incision. (From Sobczak, K., *AORN Journal, 93*(6), 707–708, 2011.)

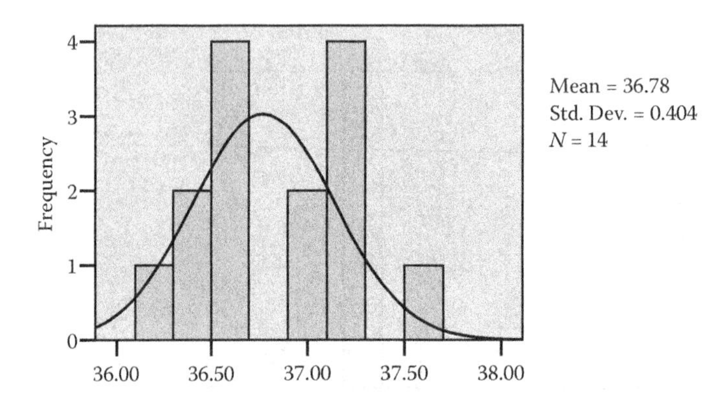

FIGURE 7.5 Device A: body core temperature at the end of surgery. (From Sobczak, K., *AORN Journal, 93*(6), 707–708, 2011.)

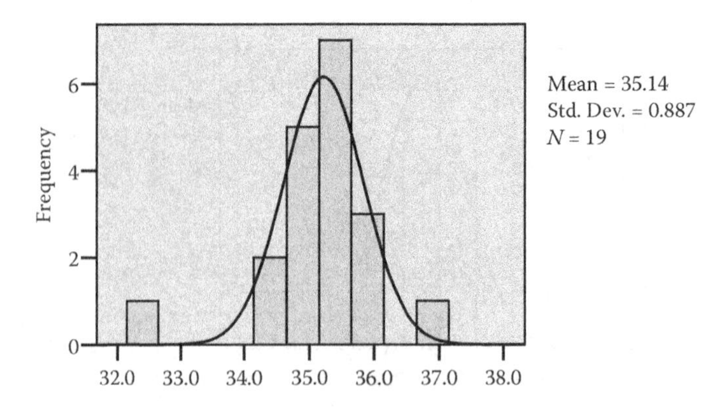

FIGURE 7.6 Device B: body core temperature at surgical incision. (From Sobczak, K., *AORN Journal, 93*(6), 707–708, 2011.)

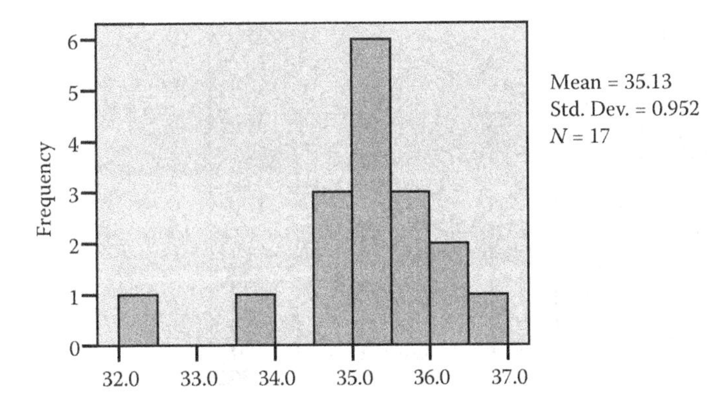

FIGURE 7.7 Device B: body core temperature at 1 h after incision. (From Sobczak, K., *AORN Journal, 93*(6), 707–708, 2011.)

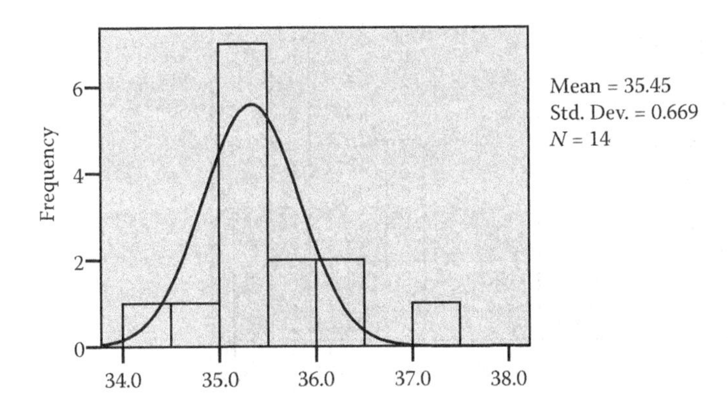

FIGURE 7.8 Device B: body core temperature at 2 h after incision. (From Sobczak, K., *AORN Journal, 93*(6), 707–708, 2011.)

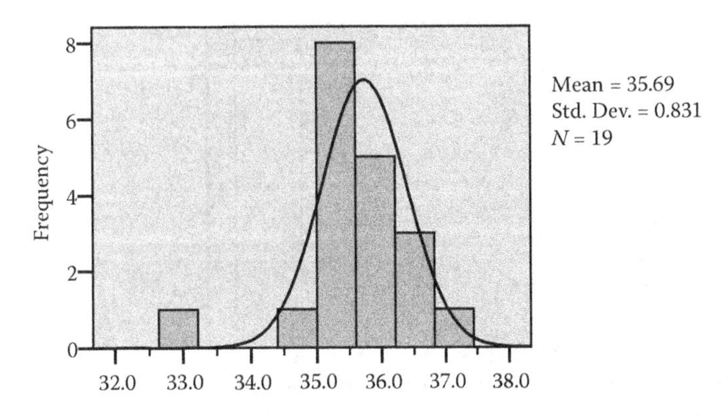

FIGURE 7.9 Device B: body core temperature at the end of surgery. (From Sobczak, K., *AORN Journal, 93*(6), 707–708, 2011.)

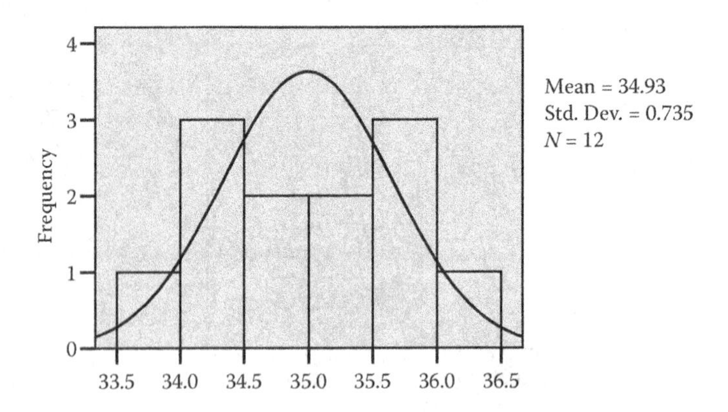

FIGURE 7.10 Device C: body core temperature at surgical incision. (From Sobczak, K., *AORN Journal, 93*(6), 707–708, 2011.)

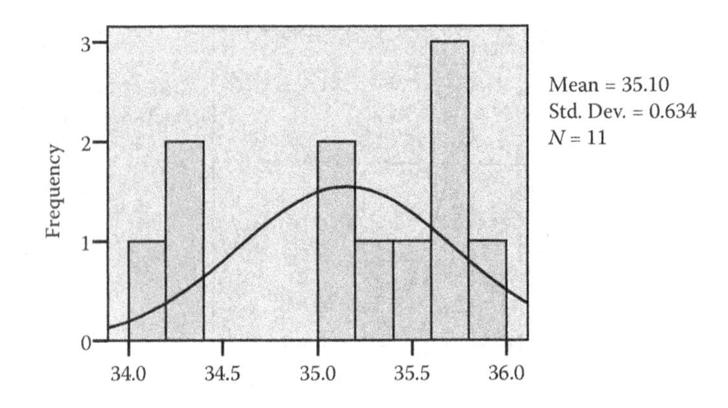

FIGURE 7.11 Device C: body core temperature at 1 h after incision. (From Sobczak, K., *AORN Journal, 93*(6), 707–708, 2011.)

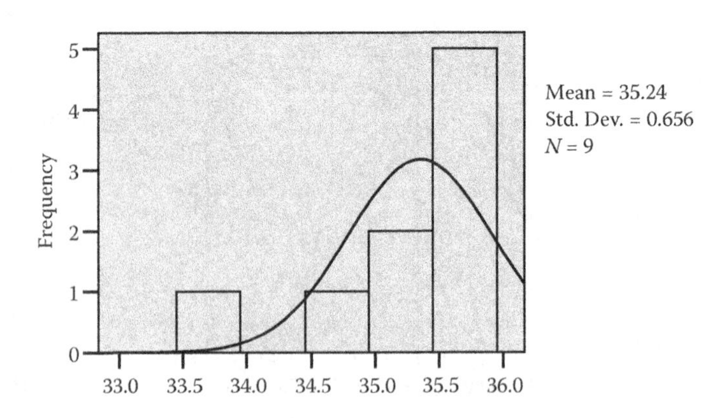

FIGURE 7.12 Device C: body core temperature at 2 h after incision. (From Sobczak, K., *AORN Journal, 93*(6), 707–708, 2011.)

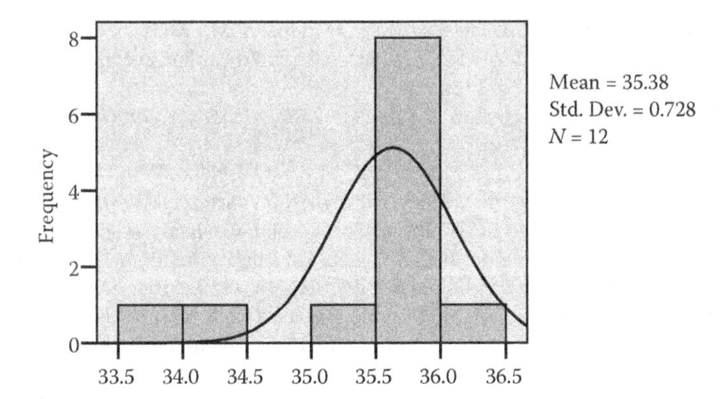

FIGURE 7.13 Device C: body core temperature at the end of surgery. (From Sobczak, K., *AORN Journal, 93*(6), 707–708, 2011.)

LIMITATIONS

It would be difficult to generalize these results because the study took place in a single facility, thus limiting the study's external validity. In addition, this nonexperimental design evidence-based project allowed for no manipulation or control over variables and no random assignment to groups. Therefore, the influence of confounding variables could have affected the study's results. The greatest threats to internal validity of the study design included selection bias because subjects were not randomly assigned to groups.

CONCLUSIONS

In this project study, forced-air warming device A (combination of upper- and lower-body warming blanket) maintained normothermia (>36.0°C) in the treated surgical patients. Patients who used device C (upper-body blanket) and device B (lower-body blanket) had body core temperatures approximately 1.1°C below normal. Perioperative hypothermia with device C (35.3°C) and device B (35.6°C) persisted until the end of surgery. Adding a combination of forced-air body warming blankets (device A, a combination of upper and lower blanket) to the thermal protocol significantly increased body core temperature in surgical patients. These results show the need for extra body warming methods to maintain normothermia in surgical patients. Thus, reviewing evidence-based medical practices (Beck and Polit 2008), and acting upon the evidence to effect positive change (such as ensuring that hypothermia is avoided in surgery), can lead to greatly improved surgical outcomes (both short and long-term) for patients, as well as significant benefits for health care organizations.

REFERENCES

Adams, T. L., and O'Neil, H. E. (2008). Nurse Executive: *The Four Principles of Management*. New York: Springer.

Akca, O., Arkilic, F. C., Kurz, A., Sessler, I. D., and Taguchi, A. (2000). Temperature monitoring and management during neuraxial anesthesia: An observational study. *Anesthesia and Analgesia Journal, 91*, 662–666.

Alfonsi, P., Nourredine, K., Adam, F., Chauvin, M., and Sessler, I. D. (2003). The effect of postoperative skin-surface warming on oxygen consumption and shivering threshold. *Anesthesia Journal, 58*(12), 1228–1234.

Ali, B., Leaper, J. D., Melling, and Scott, M. E. (2001). Effects of perioperative warming on the incidence of wound infection after clean surgery: A randomized controlled trial. *The Lancet, 388*, 876–880.

Beattie, C., Breslow, J. M., Fleischer, A. L., Frank, M. S., Higgins, S. M., Kelly, S. et al. (1997). Perioperative maintenance of normothermia reduces the incidence of morbid cardiac events. *Journal of the American Medical Association, 14*(277), 1127–1134.

Beck, T. C., and Polit, F. D. (2008). *Generating and Assessing Evidence for Nursing Practice* (8th ed.). Philadelphia, PA: Lippincott.

Brauer, A., Braun, U., English, J. M., Perl, T., Uyanik, Z. M., and Weyland, W. (2004). Perioperative thermal insulation: Minimal clinically important differences? *British Journal of Anesthesia, 92*(6), 836–840.

Brauer, A., Braun, U., Mielck, F., Perl, T., Timmermann, A., and Weyland, W. (2003). Differences among forced-air warming systems with upper body blankets are small. A randomized trial for heat transfer in volunteers. *Acta Anaesthesiologica Scandinavica Journal, 47*, 1159–1164.

Chan, H. Y., Lim, Y. P., Loh, H. K., Ng, F. S., Ong, C. B., and Oo, S. C. (2003). A comparative study of three warming interventions to determine the most effective in maintaining perioperative normothermia. *Anesthesia and Analgesia Journal, 96*, 171–176.

Cooper, S. (2006). The effect of preoperative warming on patient's postoperative temperatures. *Journal of the Association of Perioperative Registered Nurses, 83*(5), 1073–1076, 1079–1088.

Kumar, S., Leaper, J. D., Melling, C. A., and Wong, F. P. (2005). Effects of perioperative hypothermia and warming in surgical practice. *International Wound Journal, 2*, 193–203.

Kurz, A. (2007). Thermal care in the perioperative period. *Best Practice and Research Clinical Anesthesiology Journal, 22*(1), 39–62.

Kurz, A., Lenhardt, R., and Sessler, I. D. (1996). Perioperative normothermia to reduce surgical-wound infection and shorten hospitalization. *New England Journal of Medicine, 19*(334), 1209–1216.

Lenhardt, R. (2003). Monitoring and thermal management. Best practice and research. *Clinical Anesthesiology, 17*(4), 569–581.

Leslie, K., and Sessler, I. D. (2003). Perioperative hypothermia in the high-risk surgical patient. *Best Practice and Research Clinical Anesthesiology Journal, 17*(4), 485–498.

Mahoney, B. C., and Odom, J. (1999). Maintaining intraoperative normothermia: A meta–analysis of outcomes with costs. *American Association of Nurse Anesthetists Journal, 67*(2), 155–164.

Paulikas, A. C. (2008). Prevention of unplanned perioperative hypothermia. *The Journal of Association of Perioperative Registered Nurses, 88*(3), 358–365.

Sessler, I. D. (1997). Mild perioperative hypothermia. *New England Journal of Medicine, 336*(24), 1730–1737.

Sobczak, K. (2011). Using a forced–air system to prevent hypothermia in surgical patients. *AORN Journal, 93*(6), 707–708.

Wolf, G. A., Hayden, M., and Bradle, J. A. (2004). The transformational model for professional practice: A system integration focus. *Journal of Nursing Administration, 34*(4), 180–185.

8 CaseView as a Web-Based Resource Technology to Expedite Patient Care and Cost Saving for an Operating Room

Kaz Sobczak, Joyce Nacario, and Ho Lom Lee

CONTENTS

INTRODUCTION

As the world is now in an era of scientific discovery and application, it is imperative that health care professionals interact with and draw upon the skills and knowledge of experts in many disciplines. Health care is the largest industry in the United States, currently employing more than 13 million individuals, with an expected growth of approximately 25% annually from 2011 through 2020 (Borkowski 2005). Each segment of the health care industry has various social, technological, economic, competitive, and political structures that might have an impact on organizations in the future. As such, today's health care organization needs to possess the skills to communicate effectively with, motivate, and lead diverse groups of people within a large, dynamic, and complex external and internal environment.

Although the benefits of new medical and information technologies are clear to most, the challenge for health care providers lies in selecting cost-effective, comprehensive, and efficient platforms and reliable vendors with deep insights and insider experience in the health care field, gaining acceptance of new approaches, and findings ways to pay for the high costs of information technology. According to Saba and McCormick (2006), there are 2.7 million nurses in the United States. Nurses constitute a significant group of professionals who directly provide clinical care to patients and act as consumer advocates. Recent literature data and surveys indicate that a growing number of nurses are also qualified as information specialists. They may serve as administrators, researchers, educators, and community health care professionals. Moreover, they may work as computer information officers, corporate executives in vendor companies, implementers of information technology, developers of systems, and consultants (Saba and McCormick 2006).

Perioperative medical staff (OR administrators, anesthesia, surgeons, nurses, programmers) are fortunate enough to be able not only to witness the health care marvels created by constantly evolving information technology but also to participate actively in designing new Web-based resources (WBRs) for health care organizations. Joyce Nacario, who is a nursing informatics expert and clinical nurse IV at the University of California San Francisco (UCSF) medical center operating room (OR), helped design and run a WBR called CaseView with the cooperation of Ho Lom Lee, a programmer, and Dr. Kulli, a perioperative medical director (Ho et al. 2007). CaseView was developed using existing computer infrastructure, Microsoft .NET, information processes, and information technology. It is a useful tool for medical staff, administration, supporting medical units, postanesthesia care units (PACUs), and preoperative (preop) and perioperative staff. CaseView provides an active dynamic and sophisticated display of information using text, still images, and various Web design techniques without requiring *any* external cost. Using *internal* resources, this WBR technology is supported by *one* informatics nurse and *one* programmer. OR administrators acknowledge that the efficiency of the OR directly correlates to the facility's bottom line. The majority of the revenue generated by a hospital can be attributed to the OR issues surrounding time management, scheduling, payroll, billing, quality assurance, budgeting, and utilization, which can be tracked and monitored by CaseView as an active technology display to eliminate inefficiency.

Purpose

The effective management of surgical ORs is crucial in ensuring safe, high-quality, and cost-effective patient care. However, such management can be very challenging due to the number of staff, patients, and rooms involved, in addition to the necessary coordination of equipment, medications, and supplies. To address this important issue, the UCSF OR has developed a novel, easy-to-use tool to visualize and manage OR use in real time. The purpose of this project was to examine cost-effectiveness and efficiency levels resulting from the use of CaseView as an active WBR technology display to expedite patient care and cost saving for an OR in clinical practice versus a standard passive practice display (passive display includes whiteboards and printed schedules).

BACKGROUND

CaseView is a Web-based application developed using Microsoft .NET. It provides a daily view of past, present, and future OR cases on a user-friendly visual interface. It also provides a complete top-level view of all cases, in all rooms, for an entire day, at the times when they occurred or are scheduled to occur. Detailed case information can then be obtained by clicking on an individual case. The display is automatically updated every few minutes from data stored in the Patient Client Information System (PICIS) Operating Room Manager database, allowing real-time management of the OR scheduling. Other innovative features of CaseView include one-click access to preference cards, a visual comparison of the scheduled versus actual times of each case, blinking screen elements to alert the user to conditions requiring attention, and an overview of the current perioperative stage of each OR patient. As CaseView was built specifically as an information tool for medical, administrative, and perioperative staff, this study investigated the effectiveness and usefulness of the current CaseView information technology available to all staff members at the UCSF OR about the OR efficiency, turnover time, cost-effectiveness, scheduling of cases, and other logistic information.

CaseView Feature

CaseView displays information of the OR schedule as it exists in the database (actively and in real time) (Figure 8.1). Information is available for past, current, and future cases, sometimes, the OR schedule is updated as needed. The clinic enters the patient's name, procedure, surgeon, duration of the case, facility, OR number, and day of surgery.

Stakeholders

A number of individuals outside the OR may have an interest in the status of surgical cases. Auxiliary staff at the surgical waiting area must provide updates to families of patients who are having surgery. Clinical administrators need to know the whereabouts of their surgeons. The admitting department needs to verify that patients having surgery are properly admitted into the system. Nurses in the units need to be able to anticipate when their patients are due back from surgery.

Usefulness

CaseView active display addresses information and management needs by providing a powerful yet intuitive visual interface to data that already exist in clinical databases but that are otherwise difficult to obtain and interpret. It allows interested individuals to make decisions on the basis of the most up-to-date information on cost-effectiveness and efficiency of utilizing OR time.

Data Elements

CaseView connects to and queries information every few minutes. It then dynamically constructs a visual representation of these data, which are accessible via a Web browser. The data elements used to build the CaseView display include the following (Figure 8.1):

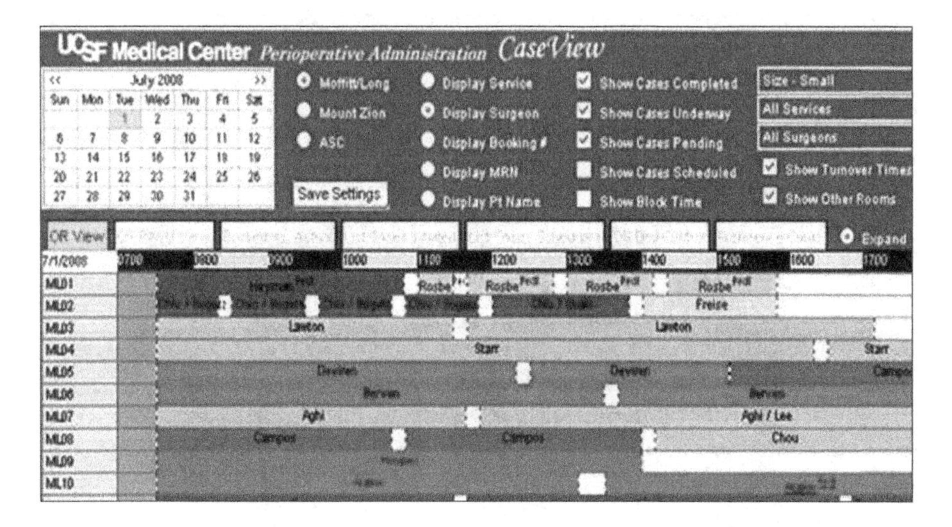

FIGURE 8.1 CaseView data elements.

- Room (OR) number
- Name of surgeon
- Name of anesthesiologist
- Name of surgical procedure
- Name of patient
- Scheduled case start time
- Scheduled case duration
- Actual case start time
- Actual case duration

Preop Phase

On the day of surgery, the patient is assessed for preop readiness in the preop area. The patient's name, family status, age, allergies, a list of personal belongings, and vital signs (blood pressure, temperature, heart rate, respiratory rate, glucose, etc.) are some of the preop checklist items that are documented into the database application (Figures 8.2 and 8.3).

Case Information

Moving a mouse over a case displays additional information about it. Clicking on a case opens a pop-up window with even more detailed information about the case. Multiple "views" are available, including a "booked" versus "actual" view, which provides a visual comparison of cases' scheduled start and end times with actual start and end times.

Real-Time Information

The OR and PACU display is shown in Figure 8.4. The color purple indicates that the patient has arrived at the surgical waiting area. Green indicates that the patient is in the preop area. Blue indicates that the patient is in the OR. Pink indicates that patient is in PACU. Yellow indicates that the patient is still going to the OR (pending cases). As can be seen in Figure 8.3, the vertical line indicates the "real-time" top layer.

Intraoperative Phase

The charge nurse looks at the OR capacity to ensure staffing coverage throughout the shifts of the day. Data on when the patient is in the OR, when the procedure began, when the procedure ended,

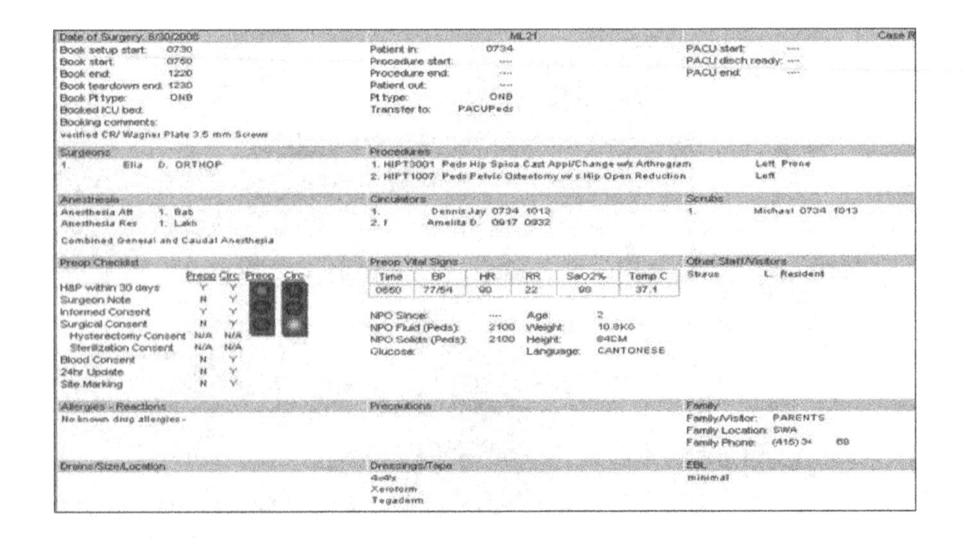

FIGURE 8.2 Preoperative phase.

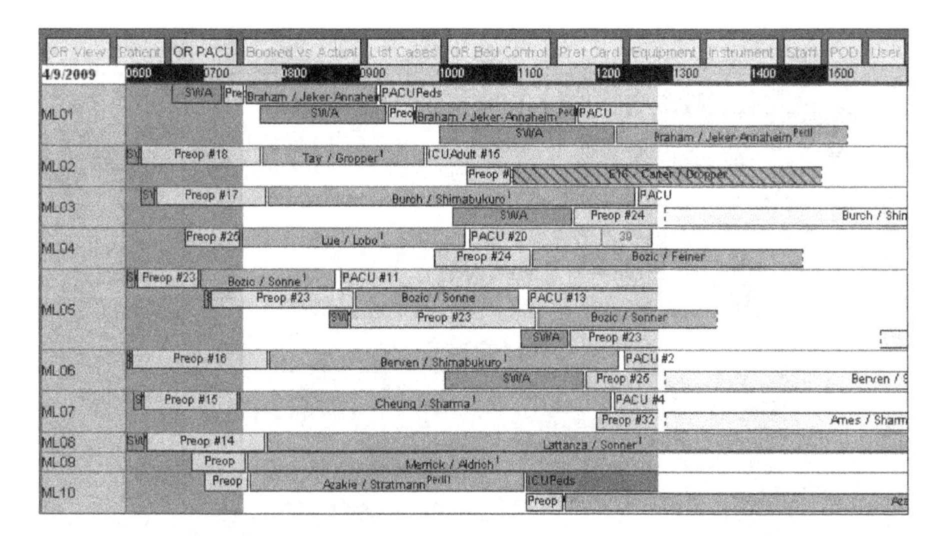

FIGURE 8.3 Preoperative phase.

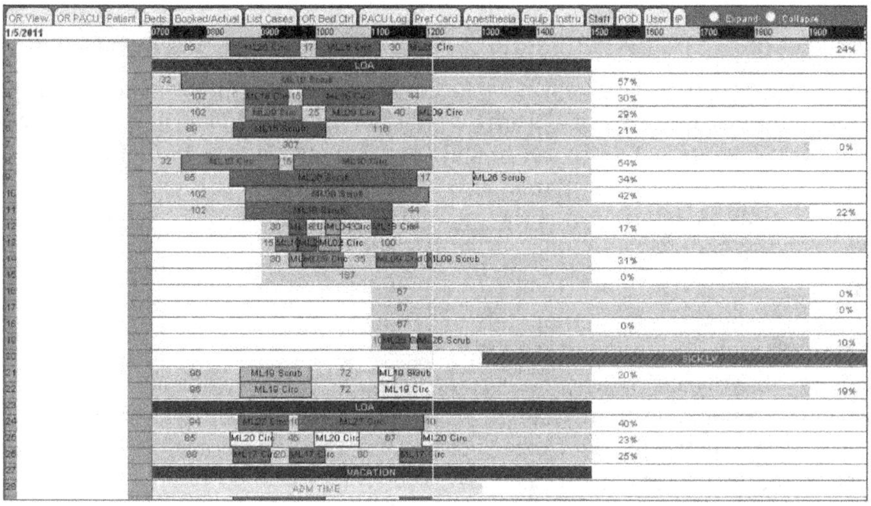

FIGURE 8.4 Real time intraoperative phase.

and when the patient is transferred to another department are being documented in the database application. These elements are then displayed in CaseView in a unique "real-time" top-layer display, which updates the existing data elements as scheduled. As one can see in Figure 8.4, presentation of all this information is done on a highly intuitive interface.

Start and End Times

Case start and end times are displayed according to the minute rather than in 15- or 30-min increments. This allows turnover times (gaps between cases) to be accurately represented on the display. Cases completed, underway, and upcoming are all shown on the same screen. Completed cases are depicted using actual start and end times, cases underway have actual start times and projected end times, while upcoming cases have projected start and end times. If a case runs longer than scheduled, all following cases are automatically pushed back and assigned new projected start and end times (Figure 8.4).

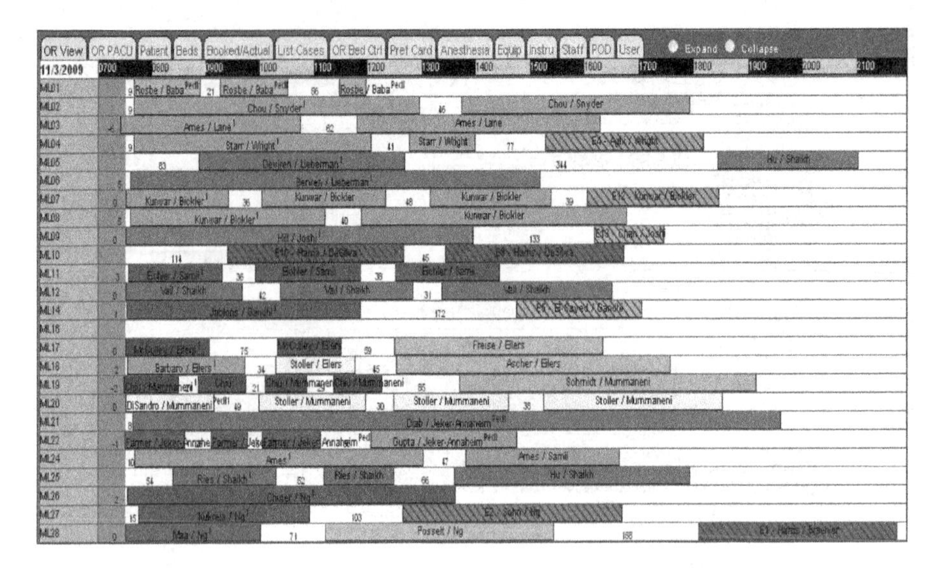

FIGURE 8.5 Turnover time.

Turnover Time

In the transition from one case to another, the turnover process needs to be as smooth as possible. Several groups of people are involved in a turnover, including the environmental services, case cart, preop, surgical waiting area, recovery room, other nursing units, and so forth. CaseView displays all of the elements necessary for the transition. A blinking indicator alerts everyone to the pending turnover, which results in improved efficiency and situational awareness (Figure 8.5).

Supplies Features

Preference Cards

Preference cards are assigned to each procedure. Each preference card contains nursing notes, preop notes, positioning, supplies, instrumentation, medications, and sutures that are necessary to set up a particular procedure in the OR. The preference card has crucial information that is passed on to several departments working closely with the OR, as discussed below.

- Case cart: The preference card is used when the case cart is being assembled by the sterile processing department (SPD) and materials services. On the day before surgery, SPD picks supplies and instruments for the surgery.
- Nurses/techs: The preference card is used by the OR circulating nurse to refer to the nursing notes, surgeon preferences, specific patient care deliverables, medications needed, sutures, instruments, and proper positioning.
- Billing: The preference card is used as the template for billing supplies used during the case. Nurses, surgeons, and surgical techs update and edit the preference cards daily to ensure accuracy (Figure 8.6).

Preop and PACU Bed Control

With a push of one button, CaseView allows health care staff to determine the availability of beds in preop and PACU. This novel active technological display information is essential to cost-effectiveness and efficiency of not only the OR but also other medical units in a health care organization. This tool visually presents the availability and occupancy of beds in preop, PACU, and the OR, which allows any administrator to utilize preop, PACU, and OR time in a cost-effective way. In turn, this

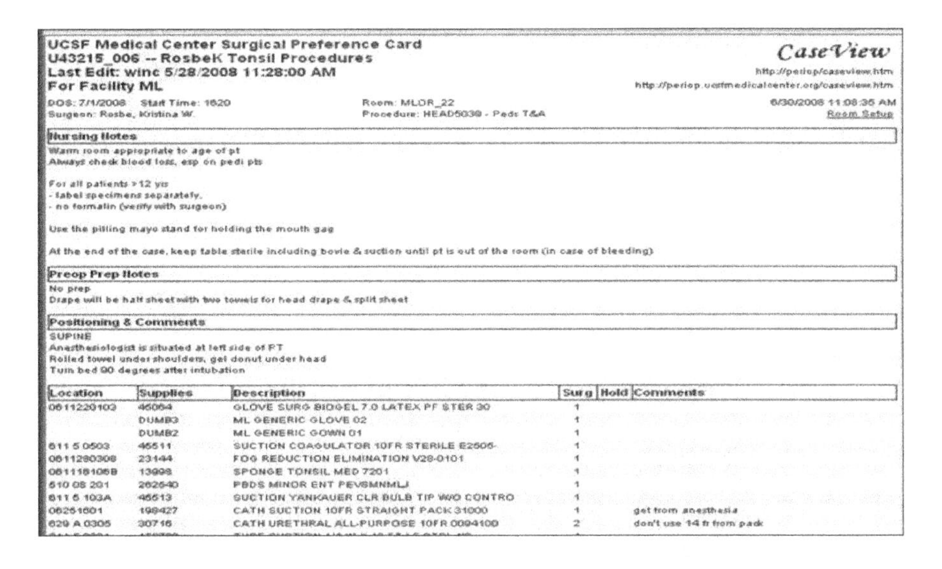

UCSF Medical Center Surgical Preference Card
U43215_006 -- RosbeK Tonsil Procedures
Last Edit: winc 5/28/2008 11:28:00 AM
For Facility ML

CaseView
http://periop/caseview.htm
http://periop.ucsfmedicalcenter.org/caseview.htm

DOS: 7/1/2008 Start Time: 1620
Surgeon: Rosbe, Kristina W.
Room: MLOR_22
Procedure: HEAD5039 - Peds T&A
6/30/2008 11:08:35 AM
Room Setup

Nursing Notes

Warm room appropriate to age of pt
Always check blood loss, esp on pedi pts

For all patients >12 yrs
- label specimens separately.
- no formalin (verify with surgeon)

Use the pilling mayo stand for holding the mouth gag

At the end of the case, keep table sterile including bovie & suction until pt is out of the room (in case of bleeding)

Preop Prep Notes

No prep
Drape will be half sheet with two towels for head drape & split sheet

Positioning & Comments

SUPINE
Anesthesiologist is situated at left side of PT
Rolled towel under shoulders, gel donut under head
Turn bed 90 degrees after intubation

Location	Supplies	Description	Surg	Hold	Comments
0611220103	46064	GLOVE SURG BIOGEL 7.0 LATEX PF STER 30	1		
	DUMB3	ML GENERIC GLOVE 02	1		
	DUMB2	ML GENERIC GOWN 01	1		
611 5 0503	46511	SUCTION COAGULATOR 10FR STERILE E2505-	1		
0611280308	23144	FOG REDUCTION ELIMINATION V28-0101	1		
061118106B	13998	SPONGE TONSIL MED 7201	1		
510 08 201	262640	PBDS MINOR ENT PEVSMNMLJ	1		
611 5 103A	46513	SUCTION YANKAUER CLR BULB TIP W/O CONTRO	1		
06251601	198427	CATH SUCTION 10FR STRAIGHT PACK 31000	1		get from anesthesia
629 A 0305	30716	CATH URETHRAL ALL-PURPOSE 10FR 0094100	2		don't use 14 fr from pack

FIGURE 8.6 Preference card.

equates to annual savings of millions of dollars. Figure 8.7 shows the bed control display. Green and red colors indicate the status of preop preparedness in preop units. They also define the patient's location in preop areas and length of stay. Pink indicates the patient's location and length of stay in PACU, while the red line indicates length of overstay in PACU.

Blood Bank

Quick access to the blood bank data regarding the blood status of the patient is imperative for an impending surgical procedure. CaseView gives any provider a quick visual of this information. As a result, patients receive the best, safest care within the best available practice. This unique application allows for quick accessing of blood product ready or ordered (Figure 8.8).

Vital Signs

As hospitals use the best existing evidence-based practice, quality of patient care and patient safety improves. CaseView allows any provider or researcher to conduct research on many variables,

FIGURE 8.7 Bed control.

Case Record | Events Log | Vital Signs | ORM Audit Trail | Supplies | History | Blood Bank

FOR BLOOD BANK USE ONLY

Blood products ordered (by Physician or MSBOS): 2 RBCs | 0 FFP | 0 platelets

Are blood products ready? Yes

Blood Bank requires: No specimens required

Note: if 2 tubes required, they must be from separate phlebotomies

Specimen(s) required to:

Blood products ready: 2 RBCs | 0 FFP | 0 platelets

Autologous RBCs: 0 RBCs Directed Donor RBCs: 0 RBCs

Other comments: TYPE AND SCREEN DONE

Insert 'TYPE AND SCREEN DONE'

Insert 'RBCs - 2 washed, 2 < 5 days, 2 < 10 days'

Insert 'Patient has antibodies to red cell antigens. Please call blood bank about blood availability.'

FIGURE 8.8 Blood bank.

which can be accessed with one click within seconds. This novel application enables a health care organization to improve health care quality provided to its patients (Figure 8.9).

Supplies

It is estimated that $200 million worth of prepared materials are discarded unused in ORs in the United States each year (Ho et al. 2007). Although some of these materials are successfully recovered for overseas donation, they nevertheless constitute an undesirable burden on health care efficiency. This situation has prompted a reevaluation of the procedures that result in the overpreparation of surgical supplies, in the hope of reducing hospital, patient, and third-party payer expenditures. CaseView tracks all supplies used in the OR at UCSF Medical Center Hospital, and it is now

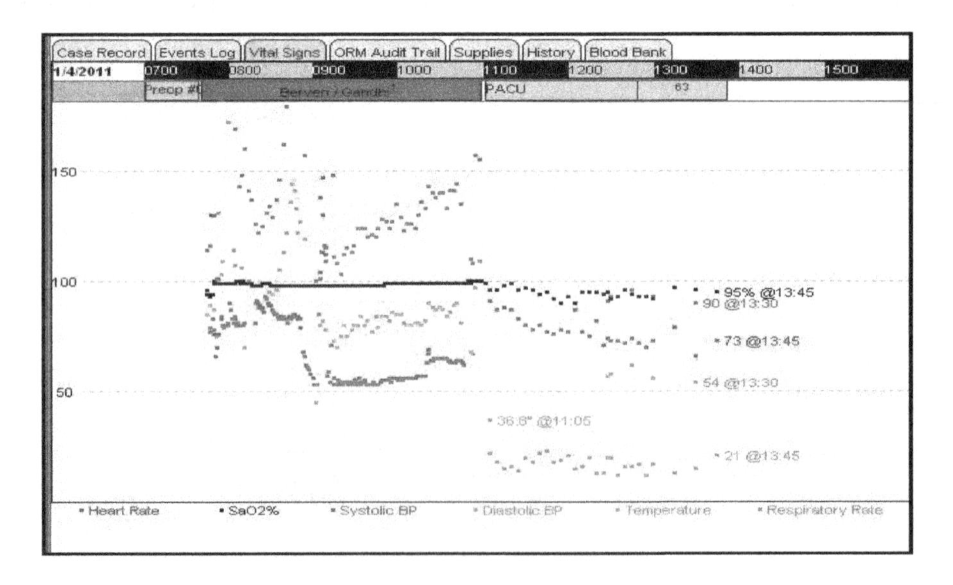

FIGURE 8.9 Vital signs.

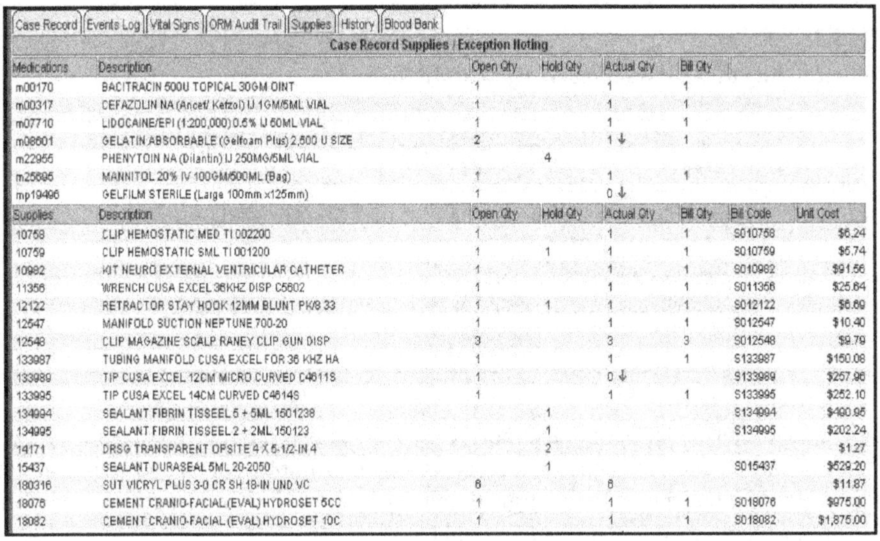

Case Record | Events Log | Vital Signs | ORM Audit Trail | Supplies | History | Blood Bank

Case Record Supplies / Exception Noting							
Medications	Description	Open Qty	Hold Qty	Actual Qty	Bill Qty		
m00170	BACITRACIN 500U TOPICAL 30GM OINT	1		1	1		
m00317	CEFAZOLIN NA (Ancef/ Kefzol) IJ 1GM/5ML VIAL	1		1	1		
m07710	LIDOCAINE/EPI (1:200,000) 0.5% IJ 50ML VIAL	1		1	1		
m08601	GELATIN ABSORBABLE (Gelfoam Plus) 2,500 U SIZE	2		1 ↓	1		
m22955	PHENYTOIN NA (Dilantin) IJ 250MG/5ML VIAL		4				
m25695	MANNITOL 20% IV 100GM/500ML (Bag)	1		1	1		
mp19496	GELFILM STERILE (Large 100mm x125mm)	1		0 ↓			
Supplies	Description	Open Qty	Hold Qty	Actual Qty	Bill Qty	Bill Code	Unit Cost
10758	CLIP HEMOSTATIC MED TI 002200	1		1	1	S010758	$6.24
10759	CLIP HEMOSTATIC SML TI 001200	1	1	1	1	S010759	$5.74
10982	KIT NEURO EXTERNAL VENTRICULAR CATHETER	1		1	1	S010982	$91.56
11356	WRENCH CUSA EXCEL 36KHZ DISP C5602	1		1	1	S011356	$25.64
12122	RETRACTOR STAY HOOK 12MM BLUNT PK/8 33	1	1	1	1	S012122	$6.69
12547	MANIFOLD SUCTION NEPTUNE 700-20	1		1	1	S012547	$10.40
12548	CLIP MAGAZINE SCALP RANEY CLIP GUN DISP	3		3	3	S012548	$9.79
133987	TUBING MANIFOLD CUSA EXCEL FOR 36 KHZ HA	1		1	1	S133987	$150.08
133989	TIP CUSA EXCEL 12CM MICRO CURVED C4611S	1		0 ↓		S133989	$257.96
133995	TIP CUSA EXCEL 14CM CURVED C4614S	1		1	1	S133995	$252.10
134994	SEALANT FIBRIN TISSEEL 5 + 5ML 1501238		1			S134994	$490.95
134995	SEALANT FIBRIN TISSEEL 2 + 2ML 150123		1			S134995	$202.24
14171	DRSG TRANSPARENT OPSITE 4 X 5-1/2-IN 4		2				$1.27
15437	SEALANT DURASEAL 5ML 20-2050		1			S015437	$529.20
180316	SUT VICRYL PLUS 3-0 CR SH 18-IN UND VC	6		6			$11.87
18076	CEMENT CRANIO-FACIAL (EVAL) HYDROSET 5CC	1		1	1	S018076	$975.00
18082	CEMENT CRANIO-FACIAL (EVAL) HYDROSET 10C	1		1	1	S018082	$1,875.00

FIGURE 8.10 Supplies.

being applied to map out a cost-effective approach to supply usage. This report summarizes the application designed for specific procedures and physician-prespecified supply lists (Figure 8.10).

SETTING

The project of CaseView was conducted in the OR unit of an academic medical center, UCSF.

METHODOLOGY

This project survey took place from February to June 2010 (Ho et al. 2007). All OR staff members were invited by e-mail to participate in a Web-based survey asking them to evaluate CaseView's effect on the health care team's efficiency, knowledge, retention, team building, leadership skills, professional development, knowledge management, evidence-based practice, and most importantly, patient safety. The convenience sample ($N = 119$) was selected from an academic medical center, UCSF, over a 3-month period and consisted of surgical staff using CaseView. The survey consisted of open-ended and close-ended questions; multiple-choice questions; matrix questions; and questions that could be answered with "yes," signifying positive, and "no," signifying negative effects.

RESULTS

Descriptive statistics was used to determine if differences reached statistical significance and to draw conclusions. Significant results based on outcomes were identified in favor of CaseView. Over 90% of the respondents expressed the opinion that the CaseView active Web-based technology display expedites superior patient care and cost saving in clinical practice versus a standard passive display (passive display includes whiteboards and printed schedules). It was also described as environmentally friendly, as it eliminates the need to print out hundreds of pages on a daily basis for paperwork.

Over 90% of the respondents surveyed also said that the CaseView active WBR display positively affected the health care team's efficiency, knowledge retention, team building, leadership skills, professional development, knowledge management, evidence-based practice, and patient safety (Table 8.1). CaseView promotes the best existing evidence-based practice to improve overall quality

TABLE 8.1
Web-Based Survey Results (Descriptive Statistics)

	Excellent	Good	Average	Below Average	Poor	Rating Average	Response Count
Quick and easy to use	71.0% (71)	25.0% (25)	3.0% (3)	0.0% (0)	1.0% (1)	4.65	100
Real-time data	67.7% (67)	27.3% (27)	4.0% (4)	1.0% (1)	0.0% (0)	4.62	99
Informative	71.0% (71)	25.0% (25)	4.0% (4)	0.0% (0)	0.0% (0)	4.67	100
Increases efficiency	69.0% (69)	25.0% (25)	5.0% (5)	1.0% (1)	0.0% (0)	4.62	100
Enables situational awareness	68.5% (63)	27.2% (27)	4.3% (4)	0.0% (0)	0.0% (0)	4.64	92

of care and patient safety at UCSF. For example, CaseView allows monitoring of patients' body core temperature at incision and at the end of surgery as a quality improvement and statstical data tool. The goal is to join our patients in their surgical journey by providing intraoperative and postoperative body core temperature at >36.0°C, unless indicated otherwise.

CaseView improves efficiency and creates a culture of patient safety within clinical practice of the OR by regular virtual rounds. With one press of a button, CaseView allows users to search quickly for right staff skills for a specific surgical procedure to deliver effective and efficient patient care.

CONCLUSION

CaseView Web-based technology addresses health care providers' need for time and cost efficiency by providing a powerful yet intuitive visual interface for data that already exist in clinical databases. It allows stakeholders to make decisions on the basis of the most up-to-date information and promotes efficiency as well as situational awareness regarding the OR. Results of the survey show that, without CaseView, monitoring data and recording the business of providing health care would be inefficient and would further strain the hospital's resources. The current pace of change in health care knowledge and information technology continues to accelerate. CaseView offers tools for health care providers to manage this exponential increase in information technology so that the most cost-effective treatment decisions can be made that will result in optimal outcomes for patients.

REFERENCES

Borkowski, N. (2005). *Organizational Behavior in Healthcare*. Sudbury, MA: Jones and Bartlett Publishers.

Ho, L. L., Kulli, J., and Nacario, J. (2007). Technology disclosure. UCSF office of technology management: CaseView. Unpublished draft.

Saba, K. V., and McCormick, A. K. (2006). *Essentials of Nursing Informatics*. New York: McGraw-Hill.

9 Technical, Ecological, and Social Aspects of Nanotechnologies

Pawel Szewczyk

CONTENTS

INTRODUCTION

The field of nanotechnologies is one of the fastest-growing and most important scientific developments in the last quarter of a century. The discoveries and innovations bridge length scales from atoms to our everyday experience and will continue to many new and unanticipated developments, as well as products such as lightweight materials, self-cleaning clothing, protective coatings, and so forth.

Nanotechnologies are the first technologies that are developing in the Internet era, and as such, they became subjects of thousands of nontruthful Internet pages, discussion forums, and other Internet means of communication. In other words, a lot has been written, but not everything that was written is true. The press had, as well, its contribution as far as the popular picture of nanotechnologies is concerned. The important achievement of the press concerning nanotechnologies is turning the attention to potential threats of any new technology and, on the other hand, asking important questions on the possible influences of different aspects of nanotechnologies on society.

This chapter is a short presentation of up-to-date concepts of nanotechnologies and a description of their progress potential in the field of quality of life, taking into account the emerging health and environmental risks. Taking advantage of the latest publications in this field, the technical, environmental, and social aspects of nanotechnologies will be presented.

CONCEPTS AND SHORT HISTORY OF NANOTECHNOLOGIES

According to a commonly used definition, nanotechnology means manipulation of atoms, molecules, and materials to form structures on the scale of nanometers (10^{-9} m) leading to manufacture of materials and devices. Nanotechnology is highly interdisciplinary, involving physics, chemistry, biology, materials science, and the full range of the engineering disciplines. The word *nanotechnology* is widely used as shorthand to refer to both the science and the technology of this emerging field. Narrowly defined, nanoscience concerns a basic understanding of physical, chemical, and biological properties on atomic and near-atomic scales. Nanotechnology employs controlled manipulation of these properties to create materials and functional systems with unique capabilities (Picraux 2010).

Hodge et al. remark in a recently edited book (Hodge 2010), "Despite its ubiquity and the undoubted importance of nanotechnology over the coming decades, the 'nanotechnology phenomenon' is itself an enigma. Its definition, meaning and historical origins continue to be a subject of contest, so that the degree to which it is really a 'new' scientific frontier requiring fresh thinking remains unclear." In the same book, Toumey (Hodge 2010) published a paper, "Tracing and Disputing the Story of Nanotechnology," in which he observes "... that there are multiple ways to tell the story of nanotechnology." After presenting and discussing a variety of stories, he concludes that "... nanotechnology can be seen as an interesting form of technology which is

- Broadly defined by scale, thereby embracing everything that deals with molecular-scale matter;
- An emerging technology, limited not to what it does now, but rather evoking multiple visions of what it could or should do at a later date; and
- A generalized technology platform, which also opens up many more possibilities than it closes."

In contrast to recent engineering efforts, nature developed "nanotechnologies" over billions of years, employing enzymes and catalysts to organize with exquisite precision different kinds of atoms and molecules into complex microscopic structures that make life possible. These natural products are built with great efficiency and have impressive capabilities, such as the power to harvest solar energy, to convert minerals and water into living cells, to store and process massive amounts of data using large arrays of nerve cells, and to replicate perfectly billions of bits of information stored in molecules of deoxyribonucleic acid (DNA) (Picraux 2010).

There are two principal reasons for qualitative differences in material behavior at the nanoscale (traditionally defined as less than 100 nm). First, quantum mechanical effects come into play at very small dimensions and lead to new physics and chemistry. Second, a defining feature at the nanoscale is the very large surface-to-volume ratio of these structures. This means that no atom is very far from a surface or interface, and the behavior of atoms at these higher-energy sites has a significant influence on the properties of the material. For example, the reactivity of a metal catalyst particle generally increases appreciably as its size is reduced—macroscopic gold is chemically inert, whereas at nanoscales, gold becomes extremely reactive and catalytic and even melts at a lower temperature. Thus, at nanoscale dimensions, material properties depend on and change with size, as well as composition and structure.

Using the processes of nanotechnology, basic industrial production may veer dramatically from the course followed by steel plants and chemical factories of the past. Raw materials will come from the atoms of abundant elements—carbon, hydrogen, and silicon—and these will be manipulated into precise configurations to create nanostructured materials that exhibit exactly the right properties for each particular application. For example, carbon atoms can be bonded together in a number of different geometries to create, variously, a fiber, a tube, a molecular coating, or a wire, all with the superior strength-to-weight ratio of another carbon material—diamond. Additionally, such material processing need not require smokestacks, power-hungry industrial machinery, or intensive human labor. Instead, it may be accomplished either by "growing" new structures through some combination of chemical catalysts and synthetic enzymes or by building them through new techniques based on patterning and self-assembly of nanoscale materials into useful predetermined designs. Nanotechnology ultimately may allow people to fabricate almost any type of material or product allowable under the laws of physics and chemistry. While such possibilities seem remote, even approaching nature's virtuosity in energy-efficient fabrication would be revolutionary (Picraux 2010).

CHARACTERIZATION OF NANOMATERIALS AND NANOSTRUCTURES

"Seeing is believing." Imaging of nanomaterials, meaning materials with at least one dimension in the nanoscale regime (1–100 nm), means not just "to create an image" but also to understand its meaning. Nowadays, there is an access to a variety of instruments that allow us to see objects at the nanoscale. In general, two fundamental types of characterization methods exist: imaging by microscopy and analysis by spectroscopy. Recent developments in nanoscience and technology cover an extremely wide field of applications and respective needs for measurement capabilities. Table 9.1 describes measurement challenges for different fields of nanotechnology.

TABLE 9.1
Measurement Challenges for Different Fields of Nanotechnology

Nanocharacterization of Fundamental Material Properties	Structural Nanomaterials	Functional Nanomaterials	Nanofabrication
Realization of nanoscale 3-D imaging capabilities	High-throughput automated nanomechanical measurements: multifunctional, high spatial resolution, rapid nanomechanical measurement and analysis	Nanoelectronics Instrumentation and metrology for advanced CMOS	Instrumentation for nanofabricated structures; mass production; reliable, reproducible, fast, and accurate measurement technology for production applications
Rapid acquisition of nanoscale data	Integration of multiple testing techniques in nanomechanics	Instrumentation and metrology for emerging novel devices	Interconnectivity of macroscopic and atomic length scales
Sample preparation and handling		Nanophotonics synthesis for nanophotonic devices	Control of the 3-D synthesis of nanostructures
Characterization of surface and interface phenomena—also in situ	Nanomechanical measurements: rapid, accurate, representative of the device or system environment in real time and length scales	Nanomagnetics	Real-time decision support for nanomanufacturing
Measurement of complex structures with compositional heterogeneity	Testing under industrial simulative environmental condition	Measurements of the magnetic properties of a cubic nanometer of material	"Real device" inspection with nanometer resolution
Combination of measurement capabilities: compositional and performance parameters to be quantitatively and reproducibly measured on the nanoscale	Standard test methods and calibration (relevant to all areas, not just nanostructured materials)	Imaging of spin dynamics	Metrology for liquid-phase manufacturing of nanomaterials
Qualitative measurements of dispersion of nanoscale substances in a matrix	Qualitative measurement of relevant material characteristics—electrical, magnetic, chemical, mechanical	Probes that decouple the measurement from the phenomenon	Production-ready standards and metrology

Source: Gennesys 2009: White Paper, A New European Partnership between Nanomaterials Science and Nanotechnology and Synchrotron Radiation and Neutron Facilities, Max Planck Institute fuer Metallforschung, Stuttgart, Germany, 2009.
Note: CMOS, complementary metal oxide semiconductor.

TABLE 9.2
Techniques Most Commonly Used for Microscale and Nanoscale Imaging

Type of Probe	Optical Probes (VIS-UV-EUV-X-ray)	Electron Probes, TEM	Mechanical or Force-Based Probes (SFM)
Dimension	2-D to 3-D	2-D to 2.5-D	2.5-D to 3-D
Advantages	– Well-understood, well-defined interaction of photons – Fast imaging technique – High penetration depth possible, depending on wavelength – Investigation of ensembles of particles	– High resolution down to atomic scale – Large interaction volume (SEM), small (TEM), smallest (STM) – Investigation of single particles	– High resolution down to atomic scale – Type of probe–sample interaction used for imaging can be selected – Very short interaction length – Investigation of single particles
Issues	– Limited resolution with optical/UV light – Unknown refractive index in UV/EUV range – Spatial resolution in elemental analysis (XPS, XRF, etc.) – Reduction of exposure time for biological materials – Reduction of carbon contamination if high-energy light source is used	– High or ultrahigh vacuum environment – Application to biological systems – Strong electron–sample interaction, partly difficult to model – Sample preparation only on surface – Elemental analysis (AES, TEM, EDX, EELS, etc.) – Carbon contamination	– Strong probe–sample interaction – Difficult to model only on surface – Slow scanning technique – Limited scan ranges
Further Developments	– Imaging x-ray optics – X-ray tomography of single biological cells or nanoparticles or nanostructures	– Improved modeling of electron–sample interaction	3-D probing of surfaces of nanostructures

Source: Gennesys 2009: White Paper, A New European Partnership between Nanomaterials Science and Nanotechnology and Synchrotron Radiation and Neutron Facilities, Max Planck Institute fuer Metallforschung, Stuttgart, Germany, 2009.

Note: AES, Auger electron spectroscopy; EDX, energy-dispersive x-ray spectroscopy; EELS, electron energy loss spectroscopy; EUV, extreme ultraviolet; SEM, scanning electron microscopy; SFM, scanning force microscopy; STM, scanning tunneling microscopy; TEM, transmission electron microscopy; UV, ultraviolet; VIS, visual; XPS, x-ray photoelectron spectroscopy; XRF, x-ray fluorescence.

Optical measurements using photons as probes are used in almost all high-accuracy dimensional measurements. This is mainly due to the fact that photons are not deflected by electromagnetic fields (like, for example, electrons) and that the photon–sample interaction is well understood. The whole measurement can be well described, permitting reliable and traceable dimensional measurements. The disadvantage is that the resolution for classical microscopy using visible and ultraviolet light is limited by the wavelength. For high-resolution applications, synchrotron radiation, neutrons, electron probes, and more recently, scanning probe microscopy are used. Table 9.2 gives an overview of common techniques using photon, electron, and force-based techniques.

TABLE 9.3
Examples of Nanoscale-Based New Materials and Innovative Products

Information and Communication Technology	New optoelectronic and molecular electronic devices, new computer concepts (quantum computer) Advanced microelectronic (nanoelectronic) devices Displays, data storage
Engineering Materials	Nanostructured materials: metals, ceramics, intermetallics, nanoparticle-loaded/strengthened polymers (composites), carbon nanotubes as strengthening components
Surface Coatings	Surface functionality and improvement, including paints and adhesives
Energy Conversion and Use	Photovoltaics, thermovoltaics, fuel cells, hydrogen storage materials, batteries/rechargeables, propellants, additives, lubricants
Sensors/Actuators	Materials and devices to generate, transduce, receive, and transform mechanical, electrical, optical, chemical, and other signals
Catalytic Synthesis	Catalysts, photocatalysts, catalyst substrates, nanoreactors, filters, adsorbents, ion exchangers
Health and Cosmetics	Diagnostic and therapeutic systems (biochips, contrast agents, drug delivery), improved implants, biological decontamination agents, cosmetics

Source: Gennesys 2009: White Paper, A New European Partnership between Nanomaterials Science and Nanotechnology and Synchrotron Radiation and Neutron Facilities, Max Planck Institute fuer Metallforschung, Stuttgart, Germany, 2009.

IMPORTANCE OF NANOTECHNOLOGY

According to the GENNESYS White Paper (Gennesys 2009), "The field of nanotechnology is one of the fastest-growing and most important scientific developments in the last quarter of a century. In view of its novelty and complexity, the importance for industry to engage in partnership with the scientific community is obvious, requiring highly skilled workforce and the most advanced test facilities to complement in-house research and development activities.

Extraordinary advances in instrumentation and powerful new experimental tools for research, provided by both national and European synchrotron x-ray and neutron facilities, are opening up the window to gain knowledge on how to control and manipulate atoms individually, observe and simulate collective phenomena on nanometer scale, probe complex material and biological systems. Applications range from nanofabrication of electronic devices to probing the secrets of nanomechanics and nano-bio folding. The discoveries and innovations bridge length scales from atoms to our everyday experience and will continue to many new and unanticipated developments, as well as products such as lightweight materials, self-cleaning clothing, protective coatings, etc." (See Table 9.3.)

Nanotechnology is driven by the potential that nanostructured materials or systems have properties that surpass, often by a large margin, those of any existing materials. This translates into cost savings as well as improvements of performance and functionality and applies to the low-tech branches (construction and textile industries), the so-called core branches of national industries (automotive and chemical industries), as well as the high-technology industrial sectors (telecommunications, transportation, defense, energy industry, pharmacy, medical engineering, and consumer goods). The development of advanced materials that are lighter, stronger, cheaper, and more versatile than existing ones will be an important asset provided by nanomaterials discoveries (Gennesys 2009).

STANDARDS FOR NANOTECHNOLOGY

The role of developing appropriate standards for nanotechnology was well presented by Miles in a recently published paper, "Nanotechnology Captured" (Hodge 2010), where he remarked: "The emergence of nanotechnologies that truly enhance the quality of life and the development of regulations to protect the public rely vitally on the development of internationally accepted documentary standards for terminology and nomenclature and an effective international infrastructure for metrology at the nanoscale." In conclusion, he finally stated: "Documentary standards that anticipate the evolution of nanotechnology are being developed by international and national standards organizations, in particular the International Organization for Standardization. ISO, through its Technical Committee 229—Nanotechnologies, plans to produce a series of standards covering Terminology, Nomenclature and Materials Specifications, Measurements and Characterization, and Health, Safety, and Environmental Aspects of Nanotechnologies. TC 229 has chosen not to adopt the traditional market-driven bottom–up approach of developing international standards but rather a more considered, planned approach where the present and future needs of the world community are identified and addressed in a harmonized fashion."

SOURCES OF ENVIRONMENTAL IMPACTS

Concerns about societal, ethical, environmental, and health implications have arisen as nanotechnology has developed and as nanotechnology-enabled products have proliferated in the marketplace. One big controversy related to the development of nanotechnology deals with the risk associated with its highly beneficial aspects. Table 9.4 contains an overview of the potential risks of nanomaterials.

TABLE 9.4
Overview of the Potential Risk of Nanomaterials

Nanotechnology and Nanomaterials

Potential Risks for Health

Toxicology of engineered nanomaterials and by-product nanoparticles:

- Ability to enter into the human body
- Impact of biochemical processes

Harmful effects of "beneficial" nanomaterials ingested in the human body:

- Impact of nanoparticles used to destroy cancer cells on the whole of the human body
- Photochemical reactions of nanoparticles used in sunscreens and cosmetics

Potential Risks for the Environment

Toxicology of engineered nanomaterials and by-product nanoparticles:

- Accumulation and transportation in water, soil, and the atmosphere

Adverse impact of "beneficial" nanomaterials for the "food chain":

- Further impact of nanoparticles transported or transformed via microorganisms such as bacteria and protozoa

Potential Risks for Safety, Security, and Ethics

Invasion of privacy
Spread of spying sensors
Nanorobotics or other bionanotechnology ambitious applications

Source: Gennesys 2009: White Paper, A New European Partnership between Nanomaterials Science and Nanotechnology and Synchrotron Radiation and Neutron Facilities, Max Planck Institute fuer Metallforschung, Stuttgart, Germany, 2009.

SOCIAL DIMENSION OF NANOTECHNOLOGIES

Already, in the year 1996, the European Parliament initiated a discussion on sustainable development and technology (Szewczyk 2010). Weber and Fahrenkrog (1996) prepared a discussion paper for the seminar, "Community Policies on Research and Sustainable Development," in which it was stated that in the last period of time, discussion on sustainable development was focused on relating this concept to environmental issues. Many other important elements and conditions, essential from the point of view of achieving sustainable development, were omitted, for example, the social and economic dimension of this concept, which represents challenges in the area of joblessness, North–South relations, and the creation of wealth. The above mentioned authors discussed the role of new technologies as driving forces of sustainable development. In conclusion, they identified four general policy strategies that support sustainable development by targeted stimulation of technological change. Basing only on scientific and technological progress is, in their opinion, not an appropriate approach.

In the time of dramatic increase of nanotechnology applications, what is expressed in everyday announcements on the Internet (see, for example, www.nanowerk.com) is that all actors active in this area of science and economy should recognize the necessity of complying with principles of sustainable development in order to achieve lasting market success. The European Commission states in its communication titled "Towards a European Strategy for Nanotechnology" (EU Communication 2004) that technological progress has to be accompanied by scientific assessment of possible undesired effects on human health, societal injustice or environmental risks. This so-called "integrated, safe, and responsible approach" has since become the core of the EU policy for nanotechnology.

A CHALLENGE FOR NANOREGULATORS AND INDUSTRY

Consumers want to know what they buy, retailers want to know what they sell, and processors and recyclers need to know what they handle. However, the relevant nanospecific information often does not reach these recipients because there is no obligation for transfer of nanospecific information. Nanomaterials are likely to become "black boxes" in terms of information; as a consequence, consumer confidence decreases, and politicians, nongovernmental organizations, and consumer advocates are calling for transparency, declaration, and labeling.

The European parliament unmistakably stated in its 2009 report on regulatory aspects of nanomaterials: "No data, no market" (Berger 2009). Members of the European parliament wanted manufactured nanomaterials to be treated as new substances, requiring extensive safety testing and mandatory labeling. Registration, Evaluation, Authorisation, Restriction of Chemicals (REACH) and other European Chemicals Acts clearly shift the responsibility to ensure safe products to manufacturers and those who put them on the market. However, it remains unclear which kind of nanospecific information is needed at the different stages in the product life cycle and how it should be delivered.

Recently, in December 2010, the *International Handbook on Regulating Nanotechnologies* finally appeared (Hodge 2010), which contains "26 chapters capturing the last decade of commentary and policy perspective regarding nano-related environmental health and safety regulatory issues" (citation from the back cover).

REFERENCES

Berger M. 2009: Nanowerk LLC, Fishing for nanotechnology risks with the wrong type of fishing net: European Parliament calls for sweeping review, http://www.nanowerk.com/news/newsid=10985.php.

EU Communication 2004: Towards a European strategy for nanotechnology. Communication from the Commission COM(2004) 338 Final. Office for Official Publications of the European Communities, Luxemburg.

Gennesys 2009: White paper. In: *A New European Partnership between Nanomaterials Science and Nanotechnology and Synchrotron Radiation and Neutron Facilities*, Dosch H. and Van de Voorde M.H. (editors). Max Planck Institute fuer Metallforschung, Stuttgart.

Hodge G.A. 2010: *International Handbook on Regulating Nanotechnologies*, Hodge G.A., Bowman D.M., and Maynard A.D. (editors). Edward Elgar Publishing Ltd, Cheltenham.

Picraux S.T. 2010: Nanotechnology—Britannica online encyclopedia, http://www.britannica.com/EBchecked/topic/962484/nanotechnology.

Szewczyk P. 2010: Nanotechnology and the risk society or the return of care. In: *Current Trends in Commodity Science. Selected Quality Problems*, Wybieralska K. (editor). Scientific Papers No. 160. Economic University Poznan, Poznan, Poland, pp. 82–90.

Weber M. and Fahrenkrog G. 1996: Sustainability and technology. A framework for discussion, Report EUR 16457 EN, Brussels.

10 Wayfinding by Colors in Public Buildings

Agnieszka Kowalska-Styczeń, Joanna Bartnicka, Christophe Bevilacqua, and Geoffrey Lepoutre

CONTENTS

INTRODUCTION

Many people have problems finding their way around public buildings such as airports, hospitals, offices, or university buildings. The problem may partially lie in their spatiocognitive abilities, but also in the architecture. Orientation and movement round the building facilitate wayfinding. We define wayfinding as "the information-gathering and decision-making processes that people use to orient themselves ad move through space; how people get from one place to another" (Billger 2000). Wayfinding can be implemented in different ways and using different strategies. Three specific strategies for navigation in multilevel buildings were compared in the work of Holscher et al. (2006). The central point strategy relies on well-known parts of the building; the direction strategy relies on routes that first head towards the horizontal position of the goal, while the floor strategy relies on routes that first head towards the vertical position of the goal. In the work of Holscher et al. (2009), it was tested whether the standard wall-mounted floor maps found in the majority of public buildings can help to navigate in a complex unknown environment. The types of descriptive features contained in wayfinding descriptions—sense of direction, wayfinding strategies, and gender—relate to wayfinding in an everyday, indoor environment were studied by Hund and Padgitt (2010). Those with a good sense of direction showed better wayfinding performance than those with a poor sense of direction.

Differences in wayfinding strategies between good and poor sense of direction were studied by Katoa and Takeuchi (2003).

In addition, gender is also important in the process of wayfinding, which has been examined by Chebat et al. (2005) and Chien-Hsiung et al. (2009).

WAYFINDING BY COLORS

The use of colors is very important and opens new perspectives in wayfinding design. Adding a contrasting color on a ceiling or on wall can help to "develop the definition of an architectural environment by reinforcing the hierarchy of spaces and landmarks" (Helvacioglu and Olguntürk 2011).

The colors can also act as signage through the building. We can use them to identify an area or a floor, and these colors can act to "benefit wayfinding, in contrast to neutral-colored elements"

(Holscher et al. 2006). Studies on the use of color in wayfinding can be found in the work of Dalke et al. (2006) and Helvacioglu and Olguntürk (2011).

We have just seen that colors can structure an environment and help one to find his/her way through a building, but we must consider the perception of those colors and integrate color blindness and illumination during the definition of the final shades.

Concerning color blindness, we simulated the perception of colors for the three following types of color deficiencies (Cassin and Solomon 1990):

- Deuteranopia: color deficiency affecting red–green hue discrimination
- Protanopia: mild color vision defect in which an altered spectral sensitivity of red retinal receptors (closer to green receptor response) results in poor red–green hue discrimination
- Tritanopia: hereditary color vision deficiency affecting blue–yellow hue discrimination

More information on color blindness can be found in the work of Sharpe and Jägle (2001) and Sharpe et al. (1999).

Selecting the shades, the issue of color blind people should be taken into consideration. Those people will not see the same colors as "normal" people, but they will differentiate them from each other and identify the indications they provide.

To optimize the use of color in wayfinding, we must consider the illumination of the environment because a color appears different according to lighting conditions. A change of light can affect the perception of colors (Billger 2000).

Creating Accessible Space in a Building by Color

According to the Polish law of regulation in the area of Infrastructure, a public building is a building intended for public administration; justice; culture; a religious cult; education; higher education; science; health care and social services; banking services; trade; catering; services; tourism; sport; road, air, water, or rail transport; a post office; or telecommunications. Other communal buildings that are designed to perform similar functions, such as an office or social building, are also considered public.

In a multifloor building, colors should be selected to highlight and contrast the various floors of the building. The choice of colors is very important because it determines the scope of services on a given floor and its destiny. The selected colors can also be used to identify individual rooms on the floor, Web sites, and so forth. It is therefore important that the colors are distinguishable by all users, including people with disabilities. The fact that the colors that are too fair are hard to identify and distinguish for people with low vision should be taken into consideration. Color is an ideal tool to simplify movement around the building and orientation. It can be used to distinguish the various floors of the building or to highlight areas on the same level, so it should be applied in a consistent manner throughout the building.

Most visible colors are those that are bright and vivid. If we want the user to find a passage easily, it is necessary to make it stand out by choosing the correct lighting and color reflecting light. There is evidence that it is easier to move through corridors where the walls are painted in a manner that gives the impression of movement and dynamics by varying in color from fair to dark.

Handling colors skillfully is a great way to show the way out of the building or its other key points. To emphasize or highlight certain elements in the hallway, we can paint their outline (e.g., fire extinguisher, resting area). However, in order to guide the user to lifts, it is recommended to paint one wall near the lift a more eye-catching color.

Yet, be very careful choosing colors that are to be used in the space hallway or stairway because some colors are reserved for specific signs and symbols. An example of such a color is red. In darkness, red is seen almost as black (so it can raise anxiety). Large expanses of that color in the stairways should be avoided because a black wall gives the impression of emptiness (black hole); it does

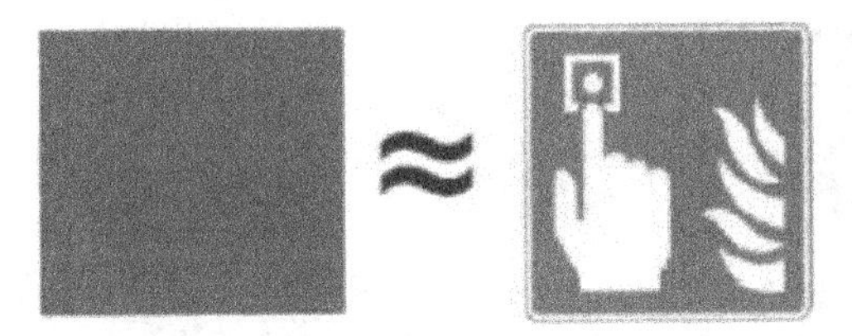

FIGURE 10.1 Fire alarm.

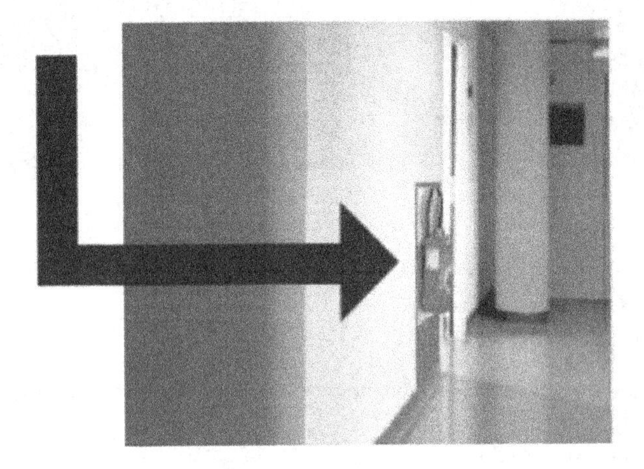

FIGURE 10.2 Identification of red fire extinguisher.

not encourage movement and does not provide psychological comfort. What's more, red is usually reserved for the prohibition of persons, information about the danger of fire, or the place of putting the fire fighting equipment (Figures 10.1 and 10.2).

An important element of the communication system in a building, including the evacuation system, is the stairways. Stairways should contain such color solutions that ensure safe movement of all persons, including persons with disabilities. It is recommended to mark the edge of the first and last step with a color contrasting with that of the floor.

Handrails in the hallway and stairwell should be of a color contrasting with the walls and floor. Similarly, stairs cheeks should be highlighted by a color. Likewise, tread and forehead tread. Such a solution would make the stairway identifiable and easily visible in the space of the building.

Example of Adaptation of Wayfinding in a Public Building

The results of research into the use of wayfinding by color solutions in a selected public building are shown below.

A building is in the process of modernization. It contains five floors: the basement, ground floor, and three floors.

Solutions in the field of color orientation and mobility were suggested in the hallway and on each floor in the main staircase and lift area.

The study used the following flow diagram:

- Analysis of project documentation on the modernization of the building
- Fieldwork and elaboration of gained documentation
- Conducting research in order to select color, taking into account color-blind people
- Suggesting solutions for wayfinding by the color of the building

(a) (b)

FIGURE 10.3 Views (a) from the ground floor to the main hall and (b) surrounding the lift and the staircase on the first floor.

Figure 10.3 shows the view from ground floor to the main hall (a) and around the lift and the staircase on the first floor during renovation.

The choice of colors proceeded in two stages:

1. Stage I: evaluation of the colors proposed in the modernization project and exclusion of the colors that do not meet the requirements of accessibility for color-blind people
2. Stage II: trials of colors tests with the following assumptions:

    - Each floor of the building is differentiated by various colors.
    - Colors assigned to each floor form a harmonious composition of colors (rainbow colors).
    - The most common basic color in the space of the building is gray.

In order to find an adapted color set, we use several tools.

- The first tool is a color blindness simulator, which allows verification of the color schemes. The objective is to check that people with color vision impairment can distinguish unambiguously the colors used to differentiate and to structure architectural elements (floors, areas, departments).
- We also use a light meter to measure the contrasts. We can use it on color samples or directly in a building. (The most relevant measures are those achieved in buildings because they take into account the lighting conditions.) In the literature, a luminance contrast of 70% between the color item and the background is recommended in order to reach the readability required by people with visual impairments.
- To adapt the wayfinding signage, we use a contrast analyzer software to validate that we achieve 70% color contrast between the text and the background of the sign and between the background of the sign and the walls for the three types of color deficiencies.

Figures 10.4 and 10.5 show the results of color analysis, which include the needs of color-blind people.

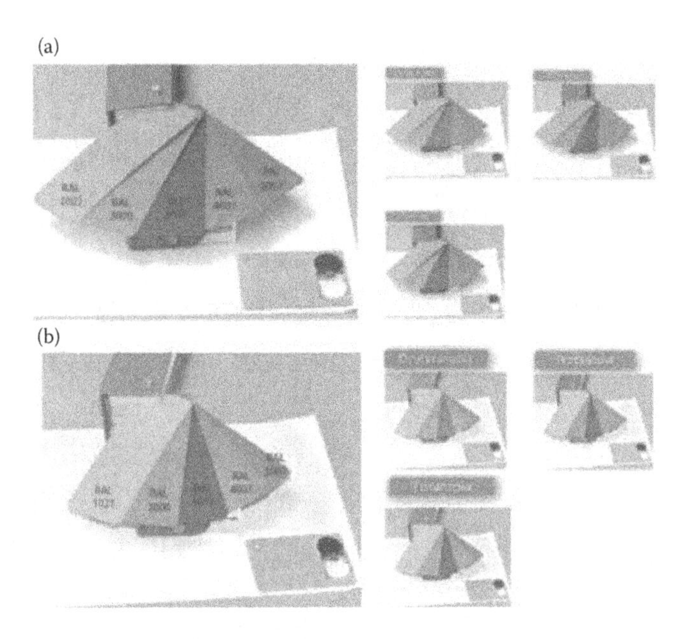

FIGURE 10.4 Tests with different colors. (a) The colors proposed by the architects and (b) RAL 1007 was changed into RAL 1021.

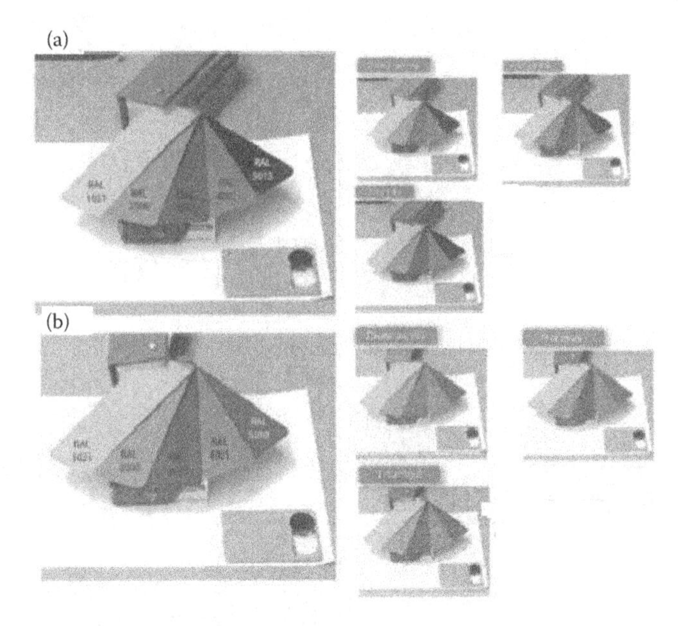

FIGURE 10.5 Tests with different colors. (a) RAL 5007 was replaced with RAL 5013 and (b) RAL 5013 was changed into RAL 5010.

There is a problem with the first and the last colors, which are not very distinguishable for people with deuteranopia and protanopia; that is why we have made a change concerning these colors.

The tests showed that only the latter proposal (Figure 10.5b) is an example of colors adapted to color-blind people.

The best choices are

- –1 floor, RAL 1021 color
- 0 floor, RAL 2000 color
- 1 floor, RAL 3013 color
- 2 floor, RAL 4001 color
- 3 floor, RAL 5010 color

The tests results were used to develop the concept of interior color, taking into account the following needs:

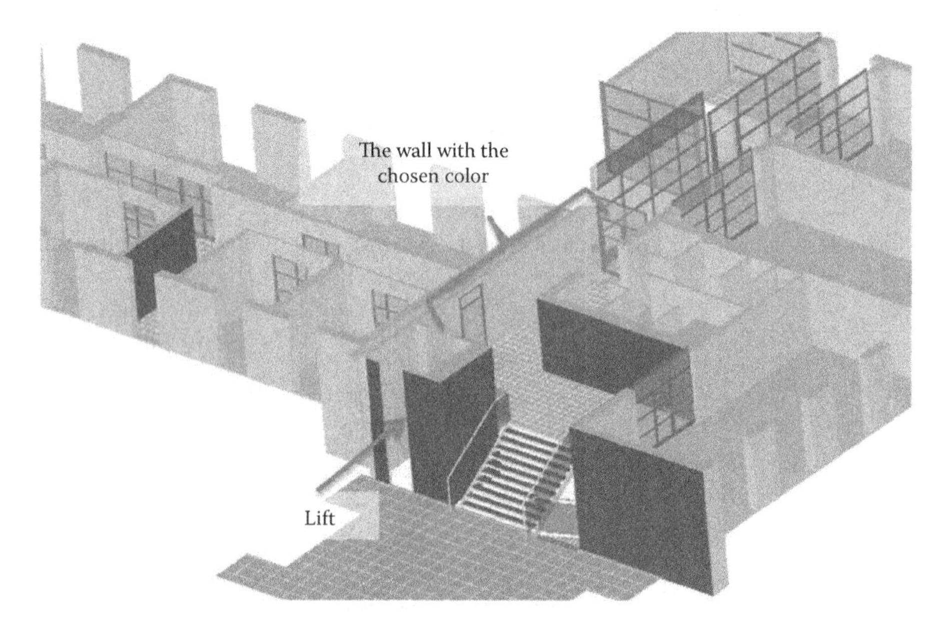

FIGURE 10.6 Visualization of the main hall and the ground floor.

FIGURE 10.7 The ground floor with RAL 2000 used in the stairs and lift area.

- After exiting the lift, or being on the top of the stairs, people are informed of the number of the floor.
- The color of floor number is identical with the color of the walls assigned to the floor.

Figure 10.6 shows the visualization of the main hall, along with an indication of the ground floor and the walls of the stairwell and lift shaft.

In Figure 10.4, the wall that should be highlighted with the chosen color for each floor is marked.

Proposed solutions for the interior wall color in the lift area are shown in the example of floors 0 and 3 (Figures 10.7 and 10.8).

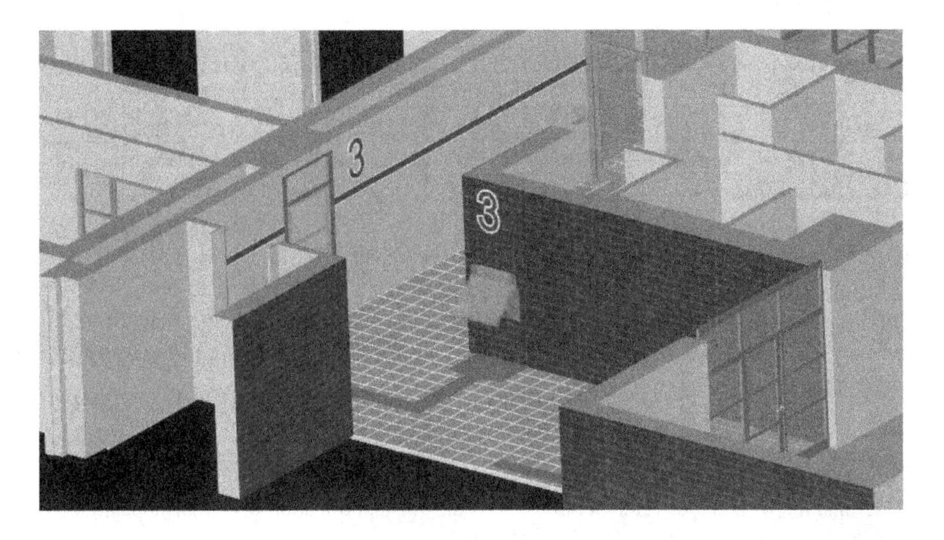

FIGURE 10.8 Third floor with RAL 5013 used in the stairs and lift area.

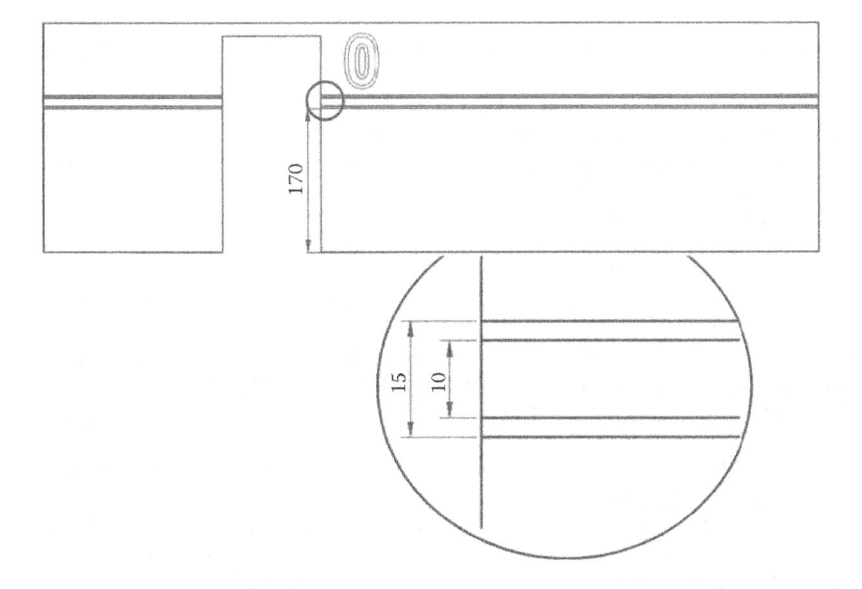

FIGURE 10.9 The scheme of painting the walls in the lift area. The dimensions are in centimeters.

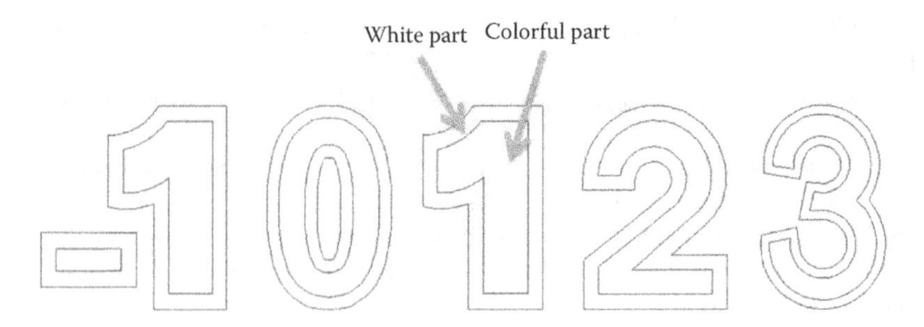

FIGURE 10.10 Model of wall plates' numbers.

We suggest painting the walls as follows:

- It is recommended to paint a stripe of a chosen color on the wall near the exits of the vertical circulations.
- The stripe should be 10 cm wide and should be placed 170 cm above the flooring.
- To emphasize the contrast effect, the colored stripe should be separated from the gray wall (RAL 7047 proposed by the architects) with 2.5 cm wide white strips.

Figure 10.9 shows the scheme of painting the colored stripes in the lift area.

An additional indication of floor colors is the number of floor. It is placed on two walls: next to the lift and on the clinker-tiled wall, which is opposite the stairway. The number of the floor is the same color as the stripe. To increase the contrast, the number of the floor should have a white border (Figure 10.10).

REFERENCES

Billger M. Evaluation of a colour reference box as an aid for identification of colour appearance in rooms. *Color Research and Application* 25, 214–225, 2000.

Cassin B., Solomon S. *Dictionary of Eye Terminology*. Gainsville, FL: Triad Publishing Company, 1990.

Chebat J.-Ch., Gelinas-Chebat C., Therrien K. Lost in a mall, the effects of gender, familiarity with the shopping mall. *Journal of Business Research* 58, 1590–1598, 2005.

Chien-Hsiung Ch., Wen-Chih Ch., Wen-Te Ch. Gender differences in relation to wayfinding strategies, navigational support design, and wayfinding task difficulty. *Journal of Environmental Psychology* 29, 220–226, 2009.

Dalke H., Littlea J., Niemanna E., Camgoza N., Steadmana G., Hilla S., Stott L. Colour and lighting in hospital design. *Optics and Laser Technology* 38, 343–365, 2006.

Helvacioglu E., Olguntürk N. Colour contribution to children's wayfinding in school environments. *Optics and Laser Technology* 43, 410–419, 2011.

Holscher Ch., Buchner S., Meilinger T., Strube G. Adaptivity of wayfinding strategies in a multi-building ensemble: The effects of spatial structure, task requirements, and metric information. *Journal of Environmental Psychology* 29, 208–219, 2009.

Holscher Ch., Meilinger T., Vrachliotis G., Brosamle M., Knauff M. Up the down staircase: Wayfinding strategies in multi-level buildings. *Journal of Environmental Psychology* 26, 284–299, 2006.

Hund A. M., Padgitt A. J. Direction giving and following in the service of wayfinding in a complex indoor environment. *Journal of Environmental Psychology* 30, 553–564, 2010.

Katoa Y., Takeuchi Y. Individual differences in wayfinding strategies. *Journal of Environmental Psychology* 23, 171–188, 2003.

Sharpe L. T., Jägle H. Ergonomic consequences of dichromacy; I used to be color blind. *Color Research and Application* 26, 269–272, 2001.

Sharpe L. T., Stockman A., Jägle H., Nathans J. Opsin genes, cone photopigments, color vision, and color blindness. In Gegenfurtner K. R. and Sharpe L. T. (editors), *Color Vision: From Genes to Perception*, Chapter 1. Cambridge: Cambridge University Press, pp. 1–51, 1999.

11 Human Classification Process
Theory, Simulation, and Practical Applications

Valery D. Magazannik

CONTENTS

STATEMENT OF THE PROBLEM

The main goal of this work is to develop and simulate the psychological mechanism by which the hierarchical structure of information in human memory functions. It was believed that the process depends on the information perceived, order of learning, last experience of the subject, and level of excitement (emotional stress). The applied goal was to develop the computerized simulation of human decision making for classification of perceived information. This simulation was expected to predict quantitative probabilistic data of human free-recall and human recognition (classification in a wide meaning) times and errors.

Characteristics of human performance are determined by external stimuli, primarily the comprehension of a situation. This is based upon the experience of the individual. Comprehension is the result of the specific semantic structure in an individual's memory. Such structure is the hierarchical grouping of information, in other words, the system of categorizing information at different levels (Bartlett 1932; Neisser 1976).

It is widely regarded that information expands during human learning through the formation of generalized integral features. However, a number of questions are not answered by previous investigations. These include the following:

1. What aggregation hierarchy will occur if the primary set of stimuli is known?
2. What are the determining factors of this hierarchy, and what rule (logic) is used to associate stimuli with each other? What are the quantitative dynamics of initial functioning of different levels of generalizations, and what are the determinants of this process?

The following are the discussions and the answers for the above questions.

The Silent Theoretical Points

The Essence of Repetition and the Role of the Space–Time Continuum

The doctrine of associationism, as old as ancient Greece, as formidable as to dominate 19th century psychology, and as recent as 20th century behaviorism, held that repetition was a critical element in forming associations—especially if followed by a reinforcing state of affairs. The repetitions can be of various types. An internal repetition may occur because of either repeated presentations of an identical object in the same locus of space or one-time presentation of a sequence of identical objects. In the latter case, the objects of the external set are repeated in different locuses of space. Associations arise in both cases. However, in the latter case, the recurrence may or may not be noticed by the observer. If it is noticed, then the association arises (or is strengthened if it already exists); if not, then the association does not arise (or it is not strengthened). Otherwise, any repetition may or may not be noticed by a subject, and the time and proximity of the repeated element affects whether or not the repetition will be noticed.

What does the term "repetition" mean? A dominant role of repetitions in the process of association strengthening is one of the main principles of associationism. What, in fact, is repeated when the sequence of objects has been read once? We do not refer now to the subject repeated verbally or in any other way. First of all, the sequence (i.e., interconnection of the object itself and its locuses in this chain) is repeated. If these objects are organized in a different sequence than in the original presentation, it is not the same objects that were read before (Rabbitt and Vyas 1973; Paivio 1974). In the broadest understanding, an object can dramatically change its sense, properties, and destination in different environments, appearing not at all "the same" to us. Therefore, something is repeated even if it looks like nothing has been repeated (if the objects are different at all).

In an extremely general case, a common thing that is able to serve as the basis of repetitions could be a span (an interval) of disturbance of a space–time continuum (i.e., the disturbance of an initial monotony). It is known that attempts to restore information in memory are accompanied by a knowledge of the existence of something that has to be restored ("something must be here, but I cannot remember what it is"), also calling feeling of knowing (FoK). Evidence of this was given by Bartlett (1932), Neisser (1976), and others. All data in this respect make us conclude that special associations are certain to link some "shadows" of objects; such associations depend on arrangement of objects in space–time rather than on the object features themselves. As to object-to-object links, they might exist as secondary ones, on the basis of such *primary* links.

It is known from the free-recall experiments that if the arrangements of the first and second lists are the same, the results of their reproduction are much better. Why? What could be repeated in such cases? We consider that the unique thing that would ensure some kind of repetition is the disturbance of the space–time continuum. In remembered objects that have nothing in common, as it appears, only their space–time dimensions can be compatible. It could be concluded that objects are associated with each other through their locations in the space–time continuum rather than directly. The latter is basic for any event, object, or situation. Such understanding agrees with the concept of the scheme (cognitive maps), in this instance, as interpreted by Neisser (1976). *The scheme* serves as a skeletal structure of a subjective image, the basis and condition of image forming. "The scheme" is the fundamental basis for repetitions.

The scheme may contain information about movement when the path of object movement is longer than the object itself (in the case of mechanical movement); in this case, then, the perception of the path is primary, and the object itself is secondary. This may be illustrated by dancing figures. We should conclude that human perception bases oneself, rather unconsciously, upon space–time estimation.

The role of semantics of some space in its subjective structuring was emphasized by Kurt Lewin back in 1943. For example, the perception of space extent is determined by the degree of familiarity with its objects; less familiar ways of moving are perceived as longer than more familiar ones. The point is that the value of subjective extension reflects the number of *segments* actualized as *parts* of the route, but the quantity of such subjective unified units (*parts*) depends, in agreement with our concept, on the strength of associations, (i.e., on the degree of familiarity with space). It is of great interest that not only the space occupied by an object has an influence on its semantic features, but also, on the contrary, in the presence of a high degree of familiarity, the semantics of the object essentially affects the perception of its space–time features; there are many illustrations of this thesis, especially in the fine arts. In a broad sense, this idea goes back as far as Bernhard Riemann (1826–1866) and William Clifford (1845–1879). Riemann (2004) pioneered the assumption that the nature of physical space had to contain some information of about what occurred in this space. This assumption meant that the properties of space might be different in different points of this space; afterwards, this idea evolved into differential geometry (as opposed to geometry of a whole, united space, Euclidean as well as non-Euclidean).

Janet Pierre (1859–1947) drew attention to a close interrelation between memory and movement and between representation of space and duration (Pierre 1929). It is known that the scheme of body movement is the measure of space relations, and speed of body movement determines peculiarities of perception of *length*, *height*, and so forth.

Thus, development of the concept of the space–time continuum together with semantic, syntactical, and other aspects has led to a more precise understanding of the concept of *the scheme* (or cognitive map).

For the reason that objects are connected not directly but through their loci, the cases become clear when the space–time information is restored but the objects in it are not restored, or when the objects are restored from different places of such organization. The fact of better reproduction if the scheme is visualized also becomes clear. "Emptiness" (holes in space) that appear while the initialization process is going on shows directions (and ways) of information gathering; such gathering produces initialization of surrounding associations and potentially the appearance of other "emptiness"; at that, the retrieved information is increased.

Also, the concept of repetition is expected to account for the level of hierarchy, the location of the object within and among other groups. It is necessary to understand the rules of association strengthening in a hierarchy. The main idea of those rules is to take into account the repetitions not only in the object's features themselves, nor in just the space–time dimensions, but also those in sequences and degrees of changes of these characteristics. The latter allows understanding of the movement (in a broad meaning), that is, the process of any changes in time and space. It is of great interest to discover the limits of the relationship between changed and unchanged parts, which ensures the perception of change as a change itself, not as a new object.

The essence of the elaborated mechanism can be accumulated in the following rules of association strengthening. The association strengthening takes place only if the following conditions are met: (1) This association has been aroused. (2) As the result of such arousing, some repetitions have been brought to light. Then the associations that link repeated elements are strengthened. At that, the repeated elements can be not only the parts of units or complete units themselves but also the size of groups if they are at the same levels and in the same serial positions in sequence of another group; at the terminal (lowest) level, the content of groups can also be repeated. Incoming information interacts with information that is in the focus of attention, that is, with the actualized set.

The rules assume that repetition of something takes place only if this *something* is located in the same serial group position, those groups containing the same quantity of objects (dimensions, elements) and being disposed of at the same level of hierarchy. This principle accounts for repetitions in changes, that is, to simulate movement or any transformations of an object, in other words, any kind of passage, both abrupt and gradual, light ones. Ordering and movement are closely related. In other words, ordering is a movement in *above-mechanical* spheres. Also, the principle facilitates the understanding of the process of discovering regularities (Magazannik 1997; Магазанник 1997).

We consider the process of discovery of regularities as the process of strengthening links of repetitive elements of the hierarchy, and it leads to increasing the probability of retrieving the information that contains such repetitions. For example, let us analyze rows of digits; remembering them and making mutual comparisons are equivalent to the comparison of the difference in magnitude in quantized continuums. Here, as well as in other cases, the higher the level of repetitiveness changes, the more, on the average, the necessary number of repetitions of the sequence to reach a definite probability of retrieving. The sequence of rows in which the levels of repetitiveness are increased one after another (minimum intervals ensuring repetition increase by one in such rows) establishes, in particular, lateral sides of the so-called Pascal triangle. All such rows are the well-known dependences; say the fourth row is the power dependence, and so forth. Here is an example of rows, where each following row contains the members of the previous one as differences.

0, 0, 0, 0, 0, 0
1, 1, 1, 1, 1, 1,...
1, 2, 3, 4, 5, 6,...
1, 3, 6, 10, 15, 21,...
1, 4, 10, 20, 35, 56,...

The row interval, which is necessary and sufficient, from the formal standpoint, for regularity demonstration, is, in reality, only the minimal interval, which permits the strengthening of links to differences (the change has been repeated) immediately following the first reading, surely, if the hierarchies (the subjective structures of the rows) match. Remember, the row interval actualized by the presented object depends on the characteristics of hierarchy and on the strength of links, but the probability of retrieving the entire sequence by one actualization is rather small if the length of the row is great enough. The mathematical expectation of the quantity of simultaneously actualized objects has some mean value; that is, the area of real consideration is less, as a rule, than the whole presented (formally possible) area. Therefore, the disclosure of regularity is more or less a long process of increasing the probability of retrieving material by taking from internal repetitions.

During training, sameness of changes produces more and more strong associations, and the probability of retrieval nears the case when changes are absent; that is, the two cases might appear as psychologically equivalent because changes (differences) are no longer perceived as such. New uniformity has been established, and a stable automatism has resulted from it. In other terms, we could say that a new gestalt has been formed.

In space, the constancy of distance differences is shaped into a straight line, and the absence of differences is shaped into a point. Equivalency of a straight line and a point can be observed when strong links have been formed among the segments of the line; therefore, the straight line is perceived as an entire object. In physics, this thesis is expressed in the first law of Newton, where the rest state and the rectilinear uniform movement are identical to each other on the basis of an absence of influence on mass. In agreement with the principle of relativity, it is impossible to discover rectilinear uniform movement while being within; that is, it is perceived as a rest. Developing this idea, it is possible to identify rectilinear uniform movement with uniformly accelerated movement on the basis of the constancy of influence. In other words, our thesis is that both the sameness of the features of objects and the sameness of the changes of the features of objects produce strong associations soon enough (after a few internal repetitions, though in the second case, the quantity of such repetitions must be greater than in the first case) and transform into subjective unified units or into a gestalt object.

Subjective Unified Units: Information Ordering Index

Any grouping, as well as intellectual activity as a whole, is based on the assumption that active memory capacity is limited. The detailed character of such hierarchies is determined by the actual meaning of active memory space. We start from the position that grouping of information by a human (in other words, subjective structuring of information) is rational and ensures the storage and retrieval of any volume of information. It can be achieved only by way of generalization, that is, through essential reduction of subjective independent chunks of information. The quantity of the subjective units that have been activated (i.e., retrieved in short-term [ST] memory) totally determines the decision time, at least if one must make a choice decision (classification tasks). The experimental data show that the recognition time depends linearly upon the quantity of subjectively independent integral units (but the quantity of the units is limited by the capacity of ST memory; that is, it cannot be more than seven to eight). Under some conditions, the capacity of ST memory can be changed; for instance, the emotional stress decreases this capacity, and it leads to certain distortion of decisions and behavior. Otherwise, some psychopharmacological drugs are able to increase it (a little). When a new object is perceived, the excitation of the subjective structure occurs, and it is possible to restore significantly more information than an object itself presented. In psychological terms, the initiation of subjective structure is actualization.

Actualization puts information in the focus of attention, where this information is repeated, compared, circulated, and quantized; where the clarification of identities and differences takes place, the clarification of identities in differences takes place, and so on; that is, the clarification of regularities takes place. The stronger the associations, the more probable that the objects will be activated as the integral union, the entire group. The formation of integral unions, in turn, can take place through the appearance of new associations and strengthening of existing ones.

What measure can be suggested for such an intuitively clear term as "association strength?" As it follows from the associative paradigm, the most suitable measure, we consider, is the value of conditional probability of one object (element, feature, situation) recalled if the other has already been recalled (actualized). This value depends on the quantity of internal repetitions.

The subjective cost or mental resources (figuratively, the "energy of the process") is neither the quantity of internal repetitions nor the quantity of readings (for the reason that a single reading can contain a few or many repetitions; it depends on the scanned information features). This cost is the quantity of elements (objects and so forth, in short, internal units) that have been read through. The degree of information ordering (quality of grouping) can be estimated as the ratio of the probability of complete stimulus set reproduction by a man at a fixed quantity of actualization to the number of elements that have been read (scanned). This index shows the measure of information ordering as well as the grouping quality. It may be presented as the increasing function of the probability of complete set recall by a person at a given quantity of trials; that is, the quantity of actualization (take this quantity as eight) and the decreasing function of the quantities of elements have been read (internally scanned). In broad meaning, the last value reflects the frequency of meetings or degree of training. Surely, the index can be standardized (to be about optimum grouping) by means of its ratio to the measure that corresponds to the best grouping of the same information, that is, to the possible maximum meaning (that is in optimum) of this index. This measure makes it possible to formulate a criterion for optimum arrangement of new information.

It is clear that the index of the degree of ordering takes its maximum value when the sizes of groups (the number of elements in the group) are relatively small and equal at all levels of hierarchy; the value of this index is a little less when the sameness of differences between groups take place, decreases gradually as the level of repetition increases, and takes its minimal value when a random sequence takes place. (This index could essentially be used as a measure of the degree of randomization of a limited sequence of items.) Such an index also shows that the number of elements read (training, in reality) has a great impact on the index of ordering.

REFERENCES

Bartlett F.C. (1932) *Remembering: A Study in Experimental and Social Psychology*. London: Cambridge Univ. Press.

Lewin K. (1943) Defining the "field at a given time." *Psychol. Rev.* 50: 292–310. Republished in *Resolving Social Conflicts and Field Theory in Social Science*. Washington, D.C.: American Psychological Association, 1997.

Magazannik V. (1997) Human Information Chunking While Classifying Decision Making—The 13th Triennial Congress of the International Ergonomics Association, Tampere, Finland, v. 3.

Магазанник В.Д. (1997) Структурирование информации человеком при принятии решений. Психологический журнал, #1.
Neisser U. (1976) *Cognition and Reality: Principles and Implications of Cognitive Psychology*. WH Freeman.
Paivio A. (1974) Spacing of repetitions in the incidental and intentional free recall of pictures and words. *J. Verb. Learn. Verb. Behav.* 13(5): 497–511.
Pierre J. (1929) L'évolution psychologique de la personnalité. Paris: Chahine.
Rabbitt P.M.A., Vyas S. (1973) What is repeated in the "repetition effect"? In *Attention and Performance*, IV, Ed. S. Korblum. N.Y.: Acad. Press, pp. 327–342.
Riemann B. (2004) Collected papers. In *Math. Reviews*, Heber City, U.T.: Kendrick Press.

12 Continuous Quality Improvement in Neonatal Care

Experience of Eastbay Newborn Services Group, Northern California, USA

Anna G. Usowicz and Golde Dudell

CONTENTS

INTRODUCTION

How do you become a good doctor in the 21st century, in the era of information technology, where you need to stay current with published literature and use it to improve patient outcomes?

The American Board of Pediatrics (ABP), the American Academy of Pediatrics (AAP), and the Accreditation Council for Graduate Medical Education (ACGME) work together to develop measures to assess competence of physicians during residency programs and maintain this competence during years of medical practice through lifelong learning. Since the 1990s, these organizations have attempted to evaluate medical professionals systematically; to find reasons for unexplained variations in medical practice, health outcomes, inadequate care, and medical errors; and to make system improvements [1]. They also defined six general competencies for physician's excellence: patient care, medical knowledge, professionalism, systems-based practice, practice-based learning and improvement, and interpersonal and communication skills. The implementation of these competencies brought a change in the education of residents and the recertification of practitioners [2,3].

ABP was founded in 1933 to establish standards of practice and certify competence of physicians. ABP has certified over 80,000 pediatricians and subspecialists. It awarded permanent certificates until 1988 when, in order to encourage self-learning, certificates became limited to 7 years. The first recertification exam was offered in 1994. In response to professional and public concerns about safety and quality of medical care, ABP created the Maintenance of Certification (MOC) program. Specialty board certifications and maintenance of certification became the most important factors in assuring quality of care and professional development for physicians. By 2004, 86% of general pediatricians and 88% of subspecialists participated in this program [4]. MOC consists of four parts: Part 1 involves maintaining a valid license to practice; part 2 involves demonstrating lifelong learning through ongoing self-assessment; part 3 involves passing a secure examination; and part 4 involves demonstrating the ability to assess and improve quality of practice [5]. ABP requires all physicians to compare their quality of care with results of their peers and benchmarks in other centers in the country and to change practices to improve care.

AAP endorsed the development and implementation of health information technology to improve child health, like electronic health records, registries, newborn screening programs, immunization information systems, trauma registries, and other databases used for public health records [6]. Traditional continuing medical education (CME), which concentrated mainly on attending formal conferences, has been evolving into continuous professional development (CPD) that involves self-directed learning appropriate for the individual physician's needs, use of Web sites like PediaLink, and telemedicine [7,8].

REVIEW

How is health information technology used to support quality improvement measures undertaken by American physicians in the subspecialty of neonatal and perinatal medicine? Among the most important organizations to influence everyday practice in neonatal–perinatal medicine is the Vermont Oxford Network (VON) nationally and the California Perinatal Quality Care Collaborative (CPQCC) in California.

VON is a voluntary collaborative group of health professionals established in 1989 to improve medical care through a coordinated program of research, education, and quality improvement projects. It is a repository for data on patients with birth weights between 401 and 1500 g. The data include demographics, medical interventions, and outcomes of these patients at member hospitals. In 2008, 576 hospitals from the United States and 174 hospitals from Europe, Asia, South America, and the Middle East countries participated in the process, and data on 106,000 infants were collected [9]. VON provides members with confidential reports that allow each center to compare their results with those of hospitals around the world. Members can also compare their results with a subgroup of hospitals with characteristics similar to their own. VON gives its members an opportunity to identify areas for improvement and track changes over time. Examples of key performance measures include mortality, length of stay, incidence of chronic lung disease (CLD), nosocomial infection, intraventricular hemorrhage, cystic periventricular leukomalacia, retinopathy of prematurity (ROP), and necrotizing enterocolitis (NEC). These conditions are of interest because of wide intercenter differences in their occurrence risk.

The VON research program includes outcome research and randomized trials that are relevant to everyday practice and encourages a habit for change, a habit for systems thinking, a habit for collaborative learning, and a habit for evidence-based practice [10]. The VON Manifesto was adopted by all members, who pledge to provide care that is family centered, safe, effective, equitable, timely, efficient, and socially and environmentally responsible. One of the specific aims is to foster a worldwide community of practice for newborn care, in which knowledge, tools, and resources for improvement are developed, managed, shared, and applied [9].

In the last 15 years, VON initiatives included six intensive multidisciplinary neonatal intensive care quality (NICQ) collaboratives, and a series of nine Internet-based NICQ collaboratives. The first quality improvement collaborative was initiated in 1995 and achieved a significant reduction in the incidence of nosocomial bacterial sepsis and a decrease in the cost of hospitalization in participating centers. Other collaborative projects involved family-centered care and a leadership series for neonatologists in administrative roles.

VON has conducted many large randomized controlled trials of various standards-of-care practices to evaluate their effect on clinically important outcomes. Data collected in hundreds of hospitals on thousands of patients provide a unique opportunity to conduct large epidemiologic studies. The trials involved the use of surfactant in respiratory distress syndrome, postnatal use of dexamethasone, and the effect of prophylactic emollient ointment on the rate of nosocomial sepsis and skin integrity. Based on these trials, recommendations were made. One of the largest trials enrolled 6196 very-low-birth-weight infants (VLBW) to identify characteristics associated with severe disability. Severe disability was found in 35.8% of survivors, and the most highly associated factors were periventricular leukomalacia, intraventricular hemorrhage, and congenital malformation. The cost of participation in VON is approximately $30,000 per hospital, which maybe a limiting factor for many hospitals.

As VON is a national perinatal quality improvement organization, the CPQCC is a regional hospital-based and community-based initiative in California. The CPQCC was established in 1997 by the California Association of Neonatologists (CAN); the state of California, with its 350 birthing hospitals; the California Children's Services which is one of the state's largest payers; the L and D Packard Foundation; and Stanford University. The CPQCC collects clinical data on 90% of all neonates receiving intensive care in California and has 129 member hospitals. In 2008, data were collected on 18,671 infants, of whom 6677 were VLBW. In 2007, an acute infant transport database was developed, and in 2009, the statewide High-Risk Infant Developmental Follow-up program was added [11]. The CPQCC outreach is based on tool kits that include academic presentations, workshops, and webcasts. Many CPQCC initiatives have been effective in improving outcome in participating units. Among them were antenatal steroids administration, postnatal steroids, hospital acquired infection prevention, improving initial lung function, nutritional support of the VLBW infant, perinatal group b streptococcus prevention, hyperbilirubinemia prevention, perinatal HIV prevention, delivery room management of the VLBW infant, and the late preterm infant management. The cost of participating in a CPQCC initiative ranges from $5000 to $7000.

DISCUSSION

How are these national and state efforts for quality improvements translated into everyday practice and care delivered to every single patient?

The Group Practice

Our group of 11 neonatologists, 3 neonatal nurse practitioners, and 7 pediatric hospitalists provides neonatal care in the East Bay Area of Northern California. Our practice was established about 50 years ago as the neonatal subspecialty started to emerge in 1960s and expanded to what it is now—one of the biggest private neonatology groups in the United States. We serve a population of 2.5 million with 28,000 births annually. We provide prenatal consultations for expectant parents and complex medical care for approximately 100 ill neonates a day in three major hospitals: Children's Hospital and Research Institute in Oakland, Alta Bates Summit Medical Center in Berkeley, and John Muir Medical Center in Walnut Creek. Our patients come from the poor neighborhoods of Alameda County as well as affluent areas of Contra Costa County; they come from diverse racial, cultural, and religious backgrounds, from families of US citizens to families of recent immigrants from all over the world. We work closely with many medical subspecialists such as perinatologists, pediatric surgeons, cardiologists, pulmonologists, neurologists, gastroenterologists, and geneticists. Our High-Risk Follow-up Clinic tracks neurodevelopmental outcomes of patients for first 2 years of life. We also provide training for 60 pediatric residents.

Developed Guidelines

Our group has developed measures to use information technology to optimize patient care, based on self-improvement and lifelong learning. Specifically, we developed measures for every member of our group to perform at the highest professional level through the following:

- Enrollment in the MOC program developed by the AAP
- Participation in international webinars
- Development of professional Web site with clinical practice guidelines

According to current guidelines of AAP, every member of our group is enrolled in the MOC program, which has a 4-year cycle. MOC requires passing a secure subspecialty examination every 7–10 years and demonstration of lifelong learning through participation in self-assessment activities. These activities are available on the AAP and ABP Web sites; participants are awarded credits after taking tests based on reading relevant references. One of the excellent learning tools is the NeoReviews-plus Internet site, which provides questions and answers developed by experts in neonatology.

MOC also requires a professional assessment of the physician and his or her interpersonal and communication skills by the parents of patients. Another MOC requirement is participation in quality improvement projects.

On a regular basis, our group has been participating in webinars, which provide an interactive forum to discuss controversial topics with neonatologists from different medical centers in the United States and abroad. Participating hospitals pay a cost of $3000 per session. In preparation for webinars, we review the relevant medical literature and answer questions based on our current practice. Afterwards, we may implement changes with the goal of practice improvement. We have eliminated established practices that have no proven efficacy or are even harmful. Last year, we reviewed the practice of red cell transfusion, noninvasive mechanical ventilation, the use of oxygen in the delivery room and the use of nitric oxide in the preterm in the NICU, and treatment of gastroesophageal reflux.

In our busy practice, we strive for excellence, and in order to achieve better outcomes for patients, we want to provide uniform care by all practitioners. We developed a corporate Web site where clinical practice guidelines are readily available to all members of our group. The guidelines are consensus statements based on the most recent recommendations of the AAP, CPQCC, and review of literature by individual members of our group. Pertinent articles are posted as well.

Our group is one of the first in the country to establish formal practice guidelines. Each member of our group is actively involved in this process. Every month, we meet for an evidence-based medicine meeting and discuss recent literature and develop consensus for practice. We invite experts from other centers to share their experience. After the meeting, guidelines are developed and placed on the Eastbay Newborn Specialist corporate Web site. Guidelines are designed to provide consistent management in different nurseries by different practitioners. These guidelines are simple and concise and easy to follow. They are intended to be guides, not strict rules. Minor variations in practice are allowed in each major hospital, depending on resources and patient populations. Every practitioner is expected to follow guidelines for routine management with the understanding that modifications may be needed based on patient response. Guidelines are discussed with residents, nurses, and respiratory therapists during teaching conferences and on work rounds. Our Web site is accessible in every hospital and office and at home. Members of our group periodically audit compliance with guidelines, in order to determine what can be improved, and incorporate those improvements into subsequent versions.

We developed numerous guidelines in many areas of our practice: respiratory care; congenital heart disease; use of blood products; nutrition; use of antibiotics, diuretics, steroids, and nitric oxide; and indications for surgeries. We developed criteria for the use of specialized interventions such as extracorporeal membrane oxygenation and brain cooling for hypoxic–ischemic encephalopathy.

Quality Improvement Projects

Our group has worked with the CPQCC in implementing various quality improvement projects, especially for VLBW infants with birth weight between 401 and 1500 g. One of these projects involved increasing the use of human milk (maternal and donor bank milk) and increasing the rate of breast-feeding for premature infants. It was developed by physicians, nurses, and lactation specialists. The benefits of breast-feeding are discussed with mothers in prenatal consultation. Every mother is provided with a breast milk pump and given instructions for collection and storage of colostrum and breast milk in the postpartum period. During convalescence of the premature baby, the mother is encouraged to hold the baby by the "kangaroo" method and to start "nonnutritive" breast-feeding as soon as tolerated. Colostrum is used for oral care and to help to establish a normal gastrointestinal flora. Hospitals have meals for lactating mothers, and they are given help to cover cost of transportation. Breast-feeding mothers are given preference over non–breast-feeding mothers for the limited number of sleep rooms, which allow them to stay overnight close to their babies. As a result, the breast-feeding rate was increased from 54% to 75% at time of discharge. Sixty-five percent of "small" babies (<1500 g) and 80% of "big" babies (>1500 g) received human milk at discharge from the hospital in 2009. Using human milk and developing a standardized feeding protocol, which ensures that enteral feedings are advanced in a uniform way by every physician, had an effect on decreasing the rate of NEC. In our practice, the incidence of NEC decreased from 9% in 2003 to less than 2% in 2010 for "small" babies.

Postnatal growth failure is a common and challenging problem of VLBW infants, with 90% of them falling below 10% for weight at 36 weeks of gestation. Postnatal growth restriction can affect neurodevelopmental outcome. According to CPQCC guidelines [12], we developed protocols to improve postnatal growth. These protocols recommend aggressive nutritional support for VLBW infants with early initiation of parenteral nutrition on the day of birth and stepwise advancement of intake of protein, glucose, lipids, and minerals in order to prevent early nutritional deficiencies. We use computerized programs for parenteral nutrition to assure safety and avoid human errors. We introduce enteral feedings early and advance them per protocol. In every nursery, we have a dedicated dietician and lactation specialist on staff to follow each patient's growth and nutritional intake during hospitalization and make recommendations to the neonatologists based on these parameters. They also assist in establishing individualized feeding plans after discharge.

Another CPQCC quality improvement initiative is aimed at decreasing the rate of ROP, the rate of surgeries for ROP, and the number of infants with vision impairment [13]. This initiative was very successful, and its goal was achieved through educational presentations to staff about oxygen toxicity and signs at bedside that identify infants who need oxygen levels maintained within the prescribed limits. The limits of acceptable oxygen saturation were lowered; nurses were instructed to keep strict control of delivered oxygen at all times and especially to avoid hyperoxia as well as repeated episodes of hyperoxia–hypoxia in VLBW infants. The alarm limits for pulse oximetry are set based on the current evidence, and the nursing staff was instructed to increase or decrease the oxygen level in small increments to avoid wide swings in the infant's oxygen saturation. Audits were designed to estimate the amount of the time that infants spent outside of the desired range, especially at saturations greater than 95%. These measures brought a striking improvement in our practice and helped to decrease the rate of severe ROP and ROP surgery from 11.4% in 2005 to 5.8% in 2010.

Reducing the rate of nosocomial infections and central line–associated bloodstream infections (CLABSI) with coagulase-negative *Staphylococcus*, fungus, and other bacterial pathogens is a major quality improvement effort required by state and third-party payers. These infections contribute to significant mortality and morbidity, cause injury of the white matter of the brain, affect neurodevelopmental outcome, and increase the cost of hospitalization [14]. In order to effectively decrease the rate of CLABSI, we introduced many evidence-based interventions simultaneously—care "bundles." One intervention was strict hand hygiene by all caretakers to reduce transmission of pathogenic bacteria from the hands of caretakers to our vulnerable patient population. Others included placing central lines and changing intravenous solutions using a strict sterile technique, using closed vascular access systems, needleless connectors, thorough hub decontamination prior to entry, and discontinuation of central catheters as early as possible. We were successful in decreasing the rate of nosocomial infections for "small" babies from 17.1% in 2005 to 2.5% in 2009 and the rate of nosocomial infections for "big" babies from 2.8% in 2005 to 0.0% in 2009. Our ultimate goal is elimination of all hospital-acquired infections altogether.

CONCLUSION

Implementation of quality improvement measures requires teamwork and close collaboration between all hospital staff working in the unit. Periodic audits of the practices are important to assure compliance and assess progress. Positive results are displayed as graphs to provide encouragement.

We also identified areas of our practice that need improvement. CLD remains a major cause of morbidity for VLBW babies. Despite enormous improvements in neonatal care, the incidence of CLD has remained constant through the last 20 years at 25.5% [15]. We undertook numerous measures to decrease the rate of CLD. We introduced various modes of noninvasive ventilation; we practice aggressive weaning from mechanical ventilation and prophylactic use of surfactant in high-risk infants. We avoid intubation in the delivery room for babies at >26 weeks of gestation. We use vitamin A, optimal nutrition, and optimal oxygen therapy. Although the rate of CLD, defined as an oxygen requirement at 36 weeks adjusted age, is below the national mean, it has remained fairly constant for the last 5 years at about 16%–22%.

Although the use of postnatal steroids was shown to improve lung mechanics and decrease CLD, their use is strongly discouraged because of toxic effects on immature brains and growth and the risk of gastrointestinal hemorrhage, infections, spontaneous bowel perforation, and hypertrophic cardiomyopathy. The AAP recommends against the routine use of steroids to prevent CLD. We placed protocols restricting the use of steroids to severe cases only. The rate of use of postnatal steroids for CLD in our practice remains at 4.3%–9.3%, compared to 8.5%–9% in other centers.

Another area for improvement is the need for a reduction of the rate of patent ductus arteriosus (PDA) ligations, which is 7.9% compared to 8.5% nationally. Surgical ligations are associated with an increased rate of CLD and adverse neurodevelopmental outcomes. We reviewed indications for pharmacologic and surgical treatment of symptomatic PDA with the cardiology department and introduced more restrictive criteria.

In summary, we have been able to improve patient outcomes through lifelong learning, introducing changes in medical practice based on scientific evidence, education of all members of the hospital staff, and collaboration between different medical centers in and outside the country.

REFERENCES

1. Dougherty D and Simpson LA. Measuring the quality of children's health care: A prerequisite to action. *Pediatrics* 2004; 113; 185–198.
2. Carracio Brown HJ, Miles PV, Perelman RH and Stockman JA III. A continuum of competency assessment: The potential for reciprocal use of the Accreditation Council for Graduate Medical Education toolbox and the components of the American Board of Pediatrics Maintenance of Certification program. *Pediatrics* 2009; 123; S56–S58.
3. Carracio C, Englander R, Wolfsthal S, Martin C and Ferentz K. Educating the pediatrician of the 21st century: Defining and implementing a competency-based system. *Pediatrics* 2004; 113; 252–258.
4. Freed GL, Dunham KM, Althouse LA and American Board of Pediatrics, Research Advisory Committee. Characteristics of general and subspecialty pediatricians who choose not to recertify. *Pediatrics* 2008; 121; 711–717.
5. Miles P. Health information systems and physicians quality: Role of the American Board of Pediatrics maintenance of certification in improving children's health care. *Pediatrics* 2009; 123; S108–S110.
6. Fairbrother G and Simpson LA. It is time! Accelerating the use of child health information systems to improve child health. *Pediatrics* 2009; 123; S61–S63.
7. Sectish TC, Floriani V, Badat MC, Perelman R and Bernstein HH. Continuous professional development: Raising the bar for pediatricians. *Pediatrics* 2002; 110; 152–156.

8. Gonzalez-Espada W, Hall-Barrow J, Hall RW, Burke BL and Smith CE. Achieving success connecting academic and practicing clinicians through telemedicine. *Pediatrics* 2009; 123; e476–e483.
9. Horbar JD, Soll RF and Edwards WH. The Vermont Oxford Network: A community of practice. *Clin Perinatol* 2010; 37; 29–47.
10. Horbar JD. The Vermont Oxford Network: Evidence—Based quality improvement for neonatology. *Pediatrics* 1999; 103; e350.
11. Gould JB. The role of regional collaboratives: The California Perinatal Quality Care Collaborative model. *Clin Perinatol* 2010; 37; 71–86.
12. CPQCC. Quality Improvement Toolkit (rev 2008). Nutritional Support of the Very Low Birth Weight Infant. Available at www.CPQCC.org.
13. Elisbury DL and Ursprung R. Comprehensive oxygen management for the prevention of retinopathy of prematurity: The pediatrix experience. *Clin Perinatol* 2010; 37; 203–215.
14. Powers RJ and Wirtschafter DW. Decreasing central line associated bloodstream infection in neonatal intensive care. *Clin Perinatol* 2010; 37; 247–272.
15. Pfister RH and Goldsmith JP. Quality improvement in respiratory care: Decreasing bronchopulmonary dysplasia. *Clin Perinatol* 2010; 37; 273–293.

13 A Model and Methods to Solve Problems of Accessibility and Information for the Visually Impaired

Gerard Uzan and Peter Wagstaff

CONTENTS

INTRODUCTION

Over the last 10 years, all the countries that are members of the European Community have initiated programs to ensure that public and private spaces and buildings, public transport, and associated infrastructure are rendered accessible. Each country has defined a legislative framework and regulations with a corresponding system of verification and centers of expertise in order to ensure that this program can progress.

Thus by law or by political initiative, many states or local authorities, responsible authorities (Transport, Highways), have implemented public consultation procedures, diagnosis, and programming work. Our purpose here is not to make a presentation of these but to identify some models that facilitate the understanding of approaches to accessibility. After dealing with some of the basic characteristics of accessibility models, we explore the requirements of mobility starting initially with the problem of mobility "without vision" and illustrating the impact of such an approach for the characterization and direction of technological choices of assistance—for all.

GENERAL PRINCIPLES

Beyond the economic and political considerations that we will not discuss here, one of the first elements is the transformation of understanding achieved by the World Health Organization (WHO) through the renewal of the International Classification of Disability: Applying the model of Wood (1975), the International Classification of Disability of the WHO in 1980 defined the situation of disability as an individual characteristic, independent of situations and resulting from the interaction between a disability (injury or impairment of an organ), incapacity (difficulty of movement, thought, or decision) and handicap (social disadvantage difficulties in completing tasks "like others" or "shared with others"). Based on an analysis of actions, interactions, and activities of daily living, the International Classification of Functioning, Disability and Health in 2001 adopted a model defining disability no longer as the personal difficulty of an individual but as the result of the combination, for one or several individuals, of a context of activity and the tasks to be performed.

To reduce or eliminate the situation of disability, it is important to act on the context and conditions to carry out tasks. The initiative should not only be left to individuals acting alone or united in nongovernment organizations but be an aim of society and concern all those involved in designing or organizing the framework of our daily living and social life. The availability of equipment (devices) and private or public premises and infrastructure (buildings, highways, transportation) are central to these triple objectives:

- The creation of norms and obligations through regulations
- The development of solutions for the accessibility of public places (buildings, highways, transportation) and equipment
- The will to apply the concept of universal design

ACCESSIBILITY

The principle of accessibility applies to two essential spheres of daily life: digital space (documents and applications on computers, smartphones, Web sites, interactive terminals, and various types of distributors) and physical spaces (vehicles, buildings, infrastructure and spaces open to the public, highways, and public transport) in the framework of personal, professional, or health requirements.

Standardization work, in addition to its territorial implications (national, continental, or international), is the result of a compromise between organizations, companies, and local and national governments, whose cultural characteristics and deployment of solutions position each actor.

We should recall that standardization has even more impact on policies and decision making since it is inscribed in national human rights associated with delays of implementation and enforcement and sanctions for failure, but this integration presupposes the feasibility of practical applications in technical and economic terms.

We therefore distinguish normative and regulatory or contractual accessibility integrating effective accessibility on one hand and the constraints of feasibility and the reality of access resulting from the measures taken on the other hand: For architects and engineers (civil or computer engineers), normative accessibility translated into law poses the question "does it exist?" Effective accessibility poses a sequence of more specific questions among the feasible solutions, which allow the following:

- Real access for people in difficulty (criterion of effectiveness)
- Completion in a reasonable time in sociocultural or professional terms (efficiency criterion)
- Nondiscriminatory and acceptable solutions (fit criterion)
- Permits the site or the digital space to be potentially accessible for all
- Is technically feasible given the state of the art and techniques

Technical and economic feasibility may be classified into three types:

- An ambient device useable by anyone
- An ambient device supplemented by a personal technical aid
- Strictly personal technical aids

The significance is the risk of overlooking solutions that are based on universal design or claiming that the first two cases are infeasible, thus benefitting the third institutional decision.

DEFINING A MODEL

Our model should include four essential elements or tasks that apply to each of the levels or stages of a sequence using a device, piece of equipment, or place: These are (1) enter, (2) act, (3) exit, and, in parallel at any time, (4) evacuate. Our experience in the field shows that it is relatively easy to include the actions of entering and evacuation. However, the notion of going out and acting are more difficult to grasp. Indeed, the concept "act" obliges architects, designers, and project leaders to adopt a common systemic approach. A major difference in accessibility between digital and physical spaces is sustainability and their modification: A Web site or application is "live" because it is renewed frequently in the course of time, often several times a year, a month, or a week, while physical sites are the object of relatively long-term programs of updating or modification over a period of 10, 30, or 50 years or sometimes more. The constraints associated with the preservation and conservation of assets, maintenance, and renewal are also very different.

In order to clarify the concept of physical accessibility, we will describe the part of our model dealing with the problems of public transport developed after multiple studies and analysis of the needs of passengers, particularly those of the visually impaired. The constraints of the site management for the design and exploitation and the validation of the personal or general systems of information for the voyager (SIVs) are crucial.

Functional specifications for human machine interfaces providing assistance for localization and guidance devices have been devised for pedestrians in urban areas following work on the mobility of pedestrians, more particularly visually impaired pedestrians. An analysis of the requirements of voyagers for guidance information to intermediate points and to information clusters and transportation data (destinations, schedules, disruptions) resulted in a categorization of requirements. Different project reports dealt with the different modes of transport and their associated areas:

- Bus stops on city streets, dedicated and general roads, and shared spaces (RAMPE 2004–2006, Infomoville 2007–2010) (Baudoin et al. 2005; Pretorius et al. 2010)
- Trains and train stations (city guide 2007–2008)
- Metro and underground stations (DANAMA 2008–2010), research on systems of information for voyagers (SIV 1 and 2 for the RATP), trams (Infomoville 2010)

In these studies and projects, we compared different technologies of localization (GPS, wireless footprints, inertial navigation, visual landmarks, and radio frequency [RF] tags) and various associated interfaces (voice, visual) on personal devices (smartphones, remote controls) or ambient systems (visual and/or sound broadcasting). Localization technologies were not the focus of the project TICTACT, which interrogates the human–machine interface (HMI) through the use of haptics as an addition, extension, or substitution to visual or vocal communication. We can, however, note the following as regards technologies of guidance and localization:

- Most of the technologies of localization do not take into account the polar ego-centered situation of the user but present his/her position in a rectangular Euclidean reference. In this context, new criteria (humility, cohesion of representations) must be added to the perceptual criterion (visibility/audibility, clarity/intelligibility) or interaction (Bastien–Scapin criteria, for example) for the choice of interactions and indications of orientation provided to guide the pedestrian or passenger or exploited as metadata.
- The pedestrian complies more or less with social codes, or he/she does not follow the signs or stops for a phone call, an idea, or a meeting. In the middle of his/her journey, he could turn back without warning. He/she has a much wider choice of "possible actions" than the driver of a vehicle, for example, for road applications, where the highway code or the kinematics of forward and/or corrected trajectories that affect representations, decisions, and actions of the user and representations and actions proposed by the HMI are more limited than for pedestrians.
- Any HMI should behave differently depending on the variability of the reliability of the data it uses. It can, of course, do so by explicit transmission of the reliability of the results but also by these modes of communication or interaction.

In transport systems (breakpoints, interchanges, and vehicles), an HMI can present, communicate, stay alert, protect, warn, share, exchange, and act not only with local or remote equipment but also using previous information. Its behavior will be closely linked to the needs of the user, the kind of information that is contextually appropriate or available, the available sources, and the requirements that will be detailed in the following section.

OUR MODEL

Our model is focused on the pedestrian/user information and his/her assimilation of information for multimodal mobility. The model takes into account each element of the chain of mobility, the acquisition of information, the motive for the journey, and the "landmarks" of the trip, plus the zones of the journey, centered on the needs of the user and the representation of the relevant information to communicate or exchange pertinent data.

CHAIN OF MOBILITY

The displacements of a pedestrian are a succession of trajectories where the user is either moving on a stationary infrastructure (walking) or almost immobile on a moving infrastructure (any vehicle, elevator, conveyor belt, or escalator). The situation can also be applied for a "pedestrian" who is in a wheelchair, driving it, or being pushed by an assistant. This has an impact on vigilance (sustained and lowered, respectively); attention (respectively saturated or shared and available); the susceptibility of the individual to stress (low and high, respectively); and the hierarchy and priority of the need for information.

- He/she moves on a "journey" (a topographic representation with "landmarks" and more or less relevant associated topologies). The journey represents the content of a representational and semantic memory, which can be internal to the subject (mental image map) or external (map, network or neighborhood, gauge line, etc.).

- He/she follows the route marked out by successive landmarks, each one in turn consisting of a goal (destination) and a point of exchange for new information (new beginning). Each milestone is a stage on the itinerary and thus a "mental switch" for the collection of information. It is a point of arrival and also a point for a new beginning. The mental attitude (cognitive and possibly affective) is different in terms of the treatment (e.g., verification of the landmark and then switching to the identification of a new segment of the route), with an "emotion" connected with the discovery or recognition of each new landmark on the itinerary.

Waiting for a vehicle is not considered here as "immobility" but is part of the pedestrian's route with an "intermediate period," which, according to its duration, could be used for other integrated activities. Immobility is considered here as a period of total inactivity.

Motives of the Journey

The displacements of a pedestrian can be grouped into four categories of motive (cause and purpose of travel):

1. Physical activity (e.g., jogging, walking, shopping, "the half-hour daily walk," etc.)
2. Social relations with others (doing things together, walking, discussion/reunions/walking/going somewhere, etc.)
3. Knowledge of the environment (including knowledge of a network, site, area, building, landmark, architecture, tourist attractions, business premises, and equipment)
4. Reaching a place, a person or object

If the place is usually territorially stable, one of the individuals or objects is potentially mobile and can also be "portable," that is to say, they move.

During a displacement, these motives can be unique or combined according to all possible configurations and can, intentionally or not, appear or disappear in preparation or during the trip. They can also be a priority or secondary during the preparation of the trip, and the priorities can change during the trip.

In terms of transport, we have focused on the final destination in the analysis of needs without neglecting others, the importance of which will increase with the evolution of intelligent transport systems.

Needs

We have regrouped the needs of the voyager into five categories.

Safety

- Avoiding falls. Often referred to by its causes: impaired perceptual or saturation of attention (vestibular imbalance, divided attention, poor distribution, etc.); the context of sources of perceptual illusions (poor lighting, visual illusions, alluring sounds, etc.); missing alarms indicating the gaps in a staircase; partially open trapdoors; irregular surfaces; and so forth. A fall is more dangerous for physical integrity than a shock or collision in a crowd.
- Avoiding shock (rough contact with elements that have a mass, shape, or material that might affect physical integrity walls and partitions, posts and poles, deposited items [bags, cardboard boxes], salient objects [pipes, steps signs]). The shock of collisions is distinguished in that it does not constitute a conflict of intent between persons whether or not objects are interposed.
- Avoiding the risk of collision, which concerns the resolution of conflicts between users (pedestrians/walking/crowds, pedestrians, and drivers).

- Avoiding the risks related to security (or the perceived risk of theft or assault).
- Ability to evacuate (rapid access to exits or to a safe area in case of an incident, accident, disaster, emergency, or crisis (fire, flood, suspicious packages, etc.).

Localization

We can distinguish between

- Ego location: location of self in a given environment: room, station, line, network, neighborhood, city, etc., to answer the question, "Where am I?"
- Halo location: knowledge of spaces and objects in their spatial relationships (topology, morphology, position, and relative dimensions), to answer the question, "What is there around me?" (in which provision, arrangement, or dimension, with what opportunities for mobility or immobility?).

Guidance

- Maintaining a straight path
- Respect of a trajectory or path
- Memorization and keeping to a destination or intermediate stage
- Memory retention and following a route
- Development of an alternative route (decision induced by the state of the environment [disturbance, roadwork, etc.]) or one's personal state (change in final destination, final detour, etc.)

Search for Information

- Information on transport (pedestrian circulation and access, destinations, routes, schedules, disruptions, availability of services)
- Information on the environment and peripheral activities excluding transport (architecture, tourism, landmarks, shopping/entertainment/culture, etc.)

Physical Mobility

- To avoid the difficulties of crossing barriers or paths with difficult accessibility.
- Indications of difficulties due to the physical characteristics of travelers (e.g., older people) or the specificities of infrastructure, vehicles, or discontinuities.
- We need to distinguish between people on foot, with walking sticks or on wheels, and those who are "overloaded."

Zones of Displacement

Articulated around the treatment of the information carried by the pedestrian, the zoning of our model does not exactly follow that derived from the functional organization of the networks and lines (shopping area/access/platform/vehicle) but remains consistent with it.

Zone at the Surface

It is characterized by the following elements:

- It is polymorphic (a diversity of configurations).
- It is multiactive: It includes commercial, cultural, and residential activities, one of which is transport.
- It is multifunctional: The pedestrians who are there are not in majority those who are simply using the transport system.

Access Zone

This is a zone where the main activity of people inside it is to move from the surface zone to the staging zone (described below) or, conversely, from a vehicle to the surface zone. This zone may or may not be dedicated specifically to transport. It can be crossed by a significant number of people or crowds of commuters hurrying to their destinations during rush hours.

Transfer Zone

This includes both the platform or curb at the bus stop or tram stop and the zones at the interior of vehicles that are opposite the doors dedicated to entry, exit, or both. It supports dual transfer:

- The arrival and the departure of the vehicle
- The transfer of passengers from the platform to the vehicle and vice versa

This is an area of time pressure on the traveler, where the need for security is a shared priority with guidance (reinsurance).

Zone of "Transportation"

It is inside the vehicle. In this area, the traveler is relieved of the responsibility of performing physical movements and delegates this task to the transport system. Cognitively, he/she is in a situation similar to that of a vehicle waiting at a bus or tram stop but is expecting not the arrival of a vehicle but the arrival of his/her vehicle at his/her destination.

Classification of Information

Information is provided in response to needs. We can classify them into security, location and orientation, transport information and transport expenses, and finally, availability of information services related to physical movement. They may also be classified according to their quality as synchronous or asynchronous to the movement.

There is, therefore

- Structural information or planned asynchronous information and synchronous updated planned information.
- Cyclical information, which is the information linked to part of a temporary duration (working away, holidays, temporary closure, and so forth). This has characteristics similar to structural information but is unique and temporary.
- Event information, which is short (interruption of traffic, a suspicious package at the station, arrival of a bus or train).

Dissemination of Information and Synchronization

These were initially defined for vocal interfaces for the location and orientation of blind people in bus interchanges or subway stations. We have identified three key instances for the dissemination of information: (1) at entry in the relevant zone, (2) when there is a useful change of state, and (3) when there is an intentional reminder of the traveler.

CONCLUSIONS

The increasing use of personal devices for localization, information, and orientation (guides and/or descriptions) and the provision of information on transport (buses, tramways, schedules, disruption, services) have led to a clear structure for the specifications of HMIs.

This model has allowed us to define a grammar and a lexicon of the specific requirements for vocal HMI devices to indicate the location and orientation of a pedestrian and for its automatization by sharing Internet information to map locations using different technologies (GPS, RF tags, wireless, inertial guidance) that are essential for the visually impaired traveler. These technologies have proved to be useful for all types of users by combining the information on location and orientation with those on transport schedules and infrastructure.

Accessibility therefore requires modeling the requirements in advance at the design stage and downstream by updating the information during the journey. The applications are currently focused on three areas: simulation of sites focused on travel and accessibility (AccesSim), guidance devices taking into account the specific characteristics of the voyager (AccesSig), and the professional constraints of technical services and operating personnel; in terms of HMI requirements, after studying vocal and visual interfaces, we are currently exploring haptic interfaces. AccesSim and AccesSig are systems currently being developed in collaboration with the EDF and the town of St. Quentin en Yvelines, respectively, plus other partners.

REFERENCES

G. Baudoin, O. Venard, G. Uzan, A. Rouseau, Y. Benabou, A. Paumier, J. Cesbron. "How can the blind get information on public transport using a PDA? The RAMPE auditive man machine interface," Proc 8th European Conf. Advancement of Assistive Technology in Europe AAATE 2005, Lille, Sept 2005.

S. Pretorius, G. Baudoin, O. Venard. "Real time information for visual and auditory impaired passengers using public transport—technical aspects of the Infomoville project," 6th. Conference Handicap 2010, IFRATH, Paris, 2010.

WHO. International classification of impairments, disabilities, and handicaps. Geneva: World Health Organization, 1980.

P.H.N. Wood. "Classification of impairments and handicaps," Document WHO/1 CDO/REV.CONF/75.15, World Health Organization, Geneva, 1975.

14 Visualization of Multicriterial Classification

A Case Study

Péter Volf, Zoltán Kovács, and István Szalkai

CONTENTS

INTRODUCTION

Management of maintenance materials and parts (maintenance, repair, and operating [MRO] inventory) is a special field of material management concerning the nature of the role it plays in effective and efficient process operation. It is especially important because the forecast of the consumption of such materials is limited. MRO has a special feature, namely, the strong relationship between safety (reliability) and costs. Optimum operation can be achieved by a trade-off between them, based on systematic analysis.

Kennedy et al. (2002) differentiated the MRO inventories from the other types of manufacturing inventories such as works in process and finished products, based on their functionalities. They highlighted the function of MRO inventory, which is assigned not to be sold to a customer but to ensure the proper condition of equipment. That is why the nature of maintenance material management depends on the maintenance strategy, which is mostly a combination of following:

- Failure based: operation until a failure occurs, and then repair.
- Time based: examination and repair after a certain operational time or output.
- Condition based: repair is scheduled using information from condition monitoring.
- Maintenance prevention: construction and usage allows avoidance maintenance (maintenance-free operation).

The last three are preventive strategies. Forecast of material requirements is the most difficult in the case of failure-based strategy. When the cause of failure comes from a systematic effect like deterioration, the expected time between failures can be modeled by a normal or normal-like Weibull distribution. It allows the estimation of the increasing probability of failure as a function of the time since the last repair.

When failures are the results of nonsystematic effects, the distribution of failure-free operational times is (near) exponential, and failure can happen any time (memoryless process). This makes the demand forecast practically impossible. The number of failures during a long time interval might provide information about the expected demand.

Dhillon (2002) separates the MRO inventories in the above-mentioned way. There are MROs for routine and nonroutine maintenance tasks. However, the MRO inventories must be controlled in the most effective way; continuous availability is important in the first case. When controlling MRO inventories, there are four sorts of information needed:

- Importance of the inventory item
- The way it should be controlled
- Quantity to be ordered at one time
- Specific point in time to place an order

The answers to the first two questions are provided by the ABC analysis, which is one of the most frequently used methods to support the decision-making procedure in inventory management. It has many advantages, like easy usage and understandability. Its base is the famous Pareto observation, the so-called 80–20 rule. This rule serves as a basis for finding the cutoff figures between the groups according to the relative importance of the items (Deis 2008). The Pareto chart provides a perspicuous way of representing the three item categories—A, B, and C (Harry et al. 2010). This approach can be applicable only for classification based on one criterion, which is its most significant shortcoming.

Recognizing this problem, numerous methods have been developed to relieve the complex multicriterial decision-making situation in the last decades. These models, like the weighted linear optimization (Ramanathan 2006), the genetic algorithm (Guvemir and Erel 1998), the analytic hierarchy process (AHP) (Partovi and Burton 1993), and the fuzzy set theory (Chu et al. 2008), are capable of taking more than one criterion into account at the same time, enabling a more sophisticated analysis framework.

Alongside the development of new multicriterial models is a dramatic increase in their complexity. The way of improving these models focuses on their applicability, ignoring their understandability, which is one of the most important aspects of selecting a model for analysis. The proper visualization of each phase of the analysis and its results can relieve the implementation of complex models as well.

The visualization technique introduced in this chapter is based on the case-based distance model developed by Chen et al. (2008). In this sense, the present research is an extension of this model, improving the interpretation of its results. Hence, the main aim of this chapter is to demonstrate how visualization can help in understanding the structure of a complex inventory analysis model step by step.

The motivation was provided by Cho and Parlar (1991), who were among the first formulating critics on the lack of analysis of creation and application of new mathematical models and theories. Additionally, Scarf (1997) confirmed that too much emphasis is on developing new models and too little on application. That is why the results will be supported by a case study to facilitate the application of this mathematical model.

LITERATURE REVIEW

First, Flores and Whybark (1987) recognized that the classification of items is determined not only by one criterion but also by more specific characteristics like criticality, reparability, scarcity, substitutability, stockability, lead time, annual dollar usage, and commonality (Ramanathan 2006). However, increasing the number of dimensions in the analysis raises difficulties in interpretation of the results. This problem can be relieved by framing the decision situation and the problem embedded in it clearly through data visualization.

Lee et al. (2003) established a psychological framework to investigate the connection between the information coming from an artificial system and the human cognitive processes. The predecessor of this research had been conducted by Lee et al. (2003), who stated that viewing data visualization works as a channel conveying information between the two systems involved in their psychological framework. In particular, "data visualization has the potential to assist humans in analyzing and comprehending large volumes of data, and to detect patterns, clusters and outliers that are not obvious using non-graphical forms of presentation" (Lee et al. 2003).

Raising the number of factors (criteria) in inventory analysis leads to large, complex databases, whose elements are multivariate data. According to the Ward's (2008) definition, multivariate data are n-dimensional data consisting of more records and represented by an $m \times n$ matrix, where m is the number of data records and n is the value belonging to each variable or dimension.

There are many ways to represent multivariate data and the results of their analysis. Among these are the glyphs. Glyphs are visualization tools (Lee et al. 2003) that can show specific clusters or point at the interaction between variables. Starfield is another example for visualizing high-dimensional data, which is a two-dimensional scatter plot with the option to expand for more dimensions. Akcay et al. (2012) applied this tool for representing the two-dimensional results of a DEA model. Besides starfield, they implemented a tie graph for representing the same results. They argue that it is an effective tool for analysts to classify the data according to a criterion and uncover unrecognized patterns in the data envelopment analysis (DEA) results. Adler and Raveh (2008) conducted a research also in the topic of visualization opportunities of DEA approach. As a result they introduced a Co-plot method that is applicable for separating efficient units from outliers.

In general, Kim et al. (2008)—like Russel et al. (2008)—claimed that visualizing data provides important aids for decision makers to improve the decision-making process. Keim (2002) stated that the field of visualizing data is an important part of the research concerning computer science to amend the understandability of the data visually.

Concerning the inventory analysis in the field of inventory management, one of the main problems is that the large number of units can lead to an impracticable data set, which can make the application of the given model more complicated and can hinder the appropriate interpretation of the results. In this sense, the case-based distance model developed by Chen et al. (2008) excels compared to other models considering the structure of the model, because it involves an untapped opportunity in the view of interpretation of results. Namely, the process of determining the parameters of surfaces separating each group of storage keeping units (SKUs) is built in the model, and visualizing these surfaces can help to discover the magnitude of groups and the relationship between them.

MODEL CONSTRUCTION

The main aim of the case-based distance model is to classify the SKUs into categories according to the set Q of criteria. Let T be the set of the units (A^i) being analyzed, where $A^i \in T_g \subset T \mid g = A, B, C$ based on $q_j \in Q \mid j = 1, 2, \ldots, m$.

The minimum $\left(c_j^{\min}\right)$ and the maximum $\left(c_j^{\max}\right)$ values in positions A^- and A^+ form an interval (Chen et al. 2008). The Euclidean distances taken from these points determine the position of the given unit on criterion q_j.

Let the set $Z_{repg} \subseteq T_g \mid g = A, B, C$ be a subset of T_g that involves the units that principally represent the set T_g and $z_{repg}^r \in T_g \mid g = A, B, C$ is one of these units, where $c_j^{\min}(A^-) \leq c_j\left(z_{repg}^r\right) \leq c_j^{\max}(A^+)$ (Chen et al. 2008).

The case-based distance model is constructed for n dimensions, which, adopted on three criteria, becomes capable to be plotted as a *three*-dimensional coordinate system defined by the three elements of Q, where, respectively, radius vector $\underline{q_1}$ represents the x, $\underline{q_2}$ the y, and $\underline{q_3}$ the z axis.

After launching the normalization factor $d_j^{\max}$, the distance between the lower and upper bounds of the intervals changes (Chen et al. 2008).

$$d_j^{\max}(A^+; A^-) = \frac{\left(c_j^{\max}(A^+) - c_j^{\min}(A^-)\right)^2}{d_j^{\max}} = 1 \mid j = 1, 2, 3 \tag{14.1}$$

Since the distance between the lower and upper bound of the interval is equal to the absolute value of the radius vector on criterion j, the Euclidean distances taken from the points of comparison can be identified as the length of the radius vectors belonging to each unit. So the distance of $z_{repg}^{r} \in T_g$ from the upper limit of the interval on the second criterion is

$$d_2^{+}\left(A^{+}, z_{repg}^{r}\right) = \frac{\left(c_2^{\max}(A^{+}) - c_2\left(z_{repg}^{r}\right)\right)^2}{d_2^{\max}} = \frac{\left(\left\|\underline{q}_{r2}^{+}\left(0; c_2^{\max}(A^{+}) - c_2\left(z_{repg}^{r}\right); 0\right)\right\|\right)^2}{\left(\left\|\underline{q}_2^{\max}\left(0; c_2^{\max}(A^{+}) - c_2^{\min}(A^{-}); 0\right)\right\|\right)^2}$$

$$= \frac{\left(\sqrt{0^2 + \left(c_2^{\max}(A^{+}) - c_2\left(z_{repg}^{r}\right)\right)^2 + 0^2}\right)^2}{q_2^{\max}} = \frac{\left(c_2^{\max}(A^{+}) - c_2\left(z_{repg}^{r}\right)\right)^2}{q_2^{\max}}$$

where

$$d_2^{\max} = \left(c_2^{\max} - c_2^{\min}\right)^2 = \left(c_2^{\max}(A^{+}) - c_2^{\min}(A^{-})\right)^2 = \left(\left\|\underline{q}_2^{\max}\left(0; \left(c_2^{\max}(A^{+}) - c_2^{\min}(A^{-})\right); 0\right)\right\|\right)^2 = \left(\left\|\underline{q}_2^{\max}\right\|\right)^2 = q_2^{\max} \quad (14.2)$$

Based on the work of Chen et al. (2008), the weighted aggregation formulas of distance are

$$D^{+}\left(z_{repg}^{r}\right) = D^{+}\left(A_j^{+}; z_{repg}^{r}\right) = \sum_{j \in Q} w_j^{+} \cdot d_j\left(z_{repg}^{r}\right)^{+} = \sum_{j \in Q} w_j^{+} \cdot \frac{\left(\left\|\underline{q}_j^{+}\left(x\left(z_{repg}^{r}\right); y\left(z_{repg}^{r}\right); z\left(z_{repg}^{r}\right)\right)\right\|\right)^2}{q_j^{\max}} \quad (14.3)$$

and

$$D^{-}\left(z_{repg}^{r}\right) = D^{-}\left(z_{repg}^{r}, A_j^{-}\right) = \sum_{j \in Q} w_j^{-} \cdot d_j\left(z_{repg}^{r}\right)^{-} = \sum_{j \in Q} w_j^{-} \cdot \frac{\left(\left\|\underline{q}_j^{+}\left(x\left(z_{repg}^{r}\right); y\left(z_{repg}^{r}\right); z\left(z_{repg}^{r}\right)\right)\right\|\right)^2}{q_j^{\max}} \quad (14.4)$$

respectively.

Note that depending on the point of comparison, we get two different sorting problems. Based on this fact, the visualization has to be separated likewise:

- If $c_j^{\min}(A^{-})$ is the point of comparison, then we talk about *minimum transformation* in the minimum (distorted) space.
- If $c_j^{\max}(A^{+})$ is the point of comparison, then we talk about *maximum transformation* in the maximum (distorted) space.

In this sense, the attribute "distortion" means that the values created by the value function $d_j\left(z_{repg}^{r}\right)$ from the original values are linear and nonlinear distorted, and this value transformation affects the sets and the shape of their separating surfaces, too.

Based on the linear and nonlinear transformations and their effects on the visualization, four different spaces ("worlds") can be identified:

1. Original space (Space 0)
2. Unit cube placed in "o" origin (Space 1)
3. Minimum (distorted) space (Space 2)
4. Maximum (distorted) space (Space 3)

Space 0: Let $P^i \in R^3$ be the point belonging to an SKU with the coordinates $P^i = \left(p_1^i, p_2^i, p_3^i\right)$, where $i = 1,2,...,D$ and

$$\begin{cases} m_1 \le p_1^i \le M_1 \\ m_2 \le p_2^i \le M_2 \\ m_3 \le p_3^i \le M_3 \end{cases}$$

where (m_1, m_2, m_3) and (M_1, M_2, M_3) are the minimum and maximum values of the original coordinates, that is, $c^{\min}\left(c_1^{\min}; c_2^{\min}; c_3^{\min}\right)$ and $c^{\max}\left(c_1^{\max}; c_2^{\max}; c_3^{\max}\right)$.

Space 1: Linear transformation: $E = (\varepsilon_1, \varepsilon_2, \varepsilon_3) = \mathrm{T}(P)$, where

$$\varepsilon_1 = \frac{p_1 - m_1}{M_1 - m_1}, \varepsilon_2 = \frac{p_2 - m_2}{M_2 - m_2}, \varepsilon_3 = \frac{p_3 - m_3}{M_3 - m_3}$$

that is, in the case of $i = 1,2,...,D$,

$$\varepsilon_1^i = \frac{p_1^i - m_1}{M_1 - m_1}, \varepsilon_2^i = \frac{p_2^i - m_2}{M_2 - m_2}, \varepsilon_3^i = \frac{p_3^i - m_3}{M_3 - m_3} \tag{14.5}$$

Consequently, the linear distorted coordinates are

$$E^i = \left(\varepsilon_1^i, \varepsilon_2^i, \varepsilon_3^i\right) = \mathrm{T}(P^i)$$

As a result, we get a unit cube placed in "o" origin.

Space 2: Minimum world (nonlinear transformation): $\xi = (\xi_1, \xi_2, \xi_3) = \Xi(E)$

Let

$$\begin{cases} \xi_1^i = \left(\varepsilon_1^i\right)^2 = \dfrac{\left(p_1^i - m_1\right)^2}{(M_1 - m_1)^2} \\ \xi_2^i = \left(\varepsilon_2^i\right)^2 = \dfrac{\left(p_2^i - m_2\right)^2}{(M_2 - m_2)^2} \\ \xi_3^i = \left(\varepsilon_3^i\right)^2 = \dfrac{(p_3^i - m_3)^2}{(M_3 - m_3)^2} \end{cases} \tag{14.6}$$

so in the case of $i = 1,2,...,D$,

$$\xi^i = \left(\xi_1^i, \xi_2^i, \xi_3^i\right) = \Xi(\mathrm{T}(P^i))$$

Note that for $i = 1,2,...,D$,

$$0 \le \xi_1^i \le 1, 0 \le \xi_2^i \le 1, 0 \le \xi_3^i \le 1$$

that is, the transformed points $\xi^i = d_j^-\left(z_{repg}^r\right)$ are in the *same* unit cube.

Space 3: Maximum world (nonlinear transformation): $\eta = (\eta_1, \eta_2, \eta_3) = \mathrm{H}(E)$

Let

$$\begin{cases} \eta_1^i = \left(1-\varepsilon_1^i\right)^2 = \dfrac{\left(M_1 - p_1^i\right)^2}{(M_1 - m_1)^2} \\ \eta_2^i = \left(1-\varepsilon_2^i\right)^2 = \dfrac{\left(M_2 - p_2^i\right)^2}{(M_2 - m_2)^2} \\ \eta_3^i = \left(1-\varepsilon_3^i\right)^2 = \dfrac{\left(M_3 - p_3^i\right)^2}{(M_3 - m_3)^2} \end{cases} \tag{14.7}$$

so in the case of $i = 1,2,...,D$,

$$\eta^i = \left(\eta_1^i, \eta_2^i, \eta_3^i\right) = \mathrm{H}(\mathrm{T}(P^i))$$

Note, that in the case of $i = 1,2,...,D$,

$$0 \le \eta_1^i \le 1, 0 \le \eta_2^i \le 1, 0 \le \eta_3^i \le 1$$

that is, the transformed $\eta^i = d_j^+\left(z_{repg}^r\right) \mid i = r$ are also in the unit cube.

It is easy to state that the relation between Space 2 and 3 is

$$\sqrt{\xi_1} + \sqrt{\eta_1} = 1, \sqrt{\xi_2} + \sqrt{\eta_2} = 1, \sqrt{\xi_3} + \sqrt{\eta_3} = 1 \tag{14.8}$$

Depending on which space we are analyzing for the position of the points P^i, the parameters searched are the following (Chen et al. 2008):

Space 2: $\xi^i = \left(\xi_1^i, \xi_2^i, \xi_3^i\right) = \Xi(\mathrm{T}(P^i))$

Weight vector: $w^- = \left(w_1^-, w_2^-, w_3^-\right)$;

Radii: R_B^- and R_C^-.

These parameters are given by the optimization model A^-MCABC expressed by the introduced transformation (Ξ) based on the work of Chen et al. (2008):

$$\min ERR = \sum_{r=1}^{n_A} \left(\alpha_C^r\right)^2 + \sum_{r=1}^{n_B} \left[\left(\alpha_B^r\right)^2 + \left(\beta_B^r\right)^2\right] + \sum_{r=1}^{n_C} \left(\beta_A^r\right)^2 \tag{14.9a}$$

where the conditions are

$$\sum_{j=1}^{3} w_j^- \cdot \xi_j^r + \alpha_C^r \le R_C^- \mid r = 1; 2; ...; n_C \tag{14.9b}$$

$$\sum_{j=1}^{3} w_j^- \cdot \xi_j^r + \alpha_B^r \le R_B^- \mid r = 1; 2; ...; n_B \tag{14.9c}$$

$$\sum_{j=1}^{3} w_j^- \cdot \xi_j^r + \beta_B^r \leq R_C^- \mid r = 1; 2; \ldots; n_B \tag{14.9d}$$

$$\sum_{j=1}^{3} w_j^- \cdot \xi_j^r + \beta_A^r \leq R_B^- \mid r = 1; 2; \ldots; n_A \tag{14.9e}$$

where

$$0 < R_C^- < 1, 0 < R_B^- < 1, R_C^- < R_B^-$$

$$-1 \leq \alpha_B^r \leq 0, \quad 0 \leq \beta_A^r \leq 1$$
$$-1 \leq \alpha_C^r \leq 0, \quad 0 \leq \beta_B^r \leq 1$$

$$w_j^- > 0; \sum_{j \in Q} w_j^- = 1 \tag{14.9f}$$

The model description of A^+MCABC in Space 3—$\eta^i = \left(\eta_1^i, \eta_2^i, \eta_3^i\right) = \mathrm{H}(\mathrm{T}(P^i))$—can be given in a similar way, where the parameters are
Weight vector: $w^+ = \left(w_1^+, w_2^+, w_3^+\right)$
Radii: R_A^+ and R_B^+ (Chen et al. 2008).

Visualization Method of Linear and Nonlinear Separating Surfaces

The nonlinear transformation implied by the value function $d_j\left(z_{repg}^r\right)$ on $z_{repg}^r = P_{repg}^i \mid i = r$ transformed the ellipsoids separating the set g into planes. In this way, the planes separating the sets in the distorted spaces can be given by the following formula:

Space 2: $\xi^i = \left(\xi_1^i, \xi_2^i, \xi_3^i\right) = \Xi(\mathrm{T}(P^i))$

$$\xi_3 = \frac{1}{w_3^-}\left(R_g^- - w_1^-\xi_1 - w_2^-\xi_2\right) \tag{14.10}$$

where $g = B^-, C^-$.

Space 3: $\eta^i = \left(\eta_1^i, \eta_2^i, \eta_3^i\right) = \mathrm{H}(\mathrm{T}(P^i))$

$$\eta_3 = \frac{1}{w_3^+}\left(R_g^+ - w_1^+\eta_1 - w_2^+\eta_2\right) \tag{14.11}$$

where $g = A^+, B^+$.

Note that due to the transformations above, the assumed relation between Space 2 and 3 cannot be assessed in distorted spaces. The design maps (DMs) have to return back to the original space in order to determine the interaction of the two classifications in the same space.

To get the three groups of SKUs and the ellipsoids separating them in the original space, execution of the inverse transformation is needed. Since $\xi^i = \left(\xi_1^i, \xi_2^i, \xi_3^i\right) = \Xi(\mathrm{T}(P^i))$ and $\eta^i = \left(\eta_1^i, \eta_2^i, \eta_3^i\right) = \mathrm{H}(\mathrm{T}(P^i))$ are monotone transformations of P^i, we can determine the ellipsoids separating the sets in the following way:

Space 2: $\xi^i = \left(\xi_1^i, \xi_2^i, \xi_3^i\right) = \Xi(\mathrm{T}(P^i)) \rightarrow \Xi(\mathrm{T}(P^i))^{-1} = \left(p_1^i, p_2^i, p_3^i\right) = P^i$.

The equation of planes dividing the sets in Space 2 is

$$\xi_3 = \frac{1}{w_3^-}\left(R_g^- - w_1^-\xi_1 - w_2^-\xi_2\right)$$

which can be reformed using Equation 14.6 into following formula:

$$\frac{(p_3 - m_3)^2}{(M_3 - m_3)^2} = \frac{1}{w_3^-}\left(R_g^- - w_1^- \frac{(p_1 - m_1)^2}{(M_1 - m_1)^2} - w_2^- \frac{(p_2 - m_2)^2}{(M_2 - m_2)^2}\right)$$

Expressing p_3, we can find the equation of the ellipsoids:

$$p_3 = m_3 + \sqrt{\frac{(M_3 - m_3)^2}{w_3^-}\left(R_g^- - w_1^- \frac{(p_1 - m_1)^2}{(M_1 - m_1)^2} - w_2^- \frac{(p_2 - m_2)^2}{(M_2 - m_2)^2}\right)} \tag{14.12}$$

where $g = B^-, C^-$.

In a similar way, in Space 3, the equation of the ellipsoids is

Space 3: $\eta^i = \left(\eta_1^i, \eta_2^i, \eta_3^i\right) = \mathrm{H}(\mathrm{T}(P^i)) \rightarrow \mathrm{H}(\mathrm{T}(P^i))^{-1} = \left(p_1^i, p_2^i, p_3^i\right) = P^i$

$$p_3 = M_3 - \sqrt{\frac{(M_3 - m_3)^2}{w_3^+}\left(R_g^+ - w_1^+ \frac{(M_1 - p_1)^2}{(M_1 - m_1)^2} - w_2^+ \frac{(M_2 - p_2)^2}{(M_2 - m_2)^2}\right)} \tag{14.13}$$

where $g = A^+, B^+$.

Note that in this case, the center of ellipsoids is the upper limit of the available values, that is, the *maximum value*.

Since the position of the points P^i representing the units in n-dimensional space is determined by the same criteria, the sets are able to be plotted in the same space, where their sections can be identified as the *nine sets before the reclassification*.

CASE STUDY

Data from analysis to be presented are from a heating power plant during the time period of June 1998–May 2011. During that time, 1595 items were in the MRO inventory. Recorded data were

- Volume (inv_vol)
- Time expired from the last outbound (last_out)
- Standard cost price (SCP)

After data screening (first eliminating items with 0 volume and then avoiding problems from different measure units), the sample size remained at 1064.

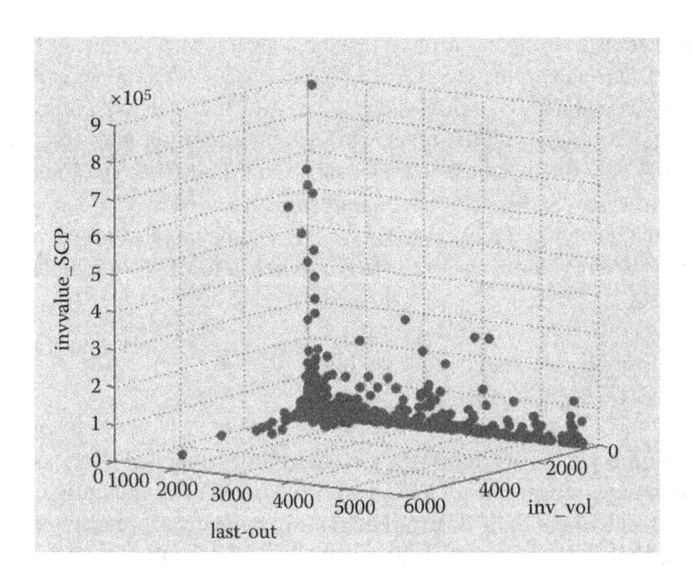

FIGURE 14.1 Distribution of item properties in 3-D coordinate system.

Figure 14.1 shows the positions of items in a 3-D coordinate system. The reason for 3-D visualization is that all analyses were carried out in a using multiple attributes, including correlation and cluster analysis. SPSS 18 statistics software has been applied to execute the systematic analyses.

Correlation and Cluster Analysis

Correlation analysis has been conducted to analyze the three classification-determining variables in terms of whether they fulfill the prerequisites of cluster analysis concerning the dependency.

The Spearman correlation coefficients show high values in two cases (Figure 14.2):

1. inv_vol – SCP
2. invvalue_SCP – SCP

Correlations			inv_vol	SCP	last_out	invvalue_SCP
Spearman's rho	inv_vol	Correlation coefficient	1.000	−.601**	−.271**	.017
		Significance (two-tailed)		.000	.000	.570
		N	1063	1063	1063	1063
	SCP	Correlation coefficient	−.601**	1.000	.073*	.755**
		Significance (two-tailed)	.000		.017	.000
		N	1063	1063	1063	1063
	last_out	Correlation coefficient	−.271**	.073*	1.000	−.114**
		Significance (two-tailed)	.000	.017		.000
		N	1063	1063	1063	1063
	invvalue_SCP	Correlation coefficient	.017	.755**	−.114**	1.000
		Significance (two-tailed)	.570	.000	.000	
		N	1063	1063	1063	1063

** Correlation is significant at the 0.01 level (2-tailed).
* Correlation is significant at the 0.05 level (2-tailed).

FIGURE 14.2 Correlation coefficients indicating the dependency between the classification-determining variables.

In the first case, there is a strong negative correlation, which means the higher the value of an item, the less of its volume is on hand. In the second case, since the value of inventory at SCP (invvalue_SCP) is a derived variable—a combination of SCP and volume— the high value of correlation is reasonable.

The high correlation of SCP with the other two variables does not allow us to apply the SCP in the analysis. There is only one variable that can substitute the SCB—the total value of a certain item on inventory (invvalue_SCP)—since it contains the effect of SCP (positive high correlation). Furthermore, since the value has a low correlation with volume and last move, instead of the initial three variables, the following three variables are the dimensions of cluster analysis:

- Volume (inv_vol)
- Time expired from the last outbound (last_out)
- Value (invvalue_SCP)

Due to the high number of MRO items, the *k*-means cluster analysis was reasonable. Keeping in mind the 80–20 rule, the *k*-means cluster analysis was repeated systematically in order to eliminate the outliers, of which classification is determined mainly by one variable. (Intermediate steps gave results with extreme shares, where small groups could be seen as outliners from a large group rather than independent groups.) As the result of this process, there remained a subset containing 531 items in the splitting shown in Figures 14.3 and 14.4.

The categories A, B, and C can be identified based on the characteristics of the previously formed groups by cluster anlysis. The key factor during the cluster analysis was the last_out; hence, the time elapsed from the last outbound determined the classification process. Its figures (minimum

Number of cases in each cluster		
Cluster	1	283.000
	2	100.000
	3	148.000
Valid		531.000
Missing		0.000

FIGURE 14.3 Cluster membership distribution in the screened subset.

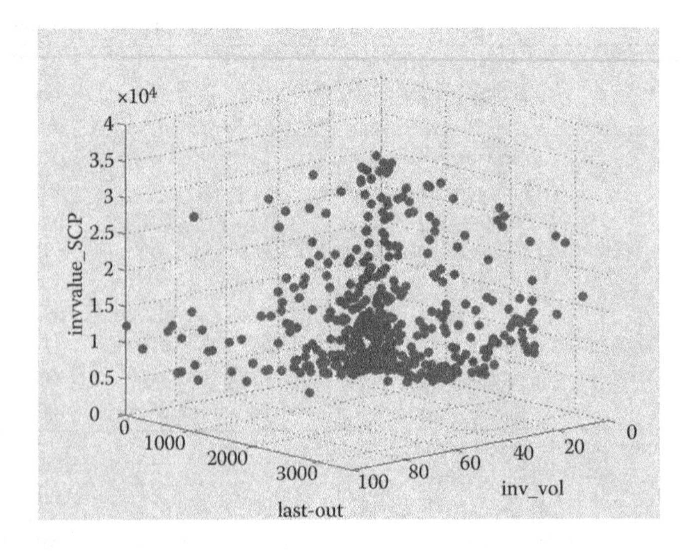

FIGURE 14.4 Distribution of item properties of the screened subset in 3-D coordinate system.

Group A
Descriptive statistics

	N	Minimum	Maximum	Sum	Mean	Standard deviation
inv_vol	100	1	89	792	7.92	12.303
last_out	100	210	3198	161,101	1611.01	675.180
invvalue_SCP	100	216	25,685	699,171	6991.71	6102.529
Valid *N* (listwise)	100					

Group B
Descriptive statistics

	N	Minimum	Maximum	Sum	Mean	Standard deviation
inv_vol	148	1	100	2175	14.70	18.596
last_out	148	6.5017	1940.5311	53,556.0908	361.8654	434.2664
invvalue_SCP	148	8392.15	30,060.00	2,495,924.59	16,864.3553	6416.1419
Valid *N* (listwise)	148					

Group C
Descriptive statistics

	N	Minimum	Maximum	Sum	Mean	Standard deviation
inv_vol	283	1	94	3329	11.76	16.347
last_out	283	0.5922	886.2879	39,176.2599	138.4320	166.7914
invvalue_SCP	283	49.2800	9218.0400	858,107.5500	3032.1821	2419.6868
Valid *N* (listwise)	283					

FIGURE 14.5 Descriptive statistics of variables in each cluster.

and maximum values, mean, deviation—Figure 14.5) are continuously decreasing from group A to group C, while the difference between the figures of volume (inv-vol) in each groups are not significan,t and no trend can be recognized in the case of value (invvalue_SCP).

Visualization of the Classification of MRO

Correlation analysis provided the three independent variables that can be taken into account to determine the categories of MRO items. Cluster analysis as a part of the screening process has been used to make an estimation for the groups of items advanced. As a result, the original set has been reduced by eliminating the outliers to come up with a subset consisting of 531 items. Applying the $A^{-}MCABC$ and $A^{+}MCABC$ models, we can obtain the weights for each criterion and the cutoff values for radii in both spaces:

Space 2 *(minimum world):* $\xi^i = \left(\xi_1^i, \xi_2^i, \xi_3^i\right) = \Xi(\mathrm{T}(P^i))$
Weight vector: $w^- = (0.243331; 0.01; 0.746669)$;
Radii: $R_B^- = 0.001406478$ and $R_C^- = 0.0001781243$.

Space 3 *(maximum world):* $\eta^i = \left(\eta_1^i, \eta_2^i, \eta_3^i\right) = \mathrm{H}(\mathrm{T}(P^i))$
Weight vector: $w^+ = (0.1620604; 0.08130314; 0.7566364)$;
Radii: $R_A^+ = 0.982917$ and $R_B^+ = 0.9410705$.

Lingo and MATLAB® software were used to calculate the necessary model parameters and to plot the surfaces.

Since the linear and nonlinear transformations affect not only the coordinates but also the separating surfaces, substituting into Equations 14.10 and 14.11 can confirm that the previously assumed nonlinear surfaces (ellipsoids) are planes in the distorted spaces (Figures 14.6 and 14.7).

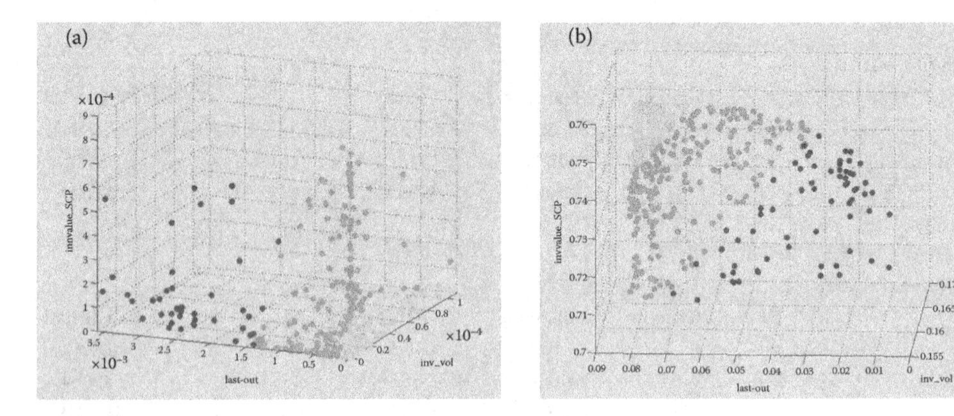

FIGURE 14.6 Visualization of groups in Space 2 and 3 (black, group A; gray, group B; light gray, group C).

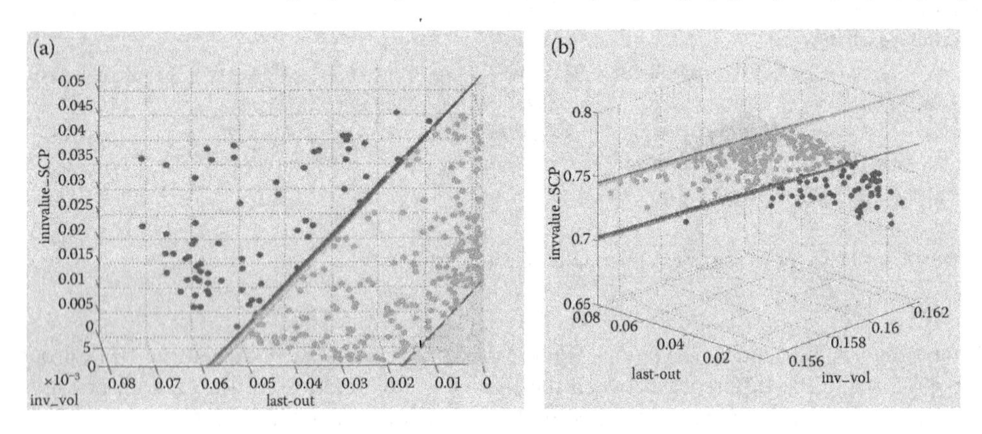

FIGURE 14.7 Visualization of groups and planes separating them in Space 2 and Space 3.

The inverse transformation ensures the feasibility of representation of the real categories of items and their separating nonlinear surfaces (substituting into Equations 14.12 and 14.13) in the same space. Figures 14.8 and 14.9 demonstrate the positions of the groups in the three-dimensional coordinate system. The relative proportion of magnitudes and density of groups can provide the same information about classification as the distribution in the case of Pareto analysis.

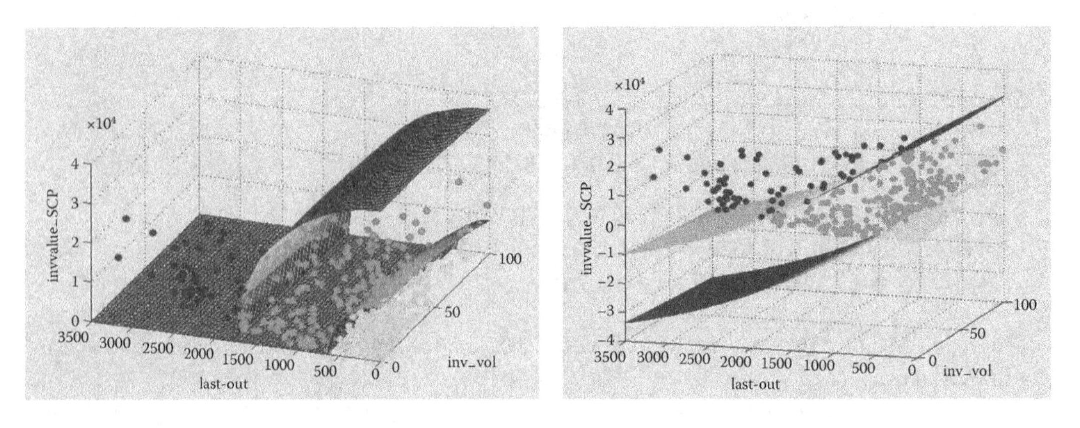

FIGURE 14.8 Visualization of groups of SKUs and ellipsoids bounding them in the minimum and maximum world using the original coordinates.

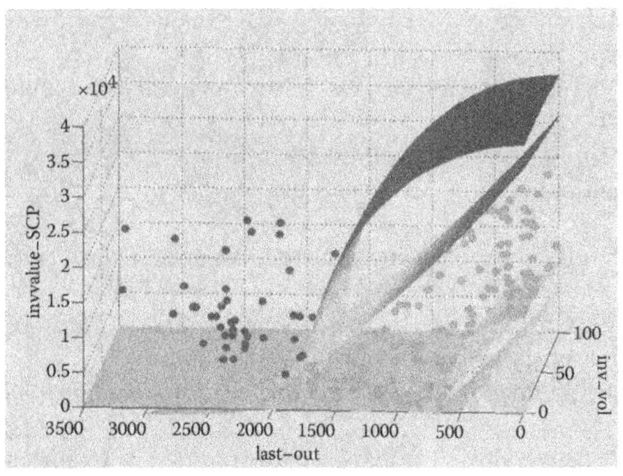

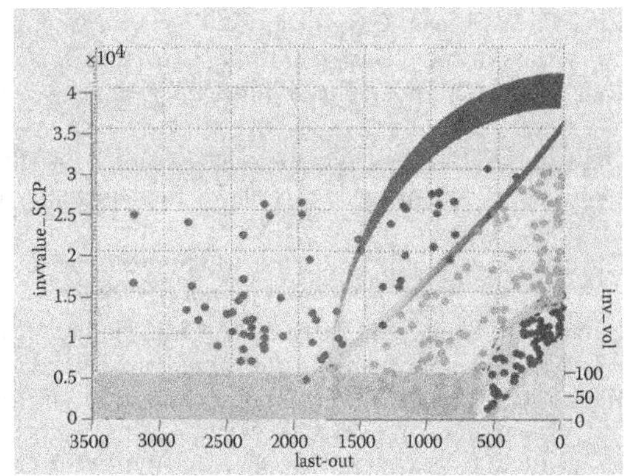

FIGURE 14.9 Visualization of classification results in one space (Space 0).

As a result, five categories have been identified (right side of Figure 14.9). Three of them are unambiguous—having the same classification results irrespective of the points of comparison (AA, BB, CC—left side of Figure 14.9). However, there are two item categories whose positions raise difficulties relating to the interpretation of classification results. Two principles can support the decision in this case. First, if the basic task is to identify the "best-performing" items (with the highest values), then the classification results of the maximum world have to be considered, and conversely, if the basic task is to identify the "worst-performing" items (with the lowest values), then the classification results of the minimum world have to be considered, which is the second principle.

CONCLUSION

The case study pointed out that the analysis of MRO inventories requires more serious methodology than analysis of other inventories does. The combination of systematic and nonsystematic effects raised difficulties in defining groups of material. Taking more properties into consideration increases the complexity of the problem. However, to alleviate the decision process or situation in the case of a sorting problem, visualization of multicriterial ABC analysis requires a sophisticated mathematical apparatus, for which an example has been provided.

In the presented case, a strong correlation was found between the unit cost (originated from purchase price) and volume.

It was also our experience that cluster analyses can be applied in an iterative way in order to make a distinction between outliers and groups.

After implementing statistical methods such as correlation analysis and cluster analysis, the presented case study gave an example of the application of visualization of a complex multicriterial classification technique (case-based distance model).

Next, researches will discover the type of inventories for which the presented methods (MCABC, cluster analysis, visualization) are applicable.

REFERENCES

Adler, N., Raveh, A. (2008): Presenting DEA graphically. *Omega* 36, 715–729.

Akcay, A.E., Gurdal Ertek, G., Buyukozkan, G. (2012): Analyzing the solutions of DEA through information visualization and data mining techniques: SmartDEA framework. *Expert Systems with Applications* 39, 7763–7775.

Chen, Y., Li, K.W., Kilgour, D.M., Hipel, K.W. (2008): A case-based distance model for multiple criteria ABC analysis. *Computers and Operations Research* 35, 776–796.

Cho, D.I., Parlar, M. (1991): A survey of maintenance models for multi-unit systems. *European Journal of Operational Research* 51, 1–23.

Chu, Ch.W., Liang, G.S., Liao, Ch.T. (2008): Controlling inventory by combining ABC analysis and fuzzy classification. *Computers and Industrial Engineering* 55, 841–851.

Deis, P. (2008): *Production and Inventory Management in the Technological Age*. Essex of Oak Park, Agoura Hills, CA.

Dhillon, B.S. (2002): *Engineering Maintenance: A Modern Approach*. CRC Press LLC, FL.

Flores, B.E., Whybark, D.C. (1987): Implementing multiple criteria ABC analysis. *Journal of Operations Management* 7(1–2), 79–85.

Guvemir, H.A., Erel, E. (1998): Multicriteria inventory classification using a genetic algorithm. *European Journal of Operational Research* 105, 29–37.

Harry, M.J., Mann, P.S., De Hodgins, O.C., Hulbert, R.L., Lacke, C.J. (2010): *Practitioner's Guide to Statistics and Lean Six Sigma for Process Improvements*. John Wiley and Sons, Hoboken, NJ.

Keim, D.A. (2002): Information visualization and data mining. *IEEE Transactions on Visualization and Computer Graphics* 8(1), 1–8.

Kennedy, W.J., Patterson, J.W., Fredendall, L.D. (2002): An overview of recent literature on spare parts inventories. *International Journal of Production Economics* 76, 201–215.

Kim, Y.G., Suh, J.H., Park, S.C. (2008): Visualization of patent analysis for emerging technology. *Expert Systems with Applications* 34(3), 1804–1812.

Lee, M.D., Butavicius, M.A., Reilly, R.E. (2003): Visualizations of binary data: A comparative evaluation. *International Journal of Human-Computer Studies* 59, 569–602.

Partovi, F.Y., Burton, J. (1993): Using the analytic hierarchy process for ABC analysis. *International Journal of Production and Operations Management* 13, 29–44.

Ramanathan, R. (2006): ABC inventory classification with multiple-criteria using weighted linear optimization. *Computers and Operations Research* 33, 695–700.

Russell, S., Gangopadhyay, A., Yoon, V. (2008): Assisting decision making in the event-driven enterprise using wavelets. *Decision Support Systems* 46(1), 14–28.

Scarf, P.A. (1997): On the application of mathematical models in maintenance. *European Journal of Operational Research* 99, 493–506.

Ward, M.O. (2008): Multivariate data glyphs: Principles and practice. *Handbook of Data Visualization*. Springer, Berlin, 179–198.

15 Toward 2020 with SSAT

Hansjörg von Brevern and Kateryna Synytsya

CONTENTS

OUTLOOK INTO 2020

A European OEM executive claimed that in "the next 10 years, we will experience more change than in the 50 years before" (Rishi et al. 2008). According to an automotive study, the global infrastructure and changes in the workforce worldwide will set new challenges for the industries. A flow of technological innovations will be aimed not only at the enhancement of system performance but also at better care of the user, providing personally oriented information and assistance as outlined in the i2010 Intelligent Car Initiative (Europe's Information Society Thematic Portal 2010). Individualization, adaptation, guidance, and assistance in problem solving are just examples of what "intelligent" technologies promise for their customers in 2020.

TODAY'S REALITY

Future challenges toward human labor, intelligence, greater personalization, dynamic guidance, situated assistance, adaptation, and information to personal habits to enhanced technical functionality of computer systems (Rishi et al. 2008) let us query where we stand today in information technology (IT)–driven organizations vis-à-vis predictions from safety environments.

In the following, we will adduce investigations in an international bank in 2011 (A) and a telecom firm in 2005 (B) and will ultimately draw interesting analogies.

AN INTERNATIONAL BANK IN 2011

This section analyzes A's contextual business, IT, and system impacts on the work activity of back-end office operators with transaction systems.

Banks are typically IT driven to ensure competitiveness, efficiency, reliability, security, stability, speed to market, and compliance with laws and regulations, which is particularly relevant for treasury (e.g., trading with foreign currencies, derivatives, and even positions at the end of the business day). The following looks at back-end office investigations of A. Investigations are part of back-end office operations within A's business organization. Work processes are divided into product classes. Internal clients are traders and mid-office teams; external clients are correspondent banks and large organizations with their own in-house trading. Investigations handle open and unsettled trades and positions of complex products. Their daily work with complex products and within an extensive system landscape is stressful, subject to high time pressure from different market opening and closing times, time zones, and changing exchange rates, and contains routine and nonroutine tasks.

Overall, contrary to A's business organization, IT is driven by the trendy process reference model (PM) capability maturity model (CMMi), which is a management improvement approach and framework to handle software projects (Campo et al. 2009; Gefen et al. 2006). While CMMi still lacks "publicized, scientific empirical validation of... effects on company performance... and user satisfaction" (Bellini 2006), only engineering can provide methods and pragmatic solutions for more intelligent, personalized, and adaptive software by 2020 to serve goal-oriented human work activity.

Despite A's eagerness to advance in CMMi levels, reform its internal PM processes, and launch it internationally, there is much room for alignment within business and IT, and between business and IT. For example, business does not follow CMMi; only sponsors and senior management decide on goals of work processes on an abstract level, business needs, software systems, system changes, and new requirements; there is no common language for business, software developers, and system analysts to describe system requirements and approve specifications. Moreover, a "major problem in (requirements engineering [RE]) is obtaining requirements that address the concerns of multiple stakeholders" so that "fairly vague expressions of users' information processing needs are translated by system analysts into a set of product specifications (sometimes very precise, sometimes fairly general, but often inaccurate in important respects)" (Boehm et al. 1995). As a result of short-term PM perspectives, the software product may have been realized successfully in time and on budget in its isolated universe. Nevertheless, the human operator and maintainer may end up with unaligned and ill-specified systems (DeMarco 1996) to suit daily operational needs. So, in the long run, the loser will be the user.

Ergo, we posit needs for collaborative horizontal and vertical alignment, shared requirements negotiation, and decision making that give the expert operator equal rights to voice and decide. A's PM implementations and management approach focus on high-level management goals and (rapid) system changes without looking beyond project boundaries and disregard low-level human work activity. PM unfortunately attempts to inappropriately constrain solid engineering where they should not and thus fail to build one horizontally and vertically aligned ground from the start. Moreover, in large organizations with considerable product diversity and technologies, "it seems impossible to achieve a unified way of developing systems" (Carlshamre and Rantzer 2001). A's IT strategy is to share and scatter organizational knowledge using pool organizations, yet management not wanting to negotiate requirements and share decision making and developers' opposition to safe-guard "their" freedom of design altogether support our argument. Such factors can lead to losing specific domain knowledge ad so forth and thus inhibit engineering.

Trading floors and mid-office have their own complex IT systems, which are mostly proprietary commercial-off-the-shelf (COTS) software, difficult to change, and often have unfriendly user interfaces yet claim to comply with common practice of work processes. Back-end office operations support the front office and mid-office and consist of investigations, processing, and trade reconciliation and confirmation teams per product class. They use legacy mainframe host systems that have been existing for >20 years, hypermedia (Web/desktop/server-side) applications (HM) ranging back to the mid 1990s, and customized and noncustomized COTS. Yet, critical legacy systems cannot be unplugged.

Nonetheless, COTS constrains interdependent business and work processes and human work activity and impede new system requirements. So, it is often easier and less expensive to build in-house reconciliation engines and add system interfaces to underlying mainframe host or Web applications. In turn, this approach multiplies the complexity of the system landscape because it stimulates additional requirements for system interfaces and underlying functionalities apart from, for example, business, organizational, legislative, regulatory, functional, and nonfunctional needs alike.

An operator needs to know critical stakeholders, and where and how to search for information across domains. Unfortunately, applications (and their help systems) are not self-explanatory, intelligent, or adaptive to human work activity. Without training, guidance, moderation, or the assistance from a more experienced peer, an operator often cannot independently explore, effectively perform, or resolve tasks with the system(s).

Exponential growth of hidden system functionalities with each of the four annual software releases complicates the blended system environment, nourishes heterogeneity, makes reengineering costly, and ultimately obstructs operators' daily work activity. Operators who are not engaged in requirements negotiation and decision-making, work, and system design processes are "virtually powerless in system development, and... 'captive' in that they have no choice but to use what is installed on their computers" (Boivie et al. 2006).

Unsurprisingly, a newcomer to investigations needs more than 3 months of training to comprehend the complexity of product class basics, work flows, critical contexts, systems, and stakeholders. The knowledge area is broad, highly complex, and mentally demanding so that it takes more than 1 year of training to become a professional (cf. Figure 15.1).

Investigations resolve incomplete trades from non-straight-through processing (non-STP). Out of the 5% of non-STP cases, 60% need individual settlement. To independently resolve individual settlements, a member of staff of investigations needs profound previous training for several months, must be able to demonstrate applied working experience and handle manual intervention cases carefully, and must have social and analytical interaction skills. Non-STP cases are the most demanding tasks because they are time-consuming, are stressful, demand a high degree of concentration and attention, and necessitate the mental ability to switch between different tasks and work processes. Threats to complex problem solving include internal and external distractions, time constraints, the pressure of jeopardizing organizational reputation, and financial loss. Learning and training (L&T) are costly for both the operator and the organization and entail continuous motivation; altogether, these constraints shape man's internal work activity and challenge task switching.

A professional operator works daily with >30 HM applications, >5 mainframe host applications, >20 mainframe host transactions, and >20 mainframe host screens.

A mainframe differs substantially from HM, which also affects work processes, design, and human work activity. During the first year, the learning curve is steep but stagnates after about 1.5 years. However, knowledge about changing market needs, habits, regulations, products, and

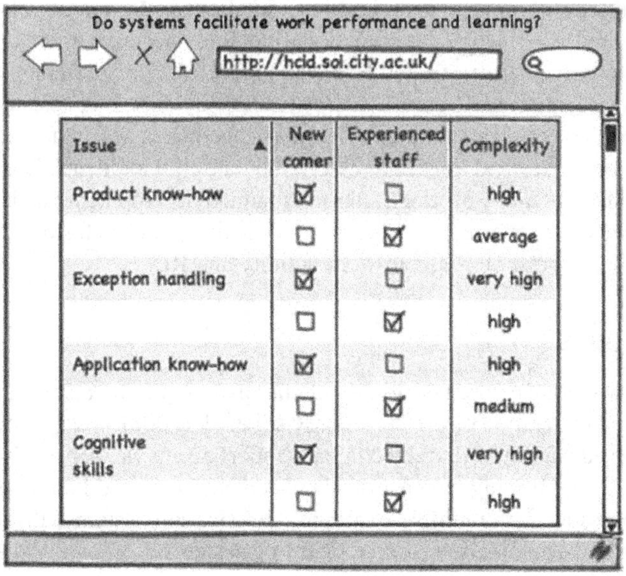

Do systems facilitate work performance and learning?

http://hcid.soi.city.ac.uk/

Issue	New comer	Experienced staff	Complexity
Product know-how	☑	☐	high
	☐	☑	average
Exception handling	☑	☐	very high
	☐	☑	high
Application know-how	☑	☐	high
	☐	☑	medium
Cognitive skills	☑	☐	very high
	☐	☑	high

FIGURE 15.1 Newcomer versus professional.

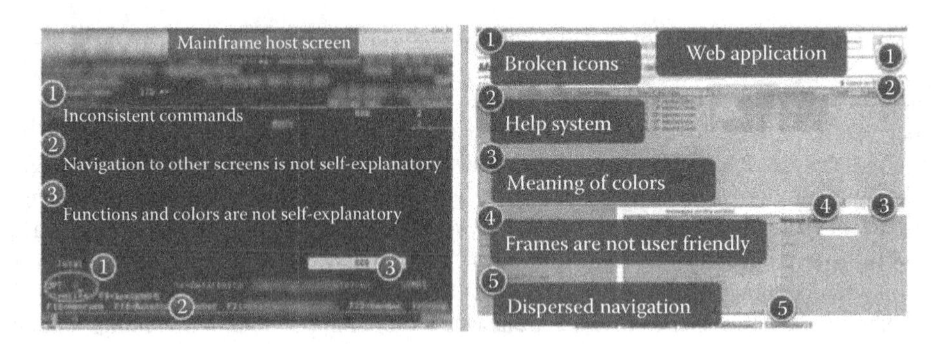

FIGURE 15.2 Comparative TUI and HM screen shots (anonymized).

services must at all times be kept up-to-date. On the average, an operator remains approximately 1.5–2 years on the job.

In mainframe host systems, all application processes eventuate on the mainframe computer except for text-based user interface (TUI) controls, which "take place on dumb terminals that simply echo text from the mainframe" (Nguyen et al. 2003). Contrary to HM graphical user interfaces (GUIs), TUIs do not recognize graphical metaphors, pictures, drag-and-drop, pull- or drop-down lists, functional buttons, boxes, fields, limit mouse events, and so forth. Hence, TUI and GUI computer-based tasks substantially differ.

Mainframe host and HM concepts cannot be merged. Indeed, work activity that includes both concepts reveals inconsistencies from the view of human-centered design. If for some reason, a new mainframe host screen is not linked properly in a linear and sequential fashion along a work flow, the operator will not be able to find it without external help. On the other hand, navigating through HM can lead to loops or getting "lost in space." Altogether, if not planned, designed, aligned between mainframe and HM systems, and based on and derived from goal-oriented human work activity, inconsistencies increase cognitive overload; cause breakdowns (i.e., "forced changes of [human] strategies of action caused by their evaluation of an unacceptable divergence between the actual results of actions and the[ir] conscious goals" [Harris 2004]); affect reliability of task performance; and reduce motivation and productivity.

Taken from A's high-level management decisions; unaligned PM; history of systems; and many uncountable, hidden, and dispersed system functionalities and interfaces, there is misalignment with goal-oriented work activity, mainframe and HM (see Figure 15.2), mainframe, and HM systems.

Altogether, A's heterogeneous systems reveal poor usability (e.g., inconsistent commands, menu navigation, metaphors, colors, looks and feels, help, and feedback, see Figure 15.3); are deficiently customized to work processes and human activity; and cause breakdowns. Breakdowns of human work activity can lead to cognitive overload, suppress human development and learning, demotivate work activity, frustrate task performance, decrease productivity, and so forth. Usability engineering (UE) is thus critically important for human work activity and RE.

A Telecom Firm in 2005

In the following, we elaborate on the essentiality of cognitive requirements from goal-oriented work activity as a pragmatic realization of the vision of Rishi et al. (2008).

In 2005, we observed and interviewed technical support agents (TSAs) at B's hotline call center. The primary focus was on missing connections between goal-oriented human work activity, L&T, and information processing with multiple unaligned information systems (ISs), customer relationship management systems, enterprise resource planning systems, L&T, and caller systems.

TSAs are a window between end consumers and represent the company's image. A friendly, professional, client-focused, and efficient problem-solving service is critical to building and retaining

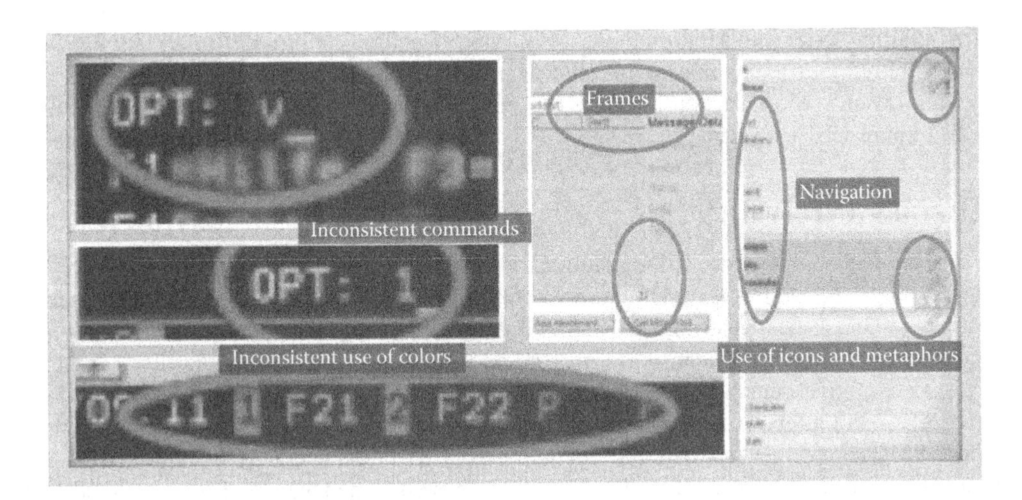

FIGURE 15.3 Comparative HM GUIs and TUIs screen shots (anonymized).

loyal client relationships. TSAs need good communication skills, to be able to resolve technical issues, and to know current products and services, customer bases, corporate policies and procedures, and how to use computer systems (Synytsya and von Brevern 2010). Nevertheless, TSAs suffer from continual stress that requires them to rest after 2–3 days of labor. Stress and job dissatisfaction cause high job fluctuation, which is always costly. Yet, TSAs are not paid well and cannot influence system changes for better assistance and productivity, nor are decision makers sufficiently envisioning investments for better-performing systems from actual TSA activity perspectives. Nonetheless, management constantly observes TSAs for better productivity.

Unlike A's operators, TSAs are typical information and knowledge workers who heavily depend on and interact with sundry systems, which should adaptively assist and return meaningful and significant material. This is equally true when a TSA may be preoccupied with another nonroutine task whose problem solution may have changed or been forgotten, and so forth (von Brevern and Synytsya 2006). Ill-structured situations like callers' unclear problem descriptions, absence or nonaccessibility of situated information in and from systems, the need to analyze assumptions (Synytsya and von Brevern 2010), and so forth can cause emotional stress. So, the more complicated and indeterminate the situation, the more mental mechanisms of imagination and image manipulation are included (Bedny and Meister 1997). The constantly changing "diversity of information about new, altered, or obsolete products, services, promotions, etc. needed in responding to telephone enquiries makes [TSAs] heavily rely on IS.... Thus, the job of [TSAs] requires them to spatiotemporally process information while answering customers' queries or betrothal in problem solving" (von Brevern and Synytsya 2006). Stress factors include cognitive overload, cognitive dissonance, challenging nonstandard tasks, and emotions, which according to Norman (2004) "modify perception, decision making, and behavior" and can "change the parameters of thought."

The nature of the job of TSAs hence requires daily L&T "to create meaning and resolve inconsistencies by assimilating new knowledge or adapting it to the existing knowledge. In this sense, knowledge and meaning are not fixed, but they are constructed by the individual within the context of meaningful learning" (Synytsya and von Brevern 2010). Thus, the issue on L&T for knowledge workers cannot just be restricted to L&T per se but tacitly penetrates business and work processes, the goal-oriented human activity system, and also RE. Notably, "meaningful learning is also a result of practical problem solving. Moreover, an active and engaged problem-based learning process continuously forms corporate knowledge" (Synytsya and von Brevern 2010).

In all, TSAs "share tasks with technology systems so that the efficiency of their work depends on the capabilities of human–computer system, justifiable task distribution, and the quality of the interaction" (Synytsya and von Brevern 2010). Hence, the activity system of information and knowledge workers requires systems to support cognitive activity, to adaptively provide timely assistance and responses, and to return meaningful and significant content to the individual in situ. These facets imply a rethink and redesign toward systems (help and GUI) to appease cognitive task requirements from the human activity system.

ANALOGIES AND FINDINGS

Both case studies demonstrate striking similarities in view of effects on an operator's daily work activity and job satisfaction vis-à-vis causes. Sources of effects are unaligned systems; heterogeneous system landscapes; unstructured and unaligned TUIs and GUIs; nonadaptive, nonsituated, nonmeaningful, and insignificant information and content provisioning; negative reinforcements onto cognitive demands; cognitive overload; cognitive dissonance; and so forth. Causes are exhaustive and intersect with effects and organizational, business, social, and environmental contexts; PM; management's excluding operators from participatory negotiation, decision making, and design of work processes, requirements, and systems; management's ignorant disdain for human work activity, the importance of engineering, UE, and L&T; and so forth. Our pragmatic findings of two independent organizations with different business types reveal that today's organizational, management, PM, and engineering procedures cannot meet 2020's claims for more intelligent, adaptive, and dynamically assisting systems according to greater personalization and individualization.

The comparison between system functionalities and actual goal-oriented human work activity discloses that present systems are incapable of coping with cognitive task requirements. The level of cognitive task requirements for "better" systems is the understanding of the specifics of human problem solving; physical, cognitive, and social interactions with computers; human goals; development; learning; and so forth. The key to this level is human goal-oriented work activity per se. Today, both organizations do not envision the importance of this key to derive cognitive task requirements for work process design and RE.

Regrettably, research literature and engineering are primarily concerned with functionality in systems and their requirements (e.g., "[s]pecifying stakeholder views on system requirements, followed by their negotiated integration..." [Robinson et al. 2003]), while they forget that systems must serve man. The emphasis is on man and not impersonal users or stick figures, which have no individual intention, emotion, thought, development, skills, and so forth. Obviously, ignorance of cognitive task requirements results in inaccurate and incomplete requirements. Inaccurate and incomplete requirements affect requirements negotiation, decision making, and design and, as such, are one major cause for ill-structured systems that commences at the threshold of software engineering (SE), that is, RE. Doing so leads to systems "that are far less usable than they should be" (Kaindl et al. 2008).

Greater computer-aided assistance interlinks with personalization, individualization, and human development. Human development therefore links to L&T and with it, the question to which extents a computer can and should provide individual and personalized assistance and help during problem solving and human performance.

Research reveals that "[l]earning processes and outcomes are shaped by previously acquired theoretical concepts and learning means that the teaching activity must provide a learner with the opportunity to incorporate the material to be learnt into existing knowledge and skill systems" (von Brevern and Synytsya 2005). From this follows that intelligent computer systems should yield "instructional assistance" based on individual human goals, tasks, and actions. More pragmatically, computer systems should continuously observe and analyze a human operator's "intrinsic and extrinsic motivation in the course of... activity... because of the possible discord between action intentions and actual actions" (von Brevern and Synytsya 2005). Ideally, they should adaptively be able to help situated problem solving or assist in moving to the next action of activity where needed. Based on sociocultural theory, we disagree with completely replacing teachers with computers because teachers "should not only moderate students' learning, but they must also shape it" and "help what a student cannot do himself" (von Brevern and Synytsya 2005).

Nevertheless, this discussion raises a further research endeavor for computer-aided assistance: to promote human development, which could be seen from developmental teaching "with its emphasis on the role of appropriate forms of content" or scaffolding that emphasizes "on the creation of a pedagogic context in which combined teacher and learner effort results in a successful outcome" (Daniels 2007). Both concepts have substantial differences that increase complexities of task, goal-directedness, meaningful context inquiry, collaboration, and so forth, albeit not further discussed.

Human–computer interaction (HCI) has brought forth >1000 UE guidelines (Nielsen and Loranger 2006), which, if combined with research from instructional design, can be a strategic tool for human development (cf. Figure 15.5). For example, instructional design principles teach us

- To avoid splitting attention between multiple sources of mutually referring information
- To present verbal information "auditorily as speech rather than visually as on-screen text both for concurrent and sequential presentations"
- To avoid redundancy
- To physically integrate both on-screen text and visual material
- To temporally synchronize verbal and visual materials
- To exclude extraneous material and so forth (Moreno and Mayer 2000)

These principles take three memory systems into account, like visual and auditory sensory memories, working or short-term memory, and long-term memory (Clark and Harrelson 2002). As such, we should strive to design adaptive applications based on UE (cf. Figure 15.4) and instructional design principles into one homogeneous system landscape.

We postulate that UE and instructional design should no longer be separate parts to overcome their unclear positioning as often discussed in literature. Instead, we regard UE and instructional design as inseparable units from RE (cf. Figure 15.5). This gives rise not only to early application of engineering techniques but to equally early embedding of RE together with instructional design, UE, and collaborative negotiation and decision-making techniques into business processes.

So, what could companies do toward customized, individual, personalized, human user-evolutionary, intelligent, and adaptive systems for better productivity, which demand systems to be "more flexible, more adaptable, more productive, more cost efficient, more schedule efficient, and more quality driven" (Layer et al. 2009)? The answer is as follows:

To place *human* goal-oriented work activity into the center.

Doing so necessitates horizontal and vertical alignment across the organization, methods for requirements negotiation and decision-making processes including both management and human operators, and studying goal-oriented work activity. This paves the way to engineering human-centered computer systems that seamlessly integrate RE with UE and instructional design. Altogether, this calls not only calls for a change of direction toward human-centered engineering based on solid engineering methods but for equally collaborating and embedding them into business processes for more efficient and reliable design solutions.

This contrasts with conventional engineering, which attempts to model systems on mere "stimulus–response" behaviors and thus "ignores cognitive mechanisms of humans and with them the embedded roles, context, tasks, and goals of technical systems" (von Brevern and Synytsya 2006).

Results of the research of Layer et al. (2009) "indicate that the context of the task does indeed affect the relationship[s]" between latent variables such as "cognitive demands," "quality of work life," and "human performance." Index variables of cognitive demands are complexity, adaptability, workload, and motivation; those of the quality of work life are learning, job satisfaction, empowerment, and supervision (Layer et al. 2009). Interestingly, the authors demonstrate that a positive reinforcement of cognitive demands improves individual performance as opposed to negatively

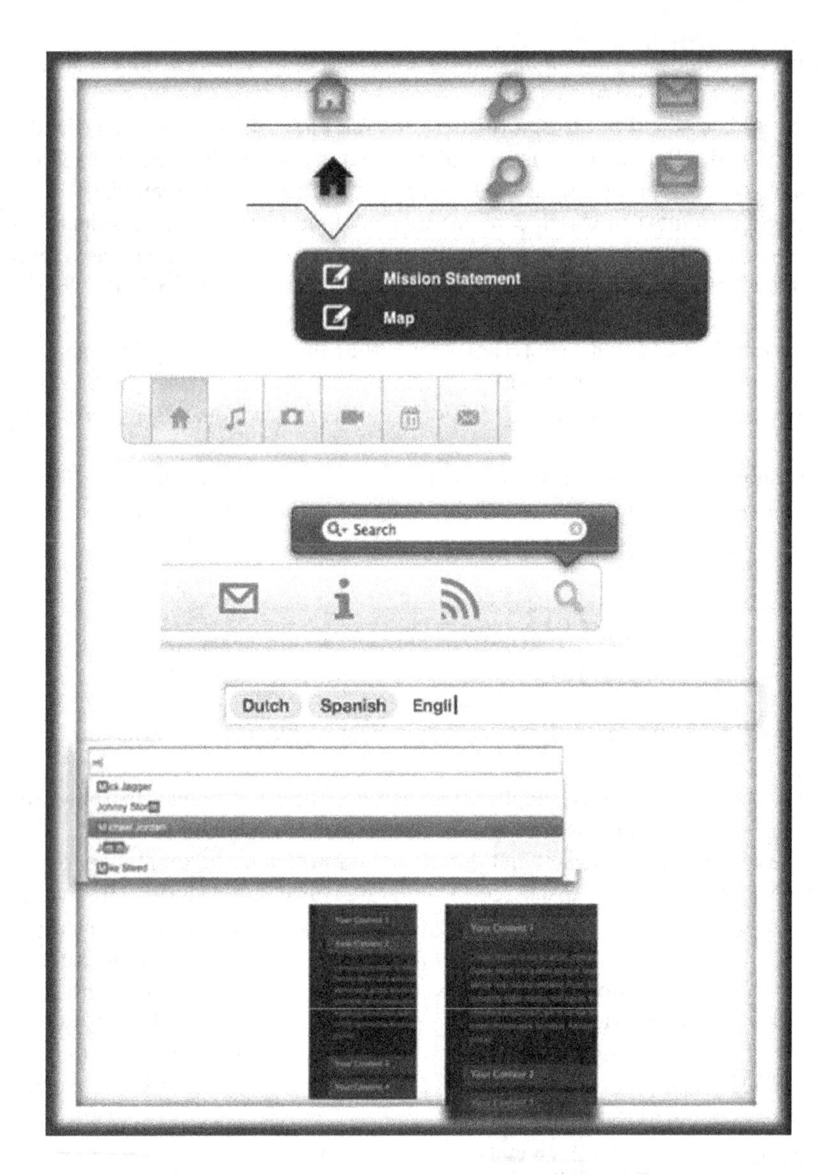

FIGURE 15.4 Examples of human-centered design elements according to UE.

reinforced cognitive demands, which decrease activity with increasing demand. Hence, the more a system supports positively perceived cognitive demands by an individual, the higher his/her performance. Furthermore, Layer et al. (2009) argue that occupational context and task specifics constrain positive effects of cognitive demands. We therefore abstract that individual personality, capabilities, skills, professional patterns, roles (Zarakovsky and Karwowski 2010), thought, perception, and so forth relate to the degree of individual acceptance of cognitive demands.

Numerous research studies argue that cognitive load (i.e., demands placed on working memory) leads to strongly reduced event detection. Nevertheless, a study by Engström et al. (2005) study reveals that cognitive load can conditionally cause positive reinforcement. Likewise, they discovered different effects between visual and cognitive load, while "visually and cognitively loading secondary tasks have radically different effects on [human] performance" (Engström et al. 2005). They conclude that visual time-sharing and cognitive load cohere as a combination of both effects. (New) task elicitation is therefore highly challenging. Consequently, engineers should not design computer-based tasks from assumptions without measuring and assessing task performance,

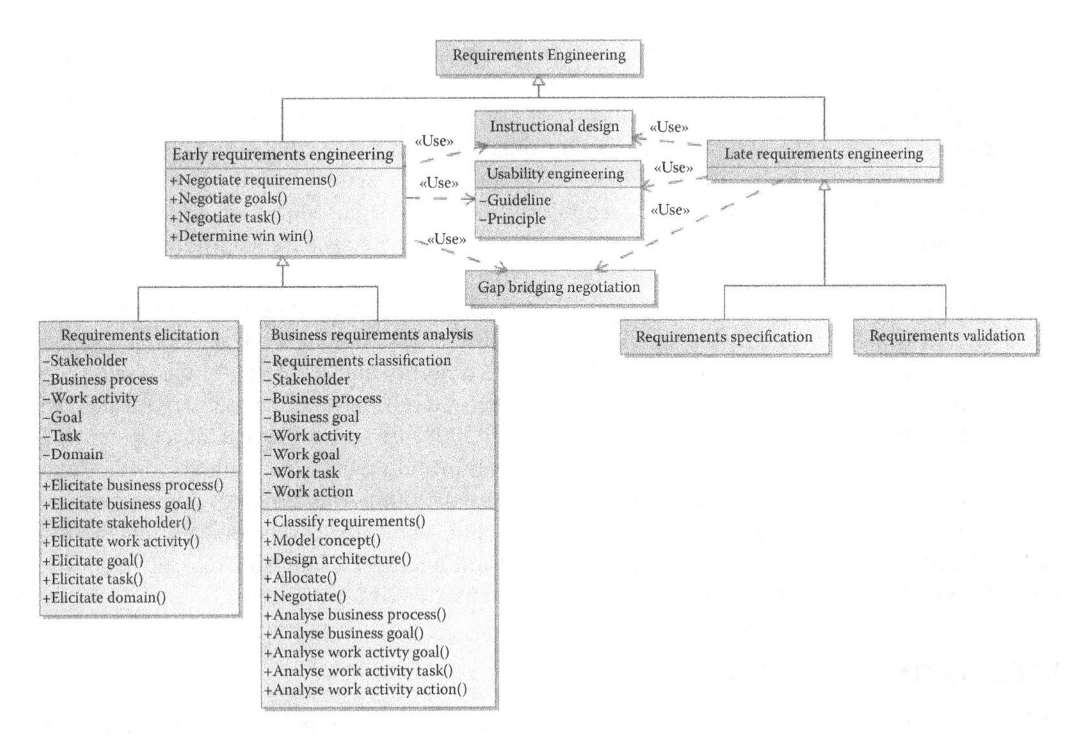

FIGURE 15.5 Extended view of RE.

because the combination of effects from human actions can vary. Likewise, human information processing, memory load, and brain functions demonstrate that L&T are critical elements of computer-based tasks for computer-aided assistance, guidance, help, individualization, customization, and so forth regardless of the type of system.

Cognitive task requirements from individual and situated human goal-oriented work activity are the answer to design greater personalization, focused assistance for personal habits, dynamic guidance and navigation, L&T, individual development, and information provisioning for present and future computer systems. The solution entails investigation and apprehension of individual human goal-oriented work activity and systems to present modular content so "that employees are always au courant with what was learnt earlier, the latest lessons learnt in customer care-taking, the latest product information, promotion details, repair processes, etc." (von Brevern and Synytsya 2006). As such, cognitive task requirements originate from at least two intersecting spheres, that is, between the human subject where, for example, sociocultural, sociohistorical, dynamic psychic, and psychological aspects play a major role, and individual interaction with physical and mental artifacts during goal-directed and self-regulative work activity. Such intersections under the umbrella of the human activity system necessitate well-grounded research for performance and better productivity for human-centered systems in RE, HCI, L&T, instructional design, and human factors.

According to Bedny et al. (2008), "any change in the physical configuration of the equipment also changes in probabilistic manner the structure of activity. Such changes cannot be described and accounted for without the concept of… human cognitive and behavior actions." Both organizations largely neglect and therefore exhibit discrepancies between (changing) computer systems and the human activity system. These disagreements impact performance and with it, reliability and accuracy of human work activity. As such, accuracy "characterizes the precision with which the goal of task is achieved. Reliability refers to failures of performance and how probability of failure can change over time or in stressful situations… human performance can be precise but not reliable, and not all errors can be considered as system failures" (Bedny et al. 2010). The structure of human activity with its tasks should therefore always be the emergence for any system changes. Sadly enough, this critical issue is largely ignored, as our case studies have revealed.

From SE perspectives, it is too late to allocate the problem space to the design phase. Instead, it belongs to RE, which is the first phase in SE. Our situation today proves that most known RE techniques like interviewing, walk-throughs, and so forth, fail to satisfactorily capture human cognition and behaviors and do not envision human activity as a holistic system. Moreover, reliance on interviews and walk-throughs are too vague because "there is often a considerable difference between what people say and what they do" (Diaper and Stanton 2004).

To elicit, analyze, negotiate, specify, and validate (IEEE 2013) meaningful requirements, engineers need models for a better understanding of the holism behind human work activity and formal methods for analyzing and decomposing it. To evaluate usability and reliability of GUI, engineers need methods to be able to analyze "the relationship between the structure of activity during task performance and interface characteristics" (Bedny et al. 2008). To support human development and learning, engineers need psychological and pedagogically valid models and methods for the study of the dynamic participant-object relationship (Bedny et al. 2008). The design of human-centered systems as envisioned by (Rishi et al. 2008) requires a philosophically, psychologically, and holistically valid approach that is equally suitable for RE. Built upon AT's long-standing and extensive scientific pillars, the systemic principles of SSAT can satisfy engineering needs.

BACK TO THE FUTURE WITH SSAT

This chapter focuses on morphological analysis in SSAT, which is a formal method of TA. SSAT is applied activity theory (AT) with an evolution in research of more than one century and is yet avant-garde for the future, albeit we will only present relevant key concepts.

AT is a philosophical and psychological "paradigm that was a foundation for the study of work behavior..." (Bedny et al. 2000). Activity includes human subjects, artificial or physical objects, sign tools, norms, and procedures. As such, activity is a complex and holistic construct within a historical–sociocultural sphere (Bedny and Karwowski 2007).

Subjects cooperatively mediate on physical or mental objects of activity toward a result by means of physical, mental, or cultural artifacts such as tools, signs, and speech. On one hand, artifacts "shape possibilities for thought and action and, in turn, are shaped by those who use them" (Daniels et al. 2007). On the other, an active subject "constantly changes the objective world and culture" through mediating artifacts and, "based on these, changes himself" (Bedny and Karwowski 2007). Social aspects "of experience are integrated into individual activity.... In the same social environment different individuals act differently and are impacted by the social environment in different ways" (Bedny and Karwowski 2007). The mental reflection of an image during activity directs the human subject in the objective (physical and artificial) world (Leontyev 1977) so that the internal condition directly affects the subject (Bedny and Karwowski 2007). Activity is therefore a dynamic and reciprocal interchange between "opposite poles, subject, and object" (Leontyev 1977). Object-oriented activity therefore forms an inseparable part of social interaction (Bedny and Karwowski 2007). Outside of these relations, human activity does not exist.

Leontyev (1979) defines activity as "a molar, not an additive unit of the life of the physical, material subject. In a narrower sense, that is, at the psychological level, it is a unit of life, mediated by psychic reflection, the real function of which is that it orients the subject in the objective world... [A]ctivity is not a reaction and not a totality of reactions but a system that has structure, its own internal transitions and transformations, its own development." Consequently, Leontyev (1979) argues that the concept of activity metamorphoses "the whole conceptual system of psychological knowledge." The psychological framework of the activity system requires acceptance "with its most important dependences and determinations" of the entire union of "its structure and in its specific dynamics, in its various aspects and forms" (Leontyev 1979). In the foci of human psychology and AT is always the individual human subject (Leontyev 1979).

However, general AT is only a psychological and philosophical framework "for studying human performance, although sometimes, it helps in the discovery of a nontraditional method of solving a practical problem" (Bedny and Karwowski 2007). From the philosophical discovery of Marx, AT inseparably incorporates human cognition and behavior: "The reflection of reality arises and develops in the process of the development of real ties of cognitive people with the human world surrounding them; it is defined by these ties and, in its turn, has an effect on their development" (Leontyev 1978).

As such, human activity is an uninterrupted, never preplanned nor predetermined, but functionally dynamic process that forms a holistic, organized, and hierarchical system (Bedny and Karwowski 2007). This formulates the point of departure in SSAT for formal methods of (informational or cognitive, parametric, morphological, and functional) analysis "of the separate elements of activity into sequential decompositions of activity" (Bedny and Karwowski 2007) under spatial–temporal considerations. The scientific works of, for example, Anokhin (1962), Bernshtein (1966), Landa (1983), and Gal'perin (1969) build the grounds of the systemic principles of activity in SSAT (Bedny and Karwowski 2007). SSAT particularizes Leontyev's (1979) definition of activity as a "goal-directed system, in which cognition, behavior, and motivation are integrated and organized by a mechanism of self-regulation toward achieving a conscious goal" (Bedny and Karwowski 2007).

AT's perspectives radically differ from the theory of behaviorism, which unfortunately still dominates today's SE. Although today's systems are highly capable of resolving technical functionality, they are regrettably largely unresponsive to the dynamics of the unity of human cognition and behavior. This lack not only is an issue of the study of the GUI in HCI but also encompasses the use and specific meaning of a software system during human work activity. SSAT helps resolve disparities between the mediating roles and behaviors among subjects, psychological and physical tools, objects, goals, and so forth and cognitive task requirements to design systems today and toward 2020.

From an AT perspective, to accomplish a computer-based task during work activity, a subject uses mental and motor actions to present, create, or modify new or existing artificial objects toward which actions are directed (Bedny and Karwowski 2007). More practically, a subject uses a computer as a mediating means to present (information, learning, help) content or to create artificial tools like a trade. Importantly, a tool is a "mediating symbol to the subject and between the subject's work performances" (Bedny and Meister 1997) and thus realizes a mediating role through the process of internalization. Under this aspect, it is incorrect to generically categorize the computer as a tool. The use of computers depends on its subjectively and objectively goal-oriented, meaningful, and situated physical or psychological functions. Likewise, the object of study determines which components of a computer are mental, artificial, or physical tools, or form the object of activity per se. A tool is, eo ipso, tightly coupled with the task of activity, its motor or mental actions, the subject and object of activity, the object of study, and indirectly with the result of activity. The dynamic interrelationship between

$$(\forall(S)\wedge\forall(O))\exists(T) := S \leftrightarrow T \leftrightarrow O \mid S \equiv Subject,\ O \equiv Object,\ T \equiv Tool$$

formulates the basic triadic schema; the extended triadic schema connects the object of activity with the result of activity from which feedback and feed-forward (control) loops lead to/from the subject of activity (Bedny and Harris 2005; von Brevern and Synytsya 2005, 2006). If we hence understand its interconnected, organized, and systemic relationships, we can deduce goals of activity, tasks, actions, subject(s), and objects of activity. Consequently, morphological analysis reveals goals of tasks and actions as well as tools and the object of activity. Likewise, the object of study is critically important to depict whether, for example, content provisioning, a UE principle, learning, the mouse, its use, or a GUI element is one of the mediating tools to the subject or the object of activity. As such, morphological analysis helps overcome another challenge in TA, where

Diaper (2004) contends, "in more than 20 years of doing [TA] he has never got the [TA] data representation of activity lists… correct the first time."

Changing goals of strategies, plans, emotions, objects of study, and so forth may affect goals of tasks and actions, or even an entire triadic schema. Nevertheless, the goal of activity is relatively stable. Goals of tasks and action are equally critically important to understand mediating tools and the object of activity in time and space. In all, the same principles apply to learning processes and design (Bedny and Karwowski 2007; Synytsya and von Brevern 2010). However, a further in-depth discussion of AT's concepts exceeds the extent of this paper.

Morphological analysis, which decodes and describes the constructive features of activity (Bedny et al. 2000), is suitable for TA to abstract cognitive task requirements. Morphological analysis consists of a qualitative and an algorithmic description of the task. The concern of qualitative analysis (QA) is an iterative process to decompose a task into members of algorithm, conscious, unconscious, mental, and motor actions using logically organized steps, procedures, and stages of analysis. Algorithmic analysis (AA) describes "human performance in a probabilistic manner and uncovers the constraints of the work process" (Bedny and Karwowski 2007).

QA consists of two subprocesses, that is, informal information gathering to identify the object of study and tasks within their context and to classify and assimilate the task with its members of algorithm and actions according to task and action taxonomies. QA therefore fits into RE elicitation.

The first subprocess may include informal discussions with stakeholders, observations, (literature) reviews, function refinement trees to understand the subject domain, and so forth and may lead a preliminary hierarchical TA (Annett 2004). The second subprocess decomposes activity into the task, its members of algorithm, and discrete cognitive and behavioral actions that are directed to achieve an intermediate and conscious goal of activity during task performance (Bedny et al. 2010). Task decomposition prepares for AA.

QA per se overcomes challenges of general TA. Human goals, motivation, emotions, and so forth are individual, dynamic, and changeable. Moreover, a "subject can evaluate the same task as being more or less difficult depending on the complexity of the task, past experience, differences between individuals, and even the person's temporal state" (Ponomarenko and Bedny 2010). Therefore, "list actions without understanding what they are for" (Annett 2004) is not meaningful. Similarly, system processes and functional system requirements that do not reflect mental and motor interactions on the tool, object, and result of activity cause cognitive dissonance, negative reinforcement of cognitive demands, breakdowns, and so forth. While TA "might be very concrete, describing specific things and behaviors in a detailed way" when analyzing behaviors of an existing system, TA "must be at a higher level of abstraction because the design has not yet been fleshed out" (Diaper 2004). The lowest level of abstraction of SSAT's hierarchical system of activity is the member of algorithm consisting of [1..4] homogenous actions. The member of algorithm is the unit of analysis of goal-oriented actions and is part of its superordinate task.

Unlike many programming languages in SE, which allow lists of sequential actions, human action processing embeds mechanisms of self-regulation. According to Rubinstein (1966), every "mental process is included in man's interaction with the world and takes part in the regulation of his actions, his behaviour; every mental phenomenon is both a reflection of being and a link in the regulation of behaviour, of the actions of people." He discovered feed-forward and feedback mechanisms and developed the model of self-regulation of conditioned reflex. He argued that the "end of an action can never serve as the beginning of another action. The end of an action in one link is the source of return afferentation which is directed to the centres of the just developed reflex and only after that, and depending on the results which this return afferentation will have in the nerve centres, will the next stage of the chain reflex begin to form" (Anokhin 1966). In this sense, Bedny and Karwowski (2007) draw an analogy between words or speech and actions, which "possess semantic, syntactic, and pragmatic features analogous to words. Syntactic features of actions are determined by their rules or organization into a system. Semantic features of action may be discovered through the relationship of an action to its objects or to other actions. Pragmatic features of actions can be determined by their role for the subject and particularly in their relation to motivation." In other words, each human action is a reflex with feedback and feed-forward control mechanisms of self-regulation and, as such, differs substantially from today's computer programming languages. Today's SE does not account for such a critical human behavior.

The research of Bernshtein (1966) has impacted Western theories of motor learning. According to Bernshtein (1966), a human subject interacts on the context according to needs. Execution of an act suggests functional mechanisms and states in the neural system that continuously allows encoding, comparing, programming, and correcting movements with the desired future. Bernshtein's (1966) levels of regulation of movements and actions explain transitions from unconscious performance at lower levels to conscious performance at higher levels. These levels can be manipulated and may change in a human subject with purposeful training, cognitive demand, and so forth.

Anokhin's (1961) model of the afferent apparatus of the conditioned reflex and Bernshtein's (1966) model of the self-regulation of movement "are based on operational and functional constructs of behaviour and their causal interrelationships" (Bedny and Karwowski 2007). Both models are critically important and integral to the concept of self-regulation in SSAT.

For example, results from our QA of B disclosed the following cognitive issues during computer-based work activity:

1. The task to understand ill-structured problems
2. The task to search for specific information within time constraints
3. Task switching
4. The task to classify information
5. The task to master skills consciously (von Brevern 2005)

The task builds a situated entity of activity and is a logically organized system of cognitive and mental actions that emerge as the primary unit for morphological analysis (Bedny et al. 2008). von Brevern and Synytsya (von Brevern and Synytsya 2006; Synytsya and von Brevern 2010) provide semiformal examples of the decomposition of activity during task performance, the hierarchical structure of activity, the hierarchically organized units of analysis, task taxonomy, and task implementation. Depending on the nature of objects, actions, and the object of study, SSAT recognizes more than one classification of actions.

AA formally transforms the verbal task description of QA into an algorithm in a language with its own syntax and symbols. This language "provides a fundamental basis for the creation of models to design any object" (Bedny and Karwowski 2007).

The logically organized steps, procedures, and stages of analysis of an algorithm provide a means to describe (a self-regulated) human performance in detail and uncover constraints of a work process (Bedny and Karwowski 2007).

We expect that statistical methods and probability theory applied to individual task performance may be useful for modeling typical and collaborative work performance and depicting the optimal algorithm of a task. Furthermore, morphological analysis leaves research space to discover cognitive task requirements from the angles of actions as typical elements of the task (i.e., technological units) or typical elements of activity (i.e., psychological units).

QA and AA are not strictly linear and can cause iterations. Altogether, unlike SE, SSAT never uses isolated techniques or procedures (Bedny and Karwowski 2007).

Nevertheless, today's SE and PM seek quick and inexpensive software implementations, which seem to contradict the granular and yet holistic levels that morphological analysis provides. The discovery of cognitive task requirements requires comprehending the complexity of task performance, which may lead to further deeper evaluations of tasks, particularly when it comes to current systems evaluation (Diaper 2004). Hence, TA should clearly be a part of RE. Altogether, the sacrifices discussed cohere with those of conventional TA, which agrees, "only a small part of a system can ever be observed" (Diaper 2004).

CONCLUSION AND FUTURE RESEARCH

Numerous environmental and internal business stimuli, PM that does not look beyond project boundaries, and so forth increase system dependence, impact the probability of the need to change work processes, and can decrease work performance. Subsequent issues altogether question our engineering ability toward computer systems more responsive to human work activity that equally support human development and provide situated assistance. The inquiry hence intended to analyze barriers of complex and sophisticated computer systems in raising efficiency of human-centered problem solving during low-level goal-oriented human work activity and to explore suitable engineering methods as a base for requirements negotiation, decision making, and system design.

To understand our stance today, we inspected low-level operators' work activity in their context within two independent IT-driven organizations in 2005 and 2011. This study differs from most studies that envisage corporate management or high-level business strategies and behaviors.

Despite the time span between both studies, we disclosed compelling environmental and behavioral similarities, constraints, and issues. Regardless of ostensible dissimilarities among both corporations, their operators need better technical systems today that cater to greater personal assistance, individualization, customization, dynamic guidance, and adaptation, and allow personal development.

In both cases, only senior management decides on processes, requirements, and design without operators' participation. This high-level tactic creates unaligned gaps between systems that were decided at management levels and low-level spatial–temporal problem solving during an operator's goal-oriented work activity. Indeed, systems are largely unfriendly, unresponsive, and nonadaptive to individual human goals, changeable tasks, and actions during work activity; cannot provide situated or timely assistance and help; do not enable human development and learning in spite of growing business domain knowledge, information, product complexity, and system complexity; and so forth.

Cognitive tasks originate from human goal-oriented work activity. During spatial–temporal human task performance, human subjects execute cognitive (thinking, decision making, etc.) and motor actions. We discovered that L&T is inseparable from professional activity, human development, individualization, and customization. Indeed, the negative causes discussed nourish the number of heterogeneous, unaligned, unmanageable, and unfriendly systems, system interfaces, functionalities, GUIs, and TUIs. These factors altogether have a negative reinforcement on operators' work performance. One of the causes is that human work activity must adapt to systems.

We hence need clever engineering that fits with early RE, UE, instructional design, and L&T simultaneously; aligns with business, management, and SE; places the human subject and goal-oriented activity at the center; understands the holism of human activity; can cope with the challenge of task discovery; is able to decompose tasks and identify mental and motor actions; and can abstract human needs. The discovery of these granular elements during human task performance provides pragmatic bottom–up arguments for requirements negotiation, decision making, and human-centered system design.

As a viable bottom–up alternative, we postulated SSAT's morphological analysis with its two parts, that is, QA and AA.

SSAT is rooted in AT. Its philosophical and psychological framework helps us to apprehend the social boundary of human activity, that is, the triadic schema. Contrary to behavioral theories that are still common in SE, the unity of human cognition and behavior is one of AT's critical principles. A physical (e.g., mouse) or psychological (e.g., signs, symbols) tool is a mediating artifact in the subject–object relationship. The understanding of the triadic schema within holistic human goal-oriented activity is valuable to model responsive computer-based tasks and their meaningful responses. The study of the dynamic and changeable human activity system hence requires a multidimensional, all-inclusive, and coherent analysis of task performance. From these, it follows that RE needs to equally integrate UE, instructional design, and negotiation techniques. Also, contrary to common practice, PM should no longer constrain solid engineering.

Although we presented advantages and some results from QA, it is AA that transforms a task description into a basic algorithm to depict cognitive task requirements for system design. In all, any method of analysis in SSAT resides under the holistic umbrella of human activity.

However, it may be argued that TA and morphological analysis are time-consuming and, as such, oppose quick software development. After all, if today's systems were flawless and human behavior plain, we could continue on the same route. Our studies through operators' eyes revealed the need for change today.

Future research seeks to formally discover cognitive task requirements using morphological analysis, link to computer-based tasks, elaborate typical and collaborative methods toward task discovery, and embed requirements negotiation techniques.

REFERENCES

J. Annett, "Hierarchical task analysis" in *The Handbook of Task Analysis for Human–Computer Interaction.* Lawrence Erlbaum Associates, Mahwaw, New Jersey; London, 2004.

P. K. Anokhin, "Features of the afferent apparatus of the conditioned reflex and their importance for psychology," *Recent Soviet Psychology*, 1961.

P. K. Anokhin, *The Theory of Functional Systems as a Prerequisite for the Construction of Physiological Cybernetics.* Academy of Sciences of USSR, Moscow, 1962, pp. 74–91.

P. K. Anokhin, "Special features of the afferent apparatus of the conditioned reflex and their importance to psychology," in *Psychological Research in the U.S.S.R.* Progress Publishers, Moscow, 1966, pp. 67–98.

G. Bedny, and W. Karwowski, *A Systemic-Structural Theory of Activity: Applications to Human Performance and Work Design.* Taylor & Francis, Boca Raton, FL, 2007.

G. Bedny, and D. Meister, *The Russian Theory of Activity: Current Applications to Design and Learning.* Lawrence Erlbaum, Mahwah, 1997.

G. Z. Bedny, and S. R. Harris, "The systemic-structural theory of activity: Applications to the study of human work," *Mind, Culture and Activity*, 12(2), 2005, pp. 1–19.

G. Z. Bedny, W. Karwowski, and T. Sengupta, "Application of systemic-structural theory of activity in the development of predictive models of user performance," *Application of Systemic-Structural Theory of International Journal of Human–Computer Interaction*, 24(3), 2008, pp. 239–274.

G. Z. Bedny, M. H. Seglin, and D. Meister, "Activity theory: History, research and application," *Theoretical Issues in Ergonomics Science*, 1(2), 2000, pp. 168–206.

I. S. Bedny, W. Karwowski, and G. Z. Bedny, "A method of human reliability assessment based on systemic-structural activity theory," *International Journal of Human–Computer Interaction*, 26(4), 2010, Taylor & Francis, pp. 377–402.

C. G. P. Bellini, *METRICS, Model for Eliciting Team Resources and Improving Competence Structures: A Socio-technical Treatise on Managing Customer Professionals in Software Projects for Enterprise Information Systems.* Doctoral Dissertation, Universidade Federal do Rio Grande do Sul, Brazil, 2006.

N. A. Bernshtein, *The Physiology of Movement and Activity.* Medical Publisher, Moscow, USSR, 1966.

B. Boehm et al., *Software Requirements Negotiation and Renegotiation Aids: A Theory-W Based Spiral Approach.* ACM, New York, Seattle, Washington, 1995, pp. 243–253.

I. Boivie, J. Gulliksen, and B. Göransson, "The lonesome cowboy: A study of the usability designer role in systems development," *Interacting with Computers*, 18(4), 2006, pp. 601–634, available: http://linkinghub.elsevier.com/retrieve/pii/S0953543805000950.

M. Campo, G. Draper, and J. L. Dutton. *The Economics of CMMI* [Online]. 2009, available: http://www.ndia.org/Divisions/Divisions/SystemsEngineering/Documents/Committees/CMMI%20Working%20Group/The_Economics_of_CMMI.pdf.

P. Carlshamre, and M. Rantzer, "Business: Dissemination of usability: Failure of a success story," *Interactions*, 8(1), 2001, ACM, pp. 31–41.

R. Clark, and G. L. Harrelson, "Designing instruction that supports cognitive learning processes," *Journal of Athletic Training*, 37(4 suppl.), 2002, pp. S-152–S-159.

H. Daniels, "Pedagogy" in *The Cambridge Companion to VYGOTSKY*. Cambridge University Press, New York, 2007, pp. 307–331.
H. Daniels, M. Cole, and J. V. Wertsch, (eds.), *The Cambridge Companion to VYGOTSKY*. Cambridge University Press, New York, 2007.
T. DeMarco, *Requirements Engineering: Why Aren't We Better at It*? IEEE Computer Society, Colorado Springs, 1996, pp. 2–3.
D. Diaper, "Understanding task analysis for human–computer interaction" in *The Handbook of Task Analysis for Human–Computer Interaction*. Lawrence Erlbaum Associates, Mahwaw, New Jersey; London, 2004, pp. 5–47.
D. Diaper, and N. A. Stanton, (eds.), *The Handbook of Task Analysis for Human–Computer Interaction*. Lawrence Erlbaum Associates, Mahwaw, New Jersey; London, 2004.
J. Engström, E. Johansson, and J. Östlund, "Effects of visual and cognitive load in real and simulated motorway driving," *Transportation Research Part F: Traffic Psychology and Behaviour*, 8(2), 2005, Elsevier, pp. 97–120.
Europe's Information Society Thematic Portal, *i2010 Intelligent Car Initiative* [Online]. 2010, available: http://ec.europa.eu/information_society/activities/intelligentcar/index_en.htm.
P. Y. Gal'perin, "Stages in the development of mental acts," in *A Handbook of Contemporary Soviet Psychology*. Basic Books, New York, 1969, pp. 249–273.
D. Gefen, M. Zviran, and N. Elman, "What can be learned from CMMi failures?", *Communications of the Association for Information Systems*, 17(36), 2006, pp. 1–28.
S. R. Harris, *Systemic-Structural Activity Analysis of HCI Video Data*. Aarhus University Press, Copenhagen, Denmark, 2004, pp. 48–63.
IEEE, *Guide to the Software Engineering Body of Knowledge (SWEBOK)* [Online]. 2013, available: http://www.computer.org/portal/web/swebok.
H. Kaindl et al., *How to Combine Requirements Engineering and Interaction Design?* IEEE Computer Society, Barcelona, Spain, 2008, pp. 299–301.
L. N. Landa, "The algo-heuristic theory of instruction," in *Instructional-Design Theories and Models: An Overview of their Current Status*. Lawrence Erlbaum, Hillsdale, 1983.
J. K. Layer, W. Karwowski, and A. Furr, "The effect of cognitive demands and perceived quality of work life on human performance in manufacturing environments," *International Journal of Industrial Ergonomics*, 39(2), 2009, Elsevier, pp. 413–421.
A. N. Leontyev, *Activity, Consciousness, and Personality*. Prentice Hall, Englewood Cliffs, 1978.
A. N. Leontyev, "Activity and consciousness," in *Philosophy in the USSR: Problems of Dialectical Materialism*. Progress Publishers, Moscow, 1977, pp. 180–202.
A. N. Leontyev, "The problem of activity in psychology," in *The Concept of Activity in Soviet Psychology*. M. E. Sharp, Armonk, NY, 1979, pp. 37–71.
R. Moreno, and R. E. Mayer, "A learner-centered approach to multimedia explanations: Deriving instructional design principles from cognitive theory," *Interactive Multimedia Electronic Journal of Computer-Enhanced Learning*, 2(2), 2000, pp. http://imej.wfu.edu/articles/2000/2/05/index.asp.
H. Q. Nguyen, B. Johnson, and M. Hackett, *Testing Applications on the Web: Test Planning for Mobile and Internet-Based Systems*. Wiley Publishing, Indianapolis, 2003.
J. Nielsen, and H. Loranger, *Prioritizing Web Usability*. New Riders Publishing, Thousand Oaks, CA, 2006.
D. A. Norman, *Emotional Design: Why We Love (or Hate) Everyday Things*. Basic Books, New York, 2004.
V. Ponomarenko, and G. Bedny, "Characteristics of pilots' activity in emergency situations resulting from technical failure, #9," in *Human–Computer Interaction and Operators Performance—Optimization of Dork Design, Activity Theory Approach*. Taylor & Francis, Boca Raton, FL, 2010, pp. 223–252.
S. Rishi, B. Stanley, and K. Gyimesi, *Automotive 2020: Clarity beyond the Chaos*. IBM Global Business Services, IBM Global Services, Somer, NY, 2008, pp. 1–28.
W. N. Robinson, S. D. Pawlowski, and V. Volkov, "Requirements interaction management," *ACM Computing Surveys*, 35(2), 2003, ACM, New York, pp. 132–190.
S. L. Rubinstein, "Problems of psychological theory," in *Psychological Research in the U.S.S.R*. Progress Publishers, Moscow, 1966, pp. 46–66.
K. Synytsya, and H. von Brevern, "Information processing and holistic learning and training in an organization: A systemic-structural activity theoretical approach, #15," in *Human–Computer Interaction and Operators Performance: Optimization of Work Design, Activity Theory Approach*. Taylor & Francis, Boca Raton, FL, 2010, pp. 385–409.

H. von Brevern, and K. Synytsya, *Systemic-Structural Theory of Activity: A Model for Holistic Learning Technology Systems*. IEEE Computer Society Press, Kaohsiung, Taiwan R.O.C., 2005, pp. 745–749.

H. von Brevern, and K. Synytsya, "A systemic activity based approach for holistic learning & training systems," *Special Issue on Next Generation e-Learning Systems: Intelligent Applications and Smart Design. Educational Technology and Society*, 9(3), 2006, pp. 100–111.

H. von Brevern, *Support of Cognitive Processes for Corporate Learning and Training*. International Research and Training Center for Information Technologies and Systems, Kiev, Ukraine, 2005, pp. 32–37.

G. Zarakovsky, and W. Karwowski, "Real and potential structures of activity and the interrelationship with features of personality, #13," in *Human–Computer Interaction and Operators Performance—Optimization of Work Design, Activity Theory Approach*. Taylor & Francis, Boca Raton, FL, 2010, pp. 329–360.

16 The Problem Statement
Issues, Importance, and Impacts on Software Development

Hansjörg von Brevern

CONTENTS

UNDERLYING ISSUES

DeMarco (1996) observed that "ill-specified systems are as common today as they were when we first began to talk about requirements Engineering." One major problem child at the threshold from business to software development is the form of requirements and the manner of how well they are formulated by business for software engineering. Subsequently, ambiguous descriptions foster software failures. Throughout the various stages of software development, different stakeholders want different levels of abstraction and views, which, altogether, a software development approach and its models need to reflect. The very essence is the problem statement (PS) that is echoed within multifarious layers of software engineering, encapsulates notions at variform degrees of granularity, and is thus central to business and engineering layers. The PS is a textual description in the form of a narrative like a well-crafted, well-focused, nonverbose, and grammatically correct essay written in well-structured language to capture the subtlety of the problem in detail. Contrary to the existing practice in the real world, engineering processes necessitate meaningful and undisguised descriptions. The almost forgotten PS as known since the 1970s (cf. Nunamaker et al. 1976) and especially when coupled with today's object-oriented (OO) software development concept reproduces a systemic rather than the conventionally known functional paradigm. As a result, such a dyadic approach at the outset of software development will contribute to truly effective and desired reactive, goal-oriented, and user-centered systems. Therefore, the PS needs to be revivified in software engineering.

BUSINESS CHALLENGES AND STIMULI

The competitive marketplace and the extremely fast-moving pace of today's economy—affecting changing goals, time planning, and costs—challenge commercial organizations greatly. These factors ultimately impact software analysis and design. Practice shows that sudden and rapid modifications befall the quality of requirements descriptions. In terms of software engineering, abstract business requests are too high level, unspecific, and meaningless. Typically for projects, requirements are dashed off unsystematically. For example, one sole (and often overstressed) business operator, a project manager who has never practically operated the system, or a requirements engineer who must meet time constraints to keep everything "green" may even have to write them.

At another instance, we may find organizations that have realized the deficiency of poor requirements and learnt from failed software projects resulting in high cost. As a countermeasure, business strategists overemphasize project management, presuming it to be in control of accuracy, cost, and timely delivery. Nevertheless, it is all about accuracy of requirements formulation where project management misses the mark and largely ignores refined specifications. Yet, the latter reside at the very interface with the software engineering (requirements engineering) at the first stage. (Throughout this chapter, the terms requirements engineering, software engineering, and software development are blurred.) Maybe, it is because project management is closer to management methods that may appear to be more tangible than the seemingly "intangible" software development methods and models. Despite the overdose of meetings that most project management methods stimulate, requirements specifications still remain poor, contain unaddressed vague assumptions, suffer from constant changes, leave insufficient time for proper requirements engineering processes, and the like. Contrary to expectations towards project management, its exaggerated utilization not only produces more documents at management level and increases costs but also imposes more stress on software development life cycles and people—unfortunately, often at the cost of the time needed for requirements elicitation.

So, Taylor and van der Hoek (2007) argue, "Designing a software application involves designing its structure as well as its user-observable properties, functional and nonfunctional alike. By removing the counterproductive boundary between requirements and design, a holistic view of product conception emerges." Hence, for the reason that software development is high risk, as the countless documented reports of software project failures prove, it is time for corporations to make a shift of paradigm by investing time and dedicated engagement in well-formulated PSs to be used by software engineering processes.

ENGINEERING CONTEXT OF THE PS

The software development method utilized determines the viewpoint, use, language, and the like of the PS, as we will argue in the forthcoming sections. So, we need to contemplate deltas between the traditional structured approach that still pertains in the minds of most organizations and engineers vis-à-vis today's OO approach.

The popular structured approach of the 1970s was supported by procedural programming languages, is centered on the system's functional views, and uses different models at various stages of the development process. Typically, the structured approach considers feature and requirements lists to exemplify system needs. Contrary to a PS that entails an entire problem domain, an individual feature and requirement is rather a context-independent static snapshot in time to capture intended system behavior. This, in return, gives rise to a requirements traceability matrix. Hence, each single change of a feature or requirement necessitates careful requirements management with its documents and simultaneously risking severely impacting layers of the typical waterfall model. In the worst case, changing features and requirements can put software development at stake because the conventional waterfall model lacks iterations. Likewise, changing features and requirements bear the danger of overemphasizing detailed features so that the contextual view drifts out of sight—notwithstanding that neither feature and requirements lists nor requirements traceability matrices are capable of modeling the context from a mere user perspective. In terms of large system development, a rapidly growing feature and requirements list can easily become unmanageable, laborious, and unfocused.

Contrary to the functional approach, the starting key focus from an OO perspective seeks to identify what the system under development (SuD) will do, but not how it will do it; technical details used to implement the tasks do not need to be specified until the design stage.

TABLE 16.1
Comparison between "Functional Requirements and Features" and Parts of a PS

#	List of Features and Functional Requirements	Extracts of a PS
1	The ability to generate reports	The store manager will be able, at any time, to print a summary report of sales in the store for a given period, including assignment of sales to sales assistants in order to calculate weekly sales bonuses.
2	Cannot be derived	A league is a group of teams that compete against each other. Each team recruits members to participate in the contests.
3	Supports basic internet-based purchasing functionality	Cannot be derived.

More practically, Table 16.1 contrasts examples of possible independent feature and functional requirements lists and extracts of a PS. Notably, Table 16.1 illustrates only extracts of a PS since the inclusion of a whole PS that describes one problem domain would exceed the boundaries of this chapter. #1 of Table 16.1 illustrates that a feature and functional requirement can be derived from the extract of the PS, which, on the contrary, cannot be derived from a feature or functional requirement, as shown in #3 of Table 16.1. From the extract of a PS of #2 of Table 16.1 neither a feature nor a requirement can be defined. The description of this possible extract of a PS is, however, necessary because it describes dependencies within a given context, which altogether is critical to derive classes during analysis and modeling of the static structure.

Therefore, the OO approach eliminates major weaknesses of the structured approach and its waterfall model where changing functions require adaptations within all sequential layers along software development and its models and also no dynamic view exists. For the reason that OO models a software as a collection of collaborating objects, it allows, for example, interaction with other objects and the development of consistent models; it is thus more cost-efficient than the structured approach. With a PS in situ and embedded into the OO paradigm, conventional feature and requirements lists and traceability matrices become redundant. Procedural design, its feature and requirements lists, and traceability matrices aim at global system knowledge and control. The same phenomenon is reflected in the area of human computer interaction (HCI) where "most of the approaches… typically concentrate on the systematic description of the human–computer interface but not on human performance itself as a system" (Bedny and Karwowski 2003). In brief, procedural analysis and design, its feature and requirements list, and traceability matrices are mainly system solution oriented. Moreover, the structured approach allows system requirements as context-independent "islands." This, however, contrasts with cognitive thinking processes because the "… thinking process reconstructs reality into a dynamic model of the situation" (Bedny et al. 2003). So, the "… pitfall of attempting to resolve human activity by conventional engineering is that it attempts to model systems on 'stimulus–response' behaviors. This approach largely ignores cognitive mechanisms of humans and with them the embedded roles, context, tasks, and goals of technical systems. Instead, the answer of how to envision, understand, and model technical systems should be explored in human activity and its manifestation. From this follows that technical systems ought to present content according to how humans process it during work.... Ergo, we need a methodology that helps us formally analyze and decompose work… as an embedded scrutiny, while an engineering methodology that enables us to extract identifiable (technical) system events and responses to actions from the context of the higher-order activity system is equally indispensable, followed by… engineering methods to ultimately model technical solutions" (von Brevern and Synytsya 2006). Such holistic views leave research space and issues of practical applications on, for example, interconnecting human activity systems with machine systems, verbalized and nonverbalized task descriptions, task analysis, and engineering methods that should be adaptive to, responsive to, integrate, and model human activity. On the one hand, the systemic-structural theory of activity (SSTA) is key to analyzing cognitive user tasks; on the other, from an engineering perspective, only OO analysis (OOA) and OO design (OOD) together with the PS in response to contextual task description can take on a user's stance—as discussed earlier. Altogether, the aspects discussed are systemic, holistic, user oriented, contextual, and domain dependent, although this chapter focuses only on OOA/OOD in combination with the verbalized PS.

RATIONALE OF THE PROBLEM DOMAIN

Understanding the problem domain is critical because its goal is to determine required SuD functionalities and the ways in which they will be used. This understanding is critical regardless of the type of SuD.

The problem (or subject) domain is the surroundings in which the proposed system will operate. "The subject domain of a reactive system is the union of the subject domains of all messages that cross its interface. To find out what the subject domain of a system is, ask which entities and events the messages sent and received by the system are about. To count as elements of the subject domain, these entities must be identifiable by the system" (Wieringa 2003).

The goal of the requirements engineering phase during software development is to close and specify the gap between requirements of the outer world and the inner world, that is, software design by circumferencing the problem domain. A mutually shared representation between business and software development requires a preliminary understanding of both software development methods and analysis of the problem domain to specify explicit functionalities, inclusions, and exclusions of the SuD. Analysis of the problem domain includes business work flow analysis and the written PS. Therefore, Tsang et al. (2005) demonstrate the need for the PS during OOD as follows: "Since the objective of domain analysis is to develop a class model that can be reused in other applications to solve problems in the same domain, it will be expedient that the problem statement describes the general requirements of the domain rather than the requirements of a specific application. The problem description should, therefore, focus on the description of the objects and their relationships in the domain rather than the specific procedures of the problem domain, since the procedures for carrying out tasks would not be the same for every organization" (Tsang et al. 2005). Also, objects form oneness: "One of the distinguishing features of object design is that no object is an island. All objects stand in relationship to others, on whom they rely for services and control" (Beck and Cunningham 1989).

Importantly, the problem domain does not attempt to meet mythical future needs but extends Wieringa's (2003) argument that "only the union of the subject domains of all messages that presently cross its interface" respectively "presently are to cross its interface."

RATIONALE OF THE PS

According to Moreno (1997), one of the main limitations attributed to OO by software engineers is the immaturity of OOA processes. Unfortunately, most of today's conventional software development literature ignores the importance of the PS and fails to clearly embed it within software development processes, link it to subsequent software engineering processes and their viewpoints, and oversee the ideal symmetry between the PS and OO. Yet, once we know the boundaries of the problem domain, we will be able to draft an unequivocal PS, which, in turn, will enable us to conduct use case modeling, textual analysis for static structure modeling, and behavioral modeling.

As seen earlier, from an OO perspective, the PS is to reflect a user's goals entailing user–system interactions that address "what" the SuD should be able to accomplish from a mere user perspective as opposed to "how" the SuD should be built. On the basis of our previous discussion and according to the argument by Beck and Cunningham (1987), a PS consists of at least one (or more) concise statement(s) of a problem written in the jargon of the problem domain, a summary of circumstances creating the problem written, and a user-oriented (and not functional system-oriented) solution that works in those circumstances. Once the problem domain has been identified, selected OO teams consisting of at least a business domain expert, an analyst, a facilitator, and a copy editor transcribe the PS on which software development depends, as illustrated in Figure 16.1.

Making a PS requires effort and may take a substantial amount of time (e.g., up to several months) because of iterative cycles. The PS of the problem domain directly impacts use case modeling and static modeling and indirectly impacts behavioral modeling, as illustrated in Figure 16.2. The PS is thus an underlying and critical tool throughout various phases of software development; for example, the PS greatly facilitates deriving but also counterevaluating test case scenarios.

In addition, Figure 16.2 reveals, the more indeterminate domain modeling has been, the more iterations may evolve, so the longer and costlier software engineering processes will become.

Hence, the PS is extremely critical in view of domain analysis and subsequent software modeling. This also means that we find varying degrees of detail of the PS because the lower the level of software engineering, the more granularity is required. Therefore, disentangling implicit and explicit behaviors of the problem during analysis and design is crucial. This, in turn, quests for methods of how to craft a PS on the one hand and textual analysis on the other.

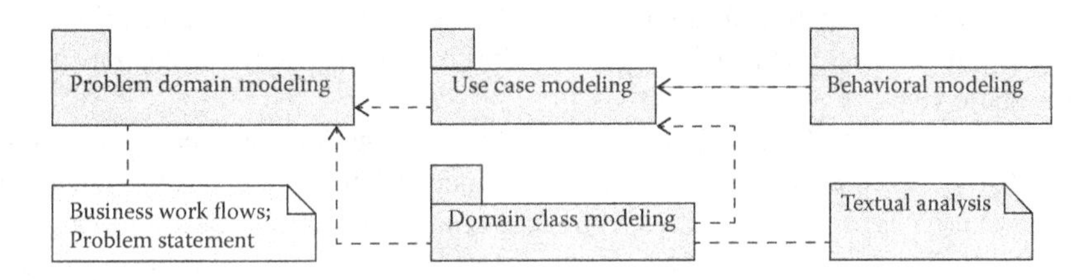

FIGURE 16.1 Problem domain modeling and its dependents.

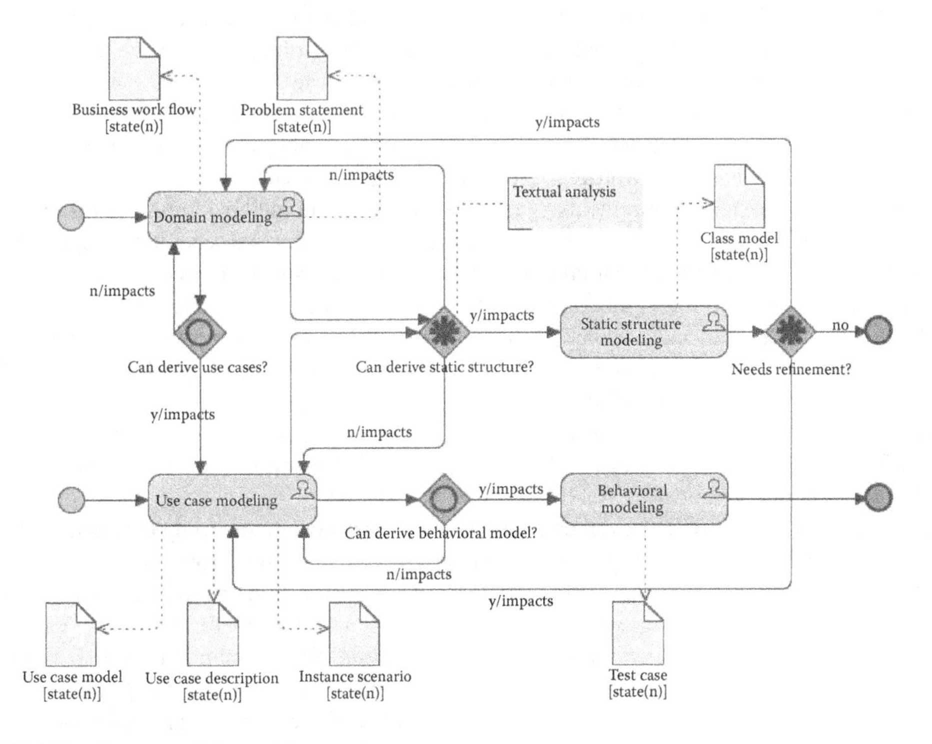

FIGURE 16.2 Impacts of the problem statement.

PS APPLIED

According to Conger (1994), the first four steps in OOA include the following:

1. Develop a summary paragraph that consists of noun–verb, noun–verb–object, and verb–object variations
2. Identify objects of interest by underlining all nouns in the summary paragraph
3. Identify processes by circling all verbs in the summary paragraph
4. Define attributes of objects by identifying adjectives or adjectival phrases and describing nouns that are objects in the solution space

More precisely, Cockburn (2000) details, "Noun... verb... direct object... prepositional phrase... I mention this simple matter because many people leave off the first noun, when they start writing use cases, write grammatically incorrect sentences. If you leave off the first noun, it is no longer clear who has the ball... If your sentence is incorrect, the story gets hard to follow. A good use case is easy to read and follow" (Cockburn 2000). Conducting textual analysis by deriving nouns, pronouns, and noun phrases during OOA/OOD from information specifications is thus not a new idea. It was posited by Abbott (1983) and then popularized by Booch (1986). Textual analysis falls into place when, for example, it comes to analyzing the PS and/or using case descriptions to perform object identification, as elucidated by Tsang et al. (2005). Textual analysis thus has an important impact on the static view.

Research done within the fields of linguistic analysis and natural language processing (NLP) seeks to formally parse sentences written in natural language. These fields categorize text input for requirements into use case scenarios, text specifications, and unrestricted descriptions.

Use case descriptions and their scenarios are part of the most simplified form of natural language. Although use case descriptions and their scenarios are not exclusive OO artifacts, they are integral to the iterative OO processes of the Rational Unified Process. The algorithmic and simple style of the stories is used to represent exact sequences of activities based on simple sentences. Often, scenarios consist of "if... then" sentence clauses. The volume of one scenario should not exceed 10–15 sentences. Software specialists (like analysts) write scenarios for other software specialists (like programmers) along the value chain of software development.

Text specifications are considered to be the most accurate natural language because they are cleaned out from unnecessary wording, are grammatically correct, and both logically and consecutively describe system requirements. Their scientific style should be comprehensible and exact, based on a simplified syntax and unobstructed semantic references, and encompass approximately 50 sentences for a small system. Domain experts write them for software specialists.

Unrestricted descriptions are not constrained by any particular style and restrictions. Often, they are the result of interviews and are characterized by repetitions, redundant wording, synonyms, and fragmentations. Sentence syntax is arbitrarily complex and long, in the range of approximately 100 sentences. Authorship and readership thus vary.

More formally, methods exist to formalize the OOA process based on analyzing linguistic information from informal specifications. Information deduced is composed of words, which, in turn, denote elements for OO modeling, such as classes, objects, properties, actions, behaviors, actors, a data dictionary, and the like. For example, Moreno (1997) proposes a mapping between the linguistic and conceptual worlds by having a set of predetermined structures from natural language to a set of grammar with defined syntax and semantics. Reasons for doing so are the potential unlimited ambiguities of natural language, vague assumptions, and unclear formulations. Various formal methods of textual analysis exist, like linguistic analysis, NLP, semantic networks, ontologies, knowledge discovery, and mining techniques, albeit extremely complex and complicated to use.

Influenced by the underlying work in linguistics and NLP, heuristics were developed that OO has adopted to alleviate domain and use case analysis based on textual analysis (cf. Rumbaugh et al.

1991; Tsang et al. 2005). Yet, contrary to formal methodologies, mere heuristic methods do not primarily aim at eliminating ambiguities of the natural language. An enticement of formal methods is, however, to commence with a PS and conclude it prematurely. Doing so tends to hide arguments and ignores the importance of iterative cycles (cf. Figure 16.2). Moreover, a method may restrict discoveries during elicitation, leading to possible misconceptions that only reflect the method and starting positions.

Probably one of best-known informal methods that have been at the forefront for over two decades has been the class, responsibilities, and collaboration (CRC) cards of Beck and Cunningham (1989). CRC helps to identify the role of an object on three dimensions; this method can be applied during static structure modeling. Although the authors of the CRC method did not address the PS, the importance of the context, naming conventions, and correctness of language cannot be missed.

"The class name of an object creates a vocabulary for discussing a design... object design has more in common with language design than with procedural program design... We urge... to find just the right set of words to describe our objects, a set that is internally consistent and evocative in the context of the larger design environment" (Beck and Cunningham 1989). Our extract of a PS as in #2 of Table 16.1 would therefore result in the candidate class "league," "team," "member," and "contest."

"A responsibility serves as a handle for discussing potential solutions. The responsibilities of an object are expressed by a handful of short verb phrases, each containing an active verb. The more that can be expressed by these phrases, the more powerful and concise the design. Again, searching for just the right words is a valuable use of time while designing" (Beck and Cunningham 1989). In our example #2 of Table 16.1, one responsibility of the object "team" is to recruit members.

And last but not least, "We name as collaborators objects which will send or be sent messages in the course of satisfying responsibilities" (Beck and Cunningham 1989). Therefore, our object "team," as in #2 of Table 16.1, collaborates with the object "member."

The arguments presented on classes, responsibilities, and collaborators cohere with our knowledge of the problem domain as discussed earlier. None among feature lists, functional requirements lists, traceability matrices, and uncoupled requirements descriptions as known by the functional approach can satisfy those needs for CRC analysis or those by the problem domain.

Nonetheless, there is hardly any commercial organization that really exercises heuristics and, even much less, that applies formal methods for textual analysis during OOA/OOD. Despite the undeniable importance of textual analysis, it is extremely rare to find any organization that employs well-formulated and structured narratives. Introducing this at a corporate level requires business to be actively integrated into requirements engineering processes by equally bearing responsibilities to deliver well-formulated and structured narratives based on formal and informal methodologies. Consequently, this postulates training for business and engineering.

Notwithstanding that there are no ready-made templates for OOA, OOD, and the PS that could be used anywhere at any time to design a well-defined system, we need to start the software engineering process by seeking ways of how we describe the problem domain clearly in the form of a PS, unambiguous, well structured, well formulated, well focused, and grammatically correct. These arguments are critical because even if the PS were complete and consistent, this would "... not guarantee that the problem was defined correctly, i.e., the wrong requirements may have been stated properly" (Nunamaker et al. 1976).

To therefore antagonize incorrect PSs at their root of origin, software engineering paradigms and their software development methodologies as well as, probably, project management methods need to follow new paths that include formal methods of human activity systems and cognition. One aspect needs to look at how to evolve holistic reactive systems from given systems; another challenging aspect needs to look at building new systems. While the first aspect is evolutionary, the second one is revolutionary. Yet, both share the same central challenges, that is, task description and task analysis; both aspects are ideal candidates for explorative research applying SSTA. Our holistic and systemic field of enquiry is paramount and starts at the forefront: "The formulation of a problem is often more essential than its solution, which may be merely a matter of mathematical or experimental skill" (Moreno 1997).

CONCLUSION AND FUTURE RESEARCH

Commonly, ambiguous requirements, the lack of business for a holistic view of software development processes, and the immaturity of OOA processes foster software failures. While business and project management seek to cope with rapid environmental changes and management issues, they are ignorant of software development approaches, their models, dependencies, and stimuli from unsatisfactory requirements that critically influence software analysis and design.

The often lacking or vague PS yet is key to counteract ill-specified systems at the doorstep from business to software engineering, whose processes necessitate meaningful and undisguised descriptions to effectively design desired reactive, goal-oriented, and user-centered systems. Although the PS is not a new idea to software engineering, conventional software engineering literature still ignores its importance and fails to clearly embed it within software development processes and link it to the various specification and software development processes within the software engineering life cycle.

The traditional structured approach produces features and functional requirements lists. Contrary to the meaningful and contextual descriptions of the PS and as needed by OOD, features and functional requirements list are not meaningful. Moreover, features and functional requirements can be isolated from each other so that traceability matrices have become the norm of project management methods that use the traditional approach. However, with a PS in situ embedded into today's OOA/OOD, conventional features and requirements lists as well as traceability matrices become redundant because OOA/OOD does not know isolated objects and for the many reasons that we have discussed in this chapter about the comparisons between yesterday's functional approach versus todays OO software development. For the reason that the software development method determines the viewpoint, use, language, and the like, we have illustrated the impacts and benefits when coupling OOA/OOD with the PS. While the problem domain is the very and only extent of each PS, the PS per se will be echoed at different degrees of granularity within the layers of OOA/OOD. Therefore, the PS must be well crafted, well focused, grammatically correct, and written in a well-structured language to capture the subtlety of the problem in detail.

Within the engineering context of software development, we have debated informal methods and heuristics of textual analysis, illustrated examples, and addressed the existence of formal methods. Unfortunately, corporate environments hardly benefit from informal methods and heuristics, while formal methods of textual analysis are almost not applied.

This chapter would be incomplete if we had not reasoned about the impossibility, from an OOA/OOD engineering standpoint, of fully being able to guarantee that the PS is correct with regard to what users expect the SuD to be. Hence, we conclude that we need to shift to holistic and systemic software development—and probably also project management—to integrate cognitive aspects of the human activity system. We propose the formal methods of task analysis and task description of SSTA in view of system evolution one the one hand and system revolution on the other.

REFERENCES

Abbott, R. J. (1983). Program design by informal English descriptions. *Communications of the ACM*, 26(11), 882–894.

Beck, K., and Cunningham, W. (1987). Using pattern languages for object-oriented programs. Object Oriented Programming, Systems, Languages, and Applications (OOPSLA 1987); Workshop on Specification and Design for Object-Oriented Programming, CR-87-43.

Beck, K., and Cunningham, W. (1989). "A laboratory for teaching object oriented thinking." Paper presented at the Object-Oriented Programming, Systems, Languages, and Applications (OOPSLA 1989), New Orleans, Louisiana.

Bedny, G., and Karwowski, W. (2003). A systemic-structural activity approach to the design of human–computer interaction tasks. *International Journal of Human–Computer Interaction*, 16(2), 235–260.

Bedny, G. Z., Karwowski, W., and Jeng, O.-J. (2003). "Concept of orienting activity and situation awareness." Paper presented at the Ergonomics in the Digital Age, IEA Congress 2003, Seoul.

Booch, G. (1986). Object-oriented development. *IEEE Transactions on Software Engineering*, 12(2), 211–221.

Cockburn, A. (2000). *Writing Effective Use Cases*. The Crystal Collection for Software Professionals. Reading, MA.

Conger, S. A. (1994). *The New Software Engineering Course*. The Wadsworth Series in Management Information Systems. Belmont, CA: Wadsworth.

DeMarco, T. (1996). "Requirements engineering: Why aren't we better at it?" Paper presented at the 2nd International Conference on Requirements Engineering (ICRE 1996), Colorado Springs.

Moreno, A. M. (1997). "Object-oriented analysis from textual specifications." Paper presented at the Ninth International Conference on Software Engineering and Knowledge Engineering (SEKE 1997), Madrid.

Nunamaker, J. F., Jr., Konsynski, B. R. J., Ho, T., and Singer, C. A. (1976). Computer-aided analysis and design of information systems. *Communications of the ACM*, 19(12), 674–687.

Rumbaugh, J., Blaha, M., Premerlani, W., Eddy, F., and Lorensen, W. (1991). *Object Oriented Modelling and Design*. New Jersey: Prentice Hall.

Taylor, R. N., and van der Hoek, A. (2007). "Software design and architecture—The once and future focus of software engineering." Paper presented at the 2007 Future of Software Engineering, Washington, D. C.

Tsang, C. H. K., Lau, C. S. W., and Leung, Y. K. (2005). *Object-Oriented Technology: From Diagram to Code with Visual Paradigm for UML*. Singapore: McGraw-Hill.

von Brevern, H., and Synytsya, K. (2006). A systemic activity based approach for holistic learning and training systems. Special issue on next generation e-learning systems: Intelligent applications and smart design. *Educational Technology and Society*, 9(3), 100–111.

Wieringa, R. J. (2003). *Design Methods for Reactive Systems: Yourdon, Statemate, and the UML*. San Francisco: Morgan Kaufmann.

17 The Influence of Individual Personality Features on Human Performance in Work, Learning, and Athletic Activity

Fred Voskoboynikov

CONTENTS

INTRODUCTION

There are two ways to ensuring the effectiveness of human performance. One is by professional selection, the so-called "screening out" of individuals with specific attributes. The other one is individual training methods directed toward the formation of individual strategies of activity based on features of personality of the individual in the process of adaptation to the objective requirements of activity. While in the West, the selection method was used more intensive, in the former Soviet Union, the attention mostly was directed toward the development of methods for individual training. The concept of individual style of activity was first introduced by Merlin (1964, 1986) and Klimov (1969). They were able to establish that different individuals can perform the same work with equal efficiency through the use of their own individual style of performance, which is more suitable to their personality features. People attempt to compensate for individual weaknesses with their personal strengths in a given task situation; that is, by implementing the individual style of activity, they diminish the impact of their negative features of personality on performance. Among other authors who studied the effect of different personality features on performance were Bedny and Voskoboynikov (1975), Merlinkin (1977), Bedny and Seglin (1999), and others.

The study of personality and individual differences is a critically important area of activity theory. The central notion in this area of study is the individual style of activity that connects features of personality with mechanisms of self-regulation and strategies of performance. According to the systemic–structural activity theory, human activity is based on principles of self-regulation, which occurs at the conscious and unconscious levels (Bedny and Karwowski 2004). Both levels are tightly interconnected and transform from one to another. Self-regulation of activity manifests itself in the way people, through trials, errors, and feedback corrections, create strategies of performance suitable to their individuality. Thus, from the point of view of systemic–structural activity theory, individual style should be considered as strategies of performance deriving from the mechanism of self-regulation, which depend on personality features (Bedny and Voskoboynikov 1975; Bedny and Seglin 1999).

Though individual style is the most effective method of adaptation of an individual to the objective requirements of activity, it has its limitations. The limitations take effect when individual features of personality are not up to the requirements of activity. In such cases, the only way to connect activity to the individual is by professional selection. In this work, we will demonstrate, based on our experimental studies, how different personality features affect the mechanism of adaptation to the objective requirements of activity.

ADAPTATION OF INDIVIDUALS TO THE OBJECTIVE REQUIREMENTS OF ACTIVITY

Individual style of performance may be shaped both consciously and unconsciously. It is important to identify the individual style of activity of a particular individual who interacts with the task situation. Through individual style, the subject can adapt to a situation more efficiently. It should be distinguished, however, that individual style of activity and methods of performance are not the same. The latter is dependent not upon individual features of personality but rather upon organizational factors, imposed supervisory procedures, and so forth. Sometimes, these methods of performance that derive from organizational factors may contradict with the individual features of personality, and that is not desirable. In cases of inadequate training, which ignores individual features of personality, the subject acquires methods of performance that contradict with his/her individuality. It may negatively affect satisfaction on the job. Implementing individual style of performance is important not only in different production situations but also in learning and training. Through individual style of performance, students comprehend situations better and acquire new knowledge and skills.

In studying the individual style of activity, it is important to observe how people with different individual characteristics acquire the same knowledge and skills. On the other hand, it is as important to identify how subjects disintegrate into distinct groups with respect to their ability to acquire skills. Such disintegration takes place as the capacity of some individuals to adjust to the requirements of activity is reduced due to the increased task complexity. In simple situations, individuals exhibit similar levels of achievement regardless of their individual style of performance. In such situations, it is hard to identify individual differences; however, when the task becomes more complicated, individuals begin to vary more in their performance. For example, an average gymnast may perform comparably with a high-level gymnast in a relatively easy task; however, when they attempt to perform a more complex task, the average gymnast will not be able to maintain the task at the same level of efficiency as the high-level gymnast would. In the experiments described below (Bedny and Voskoboynikov 1975), we will illustrate the aforementioned statements.

One difficulty in the study of this phenomenon in the experimental setting is ensuring that participants are not aware of what the experiment is trying to measure. Considering this factor, we chose for our experiment elementary school students, who we felt would be less attuned to the goal of the experiment. We selected 10 students who had completed the first grade and, according to the program, were supposed to master addition and subtraction of single-digit numbers. Upon their teacher's evaluations of their mathematical ability, they were split into three subgroups: superior (three exceptionally gifted students), average (four students), and poor (three weak performers, who were being considered for special educational placement). The question at issue was how quickly they would perform in repetitive mathematical calculations. For this purpose, we used the simplest mathematical tasks presented in a written form—additions and subtractions of single-digit numbers. The subjects were instructed to alternate their tasks from additions to subtractions. The experimenter worked with each subject individually in order to make careful behavioral observations. If the subject made errors, the experimenter signaled that, and the subject was to correct them. Time of task performance was measured.

The students performed each task three times per day over a period of 3 days. In total, they performed 30 tasks. Based on the obtained data, we generated learning curves for each individual. At the beginning of the experimental study, weaker subjects spent up to 40 min to complete the task, whereas superior subjects spent less than 4 min, and average students spent between 7 and 29 min. Based on the results of the first days of experiment, we assumed that these differences would be sustained, which could be explained by the significance of their individualities. However, the overall experimental results were contrary to our expectations. In the following days, weaker subjects sharply increased their performance speed. Subjects with superior ability exhibited a much flatter learning curve, showing only a slight improvement, suggesting that they were near the ceiling at the outset. They reached their best stabilized results at 13 trials. Average and weaker subjects continued to reduce their task performance time, approaching superior students' performance, but by the 18th and 19th trials, both groups stabilized the times of their performance, and no further improvements were noted from the 19th to the 30th trial. The initial range of the times for all students at 37 min, 10 s, to 3 min, 30 s, was substantially reduced to 3 min, 40 s, at the completion of the experiment.

Next, our task was to explain how, through training, subjects with such substantial differences in their mental development improved their performance. We compared our observational data, analyzed the results, and discussed them with students. Our observation demonstrated that at the first stages, subjects with weaker abilities used their fingers as well as external speech. It meant that their mental operations depended on external practical actions. Only after multiple executions of the task with the experimenter's help did they start to rely exclusively on internal mental operations. Subjects with average ability also had an initial tendency to rely on external practical actions to perform the task. Their calculations were accompanied by whispers that were barely audible; that is, they facilitated slow mental operations with some external actions. The students with superior abilities from the outset started their calculation without any relations to external practical actions.

To continue analyzing the data, we conducted a brief control experiment. Subjects were instructed to perform a similar task with new numbers, four trials in a single day. The results demonstrated that the time of performance for the weaker students and relatively for the average students "jumped" back quite high. Only the superior students demonstrated practically the same results as in the last trial of the first experiment. Upon the results of our brief experiment, we concluded that the students of the weaker group utilized rote memorization in perfecting their performance. For average students, adaptation manifested itself in the refinement of external and internal calculative mental actions. The superior students immediately focused on consolidating their internal mental operations. It is noteworthy that memorization at different degrees occurred in all the subjects' performance during the first series of the experiment.

Taking the results into consideration, we decided to test the strategies of the subjects by increasing the complexity of the task, thus reducing the viability of memorization. In our third experiment, we asked the subjects to perform addition and subtraction of two pairs of numbers. If the preliminary result of the first operation was greater that the results of the second operation, the subjects then were to subtract the second from the first. If the second was greater than the first, then they were to add them. In order to frustrate rote memory, the same numbers were used in different positions across several versions of the task. The total number of trials was 30. The results of this experiment demonstrated that adaptation abilities were more restricted compared to the first experiment. It was seen in the analysis of the curves of acquisition. That is, the subjects of weaker and average groups needed the experimenter's assistance during first few trials of performing the task. Only the superior students were immediately able to perform the task without any assistance. They demonstrated a high level of performance from the start and within 15 trials stabilized the times of their performance at 1 to 1.5 min. Subjects of the average group stabilized the time of their performance at 4 to 6 min within 20 trials. The weaker students also stabilized the time of their performance within 20 trials, at 8 to 10 min. After 15 trials for superior students and 20 trials for average and weaker ones, there were no improvements in any group.

Compared to the first experiment, where all subjects converged to a similar level of performance, this experiment showed three diverse groups with different levels of performance. Thus, we could observe that in performing relatively simple tasks, subjects were able to reduce differences in performance quite noticeably, whereas increased complexity of the task requirements resulted in distinct groups with distinct levels of performance.

STUDY OF THE INDIVIDUAL STYLE OF ACTIVITY

All kinds of human activity present more than one objective requirement to people in order to perform; this allows different individuals to rely on their personal strength to compensate for individual weaknesses. Such a strategy occurs at the conscious and unconscious levels and is based on principles of self-regulation. The process of self-regulation manifests itself in the formation of desired goals, in developing a program of actions that corresponds with these goals, in conditions for achieving these goals, and in the person's individual abilities (Tomaszewski 1975). When studying the individual style of activity, one should distinguish between the microstructural or analytical and macrostructural or systemic approach (Bedny and Seglin 1999). A microstructural or analytical approach involves the study of isolated features of personality and how these features influence individual aspects of performance if they are critically important for a particular task. A macrostructural or systemic approach is one in which a more salient relationship among different features of personality and individual style of activity is discovered. To illustrate a macrostructural approach in the study of individual style of activity, we will first analyze an experiment by Merlinkin (1977) where he used a microstructural approach according to our terminology. He studied the comparable individual styles of performance of gymnasts with "flexible" or "inertial" nervous systems, qualities of the nervous system that represent two opposite dimensions of mobility. Mobility is associated with the "speed of reconditioning" as a reaction of an individual to changing stimuli. The opposite of "mobility" is "inertness," a quality of reduced mobility. In his experiment, gymnasts of both groups were asked to perform three consecutive forward rolls, finishing at the upright position. He discovered that both groups performed the task correctly, but each group utilized a distinct individual style of activity. Gymnasts with a "flexible" nervous system performed all three forward rolls at the same speed and immediately stopped. Those with an "inertial" nervous system performed each forward roll at a different speed. The first forward roll they performed quickly; the second one relatively slower; and on the third one, they were visibly slowing down in order not to miss the required finish into the upright position. The demonstrated different individual styles of performance were done unconsciously.

Bedny and Seglin (1999) emphasized that the study of individual style of activity must be from the position of the systemic (macrostructural) approach, where not only isolated features of personality but also their relationship influence individual style of performance. From this point of view, we can observe that such features of personality as ability, past experience, level of motivation, anthropometrical peculiarity, and so forth may overshadow features of the nervous system and temperament. It should be noted that temperament is a quality of an individual that depends on features of nervous system. With that approach, we conducted an experiment similar to Merlinkin's but with some important differences. Two groups of subjects were selected. One group included 16 highly experienced gymnasts and tumblers with an "inertial" nervous system (hereinafter referred to as "gymnasts"). The other group consisted of 18 track-and-field sprinters with a "flexible" nervous system (hereinafter referred to as "nongymnasts"). The latter had some experience with forward rolls from general physical education classes. The subjects' features of the nervous system—"flexible" or "inertial"—were identified through our long-time observations and through Hilchenko's (1966) experimental method.

Both groups were asked to perform three different tasks: three forward rolls at their own speed, three forward rolls into a precise stop, and three forward rolls into a precise stop blindfolded. Each of them had to perform each task three times, and then the results of each subject were averaged.

All their performances were recorded with a 16-frame/second movie camera. It was discovered that in all the series, gymnasts with an "inertial" nervous system performed all forward rolls with maximum speed. At the same time, nongymnasts with a "flexible" nervous system slowed down the second and particularly the third forward roll. The most significant slowing down of the last forward roll in the nongymnasts' performance was observed in the "blindfolded" conditions. The comparison of this study with Merlinkin's demonstrated that both "his" gymnasts with an "inertial" nervous system and "our" nongymnasts with a "flexible" nervous system used similar individual style in performing the same task. Thus, we could conclude that, if in his experiment, the slowdown of the final forward roll was caused by the "inertial" nervous system, in our experiment, it was caused by the lack of past experience.

To further demonstrate that the individual style of performance might be based on other features of personality rather than on the features of the nervous system, we will consider the following example. There is a gymnastics skill that is performed on the rings called *a kip mount* (Figure 17.1). This skill requires a certain level of physical strength and mobility of the nervous system. Short gymnasts with much-developed physical strength and tall gymnasts who are not that physically strong relative to their body size are able to perform the skill equally successful by using different techniques. Below is a description of the basic technique to perform the skill.

A gymnast, from an inverted straight body hang, quickly descends into an inverted pike hang and then powerfully kicks his legs up toward the rings and slightly out and at the same time pulls the rings strongly to his hips and begins to lift his torso up by pushing strongly down on the rings with his arms and finishes to an arm support. Each of the three elements—kicking, pulling, and pushing—blends together seamlessly as one continuous movement, and it really looks that way to the naked eye. However, a shorter gymnast with developed strength begins pulling down the rings with his arms at the same time that he kicks his legs up toward the rings. A taller gymnast who is not that physically strong kicks his legs up first and then begins pulling the rings down with his arms. In other words, he tries to create some momentum by his leg movement, which makes it easier for him to bring his torso up into the arm support position. We would like to underline that both short gymnasts and tall gymnasts use the described individual styles of performance unconsciously by utilizing their different qualities.

The described example shows that gymnasts with different levels of physical strength and body size perform the skill equally successful by using different individual styles of performance. That is, in this example, the individual style of activity depends on other features of personality rather than on the features of the nervous system.

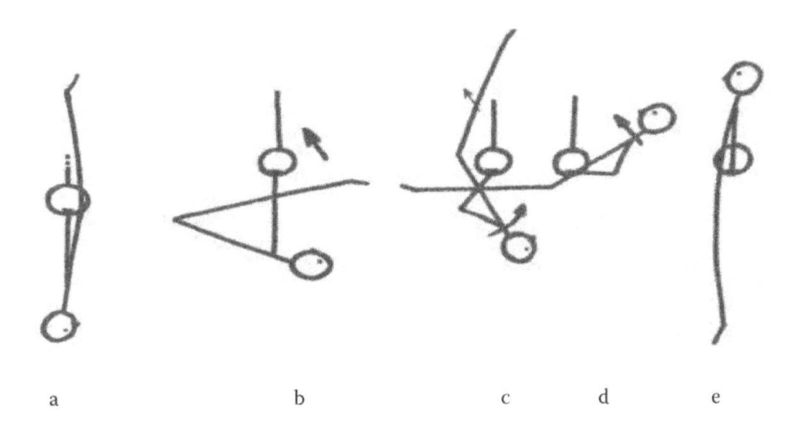

FIGURE 17.1 Kip mount technique on the rings.

CONCLUSION

The presented material demonstrates that the individual style of activity is the most efficient method of adaptation to the objective requirements of the task performance. From the point of view of self-regulation of activity, the individual style of activity can be considered as strategies of performance that derive from the individual features of personality. The individual style of activity can be formed consciously or unconsciously. In most cases, it forms unconsciously and involuntary. Often, the efficiency of this style of performance is not sufficient; therefore, one of the important duties for practitioners should be discovering individual features of personality and developing individual strategies of performance suitable for the particular individual. Strategies of activity depend not only on separate features of personality but also on their structural relationship. The individual style of activity should not contradict the normative requirements of performance in a production environment. Individual strategies of activity could be optimized on the basis of qualitative analysis or quantitative evaluation of the task complexity.

REFERENCES

Bedny, G.Z. and Karwowski, W. (2004). Activity theory as a basis for the study of work. *Ergonomics*, 47, 134–153.

Bedny, G. and Seglin, M. (1999). Individual style of activity and adaptation to standard performance requirement. *Human Performance*, 12, 59–78.

Bedny, G.Z. and Voskoboynikov, F. (1975). Problems of how a person adapts to the objective requirements of activity. In Aseev, V.G. (Ed.), *Psychological Problems of Personality*, 2. Irkutsk, Russia: Irkutsk University Press, 18–30.

Hilchenko, A.E. (1966). Study of the mobility of neural processes in man. *Physiology Journal*, 6, 67–81.

Klimov, E.A. (1969). *Individual Style of Activity*. Kazan: Kazansky State University Press.

Merlin, V.S. (1964). *Outlines of a Theory of Temperament*. Perm, Russia: Perm Pedagogical Institute.

Merlin, V.S. (1986). *Outlines of Integral Study of Individuality*. Moscow: Pedagogy.

Merlinkin, V.P. (1977). Some neuropsychological differences in development novice tumblers' skills. In Klimov, E.A. (Ed.), *Neuropsychological Study of Personality*. Perm, Russia: Perm University Publishers, 72–87.

Tomaszewski, T. (Ed.), (1975). *Psychology*. Warsaw: PWN.

18 Activity Theory as the Initial Paradigm for a Hypothesis about the Collective Goal Orientation of the Macroevolution of Mankind

Georgy M. Zarakovsky

CONTENTS

INTRODUCTION

Over recent years, the general psychological theory of activity initiated by A.N. Leontev and S.L. Rubinshtein has been successfully developing in an operational and structural direction with reference to activity of an individual (Bedny and Karwowski 2007). But this theory has good potential for developing population direction as well. According to Brushlinsky et al. (2000), a certain community can be considered as a *collective subject*. We will call a goal for such a collective subject a collective goal. This term is, to some degree, similar to the collective goal introduced by praxiologist K. Marek (1993). The difference is that Marek utilized this term to study cooperation of small groups. We utilize this term to study the country population. Hence, to analyze the collective goal-oriented activity of a given individual, we may use basic provisions of activity theory. The first such proposition is that activity is initiated by certain domineering needs. There are reasons for considering the whole mankind as the ultimate collective subject. This subject is comprised of different individuals at different times, but basic needs of such individuals remain unchanged. This consideration might help to develop a hypothesis of a collective goal-oriented *mega-evolution* of mankind, which would be better scientifically grounded than the existing hypotheses. It is relevant because today, the mentality of people is exposed to the influence of a considerable number of inconsistent and often destructive ideas and worldviews. And it is becoming dangerous for sustainable development of mankind as a whole. Friedrich Nietzsche (1990, p. 43) had a reason for saying: "A thousand goals have there been hitherto, for a thousand peoples have there been. Only the fetter for the thousand necks is still lacking; there is lacking the one goal. As yet humanity hath not a goal."

Numerous philosophical, historic, and theological works are devoted to searching for this goal. Some of them are especially important in substantiating this hypothesis.

In 1795, Antoine de Condorcet (1909) wrote *Outlines of an Historical View of the Progress of the Human Mind*. At that time, the European public conscience was shaped by the religious (Christian) idea of finiteness of the world (history was expected to end at the Last Judgment). By contrast, in his book, de Condorcet set forth the idea of eternal progress and development of humanity towards the truth and happiness. The progress is determined by discoveries and development of engineering and technologies. But there are three conditions that should be met: inequality among nations should be eliminated; inequality among various classes should be eliminated; and human nature should generally improve.

Georg Wilhelm Friedrich Hegel (1993) believed that history is the instrument and tool of the World-Spirit and that world history's ultimate goal is "comprehension by spirit of its freedom." The way to it lies through a great ordeal and struggle between the good and the evil. S.L. Frank (1992) reflected a great deal on the meaning of human life and came to the conclusion that life is eternal, but only in the spiritual sense, which is a question of the immortality of human souls.

The concept of historical development by K. Jaspers (1991) requires special attention. It is based on the idea of the Axial Age. The meaning of this concept is to explain the unity of the global human history. The way Jaspers treats the unity of history is transcendental and is connected with the hidden meaning of history. This meaning is revealed in the Axial Age; it is oriented to the future, which might be infinite. At the same time, Jaspers claims that each individual is responsible for everything. He also believes that every personal action has a transcendental reference point, as a lighthouse "shining from the depth of the future and from the depth of a person." In Jaspers's narrative, there are two moments, which are particularly important for his hypothesis. First, the presence of a certain meaning of the historical process is postulated, and this process can be infinite. Second, this meaning, or purport (transcendental reference point), somehow directs people's activity. These two moments are responsible for the future of humanity.

To my mind, the search for the transcendental reference point (which is not consciously realized by people), and hence, the search for the sensible collective goal orientation of life of mankind, should start with the analysis of intentions, which are initially set and lie at the heart of all people's activities.

INTENTION AND TECHNOLOGY

There are various classifications of the intentions peculiar to a person (requirements, motives, values, aspirations). According to the self-determination theory, which is rather popular today, an individual is characterized by three basic requirements beyond the biological necessities level: autonomy, competence, and contact with other people (affiliation motive) (Gordeyeva 2010).

The need for autonomy is the need for freedom of choice and self-determination in a person's behavior. It is a universal need of a person to feel like an actor, initiator, and the reason for his/her own life in harmony with his/her integrated "ego."

The need for competence means aspiration to be effective, to cope with problems of an optimal level of difficulty as a reaction to environmental challenges.

The need for contact with other people means the desire to have a reliable connection with significant people and to be understood and accepted by them.

The above listed needs are not the only ones in the list of basic nonbiological requirements of the person. They belong to the mental realities, which constitute people's psychological potential (Zarakovsky 2009). And, according to activity theory, a person should have basic requirements for realization of his/her potential. This group of requirements may be described as *activity intentions* that constitute the motive force vector of people's lives. I have replaced the term *requirement* or *need* with the term *intention* for a reason. The basic psychological realities, these unusual requirements that initiate and support a person's active behavior (first of all, goal-oriented activity), are most similar to instincts, which stimulate activity of a certain orientation at a subconscious level.

I, therefore, offer a classification of fundamental basic intentions that determine the general orientation of people's behavior and activity.

The first basic intention is the aspiration for immortality. It is reflected in such motivational tendencies as

- Searching for the maximum duration of a person's own life, increasing its security and health preservation in various ways
- Biological continuation of one's existence—through descendants
- Virtual continuation of one's existence through providing descendants with one's visual, acoustic, or mental shape: photos, video and audio recordings, memoirs, monuments, and so forth
- Leaving behind the legacy of one's creativity in the form of pieces of material or spiritual culture and transferring one's experience to children and disciples
- Searching for subjective virtual immortality in the form of religious belief in the soul's afterlife

The second basic intention is the aspiration to learn more about the surrounding world. It is reflected in the following aspirations:

- To novelty of impressions
- To establish cause-and-effect connections
- To learning more about the world's structure and its laws
- To artistic (aesthetic) cognition of the world

The third basic intention is the aspiration to creative transformation of the environment. It is reflected in

- Inventive and design activity
- Material and information productive activity that is a realization of inventions and projects

There are some other basic intentions such as self-actualization, self-assertion, personal identification with a certain society, and so forth, but they have no direct connection with the problem in question.

Once the *Homo sapiens* species acquired these three specific (compared to other fauna species) basic intentions, people got a real opportunity to direct their activity to achieving the most important need, that is, objective, such as immortality. And this opportunity is really being exploited, though society may not qualify it as a development vector of mankind. The course of this actualization in the form of the activity initiated by aspiration to knowledge and creative transformation of the world is traced throughout centuries. As a result, people's lives are becoming more and more independent of the processes that take place in primordial nature. And over the past two centuries, the increase in this independence has dramatically accelerated.

People now can live comfortably in any environmental conditions at the expense of life-support systems in inhabited and working premises and in any settlements in general. Infectious epidemics became fairly rare. The population is securely provided with food of natural origin despite fluctuations of agricultural productivity. The dangers connected with earthquakes have been considerably reduced now that earthquake-proof houses and early warning systems have appeared. Modern ways of preventive maintenance and treatment of illnesses and the development of essentially new medical equipment allow us to reduce the death rate and to increase the duration of people's active lives. Developments in the area of renewed energy sources that raise the energy security of the inhabitants of the planet are carried out rather successfully.

Recently, real possibilities to secure the Earth from collision with space objects, such as comets and asteroids, have appeared. Telescopes and other equipment that allow the location of dangerous objects at a considerable distance have been created. This allows accepting necessary measures of protection. Real technological facilities to influence space objects have been developed. It is now possible to influence them; for example, tractor spaceships can change their trajectory. It is possible to blow up an object with a nuclear explosive.

The next step may be adopting the Moon or other planets and their satellites. Thanks to the creation of a new tool kit for studying space bodies, it has become possible to find water on the Moon, on Mars, and on a satellite of Jupiter. It has encouraged the development of techniques and technologies that allow the creation of manned bases on the mentioned space bodies. Thus, scientific and technological progress really makes humans less dependent on "whims" of nature, thereby raising the probability of eternal existence of mankind. Certainly, scientific and technical progress generates new dangers as well, but these dangers are of a known origin. Ways of addressing them are also known. And thus, it becomes a matter of politics, sociology, and psychology (this will be discussed below).

As for a philosophical vision of scientific and technological progress, we first need to turn to the works of M. Heidegger (1986) and Gasparski and Kozminski (2007). Heidegger views technology as a way of self-actualization of mankind. He shows that consequences of technical intrusion are diverse and, in the long run, are even difficult to predict. In Heidegger's philosophy, technology appears not just as a means for achieving a certain purpose or a tool for solving pragmatic tasks. It represents a universal value of universal proportion. The mentioned positions of Heidegger's conception suggest that his idea is close to mine, but he does not completely unfold the notion of technology as a "value of universal proportion."

Gasparski and Kozminski consider the praxiological concept of technology. They pay attention to an action-based definition of technology. In their study, the term *action* is close to the term *activity*. Activity theory and praxiology are considered as a means of studying human practice.

DEVELOPMENT OF CIVILIZATION

One more aspect of the issue discussed above is connected with civilization development of humanity. Stepin (2007) writes that in the development of civilization, a critical era has just begun. Toynbee (1934) offered a famous conception of world "civilization." He allocated and described 21 civilizations that have existed in human history. They can be classified into two big categories: traditionalist civilizations and so-called Western, civilizations named after the region of their occurrence. However, now it is presented not only by the Western countries. V.S. Stepin, therefore, prefers to describe the latter as technogenic as in its development, the crucial role belongs to constant search and application of new technologies—not only industrial ones, which provide economic growth, but also technologies of social management and social communication.

Traditional cultures have never viewed world transformation or establishment of human power over nature as their goal. In technogenic cultures, this goal dominates. Among the basic values of technogenic cultures, it is also possible to highlight the understanding of nature as an inorganic world, which is a material resource for human activity. In contrast to this approach, traditionalism regards nature as a living organism, of which humans are a small part.

In these two types of civilization, the understanding and treatment of a person's value is considerably different. If in traditional cultures, the person is defined, first of all, through his/her inclusiveness into strictly determined (and often set since his/her very birth) relations within a family, clan, caste, or class; in technogenic civilization, the essential value is the ideal of a free person who can join various social groups on an equal basis with others. Hence, there is no priority of individual freedoms and human rights in traditionalist cultures.

In a technogenic civilization's system of domineering vital concepts, a special place belongs to the value of innovation and progress, which is again different in traditionalist societies. Technogenic societies at a certain stage of maturity begin to influence traditional civilizations, making them change. Sometimes, these changes are results of military capture and colonization, but more often, they are a result of "catch-up modernization," which traditional societies have to carry out under pressure from technogenic civilizations.

Having entered the postindustrial phase, the technogenic civilization started a new cycle of expansion to various countries and regions of the planet. The technogenic type of development unifies public life much more than the traditionalist. Science, education, technological progress, and the expanding market generate a new mentality and lifestyle, transforming traditional cultures. And what we call globalization today is a product of expansion of the technogenic civilization. It gets involved in various regions of the world, first of all through technological expansion, causing the accelerated modernization of traditional societies, putting them on the rails of technogenic development.

The technogenic civilization has given mankind quite a number of achievements. Scientific and technological progress, as well as economic growth, has led to a new quality of life, provided an increasing level of consumption and health care services, and increased average life expectancy. Most people associate progress of this civilization with hope for a better future. Half a century ago, very few people believed that a technogenic civilization would lead mankind to global crises, when it would literally find itself on the threshold of self-destruction. Environmental crisis, anthropological crisis, a broadening process of alienation of social groups, invention of new means of mass destruction threatening to destroy the whole mankind—all these are side effects of technogenic development. And consequently, now there is a question of whether it is possible to overcome these crises without changing the basic system of values of technogenic culture.

The analysis of the structure of social reality and the internal logic of civilization development allows the conclusion that the system of values should be changed. Overcoming global crises will require a change in the purposes of human activity and its ethical regulators. And in this respect, the postindustrial stage may become the beginning of a transition into a new type of civilization development.

There are two approaches regarding the development of a postindustrial society. Usually, it is considered as a simple continuation, a special stage of technogenic development. In this case, the change in basic values is not taken into consideration, and the approach discusses only the changes that bring new technologies into human life, social communications, and relations between the states. This approach understands sustainable development as a prolongation of today's technological progress with some nature-protective restrictions. In view of today's tendencies of globalization, this direction of development will not lead to radical change in the destructive effect of a technological civilization.

The second approach to the development of a postindustrial society postulates that this development is not a simple continuation of a technogenic civilization. It can be interpreted as the beginning of a historically new third (in relation to traditional and technogenic) type of civilization development. But then, formation of a postindustrial civilization should be connected not only with technological revolution but also with spiritual reform, criticism, and revision of some former basic values of technogenic culture (its relation to nature, the cult of force as the basis of transformation activity, ideals of the consumer society based on growing consumption of things and energy, etc.).

There are no reasons to think that the future human attitude to nature will be reduced to contemplation and self-adaptation to it. Humans will keep transforming nature. It is rather probable that overcoming the environmental crisis will be connected not with preservation of wild nature on the global scale (which is impossible today without sharp reduction of the population of the Earth) but with expansion of cultivated areas. In this process, an important role will belong not only to protection of nature directed to preservation of local natural ecosystems but also to artificial creation of new biogeocoenoses aimed to provide the necessary level of their variety as a condition for the sustainability of the biosphere. It is quite probable that in this favorable-for-mankind scenario, the environment will be more and more similar to an artificially created park or a garden, which cannot be reproduced without purposeful activity of people. This will be the mission of humankind, who has changed the shape of the planet so much that people have become the real force defining the preservation of the biosphere.

What V.S. Stepin has said about the critical era in the human civilization development is mostly agreeable, with one exception: He claims that the mission of humans will be to artificially maintain the environment. It is a principal mistake. Using the results of technical progress for preservation of what it changes cannot be the goal. Any goal is something external in relation to a system. This goal is beyond self-maintenance of mankind in conditions set by nature (life on Earth), and it really exists. As for V.S. Stepin's idea, it is not the general goal but a "subpurpose," which is designed to maintain the conditions for achieving the general goal at this stage of evolution of the solar system and the Earth, in particular.

From the positions of my hypothesis about the *mega-evolutionary* sense of scientific and technical progress, the technological civilization is a way of humanitarian development that is based on the fundamental activity intentions of humans. Perhaps, not every individual actualizes these intentions but only those people who are "urged" to work in order to carry out this mission of mankind, which is the achievement of the infinity of its life. Of course, it is impossible to tolerate that natural and social conditions were beyond the necessary corridor of the current life of each generation. This will certainly require transformation of the classical technogenic civilization into the postclassical technogenic civilization, which V.S. Stepin and many others have pointed out.

Now that I have provided my arguments, I can formulate the whole hypothesis of collective goal-orientated macroevolution of mankind, which is basically a development of the conception of "the phenomenon of man" by Pierre Teilhard de Chardin (1965).

The essence of life of mankind as a whole, particularly the immanently inherent goal orientation of the historical process, is expressed by Teilhard de Chardin in theological terms. But upon thorough consideration, it becomes clear that he has developed a quite scientifically grounded theory of spiritual and other laws of development of human society on the basis of generalization of paleontological, biological, and historical facts, as well as theological (mostly psychological and sociopsychological) ideas about consciousness and behavior of people. Teilhard de Chardin has constructed the conception of two-vector dynamics of "the phenomenon of man." One vector is the tangential development; the second is radial development. The tangential vector is essentially the adaptation and homeostatic component of human activity aimed at preservation of life as such, in a biological and sociopsychological sense. The radial vector represents "ascending" development of mankind targeted at the "Omega Point." This is the point where human merges with God as a result of spiritual growth, leading to the ideal and complete self-knowledge.

Thus, the radial vector represents a process of self-improvement of people, their spiritual growth, which is the development of the higher mental functions peculiar to the human. If we turn to Alexander Men's (2001) vision of spirituality (he tended to appreciate Teilhard de Chardin's views), which in particular he regards as creative activity, we will be able to conclude that "radial development" should include not only spirit perfection but also its objective display in the form of creative material activity. God, according to both thinkers, "God the Creator," is some being (or some essence) that created the World, including mankind, first in its primordial version, and later continued the creation in a great measure by means of minds, feelings, and hands of the people created by Him. Hence, the "radial dynamics" is not only spiritual development but also the development of the whole world in all its forms by means of people's activity.

Based on all these considerations and the known empirical and scientific data about target orientation of people's activity and development of technologies over different historical periods, I suggest the following hypothesis.

The general direction of human activity as a whole is expansion to the Universe, aimed at achievement of real immortality of mankind. It is expressed in cognition and adaptation of nature, which leads to increasing independence of mankind from natural changes on the Earth and, in the long run, may lead humanity beyond its framework.

There are two conditions for actualization of movement to this real nonmythical immortality of mankind. The first is development of science and technologies. The second condition is constant maintenance of parameters of natural, anthropogenic, and social environments within the framework that is necessary for the biological existence of people and their productive social life (which may be treated as a tangential vector of the development of mankind). Thus, it is necessary to bring some amendments into Teilhard de Chardin's conception: "Omega Point" is not the fixed ultimate goal but the goal focused on infinity. It directs mankind's activity. This point is, in fact, a binary system from two connected "points" with two *subpurposes*. The first is cognition and adaptation of nature, which means mastering it and, as a consequence, increasing the territory of people's dwellings, as well as increasing the stability of its existence in a broader variability of environmental parameters, including the influence of extraterrestrial factors. It is the process of intellectual and material expansion carried out by means of scientific and technical progress.

The second *subpurpose* is achieving the spiritual ideal of mankind in the sense of morality and the ability to carry out productive, nondestructive interaction of societies. This is a process of self-knowledge and perfection of the people forming societies. It has adaptation and homeostatic meaning, which is maintenance of the stable existence of mankind not only in a biological but also in a sociopsychological respect. Harmonization, that is, coordination of radial and tangential processes, is the primary challenge facing mankind in the near future. Without such harmonization, mankind's movement towards the general goal may be discontinued as a result of either exceeding the threshold of social contradictions, or exhaustion of natural resources, or the lack of technological facility to protect mankind against natural cataclysms.

Thus, the general orientation of all generations' activity seems to be the actualization of the main basic intention, that is, the aspiration for temporal infinity of the human race, while at the same time increasing quality of life for all people. This is not religious but, rather, scientifically grounded understanding of the ultimate essence of people's lives and mankind's life as a whole. It is the process of expansion aimed at preservation and development of the life of mankind within broadening environmental borders by means of scientific and technological progress, personal perfection, and overcoming social confrontations.

The suggested hypothesis about a general orientation of development of mankind is, in fact, a rather plausible interpretation of the impulse of the "Axial Age" of history, which K. Jaspers considers as transcendental. The hypothesis differs from known philosophical interpretations and forecasts of scientific and technical progress in one basic respect: Describing this progress, it highlights its fundamentally humanistic meaning of actualizing the people's immanently inherent aspiration to infinite life. It is important that the basic intention to achieve infinite life is being actualized not in respect to a single individual but to mankind as a whole. This presupposes a shift of the "immortality motive" from an individual to the whole *Homo sapiens* species. Individual immortality is impossible (those who cannot live without hope for it may find subjective immortality in various religious beliefs), but immortality of mankind is possible.

CONCLUSION

There are several practical conclusions that can be made from the presented hypothesis.

1. In governmental practice of regulating the quality of life of the population, subjective standards (criteria) of quality of life are less important than the objective ones. It is necessary to realize that objective standards are not only material but also spiritual and psychological. The choice of such target standards should be based first of all on the "ultimate all-mankind" criterion. Target standards should promote positive social contribution to scientific and technological progress. Therefore, sponsors' investments in education, science,

or new technologies should be estimated by power structures and society as considerably more important than sponsors' contributions in sports, show business, other sorts of charities, and so forth.

2. It is expedient to develop and launch additional indicators into estimation of comparative success of different countries. Not only social and economic development and development of human potential but also the contribution by these countries to universal civilization development should be considered. These may be indicators characterizing the share of expenses for fundamental scientific research (first of all, in physics, astronomy, biology) and also for innovative activity directed to realization of the results of these pieces of research into practice.
3. The hypothesis should be transformed into a modern "philosophical belief," that is, into a certain constructive ideology for uniting people. To begin with, it is advisable to form a system of values where the importance of science and technologies will prevail over other directions of social and economic development. This system should be possibly built into public vision (and especially into the worldviews of the elite).
4. The previous three considerations may be actualized at observance of frame conditions: preservation of key parameters of natural properties of the Earth, maintenance of parameters of the population's activity at the level providing progressive reproduction of human potential, and competitive behavior of different societies without transition to catastrophic conflicts. In order to encourage such development, big efforts in the sphere of solving political, economic, and sociocultural issues of the modern society on the global scale are required.
5. It is impossible to agree with the position widely transmitted by mass media that the rights and interests of each separate person have priority over the rights and interests of a society (the collective subject) as a whole. These interests should be balanced by a criterion of a suboptimum parity between current interests of different social groups and the interests aimed at new civilization development of mankind.

The general conclusion is that the meaning of life of mankind, the *Homo sapiens* species, consists of cognition of the natural world and creation of an artificial world providing progressive independence of human life from spontaneous processes in the environment. It creates conditions for infinite life of mankind. It is, however, necessary to make sure that intergroup cataclysms in a society and uncontrollable exploitation of natural resources do not lead to the termination of this life.

REFERENCES

Bedny G. Z., Karwowski W. (2007). *A Systemic-Structural Theory of Activity: Application to Human Performance and Work Design*. Boca Raton, London, New York: Taylor & Francis.

Brushlinsky A. V., Volovikova M. I., Druzhinin V. N. (Eds.) (2000). *A Problem of the Subject in a Psychological Science*. Moscow, Russia: Academic Project Publishing Co.

de Chardin P. T. (1965). *Phenomena of the Human*. Moscow, Russia: Progress.

Frank S. L. (1992). *Spiritual Bases of a Society*. Moscow, Russia: Republic.

Gasparski W. W., Kozminski L. (2007). The praxiological concept of technology. In W. W. Gasparski, T. Airaksinen (Eds.), *Praxiology and the Philosophy of Technology. Praxiology: The International Annual of Practical Philosophy and Methodology*. New Brunswick (USA), London (UK): Transaction Publishers, pp. 45–54.

Gordeyeva T. O. (2010). Theory of self-determination: The present and the future. A part 1: Problems of development of the theory. *Electronic Resource/Psychological Researches: Electronic Journal* N 4 (12). http://psystudy.ru, reference date September 20, 2010.

Hegel G. W. F. (1993). *Philosophy of History*. St. Petersburg: Nauka.

Heidegger M. (1986). *The Question about the Technician. The New Technocratic Wave in the West*. Moscow, Russia: Progress.

Jaspers K. (1991). *Sense and History Appointment*. Moscow, Russia: Politizdat.

Marek K. M. (1993). Collective action: A tentative model. In T. Airaksinen, W. W. Gasparski (Eds.), *Practical Philosophy and Action Theory*. New Brunswick (USA), London (UK): Transaction Publishers, pp. 103–118.

Men A. (2001). *Origins of Religion*. Moscow, Russia: Fund of Alexander Men's.

Nietzsche F. (1990). Selected works in 2 v., V. 2/Composer, ed. and notes K.A of Svastjan. Moscow, Russia: Thought Publisher.

Stepin V. S. (2007). Section of philosophy, sociology, psychology and the right of branch of social studies of the Russian academy of sciences. In V. S. Stepin (Ed.), *Types Civilization Developments/Russia in the Globalized World: World Outlook and Socio-Cultural Aspects*. Moscow, Russia: Science, pp. 3–13.

Toynbee A. (1934). *A Study of History*. Oxford University Press.

Zarakovsky G. M. (2009). *Quality of Life of the Population of Russia: Psychological Components*. Moscow, Russia: Sense.

19 Task and Its Complexity Assessment

Gregory Z. Bedny, Waldemar Karwowski, and Inna S. Bedny

CONTENTS

INTRODUCTION

A famous Polish scientist, Tadeusz Kotarbinski (1965), developed praxiology as a philosophical framework. Praxiology is the study of human actions in terms of their efficiency. An action in praxiology is defined as intentional, goal-directed behavior, which is specific to a human beings.

German action theory defines an action as follows: "Action is goal-oriented behavior that is organized in a specific way by goals, information integration, plans, and feedback and can be regulated consciously or via routines" (Frese and Zapf 1994). German action theory has been influenced by activity theory in general and by Rubinshtein (1968), Leont'ev (1978), Vygotsky (1962), and so forth in particular, and some important ideas have derived from Polish praxiology (Tomazevski 1978).

The definition of action in praxiology and German action theory is similar to a concept of activity (*deyatel'nost'*) in activity theory. From systemic-structural activity theory (SSAT) perspectives (Bedny and Karwowski 2007), activity can be defined as a goal-directed system, in which cognition, behavior, and motivation are integrated and organized by mechanisms of self-regulation toward achieving a conscious goal. Action in activity theory is one of the main units of analysis. Activity during task performance can be divided into a system of logically organized cognitive and behavioral actions. Hence, actions from activity theory perspectives can be cognitive and behavioral. Cognition in activity theory is described as a cognitive process or a system of cognitive actions (Bedny and Karwowski 2007).

From activity theory perspectives, cognition can be presented as unconscious mental operations automatically unfolding over time or voluntarily performed conscious cognitive actions. These two levels of information processing are interdependent and influence each other. Activity is a goal oriented system. The goal of activity is a conscious mental representation of a future desire result. It includes verbally logical and imaginative components. In action theory and cognitive psychology, the goal includes cognitive and motivational mechanisms. According to German action theory (Frese and Zapf 1994), "...the action is 'pulled' by the goal (the motivational aspect)." In contrast, in activity theory, the goal is a purely cognitive mechanism, and motivation is an energetic component of activity. Motives or motivation create a vector motive(s)—a goal that gives activity goal-directed character. The more intensive motives are, the more efforts a subject will expend to reach the goal.

Motivation pushes activity. Sometimes, there is a contradiction between motives or a conflict of motives.

There are a number of current publications on activity theory in the West. However, translation and interpretation of some basic concepts of activity theory very often are incorrect. As an example, Kaptelinin and Nardi (2006) wrote, "*A way to understand* objects *of activities is to think of them as objectives*...," mixing concepts of object with objectives. In activity theory, an object that can be material or mental cannot be considered as an objective. An object of activity is something that can be modified according to the activity goal.

Through the use of mental or external tools, the object is modified in accordance with the activity goal. In order to study human work, some basic concepts such as subject, object, goal, action, product, result, self-regulation, strategies, efficiency, and so forth have been developed. In this chapter, we, in an abbreviated manner, describe basic principles of task complexity assessment.

TASK COMPLEXITY

From SSAT perspectives, a task is a system that consists of cognitive and motor actions, cognitive operations, and processes required to achieve a task's goal. The most important attribute of task as a system is complexity. Complexity is determined by the number of elements in the system, the specificity of each element, the manner in which these elements interact, the number of modes in which the system can function, and so forth. In this work, we do not consider the complexity of the technical systems, including a "man–computer system." The object of this chapter is complexity of human activity during task performance. We are interested in the complexity of the activity during task performance because one of the main purposes of ergonomic design is to reduce the cognitive demands of the task. The basic, as well as the most general, characteristic of cognitive demands is task complexity. The more complex the task is, the higher are the cognitive demands of the task performance. The concept of task complexity is important for the study of human labor by such professionals as industrial engineers, economists, philosophers, and scientists. Evaluation of task complexity is also fundamental to the study of human-computer interaction (HCI). Hence, task complexity evaluation has an important theoretical and practical meaning. In this work, we introduce general principles of task complexity evaluation and how these principles can be used for evaluation and optimization of human performance. Quantitative evaluation of task complexity facilitates more accurate evaluation of performance and its optimization based not only on qualitative but also on quantitative criteria. Task complexity is an important characteristic not just of cognitive but also motor components of activity because motor actions include cognitive components responsible for regulation of motor actions and motions. Hence, the concept of complexity can be applied not only to the cognitive components of activity but also to the motor component, or more specifically, to mental aspects of motor actions' regulation. For example, the more precise the motor action is, the more concentration of attention it requires, and therefore, more mental efforts are required for its execution. The importance of task complexity is in its relation to design. Design is the creation of models of artificial objects in accordance with requirements and characteristics with the purpose of materializing these objects. Measures of complexity can be used as criteria for evaluating efficiency of hardware or software design. Description and analysis of activity structure is an important stage of task complexity evaluation. In cognitive psychology, where the major concept is process, it is impossible to describe the structure of activity and, hence, evaluate task complexity.

ALGORITHMIC DESCRIPTION OF TASK PERFORMANCE

Activity during task performance often has very complex logical organization. Description of logical organization of activity is performed by utilizing a concept of human algorithm (Bedny and Karwowski 2007). Members of human algorithm, cognitive and behavioral actions, and operations are major units of analysis during algorithmic description of activity. Members of an algorithm are made of actions with their associated subgoals integrated through members of algorithm supervening goals. Each member of the algorithm is designated by a special symbol. For example, operators are represented by the symbol "*O*" and logical conditions by the symbols "*l*" or "*L*." A member of the algorithm usually can include up to four to five actions (limited by capacity of working memory).

All operators involved in receiving information are categorized as afferent operators and are designated by a superscripts α, as in "O^{α}" If the operator is involved in extracting information from long-term memory, the symbol μ is used, as in O^{μ}. Am operator with the symbol $O^{\alpha th}$ is involved in performing thinking actions with support of visual information. The symbol $O^{\mu w}$ is associated with keeping information in working memory. Operators with the symbol O^{ε} are efferent operators associated with execution, for example, moving a gear. In deterministic algorithms, the logical conditions designated "*l*" have two values, 0 or 1. In some cases, logical conditions can be a combination of simpler ones. Such simple logical conditions are connected by "and," "or," "if–then" rules and so forth. Complex logical conditions are designated by a capital "*L*," while simple logical conditions are designated by a lowercase "*l*." The order number of each member of the algorithm is depicted by a subscript (for example, symbol O_4^{ε}, where number 4 indicates the order number of this particular member of the algorithm). Operators and logical conditions have independent numbering.

In a probabilistic algorithm, logical conditions may have two or more outputs with a probability between 0 and 1. As a simple example, suppose an algorithm has logical conditions with three outputs with distinct probabilities of occurrence. In such a case, the logical condition can be designated as $L_1 \uparrow^{1(1-3)}$, which possesses not two potential values but three. In this case, there are three versions of output. $\uparrow^{1(1)}$, $\uparrow^{1(2)}$, and $\uparrow^{1(3)}$, with different probabilities (for example, the first output has the probability 0.3, the second 0.2, and the third 0.5). Knowledge of the probability of the output may be taken into consideration when studying probability of performance of various actions, strategies of performance, calculation of performance time of the algorithm or components of the algorithm, and evaluation of task complexity. Frequently, in an algorithmic description, an always-false logical condition is used, which is defined by the symbol "ω." This logical condition is introduced only to make it easier to write the algorithm and does not designate real actions performed by the subject. It always defaults to the next member of the algorithm as indicated by the arrow included in the specification of this always-false logical condition. An arrow designates the logic of transition from one member of an algorithm to another. Thus, the algorithm exhibits all the possible actions and their logical organization and, therefore, constitutes a precise description of human performance. It describes activity of a subject in terms of actions through which the subject attains the goal of activity.

Let us consider, in an abbreviated manner, a fragment of the experimental task "installation of pins." During a laboratory experiment, a subject was asked to fill a pinboard with 30 pins. Ten of the pins had a clearly visible flute. The pins were inserted into the holes in the pinboard according to the position of the flute in right and left hands. The task performance included receiving information, interpreting it, making a decision, and performing motor actions according to logical conditions. Cognitive actions were combined with motor actions.

The pins were put in the holes according to specific rules.

1. If a pin has no flute, it can be installed in any position.
2. If a fluted pin is picked up by a subject's left hand, it must be placed so the flute is placed inside the hole.
3. If a fluted pin is picked up by a subject's right hand, it must be placed so the flute is above the hole.

The left column of Table 19.1 depicts standardized symbolic descriptions of members of the algorithm, and the right column contains verbal descriptions of members of the algorithm.

TABLE 19.1
Algorithmic Description of the Operation "Installation of Pins" (Fragment)

Members of Algorithm	Description of Member of Algorithm
$_{\omega 1}\!\downarrow O_1^{\varepsilon}$	Move both hands to the pin box and grasp two pins.
O_2^{α} $\left({}^1O_2^{\alpha} - {}^4O_2^{\alpha}\right)$	Determine the type and position of the pins: ${}^1O_2^{\alpha}$—both pins have no flutes; ${}^2O_2^{\alpha}$—pin in the left hand has a flute; ${}^3O_2^{\alpha}$—pin in the right hand has a flute; ${}^4O_2^{\alpha}$—both pins have flutes.
$\mathbf{L}_1 \uparrow^{1\,(1\text{–}4)}$	If a flute, in either the right or the left pin is absent ($l_l = 0$ and $l_r = 0$), perform O_3^{ε}. If this condition is not observed ($L_1 = 0$), transfer to L_2. If L_2 is also not noticed ($L_2 = 0$), transfer to L_3. If L_3 is not observed ($L_3 = 0$), transfer to L_4.
${}^{1\,(2)}\!\downarrow \mathbf{L}_2 \uparrow^{2\,(1\text{–}3)}$	If the left pin is fluted and the right pin is not ($l_l = 1$ and $l_r = 0$), a decision should be made to install the right pin in any position and to turn the *left pin* so the fluted side would be placed *inside a hole*, and perform O_4^{ε}. If the right pin has a flute, not the left one ($L_2 = 0$), transfer to L_3. If L_3 is also 0, then transfer to L_4.
$\downarrow^{1\,(3)} \downarrow^{2\,(2)} \mathbf{L}_3 \uparrow^{3}$	If the left pin is not fluted but the right pin is ($l_l = 0$ and $l_r = 1$), a decision must be made to put the left pin in any position and to turn the *right pin* so that fluted side would be placed *outside a hole*, and then perform O_5^{ε}. If L_3 is also not observed ($L_3 = 0$), transfer to L_4.

TASK COMPLEXITY AND DIFFICULTY

The concept of task complexity goes hand in hand with the concept of task difficulty. Although in most cases, complexity and difficulty are considered synonymous, they should be differentiated. Complexity is an objective characteristic of the task, and difficulty is the performer's subjective evaluation of the task complexity. Depending on the skills and the individual features of the subject, the same task might be evaluated by her/him as more or less difficult. For instance, an experienced computer user might find certain tasks easy, whereas an inexperienced user would need a lot of assistance to perform the same tasks. Complexity does not have a subjective component. Therefore, the performer cannot experience complexity by itself but, rather, perceives it as a subjective difficulty. When complexity of a task is higher, the probability that performance requires more cognitive effort and motivational mobilization increases. Quantitative measures of complexity include a combination of indexes that describe intensity and frequency of cognitive efforts during performance of qualitatively various activity elements.

Complexity is a multidimensional concept that requires multiple measures for its evaluation. For each specific task, some measures might be more important than others. Some of these measures can be of zero value if they are not important for a particular task. Such concepts as level of attention concentration, level of activation of neural centers, and level of wakefulness are important as a theoretical basis for evaluation of mental efforts and task complexity.

SSAT developed five categories of complexity depending on the level of concentration of attention during performance. Each category represents a range of complexity, which means that complexity varies inside an interval but that such variation can be ignored. Category 1 is the simplest, and category five is the most complex. The presented five-point scale of complexity can be applied with sufficient precision to the complexity evaluation of various elements of activity. Complexity of activity elements also depends on specificity of combining performance of activity elements in time.

Multiple attempts were made to develop a quantitative method of task complexity evaluation. Every measurement procedure requires selection of adequate units of measure. It was suggested to utilize different units of measure such as number of controls and indicators, number of actions and alternatives in multiple-choice tasks, and so forth. For measuring computer-based tasks, such measures as task solving time, number of transitions, and total number of system's states were suggested by Rauterberg (1996). However, the suggested units of measure are inadequate from a mathematical point of view because they are noncommensurable units of measure. A complicated task can be performed at the same time as a simpler one. The subject can spend more mental effort performing a task in a short time or with fewer transitions during performance. Manipulation of one control can be more complex than manipulation of several controls. Similarly, one cannot just calculate the amount of actions performed by an operator during task performance for task complexity evaluation because one motor action or instance of decision making can be more complicated than several simple ones, and so forth. The above listed examples demonstrate an attempt to utilize noncommensurable units of measure. SSAT developed units of measuring complexity, procedures, and measures for evaluation of task complexity (Bedny and Karwowski 2007; Bedny and Karwowski 2008).

BASIC PRINCIPLES OF SSAT TASK COMPLEXITY ASSESSMENT

The activity as a system calls for multiple stages and levels for its description. These stages and levels are qualitative analysis, analysis of logical organization of activity (algorithmic description of activity), analysis of activity time structure, and quantitative description of activity. These stages have loop structure organization. This means that the next stage or level of analysis might require reconsideration of the previous stage. These stages of task analysis were described by Bedny and Karwowski (2007), Sengupta and Bedny (2011), Bedny et al. (2010), and Bedny and Bedny (2011). Below, we will consider quantitative analysis. Quantitative assessment of task complexity is related to the last stage of activity analysis. It can be done only after algorithmic and time structure description of activity is completed. Algorithmic description of activity depicts logical organization of activity. Time structure represents a logical sequence of activity elements, their duration, and the possibility of their simultaneous or sequential performance.

In the description of an activity, one can outline two types of units of analysis. One of theses types is the "typical task element" or "technological element," and the other is referred to as the "typical activity element" (standard description of activity elements). For example, "move lever," "press button," and "take reading from instrument" are all examples of typical task elements (technological units). Depending on the specificity of their performance, they can involve various activity elements. A task element such as "move lever" can be performed throughout different operations (motions) depending on distance, amount of effort, and precision of movement. Examples of standard elements of activity are "detection," "decision making at sensory perceptual level," "reach-R30C (reach an object, distance 30 cm, object jumbled with others so that searching and selection also take place)," "eye focus," and so forth.

It is interesting that leading specialists in system methods-time measurement-1 (MTM-1) (Karger and Bayha 1977), did not distinguish between these two units of analysis. At the algorithmic stage of analysis, both psychological (typical elements of activity) and technological (elements of task) units of analysis are utilized. Only psychological units of analysis are utilized at the stage of time structure development and quantitative complexity assessment. Failure to distinguish between these two unit types causes the practitioner to confuse the time line chart with the time structure of activity during task performance. In our further discussion, we consider only how time structure of activity can be used for quantitative analysis. In our example, the left column of Table 19.1 utilizes psychological units of analysis. In the right column, technological units of analysis are used. Below, we will describe the time structure of activity and complexity assessment of one member (O_5^ε) of an algorithm in the above considered task.

At the previous stage, this task was described algorithmically (Table 19.1). At the next stage, the time structure of activity during performance of this task was developed. At this stage of analysis, typical elements of task were transformed into typical elements of activity. Figure 19.1 graphically displays the time structure of one member of the algorithm, O_5^ε, when one out of two pins is fluted. This requires making a decision on how to turn the pin with the flute.

This member of an algorithm is an efferent operator according to the algorithmic description terminology. One member of an algorithm can usually combine from one to four actions. The capacity of working memory limits a subject's ability of combining a larger number of activity elements into an integral structure of activity element. Thus, in our task analysis, we utilize such units of analysis

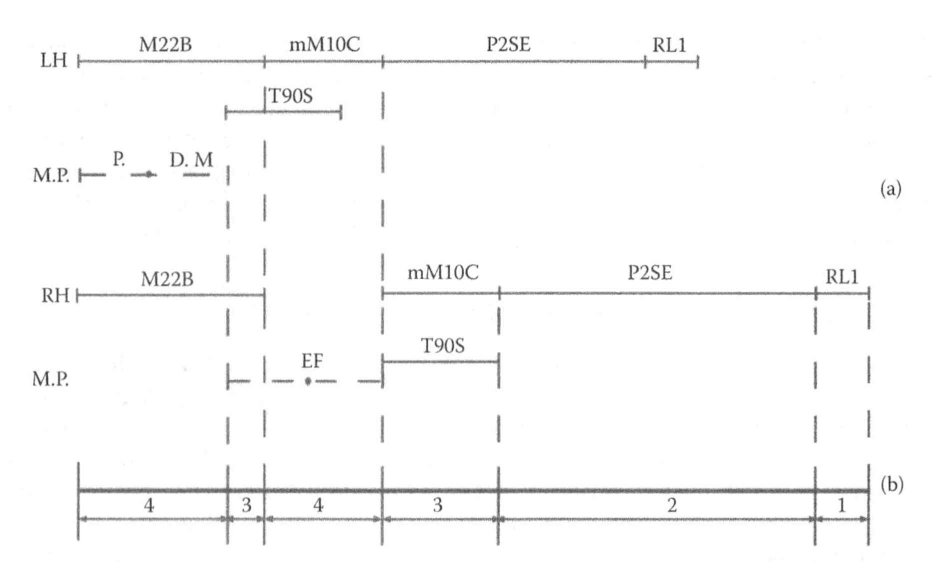

FIGURE 19.1 Time structure and category of complexity of activity elements during filling pinboard when one pin has a flute. (a) Time structure; (b) category of complexity.

as a member of algorithm (mode of performance of some elements of task), cognitive and behavioral actions, and operations that are elements of the member of the algorithm. Figure 19.1 depicts the time structure of one member of the algorithm, when a subject, after grasping two pins, moves them simultaneously into the holes when one pin is fluted. This requires decision making on how to turn the pin with the flute. In Figure 19.1, the horizontal segments index the duration of the various elements of activity. The length of the segments in the model reflects the duration of elements of activity (duration of separate operations that are used for action performance). LH and RH mean left and right hand actions. MP designates mental action. The elements or operations of activity are specified by symbols above the segments. Most elements are specified utilizing MTM-1 system symbols. EF is used to describe perceiving signals and simple decision making that includes a "yes–no" or "if–then" decision. According to SSAT, such elements are simple decision-making actions at the sensory–perceptual level. Elements of activity with symbols P and DM designate more complicated decision-making actions performed based on information extracted from memory. Such decision-making actions are more complicated than EF decision-making elements in MTM-1. Left hand motor action includes the following operations or elements: M22B—moving an object 22 cm with average level of concentration of attention; mM10C—moving an object 10 cm with a high level concentration of attention; P2SE—positioning or installing a symmetric object (pin) into the hall with light pressure and easy handling; RL1—release by unfolding the fingers; and T90S—turn object with light weight 90°. Similar elements or operations include right hand action. It is important to note that SSAT utilizes hierarchically organized units of analysis such as a member of an algorithm (mode of performance of some element of task), cognitive and behavioral actions, and operations. MTM-1 has only a concept of motion and does not utilize such important concepts as time structure of activity and so forth. In SSAT, there are rules of simultaneous and sequential performance of cognitive and behavioral components of activity. Elements of activity are also classified in according to their complexity depending on the level of concentration of attention on separate elements of activity and specificity of combining activity elements in time. These rules can be found in the work of Bedny and Karwowski (2007). Line B in Figure 19.1 demonstrates the category of complexity of various time intervals during performance of the considered member of the algorithm. Such a method permits the development of the system of measures for complexity evaluation of work activity during performance of different types of tasks.

CONCLUSION

SSAT has developed principles of algorithmic and time structure description of activity. This allows a quantitative assessment of task complexity. Hierarchically organized units of analysis such as members of an algorithm, cognitive and behavioral actions, and operations have been developed for this purpose. For quantitative assessment of complexity, psychological and technological units of analysis are distinguished. Time intervals for various components of activity are used as units of complexity measurement. The complexity of the time intervals for elements of activity depends on the level of concentration of attention during these intervals, the way these elements of activity are combined, the probability of the performer utilizing them, and the existence of emotional stress at the time of performance.

REFERENCES

Bedny, G. Z., Karwowski, W. (2007). *A Systemic-Structural Theory of Activity. Application to Human Performance and Work Design*. Boca Raton, London, New York: Taylor & Francis.

Bedny, G. Z., Karwowski, W. (2008). Application of systemic-structural theory of activity to design and management of work systems. In W. W. Gasparski, T. Airaksinen, (Eds.). *Praxiology and the Philosophy of Technology*. New Brunswick and London: Transaction Publishers, pp. 97–144.

Bedny, I., Karwowski, W., Bedny, G. (2010). A method of human reliability assessment based on systemic-structural activity theory. *International Journal of Human–Computer Interaction*, 26 (4), 377–402.

Bedny, I., Bedny, G. (2011). Abandoned actions reveal design flaw: An illustration by a web-survey task. In G. Bedny, W. Karwowski (Eds.). *Human-Computer Interaction and Operators' Performance. Optimizing Work Design with Activity Theory*. Taylor & Francis, pp. 149–184.

Frese, M., Zapf, D. (1994). Action as the core of work psychology: A German approach. In H. C. Triandis, M. D. Dunnette, L. M. Hough (Eds.). *Handbook of Industrial and Organizational Psychology*, 2nd ed., vol. 4. Palo Alto, CA: Consulting Psychologists Press, pp. 271–339.

Kaptelinin, V., Nardi, B. A. (2006). *Acting with Technology. Activity Theory and Interaction Design*. Massachusetts: Cambridge.

Karger, D. W., Bayha, F. H. (1977). *Engineering Work Measurements*, 3rd ed. New York: Industrial Press, Inc.

Kotarbinski, T. (1965). *Praxiology*. Oxford: Pergamon Press.

Leont'ev, A. N. (1978). *Activity, Consciousness, and Personality*. Englewood Cliffs, NJ: Prentice-Hall.

Rauterberg, M. (1996). How to measure cognitive complexity in human-computer interaction. In R. Trappl (Ed.). *Cybernetic and Systems*. Austrian Society for Cybernetic Studies, pp. 815–820.

Rubinshtein, S. L. (1968). *Foundations of General Psychology*. Berlin: Verlag deutscher Wissenschaftten.

Sengupta, T., Bedny, I. (2011). Microgenetic principles in the study of computer-based tasks. In G. Bedny, W. Karwowski, (Eds). *Human–Computer Interaction and Operators' Performance. Optimizing Work Design with Activity Theory*. CRC Press, Taylor & Francis, pp. 117–148.

Tomazevski, T. (1978). *Activity and Consciousness*. Weinheim, Germany: Beltz.

Vygotsky, L. S. (1962). *Thought and Language*. Cambridge, MA: MIT Press.

20 Virtualization of Hospital Processes in Forming the Knowledge-Based Organization

Joanna Bartnicka and Teodor Winkler

CONTENTS

INTRODUCTION

Hospital processes are characterized by a high degree of complexity, which is connected with the complexity of the structure of the participants in the processes; the complexity of the technical infrastructure, including specialized equipment and medical instruments; and most of all, the complexity of information resources and knowledge needed for them to be carried out effectively and efficiently. Here, we take account of the key media of storage of such resources, which is the hospital staff. Therefore, the postulate is true that the improvement of patient treatment and patient care is directly proportional to the intellectual resources of the hospital (Sharma et al. 2005). However, the other major media of storage of hospital knowledge are hospital databases, specialized Web sites, documents, and written and unarticulated principles and procedures. They also undergo dispersal and have different forms—from unstructured knowledge to ordered knowledge that is easy to share.

In spite of the great importance of knowledge and their media of storage in effective and efficient hospital processes, these resources are often lost from the hospital through such factors as rotation and movement of workers, cost savings, or incorrectly prepared documentation. It also happens that the hospital is not aware of having a part of the knowledge acquired during years of activity (Chase 1998). Knowledge that is disordered and difficult to locate impedes access to it and the use of it in the process of treatment and patient care, which, in turn, translates negatively to the work of individual employees, teamwork, and, thus, the quality and correctness of the whole process.

Disquieting is the fact that apart from problems of access and use of knowledge, there have been a number of medical errors with negative effects on health and even on the lives of patients. Research conducted in the United States showed that 44,000–98,000 people died as a result of medical errors, and an estimated million were injured (Warner 2004; IOM 1999).

In the context of the consideration described above, there is no doubt that one way of improving the process of treatment and patient care is improving the process of knowledge management in the hospital. As reported in the literature, more effective sharing of knowledge will result in benefits such as reducing the time of performing of hospital processes, cost reduction, higher satisfaction index, better medical care, and better medical and education levels (Antrobus 1997).

To address the problem of knowledge management in health care, work on elaboration of complex systems supporting access to necessary knowledge resources by hospital staff during realization of hospital processes was undertaken at the Faculty of Organization and Management in the Silesian University of Technology. System elaboration is a part of a research development called "Knowledge-Based Enhancement of the Work Conditions in Health Care Organizations"—the virtual hospital is financed by the National Research and Development Center (Virtual Hospital 2009–2011).

KNOWLEDGE MANAGEMENT AND VIRTUALIZATION OF HOSPITAL PROCESSES

Wanting to build knowledge-based hospital processes, you should first identify and review the resources, which are involved in all stages of these processes. The problem connected with such action is the identification and access to knowledge that is located in the human mind, further dispersal of knowledge, and lack of structured knowledge located in other media storage. Activities related to the identification and acquisition of knowledge should include three main categories of knowledge (Nonaka and Nishiguchi 2001):

- Tacit (silent) knowledge resources—to which access is not direct, but through the observation of the behavior of people who possess such resources. This is noncodified knowledge.
- Implicit knowledge resources—to which access is not direct but is revealed by asking questions or discussions. Therefore, such knowledge is also noncodified.
- Explicit knowledge resources—easily available, documented in a formal medium of storage with knowledge. This kind of resource can be well ordered and organized with the help of informatics technology (IT) tools such as electronic patient records, medical systems, specialized portals, and so forth. They can also take the form of paper documentation.

Knowledge resources that are engaged in the hospital can be included for each listed category. The authors of this chapter use such useful tools for process identification as process mapping and detailed process analysis, whereas the helpful methods in gaining tacit and implicit knowledge are participant observations, photographic and video recordings, as well as panel discussions and interviews conducted with stakeholders.

Identification of hospital processes and, related with them, knowledge resources is the first stage in the process of knowledge management. Subsequently, there are such activities as codification of knowledge, organization, gathering, and finally, sharing of knowledge. These types of operations are necessarily supported by tools and technologies in the field of IT and information and communication technology (ICT). Codification of knowledge is, in particular the creation of representations of so-called knowledge objects. The attribute of knowledge objects (Shepherd 2004) is their repeated use and combination, thus creating a shared knowledge, depending on the contextual situation. They represent skills, good practices, models and frameworks, methods and techniques, tools, concepts, practical experience, presentations, and so forth. The knowledge objects can thus take the form of the following:

- Text descriptions
- Animation and computer simulations
- Drawings, schemes
- Computer models
- Tables
- Movies

- Internet links
- Checklists, procedures, catalogues

Process mapping allows for a graphical representation of interrelated activities on the background of organizational units (Rummler and Brache 2000). In addition to identification of activities, knowledge resources and their medias of storage needed to carry out specific activities are also identified.

Empirical studies of authors in area of the functioning of public hospitals allowed for the development a model of hospital process map (Figure 20.1), which can be called a traditional model. The model shows, in particular, the high degree of scattering of knowledge, both in the hospital, in various hospital units, as well as in an external environment outside the hospital, which are the economic, political, scientific research, and social environments.

Research has shown that such a high degree of scattering of knowledge resources and diversity of media of storage can cause the following problems:

- Knowledge needed for the implementation of the steps in the process may not be provide on time.
- There may be gaps in knowledge, for example, due to an unknown location of the current media of storage of knowledge.
- There are difficulties in the transfer and sharing of knowledge, especially tacit and implicit knowledge.
- The same information resources are generated at different locations during the process; for example, the same medical records are generated.

To address the identified problems, a way of gradual transforming the traditional functioning of health care units (according to the diagram in Figure 20.1) in knowledge-based organizations suggested. This suggested method is the virtualization of processes involving the digitization and augmentation of processes in the hospital knowledge resources, which are needed for their effective and efficient implementation. Virtual representation of these processes is a network of relationships between knowledge objects, whereas the representation is defined on the basis of knowledge of the

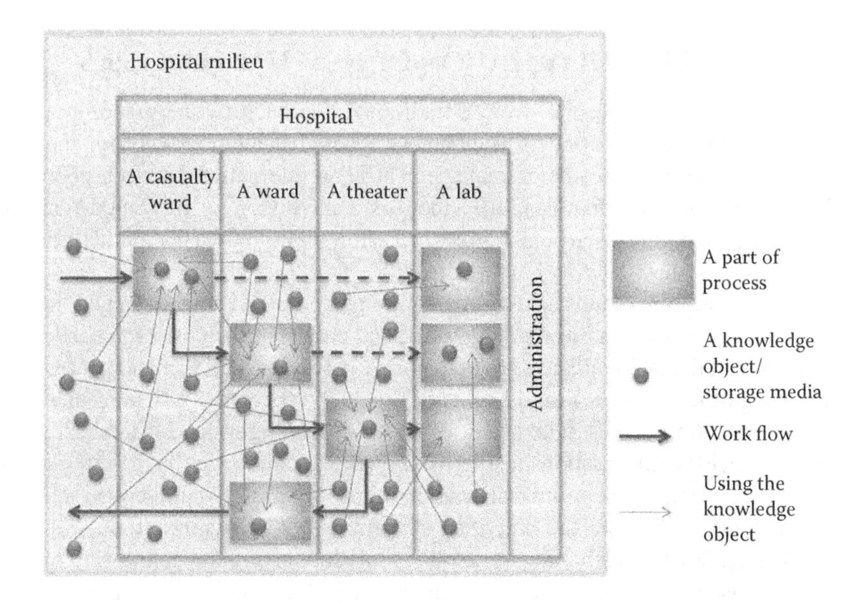

FIGURE 20.1 A model of a process map in a hospital with a traditional approach.

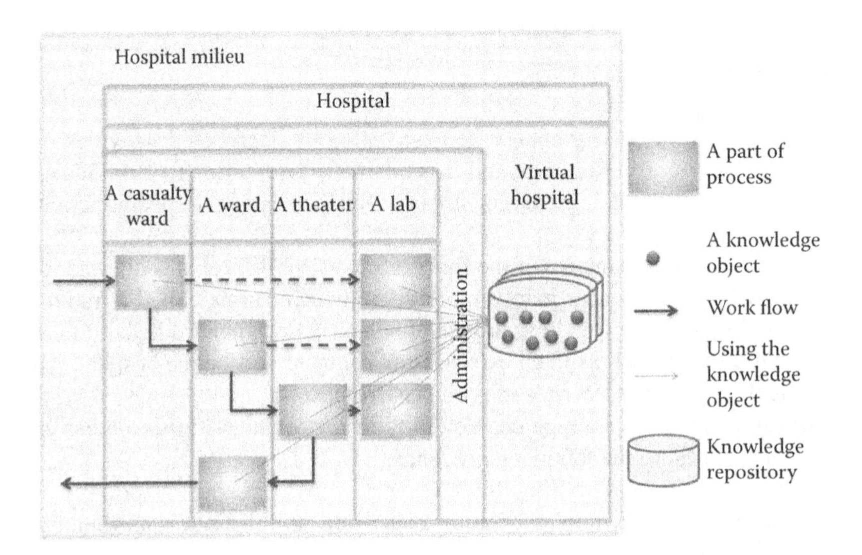

FIGURE 20.2 A model of a process map in a hospital after virtualization.

relationship occurring in the real process in the hospital. The network of relations is then implemented into a Web application. Consequently, the virtualization of processes leads to emergence of a computerized knowledge management system, which possesses structured, standardized, and repeatedly used knowledge objects, shared with the principle of "just enough" and "just in time," where the goal is to provide the right information, to the right people, at the right time (Kerschberg and Jeong 2005). The structure of such a system is based on thematic repositories of knowledge, whose content is made available to real users in real running processes using ICT tools.

Figure 20.2 shows the process map model of a hospital after virtualization. The background of the processes performed in a hospital in this case is the knowledge gathering in the repositories. Therefore, the problem of scattering of resources and media of storage of knowledge disappears. Participants in hospital processes will be able to have unfettered access to the knowledge, which they will be need at any given moment.

A METHODOLOGY OF VIRTUALIZATION OF HOSPITAL PROCESSES

In changing the activity of the hospital from a traditional approach to one based on a knowledge organization's approach, there is a need to elaborate a detailed methodology for the procedure. Keeping in mind the fact that knowledge management is an interaction between people, process, and technology (Oduwole and Onatola 2010), there is a methodology presented that takes into account these three aspects in this article. Figure 20.3 shows the links between these three aspects from the study's point of view.

In the study, there are four groups of users who have personalized access to specific knowledge resources: medical staff, nurses, administrative staff, and patients. Each user is a participant in the processes performed in the hospital and a participant in knowledge management processes, such as creating and sharing of knowledge. Finally, all issues related to knowledge management, including virtualization of hospital processes, take place with the using of high-tech IT and ICT.

The literature provides information on the use of such technologies in health care in terms of knowledge management. It pays, inter alia, attention to the tools for the construction of medical libraries and tools supporting the activities of trainings of medical staff in work processes (Oduwole and Onatola 2010). It also indicates mobile tools supporting the monitoring of patients by health workers, in this case, the midwives (Chib et al. 2008). It emphasizes the need for permanent

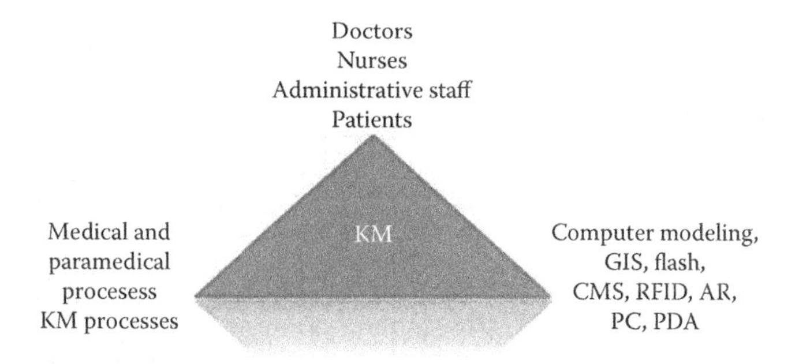

FIGURE 20.3 A model of knowledge management in a hospital.

improvement of skills and knowledge that can be done through technology and the Internet/intranet (Jacko et al. 2001; Mac et al. 2001).

In addition to the construction of knowledge repositories, tools for automatic data reading using radio waves are proposed. Discussion about the above theme concerns, inter alia, using radio-frequency identification (RFID) technology to enhance patient safety, for example, in terms of orientation in the space of the hospital (R-Moreno et al. 2011); effective treatment and reflection on economic issues related with implementation of a system based on RFID (Wicks et al. 2006); and the use of RFID in the form of implants of so-called Veri-Chips (Smith 2008). The literature reports that RFID has potential use all along the patient supply chain, particularly as a clearinghouse of information regarding medications, testing, treatments, patient tracking, equipment location, inventory, timing, and many other applications in ancillary care, pharmaceuticals, direct patient care, medical supplies, and work flow (Revere et al. 2010; Bartnicka and Smolorz 2010).

Training plays an important role in forming the knowledge-based hospital organization. An environment that supports preparation of training materials is a virtual environment. It uses technology based on both virtual reality (Keskitalo 2011; Rosen 2008) as well as augmented reality (AR) (Sielhorst et al. 2004).

In addition to technologies that support the activities of hospitals mentioned in the literature, the authors of this chapter propose a new solution based on geographic information systems (GISs). A GIS is a complex computer software for saving and presentation of graphic data of spatial character (Longley et al. 2006; Gotlib et al. 2007).

Taking into account the opportunities posed by modern informatics and ICT, a model of creating knowledge-based virtualization of hospital processes was elaborated.

In frame of the study, virtualization of the process of arthroplasty of the knee joint was done. The conversion from a traditional model to a knowledge-based model is illustrated in Figure 20.4.

The particular stages were the following:

1. Identification and formal recording of a process using such methods as interviews and expert panels with the participation of hospital staff. At this stage, both the procedures encoded in the form of documentation, as well as the unwritten procedures, which represent the customary behaviour of staff (implicit knowledge), were taken into account.
2. Identification and acquisition of knowledge involved in the process. The methods in this stage were participating in observations, video and photographic recordings, interviews, and analysis of the literature. Recording of activities described in the map process, including some surgical operations, hospital infrastructure, and service equipment, was done. To a large extent, the acquired knowledge is tacit and implicit.

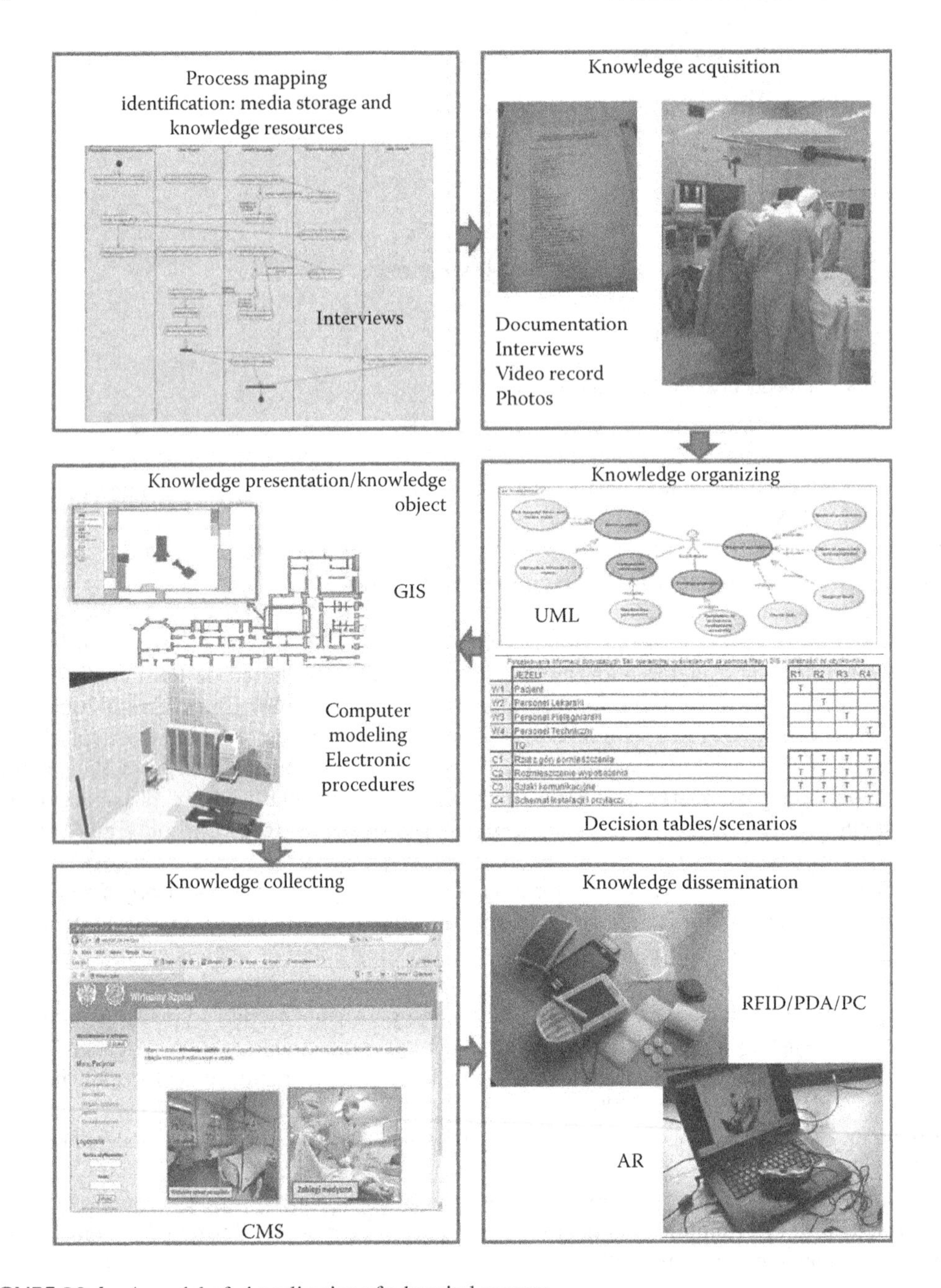

FIGURE 20.4 A model of virtualization of a hospital process.

3. Organizing knowledge assets using methods such as decision tables, scenarios of possible variants of processes, and unified modeling language (UML) diagrams.
4. Knowledge presentation using computer methods, such as computer modeling methods based on computer aided design (CAD) systems, GISs, and electronic documents. At this stage, training movies based on computer simulation were prepared.
5. Gathering resources using content management systems content management system (CMS) and the creation of a Web application.
6. Sharing knowledge via such technologies as RFID and AR.

Selection of tools and technologies for knowledge sharing for groups of users defined above is intended to provide contextual knowledge in a given place and time without needing to search for the right resources in the repository.

CONCLUSION

Virtualization of hospital processes is the first step in building a knowledge-based organization. The difficulties associated with this phase are primarily the acquisition and the codification of knowledge, especially tacit and implicit knowledge, as well as the way of ordering and the selection of key resources for the process. Subsequently, it becomes problematic to choose the appropriate size of the knowledge objects so as to be able to reuse them in multiple contextual situations.

Another step, perhaps more difficult to achieve, is to implement a computerized system based on the virtualization of processes in the real activity of hospital processes and activity of their participants. Changing habits, although justified, has repeatedly met resistance from workers, all the more in frame of the application of modern ICT technology. This problem is present in several publications in the field of knowledge management and development of knowledge-based organizations.

ACKNOWLEDGMENT

This work was elaborated in the frame of the development project (contract no. N R11 0026 06/2009) called "Knowledge-Based Enhancement of the Work Conditions in Health Care Organizations" financed by the National Research and Development Center (project leader: Professor Teodor Winkler, Silesian University of Technology).

REFERENCES

Antrobus, S., Developing the nurse as a knowledge worker in health—Learning the artistry of practice. *Journal of Advanced Nursing*, 25(4), 823–829, 1997.

Bartnicka, J. and Smolorz, M., Zastosowanie technologii RFID w zarządzaniu zasobami w placówkach opieki zdrowotnej. In: R. Knosala (ed.), *Komputerowo zintegrowane zarządzanie*. Tom I, Oficyna Wydawnicza Polskiego Towarzystwa Zarządzania Produkcją, Opole, 2010.

Chase, R. L., Knowledge navigators. *Information Outlook*, 2(9), 18, 1998.

Chib, A. et al., Midwives and mobiles: Using ICTs to improve healthcare in Aceh Besar, Indonesia. *Asian Journal of Communication*, 18(4), 348–364, 2008.

Gotlib, D. et al., *GIS. Areas of Applications*. PWN, Warsaw, Poland, 2007.

Institute of Medicine (IOM), *To Err is Human: Building a Safer Health System*. National Academy Press, Washington, DC, 1999.

Jacko, J., Sears, A. and Sorensen, S., Framework for usability: Healthcare professionals and the Internet. *Ergonomics*, 44(11), 989–1007, 2001.

Kerschberg, L. and Jeong, H., Just-in-time knowledge management. Proceedings of Wissensmanagement '2005, 1–18, 2005.

Keskitalo, T., Teachers' conceptions and their approaches to teaching in virtual reality and simulation-based learning environments. *Teachers and Teaching*, 17(1), 131–147, 2011.

Longley, P. A. et al., *GIS. Theory and Practice*. PWN, Warsaw, Poland, 2006.

Mac, H.-Y., Mallard, A. and Kwok, J., Designing a hospital intranet nurse learning system for improving continuing nursing education. *Innovations in Education and Teaching International*, 38(4), 397–405, 2001.

Nonaka, I. and Nishiguchi, T., *Knowledge Emergence*. Oxford University Press, Oxford, 2001.

Oduwole, A. A. and Onatola, A. D., Electronic technology tools for knowledge management by health information professionals. *Journal of Hospital Librarianship*, 10(3), 305–314, 2010.

Revere, L., Black, K. and Zalila, F., RFIDs can improve the patient care supply chain. *Hospital Topics*, 88(1), 26–31, 2010.

R-Moreno, M. et al., Multi-agent intelligent planning architecture for people location and orientation using RFID. *Cybernetics and Systems*, 42(1), 16–32, 2011.

Rosen, K., The history of medical simulation. *Journal of Critical Care*, 23, 157–166, 2008.
Rummler, G. A. and Brache, A. P., *Improving the Efficiency of Organization*. PWE, Warsaw, Poland, 2000.
Sharma, S., Wickramasinghe, N. and Gupta, J., Knowledge management in healthcare. In: N. Wickramasinghe, J. Gupta and S. Sharma (eds.), *Creating Knowledge-Based Healthcare Organizations*. Idea Group Publishing, USA, UK, 2005.
Shepherd, C., Objects of interest. TAFE Innovative Project, http://fastrak-consulting.co.uk/tactix/features objects/objects.htm, 2004.
Sielhorst, T. et al., An augmented reality delivery simulator for medical training. Workshop AMI-ARCS 2004, held in conjunction with MICCAI '04, September 30, 2004, Rennes, France, http://ami2004.loria.fr/PAPERS/26obetoebiel.pdf, 2004.
Smith, A., Evolution and acceptability of medical applications of RFID implants among early users of technology. *Health Marketing Quarterly*, 24(1), 121–155, 2008.
Virtual Hospital, Research development "working conditions in health care organizations improvement based on knowledge" called: "virtual hospital" (contract no N O11 0026 06/2009), 2009–2011.
Warner, M., Under the knife. *Business 2.0*, 5(1), 2004.
Wicks, A., Visich, J. and Li, S., Radio frequency identification applications in hospital environments. *Hospital Topics*, 84(3), 3–9, 2006.

21 Advanced Technologies in the Aerospace Industry

Marek Goral, Jan Sieniawski, Krzysztof Kubiak, and Arkadiusz Onyszko

CONTENTS

INTRODUCTION

During the last decade, the global air traffic has been growing rapidly. This caused the growth in the production of modern liners. There is a tendency to reduce the exploitation costs, including the amount of fuel being used. Simultaneously, more restricted requirements in the area of environment protection are being introduced, which involves a reduction of the amount of noxious substances formed during combustion of fuel. One of the essential development directions in aircraft engine construction is raising the combustion temperature, which increases the efficiency, and reduction of the amount of noxious substances being created. This determines the usage of high temperature–resistant and creep-resisting materials with complex chemical composition and sophisticated microstructure, for elements of the hot parts of aircraft engines, for example, turbine blades.

The elevation of aircraft engines' service temperature is possible by using turbine blades made of nickel-based superalloys with equiaxed microstructure (EQ), directionally solidified (DS), or with single-crystal (SC) microstructure. The third generation of SC alloys used in turbine engines is currently used. At present, there is also development work on the fourth generation of those materials containing platinum metal alloy additions being done. High temperature–resistant coatings are being deposited on the surface of turbine blades. They have a function of corrosion protection, oxidation resistance, and thermal isolation. At present, the thermal barrier coatings are used for this purpose.

In the Research and Development Laboratory for Aerospace Materials at Rzeszow University of Technology, the production technologies of advanced materials and coatings, used in most modern aircraft engines in the world, are being developed. The concept of work of the laboratory is connected with development of technology for production of elements in each stage—from the computer aided design (CAD) model to the final product. The end product can be, for example, a single-crystal blade, after a heat treatment process, with a thermal barrier coating deposited after final machining. Thus, it is necessary to understand the metallurgy of the blade material with respect to the microstructure, chemical and phase composition, and mechanical and corrosion properties. The laboratory is also certified by NADCAP for conducting research in the area of material characterization and consistency of conducted research with the ISO 17025 standards.

NUMERICAL SIMULATION OF THE CRYSTALLIZATION PROCESS CASTED BY THE BRIDGMAN METHOD

The application of dedicated software for creating cast geometry and for casting process modeling during the design process of cast manufacturing technology using the lost-wax casting method (cire perdue method) is a huge convenience.

The Research and Development Laboratory for Aerospace Materials is equipped with ProCAST software for numerical simulation of the casting process. This program enables numerical simulation of processes: filling a mold with liquid metal, directional crystallization, volume crystallization of casting alloys in a casting mold during the cooling process, and creation of stresses in casts and molds during the crystallization process. Its functionality includes modeling the size and shape of grains and their crystallographic orientation and prediction of cast defects. There is also a possibility to determine, during so-called inverse modeling, the thermophysical parameters of materials and boundary conditions.

According to the direction of conducted research, one can use suitable modules of ProCAST to simulate of some casting processes, that is, MeshCAST, FlowSolver, thermal solver, stress solver, inverse modeling, and CAFE module (Szeliga et al. 2009).

Figure 21.1b shows an example of temperature distribution in a ceramic mold and of a furnace heating chamber for directionally crystallized casts made by Bridgman method. As a result of shifting of the ceramic mold through the cooling area (cooling ring), the temperature distribution in the ceramic mold changes according to time of duration of crystallization process.

The CAFE module is based on a combination of cellular automats (CAs) and the finite element method (FEM). It allows us to predict a type, size, and grain amount in a cast (Figure 21.2). This

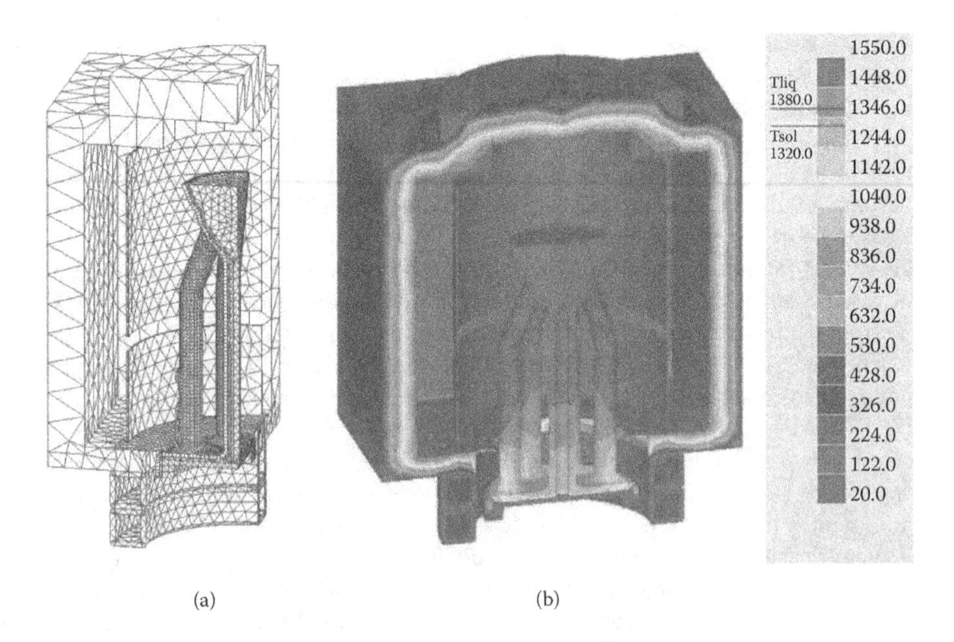

FIGURE 21.1 The finite elements of mesh and temperature distribution of ceramic mold layers and furnace heating chambers (a) after shifting of the mold at a distance of 67 mm (b).

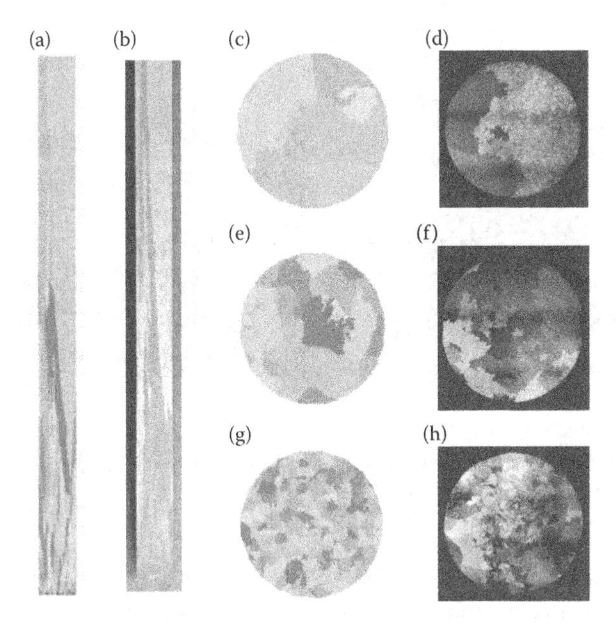

FIGURE 21.2 Predicted (a) and actual (b) shape of grains in cross section and longitudinal section of bar cast at a height of 140 mm (c and d), 60 mm (e and f), and 13 mm (g and h).

makes it possible to reduce or even entirely eliminate, at a certain stage, the necessity to create expensive test casts. A differential equation describing the temperature field is solved with the use of FEM, whereas a microstructure development with release of phase transition heat during the crystallization process is analyzed using CA. The model area in entirely discretized with overlapping FEM and CA meshes.

TECHNOLOGY OF SINGLE-CRYSTAL BLADE PRODUCTION

The Research and Development Laboratory for Aerospace Materials has conducted, for several years, a research project within many research projects, on development casting techniques for an aircraft engine turbine element made of nickel-based single-crystal superalloys (e.g., CMSX-4). The process of blade production with the same structure is a multistage process and includes

- Making a blade wax model followed by preparation of a packet-form blade set
- Creating a ceramic mold and melting the wax in an autoclave
- Creating a mold
- Casting the material into the mold
- Disassembling the mold and cutting the blades off
- Heat treatment of blades in a high vacuum furnace

The first stage of creating the single-crystal turbine blades is making wax model (blades, gating systems, deaeration systems, pans) with the use of an injection molding method. Elements made of waxes with different physical, chemical, and mechanical properties are joined into a model set.

The model set is, after degreasing, coated with the first sand layer (so-called facing sand). Afterward, the first layer is lavished with ceramic mix (mostly used are zirconium orthosilicate [$ZrSiO_4$], Al_2O_3, ZrO_2, mullite, and Y_2O_3) with suitable granularity (200–300 mesh). After the coating of a proper amount of layers and final drying of the model set, the wax melt-out is carried out.

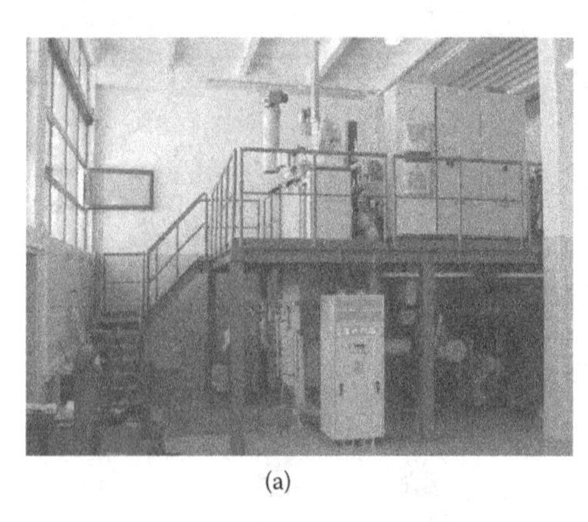

(a)

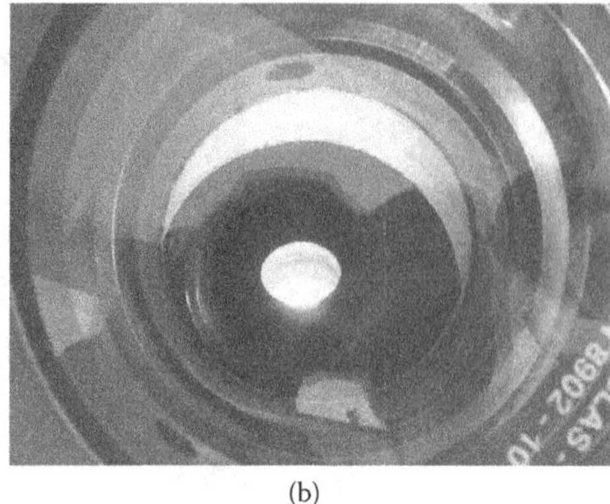

(b)

FIGURE 21.3 Furnace for melting and casting of SC, DS, and equiaxed nickel- and cobalt-based superalloys (a) and casting process of single-crystal turbine blades (b).

The process is run in autoclave at a temperature of 180°C and under about 8 atm pressure of water vapor. The next stage is drying and annealing of molds in an electric furnace at a temperature of 1300°C. After annealing, molds are submitted to a leak test, which allow us to eliminate all pieces with fractures. Single-crystal turbine blades of aircraft engines are made of CSMX-4 and CMSX-6 alloys in a vacuum furnace (manufactured by ALD) for melting and casting of elements with equiaxed microstructure, directionally crystallized, and monocrysta line (Figure 21.3).

The casting process is based on Bridgman's method, which includes gradual alloy crystallization and isolation of a single particle in a spiral-shaped selector. Besides the selector, the cast should consist of a single particle. Temperature control and mold feeding speed are critical factors, determining the cast quality. The typical temperature gradient is 80°C–170°C, and the shift speed is about 30 cm/h. It is thought that the higher the temperature gradient, the better the cast quality is. The significant difference between the CS and DS casting methods is the manner of selection of a single particle with suitable orientation. In most of the used technologies, the special section with a spiral-shaped section is located between the cooling plate and casting mold. This helix or the spiral-shaped particle selector works as a filter and allows only one particle to move in a certain direction. Because the superalloys solidify with a negative temperature gradient, the dendritic growth can occur in all mutual perpendicular directions. The constant change of helix directionality limits the solid-phase growth to one particle, until the moment of appearance in the mold. Obtained blade casts—for purposes of maintaining the proper mechanical properties—are heat treated in a high vacuum furnace designed to obtain a high cooling rate during the solutionizing process.

HIGH TEMPERATURE–RESISTANT COATINGS ON TURBINE BLADE ELEMENTS

The process of creating of single-crystal turbine blades is always connected with deposition of high temperature–resistant coatings on their surface. The application of this process allows us to increase the durability and service temperature of modern aircraft engines. The function of coatings is to protect the surface from high temperature and the aggressive environment of gases created in the combustion chamber of an aircraft engine. The most advanced coatings are thermal barrier coatings, which have an influence on increasing the durability of blades through

- Protection of base material from oxidation and corrosive influence of the fuel component, for example, sulfur
- Decreasing the elements' surface temperature function of thermal isolation

Thermal barrier coatings (TBC) consist of, at least, two layers, each of them having one of two functions described above. A bond coat is a metallic coating consisting of elements that increase heat resistance of the material surface. There are two kinds of bond coats that are currently used:

- Multicomponent alloys containing nickel, chromium, and aluminum (MCrAlY)
- Diffusion, platinum-modified coatings, which are based on the NiAl phase (Feuerstein et al. 2008)

ALUMINIDE COATINGS

Diffusion aluminide coatings are based on the intermetallic β-NiAl phase and serve as a bond coat mainly for ceramic coatings deposited with the electron beam physical vapour deposition (EB-PVD) method (Haynes et al. 2001). Until now, the main process used to create this coating—with regard to low costs and ability of coating many small elements in one process—was the pack cementation method. Elements were placed in the container with aluminum oxide powder (filler), aluminum (coating-forming material), and halide activator. Deposition was performed in a temperature range between 800°C and 1000°C as a result of halide reaction with the element's surface, causing the deposition of aluminum followed by formation of the NiAl phase in a process of diffusion and chemical reactions. More advanced aluminizing methods are "over-the-pack methods," vapor phase aluminizing, and chemical vapor deposition (CVD). The described processes allow us to control aluminum halide flow to chosen elements areas, which are to be coated and are usually applied for coating of inner elements, for example, blades with internal cooling channels (Warnes and Punola 1997).

According to aluminum activity, one can distinguish two basic types of aluminizing process:

- High temperature, low activity (HTLA), conducted in the temperature range of 1050°C–1100°C, in which the NiAl phase is formed during outward nickel diffusion
- Low temperature, high activity (LTHA), conducted in the temperature range of 700°C–950°C, which results in Ni_2Al_3 are being created (formation of the NiAl phase occurs as a result of inward aluminum diffusion). To obtain a homogeneous NiAl coating, the additional diffusion treatment in the vacuum of a protective atmosphere environment is performed for this kind of coating.

In the Research and Development Laboratory for Aerospace Materials, research is being conducted on the development of technology for obtaining diffusion aluminide coatings by CVD method (Figure 21.4). The deposition process is performed in a device consisting of

- External generators of coating-forming gases
- A carrier and reaction gas line with a gas flow controlling system and a vacuum pump system
- A top hat furnace with retort, in which coated elements are placed, with a multizone temperature regulation system

As a result of the chemical reaction between hydrogen chloride and aluminum granules taking place in the outer generator, the aluminum chloride is created, which, after feeding inside the retort, reacts with the blade surface made of a nickel-based superalloy and forms the NiAl phase. The process is conducted under low pressure—from 50 to 500 mbar—in the temperature range between 800°C and 1000°C and takes from 2 to 8 h.

Further increase in oxidation resistance of aluminide coatings can be obtained by platinum addition. The influence of platinum is connected with

- A rise in the degree of aluminum diffusion into the base material during diffusion aluminizing
- A significant increase in oxidation resistance of platinum-modified aluminide coatings

FIGURE 21.4 The CVD laboratory system for aluminide coating production.

The typical coating process includes electroplating of 5–10 μm platinum followed by diffusion aluminizing in low- or high-activity processes (Zhao and Xiao 2008). The microstructure of platinum-modified coating can have a few characteristic structure types, due to platinum amount and process type. In a single-phase structure, platinum dissolves in a solid solution, creating (Ni, Pt) Al. In double-phase coatings, the $PtAl_2$ phase separation on a NiAl phase base is created. It can be observed in the form of many separations, which can form even a continuous zone. The fundamental disadvantage of double-phase coatings is their brittleness (Angenete and Stiller 2001).

High costs of platinum during recent years caused demand for research conducted on application of other elements that have similar properties and allow us to obtain cheaper coatings. The two most important elements that can replace Pt are hafnium and palladium, introduced with the use of physical, diffusion, and galvanic methods (Hong et al. 2009).

In the Research and Development Laboratory for Aerospace Materials, research is being conducted on obtaining platinum-modified coatings and palladium-modified coatings formed in a process of depositing galvanic Pt and Pd coatings, followed by diffusion treatment and CVD aluminizing. Coatings created in such a way as to form a bond coat for the most advanced TBC coatings obtained during, for example, the EB-PVD method. Obtained coatings play a role in increasing high temperature resistance of IN-738 and CMSX-4 alloys, which was confirmed by cyclic oxidation tests.

For further rise of oxidation resistance of aluminide coatings and extending of engine exploitation time, the additional modifying elements like hafnium and zirconium during the CVD process to the coating are introduced. It is possible to obtain through mounting of additional high-temperature reactive gas generators. A small addition of those elements is introduced to aluminide coatings and allows us to increase the high temperature resistance. This is a relatively simple treatment implemented during a typical aluminizing process. Moreover, it is possible to perform a diffusion chromium plating, which allows us to increase significantly hot corrosion resistance, caused by sulfur compounds.

The CVD technology for manufacturing coating does not apply only to high temperature–resistant coatings. With the use of a device installed in the laboratory, it is possible to perform a process to obtain hard, multilayer coatings such as TiN, TiCN, and Al_2O_3 hard ceramics, which are commonly used for increasing cutting tools' durability, for example, milling cutters, drills, and wear plates. There is a possibility of conducting development work for many kinds of coatings used for increasing the wear resistance. Hard coatings are also utilized for protection of compressor blades' surface against the influence of erosion factors, for example, in desert conditions. However, it is necessary to develop a multilayer coating, which protects not only against erosion but also against corrosion of the compressor blade, for example, during flights over maritime areas.

Besides the conventional thermal activated CVD method, there are plans to conduct research on CVD processes with the use of glow discharge. The use of this process makes it possible to obtain a coating at a lower temperature and with more complex chemical composition. A machine enabling carburizing, glow-discharge nitriding with oxidation will be utilized. Obtained coatings can be used not only for compressor blades but also for increasing the durability of machine components, for example, gear wheels, and so forth.

MeCrAlY BOND COATS

The first commercial bond coat was developed by the Pratt Whitney company in the late sixties. At first, those kinds of coatings were created, that is, with the EB-PVD method; later, however, there were also other technologies, that is plasma spraying (air pressure plasma spraying [APPS], low pressure plasma spraying [LPPS], vacuum plasma spraying [VPS]) and high velocity oxy fuel (HVOF) spraying. Praxair company developed a technology for galvanic deposition of Ni or Co with the addition of CrAlY followed by diffusion annealing. Application of MeCrAlY coatings was expanded to coating of many turbine engine elements, that is, casing, abrasive coating of blades, and bond coats for TBC coatings. In comparison to diffusion coatings, these coatings are independent from the base material chemical composition and have higher elasticity. The chemical composition of a coating can be matched precisely depending on the predicted degradation mechanism. MeCrAlY coatings usually contain at least four elements. Chromium allows us to obtain high corrosion resistance and oxidation resistance at the same time. After coating, it is necessary to perform additional heat treatment in vacuum for adhesion improvement. Those coatings are characterized by a double-phase structure ($\alpha + \beta$) (NiAl + Ni). The NiAl phase in a coating serves as an aluminum reservoir. With the increase of the thermally grown oxides (TGO) zone, the thickness of phases with a high content of aluminum in the outer area of coating is decreasing, and the γ phase becomes predominant. The depleted area in NiAl can cause the creation of oxidation pits. Those pits are created sometimes as a result of reactive elements, that is, zirconium and hafnium. Those oxides are called "pegs" in the literature, and they are one of many mechanisms responsible for the increase in oxide scale adhesion. The performance of MeCrAlY coatings is a result of thermal and chemical consistency with the base material and minimal interaction with base material properties. The efficiency of MeCrAlY coatings is a result of their ability to form a dense, protective scale, which moderates all reactions between the base material and hot corrosive environments. Aluminum oxide is the main protective scale; any other elements forming oxides do not have good protective properties like aluminum, although their creation does not favor further surface alloying of MeCrAlY.

The formation of scale is connected with aluminum activity and its diffusivity in an alloy. The increase of aluminum activity is possible through the introduction of 17%–25% of chromium, which causes the decrease of aluminum content necessary to create and maintain the oxide layer and results in high corrosion resistance.

Aluminum used in coatings in the amount of 8–12 wt.% forms an oxide layer, is responsible for the creation of a slowly growing adhesive oxide scale (TGO), and has the function of aluminum reservoir, completing the formation of the oxide layer. Yttrium, hafnium, and zirconium are reactive oxides that have a positive influence on scale adhesion. MeCrAlY coatings contain usually 1% Y (or less). Addition of hafnium and zirconium plays the same role; however, it also protects from scaling of TGO.

Research was conducted to introduce those elements to TBC coatings. Addition of rhenium results in an increase in cyclic and isothermal oxidation and thermal fatigue resistance. Silicon

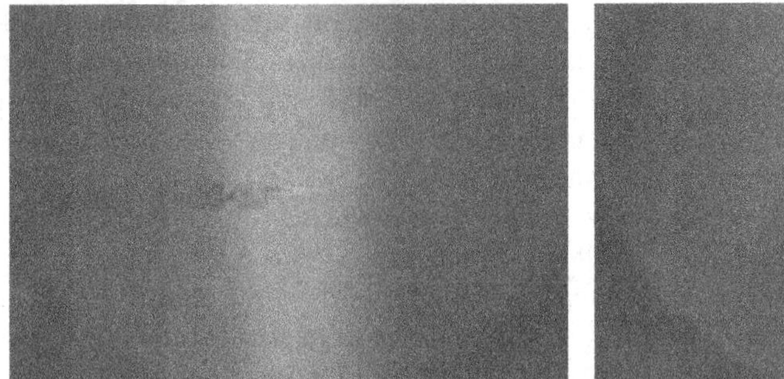
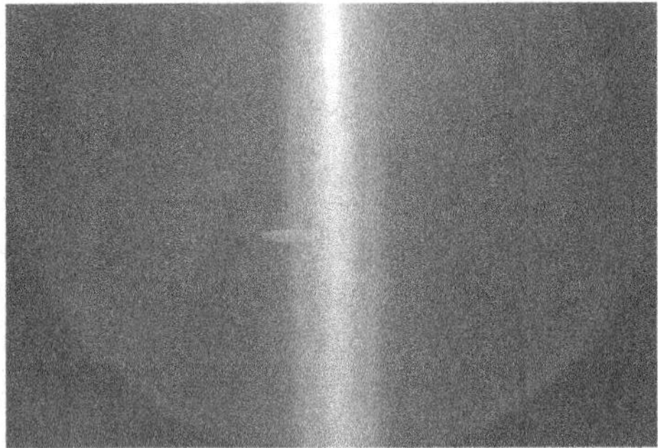

FIGURE 21.5 The plasma spraying process using LPPS thin film technology.

influences, in lower quantity, positively on aluminum oxide scale adhesion, and introduction of high-melting tantalum also improves high-temperature properties and sulfur corrosion resistance.

The Research and Development Laboratory for Aerospace Materials uses plasma spraying technology under low pressure (LPPS thin film) for deposition of MeCrAlY bond coats (Figure 21.5). In this method, the metallic powder is partially melted and sprayed on the blade surface by plasma gun. For the first time in Poland, it is possible to perform such a process in vacuum conditions. It ensures perfect properties of the coating, containing no oxides, which occurs in processes of plasma spraying under atmospheric pressure (APS). Moreover, the obtained coatings are characterized by lesser thickness in comparison to those obtained by APS, as a result of using thin film technology. Multicomponent MeCrAlY coatings protect the surface of blades, which are under particularly large influence of sulfur compounds (hot corrosion). Resistance is ensured by the presence of chromium in a coating. Additionally, aluminum protects the base material from oxidation. Currently, there is research being conducted on spraying of powders with different chemical compositions aimed at determining coating process parameters providing the best properties. One of the more promising processes is connected with developing multilayer coatings with gradient structure, obtained during mutual LPPS plasma spraying and CVD aluminizing. It enables us to create coatings with higher aluminum content on the surface and better corrosion resistance and characterized by chemical composition independent from the base material.

LPPS technology has its applications not only in aviation but also in biomedicine (e.g., hydroxyapatite spraying on titanium implants) and fuel cell manufacturing technology (ceramic layers).

The technology of plasma spraying under low pressure is also used to create outer ceramic TBC coatings. The unique function of the LPPS thin film device is the ability to obtain low pressure in a vacuum chamber (below 1 mbar), which allows the creation of a high-energy plasma jet, which enables evaporation of ceramic material (plasma spray–physical vapor deposition). The formed ceramic layer can have complex columnar crystal microstructure—close to those obtained with the EB-PVD method. This technology is unique not only in scale of the country but also in scale of the Europe and is a subject of interest of leading aircraft engine manufacturers (Refke et al. 2010).

CERAMIC COATINGS

The outer TBC coating is made of ceramic material—yttria-stabilized zirconia oxide—providing thermal isolation. Electron beam physical vapor deposition is the most advanced method of ceramic coating deposition on particularly loaded single-crystal turbine blades of aircraft engines. EB-PVD technology is used by the world's leading aircraft engine manufacturers. The EB-PVD process consists in evaporating in high vacuum conditions of ceramic material, in the form of a ceramic ingot, with the use of a high-power electron gun "scanning" its surface. Vapor sediments on the blades' surface form columnar crystals and create a compact coating with a thickness of about 0.1 mm. These kinds of coatings are used only for the first stages of an aircraft engine and ensure a decrease in surface temperature. EB-PVD technology enables the obtaining of blade surface of high smoothness and with variable thermal stress resistance (Singh et al. 2002).

The prototype devices for deposition of this type of coating was installed in the Research and Development Laboratory for Aerospace Materials at the Rzeszow University of Technology and will be used for the deposition of ceramic coatings on bond coats created with the methods described above: CVD and plasma spraying LPPS thin film. There is also research being conducted on creating new ceramic materials, which are characterized by lower thermal conductivity than the standard values for this kind of application of yttrium-oxide–stabilized zirconium oxide (YSZ). There is also a plan to deposit ceramic coatings with complex crystal structure, for example, "zigzag," with better properties and to undertake an investigation of a numerical model describing coating formation (Schultz et al. 1997; Lugscheider et al. 1998).

LASER TECHNOLOGIES

Creating turbine blades for advanced aircraft engines with TBC coatings requires making, after coating, small holes, which enable supplying of cooling air. They have a diameter of about 1 mm and have to be created in strictly defined places. An additional difficulty is the ceramic coating, which is characterized by high embitterment and thermal resistance. The only used method of creating holes is by application of laser techniques. It allows us to obtain holes without significant influence on material properties and preserves the good properties of TBC coating in its surroundings.

In the laboratory, research will be conducted on creating holes in TBC coatings obtained in the EB-PVD and plasma spraying process. Moreover, a laser device will be available, which will enable the conducting of work on cutting and welding of different kinds of materials used in aviation.

CONCLUSION

Intensive development of the aviation industry, including engine construction, requires the usage of modern technologies of creating of materials and coatings. As a result, it is necessary to develop new kinds of materials and coatings and to test currently used technologies for new engines on a laboratory scale. This generates large costs, which must be covered by aerospace companies; therefore, it is required for the manufacturers to cooperate with research institutes to enhance a range of conducted investigations and introduce completely new technologies. The Research and Development Laboratory for Aerospace Materials is able to conduct research and development in every aspect of aircraft engines elements manufacturing, from computer models to the end product, for example, aircraft engine blades. Development of current, modern technologies, which have not been used before in the aviation industry, allows us to design better aircraft engine constructions. The equipment of the laboratory enables us to test developed materials in terms of their microstructure, chemical and phase composition, mechanical properties (for different temperatures), and corrosion resistance. Planned research will contribute to further intensive development of the Polish and world aviation industry.

REFERENCES

Angenete J., Stiller K. A comparative study of two inward grown Pt modified Al diffusion coatings on a single crystal Ni base superalloy. *Materials Science and Engineering* A316, (2001), 182–194.

Feuerstein A., Knapp J., Taylor T., Ashary A., Bolcavage A., Hitchman N. Technical and economical aspects of current thermal barrier coating systems for gas turbine engines by thermal spray, and EBPVD: A review. *Journal of Thermal Spray Technology* 17(2), (2008), 199.

Haynes J.A., Lance M.J., Pint B.A., Wright I.G. Characterization of commercial EB-PVD TBC systems with CVD (Ni,Pt)Al bond coatings. *Surface and Coatings Technology* 146–147, (2001), 140–146.

Hong S.J., Hwang G.H., Han W.K., Kang S.G. Cyclic oxidation of Pt/Pd-modified aluminide coating on a nickel-based superalloy at 1150°C. *Intermetallics* 17, (2009), 381–386.

Lugscheider U., Barimani C., Dopper G. Ceramic thermal barrier coatings deposited with electron-beam physical vapour deposition technique. *Surface and Coatings Technology* 98, (1998), 1221–1227.

Refke J., Gindrat M., von Niessen K., Damani R. LPPS thin film: A Hybrid Coating Technology between Thermal Spray and PVD for Functional Thin Coatings and Large Area Applications. Thermal Spray 2007. Global Coating Solutions: Proceedings of the 2007 Thermal Spray Conference, Beijing China.

Schultz U., Fritscher K., Peters M. EB-PVD Y_2O_3 and CeO_2/Y_2O_3—stabilized zirconia thermal barrier coatings-crystal habbit and phase composition. *Surface and Coatings Technology* 82, (1997), 259–269.

Singh J., Wolfe D.E., Singh S. Architecture of thermal barrier coatings produced by electron-beam physical vapour deposition (EB-PVD). *Journal of Materials Science* 37, (2002), 3261–3267.

Szeliga D., Suchy J.S., Sieniawski J. Symulacja numeryczna procesu krystalizacji odlewów łopatek turbinowych z weryfikacją doświadczalną [Numerical Simulation of Crystallisation Process of Turbine Blades Cast with Experimental Verification]. Conference Proceedings "XXXVII Szkoła Inżynierii Materiałowej, Kraków–Krynica 2009."

Warnes B., Punola C. Clean diffusion coatings by chemical vapour deposition. *Surface and Coatings Technology* 94–95, (1997), 1–6.

Zhao X., Xiao P. Effect of platinum on the durability of thermal barrier systems with a $\gamma+\gamma'$ bond coat. *Thin Solid Films* 517, (2008), 828–834.

Section II

Psychology

22 Leaders' Mind Types and Their Relations to Preferred Leadership Styles

Jolanta Babiak and Czeslaw S. Nosal

CONTENTS

INDIVIDUAL DIFFERENCES IN LEADERSHIP RESEARCH

In management psychology research, cognitive attributes of leaders have probably been the most frequently studied traits within the individual differences paradigm (Zaccaro, Kemp, and Bader 2004). Although trait theories have been virtually neglected for many decades (see Bass 1990 for review), the newest evidence supports the notion that, among numerous variables influencing leadership processes, broadly defined personality and cognitive traits as well as leadership behaviors have great impact on leadership outcomes (Norton 2010; Zaccaro 2007). Early meta-analysis conducted by Lord, DeVader, and Alliger (1986) asserted that some individual characteristics, such as intelligence, masculinity–femininity, and dominance, are connected to leadership perceptions. In the similar vein, more recent meta-analysis provided evidence that general intelligence is significantly related to leadership perceptions and effectiveness (Judge, Colbert, and Illies 2004). However, while the linkages of cognitive ability, for example, intelligence, to leadership have been well researched, there are fewer studies that propose other types of cognitive traits, for example, original thinking (Morrow and Stern 1990), cognitive openness, comprehensive reasoning, and acceptance of the unknown (see Norton 2010), as important leader attributes.

While organizational leaders are faced with so many new and unpredictable challenges, signified by increasing speed of change in the external environment of the organizations, world's economy near-collapse situation, and constant pressure to create more added value, it appears to be especially important to study mental attributes, other than general intelligence, which might help and enable leaders to successfully function in this fast developing and unpredictable environment. Current scholar attention to leadership and psychological mechanisms governing leaders' behavior has taken into account this general perspective of a situation being in constant flux.

In line with the current perspective, in our approach to managerial leadership presented in this article, we focus our attention on the manifestation of more compound individual differences. In our study, we concentrate on the linkages between various individual cognitive traits and managerial behaviors in a group of 323 Polish managers. Our research methodology follows two approaches, namely, variable approach, which allowed investigating relationships between cognitive preferences and leadership styles, and a typological approach, which led to identifying and examining relationships between specific mind types and leadership profiles.

STUDY PARTICIPANTS AND MEASURES USED

Participants were 323 Polish managers working in various types of business organizations. The sample was composed of 211 males (65.3%) and 112 females (34.7%). Almost one-third of the entire sample was 50 years old or older and the average age was 41.53. Top positions, such as company presidents, vice presidents, and managing directors, accounted for 20% of the entire sample. To assess the leadership styles and cognitive preferences, all participants completed two questionnaires, both in self-descriptive form.

LEADERSHIP STYLES

Leadership styles were operationalized based on a new instrument developed solely on Polish sample. The questionnaire used draws upon classical (e.g., authoritarian, task oriented, democratic, people oriented) and contemporary (e.g., visionary, transformational, charismatic, transactional) conceptions of leaders' patterns of behavior, but the pool of items was newly formulated in Polish language. The *Managerial Styles of Leading* (MSL) instrument is a 51-item measure, which was validated using confirmatory factor analysis (Babiak 2010*). All items are grouped into six dimensions, each of which represents distinct leadership style. The labels, descriptions of each style, and Cronbach's α reliabilities, which were assessed based on a larger sample of study participants, n = 477 (Babiak 2010), are presented in Table 22.1.

COGNITIVE PREFERENCES

Cognitive preferences were measured using the self-description checklist *Mind Types* (abbreviated CTU-96), developed by Polish cognitive psychologist Czesław S. Nosal (see Bajcar 2002†). Nosal's instrument is based on the assumption that individuals differ along five cognitive preference dimensions. These preferences relate to an individual person's perception and judgment. Nosal's measure, CTU-96, is concerned with self-descriptions of people's making preferences for taking information, analyzing it and judging it in five different styles of thinking. This is a 120-item instrument consisting of adjectives, which describe one's cognitive preferences. Sample items are the following: foreseeing, oscillating, superstitious, clever, sophisticated, meditative, radical, sensible, creative, penetrating, rational, receptive, pragmatic, common, inventive, ordinary, and principled. Table 22.2 provides descriptions of the five cognitive preferences together with their corresponding Cronbach's α reliabilities (see Bajcar 2002).

Analytic Procedure

First, the relationships between cognitive preferences and leadership styles were assessed in structural modeling analysis. We used structural modeling techniques in a context of model generation

---

* The *Managerial Styles of Leading* (MSL) questionnaire was originally created in Polish language (Babiak 2010). The English version of the MSL questionnaire is available upon request from the first author.

† The instrument for diagnosing mind types (CTU-96) was originally created in Polish language (see Bajcar 2002). The English version of the CTU-96 questionnaire is available upon request from the second author.

TABLE 22.1
Description of *Managerial Styles of Leading* (MSL) Scales and Cronbach's α Reliabilities

Label	Description	Cronbach's α
Error prevention	Focusing on flawless execution of tasks, encouraging more effort, addressing high-quality results, close monitoring, good orientation in followers' professional capabilities	.790
Formal discipline	Controlling and maintaining high discipline of work, focusing on implementation of tasks, emphasizing power and authority, demanding to comply with the standards during the implementation of tasks	.676
Participation	Encouraging participation and acceptance of followers in decision making, emphasizing the importance of strong commitment, maintaining close relationships with followers	.617
Effective Machiavellianism	Focusing on self-presentation techniques, manipulating information, accepting immoral and unethical ways of accomplishing tasks, authoritative manner of communicating with followers, imposing his own will	.743
Rewarding	Focusing on recognition for achievements, praising, frequent rewarding, communicating vision, mission and organizational goals, sharing expert knowledge	.642
Outcome orientation	Avoiding leadership responsibilities and duties, evasive approach to meetings with followers, ignoring importance of quality of work process, focusing merely on work outcomes, delegating all issues concerning work processes to followers	.614

TABLE 22.2
Description of Cognitive Preferences (CTU-96) Scales and Cronbach's α Reliabilities

Label	Description	Cronbach's α
Rationality	Preference for analytical, logical, insightful, explorative, effective processing of incoming information	.92
Emotionality	Preference for judgment based on emotions, moods, and personal feelings and for subjective comprehension of facts and events	.86
Creativity	Preference for innovative, open, and original evaluation of information and problems	.83
Intuition	Preference for global and comprehensive evaluation; detection and judgment based on relating facts from various and previously unconnected sources	.71
Typicality	Preference for simplistic, stereotypical, and superficial evaluation of problems and issues	.77

(Kline 2011). Based on its fit parameters, which inform how well the data support the model, we inferred causal relationships between variables. We acknowledge, however, that structural modeling techniques and analysis of paths between variables allow for quasi-causal inferences. Nevertheless, following commonly adopted terminology in quantitative data analysis, we will refer to the discussed relationships based on the derived model, as causal. In this study, cognitive preferences were considered as variables that determine specific behavioral styles manifested by managers.

Next, we conducted *k*-means cluster analysis (Romesburg 2004) to generate four distinct profiles of leaders' behaviors based on the means of manifested styles. The configuration of leadership styles of each cluster resulted in the following labels: **pseudo supervisors**, **Machiavellians**, **leaders**, and **pseudo democrats** (Babiak 2010). The aim of this analysis was to determine whether these managerial profiles are characterized by different cognitive preferences. Analysis of variance (ANOVA) models (Field 2009) yielded significant differences regarding leadership profiles and cognitive preferences. Games-Howell *post hoc* (Field 2009) comparisons were used, wherever significant differences existed.

Lastly, in order to examine the relationships between cognitive types and leadership styles, using *k*-means methodology (Romesburg 2004), we extracted three mind types based on the means of cognitive preferences. ANOVA models (Field 2009) yielded significant differences regarding leadership styles and mental types. Games-Howell *post hoc* (Field 2009) comparisons were used, wherever significant differences existed. Through contingency table and χ^2 test (Field 2009), we investigated whether certain mind types are more likely than others to dominate in identified leadership profiles.

Relationships between Cognitive Preferences and Leadership Styles—Causal Model

In Table 22.3, we reported means and standard deviations for both measures as well as correlations between dimensions of each measure.

Results in Table 22.3 indicate that cognitive preference labeled *rationality* and leadership style labeled *error prevention* exhibit the strongest correlation ($r = .41$, $p < .01$). Other correlations do not exceed the value of .35.

A structural model examining the influence patterns between five cognitive preferences and six leadership styles was built using AMOS software package. Estimation of model parameters was based on asymptotically distribution free estimator (ADF; Browne 1982), since this method allows for estimation of the relationships in cases of not normal data distribution.

Figure 22.1 presents the relationships between cognitive preferences and leadership styles. Fit indexes reveal a very good correspondence of the model to the obtained data: $\chi^2 = 52.090$; df = 31; χ^2/df = 1.680; RMSEA = 0.046; p (RMSEA $\leq$ 0.05) = .597; GFI = 0.973; AGFI = 0.942.

The relationships depicted in the model indicate that cognitive preference labeled *typicality* has the highest number of direct effects of all other cognitive preferences. *Typicality* produced the strongest direct effect on *outcome orientation* ($\beta = .32$), moderate effect on *formal discipline* ($\beta = .22$), and slight effects on *error prevention* ($\beta = -.12$) and *rewarding* ($\beta = -.11$). Given that *typicality* refers to simplified ways of assessing information and using stereotypes while evaluating events and people, these direct effects are not too surprising. Leadership styles labeled *outcome orientation* and *formal*

TABLE 22.3
Descriptive Statistics for Cognitive Preferences (CTU-96) and Leadership Styles (MSL) and Correlations between Corresponding Variables

	Cognitive Preferences				
Leadership Styles	**Rationality** $M = 135.05$ $SD = 15.69$	**Emotionality** $M = 53.57$ $SD = 10.43$	**Creativity** $M = 58.49$ $SD = 9.38$	**Intuition** $M = 42.01$ $SD = 5.22$	**Typicality** $M = 24.80$ $SD = 6.60$
Error prevention $M = 49.43$ $SD = 5.55$	.41**	–.03	.24**	.35**	–.12*
Formal discipline $M = 23.08$ $SD = 4.48$	.25**	.02	.04	.15**	.24**
Participation $M = 21.47$ $SD = 343$	.21**	.12**	.27**	.25**	–.03
Effective Machiavellianism $M = 27.54$ $SD = 3.59$	.02	.07	.09	.11	.10
Rewarding $M = 23.47$ $SD = 6.92$	.24**	.11**	.30**	.31**	–.16**
Outcome orientation $M = 22.97$ $SD = 5.11$	–.04	.05	–.07	–.06	.33**

*$p < 0.05$, two-tailed; **$p < 0.01$, two-tailed.

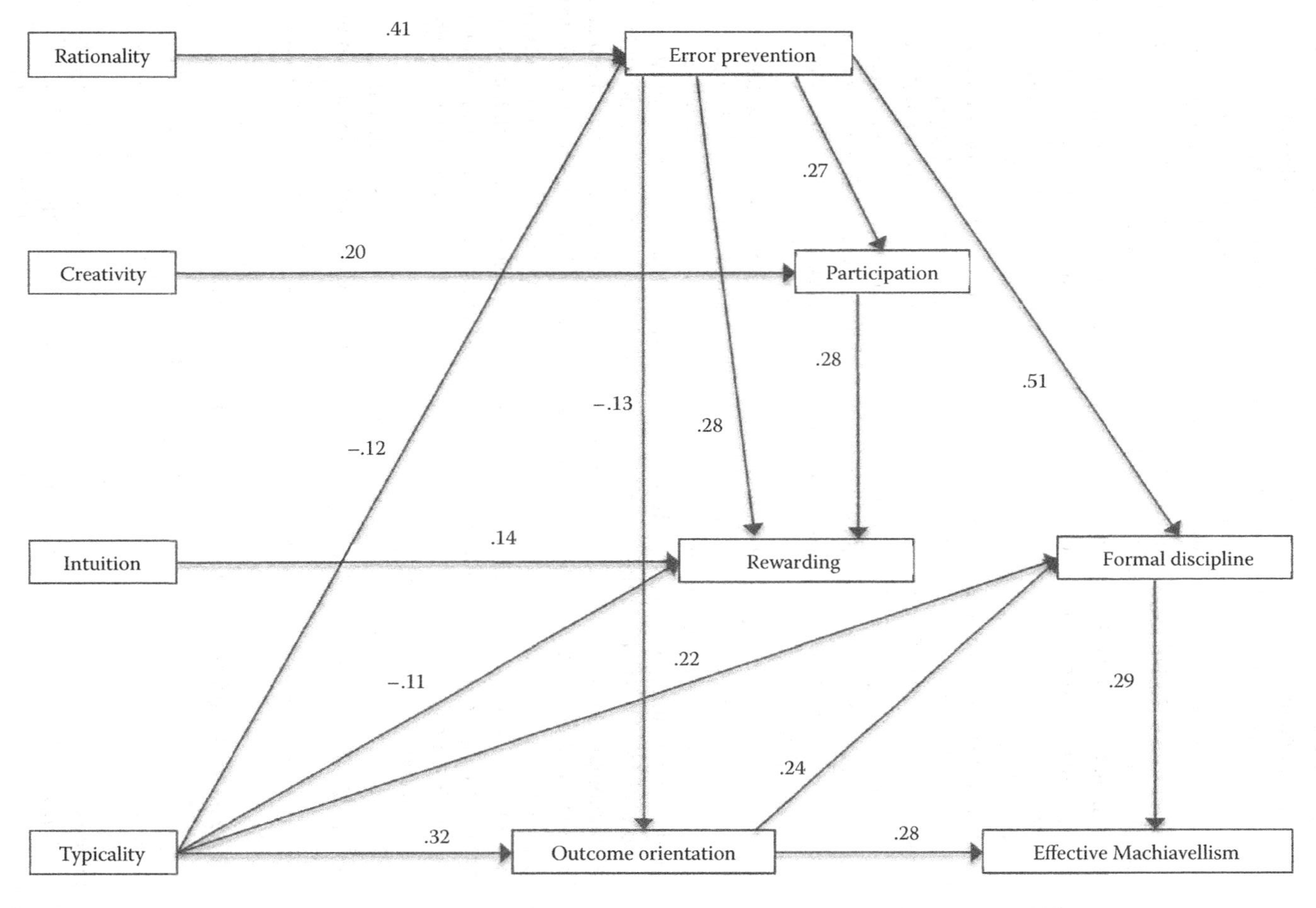

FIGURE 22.1 Structural model of causal relationships between four cognitive preferences and six leadership styles, $n = 323$.

discipline exhibit rather schematic and inelastic way of behaving, but, interestingly, they have very little in common. *Outcome orientation* can be roughly qualified as *laissez-faire* management style (Bass 1990), whereas *formal discipline* is characterized by tight control and strong interest in all activities that lead to success. However, both styles follow a pattern of uncomplicated behavioral acts characterized by clear-cut approach to managing. These two effects seem to justify the idea that *typicality* is reflected in tendencies to use unsophisticated categories in understanding the reality and making decisions, in superficial approach to power distribution, and in authoritarian way of dealing with people. The other two leadership styles influenced by *typicality* are *error prevention* ($\beta = -.12$) and *rewarding* ($\beta = -.11$). Both of these styles are distinguished by great interest in delivering high-quality results of work, concern with the well-being of followers, and good understanding of managerial duties. *Typicality* predicts slight decrease on both styles.

Table 22.4 summarizes total effects of *typicality* on all six leadership styles. Total effects were calculated as the sum of powers of all effects of one variable on another (Kline 2011).

Typicality produces weak to moderate effects across the paths, except for effect on *participation*, which can be considered as nil.

Between the five cognitive preferences, *rationality* had the strongest direct effect on *error prevention* ($\beta = .41$). This is an important finding because it shows the role of logical, analytical, insightful, and expedited style of evaluating and processing of incoming information represented by *rationality* in managing by ways of caring for subordinates' effectiveness at work, emphasizing to deliver high-quality end results, and showing interest in followers' professional shortcomings and their needs. As presented in Table 22.4, *rationality* also produced moderate effect on *formal discipline* (.20) and slight effects on *rewarding* (.15) and *participation* (.11).

The relationships depicted in Figure 22.1 indicate that *creativity*, interpreted as preferences for formulating new ideas and insightful, unconventional, and experimenting ways of approaching problems, had a direct effect on *participation* ($\beta = .20$). This managing style reflects the acceptance of collective decision making as well as tendency to rely on subordinates' opinions of manager's own plans and decisions. It can therefore be implied that open-minded approach to problems can produce welcoming grounds for subordinates' participation in decisions. Total effects summarized in Table 22.4 show that *creativity* did not predict any other substantial change on other leadership styles. Perhaps this stems from the fact that creative thinking does not substantially influence managing styles, which by their nature reflect focusing mainly on task implementation. In fact, as it was presented previously, *typicality*, that is, simplified, stereotypical, and prompt evaluation of information, in other words cognitive preference contradictive to *creativity*, displayed stronger and greater number of relationships with leadership styles.

Last of the direct effects obtained in the model in Figure 22.1 concerns preference for *intuition* on leadership style labeled *rewarding* ($\beta = .14$). *Intuition* is characterized by the ability to

TABLE 22.4
Total Effects for Cognitive Preferences on Leadership Styles

Variables	Rationality	Creativity	Intuition	Typicality	Total Effects
Error prevention	.41	n/e	n/e	−.12	.19
Formal discipline	.20	n/e	n/e	.24	.10
Participation	.11	.21	n/e	−.03	.06
Effective Machiavellianism	.04	n/e	n/e	.16	.03
Rewarding	.15	.06	.14	−.15	.05
Outcome orientation	−.05	n/e	n/e	.33	.13
Total effects	.25	.05	.02	.23	

Note: n/e means total effect = .001.

globally comprehend incoming information and comprehensive ways of understanding the reality, understanding the whole based on bits of information, and recognizing implicit or indirect sensible alternatives. It seems, therefore, that *intuition* may support more frequent display of rewarding leadership style, defined by appropriately distributing rewards, formulating vision, and willingly communicating it to subordinates. Total effects in Table 22.4 show that *intuition* did not produce any other effects on other leadership styles.

In this analysis, it was found that one of the cognitive preferences, namely, *emotionality*, had no effects on any of the leadership styles. This finding probably bears on the role of calm, realistic, and clear judgment of people, facts, and occurrences in the situation of managing. *Emotionality* reflects quite an opposite mental evaluation of information. It is determined by subjective views, affective thoughts, and employment of rather intensive emotions. In the managerial world, any such affect-driven problem resolution or temporary moods may adversely influence work outcomes.

Relationships between Cognitive Preferences and Leadership Profiles

Initially, leadership profiles were identified using k-means clustering procedure on the standardized z scores for the MSL scales as part of separate analysis ($n = 477$) (Babiak 2010). Four clusters were retained, after analyzing alternative results for a different number of clusters. Uniform or close to uniform group sizes, as well as meaningful interpretation possibility, led to choosing a particular combination. Based on the mean values of each leadership style in the particular combination, the following names were assigned: **pseudo supervisors**, **Machiavellians**, **leaders**, and **pseudo democrats** (Babiak 2010).

In the analyses provided in this article, the mean leadership profiles were formed based on entering the initial clustering centers into the new matrix. Intergroup distances between the means were recalculated, and four identified in previous study leadership profiles (Babiak 2010) were retained. Figure 22.2 presents a plot of the cluster centers for each of the four leadership profiles in relation to cognitive preferences.

ANOVA was then conducted using a general linear model (GLM), in which cognitive preferences served as dependent variables and leadership profiles served as independent variables. Games-Howell *post hoc* comparisons were then examined to see which leadership profiles significantly differed on each measure of cognitive preferences. ANOVA results presented in Table 22.5 indicate that leadership profiles account for the largest variance, 12%, in *rationality* $F(3319) = 12.04$; $p < .001$; $\eta^2 = .12$, and *intuition*, 10%, $F(3319) = 12.04$; $p < .001$; $\eta^2 = .10$. Insignificant result was found for one cognitive preference, *emotionality*.

As presented in Figure 22.2, **pseudo supervisors** scored the lowest, between the four groups, in *rationality*, *creativity*, and *intuition* and were moderate on *emotionality* and slightly above average on *typicality*. Following description of the cognitive preferences (see Table 22.2), these individuals are likely to employ simplistic and stereotypical interpretation of surrounding reality. They are not likely to be interested in critical, insightful, or global comprehension of organizational problems. They are rather sensitive and moody when judging incoming information. This pattern of cognitive preferences suggests weak mental organization and tendency to put little effort in appraising and analyzing problems. This may lead to low motivation for implementation of tasks and achievement of organizational goals, and greater aversion to undertake new activities and risks.

Machiavellians scored the highest, between the four groups, in *rationality*, *creativity*, *intuition*, *emotionality*, and *typicality*. Furthermore, **Machiavellians** were above average on *creativity*. These individuals are likely to logically, practically, economically, straightforwardly, and superficially assess information and occurrences. They seem to be sometimes inclined to see the big picture and discover its hidden elements. They also show some flexibility in their thinking, and their judgments rely on tendencies to employ emotions. This configuration of cognitive preferences paints a picture of pragmatic, achievement-driven, and control-oriented individuals, yet simplistic and stereotypical in their thinking and judging.

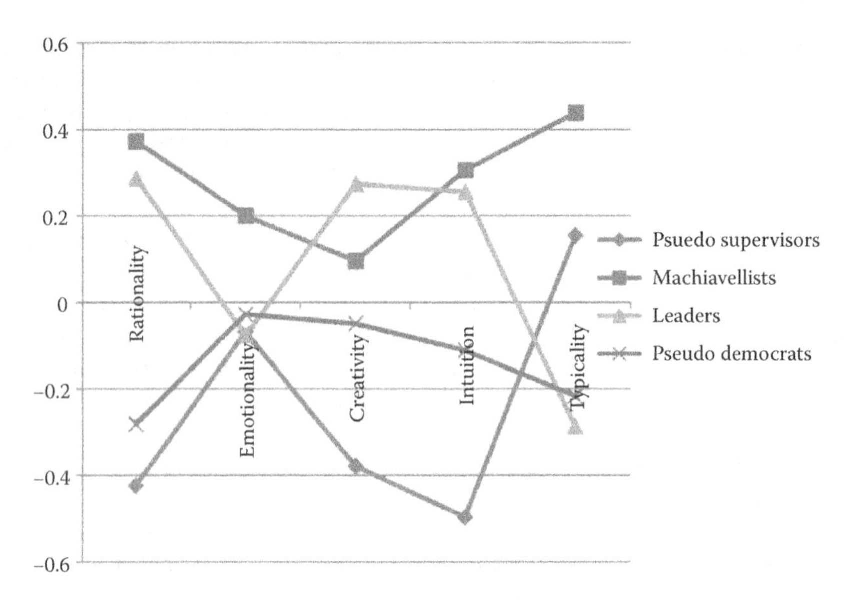

FIGURE 22.2 Relationships between cognitive preferences and leadership profiles, $n = 323$; pseudo supervisors, $n = 76$; Machiavellists, $n = 74$; leaders, $n = 94$; and pseudo democrats, $n = 79$.

TABLE 22.5
ANOVA Results for Cognitive Preferences by Leadership Profiles, $n = 323$

Cognitive Preferences	$F(3319)$	Significance	Partial η^2
Rationality	14.20	.001	.12
Emotionality	1.31	.271	.01
Creativity	6.60	.001	.06
Intuition	12.05	.001	.10
Typicality	9.91	.001	.09

The third profile, **leaders**, scored the highest, between the four groups, on their *creativity* and the lowest on their *typicality*. Simultaneously, they rated themselves as high on *rationality* and *intuition*, and as moderate on *emotionality*. Based on this configuration of cognitive preferences, we infer that members of this group show tendencies to evaluate information in sensible, novel, comprehensive, and intelligent way. These managers are likely to be logical, analytical, open-minded individuals, who are focused on analyzing the core of problems. What is important, compared to **Machiavellians**, **leaders** are more emotionally stable in evaluating information and appear to be more inclined to deal with abstract ideas in an objective, global, and candid way.

The fourth profile, **pseudo democrats**, had moderate or low scores on all cognitive preferences. As indicated by a description of the cognitive preferences (see Table 22.2), these managers would appear to employ some indistinctive, more casual strategies of information evaluation and problem solving. Fairly noticeable characteristic of these individuals is that they probably do not fully comprehend complex reality of managing. Hence, they appear somewhat reflective and lost.

Relationships between Mind Types and Leadership Profiles

Independent analyses were conducted to identify distinct groups of managers based on their specific configuration of cognitive preferences, referred to as **mind types**. Our aim was to examine which specific mind types would be dominating between the four leadership profiles. We performed *k*-means cluster analysis on the standardized *z* scores for the cognitive preferences scales of CTU-96. Figure 22.3 presents the outcome of the cluster centers for each of the three groups, which were based on the mean value of each cognitive preference they represented. We refer to these groups as **mind type 1**, **mind type 2**, and **mind type 3**.

Members of **mind type 1** obtained the lowest scores on *rationality*, *creativity*, and *intuition*. The highest score occurred on *typicality* and the average on *emotionality*. This pattern of cognitive preferences suggests orientation toward superficial, inelastic style of reasoning, low openness to making effort for new, unknown solutions, and preference for focusing on trivial and stereotypical evaluations of people and facts.

Mind type 2 managers depict a configuration consistent with a general conception of how leaders should evaluate and comprehend things, events, facts, and people. These individuals received the highest scores between the three groups on *rationality*, *emotionality*, *creativity*, and *intuition*. They scored average on *typicality*. Such configuration of cognitive preferences indicates that members of this group are likely to be logical, analytical, insightful, creative, and open-minded. Furthermore, they appear to base their thinking and decision-making on emotions and current moods, which could predispose them to think and act more spontaneously than calculative. In sum, these managers show tendencies for performing pragmatic, sensible, abstract, and innovative evaluation of problems and issues. When working on new problems and ideas, they appear to have proclivity for drawing more upon emotions than on stereotypes and simple solutions.

Mind type 3 individuals obtained the lowest scores on *emotionality* and *typicality* and were moderate on *rationality*, *creativity*, and *intuition*. This configuration of scores of the cognitive preferences suggests that these managers are moderately insightful, moderately conceptual, and innovative and are rather casual in evaluation of facts and events. In general, this pattern of cognitive preferences characterizes individuals with high rating of their own cognitive abilities, which follows pragmatic, unbiased, rather insensitive style of judging and thinking.

Mind Types by Leadership Styles

Using a GLM, we conducted ANOVA in which the mind types served as independent variables and the six styles of leadership served as the dependent measures. In general, this particular analysis

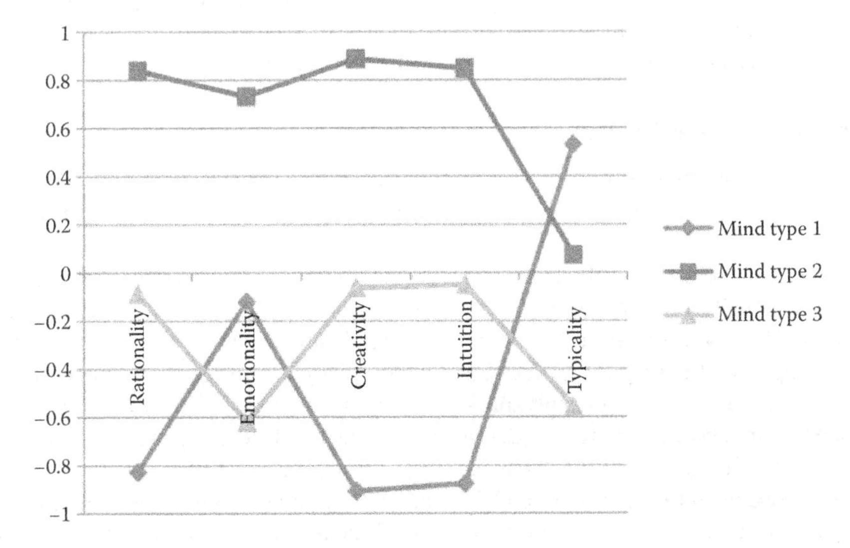

FIGURE 22.3 Cognitive preference cluster membership for mind types, $n = 323$; mind type 1, $n = 101$; mind type 2, $n = 111$; mind type 3, $n = 111$.

TABLE 22.6
ANOVA Results for Mind Types by Leadership Styles, $n = 323$

Leadership Styles	$F(2320)$	Significance	Partial η^2
Error prevention	22.70	.001	.12
Formal discipline	2.87	.058	.09
Participation	9.38	.001	.05
Effective Machiavellianism	2.04	.132	.01
Rewarding	13.20	.001	.08
Outcome orientation	3.82	.023	.02

allowed us to investigate whether individuals with different mind types display different leadership styles. For significant main effects, we examined Games-Howell *post hoc* comparisons to see which mind types significantly differed on measures of leadership style. ANOVA results presented in Table 22.6 indicate that leaders' mind types account for the largest variance, 12%, in *error prevention* $F(2320) = 22.70$; $p < .001$; $\eta^2 = .12$, and *rewarding*, 8%, $F(2320) = 13.20$; $p < .001$; $\eta^2 = .08$. The only insignificant result was found for one leadership style, *effective Machiavellianism.*

Members of **mind types 2** and **3** were significantly more frequently manifesting three leadership styles, *error prevention*, *participation*, and *rewarding*, than members of **mind type 1.** Every one of these leadership styles is beneficial from a standpoint of achieving organizational goals. These styles have a common ground, namely, high managerial activity, both in task orientation and people orientation. Thus, both groups of managers who showed high (**mind type 2**) and moderate (**mind type 3**) preference for analytical, insightful, and innovative evaluation of managerial issues and problems tend to demonstrate desirable leadership behaviors, such as focusing on delivering high-quality work outcome, allowing followers to participate in decision making, adequate rewarding, and maintaining good contact with followers. We also found significant differences between the mean scores of the members of **mind type 1** and members of **mind type 3** with respect to their preference for *outcome orientation.* This leadership style is characterized by lack of interest in quality of subordinate work process, tendency to checking off end results, and rather passive approach toward basic managerial duties. The differences in leadership styles between these two **mind types** (**1** and **3**) can be inferred from their underlying general characteristics. **Mind type 1** managers tend to evaluate information in a shallow, stereotypical, effortless manner, whereas **mind type 3** members show preferences for deeper analytical activity before reaching conclusions.

Mind Types by Leadership Profiles

We also examined the relationships between three mind types and four leadership profiles in our sample of managers. In this analysis, we used χ^2 test and contingency tables. The results are presented in Table 22.7.

Significant χ^2 result indicates that the three mind types differ with respect to identified leadership profiles.

As may be seen in Table 22.7, **mind type 1** is dominant among **pseudo supervisors** and **pseudo democrats**. Members of **mind type 1** appear to have tendency for using stereotypes in their judgment and for superficial perception of facts and events. They are likely to choose easier ways to find solutions to a problem through analyzing it in clear-cut categories. This result is not surprising because **pseudo supervisors** and **pseudo democrats** are the two leadership profiles that tend to stay away from supervising and show little interest in the work processes; they also tend to minimize occasions to make concluding decisions, and, in general, they demonstrate passiveness in managing.

Mind type 2 had substantial representation among **Machiavellians** and **leaders**. This particular mind type represents analytical, logical, innovative way of problem solving and most probably characterizes individuals who are working well with tangible problems as well as abstract ideas. Overall

TABLE 22.7
Mind Types by Leadership Profiles, n = 323

		Mind Types			
Leadership Profiles		**Mind Type 1**	**Mind Type 2**	**Mind Type 3**	**Total**
Pseudo supervisors	Frequency	37	19	20	76
	Percentage	36.6%	17.1%	18.0%	23.5%
Machiavellians	Frequency	17	35	22	74
	Percentage	16.8%	31.5%	19.8%	22.9%
Leaders	Frequency	16	40	38	94
	Percentage	15.8%	36.0%	34.2%	29.1%
Pseudo democrats	Frequency	31	17	31	79
	Percentage	30.7%	15.3%	27.9%	24.5%
Total	Frequency	101	111	111	323
	Percentage	100.0%	100.0%	100.0%	100.0%

Note: χ^2 = 31,323; df = 6; p = .001.

behavioral characteristic of **Machiavellians** and **leaders** corresponds to **mind type 2**. These two leader types are active managers. This significant relationship suggests that they are likely to be intelligent, to be economical in distributing their mental resources, to be innovative, and also to understand well the area of social interactions. The difference, however, probably lies in how all these resources are utilized by managers of these two leadership profiles. Most likely, **Machiavellians** lead through manipulation and coercive power, and **leaders** achieve their goals through successful managing tasks and leading people.

Mind type 3 is dominant among **leaders** and **pseudo democrats**. This particular mind type suggests rather analytical, inventive, intuitive, and somewhat emotional approach to problems and issues. This result is a bit surprising, since these two leadership profiles are very different with respect to their preferred leadership styles. We infer, however, that open-mindedness, which is manifest for **mind type 3**, refers to individuals who maintain good relations with people, which in turn is a common characteristic for both **leaders** and **pseudo democrats**.

CONCLUSION

The results obtained in this study have some important implications for understanding how individual differences in cognitive perception, evaluation, and judgment influence behaviors preferred by managers.

An important finding refers to a role of critical and efficient information processing in managing. As ascertained, *rationality*, as a cognitive preference, produced the strongest effect on *error prevention*, the leadership style characterized by activity, constant monitoring, and focus on delivering high-quality effects related to work. *Rationality* emphasizes critical, logical, and analytical style of information evaluating and also efficiency in cognitive processing. This finding is consistent with results of other studies regarding the role of cognitive processes in different leadership criteria, including manifested behavioral patterns. More specifically, there have been considerable research results providing evidence for the importance of divergent thinking, idea generation, and other non-IQ elements of cognitive processes and preferences of organizational leaders (e.g., Connelly et al. 2000; Vincent, Decker, and Mumford 2002) for achieving organizational goals. In general, results of these studies suggest that effective leaders are rather logical, creative, analytical, fast, and rational when processing information.

Another important finding is that *typicality*, as a cognitive preference, had the greatest number of direct effects on leadership styles. *Typicality*, defined by schematic, stereotypical, unsophisticated thinking style, exerted the greatest effect on *outcome orientation*, which is generally characterized by laissez-faire behaviors. Preference for simplistic and stereotypical evaluation of facts and details enables to reach conclusions a lot quicker than through detailed analysis of information. It is not surprising then that this type of cognitive preference has exerted the greatest effect on simplistic, schematic, and uninvolved style of leading. This relationship shows that perception and evaluation based on stereotypes and routine are represented by avoidance of active monitoring of followers' work quality and overall somewhat evasive approach to leadership responsibilities. It can be implied that facts and events encountered by managers high on *typicality* are probably categorized based on the simplest, easiest classification methods, within the framework of superficial evaluation of its structure.

The other cognitive preferences had also some direct effects on leadership styles, but their impact was lesser than that of *rationality* and *typicality*. Lastly, *emotionality* did not influence any of the managerial behaviors. This could probably be attributed to the fact that in management situations, judgment shall be made on poised, realistic, and calm perception of people, facts, and occurrences.

The analysis regarding relationships with respect to linking leadership profiles to cognitive preferences has also yielded some interesting results. **Leaders** and **Machiavellians** exhibited preference for logical, practical, novel, and comprehensive ways of evaluating information. They differed, however, on *emotionality* and *typicality*, which was a lot higher in the group of **Machiavellians**. The other two leadership profiles, namely, **pseudo supervisors** and **pseudo democrats**, scored rather low on all cognitive preferences. This indicates that these managerial groups employ rather simplistic, stereotypical, and casual strategies of information evaluation. Our explanation for this pattern of relationships is that the primary function of organizational leaders is to attain organizational goals, which could be described in terms of getting the job done. Therefore, cognitive preferences associated with logic, quick reasoning, criticism, and insight would have the greatest influence on behaviors leading to advancement in the work of followers. These behaviors are manifested by **leaders** and **Machiavellians** rather than **pseudo supervisors** and **pseudo democrats**. In other words, we see that solutions to managerial problems are generated through coherent thinking, which is then translated into real action.

Similarly, the results of the analysis regarding relationships between the three yielded mind types and leadership profiles reflect the above conclusions. **Leaders** and **Machiavellians**, the two active types of behaviors, represented rational, logical, creative, and global orientation toward information perception and judgment. Both of these groups are probably energetic thinkers, who are likely to deal with practical problems with success. On the other hand, **pseudo supervisors** and **pseudo democrats** tended to be rather stereotypical and superficial in their perception and judgment.

It is worth emphasizing that the primary objective of organizational leaders is to accomplish organizational goals, being at the same time aware that human behavior *per se* is not limited only to pursuing and reaching various goals (Nosal 1991). Earlier studies that examined the nature of leadership styles suggest that certain capacities, such as general intelligence, flexibility and fluency in idea generation, and creative problem solving, among others, influence behaviors related to organizational performance (see Bass and Bass 2008 for review). Between the six leadership styles yielded in the sample of Polish managers (Babiak 2010), *error prevention* seems to be the one that represents orientation toward goal attainment, bringing value and quality to work processes, and stimulating and encouraging more work effort. The results obtained in this study demonstrate that the preference for poised, analytical, logical, explorative thinking, labeled *rationality*, exerts the greatest effect on *error prevention*, the repertoire of goal-orientated behaviors, which reflect the primary function of organizational leaders.

REFERENCES

Babiak, J. (2010). *Różnice w profilach stylów kierowania uwarunkowane przez cechy osobowości i typy umysłów menedżerów [Differences in Profiles of Leadership Styles Determined by Personality Traits and Mind Types of Managers]*. (Unpublished doctoral dissertation). Wroclaw, Poland: Wroclaw University.

Bajcar, B. (2002). *Profile orientacji temporalnej—Wymiary i konsekwencje psychologiczne [Temporal Orientation Profiles—Dimensions and Psychological Consequences].* (Unpublished doctoral dissertation). Opole, Poland: Opole University.

Bass, B. M. (1990). *Bass and Stodgill's Handbook of Leadership. Theory, Research and Managerial Applications.* New York: The Free Press.

Bass, B. M., and Bass, R. (2008). *The Bass Handbook of Leadership. Theory, Research and Managerial Applications.* New York: The Free Press.

Browne, M. W. (1982). Covariance structures. In D. M. Howkins (Ed.), *Topics in Applied Multivariate Analysis* (pp. 72–141). Cambridge: Cambridge University Press.

Connelly, M. S., Gilbert, J. A., Zaccaro, S. J., Threlfall, K. V., and Marks, M. A. (2000). Exploring the relationship skills and knowledge to leader performance. *Leadership Quarterly, 11*(1), 65–85.

Field, A. (2009). *Discovering Statistics Using SPSS.* London: SAGE Publications, Ltd.

Judge, T. A., Colbert, A. E., and Ilies, R. (2004). Intelligence and leadership: A quantitative review and test of theoretical propositions. *Journal of Applied Psychology, 89*(3), 542–552.

Kline, R. B. (2011). *Principles and Practice of Structural Equation Modeling.* New York: The Guilford Press.

Lord, R. G., De Vader, C. L., and Alliger, G. M. (1986). A meta-analysis of the relation between personality traits and leadership perceptions: An application of validity generalization procedures. *Journal of Applied Psychology, 71*(3), 402–410.

Morrow, I. J., and Stern, M. (1990). Stars, adversaries, producers, and phantoms at work: A new leadership typology. In K. E. Clark, and M. B. Clark (Eds.), *Measures of Leadership* (pp. 419–439). West Orange: Leadership Library of America, Inc.

Norton, L. W. (2010). Flexible leadership: An integrative perspective. *Consulting Psychology Journal: Practice and Research, 62*(2), 143–150.

Nosal, C. S. (1991). Neurobiology of subjective probability. In E. Eells, and T. Maruszewski (Eds.), *Probability and Rationality* (*21*, pp. 173–180). Rodopi: Poznań Studies in the Philosophy of the Sciences and the Humanities.

Romesburg, H. C. (2004). *Cluster Analysis for Researchers.* Morrisville, NC: Lulu Press, Inc.

Vincent, A. S., Decker, B. P., and Mumford, M. D. (2002). Divergent thinking, intelligence, and expertise: A test of alternative models. *Creativity Research Journal, 14*(2), 163–178.

Zaccaro, S. J. (2007). Trait-based perspectives of leadership. *American Psychologist, 62*(1), 6–16.

Zaccaro, S. J., Kemp, C., and Bader, P. (2004). Leaders traits and attributes. In J. Antonakis, A. T. Cianciolo, and R. J. Sternberg (Eds.), *The Nature of Leadership* (pp. 101–124). Thousand Oaks: Sage Publications, Inc.

23 Advanced Theory of Political Marketing

Wojciech Cwalina, Andrzej Falkowski, and Bruce I. Newman

CONTENTS

MODELS OF POLITICAL MARKETING

Earlier theories of political marketing originated, to a large degree, from theories of marketing developed for the consumer goods market (Kotler 1975; Reid 1988; Shama 1975; Wring 1997). However, in the course of time, important differences have emerged between the practice and efficiency of marketing theories used for political and economic purposes. Political marketing, to a larger and larger extent, drew from such disciplines as sociology, political sciences, or psychology (Cwalina et al. 2008; Lees-Marshment 2003; Scammell 1999). That leads to defining political marketing as a separate branch of science, with its own subject matter and methodology of research (Lock and Harris 1996; Newman 1994).

The first model of political marketing has been proposed by Niffenegger (1988) where he tries to show the use of the classic marketing mix tools for political campaigns. He stresses that in political marketing, one can notice efforts aimed at integration within the *marketing mix* (*4Ps*) to control the voters' behaviors more efficiently. Despite the fact that it attempts to show the efficiency of using marketing strategies for political campaigns, Niffenegger's concept of political marketing is in fact a copy of the concepts used in commercial marketing. It seems, then, that it does not distinguish to a sufficient extent between consumer and political choices.

Reid's (1988) concept is also an attempt to apply some concepts from mainstream marketing to political marketing. It focuses on this element of the voting process that refers to voting understood as a buying process. Reid stresses that by looking at the problem from a consumer perspective, it appears that a broader marketing approach could make a useful contribution toward aiding a better theoretical knowledge of the "voting decision process." Reid's approach to political marketing corresponds very well to the marketing concept, which is the last stage of the evolution process in which presidential candidates have gone from campaign organizations run by party bosses to organizations run by marketing experts (Newman 1994). Its analysis is a pretty accurate reflection of the concepts developed in mainstream marketing and used for political behavior. However, this excludes a number of specific characteristics both of the political market and different strategies of running political campaigns.

Kotler and Kotler (1999) present a structured process of marketing activities related to political campaigns, which consists of six stages. The analysis of these activities creates a so-called *candidate marketing map.* Thus, a professionally planned political campaign consists of (1) environmental research, (2) internal and external analysis, (3) strategic marketing, (4) setting the goals and strategy of the campaign, (5) planning communication, distribution, and organization, and (6) defining key markets for the campaign. A candidate marketing map proposed by Kotler and Kotler is compatible to the process of planning and organizing political campaigns described by Mauser (1983). According to him, this process includes three stages: (1) the preparation process during which the candidate assesses his and his competition's strengths, (2) the process of developing a strategy of influencing voters, and (3) the process of implementing the strategy.

The comprehensive political marketing (CPM) described by Lees-Marshment (2003) is also consistent with the development of the concept of product in economic marketing. She believes that a candidate's or party's CPM should be based on six fundamental principles. Firstly, it is more than political communication only. Secondly, it applies to all political organization behaviors/activities, not only to political campaigns but also to the way in which product is designed. Thirdly, CPM uses marketing concepts and not only techniques. Fourthly, it also includes elements of political sciences to better utilize and adapt such knowledge for the purpose of marketing. Fifthly, it adapts marketing theory in such a way as to make it for the nature of politics. And, finally, it applied marketing to all political organizational behaviors including interest groups, politics, the public sector, media, parliament and local governments, as well as parties, candidates, and elections.

According to Harris (2001), the changes taking place in modern democracies, in the development of new technologies, and in citizens' political involvement significantly influence the theoretical and practical aspects of political marketing efforts. Above all, modernization causes changes from direct involvement in election campaigns to spectatorship. Campaigns are conducted primarily through mass media and citizens participating in them as a media audience. In this way, politicians more and more often become actors in a political spectacle rather than focus on solving real problems that their country faces. They compete for their voters' attention not only against their political opponents but also against talk shows or other media events. For instance, during the Polish presidential campaign in 2005, the debate between two major candidates (Lech Kaczyński and Donald Tusk) was rescheduled for another day, because otherwise it would have competed for the viewer against the popular *Dancing with the Stars.*

Newman's (1994) concept of political marketing is the most thorough model from those discussed so far describing the marketing approach in political behavior. It provides procedures for a number of concepts related to marketing activities on the voting market. It has also been the source of inspiration for a number of empirical researches expanding the theory (Cwalina et al. 2008).

In his model, Newman (1994) introduces a clear distinction between the processes of a marketing campaign and those of a political campaign. The marketing campaign helps the candidate go through the four stages of the political campaign, including everything from the preprimary stage of a politician's finding his own place in politics to his already formed political image at the general election stage. It is natural then that both campaigns are closely connected. The process of a marketing campaign is the foundation of the model because it includes all the marketing tools needed to conduct the candidate through all the levels of the political campaign.

The common element of the theories of political marketing presented here is their focus on the voter as a starting point for any actions undertaken by political consultants on the competitive voting market. An in-depth analysis of the similarities and differences between these theories may help one develop a new and advanced theory of political marketing. This new concept is the foundation of the problems and research on modern political marketing presented by Cwalina et al. (2011).

CHALLENGES FOR POLITICAL MARKETING

Political marketing campaigns are integrated into the environment, and, therefore, they are related to the distribution of forces in a particular environment (Cwalina et al. 2008; Newman 1994; Scammell 1999). In this way, changes in societies, legal regulations, or the development of new technologies force modifications of particular marketing strategies and make marketing needs regenerate as well (Harris and Rees 2000; Vargo and Lusch 2004). Each of these elements represents an area where dynamic changes have taken place in the past few decades. These changes facilitate the development of marketing research and are becoming more and more important for the election and governing processes. Thus, political marketing should include changes taking place in modern democracies, such as shift from citizenship to spectatorship, and assess and show new ways of increasing citizen support. Besides, the relations between political marketing and such areas of knowledge as practice, public relation, or political lobbying also need to be clearly defined (Baines et al. 2002; Harris 2001).

The emphasis on the processes of election exchanges cannot obscure the fact that political marketing is not limited only to the period of the election campaign. In the era of permanent campaign, in reality, there is no clear difference between the period directly before the election and the rest of the political calendar (Harris 2001). Governing involves tough endless campaigning that secures politicians' legitimacy by stratagems that enhance their credibility (Nimmo 1999). Dulio and Towner (2009, p. 93) state that modern campaigning extends to governing, and "each day is election day."

Permanent Campaign

Through the concept of the permanent campaign, political marketing has become the organizing principle around which parties and government policies were constructed. Political marketing is no longer a short-term tactical device used exclusively to win voters' support, but a long-term permanent process that aims to ensure continued governance. According to Nimmo (1999), the permanent campaign is a process of continuing transformation. It never stops. From this perspective, governing is then turned into a perpetual campaign that remakes government into an instrument designed to sustain an elected official's public popularity.

Media and Politics

Together with the political changes, a number of changes in the ways the media operate took place. These changes concerned both the legal regulations of the media market and its opening up to commercial broadcasters and to introducing new technologies and improving the quality of the broadcast (Kaid and Holtz-Bacha 2006). According to Diamond and Bates (1992), the development of television production, marketing methods, and public opinion polls led to the establishment of today's high-tech political communication.

Gamson et al. (1992) stated that a wide variety of media messages act as teachers of values, ideologies, and beliefs, and they provide images for interpreting the world whether or not the designers are conscious of this intent. It seems, however, that in relation to politics, developers of media messages are fully aware of what content and in what form they are trying to communicate to society.

Political Public Relations and Lobbying

Kotler and Keller (2006) believe that public relation is one of the six major modes of communication within *marketing communications mix*. Public relation (PR) is company-sponsored activities and programs designed to create daily or special brand-related interactions. It involves a variety of programs designed to promote or protect a company's image or its individual products. Public relations include communications directed internally to employees of the company or externally to consumers, other firms, the government, and media. According to these authors, the appeal of public relations is based on three distinctive qualities: (1) high credibility (the news stories and features are more authentic and credible to readers than ads); (2) ability to catch buyers off guard (PR can reach prospects who prefer to avoid salespeople and advertisements); and (3) dramatization (PR has the potential for dramatizing a company or product). Then, major tools in marketing PR include, according to Kotler and Keller, publications (e.g., reports, press and the web articles, or company newsletters), events (e.g., news conferences, seminars, or outings), sponsorships (sports and cultural events), news (the media releases), speeches, public-service activities (e.g., contributing money and time to good causes), and identity media (e.g., logos, stationery, business cards, buildings, or uniforms).

Political Consultants

Today, politics has become a big, profitable business to consultants who help manufacture politicians' images. Bowler et al. (1996) even talk about the emergence and dynamic development of "political marketing industry." In California, the first permanent organization devoted to political campaigning, Whitaker and Baxter's Campaigns Inc., was founded in 1930. Sixty years later, the 1990 *California Green Book* listed 161 general campaign consultants, 14 polling firms, 3 petition management companies, 22 professional fundraising firms, and 15 legal firms offering legal and accounting device to campaigns (Bowler et al. 1996). O'Shaughnessy (1990, p. 7) describes political consultants as "the product managers of the political world."

AN ADVANCED THEORY OF POLITICAL MARKETING

The processes described above show clearly the shift in the focus and range of political marketing. It is expanded to permanent strategic element of governance. These changes facilitate the development of marketing research and are becoming more and more important for the election and governing processes. They also make one develop more appropriate models of political marketing including those processes.

Many scholars, among them Newman (1994, 1999) and Cwalina et al. (2009), point to multiple possibilities and paths through which the transition of political marketing into political marketing science may occur. They are also the foundation for advanced theory of political marketing presented in Figure 23.1.

The starting point for developing the advanced model of political marketing is the model elaborated by Newman (1994). The advanced model of political marketing brings together into a single framework the following two campaigns: the permanent marketing campaign and the political marketing process. These two components are realized within a particular country's political system, and the system depends, above all, on political tradition as well as the efficiency of the developed democratic procedures. In this way, "democracy orientation" determines how the functions of the authorities are implemented and, also, who is the dominant object in the structure of government. On the other hand, democracy orientation also defines who the voters focus on during elections. From this perspective, one may distinguish four fundamental types of such orientation: candidate-oriented democracy, party leader-oriented democracy, party-oriented democracy, and government-oriented democracy.

A good example of candidate-oriented democracy is the United States, where the choice in an election is very much a function of the sophisticated use of marketing tools to move a person into contention. It is characterized by the electorate's attention shifts from political parties to specific candidates running for various offices, and, particularly, for president. The shift is accompanied by the growing importance of a candidate's individual characteristics, of which his/her image is

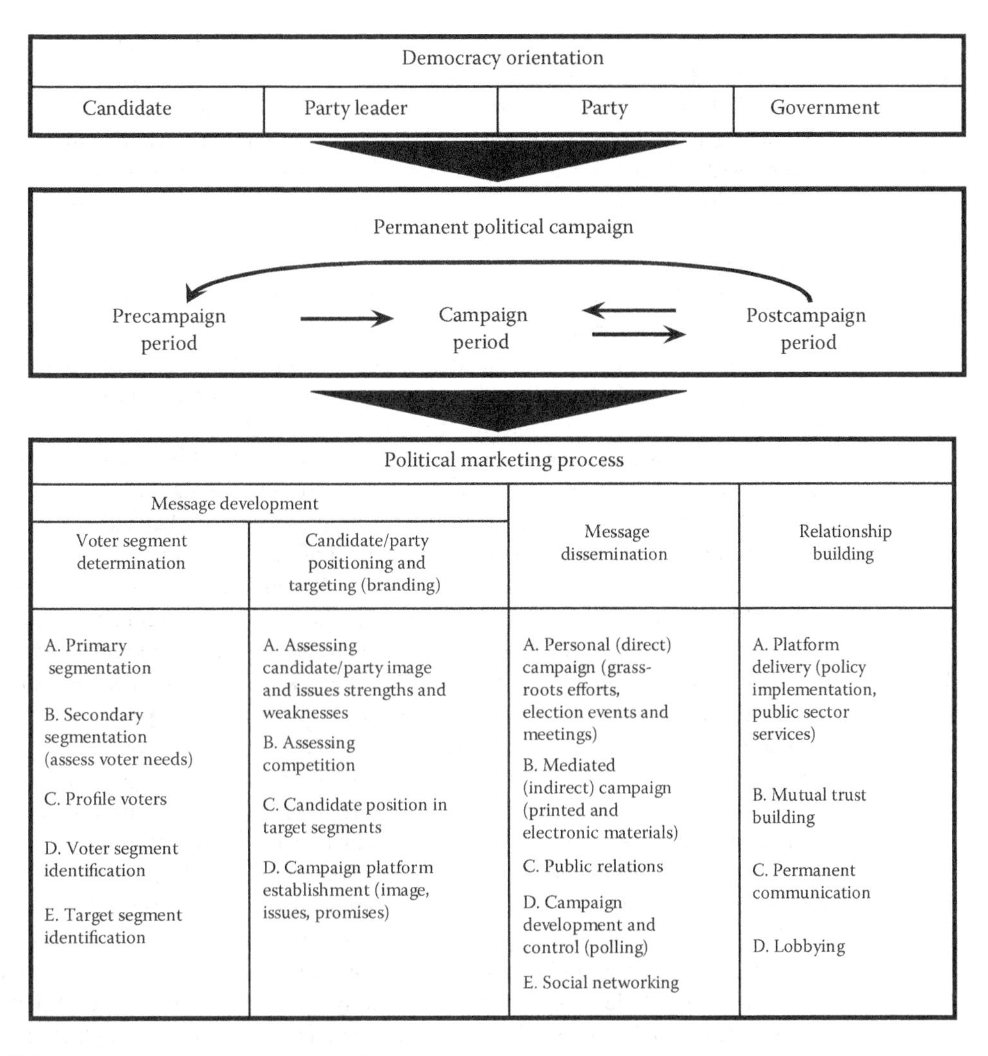

FIGURE 23.1 The advanced model of political marketing.

made up. American parties have little direct control over either candidate selection or the running of campaign. Key decisions about campaign strategy are made at the level of the individual candidate. Although the national party committees play a supportive role, candidate image, character, and policy pledges are the prime "products" to offer in elections rather than party behaviors and platforms.

Party leader–oriented democracy seems to be a characteristic of the United Kingdom or Mexico, where there is still a focus on the individual in the campaign, but the choice in an election is more a function of the "approval" of a superbody of influentials who decide who will run in an election. To a great degree, the political party is in a very powerful position, but there is still active use of marketing techniques once the party chooses who the nominee will be. British parties are ideologically cohesive and disciplined, with centralized and hierarchical national organizations, and their leaders are focused on directing the behavior of the whole party in the search for office. While analyzing the data from parliamentary elections in Britain between 1979 and 1987 collected in the archives of the Economic and Social Research Council (ESRC) and from TV polls conducted by the BBC together with Gallup, Johnston et al. (1988) stated that, all other things being equal, the most popular leaders will be those with the most popular policies. But this is not all, because some leaders are much more popular than the policies they stand for, whereas others are less popular. Thus, a voter may prefer the Labour Party's policies, but because of quality of the party's leader, the promises in those policies will not be delivered, and, as a consequence, another party may get that person's vote because it has a leader who is believed to be better equipped to fill the role of Prime Minister. To some voters, leadership may be even more important than direction in which it leads. For example, Johnston et al. found that about three-quarters of those who thought Margaret Thatcher would make the best Prime Minister voted for her at each of the three analyzed elections. Another piece of evidence confirming these results is the data from the 1987 British Campaign Study (BCS) conducted by Stewart and Harold D. Clarke (1992). They found that favorable perceptions of Thatcher as competent and responsive enhanced, while similar perceptions of other leaders reduced, the likelihood of the Conservative voting. Similar interrelations were also found for leaders of the other parties. The dominant position of the leader in relation to his/her party is also confirmed by the results of the polls conducted by MORI in the context of the 1987–2001 British general elections analyzed by Worcester and Mortimore (2005). They showed leader image to be a greater determinant of voting behavior than party image: in 1987, 35% and 21%; in 1992, 33% and 20%; in 1997, 34% and 23%; and in 2001, 32% and 24%, respectively. In this context, it seems justifiable to assume that a political party and voters' identification with it are an important factor influencing voting decisions. However, its image is to a large degree based on how its leader is perceived. It is the leader that voters focus on, and it is the leader whose promotion is the main goal of the campaign.

Party-oriented democracy is characteristic of such countries as Poland, Finland, Czech Republic, or Romania, where the political party is really presenting itself to the voters as the real choice being made. The Polish political system is based on a party system. Therefore, in the parliamentary, presidential, and local elections, candidates supported by significant political parties have a better chance of success. The politics of Poland takes place in the framework of a parliamentary representative democratic republic, whereby the Prime Minister is the head of the government and of a multiparty system. The president, as the head of state, is chosen during elections based on the majority rule and has the power to veto legislation passed by parliament, but otherwise has a mostly representative role. During general parliamentary elections, the citizens of Poland elect their representatives, who belong to various political parties. These parties then take seats in the Sejm and Senate (lower and higher chambers of parliament) depending on the number of votes they receive during an election. For a particular candidate to be elected, his/her party (or election committee) must get at least 5% of the votes across the whole country, and if he/she represents a coalition of various parties, it has to be 8% of the votes cast in the whole country. So a situation may occur (and does occur) that a candidate who won most votes in his/her constituency will not become an MP because his/her party was below the 5% or 8% threshold in the whole country. Such legal regulations lead to campaigns being mainly concentrated on political parties. Obviously their leaders are an important element of winning such support; however, even their personal success does not guarantee party's success. Besides, the winning party designates their candidate for Prime Minister, and it does not have to be its leader but only a person nominated by the party.

Government-oriented democracy seems to be characteristic of countries like Russia, China, and other countries where governing is dominated by one party. Such a system is defined on the website of the China Internet Information Center (www.china.org.cn) as "democratic centralism." The Communist Party of China (CPC) has established formal (through elections within the Party) and informal (appointed by the organization of the higher level) organizations within the Chinese government and various levels and walks of life in the country. According to the principle of democratic centralism, the individual CPC member is subordinate to a party organization, the minority is subordinate to the majority, the lower level organization is subordinate to the higher level, and each organization and all members of the whole CPC are subordinate to the Party's National Congress and the Central Committee. Furthermore, leading bodies at various levels of the party, except for their agencies and for leading party groups in nonparty organizations, are all elected, and the party prohibits personality cult in any form. Such elections are two-tier elections: direct and indirect election. Direct elections are applicable to the election of deputies to the people's congresses of the counties, districts, townships, and towns. They adopt the competitive election method, which means that a candidate wins the election when he/she receives more than half of the votes cast. Indirect elections, then, are applicable to the election of deputies to people's congresses above the county level and deputies among the armed forces at the same level and deputies to the National People's Congress (NPC) elected from special administrative regions. Candidates may be nominated by political parties or mass organizations jointly or independently or by more than ten deputies. Expenses for the election of the NPC and local people's congresses at various levels are to be provided from the national treasury. In the case of government-oriented democracy, major tasks of the political campaign focus on the communication between the government and citizens rather than direct election struggle between candidates or political parties.

Depending on democracy orientation then, political campaigns may focus on different goals and different means of reaching them. Nevertheless, in all the systems discussed above, the political campaigns are permanent. The permanent campaign is a process of continuing transformation. It never stops. Therefore, distinguishing the particular stages of the campaign (pre-campaign period, campaign period, and post-campaign period) is in a sense artificial because those particular stages often merge into one another, and there is no clear division line between them. From this perspective, governing is then turned into a perpetual campaign that remakes government into an instrument designed to sustain an elected official's public popularity.

The permanent marketing campaign is the heart of the model because it may be successfully conducted only within the political marketing process. It contains three key elements: politician/party message development and dissemination, and relationship building. Message development refers to distinguishing particular groups of voters for whom an individualized and appropriate campaign platform will be designed. Voter segment determination is a process in which all voters are broken down into segments, or groupings, that the candidate then targets with his/her message. With political marketing, one may distinguish two levels of voter segmentation: primary and secondary (Cwalina et al. 2009). *The primary segmentation* focuses on dividing voters based on the two main criteria: (1) voter party identification (particular party partisanship vs. independency) and (2) voter strength (from heavy partisans to weak partisans to floating voters). From the perspective of the whole marketing campaign, the goal of campaign should be to reinforce the decisions of the supporters and win support of those who are uncertain and whose preferences are not crystallized as well as those who still hesitate or have poor identification, for a candidate or party that is close ideologically. It is these groups of voters that require more study—*the secondary segmentation.*

After identifying voting segments, one needs to define the candidate's position in each of the multistage process of positioning. It consists of assessing the candidate's and his/her opponents' strengths and weaknesses. The key elements here include (1) creating an image of the candidate emphasizing his/her particular personality features and (2) his/her developing and presenting a clear position on the country's economic and social issues. These elements may be used jointly for positioning politicians, or separately as: *positioning via policies on issues* or *image and emotional positioning* or based on, for example, the model of "political triangle" proposed by Worcester and Mortimore (2005). The goal of message development is elaboration and establishing the campaign platform. It evolves over the course of the permanent political campaign (in the time of election period and governing). The campaign platform is defined in terms of candidate leadership, image, and issues and policies he/she advocates. It is influenced by several factors, including the candidate himself/herself, the people in his/her organization, the party, and, especially, the voters.

The established politician or party message is then distributed on the voter market. Personal (direct) campaign primarily refers to the grassroots effort necessary to build up a volunteer network to handle the day-to-day activities in running the campaign. The grassroots effort that is established becomes one information channel that transmits the candidate's message from his/her organization to the voter, and feedback from the voters to the candidate. The goal here is then not only the distribution of candidate's message but also an attempt to establish and/or enhance relationships with voters and other political power brokers. Direct marketing consists of candidate's meetings with voters and power brokers, such as lobbyists, interest groups, etc.

Mediated (indirect) campaign becomes a second information channel for the candidate. Instead of the person-to-person channel used with a direct marketing approach, this channel makes use of electronic and printed media outlets such as television, radio, newspapers, magazines, direct mail, the Internet (e.g., e-mail, websites, blogs), campaign literature (e.g., flyers, brochures, fact sheets etc.), billboards, and any other forms of promotion that are available. Political marketing also adopts new ways of communicating with the voter, mainly related with the development of new technologies such as jag social networking or mobile marketing. After the so-called "digital revolution," which was mainly related with the Internet's development and spreading, initiating convergence between traditionally separated technologies, we are currently experiencing some kind of "mobile revolution" in which all ICT and media usage seem to be going mobile. Mobile marketing is the use of wireless media as an integrated content delivery and direct-response vehicle within a cross-media marketing communications program. The wave of mobile telephony is largely behind us, but has created an environment in which every single person suddenly owns a personal mobile device and everyone is always "on" and reachable. In today's fragmented political market, it is evident that traditional mass market (and mass media) approaches need to make way for a more differentiated and personalized approach of (micro) segment targeting. In order to achieve this, and given the fact that almost everyone has a personal mobile device, the advertising and marketing sector is rigorously experimenting with a diversity of new mobile marketing paths. The ubiquity of the mobile phone extends the traditional media model of time–space. Mobile advertisers can deliver timely short message service (SMS) ads to consumers based on their demographic characteristics and geographic information. Worldwide, wireless advertisers have already integrated SMS into the media mix. SMS has started its ascent toward reaching critical mass as a direct marketing medium (Scharl et al. 2005). In political campaigning, through the use of SMS, politicians try to influence voters directly. They try to inform citizens about events they organize, to invite them to participation in campaign, and to ask for their vote. The results of research conducted by Mylona (2008) between December 2004 and January 2005 among Greek MPs suggest that 40% of them use SMS for communicating with their voters. However, younger politicians use this communication tool more than older. Mobile technology seems a potent political tool because it appeals to voters' emotions, is individualized, and reaches voters immediately.

The use of social networking by the Barack Obama campaign as both a personal and mediated information outlet in 2008 was integral to his victory. His MyBarackObama.com or MyBO site offered a wide range of Web 2.0 type functionality such as personal blogging, social networking, video, and photo sharing that moved campaign web offerings forward from the largely static "brochureware" style. Obama's innovative campaign used social network sites, text-messaging, email, and micro-targeting to raise money and reach young voters. For example, Obama had the following links to his website: Facebook, Myspace, YouTube, Flickr, Digg, Twitter, Eventful, LinkedIn, BlackPlanet, Faithbase, Eons, and Glee. The Obama campaign did not invent anything completely new. Instead, by bolting together social networking applications under the banner of a movement, they created an unforeseen force to raise money, organize locally, fight smear campaigns, and get out the vote that helped them topple the Clinton machine and then John McCain and the Republicans. For this precise reason, Obama's 2008 presidential bid became an innovative and politically revolutionary one that used media in ways the public had never seen before. Obama was able to use the Internet to not only provide information but also encourage activism. During his campaign, Obama was given over $55 million dollars in donations by linking from his social media sites and website to a MyBO (Hellweg 2011).

Obama won in 2008 in part because he made better use of the Internet and other marketing-related technologies to support his marketing efforts. Through the use of Facebook, Flickr, Twitter, and other social networking sites, the Obama strategists successfully targeted young voters who wanted change in the US political system. These technological outlets were first used by Howard Dean in 2004, even though he was unsuccessful in his bid to win the Democratic nomination. The use of social networking by the Dean campaign organization was beneficial in getting volunteer support and some fundraising, primarily as a personal, targeted approach to interested voters and citizens.

The Obama campaign's use of these same technologies was boosted to a level never before attained in a presidential campaign, with hundreds of millions of dollars raised over the Internet, and thousands of hours of support received. The funds were raised through both personal requests to individual donors who had signed up at events during the course of the campaign and through Internet channels targeted at selected segments of voters who followed the campaign over the Internet. Millions of supporters each gave under $300/person, on average, but enough to build a multimillion dollar campaign that would support the advertising that eventually worked to defeat the McCain campaign. At the same time, messages sent through mediated Internet outlets were used to allow supporters to keep track of the campaign at all times. Any interested party could tap into a website like Flickr or Twitter and both follow and be alerted to the daily activities of the campaign. Social networking at both the personal and mediated levels will continue to play an integral role in campaigns in democracies around the world in the future.

The foundation of message dissemination is organizational tasks connected with assembling staff for the campaign team, defining their tasks, and monitoring their activities where soliciting funds for the campaign plays an important role. Then, polling represents the data analysis and research that are used to develop and test new ideas and determine how successful the ideas will be. Polls are conducted in various forms (benchmark polls, follow-up polls, tracking polls) throughout the whole voting campaign and implemented by various political entities among the campaigns. One should also note the growing importance of polling specialists. The results of their analyses given to the general electorate not only reflect the electorate's general mood but also influence the forming of public opinion.

The third element of the political marketing process and the goal of political party or candidate is to establish, maintain, and enhance relationships with voters and other political power brokers (media, party organizations, sponsors, lobbyists and interest groups, etc.), so that the objectives of the parties involved are met. And this is achieved by a mutual exchange—both during election campaign and after it, when he/she is either ruling or in opposition. The integral element of the relationship building is the "promise concept." The key functions related to it include the following: to give promises, to fulfill promises, and to enable promises. Therefore, an important element of building stable relations is trust, which is a willingness to rely on an exchange partner in whom one has confidence. Trust is also the foundation of developing relationship commitment, where an exchange partner believes that an ongoing relationship with another is so important as to warrant maximum efforts at maintaining it. That is, the committed party believes the relationship is worth working on to ensure that it endures indefinitely. In order to achieve that, one also needs to establish communications channels functioning on a constant basis.

The advanced model of political marketing presented have been further developed and supported by the results of empirical research related to its particular components (Cwalina et al. 2011). The model is an attempt to include the changes taking place in modern democracies and turning political marketing into political marketing science.

REFERENCES

Baines, P.R., Harris, P., Lewis, B.R. (2002). The political marketing planning process: Improving image and message in strategic target areas. *Marketing Intelligence and Planning*, 20(1), 6–14.

Bowler, S., Donovan, T., Fernandez, K. (1996). The growth of the political marketing industry and the California initiative process. *European Journal of Marketing*, 30(10/11), 166–178.

Cwalina, W., Falkowski, A., Newman, B.I. (2008). *A Cross-Cultural Theory of Voter Behavior*. New York: Haworth Press/Taylor & Francis Group.

Cwalina, W., Falkowski, A., Newman, B.I. (2009). Political management and marketing. In D.W. Johnson (ed.), *Routledge Handbook of Political Management*, (67–80). New York: Routledge.

Cwalina, W., Falkowski, A., Newman, B.I. (2011). *Political Marketing: Theoretical and Strategic Foundations*. New York: M.E. Sharpe, Inc.

Diamond, E., Bates, S. (1992). *The Spot: The Rise of Political Advertising on Television*. Cambridge, MA: MIT Press.

Dulio, D.A., Towner, T.L. (2009). The permanent campaign. In D.W. Johnson (ed.), *Routledge Handbook of Political Management*, (83–97). New York: Routledge.

Gamson, W.A., Croteau, D., Hoynes, W., Sasson, T. (1992). Media images and the social construction of reality. *Annual Review of Sociology*, 18, 373–393.

Harris, P. (2001). To spin or not to spin that is the question: The emergence of modern political marketing. *The Marketing Review*, 2(1), 35–53.

Harris, P., Rees, P. (2000). Pictures at an exhibition: Milton, Machiavelli, Monet, Mussorgsky and marketing. *Marketing Intelligence and Planning*, 18(6/7), 386–373.

Hellweg, A. (2011). Social media sites of politicians influence their perception by constituents. *Elon Journal of Undergraduate Research in Communication*, 2(1), 22–36.

Johnston, R.J., Pattie, C.J., Allsopp, J.G. (1988). *A Nation Dividing? The Electoral Map of Great Britain 1979–1987*. London: Longman.

Kaid, L.L., Holtz-Bacha, C. (eds.) (2006). *The Sage Handbook of Political Advertising*. Thousand Oaks, CA: Sage Publications.

Kotler, P. (1975). Overview of political candidate marketing. *Advances in Consumer Research*, 12(2), 761–769.

Kotler, P., Keller, K.L. (2006). *Marketing Management*. 12th edition. Upper Saddle River, NJ: Pearson Prentice Hall.

Kotler, P., Kotler, N. (1999). Political marketing. Generating effective candidates, campaigns, and causes. In B.I. Newman (ed.), *Handbook of Political Marketing*, (3–18). Thousand Oaks, CA: Sage Publications.

Lees-Marshment, J. (2003). Political marketing: How to reach that pot of gold. *Journal of Political Marketing*, 2(1), 1–32.

Lock, A., Harris, P. (1996). Political marketing—*Vive la différence! European Journal of Marketing*, 30(10/11), 14–24.

Mauser, G. (1983). *Political Marketing: An Approach to Campaign Strategy*. New York: Praeger.

Mylona, I. (2008). SMS in everyday political marketing in Greece. *Journal of Political Marketing*, 7(3/4), 278–294.

Newman, B.I. (1994). *The Marketing of the President: Political Marketing as Campaign Strategy*. Thousand Oaks, CA: Sage Publications.

Newman, B.I. (1999). *The Mass Marketing of Politics: Democracy in an Age of Manufactured Images*. Thousand Oaks, CA: Sage Publications.

Niffenegger, P.B. (1988). Strategies for success from the political marketers. *Journal of Services Marketing*, 2(3), 15–21.

Nimmo, D. (1999). The permanent campaign: Marketing as a governing tool. In B.I. Newman (ed.), *Handbook of Political Marketing*, (73–86). Thousand Oaks, CA: Sage Publications.

O'Shaughnessy, N. (1990). High priesthood, low priestcraft: The role of political consultants. *European Journal of Marketing*, 24(2), 7–23.

Reid, D.M. (1988). Marketing the political product. *European Journal of Marketing*, 22(9), 34–47.

Scammell, M. (1999). Political marketing: Lessons for political science. *Political Studies*, 47(4), 718–739.

Scharl, A., Dickinger, A., Murphy, J. (2005). Diffusion and success factors of mobile marketing. *Electronic Commerce Research and Applications*, 4(2), 159–173.

Shama, A. (1975). Applications of marketing concepts to candidate marketing. *Advances in Consumer Research*, 2(1), 793–801.

Stewart, M.C., Clarke, H.D. (1992). The (un)importance of party leaders: Leader images and party choice in the 1987 British election. *Journal of Politics*, 54(2), 447–470.

Vargo, S.L., Lusch, R.F. (2004). Evolving to a new dominant logic for marketing. *Journal of Marketing*, 68(1), 1–17.

Worcester, R.M., Mortimore, R. (2005). Political triangulation: Measuring and reporting the key aspects of party and leader standing before and during elections. *Journal of Political Marketing*, 4(2/3), 45–72.

Wring, D. (1997). Reconciling marketing with political science: Theories of political marketing. *Journal of Marketing Management*, 13(7), 651–663.

24 Marketing Strategies in the Conceptual Age

Tymoteusz Doligalski and Jacek Wojcik

CONTENTS

ILLUSION OF A POWERFUL CONSUMER

The conceptual age is perceived as an era in which relations between companies and customers* gain a new dimension. Is it true that we are witnessing the dawn of a new era in exchange process relations on the consumer market? One cannot deny the considerable impact of information and communications technology (ICT) on market relations. However, reality does not always change according to the expectations of apologists of new ideas. In the analysis of relations between consumers and companies, one should take two aspects into consideration:

1. Who is the dominant party?
2. What is the effect of these relations?

In this context, the notion of relations was used deliberately, since the interactions between the two subjects are not limited to exchange process (i.e., the buy/sell process). The situation when there is a nonmonetary exchange between the customer and the company, and when the exchange conditions for other subjects (both customers and companies) are formulated, should be also included in the scope of consideration.

Marketing doctrine evolved from the approach where a company builds relations with a customer independently to a new approach, according to which a company should follow the lead of a modern customer, also referred to as a prosumer, a consumer 2.0, etc. Jaworski et al. (2000) distinguish two different types of market orientation—market-driven approach and driving markets approach. The first one consists of identifying and understanding customers' (or other market players') needs. The latter implies influencing the market structure and the behaviors of market players in a direction that enhances the competitive position on the market. The goal of the market driving concept is to "build new customers" by creating new needs and demand instead of maintaining servient relationship with customers. Jaworski et al. (2000) describe market driving operations as *the extent to which a firm invests resources (time, money, effort) in changing the consumer mindset in a way that*

* In this paper the authors use both the notion of customer and the notion of consumer. When using the notion of consumer, the authors refer only to the B2C market. The notion of customer will be used in the context of phenomena occurring on both B2C and B2B markets.

enhances the benefit consumer perceives from the focal product. Furthermore, the authors divide the market driving operations into two categories: the shaping of market structure and the formation of market behaviors. The market structure shaping may be achieved by

- Elimination of market players (middlemen, competitors, or suppliers)
- Building new market players, created by a new market structure
- Changing functions performed by market players

Hills and Sarin (2003) have conducted a short survey among over 100 marketing research workers on the substance of market driving actions. The obtained results can be classified from three perspectives: value building, change, and leadership. In the scope of value building, there were opinions that new markets can be created through introducing innovative solutions and identifying customers' needs that have not been previously recognized.

The idea of market driving has gained a considerable popularity in the high tech industry. Schindehutte et al. (2008), alongside market orientation, distinguish technology orientation and entrepreneurial orientation. A technologically oriented company prioritizes technology and innovations over its current customers. This kind of orientation is associated with the risk of distancing from the customers, which may lead to a situation when the products introduced by the company do not correlate with customers' needs.

Another approach to market shaping can be observed in the method developed by Kim and Mauborgne (2005), called the Blue Ocean Strategy. The authors describe the strategy as a redefinition of value for a customer, which allows the company to create and capture new demand. In consequence, competition becomes irrelevant, since the company operates in a competition-free market space. The research conducted by the authors yielded very interesting results. It appears that from the point of view of revenues and profitability, the most beneficial action for companies was to boldly redefine the composition of values offered to customers. The authors present various examples that prove that the critical success factors were to question the market standards, to introduce values that have never been offered before, and, by doing that, to establish the company's position on a market on which competitive constraints are no longer important.

It was shown that a company that takes initiative in market relation shaping can achieve success, and in many cases, such proceedings are more effective than the market-driven approach. It is worth noticing that this kind of strategy is very popular in the high-tech industry, even though new technologies are usually perceived as the customer's ally. It should be also stressed that the market driving strategy holds usually a much higher level of risk.

The concept of market-driven operations correlates with other marketing concepts. One of the most frequently quoted is the idea of one-to-one marketing. According to Peppers and Rogers (1997), in modern economy, the critical success factor is to build long-lasting relationships with customers, which help to provide them with products customized to their individual needs. The authors formulated four levels of the one-to-one marketing. These are customer identification, customer differentiation, customer interaction, and customization. The most important element of the model is customization, which consists of cooperation between the company and customers in order to adjust the company's proceedings to the customers' expectations. These steps result in improving the level of customers' satisfaction and in raising switching costs, which can help in increasing customer loyalty.

Since the 1990s, the idea of mass customization has been perceived as an extremely appealing marketing concept, enabling an individualized approach in fulfilling the needs of mass market customers. According to one of the first definitions of the phenomenon, formulated in 1993 by Pine, mass customization consists of developing, producing, marketing, and delivering affordable goods and services with enough variety and customization that nearly everyone finds exactly what they want. Kleeman and Voss (2008) consider that a mass customization is an *isolated activity of individual customers (...), not the collective activity of many individuals.* It is worth noticing that mass customization does not necessarily mean creation or modification of a product according to customer's needs. Offering a wide range of products and communicating with customers in a personalized manner, which aims at providing customers with products corresponding to their needs, are also included in the idea of mass customization. In the marketing literature, personalization of communication means departing from typical mass communication and is perceived principally as a manifestation of the customer's growing importance in relation with a company. Is it justified?

1. It should be stressed that personalization of stimuli increases the persuasive impact of promotional announcements.
2. It is of importance to determine on what kind of information is the base of the communication personalization: is the information voluntarily disclosed or is it gathered by a company without customers' knowledge? The latter results in behaviorally targeted advertising, a synonym for communication activities in Web 1.0 (O'Reilly 2005).

The personalization of persuasion can take various forms. The simplest and the most widely known is the contextual advertisement used by search engines (Web 2.0). The two types of advertising draw from different conceptions of motivation. As its name indicates, the behavioral advertising relies on the concept of behaviorism. This approach perceives the Internet user as a subject guided by the advertiser. The advertiser selects stimuli (of which the advertisement consists) in order to stimulate a desired reaction of the Internet user. The idea of Web 2.0 refers to the motivation theories formulated by the school of humanistic psychology and cognitive psychologists (our choices are independent and rational and stem from cognitive activity). Contextual advertising with the use of search engines is one of the most acceptable forms of advertising. However, lack of general approval for a given form of advertising does not necessarily have to have a negative influence on the advertiser's results. The research conducted by the Next Century Media company showed that the advertisement based on the analysis of past behaviors of Internet users is more effective, as it is more likely to draw the Internet user's attention (conclusions based on an eye tracking research). The behavioral advertisements received 17% more attention (the number of times the subjects looked and time spent looking at the advertisements). What is more, during subsequent exposures, the number of looks increased by up to 54%. In the case of behavioral targeting, the cost of reaching potential buyers was also 50% lower than with contextual targeting. Finally, behavioral targeting generated a remarkable increase in unaided brand awareness (Marcus 2006; Marketing Vox 2006).

The company BL Labs analyzed Internet users' response to over 400 million advertisements displayed online. It was proven that behaviorally targeted advertisements placed on websites with the content that did not enter in the scope of the Internet user's interests had a remarkably higher efficiency coefficient than the advertisements adapted to context (i.e., the content of a website) and the user's behavior (the click-through rate was 108% higher and the action-trough rate was 19% higher than in the case of contextually relevant advertisements) (Newcomb 2006). In the following years, companies specialized in this kind of advertising obtained even better results (Nugg.ad 2011). It seems evident that stimuli differentiation is a critical success factor for behavioral advertising, provided that its content is personalized (customized). These results do not come as a surprise for psychologists but are revealing from the point of view of marketers who for the last 10 years were focused on contextual advertising. Google company, which had promoted contextual advertisements, noticed the advantages of behavioral marketing (Baker 2005; Lee 2005). In order to accelerate the implementation of the new technology, Google purchased the company DoubleClick. Google's proceedings can be perceived as another proof of the efficiency of behavioral targeting.

Mass customization revolves around the idea of prosumer. Chronologically, it was the notion of prosumer that appeared first. In 1980, it was coined by Toffler (1980) to underline the role of consumers in product co-creation process (primarily in the scope of design works), the goal of which is to satisfy consumer's needs to the highest possible extent. In recent years, the notion assumed an increasingly important role. Google (2011) trends showed a sudden intensification in the search for the term after 2000, especially in the years 2005–2009. It seems that there is a tendency to overuse the notion of prosumer. Not every kind of consumer's participation in product creation can be qualified as a manifestation of prosumerism within the meaning of Toffler's definition. It does not include activities related to self-service, such as product assembly.

In successive years, the meaning of the term broadened. The notion of prosumer includes equally

- A professional consumer
- A proactive consumer

The above-mentioned meanings are not mutually exclusive, but they do not necessarily have to overlap. Consumer's activity does not have to manifest in willingness to participate in product creation. It can be restricted to search for information on a product, which in turn should not be always perceived as a manifestation of consumer's competence. For many authors, exhibiting activity in communication process and gathering information from the sources beyond the supplier's control, especially from other consumers, seem to be sufficient determinants of prosumerism. Researches based on the simplest methods (retrospective survey and respondents' declarations) are not able to verify the cost-effectiveness and the efficiency of consumers' actions, which are the aspects that combine consumer's activity (which is relatively easy to measure) with purposefulness (including consumer's competences, which are significantly more difficult to estimate) (Nielsen 2011). Exchanging opinions about products with other consumers for many years has been a typical practice, mentioned even by rather old publications (Katz and Lazarsfeld 1955). Dependence on the opinions of others can be perceived as a manifestation of shallow information processing. Wojciszke indicates that in the case of peripheral information processing, the recipients' reactions can be interpreted as a sign of being influenced by others (Wojciszke and Doliński 2008). It should be mentioned that in Rogers's (1962) theory of diffusion of innovations, the persons that adopt an innovation relatively late are more susceptible to the influence of others than the innovators.

Consumer's participation in the product co-creation can take place on three levels:

1. Co-creation effectuated by consumers for themselves (which can be viewed as one of customization processes).
2. Co-creation by the company and the consumer of products for other consumers (the introduction of mechanisms based on social proof to search engines, e.g., Page Rank, can be considered as one of the most advanced effects of such cooperation).
3. Co-creation of products by consumers for other consumers (individually or in group with other customers)—the consumer becomes the producer.

In the case of levels 1 and 2, the product creation process takes place in cooperation with the producer and is controlled by the company (the experts), which is not the case of the third level (new middlemen provide only a cooperation platform). The quality of products created by consumers for consumers (especially of the so-called UGCs—user-generated contents, such as Reddit or Digg) still arouses controversy. In the case of UGC products, content valorization is usually based on audience ratings. On the Internet, the audience rating started to be perceived as the crucial ranking factor. Hence, some authors do not hesitate to describe this phenomenon as collective stupidity or digital maoism (Lanier 2006). However, the adherents of the digital wisdom appear as the dominant group. It comprises researchers such as Lévy (1997) ("collective intelligence"), Rheingold (2002) ("smart mobs"), or de Kerckhove (1997) ("connected intelligence") and authors representing business approach, such as Howe (2008) ("crowdsourcing"), Tapscott (Tapscott and Williams 2008) ("wikinomics"), and Surowiecki (2004) ("the wisdom of crowds"). While evaluating such undertakings, one should be aware that only a few of them, such as Wikipedia or Linux, achieved a spectacular success. In many areas, the products co-created by customers did not gain popularity, vide: the case of the open-source search engine.

In conclusion, it seems that humanistic and common sense psychology had a considerable impact on the creation of the idea of prosumer and Web 2.0. These concepts seem to respond to our desires, a fact that is the reason for their popularity. Their market implementation, however, does not always bring expected results, because of cognitive limitations of consumers, suffering from information overload.

NETWORK EFFECTS: IS THE NUMBER OF USERS MORE IMPORTANT THAN THE PRODUCT QUALITY?

In modern economy, one of the most significant elements that influence the customer value formation process are *network effects* (*network externalities*). The notion of network effect describes a situation in which values for a customer depend on the number of a given product's users and increase with their number (Wang et al. 2010).

Network effects occur most frequently in the case of products and services related to the Internet or associated with information and communication technologies, less often in connection with traditional products. The products that use the network effect are communication products (i.e., telephones, fax machines, web messengers), online communities enabling interaction between users (i.e., online auctions, discussion forums, social networking services), and products based on a particular standard (i.e., software, data storage devices, cassettes, photographic films).

According to Shapiro and Varian (1999), pioneer companies that enter markets on which network effects are particularly pronounced can count on the so-called first-mover advantage. It is a possibility to gain a considerable number of customers (a customer base) in a short period of time, which is likely to restrict market access for competitors. Therefore, a company should aim at reaching critical mass, as it can become a serious barrier to potential rivals. In such a situation, two effects can appear: the lock-in effect, which prevents customers from leaving the company due to the lack of satisfactory alternative solutions, and the lock-out effect, which consists of eliminating competition from the market. Eventually, it leads to the situation when due to a significant number of customers using one solution, the costs of switching suppliers increase and the particular solution becomes a market standard.

In the article "Does Quality Win? Network Effects Versus Quality in High-Tech Markets," Tellis et al. (2009) disagree with the aforementioned theory. The authors conducted a research to determine which of the two—precedence in offering a product (service) or product's quality—are the critical drivers of success for the network effect-dominated markets. The problem is important not only from the point of view of marketing strategies but also from the point of view of economy, as it concerns a much discussed issue: whether the market is always effective (in this case: whether a product offering the highest values always becomes the market leader) or, quite the contrary—is it subject to hysteresis, that is, does the balance and structure of the market depend on the prior events. Therefore, the predominant value of a product does not necessarily have to translate into market predominance.

Tellis et al. (2009) quote opinions on the subject, which are divided. Katz and Shapiro claim that the markets that are driven by network effects show a tendency to get locked-in with outdated standards or technologies. Krugman doubts whether markets aim at the best possible solutions and claims that historical events on the market have an influence on its future shape (*outcome of market competition*). On the other hand, some authors claim that network effects do not protect markets from competition, hence making them efficient.

Tellis et al. (2009) research was conducted on various product categories related to personal computers, mainly software. The authors assessed the quality of products resorting to ratings and reviews from professional journals. The study yielded the following results concerning markets with the presence of network effects:

- Market leadership changes frequently and a new leader appears every 3.8 years, on average.
- Change in market leadership is generally associated with a change in quality the same year or a few years earlier.
- Both network effects and quality determine market share, but quality seems to be more important.
- Even in the presence of network effects, the market is efficient.

The results described above can be confirmed by the changes that took place in the years 1984–1997 in the word processor market, an industry being considerably influenced by network effects. Word processors provide customers with network-independent values, such as computer-based text processing, but also with a possibility to share files with other users. In the early 1980s, the industry was dominated by the WordStar program, introduced in 1978. The product offered the highest quality on the market; hence, in 1985 it had a 70% market share. However, since 1984, the program's quality relatively began to decline, entailing a decrease in market share. In 1985, WordPerfect surpassed WordStar in quality, and its market share began to rise. In 1989, WordPerfect achieved a leading position on the market. In 1991, the program that, according to analysts, had the highest quality was Microsoft Word, which in two years became the market leader, obtaining in 1997 a 90% market share.

The most important element of the above-mentioned research is, possibly, the correlation between quality and network effects. According to the authors, the two variables do not diverge but converge, since the network effect enables the quality leader to gain the market leadership in a considerably shorter period of time. The authors note that such a situation takes place when customers notice the difference in quality of the products, a network effect occurs, and switching costs are not excessively high. The study shows that the markets affected by network effects stay effective, that is, the share in market is influenced to a greater extent by quality than by the number of customers.

CUSTOMER SWITCHING COST AS THE DARK SIDE OF CUSTOMER LOYALTY

For the needs of this paper, customer loyalty will be defined as customers' intention to perform particular actions in the future, such as buying a product or recommending the company to others. Understood in that way, loyalty is determined by two factors: customer's willingness to continue the relationship with a company and supplier switching costs. The following section provides information on the latter. For the needs of this paper, after McSorley et al. (2003), we define switching costs as the real or perceived costs that are incurred when changing supplier but which are not incurred by remaining with the current supplier.

According to Shapiro and Varian (1999), identification, understanding, and calculation of switching costs, as well as proper strategy development, are the key factors of efficient competition in the modern economy. Farrell and Shapiro (1988) include switching costs in relationship-specific assets.

In the subject literature, switching costs, along with customer satisfaction, are considered to be two of the most important determinants of customer loyalty (Fornell 1992; Caruana 2004). Quite frequently, the publications on customer loyalty concentrate on particular aspects of switching costs, such as risk aversion. Gourville (2005) reports that, for a customer, a loss of a benefit is more significant than a gain of the same amount. Some authors assume that for individual consumers, the psychological barrier of switching costs is the most difficult to deal with (Caruana 2004).

McSorley et al. (2003) have established the following classification of switching costs:

- Transaction costs
- Compatibility costs
- Learning costs

- Contractual switching costs
- Uncertainty costs
- Psychological costs

High switching costs can cause the so-called lock-in effect, which is a situation when a customer repeatedly purchases the same product not because of satisfaction but because of switching costs. Shy states that existence of the lock-in effect on a market allows one to fix the price above marginal costs (Aydin and Ozer 2006). The Delta model, presented in the next chapter of this paper, reflects marketing strategies, including those based on the lock-in effect.

The influence of switching costs on the market operation and firm strategies was broadly discussed in Klemperer's works written from the perspective of *industrial organization,* a field of knowledge that studies the behavior of markets and companies, from the perspective of microeconomic analysis. Klemperer's research shows that switching costs can lead to a decrease in market competitiveness, including rise in prices (Klemperer 1987b), market entry deterrence (Klemperer 1987a), and reduction of losses resulting from price wars (Klemperer 1989). The results of the research prove that it is the company imposing switching costs on customers that draws the biggest benefit from it. The subjects most damaged by these actions are the company's customers, other customers, and potential competitors willing to enter the market.

Hence, it seems plausible to take into consideration the ethical implications of supplier switching costs. Since customers are perceived as a company's main asset, providing it with incomes and other indispensable values, is it justified from the ethical point of view to burden them with supplier switching costs?

It is worth noticing that switching cost can stem from the values offered to customers. If a service provider operating on institutional market (e.g., an advertising agency) acquires a profitable client, it is likely to employ the *client-proximity strategy*: an advertising agency will try to most accurately identify the client's needs in order to adjust to them the values represented by the final product, but also by the patterns of communication and proceedings. From the client's point of view, the situation may seem very comfortable, as the advertising agency tries to fulfill their needs. However, with every new attempt to understand client's needs and adjusting the offer, advertising agency burdens customers with supplier switching costs. From the client's perspective, finding an alternative supplier and developing a new customer–supplier relationship requires an investment of time, money, and considerable effort, without guaranteeing satisfactory final results. Thus, customers offered competitive values tend to build a relationship with a company, which in turn increases the company's competiveness and allows the company to obtain benefits described by Klemperer and other authors.

COMPETITIVE STRATEGIES IN CONCEPTUAL AGE ACCORDING TO THE DELTA MODEL

The above-described phenomena may be defined within the scope of the strategy model introduced by Hax and Wilde II (1999) called, after the Greek letter delta, the Delta model. Hax and Wilde II distinguished three areas of competitive advantage in modern (network) economy. These are (1) best product, (2) total customer solutions, and (3) system lock-in. According to the authors, all the strategies based on respective areas of the model aim at establishing bonds with a customer, but their ways to achieve the goal are completely different.

The *best product* strategy focuses on the creation of value based on product portfolio management. The bond with a customer is built by a rapid introduction of new products to the market (the rule of *first to market*) or by the products' superiority (the rule of *dominant design*). As a consequence, the bond with a client obtained through the best product strategy is the weakest. According to the authors, in the scope of the best product strategy, customers are generic, numerous, and faceless. Competitors' actions are usually the central focus of the company's operations, which include imitation strategies and price wars. This kind of behavior was exhibited by Google at the beginning of its activity. The company launched a search engine, assuming that its quality would encourage customers to use it and recommend it to others.

In the *total customer solutions* strategy, the company builds value by managing relations with customers. Products and services offered by the company are adjusted to particular customers' needs. During the process, both parties become acquainted and learn about each other, which positively influences product and service customization. According to the authors, the difference between the best product strategy and the total customer solutions strategy can be compared to the difference between a war with a rival company and love for a customer. Even if the use of the notion of love in this context may seem slightly exaggerated, the relation between the customer and the company can be described in terms of mutual cooperation. This strategy is employed, for instance, by Amazon.com. The company is not particularly attached to the goods it sells, for example, giving customers a possibility to write negative reviews, but concentrates on the relationship with clients.

In the *system lock-in* strategy, a company tries to create a situation where the switching costs are high enough to prevent loss of customers. Competitive advantage is gained not by product portfolio or customer value management but by cooperation with companies offering complementary products. From the perspective of the system lock-in strategy, the number of complementary solutions providers is one of the determinants of efficiency. It can help to eliminate competitors from the market, limiting customer's choice and in some cases even restricting the selection to a single solution. The example of Microsoft and its Windows operating system seems to be a model illustration of lock-in strategy application. Microsoft's complementors are software companies that produce applications running only on the Windows operating system. The functioning of the YouTube video service can be equally described in the context of the system lock-in strategy. In this case, the role of customers is played by viewers, and people or companies uploading videos act as complementors.

Hax and Wilde II contributed to the theory of customer relationship development by putting in evidence the different roles played by customers in company's marketing strategies. Only one of the three strategies perceives the customer as a distinguishable entity, who, according to the authors, enters in a relationship of love with a company. In other words, the company and the customer cooperate in order to fulfill the needs of the latter (the customer solutions strategy). In the two other strategies, an individual customer cannot be distinguished from the group (the best product strategy) or is treated as a hostage forced to use the services provided by the company (customer lock-in). What is more, the actions aimed at gaining competitive advantage proposed by Hax and Wilde II concentrate on imposing switching costs on customers and limiting competitors' actions.

CONCLUSION

In the scope of marketing communication, the idea of market-driven operations manifests in the Web 2.0 concept. Together with prosumerism, they may lead to product unification (conformity effect, social righteousness effect) rather than to its customization.

In the market-driven model, the servient relationship with customers shows faith in customers, in their cognitive capacities and will to cooperate. The efficiency of behaviorally targeted advertising proves that the Web 1.0 marketing communication model can be more beneficial for companies than the Web 2.0 model. The company controls the communication process and can take advantage of information asymmetry (i.e., the company has better access to information and tools enabling its processing).

Another element that gives a company advantage over a customer is the lock-in effect. The notion describes a situation where a customer, even though unsatisfied with the company's offer, continues using their services. This situation results from high switching costs, which include time and effort spent on search for a new supplier. However, examination of switching costs only from the perspective of customer's losses does not seem justified. The switching costs may be related to correct fulfillment of customer's needs. In such a case, the customer is satisfied and willing to maintain the relationship with the company, also because finding a company offering similar benefits is hampered.

Switching costs can originate from network effect, that is, a situation when a customer's benefits from using a given product depend on the number of users of the product. There was the fear that markets with network effect would be ineffective, since these markets exhibit the first-mover advantage, as a consequence of which they tend to get locked-in with outdated standards. The research shows, however, that the markets remain effective, which indicates that, in this case, the product's value is more important than the order of entering the market.

The above-described phenomena are reflected in companies' strategies. The Delta model shows that companies gain competitive advantage by imposing switching costs on customers (customer lock-in) and afterward eliminating competitors (competitors lock-out). As a consequence, not only the current clients of a company have difficulties in switching suppliers, but also potential customers are somehow restrained to a particular company's services.

REFERENCES

Aydin, S., Ozer, G. (2006). How switching costs affect subscriber loyalty in the Turkish mobile phone market: An exploratory study. *Journal of Targeting, Measurement and Analysis for Marketing, 14*, 141–155.

Baker, L. (2005). Google advertising patents for behavioral targeting, personalization and profiling. *Search Engine Journal*, 7.10.2005, http://www.searchenginejournal.com/?p=2311, [14.03.2011].

Caruana, A. (2004). The impact of switching costs on customer loyalty: A study among corporate customers of mobile telephony. *Journal of Targeting, Measurement and Analysis for Marketing, 12*, 256–268.

de Kerckhove, D. (1997). *Connected Intelligence: The Arrival of the Web Society*. Toronto: Somerville House.

Farrell, J., Shapiro, C. (1988). Dynamic Competition with Switching Costs. *The RAND Journal of Economics, 19*, 125–137.

Fornell, C. (1992). A national consumer satisfaction barometer: The Swedish experience. *Journal of Marketing, 56*, 6–21.

Google. (2011). *Google Trends*. http://www.google.com/trends.

Gourville, J.T. (2005). The curse of innovation: A theory of why innovative new products fail in the marketplace. Harvard Business School Research Paper, no. 05-06.

Hax, A.C., Wilde II, D.L. (1999). The delta model: Adaptive management for a changing world. *Sloan Management Review, 40*, 11–28.

Hills, S.B., Sarin, S. (2003). From market driven to market driving: An alternate paradigm for marketing in high technology industries. *Journal of Marketing Theory and Practice, 11(3)*, 13–24.

Howe, J. (2008). *Crowdsourcing: Why the Power of the Crowd Is Driving the Future of Business*. New York: Crown Publishing Group.

Jaworski, B., Kohli, A.K., Sahay, A. (2000). Market-driven versus driving markets. *Journal of Academy of Marketing Science, 28(1)*, 45–54.

Katz, E., Lazarsfeld, P.F. (1955). *Personal Influence*. New York: Free Press, p. 234.

Kim, W.Ch., Mauborgne, R. (2005). *Blue Ocean Strategy*. Boston: Harvard Business School Press.

Kleemann, F., Voss, G.G. (2008). Un(der)paid innovators: The commercial utilization of consumer work through crowdsourcing. *Science, Technology and Innovation Studies, 4(1)*, 5–16.

Klemperer, P. (1987a). Entry deterrence in markets with consumer switching costs. *The Economic Journal, 97*, 99–117.

Klemperer, P. (1987b). The competitiveness of markets with switching costs. *Rand Journal of Economics, 18(1)*, 138–150.

Klemperer, P. (1989). Price wars caused by switching costs. *Review of Economic Studies, 56*, 405–420.

Lanier, J. (2006). Digital maoism: The hazards of the new online collectivism. *Edge*, 30.05.2006, http://edge.org/conversation/digital-maoism-the-hazards-of-the-new-online-collectivism.

Lee, K. (2005). Search personalization and PPC Search Marketing. *ClickZ*, 15.07.2005, http://www.clickz.com/clickz/column/1714335/search-personalization-ppc-search-marketing, [14.03.2011].

Lévy, P. (1997). *Collective Intelligence: Mankind's Emerging World in Cyberspace*. Cambridge: Perseus.

Marcus, N. (2006). Behavioral targeting has more visibility. *iMedia Connection*, 1.02.2006, http://www.imediaconnection.com/content/8121.asp, [14.03.2011].

Marketing Vox. (2006). Behavioral targeting reaches more buyers, at lower cost. *Marketing Vox*, 24.04.2006, http://www.marketingvox.com/study_behavioral_targeting_reaches_more_buyers_at_lower_cost-021590/, [14.03.2011].

McSorley, C., Padilla, A., Williams, M., Fernandez, D., Reye, T. (2003). Switching costs. A report prepared for the Office of Fair Trading and the Department of Trade and Industry by National Economic Research Associates, 04.2003, http://www.oft.gov.uk/shared_oft/reports/comp_policy/oft655aannexea.pdf, [14.03.2011].

Newcomb, K. (2006). Study: Behavioral ads convert better out of context. *ClickZ*, 16.10.2006, http://www.clickz.com/clickz/news/1712836/study-behavioral-ads-convert-better-out-context, [14.03.2011].

Nielsen. (2011). *Nielsen Global Online Consumer Survey.* http://nielsen.com/us/en/nielsen-solutions/nielsen-measurement/global-consumer-confidence.html.

Nugg.ad. (2011). Case study. http://www.nugg.ad/en/index, [14.03.2011].

O'Reilly, T. (2005). What is web 2.0, design patterns and business models for the next generation of software? *O'Reilly Media Inc*, 30.9.2005, http://oreilly.com/web2/archive/what-is-web-20.html?page=1, [2011.03.14].

Peppers, D., Rogers, M. (1997). *Enterprise One-to-One, Tools for Competing in the Interactive Age*. New York: Curency Doubleday.

Pine II, J.B. (1993). *Mass Customization—The New Frontier in Business Competition*. Cambridge: Harvard Business School Press.

Rheingold, H. (2002). *Smart Mobs: The Next Social Revolution: Transforming Cultures and Communities in the Age of Instant Access*. Cambridge: Perseus.

Rogers, E.M. (1962). *Diffusion of Innovations*. New York: Free Press, p. 162.

Schindehutte, M., Morris, M.H., Kocak, A. (2008). Understanding market-driving behavior: The role of entrepreneurship. *Journal of Small Business Management, 46*, 4–26.

Shapiro, C., Varian, H.R. (1999). *Information Rules: A Strategic Guide to the Network Economy*. Boston: Harvard Business Review Press.

Surowiecki, J. (2004). *The Wisdom of Crowds: Why the Many are Smarter than the Few and How Collective Wisdom Shapes Business, Economies, Societies, and Nations*. New York: Doubleday.

Tapscott, D., Williams, A.D. (2008). *Wikinomics: How Mass Collaboration Changes Everything*. New York: Penguin Group.

Tellis, G.J., Yin, E., Niraj, R. (2009). Does quality win? Network effects versus quality in high-tech markets. *Journal of Marketing Research, XLVI*, 135–149.

Toffler, A. (1980). *The Third Wave*. New York: Bantam Books.

Wang, Q., Chen, Y., Xie, J. (2010). Survival in markets with network effects: Product compatibility and order-of-entry effects. *Journal of Marketing, 74(4)*, 1–14.

Wojciszke, B., Doliński, D. (2008). Psychologia społeczna [Social psychology]. In: J. Strelau, D. Doliński (Eds.) *Psychologia* [Psychology]. Gdańsk: Gdańskie Wydawnictwo Psychologiczne, p. 348.

25 Work Locus of Control and Burnout Syndrome

Jozef Maciuszek and Bartlomiej Melges

CONTENTS

INTRODUCTION

The paper analyzes the significance of an individual's perception of locus of control (LOC), which determines their functioning at work. It also looks at the results of our research into the relationship between work LOC and degree of occupational burnout. Numerous studies of the relationship between LOC and occupational burnout have been conducted before ours, but they focused mainly on a generalized perception of the LOC. Our research, however, investigates the relationship between individual beliefs about sources of control in the workplace and burnout syndrome.

ON PROFESSIONAL BURNOUT

One of the first researchers concerned with professional burnout was Maslach (1993), who defined its three main symptoms: (1) emotional exhaustion; (2) depersonalization, which is an indifferent, disinterested, distanced, and even dehumanized way of treating other people; and (3) a reduced sense of personal achievements, a sense of professional failure going hand in hand with reduced professional commitment. Theoretical and methodological considerations alongside a number of research reports (Maslach 1998, 2003; Maslach et al. 2001) argue for a multidimensional burnout model. The phenomenon of professional burnout refers to professional interactions such as helping, treatment, therapy, teaching, and upbringing. For this reason, most of the research refers to professions such as teacher, doctor, nurse, and social worker.

Our approach to the analysis of the role of convictions in the context of burnout is based on a few elementary assumptions. We assume the multidimensional burnout model proposed by Maslach (2003), and we presume that burnout symptoms appear in conditions of chronic social stress. We consider stress from a transactional-cognitive perspective (Lazarus and Folkman 1984) as a dynamic sequence of stressor evaluation processes (primary appraisal) and own competences and resource evaluation processes (secondary appraisal). Alongside environmental factors in burnout processes, we also recognize the important role of subjective factors (including the individual's conviction system).

Two main groups of factors affecting occupational burnout are individual variables (especially social, demographic, and personality-related) and situational variables (related to the nature of an individual's job and the characteristics of their work environment). Among individual variables, the following can be identified as predictors of occupational burnout: type A behavior model (Schaufeli and Enzmann 1998), "Big Five" model neuroticism (which is positively correlated with emotional exhaustion and depersonalization; Bakker et al. 2006), external LOC (Schaufeli and Enzmann 1998), and an emotion-oriented coping style (Lee and Ashforth 1993). In turn, certain temperament traits, especially the strength of the actuation process and mobility, have a negative correlation with burnout (Langelaan et al. 2006). Maslach and Leiter (1997) strongly emphasize the role of the work environment in the job burnout process. They claim that burnout says more about the work environment of employees than about the employees themselves, and the root causes of burnout are more likely to be found in the former than in an individual's character traits. Key causes include a mismatch between an individual's needs and requirements of their jobs. The six most frequently described types of such mismatch are (1) excessive workload; (2) a feeling of a lack of control over one's work; (3) insufficient gratification from one's contribution; (4) no sense of community in the workplace; (5) unfair treatment; and (6) conflicting values. According to this approach, job burnout is a consequence of excessive workloads, a lack of control over one's own work or professional decisions, inadequate remuneration, unfair treatment, and a clash between an employee's values and the system of values prevailing at the organization.

It is clear that Maslach and Leiter perceive a lack of control at work as a type of mismatch that can lead to occupational burnout. Lack of control is understood as a situation in which employees do not have sufficient autonomy and independence as regards the tasks they are required to perform, they are unable to take part in the decision-making process or to make relevant choices, and they have a feeling of being overwhelmed by their work responsibilities. A feeling of a lack of control arises when employees are unable to use their skills to set priorities, make choices about the right ways to act, and decide about the use of resources (as a consequence of a top-down policy approach). This feeling also ties in with the obligation to carry out orders that the employee does not agree with.

Understood in this way, a sense of a lack of control at work is seen as one of the organizational predictors of occupational burnout. This paper refers to the category of control in the context of individual variables and is closely associated with an individual's system of beliefs.

CONTROL UNDERSTANDING

"Control" is a very broad term that refers to many aspects of human functioning. In order to describe the role of convictions about control in the context of professional burnout, it is necessary to mark out the most important criteria for defining the concept of control (Sokołowska 1993). Firstly, in talking about control, we can mean either an objective state of affairs or a perceived or felt control, independent of an objective state of affairs. There has been a good deal of research on the influence of general convictions referring to control on the behavior of participants who found themselves in the same objective situation. Secondly, control is understood either as control of an environment or as control of one's own actions and their results (personal control). In the first case, there is a similarity between understood control and the power definition (influence on the behavior of others). Personal control is perceived as control by an individual over the choice of the direction, process, and results of his/her activity (Sokołowska 1993). Thirdly, personal control is defined in two ways. One is as perceived dependence between one's own activity and obtained results; one can think that the results depend on oneself or on other factors such as coincidence or other people (Rotter 1981). The other definition is as a perceived possibility to obtain desired results owing to one's own activity, perceiving one's own effectiveness or competence (Bandura 1997). Fourthly, control of one's own activity and its results can be divided into behavioral and cognitive control. Behavioral control is the management of events by one's own intentional activity (active control over environment rather than passive dependence and indulgence); it is a situation in which a particular event appearance depends on the individual, on his/her behavior. Cognitive control refers to the ability to understand events taking place (e.g., revealing the cause and effect relationships that make sense of events) and to the ability to anticipate the events happening around us (bringing order instead of chaos and uncertainty). We will make particular reference in this article to these aspects of control in which the role of the individual's convictions is revealed.

LOCUS OF CONTROL

LOC refers to the system of convictions regarding the factors determining the results of an individual's activity. People who display a greater external LOC are convinced that events do not depend on their activity but are rather defined by external forces such as chance, fate, other people, or the so-called objective situation. People with an internal LOC hold opposite convictions; they believe that the results of their activity and course of life depend on their behavior and traits.

Internal LOC is not synonymous with a feeling of self-efficacy (Bandura 1997). A feeling of internal LOC may be expressed not only in the conviction that one can achieve goals thanks to one's own effort but also in the belief that some results cannot be achieved, for example, because some skills, competences, or characteristic traits are missing. This generalized feeling of internal or external LOC is perceived as an individual's personality trait that determines their behavior, especially in new, complex, or obscure situations (Rotter 1966). According to some (cf. Judge et al. 2002), LOC is genetically conditioned; however, the prevailing view is that it arises as a result of repetitive experiences.

A large volume of research has been carried out measuring the LOC with Rotter's I-E Scale. The results obtained reveal many differences between the two representatives of conviction systems mentioned above. As expected, people who had an external LOC resigned faster from experiments on acquired helplessness, whereas those with an internal LOC were persistent in aspiring to final success and were not discouraged by initial defeats. People who have an internal LOC are more resistant to social influence and more "I" dependent, that is, less conformist. In experimental task situations, they are more focused on tasks whose results depend more on skills than on those with results depending on chance. In contrast, a generalized conviction of an external LOC goes hand in hand with more frequently choosing activities whose results depend on random factors rather than on one's own efficiency, and with greater motivation to take up such activities (Rykman 1979).

Researchers (see Zimbardo and Ruch 1977) have also observed a relation between LOC and socioeconomic variables. People from socioeconomically disadvantaged groups are convinced of an external LOC and the uselessness of individual effort, which can explain the apathy and lack of motivation for achievements characteristic of such groups. Greater idleness and fatalism during natural disasters have also been observed in the case of the "externals." The results of research carried out under Kozielecki's (1981) guidance show that a conviction of an internal LOC affects the choice of more risky activities, greater consistency in decision-making, and application of a long-term perspective strategy (taking the distant consequences of one's own activities into consideration) in complex decision-making situations.

People with external LOC have a lower achievement motivation, which becomes a drive to avoid failure rather than to achieve success. People with internal LOC function better socially; they are more inclined to establish close social relationships, better able to influence others, and have a higher sociometric status. They are characterized by a task-oriented stress-coping style, higher frustration tolerance, greater joy of life, and a preference for skill over chance-determined situations.

Although Rotter (1966) assumed that LOC was uniform, other researchers have identified more detailed dimensions within this general category. For instance, a distinction is made between an individual's beliefs about themselves (locus of personal control) and their beliefs about man in general ("ideology of control"). A distinction is also often made between a feeling of control over success and failure (Weiner and Potepan 1970). Furthermore, it is often noted that the LOC concept can be applied to describe various spheres or aspects of life, for example, control over one's own health, control in interpersonal relationships, or a feeling of control at work.

We are especially interested in the significance of work LOC in the context of the burnout syndrome. Work LOC describes an individual's beliefs about what determines the results they achieve, their rewards and punishments, promotion chances or supervisors' recognition, the general course of their career, etc. Research results have generally confirmed that people with an internal LOC function better in the work environment: they tend to be more effective as workers and are more satisfied with their jobs (Judge and Bono 2001); they feel a stronger internal motivation to work and have better potential management skills (Spector 1988). However, it is also often mentioned that the role of LOC in one's functioning at work depends on the type of work performed and its nature.

Various studies have been performed focusing on the relationship between a generalized feeling of LOC and occupational burnout. In our research, we were mainly interested in the beliefs about one's work situation and its various aspects.

LOC AND THE BURNOUT SYNDROME

The principal objective of our study was to identify the relationship, if any, between a detailed perception of LOC (individual variables) and professional burnout syndrome. Another question we asked ourselves was whether the individual variables under consideration made it possible to predict burnout. The research was carried out in the Spring of 2012 with a group of 202 music teachers (165 women and 37 men) working at nine primary schools in southern Poland.

In our study, we assumed that individual beliefs about sources of control at work can be more strongly correlated with professional burnout than the generalized LOC. Therefore, we used the "Man at Work" questionnaire developed by Matczak et al. (2009) to measure work LOC.

The Man at Work questionnaire consists of two main scales: the *LOC* scale (41 items) and the *Control Scale* (CS) (19 items). The CS measures the subject's tendency to give socially approved answers. The LOC scale looks at the subject's perception of work LOC, understood as beliefs about the factors—external (located in the environment) or internal (subjective)—that determine the effects of a person's work activities. The LOC scale consists of seven subscales that focus on various detailed aspects of LOC:

1. *Feeling of Inefficacy (FI):* describes the degree to which the subject is convinced that their professional career and work results do not depend on their competences and motivation.
2. *Sense of Fatalism (SF):* describes the degree to which the subject is convinced that their professional development and success are determined by fate, chance, or good or bad luck (a tendency to attribute sources of control to nonsocial external factors).
3. *Sense of Being Dependent on Others (SDO):* assesses the degree to which the subject perceives that the results of their work depend on the attitudes and activities of others (perceiving external social factors as sources of control).
4. *Personal Control (PC):* concerns the belief about one's ability to control one's professional career through one's own actions, abilities, and skills.
5. *Ideology of Control (I):* concerns the beliefs about LOC in "people in general," that is, to what extent people can decide about their professional lives.
6. *Successes (S):* this subscale looks at the sources of control in successfully completed activities.
7. *Failures (F):* this subscale concerns the sources of control of failures.

The questionnaire applies a four-degree assessment scale (*strongly disagree, disagree, agree, strongly agree*), and the highest-scoring answers are those indicating an external LOC.

To assess professional burnout, we resorted to the *Maslach Burnout Inventory* (MBI) questionnaire, that is, the most common and best-standardized tool used to assess this phenomenon (Maslach and Jackson 1986). It evaluates three dimensions of burnout: nine statements concern emotional exhaustion, five concern depersonalization, and eight look at a reduced feeling of personal accomplishment. Answers are given according to a seven-point scale describing the occurrence of a given feeling, ranging from “never” to “every day.”

Results

In order to answer the question about the existence of a link between LOC and professional burnout, a correlation analysis based on Pearson’s *r* method was applied. Table 25.1 lists the obtained coefficients of correlation between LOC and various dimensions of professional burnout.

The overall results of the questionnaire (the LOC scale) indicate a strong link between LOC and all dimensions of professional burnout, that is, emotional exhaustion, depersonalization, and reduced personal accomplishment. The relationship between the LOC scale and burnout consists of the fact that external LOC has a positive correlation with emotional exhaustion ($r = 0.45$, $p < 0.001$) and depersonalization ($r = 0.19$, $p < 0.19$), and a negative correlation with the feeling of personal accomplishment ($r = 0.34$; $p < 0.01$). This means that a higher general indicator of external LOC results in

- A greater feeling of an excessive emotional burden, fatigue, sentimental emptiness, lack of mental energy (i.e., greater emotional exhaustion)
- A more negative attitude toward trainees, the greater the likelihood of treating subordinates in a passive, indifferent, impersonal, instrumental, cynical, or contemptuous way (i.e., greater depersonalization)
- A lower assessment of one’s own work, professional skills, and achievements (a feeling of a lack of accomplishment)

We obtained identical results on all subscales of the questionnaire, which measured various aspects of work LOC. These subscales included the following: Feeling of Inefficacy (FI), Sense of Fatalism (SF), Sense of Being Dependent on Others (SDO), Personal Control (PC), Ideology of Control (IC), Successes (S), and Failures (F). It turns out that various aspects of external LOC are positively correlated with emotional exhaustion and depersonalization and have a negative correlation with the sense of personal accomplishment at work.

TABLE 25.1
Pearson’s *r* Correlation between the Dimensions of the Work LOC and Professional Burnout

LOC Dimensions	Emotional Exhaustion	Depersonalization	Reduced Personal Accomplishment
LOC (overall result)	.45**	.19**	–.34**
FI	.39**	.15*	–.42**
SF	.35**	.10	–.25**
SDO	.36**	.26**	–.11
PC	.42**	.17*	–.37**
IC	.44**	.19**	–.30**
S	.43**	.22**	–.30**
F	.41**	.15*	–.27**
CS	–.02	–.25**	.37**

*p <.05; **p <.01.

We also looked at which aspects of work LOC are predictors of particular dimensions of professional burnout. A regression analysis proved that exhaustion was most strongly determined by a generalized feeling that employees do not have a real impact on the course of their professional careers (measured on the Ideology of Control subscale; beta = 0.48, $p < 0.001$). Depersonalization is most strongly determined by a sense of dependence on others (beta = 0.34, $p < 0.001$). A feeling of a lack of professional effectiveness is determined by the sense of inefficacy (beta = −0.51, $p < 0.001$).

CONCLUSION

Among the subjective variables affecting stress and professional burnout, various conviction patterns play an important role. They are judgments referring to the order of things in the world, to other people, and to ourselves, which are characterized by a high degree of certainty (high subjective probability). The substance of convictions is a belief in their truth, the sense of certainty of something, certainty referring to the causes of events, the meaning of particular phenomena or facts, and what is important and what matters most.

The most important categories of convictions refer to ourselves, other people, and the order of the world. Most often, they consist of one of three types: the first refers to convictions about cause and effect relationships. We hold many convictions that one thing causes another, and according to them, we establish rules to live by. What makes one successful in his/her professional life? What is the cause of failure? What causes cancer? What makes people creative when doing their job? The conclusions we draw come from our convictions. The second type of convictions can refer to meaning. What does the behavior of our boss mean (e.g., criticizing my performance)? Does it mean that I am a bad worker? Does it mean that I have to look for another job? Does it mean that I have to change my strategy for talking to clients? Convictions referring to meaning will affect our behavior in accordance with their content. The third type of convictions is connected with what we believe is important and what really matters, revealing our values and criteria for evaluation.

An individual's belief system and especially such elements as a pessimistic style of explaining failures, a generalized sense of a lack of influence on events (Miller and Norman 1979; Pines 1993), low self-efficacy (Cherniss 1993), negative views of other people (Van Dierendock et al. 1994), and irrational beliefs about one's professional role (Sęk 2000) are all important factors affecting the job burnout process.

Internal LOC occupies a prominent place among belief patterns as a factor contributing to job burnout. The results of our studies confirmed the conclusions drawn from earlier research. There have been many studies on the relation between LOC and components of burnout. In discussing hypothetical links of burnout with LOC, we must emphasize Rotter's (1975, p. 62) statement that "there is absolutely no justification for thinking in terms of a typology." A meta-analysis of a dozen or so studies indicates that external control accounts for about 10% of emotional exhaustion variance, 5% of depersonalization variance, and 5% of lack of professional satisfaction variance (Glass and McKnight 1996). A relation between the external LOC and exhaustion, depersonalization, and reduced professional satisfaction has also been observed by Buunk and Schaufeli (1993), Capel (1987), and Pierce and Molloy (1990). A similar result was obtained by Chrzanowska (2004) in her studies on the teachers of mentally handicapped pupils; a more external LOC resulted in greater emotional exhaustion and tendency to depersonalization. The more internal the LOC, the greater the commitment to work and the lesser the threat of professional burnout syndrome. A conviction of having control affects human functioning in difficult situations, danger extent estimation, and taking up preventive activities (Sęk 1996). People with external LOC are more often convinced that they have no influence on a stressful situation and that they cannot change it. This explains why they function worse in stressful situations: they are more focused on their own emotional experiences and on an avoidance strategy than on problem-solving strategies (Wong 1993).

In our research, we implemented a new tool designed to measure work LOC (Matczak et al. 2009). The results of the study concern seven specific aspects of LOC, including such expressions of external LOC as a feeling of inefficacy, a sense of fatalism, and a feeling of being dependent on others. It is clear that all aspects of external work LOC are closely related to the three dimensions of occupational burnout, which corroborates the results obtained in earlier studies, including those investigating the relationship between general LOC and job burnout.

REFERENCES

Bakker, A. B., Van Der Zee, K. I., Lewig, K. A. and Dollard, M. F. (2006). The relationship between the big five personality factors and burnout: Study among volunteer counselors. *Journal of Social Psychology, 146*, 31–50.

Bandura, A. (1997). *Self-Efficacy: The Exercise of Control*. New York: Frejman.

Buunk, B. P. and Schaufeli, W. B. (1993). Burnout: A perspective from social comparison theory. In: Schaufeli, W. B., Maslach, C. and Marek, T. (eds.), *Professional Burnout: Recent Developments in Theory and Research*. Washington, DC: Taylor & Francis, (pp. 53–74).

Capel, S. A. (1987). The incidence and influence on stress and burnout in secondary school teachers. *British Journal of Educational Psychology, 57*, 279–299.

Cherniss, C. (1993). Role of professional self-efficacy in the etiology and amelioration of burnout. In: Schaufeli, W. B., Maslach, C. and Marek, T. (eds.), *Professional Burnout: Recent Developments in Theory and Research*. Washington, DC: Taylor & Francis, (pp. 135–149).

Chrzanowska, I. (2004). *Wypalenie zawodowe u nauczycieli* [*Teacher Burnout*]. Łódź: Wydawnictwo Uniwersytetu Łódzkiego.

Glass, D. C. and McKnight, J. D. (1996). Perceived control, depressive symptomatology and professional burnout: A review of the evidence. *Psychology and Heath, 11*, 23–48.

Judge, T. A. and Bono, J. E. (2001). Relationship of core evaluations traits—self-esteem, generalized self-efficacy, locus of control, and emotional stability—with job satisfaction and job performance: A meta-analysis. *Journal of Applied Psychology, 86*, 80–92.

Judge, T. A., Erez, A., Bono, J. E. and Thoresen, C. J. (2002). Are measures of self-esteem, neuroticism, locus of control, and generalized self-efficacy indicators of a common construct? *Journal of Personality and Social Psychology, 83*, 693–710.

Kozielecki, J. (1981). *Psychologiczna teoria samowiedzy* [*A Psychological Theory of Self-Knowledge*]. Warszawa: PWN.

Langelaan, S., Bakker, A. B., Van Doornen, P. J. P. and Schaufeli, W. B. (2006). Burnout and work engagement: Do individual differences make a differences? *Personality and Individual Differences, 40*, 521–532.

Lazarus, R. S. and Folkman, S. (1984). *Stress, Appraisal, and Coping*. New York: Springer.

Lee, R. T. and Ashforth, B. E. (1993). A longitudinal study of burnout among supervisors and managers: Comparison between the Leiter and Maslach (1999) and Golembiewski et al. (1986) models. *Organizational Behavior and Human Decision Process, 54*, 369–398.

Maslach, C. (1993). Burnout: A multidimensional perspective. In: Schaufeli, W. B., Maslach, C. and Marek T. (eds.), *Professional Burnout: Recent Developments in Theory and Research*. Washington, DC: Taylor & Francis, (pp. 135–149).

Maslach, C. (1998). A multidimensional theory of burnout. In: Cooper, C. L. (ed.), *Theories of Organizational Stress*. New York: Oxford University Press (pp. 68–85).

Maslach, C. (2003). Job burnout: New directions in research and intervention. *Current Directions in Psychological Science, 12*, 189–192.

Maslach, C. and Jackson, S. (1986). *Maslach Burnout Inventory. Manual*. Palo Alto: Consulting Psychologists Press.

Maslach, C. and Leiter, M. P. (1997). *The Truth About Burnout*. San Francisco: Jossey-Baas.

Maslach, C., Schaufeli, W. B. and Leiter M. P. (2001). Job burnout. *Annual Review of Psychology, 52*, 397–422.

Matczak, A., Jaworowska, A., Fecenec, D., Stanczak, J. and Bitner, J. (2009). *Człowiek w pracy. Podręcznik (The Man in Work. Manual)*. Warszawa: PTP.

Miller, I. and Norman, W. E. (1979). Learned helplessness in humans: A review and attribution—theory model. *Psychological Bulletin, 86*, 93–118.

Pierce, C. M. and Molloy, G. N. (1990). Psychological and biographical differences between secondary school teachers experiencing high and low levels of burnout. *British Journal of Educational Psychology, 60*, 37–51.

Pines, A. M. (1993). Burnout: An existential perspective. In: Schaufeli, W. B., Maslach, C. and Marek, T. (eds.), *Professional Burnout: Recent Developments in Theory and Research*. Washington, DC: Taylor & Francis, (pp. 35–51).

Rotter, J. B. (1966). Generalized expectations for internal-external control. *Psychological Monographs: General and Applied, 80*, 1–28.

Rotter, J. B. (1975). Some problems and misconceptions related to the construct of internal versus external control of reinforcement. *Journal of Consulting and Clinical Psychology, 43*, 56–67.

Rotter, J. B. (1981). The psychological situation in social-learning theory. In: Magnuson, D. (ed.), *Toward a Psychology of Situations. An International Perspective*. New Jersey: Lawrence Erlbaum Associates.

Rykman, R. M. (1979). Perceived locus of control and task performance. In: Perlmuter, L. C. and Monty, R. A. (eds.), *Choice and Perceived Control*. New Jersey: Lawrence Erlbaum Associates.

Schaufeli, W. B. and Enzmann, D. (1998). *The Burnout Companion to Study and Practice: A Critical Analysis*. London: Taylor & Francis.

Sęk, H. (1996). *Wypalenie zawodowe, psychologiczne mechanizmy i uwarunkowania* [*Professional Burnout. Psychological Mechanisms and Conditioning*]. Poznań: Zakład Wydawniczy K. Domke.

Sęk, H. (ed.). (2000). *Wypalenie zawodowe. Przyczyny, mechanizmy, zapobieganie* [*Professional Burnout. Causes, Mechanisms, Prevention*]. Warszawa: Wydawnictwo Naukowe PWN.

Sokołowska, J. (1993). *Przewidywania i wybory a przekonanie o własnej kontroli* [*Impact of Perceived Control on Predictions and Choices*]. Warszawa: Wydawnictwo Instytutu Psychologii PAN.

Spector, P. E. (1988). Development of the work locus of control scale. *Journal of Occupational Psychology, 61*, 335–340.

Van Dierendock, D., Schaufeli, W. B. and Sixma, H. J. (1994). Burnout among general practitioners: A perspective from equity theory. *Journal of Social and Clinical Psychology, 13*, 86–100.

Weiner, B. and Potepan, P. A. (1970). Personality characteristics and affective reactions towards exam of superior and failing college students. *Journal of Educational Psychology, 61*, 144–151.

Wong, P. T. (1993). Effective management of live stress: The resource—congruence model. *Stress Medicine, 9*, 51–60.

Zimbardo, P. G. and Ruch, F. L. (1977). *Psychology and Life*. Illinois: Scot, Foresman and Company.

26 Cultural Factors of Economic Effectiveness and Innovation

Organizational Psychology of Innovation

Damian Grabowski

CONTENTS

CULTURE, ECONOMIC EFFECTIVENESS, INNOVATIVENESS, AND INNOVATION

The aim of the article is to show relations between culture, economic growth, and innovativeness, as well as present that values and norms play the key role for the economic growth of some countries and for its lack in case of others. Landes (2000) wrote, "Max Weber was right. Studying the history of economic development you cannot stop thinking that culture virtually decides about everything." Economic success of certain immigration ethnic minorities can be taken as an example here, such as the Chinese in East and Southeast Asia or the Lebanese in West Africa (Grabowski 2010).

Culture is a commonly applied term in social sciences, such as sociology, psychology, and anthropology, in particular. Bjerke (1999), dealing with the problems of culture in reference to organizations, reviewed its definition and concept. However, these theories are observed to share some elements. Namely, the first group is behaviors and values, and the second one is conscious behaviors and values and subconscious behaviors and values. These four elements comprise various cultural concepts with which norms, attitudes, and the meaning system are connected.

It has been assumed that economic effectiveness equals obtaining economic profit, which consequently leads to gaining yet another one. This term has been replaced here with other expressions, such as prosperity, economic development, success, growth, and finally wealth.

Porter (2000) has observed that prosperity, or standard of living of a country, results from productivity, with which it manages its human, capital, and natural resources. This productivity is responsible for the level of real earnings and for the profitability of the capital, thus influencing the main factors that determine the state revenue per capita. Income per capita is the basic rate of economic growth here (Grabowski 2010).

Porter (2000), when describing prosperity macroeconomics, mentions innovativeness as one of the contributing elements. The term has a similar meaning to the word innovation, which means "the implementation of a new or significantly improved product (good or service), or process, a new marketing method, or a new organisational method in business practices, workplace organisation or external relations" (Oslo Manual 2005).

Innovation takes place when production, process, marketing, or organizational method is new or improved, which may render it more effective. An example of an organizational innovation may be lean manufacturing, which derived mostly from the Toyota Production System (Zgorzelski 2002).

Therefore, it can be assumed that innovation is strongly connected with economic growth. It can even be hypothesized that the former conditions the latter and expects that cultural factors that enhance innovativeness mark economic growth, and contrariwise, culture that conditions growth enhances innovativeness (Grabowski 2010). However, in case of growth regarded as efficiency (productivity), the culture that enhances growth has reduced innovation. Stable, bureaucratized, and tight cultures that sustain growth hinder innovation processes (Triandis 1984). Nonetheless, Taylor has introduced a lot of innovations that have increased production.

In social sciences, apart from innovation, there appear such terms as creativity, ingenuity, or creativeness. Innovation should be assumed a domain of economic practice (production, commerce), whereas creativity means new ideas that precede a product prototype being made in the process of its making or a new method being worked upon. The prototype domain is already an innovation, which both in European and American cultures have positive connotations. (These connotations are used by marketing. For example, Toshiba Corporation's tagline is "Leading Innovation" or Hewlett-Packard's tagline is "Invent.") In economy, innovations are assumed to be such new products and methods that enhance efficiency and profits. In many circumstances, innovations are linked with lowering the profits and spending money, for example, on research. Higher profits come much later. Thus, innovations are primarily associated with investments (Grabowski 2010).

This article is an attempt to answer the question on cultural factors of economic growth and innovativeness. Therefore, what values, norms, and attitudes enhance economic growth and innovativeness? What kind of meaning system can be discussed in reference to countries that have succeeded economically provided that their success is associated with innovation?

CONCEPTIONS OF ECONOMIC GROWTH AND INNOVATIVENESS FACTORS

Culture is not obviously the sole factor that determines economic growth. To answer the question concerning cultural influence, the system of many variables and their connections needs to be analyzed (Grabowski 2010). Sachs (2000) points at three groups of such factors.

First is geographic location, which is associated with some areas being more privileged than others. Among the benefits that result from a convenient location, there is access to basic natural resources, favorable conditions for farming, and a good climate. The last factor is actually significant when it comes to European countries. Landes (1998) observed that European winter is severe enough to limit the growth of pathogenic factors and plagues. According to Sachs (2000), in the temperate climactic zone (European case), contagious diseases are transmitted via direct contact between people, whereas in the subtropics, such diseases are transmitted via insects and mollusks.

Sachs described three plausible causes of continual poverty in tropical zones: factors associated with agriculture, health, and scientific background mobilization. Problems that agriculture encounters in this area start with barren soil and strong erosion, as well as soil depletion that used to be covered with rainforest; irregular rainfall; and the risk of drought in the regions that are characterized by high humidity fluctuations, a substantial number of vermin, and substantial losses during food storage, finishing with a lower synthesis capacity at night hours in the areas characterized by high temperatures. As a result, considering lower agricultural capacity, a substantial percentage of the population is accumulated in this very sector.

Another set of factors is a social system. Within this group, cultural factors may actually be placed. Sachs (2000) presents here Weber's theses on capitalistic societies. Within such systems, the state is based on the rule of law. The society is characterized by high mobility, and economic exchange takes place by means of market institutions.

The third group of variables is the so-called positive feedback. This phenomenon, whose nature is the situation in which the rich becomes even richer and more powerful, may be defined as the escalation of inequality (Tomaszewski 1982).

Triandis (1984) points at three environments, that is, factors of economic development. These environments are physical, sociocultural, and economic. Figure 26.1 shows the relationship between these environments and "human predispositions, which lead to human behaviours which lead to industrialization, which leads to wealth." Physical environments include variables such as availability and utilization of resources. Sociocultural environments include the major attributes of cultural differences such as cultural complexity, tightness, norms, roles, and values. Economic environments include variables such as gross national product, interest rates, etc.

Triandis (1984) points out that economic development is associated with the degree of given external predictability. The above-mentioned tropics may be rendered areas of low predictability. The fact that events can be predicted strengthens the conviction as to the value of economic success, which equals rewards, and these are possible provided that events are predictable. Rewards are certain predictable states, too. Lack of predictability leads to helplessness, whereas predictability enables achieving constant profits.

This variable is also linked with a higher level of trust. In societies where the level of trust is low, organizations employ two to three workers to perform the same task, which in the society characterized by a high level of trust would be performed by a single employee. "Work is organized so that one person's work is checked by several others" (Triandis 1984). Such control limits freedom of action.

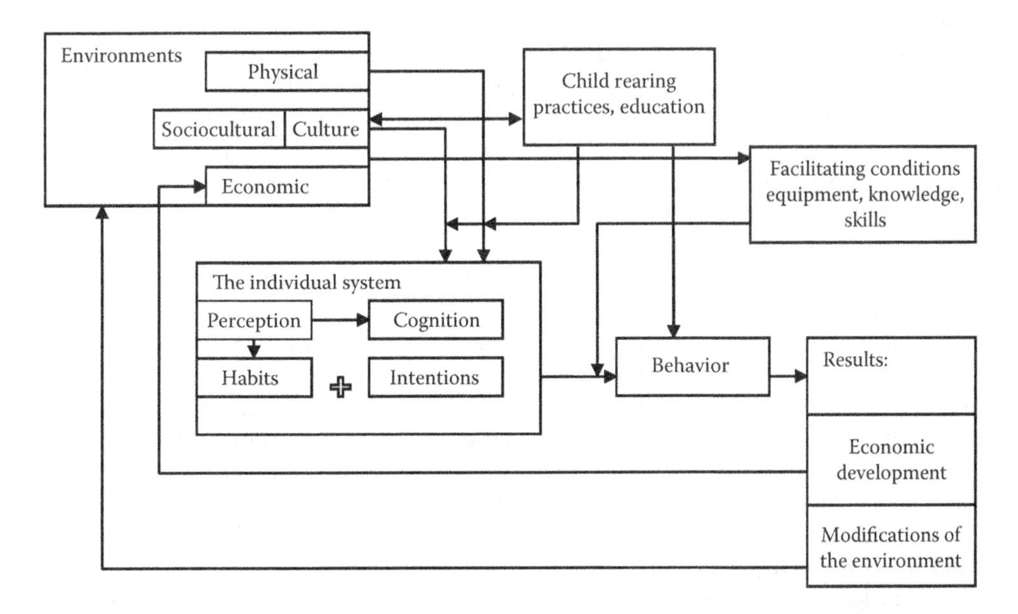

FIGURE 26.1 Relationships between environments, individual systems, behavior, and economic development (growth). (Adapted from Triandis, H. C., *International Journal of Psychology*, *19*, 79–95, 1984.)

Predictability is connected with efficient time management. This is particularly vital in the context of planning and attitude toward future. Only predictable reality enables effective time management and planning.

Stability of a given culture, tightness (little room for maneuver in terms of departure from the norm), and reliability of parental behavior are associated with predictability in the context of cultural variables, which favor industrialization.

CULTURAL FACTORS FOR ECONOMIC EFFICIENCY AND INNOVATIVENESS

Numerous variables that affect economic growth in reference to the social system can be described. Grondona (2000) described 20 cultural factors that favor economic growth and innovativeness. These factors can be placed on the continuum within two poles: (1) tightness and stability and (2) looseness and flexibility.

AROUND TIGHTNESS AND STABILITY

Pharisaic Religion

The first factor mentioned by Grondona is pharisaic religion, a system where "the rich celebrate their success as evidence of God's blessing and the poor see their condition as God's condemnation. Both the rich and the poor have a strong incentive to improve their condition through accumulation and investment." In publican religion, "the poor will feel justified in their poverty, and the rich will be uncomfortable because they see themselves as sinners" (Grondona 2000) tormented by unclear conscience.

Protestantism is an example of pharisaic religion. Weber (1958) described the so-called ascetic branches of Protestantism, that is, Calvinism, Pietism, Methodism, and sects stemming from Baptism movements. The way to achieve wealth (i.e., work) had a religious meaning. Work "was a vehicle toward economic success and economic success a sign of salvation" (Miller et al. 2002). This pattern was defined as work ethic, which is elaborated on below.

Importance of Work

One of 20 factors noted by Grondona is work ethic. Work ethic is understood as prioritizing work and putting it in the center of life. It is only reformation, especially in the Calvinist meaning, that led to a revaluation of the social scale of prestige and let put work ethic at a prominent place.

More precisely, a pattern of behavior that is referred to as "work ethic" comprises the following elements (ingredients) (Cherrington 1980; Furnham 1984, 1990; Furnham et al. 1993; Grabowski 2008, 2010; Mirels and Garret 1971; Sennett 1998; Tang 1993):

1. Perceiving work as moral obligation and virtue. Dedicated work is a desirable activity and considered as a mark of good character.
2. Spending as much time as possible on work (simply an order: devote most of your time to work). In other words: save your time, do not waste time.
3. Putting as much energy as possible into work (most of your desires should be work-oriented).
4. Delay of gratification. Analyzing this construct, a conclusion can be drawn that postponing gratification is a pattern of behavior, which is about the skill to await a reward for a long time. Commonly, this kind of behavior is interpreted as patience. Essentially, it is about saving and performing work without receiving big awards. The work in itself is a reward, and bigger ones are planned in distant future. Grondona notes that pro-development societies are less prone to consume but rather tend to save. People living in such systems concentrate on near future and so are able to plan efficiently. Consumption equals lack of investment, which is indispensable for generating innovation.

5. Reluctance toward free time, which results from a preferred order: devote most of your time to work. Anyhow, free time was treated as renewing energy, regenerating forces indispensable to perform work. Free time had a clearly assigned function. Rest after work serves a better performance of further work. Spending free time on other aims or prolonging it is reprehensible.
6. Independence—self-reliance and self-standing primarily in the economic sense, performing work on your own, reluctance to receiving help from others. Independence means acceptance of responsibility for the economic situation.
7. Conviction that such behavior leads to success and progress.
8. The cult of efficiency is not mentioned by Weber (1958) as a separate ingredient, but it results directly from the above-mentioned ingredients. This cult is connected individually with the willingness to achieve success and stems from striving for wealth, which obviously should not be consumed; efficiency is, for example, creating as many goods as possible in a short time span or more direct aiming for the realization of a given aim, that is, elimination of obsolete activities to avoid wasting energy.

Within the construct of work ethic, a notion of the so-called hard work is of great importance. This term means spending a lot of energy, as well as a long time on performing work, as well as paying a lot of attention to doing work and following orders such as work fast and earnestly. Thus, an ideal is efficient work that takes most of the 24-hour period. Practice is a kind of work that fulfills most of the day. In the modern era, such a situation was very often witnessed. Landes (1998) describes the situation of draft oxen in the industrial era, that is, Japanese female workers who were sleeping next to the machines in factories working for a pittance. On the other hand, it was in fact in this era, just later, that a 60-hour long working week was reduced to a 40-hour one. Hence, work ethic refers primarily to hard efficient work.

Tightness and Moral Imperative

Grondona (2000) mentions that the next factor for economic development is a moral imperative. In cultures that favor development, laws and norms are generally in harmony. These are not excessively demanding and so possible to execute. "Moral law and social reality virtually coincide. In development-resistance cultures, on the other hand there are two worlds that are out of touch with each other. One is the exalted world of the highest standards and the other is the real world of furtive immorality and generalized hypocrisy. The law is the utopian ideal that does little more than express what people might in theory prefer, whereas the real world, effectively out of touch with all law, operates under the law of jungle, the law of cleverest or the strongest, a world of foxes and lions disguised as lambs" (Grondona 2000).

The imperative is associated with the construct of tightness described by Triandis (1984; see also Chan et al. 1996). This word means behavior consistent with norms and rules. Tightness is a situation of high consistency of behaviors and norms, as well as a complex system of sanctions for the behavior incompatible with the pattern (Chan et al. 1996). Even in the case of the United States, which is considered to be a less tight culture, compliance with norms in the workplace is required (Bell 1976). Due to the patchwork nature of this culture (specific culture) (Trompenaars and Hampden-Turner 1998), everyone conforms to the norms and rules in the work situation, but outside it (culture and politics), they can afford freedom, which allows for creativity and ingenuity. Specific culture has a small area of privacy, which is clearly separated from public life. This culture has many personalities/sectors where they are acting and just there like socializing in clubs, associations, and companies. Moral imperative is a situation of cultures that are tight to an average degree.

Two Concepts of Wealth

An important factor in economic development and innovation, associated with delay of gratification, is also a social concept of wealth. In the societies that are in favor of wealth resulting from economic growth, it is considered to be mainly something that does not exist yet, whereas in the systems that are against it, wealth is something that exists. In other words, development is enhanced by the conviction that "wealth can be eternally multiplied because it depends on the creative potential and intuition, not on the amount of permanent resources. Wealth can be multiplied by improving productivity." This way of thinking generates behaviors aimed at raising efficiency, and consequently, the bulk to be shared is constantly increased. The conviction that wealth is a constant value, on the other hand, induces numerous groups to fight for the already existing pool, which consequently "contributes to lower productivity" (Porter 2000).

The former conviction is the situation of genuine modern capitalism described by Weber (1958). Rational methods of achieving profits may be identified with the constraint, or at least with a rational mitigation of an irrational aspiration for profit; accursed hunger for gold (auri sacra fames); or the conviction that wealth is a permanent value, which does not enhance innovation.

Around Looseness and Flexibility

Trust in the Individual, Individualism, and Collectivism

As Grondona states, the important factor of economic growth is faith in an individual. Freedom and having control over their destiny enhances risk taking and enterprise. This feeling is an essential element of individualism, that is, a variable, which has been subject of numerous researches (Hofstede 1997; Tiessen 1997; Triandis 2000, 2001; Triandis and Suh 2002).

Individualism is regarded as one of the two poles of individualism versus collectivism continuum or as a separate dimension (Robert and Wasti 2002). Individualism and collectivism are presented as cultural syndromes that consist of many norms, values, and generally speaking, behavioral patterns.

Triandis (1999) described four essential, original components of individualism and collectivism (Singelis et al. 1995).

1. Definition of ego—autonomic ego and independent on the group (individualism)—ego interdependent on others, defining oneself as part of a group or in categories concerning certain aspects of this group (collectivism).
2. The structure of the goals—individual aims do not correlate with those of a group—individual aims compatible with those of a group. Individuals put their own aims higher than those of a group, whereas collectivists put more emphasis on group aims.
3. Pressure on attitudes—norms. Individuals' behavior is determined by attitudes and individual needs, and also other internal processes, whereas collectivists' behavior is first and foremost determined by norms, duties, and commitments. Collectivists point at a weaker relationship between attitudes and behavior than individualists.
4. Emphasis on rationality—bonds. Individualism stresses appraisal of own profits and costs in the process of building relationships with other people. This is why individuals keep such contacts that are beneficial for themselves. Collectivists put higher relations with others and other people's needs, even if such relations are harmful for them. The basic form of an individualistic formation is an association, whereas a collectivistic one's a community (Triandis 1993).

In cross-cultural psychology, two types of collectivism and individualism are described. Vertical collectivism (VC) and horizontal one (HC), as well as vertical (VI) and horizontal individualism (HI). VC is a concentration on internal group cohesion, respect for norms and those in power, submissiveness to the rule, and sacrifice for the group. The horizontal version of collectivism is, on the other hand, empathy, sociability, and cooperation.

VI is the US perspective. It is the culture where competition and rivalry are strongly accentuated. You need "to be the best" to obtain the highest position in the organizational and social structure. In horizontal, individualistic cultures (Australia, Scandinavian countries), such values as self-reliance, independence on others, as well as exceptionality and uniqueness are stressed.

Based on the above characteristics, economic development can be assumed to be the perspective of vertical and HI. Japan is an interesting example here. This country is rather characterized by the culture of VC. Japan's culture pattern is in turn VC: 40%, HC: 20%, VI: 25%, and HI: 15% (Singelis et al. 1995). The indicator of VI, which is 25%, is remarkable. Hence, Japan first of all represents an individualized approach. Secondly, a vertical culture is of perpendicular nature to Western observers. Japanese hierarchy is not so much a chain of orders, but rather a series of coordinations or synchronizations reaching higher and higher. A boss is a coordinator of ideas, which are conceived at lower levels of hierarchy. He or she acts on behalf of and for the benefit of the community, making syntheses and configurations of ideas and remarks of all their subordinates (Hampden-Turner and Trompenaars 1993). A lot of Japanese organizations are units in which the employees rationalize particular workplaces. In the development policy described by Bjerke (1999), comprising imitation, intervention, improvisation, innovation, and invention, Japan epitomizes both innovation, which means Japanese products are superior to many others manufactured in the West, and invention, that is, inventing original goods such as artificial intelligence.

Therefore, the following thesis by Grondona (2000) could be referred to Japan: "to trust the individual, to have faith in the individual, is one of the elements of value system that favors development."

Summing up the reflections on individualism and collectivism and referring them to the context of innovation, a possible conclusion is that excessive individualism may mean egoism and ruthless fighting, where precious ideas are lost and excessive collectivism may mean lack of freedom for an individual. Such patterns obviously mean lower innovativeness. On the other hand, individualism is a perspective of independence and freedom, whereas collectivism is epitomized by the forces rooted in cooperation and human group. Both of these patterns are essential for innovativeness (Grabowski 2010).

The factor associated with VI described by Grondona is social acceptance for competition. To a large degree, the collective culture of Japan could seem to be devoid of this phenomenon. There is nothing more erroneous. Admittedly, coworkers cooperate with one another, but their companies compete (Hampden-Turner and Trompenaars 1993). Besides, the Japanese naturally go from one value to opposite one another: from cooperation (harmony) to rivalry (competition). The Japanese rivalry does not aim at destroying competitors. The loser can save face, and they learn from the winner a better course of action. A strong individualism leading to ruthless fighting does not lead to enhanced innovativeness.

EFFICIENCY, PRODUCTIVITY, AND FLEXIBILITY: MODERN AND POSTMODERN CAPITALISM

Efficiency or productivity is a kind of activity characterized by speed and tempo (Grabowski 2010). An efficient person is the one who performs more actions in a given unit of time. Henry Ford's practices were actually aimed at containing more in a given time and he succeeded in that. The production of the Ford model T was at some point reduced from 728 hours to 93 minutes (Stoner et al. 1997). Taylor's methods, as well as his students and successors, besides the division of work and assembly line, turned out to be placed among the factors that enhanced production.

Accelerating the pace of work or in fact production was made possible owing to a division of work and analytical way of thinking. Another factor that enhanced the pace was a sequential time orientation (Trompenaars and Hampden-Turner 1998) associated with analysis. In the cultures characterized by this orientation, only one action is performed at a time; time is graspable and measurable. Such patterns of thinking actually contributed to an increased production in modern capitalism, but at present, in the period of client- and consumer-oriented postmodern capitalism (Bauman 2004), an analysis should be combined with a synthesis (integration), and sequentionalism with synchronic orientation. A symbol of such a combination is Toyota (Hampden-Turner and Trompenaars 1993; Zgorzelski 2002) in which a flexible, "lean" production was established. Such conduct of action may be named as the one producing only a necessary number of elements. "As many as we need" is an ideal of contemporary production, which first and foremost means manufacturing as many goods as can be sold (Grabowski 2008). Overproduction is at present, that is, in the period when consumers hold power, one of the biggest problems of the managing staff. Henry Ford wanted to produce more, whereas today, as much should be produced as can be sold. Ford did not have to worry excessively about the market. On the other hand, in the present day, the sales are analyzed even before the production starts (Grabowski 2008). At present, economic behavior is adequate reacting to what is going around, that is, to the needs and the whims of customers.

METAFLEXIBLE ORGANIZATIONAL CULTURE AND INNOVATIVENESS: CULTURE OF POSTMODERN CAPITALISM

The factors of economic development described above refer both to the modern and the postmodern capitalism. Grondona has specified cultural factors for economic growth that suit modern capitalism (work ethic, moral imperative) as well as postmodern capitalism (individualism, independence, openness of the mind marked in the description on the role of heresy) and those that refer to both periods (concepts of wealth and acceptance of competition).

If economic activities are divided into such stages as planning a product, manufacturing it (production), and sale, for each of these phases, cultural factors that support them in certain activities may be pointed out. The perspective of postmodern capitalism is existence of a number of factors that ought to be combined. "Today economic activity needs creativeness, flexibility or, in short, aiming for divergence, which equals many ways of achieving an aim and possibilities, as well as discipline, i.e., striving for convergence" (Grabowski 2010). Striving for divergence is determined by independence, individualism, or generally speaking, through the freedom of action, which allows organizational flexibility. Striving for convergence, on the other hand, means the perspective of reducing the possible ways associated with control, discipline, and stability. Discipline imposes tightness, that is, action in accordance with the norms and behavioral pattern.

In creative thinking, a stage of generating ideas (increasing divergence) and the stage of its appraisal (increasing convergence), and a stage of invention and decision making can be differentiated. In reference to the economic activities described above, it may be concluded that in contemporary economic activities, the factors that condition creativeness in the planning phase (invention) play a role, as well as tightness, which is required in the process of implementing new products (innovation). Nonetheless, the latter one must also be more flexible now, which has been proven by the case of Japan and the model of production that has come into being in Toyota.

Organizational culture is a term that comprises the configuration of the cultural factors that are associated with one another and that have been described in the third part of this article. According to the conception of Cameron and Quinn (2006), two dimensions of effectiveness are vital. The first one concerns the poles, flexibility, discretion, dynamics, and freedom of action on the one hand, and stability, order, control, and durability on the other. Some organizations are efficient when they change, adapt, and are characterized by a low level of formality, whereas some others operate well when they are stable, predictable, and formalized. Another dimension is attitude to internal affairs, integration, and unity, whereas its opposite is the external orientation, differentiation, and rivalry. Some organizations are effective when they form a harmonious unity, but some others need to concentrate on cooperation or competition.

Two dimensions have allowed one to describe four types of organizational cultures: clan (flexibility and integration), adhocracy (flexibility and external focus), hierarchy (stability and integration), and market (stability and external focus). Thus, the division comprises both modern, tight cultures and postmodern, flexible ones. A symbol of modern culture is hierarchy or bureaucracy described by Weber. Long (2003) describes these cultures as role-oriented organizational cultures. "The role-oriented organizational culture emphasizes and values specialized work roles and employees who best fill those roles. This work role emphasis also reflects affinity for hierarchical and functional role differentiation" (Long 2003). Examples of this culture are bureaucracy described by Weber, scientific management formalized by Frederick Taylor, and principles of management formalized by Henri Fayol. An essential feature of these cultures is certain tightness; real behavior should be consistent with the norm. Hence, an important point is efficiency or increased productivity. Such an organization is a situation of lower equifinality (Katz and Kahn 1966), which means there are a limited number of ways to achieve the final aims. It has been assumed here that bureaucratic organizations are open systems. It even concerns the so-called closed social systems (e.g., totalitarian ones).

In the postmodern era, there have appeared organizations with flexible culture. The essential feature of such organizations is adapting to external conditions, reacting to changing circumstances. Such organizational forms are short lived; they are "'tents rather than palaces' in that they can reconfigure themselves rapidly when new circumstances arise" (Cameron and Quinn 2006). The organizational structure is thus changeable and lacks an organizational scheme. Such a situation is characterized by a higher equifinality, which means there are many ways to achieve a given aim.

The flexible culture that favors innovation is adhocracy. In this culture, an essential element that above all conditions innovative activities is freedom of action. Adhocracy is certainly a type of culture that favors innovative behavior, whereas as far as the innovations themselves are concerned, that is, implementing new methods and products, hierarchy or another bureaucratized organizational structures are indispensable too. Both flexibility and freedom of behavior may make an organization close and also bring about chaos. However, the organizational tightness may also lead to defense against novelty. Bureaucratized, hierarchical organization is a perspective of a high predictability and uncertainty avoidance. This predictability, however, concerns organizations, and emphasizing it may lead to the situation in which the organization will stop reacting to changes in the environment (conforming to organizational norms may start being more important than customers' needs, which is defined as bureaucracy in a negative sense) (Katz and Kahn 1966). Hence, the procedures and instructions need to be adjusted in critical moments.

Generally speaking, it is assumed that innovations are favored by flexible organizational structures. This thesis, nonetheless, calls for more precision. Innovative and creative (invention) activities may be said to be favored by loose, flexible cultures such as adhocracy since these are structures open to information. In case of the innovations themselves, understood as implementation of a new product, tighter and bureaucratized organizational structures are required. Innovations may be said to require a synthesis of tight solutions, stable ones done in combination with flexible ones, that would imply freedom of action. An example of such a synthesis is Japanese companies with Toyota at the front. Such a situation may be defined as metaflexibility. Sometimes an organization acts in a way typical of adhocracy and in another applies solutions typical of bureaucratic cultures (Grabowski 2010; Trompenaars and Hampden-Turner 2004), which is presented in Figure 26.2. Metaflexibility is the trait of culture of postmodern capitalism.

Therefore, it may be stated that in modern economy, innovation was associated with striving for stable tight organizations, which made efficiency possible. In the 19th and at the beginning of the 20th century, the problem was to control chaotic production activities. Striving for convergence and decreasing equifinality were typical of the work ethic and Ford's way of thinking. Hierarchical cultures introduced order and decreased uncertainty. In the postmodern, post-industrial era, first flexible cultures were emphasized to be now combined with hierarchical solutions, which may be defined as metaflexibility. The metaflexible organizational culture consists of subcultures: bureaucratized and flexible ones. Bureaucratized departments care about production, whereas flexible structures are environment-oriented and react at the right moment by sending information to the former culture. Thus, the whole organization oscillates, once getting close to the pole of tightness, which means stability, control, and avoiding uncertainty, and in another situation by acting flexibly, emphasizing freedom of action (Grabowski 2010).

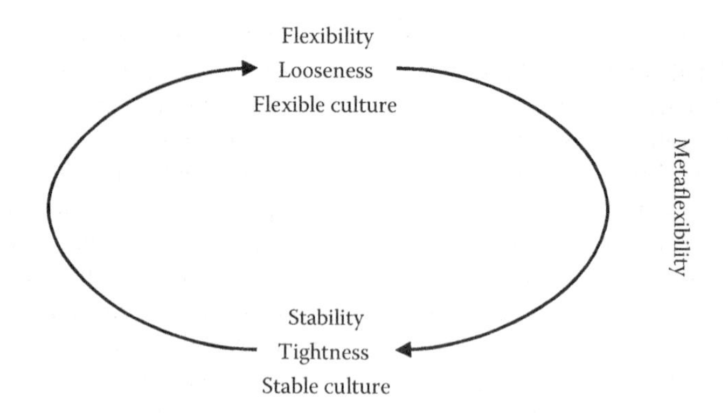

FIGURE 26.2 Model of metaflexible organizational culture.

It is worth mentioning that factors perceived in sociology as the ones associated with tightness may change the contents. For example, discipline understood as behavior compliant with a particular pattern, that is, with behavior required by a given organization, may be considered as an employee's attachment to the aims of the organizations, but such an employee has freedom of action in its realization. In the first case, discipline means emphasizing the way of action, which leads to achieving a particular aim. This discipline can be described as modern. In the second case, it is accentuating the aim. Here an employee can act differently, but it is important for them to achieve a particular final result (e.g., recruit new customers). This version of discipline can be named "post-modern discipline."

CONCLUSION

It is hard to answer the question which cultural factors or organizational culture favors efficiency and innovativeness. This article was supposed to show that both the so-called tight cultures and the flexible ones are associated with innovativeness. Nonetheless, it should be pointed out that excessive tightness, stability of organizational culture, and organization will equal hindering innovativeness, which may be observed in highly bureaucratized organizations. On the other hand, a high flexibility will equal chaos and make an organization disappear (after all, an organization is a structure). The organizations that strengthen innovativeness are thus those open to structural changes, which react to what is going on outside. Such organizations combine tightness with flexibility, bureaucratic solutions with such that emphasize freedom of action. Flexible subcultures, by reacting to the outside (world), enable changes in the production and subcultures that are stable, make this production possible, and maintain it. Organizational cultures, which combine such subcultures, may be defined as metaflexible. In such organizations, a given tight culture (hierarchy) or flexible one (adhocracy) may dominate, but if it is only one culture, the organization may disappear. In metaflexible cultures, one of them may be dominant, but solutions that are reserved for others also have their say.

REFERENCES

Bauman, Z. (2004). *Work, Consumerism and the New Poor.* Buckingham, Philadelphia: Open University Press.

Bell, D. (1976). *The Cultural Contradictions of Capitalism.* New York: Basic Books.

Bjerke, B. (1999). *Business Leadership and Culture. National Management Styles in the Global Economy.* Northampton, Massachusetts: Edward Elgar Publishing.

Cameron, K. S., Quinn, R. E. (2006). *Diagnosing and Changing Organizational Culture. Based on the Competing Values Framework.* San Francisco: Jossey-Bass, A Wiley Imprint.

Chan, D. K.-S., Gelfand, M. J., Triandis, H. C., Tzeng, O. (1996). Tightness—Looseness revisited: Some preliminary analyses in Japan and United States. *International Journal of Psychology, 31*, 1–12.

Cherrington, D. (1980). *The Work Ethic. Working Values and Values that Work*. New York: Amacom, A division of American Management Associations.
Furnham, A. (1984). The Protestant work ethic: A review of the psychological literature. *European Journal of Social Psychology, 14*, 87–104.
Furnham, A. (1990). *The Protestant Work Ethic. The Psychology of Work-Related Beliefs and Behaviours*. London and New York: Routledge.
Furnham, A., Bond, M., Heaven, P., Hilton, D., Lobel, T., Masters, J., Payne, M., Rajamanikam, R, Stacey, B., Van Daalen, H. (1993). A comparison of protestant ethic beliefs in thirteen nations. *Journal of Social Psychology, 133*, 185–197.
Grabowski, D. (2008). Etyki ponowoczesności a przyszłość psychologii pracy [Ethics of postmodernity and future of work psychology]. *Czasopismo Psychologiczne, 14*, 49–60.
Grabowski, D. (2010). Kulturowe czynniki efektywności gospodarczej i innowacyjności. Kultura, efektywność a innowacyjność [Cultural factors of economic effectiveness and innovativeness. Culture, effectiveness and innovativeness]. *Chowanna, 35*, 77–97.
Grondona, M. (2000). Cultural typology of economic development. In L. E. Harrison, S. P. Huntington (Eds), *Culture Matters: How Values Shape Human Progress* (pp. 44–55). New York: Basic Books.
Hampden-Turner, Ch., Trompenaars, A. (1993). *The Seven Cultures of Capitalism: Value Systems for Creating Wealth in Britain, the United States, Germany, France, Japan, Sweden and the Netherlands*. New York: Doubleday.
Hofstede, G. (1997). *Cultures and Organizations. Software of the Mind*. New York: McGraw-Hill.
Katz, D., Kahn, R. L. (1966). *The Social Psychology of Organizations*. New York: John Wiley and Sons.
Landes, D. (1998). *The Wealth and Poverty of Nations. Why Are Some So Rich and Others So Poor?* New York: W.W. Norton and Company.
Landes, D. (2000). Culture makes almost all the difference. In L. E. Harrison, S. P. Huntington (Eds), *Culture Matters: How Values Shape Human Progress* (pp. 2–13). New York: Basic Books.
Long, L. W. (2003). Mapping organizational culture: An integration of communication and organizational design. *Intercultural Communication Studies, XII-2*, 127–142.
Miller, J., Woehr, D., Hudspeth, N. (2002). The meaning and measurement of work ethic: Construction and initial validation of a multidimensional inventory. *Journal of Vocational Behavior, 60*, 451–489.
Mirels, H. L., Garrett, J. B. (1971). The Protestant ethic as a personality variable. *Journal of Consulting and Clinical Psychology, 36*, 40–44.
Oslo Manual (2005). *Guidelines for Collecting and Interpreting Innovation Data. Organisation for Economic Co-operation and Development*. Paris: Statistical Office of The European Communities.
Porter, M. (2000). Attitudes, values, beliefs and microeconomics of prosperity. In L. E. Harrison, S. P. Huntington (Eds), *Culture Matters: How Values Shape Human Progress* (pp. 14–28). New York: Basic Books.
Robert, Ch., Wasti, A. (2002). Organizational individualism and collectivism: Theoretical development and an empirical test of a measure. *Journal of Management, 28*, 544–566.
Sachs, J. (2000). Notes on a new sociology of economic development. In L. E. Harrison, S. P. Huntington (Eds), *Culture Matters: How Values Shape Human Progress* (pp. 29–43). New York: Basic Books.
Sennett, R. (1998). *The Corrosion of Character. The Personal Consequences of Work in the New Capitalism*. New York, London: W. W. Norton and Company.
Singelis, T. M., Triandis, H. C., Bhawuk, D. P. S., Gelfand, M. J. (1995). Horizontal and vertical dimensions of individualism and collectivism: A theoretical and measurement refinement. *Cross-Cultural Research, 29*, 240–275.
Stoner, J., Freeman, R. E., Glibert, D. R. Jr. (1997). *Kierowanie. [Management]*. Warszawa: Polskie Wydawnictwo Ekonomiczne.
Tang, T. L.-P. (1993). A factor analytic study of the Protestant work ethic. *Journal of Social Psychology, 133*, 109–111.
Tiessen, J. H. (1997). Individualism, collectivism and entrepreneurship: A framework for international comparative research. *Journal of Business Venturing, 12*, 367–384.
Tomaszewski, T. (1982). The dynamics of inequality. *Dialectics and Humanism, 2*, 103–109.
Triandis, H. C. (1984). Toward to psychological theory of economic growth. *International Journal of Psychology, 19*, 79–95.
Triandis, H. C. (1993). Collectivism and individualism as cultural syndromes. *Cross-Cultural Research, 27*, 155–180.
Triandis, H. C. (1999). Cross—Cultural psychology. *Asian Journal of Social Psychology, 2*, 127–143.
Triandis, H. C. (2000). Culture and conflict. *International Journal of Psychology, 35*, 145–152.

Triandis, H. C. (2001). Individualism—Collectivism and personality. *Journal of Personality, 69*, 907–924.
Triandis, H. C., Suh, E. M. (2002). Cultural influences on personality. *Annual Review of Psychology, 53*, 133–160.
Trompenaars, A., Hampden-Turner, Ch. (1998). *Riding the Waves of Culture: Understanding Diversity in Global Business*. New York: McGraw Hill.
Trompenaars, A., Hampden-Turner, Ch. (2004). *Managing People across Cultures*. Chichester: Capstone Publishing Ltd.
Weber, M. (1958). *The Protestant Ethic and the Spirit of Capitalism*. New York: Charles Scribner's Sons.
Zgorzelski, M. (2002). *Hamburgery ze Świętej Krowy. Szkice o amerykańskiej teorii i praktyce zarządzania. [Hamburgers from the Holy Cow. The Drafts on American Theory and Practice of Management]*. Kraków: Oficyna ekonomiczna.

27 A Short Version of the Need for Closure Scale

*Item Selection and Scale Validation**

Małgorzata Kossowska, Krzysztof Hanusz, and Mariusz Trejtowicz

CONTENTS

INTRODUCTION

Webster and Kruglanski (1994) created the Need for Cognitive Closure (NFC) construct to describe and explain individual differences in the extent to which people aim for and achieve firm belief in order to reduce uncertainty and ambiguity. Research shows that the NFC influences the way in which people create and use mental representations such as schemas, prototypes, and stereotypes, and determines the way that they think and feel about the social world and how they act within it (review in Kruglanski et al. 2009). Individuals with a high NFC prefer order, predictability, and decisions that are easy to make, and are usually close-minded, tending to feel discomfort when they encounter ambiguity. A high NFC promotes superficial analysis and leads people to search only for information that fits into their already existing belief structures. This results in a simplified picture of social situations, which is resistant to change and promotes the feeling that outcomes can be predicted and that all required information is available explicitly in their belief system. In contrast, a low NFC relates to a greater tolerance for ambiguity and creates the conditions for insightful and detailed analysis. People with a low NFC are open to new information and less likely to make rigid, unchanging, and final judgments. They perceive any given situation as complex and move beyond stereotypes and already existing understandings in order to consider alternative interpretations that allow them to assimilate new information and adapt to changes.

---

* A description of the scale construction and validation in Polish was published in *Psychologia Społeczna* (2012), *5*, 89–99.

Kruglanski and Webster (1996) posit two processes underlying the NFC: (1) *seizing* information and (2) *freezing* the seized information into already existing belief structures. People with high NFC exhibit a tendency to seize information selectively. They favor information that is coherent with their belief systems and quickly incorporate that information into existing belief structures, while ignoring information that might undermine the stability of their belief systems. According to researchers, these two sequential processes influence the formation of opinions, judgments, and beliefs, including political attitudes and orientations, such as authoritarianism, conservatism, and liberalism (Kossowska and Van Hiel 2003). In addition, it has been found that the need for closure affects the number of stereotypical judgments that people make, on their resistance to persuasion, and on their ability to deal with new situations (review in Kruglanski et al. 2009).

THE NFC SCALE

Webster and Kruglanski (1994) have suggested that the NFC manifests itself in multiple dimensions, yet the scale that has traditionally been used to measure NFC is one-dimensional. However, Neuberg et al. (1997) have shown that the original scale has a low internal coherence and a two-dimensional structure, in which one dimension focuses on decisiveness whilst the second contains the remaining items. This latter view has been confirmed by analyses on several American samples (Kruglanski et al. 1997), three European samples (Mannetti et al. 2002) and a European, Asian, and American sample (Kossowska et al. 2002). In addition, the scale structures appear to be invariable across Dutch, Italian, Croatian, and American groups (Mannetti et al. 2002), as well as Polish, Belgian, Korean, and American groups (Kossowska et al. 2002).

A two-dimensional structure of the NFC scale is widely accepted, though there is no agreement on the interpretation of the processes that underlie the NFC. Neuberg et al. (1997) suggest that the two dimensions represent the tendency to seize information (items focused on decisiveness) and the tendency to freeze information into belief structures (other items). This suggestion is rejected by other researchers (Kruglanski et al. 1997; Roets and Van Hiel 2007), who claim that both processes constitute decisiveness and that it is impossible to separate the two dimensions. Roets et al. (2006) also suggest an alternative interpretation of the two-dimensional scale structure, stating that the separation of decisiveness from the rest of the scale is an effect of operationalizing this dimension incorrectly. They emphasize that decisiveness pertains to both motivational and ability aspects of closure and propose changes to the subscale to ensure that it pertains only to the motivational aspect. The result of these changes would be a 41-item scale for measuring the motivational aspects of cognitive closure (Roets and Van Hiel 2007). Bar-Tal and Kossowska (2010) showed that original decisiveness correlates with dimensions that are usually treated as pertaining to ability and not to motivation. They proposed a new version of the subscale that pertains only to the ability aspect of NFC (Bar-Tal and Kossowska 2010).

PROPOSAL OF A NEW SHORTENED VERSION OF THE NFC SCALE

Over the last 20 years or so, the need for closure construct has been a subject of interest for researchers, and the scale has been used in hundreds of studies conducted both in Poland and across the world in the fields of psychology and business (marketing and management) (see Kossowska 2005). However, the scale is longer than most; as a result, it is common practice to use only selected items rather than administering the whole set. For example, Keller (2005) used 14 items in his research, Kemmelmeier (2010) used 26 items, and Lynch et al. (2010) used 8 items. Polish researchers are also using shortened versions of the NFC scale (e.g., Jaśko 2011; Bukowski et al. 2011). However, in none of these studies did the researchers report that they had selected items or the criteria they used to select them. In each case, the shortened version of the scale yielded the expected results. However, the use of items that are chosen on an idiosyncratic basis is problematic. Firstly, the

TABLE 27.1
Sample Sizes and Demographic Summaries for Studies Used

Study	*N*	Gender: % Women	Age (Average)	Age (SD)	Age (Min)	Age (Max)	Average Age: Men	Average Age: Women
1	52	94.2%	66.6	6.7	53	81	73.0	66.2
2	223	34.1%	41.7	20.6	21	85	33.2	58.3
3	229	52.8%	43.9	12.1	20	84	43.6	44.1
4	340	58.8%	40.9	12.5	18	80	41.5	40.5
5	76	61.8%	46.9	2.9	42	50	47.0	46.8
6	98	50.0%	67.8	2.4	65	72	68.1	67.5
7	130	42.3%	21.7	1.9	19	26	22.2	21.0
8	120	65.8%	46.6	209	21	84	49.2	45.2
9	143	54.5%	46.5	22.5	21	77	44.8	48.0
10	118	46.6%	23.6	6.2	18	59	23.3	23.9
11	56	71.4%	20.9	2.2	19	29	21.5	20.6
12	83	72.3%	21.0	1.6	19	25	21.1	21.0
13	86	77.9%	28.5	7.9	16	61	27.8	31.0
Total	*1754*	*55.6%*	*40.0*	*18.3*	*16*	*85*	*37.6*	*42.1*

arbitrary selection of items makes it difficult to compare results between samples and to replicate studies. Secondly, it is not clear whether such shortened versions represent the key aspects of the NFC and to what extent the items that are selected represent specific dimensions of the scale. Therefore, the aim of the analysis presented in this paper is the creation of a shortened yet valid version of the NFC scale.

To achieve this goal, we gathered data using the Polish version of the full 32-item scale prepared by Kossowska (2003). Data were gathered using a large, heterogenic sample of youth and adults over the course of several years by researchers in various laboratories.* We analyzed the relations between different items in order to select those items that best represent the four subscales (without decisiveness) and the whole construct. Next, we compared this shortened scale to the original scale and also checked the correlation between the new shorter scale and other scales with which the NFC scale should, in theory, be correlated.

Participants

Data from 13 studies ($N = 1680$) were used in the analysis. The final sample comprised 915 women and 765 men with an average age of 40.7 (SD = 18.3). Table 27.1 contains sample sizes and essential descriptive statistics on participant demography. Before further analysis, the *EM* method was used to impute missing data (the average number of missing data cells ranged from 0% to 4%); this method does not change variable distributions or covariances between them.

Research Materials

One of the goals of the analysis was to check the correlation of the new shortened scale with other scales that have been used to measure similar constructs. So, in addition to the original 32-item NFC

* We used data published by Kossowska (2007a, 2010) and Kossowska et al. (Cornelis et al. 2009; Bar-Tal and Kossowska 2010; Piotrowska et al. 2011; Kossowska et al. 2012) and unpublished works (Kossowska and Bar-Tal, in press; Bukowski et al., under review; Sędek and Safrończyk, unpublished report; Sędek and Gradowski, unpublished data set; Jaśko, unpublished PhD thesis; Piber-Dąbrowska, unpublished data sets; Trejtowicz, unpublished data sets).

scale adapted by Kossowska (2003), we also used scales for measuring the following: right-wing authoritarianism (Koralewicz 1987), social dominance orientation (in the experimental version prepared by Kossowska 2002), five personality factors (NEO-FFI; Zawadzki et al. 1998), and mental speed from the Wechsler Intelligence Scale (WAIS-R) (Wechsler 1997). The relation between the new shortened version of the NFC scale and other scales that measure similar constructs would indicate the high external validity of the new version.

RESULTS

Structure of the New Scale

Measurement models (confirmatory factor analysis) were built separately for all five subscales of the new shortened NFC scale. Two items in the *Closedmindedness* subscale did not meet minimal criteria for measurement quality (average variance that can be attributed to the latent factor below 5%), so they were excluded from further analysis. Next, a principal components analysis (on the rest of the 30-item version of the scale) with an imposed 2-factorial solution (with Varimax rotation) was performed. We expected that the items in the *Decisiveness* subscale would form a separate orthogonal factor in an exploratory analysis that would be free from theoretical assumptions. Four intercorrelated NFC subscales (with an exclusion of *Decisiveness*) were entered into confirmatory factor analysis, assuming one underlying factor.

Table 27.2 contains standardized regression weights from measurement models for each subscale as well as from one factor solution (where *Decisiveness* is excluded). Factor loadings from 2-factorial principal components analysis are also reported.

The items in bold are those that were selected for the shortened 15-item version of the NFC scale. In addition to statistical criteria, two additional criteria were used; items corresponding to Roets and Van Hiel (2010) shortened NFC scale were preferred and items that were semantically similar were avoided.

The Shortened NFC Scale Intercorrelations and Correlations with Other Constructs: Convergent Validity and Internal Consistency Testing

Table 27.3 presents correlations for (a) the original subscales of the NFC scale and (b) their equivalents from our shortened version, with scale sum and with other variables: age, mental speed from Wechsler's Adults Intelligence Scale (WAIS-R), Social Dominant Orientation scale, and Right-Wing Authoritarianism scale (subsample sizes for these measurements are presented in parenthesis for each of these constructs).

High correlations between different NFC subscales confirm that the shortened and original versions of the scale are empirically equivalent. The internal consistency and convergent validity both speak in favor of using the shortened scale, from which items interfering with the measurement have been excluded. Similar values for the correlations of the results for both versions of the subscales with results for other constructs provide evidence that the original and shortened versions of the scale possess similar criterion validity.

TABLE 27.2
Items Analysis of NFC Scale

Item	Stand. Weight (CFA: Measurement Models for Each Subscale)	Stand. Weight (CFA): NFC Model (4 Subscales)	Factor Loadings (PCA) - 1. Component	Factor Loadings (PCA) - 2. Component	Position Present in NFC15 (Roets and Van Hiel 2010)
		Intolerance of Ambiguity Subscale			
I don't like situations that are uncertain.	**0.448**	**0.428**	**0.469**	**0.122**	Yes
I feel uncomfortable when I don't understand why an event occurred in my life.	**0.507**	**0.405**	**0.488**	**−0.161**	Yes
I feel uncomfortable when I am not certain what to think on an important issue.	0.464	0.383	0.462	−0.209	–
I think that I would learn best in a class that lacks clearly stated objectives and requirements. R	0.198	0.115	0.153	0.294	–
I dislike it when a person's statement could mean many different things.	0.476	0.367	0.508	−0.148	Yes
I feel uncomfortable when someone's meaning or intention is unclear to me.	**0.609**	**0.405**	**0.466**	**−0.163**	–
Measurement model summary: Ambiguity Avoidance					
Chi-square = 5.40 (df = 7); p = 0.612					
AGFI = 0.997					
NNFI = 1					
RMSEA[a] = 0.025					
AVE = 22%					

(*continued*)

TABLE 27.2 (Continued)
Items Analysis of NFC Scale

Item	Stand. Weight (CFA: Measurement Models for Each Subscale)	Stand. Weight (CFA): NFC Model (4 Subscales)	Factor Loadings (PCA) - 1. Component	Factor Loadings (PCA) - 2. Component	Position Present in NFC15 (Roets and Van Hiel 2010)
		Preference for Order Subscale			
I think that having clear rules and order at work is essential for success.	0.463	0.476	0.508	0.050	–
I find that a well-ordered life with regular hours suits my temperament.	**0.688**	**0.701**	**0.672**	**0.222**	**Yes**
My personal space is usually messy and disorganized. R	0.308	0.228	0.264	0.512	–
I believe that orderliness and organization are among the most important characteristics of a good student.	0.604	0.627	0.641	0.053	–
I find that establishing a consistent routine enables me to enjoy life more.	**0.703**	**0.735**	**0.729**	**0.040**	**Yes**
I enjoy having a clear and structured mode of life.	**0.917**	**0.875**	**0.804**	**0.113**	**Yes**
I like to have a place for everything and everything in its place.	0.792	0.642	0.687	0.111	–
Measurement model summary: Order preference					
Chi-square = 11.96 (df = 6); p = 0.063					
AGFI = 0.991					
NNFI = 0.995					
RMSEA[a] = 0.044					
AVE = 44%					

Preference for Predictability Subscale					
I like to have friends who are unpredictable. R	0.341	0.288	0.312	0.397	–
When dining out, I like to go to places where I have been before so that I know what to expect.	0.502	0.489	0.541	0.024	–
I don't like to go into a situation without knowing what I can expect from it.	**0.723**	**0.581**	**0.648**	**0.092**	**Yes**
I think it is fun to change my plans at the last moment. R	0.260	0.238	0.258	0.528	–
I enjoy the uncertainty of going into a new situation without knowing what might happen. R	0.404	0.321	0.359	0.491	–
I don't like to be with people who are capable of unexpected actions.	**0.686**	**0.594**	**0.645**	**0.074**	**Yes**
I prefer to socialize with familiar friends because I know what to expect from them.	0.587	0.615	0.672	0.008	–
I dislike unpredictable situations.	**0.697**	**0.622**	**0.673**	**0.031**	**Yes**
Measurement model summary: Predictability preference					
Chi-square = 19.18 (df = 13); p = 0.117					
AGFI = 0.992					
NNFI = 0.996					
RMSEA[a] = 0.031					
AVE = 30%					
Closedmindedness Subscale					
Even after I've made up my mind about something, I am always eager to consider a different opinion. R	**0.593 (0.589)[b]**	**0.145**	**0.184**	**–0.420**	**Yes (reversely stated)**
I dislike questions which could be answered in many different ways.	0.306 (0.300)[b]	0.217	0.256	–0.163	Yes
When considering most conflict situations, I can usually see how both sides could be right. R	**0.618 (0.619)[b]**	**0.120**	**0.128**	**–0.344**	**–**

(continued)

TABLE 27.2 (Continued)
Items Analysis of NFC Scale

Item	Stand. Weight (CFA: Measurement Models for Each Subscale)	Stand. Weight (CFA): NFC Model (4 Subscales)	Factor Loadings (PCA) - 1. Component	Factor Loadings (PCA) - 2. Component	Position Present in NFC15 (Roets and Van Hiel 2010)
		Closedmindedness Subscale			
I like to know what people are thinking all the time.	Excluded (0.209)[b]	–	–	–	–
I always see many possible solutions to problems I face. R	Excluded (–0.112)[b]	–	–	–	–
When thinking about a problem, I consider as many different opinions on the issue as possible. R	**0.594 (0.598)[b]**	**0.103**	**0.131**	**–0.384**	**–**
Measurement model summary: Closedmindedness					
Chi-square = 1.50 (df = 2); p = 0.471					
AGFI = 0.997					
NNFI = 1					
RMSEA[a] = 0.044					
AVE = 29%					
Measurement model summary: NFC (1 latent construct model)					
Chi-square = 149.13 (df = 135); p = 0.191					
AGFI = 0.983					
NNFI = 0.998					
RMSEA[a] = 0.014					
AVE = 23%					

		Decisiveness Subscale			
When I go shopping, I have difficulty deciding exactly what it is I want. R	0.531	–	–0.089	0.556	N/A
When faced with a problem I usually see the one best solution very quickly.	0.277	–	0.210	0.074	N/A
I usually make important decisions quickly and confidently.	**0.532**	**–**	**0.114**	**0.212**	**N/A**
I would describe myself as indecisive. R	**0.808**	**–**	**–0.013**	**0.639**	**N/A**
I tend to struggle with most decisions. R	**0.525**	**–**	**–0.139**	**0.555**	**N/A**
Measurement model summary: Decisiveness					
Chi-square = 2.49 (df = 2); p = 0.287					
AGFI = 0.996					
NNFI = 0.998					
RMSEA[a] = 0.050					
AVE = 31%					
Variance explained (EFA):			22%	9%	

Note: R, the score should be reversed.

[a] 90% high confidence interval reported for RMSEA.

[b] Weight for measurement model of the full six-item version of *Closedminded* subscale presented in brackets (containing also two items excluded from further analysis).

TABLE 27.3
Descriptive Statistics, Internal Reliability, and Correlations for NFC Subscale (for Both Short and Original Versions)

Statistics	Intolerance of Ambiguity		Preference for Order		Preference for Predictability		Closedmindedness		Need for Cognitive Closure		Decisiveness	
	Original	Short	Original	Short	Original	Short	Original (4 Items)	Short	25 Items	12 Items	Original	Short
Descriptive Statistics												
Mean	24.7	12.7	28.9	12.1	31.0	11.1	16.6	12.7	101.2	48.6	18.5	11.0
SD	4.18	2.57	6.18	3.17	6.66	3.17	3.13	2.49	15.08	7.94	4.26	2.95
Variance coefficient[a]	0.169	0.202	0.214	0.262	0.215	0.286	0.189	0.196	0.149	0.163	0.230	0.268
Reliability												
Cronbach's alpha	0.591	0.524	0.823	0.804	0.781	0.744	0.582	0.628	0.860	0.737	0.672	0.647
AVE (CFA)	22%	27%	44%	60%	30%	49%	29%	36%	23%	28%	31%	41%
Correlations												
Between versions ($n = 1754$)	0.875***		0.918***		0.847***		0.903***		0.923***		0.924***	
Intolerance of ambiguity	1	1	0.522***	0.426***	0.517***	0.462***	0.158***	0.138***	0.751***	0.721***	−0.046	−0.025
Preference for order	0.522***	0.426***	1	1	0.637***	0.617***	0.144***	0.093***	0.865***	0.812***	0.174***	0.085***
Preference for predictability	0.517***	0.462***	0.637***	0.617***	1	1	0.000	0.009	0.845***	0.797***	0.048*	−0.023

Closedmindedness	0.158***	0.138***	0.144***	0.093***	0.000	0.009	1	1	0.310***	0.399***	0.032	–0.013
NFC	0.751***	0.721***	0.865***	0.812***	0.845***	0.797***	0.310***	0.399***	1	1	0.087***	0.012
Decisiveness	–0.046	–0.025	0.174***	0.085***	0.048*	–0.023	0.032	–0.013	0.087***	0.012	1	1
Age ($n = 1743$)	0.167***	0.171***	0.377***	0.399***	0.360***	0.420***	–0.038	–0.080***	0.352***	0.357***	0.078**	0.076**
Mental speed (WAIS-R: $n = 223$)	–0.283***	–0.334***	–0.37***	–0.347***	–0.452***	–0.444***	–0.063	0.094	–0.450***	–0.426***	–0.008	0.002
SDO ($n = 228$)	–0.030	–0.069	0.129	0.080	0.058	0.089	0.075	0.028	0.081	0.051	0.030	0.073
RWA ($n = 340$)	–0.084	–0.023	0.315***	0.326***	0.136*	0.149**	0.220***	0.148**	0.228***	0.270***	–0.025	–0.055
Neuroticism ($n = 340$)	0.069	0.070	–0.02	0.066	0.003	0.031	–0.094	–0.025	–0.009	0.060	–0.564***	–0.537***
Extraversion ($n = 340$)	–0.060	–0.008	–0.09	–0.173**	–0.274***	–0.232***	–0.174**	–0.183***	–0.225***	–0.262***	0.401***	0.464***
Openness to experience ($n = 340$)	–0.084	–0.013	–0.33***	–0.373***	–0.431***	–0.324***	–0.509***	–0.383***	–0.481***	–0.476***	0.065	0.134*
Agreeableness ($n = 340$)	0.108*	0.022	0.216***	0.214***	0.244***	0.252***	–0.021	–0.052	0.236***	0.207***	0.216***	0.200***
Conscientiousness ($n = 340$)	0.184***	0.233***	0.490***	0.433***	0.303***	0.327***	–0.055	–0.141**	0.399***	0.385***	0.425***	0.395***

[a] Standard deviation divided by mean. This coefficient allows us to make comparisons between variability of different variables (here: short and original version).
$*p < 0.05$; $**p < 0.01$; $***p < 0.001$.

DISCUSSION

Our goal was to formulate and validate a shortened version of the scale for measuring the NFC. The shortened version consists of 15 items, chosen on the basis of regression weights from each of the subscales of the original scale. We aimed to make our selection of items consistent with the choices made by Roets and Van Hiel (2010) when they prepared a shortened English version of the scale. The proposed shortened version therefore measures the NFC in a similar way to the scale suggested by these authors. The psychometric values of the new shortened scale show better qualities than the original, due to the exclusion of items that resulted in high measurement error. This new tool also has satisfactory external validity, which is indicated by the correlation between the new scale and other scales that measure similar constructs as the original NFC scale. Using the analyses presented herein as a basis, we suggest that the shortened version of the NFC scale has a two-dimensional structure, in which one dimension accurately represents the motivational aspect (Kruglanski 1989) and the second refers to the cognitive aspect of the closure processes. This scale structure is consistent with findings regarding the original NFC scale. This new tool could be successfully used in future empirical research.

ACKNOWLEDGMENT

This research was supported by a grant DEC 2011/02/A/HS6/00155 from the National Science Center awarded to Małgorzata Kossowska.

REFERENCES

Bar-Tal, Y., Kossowska, M. (2010). The efficacy to fulfill need for structure: The concept and its measurement. In: J. P. Villanueva (ed.), *Personality Traits: Classifications, Effects and Changes* (pp. 47–64). New York: Nova Publishers.

Bukowski, M., von Hecker, U., Kossowska, M. (2011). *The Role of Need for Cognitive Closure in the Construction of Social Mental Models*. Unpublished research report.

Cornelis, I., Van Hiel, A., Roets, A., Kossowska, M. (2009). Age differences in conservatism: Evidence on the mediating effects of personality and cognitive style. *Journal of Personality, 7*, 51–88.

Jaśko, K. (2011). *Psychologiczne uwarunkowania legitymizacji systemu demokratycznego* [*Psychological Determinants of Legitimization of Democratic System*]. Unpublished PhD thesis. Kraków: Uniwersytet Jagielloński.

Keller, J. (2005). In genes we trust: The biological component of psychological essentialism and its relationship to mechanisms of motivated social cognition. *Journal of Personality and Social Psychology, 88*, 686–702.

Kemmelmeier, M. (2010). Gender moderates the impact of need for structure on social beliefs: Implications for ethnocentrism and authoritarianism. *International Journal of Psychology, 45*, 202–211.

Koralewicz, J. (1987). *Autorytaryzm, lęk, konformizm. Analiza społeczeństwa polskiego końca lat siedemdziesiątych* [*Authoritarianism, Anxiety, Conformism. Analysis of Polish Society in the End of 70s*]. Wrocław: Ossolineum.

Kossowska, M. (2002). *Walidacja skali Orientacji na Dominację Społeczną* [*Validation of Social Dominance Scale*]. Unpublished research report.

Kossowska, M. (2003). Różnice indywidualne w potrzebie poznawczego domknięcia [*Individual differences in need for cognitive closure*]. *Przegląd Psychologiczny, 46*, 355–375.

Kossowska, M. (2005). *Umysł niezmienny. Psychologiczne mechanizmy sztywności* [*Unchangeable Mind. Psychological Mechanisms of Rigidity*]. Kraków: WUJ.

Kossowska, M. (2007a). The role of cognitive inhibition in motivation toward closure. *Personality and Individual Differences, 42*, 1117–1126.

Kossowska, M. (2007b). Motivation towards closure and cognitive processes: An individual differences approach. *Personality and Individual Differences, 43*, 2149–2158.

Kossowska, M. (2010). Poznawcze i motywacyjne źródła nabywania wiedzy [*Cognitive and motivational sources of knowledge formation*]. In: A. Kolańczyk, B. Wojciszke (eds.), *Motywacje Umysłu* [Motivated Mind] (pp. 15–28). Sopot: Smak Słowa.

Kossowska, M., Bar-Tal, Y. (in print). Need for closure and heuristic information processing: The moderating role of ability to achieve cognitive closure. *British Journal of Psychology*.

Kossowska, M., Jaśko, K., Bar-Tal, Y., Szastok, M. (2012). The relationship between need for closure and memory for schema-related information among younger and older adults. *Neuropsychology, Cognition and Aging*, *19*, 283–300.

Kossowska, M., Van Hiel, A. (2003). The relationship between need for closure and conservatism in Western and Eastern Europe. *Political Psychology*, *24*, 501–518.

Kossowska, M., Van Hiel, A., Chun, W. Y., Kruglanski, A. (2002). The Need for Cognitive Closure Scale: Structure, cross-cultural invariance, and comparison of mean ratings between European–American and East Asian Samples. *Psychologica Belgica*, *42*, 267–286.

Kruglanski, A. W. (1989). *Lay Epistemics and Human Knowledge: Cognitive and Motivational Bases*. New York, US: Plenum Press.

Kruglanski, A. W., DeGrada, E., Mannetti, L., Atash, M. N., Webster, D. M. (1997). Psychological theory testing versus psychometric nay-saying: Comment on Neuberg et al.'s (1997) critique of the Need for Closure Scale. *Journal of Personality and Social Psychology*, *73*, 1005–1016.

Kruglanski, A. W., Orehek, E., Dechesne, M., Pierro, A. (2009). Three decades of lay epistemics: The why, how, and who of knowledge formation. *European Review of Social Psychology*, *20*, 146–191.

Kruglanski, A. W., Webster, D. (1996). Motivated closing of the mind: "Seizing" and "freezing". *Psychological Review*, *103*, 263–283.

Lynch, J. G., Netemeyer, R. G., Spiller, S. A., Zammit, A. (2010). A generalizable scale of propensity to plan. The long and the short of planning for time and money. *Journal of Consumer Research*, *37*, 108–128.

Mannetti, L., Pierro, A., Kruglanski, A., Taris, T., Bezinovic, P. (2002). A cross-cultural study in the need for cognitive closure scale: Comparing its structure in Croatia, Italy, USA and the Netherlands. *British Journal of Social Psychology*, *41*, 139–156.

Neuberg, S. L., Judice, T. N., West, S. G. (1997). What the Need for Closure Scale measures and what it does not: Toward differentiating among related epistemic motives. *Journal of Personality and Social Psychology*, *72*, 1396–1412.

Neuberg, S. L., West, S. G., Judice, T. N., Thompson, M. M. (1997). On dimensionality, discriminant validity, and the role of psychometric analyses in personality theory and measurement: Reply to Kruglanski et. al.'s (1997) defense of the Need for Closure Scale. *Journal of Personality and Social Psychology*, *73*, 1017–1029.

Piber-Dąbrowska, K. (2010). Unpublished data set.

Piotrowska, K., Kossowska, M., Bukowski, M. (2011). Wpływ sytuacyjnie wzbudzonej potrzeby poznawczego domknięcia na różnice indywidualne w aktywizacji treści stereotypowych [*The influence of situationally evoked need for closure on individual differences in the stereotype activation*]. *Czasopismo Psychologiczne*, *17*, 229–240.

Roets, A., Van Hiel, A. (2007). Separating ability from need: Clarifying the dimensional structure of the Need for Closure Scale. *Personality and Social Psychology Bulletin*, *33*, 266–280.

Roets, A., Van Hiel, A. (2010). Item selection and validation of a brief, 15-item version of the Need for Closure Scale. *Personality and Individual Differences*, *50*, 90–94.

Roets, A., Van Hiel, A., Cornelis, I. (2006). The dimensional structure of the need for cognitive closure scale: Relationships with 'seizing' and 'freezing' processes. *Social Cognition*, *24*, 22–45.

Sędek, G., Gradowski, Ł. (2010). Unpublished data set.

Sędek, G., Safrończyk, K. (2011). Unpublished data set.

Trejtowicz, M. (2012). Unpublished data set.

Webster, D. M., Kruglanski, A. W. (1994). Individual differences in need for cognitive closure. *Journal of Personality and Social Psychology*, *67*, 1049–1062.

Wechsler, D. (1997). *Wechsler Adult Intelligence Scale—Third Edition*. San Antonio: The Psychological Corporation.

Zawadzki, B., Strelau, J., Szczepaniak, P., Śliwińska, M. (1998). *Inwentarz osobowości NEO-FFI Costy i McCrae* [*NEO-FFI Personality Inventory by Costa and McCrae*]. Warszawa: Pracownia Testów Psychologicznych PTP.

APPENDIX: THE SHORT NEED FOR CLOSURE SCALE

INSTRUCTIONS: Read each of the following statements and decide how much you agree with each according to your beliefs and experiences. Please respond according to the following scale.

1.........strongly disagree
2....moderately disagree

3...........slightly disagree
4................slightly agree
5.........moderately agree
6..............strongly agree

1.	Even after I've made up my mind about something, I am always eager to consider a different opinion. R (Closedmindedness)	1 2 3 4 5 6
2.	I don't like situations that are uncertain. (Intolerance of ambiguity)	1 2 3 4 5 6
3.	I find that a well ordered life with regular hours suits my temperament. (Preference for predictability)	1 2 3 4 5 6
4.	I feel uncomfortable when I don't understand why an event occurred in my life. (Intolerance of ambiguity)	1 2 3 4 5 6
5.	I don't like to be with people who are capable of unexpected actions. (Preference for predictability)	1 2 3 4 5 6
6.	I usually make important decisions quickly and confidently. (Decisiveness)	1 2 3 4 5 6
7.	I would describe myself as indecisive. R (Decisiveness)	1 2 3 4 5 6
8.	I tend to struggle with most decisions. R (Decisiveness)	1 2 3 4 5 6
9.	When considering most conflict situations, I can usually see how both sides could be right. R (Closedmindedness)	1 2 3 4 5 6
10.	I don't like to be with people who are capable of unexpected actions. (Preference for predictability)	1 2 3 4 5 6
11.	I find that establishing a consistent routine enables me to enjoy life more. (Preference for order)	1 2 3 4 5 6
12.	I enjoy having a clear and structured mode of life. (Preference for order)	1 2 3 4 5 6
13.	I feel uncomfortable when someone's meaning or intention is unclear to me. (Intolerance of ambiguity)	1 2 3 4 5 6
14.	When thinking about a problem, I consider as many different opinions on the issue as possible. R. (Closedmindedness)	1 2 3 4 5 6
15.	I dislike unpredictable situations. (Preference for predictability)	1 2 3 4 5 6

R – the score should be reversed

28 Psychology in Supporting and Stimulating Innovativeness

The Role of Values and Spirituality

Barbara Kożusznik and Jarosław Polak

CONTENTS

> When it comes to attracting, keeping and making teams out of talented people, money alone won't do it. Talented people want to be part of something they can believe in, something that confers meaning on their work and their lives.
>
> **John Seely Brown**
> *(In Search of Meaning in the Workplace, Linda Holbeche and Nigel Springett)*

INTRODUCTION

On the ground of Daniel Pink ideas (2005, 2009), the authors advocate that business approach based on values and spirituality can meet the requirements of conceptual age in postindustrial countries. In the face of rapidly changing technology in the global world and increasing expansion of Asian countries, our postindustrial civilization may survive provided that it makes the most of human creativity and ability to adapt to changes. We are in expectation of inevitable changes, and it is essential to activate creativity and innovativeness of an organization and its workers. Thus, psychological knowledge and skills will play a crucial role in the process of innovativeness activation, and transfer, implementation, and adoption of innovations. Considering a low innovativeness level, Poland in particular should get involved in these activities. Psychology offers a wide variety of approaches, theoretical models, and practical solutions improving innovative processes (Kożusznik 2011). It is also still looking for new sources of people's motivation and energy. However, the research results indicate that there are a number of psychological barriers to innovativeness. Probably all of them may be overcome on condition that psychology is treated as a partner by other disciplines, and psychologists themselves, through their expertise and determination, need to participate in supporting and improving innovative processes. Moreover, a number of authors agree that spirituality (most often expressed in terms of specific religious tradition) plays the important role in individual and social processes leading to innovation (Handy 1999; Biberman and Whitty 2007; Gibbons 2000; Duffy 2006) especially by strengthening cooperation and building meaningful environment within organization. Contemporary psychological look at the spirituality at the workplace is also realistic, and there may be negative effects as well as positive ones (Koenig 1997).

CONCEPTUAL AGE CHALLENGES: ERA OF INVENTION AND INNOVATIVENESS

The challenges that are to be faced by the world economy are referred to as "conceptual challenges." The notion of *Conceptual Age* was used the first time by Alan Greenspan at his speech at Connecticut University in 1997 where he stated that there has been a colossal increase in demands toward workers not only in terms of know-how, information, etc., but also regarding conceptual properties possessed by them, that is, their command of creative, analytic, and transforming skills as well as effective interaction with them. Pink (2005) clarifies the way the economy of information society based on knowledge is converting from the information era into the conceptual era where economy consists of creativity, innovation, and ability to design and change reality for the purpose of attaining market competitiveness, particularly in the global world. In this global world, developments have been occurring on a grand scale; nevertheless, there has been a shift in the emphasis from the economy focusing on the product to the economy concentrating on services and information processing (Huitt 1999a; Naisbitt 1982; Toffler 1981, 1990).

The Conceptual Age can be viewed as continuation of the Information Age. This is a consequence of widespread access to technologies applied for gaining, processing, storing, accessing, and using information, the tempo of creating new technologies and mass production, the need of simultaneous development of technologies and social skills, and easiness in transferring them across borders (Huitt 1999b). The Conceptual Age results from growing importance of distribution rather than production. It is also a consequence of decentralization and the emphasis put on training courses for people and on general development of organizations geared at improving decision-making skills and decision implementation, as well as client-oriented economy (availability of alternatives and possibilities of ordering the same product or service). We can observe changes affecting workforce: *dejobbing* (employment for a fixed time period is more common than employment on a full-time basis); people take up a few posts, different kinds of work, and different occupations within their lifetime; the phenomenon of so-called "multitude of careers" occurs. In addition, the stress on entrepreneurship and home business has been intensified, whereas in the range of health care, pension, etc., a shift to personal responsibility has been made. At the same time, what can be noticed is the increased focus on institutional assistance and a reduction in government support networks. Thanks to medical advances prolonging a human life (Canton 2006), there is a tendency indicating that an increasing number of people will live up to 100. There are noticeable economic possibilities resulting from this. Kotlikoff and Burns (2004) underline that it will cause economic tension not being experienced up to date, particularly in the scope of medical care and finances designated for pension benefits, which also lays the foundations for the Conceptual Age.

Pink (2005) supplements the above catalogue of causes leading to formation of the Conceptual Age with occurrence, apart from widespread access to information, of three considerably significant phenomena:

- *Abundance*—there are abundant alternatives for all skills, products, and services offered by individual and organization in demand on the market.
- *Asia*—all products can be delivered cheaply using Asian labor.
- *Automation*—products and services can be made or delivered faster, better, and cheaply by the computer and machines (Pryor and Bright 2011).

They constitute the main grounds for appearance of totally new demands, challenges, or simply necessary changes, which enable well-developed countries of Europe, the United States, and Japan to compete with other markets, Asian markets in the first place.

KEY ROLE OF PSYCHOLOGY IN SUPPORTING AND DEVELOPMENT OF INVENTIVENESS

In the face of the above-described intensifying phenomena, the question appears how our postindustrial world is supposed to survive and cope with competition. Pink states that it is not traditional "knowledge workers" but creators and sensitive and empathic persons who will help organizations to survive in postindustrial countries in Europe, the United States, or Japan. These organizations, in order not to cease to exist, are awaiting next transformation—knowledge workers have to be replaced or supplemented by workers of concepts, full of ideas and empathy. In the conceptual era, traditional jobs—legal, accounting, engineering ones—can be outsourced and performed by cheaper workforce. Nevertheless, Pink writes that creativity cannot be outsourced and states that the era of the left hemisphere domination is coming to an end. Organizations will be able to survive only due to activation of the right hemisphere possibilities—*inventiveness, resourcefulness, meaning*, and *empathy*—which will constitute the foundations of success of these organizations that will base on them (Pink 2006).

In the world of abundance of goods and services, new tasks to be undertaken by organizations consist of providing to people or customers such products that will embody special meaning (Miller 2005) since a prosperity period has prompted societies to ask fundamental questions with regard to the overall condition of a human life: "What do I have in common with other people? How am I connected with the world?" People started, in a far greater degree, to look for an answer to the question about things greater than themselves (Pink 2005; Seligman 2002). People become aware of a necessity of changes in consumption and production, and this awareness constitutes the major premise of emergence of the Conceptual Age. Aburdene (2005) stated that people more frequently search for values that go beyond material consumption; they search for the meaning (Frankl 1984; Handy 1999; Maslow 1971; Inglehart 2004). This tendency may have influence on society in the following manner (Aburdene 2005):

1. *Value-oriented consumption*—consumers who form the minority at present, but will become the majority in the next decades, will seek products and services corresponding to their values.
2. *Spirituality in business*—there are a number of organizations that have already laid emphasis on meaning and spiritual values. There is going to be growing demand for talented workers, for people oriented toward realization of important values. A number of organizations strive both to achieve a business success and to create social goods.
3. *Socially responsible investments*—they constitute another aspect of consumption geared at realization of values; there will be a search for investments that will correspond to values represented by investors.

Whether our European countries and the United States manage to cope with pressures of global markets will depend on whether we will be actually able to embed values into economic activity (Canton 2006) and whether knowledge, attitude, and skills indispensible in this era will be employed. It is innovativeness and creativity that will be a decisive factor conditioning whether people survive or not. Therefore, it is necessary to launch new ideas and concepts onto the today market. This entails creativity, empathy, understanding and respect, and making reference to values. All this means that psychology and its mechanisms are the most significant since innovations are devised relying on new concepts, and psychology provides us with the knowledge how these particular concepts are created and how they should be stimulated. Psychology is also a source of knowledge clarifying how new ideas and concepts are to be brought into practice (thanks to proper transfer allowing for psychosocial barriers) and how to manage innovative implementation processes by lowering psychological resistances toward changes.

Certainly, psychologists have to show direct or indirect links between the complex phenomenon of human spiritual values and virtues and innovativeness. Such research will contribute to facilitating the discussion on the topic. In the reflection on spirituality and innovativeness of civilization, an unavoidable question is the way the world economic and technological progress should follow. Should new innovations contribute to the multiplication of corporation wealth or should they ensure a better life for everyone? Should they increase the market absorbance or rationalize utilized resources and conservation? These questions need not be answered in association with spirituality, values, and religiosity, but they may be provoked by the spiritual or religious viewpoint (not necessarily associated with any religious tradition). They are, in fact, fundamental questions, and they assume a vision of origin, as well as the role values play in the world. Thus, a question should be asked not only about how to stimulate innovativeness but also why do it and which way human progress based on these innovations should follow.

If theoreticians are right and innovation really is the most important world economic engine, then efforts should be focused on it in order to recognize its conditions, also in the field of psychology. The complexity of this process largely stems from the part played by creative values and the ones that refer to the value system of human power. Perhaps it is an enormous task of modern psychology to discover these powers and processes, which may support the creation of innovation.

DOES SPIRITUALITY MATTER?

Spirituality and certain kinds of religiosity have a lot to do with either individuals or social groups taking action of business character, which is also an innovative one. Vaughan (1991) provided synthetic definition of spirituality: "a subjective experience of the sacred" (p. 105). Shafranske and Gorsuch (1984) defined spirituality as "a transcendent dimension within human experience discovered in moments in which the individual questions the meaning of personal existence and attempts to place the self within a broader ontological context" (p. 231). Spirituality and religiosity may satisfy needs for a purpose in life, anchoring a sense of what is right and wrong (Baumeister 1991). Spirituality may also influence the innovative potential of individuals and social groups. Thus, knowledge on spiritual forms and relations between them is important from a practical point of view. Perhaps it should matter for each company owner if their workers are spiritual or not and if so what kind of spiritual involvement it is. It might not be a pure coincidence that among many innovators and powerful entrepreneurs, a great many of them were Jews.* Knowledge and potential consequences of Max Weber's thesis on the link between religious values and the economic development of nations (Davie 2007) are a source of interest in religiosity in the field of economics. Undergoing discussions concern especially the Middle East countries (namely, those where dominant religions are either Hinduism or Islam), and the underlying question is whether these countries are capable of changing the economic power scheme in the world (Harrison et al. 2009). A considerable force that could contribute to it is the innovativeness of those economies, which, according to M. Weber's thesis, is dependent on religious norms that apply there.

Understandably, innovativeness in reference to human psyche is an exceptional phenomenon. It is the final effect of human creative intellectual power, as well as passion and engagement in the process of creating new things (Handy 1999). Hence, innovativeness is part of the considerations on the expression of human spirit that is engaged in the process of making the world better. Besides, the creation of a new technology that could change the whole civilization is a moral deed. And from here, there is only one step to religiosity, in which a reflection on profound and universal grounds

* According to Cochran et al.'s (2006) controversial thesis, a simple correlation between the level of intelligence and being a Jew is nonexistent, because a higher than average intelligence concerns solely Ashkenazi Jews, not all Jews.

for human activity is always present. From a more practical perspective, admittedly, the skeptics are right that a lot of modern and fast emerging product innovations have a purely marketing character and no underlying meaning should be sought. But considering Louis Pasteur, it becomes clear what subject is under discussion here. It does not mean that all big innovators were religious by being part of a religious tradition. Nonetheless, a vast number of them were known to be deeply inspired by universal values. Iconic Randy Pausch is a perfect example of a person pointing at sources of inspiration very close to spirituality. It should also be noted that almost all the Nobel Prize winners so far have been people for whom religion played an important role (Donahue and Nielson 2005).

In this part, other questions concerning psychology of spirituality and religion will be posed that are apparently associated with the processes that lead to innovation creation. These will comprise both an interpretation and empirical approach. A concept that is not very common in literature but still an interesting one is an assumption that certain religious ideas may stimulate human innovativeness (Fleissner 2004). Action mechanism reminds a little the role of Immanuel Kant's regulative ideas, which described a certain status quo but were beyond any empirical verification, concerned the problem of universe as a whole, immortal nature of the soul and freedom, and were indispensable part of human thought that allowed for full understanding of the outside world. Religious notions of omnipotence, omnipresence, and immortality have a similar status; they are unattainable states, which people are aware of thanks to religious ideas and which they still try to achieve. In renaissance, such a method was alchemy; today it is innovation. According to Peter Fleissner, all modern areas of knowledge whose aim is to free humans from restrictions imposed by their form, time, and space are governed by these ideals (unaware of their source). Just like in I. Kant's thoughts, these notions taken out of the religious context sound like fantasies, but they play a certain social function. Obviously, it is impossible to prove that they have purely religious grounds. From the psychology of religion viewpoint, in which *sacrum* (or as Rudolf Otto sees it, the idea of the holy) is interpreted as a taboo, believers will doubtfully set up to cross over it, but certain interpretations of the promise "you shall be as Gods" seem to contradict it. Summing up, religious ideas can be said to be a potential source of inspiration for innovative ideas.

Another example of a potential influence of spirituality and religiosity on innovativeness is direct ethical and practical grounds included in religious systems. Judaism and Christianity are best examples of this thesis, in which orders to subdue the earth, multiply resources, and love others may all significantly affect innovativeness. An archetypal example of the way these ideas work is British colonialism. In this ruthless system (and devoid of religious motivation), wealth was acquired from yet new colonies, worsening the situation of the subjected people. But together with the colonial expansion, the invention of print was disseminated. The activities of protestant missionaries, who followed biblical rules, contributed to this (Harrison et al. 2009). The research confirms that religious values orientation may motivate people to openness for social and physical needs of others (Idler et al. 2003, p. 327); it gives a feeling of purposefulness in life (French and Joseph 1999). A number of modern Christian groups (not only Protestant ones) base their evangelization process on fostering life in accordance with biblical rules, which must bear fruit in both material and spiritual prosperity (Davenport 2009; Oster 2009). Similar movements also appear based on Islam. Many acclaimed business authors who specialize in self-improvement, such as Stephen Covey or Zig Ziglar, base their advice concerning efficient (including creative) activities on the values that stem from religion.

Many theoreticians of responsible business, judging the way modern companies operate, simply advise their operation be based on spiritual rules. Overtones of this kind of thinking are present in books of such authors as Adrian Hodges (Grayson and Hodges 2001) and Marcello Palazzi (Palazzi and Starcher 1997, p. 8). Palazzi, who cooperates with religious organizations, proves work that gives an impression it makes sense, and that has its grounds in some values, must increase productivity, quality and involvement, and, importantly, innovativeness. Respecting other people's rights, besides being based on deeply rooted and spiritual values, provides measurable business profits. It coincides with Katz and Kahn's (1979) approach, devoid of religious reference, to finish with radical social changes in the world of organization, including organizational innovations. Katz and Kahn claim any technological changes be preceded with decisions on social aims. It is rational behavior and common in the field of management once costs that social systems have to bear to repair the aftermath of disturbances evoked by technology are considered. Researchers recommend experiments, series of attempts, and appraisals that should precede negative results. They have named their attitude pragmatism. They postulate "using such measures of organisational efficiency, in which the needs of leader and subjects, shareholders and workers, consumers and citizens would be considered" (Katz and Kahn 1979, p. 727). In result, a concept has been born of responsible organizations whose activities are morally judged and justified, and often rooted in spiritual and religious attitudes.

Noteworthy is a concept of spiritual capital of companies as an important part of individual social capital (Iannaccone and Klick 2003). Responsible value-based behavior allows companies to count on greater involvement of their workers in the realized objectives. A certain example of activities that take into account religious values are business sects, one of which is the famous Amway company. Its philosophy refers to Christian values. In literature, examples of companies that apply various methods of the so-called spirituality management may be found (Giacalone and Jurkiewicz 2010).

Another vast area of psychology of religion, which could be a potential source of hypotheses that explain psychological and social sources of individual innovativeness, are concepts that stem from Gordon Allport's hypotheses and concern religious maturity and intrinsic–extrinsic religious orientation (Allport and Ross 1967). Intrinsic religious orientation, simply speaking, is such an approach to religious dilemmas that assumes their autonomous nature. Among people with such a religious orientation, religious questions fulfill their life with motivation and meaning. On the other hand, extrinsic religious orientation means piety in economic sense, and religion is a source of interest due to its usability for own interests. This division largely overlaps with the concept of religious maturity (Francis and Pocock 2008). Gradually, a third dimension was added called Q (which stands for Quest), which means authentic search for religious values accompanied by successful coping with doubts and problems of everyday life at the same time (Batson et al. 1983). Although the discussion on the associations between a type of religiosity and practical activities is still in progress (and is not likely to end soon), the data suggest irrefutably that people with an intrinsic religious orientation are characterized by greater stability in their actions and are more consistent in aim realization that stems from religious commands (research results as well as the concept itself are discussed in Kirkpatrick and Hood 1990). Attitudes to own religiosity examined by means of *Francis Scale of Attitude towards Christianity* (based directly on Allport's concept) also show correlations between religiosity and altruism, feeling of happiness, psychological health, and readiness to fulfill social needs of other people (Hills and Francis 2003), which may also influence some variables that condition the innovative process. This association appears hypothetical and indirect, but its occurrence cannot be excluded.

PSYCHOLOGICAL MODELS OF INNOVATION PROCESS

A number of psychological models of innovation process conditions have been proposed in the literature. To recapitulate, on the basis of the analysis of these models, it can be assumed that a psychological innovative model is composed of the following elements:

- Individual and team creativity (e.g., Altshuller 1984; de Bono 1973; Whitfield 1975; Osborn 1959; Cottrell 1972; Robson 1993).
- Ability to transfer creativity to the organization (transfer and appearance of the innovation) (e.g., Rogers 1962; Argyris 1976).
- Implementation (innovation adoption by the organization; e.g., Zaltman et al. 1973; Argyris 1970; Sotiriou and Wittmer 2001).

Connections between presented model and spirituality are rather obvious in the light of Pinks demands: inventiveness, resourcefulness, meaning, and empathy. Creativity is considered the key element that enables making new ideas in the *invention process*. And supporting innovations through openness to employees' ideas is the most important factor in the innovation making process for the questioned company employees. According to Mark Bandsuch (Bandsuch and Cavanagh 2005), a specific kind of openness that is based on respect for everything that is of spiritual nature in an organization brings companies profit, and also as increased creativeness and productivity. The fact that employees can participate in creative problem solving is said to be common practice of the organizations that mind employees' spiritual needs. This could point out at the practically applied relation of spirituality and creativity (Burack 2000). Most researchers claim, however, that religious people are more conservative than nonreligious ones. Dollinger's (2007) research proves that conservatism correlates negatively with creativity. What is interesting, among those examined, the creativity of religious people is often linked with religion in terms of the contents. A religious form, nonetheless, in modern, secular societies may lower social acceptance of such creations or ideas. Thus, perhaps religious people, taking part in the innovation process, may turn out to be insufficiently creative or be subject to other (maybe more frequent) barriers against creativity. Still, some religions may require from their believers to overcome habits and ways of achieving aims applied before, as well as create the ability to distance oneself and question the status quo, which are all important elements of creativity.

The second part of the innovative process is transfer creativity to the organization, and its underpinnings are the networks and the communication that affects (among others) diffusion of innovations. According to the functional approach of Durkheim (2001) and Haidt (Graham and Haidt 2010), the first role of religion and spiritual assumptions is to bind people in groups to allow them to work together. Donahue and Nielsen (2005) point at a possible connection between social behavior and religiosity; however, this relation is not always positive. Wu (2009) discovered that among factors that affect diffusion of innovations on Facebook social networking service is religiosity. People who adapt themselves most easily have turned out to be Christians, followed by people representing other denominations to finish with Muslims. Some results may point out that in the communication process, religious people have more respect for the feelings of those they communicate with and restrain themselves from activities that could be detrimental for them (Donahue and Benson 1995). The research results of Sikorska-Simmons (2002) prove that cooperation with spiritual and religious people may be a good work satisfaction predictor. Additionally, the author examined people employed in the professions associated with helping others. Spirituality can also influence such psychological variables like job involvement and job commitment. People more involved in the innovation process probably devote more time and attention on developing an idea and disseminating it. They also tend to better deal with obstacles and setbacks in their activities. A positive correlation between job involvement and spirituality/religiosity has been shown in the research of Sikorska-Simmons (2002) and Knotts (2003). Still, in Knotts's work, the dependence solely concerns protestants and not non-protestants. Roundy (2009), on the other hand, examined a possible contradiction between religious and job involvement in laic organizations and also managed to establish a correlation there. It turned out that the bigger the religious involvement, the lower the work involvement. This has also been confirmed by Polak (2011), who, based on his own observations, states that spirituality in an organization may be the reason for lack of involvement in business purposes realization, especially among spiritual or religious people.

The major constraints in the implementation stage are deficiencies in management of the innovation process not eliminating resistances to changes, fears, and negative emotions connected witch changes. Besides, ways of coping with emotions should be linked with individual innovativeness. Duriez and Hutsebaut (2001) and Park (2005) studied a relation between religiosity and emotionality; they interpreted religiosity as a system that supports negative emotions coping. Harris et al. (2002) claimed a lower level of fear among those involved religiously and praying. Religiosity and spirituality may also support coping with stress by making things meaningful (Park 2005). An important trait that managers possess that makes it easier for companies to accept changes in the organizational systems is reflexive nature associated with both religiosity and spirituality. Nevertheless, Kazama et al.'s (2002) research proves that the ability of intellectual reflection on the strategies, targets, and processes enhances positive attitudes among workers on changes and facilitates their implementation.

The research tracks mentioned in this part have just been outlined. This scope of knowledge should definitely undergo further research. Certainly, these examinations must subtly differentiate between analyzed variables because giving a simple answer to the question of a relation between spirituality and innovativeness in a general sense must fail.

Innovation psychology does not only focus on studies on talents and creation but also deals with the whole process and all its phases. Psychological problems occur both in the creation as well as transfer and adoption stages. Unfortunately, our findings have been understated and treated superficially, and psychologists are not invited as partners to undertake, together with economists, management specialists, etc., the activities aimed at creation, transfer, and implementation of innovations. Although research findings have proved that there is a correlation between human resources management and organizational effectiveness and its inventiveness, studies into a relation between human resources management and innovativeness are rarely conducted (Selvarajan et al. 2007). Psychologists have at their disposal the research results regarding the innovation process ranging from creation through transfer to implementation of the innovation. These studies' results, nevertheless, are seldom taken into account, which frequently results in failure of change implementation, resistances, and additional costs. The proposed model shows how significant is linking a moment of innovation creation, understanding its sense and its goal for its subsequent implementation. The research results in respect of barriers and psychological problems indicate what tools should be made available for managers in the first place, and as a form of assistance in creation, transfer and better implementation of the innovation.

A ROLE OF PSYCHOLOGY IN AN INNOVATION ERA

To recap on the above reflections, it may be stated that psychology, organizations, and world economy will have to cope with a lot of important tasks including support and stimulation of innovativeness. It may be assumed that in the face of a key role played by psychology in these processes, they will not be crowned with success without contribution and active participation of psychological knowledge and skills.

Correlation between Values, Spirituality, and Economy

Psychology has to rise to new challenges posed toward the innovative era and conceptual era. The research paradigm in organizational psychology elaborated on the foundations of the knowledge-based economy should be developed on the basis of an approach focusing on conceptual factors. Further increase in competitiveness of economy of postindustrial states is conditioned by effective use of human factors—knowledge but also emotions, creativeness, empathy, and intuition (Markman and Wood 2009). The point is that innovations should be full of inventiveness and need to correspond to expectations and values of people aspiring to something beyond satisfaction of their basic needs. A novel attitude toward an organization and to ourselves will provide a stimulus to the development of products and services adapted to the needs of the conceptual era—a need of understanding and sense of meaning. Innovations may play a significant social role since it is not actually a model of success ideology but a model of coping with constant changes, acceptance of uncertainty, looking for own one's place, finding sense and deeper meaning, and acceptance of diversity that has become a dominant pattern for development.

Accounting for Psychological Factors at Each Stage of the Innovation Process

The success of innovation processes does not only depend on creativity and creative workers. It also hinges on strenuous work at all phases of the innovation process as the success of an innovative process is conditioned by participation of people in this process—their needs, values, and adaptive capabilities (Denning 2004). Significant objectives set for psychology consist of getting involved in stimulation and proper implementation of innovative processes and raising awareness of importance and a role of psychological factors in making the innovation successful. Even the most creative ideas and inventions need people's willingness to adopt them. According to Drucker (1992, 2002), they require strenuous work and participation of psychologists during the whole course of innovative processes, and the care exercised by them at each phase of this process in the range of psychological values. This work is strenuous since surmounting psychological barriers needs time, patience and determination, inventiveness, and communicative skills. The research studies indicate that psychological problems are essential to the phase of invention and stages of transfer and implementation. There is an increasing demand that managers should be equipped with appropriate psychological tools, which should be used at consequent individual phases of the innovation process. It should be emphasized that significance of psychological factors is based not only on stimulation of creativity at the level of innovation creation, as it is assumed by representatives of other discipline, but also it is particularly essential to the remaining phases of the innovation process—e.g., while forming and supporting self-managing teams, during impact reduction and breaking stereotypes (e.g., between science and practice; minority and majority groups), and when applying psychological methods at each phase of the innovative process (e.g., selection of people for group formation, choice of representatives, etc.). The disregard for a role of psychology played at these stages is often a result of necessity to incur financial costs and to devote time, energy, and expertise to follow the guidelines. This makes psychologists competitive toward representatives of other disciplines and reveals deficiencies of systems not allowing for recommendations of psychology. Unfortunately, it can be often observed not merely in respect of innovative processes but regarding processes providing safety at work.

In order to achieve the above enumerated objectives, subsequent challenges are to be posed for education and management. Other challenges to respond to are cooperation with economists and representatives of other disciplines and the need to elaborate a coherent model of innovative research and to give a new quality to psychological studies on innovations.

Challenges for Education

New tasks are emerging for education, training systems, and overall development of workers and managers. Pink (2005) describes qualities that should be improved by managers and workers in the childhood, by the school system, and in a later period. They are psychological qualities of crucial importance in the innovative era and conceptual era:

Empathy: which should supplement logical and critical thinking, allows to imagine oneself in another person's place and to look at a particular situation from the perspective of others.

Stories: the ability to communicate by discovering more profound meanings, understanding facts in contexts, and imparting emotional meaning to them, that is, spinning "stories" that are a manifestation of deeper understanding and creative thinking.

Design: implies that it is not sufficient to produce functional services or products—they must be also beautiful and emotionally involving. It is also essential to teach people to design objects, spaces, etc., in conformity with values and to adapt them to authentic people's needs (e.g., hospital spaces, drug packagings for elderly people, etc.).

Symphony: qualities and skills of perceiving the whole pictures, combining elements, perceiving them in reciprocal relations.

Play: importance of distancing oneself, a sense of humor as a tool for coping with stress, tension, pace of life, and work.

Meaning: a skill to perceive significance of matters, priorities, and things that really count. Pink states that searching for meaning and sense is one of the most important phenomena in the United States and European countries.

All these qualities constitute indispensible psychological equipment in the period when the world of Europe and the United States has to fight for survival, further development, and its own place. They should be used for creating innovations and convincing people how valuable they are.

Assistance and Hints for Managers and Management

According to Auhagen (2002) and on the basis of the interviews conducted with managers, it can be assumed that supporting, creating, and adopting innovations is successful provided that being people endowed with free will and acting in a responsible manner, we can accept one another. Creating and adopting innovations is successful if people can demonstrate their creativity and other innovative competences in a situation ensuring them a sense of security and a sense of their own effectiveness. Creating and adopting innovation is successful on condition that communication in the organization is overt, transparent, and based on exchange of information, while the control is not too restrictive. Creating and adopting innovation is successful if people or teams have a sense of responsibility for themselves, as well as for activities with regard to others. Creating and adopting innovation is successful providing workers perceive a positive sense of activity and accept clear motives and objectives (Kożusznik 2009).

It is an ethical aspect that is of so big importance in supporting innovativeness since it demands from the management a change in perception of a human being from as an individual who only aims at achieving material benefits and counts the costs of their activity by applying this criterion, to the direction of an image of a worker as a person being responsible and actively forming their life, deserving respect and subjective treatment.

REFERENCES

Aburdene, P. (2005). *Megatrends 2010: The Rise of Conscious Capitalism*. Charlottesville, VA: Hampton Roads.

Allport, G. W., and Ross, J. M. (1967). Personal religious orientation and prejudice. *Journal of Personality and Social Psychology*, *5*(4), 432–443.

Altshuller, G. S. (1984). *Creativity as an Exact Science*. Gordon and Breach Publishers Inc., Amsterdam, Netherlands.

Argyris, Ch. (1970). *Integrating Individuals and Organizations*. New York: Wiley and Sons.

Argyris, Ch. (1976). Leadership, learning and changing the status quo. *Organizational Dynamics*, *4*(3), 29–43.

Auhagen, A. E. (2002). *Psycho-soziale Faktoren von Innovationen Gruppendynamik und Organisationsberatung*, 33 Jahrg., 3/2002.

Bandsuch, M., and Cavanagh, G. (2005). Integrating spirituality into the workplace: Theory and practice. *Journal of Management, Spirituality and Religion*, *2*(2), 221–254.

Batson, C., Schoeanrade, P., and Ventis, L. W. (1983). *Religion and the Individual: A Social-Psychological Perspective*. New York: Oxford University Press.

Baumeister, R. F. (1991). *Meanings of Life*. New York: Guilford Press.

Biberman, J., and Whitty, M. D. (2007). *At Work: Spirituality Matters*. Chicago: University of Chicago Press.

Burack, E. (2000). Spirituality in the workplace. In: J. Biberman, and M. D. Whitty (Eds.), *Work and Spirit: A Reader of New Spiritual Paradigms for Organizations*. Scranton: University of Scranton Press, pp. 95–110.

Canton, J. (2006). *The Extreme Future: The Top Trends that will Reshape the World in the Next 5, 10, and 20 Years*. New York: Dutton.

Cochran, G., Hardy, J., and Harpending, H. (2006). Natural history of Ashkenazi intelligence. *Journal of Biosocial Science*, *5*(38), 659–693.

Cottrell, N. B. (1972). Social facilitation. In: C. G. McClintock (Ed.), *Experimental Social Psychology*. New York: Holt, Rinehart, and Winston, pp. 185–236.

Davenport, T. (2009). Process innovations: A "catholic" approach to process management. Retrieved August 20, 2011, from http://www.bptrends.com/publicationfiles/02-04%20COL%20Catholic%20-%20%20Davenport.pdf.

Davie, G. (2007). *The Sociology of Religion*. London: SAGE.

de Bono, E. D. (1973). *Lateral Thinking: Creativity Step by Step*. Harper & Row.

Denning, P. J. (2004). The social life of innovation. *Communication of the ACM*, *47*(4).

Dollinger, S. (2007). Creativity and conservatism. *Personality and Individual Differences*, *43*(5), 1025–1035.

Donahue, M. J., and Benson, P. L. (1995). Religion and the well-being of adolescents. *Journal of Social Issues*, *51*(2), 145–160.

Donahue, M. J., and Nielsen, M. E. (2005). Religion, attitudes and social behavior. In: R. F. Paloutzian and C. L. Park (Eds.), *Handbook of the Psychology of Religion and Spirituality*. New York: Guilford Press.

Drucker, P. F. (1992). *Innowacja i Przedsiębiorczość (Innovation and Entrepreneurship)*. Warszawa: PWE.

Drucker, P. F. (2002). The discipline of innovation. *Harvard Business Review*, *80*, 95–104.

Duffy, R. D. (2006). Spirituality, religion, and career development: Current status and future directions. *Career Development Quarterly*, *55*(1), 52–63.

Duriez, B., and Hutsebaut, D. (2001). Approaches to religion and the moods and emotions associated with religion. *Journal of Empirical Theology*, *14*(2), 75–84.

Durkheim, É. (2001). *The Elementary Forms of Religious Life*. Oxford, New York: Oxford University Press.

Fleissner, P. (2004). Can religious belief systems influence technological and social innovations? *International Journal of Information Ethics*, *2*(11), 1–11.

Francis, L. J., and Pocock, N. (2008). Personality and religious maturity. *Pastoral Psychology*, *57*(5–6), 235–242.

Frankl, V. (1984). *Man's Search for Meaning*. New York: Washington Square Press.

French, S., and Joseph, S. (1999). Religiosity and its association with happiness, purpose in life, and self-actualisation. *Mental Health, Religion and Culture*, *2*(2), 117–120.

Giacalone, R. A., and Jurkiewicz, C. L. (Eds.) (2010). *Handbook of Workplace Spirituality and Organizational Performance*. New York: M E Sharpe Inc.

Gibbons, P. (2000). Spirituality at work: Definitions, measures, assumptions, and validity claims. Paper presented at the Conference Academy of Management, Toronto.

Graham, J., and Haidt, J. (2010). Beyond beliefs: Religions bind individuals into moral communities. *Personality and Social Psychology Review*, *14*(1), 140–150.

Grayson, D., and Hodges, D. G. A. (2004). *Corporate Social Opportunity!: 7 Steps to Make Corporate Social Responsibility Work for Your Business*. Sheffield: Greenleaf Publishing.

Handy, C. (1999). *The Hungry Spirit: Beyond Capitalism: A Quest for Purpose in the Modern World*. New York: Broadway Books.

Harris, J. I., Schoneman, S., and Carrera, S. (2002). Approaches to religiosity related to anxiety among college students. *Mental Health, Religion and Culture*, *5*(3), 253–265.

Harrison, L., Kuran, T., and Woodberry, R. (2009). Religion and the Open Society Symposium: Session Three: Religion, Innovation, and Economic Progress. *Council on Foreign Relations*. Retrieved August 5, 2011, from http://www.cfr. org/publication/15887/religion_and_the_open_society_symposium.html.

Hills, P., and Francis, L. J. (2003). Discriminant validity of the francis scale of attitude towards christianity with respect to religious orientation. *Mental Health, Religion and Culture*, *6*(3), 277–282.

Huitt, W. (1999a). Success in the information age: A paradigm shift. Revision of background paper developed for workshop presentation at the Georgia Independent School Association, Atlanta, Georgia. Retrieved August 2005, from http://chiron.valdosta.edu/whuitt/col/context/infoage.html.

Huitt, W. (1999b). The SCANS report revisited. Paper delivered at the Fifth Annual Gulf South Business and Vocational Education Conference, Valdosta State University, Valdosta, GA, April 18, 1997. Retrieved March 2007, from http://chiron.valdosta.edu/whuitt/col/student/scanspap.html.

Iannaccone, L. R., and Klick, J. (2003). Spiritual capital: An introduction and literature review. Preliminary draft, prepared for the Spiritual Capital Planning Meeting, October 9–10, 2003, Cambridge.

Idler, E., Ellison, C., George, L., Krause, N., Ory, M., Pargament, K., Powell, L. et al. (2003). Measuring multiple dimensions of religion and spirituality for health research: Conceptual background and findings from the 1998 general social survey. *Research on Aging*, *25*, 327–365.

Inglehart, R. (2004). *Human Beliefs and Values: A Cross-Cultural Sourcebook Based on the 1999–2002 Values Surveys* (1st ed.). Mexico: Siglo XXI.

Katz, D., and Kahn, R. L. (1979). *The Social Psychology of Organizations*. New York: Wiley.

Kazama, S., Foster, J., and Hebl, M. (2002). Impacting climate for innovation: Can CEOs make a difference. Paper presented at the 17th Annual Conference of the Society for Industrial and Organizational Psychology, Toronto.
Kirkpatrick, L. A., and Hood, R. (1990). Intrinsic-extrinsic religious orientation: The boon or bane of contemporary psychology of religion? *Journal for the Scientific Study of Religion*, *29*(4), 442–462.
Knotts, T. L. (2003). Relation between employees' religiosity and job involvement. *Psychological Reports*, *93*(3 Pt 1), 867–875.
Koenig, H. (1997). *Is Religion Good for Your Health?: The Effects of Religion on Physical and Mental Health.* New York: Haworth Pastoral Press.
Kotlikoff, L., and Burns, S. (2004). *The Coming Generational Storm: What You Need to Know about America's Economic Future.* Cambridge, MA: The MIT Press.
Kożusznik, B. (2009). Psychological factors of innovativeness. In: L. Karamushka and B. Kozusznik (Eds.), *Work and Organizational Psychology Problems and Challenges in Poland and Ukraine.* Kiev: Kostiuk Institute of Psychology.
Kożusznik, B. (2011). Psychological aspects of innovative changes in organization and organizational development. In: L. Karamushka and B. Kożusznik (Eds.). Kiev: Kostiuk Institute of Psychology, p. 128.
Markman, A. L., and Wood, K. (2009). *Tools for Innovation.* Oxford University Press.
Maslow, A. (1971). *The Farther Reaches of Human Nature.* New York: The Viking Press.
Miller, V. J. (2005). *Consuming Religion: Christian Faith and Practice in a Consumer Culture.* New York, London: Continuum International Publishing Group.
Naisbitt, J. (1982). *Megatrends.* New York: Warner Books, Inc.
Osborn, A. F. (1959). *Applied Imagination.* New York: Scribner.
Oster, G. (2009). Christian Innovation, Descending Into the Abyss of Light. Dostęp: 10.09.2009. Tryb dostępu: [Regent Global Business Review:]. www.regent.edu/rgbr.
Palazzi, M., and Starcher, G. (1997). *Corporate Social Responsibility and Business Success.* Paris: European Bahá'í Business Forum.
Park, C. L. (2005). Religion as a meaning-making framework in coping with life stress. *Journal of Social Issues*, *61*(4), 707–729.
Pink, D. (2005). *A Whole New Mind: Moving from the Information Age to the Conceptual Age.* New York: Riverhead Hardcover.
Pink, D. (2009). Oprah talks to Daniel Pink. Interview with Oprah Winfrey. Retrieved July 30, 2009, from http://www.oprah.com/printarticlefull/omagazine/200812_omag_ocut_pink.
Polak, J. (2011). Spirituality at the workplace—Exploratory study. In: G. Rossi and M. Aletti (Eds.), *International Association for the Psychology of Religion. Congress 2011. 21st–25th August, Bari—Italy. Programme and Book of Abstracts.* Azzate/Varese: Arti Grafiche Tibiletti, pp. 19–20.
Pryor, R., and Bright, J. (2011). *The Chaos Theory of Careers: A New Perspective on Working in the Twenty-First Century.* New York: Taylor & Francis.
Robson, M. (1993). *Problem Solving in Groups.* Aldershot, UK: Gower Publishing Company Ltd.
Rogers, E. M. (1962). *Diffusion of Innovations.* New York: Free Press.
Roundy, P. (2009). Work and religion: Artificial dichotomy or competing interests? *International Journal of Human and Social Sciences*, *4*(5), 311–317.
Seligman, M. (2002). *Authentic Happiness.* New York: Free Press.
Selvarajan, T., Ramamoorthy, N., and Patrick, F. (2007). The role of human capital philosophy in promoting firm innovativeness and performance: Test of a causal model. *International Journal of Human Resource Management*, *18*(8), 1456–1470.
Shafranske, E., and Gorsuch, R. (1984). Factors associated with the perception of spirituality in psychotherapy. *Journal of Transpersonal Psychology*, *16*, 231–241.
Sikorska-Simmons, E. (2005). Religiosity and work-related attitudes among paraprofessional and professional staff in assisted living. *Journal of Religion Spirituality and Aging*, *18*(1), 65–82.
Sotiriou, D., and Wittmer, D. (2001). Influence methods of project managers: Perceptions of team members and project managers. *Project Management Journal*, *32*(3),12–21.
Toffler, A. (1981). *The Third Wave.* New York: Bantam Books.
Toffler, A. (1990). *Powershift.* New York: Bantam Books.
Vaughan, F. (1991). Spiritual issues in psychotherapy. *Journal of Transpersonal Psychology*, 23, 105–119.
Whitfield, P. R. (1975). *Creativity in Industry.* "Business and Management." London: Penguin Books.
Wu, S. (2009). A study of social influence in diffusion of innovation over Facebook. Pobrano 07 17, 2009 z lokalizacji Cornell University. Retrieved from http://www.cs.cornell.edu/~sw475/IS_breakfast_Dec_5.ppt.
Zaltman, G., Duncan, R., and Holbek, J. (1973). *Innovation and Organization.* New York: Wiley & Sons.

29 The Polish Adaptation of the Short Form of the Dutch Work Addiction Scale

Małgorzata W. Kożusznik, Anna Dyląg, and Magdalena Anna Jaworek*

CONTENTS

INTRODUCTION

Excessive and compulsive work fails to elicit the same judgment as other addictions. In fact, extended work involvement is often supported by our culture (Porter 2001), applauded by the society (Porter 1996), and valued positively by the employers.

However, workaholism seriously impairs the functioning at the workplace both at short as well as at the long term. On the one hand, one of the immediate negative outcomes of workaholism is its negative influence on personal relationships at work (Porter 1996) or work family conflict (Bakker et al. 2009). On the other hand, the long-term detrimental effects for an individual and organization (Porter 1996) are stress reactions, burnout, decrease in performance, increases in health- and accident-related expenses, and higher turnover rates (Maslach and Jackson 1981; Pines and Aronson 1988).

* The contribution of Małgorzata Kożusznik to this article was supported by a predoctoral scholarship V Sègles and a V Sègles grant to carry out a stay in the framework of the international mention of the doctoral degree, from the University of Valencia.

Workaholism

Workaholism is a disorder manifested in over-average energy exhaustion, perseveration in action, setting difficult standards, and disability to withdraw from task involvement (see Burke 2000a, b; Rohrlich 1981; Buczny and Wojdyło 2010). It means dedicating an excessive amount of time and energy to one's work while overlooking other areas of life (e.g., Buelens and Poelmans 2004; Mudrack and Naughton 2001). It means "compulsion or the uncontrollable need to work incessantly" (Oates 1971, p. 11). It is an "irresistible inner drive to work excessively hard" (Schaufeli et al. 2008a; Schaufeli et al. 2009, p. 251), which combines two dimensions: working excessively (WE) and working compulsively (WC) (Schaufeli et al. 2009). On the one hand, WE is a behavioral component of workaholism (Schaufeli et al. 2009). It refers to "time spent working or thinking about work" (McMillan and O'Driscoll 2006) or spending great deal of time on work when given discretion to do so (Scott et al. 1997). WC, on the other hand, is a cognitive component of workaholism (Schaufeli et al. 2009), which is connected to an "obsessive personal style" (McMillan and O'Driscoll 2006). It is manifested by reluctance to disengage from work and persistent thinking about work, even when not working (Scott et al. 1997).

Workaholism is often confused with work engagement. In fact, workaholism includes the aspect of work enjoyment and thus is closely related to engagement (Schaufeli et al. 2009; Beckers et al. 2004), a positive, fulfilling work-related state of mind that is characterized by vigor, dedication, and absorption (Schaufeli et al. 2002). However, it is crucial to note that these two phenomena should be differentiated as they are differently related with several indicators (i.e., excessive work, job demands, job resources, social relations, and health and organizational outcomes; Schaufeli, Taris, and Van Rhenen 2008). Following the suggestion of Schaufeli et al. (2009), we discriminate workaholism (being bad) and engagement (being good). In addition, low levels of health outcomes, particularly burnout, have been reported by workaholics (Andreassen et al. 2007; Schaufeli et al. 2008b).

Recently, conditions facilitating workaholism have appeared in Poland. In fact, the new forms of work (i.e., telework), the growing competition in increasing the quality of the products and the performance of the employees while decreasing the price of the products and services, as well as the increase in engagement at work of the employees constitute altogether a fertile ground for the development of workaholism. We need more knowledge about this phenomenon to cope with it. However, empirical studies on psychological and social factors determining workaholism are limited, which in part might be due to a lack of reliable tools available in Polish language version to measure this phenomenon.

The above limitations in the measurement of workaholism in Poland provided the initial impetus for the adaptation of the Dutch Work Addiction Scale (DUWAS; Schaufeli et al. 2009). Therefore, the objective of this study is to develop and validate the Polish version of the 10-item DUWAS (Schaufeli et al. 2009).

METHOD

Participants

Participants were postgraduate students at the Jagiellonian University in Kraków, Poland, who at the same time were employed. The sample consisted of 200 participants. The majority were women (60%). The participants ranged in age from 21 to 61, and the mean age was 32.98 (SD = 10.36). Participants were volunteers and received no credit for their participation in the study.

Measures

Dutch Work Addiction Scale

The DUWAS (Schaufeli et al. 2009) is a 10-item self-report questionnaire, which consists of two scales corresponding to two different dimensions of workaholism: WE (5 items) and WC (5 items). Instructions ask the subject to "decide how often you ever feel this way about your job." The 10 items are rated on a 4-point Likert scale from *(almost) never* to *(almost) always*. The Polish translation of the DUWAS was developed with a back-translation procedure by two independent translators. Afterward, the committee of three functionally bilingual translators, all content experts, compared the original version, the back-translated version, and the consensually derived target version and concluded that there are no significant differences between these three versions.

Maslach Burnout Inventory—General Survey

To measure burnout, a phenomenon related to workaholism (Andreassen et al. 2007; Schaufeli et al. 2008b), the Maslach Burnout Inventory—General Survey (Schaufeli et al. 1996), was used. The scale has 16 items with a response scale from 0 (never) to 6 (every day), and it reveals a good internal consistency (Cronbach's $\alpha = .86$). Burnout was considered as one factor, where the higher the score, the higher the level of burnout. This approach to burnout as one factor has already been used (Kozusznik et al. 2012).

Utrecht Work Engagement Scale

Work engagement, a phenomenon closely connected to workaholism (Schaufeli et al. 2009), was assessed with the "shorter version of the Utrecht Work Engagement Scale" (UWES-9), reduced by the authors (Schaufeli et al. 2006). The scale ranges from 0 (never) to 6 (every day). The measure applied to our population is characterized by satisfactory psychometric values (Cronbach's $\alpha = .90$). Work engagement was considered as one factor, where the higher the score, the higher the level of engagement.

Procedure

Participants completed the DUWAS-PL and other instruments in group sessions. The whole procedure lasted about 1 hour.

Data Analysis

Data analysis was structured in two steps. First, psychometric properties of the Polish DUWAS were investigated. Means, standard deviations, and reliability coefficients were calculated, and confirmatory factor analysis was conducted using LISREL 8.8 (Jöreskog and Sörbom 2006). Second, we run correlation analyses in which we tested relationships of the WE and WC scales with burnout and engagement.

RESULTS

Psychometric Properties and Factorial Structure

Cronbach's α reliability coefficients were .69 for WE and .74 for WC. Item analysis confirmed internal consistency for both scales: item–total correlations ranged from .36 to .49 for WE and from .39 to .59 for WC (Table 29.1). WE and WC were significantly correlated ($r = .62$, $p < .01$).

A *t*-test was performed to examine gender differences. However, there was no statistically significant difference found for WE scale [$t(198) = .31$, $p = ns$]. Overall means were 2.25 (SD = .60) for

TABLE 29.1
Item–Total Correlation and Confirmatory Factor Loadings for All Items

Item #	English Item Original and Polish Translation	Item–Total r	Standard Factor Loadings CFA
	Working Excessively Items		
1	I seem to be in a hurry and racing against the clock. Wydaję mi się, że ciągle pracuję w pośpiechu i ścigam się z czasem.	.49	.66
2	I find myself continuing to work after my co-workers have called it quits. Pracuję nawet wtedy, gdy pozostali współpracownicy wyszli już pracy.	.40	.62
4	I stay busy and keep many irons in the fire. Pomimo natłoku zajęć „łapię dziesięć srok za ogon".	.48	.61
6	I spend more time working than on socializing with friends, on hobbies, or on leisure activities. O wiele więcej czasu poświęcam pracy niż zabawie z przyjaciółmi, hobby czy wypoczynkowi.	.46	.66
8	I find myself doing two or three things at one time such as eating lunch and writing a memo, while taking on the telephone. Łapię się na tym, że zajmuję się kilkoma rzeczami naraz (np. jem, piszę i rozmawiam przez telefon).	.36	.46
	Working Compulsively Items		
3	It's important to me to work hard even when I don't enjoy what I'm doing. Niezależnie od tego, czy praca sprawia mi przyjemność, czy też nie – lubię ciężko pracować.	.39	.52
5	I feel that there's something inside me that drives me to work hard. Czuję, że coś mnie zmusza do wykonywania ciężkiej pracy.	.46	.65
7	I feel obliged to work hard, even when it's not enjoyable. Czuję się zobligowany do ciężkiej pracy, nawet jeśli nie znajduję w tym przyjemności.	.58	.71
9	I feel guilty when I take time off work. Czuję się winny(a) kiedy biorę wolne od pracy.	.51	.69
10	It is hard for me to relax when I'm not working. Trudno jest mi się zrelaksować, gdy nie pracuję.	.59	.75

Source: Items © by Schaufeli W. et al., *International Journal of Stress Management, 16*(4), 249–272, 2009.
Note: $N = 200$.

women and 2.23 (SD = .61) for men. No significant difference was found for the WC scale [$t(198) = .29$, $p = ns$]. Overall means were 1.94 (SD = .61) for women and 1.92 (SD = .66) for men. Next, in order to assess whether there is a relationship between the age and the score on the two scales of the DUWAS, we calculated the correlations between the age and the WE and WC scales. The results show no significant relationships between the age and the two scales ($r = .03$, $p = .70$ for WE and $r = .13$, $p = .07$ for WC).

In order to assess whether the two-factor structure of the model fitted the Polish translation, we run a confirmatory factor analysis. Maximum likelihood method (ML) was used to estimate the model parameters, and Pearson correlation matrix was used as input for the analysis. The χ^2/df ratio was below 2.0, the nonnormed fit index (NNFI) was above .95, and the comparative fit index (CFI) was above .95. The standardized root mean square residual (SRMR) as well as the root mean square of approximation were both below .08 (see Table 29.2). The standardized regression weights (factor loadings) ranged from .46 to .75.

TABLE 29.2
Fit Indices for Confirmatory Factor Analysis

χ^2	Df	SRMR	NNFI	CFI	RMSEA
67.38	34	.07	.96	.97	.07

Note: $N = 200$.

TABLE 29.3
Relation to Other Constructs

	Working Excessively	Working Compulsively
Burnout	.12	.20**
Engagement	.15*	.15*

Note: $N = 200$.
* $p < .01$.
** $p < .05$.

Correlations with Other Constructs

The results of the correlation between WE scale, WC scale, burnout, and engagement are shown in Table 29.3. As expected, WC was significantly positively correlated with burnout and work engagement. WE, by contrast, was positively correlated only with work engagement.

DISCUSSION

Our findings confirm the reliability, factor structure, and validity of the Polish adaptation of the DUWAS (Schaufeli et al. 2009). With respect to reliability, internal consistency coefficients of both WE and WC scales were comparable to those obtained by its authors using the original version. The predicted two-factor model fit well across all fit indices. Additional evidence of the validity of the Polish version of the DUWAS comes from the pattern of relations with burnout and engagement measures.

WC was positively and significantly related to burnout and engagement. It means that the person who works in a compulsive way is more likely to be more engaged in their work, but at the same time, they are more likely to suffer burnout. These results go in line with previous research, which showed the relationships between workaholism, burnout, and engagement (e.g., Andreassen et al. 2007; Schaufeli et al. 2008).

WE was positively and significantly related to engagement. This means that individuals who commonly work in an excessive way are more likely to be engaged at work. The lack of the relationships between this dimension of workaholism and burnout may suggest that WE is not enough to provoke burnout. The results support the findings that engagement is associated with overwork (Beckers et al. 2004).

This study provides evidence that the Polish version of the DUWAS is a reliable and valid self-report tool for assessing workaholism. The results indicate equivalence between the original and the adapted instrument; the two-factor structure that underlies the original DUWAS version was replicated. There were no gender differences found in case of the two scales. External correlates were also confirmed: associations were found between WE, WC, burnout, and engagement.

LIMITATIONS AND FUTURE DIRECTIONS

The results of this study require cautious interpretation due to a notable limitation. The study used a sample of postgraduate students, the majority of which being employed at the same time. Future studies will need to test the generalizability of these findings using samples from different occupations, preferably those that are most exposed to workaholism.

The study provides some practical implications and offers the Polish adaptation of DUWAS as a reliable and valid measure of work addiction. Hopefully, it will help to broaden the knowledge about workaholism in order to cope with this phenomenon.

REFERENCES

Andreassen, C. S., Ursin, H., and Eriksen, H. R. (2007). The relationship between strong motivation to work, "workaholism," and health. *Psychology and Health, 22*, 625–629.

Bakker, A. B., Demerouti, E., and Burke, R. (2009). Workaholism and relationship quality: A spillover-crossover perspective. *Journal of Occupational Health Psychology, 14*, 23–33.

Beckers, D. G. J., Van der Linden, D., Smulders, P. G. W., Kompier, M. A. J., Van Veldhoven, J. P. M., and Van Yperen, N. W. (2004). Working overtime hours: Relations with fatigue, work motivation, and the quality of work. *Journal of Occupational and Environmental Medicine, 46*, 1282–1289.

Buczny, J., and Wojdyło, K. (2010). Impulsive and reflective mechanisms of workaholism. Implicit and explicit measures in prediction of behavioral outcomes. *2nd Early Career Summer School for Advanced Work and Organizational Psychology*, Valencia, Spain.

Buelens, M., and Poelmans, S. A. Y. (2004). Enriching the Spence Robbins' typology of workaholism. *Journal of Organizational Change Management, 17*(5), 440–458.

Burke, R. J. (2000a). Workaholism and extra-work satisfaction. *The International Journal of Organizational Analysis, 7*, 352–364.

Burke, R. J. (2000b). Workaholism in organizations: The role of personal beliefs and fears. *Anxiety, Stress and Coping, 13*, 53–64.

Jöreskog, K. G., and Sörbom, D. (2006). *LISREL 8.80 for Windows* [*Computer Software*]. Lincolnwood, IL: Scientific Software International, Inc.

Kozusznik, M., Rodríguez, I., and Peiró, J. M. (2012). Cross-national outcomes of stress appraisal. *Cross Cultural Management, 19*(4), 507–525.

Maslach, C., and Jackson, S. E. (1981). *Maslach Burnout Inventory Manual* (2nd ed.). Palo Alto, CA: Consulting Psychologists Press.

McMillan, L. H. W., and O'Driscoll, M. P. (2006). Exploring new frontiers to generate an integrated definition of workaholism. In R. J. Burke (Ed.), *Research Companion to Working Time and Work Addiction*. Cheltenham: Edward Elgar, pp. 89–107.

Mudrack, P. E., and Naughton, T. J. (2001). The assessment of workaholism as behavioral tendencies: Scale development and preliminary empirical testing. *International Journal of Stress Management, 8*, 93–111.

Oates, W. (1971). *Confessions of a Workaholic: The Facts about Work Addiction*. New York: World Publishing.

Pines, A., and Aronson, E. (1988). *Career Burnout: Causes and Cures*. New York: Macmillan.

Porter, G. (1996). Organizational impact of workaholism: Suggestions for researching the negative outcomes of excessive work. *Journal of Occupational Health Psychology, 1*(1), 70–84.

Porter, G. (2001). Workaholic tendencies and the high potential for stress among co-workers. *International Journal of Stress Management, 8*(2), 147–164.

Rohrlich, J. B. (1981). The dynamics of work addiction. *Israel Journal of Psychiatry and Related Sciences, 18*, 147–156.

Schaufeli, W., Bakker, A., Van Der Heijden, F., and Prins, J. (2009). Workaholism among medical residents: It is the combination of working excessively and compulsively that counts. *International Journal of Stress Management, 16*(4), 249–272.

Schaufeli, W. B., Arnold, B., Bakker, A. B., and Salanova, M. (2006). The measurement of work engagement with a short questionnaire: A cross-national study. *Educational and Psychological Measurement, 66*(4), 701–716.

Schaufeli, W. B., Leiter, M. P., Maslach, C., and Jackson, S. E. (1996). Maslach burnout inventory-general survey (MBI-GS). In C. Maslach, S. E. Jackson, and M. P. Leiter (Eds.), *Maslach Burnout Inventory Manual*, 3rd ed., Palo Alto, CA: Consulting Psychologists Press.

Schaufeli, W. B., Salanova, M. Gonzalez Roma, V., and Bakker, A. B. (2002). The measurement of engagement and burnout: A two sample confirmatory factor analytic approach. *Journal of Happiness Studies, 3*, 71–92.

Schaufeli, W. B., Taris, T. W., and Bakker, A. B. (2008a). It takes two to tango: Workaholism is working excessively and working compulsively. In R. J. Burke, and C. L. Cooper (Eds.), *The Long Work Hours Culture: Causes, Consequences and Choices*. Bingly: Emerald.

Schaufeli, W. B., Taris, T. W., and Van Rhenen, W. (2008b). Workaholism, burnout and engagement: Three of a kind or three different kinds of employee well-being? *Applied Psychology: An International Review, 57*, 173–203.

Scott, K. S., Moore, K. S., and Miceli, M. P. (1997). An exploration of the meaning and consequences of workaholism. *Human Relations, 50*, 287–314.

30 Why Some Arguments Are More Compelling than Others

The Answers from Cognitive and Social Psychology

Robert Mackiewicz and Paweł Koniak

CONTENTS

INTRODUCTION

Both in formal and informal settings, we use different arguments intended to persuade others to believe in what we believe or to choose a particular course of action. Such efforts of persuasion can be defined as the formation or change of attitudes in response to a message about the object of this message. One of the key points in the debate about persuasion effectiveness has been to distinguish the roles of two factors: the content of the message and external or peripheral cues that can influence this message (Bohner et al. 1995). This debate can be also formulated by the use of linguistic distinction between semantics, that is, the meaning of the message, and pragmatics, that is, the relation between the message and those who "send" and "receive" it. Therefore, each attempt at persuasion depends on decoding the message and establishing what it means and making inferences about the intention, credibility, and truthfulness of the source. In other words, someone who is being persuaded has to make inferences about what is relevant in the content and context of the message (Sperber and Wilson 1995).

The aim of this chapter is to present the empirical research from two domains—cognitive and social psychology—that are intended to show how the meaning of the message can influence the interpretation of what is pragmatically relevant and how the contextual features of an utterance influence the way its meaning is understood. So far, researchers from both domains have independently studied what makes an argument valid or persuading. The cognitive approach mainly concentrates on the logical validity of arguments. As Evans (2002) points out, the researchers typically adhere to the so called "deduction paradigm," and they present the participants of psychological studies with sets of inferences and either ask them to evaluate given conclusions or to produce their own ones. In a typical study from social psychology, participants learn information about an issue and then report their attitudes toward that issue. Social psychologists therefore study what is the persuasive value of strong and weak arguments, and how they are modified by the knowledge of the source and participants' motivation to process provided information. Many theorists from both areas share the same conviction that messages can be processed by human cognitive systems in two different ways: heuristically and analytically (e.g., Sloman 1996 in cognitive psychology and Chen and Chaiken 1999 in social psychology). Heuristic processing is automatic and rapid and is similar to processing information in perception: one does not have to think if one perceives something or not. But analytic processes monitor the quality of answers given by the heuristic system and endorse them, correct, or modify (Evans 2003). For example, if you hear a politician saying that "current problems of national health care have to be resolved," you may spontaneously agree with him (heuristic processing), but after a while, you might realize that his message is ambiguous, as he does not give any clues on how he would resolve problems with the health care system (analytic processing). Therefore, the first outcome of heuristic processing is establishing the truth value of the persuasion message. As Gilbert (1991) asserts, comprehending a message entails accepting it, at least in the first stage of its processing. Gilbert draws an analogy with the perceptual system. If you see something, you just know it exists, and if you hear a proposition, you just think it is true. Any doubt comes later and is typically associated with finding counterexamples to the content of the message. Looking for counterarguments is a core faculty of the human mind (Johnson-Laird 2007), and it allows humans to go "beyond information given" and see what may happen if the propositions one started to believe in are false. We suppose that the search for counterexamples and the inspection of false possibilities are the outcomes of analytic processing. We assume that in persuasion, analytic processing is associated with different pragmatic factors that broaden the semantic meaning of the message.

There are many pragmatic cues that can influence analytic processing. One of the most important is the knowledge about the message source (Petty and Wegener 1998). A message coming from a credible source has more persuading influence mainly due to his or her expertise (Bohner et al. 2002) and truthfulness (Eisend 2006). Social psychologists have also studied such pragmatic factors as giving arguments for or against a certain issue (Bohner et al. 2003), the number of arguments (Petty and Cacioppo 1984), and the order in which they are presented (Haugtvedt and Wegener 1994). Pragmatic factors can change the meaning, that is, the semantics, of the message. For example, if you hear something from a politician whom you do not like, you may wonder why he is saying this and you may start to generate counterarguments. Pragmatic factors therefore can help in differentiating between strong and weak arguments (Priester and Petty 1995), looking for false possibilities (Gershoff et al. 2007), or activating knowledge about hidden motivations of the message source (Eagly et al. 1978).

At the heuristic stage, the understanding–accepting equality is a semantic process of assigning propositions with truth-values. When an individual hears a certain statement, he instantly creates a "possible world" in which this statement is true. According to Johnson-Laird's (1983) theory of mental models, possible worlds are mental models that are constructed in the human mind. Such mental models are iconic insofar as they can be and they correspond to the structure of what they represent. Mental models can represent relations and the relations between relations (Goodwin and Johnson-Laird 2005). So, for example, if you hear that Polish government's expenditure on health care has risen recently, you understand this by constructing the mental model that iconically represents the previous and current sum of money spent on public health care. Such a representation can be graphically presented in the way presented in Figure 30.1.

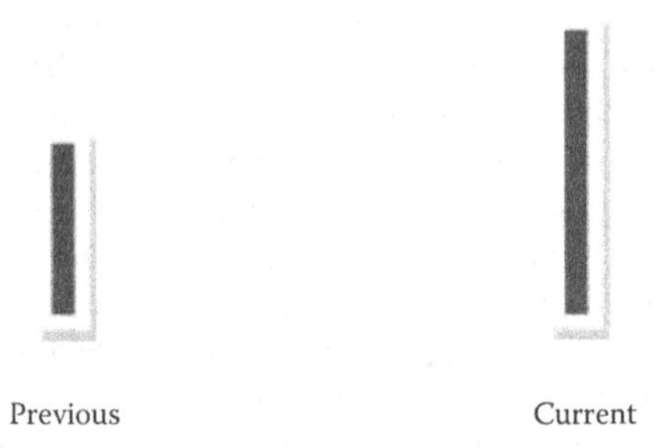

FIGURE 30.1 Mental model for the representation of the lower–higher relation.

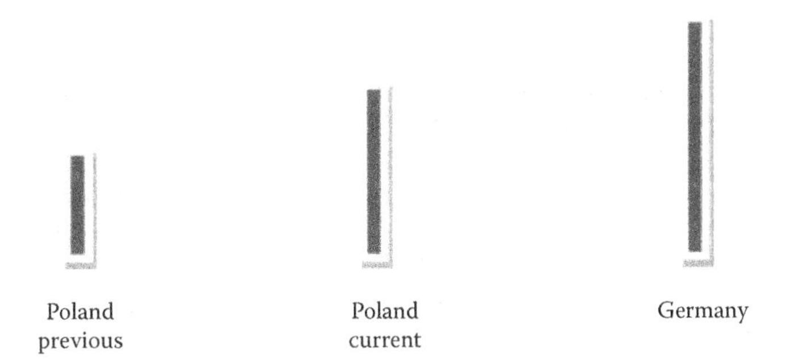

FIGURE 30.2 Mental model representing linear relation between three entities.

Of course, the human mind does not create vertical bars; rather they are used in Figure 30.1 in order to point out that the mind represents only relevant aspects of the message. Such initial mental models can be updated with new information, for example, "Current expenditure is still smaller than in Germany." Figure 30.2 provides an example of the mental model for the linear representation of such a relation.

Mental models are based on the so-called principle of truth (Johnson-Laird 2007), that is, they represent what is true and not what is false. So, for example, the message "Either the average retirement age in Europe is less than 64 or else it is less than 73" would be represented by mental equivalents of those two propositions:

Retirement age is less than 64
Retirement age is less than 73

Separate lines indicate that both propositions are different and mutually exclusive. Mental models are heuristically asserted as true possibilities, and without deeper consideration, one might think that both are equally probable. Suppose that you heard them from two politicians. At the first stage of processing, you may think that both can be right and you would have problems with choosing the one who is actually speaking the truth. But only at the analytic stage would you come to the conclusion that if the retirement age is less than 64, then it is also less than 73 and both propositions are true. As the propositions are presented within an exclusive disjunction, they cannot both be true at the same time, so one of them has to be false. Deeper inspection of false possibilities reveals that both messages are compatible with only one situation:

Retirement age is more than 64
Retirement age is less than 73

That means that only after the stage of analytic processing and taking into account false possibilities can one construct the mental model that truly represents the meaning of propositions. We hypothesize, however, that individuals normally do not think of what might be the case if an argument is false. If there are no pragmatic clues, they treat different arguments as equally probable possibilities. We tested this prediction in our first experiment.

Experiment 30.1

Twenty seven undergraduate students took part in this study (mean age: 20.26). They were divided in two groups, and half of them received a booklet with a number of general knowledge statements that were presented as messages coming from two different but unknown individuals. Here are typical examples of those statements:

An average number of letters in an email is around 139, or else an average number of letters in email is around 159.
The maximum height of trees in Poland is higher than 19 meters, or else the maximum height of trees in Poland is higher than 32 meters.

Both possibilities in the first sentences can be true, while the correct answer in the second sentence needs to take into account that one of the possibilities must be false, so the correct answer must lie between both numbers. Each participant was presented with six sentences of the first type and six of the second one. The participants from the second group were also presented with 12 pairs of statements, and in half of them, both possibilities could be true and in the other half the correct answer required thinking of what is false. All statements were described as "sentences coming from different politicians currently active in political life." Here are two examples of such statements:

The current rate of unemployment in Poland is around 8%, or else the current rate of unemployment in Poland is around 14%.
The average holiday time in EU is longer than 19 days, or else the average holiday time in EU is longer than 32 days.

Students who participated in the study wrote their answers in the space provided below each pair of sentences.

There were not many pragmatic clues both in abstract and political settings, so we predicted that the participants would quite often treat both options as equally probable. Indeed, on average, they just repeated one of the options as their own answer in 70% of cases. However, the knowledge that an utterance comes from a politician served as a pragmatic cue for some of the individuals, and those who received booklets with political statements less often choose one of the options (55%) in comparison to those who answered abstract questions (86%). Although the difference is statistically significant [$F(1,25) = 6.87$, $p < .05$], it is worth noting that the percentages of those who thought that both pieces of information are equally believable were relatively high in both groups. As in half of the trials both options were equally possible, choosing any of them should be treated as the possibly correct answers. But, in the other half, only the answers that used numbers between those provided could be treated as correct answers. The average percentages of such answers are presented in Figure 30.3.

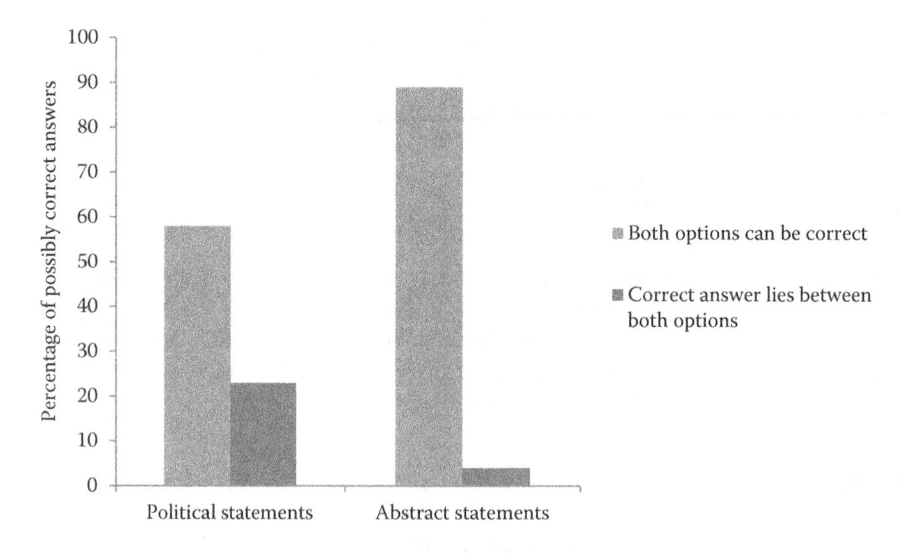

FIGURE 30.3 Average percentages of possibly correct answers for abstract and political statements in Experiment 30.1.

Only in 13% of the cases the participants gave the answer that could be correct, when it required thinking of what is false, but choosing one option in the other half of trials resulted in 73% of correct answers [$F(1,25) = 54.01$, $p < .001$]. As there were not many of those who gave correct answers in the former trials, the difference was not significant between participants from the "abstract" and "political" groups. But when both statements could be correct, the participants who received political statements less often gave the correct answers than those who received abstract statements [$F(1,25) = 8,92$, $p < .003$].

The results of Experiment 30.1 proved the prediction that statements that we hear without a broader context are heuristically tagged with "true" value. Indeed, 70% participants chose one of the pieces of information provided as the true answer. However, this "believability heuristics" gets weaker when it comes to the world of politics. The participants who knew that the statements they had read come from politicians less frequently chose them as correct answers. However, as they did not think of false possibilities, they less often gave answers that could be logically correct.

Experiment 30.2

The aim of the next experiment was to check how the semantic content of the message can influence its pragmatic value. In this study, we presented individuals with three types of political statements. Some of them were positive, for example:

> If party X wins the next local election, they will spend more money on local roads.

and some negative, for example:

> If party X wins the next local election, they will spend less money on local roads.

We also used an ambiguous version of each message, for example:

> If party X wins the next local election, they will concentrate on traffic problems.

We hypothesized that the semantically ambiguous message will have the highest pragmatic value, as it is difficult to form counterarguments to ambiguous statements. Positive messages can induce some doubts as they may trigger some caution among the voters. Assessing the possibility of negative outcomes should, however, depend on their content and political views of the voters.

In this experiment, we tested 74 participants from the general public (aged from 19 to 29, mean = 25). We conducted the study during the local election in Gdańsk in 2010. We asked the participants to assign the probabilities (on the scale ranging from 0 to 100) to different outcomes that may happen, if different political parties win the election. The participants were divided into three groups, and they received four possible outcomes described in a positive, negative, or ambiguous way. In the "positive" group, we asked to assess the probability of (1) raising the punishment for corruption, (2) extending the scope of free health service, (3) raising the teachers' wages, and (4) spending more money on local roads. The same four outcomes were presented ambiguously in the second group and were changed to negative ones (as in the example above) in the third group.

Although the estimations of probabilities were different for the four main political parties, we present here only the results for Platforma Obywatelska (Civic Platform)—the party that won the local election in Gdańsk and had the biggest number of supporters in our research group. Average estimations of probabilities for each type of outcome are given in Figure 30.4.

As predicted, the participants assigned the highest probabilities to outcomes presented in an ambiguous way (mean 45.3), and those estimates were significantly higher than estimates of

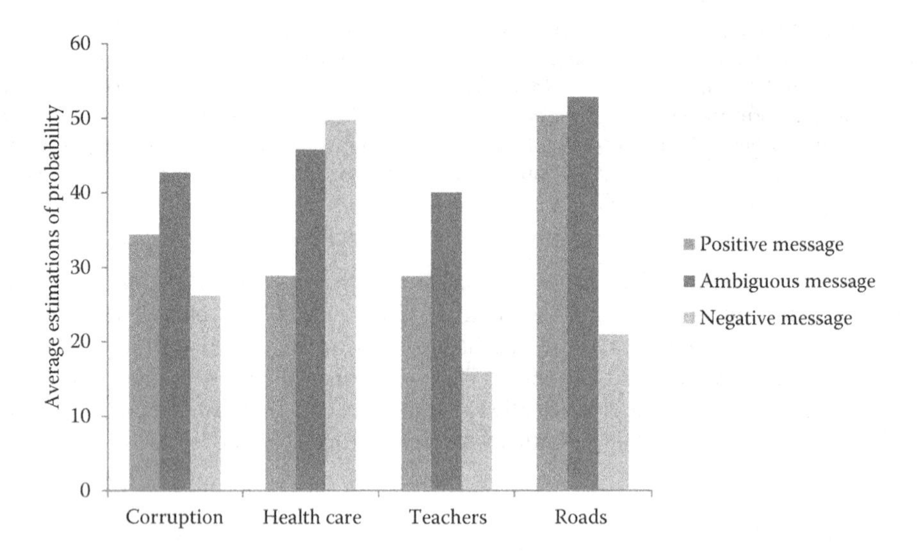

FIGURE 30.4 Average estimations of probabilities of different outcomes depending on whether the message is ambiguous, positive, or negative in Experiment 30.2.

positive outcomes (mean 35.6) and negative outcomes [mean 28.2; $F(2,71) = 3.87$, $p = .03$]. As the inspection of means from Figure 30.4 reveals, in one case (health care), individuals assigned equal probabilities to outcomes presented in ambiguous and negative ways, and in another (public roads), the probabilities of ambiguous and positive outcomes were similar. This interaction can be accounted for by the background knowledge of the participants. At the time of the election, many politicians from other parties accused Civic Platform that this party's public health reform would result in a larger number of commercial hospitals. So probably some voters thought that if Civic Platforms won the election, they would privatize some sectors of public health care, and that is why they assigned quite high probabilities to this outcome. Gdańsk is one of the cities that hosted UEFA Euro 2012 Championship; therefore, at the time of the election, there was a lot of construction work in the city. That is why, as we assume, the participants from the "positive" group assigned the highest probability to the possibility that there would be many new roads built in future. In case of other two issues (corruption and teachers' wages), ambiguous statements got bigger probability assignments than positive and negative.

Generally, the participants were not very much convinced that Civic Platform will tackle any of the presented issues (mean estimate 36.4). However, estimates for this party were still significantly higher than for other three (probability estimates ranged from 24 to 27). The messages coined in ambiguous terms got the highest probability estimates, and this means, as we hypothesized, that semantically ambiguous messages are mentally tagged as "possibly true" and they trigger less counterexamples and less critical thinking than positive or negative messages.

Experiment 30.3

The aim of the next two experiments was to investigate how more complicated persuasive messages are processed analytically. We assume that in order to find out what is the motivation of the speaker and if his or her arguments are valid, one has to pay more effort in the processes of comprehension. Hence, we predicted that such efforts would be present mainly at the analytic stage, and pragmatic aspects of the message could also influence its semantic understanding (Wyer and Gruenfeld 1995). Individuals may be able to pay more attention to the message content when they can compare it with the credibility of the source and potential gains and losses that he or she might attain (Priester and Petty 1995). Therefore, in Experiment 30.3, we presented information about the credibility of the source and his personal view on the issue in question. In this experiment, the speaker presented equally strong arguments for and against equaling the numbers of male and female Members of Polish Parliament (the issue seemed fictitious at the time of the study). We predicted that the participants would be more likely to agree with the speaker, if he is credible, as his arguments are valid and generate fewer searches for counterarguments. We also predicted that the force of negative arguments would be stronger when the speaker is personally against a particular issue. Generally, in judgment and choice, negative personal experiences are weighed more heavily than positive ones (Folkes and Kamins 1999). In everyday social interaction, people often inform us what they like and why. We often treat that as a conventional way of describing experiences. On the other hand, if someone expresses his or her negative feelings, he or she usually does it for a reason (Gershoff et al. 2007). Therefore, we assumed that "being against" would activate more pragmatic search for the speaker's motivation and result in a bigger change in the readers' attitudes to the parity issue.

The detailed results of this experiment and Experiment 30.4 have been published elsewhere in Polish (Koniak and Cwalina 2008), so we only here summarize the main points. One hundred four undergraduates took part in the study, and their mean age was 20.1 years. Half of the participants were informed that Andrzej Nowak—a fictitious MP member who presented the issue—personally supports the view that female MP's should have parity with their male colleagues, and the other half were informed that he is against it. Each group was divided into two halves: one half was informed that he was not very sincere in his previous political activities and the other half read two examples showing his honesty in political life. We measured the participants' attitudes to the parity issue twice: before and after the presentation of the speaker's point of view. In calculating the results, we measured the difference between the second and the first score. Positive numbers mean that participants expressed more support for the stand taken by the speaker. They were more for parity after reading his arguments than at the beginning of the experiment if the speaker was also for, and were more against the parity if the speaker was also against. A shift of attitudes against a speaker's point of view was coded as a negative number. The averages of such shifts of attitudes are presented in Figure 30.5.

The participants were quite reluctant to change their minds after hearing the arguments for and against parity in parliament: most of the scores were around zero. But there was one exception: they agreed more eagerly with someone who was presented as truthful and was against the issue [$t(26) = 3.22$, $p = .003$]. Therefore, personal views of the sender are more convincing because his or her personal point of view provides a very strong counterexample to the message's topic. If he or she is for a certain issue, his or her personal views only back up initial heuristic interpretation of

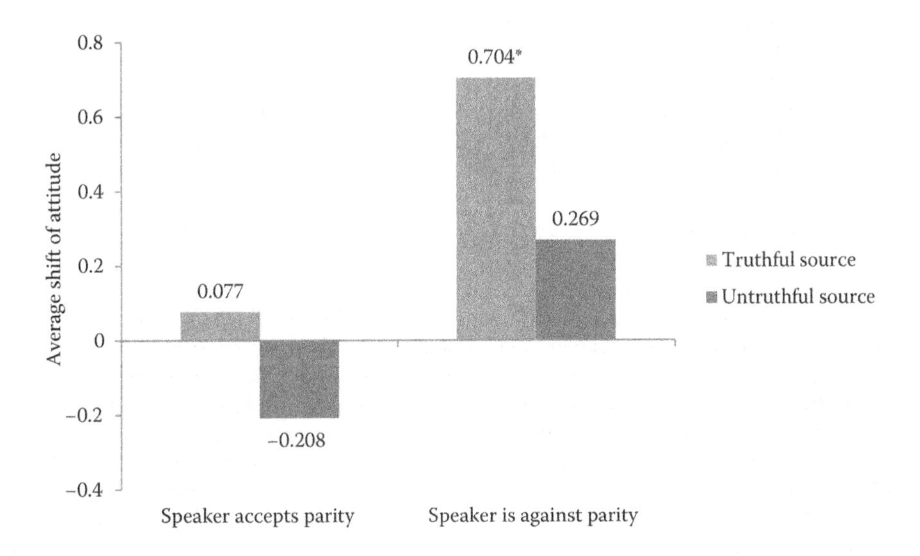

FIGURE 30.5 Average shift of attitudes in the same direction as the speaker's point of view or in the opposite direction in Experiment 30.3. (* – significant at $p < .05$.)

positive arguments. But by the virtue of the principal of truth, such arguments would be regarded as positive, and the positive stand of the sender could not change the semantic value of arguments. Someone who is for an issue just does not trigger pragmatic analysis: Individuals concentrate only on arguments and do not ask why the sender presents both sided argumentation. On the contrary, when someone is against an issue and presents also positive arguments, the receivers of the message tend to ask why he or she presents arguments contrary to his or her own stand. If he or she is trustworthy, his or her audience might think that his or her presentation reflects different points. However, a speaker's negative attitude triggers more pragmatic thinking about his or her motivation.

Experiment 30.4

The results of the previous experiment showed that the personal stand of the speaker matters, but only when he or she is against something. Being positive about an issue is treated as a more neutral or conventional way of expressing one's own opinions, and so, regardless whether you are credible or not, your personal view does not matter much. But if you are against something, the listeners will take that into account, only if you are credible. On the level of analytic processing, your negative stand may be treated as a very good counterexample to all positive arguments that you have decided to argue in favor of. Is that really so in all circumstances? Would it be enough to be trustworthy and against something to convince people to change their attitudes? We devised Experiment 30.4 in order to give answers to those questions. One hundred twenty two participants from the same population as in Experiment 30.3 took part in the study. In this experiment, half of the participants read only positive and half only negative arguments for eliminating the concept of "No-fault divorce" from the Polish legal system. Half of the participants also read that the message sender is a professor of law school at a major university, and the other half were informed that the arguments come from a person who works in a car repairing garage. As in the previous study, only the arguments from the competent source mattered, so we concentrated only on the results from the participants who were reading the statements provided by the competent expert in the field.

There were two more variables that we used to divide the participants into subgroups. Some of them were informed that the speaker is personally for the elimination of the no-fault divorce, and he or she presented either strong or weak arguments for the elimination, for example:

> It is estimated that about 60% (in the strong arguments group)/2% (in the week arguments group) of no-fault divorces are fictitious and the divorcees aim to get extra welfare support for their families.

The other two groups were also presented with strong and weak arguments, but they were all against the elimination of no-fault divorce and this was also the personal stand of the speaker. Here are typical arguments of both types:

> The elimination of no-fault divorces resulted in a rise of 30% (strong version)/1% (week version) of domestic violence.

We started the experiment with measuring the initial opinions on the issue that were quite varied: the mean was around the middle of the five-point scale. Then, we checked if the participants understood the experimental statements in the same way as we did, so we asked them to evaluate the trustworthiness of the speaker and the quality of the arguments. As the manipulation checks yielded positive results, we asked the participants to express their own attitudes again, taking into consideration the arguments that they had just read. Similarly to the previous study, we calculated the attitude shift score by subtracting the numerical estimation from the first measure from the estimation of the second one. Positive numbers meant that someone changed their view in accordance with the speaker, and negative meant that whatever his or her point was before now he or she is against the speaker. The average numbers of attitude change in different groups presented with arguments from a competent speaker are given in Figure 30.6.

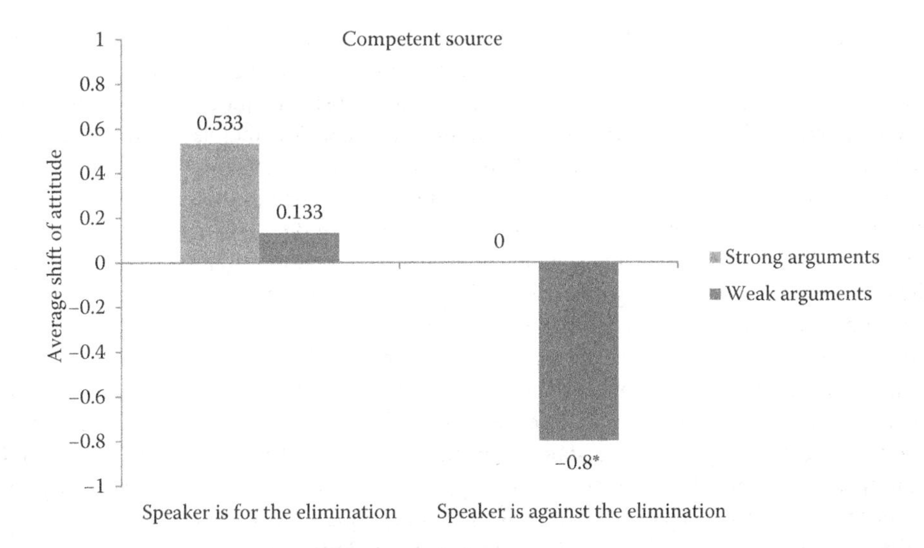

FIGURE 30.6 Average shift of attitudes in the same direction as the speaker's point of view or in the opposite direction depending on the speaker's own point of view and the quality of his or her arguments in Experiment 30.4.

Again, the message got pragmatic interpretation and influenced the receivers' attitudes only in the group presented with arguments against the issue. But this time, the only statistically significant shift could be observed among students who were presented with weak arguments [$t(14) = 2.26$, $p = .04$]. This result can also be explained by pragmatic factors operating at the analytic stage of message interpretation. If a competent speaker presents weak arguments for his or her thesis, it is hard to believe that he or she adheres to what he or she says. Individuals could be more likely to look for counterexamples to the speaker's point of view and easily access better arguments from their long-term memories. Hearing weak arguments from a competent person may also trigger some doubts about his or her motivation. After all, if he or she believed in his or her contentions, he or she would give much better arguments.

CONCLUSIONS

Many cognitive and social psychologists agree that humans can process information in two ways: heuristically and analytically. There is some discussion whether both systems work in parallel or whether the first stage is usually heuristic and the analytic system can only change the outcome of heuristic processing (Evans 2008). We argue that at the first stage of understanding, the persuasive message, the heuristic processes come into play. They are automatic and rapid and require low effort. Heuristic processing results in understanding what the message is all about, that is, in understanding its semantics. The fact that semantic analysis is a part of the heuristic system can be also backed up by the studies of automatic semantic activation of concept knowledge. Following Allport (1977), we argue that understanding meaning is a rapid and automatic process. We suppose also that understating at the heuristic stage equates with believing or accepting the message (Gilbert 1991). Contrary to some researchers from social psychology (cf. Ranganth et al. 2010), we believe that pragmatic factors operate in the stage of analytic processing. It is at this stage that we can start to doubt what we have just heard. Analytic processing is controlled and slow and requires cognitive effort, and the same features are associated with understanding the message in particular context. Only at the analytic stage can individuals ask the questions "Why is the speaker saying this?", "Is he trustworthy?", and "What could have motivated him to say so?". Hence, analytic processing results in looking for false possibilities to provided information.

We reported four experiments that were designed to test our predictions. In the first experiment, we presented the participants with pairs of different opinions on general knowledge questions and different political issues. Each pair was presented as an exclusive disjunction and so the only one could be true. We predicted, following the mental models theory's principle of truth, that individuals would understand all statements as equally probable and would just choose one of the options as their answers. Indeed, the vast majority of participants repeated one of the statements as their own answers and that led to erroneous conclusions when the truth of one sentence excluded the truth of the other one. The second experiment intended to check what is the pragmatic status of political promises presented in positive, negative, or ambiguous ways. The results corroborated our prediction that ambiguous messages lead to less counterfactual thinking, so the participants rated ambiguous promises as more probable than positive or negative ones.

Our next two experiments confirmed our prediction that pragmatic information about speaker's credibility and the types of arguments he or she puts forward can influence heuristic processing and in consequence can change the understanding of the message. In Experiment 30.3, we found that when the speaker presented arguments for and against a certain issue, individuals may agree with him or her only if he or she is personally against the issue in question. We suppose that being against something gives better access to counterfactual thinking, and the competent speaker's point of view provides a strong argument against a certain issue. But, as we proved in Experiment 30.4, when the source of persuasion is a competent expert who presents weak arguments, those arguments are treated as counterexamples against his or her own point of view. The receivers of a persuasion message are more likely to think about his or her hidden motivation and will not agree with the point that he or she represents.

In many real-life situations, we are persuaded to accept someone's point of view. At the face value, we quite often treat such arguments as true or valid. The corner stone of human intelligence is the ability to envisage false possibilities and to search for counterexamples. We can do so if we try to think critically of arguments that we have been presented with. Pragmatic knowledge about the context of the message and characteristic features of the one who utters it can only help us in inducing this critical thinking.

REFERENCES

Allport, D.A. (1977). On knowing the meaning of words we are unable to report: The effects of visual masking. In S. Dornic (Ed.), *Attention and Performance* (vol. VI, pp. 505–533). Hillsdale, NJ: Erlbaum.

Bohner, G., Einwiller, S., Erb, H.-P., Siebler, F. (2003). When small means comfortable: Relations between product attributes in two-sided advertising. *Journal of Consumer Psychology, 13*, 454–463.

Bohner, G., Moskowitz, G.B., Chaiken, S. (1995). The interplay of heuristic and systematic processing of social information. In W. Stroebe, M. Hewstone (Eds.), *European Review of Social Psychology* (vol. 6, pp. 33–68). Hoboken, NJ: Wiley.

Bohner, G., Ruder, M., Erb, H.P. (2002). When expertise backfires: Contrast and assimilation effects in persuasion. *British Journal of Social Psychology, 41*, 495–519.

Chen, S., Chaiken, S. (1999). The Heuristic-Systematic Model in its broader context. In S. Chaiken, Y. Trope (Eds.), *Dual Process Theories in Social Psychology* (pp. 73–96). New York: Guilford Press.

Eagly, A.H., Wood, W., Chaiken, S. (1978). Causal inferences about communicators and their effect on opinion change. *Journal of Personality and Social Psychology, 36*, 424–435.

Eisend, M. (2006). Source credibility dimensions in marketing communication—A generalized solution. *Journal of Empirical Generalisations in Marketing Science, 10*, 1–33.

Evans, J.St. (2002). Logic and human reasoning: An assessment of the deduction paradigm. *Psychological Bulletin, 128*, 978–996.

Evans, J.St. (2003). In two minds: Dual-process accounts of reasoning. *Trends in Cognitive Sciences, 7*, 454–459.

Evans, J.St. (2008). *Hypothetical Thinking: Dual Processes in Reasoning and Judgment*. Hove: Psychology Press.

Folkes, V.S., Kamins, M.A. (1999). Effects of information about firms' ethical and unethical actions on consumers' attitudes. *Journal of Consumer Psychology, 8*, 243–259.

Gershoff, A.D., Mukherjee, A., Mukhopadhayay, A. (2007). Few ways to love, but many ways to hate: Attribute ambiguity, and the positivity effect in agent evaluation. *Journal of Consumer Research, 33*, 499–505.

Gilbert, D.T. (1991). How mental systems believe. *American Psychologist, 46*, 107–119.

Goodwin, G., Johnson-Laird, P.N. (2005). Reasoning about relations. *Psychological Review, 112*, 468–493.

Haugtvedt, C.P., Wegener, D.T. (1994). Message order effects in persuasion: An attitude strength perspective. *Journal of Consumer Psychology, 21*, 205–218.

Johnson-Laird, P.N. (1983). *Mental Models. Toward a Cognitive Science of Language, Inference, and Consciousness.* Cambridge, MA: Cambridge University Press.

Johnson-Laird, P.N. (2007). *How We Reason.* Oxford: Oxford University Press.

Koniak, P., Cwalina, W. (2008). Zmiana wrażenia o nadawcy a kształtowanie postawy wobec obiektu perswazji. (The change of the image of the message sender and the formation of attitudes towards the object of a persuasive message.) *Studia Psychologiczne, 46*, 43–56.

Petty, R.E., Cacioppo, J.T. (1984). The effects of involvement on responses to argument quantity and quality: Central and peripheral routes to persuasion. *Journal of Personality and Social Psychology, 46*, 69–81.

Petty, R.E., Wegener, D.T. (1998). Attitude change: Multiple roles for persuasion variables. In D. Gilbert, S. Fiske, G. Lindzey (Eds.), *The Handbook of Social Psychology* (pp. 323–390). New York: McGraw-Hill.

Priester, J.R., Petty, R.E. (1995). Source attributions and persuasion: Perceived honesty as a determinant of message scrutiny. *Personality and Social Psychology Bulletin, 21*, 637–654.

Ranganth, K.A., Spellman, B.S., Joy-Gaba, J.A. (2010). Cognitive "category-based induction" research and social persuasion research are each about what makes arguments believable: A tale of two literatures. *Perspectives on Psychological Science, 5*, 115–122.

Sloman, S.A. (1996). The empirical case for two systems of reasoning. *Psychological Bulletin, 119*, 3–22.

Sperber, D., Wilson, D. (1995). *Relevance: Communication and Cognition*. Oxford: Blackwell Publishing.

Wyer, R.S., Gruenfeld, D.H. (1995). Pragmatic information processing in social contexts: Implications for social memory and judgment. In M.P. Zanna (Ed.), *Advances in Experimental Social Psychology* (vol. 27, pp. 49–92). San Diego, CA: Academic Press.

31 Job Attitudes of IT Sector Specialists

Beyond Stereotypes: What Is the Challenge?

Jerzy Rosinski

CONTENTS

INTRODUCTION

Modern organizations experience a change that is not only semantic in nature: Employees are no longer human capital components—they become owners and investors of the capital (Morawski 2006). The change in the importance of employees and their intellectual capital is related not only to the processes forming megatrends; the key role of specialists is connected, *inter alia*, with the ever-narrowing specialization, a rapid technical development, IT development creating new opportunities for communication and transfer of knowledge. The consequence for enterprises is that they perceive results achieved in the area of innovation as their strategic objectives. This change has not only pushed routine or problem-solving processes into the background, but it has made the success of a company conditional upon the performance of employees.

Due to the changes that are taking place, the specialists' own perception of organizations employing them has changed, too. They do not identify themselves with any specific organization but rather with their professional milieu, believing that they work "on a project" and not "for a company."

This paper deals with selected economic and organizational background processes. It includes a presentation of results of a survey conducted among high-tech and low-tech sector employees. The comparison of the survey results for each group shows the characteristic features typical of employees in the post-industrial era.

IT SPECIALIST'S ENVIRONMENT

Current economic changes, referred to at the end of the 20th century as the third wave (Toffler and Toffler 2007), the post-capitalist economy (Drucker 1999), or the third industrial revolution (Warnecke 1999), are only a stage in a longer process. Similarly, the specificity of IT specialists' operation results from numerous earlier changes; interestingly enough, a similar phenomenon (the emergence of a distinguishable social stratum of well-paid specialists) had already been observed in Medieval Europe. In this paper, the analysis of "background processes" related to a high tech-based economy will be followed by the presentation of several features of IT specialists themselves. Hopefully, the analogies with past processes will enhance our understanding of the status quo.

ECONOMY

A knowledge-based post-capitalist economy is rooted in the (first) industrial revolution of the mid-18th century (Morawski 2006). One could say that the transformations of the last three centuries, despite numerous periods of acceleration referred to as a "revolution," are, in fact, evolutionary (Toffler and Toffler 2007) and characterized by stable dynamics of development within the so-called "limits to growth" (Senge et al. 2008). Regardless of names given to individual landmarks, we not only deal with the same process dynamics that have the form of development within the limits of growth (see Figure 31.1), but, in each case, the development began with centrally implemented solutions evolving toward decentralized ones (Warnecke 1999). The acceleration observed at the moment is consequent upon IT development, yet the same growth has been previously characteristic of the transformations caused by the steam engine and subsequently by the railway, telegraph, telephone, power generator, combustion engine, automation, and establishment of production process chains (Carr 2003; Warnecke 1999).

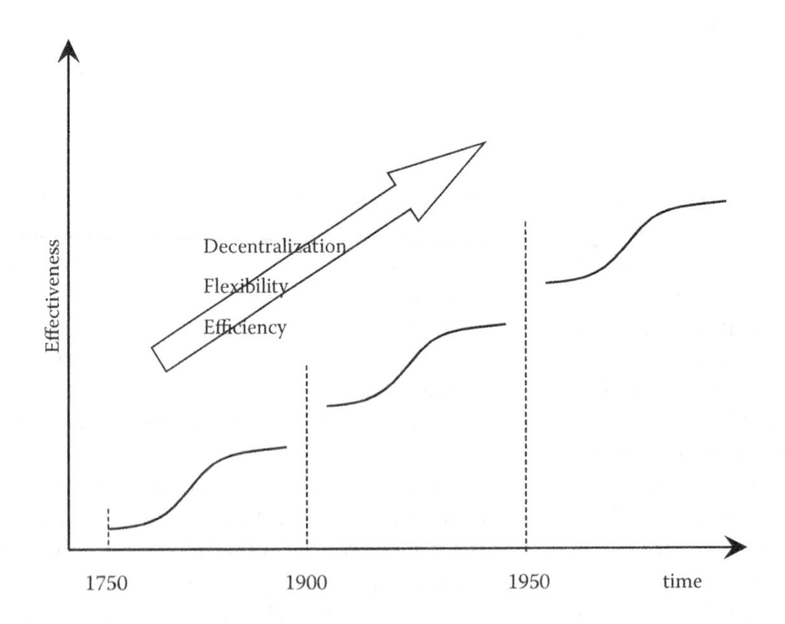

FIGURE 31.1 Efficiency growth during each industrial revolution. (From own analysis based on Warnecke, H.J., *A Revolution in Corporate Culture: The Fractal Company* [Polish language edition], Wydawnictwa Naukowe PWN, Warszawa, 1999. With permission.)

The acceleration brought about by the data collection and processing technology—the stage of "data and information consumption"—is coming to an end (Perechuda 2005). The need to facilitate the creation of knowledge within an organization is becoming more and more urgent, which means the transition toward effective management of knowledge workers as was the case with the management of physical labor and material services in the 20th century.

As a result of transformations in the global economy, the traditional production factors (capital, land, labor) have retained their importance; however, compared to knowledge, they have become secondary (Drucker 1999; Sikorski 2002); also the traditional competitive advantage factors lost their primary significance (Perechuda 2005). The changes affected also the key success factors; chronologically, these were compulsion in the pre-industrial age, wealth in the industrial era, and knowledge in the post-industrial era (Morawski 2006). According to Drucker (1999), "The change in the significance of knowledge, which began 250 years ago, has transformed society as well as the economy. ... Knowledge is the only meaningful resource today." The traditional factors of production—land (i.e., natural resources), labor, and capital—have not disappeared, but they have become secondary. They can be obtained and obtained easily, providing there is knowledge. "Knowledge ... is a utility, the new means to obtain social and economic results." Currently, knowledge is the basic resource in a post-capitalist society while the traditionally understood labor and capital are being sidelined (Toffler and Toffler 2007). What matters is where innovations originate from—where a product has been invented—while the place of manufacture is of secondary importance. Also, to a greater and greater extent, capitalism happens without capitalists; a capitalist is not a person, a pension fund with a capital exceeding the funds gathered by persons perceived by us as reach capitalists (Drucker 1999).

Technology

Presently, as never before, technological development is linked with changes in the economy. On the one hand, technology causes virtualization of many fields of activity (e.g., installation designing or sale of music pieces), pushing whole sectors into oblivion (e.g., amateur analog photography market). On the other hand, as Bill Gates claims, "The omnipresent and equal access to information" ensured by technological changes "will bring us as close as possible to the Adam Smith's perfect market." Hence, technology and economy not only affect each other; paradoxically enough, due to technological advancement, we can go back to the "source" of entrepreneurship.

The dynamics of development within the limits of growth discussed above (Figure 31.1) may be also considered from the perspective of technological advancement and, consequently, from the perspective of the IT sector (Figure 31.2). What we deal with here are the processes characteristic of the whole knowledge-based economy, such as transition from centralized solutions to decentralized ones.

The current level of development in the high-tech sector may be the end of the dynamic efficiency growth based exclusively on hardware capabilities. For instance, as regards the efficiency of IT tools enabling automatic classification and elaboration of systematics, the current efficiency level is approximately 85%, which, in many cases, is sufficient. Thus, from the perspective of a long-term trend (Figure 31.2, dotted line), the chart may be interpreted as depicting the situation in which capabilities of tools for gathering and processing overt knowledge ("know what" type of knowledge) are nearly exhausted. In this case, the limit to growth is the lack of tools for covert knowledge ("know how" type of knowledge). The crossing of the limit to growth and entering another phase of increased efficiency of tools facilitating knowledge management will be possible when new effective tools are created, facilitating the transformation of covert knowledge into overt knowledge. This means, *inter alia*, that the aspects of management that are connected with the understanding of relationships between entities at the descriptive level offered by the psychology or sociology of an organization will be more and more useful to new technology sectors.

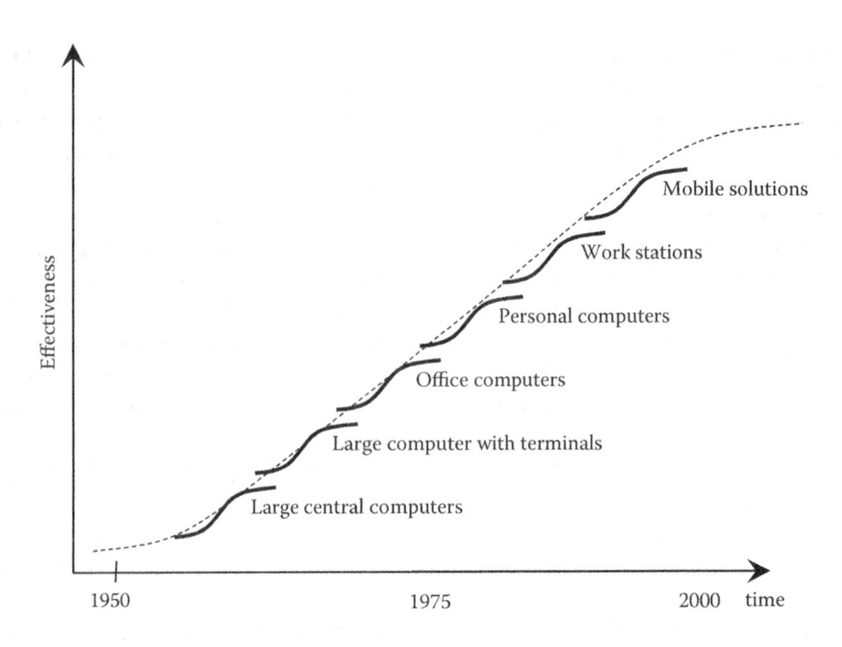

FIGURE 31.2 Efficiency growth of information processing in the IT sector—development stages related to technological changes. (From own analysis based on Warnecke, H.J., *A Revolution in Corporate Culture: The Fractal Company* [Polish language edition], Wydawnictwa Naukowe PWN, Warszawa, 1999.)

IT Sector

As Peter Drucker points out, so far, the IT sector has focused on technological aspects: data gathering, storing, transmitting, and presenting. In other words, the information technology sector has been focusing on technology alone. Presently, the information is gaining ever-greater importance. Thus, from the issue of efficiency of storage and aggregation, we move to the question about the significance of information, namely "what is the purpose of information obtained from the system?" (Drucker 2000). Hence, the focus in the IT sector is shifted from technology to information, not only with respect to business management applications, but also to other sectors of economy (Drucker 2000). Naturally, IT is indispensible in knowledge management: Adequately working hardware, networks, and software are necessary in management, but they are just tools. One should remember that knowledge is inextricably linked with people; it cannot be created without their involvement (Morawski 2006).

When discussing the changes in and development of perspectives in the IT sector, worth attention is an extremely interesting analogy with the information revolution of the 15th century. It was related to the invention of the printing press (Drucker 2000). Within 50 years after the printing press was invented (1456–1500), in Europe printers replaced monks-copyists because they were more than a thousand times more efficient. Not only did the advent of the printing press cause a dramatic decrease in book prices but also cultural changes (printing of books in mother tongues, taking up issues other than theology and philosophy), social changes (printing of Martin Luther's theses brought about Reformation—before the age of the printing press, similar social movements were quickly suppressed and had local impact), and economic changes (dissemination of information about discoveries of sailors, printed maps). The revolution was happening very fast in those days: In 1455, the occupation of a printer did not exist, yet 25 years later, printers were indispensible. Printers were held in high regard, like princes, and soon became extremely wealthy. In the 16th century, printers were preoccupied with innovations brought about by the printing technology; trying to earn more and more, they focused on the technical aspects of printing and stopped performing the function of publishers. Consequently, already in 1580, printers became mere craftsmen, thus losing their wealth and status. The social and economic role of printers was taken over by publishers: They started to attract capital and earn ever-growing profits. The change that took place around 1580 was consequent upon the technology reaching its "saturation point"—information management gained importance, and so printers became technicians while publishers, focused on the dissemination of information meaningful to the recipients, were gaining respect.

A similar situation can be observed in the IT sector: leading high-tech companies are developing very fast, and the market leaders enjoy respect and wealth. The development is focused on technology; however, it seems that users perceive the current level of development as "sufficient," and often an average user of a text editor, spreadsheet package, or an e-mail application does not use new functionalities offered by subsequent versions of the application, so hardly ever does a user (except for passionate game players) need upgraded hardware. This means that the technological component is reaching the saturation point, and the information aspect is growing in importance. Therefore, the leadership in the IT sector will be available to entities providing tools responding to users' information needs, and specialists will gradually become technicians.

Current information technologies coupled with management models perfectly replace employees in simple tasks and may significantly reduce the importance of red tape. However, by replacing mechanical tasks performed by humans with IT operations, we do not replace a human being as such. It is nothing more than what we saw during the industrial revolution, namely the replacement of simple work performed by humans with machine work. At present, cyclic and repetitious simple intellectual tasks are being taken over by "digital machines" (Biernat 2006; Błotnicki and Wawrzynek 2006). But commitment and creativity contributed by an employee to the knowledge possessed by an organization is irreplaceable. Hence, excessive spending on information technology does not translate into better knowledge management and, consequently, does not result in achieving a competitive advantage. It seems that the "development of a communication technology system serving as a tool for information technology" is becoming crucial (Perechuda 1997; Butryn 2006).

Therefore, it can be concluded that also narrower analyses concerning the situation in the IT sector show that the increase of efficiency in knowledge management is not so much (if at all) dependent on developing hardware and modifying organizational structure, but rather on the attitudes and behaviors of employees. To effectively modify attitudes, one has to first understand them; hence the necessity of describing the knowledge workers (*inter alia* their habits and views) in the high-tech sector able to cause positive and lasting changes of attitudes.

SPECIFICITY OF PROFESSIONAL FUNCTIONING OF IT SPECIALISTS: A SURVEY

Survey Procedure and Tools

A group of 470 persons was surveyed, comprised of employees and lower-level managerial staff (immediate superiors) in organizations within and outside the IT sector. The IT sector group of employees consisted exclusively of persons directly involved in developing, testing, and implementing applications and of persons managing project groups. Surveyed were 228 employees of the IT sector. The surveyed group of employees from outside the IT sector was composed of salespersons from the fast-moving consumer goods sector and local government employees. The group totaled 242 persons. The surveyed persons have been asked to fill in a couple of research tools; however, the present paper demonstrates only the results of the questionnaire survey. The questionnaire dealt with the way employees perceive their organization and the internal relations in the organization. The questionnaire was constructed on the basis of individual interviews with IT sector employees, and its purpose was to explore the area revealed during qualitative surveys without an *a priori* reproduction of any theory describing the culture of an organization.

According to the constructive principles of the tool, the questionnaire on the perception of the internal relations in an organization referred to the way employees perceive the following aspects:

the employing organization and internal relations in the company (Part 1 of the questionnaire), material aspects connected with the work for a given organization (Part 2), and personal professional functioning (Part 3). The assumptions made as to the three aforementioned aspects proved correct although Part 1, in particular, required additional statistical analyses.

For Part 1 of the questionnaire, after collecting the survey results, the factor structure of the questionnaire indicators was investigated.

The resultant factors were characterized by satisfactory factor loadings (above 0.5) and enabled sensible—in terms of subject matter—synthesis of survey questions making up a given factor. Their meanings are discussed below.

Factor 1: Appreciation by the company—it may be described as a "general acceptance of the company." The more intensive is the factor, the more the company concerned cares for the improvement of skills and competence of its employees; employees receive recognition from their managers, and managerial staff do not patronize employees or show their superiority. It refers rather to "acceptance of the organization" in terms of its formal functioning (career, development, employee appraisal).

Factor 2: Familiarity in the organization—detachment of managers and relationships among employees. The more intensive the factor is, the less detached is the managerial staff and the closer are the relationships. This issue is related rather to "acceptance of organization" in terms of its informal functioning (everyday situations, contacts at work and during breaks at work).

Factor 3: Identification with tasks—the urge to succeed in this particular company, which, however, doesn't have to mean identifying oneself with the organization's values; a calculative model of organizational involvement is possible here (Etzioni 1985). This means that an employee is strongly committed to carry out tasks for the organization; however, this attitude is the result of a conviction that this particular organization is the best place to achieve personal objectives and satisfy one's own needs.

Factor 4: Sense of responsibility—fulfilling obligations toward persons within ones' own team (most often a project team) as well as the organization (the so-called internal customer).

Factor 5: Formalization of interpersonal relations—the quality of collaboration, contacts, being together when carrying out organizational tasks; the more intensive the factor, the less eager are employees to collaborate. High values of this factor may be also a sign of very formal relations between the "IT specialists" and "business" or "administration"—in other words, formality of relations may be related to division of the organization into various departments (see also Item 2 in Factor 5); Factor 5 does not describe only the relationship between boss and subordinate (detachment of managers is also covered by Factor 2), but also the relationships among colleagues.

Factor 6: Tangible incentives—how important material incentives are to employees (both salary-related and other); the more intensive the factor, the greater significance employees attach to material incentives offered by the company.

Factor 7: Individualism—positive evaluation by employees of autonomy in work understood as emotional detachment in contacts with superiors and colleagues as well as nonconformity as regards selected company rules.

Both the results of Part 1 of the questionnaire (including these seven factors) and the results of Part 2 have been subject to comparison in relation to the following independent variables:

- Group: IT personnel—*high-tech* (referred to as the IT people in Table 31.1) and non-IT personnel—*low-tech* (referred to as non-IT people in Table 31.1)
- Work experience: employees with work experience up to three years or with four years or more
- Role: employees whose dominant role is specialist according to M. Belbin or employees having other team roles (described in the same Belbin's typology)

TABLE 31.1
Dependencies between Three Independent Variables and Separated Factors

Respondents' Subgroups Selected on the Basis of the Independent Variable												
Group				Work Experience				Role				
IT Persons		Non-IT Persons		Little Work Experience		A Lot of Work Experience		Specialists		Non-Specialists		
Mean	Standard Dev.	Mean	Standard Dev.	Mean	Standard Dev.	Mean	Standard Dev.	Mean	Standard Dev.	Mean	Standard Dev.	p
Factor 1—Appreciation by the Company												
1.67	0.46	2.10	0.60	–	–	–	–	–	–	–	–	< 0.001
–	–	–	–	1.74	0.55	1.97	0.57	–	–	–	–	0.023
–	–	–	–	–	–	–	–	1.90	0.56	1.87	0.59	0.975
Factor 2—Familiarity of the Organization												
1.26	0.42	2.41	0.76	–	–	–	–	–	–	–	–	< 0.001
–	–	–	–	1.54	0.72	2.01	0.85	–	–	–	–	< 0.001
–	–	–	–	–	–	–	–	1.92	0.90	1.78	0.78	0.343
Factor 3—Task Identification												
2.39	0.52	1.90	0.47	–	–	–	–	–	–	–	–	< 0.001
–	–	–	–	2.27	0.53	2.08	0.55	–	–	–	–	0.217
–	–	–	–	–	–	–	–	2.18	0.56	2.12	0.54	0.460
Factor 4—Sense of Responsibility												
1.46	0.51	1.49	0.51	–	–	–	–	–	–	–	–	0.566
–	–	–	–	1.44	0.52	1.49	0.50	–	–	–	–	0.546
–	–	–	–	–	–	–	–	1.51	0.50	1.45	0.51	0.208

(continued)

TABLE 31.1 (Continued)
Dependencies between Three Independent Variables and Separated Factors

Respondents' Subgroups Selected on the Basis of the Independent Variable												
Group				Work Experience				Role				
IT Persons		Non-IT Persons		Little Work Experience		A Lot of Work Experience		Specialists		Non-Specialists		
Mean	Standard Dev.	Mean	Standard Dev.	Mean	Standard Dev.	Mean	Standard Dev.	Mean	Standard Dev.	Mean	Standard Dev.	p
Factor 5—Formalization of Interpersonal Relationships												
3.27	0.48	2.82	0.55	–	–	–	–	–	–	–	–	< 0.001
–	–	–	–	3.11	0.50	3.00	0.59	–	–	–	–	0.791
–	–	–	–	–	–	–	–	3.01	0.62	3.06	0.51	0.759
Factor 6—Tangible Motivators												
1.74	0.65	1.68	0.66	–	–	–	–	–	–	–	–	0.111
–	–	–	–	1.66	0.61	1.74	0.68	–	–	–	–	0.094
–	–	–	–	–	–	–	–	1.73	0.70	1.69	0.62	0.841
Factor 7—Individualism												
2.90	0.47	2.64	0.48	–	–	–	–	–	–	–	–	< 0.001
–	–	–	–	2.82	0.48	2.74	0.50	–	–	–	–	.630
–	–	–	–	–	–	–	–	2.78	0.52	2.76	0.47	.371

Source: Own study.

TABLE 31.2
Collective Results with Regard to the Significance of Differences Occurring for the Variables Group, Role and Work Experience in Relation to the Goods Employees Received from the Company

	Laptop	Mobile Phone	Company Car	Shares/Stocks	Other
Group	$p_1 = .957$	$\mathbf{p_1 = < .001}$	$\mathbf{p_1 = < .001}$	$p_1 = .085$	$\mathbf{p_1 = < .001}$
	$p_2 = .999$	$\mathbf{p_2 = < .001}$	$\mathbf{p_2 = < .001}$	$p_2 = .118$	$\mathbf{p_2 = < .001}$
Role	$\mathbf{p_1 = .016}$	$\mathbf{p_1 = .018}$	$\mathbf{p_1 = .023}$	$p_1 = .538$	$p_1 = .403$
	$\mathbf{p_2 = .020}$	$\mathbf{p_2 = .022}$	$\mathbf{p_2 = .027}$	$p_2 = .564$	$p_2 = .435$
Work Experience	$p_1 = .143$	$\mathbf{p_1 = < .001}$	$\mathbf{p_1 = .029}$	$\mathbf{p_1 = .045}$	$p_1 = .983$
	$p_2 = .161$	$\mathbf{p_2 = < .001}$	$\mathbf{p_2 = .031}$	$\mathbf{p_2 = .065}$	$p_2 = .999$

Source: Own study.
Note: p_1, results achieved using a chi-square; p_2, results achieved using p Fisher's exact test.

Results related directly to the specificity of IT people operation are ratios connected with the variable the group.

The differences occurring in the interaction between the aforementioned variables have also been subject to analysis.

Table 31.1 shows the distribution of basic data for Part 1 of the questionnaire.

The first part of the questionnaire for the respondents was connected with the way employees perceive an organization and internal relations in the company (between their coworkers and superiors). The second part, however, refers to the elements defining a relationship with the organization in a more measurable way; that is in terms of goods received from the company, the work space, etc. While Part 1 of the questionnaire focuses on the subjective perception of the employee-organization relationship, Part 2 is connected with elements material in character: The respondents were asked about goods received from the company, such as company laptops, mobile phones, cars, or shares.

The statistical analyses of this part were done using a chi-square test and Fisher's exact test (Table 31.2).

Parts 1 and 2 of the questionnaire refer to the context of organizational functioning, and Part 3 focuses, to a greater extent, on the private aspect of professional functioning, bearing no direct connection with any specific organization. In Part 3, the focus is on the subjective evaluation of private life satisfaction and on the subjective evaluation of professional knowledge in contrast to other employees of the same sector.

Private Life Satisfaction

Table 31.3 shows the results obtained after variance analysis for the independent variables: Group, Work Experience, and Role with regard to the dependent variable, personal satisfaction. The interaction between the variables Group, Work Experience, and Role has not been presented owing to the fact that it does not occur at the statistically significant level.

Another table (Table 31.4) shows the results of the respondents' subjective evaluation of their professional knowledge, the professional knowledge of other IT specialists, and the breakdown of the two kinds of data in question (the difference between one's own estimated knowledge and the knowledge of other IT specialists). The evaluation of the level of knowledge could be expressed by values 0–100, and the respondents marked the evaluation of their knowledge on a line labeled with numbers. Originally, only the latter kind of data appeared to be the clearest for interpretation (the difference between one's own level of knowledge and that of other IT specialists); however, also the intermediate results seem to be a source of interesting information. Table 31.4 shows results for independent variables Group,

TABLE 31.3
Respondents' Private Satisfaction (Distribution according to Independent Variables)

			Subgroups of Respondents Selected on the Basis of the Independent Variable											
			Group				Work Experience				Role			
			IT People		Non-IT People		Little Work Experience		A Lot of Work Experience		Specialists		Non-Specialists	
Independent Variables	*F* (1.454)	p	Mean	Standard Deviation	Mean	Standard Deviation	Mean	Standard Deviation	Mean	Standard Deviation	Mean	Standard Deviation	Mean	Standard Deviation
Group	3.81	0.052	79.34	20.06	76.02	17.62	–	–	–	–	–	–	–	–
Work Experience	6.22	0.013	–	–	–	–	76.02	21.77	79.34	16.99	–	–	–	–
Role	1.45	0.228	–	–	–	–	–	–	–	–	77.30	19.68	78.87	18,18

TABLE 31.4
Subjective Evaluation of the Respondents' Level of Professional Knowledge and the Level of Professional Knowledge of Other IT Specialists (Distribution according to Independent Variables)

Evaluation of One's Own Professional Knowledge

			Subgroups of Respondents Selected on the Basis of the Independent Variable											
			IT People		Non-IT People		Little Work Experience		A Lot of Work Experience		Specialists		Non-Specialists	
Independent Variables	***F* (1.454)**	**p**	**Mean**	**Standard Dev.**	**Mean**	**Standard Dev.**	**Mean**	**Standard Dev.**	**Mean**	**Standard Dev.**	**Mean**	**Standard Dev.**	**Mean**	**Standard Dev.**
Group	0.00	0.994	62.88	20.23	68.79	18.74	–	–	–	–	–	–	–	–
Work experience	48.84	< 0.001	–	–	–	–	58.13	21.76	70.19	17.03	–	–	–	–
Role	0.00	1.000	–	–	–	–	–	–	–	–	67.87	20.30	64.30	19.06

Evaluation of Other IT Workers' Professional Knowledge

			Subgroups of Respondents Selected on the Basis of the Independent Variable											
			IT People		Non-IT People		Little Work Experience		A Lot of Work Experience		Specialists		Non-Specialists	
Independent Variables	***F* (1.454)**	**p**	**Mean**	**Standard Dev.**	**Mean**	**Standard Dev.**	**Mean**	**Standard Dev.**	**Mean**	**Standard Dev.**	**Mean**	**Standard Dev.**	**Mean**	**Standard Dev.**
Group	12.29	< 0.001	66.89	17.37	60.50	21.25	–	–	–	–	–	–	–	–
Work experience	1.06	0.303	–	–	–	–	66.45	19.26	62.05	19.79	–	–	–	–
Role	0.25	0.619	–	–	–	–	–	–	–	–	64.74	19.82	62.68	19.59

(continued)

TABLE 31.4 (Continued)
Subjective Evaluation of the Respondents' Level of Professional Knowledge and the Level of Professional Knowledge of Other IT Specialists (Distribution according to Independent Variables)

			Comparison of One's Own Knowledge with the Knowledge of Other IT Workers											
			Subgroups of Respondents Selected on the Basis of the Independent Variable											
			IT People		Non-IT People		Little Work Experience		A Lot of Work Experience		Specialists		Non-Specialists	
Independent Variables	*F* (1.454)	p	Mean	Standard Dev.	Mean	Standard Dev.	Mean	Standard Dev.	Mean	Standard Dev.	Mean	Standard Dev.	Mean	Standard Dev.
Group	10.19	0.002	–4.01	20.91	8.29	23.50	–	–	–	–	–	–	–	–
Work experience	48.40	< .001	–	–	–	–	–8.31	20.62	8.13	22.31	–	–	–	–
Role	0.21	0.648	–	–	–	–	–	–	–	–	3.13	21.60	1.62	24.26

Note: The "–" sign, appearing repeatedly in this part of the table, comes from the comparison of two values and pointing to the result of that comparison (the result of the difference). The "–" sign means that, in this case, IT workers evaluate their knowledge below the average observed in their sector and that people with little work experience evaluate their knowledge below the estimated average level of knowledge of the people with equal work experience.

Work Experience, and Role with regard to the dependent variable "subjective evaluation of one's own professional knowledge," "subjective evaluation of one's own professional knowledge by the respondents' professional counterparts."

FUNCTIONING OF IT SPECIALISTS

The research data is a source of information, *inter alia*, about the specificity of the functioning of the IT specialists. It is worth noticing that the original labeling trend (specificity of IT) coexists with numerous pieces of other information showing the disappearing differences, for example, as a result of the professionalization of the labor market and removing the differences between high-tech and low-tech workers. The data for this part of the final results drawn come from the data obtained from a questionnaire prepared by the author of this paper and using the personality test—both methods being analyzed with regard to the variable Group and interaction between the variables Role and Group.

CONSIDERABLE SIMILARITIES

An essential conclusion, concerning all the results, about the specificity of the functioning of the IT specialists seems to be the interpretation of the fact that all differences concerning the IT sector have been revealed on the basis of the questionnaire. The lack of differences between IT and non-IT is shown by means of methods created to measure the personality factors (self-image, team role). This could be due to the character of the tools applied; however, there is an interpretation worth considering, namely, that the differences between IT and non-IT workers might result from social (environmental) factors and not individual ones (personality-related). Such interpretation is confirmed by the results presented below, diverse for the IT sector, connected with the impact of the length of work experience (thus being in a certain environment).

No differences regarding the subjective evaluation of the level of one's own professional knowledge have been observed. Both professional groups place their professional knowledge at a similar level, and, simultaneously, above 50, which is the average value possible to choose from on the scale (62.88 and 68.79 on the 0–100 scale). A suggested interpretation of that result is work professionalization—people with professions outside new technology may also be in need of the advance knowledge necessary to complete tasks at work. Interpretations relating to personality factors (such as the absence of a feeling of superiority among IT workers) failed to be positively verified by the results achieved by the application of methods that research personality (the presence of a feeling of professional pride, willingness to come in contact solely with people with an adequate level of specialist competence).

Only minor differences among individuals in the role of Specialist employed in and outside the IT sector have been observed:

- There are no differences in the main analyses conducted by applying methods that measure personality factors.
- There are differences in the interactions between variables (Group and Role); merely two differences become apparent (greater fear about the future, less conscientiousness among IT specialists). The two differences listed are also observed in the questionnaire results; therefore, a question arises as to the origin of the changeability: whether it is individual (personal) or social (environmental).

The interpretations of the observed differences are ascribed to environmental factors (such as the rapid increase of knowledge in IT) and not the construct (comprised of personality, cognitive schemata, and habits in behavior) underlying the typology of team games. Due to only minor differences (which can also be explained by factors other than team role), an interesting paradox emerges concerning team roles according to Meredith Belbin's typology. The role of Specialist is stemmed in connection with the development of new technologies; classic elements of the description of the Specialist are encountered nowadays equally often among IT workers as new technologies outside.

Lack of differences in terms of descriptive components associated with the Specialist's team role may show the consequences of work professionalization at the level described by psychology (behavior and attitudes). Workers outside the IT sector begin to exhibit features of Specialists. In other words, changes at the level of economy may push the workers outside the IT sector into behaving the way that not so long ago—in the 1990s—was ascribed only to workers in the IT sector or, more widely, in companies within new technologies (high tech).

OBSERVED DIFFERENCES

The differentiations existing between the groups of IT workers and respondents working outside the IT sector imply that the respondents representing the IT sector, in contrast to workers outside the IT sector, may exhibit specific features of functioning within the following three areas:

1. Perception of the organization employing them
2. Relationships with coworkers
3. Perception of their career

The conclusions concerning the functioning of the IT workers in the three areas in question shall be presented respectively.

In terms of perception of the employing organization, an unusually negative picture of the present employer emerges. A typical function of IT workers in the employing organization seems to be based on calculative participation; IT workers see the companies at which they are employed as systems

- With a rather unfriendly atmosphere, a considerable distance in relationships with superiors, and an unpleasant atmosphere among coworkers
- With a lack of feedback from superiors
- In which the employees' competence is not enhanced

In terms of how the organization's familiarity is perceived, IT workers seem to differ considerably from people from other sectors, regardless of the length of employment. Among the representatives of non-IT sectors, a negative picture of the organization appears to change in time: The longer you work for a certain organization, the more employee-friendly it appears. However, IT workers, regardless of the length of employment, perceive the employing organizations as rather unfriendly; both workers with shorter and longer periods of employment express a negative evaluation of their employers. It is true, though, that, with time, negative opinions of the employer seem to recede; however, both workers with a shorter and longer length of employment perceive their employers much worse than people from non-IT sectors perceive theirs.

The aforementioned negative components may be reflected in the systems of evaluation, remuneration, competence development, or quantity of equipment and friendliness of workplace. It could be more of a cognitive explanation of the adopted attitude of personalization with a certain professional milieu but with a simultaneous lack of personalization with the employing organization. This means that an employee tries to see reason in the situation they have found themselves in and is searching for a point of reference not among their colleagues in the company, but among individuals of the same profession. If it is the professional group they find significant, they will apply a greater value to it, thus diminishing the value of the present employer. Such interpretation may be supported by the need to be successful in that particular organization, which coexists with the negative perception of it.

While the attitude itself may originate in the labor market, the ease of providing services in the IT sector for often virtual or chain organizations worldwide—the consequences at the level of functioning of an individual—can be described by means of tools from the area of psychology, for example, connected with the reduction of cognitive dissonance and retaining coherence.

In terms of relationships with coworkers, certain differentiating elements can be recognized among IT workers in comparison with the perception of such relationships by people employed outside the IT sector. The differentiating elements among IT workers, as far as the perception of relationships with coworkers is concerned, are as follows:

- Perception of the atmosphere among coworkers as unfriendly. Such a conception may result from fierce competition with other IT workers (on the open labor market for IT specialists) when others are seen as potential competition and, hence, the description of the atmosphere at work as unfriendly.
- Reluctance of IT workers and specialists as team leaders to work in a team and to cooperate in performing tasks with other departments; this reluctance tends to become stronger with the length of employment.

  With the length of employment, specialist competence increases, and so does an employee's value on the labor market. The reluctance to cooperate with others may spring from the fear of losing the outgoing status resulting from having unique knowledge and specialist competence It may also be a sheer "business calculation" appearing in the form of the idea: "it's just not worth working with this group of people; I won't increase my level of competence significantly, but at the same time, the group will learn a lot from me."
- A very good evaluation of the knowledge of other IT workers; in fact, better than that of one's own professional knowledge. Such belief may be a reaction to the requirement (included in the professional ethos) of competition, based on up-to-date, highly specialized knowledge, which leads to making more frequent and stricter comparisons with the membership group in order to make certain that one still belongs to that (elite) group. It is also possible that this outcome derives from a much simpler mechanism; namely, it may be a result of perceiving others as potential competitors on the labor market.

In terms of relationships with coworkers, we also deal with elements that do not differentiate the IT sector from other organizations outside this line of business. These include the following:

- Evaluation of one's own professional knowledge
- Attaching importance to customer relations
- Attachment to material incentives

Attaching importance to relationships with customers and evaluation of one's own professional knowledge—both similar for IT and non-IT workers—may be further indicators of work professionalization and increasing the quality of customer relations, regardless of the line of business. Similarly, as in other lines of business, attachment to material incentives may be connected with processes in the macroscale; for example, individualization of consumer styles results in an employee's interest in material elements as a possibility to make individual choices and to satisfy their own specific needs, thanks to the goods received from the employer.

In terms of perceiving one's own career, the following elements seem to differentiate IT workers from non-IT workers:

- Being strongly orientated toward individual success with no apparent attachment to the present organization (it is treated as a means of achieving a higher value on the labor market) and perceiving other employees as potential competitors. The discrepancies in being orientated toward individual success may be explained in categories based on individual differences, strengthened by formal education and the initial years of employment.

- IT workers' high level of private life satisfaction (on the scale adopted in the research); in comparison with workers outside IT, IT workers claim a higher level of satisfaction. The high level of private life satisfaction was accounted for in categories of placing professional work quite high in the hierarchy of private values and job satisfaction ("self-realization") connected with having a profession that wins public recognition (according to *CBOS* – Polish Public Opinion Research Centre).
- An attitude externally perceived as "lack of conscientiousness" or "being unconcerned with anything." It may be perceived by highly qualified IT people themselves as defending their freedom and independence as organizational experts. Discrepancies in pessimism about the future, deriving from a necessity to be up to date with technological advances, may be explained by variables social in character (described both by the macroeconomic processes and processes identified by social psychology).
- Awareness of a necessity of continual development of specialist competence and, accordingly, pessimism about the evaluation of one's own professional knowledge, increasing with the length of employment. Discrepancies in terms of lack of conscientiousness (or, as IT workers themselves put it, "greater independence") stem from the extent of individual factors.

Additional information for conclusions drawn in relation to how IT workers perceive their career can be found below.

The following partial conclusions repeated in the questionnaire may be supportive of the individual career success orientation:

- High individualism of performance
- A strong, unambiguous attitude connected with a desire to achieve success in a particular organization
- Strong task orientation
- A high evaluation of one's professional autonomy
- "Cosmopolitan" professional orientation
- Perceiving other employees as potential competition (on the global market)
- High level of expectations (even staking claims) toward the organization's scope of action with regards to its employees' development
- Possible preference to work on one's own equipment as a reaction against "limiting the freedom" by the employer (e.g., lack of permission for changes in the company computers' software)
- Fulfillment of needs connected with status presentation through higher pay expectations or financing costly elements of professional development (which could be turned into financial benefits in the long run)

The phenomenon "lack of conscientiousness" is a symptom recognized by the environment; an employee's conviction—revealing private freedom as an expert—has been identified for analyses conducted for a Specialist's team role and comparison of the functioning of people in the team role in and outside the IT sector. However, due to the specificity of functioning of this team role (professional development orientation, focus on specialist knowledge) and its origin (due to development of new technologies' lines of business), it seems reasonable to provide data appropriate for IT workers functioning in a specific team role as typical for a wider group of IT workers. Such course of action appears to be supported by the common belief connected with defending one's freedom/an expert's independence found in other research data, which relate to the whole group of IT workers (see factor analysis for the questionnaire).

Along with increasing work experience, pessimism in perceiving one's own professional knowledge among IT workers also seems to grow:

- IT workers with little work experience evaluate their professional knowledge considerably better than non-IT workers asked to carry out the same sort of self-evaluation.
- IT workers with long work experience evaluate their professional knowledge considerably worse than their counterparts outside the IT sector (non-IT workers with long work experience) asked to carry out the same sort of self-evaluation.

The existing discrepancies among people with short work experience may be explained by factors individual in character (being more optimistic about the future) or variables social in character (university as a factor selecting candidates for diverse faculties, an IT specialist being a prestigious profession).

The differences for people with a lot of work experience tend to be explained solely in terms of social factors—a possible factor social in character is making comparisons of records of achievement with people in the same line of business, which, with the rapid updating of knowledge in IT, can be a source of negative comparative results for the people in IT with more work experience. It is true that IT workers with a lot of work experience still enjoy public recognition of their profession; however, with a strong identification with their professional milieu, the comparison with other (younger) people from the same line of business may be a reason for a less optimistic perception of their own knowledge.

The line of interpretation in terms of social factors is also supported by the results obtained for the evaluation of the knowledge of colleagues from IT (connected with rapid disintegration of up-to-date knowledge) and analysis results of the interaction between the variables Role and Group for data obtained in a personality test—interpreted as anxiety about the loss of position of an organizational expert in connection with an ongoing necessity to be familiar with up-to-date technological innovations in IT.

Regardless of the line of business, the comparison of one's own knowledge at the onset of one's professional career and after a period of four years is identical in character. With increasing work experience, the difference between the evaluation of one's own knowledge and the knowledge of others in the same line of business tends to diminish. In other words, with greater work experience, we encounter a subjective increase of one's own knowledge in comparison with our coworkers. It would appear that it is quite a clear process related to a subjective increase of one's competence, connected with the increase of specialist competence.

CONCLUSION

Functioning of IT Workers: Summary

Research data gathered about the functioning of IT workers provides information that indicates the coexistence of processes of disappearing differences between high technology and low technology as a result of professionalization of the labor market and leveling discrepancies between workers. At the same time, we deal with a multi-aspect differentiation relating to the functioning of IT workers.

Calculative participation, which is characteristic for IT persons, may be an element specific mostly to virtual organizations (Sikorski 1998; Gableta and Pyszczek-Pietroń 2004). Still, elements characteristic for IT persons should be considered as features that may refer to a broader group: a group of highly specialized knowledge workers. Descriptive elements of these vocational groups, disclosed in the survey, should be analyzed:

- Absence of strong bonds formed between an employee and his organization combined with simultaneous appreciation of professional independence
- Perception of the organization and team as a highly competitive environment
- Focus on personal professional development and individual success
- A possible contamination of perceiving one's private life through one's professional life

These elements, together with the above-mentioned calculative participation, may be interesting for management practitioners working for organizations whose market success is dependent on the effectiveness of their high-class specialists. They may guide modifications of incentives, motivation systems, and career paths. Furthermore, they are helpful when trying to understand the behavior of specialists both at the level of an organization and a project team and, for this reason, may inspire project leaders or line managers.

REFERENCES

Belbin M., 2003, *Twoja rola w zespole*, GWP, Gdańsk.

Biernat J., 2006, *Model zarządzania a informatyzacja w gospodarce opartej na wiedzy (GOW)*, [in:] Binsztok A., Perechuda K. (edit.) Koncepcje, modele i metody zarządzania informacja i wiedzą, Wydawnictwo Akademii Ekonomicznej im. Oskara Langego we Wrocławiu, Wrocław.

Błotnicki A., Wawrzynek Ł., 2006, *Od porządkowania danych do business intelligence – jak uświadomiona wiedza staje się elementem konkurencyjności organizacji*, [in:] Binsztok A., Perechuda K. (edit.) Koncepcje, modele i metody zarządzania informacja i wiedzą, Wydawnictwo Akademii Ekonomicznej im. Oskara Langego we Wrocławiu, Wrocław.

Butryn B., 2006, *Technologie informacyjno – komunikacyjne a zarządzanie sieciami partnerskimi,* [in:] Binsztok A., CBOS 2009 *Prestiż zawodów. Komunikat z badań*, BS/8/2009, CBOS, Warszawa, s. 2–3. Perechuda K. (edit.) Koncepcje, modele i metody zarządzania informacja i wiedzą, Wydawnictwo Akademii Ekonomicznej im. Oskara Langego we Wrocławiu, Wrocław.

Carr N. G., 2003, *IT się nie liczy*, "Harvard Business Review Polska," Issue 9 (9) – November 2003, pp. 84–93.

Drucker P. F., 1999, *Społeczeństwo pokapitalistyczne*, Wydawnictwo Naukowe PWN, Warszawa.

Drucker P. F., 2000, *Zarządzanie w XXI wieku*, Wydawnictwo Muza S.A., Warszawa.

Etzioni A., 1985, *Władza, uczestnictwo i uległość w organizacjach* [in:] Marcinkowski A., Sobczak J.B. (edit.) Wybrane zagadnienia socjologii organizacji, Uniwersytet Jagielloński, Kraków.

Gableta M., Pyszczek-Pietroń A., 2004, *Funkcjonowanie pracownika w wirtualnych warunkach gospodarowania,* [in:] Wiśniewski Z., Pocztowski A. (edit.) Zarządzanie zasobami ludzkimi w warunkach nowej gospodarki, Oficyna Ekonomiczna, Kraków.

Morawski M., 2006, *Zarządzanie wiedzą. Organizacja – system – pracownik, Wydawnictwo Akademii Ekonomicznej im. Oskara Langego we Wrocławiu, Wrocław.*

Perechuda K., 1997, *Organizacja wirtualna, Ossolineum, Wrocław.*

Perechuda K., 2005, *Interakcyjne łańcuchy transferu wiedzy*, [w:] Galant V., Perechuda K. (edit.) Modele i metody zarządzania informacja i wiedzą, Wydawnictwo Akademii Ekonomicznej im. Oskara Langego we Wrocławiu, Wrocław.

Senge P. M., Kleiner A., Roberts Ch., Ross R. B., Smith B. J., 2008, *Piąta dyscyplina. Materiały dla praktyka. Jak budować organizację uczącą się*, Wolters Kluwer Polska - OFICYNA, Kraków.

Sikorski Cz, 1998, *Ludzie nowej organizacji. Wzory kultury organizacyjnej wysokiej tolerancji niepewności*, Wydawnictwo Uniwersytetu Łódzkiego, Łódź.

Sikorski Cz., 2002, *Zachowania ludzi w organizacji*, Wydawnictwo Naukowe PWN, Warszawa.

Toffler A., Toffler H., 2007, *Rewolucyjne bogactwo*, Wydawnictwo Kurpisz, Poznań.

Warnecke H. J., 1999, *Rewolucja kultury przedsiębiorstwa. Przedsiębiorstwo fraktalne,* Wydawnictwa Naukowe PWN, Warszawa.

32 The Effect of Comparative Advertising on the Recall of Brand Opinion

Empirical Research in the Backward Framing Paradigm

Alicja Grochowska and Dorota Szablisty

CONTENTS

Different advertising strategies are undertaken to enhance memorizing the brand and to form attitudes toward it. Some marketers believe that comparative advertising is an efficient tool in building and/or strengthening brand image and brand awareness. This paper shows the maleficent role of comparative advertising in forming an associative structure of the brand and demonstrates that this format of advertising can distort the memory of the brand. Literature in the area of consumer research shows that comparative advertising has not been investigated in this approach.

ADVANTAGES AND DISADVANTAGES OF COMPARATIVE ADVERTISING

Studies on the effectiveness of comparative advertising provide evidence that using comparative arguments (versus noncomparative) results in a better elaboration of the ad content (Dröge 1989; Hill and King 2001). Comparative advertisement content is more likely to be processed deeply, via a central route as compared to noncomparative ads, which are processed more superficially, via a peripheral route (Hill and King 2001; Pechmann and Stewart 1990; Cacioppo and Petty 1984). In addition, Jain et al. (2000) found comparative ads to be associated with higher counter argumentation, more negative attributions, fewer positive attributions, and lower claim believability as compared to noncomparative ads. Comparative argumentation provides consumers with more information and can, therefore, lead to more effective decision-making in the consumption process (Barry 1993). Moreover, comparative advertising can effectively encourage consumers to engage in relative judgments, generating either an association or differentiation effect (Chang 2007). Some studies suggest that this format of ads appeals to a wider audience and enhances the sponsor's brand identification, message persuasiveness, and market share (Muehling et al. 1990). Comparative advertisements attract consumers' attention—users of the sponsored brand as well as users of the comparison brand (Pechmann and Stewart 1991). This format of advertising is better recalled and generates more purchases than noncomparative ads (Grewal et al. 1997; Pechmann and Stewart 1990). The novelty of comparative ads plays an important role for consumers in countries in which they are not widely used or not used at all (Donthu 1998). Thus, comparative ads grab more attention, are deeply elaborated, and better remembered and, in this sense, can be more efficient than noncomparative ads.

However, better elaboration may actually have the opposite effect than marketers expected. The persuasion effects of comparative advertising are not always superior to those of noncomparative advertising. This format of advertising is less believable and generates less favorable attitudes toward the ad than noncomparative advertising (Grewal et al. 1997). Comparative ads increase consumer awareness of competitors' brands, decrease claim credibility, and produce confusion rather than effective communication (Muehling et al. 1990). Moreover, comparative ads elicit negative thoughts about a sponsored brand (Golden 1979; Goodwin and Etgar 1980; Shimp and Dyer 1978; Thomson and Hamilton 2006). Williams found that comparative advertising actually leads to consumer preference to the named competing brand (1978, c.f. Gnepa 1993). Comparative advertising might enhance memorizing the name of a competitor's brand and, in addition, consumers may not remember which brand was being advertised (Pechmann and Stewart 1990). This format of advertising elicits more negative cognitive responses and more negative emotions than noncomparative ads (Belch 1981).

Cross-cultural studies showed that consumer attitudes toward comparative advertising can be determined by a mere exposure effect. Consequently, consumers in countries in which comparative advertisements are commonplace have been exposed to comparative ads often and, hence, may have a more positive attitude than consumers in other countries where such advertisements are unfamiliar (Donthu 1998). We investigated the effects of comparative advertising in Poland—a country where such ads are rarely used.

It is worth noting that the content of comparative ads and evaluations of the brand are linked to the knowledge of the brand that consumers hold in memory and are essential elements of brand knowledge. Keller (1993) stated that brand knowledge is conceptualized according to an associative network memory model in terms of two components, brand awareness and brand image (i.e., a set of brand associations). Research on comparative advertising showed that this format of advertising facilitates better elaboration of ad content and elicits more negative evaluations and emotions than noncomparative ads. These newly learned pieces of information and negative evaluations are being joined to the associative structure of the brand. Several studies showed that people forget the *source* of information, but the content of information is stored in the memory (Braun 1999; Garry et al. 1996; Johnson et al. 1993). In this sense, comparative advertising can be maleficent for a sponsored brand.

An associative structure of the brand is a result of a network structure of the memory. This structure is dynamic and susceptible to distortions. Therefore, comparative advertisements, which are better elaborated than noncomparative ads, may become a source of memory distortions about a sponsored brand. In our studies, we investigate the comparative advertisement as information, which may distort the memory of the brand and affect its associative structure. This approach to comparative advertising and memory distortion is particularly significant for new and unknown brands and for low-involvement products. Consumers have a variety of different types of associations for familiar brands, but not for unfamiliar brands because they have not had any experiences with them. Information is actively searched, better remembered, and critically evaluated for high-involvement products but not for low-involvement products. Thus, in our studies, brand familiarity and product involvement are considered.

ASSOCIATIVE STRUCTURE OF THE MIND AND MEMORY DISTORTIONS

The theoretical basis for our studies are network models of the mind in which encoded information is stored in memory as a network structure, consisting of nodes representing concepts and links representing associations among concepts. When new information is acquired or when internal information is retrieved from memory, a node containing this information can be a source of activation for other nodes. Activation can spread from this node to other linked nodes in memory. The stronger the associations among nodes, the easier connected pieces of information are retrieved from memory (Anderson 1983; Keller 1993; McClelland 1995; McClelland and Rogers 2003). Processes of acquiring new information about the brand from different sources, also from comparative advertising, and mechanisms of memory distortions can be explained in this approach. Consistent with an associative network memory model, brand knowledge is conceptualized as consisting of a brand node in memory to which a variety of associations are linked (Keller 1993). Information from comparative advertisements is added to the associative structure of the brand. Manning and colleagues (2001) noted that association is the general effect of comparative advertising.

On the other hand, memory is an active constructive process. The brain's structure is plastic and dynamic and can be changed by new experiences, and mental representations of those experiences are also plastic and changing (Braun and Zaltman 1997; LeDoux 1996). According to Edelman (1992), the brain is an active system in which shifting is constant, and encoded material is reprocessed and updated continually (see: Braun and Zaltman 1997). Memories are constructed from fragments of information that are distributed across different brain regions and depend on influences operating in the present as well as the past. The areas involved in associative learning and memory consolidation are highly dependent on the hippocampal system in which associations become "glued" together (Braun and Zaltman 1997; Schacter 1996). This means that pieces of information coming from different sources are joined in the memory and form an associative structure of a given object. It is easier to join newly learned information when the knowledge about the given object is not well consolidated in the memory (Schacter 2001). Thus, mental representations of unfamiliar brands (as compared to familiar ones, which are better consolidated in memory) and low-involvement products (as compared to high-involvement products, information about which is more deeply processed) can be more susceptible to memory distortions.

Newly learned information can influence the memory both when it precedes and when it follows an experience. When new information is joined to the previous experience, people are biased to "see what they expect to see." This process is called "forward frame." On the other hand, when past information is altered in the memory by information learned *after* it was encoded, we call that a "backward frame" (Braun and Zaltman 1997; Braun-LaTour and LaTour 2005). Comparative advertisements often act as a "backward frame" when competitors try to depreciate a competitive brand. In our studies, a backward framing paradigm was applied to investigate the effects of comparative advertising.

BACKWARD FRAMING AS A FORM OF MARKETING CONTROL

Research conducted in the backward framing paradigm showed that information acquired *after* an experience can transform the memory of that experience. The aim of our research is to investigate how comparative advertisements can distort the memory of the previous experience (i.e., brand evaluation), considering brand familiarity and product category. Past studies showed that the advertising following a direct product experience (tasting a bad mixture of orange juice) transformed consumers' memory of the original experience and resulted in more favorable product evaluation (Braun 1999; Braun-LaTour and LaTour 2005). However, we can expect that a comparative advertisement acting as a "backward frame" can transform the memory of product and brand evaluations to be *less* favorable. Braun and Zaltman (1997) demonstrated that a negative critic review affected consumers' evaluation of a product (trailer of a movie), and a previous evaluation was recalled as less favorable. Research on the memory processes and studies in the backward framing paradigm prove that the better an experience is consolidated in the memory, the less it is distorted by post-experience information.

Furthermore, the better new information (after-the-fact experience) is elaborated, the stronger it can distort a previous experience in the memory (Braun et al. 2002; Braun and Loftus 1998; Braun and Zaltman 1997; Braun-LaTour et al. 2004; Loftus 1996, 2005; Loftus and Pickrell 1995; Schacter 2001). Comparative advertisements are better elaborated than noncomparative ads. Thus, a comparative ad acting as after-the-fact information can strongly distort the memory of a previous experience with the brand. Gunasti et al. (2008) note that explanations of post-experience advertising effects have implicated cognitive factors, such as consumers' knowledge and ability, as the reason for memory distortions caused by marketing communications. However, motivational factors should be considered as well. Motivation not only biases what a person recalls from memory (biased memory search), but also how the contents of memory are interpreted (biased memory reconstruction) (cf Gunasti et al. 2008). Motivation acts as an internal context for processing—perceiving and memorizing—products. Brand familiarity and product involvement can be understood as motivational factors in processing information on a product and increasing the cognitive ability of the consumer to reduce the misleading effects of advertising (Cowley and Janus 2004; Gunasti et al. 2008).

When competitors try to depreciate a competitive brand, they often use comparative advertising. The effects of depreciation can be particularly severe for new/unfamiliar brands and for low-involvement products. The role of comparative advertising acting as a backward frame has not been examined in past studies conducted in the backward framing paradigm. Brand familiarity and product involvement determine how well a previous experience is consolidated in the memory. Thus, these factors are considered in our research.

BRAND FAMILIARITY IN PROCESSING COMPARATIVE ADVERTISEMENTS

Knowledge of the brand, stored in the consumer's mind, creates an internal context for perceiving the brand and its properties. Research shows that advertisements for familiar brands are more easily detected in advertising clutter and are better evaluated than ads for unfamiliar brands. Snyder (1989) found that familiarity with the advertised brand's name played an important role in consumers' responses to advertising claims. Results of her research showed that claims for familiar brands were more believable and led to higher quality ratings as well as to more interest in trial than claims for fictitious brands. Moreover, familiar brands are less affected by competing claims from other brands and have a more persuasive power of sources of claims (Chattopadhyay 1998; Kent and Allen 1994; Machleit et al. 1993; Pechmann and Stewart 1990). Familiar brands elicit more associations and more personal associations (Dahlén and Lange 2005; Low and Lamb 2000). On the other hand, ads for unfamiliar brands are processed more thoroughly than ads for familiar brands. Consumers have stored knowledge in memory for familiar brands and, thus, are likely to process ads for familiar brands less extensively than those for unfamiliar brands. Familiar and unfamiliar brands differ in terms of the knowledge regarding the brand that a consumer has stored in memory. Consumers tend to have a variety of different types of associations for familiar brands—but not for unfamiliar brands because they have not had any experiences with them. Consumers have well-established brand schemas for familiar brands, which lead them to expect a certain kind of communication from the brand. Familiar brands have a well-developed brand schema, are cognitively available, and activate a wide area of an associative network in contrast to unfamiliar brands (Campbell and Keller 2003; Dahlén and Lange 2004). Therefore, one can expect that the effect of comparative advertisements on memory distortions is modified by brand familiarity.

Comparative advertising is more often and more willingly used by new, unknown brands, which try to attract consumers' attention or to highlight similarity by comparing themselves to a leader brand. In a survey of advertisers conducted by Muehling et al. (1989), "allowing a small unknown firm to successfully compete with much larger firms" was often mentioned as one of the greatest values of comparative advertising (Muehling et al. 1990). According to Gotlieb and Sarel (1991), comparative advertising is an effective method of communicating benefits of new brands. The authors found that when higher involvement is activated and a source of higher credibility is included in the advertisements, comparative advertising for a new brand has a more positive effect on purchase intentions than noncomparative advertising. This format of advertising is also used by competing brand leaders (e.g., Coca-Cola vs. Pepsi). Thus, brand familiarity is connected with using comparative advertising. On the other hand, the strength of memory distortion is determined by brand familiarity.

Therefore, a comparative advertisement acting as a "backward frame" is processed more thoroughly and may cause stronger memory distortions for unfamiliar than familiar brands. Furthermore, an associative structure of an unfamiliar brand is not well consolidated in the memory and more susceptible to memory distortions than an associative structure for a familiar brand.

PRODUCT INVOLVEMENT AND PROCESSING OF COMPARATIVE ADVERTISEMENTS

The memory of a brand and the strength of memory distortion can also be modified by the category of the product advertised. A primary criterion of how products are categorized is the level of involvement in processing information about them. The marketing literature distinguishes involvement into *product involvement* and *consumer involvement*. Consumer involvement refers to a person's perceived relevance of the object based on inherent needs, values, and interests (Zaichkowsky 1994). To the contrary, product involvement refers to product classes, which evoke interest or internal drive influencing consumers' cognitive and behavioral responses to marketing stimuli (Tsiotsou et al. 2010).

In our studies, product involvement is taken into consideration. The level of involvement (low versus high) differs according to product importance, utility, and value and perceived risk (financial, physical, and psychosocial). Consumers spend time and energy to make a decision in purchasing high-involvement products whereas they spend less time and effort for low-involvement products. For high-involvement products, brand evaluation is clear and distinct, and it is vague and general for low-involvement products. Information is actively searched and critically evaluated for high-involvement products and ignored or accepted without evaluation in the case of low-involvement products (Boriboon et al. 2010; Chen and Wang 2010; Tsiotsou et al. 2010). Thus, fast-moving goods (e.g., food, household products) can be classified as low-involvement products, and durable goods (e.g., cars, electronic appliances, and services) are high-involvement products.

Advertisements for high-involvement products as compared to ads for low-involvement products are processed more actively, demand a greater cognitive effort, and focus consumers' attention on brand claims (Celsi and Olson 1988; Warrington and Shim 2000). This is a consequence of Cacioppo and Petty's Elaboration Likelihood Model (1984), according to which a greater elaboration refers to high-involvement products, and a lower elaboration refers to low-involvement products. Ads for low-involvement products are processed via a peripheral route, and consumers' attention is focused on peripheral cues (ad execution cues) whereas ads for high-involvement products are processed via a central route, and attention is focused on the product claims. Ads for high-involvement products are processed more thoroughly; consumers closely monitor marketing messages and dedicate greater attention and effort to decision making via a central route as compared to ads for low-involvement products.

Generally, the higher the product involvement level, the more information consumers look for about the product, for example, when comparing advertisements or different brands' products (Coulter, Price and Feick 2003). Research shows that product involvement can influence a purchasing decision-making process and attitude toward the product as well as perception of competitors (Te'eni-Harari et al. 2009; Celsi and Olsen 1988; Brisoux and Chéron 1990).

The level of involvement in relation to certain products is dependent on the needs and values of the consumer. However, there are products that are, by definition, high or low involving (Warrington and Shim 2000). When buying a low-involvement product, product satisfaction is the most important factor in decision-making. However, with high-involvement products, consumers are more likely to focus on external factors, such as brands' prestige and image (Suh and Yi 2006).

According to Muehling et al. (1990), while buying low-involvement products consumers focus more on peripheral cues and pay less attention to the products' attributes. On the other hand, high-involvement products tend to make consumers think in product-related categories and increase the probability of elaboration; as a result, this information is processed through the central route (Torres and Briggs 2007). High-involvement products are more related to the values of consumers, who are more loyal to the product's brand and are more likely to invest more in the product than when purchasing low-involvement products. Despite marketing campaigns, consumers do not differentiate between low-involvement products in terms of brand and do not demonstrate any clear preferences between them. However, when purchasing a high-involvement product, a consumer is more likely to make a decision through cognitive elaboration, and low-involvement products are purchased without any reflection (Muehling et al. 1990; Warrington and Shim 2000).

Putrevu and Lord (1994) note that the relative effectiveness of comparative versus noncomparative advertising may vary as a function of product characteristics. Pechmann and Esteban (1993) investigated the persuasion processes (routes) associated with comparative and noncomparative advertising by manipulating argument strength, and they also examined the moderating role of involvement. In a high-involvement condition, argument strength influenced brand attitude and purchase intent regardless of ad type (central route persuasion) whereas, in a low-involvement condition, there were no effects for argument strength (peripheral-route persuasion).

HYPOTHESES

We can expect that distortions of memory for low-involvement products are going to be greater than for high-involvement products. Information about high-involvement products is processed more thoroughly; thus, as a result, the memory of previous experience with high-involvement products will be less susceptible to distortion in comparison to low-involvement products. Information about low-involvement products is processed more superficially; therefore, details about these products are more prone for distortion by further information. The degree of memory distortions of high- versus low-involvement products is modified by brand familiarity: It is more difficult to distort the memory about a familiar brand in comparison to an unfamiliar brand. Knowledge about familiar brands is well consolidated in memory. This is why it is expected that product involvement effects will occur in unfamiliar brands. In the case of familiar brands, memory distortions created by comparative advertisements should not be expected.

The memory distortion effects through comparative advertising can be modified through different types of measurement. Manning et al. (2001) note that regarding the comparative advertising associative effect, using unstructured, open-ended measures of memory is reasonable. Furthermore, these measures possess greater correspondence to the representations residing in long-term memory. On the other hand, using structured measures (anchored scales) for assessing comparative advertising effects explores a more superficial nonsemantic level of the comparative advertising effects. Thus, one can expect that the effects of memory distortions through comparative advertising will be more visible at an associative level than measured with anchored scales.

H1: In accordance with the backward framing paradigm, brand evaluation recall will be influenced by a comparative advertisement. The memory of the primary evaluation will be changed, and as a result, the final evaluation will be more negative.
H2: Evaluations of market leading brands will be more resistant to memory distortions than evaluations for unfamiliar brands. Comparative advertisements will distort the memory of previous evaluations to be more negative.
H3: Evaluations of low-involvement products will be more susceptible to memory distortions than evaluations of high-involvement products.

METHOD

Design of the Experiment

A 2 (product involvement: high vs. low) × 2 (brand familiarity: familiar vs. unfamiliar) design was used. Participants were randomly assigned to one of the four conditions:

1. Play (familiar brand, high-involvement product)
2. Gravity (unfamiliar brand, high-involvement product)
3. Pepsi (familiar brand, low-involvement product)
4. IQ (unfamiliar brand, low-involvement product)

Research was designed in the backward framing paradigm (Braun and Zaltman 1997; Braun-LaTour and LaTour 2005). After reading information about the brand (familiar/unfamiliar, for high-/low-involvement product), participants evaluated the brand, and then they were presented with a comparative advertisement. Further, they recalled their previous evaluations. The advertisement acted as a "backward frame," altering how consumers remembered their previous evaluations of the brand.

Participants

Sixty undergraduates, from different faculties, aged 18–57 (M = 27.54; SD = 12.32), participated in the experiment.

Stimuli

Information about the Brand

According to the backward framing paradigm, information about the brand acted as a "previous experience" with the brand. It has been designed to have the same structure regardless of the brand that it is related to. In addition, information concerning the corresponding ads, Play-Gravity and Pepsi-IQ, was the same. The purpose of those materials was to provide participants with the same basic information about each brand. An example of information: "Gravity is one of four mobile providers operating on the Polish market (since 2006). According to the international market research agency Synovate, Gravity has earned a distinction for its wide network, which they managed to achieve in a very short time frame. According to Synovate, attention should be drawn to extremely favourable rates, which forced competing networks such as Orange, Plus, Era to lower their prices. Beata Czerniakiewicz, a market analyst from Synovate quotes: 'Gravity has a very efficient and courteous customer service, which can help at any time of day or night.' With high efficiency, Gravity is also introducing new products into the Polish market."

Advertisements

Four comparative print advertisements were used. An ad acted as a "backward frame" in the experiment. Advertisements for familiar brands were selected from a set of several comparative ads. For high-involvement products, an ad for the mobile provider Play (versus Era) was used, and for low-involvement products, a Pepsi (versus Coca-Cola) ad was used. Advertisements for unfamiliar brands (mobile provider Gravity and energy drink IQ) were constructed for the purpose of the experiment. Brand names were selected in a pilot study. Layouts of advertisements for unfamiliar brands were constructed in a way not to resemble familiar brands' ads but, at the same time, to have the same structure as familiar brands' ads. All of them included the advertised product and a headline referring to a competitor brand.

MEASUREMENT OF DEPENDENT VARIABLES

Brand evaluations were measured with anchored, graphic scales (feelings thermometer, face scale) and with associative measures (brand evaluation scale, description of the brand in the participant's own words and in three adjectives). All measures were implemented twice: (1) after reading the description of the brand and then (2) after viewing a comparative ad, which acted as a backward frame. At the second time, participants were asked to *recall* their previous evaluations and descriptions.

Feelings Thermometer

A feelings thermometer was originally used by Braun-LaTour and LaTour (2005). The scale consists of temperatures from 0°C, very cold and unfavorable feeling, to 100°C, very warm and favorable feeling. Low temperatures refer to negative feelings toward the brand. On the other hand, high temperatures refer to positive feelings toward the brand. Respondents were asked to choose the temperature that describes best their feelings toward a brand.

Face Scale

The face scale consists of seven schematic faces from very sad to very happy (Braun-LaTour and LaTour 2005). The participants' task was to choose the face that best described his/her feelings toward the brand. Graphical scales were coded from one for very negative feelings to seven for very positive feelings.

Brand Evaluation Scale

Brand evaluation was assessed on four seven-point semantic differential scales (e.g., favorable-unfavorable, good quality–bad quality). The four measures were averaged to form overall measures of ads; Cronbach-alpha was .94.

Description of the Brand Using Three Adjectives and Participants' Own Words (Associative Measures)

Participants were asked to describe the brand with three adjectives that first came to mind and then to describe the brand in their own words, i.e., to write two to three sentences. Descriptions have been divided into smaller units, in accordance with the principle that one unit carries one opinion: positive, negative, or neutral. For example, the sentence "It forces other mobile operators to change their offer, is ahead of the markets, takes care of customers" was split into three units: (1) It forces other mobile operators to change their offer, (2) is ahead of the markets, (3) takes care of customers. Three independent judges coded the units of descriptions and three adjectives to evaluate an emotional valence of associations on the five-point scale (–2 very negative, –1 negative, 0 neutral, +1 positive, +2 very positive). Judges' evaluations were summed up (positive, negative, and neutral) for each participant. This way, four variables were acquired: judges' sums of brand evaluation obtained on the basis of descriptions, before and after viewing a comparative advertisement, and, similarly, sums for adjectives. Judges' interrater reliability was 0.89, and sums were used for analyses.

DISTRACTION TASKS

In order to eliminate the short-term memory effect, after the presentation of information about a brand, participants completed a distraction task. They were asked to find differences between attached pictures. The average time for solving the task was five to seven minutes. The second distraction task was applied after the presentation of a comparative advertisement. It was a survey on consumer behavior containing 81 items, which had no relationship to the present experiment. The average time of filling the survey was about 10 minutes.

PROCEDURE

Participants were randomly assigned to one of the four experimental groups (Play, Gravity, IQ, Pepsi). After reading information about the brand, they evaluated the brand on the feelings thermometer, face scale, and brand evaluation scale and described the brand with three adjectives and in their own words (two to three sentences). Then distraction task "find the difference" was implemented. Next, they were presented with a comparative advertisement. The ad acted as a backward frame, altering how they remembered their own evaluations. To eliminate short-term memory effects, the distraction task "consumer behavior" was used. After that, they were asked to recall their previous answers on scales and in descriptions. At the end of the experiment, participants were debriefed and informed about the aim of the research.

RESULTS

Semantic and Nonsemantic Measures

According to Manning et al. (2001), regarding comparative advertising's associative effect, using unstructured, open-ended measures of memory is reasonable. It could be expected that anchored numerical measures differ from associative measures in obtaining results of memory distortions. Thus, two scales were calculated: nonsemantic: a sum of feelings thermometer + faces scale (alpha = 0.95) and semantic: a sum of brand evaluation scale + description + three adjectives (alpha = 0.74).

Affinity Index as a Measure of the Resistance to Memory Distortions

Kleine and Kernan's (1988) method of measuring "consumption objects meaning" (affinity index) has been applied to analyses of the resistance of memory to distortions (the stability of memory).

TABLE 32.1
Affinity Indices and Diagonal Values for the Four Experimental Conditions

Product Involvement	Brand Familiarity	Brand	Diagonal Value *before* Viewing a Comparative Advertisement	Diagonal Value *after* Viewing a Comparative Advertisement	Total Value	Affinity Index
High	Familiar	Play	133	129	234	234/(133 + 129) = **0.893**
High	Unfamiliar	Gravity	101	88	174	174/(101 + 88) = **0.921**
Low	Familiar	Pepsi	110	108	210	210/(110 + 108) = **0.963**
Low	Unfamiliar	IQ	100	90	143	143/(100 + 90) = **0.753**

Kleine and Kernan measured the strength of associations between particular consumption objects. In our studies, affinities between descriptions of the brand generated *before* viewing a comparative advertisement and recalled *after* viewing a comparative ad were measured with the affinity index. Phrases of descriptions are regarded as associations, e.g., "a good brand, ecological, helps Africa, ... emotional, expensive, my friends do not use it." Indices were calculated for each of the four experimental conditions. The method of continued associations (Szalay and Deese 1978) provides a foundation from which to build such a measure. The affinity index value for each pair of descriptions in each experimental condition was a measure of the stability of memory (the resistance to memory distortions). Affinity is operationalized as the amount of overlap between two response lists (the number of associations two objects have in common). These associations that first come to mind are the most readily available in memory and the most important for a subject (Tversky 1977). Because a participant's first responses are assumed to be more dominant (i.e., salient), each response is assigned a dominance score that is a measure of its relative salience. These scores were assigned according to Szalay and Deese's (1978) method: six to the first response produced by a participant, five to the second response, four to the third response, three to the fourth through seventh responses, two to the eighth and ninth responses, and one to each subsequent response. Dominance scores for common responses were summed up across subjects for each phrase of a description. This way, two lists of associations were prepared for each experimental condition: associations (phrases from descriptions) generated *before* and recalled *after* viewing a comparative advertisement. The sums of dominance scores for each description (diagonal values) were used for calculating the affinity index. Calculation of the inter-object affinity index involves summing up the dominance scores across the overlapping elements and across stimuli. This total is then divided by the sum of the total dominance scores of the objects being compared. The resulting index value is the proportion of the combined total dominance scores accounted for by the affinial relations. The index values could vary between zero and one and increase in value as inter-object affinity increases. The higher value of the index, the better stability of the memory (resistance to memory distortions). Affinity indices and diagonal values for the four experimental conditions are presented in Table 32.1.

VERIFICATION OF HYPOTHESES

Effects of the Comparative Advertisement on the Memory Distortion of Brand Evaluation

To verify hypotheses, a repeated measures ANOVA was used. Measures of feelings temperature, feelings expressed on the face scale, and brand evaluation scale and evaluations obtained from descriptions and three adjectives were used as dependent variables (DV1: evaluation before viewing the comparative advertisement; DV2: memory of evaluation after viewing the comparative advertisement). Product category (involvement) and brand familiarity were independent variables (in analyses referring to hypotheses 2 and 3).

Hypothesis 1 stated that comparative advertisements distort the memory of brand evaluations and evaluations are recalled as more negative than before viewing a comparative ad. It was also expected that effects of memory distortions are more distinctive in the case of semantic (associative) measures than nonsemantic (anchored) measures. Means and standard deviations for particular measures before and after viewing a comparative advertisement are presented in Table 32.2.

ANOVA for repeated measures has shown the effect of the comparative advertisement on the memory distortion of brand evaluation. As predicted, participants recalled their own evaluations as more negative than they were before viewing a comparative ad. Effects were statistically significant for the brand evaluation scale [$F(1, 59) = 8.98, p < .01$], brand evaluation from the three adjectives [$F(1, 34) = 5.36, p < .05$], brand evaluation from the descriptions [$F(1, 37) = 5.01, p < .05$] and semantic measures [$F(1, 34) = 13.0, p < .001$]. There were no such differences observed for anchored measures: feelings thermometer, face scale, and nonsemantic measures were not significant.

It has been found in previous studies that when an advertisement is received post-experience, it can be used to help consumers reconstruct their prior experience. In previous experiments (Braun 1999; Braun-LaTour and LaTour 2005), advertisements that acted as a backward frame caused more positive thoughts and more positive evaluation of the product advertised. In contrast, the results of our experiment showed that previous experience can be transformed to be more negative. Comparative advertisements acting as a backward frame transformed evaluations of the brand to be more negative. Participants recalled their own thoughts and evaluations as more negative than they were before viewing a comparative advertisement.

The effects of memory distortions were significant for semantic, but not for nonsemantic measures. Similarly, differences between associative and anchored measures have been found in Manning and colleagues' (2001) research.

Our results are in accordance with network models of memory. That is a semantic network that is sensitive to changes and distortions. Furthermore, consumers are not conscious of the influence of advertisements and other marketing stimuli on their memory. Pieces of information from a comparative advertisement are joined to the memory network and are recalled as consumers' own knowledge about the brand (see: Johnson et al. 1993; Schacter 2001). This mechanism has its consequences for the associative structure of the brand stored in the consumer's mind. Negative

TABLE 32.2
Means and Standard Deviations for Evaluations and the Memory of Evaluations (Before and after Viewing a Comparative Advertisement)

	Before Viewing a Comparative Ad		After Viewing a Comparative Ad				
Measure	**M**	**SD**	**M**	**SD**	**n**	**F**	**p**
Feelings thermometer	62.47	13.60	62.85	13.85	60	$F(1, 59) = 1.999$	.163
Face scale	5.27	1.10	5.20	1.07	60	$F(1, 59) = 2.744$	.103
Brand evaluation scale	4.87	1.12	4.66	1.09	60	$F(1, 59) = 8.978$	.004
Evaluation from description	5.45	5.58	4.37	5.54	38[a]	$F(1, 37) = 5.009$	.031
Evaluation three adjectives	7.77	5.24	6.46	4.73	35[a]	$F(1, 34) = 5.359$	.027
Nonsemanic measures	67.70	14.45	68.05	14.65	60	$F(1, 59) = 1.320$	.256
Semantic measures	32.46	10.36	29.26	9.24	35[a]	$F(1, 34) = 12.997$	.001

[a] Several participants did not write associations

evaluation is joined to the associative structure of the brand and can result in the depreciation of the brand. And this is a maleficent effect of comparative advertising.

BRAND FAMILIARITY AND MEMORY DISTORTIONS

According to hypothesis 2, evaluations for familiar brands are more resistant to memory distortions than evaluations for unfamiliar brands. Previous evaluations are recalled as more negative than before viewing a comparative advertisement.

Analysis of variance with repeated measures revealed the effects of brand familiarity on the memory distortion. When comparative advertisements acted as a backward frame, the memory of evaluations for unfamiliar brands was more susceptible to memory distortions than the memory of evaluations for familiar brands. Significant effects occurred for the face scale [contrast: $F(1, 56) = 3.0$; $p < .1$], brand evaluation scale [contrast: $F(1, 56) = 27.57$; $p < .00001$], brand evaluation obtained from descriptions [contrast: $F(1, 34) = 8.48$; $p < .01$], and semantic measures [contrast: $F(1, 33) = 13.35$; $p < .001$] (Figure 32.1). Results in feelings thermometer, nonsemantic measures, and brand evaluation obtained from the three adjectives were not significant.

Calculated affinity indices, as measures of the resistance to memory distortions (the stability of memory) showed that familiar brands are more resistant to memory distortions than unfamiliar brands (mean affinity indices: .928 and .837, respectively). In the case of low-involvement products, comparative advertisements distorted the memory descriptions—more for unfamiliar (IQ = .753) than familiar (Pepsi = .963) brands. However, in the case of high-involvement products, differences in the stability of the memory between familiar and unfamiliar brands were smaller and an unfamiliar

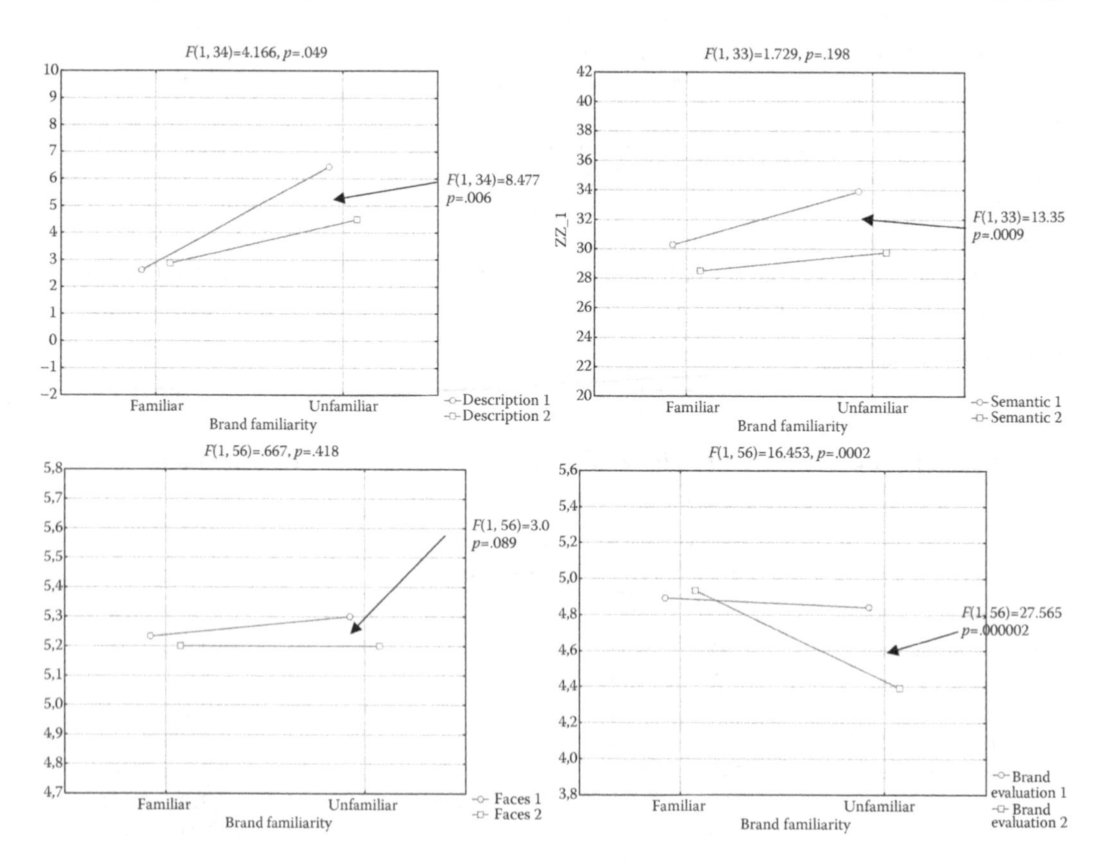

FIGURE 32.1 Effects of comparative advertisement on memory distortions for familiar and unfamiliar brands.

brand (Gravity = .921) was more resistant to memory distortions than a familiar brand (Play = .893). It is worth noting that a familiar brand, Play, elicited more negative associations than an unfamiliar brand, Gravity, (means of judges' evaluation: $M = +4$ and $M = +7$, respectively), and this could be a reason for memory distortions. In addition, a new, unknown brand, Gravity, could cause vigilance and interest and, for this reason, be better remembered. Furthermore, there were more associations for familiar ($M = 120$) than unfamiliar brands ($M = 94.75$), calculated from diagonal values (Table 32.1). The memory network for a familiar brand is better developed than for an unfamiliar brand.

The presented results support hypothesis 2. When a comparative advertisement acts as a backward frame, it is easier to distort the memory of evaluations for unfamiliar than familiar brands. Memory distortions occur rather at the semantic (associative) level than at the nonsemantic level (when measured at anchored numerical scales). An associative structure for a familiar brand is better developed and better consolidated in the memory and, therefore, more resistant to memory distortions. On the other hand, an associative structure for an unfamiliar brand is not well developed and not well consolidated in the memory; thus, it is more susceptible to memory distortions and information acquired after an experience can transform the memory of that experience. The maleficent role of comparative advertising appears especially in the case of unfamiliar, new brands. Thus, new brands shouldn't use comparative advertising. Negative evaluations are joined to the memory network of a sponsored brand, not of a comparison brand.

PRODUCT INVOLVEMENT AND THE EFFECT OF COMPARATIVE ADVERTISING ON MEMORY DISTORTION

It was expected that comparative advertisements acting as a backward frame could distort the memory of previous evaluations of a brand—more easily for low- than high-involvement products. Effects of comparative advertisements on memory distortions of low- versus high-involvement products could be observed only when brand familiarity was taken into consideration. Effects occurred for brand evaluation scale, brand evaluation obtained from descriptions and semantic measures (Figure 32.2) whereas for feelings thermometer, the face scale, and for brand evaluation obtained from the three adjectives results were not significant.

In the brand evaluation scale, the effect of memory distortions occurred for unfamiliar brands and was stronger for low-involvement products [contrast $F(1, 56) = 20.59$, $p < .0001$] than for high-involvement products [contrast $F(1, 56) = 8.34$, $p < .01$]. Similarly, in brand evaluations obtained from descriptions, the effect of memory distortions was a little stronger for low- than high-involvement products [contrasts: $F(1, 34) = 4.83$, $p < .05$ and $F(1, 34) = 3.78$, $p < .10$, respectively]. In the case of semantic measures, stronger memory distortions occurred for low- [contrast $F(1, 31) = 7.87$, $p < .01$] than high-involvement products [contrast $F(1, 31) = 7.17$, $p < .05$; this difference was very small]. This means that an associative network for a familiar brand is well consolidated in the memory, and the effect of product involvement is of no significance here. The effect of product involvement on the memory distortion could be observed only in the case of unfamiliar brands: A comparative advertisement distorted the memory of consumer's opinion stronger for low- than high-involvement products. Therefore, hypothesis 3 has been supported only for unfamiliar brands. In marketing practice, new, unfamiliar brands often use comparative advertising. Apparently, using comparative advertising for low-involvement products by new, unfamiliar brands can have the opposite effect from what marketers intended. Brand evaluations are distorted in the consumers' memory and are recalled as lower than before viewing a comparative advertisement.

Affinity indices as measures of the resistance of memory to distortions also support hypothesis 3. These measures showed that associations (descriptions) for brands of high-involvement products are more resistant to memory distortions as compared to low-involvement products (mean affinity indices: .907 and .858, respectively). This result confirms that the memory is distorted at the level of its associative structure. And comparative advertising is particularly maleficent for low-involvement products.

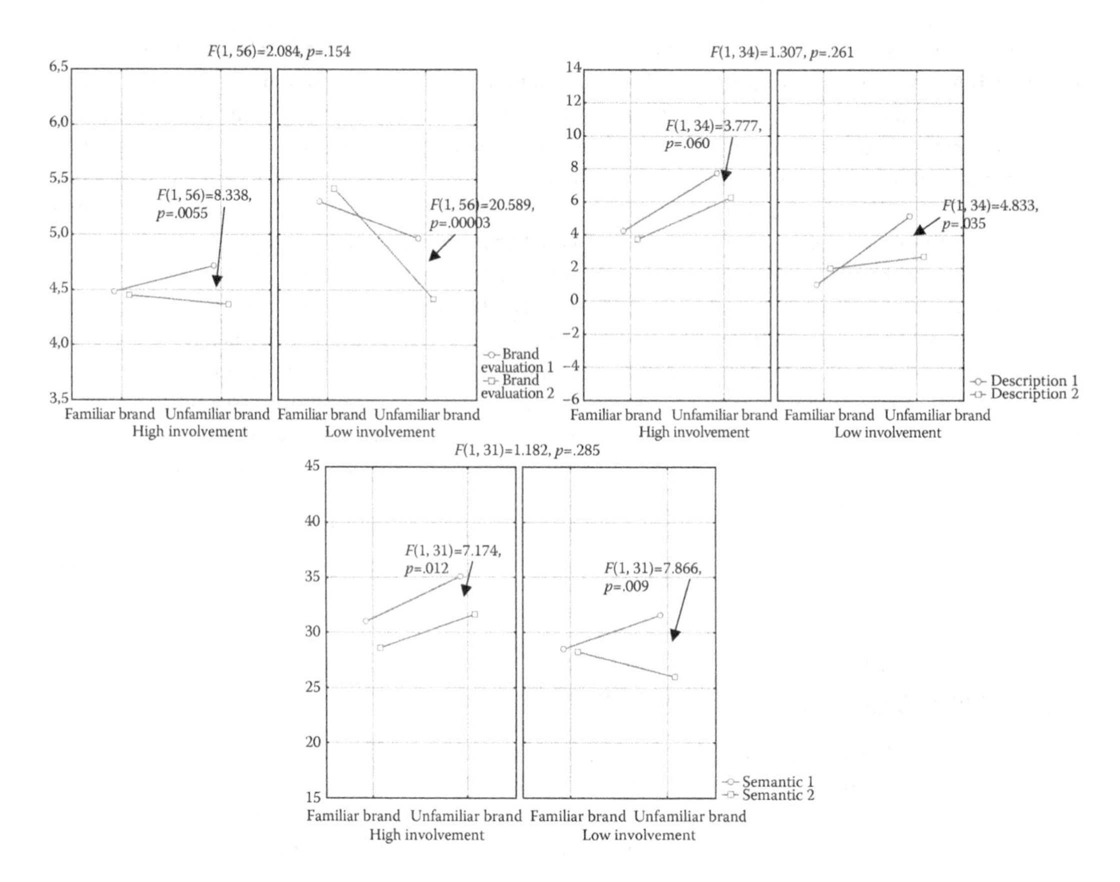

FIGURE 32.2 Effects of comparative advertisement on memory distortions for high- and low-involvement products.

CONCLUSION

Our research showed that the associative and network structure of the brand is sensitive to comparative advertising. The effects of memory distortions were significant for semantic but not for nonsemantic measures. Negative evaluation, joined to the associative structure of the brand, can result in the depreciation of the brand.

A maleficent impact of comparative advertising on memory distortions could be observed particularly for unfamiliar brands. Our research showed that negative evaluations are joined to the memory network of a sponsored brand but not of a comparison brand. Thus, marketers should avoid using comparative advertisements for new, unfamiliar brands.

The effect of product involvement on memory distortions appeared only for unfamiliar brands: A comparative advertisement distorted the memory of consumer's evaluations stronger for low- than high-involvement products. It is possible that semantic network for familiar brands is better consolidated, and it is more difficult to distort it in the memory.

REFERENCES

Anderson, J. R. (1983). *The Architecture of Cognition*, Cambridge, MA: Harvard University Press.

Barry, T. E. (1993). Comparative advertising. What have we learned in two decades? *Journal of Advertising Research, 33*(2), 19–29.

Belch, G. E. (1981). An examination of comparative and noncomparative television commercials: The effects of claim variation and repetition on cognitive response and message acceptance. *Journal of Marketing Research, 18*, 333–349.

Boriboon, N., Alsua, C. and Suvanujasiri, A. (2010). High and low involvement products and their relationship with purchase intentions of Thai consumers. *Journal of International Business and Economics, 10*(2), 37–54.
Braun, K. A. (1999). Postexperience advertising effects on consumer memory. *Journal of Consumer Research, 25*, 319–334.
Braun, K. A., Ellis, R. and Loftus, E. F. (2002). Make my memory: How advertising can change our memories of the past. *Psychology and Marketing, 19*(1), 1–23.
Braun, K. A. and Loftus, E. F. (1998). Advertising's misinformation effect. *Applied Cognitive Psychology, 12*, 569–591.
Braun, K. A. and Zaltman, G. (1997). Backward framing through memory reconstruction, Report No. 98-109 for Marketing Science Institute.
Braun-LaTour, K. A. and LaTour, M. S. (2005). Transforming consumer experience. When timing matters. *Journal of Advertising, 34*, 19–30.
Braun-LaTour, K. A., LaTour, M. S., Pickrell, J. E. and Loftus, E. F. (2004). How and when advertising can influence memory for consumer experience. *Journal of Advertising, 33*, 7–25.
Brisoux, J. E. and Chéron, E. J. (1990). Brand categorization and product involvement. *Advances in Consumer Research, 17*, 101–109.
Cacioppo, J. T. and Petty, R. E. (1984). The elaboration likelihood model of persuasion. *Advances in Consumer Research, 11*, 673–675.
Campbell, M. C. and Keller, L. K. (2003). Brand familiarity and advertising repetition effects. *Journal of Consumer Research, 30*, 292–304.
Celsi, R. L. and Olson, J. C. (1988). The role of involvement in attention and comprehension processes. *Journal of Consumer Research, 15* (September), 210–224.
Chang, C. (2007). The relative effectiveness of comparative and noncomparative advertising. *Journal of Advertising, 36*(1), 21–35.
Chattopadhyay, A. (1998). When does comparative advertising influence brand attitude? The role of delay and market position. *Psychology and Marketing, 15*, 461–475.
Chen, Y. and Wang, Y. (2010). Effect of herd cues and product involvement on bidder online choices. *Cyberpsychology, Behavior, and Social Networking, 13* (August), 423–428.
Coulter, R. A., Price, L. L. and Feick, L. (2003). Rethinking the origins of involvement and brand commitment: Insights from postsocialist Central Europe. *Journal of Consumer Research, 30*(2), 151–169.
Cowley, E. and Janus, E. (2004). Not necessarily better, but certainly different: A limit to the advertising misinformation effect on memory. *Journal of Consumer Research, 31* (June), 229–235.
Dahlén, M. and Lange, F. (2004). To challenge or not to challenge: Ad-brand incongruency and brand familiarity. *Journal of Marketing Theory and Practice, 12* (Summer), 20–35.
Dahlén, M. and Lange, F. (2005). Advertising weak and strong brands: Who gains? *Psychology and Marketing, 22*(6), 473–488.
Donthu, N. (1998). A cross-country investigation of recall of and attitude toward comparative advertising. *Journal of Advertising, 27*, 111–122.
Dröge, C. (1989). Shaping the route to attitude change: Central versus peripheral processing through comparative versus noncomparative advertising. *Journal of Marketing Research, 26*(2), 193–204.
Edelman, G. (1992). *Bright Air, Brilliant Fire*. New York: Basic Books.
Garry, M., Manning, C. G., Loftus, E. F. and Sherman, S. J. (1996). Imagination inflation: Imagining a childhood event inflates confidence that it occurred. *Psychonomic Bulletin and Review, 3*, 208–214.
Gnepa, T. J. (1993). Observations: Comparative advertising in magazines: Nature, frequency, and a test of the "underdog" hypothesis. *Journal of Advertising Research, 33*(5), 73–75.
Golden, L. L. (1979). Consumer reactions to explicit brand comparisons in advertisements. *Journal of Marketing Research, 16*(4), 517–532.
Goodwin, S. and Etgar, M. (1980). An experimental investigation of comparative advertising: Impact of message appeal, information load, and utility of product class. *Journal of Marketing Research, 17*(2), 187–202.
Gotlieb, J. D. and Sarel, D. (1991). Comparative advertising effectiveness: The role of involvement and source credibility. *Journal of Advertising, 20*(1), 38–45.
Grewal, D., Kavanoor, S., Fern, E. F., Costley, C. and Barnes, J. (1997). Comparative versus noncomparative advertising: A meta-analysis. *Journal of Marketing, 61*, 1–15.
Gunasti, K., Baumgartner, H. and Ding, M. (2008). A reexamination of post-experience advertising effects: The moderating role of accuracy motivation. *Advances in Consumer Research, 35*, 864.
Hill, M. E. and King, M. (2001). Comparative vs. noncomparative advertising: Perspectives on memory. *Journal of Current Issues and Research in Advertising, 23*(2), 33–52.

Jain, S. P., Buchanan, B. and Maheswaran, D. (2000). Comparative versus noncomparative advertising: The moderating impact of prepurchase attribute verifiability. *Journal of Consumer Psychology, 9*(4), 201–211.

Johnson, M. K., Hashtroudi, S. and Lindsay, D. S. (1993). Source monitoring. *Psychological Review, 114*, 3–28.

Keller, K. L. (1993). Conceptualizing, measuring, and managing customer-based brand equity. *Journal of Marketing, 57* (January), 1–22.

Kent, R. J., and Allen, C. T. (1994). Competitive interference effects in consumer memory for advertising: The role of brand familiarity. *Journal of Marketing, 58* (July), 97–105.

Kleine, R. E. and Kernan, J. B. (1988). Measuring the meaning of consumption objects: An empirical investigation. *Advances in Consumer Research, 18*, 311–324.

LeDoux, J. E. (1996). *The Emotional Brain: The Mysterious Underpinnings of Emotional Life*. New York: Simon and Schuster.

Loftus, E. F. (1996). Memory distortion and false memory creation. *Bulletin of the American Academy of Psychiatry and the Law, 24*, 281–295.

Loftus, E. F. (2005). Planting misinformation in the human mind: A 30-year investigation of the malleability of memory. *Learning and Memory, 12*, 361–366.

Loftus, E. F. and Pickrell, J. (1995). The formation of false memories. *Psychiatric Annals, 25*, 720–725.

Low, G. S. and Lamb, C. W., Jr. (2000). The measurement and dimensionality of brand associations. *Journal of Product and Brand Management, 9*(6), 350–368.

Machleit, K. A., Allen, C. T. and Madden, T. J. (1993). The mature brand and brand interest: An alternative consequence of ad-evoked affect. *Journal of Marketing, 57*, 72–82.

Manning, K. C., Miniard, P. W., Barone, M. J. and Rose, R. L. (2001). Understanding the mental representations created by comparative advertising. *Journal of Advertising, 30*(2), 27–39.

McClelland, J. L. (1995). Constructive memory and memory distortions: A parallel distributed processing approach. In Daniel Schacter (ed.), *Memory Distortion* (pp. 69–90). Cambridge, MA: Harvard University Press.

McClelland, J. L. and Rogers, T. T. (2003). The parallel distributed processing approach to semantic cognition. *Nature, 4*, 310–322.

Muehling, D. D., Stem, D. E., Jr. and Raven, P. (1989). Comparative advertising: Views from advertisers, media, and policy makers. *Journal of Advertising Research, 29*(5), 38–48.

Muehling, D. D., Stoltman, J. J. and Grossbart, S. (1990). The impact of comparative advertising on levels of message involvement. *Journal of Advertising, 19*, 41–50.

Pechmann, C. and Esteban, G. (1993). Persuasion processes associated with direct comparative and noncomparative advertising and implications for advertising effectiveness. *Journal of Consumer Psychology, 2*(4), 403–432.

Pechmann, C. and Stewart, D. W. (1990). The effects of comparative advertising on attention, memory, and purchase intentions. *Journal of Consumer Research, 17* (September), 180–191.

Pechmann, C. and Stewart, D. W. (1991). How direct comparative ads and market share affect brand choice. *Journal of Advertising Research, 31*(6), 47–54.

Putrevu, S. and Lord, K. R. (1994). Comparative and noncomparative advertising: Attitudinal effects under cognitive and affective involvement conditions. *Journal of Advertising, 23*(2), 77–91.

Schacter, D. (1996). *Searching for Memory*. New York: Basic Books.

Schacter, D. (2001). *The Seven Sins of Memory: How the Mind Forgets and Remembers.* Boston: Houghton Mifflin.

Shimp, T. A. and Dyer, D. C. (1978). The effects of comparative advertising mediated by market position of sponsoring brand. *Journal of Advertising, 7*, 13–19.

Snyder, R. (1989). Misleading characteristics of implied-superiority claims. *Journal of Advertising, 18*(4), 54–61.

Suh, J. C. and Yi, Y. (2006). When brand attitude affect the customer satisfaction: Loyalty relation: The moderating role of product involvement. *Journal of Consumer Psychology, 16*(2), 145–155.

Szalay, L. B. and Deese, J. (1978). *Subjective Meaning and Culture: An Assessment Through Word Associations.* Hillsdale, NJ: Erlbaum.

Te'eni-Harari, T., Lehman-Wilzig, S. N. and Lampert, S. I. (2009). The importance of product involvement for predicting advertising effectiveness among young people. *International Journal of Advertising, 28*(2), 203–229.

Thompson, D. V. and Hamilton, R. W. (2006). The effects of information processing mode on consumers' responses to comparative advertising. *Journal of Consumer Research, 32*, 530–540.

Torres, I. M. and Briggs, E. (2007). Identification effects on advertising response: The moderating role of involvement. *Journal of Advertising, 36*(3), 97–108.
Tsiotsou, R. H., Rigopoulou, I. D. and Kehagias, J. D. (2010). Tracing customer orientation and marketing capabilities through retailers' websites: A strategic approach to Internet marketing. *Journal of Targeting, Measurement and Analysis for Marketing, 18* (June), 79–94.
Tversky, A. (1977). Features of similarity. *Psychological Review, 84*(4), 327–352.
Warrington, P. and Shim, S. (2000). An empirical investigation of the relationship between product involvement and brand commitment. *Psychology and Marketing, 17* (September), 761–782.
Zaichkowsky, J. (1994). The personal involvement inventory: Reduction, revision, and application to advertising. *Journal of Advertising, 23*, 59–70.

33 Brand Depreciation in Indirect Comparative Advertising

The Legal Approach and Psychological Empirical Research on Image Shaping and Brand Positioning on the Polish Advertisement Market

Agnieszka Woźnica-Kowalewska

CONTENTS

INTRODUCTION

In today's reality, we are accompanied by advertisements at all times. Advertising communication reaches us in huge quantities. Among them, there are various types of comparative advertisements, which directly or indirectly compare two or more products or services and which state or imply that the advertised product remains in a relationship with a comparable one irrespective of whether the competitor is mentioned by name (Falkowski and Woźnica 2008a, b).

A comparative advertising communication can take different forms, including direct comparative advertising and indirect comparative advertising. The former refers to a sales promotion message that demonstrates the merits of the advertiser's goods in comparison with a particular product offered by the competitor identified by the name or trademark whereas the latter does not name the competing brand. Indirect comparative advertisements imply some relationship between the two competing products, but do not openly point out exactly which brand is involved in the comparison.

For many years, both types of comparative advertising have been "the number 1" in stirring up global disputes and controversies virtually on all aspects of such messages. A substantial consideration is given to their psychological persuasiveness, marketing efficiency, economic profitability, and ethical relevance and also legal characteristics.

However, even though comparative advertising keeps provoking many contentions and opposing views, its popularity tends to grow continuously. This tendency can be perceived particularly in Europe, Poland included. As the Polish show more and more demand for comparative advertising, social psychologists in our country also become more and more interested in its influence on consumers. The aim of my research is (apart from verifying the persuading effectiveness of comparative advertisements versus conventional ones) carrying out an analysis of the effectiveness of comparative advertising in connection with the ad form (direct and indirect advertisements) and the product kind (economic and political advertisements).

In this paper, I would like to present some of the analyses and findings concerning indirect economic comparative advertising. The study was based on TV commercials about the insurance market in Poland. The comparative advertisement was emitted by the company *Link 4*. The research focused on verifying the influence of the negative message in the comparative commercial made by the insurance company *Link 4* on the changes in the consumer's attitude toward the advertised brand and its opponents as well as the whole professional community of insurance agents. The commercial underlined that the insurance company *Link 4* does not employ agents, which means that it saves on provisions that traditional companies pay to their agents and thus, consequently, it can offer much lower insurance rates to its clients. Therefore, these TV ads make the viewers infer indirectly about the competition of *Link 4*. If the agent inspires the feelings of pity and discouragement, these two emotions put together denigrate his image and—by extension—the image of the company that he is a representative of. Accordingly, the consumer makes the following causal connection: "If the agent is useless (worthless), the company that he represents is useless (worthless) as well."

COMPARATIVE ADVERTISING: LEGAL ASPECTS

Before moving on to the question of the influence exerted upon consumers by comparative advertising, I would like to rest for an indispensable while on the legal context of this type of commercial advertisements. It is worth noting that legal regulations treat comparative advertising with a special consideration. On the one hand, a comparative message is very important in regards to protecting the interests of consumers, for whom commercial comparisons might be the source of necessary market information as well as to the fact that comparisons of commercial goods and services conduce to improving market transparency and stimulating competition. On the other hand, however, this type of advertising involves a huge risk of transgressing the rules of honest competition by creating the impression that the competitor's products are "worse than" or at least "not as good as" the advertiser's ones, which may result in the competitor's good will being harmed.

The Polish jurisdiction governs the activities of comparative advertising on the basis of article 16(3) of the Act of Combating Unfair Competition (*Journal of Laws of the Republic of Poland 2003*, no 153, item 1503). Pursuant to the above mentioned article, comparative advertising is prohibited as an unlawful act of unfair competition if it is *contra bonos mores* (i.e., against good morals). Accordingly, in order to be permitted, it has to fulfill many conditions. First of all, the comparison must not be misleading. In other words, it has to relate to objective, verifiable features of the products that satisfy the same needs. Price is a good example of such a feature. Second, it is not permissible that comparative advertisements aim to discredit the competitor, provoke mistakes between the competitor and the advertiser, or unfairly use the reputation of the other company.

There are therefore strict regulatory restrictions relative to comparative advertising. One may even have an impression that in order to create a proper comparative message without breaching any laws, he or she would have to give it the form of honest, uninteresting, and not attractive information. And yet, in spite of all those restrictions, comparative ads continuously grow in popularity, and many advertisers prefer exactly this sort of promotion of their goods. The questions remain whether this form of commercial message is indeed so effective and what conditions have to be met to validate the statement that comparative advertising is better than conventional one. The analysis of these theses is the subject matter of this study.

BRAND EQUITY

The marketing success of a brand and its reputation is determined by what consumers know about it. In order to make this knowledge a part of the consumers' cognitive structure, it is essential to implement an appropriate brand-building process. The process is based on making advantageous associations with the brand name through various marketing activities, including advertisements.

Prior to bringing up the subject of brand depreciation (that is, of a decrease of its value in the view of consumers), it is necessary to define the notion of brand itself. In this research, I have adopted the definition proposed by Kelvin L. Keller (1993) because his concept highlights the role of the consumer and his reactions to marketing procedures, which are the key factors in analyzing the question of brand depreciation in comparative advertising. Keller's approach to brand equity can also be regarded as the most synthetic because he takes into consideration the roles of the brand owner and the consumer as well as the relationship between what he knows about the brand and how it is advertised. According to Keller, brand equity is based predominantly on prior brand knowledge because it determines the consumer's opinions on this matter. Brand knowledge is described as the brand awareness combined with the brand image, which was created and preserved in the consumer's mind.

Brand awareness is the consumer's ability to identify a given trademark among other ones that are offered in the marketplace. It refers to the strength of its memory trace and the ability to recognize it irrespective of changing circumstances. Brand awareness is measured with the extent to which the brand is recognizable. Widely known brands are well established in consumers' minds and thus easily recalled and recognized.

Brand image is defined as the set of associations created within consumers' minds and reflected in their experience relating to the brand in question.

Keller's whole concept of brand equity is showed herein below in Figure 33.1.

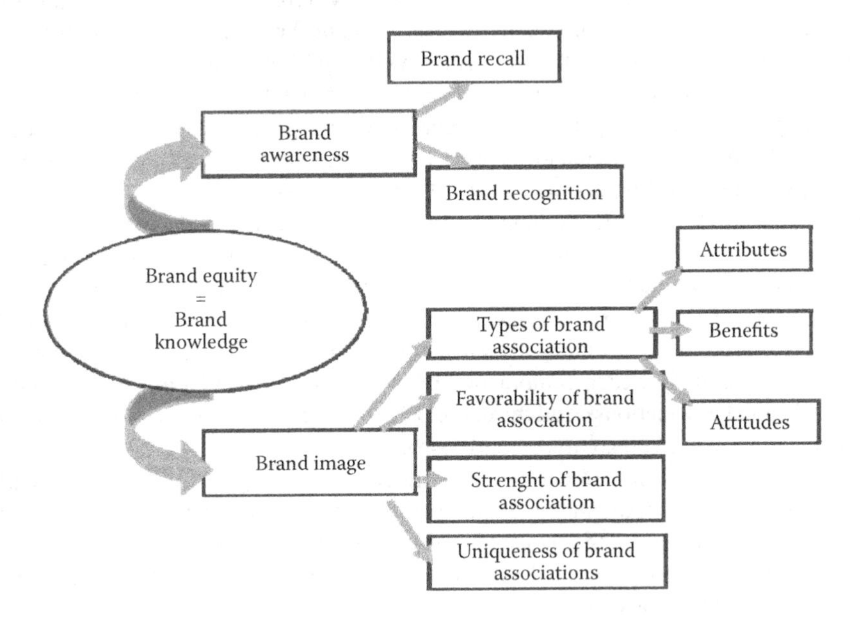

FIGURE 33.1 Dimensions of brand knowledge.

In accordance with the presented guidelines, brand equity is characterized with the strength and number of positive and negative associations with the given brand. Consequently, a high-equity brand is conditional on high brand awareness and positive brand image.

To enhance their brand equity, companies use many types of marketing communications, including advertisements devised to boost positive brand associations in the target audience.

However, comparative advertising has, by definition, yet another task to fulfill, namely depreciation of the competitor's brand, i.e., diminishing its worth by means of the message communicated in the ads. The comparison is made with a view toward denigrating the assumed quality of the compared product by building up negative associations with the competitor's brand. The findings of the empirical research that has been carried out so far on consumer behavior point out that the lower the assumed quality of a product is, the more negative associations the product evokes (Aaker and Keller 1990; Keller 1993).

Thus, negative comparative advertising disturbs the rival brand's structure of associations by provoking negative connotations and simultaneously eliminating positive ones.

ASSOCIATIVE SIMILARITY INDEX

As explained above, brand equity depends upon what consumers know about the brand and how they perceive it. In other words, it refers to a structure of brand associations in the customer's memory. The structure has a cognitive character and determines the connotative meaning of the brand containing the cognitive-affective experience as of yet (Falkowski et al. 2009a).

Still, the crucial question arises concerning how to empirically specify the brand knowledge and to measure brand depreciation caused by negative messages in comparative advertising.

The contemporary professional literature provides many methods of studying connotative meaning, yet it seems that the most effective one is the so-called free association task put forward by Szalay and Deese (1978). The accuracy of this method lies in that the respondent is not presented in advance with the describing characteristics of products or brands but, on the contrary, is expected to describe them himself at will according to his views and preferences. Thanks to this method, it is possible to determine the strength of the associations evoked from the memory into the consciousness with regard to various brands. Szalay and Deese worked on the assumption that the strength of associations with a given object is directly proportional to the order in which they were evoked. The association that came first is more strongly related to the object than the ones that came later on. On this basis, there appeared the descending function of the association strength, which, in turn, inspired further research. The findings led to precise computations, which enabled them to determine the associative structure of the connotative meaning of objects. Finally, the first association was assigned the number six; the second one, the number five; the third one, four; the fourth up to the seventh one got the number three; the eight and ninth one, two; whereas all the others were assigned the number one.

The associative structure of the connotative meaning of objects was later used by Kleine and Kernan (1998), who devised a method of measuring the similarities in connotative meaning of two objects with aid of the associative index. This method has proved critical in analyzing comparative commercial messages. According to this method, the total value of the similarity index is obtained in the following way: First, the values of associations measured for each person and each object need to be added. Second, for each combination of two pairs of objects, one needs to add up only the values of identical associations. And third, the outcome needs to be divided by the total number of all the associations concerning two compared objects.

This study was undertaken to focus on comparative advertising in the insurance market and poses a good example of how the method can be employed. Herein below there are presented the associative similarity indices for the two analyzed insurance companies (*PZU* and *Link 4*, respectively). The indices were worked out on the basis of the responses by a group of people who watched comparative ads created by the company *Link 4*.

	Link 4	PZU	Allianz	Agent
Link 4				
PZU	0.24			
Allianz	0.28	0.37		
Insurance agent	0.09	0.13	0.09	
Ideal company	0.16	0.26	0.15	0.27

FIGURE 33.2 Matrix of the associative similarity index between insurance companies, an insurance agent, and an ideal company (pretest for the group with the *Link 4* ad).

In the case of the pretest: $\frac{(186+104)}{1226}=0.24$.

In the case of the post-test: $\frac{(132+50)}{1070}=0.17$.

The similarity indices for two particular items go to the making of the triangle matrix of similarity indices (Figure 33.2), which is the basis of multidimensional scaling (to be discussed next).

MULTIDIMENSIONAL SCALING

Multidimensional scaling is a technique for taking consumers' perceptions of items (brands) and representing them on a perceptual map, which makes it possible to visually display exact similarities between the objects being compared. Displaying respondents' perceptions of related products on a visual grid can be of a perceptive type (when the map shows real products available on the market) or of a preferential one (when the map includes also a point of maximum preference representing an ideal brand). The closer to this point of maximum preference a given brand is located, the better opinion it has, and the more liked it is by consumers.

In conclusion, thanks to multidimensional scaling it is possible to check whether comparative advertisements influence consumers' opinions about the advertised brand as well as about the rival brands.

METHOD, PROCEDURES, AND MEASURES

The influence of indirect negative comparative TV commercials created by the insurance company *Link 4* on perceiving insurance agents and rival brands was researched in accordance with the experimental procedure of a pretest and a post-test (meaning that measurements were taken before and after a stimulus exposure). The research was carried out according to the below-presented scheme:

In the research sample, there were 88 people with secondary or university education and aged 19–55 (on average, 26.7 years old). The respondents were randomly allotted to one of the three groups described in Figure 33.3.

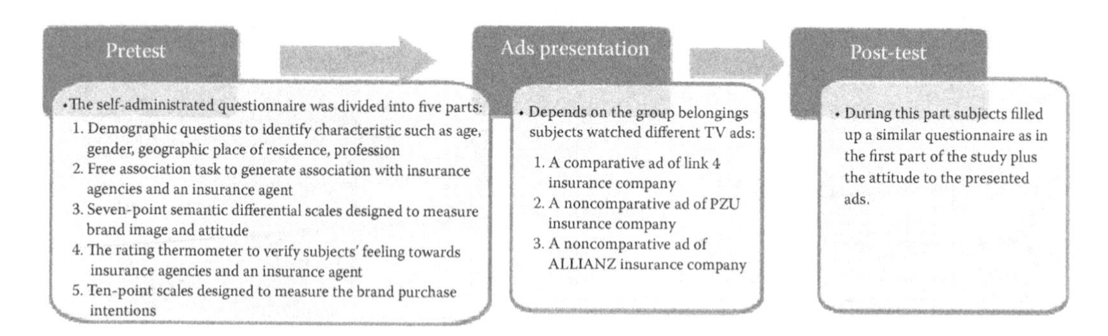

FIGURE 33.3 Research scheme.

RESEARCH FINDINGS

The research findings corroborated the initial assumptions concerning the results of the indirect negative comparative advertising. It turned out that the presented commercials indeed increased the brand equity of the advertiser (i.e., *Link 4*) and, moreover, had an unfavorable effect on a few elements that build the brand equity of rival companies.

1. The influence of indirect comparative advertising by Link 4 on the insurance agent's image.
   a. After the ad exposition, a decrease in positive opinions about an insurance agent's image was clearly observable.
   b. The agent's image depreciation after watching the negative *Link 4* commercial consisted of him being perceived as unprofessional, disagreeable, useless, old-fashioned, inefficient, unapproachable, and not giving a sense of security (Falkowski et al. 2009b).
   c. There was also a change in his general assessment. The research findings clearly show that after watching the ads, the viewers' evaluation of an insurance agent was negative. The change in opinion is statistically relevant ($t = 2.51$; $p < 0.02$) (Figure 33.4).
   d. The *Link 4* comparative ad presentation also visibly disinclined the respondents from making use of the insurance agent's services ($t = 3.03$; $p < 0.01$) (Figure 33.5).
2. The influence of indirect comparative advertising by Link 4 on the brand images of the rival insurance companies.

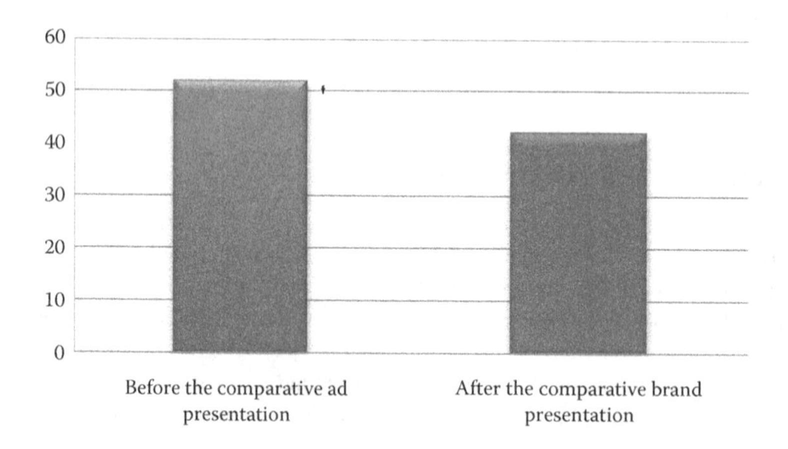

FIGURE 33.4 General evaluation of an insurance agent.

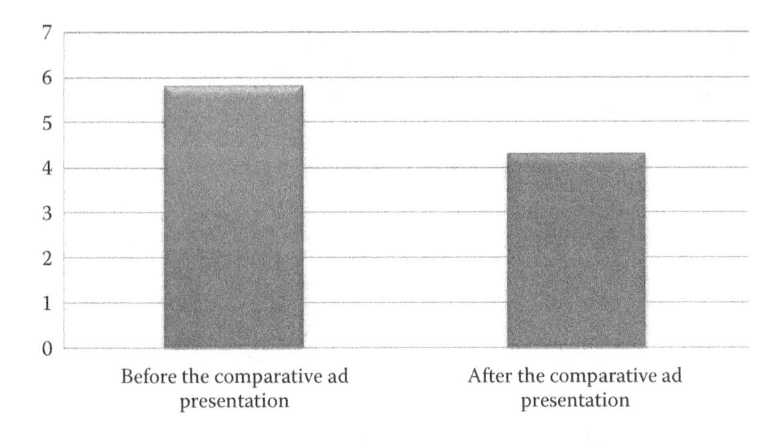

FIGURE 33.5 Intention (make use of insurance agent services).

The presentation of the *Link 4* comparative commercials changed also the respondents' approach toward other insurance companies. It was observed that they started regarding the brand images of the competitors (such as *PZU* and *Allianz*) as inferior to the one of the advertised company, which, in addition, gained in marketability. It is worth underlying that the brand depreciation is noticeable especially with reference to the company *Allianz*, which confirms the current research on the influence of advertising activities on brand evaluation (Figure 33.6). Widely known brands are rather resistant to advertising, and their image changes to a very small extent (*PZU*) while the customers' opinion on not so familiar brands is much more prone to considerable changes (*Allianz*) (Campbell and Keller 2003).

Figure 33.7 shows some changes in the general evaluation of the insurance companies before and after the comparative ad presentation. Here, the gain of the advertised company is not so spectacular as in the case of the brand image, and we can also observe a rather small decrease in the evaluation of *PZU*. All the same, being the least familiar to the respondents, the company *Allianz* was clearly evaluated as much worse.

1. An analogous result was obtained with regard to consumer behavior toward the insurance companies. The *Link 4* advertisements inclined the respondents to use the services of this company while making them feel discouraged from choosing *PZU*, let alone *Allianz* (Figure 33.8).

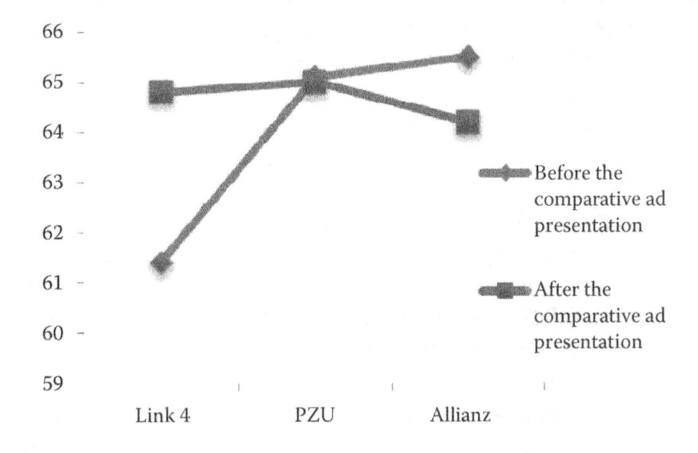

FIGURE 33.6 The brand image of different insurance companies (14 of seven-point semantic differential scales).

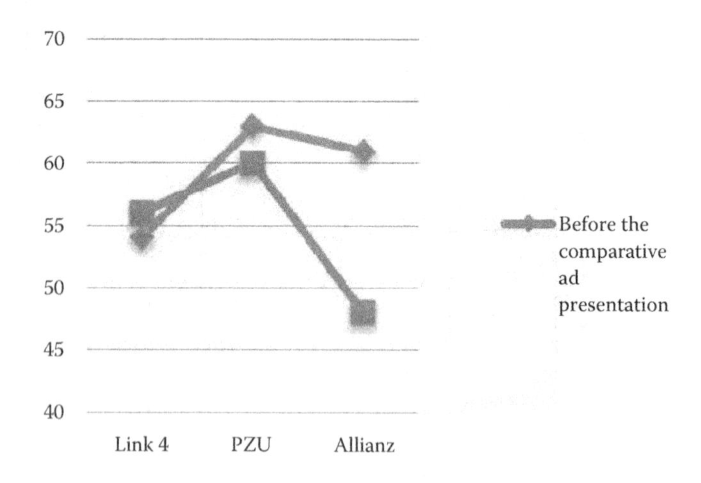

FIGURE 33.7 General evaluation of insurance companies.

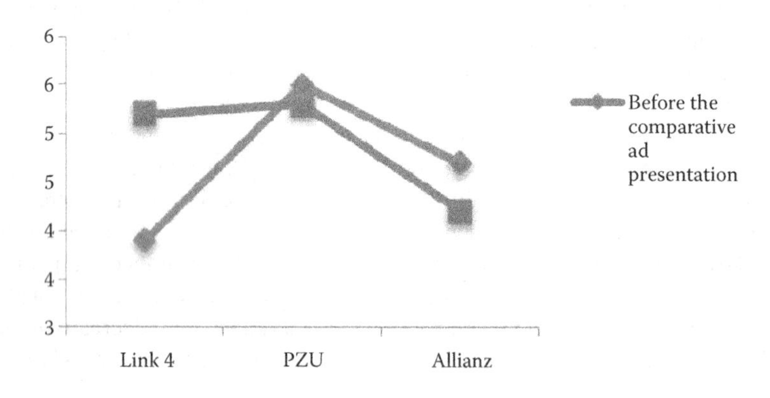

FIGURE 33.8 Intention (make use of insurance companies services).

BRAND DEPRECIATION RESULTING FROM NEGATIVE INFORMATION IN COMPARATIVE ADVERTISING

What is particularly interesting is the analysis of the customers' associative structures of perceptions concerning the insurance companies and the changes in their views after watching negative comparative TV commercials for *Link 4*. Negative comparative advertising has a harmful effect on the associative structure of the competitor's brand in the sense that it provokes associations of inferiority while eliminating the positive elements of the image. In contrast, the advertised brand (here the insurance company *Link 4*) enjoys the opposite effect. Namely, this phenomenon is the core of brand depreciation.

The way to use the method of free associations and how to follow the procedure of creating associative similarity indices has already been presented in this paper. In this analysis, the following steps were taken: First, the respondents were asked to provide as many associations with the given insurance companies (*Link 4, PZU, Allianz*), an insurance agent, and an ideal insurance company as possible (in both situations, before and after the ads presentation). Second, there were calculated the similarity indices with regard to each insurance company, the insurance agent and the ideal insurance company. Then, the lists of associations were compared in search of the ones common for two items. And finally, there was implemented the method of multidimensional scaling with the use of the MINISSA program in order to precisely illustrate, on the one hand, which insurance companies are close to one another or the agent or the ideal insurance company (similarities), and on the other hand, which are apart (dissimilarities). The scaling pointed out which items were recognizable to the viewers and which were unfamiliar, and it also illustrated the influence of the comparative ad presentation on the disposition of the elements on the map.

Figure 33.9 shows the representations of the insurance companies and of the agent and their disposition on the scaling graph for the group that watched the *Link 4* comparative advertisement. The closer the distance between two elements, the more similar they appeared to the respondents.

On the graph above, there are presented spatial maps of preferences in terms of the insurance market as of before and after the comparative advertisement presentation. It is noticeable that before the comparative advertising got involved, the insurance companies and the agent are rather distant from one another and also quite far away from the ideal insurance company. It seems justified, therefore, to conclude that despite all those brands representing the same category—"insurance services"—consumers see them as clearly different. Yet, they do not show a strong preference for one particular brand, which makes them rather tolerant in their choice as they do not exclude any of the presented items. The distances from the ideal insurance company (i.e., the maximum preference) are as follows: the brand *Link 4* (d = 1.51), the brand *PZU* (d = 1.00), the brand *Allianz* (d = 1.51). The closer to the ideal company the given brand is located, the distance d between them gets smaller. Inversely, the farther they are from each other on the graph, the distance d is greater. Accordingly, it can be clearly seen that before watching the *Link 4* comparative ads, the consumers considered the brand *PZU* as relatively the most similar to the ideal insurance company (on the left side of the graph, *PZU* is at the smallest distance to the ideal insurance company).

Interesting changes in the spatial representation of the brands were observed after the insurance company *Link 4* launched its indirect negative comparative advertising campaign. The presentation of the *Link 4* advertising messages changed the map of preferences on the Polish insurance market in a significant way. It actually did not affect *PZU*, but it weakened *Allianz* and strengthened brand equity of the advertiser. Thanks to its comparative advertisements, the brand *Link 4* came significantly closer to the ideal insurance company (d = 0.95). Having seen the ads, the respondents started perceiving *Link 4* as definitely much similar to the ideal insurance company. As far as the rival brands are concerned, the comparative advertising changed the customers' attitude toward *PZU* to a tiny extent—the distance between PZU and the ideal insurance company slightly diminished (d = 0.75). In this context, there was observed no depreciative effect of the comparative advertising with reference to *PZU*. However, the conclusions concerning the firm *Allianz* are completely to the

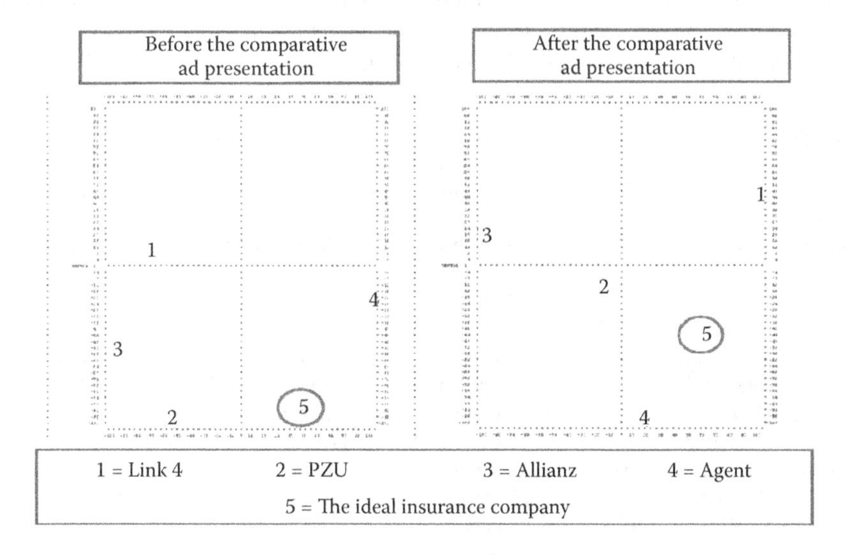

FIGURE 33.9 The association structure in the perception of insurance companies and an insurance agent.

contrary. The negative indirect comparative advertising presented by *Link 4* changed the way in which *Allianz* was perceived: The respondents thought it less similar to the ideal insurance company than before watching the ads ($d = 1.71$). In this case, therefore, the comparative advertising exerted the intended effect by depreciating *Allianz* as the rival brand and making it less popular among customers. The findings of this research corroborate the already mentioned principle of high resistance that well-established brands (*PZU*) show toward marketing strategies and susceptibility of a less known brand (*Allianz*) to depreciation present in a competitor's comparative advertising (*Link 4*). Only widely known brands have a steady place in consumers' minds, and this makes them fairly resistant to such advertising messages.

CONCLUSION

The research analysis presented in this article confirmed the initial assumptions concerning the effects of negative indirect comparative advertising. The findings achieved by means of the semantic differential analysis (the influence of comparative advertising on the brand image of each insurance company) as well as of the associative similarity index and multidimensional scaling (the influence of comparative advertising on how customers perceive each brand and which one they prefer) showed that indirect comparative advertising effectively enhances the brand equity of the advertised company (*Link 4*) and significantly contributes to brand depreciation of the competitors, especially the ones that are not very well known on the market (*Allianz*).

The next step in conducting research on comparative advertising will be an analysis of its effectiveness in exercising influence not only in the domain of marketing, but also in the area of politics with the two form variables taken into account: directness and indirectness.

Another remarkably interesting aspect of the research will be the analysis of the very contents and direction of the associations constituting the brand equity. According to the above-presented assumptions, the brand equity is determined by the strength and number of positive and negative brand associations. The more positive associations there are generated and highlighted by advertising, the higher the brand equity becomes; consumers view the advertised brand as closer to the ideal than other brands. Brand depreciation consists of a decrease of positive brand associations or an increase of negative ones, and this is the aim of comparative advertising. However, the question arises whether depreciation of a brand depends in its strength more on a decrease in positive associations with the brand or rather an increase in negative associations with it. It can be assumed that the strength of negative assumptions will be more visible in this particular type of advertising message. Indeed, it turned out that adding negative associations contributed more to the competitor's brand depreciation than eliminating positive associations.

REFERENCES

D. A. Aaker (1991). *Managing Brand Equity: Capitalizing on the Value of a Brand Name*, The Free Press. New York.

D. A. Aaker, and K. L. Keller (1990). Consumer evaluation of brand extensions, *Journal of Marketing, 54*, s. 27–41.

M. C. Campbell, and K. L. Keller (2003). Brand familiarity and advertising repetition effects, *Journal of Consumer Research, 30*, s. 292–304.

A. Falkowski, A. Dermont, and J. Szymaniak (2009). *Percepcja polityków w świetle wyobrażeń wyborców. Pozycjonowanie według skojarzeniowego indeksu podobieństwa*, w: J. Miluska (red.) *Polityka i politycy. Diagnozy, ocena, doświadczenia*, Wydawnictwo UAM, Poznań, s. 279–293.

A. Falkowski, M. Szymański, and A. Woźnica (2009). Deprecjacja marki w reklamie porównawczej. Psychologiczne badania na rynku ubezpieczeniowym, *Marketing i Rynek, nr 1*, s. 15–21.

A. Falkowski, and A. Woźnica (2008a). Pułapki reklamy porównawczej. *Marketing i Rynek, 15 (1)*, 17–23.

A. Falkowski, and A. Woźnica (2008b). Reklama porównawcza w praktyce. *Marketing i Rynek, 2*, s. 25–32.

K. L. Keller (1993). Conceptualizing, measuring, and managing customer-based brand equity, *Journal of Marketing, 57*, 1–22.

R. E. Kleine, and J. B. Kernan (1998). Measuring the meaning of consumption objects: An empirical investigation, *Advances in Consumer Research, 15*, 498–504.

L. B. Szalay, and J. Deese (1978). *Subjective Meaning and Culture: An Assessment through Word Associations*, Lawrence Erlbaum Associates, New York.

34 The Effects of Learned Helplessness and Message Framing on the Processing of Verbal and Visual Information in Advertisement

Magdalena Gasiorowska

CONTENTS

The aim of the presented research is to show relationships between learned helplessness and information processing. Helplessness is associated with depression (Seligman 1975; Sedek et al. 2010), passivity, lack of initiative, a pessimistic style of thinking, loss of intrinsic motivation, and deterioration of functioning. In several studies the causes and consequences of learned helplessness, especially negative ones, have been explored. It has been shown that as a result of an experience with an uncontrollable situation, a group of cognitive, emotional, and motivational deficits arises. Individuals who experience helplessness cope with new problem situations less effectively, engage in action less often, make more mistakes, and focus on themselves and not on the task. These effects have been investigated in many domains, such as depressive disorders; educational attainment; and substance abuse, treatment, and hospitalization (e.g., Ozment and Lester 2001;

Sutherland and Singh 2004; Chovil 2005; Gottheil et al. 2002; Bodner and Mikulincer 1998; Cemalcilar et al. 2003). Studies on a cognitive model of learned helplessness by Sedek and Kofta (1990, 1993) show that the deficits affect mainly more sophisticated levels of processing. In uncontrollable surroundings, intensive cognitive effort in task solving cannot lead to real progress and results in an altered psychological state, which Kofta and Sedek (1993) term cognitive exhaustion. In this state, simple ways of thinking are used, and complex cognitive processes (such as information-gathering strategies) are blocked.

Consequently meaning helpless people have limited cognitive resources. Several studies suggest that they process information less thoroughly, peripherally rather than centrally, not paying attention to major aspects of messages addressed to them (Amichai-Hamburger et al. 2003). For this reason, helpless subjects were not able to perceive, process, and remember information requiring high cognitive resources. Verbal stimuli are an example of stimuli demanding great cognitive effort to be processed in contrast to pictorial stimuli, which is processed with much greater ease. Words require deeper and more precise processing, and images, on the other hand, are processed more shallowly and superficially (Kosslyn 1990; Braun and Loftus 1998; Grochowska and Falkowski 2012a). Therefore, due to cognitive deficits, helpless people should process verbal and pictorial information differently.

On the one hand, learned helplessness is associated with cognitive deficits but, on the other, with emotional dysfunctions. For this reason, the emotional aspect was included in the presented studies. Helpless people more often experience negative emotions, such as anxiety, anger, sadness, and self-dissatisfaction (Sedek and Kofta 1990). Thus, due to these emotional deficits, they should perceive emotional information in a different way as compared to individuals who do not experience helplessness. There is a stream of research referring to the prospect theory proving that positively framed messages (emphasizing gains) are processed differently than negatively framed messages (emphasizing losses). However, so far, there are no studies undertaken to combine the perception of gains and losses with learned helplessness syndrome. These relationships seem to be important because of the specific function of negative emotions in learned helplessness. It is known that negative emotions contribute to analytical and precise processing (Ashby et al. 1999; Falkowski and Grochowska 2009). However, this is not the case with people experiencing learned helplessness. Negative emotions co-occur with cognitive deficits, which manifest in shallow levels of information processing. In this article, the relationships between cognition and emotion in this meaning are examined.

In the presented research, print ads have been used for several reasons. First of all, some studies suggest that elements of print advertisements may be understood as elements of an ad hoc category organizing into a whole by a common goal: to encourage the customer into purchasing the product (Grochowska and Falkowski 2010). These elements are brand name, headline, photograph, product illustration, and brand claims (Keller 1987). Second, an advertisement is a material that can be divided into verbal and pictorial elements and can be positively framed (emphasizing the advantages or benefits the product can bring to the consumer) or negatively framed (emphasizing the losses or disadvantages consumers may suffer for not using the product) (Maheswaran and Meyers-Levy 1990). Finally, there are a growing number of studies conducted on customer dissatisfaction from products or services offered by companies and complaints submitted by clients. The issue of helplessness is increasingly considered in this context (La Forge 1989; Wells and Stafford 1995; Sharad and Hardik 2008; Gelbrich 2009; Krishnan 2010). For instance, Gelbrich (2009) shows which methods are effective in maintaining customer satisfaction and brand loyalty. It turns out that instrumental support is effective in gaining service recovery when customers experience a lack of control. Moreover, the awareness that opportunity to assert one's rights exists may reduce feelings of helplessness (Wells and Stafford 1995), and conversely, learned helplessness can become a cause of passivity in taking the complaining action (La Forge 1989). According to the assumptions presented above, a feeling of helplessness appears in a consumer's life, and attempts to control it are made. However, in the marketing area, there is a gap in the research into the way in which persuasive messages are processed by helpless consumers. In addition, the ads are a regular part of everyday life, and purchasing decisions are made continuously. Therefore, it becomes interesting to investigate the mechanisms supporting the processing of advertisements considering helplessness.

THEORETICAL BACKGROUND

Learned Helplessness in Information Processing

The phenomenon of learned helplessness (LH) was demonstrated for the first time in experiments in animals (Seligman and Maier 1967; Overmier and Seligman 1967). In Seligman's experiments, exposure to inescapable shocks caused dogs that were not able to produce an adequate response to reinforcements to a new problem. In other words, prior exposure to control deprivation impairs performance on a subsequent task. The same mechanism was soon observed in humans. Prolonged deprivation of control led to a number of changes in human behavior called the learned helplessness syndrome (Seligman 1975; Hiroto and Seligman 1975; Maier and Seligman 1976). This consists of motivation decline toward solving a problem (motivational deficit), depressed mood, and a tendency to experience negative emotions (emotional deficit) and, finally, difficulties in learning effective reactions (cognitive deficit) (Seligman 1975; Sedek 1995).

According to classical LH theory (Seligman 1975), during exposure to uncontrollability people learn that there is no contingency between action and outcome and form a generalized expectation of noncontingency in the future, which is directly responsible for LH deficits. Therefore, in the original theory, response-outcome noncontingency is considered as the crucial aspect; this means that the source of helplessness is located in a lack of control over reinforcements. Under experimental conditions, participants are deprived of influence on the course of events, which is called the helplessness training (Sedek and Kofta 1990). There were several methods commonly used to induce a control deprivation: inescapable noise (Hiroto 1974; Hiroto and Seligman 1975; Hatfield et al. 2002), electric shocks (Seligman and Maier 1967; Overmier and Seligman 1967), and unsolvable tasks (Sedek et al. 1993; Amichai-Hamburger et al. 2003). The last method is currently the most popular one, and the effectiveness of such training appears to be well explained by the cognitive model of helplessness proposed by Sedek and Kofta (1990, 1993).

According to this model, in an uncontrollable situation people engage in intensive but unsuccessful cognitive work. Meaningless cognitive involvement leads to the state of cognitive exhaustion in which constructive and integrative mental processing are impaired. It has been found that cognitively exhausted individuals use simple ready procedures instead of building effective programs of action. One might say that they cognitively froze. Generative thinking is blocked; intrinsic motivation fades. As a consequence, the global impairment in the performance of new tasks can be observed (Kofta and Sedek 1993). However, a lower level of performance does not mean completely chaotic behavior. An individual just switches to an easier cognitive style requiring less effort. One may say, the cognitive system works in a "lazy" way. In uncontrollable situations, this might be a manifestation of adaptive behavior to such uncontrollability. Furthermore, a reduced level of performance does not apply to all tasks and all cognitive functions. The impairment concerns especially the higher cognitive functions that require the intensive cognitive effort (McIntosh et al. 2005). According to this view, such adverse consequences are likely to occur in tasks that are complex and cognitively demanding while the performance of simple tasks remains unchanged (Sedek and Kofta 1990; Sedek et al. 2010). In other words, helpless individuals do not have problems in using lower cognitive strategies, and cope with simple tasks. However, they might have difficulties in using more sophisticated strategies. Constructive and integrative mental processing are especially impaired. Deficits concern mainly tasks that require the integration of partial information into coherent mental representations. Harris and Highlen (1982) show a relationship between helplessness and conceptual complexity. In their research, conceptually complex participants (with higher capacity to integrate stimulus input) were more resistant to the helplessness training in comparison to conceptually simple participants.

Cognitive deficits in the functioning of helpless people, identified above, might also affect their consumer experiences and decisions. Experience with different companies, brands, products, and lots of information about them, coming from different sources, often inconsistent, could be a kind of helplessness training. A consumer plunged into helplessness as a result of inconsistent activities undertaken by some companies or as a result of receiving conflicting and contradictory information about specific products or brands, may have difficulties in processing the numerous, complex, and ambiguous marketing stimuli. In the presented research, a print advertisement is examined because consumers in a state of helplessness may have trouble in understanding the text contained in the ad. They could also process positive and negative messages in different ways. These problems will be discussed in the following parts of the article.

Learned Helplessness in Processing of Verbal and Pictorial Information in the Advertisement

Pictures are remembered better than words. The picture-superiority effect was presented in several studies (Nelson et al. 1976; Kosslyn 1990; Braun and Loftus 1998). Pictures are remembered more easily and are more persuasive. However, they could be distorted in memory more easily (Kosslyn 1990; Braun and Loftus 1998). In contrast, verbal stimuli are much less susceptible to distortion but demand greater cognitive effort in a coding phase. In other words, more cognitive resources are needed to memorize the verbal message, and this message is better established in the memory. However, a pictorial message is processed with less effort but is more easily distorted in the memory. According to Craik and Lockart's (1972) levels-of-processing approach, the level of processing semantic information (using verbal code) is deeper than the level of processing graphemic information (using pictorial code). In a classic investigation of reconstructive memory processes, it has been found that the verbal label affects the way the pictorial object was encoded and later reconstructed in the memory. When participants were given verbal labels for the presented unambiguous pictorial objects, pictures were reconstructed in the memory according to the verbal label (Carmichael et al. 1932; Neisser 1967). To sum up, verbal information requires deep and precise processing, and pictures are processed in a shallow and superficial way (Kosslyn 1990; Braun and Loftus 1998; Grochowska and Falkowski 2012a). On the basis of the assumptions presented above, we can expect that individuals who have insufficient cognitive resources (lack of motivation, lack of ability, tiredness, helplessness) may have difficulties in remembering verbal stimulus, but they should not have a problem with memorizing a pictorial one. Individuals who have appropriate resources to process the text are more likely to focus their attention on the text and remember it better than pictorial information.

An advertisement is a specific kind of stimulus. A picture is an integral part of the ad, and pictorial ads are much more attractive and more easily perceived than verbal ads (Chang 2006; Keller and Block 1997). However, it is the verbal information that is important in a proper communication with the consumer. Similarly to other verbal stimuli, verbal ads are processed with more difficulty, but again—if a consumer has already put effort into their processing—they are being less often distorted (Braun-LaTour et al. 2004). The problem arises when a consumer is not motivated to process information or, more important from the perspective of the discussed issues, if the consumer does not have sufficient cognitive resources for deep processing as is the case of helpless people. For example, Amichai-Hamburger et al. (2003) found that individuals after helplessness training were more influenced by the attractiveness of the figure that appeared in an advertisement (thus referring to the picture), whereas participants without helplessness training were more affected by the quality of the arguments in an advertisement (thus referring to verbal information). On this basis, one can say that people under helplessness training switch their processing of advertisements from central to peripheral.

To sum up, helplessness is closely associated with impaired cognitive functions. A helpless individual is unable to engage in deep, effort-demanding and, more complex cognitive processes (von Hecker and Sedek 1999); however, he or she is capable of processing information at a shallower level, which does not require much effort (McIntosh et al. 2005). And if the text, in comparison to a picture, needs deeper processing, it appears likely that helpless individuals should process verbal information in an advertisement less thoroughly, but they should have no difficulties with processing pictorial information.

Hypothesis 34.1

Participants in the failure condition (unsolvable tasks) are less affected by the verbal message than the pictorial message whereas participants in the no-failure condition (solvable tasks) are less affected by the pictorial message than the verbal message. Helpless participants cannot afford to process the text, so they focus their limited resources on the picture. Participants without training engage their efforts in the text processing. ■

Learned Helplessness in Perception of Gains and Losses: An Emotional Aspect of the Prospect Theory

In the original form of the prospect theory, Kahneman and Tversky (1979) explain decision making under risk. They show that the relationship between subjectively perceived gain (loss) and an objective monetary value is not linear. Small monetary values are overestimated in the subjective evaluations in comparison to large values, and, what is more important: large monetary values are underestimated in the subjective evaluations, compared to small values. Moreover, the value function is steeper for losses than for gains, which means in practice that "losses loom larger than gains." In this context, the prospect theory is considered only in terms of cognition. However, it is worth noting that emotional processes are engaged in perceiving gains and losses. The feeling of losses causes negative emotions and conversely, the feeling of gains causes positive emotions. Therefore, in this paper, the prospect theory was enriched with the emotional content.

There is a stream of research emphasizing the emotional aspect in perception of gains and losses in consumer behavior. With a view to the undertaken study, attention should be paid to two paths of research conducted in consumer behavior topics, referring to the prospect theory in this emotional approach. The first concerns the consumer motivation and distinction of products purchased for pleasure (promotion) from products purchased out of necessity (prevention) (Rossiter et al. 1991; Zhou and Pham 2004). According to regulatory focus theory, people differently process information when they strive for achieving gains (promotion focus) or avoiding losses (prevention focus) (Higgins 1997). The second stream seems to be especially interesting in as much as it is directly associated with emotions. This concerns the issue of positive versus negative message framing effectiveness. The feeling of gains from the purchase or the feeling of losses from not having the product could be created by an advertisement communication. The photograph, brand claims, or the headline can emphasize the advantages or benefits the product can bring to the consumers (for example, "You keep beautiful and young looking skin for a long time") or can emphasize the losses or disadvantages consumers may suffer for not using the product (for example, "You will have wrinkles and flabby and tired skin"). This way of meaning to the prospect theory is the subject of numerous studies (e.g., Block and Keller 1995; Chang 2007; Cox et al. 2006; Lee and Aaker 2004).

Effects of positive and negative message framing have been investigated in several empirical studies, but the results are inconclusive. For example, research showed that in the case of health-related products, negatively framed messages are more effective than positively framed messages because advertisements of medical services and medicines emphasizing losses are analyzed more thoroughly in comparison to ads emphasizing gains (Smith and Petty 1996; Tsai and Tsai 2006).

The effect is observed mostly for high-risk products whereas for low-risk products a positively framed message is more effective (Chang 2007). Chang (2007) found that positive frames are more effective than negative frames for unfamiliar brands of medicine. However, in the case of transformative products, positive message framing produces usually the best advertising effect and is more persuasive (Eagly and Chaiken 1993; Levin and Gaeth 1988; Buda and Zhang 2000). Moreover, according to Donovan and Jalleh (1999), the involvement plays an important role for the effectiveness of these two types of communication. Positively framed messages are more effective in a low-involvement condition, and negatively framed messages are more effective in a high-involvement condition. Similar results were obtained by Martin and Marshall (1997) and Maheswaran and Meyers-Levy (1990). It has also been found that mood moderates processing positively and negatively framed messages. In the positive mood, it is more effective to appeal to the losses, and in the negative mood to the gains (Keller et al. 2003; Chang 2007). On this basis, one may say that the effectiveness of message framing is determined by situational factors, such as involvement and emotions. It is worth looking at the emotional one.

Previous studies suggest that negative emotional information is processed more thoroughly with greater attention and accuracy (Ashby et al. 1999; Falkowski and Grochowska 2009). The threat increases vigilance and concentration on the threatening stimulus. On the other hand, positive emotional stimuli are processed widely but superficially. Talarico, Berntsen and Rubin (2009) showed that a positive affect enhances the recall of peripheral details. Other research showed that stimuli incompatible with mood can become more persuasive (cf. Agrawal and Duhachek 2010; Keller et al. 2003). Grochowska and Falkowski (2012b) found that consumers are more sensitive to messages that violate their expectances. Therefore, if a consumer expects positive communication in the ad, a greater sensitivity will be observed when negative stimuli appear. And conversely, if the negative communication is expected, the individual will react with greater sensitivity to positive ads. Additionally, Buda and Zhang (2000) showed that when information in a message violates expectations, it is processed more thoroughly. One can assume that when people expect the message to be framed positively, the negatively framed message will be processed more thoroughly. However, negatively framed messages are analyzed more extensively only when people expect the message to be positively framed. Although message framing expectations in advertising are mostly positive, in some instances they might be negative. For example, from medical or health advertisements, people expect negative information (Buda and Zhang 2000; Grochowska and Falkowski 2012b), such as a running nose or a back pain. Thus, positively framed ads of health products should be more accurately analyzed because they are unusual and unexpected. The processing of incompatibility between message framing and expectations to the ad requires considerable cognitive resources. Therefore, helpless consumers might have difficulties in perceiving the inconsistency and process a negative message more deeply concentrating on a direct threat (without a correction for their expectations). There is some research into depression proving that depressed people exhibited significantly greater difficulty in identifying the contradictions in the text than those in a normal mood (Ellis et al. 1997; cf. Sedek et al. 2010). Relationships between helplessness and depression are very close and were emphasized in previous parts of the article.

On the basis of the presented assumptions, it seems that helpless people are less sensitive to messages in the advertisements due to their limited cognitive resources. This will be especially evident for positively framed medicine or health ads that violate consumer's expectations. People who do not experience the helplessness can dispose their cognitive resources, allowing for more accurate processing of the message, particularly that which is contrary to expectations, such as a positive framing in medicine advertisements.

To sum up, the prospect theory is used in this article to examine the role of helplessness in gains and losses perception. A helpless consumer is a person with impaired cognitive functions who processes information with lower involvement in comparison to an individual who does not experience helplessness. If the involvement is low and ad processing is peripheral, one may expect that the consumer will be less sensitive to the ad. On the other hand, consumers who are not helpless administer a larger pool of resources and process the message more thoroughly, and we can assume that they will be more sensitive to advertisements. Because consumers expect messages emphasizing losses for advertisements of medical and health products, they will be more sensitive to the ads emphasizing gains. The medicine advertised in a positive way might cause a sort of dissonance, resulting in a greater sensitivity to the ad. Therefore, it is expected that consumers will be less sensitive to positively than negatively framed messages. However, sufficient resources are essential to process that incompatibility, thus, helpless people do not perceive it and react with a higher vigilance to the losses (according to the "losses loom larger than gains" assumption) without taking into account that this is a health care product advertisement.

Theoretical Model

To test hypotheses, a theoretical model for print advertisement is proposed (Figure 34.1). A structural equation modeling method is used to explain relationships between emotional and cognitive determinants of processing information of the ad. Cause-effect relationships between emotions, ad evaluation, product evaluation, purchase intention, and anxiety are included in the model to understand the entire effectiveness of an advertisement. The efficiency is defined here as a sensitivity to the advertisement (vigilance), and the fit of the empirical data to the theoretical model is treated as a measure of such sensitivity. Therefore, the model can be interpreted as the emotional-cognitive network, sensitive to the content of the ad.

The presented model is based on Burke and Edell's (1989) research, which examined relationships between emotions elicited by an advertisement and evaluation of the ad and brand. One may say, emotions elicited by the ad influence the ad and product evaluation and, indirectly, the purchase intention (Grochowska 2009). Moreover, trait anxiety and state anxiety have been included in the model. The processes of perception are moderated by personality traits (Bruner 1957). Research conducted into consumer behavior also indicates that there are differences in the ad perception depending on personality traits. Among them, a role of the anxiety was mentioned several times (Hill 1988; Jones and Murray 1996). For example, risk perception is determined by anxiety (Kallmen

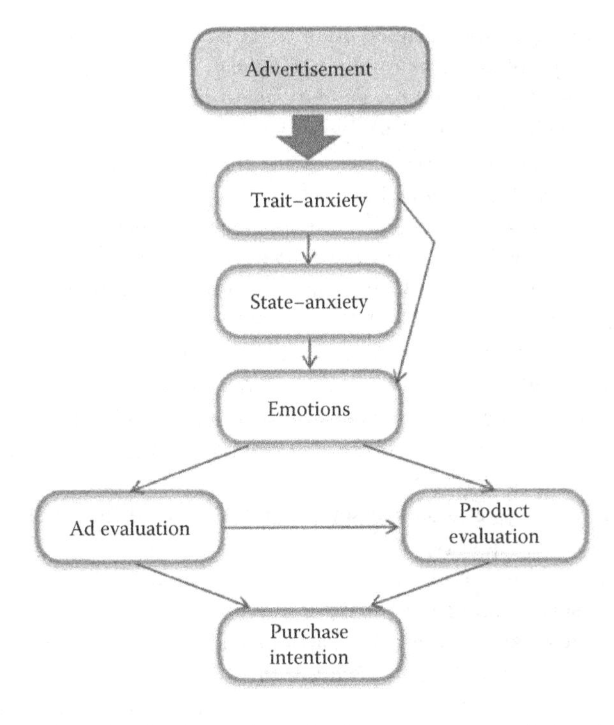

FIGURE 34.1 Theoretical model: Anxiety and emotion in the perception of advertisements.

2000; Turner et al. 2006). Thus, it can be assumed that in ad perception, the anxiety will moderate the feeling of gains and losses associated with the product.

As it was noted, helpless consumers have limited cognitive capabilities and a lower involvement in information processing. Thus, it can be expected that such consumers will be less sensitive to advertisements, which results in a worse fit of the theoretical model to the empirical data. On the other hand, consumers who do not experience helplessness have a greater cognitive capacity, process messages more thoroughly and, thus, it can be expected that they will be more sensitive to the advertisements (which results in a better fit of the model). Moreover, these effects will be moderated by positively and negatively framed advertisements.

Hypothesis 34.2

A. Participants in the failure condition (unsolvable tasks) are less sensitive to an advertisement than participants in the no-failure condition (solvable tasks).
B. Participants in the failure condition (unsolvable tasks) are more sensitive to negatively than positively framed messages whereas participants in the no-failure condition (solvable tasks) are more sensitive to positively than negatively framed messages. ■

METHOD

Participants

One hundred and fifty-five undergraduate students (Males = 55, Females = 100), age 18–48 ($M = 23.45$), took part in the study. Participants were volunteers from universities and colleges in and around Warsaw.

Materials

Stimuli. The pain reliever Apap was selected as an advertised product. Two print advertisements of Apap were used in the experiment: containing a positively framed message using phrases such as "*With Apap you can return to a perfect image and appearance to others*" and a negatively framed message using phrases such as "*With Apap you avoid having a bad image and appearance to others.*" Each ad consisted of the following elements: the photograph, product illustration, brand claims, brand name, and headline (Keller 1987).

The State-Trait Anxiety Inventory (STAI) by Spielberger, Gorsuch, and Luschene (1970/adapted by Wrzesniewski and Sosnowski 1987) was used as a measure of the anxiety.

The Emotional Network Scale by Falkowski and Grochowska (2009) was used to measure emotions elicited by the advertisement. There were 13 items for positive emotions and eight items for negative ones. The scale allows for three indices: sum of positive emotions, sum of negative emotions, and net affect (the difference between positive and negative emotions). In structural equation modeling, two measures were included: positive emotions and negative emotions.

Ad Evaluation was assessed on four seven-point semantic differential scales (e.g., good-bad, interesting-uninteresting).

Product Evaluation was measured on four seven-point semantic differential scales (e.g., favorable-unfavorable, good quality–bad quality).

The Purchase Intention Scale: Participants marked answers to the question "Would you like to buy this product?" on a seven-point Likert scale.

The Memory Test: Free recall test was used. Participants were asked to list what they could recall from the ad as accurately as possible. Then, the information listed by the participants was categorized into verbal information (brand claims, brand name, and headline) and pictorial information (photograph, and product illustration).

Helplessness training by Sedek and Kofta (1993) was used as a control deprivation method. It was adapted by the authors according to the training used previously in the studies of learned helplessness (Hiroto and Seligman 1975). Participants performed 10 discrimination problems with eight trials for each problem. Every trial was a combination of five dimensions: size of the figure (small or large), shape of the figure (triangle or circle), color of the figure (light or dark), position of line (at the top or bottom of the figure), and size of the letter "r" in the middle of the figure (small or large). The participant's task was to discover the figure feature from the 10 possibilities listed above. In every trial, participants were informed if the feature they were looking for is in this trial. After seeing the eight trials, they were asked to give the correct answer (for example: a triangle figure). In the control group, participants were given 10 solvable problems whereas in the experimental condition participants received 10 unsolvable tasks.

The control questions: Ad familiarity, feeling of control during solving figure tasks, and involvement in ad processing were measured on seven-point Likert scales.

Procedure

The study was conducted in laboratory conditions. At first, participants took part in the training phase. They solved 10 discrimination problems (solvable or unsolvable depending on the experimental condition). Then, the STAI was implemented. Further, they participated in the test phase in which they were asked to view one of the two ads for one minute (positively or negatively framed). After that, the participants were asked to fill out the Emotional Network Scale, Ad Evaluation Scale, and Product Evaluation Scale. Then, they were asked to recall everything that they remembered from the advertisement. Finally, purchase intention scale and control scales were filled out.

RESULTS

Manipulation Check

To examine the validity of the experimental manipulation, an analysis of variance (ANOVA) was conducted. The ANOVA showed that participants in the failure condition reported a lower level of control ($M = 3.84$) than participants in the control group ($M = 4.73$): $F(1, 149) = 11.59$; $p < 0.001$. Pearson's correlation coefficient between the feeling of control and a number of correctly solved tasks in the training phase was also calculated: $r = 0.385$; $p < 0.01$. The more tasks the subjects solved correctly, the higher feeling of control they declared.

Memory of Verbal and Pictorial Information

Repeated measures analysis of variance of text and picture memory was made for the type of the advertisement (positively versus negatively framed) and the informational training (solvable versus unsolvable tasks). The number of recalled elements was analyzed.

The results support hypothesis 1: As we can see in Figure 34.2, individuals after helplessness training do not differentiate the text included in the ad depending on message framing. They remembered a similar number of elements for positively ($M = 1.38$) and negatively framed messages ($M = 1.51$). Instead, there are differences in memory of pictorial information. In the failure condition, participants remembered more pictorial elements for the positively framed ad ($M = 1.55$) than the negatively framed ad ($M = 1.21$): $F(1, 151) = 4.40$, $p = 0.038$. On the other hand, in the solvable task conditions, the differences appeared for memory of verbal information. Participants in the control group remembered more verbal elements for the positively framed ad ($M = 1.76$) than the negatively framed ad ($M = 1.15$). However, there were no differences for pictorial elements. They remembered a similar amount of elements for positively framed ($M = 1.26$) and negatively framed ads ($M = 1.36$).

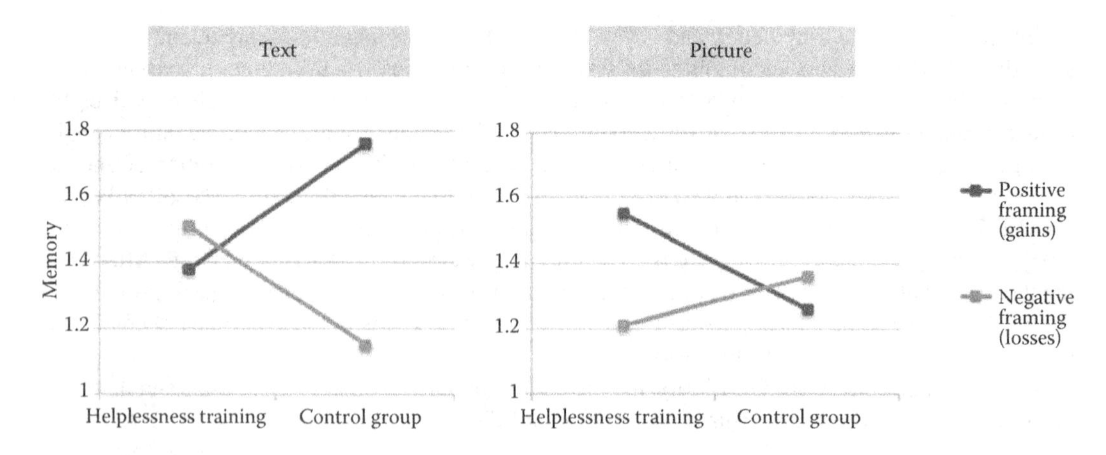

FIGURE 34.2 Effects of helplessness training on memory of verbal and pictorial information for positively and negatively framed advertisements.

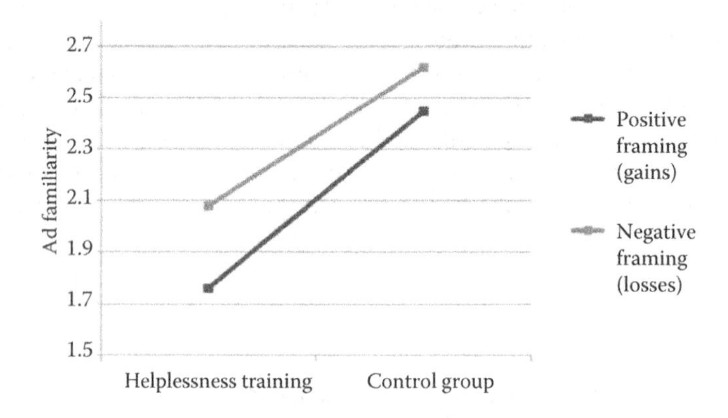

FIGURE 34.3 Effects of helplessness training and message framing on ad familiarity.

Effects of Ad Familiarity

The ANOVA for helplessness training and message framing revealed an interesting effect on ad familiarity.

As we can see in Figure 34.3, for helpless participants, the advertisement seems to be less familiar ($M = 1.92$). By contrast, people in the control group indicate a greater ad familiarity ($M = 2.53$): $F(1, 150) = 4.20$, $p = 0.042$. This effect is opposite to the intuition but seems to be interesting. Advertisements were prepared for purposes of research; thus, participants could not see them before. This means that helpless consumers recognize the advertisement more accurately as an unknown. For people who did not experience helplessness, the advertisement seems to be incorrectly more familiar.

Indices of Fit and Path Coefficients

It was expected that the structure of relationships between anxiety, emotions elicited by the ad, evaluations of the ad and product, and the purchase intention were modified by the learned helplessness and message framing. To examine these expectations, the structural equation modeling was used. The theoretical model (Figure 34.1) was tested for the data obtained for positively or negatively framed ads for the helpless or control participants. Additionally, analyses were conducted separately for the models including positive emotions and, negative emotions. Thus, in the following analyses, the eight empirical models have been tested. Structural equation models, including positive emotions are presented in Figure 34.4a, and negative emotions in Figure 34.4b.

Figure 34.4 presents cause-effect relationships for the helpless or the control participants for positively or negatively framed ads, separately for the model including positive emotions (Figure 34.4a) and the model including negative emotions (Figure 34.4b). Path coefficients are indicated above the arrows, and significant paths are bolded. The chi-square test (χ^2) is used as a measure of fit.

State anxiety and trait anxiety determined the fit of the entire model. The analysis of particular paths showed the effect of state anxiety on emotions elicited by the ad only in helplessness training groups. This means that in these groups, the greater the anxiety, the stronger the emotions. Path coefficients are positive for all of the four conditions although the paths are significant only for the negatively framed ad in the model for positive emotions (Figure 34.4a, model 4) and for the positively framed ad in the model for negative emotions (Figure 34.4b, model 3). This means that in the case of helpless individuals, the greater the anxiety, the stronger the positive emotions that are elicited by the negatively framed ad, and the greater the anxiety, the stronger the negative emotions that are elicited by the positively framed ad. However, results in the groups that did not experience helplessness are different. In these groups, the effect of state anxiety on emotions elicited by the ad was not observed (Figure 34.4a, model 1–2) or the effect was inverse in the case of negative emotions (Figure 34.4b, model 1–2, but not statistically significant).

It is also worth noting that emotions elicited by the ad affect ad evaluation in each condition. This means that on the one hand the more positive emotions induced by the ad, the better the ad evaluation. The effect was observed for both the helplessness and control participants and for both advertisements (positively and negatively framed). Path coefficients are statistically significant for each of the four conditions (Figure 34.4a). On the other hand, the more negative emotions induced by the ad, the worse the ad evaluation. Similarly, this effect was also obtained for all of the four conditions. All path coefficients are negative although they are significant only for the control participants (Figure 34.4b).

Another result shows that the ad evaluation affects the purchase intention but only for the helpless participants and only when the advertisement emphasizes the losses. The effect was observed for both the positive emotion model (Figure 34.4a, model 4) and the negative emotion model (Figure 34.4b, model 4).

Sensitivity to the Advertisement

It was hypothesized that there are effects of the learned helplessness and message framing on the sensitivity to the advertisement. To test these expectancies, the indices of fit for the four conditions were compared separately for the model for positive and negative emotions. Thus, in the following analyses, the eight empirical models described above have been tested. The fit of the empirical data to the theoretical model (χ^2) was a measure of sensitivity to the advertisement: The better the fit (i.e., the lower value of the chi-square test), the higher the sensitivity to the advertisement. The chi-square tests were used to estimate the fit of the empirical data to each theoretical model. The results are shown separately for the positive emotion model (Figure 34.5a) and the negative emotion model (Figure 34.5b).

The results support hypothesis 2A for both models and hypothesis 2B for the positive emotion model and partly for the negative emotion model. As we can see in Figure 34.5a, the fit of the model to the empirical data was better for the control group (lower values of the chi-square test) in comparison to the helplessness training group (higher values of the chi-square test). The outcomes suggest that in the case of positive emotions people who did not experience helplessness are more sensitive to the advertisement than helpless consumers. In other words, helpless consumers are less vigilant and process the message less thoroughly.

We can also observe in Figure 34.5a that the fit of the model for positive emotions to the empirical data was modified by the message framing. For helpless participants, the fit was better in the

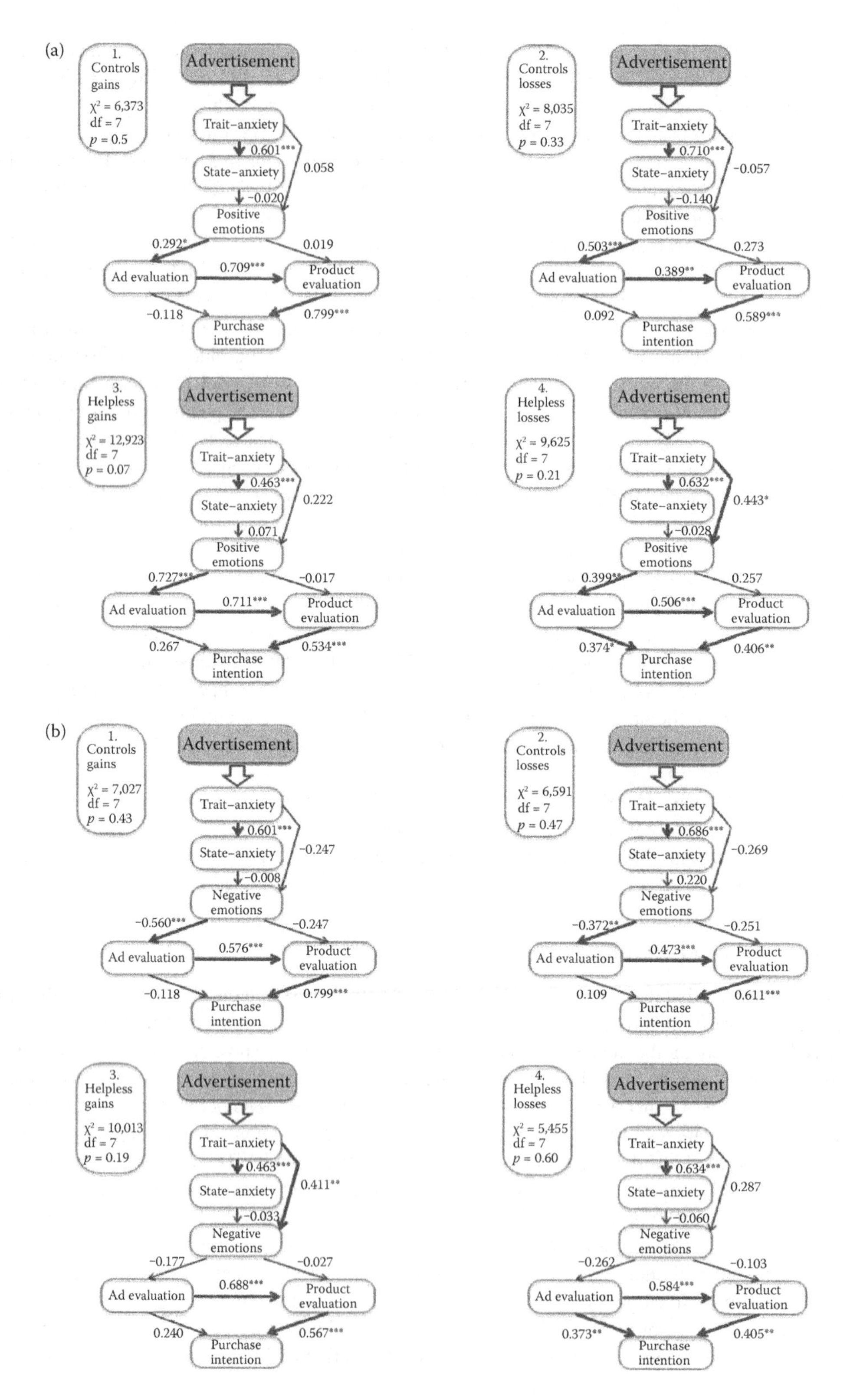

FIGURE 34.4 Structural equation models for positive emotions (a) and negative emotions (b). *, significant at the 0.05 level; **, significant at the 0.01 level; ***, significant at the 0.001 level.

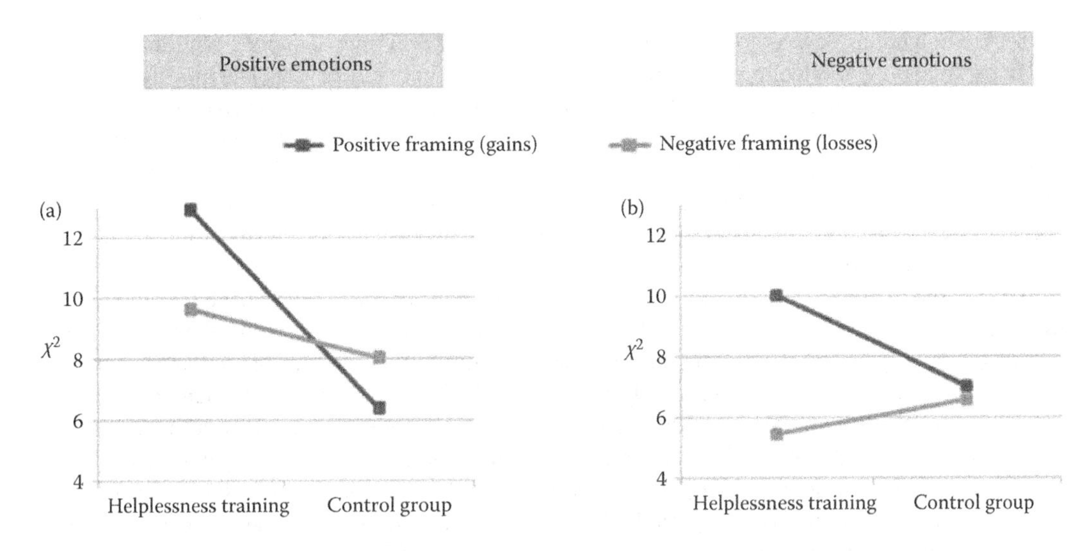

FIGURE 34.5 Values of fits (χ^2) for models for positive emotions (a) versus negative emotions (b).

case of the negatively framed than the positively framed advertisement. For individuals who did not participate in the helplessness training, the better fit was observed if the advertisement emphasized the gains, not the losses. The results support hypothesis 2B, indicating the greater sensitivity to the positively framed ad for consumers who do not experience helplessness and the greater sensitivity to the negatively framed ad for helpless consumers. These outcomes can be explained by the two mechanisms. First, consumers are more sensitive to messages that violate their expectances (Buda and Zhang 2000). Therefore, the health care product that is positively advertised causes a dissonance, resulting in a greater vigilance (Grochowska and Falkowski 2012a). Second, helpless individuals react to a direct threat. They do not have sufficient resources to process the inconsistency. Thus, it leads to a greater sensitivity to the losses regardless of the category of the advertised product.

Furthermore, the results show different functions of positive and negative emotions. As we can see in Figure 34.5b, the fit of the empirical data is much better to the negative emotion model than to the positive emotion model (Figure 34.5a). The fit of the negative emotion model is better, especially for the helpless participants and (among them) particularly for the negatively framed ad. On the other hand, for participants who did not experience helplessness, the fit of the empirical data to the theoretical model is the same for positively and negatively framed ads. One may say the dissonance to the positively advertised pain reliever disappears when negative emotions are activated.

CONCLUSION

This research showed that helpless consumers, in comparison to the consumers who do not experience helplessness, process persuasive messages differently. A helpless person processes an advertisement at a lower level, which results in a better differentiation of pictorial information (requiring fewer cognitive resources) and in a worse differentiation of verbal information (requiring more cognitive resources). Moreover, the impaired cognition functions of a helpless consumer cause a lower sensitivity to the ads incompatible with his or her expectations. This could be explained by the inability to perceive inconsistency between the message framing and one's expectations to the ad in helpless individuals. Finally, for helpless consumers, the advertisement seems to be less familiar. This result is in accordance with network models of memory. A wide area of the cognitive network might be activated in the case of individuals who did not participate in the helplessness training. As a consequence, connections between the presented message and any prior knowledge are formed and lead to a feeling of familiarity. Therefore, the new task for consumers who did not experience helplessness is not as new as for the helpless clients. That is why they can cope with the task better. In this meaning, the feeling of familiarity is a kind of adaptive behavior. The familiarity result might be well explained also in the light of the limited cognitive resources characteristic to the helpless individuals. Control deprivation leads to the impairment of complex cognitive functions, including the integrative processing. Thus, helpless consumers integrate elements of information worse than other consumers, which results in a weaker integration of incoming messages with any previous knowledge. On the other hand, people who do not experience helplessness integrate information better and efficiently connect it to previous knowledge.

Further, the obtained findings exhibit the various functions of positive and negative emotions in the context of learned helplessness. Higher sensitivity to the advertisement observed under negative emotions is particularly evident for helpless individuals. The feeling of vigilance increases then both for the positively and negatively framed ads. This effect supports the statement, which is well established in the literature, that activation of negative emotions leads to a more accurate analysis of the presented stimuli. Negative emotions are responsible for alerting individuals to be careful, especially in the case of individuals with learned helplessness syndrome. They become even more sensitive to negatively framed advertisements. Sensitivity to the ad in the case of consumers who do not experience learned helplessness is only slightly higher when negative emotions are activated. However, the ability to perceive the inconsistency between the expectations to the ads (expected negative messages in a medical advertisement) and the given message (referring to gains) disappears. On the other hand, the inconsistency might be perceived when positive emotions are activated, which results in a greater sensitivity to the positively framed ad of a health care product whereas, under negative emotions, people who are not helpless react with the same level of vigilance to both types of advertisements. On this basis, one may say that the perception of incompatibility requires flexible multifaceted thinking, and that is why positive emotions are helpful in this process. In contrast, negative emotions are helpful in narrow information processing, which results in focusing on the threatening stimulus, such as the negatively framed message in the ad.

In conclusion, negative emotions and limited cognitive functions coexist in people with learned helplessness syndrome. This is reflected by cognitive deficits, particularly in situations when stimulus has to be processed at a higher level of abstraction (how this occurs when the consumers must examine their own expectations and the message targeted at them). Sufficient cognitive resources and flexibility of mind are essential in this process. Therefore, individuals who were not deprived of control cope better with this task, especially when positive emotions are activated. Helpless consumers are unable to handle this. They do not perceive the inconsistency even when positive emotions are activated. However, they are not confused in response to the threatening stimuli. At the time, activation of negative emotions leads to higher sensitivity to the negatively framed advertisement. Moreover, under negative emotions, helpless consumers react to advertisements with much greater sensitivity, which does not differ significantly from the sensitivity of people without learned helplessness syndrome.

Additionally, the results suggest some practical implications for media communication. For example, for helpless consumers, pictorial elements of an ad are more persuasive and better remembered. Further, in the case of such consumers, the use of contradictory elements in advertisements should be reduced. Moreover, for helpless consumers, the negatively framed ad of a health care product is more persuasive, and conversely, for the consumers who do not experience helplessness the positively framed ad is more persuasive.

REFERENCES

Agrawal, N., Duhachek, A. (2010). Emotional compatibility and the effectiveness of antidrinking messages: A defensive processing perspective on shame and guilt. *Journal of Marketing Research, 47 (2),* 263–273.

Amichai-Hamburger, Y., Mikulincer, M., Zalts, N. (2003). The effects of learned helplessness on the processing of a persuasive message. *Current Psychology: Developmental, Learning, Personality, Social, 22 (1),* 37–46.

Ashby, F. G., Isen, A. M., Turken, A. U. (1999). A neuropsychological theory of positive affect and its influence on cognition. *Psychological Review, 106*, 529–550.

Block, L. G., Keller, P. A. (1995). When to accentuate the negative: The effects of perceived efficacy and message framing on intentions to perform a health-related behavior. *Journal of Marketing Research, 32*, 192–203.
Bodner, E., Mikulincer, M. (1998). Learned helplessness and the occurrence of depressive-like and paranoid-like responses: The role of attentional focus. *Journal of Personality and Social Psychology, 74 (4)*, 1010–1023.
Braun, K. A., Loftus, E. F. (1998). Advertising's misinformation effect. *Applied Cognitive Psychology, 12*, 569–591.
Braun-LaTour, K. A., LaTour, M. S., Pickrell, J. E., Loftus, E. F. (2004). How and when advertising can influence memory for consumer experience. *Journal of Advertising, 33 (4)*, 7–25.
Bruner, J. S. (1957). Going beyond the information given. In J. S. Bruner, E. Brunswik, L. Festinger, F. Heider, K. F. Muenzinger, C. E. Osgood, D. Rapaport (Eds.), *Contemporary approaches to cognition* (pp. 41–69). Cambridge, MA: Harvard University Press.
Buda, R., Zhang, Y. (2000). Consumer product evaluation: The interactive effect of message framing, presentation order, and source credibility. *Journal of Product and Brand Management, 9*, 229–242.
Burke, M. C., Edell, J. E. (1989). The impact of feelings on ad-based affect and cognition. *Journal of Marketing Research, 26 (1)*, 69–83.
Carmichael, L., Hogan, H. P., Walter, A. A. (1932). An experimental study of the effect of language on the reproduction of visually perceived form. *Journal of Experimental Psychology, 15 (1)*, 73–86.
Cemalcilar, Z., Canbeyli, R., Sunar, D. (2003). Learned helplessness, therapy, and personality traits: An Experimental Study. *The Journal of Social Psychology 143 (1)*, 65–81.
Chang, C. T. (2006). Is a picture worth a thousand words? Influence of graphic illustration on framed advertisements. *Advances in Consumer Research, 33*, 104–112.
Chang, C. T. (2007). Health-care product advertising: The influences of message framing and perceived product characteristics. *Psychology and Marketing, 24*, 143–169.
Chovil, I. (2005). Reflections on schizophrenia, learned helplessness-dependence, and recovery. *Psychiatric Rehabilitation Journal 29 (1)*, 69–71.
Cox, A. D., Cox, D., Zimet, G. (2006). Understanding consumer responses to product risk information. *Journal of Marketing, 70*, 79–91.
Craik, F. I., Lockhart, R. S. (1972). Levels of processing: A framework for memory research. *Journal of Verbal Learning and Verbal Behavior, 11 (6)*, 671–684.
Donovan, R. J., Jalleh, G. (1999). Positively versus negatively framed product attributes: The influence of involvement. *Psychology and Marketing, 16*, 613–630.
Eagly, H. A., Chaiken, S. (1993). *The Psychology of Attitudes*. Fort Worth, TX: Harcourt Brace Jovanovich, Inc.
Ellis, H. C., Ottaway, S. A., Varner, L. J., Becker, A. S., Moore, B. A. (1997). Emotion, motivation, and text comprehension: The detection of contradictions in passages. *Journal of Experimental Psychology: General, 126*, 131–146.
Falkowski, A., Grochowska, A. (2009). Emotional network in control of cognitive processes in advertisement. In A. L. McGill, S. Shavitt (Eds.), *Advances in Consumer Research,* 36, pp. 405–412.
Gelbrich, K. (2009). Beyond just being dissatisfied: How angry and helpless customers react to failures when using self-service technologies. *Schmalenbach Business Review, 61*, 40–59.
Gottheil, E. M. D., Thornton, C., Weinstein, S. (2002). Effectiveness of high versus low structure individual counseling for substance abuse. *The American Journal on Addictions, 11*, 279–290.
Grochowska, A. (2009). *Procesy kategoryzacji i zniekształcenia pamięciowe w reklamie* [*Processes of categorization and memory distortion in the advertisements*]. Gdańsk: Gdańskie Wydawnictwo Psychologiczne.
Grochowska, A., Falkowski, A. (2010). Conceptual structure of advertisement: Methodological analysis according to ad hoc and common categories. *Ergonomia, An International Journal of Ergonomics and Human Factors,* 32, 1, 13–36.
Grochowska, A., Falkowski, A. (2012a). Message framing in the ads for health-related products: Verbal and visual information for prevention versus promotion products. *The Proceedings of the 2012 Conference of the American Academy of Advertising*, 70–80.
Grochowska, A., Falkowski, A. (2012b). Construction of print advertisement in the context of conceptual coherence and memory distortion. *Advances in Consumer Research, 38*, 289–295.
Harris, R. M., Highlen, P. S. (1982). Conceptual complexity and susceptibility to learned helplessness. *Social Behavior and Personality, 10 (2)*, 183–188.
Hatfield, J., Job, R. F., Hede, A. J., Carter, N. L., Peploe, P., Taylor, R., Morrell, S. (2002). Human response to environmental noise: The role of perceived control. *International Journal of Behavioral Medicine, 9 (4)*, 341–359.

Higgins, E. T. (1997). Beyond pleasure and pain. *American Psychologist, 52*, 1280–1300.
Hill, R. P. (1988). An exploration of the relationship between AIDS-related anxiety and the evaluation of condom advertisements. *Journal of Advertising, 4,* 35–42.
Hiroto, D. S. (1974). Locus of control and learned helplessness. *Journal of Experimental Psychology, 102 (2)*, 187–193.
Hiroto, D. S., Seligman, M. E. P. (1975). Generality of learned helplessness in man. *Journal of Personality and Social Psychology, 31*, 311–327.
Jones, D. B., Murray, K. B. (1996). State-trait communication apprehension dimensions and potential client's responses to help offering ads. *Psychology and Marketing, 13,* 1–17.
Kahneman, D., Tversky, A. (1979). Prospect theory: An analysis of decision under risk. *Econometrica, 47,* 263–291.
Kallmen, H. (2000). Manifest anxiety, general self-efficacy and locus of control as determinants of personal and general risk perception. *Journal of Risk Research, 3*, 111–120.
Keller, K. L. (1987). Memory factors in advertising: The effect of advertising retrieval cues on brand evaluations. *Journal of Consumer Research, 14*, 316–333.
Keller, P. A., Block, L. G. (1997). Vividness effects: A resource-matching perspective. *Journal of Consumer Research, 24 (3)*, 295–304.
Keller, P. A., Lipkus, I. M., Rimer, B. K. (2003). Affect, framing, and persuasion. *Journal of Marketing Research, 40*, 54–64.
Kofta, M., Sedek, G. (1993). Wyuczona bezradność: Podejście informacyjne [Learned helplessness: Informational approach]. In M. Kofta (Eds.), *Psychologia aktywności: Zaangażowanie, sprawstwo, bezradność* (pp. 171–223). Poznań: Nakom.
Kosslyn, S. (1990). Mental imagery. In D. N. Osherson, S. M. Kosslyn, J. M. Hollerbach (Eds.), *Visual cognition and action: Vol 2. An invitation to cognitive science* (pp. 73–97). Cambridge, MA: Massachusetts Institute of Technology Press.
Krishnan, P. (2010). Consumer alienation by brands: Examining the roles of powerlessness and relationship types. *Dissertation Abstracts International Section A: Humanities and Social Sciences, 70 (7–A)*, 2618.
La Forge, M. C. (1989). Learned helplessness as an explanation of elderly consumer complaint behavior. *Journal of Business Ethics, 8*, 359–366.
Lee, A. Y., Aaker, J. L. (2004). Bringing the frame into focus: The influence of regulatory fit on processing fluency and persuasion. *Journal of Personality and Social Psychology, 86 (2)*, 205–218.
Levin, I. P., Gaeth, G. J. (1988). How consumers are affected by the framing of attribute information before and after consuming the product. *Journal of Consumer Research, 15 (3),* 374–378.
Maheswaran, D., Meyers-Levy, J. (1990). The influence of message framing and issue involvement. *Journal of Marketing Research, 27 (3),* 361–367.
Maier, S. F, Seligman, M. E. P. (1976). Learned helplessness: Theory and evidence. *Journal of Experimental Psychology: General, 105*, 3–46.
Martin, B., Marshall, R. (1997). The interaction of message framing and felt involvement in the context of cell phone commercials. *European Journal of Marketing, 33,* 206–218.
McIntosh, D. N., Sedek, G., Fojas, S., Brzezicka-Rotkiewicz, A., Kofta, M. (2005). Cognitive performance after preexposure to uncontrollability and in depressive state: Going with a simpler "plan B". In W. R. W. Engle, G. Sedek, U. von Hecker, D. N. McIntosh (Eds.), *Cognitive Limitations in Aging and Psychopathology* (pp. 219–246). New York: Cambridge University Press.
Neisser, U. (1967). *Cognitive psychology*. New York: Appleton.
Nelson, D., Reed, V., Walling, J. (1976). Pictorial superiority effect. *Journal of Experimental Psychology: Human Learning and Memory, 2*, 523–528.
Overmier, J. B., Seligman, M. E. P. (1967). Effects of inescapable shock upon subsequent escape and avoidance learning. *Journal of Comparative and Psychological Psychology, 63*, 23–33.
Ozment, J. M., Lester, D. (2001). Helplessness, locus of control and psychological health. *Journal of Social Psychology, 141 (1)*, 137–138.
Rossiter, J. R., Percy, L., Donovan, R. J. (1991). A better advertising planning grid. *Journal of Advertising Research, 31 (5)*, 11–21.
Sedek, G. (1995). *Bezradność intelektualna w szkole* [*Intellectual helplessness in school*]. Warszawa: Wydawnictwo Instytutu Psychologii PAN.
Sedek, G., Brzezicka, A., von Hecker, U. (2010). The unique cognitive limitation in subclinical depression: The impairment of mental model construction. In A. Gruszka, G. Matthews, B. Szymura (Eds.), *Handbook of individual differences in cognition: Attention, memory, and executive control* (pp. 335–352). New York: Springer.

Sedek, G., Kofta, M. (1990). When cognitive exertion does not yield cognitive gain: Toward an informational explanation of learned helplessness. *Journal of Personality and Social Psychology, 58 (4)*, 729–743.

Sedek, G., Kofta, M. (1993). W poszukiwaniu uniwersalnych wyznaczników zjawiska wyuczonej bezradności: Przegląd klasycznych wyników eksperymentalnych i test empiryczny koncepcji egotystycznej [Looking for universal determinants of learned helplessness: Research overview on egotism approach]. In M. Kofta (red.), *Psychologia aktywności: Zaangażowanie, sprawstwo, bezradność* (pp. 133–170). Poznań: Nakom.

Sedek, G., Kofta, M., Tyszka, T. (1993). Effects of uncontrollability on subsequent decision making: Testing the cognitive exhaustion hypothesis. *Journal of Personality and Social Psychology, 65 (6)*, 1270–1281.

Seligman, M. E. P. (1975). *Helplessness: On depression, development and death.* San Francisco: Freeman.

Seligman, M. E. P., Maier, S. F. (1967). Failure to escape traumatic shock. *Journal of Experimental Psychology, 74*, 1–9.

Sharad, S., Hardik, S. (2008). Effect of organizational culture on creating learned helplessness attributions in R&D professionals: A canonical correlation analysis. *Vikalpa, 33 (2)*, 25–45.

Smith, S. M., Petty, R. E. (1996). Message framing and persuasion: A message process analysis. *Personality and Social Psychology Bulletin, 22*, 257–268.

Spielberger, C. D., Gorsuch, R. L., Lushene, R. E. (1970). *Manual for the State-Trait Anxiety Inventory*. Palo Alto, CA: Consulting Psychologists Press.

Sutherland, K. S., Singh, N. N. (2004). Learned helplessness and students with emotional or behavioral disorders: Deprivation in the classroom. *Behavioral Disorders, 29 (2)*, 169–181.

Talarico, J. M., Berntsen, D., Rubin, D. C. (2009). Positive emotions enhance recall of peripheral details. *Cognition and Emotion, 23 (2)*, 380–398.

Tsai, C. C., Tsai, M. H. (2006). The impact of message framing and involvement on advertising effectiveness — the topic of oral hygiene as an example. *The Journal of American Academy of Business, 18*, 222–226.

Turner, M. M., Rimal, R. N., Morrison, D., Kim, H. (2006). The role of anxiety in seeking and retaining risk information: Testing the risk perception attitude framework in two studies. *Human Communication Research, 32*, 130–156.

von Hecker, U., Sedek, G. (1999). Uncontrollability, depression, and the construction of mental models. *Journal of Personality and Social Psychology, 77*, 833–850.

Wells, B. P., Stafford, M. R. (1995). Service quality in the insurance industry: Consumer perceptions versus regulatory perceptions. *Journal of Insurance Regulation, 13 (4)*, 462–477.

Wrzesniewski, K., Sosnowski, T. (1987). *Inwentarz Stanu i Cechy Leku (ISCL). Polska adaptacja STAI. Podrecznik* [*Manual for the State-Trait Anxiety Inventory. Polish adaptation*]. Warszawa: Laboratorium Technik Diagnostycznych.

Zhou, R., Pham, M. T. (2004). Promotion and prevention across mental accounts: When financial products dictate consumers' investment goals. *Journal of Consumer Research, 31*, 125–135.

35 Perceived Risk–Return Relationship of Investments and Other Risky Actions

Joanna Sokolowska

CONTENTS

INTRODUCTION

In the last four decades, it has been reported in many studies that people perceive risk and return as negatively correlated. For example, Shefrin (2001) argues that investors expect higher returns from safer stocks. Such reasoning contradicts basic assumptions in theory of decision-making, economy, and finance. A positive relationship between risk and return lies at the heart of these theories. It is assumed that decision makers should think about the expected return and variance of returns (or other measures of risk) and make their choices under the assumption that the higher the risk, the higher the expected return. This holds true for both financial investments and other risky actions, e.g., adopting new technologies or medical procedures.

Because risk brings profits, it was traditionally considered to be a necessary condition of a free market economy, in particular, a necessary condition of "entrepreneurship spirit" (e.g., Knight 1921). This assumes trade-offs between dangers and benefits. The past proved human ability to take a challenge and accept such trade-offs in the name of scientific progress and economic development.

Since the mid-'70s, however, there has been the growing social demand for the "zero-risk society" and a negative attitude toward risk. As Wildavsky and Dake (1990) have noticed, traditionally high regard for risk among Americans no longer holds. According to Mary Douglas (1990), risk has changed its meaning from "chance" to "danger." This is consistent with the social demand for the "zero-risk society." In some instances, it is possible to eliminate a given hazard entirely. For example, if all nuclear power plants were closed, there would be no risk of nuclear meltdown and radioactive fallout. However, by doing so without reducing consumption of electricity, the amount of greenhouse gases in the atmosphere would increase to potentially harmful levels. Are people not aware that the cost-effectiveness function of risk reduction decreases sharply as the risk level falls?

Some empirical findings point at the lack of such awareness. For example, Alhakami and Slovic (1994) found that people perceive an inverse relationship between risk and benefit. This means that hazards judged as very high in danger are also considered very low in benefits whereas hazards judged as low in danger are judged as high in benefits.

Although risk and benefit are positively correlated in the world, empirical results show that they are negatively correlated in people's perception. One possible reason might be that positive and negative consequences of risky activities are rarely symmetric and refer to different domains. For example, online shoppers find a broader selection of goods at lower prices, save time, and so on. On the other hand, they become exposed to possible financial fraud. Another explanation relies on the finding that the inverse relationship between perceived risk and benefits is linked to a person's overall evaluation of an activity or technology (Alkahami and Slovic 1994). On this basis, it has been concluded that confounding benefits and losses in people's minds steams from affect. Researchers have called this process affect heuristic (e.g., Finucane et al. 2000; Slovic et al. 2002a, b, 2004).

Both explanations are rooted in dual-process theories of cognition in which two different modes of cognition are assumed: System 1 and System 2 (e.g., Epstein 2003; Evans 2006; Sloman 1996; Kahneman and Frederic 2002, 2005; Stanovich and West 2000). In most general terms, System 1 is automatic, fast, and experience and affect based whereas System 2 is analytic, deliberative, and reason based. In line with the first explanation, the complexity of relationships between different consequences in the real world requires abstract, rule-based analysis, which is possible only within System 2. Unfortunately, System 1 is activated first, and operations within this system cannot lead to adequate understanding of a situation. In line with the second explanation—the affect heuristic—relying on affect means solving problems within System 1.

In the presented experiments, accuracy of the above theoretical framework for the inverse relationship was checked through variations: (1) in familiarity and knowledge about a risky activity, thus in the ability to use System 2 during the evaluation of positive and negative consequences (using laymen and experts as respondents) and (2) in the extent to which a given risky activity evokes affect (using risky activities related to finance or health and, additionally, being involved in social and moral controversies). One additional issue studied here is the relative impact of cognition and affect on the inverse relationship in evaluating positive and negative consequences of risky activities.

INVERSE RELATIONSHIP AND INFORMATION PROCESSING WITHIN SYSTEM 1

System 1 is considered to be the primary and more basic mode of cognition. For example, Epstein (2003) claims that System 1 is common to humans and higher animals. Basic principles of operation within this system are similarity and contiguity (e.g., Sloman 1996), and the content of these operations are precepts, current stimulation, concrete and generic concepts, images, features, stereotypes, and so on (e.g., Kahneman 2003; Kahneman and Frederick 2002, 2005). Epstein (2003) argues that representations in System 1 are in the form of prototypes, metaphors, and narratives. These characteristics of System 1 make it impossible to capture correctly the description of complex relationships based on hierarchical or logical rules (e.g., not symmetric gains and losses). This could explain the mentioned lack of awareness that the cost-effectiveness function of risk reduction decreases sharply as the risk level falls. In another example, Shefrin (2001) asked respondents to specify the return expected for stocks of eight companies over the next 12 months. Respondents also rated perceived risk for each stock. The return expectations of respondents were consistently negatively correlated with their risk perceptions. Shefrin (2001) explained this result in terms of using the stereotype of a good company instead of considering information about stocks. He supported this explanation with analysis of data from the annual reputation survey conducted by *Fortune* magazine (Shefrin and Statman 1995), which showed that investors considered as good investment stocks of good companies (judged by, e.g., quality of management, quality of products, financial soundness, and so on) with the correlation coefficient of 90%. In line with this stereotype, stocks of good firms are not risky.

Despite almost common agreement that System 1 is the primary one and activated first (e.g., Epstein 2003; Evans 2006; Darlow and Sloman 2010), some authors think that under some circumstances one can overcome this characteristic and activate System 2. For example, Epstein (2003) illustrates this with an example of people's ability to overcome irrational reactions to unfavorable arbitrary outcomes. He modified an experiment by Tversky and Kahneman (1983) in which they found that among people who were late at the airport and missed their flight, those who barely missed the flight were more upset than those whose flight was on time. Epstein's modification was to make participants respond from three perspectives: how they believed most people would react, how they would react, and how a completely logical person would react. The findings supported the hypothesis that, taking the last perspective, people did not report irrational reactions. This suggests that under favorable circumstances people are capable of reflective analysis of a situation.

One factor that may favor activation of reflective thinking is familiarity. Indeed, Ganzach (2000) found that participants (finance majors in an M.B.A. program) generally judged risk and return for unfamiliar stocks either as having high return and low risk or as having low return and high risk. For familiar stocks, however, judgments of risk and return were derived from their values in the financial markets and, therefore, were positively correlated.

One may expect that expertise should cause a similar effect. Thus, in the presented experiments, two groups of respondents were used: laymen and experts. If the reason for the inverse relationship in evaluation of gains and losses is that information is processed and judgments are made within System 1, the relationship should be present among laymen but not among experts.

INVERSE RELATIONSHIP AND AFFECT

In line with Epstein (2003), System 1 is intimately related to experience of affect and is used to facilitate positive affect and avoid negative one. Kahneman (2003) argues that people use attributes that are easy to access. Because affect can be easily substituted for other attributes hard to process (Kahneman and Frederick 2002, 2005), it is extensively used for judgments within System 1.

The second explanation of the inverse relationship in terms of affect is through the mechanism of the affect heuristic. Alhakami and Slovic (1994) found that the inverse relationship between perceived risk and perceived benefit of an activity (e.g., using pesticides) was linked to the strength of positive or negative affect associated with that activity. If people liked an activity, they judged its risks as low and benefits as high. Oppositely, if they disliked it, they judged its risks as high and benefits as low. This implies that people base their judgments of an activity or a technology on what they feel about it, and then the affect heuristic guides perceptions of risk and benefit. This implies that providing information about risk should change the perception of benefit and vice versa. For example, information stating that risk was low for some technology should lead to more positive overall affect that would, in turn, increase perceived benefit. Indeed, Finucane et al. (2000) confirmed these predictions.

If the reason for the inverse relationship is that judgments of risk and benefit are determined by the overall affective evaluation, the inverse relationship should not be present in evaluating risky activities that do not evoke affect. Thus, in the presented experiments, three risky activities were investigated, which differed in the extent to which they evoke affect. It might be also expected that evaluating consequences delayed in time evokes weaker affect than evaluating current consequences. Thus, in one experiment, respondents evaluated both current and future benefits and losses.

RELATIVE INPUT OF COGNITION AND AFFECT INTO INVERSE RELATIONSHIP

The possible interplay between cognition and affect has been prominent in dual-process theories of cognition. Such interplay was studied in decision making with no clear conclusions. In some studies, very strong impact of affect was demonstrated, even among experts. For example, Slovic et al. (1997) surveyed members of the British Toxicological Society and found that their judgments of risk and benefit exhibited the inverse relationship. As expected, the strength of this relationship was found to be mediated by the toxicologists' affective reactions toward the hazardous items being judged. In another study, these same toxicologists were asked to rate 30 chemicals on an affect scale (bad-good). Next, they were asked to judge the degree of risk associated with the exposure to levels of chemicals markedly below those considered as dangerous by the regulatory agency. For such low exposures, one might expect these risk judgments to be negligible, which means that there should be no correlation between the ratings of risk and affect. Instead, such a correlation was very strong. However, Slovic et al. (1997) also reported individual differences in the extent to which toxicologists relied on affective processes in their judgments of chemical risks.

Berndsen and Van der Pligt (2005) examined the impact of affective versus cognitive focus on risk perception and acceptance of meat consumption by presenting participants with risk descriptions in bogus newspaper articles phrased in either more affective or more cognitive terms. They expected that exposure to cognitive or affective stimuli would affect perceptions of health risks and found that affective focus had a stronger impact than cognitive focus. However, Van Gelder et al. (2009) demonstrated that both cognition and affect had impact on risky choices.

In this study, the relative importance of cognition and affect on the presence of the inverse relationship was investigated through surveying experts focused on either technical or ethical aspects of a hazard. They judged not only current but also future benefits and losses at different time perspectives. It was expected that the inverse relationship among experts would be present in judging ethical but not technical aspects. It was also expected that the inverse relationship would become weaker as the time perspective increases.

EXPERIMENTAL DESIGN

Overview

The purpose of this research was to investigate whether the inverse relationship in evaluation of gains and losses results from the following:

1. Low-level analysis of information, which, as a consequence, should be absent in judgments of experts
2. Affect evoked by a risky activity, which, therefore, should be absent in judgments of emotionally neutral activities that are not socially and morally controversial

Additionally, the interplay between cognition and affect was studied using judgments of benefits and losses of a highly controversial hazard. These judgments were (a) made for different time perspectives and (b) done by experts focused on either technical or ethical aspects of the hazard.

In three experiments, respondents evaluated gains and losses, and declared their attitudes toward risky activities related to finance (Experiment 1), health (Experiment 2), and scientific research on stem cells (Experiment 3). It was assumed that the first activity was emotionally neutral whereas the other two were controversial and high in affect. In all experiments, participants were laymen and experts. In Experiment 3, there were two groups of experts focused on either the technical or ethical aspects of the activity. In this experiment, future benefits and losses in five, 10, and 15 years were also judged.

Participants: Laymen and Experts

In Experiment 1, 42 persons working for banks in Warsaw were surveyed as experts and the others—43 students and 90 adults (no unemployed persons)—were laymen. In Experiment 2, experts were 40 medical interns and young physicians; laymen were 160 students who did not study biological or life sciences. In Experiment 3, one group of experts consisted of 27 physicians and biologists who professionally deal with stem cell research. The second group of experts consisted of 28 persons, mostly priests, graduated in humanities and working for or associated with Polish Catholic popular science journals and covering the stem cell research topic. These two groups of experts were selected on the basis of an assumption that they differed with respect to their focus on technical or ethical aspects of stem cell research and therefore might have different overall attitudes toward this hazard.

Selection of Risky Activities

On the basis of the Internet and mass media search, a preliminary list of risky activities was created, including such activities as nuclear plants, production of genetically modified food, drugs and alcohol abuse, *in vitro* insemination, abortion, applications of stem cell research, stock market investments, individual retirement accounts, and extreme sports. Subsequently, three judges evaluated the affect and controversy invoked by these activities. On this basis, three activities were selected: Individual retirement accounts as evoking relatively low affect and controversy and abortion and stem cell research applications as involving affect and controversy related to ethical/religious views. The last hazard was perceived as the most controversial activity.

Creating Lists of Positive and Negative Consequences for the Selected Activities

Individual retirement accounts (IRA): In order to create a list of positive and negative consequences of IRA, information published in mass media, on the Internet, and delivered by financial institutions offering such accounts was analyzed. On the basis of this information and the evaluation by three judges, a list of advantages and disadvantages was created. Some similar consequences, e.g., tax deductions related to capital gain tax and to inheritance tax were grouped. The final list of consequences consisted of six positive items and seven negative items. Subjects evaluated all selected consequences—positive and negative—on an 11-point response scale: from 0: "very small/insignificant" to 10: "very high."

The reliability of the scale for all items was low (*Cronbach* $\alpha = 0.611$). This is in line with the assumption that the scale contains heterogeneous items related either to positive or to negative consequences. Thus, the reliability of these two subscales was checked and a better *Cronbach* α was obtained for each subscale (0.771 and 0.668 for positive consequences and negative consequences, respectively). Removing one item from the disadvantages (i.e., no possibility to have more than one IRA) shifted *Cronbach* α to 0.696.

The two-subscale structure was additionally supported by factor analysis, principal component method with varimax rotation, which yielded two factors accounting for 46% of the variance. The first factor included six positive consequences and the other one included six negative ones.

The factor analysis was repeated for laymen and expert groups. For laymen, the same structure emerged. For experts, however, the best structure was a two-factor solution in which one item (benefits from flexible schedule of payment) was removed from the list of positive consequences due to low factor loading.

Abortion: A preliminary list of positive and negative consequences of abortion was created on the basis of mass media publications and information from medical sites on the Internet. This list was next evaluated by three judges. On the basis of this information and the evaluation by the judges, the final list of consequences consisted of four positive and five negative items. Subjects evaluated all selected consequences—positive and negative—on an 11-point response scale: from 0: "very small/insignificant" to 10: "very high."

The reliability subscales for positive items (*Cronbach* $\alpha = 0.785$) and for negative items (*Cronbach* $\alpha = 0.785$ and 0.715, respectively) was higher than for all items (*Cronbach* $\alpha = 0.542$). One positive (benefits for a position of a woman in the family and in the society) and one negative item (minor immediate negative health consequences) were removed to improve the reliability. *Cronbach* α increased to 0.837 and 0.757, respectively.

The factor analysis performed on nine items, principal component method with varimax rotation, gave the two-factor solution, which accounted for 57% of the variance. The two factors are labeled "positive consequences" and "negative consequences." However, two items had low factor loadings—one positive and one negative—the same which lowered *Cronbach* α. Thus, these items were removed from further analysis. The same structure was obtained for all respondents and for both the laymen and expert groups.

Stem cell research and its applications: To create a list of positive and negative consequences of stem cell (SC) research and its applications, mass media publications were used, and interviews were carried out with professionals either dealing with such research (e.g., working for stem cell banks) or responsible for information about such research in Catholic publications. The list of eight benefits for science and health care (e.g., scientific progress, progress in treatment of genetic diseases and cancer) and seven negatives (e.g., undesirable cell mutations or adoption of malfunctions of cancer cells) were created. Subjects evaluated all selected consequences—positive and negative—on an 11-point response scale: from 0: "very small/insignificant" to 10: "very high."

The reliability of subscales for positive consequences and negative consequences (eight items *Cronbach* $\alpha = 0.939$ and seven items *Cronbach* $\alpha = 0.951$, respectively) was higher than for all items (*Cronbach* $\alpha = 0.554$). The structure was supported by the factor analysis performed on all 16 items, principal component method with varimax rotation. The two-factor solution emerged, which accounted for 76% of the variance. The factors were labeled "positive consequences" and "negative consequences." The same structure that was found for all respondents was also found for the laymen group and for each of the two expert groups.

Index of the Inverse Relationship between Positive and Negative Consequences

According to Alhakami and Slovic (1994), the index of the inverse relationship is negative linear correlation (e.g., *Pearson's r*) between evaluations of gains and losses for a given risky activity. The absence of such correlation or positive correlation points to the lack of the inverse relationship. Unfortunately, positive and negative consequences of real risky activities are rarely symmetric, which makes it impossible to estimate directly such correlations. This can be done only (1) when symmetric consequences are correlated (e.g., benefits from cancer treatment due to stem cell research versus losses resulting from adoption of cancer cell malfunctions by stem cells), or (2) averages for positive and negative consequences for each activity are correlated.

Attitudes and Acceptance

Individual retirement accounts: First, respondents were asked whether they had such accounts. The majority (118 out of 175) did not have an IRA. However, when students were excluded from the analysis, the proportion was 54 versus 78 with no difference in frequency distribution between laymen and experts ($\chi^2 = 0.005$, $df = 1$, $p = .945$).

Abortion: Respondents rated their attitudes toward unconditional legalization of abortion in Poland on an 11-point response scale from 0: "strongly no" to 10: "strongly yes."

Stem cell research and its applications: Respondents rated their attitudes toward stem cell research on a 11-point response scale from 0: "very negative" to 10: "very positive."

RESULTS

Relationship between Current Overall Positive and Negative Consequences due to Expertise and Affect Invoked by a Hazard

In Figure 35.1, the difference in evaluating benefits and losses for the investigated risky activities is presented. This difference was calculated for averaged evaluations of positive and negative consequences.

As can be seen in Figure 35.1, all respondents evaluated the advantages of an IRA slightly higher than its disadvantages. There was no difference in evaluation of consequences between laymen and experts. In contrast, the negative consequences of abortion were evaluated higher than positive ones. This difference is significantly smaller in the expert group than in the layman group. For applications of stem cell research, there was significant difference in evaluating benefits and negative consequences by Catholic priests and by all other respondents. Students, biologists, and physicians evaluated benefits higher than losses. In contrast, Catholic priests evaluated losses higher.

Taking into account a small difference in perception of positive and negative consequences for IRA, one might expect no inverse relationship for this activity. In contrast, one might expect the inverse relationship in evaluating positive and negative consequences of the other two hazards, in particular for laymen and Catholic priests.

Indeed, for IRA the inverse relationship was found neither for laymen (*Pearson* $r = -0.148$; $p = 0.094$, $N = 127$) nor for experts (*Pearson* $r = -.070$; $p = 0.663$, $N = 39$). For abortion, no inverse relationship was found in the evaluation done by medical experts (*Pearson* $r = -0.025$; $p = 0.883$,

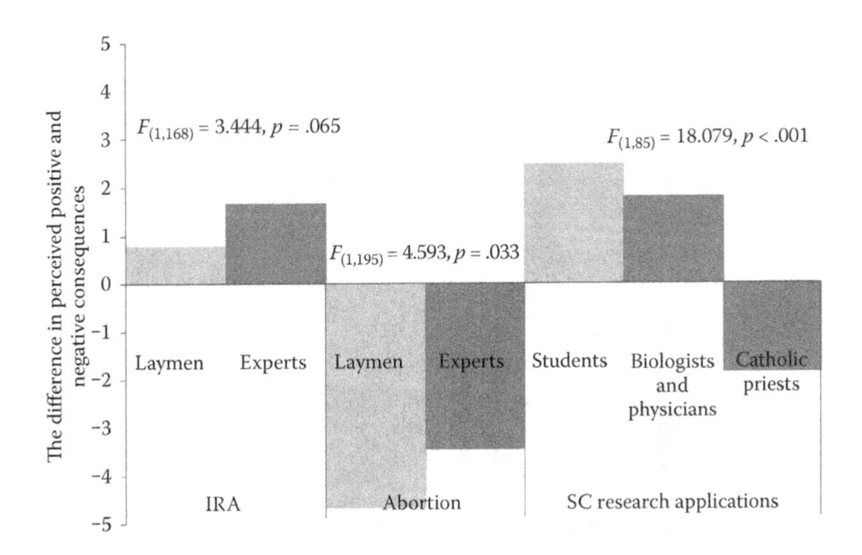

FIGURE 35.1 The difference in evaluation of positive and negative consequences of the investigated risky activities.

$N = 38$). In contrast, such a relationship was observed for laymen (*Pearson* $r = -0.265$; $p < 0.01$, $N = 159$). The significant inverse relationship was also observed in evaluation of SC research applications done by all experimental groups (*Pearson* $r = -0.526$; $p = 0.02$, $N = 33$ for laymen; *Pearson* $r = -0.432$; $p = 0.022$, $N = 28$ for Catholic priests; and *Pearson* $r = -0.06$; $p = 0.977$, $N = 27$ for medical experts and biologists).

These findings are in agreement with the thesis that there is no inverse relationship in judgments of benefits and losses for emotionally neutral risky activities. In contrast, the inverse relationship is observed for hazards that are controversial and evoke affect. Expertise protects against simplified, black-and-white perceptions of hazards (medical experts' evaluations of abortion). However, when affect is strong, its impact on the evaluation of hazards is stronger than the impact of expertise (evaluation of SC research applications by both expert groups).

Relationship between Gains and Losses and Time Perspective

It was possible to determine almost symmetric gains and losses for two applications of SC research, i.e., (1) benefits from genetic disease treatment versus loss from undesirable stem cell mutations and (2) benefit from cancer treatment versus loss from adoption of the cancer cell malfunctions by stem cells. For these two applications, the judgment of future consequences was investigated. The difference in evaluation of future benefits and losses was checked in all groups for three time perspectives: 5, 10, and 15 years. The results are presented in Figure 35.2.

As can be seen in Figure 35.2, students always evaluated future benefits higher than future losses. Biologists and physicians evaluated benefits lower than losses in the five-year perspective, but for longer time perspectives evaluated benefits higher than losses. Catholic priests evaluated benefits higher than losses only in the 15-year perspective.

The results indicate that overall evaluation of hazards became more positive with time. For students, biologists, and physicians, the longer time perspective, the higher observed favorable differences in evaluating positive and negative consequences. Following the same trend, the unfavorable difference observed for Catholic priests diminished with time and eventually becomes positive.

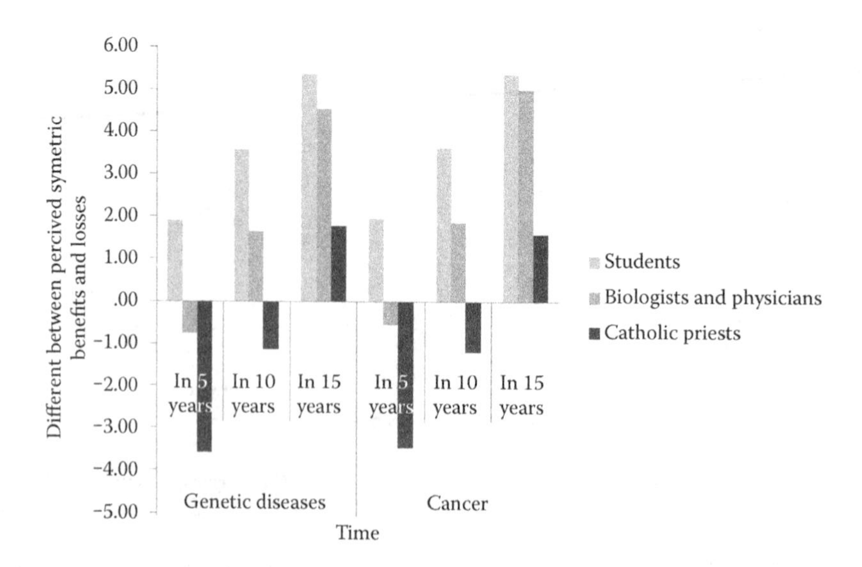

FIGURE 35.2 The difference in evaluation of future symmetric benefits and losses for two SC research applications.

INVERSE RELATIONSHIP AND ATTITUDES/ACCEPTANCE

As mentioned in the introduction, Alhakami and Slovic (1994) found that if people liked an activity, they judged its risks as low and benefits as high. Oppositely, if they disliked it, they judged its risks as high and benefits as low. They concluded that the inverse relationship between perceived risk and perceived benefit of an activity was linked to the strength of a person's overall positive or negative evaluation of an activity. Thus, one may think that the difference in evaluation of benefits and losses is related with a sign of such overall evaluation (general attitude) whereas the absolute difference is related to the strength of general attitudes. Then, the absolute difference in evaluation of positive and negative consequences is considered in this section.

In the case of IRAs, a behavioral index of acceptance was used, i.e., having an IRA was treated as high acceptance and not having such an account as low acceptance. The analyses were performed for 42 persons working for banks (experts) and for 90 regular members of the society, having regular jobs. The difference and the absolute difference in evaluating positive and negative consequences for both layman and expert group are shown in Figure 35.3.

As can be seen from Figure 35.3, those who had an IRA evaluated its benefits higher than losses in comparison with those who did not have such an account. However, the absolute difference in perceived overall benefits and losses was similar in both groups. Thus, not surprisingly, no inverse relationship between the averages for perceived benefits and losses was found in either group (had IRA account: *Pearson* $r = -0.151$; $p = 0.281$, $N = 53$; no IRA account: *Pearson* $r = -0.028$; $p = 0.808$, $N = 76$).

For abortion and SC research, the absolute differences in evaluating positive and negative consequences were analysed in relation to the attitudes toward these hazards. This is shown in Figure 35.4.

From Figure 35.4, one can conclude that the distance between evaluation of positive and negative consequences for abortion was higher than for applications of SC research, and the attitudes toward this hazard were less positive than the attitudes toward SC research.

The statistically significant Pearson correlations between these two factors were found for laymen in Experiment 2 ($r = -0.410$, $p < .001$, $N = 159$) as well as for laymen and Catholic priests in Experiment 3 ($r = 0.499$, $p = .009$, $N = 33$; $r = -0.598$, $p = .031$, $N = 28$, respectively). The distance

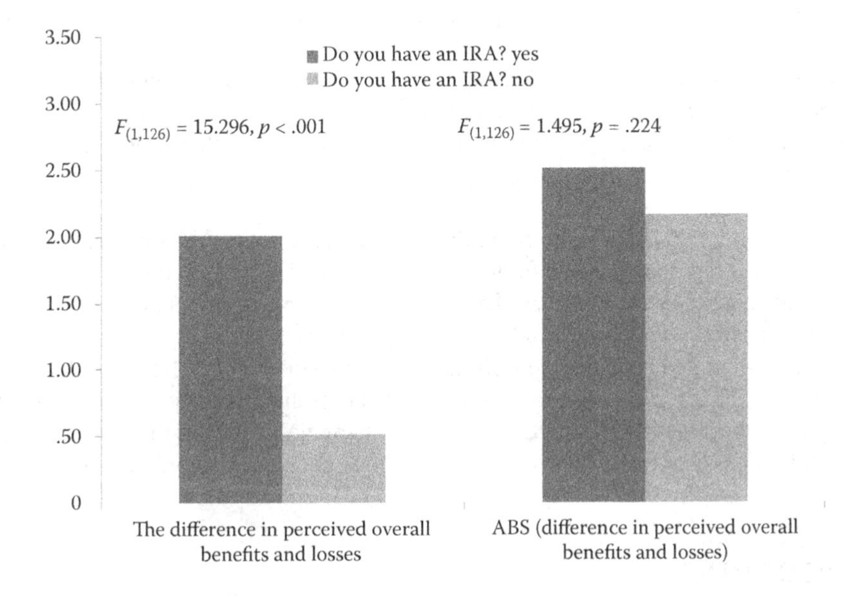

FIGURE 35.3 The absolute difference in evaluation of positive and negative consequences of IRA by persons having or not having such an account.

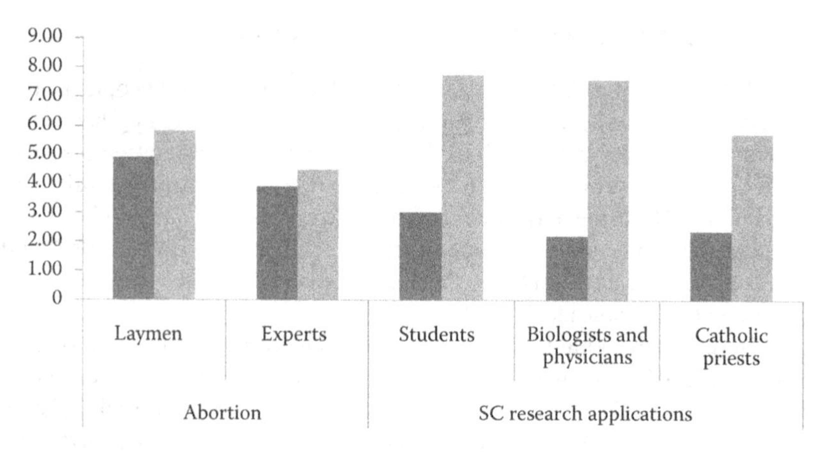

FIGURE 35.4 The absolute difference in evaluation of benefits and losses and the general attitudes toward abortion and stem cell research applications.

in evaluation of positive and negative consequences for abortion and for SC research applications by medical experts and biologists is not related to their attitude toward these hazards. This suggests that larger distances in perceived positive and negative consequences are related to strength of attitudes among respondents with no technical expertise (laymen and Catholic priests). In contrast, for technical experts, such distances are uncorrelated with general attitudes.

CONCLUSION

The main findings from the reported study are the following:

1. The inverse relationship was found in judgments of benefits and losses only for the hazards that evoke affect.
2. For those hazards, the inverse relationship was observed either among laymen only and not for experts (Experiment 2) or for all respondents (Experiment 3).
3. Judgements of benefits and losses far in time were more positive than judgments of consequences in shorter time perspectives.

In general, these findings support the claim that the affect heuristic guides perception of benefits and losses. However, expertise focused on technical aspects of hazards helps to avoid the simplified judgments of consequences. However, the question what characteristics of a situation reinforce either cognition or affect needs more studies. As suggested by Van Gelder et al. (2009), one line of research worth pursuing is to compare well-defined and vague risky activities. In Experiment 3, it has been observed that for well-defined symmetric benefits and losses for the application of SC research in genetic diseases and cancer treatment no systematic pattern of the inverse relationship was observed. It is possible that in well-defined situations people rely more on cognition than on affect.

ACKNOWLEDGMENT

This research was supported by National Center for Science Grant 3637/B/H03/2011/40.

REFERENCES

Alhakami, A. S., Slovic, P. (1994). A psychological study of the inverse relationship between perceived risk and perceived benefit. *Risk Analysis*, 14 (6), 1085–1096.

Berndsen, M., Van der Pligt, J. (2005). Risks of meat: The relative impact of cognitive, affective and moral concerns. *Appetite*, 44, 195–205.

Darlow, A. L., Sloman, S. A. (2010). Two systems of reasoning: Architecture and relation to emotion. *Cognitive Science*, 1 (1), 1–11.

Douglas, M. (1990). Risk as forensic resource. *Daedalus — Journal of the American Academy of Arts and Sciences*, 119, 5–16.

Epstein, S. (2003). Cognitive-experiential self-theory of personality. Millon, T., Lerner, M. (eds.), *Comprehensive Handbook of Psychology*, Vol. 5: *Personality and Social Psychology* (pp. 159–184). Hoboken, NJ: Wiley and Sons.

Evans, J., St., B. T. (2006). The heuristic-analytic theory of reasoning: Extension and evaluation. *Psychonomic Bulletin and Review*, 13 (3), 378–395.

Finucane, M. L., Alhakami, A., Slovic, P., Johnson, S. M. (2000). The affect heuristic in judgments of risks and benefits. *Journal of Behavioral Decision Making*, 13 (1), 1–17.

Ganzach, Y. (2000). Judging risk and return of financial assets. *Organizational Behavior and Human Decision Processes*, 83 (2), 353–370.

Kahneman, D. (2003). A perspective on judgment and choice: Mapping bounded rationality. *American Psychologist*, 58 (9), 697–720.

Kahneman, D., Frederick, S. (2002). Representativeness revisited: Attribute substitution in intuitive judgment. T. Gilovich, D. Griffin, and D. Kahneman (eds.). *Heuristics and Biases: The Psychology of Intuitive Judgment* (pp. 49–81). Cambridge University Press.

Kahneman, D., Frederick, S. (2005). A model of heuristic judgment. Holyoak, K. J., Morrison R. G. (eds.) *The Cambridge Handbook of Thinking and Reasoning* (pp. 267–293). Cambridge University Press.

Knight, F. H. (1921). *Risk, uncertainty and profit*. New York: Houghton Mifflin.

Shefrin, H. (2001). Do investors expect higher returns from safer stocks than from riskier stocks? *Journal of Behavioral Finance*, 2 (4), 1–15.

Shefrin, H., Statman, M., (1995). Making sense of beta, size and book-to market. *The Journal of Portfolio Management*, 21 (2).

Sloman, S. A. (1996). The empirical case for two systems of reasoning. *Psychological Bulletin*, 119 (1), 3–22.

Slovic, P., Finucane, M. L., Peters, E., MacGregor, D. G. (2002a). The affect heuristic. Gilovich, T., Griffin, D., Kahneman, D. (eds.), *Heuristics and Biases: The Psychology of Intuitive Judgment* (pp. 397–420). New York: Cambridge University Press.

Slovic, P., Finucane, M. L., Peters, E., MacGregor, D. G. (2002b). Rational actors or rational fools: Implications of the affect heuristic for behavioural economics. *Journal of Socio-Economics*, 31, 329–342.

Slovic, P., Finucane, M. L., Peters, E., MacGregor, D. G. (2004). Risk as analysis and risk as feelings: Some thoughts about affect, reason, risk, and rationality. *Risk Analysis*, 24 (2), 311–322.

Slovic, P., MacGregor, D. G., Malmfors, T., Purchase, I. F. H. (1997). Influence of affective processes on toxicologists' judgments of risk. Unpublished Study. Eugene, OR: Decision Research.

Stanovich, K. E., West, R. F. (2000). Individual differences in reasoning: Implications for the rationality debate? *Behavioral and Brain Sciences*, 23 (5), 645–665

Tversky, A., Kahneman, D. (1983). Extensional versus intuitive reasoning: The conjunction fallacy in probability judgment. *Psychological Review*, 90 (4), 293–315.

Van Gelder, J.-L., de Vries, R. E., Van der Pligt, J. (2009). Evaluating a dual-process model of risk: Affect and cognition as determinants of risky choice. *Journal of Behavioral Decision Making*, 22, 45–61.

Wildavsky, A., Dake, K. (1990). Theories of risk perception: Who fears what and why? *Daedalus Journal of the American Academy of Arts and Sciences*, 119, 41–60.

36 Creativity Barriers, Creative Behaviors in Organization

Research in the Polish Media Sector

Roksana Ulatowska and Tadeusz Marek

CONTENTS

CREATIVITY BARRIERS IN ORGANIZATION

Organizations as a kind of social system have a long tradition. They can be rigid, friendly, innovative, or conservative; they are complex and difficult to comprehend. But the last time the research that concerns analysis of organizations is absolutely necessary as the understanding of creativity's nature is indispensable is in problem solutions in organizations, firms, and institutions.

It is commonly known that the configuration of a firm's organizational factors, style of management, motivational system, communication style, or climate of organization helps or inhibits the creative invention of its participants. Creativity is not only the basic element of a mission at many organizations and firms, but it is also an indispensable condition for their existence. That's why recognition and understanding of an inner organization's essence, which influences action's proficiency of its participant, is so important, and that's why so much research concerns preparing instruments for examining social context in an organization. (Alencar 2012; Isaksen et al. 2001; Gamache and Kuhn 1998).

The literature of this domain enumerates a lot of factors which block creative thinking in an organization. They are connected with economical obstacles, with payment systems, with attitudes toward creative thinking in the work place, and with the organization's climate (Amabile et al. 1990). These factors are, e.g., excessive bureaucracy; an over-developed, centralized structure; inappropriate motivational systems; incorrect information fluency; or too autocratic a style of management.

STYLES OF CREATIVE BEHAVIOR

Understanding what characterizes creative behaviors in an organization requires a holistic approach. The model of the Style of Creative Behavior (SCB) (Strzałecki 1993) and the tool that is the operating version of the model allows a synthetic view of many areas of creative behavior.

The style of creative behavior definition comprises the mechanisms determining an individual's preferred and characteristic ways of thinking, problem solving, perceiving, remembering, and responding to stimuli. Strzałecki adopts Royce's definition of cognitive style as a multidimensional, organized subsystem of processes by means of which a person manifests cognitive and/or affective phenomena. Creativity is considered from the point of view of action and draws our attention to the fact that the transformation of a vision into reality that occurs in this area of human activity depends on the structure of ability and personality as well as on motivation and skills. The tool developed by this author makes it possible to define creative problem-solving styles and identify the relationship between the qualities of creative persons and the very process of creating action programs. The model is based on the author's belief that creativity is not a one-dimensional phenomenon and a number of spheres that influence it should be distinguished in understanding it. This model demonstrates the significance of different areas of a person's functioning: temperamental, personality, axiological, and contextual areas. Table 36.1 contains a description of five dimensions of the questionnaire. The questionnaire has 120 items ordered in five factors with three points for evaluation: yes, no, and do not know.

Strzałecki's research (1993) conducted in the context of creative enterprises and problem solving indicates a triad of the *Strength of the Ego*, *Self-Realization*, and *Internal Locus of Control* as the specific configuration of the qualities of creative persons in different areas of their activity.

TABLE 36.1
Description of Five Dimensions of the SCB Questionnaire

Factors	Interpretation
Affirmation of life	Willingness to enjoy life despite failures; ability to make independent decisions and to follow one's system of values
The strength of ego	Concentration on problem solving, ability to cope with anxiety due to the situation of the problem, ability to make decisions
Self-realization	Accomplishment tendency, strong motivation for solving distant problems during long periods, ambition
Flexibility of cognitive structures	Elasticity in the use of problem solving strategies; ability to combine ideas from remote fields; originality and innovativeness; fluency in data analysis and synthesis; ability to find analogies; ability to undertake difficult tasks and to generate a high number of solutions
Internal locus of control	Readiness to be guided according to internal value system, flexibility of problem analyzing, independence of external pressure

CREATIVITY BARRIERS QUESTIONNAIRE

The pattern for the questionnaire was the American tool *Work Environmental Inventory* (Amabile and Gryskiewicz 1989). The preparation of CBQ had two phases. During the first one, subjects answered four questions in writing, connected with situations in which they quit, to claim, to describe, and to realize some new ideas. Using open questions helped to discover "critical episodes" and also to generate and to describe creativity barriers present in the work place. Additionally, for extending these data, interviews were done in which subjects were asked about perceived creativity inhibitors connected with generating and initiating new ideas. In this way, employees of investigated firms were given opportunities to openly talk and freely express their opinions, and the author could recognize factors that are perceived as obstacles in the creative process. During the second phase, on the basis of all this information, 10 categories were established and defined. The next step was the preparation of questions that would examine creation obstacles in an organization and could be included to future questionnaire. The questions were ordered in categories. From 202 questions, the competent jury chose those that, in their opinion, were mostly accurate. The author, on the basis of the jury's agreement, accepted six questions in each category. The result was the tool with 60 items.

The Creativity Barriers Questionnaire is the tool for diagnosis of creativity inhibitors in an organization that are connected with the organizational climate. Each member of the organization experiences and describes it from his or her point of view. The climate perception is psychologically important because it is the basis for the characterization of the organizational system.

The questionnaire has 60 items ordered in 10 categories with five points for evaluation, in which 1 means "I strongly disagree with this opinion," 2 means "I rather disagree with this opinion," 3 means "it's difficult to say." 4 means "I rather agree with this opinion," and 5 means "I strongly agree with this opinion." It is established that answers 1 and 2 mean that barriers are rather low; 3 means medium barriers, and 4 or 5 mean a high level of barriers. By the way, the answer "it's difficult to say" is difficult for interpretation but, in spite of all, is diagnostic. The subject evaluates the accuracy of each item comparing it with his or her own feelings, experiences, and opinions using the five-point scale. For example,

Our boss inspects and controls us strongly

1 2 3 4 5

I strongly disagree with this opinion   I strongly agree with this opinion

CBQ identifies factors connected with commitment, feeling of freedom, general atmosphere, style of management, communication, approachable resources, motivating system, status, and bureaucracy. These areas are strongly responsible for important creativity barriers. Detailed characteristics of the 10 categories are presented in Table 36.2.

The Creativity Barriers Questionnaire gives information about the level of creativity barriers perceived by each employee in an organization (according to his or her standing—higher or lower) and can be used in recognizing their condition. This tool can help also in improvement of the climate in Polish organizations and influence their workers' creativity increasing. It is helpful everywhere when we would like, using simple measures, to find the cue describing an organization's atmosphere and to determine its influence on employees. Psychological intervention like that can give a lot of interesting information.

TABLE 36.2
Characteristics of 10 Categories of CBQ

Freedom (taking risk)	Independence of actions in organization. In the climate of freedom, people have autonomy and are able to make decisions concerning their work. They have an opportunity to take the initiative over in order to achieve "something"; they share their work and results with each other. Otherwise, people work according to strict rules and instructions. They perform their work according to previously set schemes with tolerance to ambiguity and insecurity at work. In a situation where a company accepts employees' risky decisions, daring ideas are realized. In a climate of not taking risks, carefulness and indecisiveness dominate. Employees try to find themselves on a safe side, separating themselves from responsibility and initiative.
Change and engagement	The level in which employees are engaged in their everyday work, long-term goals, and visions. If the level of engagement is high, employees feel motivated and obliged to contribute. Climate is then dynamic, electrifying and inspiring. Employees find joy and purpose in their actions. Otherwise, when they are not engaged, they feel apathy and are alienated. They are not interested in success.
Resources	Free access to proper number of resources, including people, materials, time, facilities, and information.
Style of management (supervision and control)	Superior's style of management. When a director allows free expression of thoughts, encourages employees to propose ideas, uses constructive criticism, and does not perform strict supervision and control, employees undertake creative actions. Otherwise, not having a sense of their efficacy, afraid of criticism, they give up on any initiative.
Atmosphere (debates, talks, supporting new ideas)	Spontaneity and relaxation; free manifestation in the place of work. Careless and light climate, atmosphere of play. Otherwise, seriousness. Atmosphere is stiff, depressing, and "heavy." Jokes and laughter are out of place and are not tolerated. Here, a way of approaching new ideas is important. In a supportive climate, people listen to each other and encourage each other when undertaking the initiative. Possibilities to test creative ideas are created. When new ideas are discussed, the atmosphere is positive and constructive. During a talk, many opinions appear, people think about different ideas from different points of view. Otherwise they submit to authoritarian patterns without asking any questions. When support is low, a "no" reaction appears automatically. The typical answer to proposed ideas is to search for blame and obstacles.
Communication (information flow)	Efficient information flow—everyone can learn about everything—employees willingly share their results, do not hide from each other anything that concerns their work. When communication is blocked, information flows only downward from management to subordinates; access to the information source is hampered.
Conflict/competition	Personal and emotional stress. If the level of conflict is high, people and groups "hate" each other. Climate is characterized by an inner war. People gossip, slander each other, do not control their impulses.
Bureaucracy	All kinds of formal obstacles: regulations, rules, standards, procedures, formal requirements, etc.
Motivation	A system of motivating and supporting employees, understood not only as a salary, but including also immaterial benefits, e.g., the possibility of trips to fairs and training.
Status/evaluation	Manifesting creative activity and its evaluation depends on employee's status. The higher up in the hierarchy, the bigger possibility that creative behaviors appear.

PSYCHOMETRIC CHARACTERISTICS OF THE QUESTIONNAIRE

Reliability

The questionnaire's reliability has been proven on a group of 230 people, members of five different media organizations. The reliability ratio has been calculated for each question (estimated by means of Cronbach's α coefficient). The in-depth reliability analysis has been presented in the chart. The scale, consisting of 60 positions is characterized by good reliability. Cronbach's α coefficient is 0.943, which proves the questionnaire to be a good creativity barrier measuring tool. Respective scales characterized by the following reliability indicators are presented in Table 36.3.

Correctness

Due to the difficulties in stating the criteria's correctness (because of the lack of the Polish tools to measure creativity barriers in an organization). I have decided to concentrate on the theoretical correctness, to which determination belonged an analysis of the connection between theoretical constructs lying at the bottom of the method and the content of the individual questions. During the construction of the questions, a couple of assumptions have been taken into account.

The first assumption was that the tools to measure creativity barriers exist. It seems to be almost a common sense assumption, and it does not require any further explanation. Second, I assume that creativity barriers in an organization do exist. In the tool's theoretical model, there has been given a detailed description of it. Third, creativity in an organization manifests itself in a form of perceptible behavior, a kind of expression, as a condition of stating the creativity, leading to the origin of new and valuable products.

Psychometric attributes of the questionnaire seem to be sufficient to be considered a good tool to measure creativity barriers in an organization. Studying climate of a company or an institution considering novelty and innovation appears to be of particular importance to identify any critical behavior and conditioning determining employees' creativity and also to diagnose the features, which can be required for improving the way of stimulating creativity. The CBQ (Creativity Barriers Questionnaire) identifies creativity barriers not only there, where creativity leads directly to a new product, but also in places where the creative approach to work itself is important. The questionnaire can be used among all the employees of a company, institution, or an organization. The research should include regular employees, administrative employees and their managers and principals, which enables the insight into both professional groups. The perception of an organization's climate depends also on the position of an employee in its hierarchy.

TABLE 36.3
Reliability of Questionnaire

Freedom (taking risk)	0.938
Change and engagement	0.949
Resources	0.938
Style of management (supervision and control)	0.940
Atmosphere (debates, talks, supporting new ideas)	0.938
Communication (information flow)	0.941
Conflict/competition	0.949
Bureaucracy	0.910
Motivation	0.941
Status/evaluation	0.942

Nowadays similar tools are used in many countries. The most popular are the following:

WEI: *Work Environmental Inventory*, (Amabile and Gryskiewicz 1989)—it identifies and diagnoses stimulators and obstacles of creativity in the work environment.
SOQ: *Situational Outlook Questionnaire*, (Isaksen, Lauer, Ekvall, and Britz 2001)—it measures perception of the character of life within organization. The main purpose is to estimate how attitudes and behaviors support creativity and change.

THE POLISH MEDIA SECTOR: CREATIVITY BARRIERS RESEARCH

In the research, the created Creativity Barriers Questionnaire was used. The participants were 230 people, representing two national Polish newspapers, a private radio station, and private and public television stations. They were regular and administrative employees as well as the employees performing executive duties.

The analysis of the obtained results suggests that making innovations is fundamentally determined by the model of a given organizational structure. The more centralized, hierarchic the structure, the higher the level of barriers in a company. Overdeveloped, it creates a network, in which exists a lot of levels of mutually subordinate positions, and therefore, an organizational climate is characterized by a severe discipline in the performance of strictly defined instructions. The basic determiners of behavior are then power and authority and unquestionable dependence upon a superior. The structure of the company's superiority as a given way of controlling and directing, supported also by a source of the system's norms and standards, developed bureaucracy, and the lack of a free flow of information and knowledge destroy initiative, making the employees only obeying instructions and performing their direct duties. That kind of climate is conducive to forming barriers in making a creative activity.

RESEARCH RESULTS

The results obtained for each organization are discussed briefly below. Each particular media differs from each other in the level of barriers $F(4, 23) = 48.58$; $p = 0.000$). Analysis of variance showed a statistically important influence of variation organization on the result obtained by the people under research (newspaper 1 = 155.24; newspaper 2 = 165.11; TV private = 125.31; radio = 151.02; TV public = 185.84). Newman-Keuls's post hoc test showed that the biggest differences were between the results obtained in public television ($p > 0.05$) and the results of the other media (Figure 36.1). The results of the employees from newspaper 1 and radio are not considerably different from each other ($p = 0.36$), but the remaining differences are statistically significant.

In radio and both newspapers, 145 people took part in the research (46 people from radio 1, 49 people from newspaper 1, and 50 people from newspaper 2). Collected results suggest that there occurred an average level of creativity barriers. Companies are characterized by a large span of management and a low level of formalization and departmentalization. It is composed of small groups, from which each one has a lot of autonomy and freedom of actions. When divided into small groups, organizations' employees act rather under general than individual supervision and control of people with the highest authority. However, although the organizational structure of these three organizations is conducive to creativity, the results of some research do not confirm it because the climate of the organization generates some barriers. Referring to the barriers, we can assume that maybe besides the formal organizational structure, during social interaction originated another, alternative, secret organization's system and led to the creation of different cultures (Schein 1984). If, however, an organization is composed of subgroups that are constant in relationship to function, work division, and importance hierarchy, then it can contain within its framework many cultures, which often are in conflict with each other. On the example of the radio and both newspapers, we can see in what way the basic character of an organization comes not inasmuch from its formal organizational scheme and code of practice as from its culture.

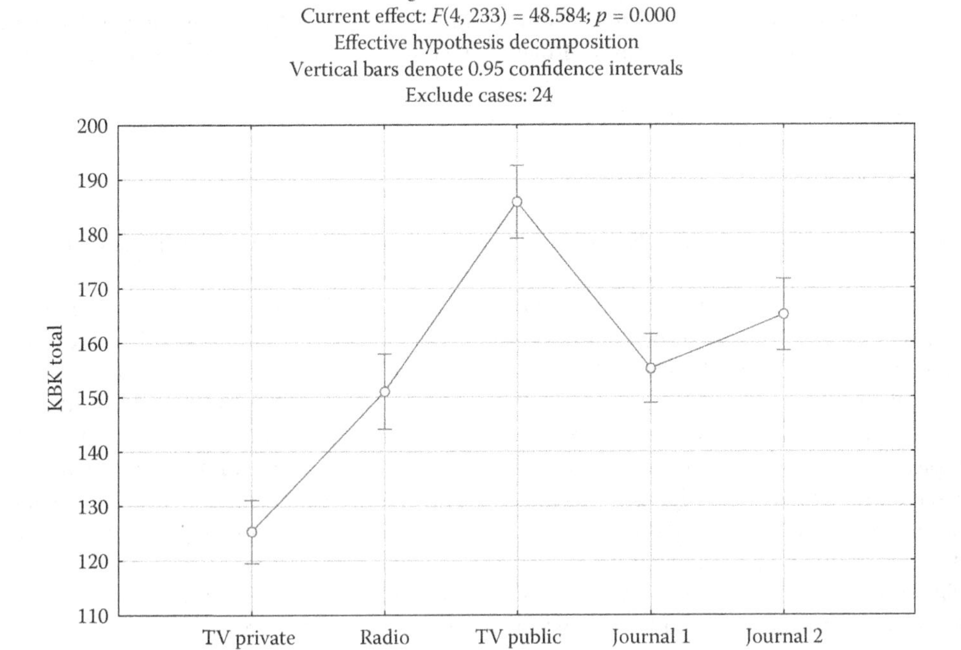

FIGURE 36.1 Mean values of creativity obtained in respective organizations.

Another organization that took part in the research was the Krakow department of public television. The results obtained from 44 employees indicate that, in this organization, the highest level of creativity barriers can be found. The organizational structure of the television is characterized by the highest level of centralization and formalization. Overdeveloped and hierarchic, it creates a network of many different levels of positions subordinate to each other, and this is why the organizational climate is of severe discipline in performing strictly defined instructions.

Legislative and executive functions are performed there by one system, represented by the board, whose directives are performed by lower positions of employees. Basic determinants are power and authority and unquestionable dependence upon the superior. Television's structure of superiority, as a particular way of control and management, and supported also by the source of the system's norms, developed bureaucracy and a lack of a free flow of information and knowledge to destroy the initiative, making employees rather obeying instructions and performing their close duties and responsibilities. Such perception of the climate is conducive to the creation of barriers in undertaking creative activity.

Participants of the last examined organization were 41 employees of a Polish nationwide television station. Private TV's organizational structure is also flat, informal, and decentralized. The level of barriers is perceived on the lowest level in comparison with the other companies. The organizational structure of that company is conducive to the creation of a creativity-friendly climate. The free flow of information and cooperation between departments is conducive to innovative, flexible, and adaptive behaviors. An open system of superiority, lack of strict control, autonomy, and freedom of employees and no stress on formal rules and procedures cause the fact that the fewest creativity limits of the employees appear.

Due to the working character of the 10 initially distinguished categories, after collecting the data, it was decided that the results would be subject to factor analysis. It was conducted in order to identify common factors and to simplify data description by means of reducing the number of the needed dimensions. As a result of factor analysis with oblique rotation *oblimin*, a three-factor structure has been obtained, which explains about 31% of the general variance. The result is not a strong rate and needs further research. The results of the analysis indicate that major components of the climate of Polish organizations, a climate unfavorable to creative initiatives, are three factors:

- Authoritarianism and limitation of free acting
- Inefficient communications
- Lack of resources

Table 36.4 shows some exemplary questions highly loading a given factor in load factor value order.

The first factor—*authoritarianism and limitation of free acting*—is connected with noticing by some employees the pressure brought by management. As an element of the working environment, it causes a feeling of losing the sense of autonomy in performed work and independence in decision-making, the opportunity of free thinking and acting, and using particular abilities. In a hampering creativity climate, directors use an authoritarian form of management, of which basic elements are lack of trust toward employees and cooperation in solving problems. Authoritarian management is based on power, strict division into rulers and ruled, and giving employees no access to making decisions. A director, who strengthens his authority by means of monopolizing information, causes obstruction of initiative and creativity.

The second factor—*inefficient communication channels*—is related to the lack of information flow, in which every person can learn what they want, and employees reluctantly share their work results. In the climate of an organization with an inefficient communication system, employees hide from each other information concerning their work, which consequently negatively affects

TABLE 36.4
Statements Highly Loading a Given Factor

Factor 1 Authoritarianism and Limitation of Free Acting
25. Even an insignificant change in our company is introduced only by means of a decision issued by a superior: *.719*
31. I don't take initiatives at work because usually they end up as failures: *.717*
40. Too many people in our company strictly observe the rules: *.704*
32. An atmosphere of indecisiveness and carefulness dominates in my work: *.695*
12. Our boss exercises strict supervision and control over us: *.694*
4. In my company, for being creative you can get, for example, prizes, words of praise and approval, letters of commendation, a possibility of trip abroad, participation in fairs and trainings[a]: *.669*
Factor 2 Inefficient Communications
13. Information flow in our company is very efficient[a]: *.705*
15. In our company nothing can be arranged by word of mouth, everything has to be presented in writing: *.695*
21. Management needs too much time to take an attitude towards presented ideas: *.693*
43. There are problems with communication in our company: *.675*
27. Our superior often speaks with people, asks them what functions incorrectly, what they don't like, and what they would like to change[a]: *.659*
45. Management doesn't inform us about their actions: *.650*
28. I'm discouraged to realize ideas because of the formalities connected with it: *.650*
Factor 3 Lack of Resources
41. I have access to facilities and information I need to perform my duties[a]: *.778*
59. Resources to which I have access at work are not sufficient to realize even small ideas: *.778*
20. There are often conflicts between the staff of our company: *.590*
11. I have sufficient resources to realize ideas in work[a]: *.539*

[a] Recoded positions/reverse-coded.

undertaking creative actions. When the communication is blocked, information flows only downward from the management to subordinates, and the access to its source is hampered.

It seems interesting that the third factor—*lack of resources*—which relates to the accessibility of resources, contains also an element of conflict and competition. It can mean that in Polish organizations the lack of material resources and information and time pressure intensify conflict formation and are conducive to the emergence of unhealthy competition.

An analysis of variance was also conducted, with respect to those three dependent variables. Significant effects were observed in all subscales of the CBQ (Creativity Barriers Questionnaire). Further analysis was carried out to explain the specific character of these effects.

Effect of the variable of the media company on factor 1—*authoritarianism and limitation of free acting*—proved to be significant, $F(4, 228) = 69.7$; $p = 0.000$) (Figure 36.2). Multiple comparisons exposed differences between all means: TV public, TV private, radio, and both newspapers (TV private = 27.719, TV public = 46.302, newspaper 2 = 37.370, newspaper 1 = 34.429, radio = 34.132; $p = 0.05$). The highest scores, substantially higher than all the other groups, were reported by TV public. The lowest scores, lower than all the other groups, were observed in TV private. Intermediate results were reported by the radio and the press, but the staff of journal 2 had substantially higher scores than the employees of journal 1 and the radio.

The impact of the variable organization on factor 2—*inefficient communications*—proved important, $F(4, 229) = 35.71$; $p = 0.00$) (Figure 36.3). The following relationships were observed: the lowest scores occur in TV private (TV private = 14.127), the highest in TV public and newspaper 2 (their results do not differ significantly from each other); then there are intermediate scores achieved by the radio and newspaper 1 (it turned out that the variable) (TV public = 20.628, newspaper 2 = 20.109, radio = 18.293, newspaper 1 = 18.30).

Organization also has an significant impact on factor 3—*lack of resources*—$F(4, 232) = 48.90$, $p = 0.00$) (Figure 36.4). The lowest results were achieved in TV private, followed by the radio (TV private = 26.649, radio = 31.902, $p = 0.00$) with the two newspapers slightly above it and TV public,

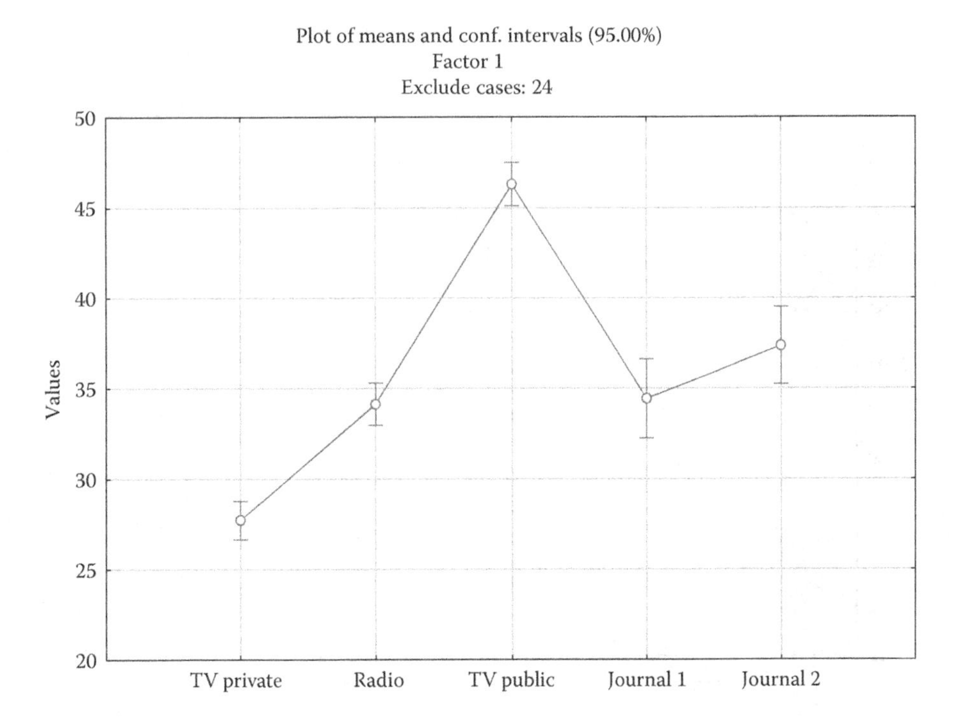

FIGURE 36.2 Mean values of creativity obtained in respective organizations—factor 1.

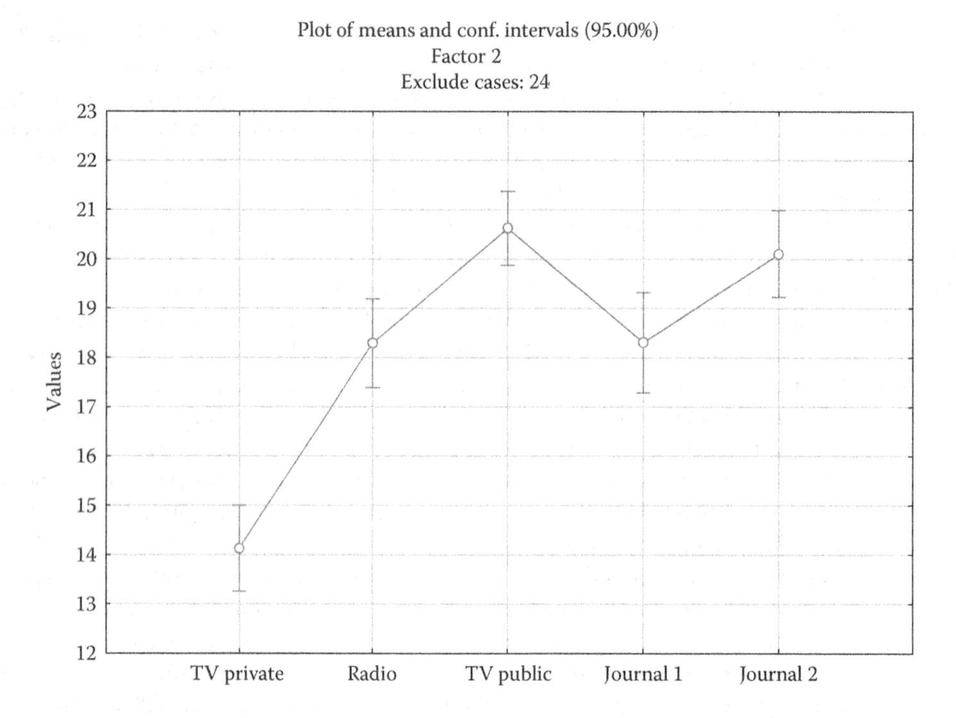

FIGURE 36.3 Mean values of creativity obtained in respective organizations—factor 2.

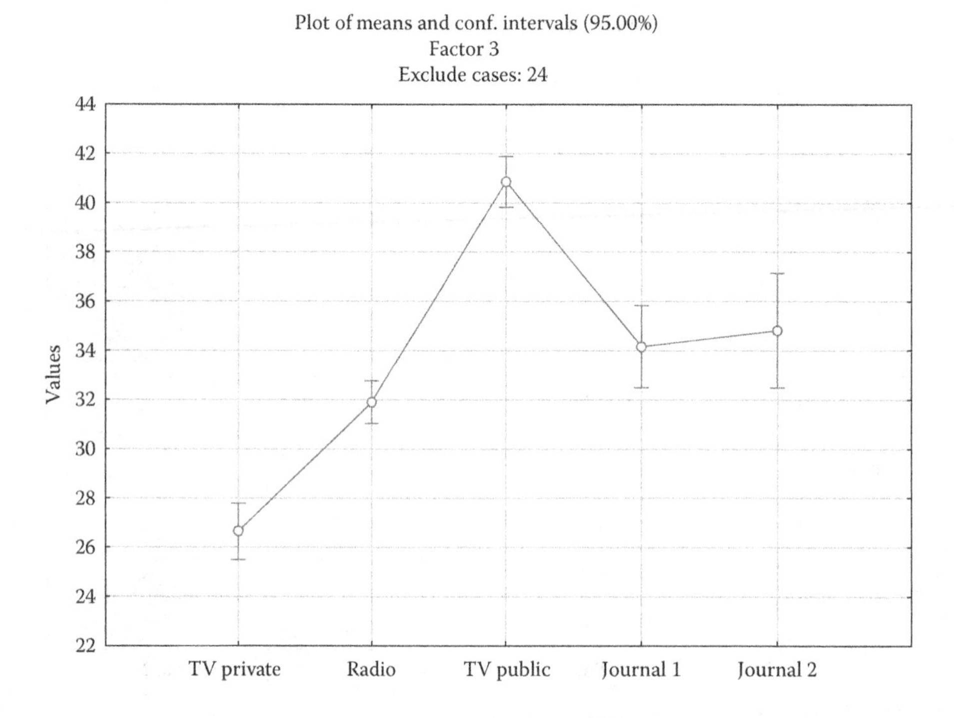

FIGURE 36.4 Mean values of creativity obtained in respective organizations—factor 3.

which had the highest scores, (differences between these averages are not statistically significant) (newspaper 1 = 34.163, newspaper 2 = 34.826, TV public = 40.864).

The study also verified the hypothesis that the staff of the media where barriers to creativity are low show the characteristic configuration of the behavioral qualities of creative people. Journalists of the media that scored low in the CBQ are going to receive higher scores for all five factors of the SCB. All factors of the SCB had high and significant negative correlations with the CBQ (the highest: *Strength of the Ego*, 0.4, $p < 0.001$; *Self-Realization*, 0.36, $p < 0.001$; and *Internal Locus of Control*, 0.14, $p < 0.05$). The SCB model seems to be a good construct for explaining the mechanisms of creativity.

CONCLUSION

Separated factors: *authoritarianism and limitation of free acting*, *inefficient communications*, and *lack of resources* are the main components of the climate of Polish organizations blocking any creative initiatives. The growth of the employees' creativity, which enhances the organization's efficiency and productivity will appear when a proper working climate is secured, when barriers related to an authoritarian style of management, connected with restricting the freedom of action and the lack of resources consisting of financial means, time, materials, knowledge, and coworkers' competence are eliminated. Unfortunately, there are still in Poland many enterprises with an autocratic management style, in which plurality of rules and procedures restrict employees' creative ideas. It has to be realized that innovative ideas often come unexpectedly, by surprise, and therefore, creative skills cannot be squeezed into formal organizational structures. Highly formalized organizations can have problems with developing creative thinking of their employees.

Directing people by supporting creative actions in the work process as well as by providing them resources understood as economical, social means and time, certainly makes it easier for them to undertake creative initiative and innovation. Necessary for creativity in working place are a low level of formalization, a liberal style of management, openness, changeable and flexible information channels, and the possibility of free communication, which is without prior adjustment with superiors, cash administration for financing the works and awarding employees. Companies' orientation toward the innovative activity should be also connected with constant regulation and improvement of the organizational structure. Delegation of authority, decentralization of decision making, making the structures more flexible, supporting flat communication, and formation of the mobile working groups not subordinate to hierarchic networks are various ways of activity that can lead to creation of good conditions for the development of creativity and initiative and proposing innovations in all the fields of company's activities.

REFERENCES

Alencar, E. M. L. S. (2012). Creativity in Organizations: Facilitators and Inhibitors. In: Mumford, M. D. (ed.), *Handbook of Organizational Creativity*. New York: Elsevier, pp. 300–320.

Amabile, T. M. and Gryskiewicz, D. (1989). The Creative Environment Scales: Work Environment Inventory, *Creativity Research Journal*, 2, pp. 231–252.

Amabile, T. M., Goldfarb, B. and Brackfield, S. C. (1990). Social Influences on Creativity: Evaluation, Co-action, and Surveillance, *Creativity Research Journal*, 3, pp. 6–21.

Ekvall, G. and Tangeberg-Anderson, Y. (1986). Working Climate and Creativity: A Study of an Innovative Newspaper Office, *Journal of Creative Behavior*, nr 3.

Ekvall, G. and Ryhammar, L. (1999). The Creative Climate: Its Determinants and Effects at a Swedish University, *Creativity Research Journal*, 12, pp. 303–310.

Gamache, D. R. and Kuhn, R. L. (1998). When Creativity Starts but Stops: Curses of Magic Ideas and Big Fallacies (pp. 27–41). In: *The Creativity Infusion. How Managers Can Start and Sustain Creativity and Innovation*. New York: Harper and Row, Publishers.

Isaksen, S. G., Lauer, K. J., Ekvall, G. and Britz, A. (2001). Perceptions of the Best Worst Climates for Creativity: Preliminary Validation Evidence for the Situational Outlook Questionnaire, *Creativity Research Journal*, 13, pp. 171–183.

Kilbourne, L. M. and Woodman, R. (1999). Barriers to organizational creativity. In: Purser, R. E. and Montuori, A. (eds.), *Social creativity.* Vol. II. Hampton, Creskill, NY, pp. 125–150.
Schein, E. H. (1984). Coming to a New Awareness of Organizational Culture. *Sloan Management Review*, 19, pp. 3–16.
Strzałecki, A. (1993). Styles in Creative Problems Solving. In: T. Marek (ed.), *Psychological Mechanism of Human Creativity: The Temptation for Reassessment*, Eburdon, Delft, pp. 7–24.

37 Trauma in Modern Society

Luiza Seklecka, Barbara Wachowicz, Koryna Lewandowska, Tadeusz Marek, and Waldemar Karwowski

CONTENTS

INTRODUCTION

Because the link between the culture, individuals, and the brain began to be investigated, it is known that many factors related to cultural variables may affect both the causes of post-traumatic stress disorder (PTSD) and its clinical manifestation (Marsella 2010). In the time of globalization and extensive migration, it is particularly important to understand that not only Western economy, politics, culture, and psychology, but also non-Western and intercultural approaches constitute grounds for trauma-related disorders and their therapy. At the same time, it is noteworthy that research in the field of culture and educational neuroscience indicated that cross-cultural differences between human cognitive function may be explained in the light of brain plasticity caused by different socio-cultural and educational patterns (Ansari 2012). Similarly, also the individual experience may affect brain changes (Cozolino 2002), influencing a person's ability to function in the environment and, thus, determining his or her vulnerability to trauma-related disorders. Taking into consideration the rapid development of neuroscience and the efforts to implement this knowledge in the field of psychotherapy, it seems particularly important to understand the interdependence between cultural and psychological factors and the underlying brain mechanisms.

CULTURAL ASPECTS OF TRAUMA

The continuous flux of changes in the societies and cultures of Western civilization has made a substantial impact on the psychological well being of individuals. It is well known that this transition of the social system and simply living in today's world is burdened with stress caused not only by the increasing noise, traffic, rush, or new technologies, but also by the sense of uncertainty and insecurity, which is related to the transformation of economy, work, and family models as well as to a reorganization of cultural values and meanings. What is more, modern societies are also struggling with effects of traffic accidents, crimes, acts of terrorism, and natural disasters. All of these enlisted factors can lead to development of stress-related disorders in the population, and none of them can happen in the social vacuum. Surprisingly, trauma—a potential cause of the PTSD syndrome—is very rarely considered in the social and cultural context and is often treated only as an individual experience.

However, there are some attempts to describe trauma at collective level. As sociologist Piotr Sztompka (2000) explains, traumatogenic transition in society can be characterized by four traits.

The first one is related to the specific speed of changes and its relativity. The most obvious example of sudden event shock is social reaction to revolution or to the terrorist attack of September 11, 2001. But collective trauma can emerge not only as a result of such events, but also as an effect of the sudden awareness of a prolonged process, which has started some time ago. A good example for this can be growing poverty or ecological threats. The second trait is related to the comprehensiveness of changes, which affect many aspects of social and personal life. Their character is fundamental (what stands for the third trait), and they lead to reconstitution of society as in the case of a shift in values or social hierarchies. As they are also surprising and unexpected (fourth trait), such changes elicit a reaction, which Sztompka describes in terms of "unbelieving mood." What is important, when this kind of transition is happening, is that the continuity of tradition and collective identity is challenged or even destroyed. This "cultural trauma" results in aura of anxiety, insecurity, helplessness, passiveness, and distrust toward other people and institutions. It is worth noting that the described social climate is quite similar to Beck's (2006) concept of modern "risk society."

A quite similar point of view is presented by Sandra Bloom (2006). As she argues, the same stress-related symptoms that can be observed in individuals also appear at the organizational and social level. In case of prolonged problems or crises, whole groups might become "chronically hyper-aroused and stressed" due to remaining in an emergency state. In such a situation, while the controlling and punitive attitude of leaders is escalating, the level of irritability and tension increases as well as helplessness and passive-aggressive behaviors of subordinates. Bloom also claims that in the case of traumatic events that affect the whole organization or society, collective memories may be distorted or forgotten similarly to individual memories. The term "corporate amnesia" refers to this phenomenon, which may manifest itself in omitting and/or denying disagreeable facts. Groups can also idealize the past in contrast to the present time with a sense of grief and loss. As Bloom concludes, such processes can be resistant to changes, which can lead a whole society to a prolonged repeating of ineffective strategies without improving the situation.

At this point, an important question appears: May the social and cultural aura increase vulnerability to stress-related disorders in individuals? Young and Erickson (1988) were observing the cultural transition in American society (especially a shift from a focus on family and community to self-realization and individual happiness) and its relationship to the PTSD syndrome. They emphasized culture as a basis of identity, providing frames for a sense of continuity of significant others, continuity with the past, and continuity with an anticipated future self (Kilpatrick 1975). Trauma experience can perturb these elements of identity continuity, which may be additively undermined by chronic changes ongoing in the social environment. In other words, the individual may be dealing with an experience that he or she cannot explain or understand while the meanings and symbols of culture—the basis for individual beliefs and assumptions—are no longer seen as permanent or lasting. Thus, the aura related to such a transition might be relevant for a reaction to a traumatic event. As the authors have claimed, it may exaggerate the feeling of desensitization, distance, numbing, and isolation—feelings that are also common in PTSD. Especially the weakening of social networks, which are so important for trauma victims, may result in an increasing sense of alienation.

Results of conducted studies seem to confirm the impact of social and cultural context on the development of post-traumatic stress disorder. However, it is important to note that this area of research is still rather poorly studied. The main problem may be the fact that, due to specifics of every culture, conclusions, in most of cases, can be applied only to the population that was studied. What is more, investigation of social factors that are associated with the occurrence of PTSD symptoms seems to strongly depend on the political, economical, and historical moment. A good example of this relationship is the study of Silove et al. (2010), which was conducted in Timor-Leste, a country in southeast Asia with a recent history of war and outbreaks of violence. Results of data analysis revealed that higher education, which is thought to be one of the most important factors preventing PTSD (DiGrande et al. 2008), in this specific context, can lead to higher levels of exposure to trauma. In Timor-Leste, highly educated individuals were often members of a resistance movement, which explains this relationship. The earlier study of DiGrande et al. (2008), conducted after the terrorist attack on the World Trade Center, showed quite the opposite; Among the residents of lower Manhattan, low education and income were associated with probable PTSD as well as Hispanic ethnicity, divorce, female gender, and older age. In Timor-Leste, the situation was different also in the last aspect; younger people were experiencing psychological symptoms more frequently (Silove et al. 2010).

The impact of culture, particularly the dimension of individualism-collectivism, on the character of the PTSD was studied by Laura Jobson and Richard O'Kearney (2008). As they have shown, participants with PTSD from an independent culture (in other words, classified as an individualistic culture, e.g., Australian) reported more trauma-related self-defining memories, self-cognitions, and goals than trauma survivors without PTSD. There was no such difference for individuals from an interdependent culture (classified as a collectivistic culture, e.g., Chinese). In this study, self-cognitions were understood as answers to the "Who am I?" question and self-defining memories were described as ones that are remembered clearly, are important, and lead to strong feelings. Researchers also asked participants to provide 15 goals that were important to them. In the second study (Jobson and O'Kearney 2011), the main concern was how the culture affects autonomous orientation ("the tendency to express autonomy and self-determination") in autobiographical remembering. Following the authors' assumptions, in individualistic cultures, personal control, uniqueness, and autonomy are affirmed, and in collectivistic cultures, these features are rather seen as undermining group harmony. Results of the study revealed that trauma survivors with PTSD from individualistic cultures had lower levels of autonomous orientation than those without PTSD. In collectivistic cultures, this relationship was the opposite. Thus, both studies have confirmed cultural differences in the impact of trauma.

Another interesting study was conducted by Lior Oren and Chaya Possick (2010). Researchers were investigating the relationship between ideology understood as shared political, moral, and religious belief systems and the severity of PTSD symptoms in Jewish residents who were evacuated from the area of Gaza by the Israeli authorities. It was found that those who were forced to leave the settlement reported higher levels of symptom severity and were more strongly convinced about the necessity to stay in Gaza than those who were voluntarily evacuated. As the authors concluded, such a result suggests that threatening one's worldview and values causes greater loss in more ideologically driven individuals.

However, measuring the severity of PTSD symptoms may be problematic itself due to culture differences, which brings a risk of underestimating a group-specific way of experiencing trauma. For example, a study of Tibetan refugees in India (Terheggen et al. 2001) revealed that the most traumatic events for this group are different from those that are estimated as highly traumatizing in Western societies. Ninety-five percent of participants from Tibet designated "witnessing the destruction of religious signs" not only as traumatic but also as the "worst possible event that could happen." What is more, the three most traumatic possible events were related to religion in the Tibetan group, not to personal danger as it is in Western culture. Interestingly, respondents also scored higher on somatically phrased items than those formulated in a psychological way. This result was consistent with the findings of Carlson and Rosser-Hogan (1994), who observed a similar type of symptom presentation in a Cambodian group. As it can be seen, the proper cultural adaptation of reaction to trauma measures is one of the most important factors in such studies.

PSYCHOLOGICAL ASPECTS OF TRAUMA

Human behavior is driven by various factors, including motivational, emotional, and social ones. There are different models defining the construction of mind (the term referred to as the mental cognitive plane) and explaining the rules for its operation as well as the underlying processes. One of the approaches based on parallel and distributed processing includes as underlying assumptions for the theory of cognitive network self-organization and self-control. Its next idea is that processing of information (in the cognitive network) is parallel and that this process is fragmented. Such a system is supposed to have different layers, which interact with each other, and their functioning is regulated by feedback. The model of cognitive networks of the mind seems to depict the neural networks on a biological level.

However, it is impossible to refer to the term of "mind" or the functioning of the unit without referring to the personality psychology. In accordance with the previously mentioned theory we have the theory of personal constructs, which refers to the way a person is experiencing, constructing, and interpreting reality. It is important to remember that each construct, according to this theory, always has two poles (Kelly 1955). Human personality is formed by the network of so-called cognitive constructs, which can be simple or more complex with one- or multi-dimensional structure. Kelly's concept of construct describes a scheme of representations, which can be an analog to the belief system in a cognitive-behavioral psychotherapy. This belief system can be regarded as related to automatic thoughts and, thus, can be a crucial determinant of the PTSD development (Curwen et al. 2000).

Horowitz (1997) proposed that cognitive factors play a significant role in the persistence of PTSD. They have linked the severity of PTSD to specific maladaptive beliefs about the self and the world, to the nature of the traumatic memory, or to both of them. A recent synthesis of cognitive factors maintaining PTSD is presented by Ehlers and Clark (2000). Preexisting cognitive vulnerabilities, cognitive processing, and anxiety during a traumatic event but also a negative response to trauma sequelae are core elements of cognitive models of PTSD (Laposa and Rector 2012). Attributional style, anxiety sensitivity, and looming maladaptive style in relation to PTSD are proposed cognitive vulnerabilities that are specific to PTSD. According to Foa and colleagues (Foa and Riggs 1993), there are three main factors that may determine susceptibility to the PTSD: the trauma, preexisting individuals' schemas about the world and self, the memory records of the trauma, and post-traumatic experiences. In other words, a maladaptive or pathological association formed in network of cognitive beliefs ("pathological fear structure") seems to be the crucial mechanism responsible for PTSD that arises as a result of a life-threatening event or violation of an individual's integrity. The aim of PTSD therapy is to adjust or rebuild a network of cognitive beliefs so that the experience of trauma can be incorporated into the newly constructed or reconstructed system of beliefs. Therefore, it becomes adequate to the existing individual's situation, and due to this, contributes to an improvement of the patient's daily functioning. Continuous change of cognitive schemas or held beliefs is an essential mechanism of post-traumatic growth or, going further, mechanisms responsible for an individual's personality integration on higher levels because it relates to a wider and more complex change of mind structure (see the previously mentioned cognitive network model of mind). This fact stays in line with results of PTSD therapies that are grown on cognitive behavioral ground and are the most effective according to the latest research (Bisson and Andrew 2009).

Following the aforementioned point of view, it is impossible to describe the phenomenon of trauma in today's society with the exclusion of the cognitive field, which is specific for a person for whom the occurrence of PTSD had been diagnosed. This issue can be considered from many different perspectives. First, we can identify cognitive patterns. The questions that should be answered are the following: What categories characterized individuals who, later on, have a diagnosis of PTSD? Is there any type of cognitive category, schemas, or features that can predispose or enhance the probability of future development of PTSD? In the literature, there are 15 cognitive variables quoted (see Dunmore et al. 2001) associated with PTSD severity: mental defeat, mental confusion, detachment, negative appraisal of initial symptoms, negative appraisal of other's responses, perceived permanent change, avoidance or safety seeking, negative beliefs after assault, negative appraisal of emotions during assault, positive perception of other's responses, undoing, negative beliefs before assault, change in beliefs, mental planning, and negative appraisal of actions. However, it should be noted, that only eight initially mentioned variables are correlated with PTSD at its beginning, six months' follow-up, and nine months' follow-up. The subsequent five variables are correlated with

PTSD in one- or two-time follow-up timeframes, and the remaining two variables failed to correlate with PTSD severity at any point. Enlisted factors can be modulated by gender, previous history of PTSD, and experienced severity of symptoms (Dunmore et al. 2001). This line of research, focused on the role that cognitive categories play in the development of disorders, is not new. Previously, it has been studied with regard to such mental disorders as compulsive-obsessive disorder or anxiety disorder (Enright and Beech 1993; Mitte 2005). In the case of PTSD, it has been studied in relation to attention (e.g., the Stroop interference), memory, verbal fluency, and intelligence (Michael et al. 2005; Bremner and Brett 1997).

Now, we stand at the point to find a reference between culture- and social-dependent changes in an individual's cognitive functioning and its role in modern society. A cognitive system seems to be built with categories of more uniform character as a result of environmental or cultural identification with the organizations where we work. Therefore, an individual is more susceptible to the development of PTSD due to the homogeneity of cognitive categories. According to the theory of cognitive constructs and the bipolarity of Kelly's categories, the more specific the cognitive categories, the closer they are within cognitive networks (Kelly 1995). The bigger discrepancy between experience and the cognitive schemes, the stronger the perceived dissonance and a sense of helplessness. This might indicate a weaker ability to rebuild those networks or stay in a state of homeostasis. In other words, it may impact on an individual's resilience. While speaking about PTSD, we have to remember that the cognitive structure is suddenly damaged and cannot be restored in its earlier form. However, another aspect that needs further investigation in the context of PTSD is the potential of the already existing individual's cognitive networks to generate new categories and connections between them. This aspect may play a significant role in the return to the state of homeostasis while facing trauma and in the process of auto-therapeutic activation. Both the codependency within categories of cognitive networks and the potential to rebuilt it by adding new categories or connections can serve as means to the functioning betterment.

Traumatic experience must be incorporated again in the new structure in its revised form. What is unknown is what will happen if the structure explaining the occurrence of the event exists, so it doesn't call for change, but still feelings of horror and helplessness associated with traumatic experience occur. If this is the case, we suspect fear-related connections in the structure will incorporate different categories as well. Probably then, psychological resources play the most significant role in the appearance of disorder. It seems a probable scenario that an individual will encounter such persistent thoughts returning or experience them in dreams, but the number of symptoms may not be sufficient to make a diagnosis of PTSD. This is an area that requires examination.

According to the object relations theory, perception of an object only as good or only as bad without combining those two features in the one object is the basis for the future psychotic disorders (Edward and Morris 1984). In such case, the susceptibility to multiple trauma increases via cognitive sensibility path (Laposa and Rector 2012). In this light, handling traumatic events may be seen as a critical turning point on the way to reaching the higher development stages in accordance with Kelly's theory of personality and connectionism (Marcus 2009). A final result—personality structure consisting of the key beliefs about the self—would be replaced after experiencing a traumatic event and changed to a more adaptive structure at a higher level of integration. Attitudes or characteristics of significant persons, art, literature, etc. may provide the source of information about the inner state of the individual or other people. Such a valuable source of corrective cognitive restructure may be self-controlled by cogitation. This can be a way to gather corrective experiences that occurs naturally and may be considered as an alternative to psychotherapy. This process—reverse to schizoid-paranoid position—stays in line with the salutogenic health model (Antonovsky 1987) and can describe an individual's development achieved through crises, and therefore, it may be also associated with the occurrence of traumatic events.

Facing trauma is definitely one of the most difficult experiences associated with negative circles of depletion of energy. Certain factors can lead to launching of development processes that help to grow by overcoming multiple crises. This is the reversed but still positive side of trauma.

NEURAL CORRELATES OF TRAUMA

According to Kolassa and Elbert (2009) multiple traumatic events (e.g., related to domestic violence) influence the development of PTSD by increasing the fear network activity (block-effect) and therefore may affect the severity of the trauma. Moreover, it is assumed that a stressful, uncertain, or impoverished family environment during childhood may have an influence on the decreased integrity between emotional (i.e., limbic) and cognitive (i.e., cortical) structures of the brain (Cozolino 2002), which may be one of the predictors of susceptibility to PTSD. Stress-related brain plasticity, particularly in the early stages of child development, is also considered as one of the factors influencing neurobiological vulnerability to PTSD later in life although the role of the genetic predispositions may be also important (see Heim and Nemeroff 2009).

It is known that the severe stress related to emotional events, irrespective of the age of an individual, may lead to structural changes at the cellular level, and thus, an emotional problem related to the traumatic event may turn into chronic anxiety disorders (see Roozendaal et al. 2009). Such changes mainly affect the hippocampus and the amygdala (e.g., Vyas et al. 2002)—two structures that play a crucial role in the PTSD neuro-circuitry system. The results obtained in various studies on the hypothalamic-pituitary-adrenal (HPA) axis' stress-related activity and its interaction with the amygdala, hippocampus, and medial prefrontal cortex (MPFC; the third main component of the PTSD neuro-circuitry system) (see Heim and Nemeroff 2009) also accentuate the role of neurodegenerative changes in stress-related disorders. It is difficult to evaluate to what extent these changes can be modulated by cultural factors. It seems that culture and society may determine what events are considered as stressful as well as their influence on revealed symptoms (Marsella 2010), whereas the character of stress-related neurobiological changes may be considered as more universal.

The current state of knowledge about the neural correlates of PTSD indicates both structural and functional changes in the aforementioned brain regions (e.g., Kolassa and Elbert 2007). The first of these structures, the amygdala, is a part of the limbic system engaged in the emotional modulation of memory, including fear conditioning as well as in the fight-flight-freeze response to threatening or stressful stimuli (see LeDoux et al. 2009). It is assumed that amygdala dysfunction may be the cause of such PTSD symptoms as hyper-arousal, anxiety, hyper-vigilance in the presence of a potential threat, and increased startle reactivity (see Shin et al. 2006). Numerous studies indicate a smaller amygdala volume in patients with chronic PTSD or a history of PTSD in the past (e.g., Rogers et al. 2009) but not in children with the diagnosis of PTSD (for the meta-analyses, see Woon and Hedges 2008). Interestingly, the recent research of Kuo et al. (2012) suggests that the volume of the amygdala may previously increase and then decrease in the further development of this disorder. Noteworthy, most of the existing functional brain studies (SPECT, PET, and fMRI) have reported exaggerated amygdala activity in PTSD patients, regardless of the duration of symptoms (e.g., Shin et al. 2005; for review, see Shin et al. 2006). What is more, PTSD patients reveal greater amygdala activation during visual imagery of an emotionally negative situation than during its perception whereas healthy subjects reveal an opposite tendency (Shin et al. 1997). It has been proven that amygdala hyper-responsiveness is positively associated with the severity of PTSD symptoms (e.g., Armony et al. 2005). Such generalized amygdala hyperactivation may indicate greater emotional reactivity in PTSD and, due to its greater activation during visual imagery, may facilitate intrusions.

Most of the current neuroimaging findings indicate a reciprocal relationship between amygdala and medial prefrontal cortex activity in PTSD (see Kolassa and Elbert 2009). The study of Shin et al. (2005) has reported not only a negative correlation between activity of the amygdala and MPFC in PTSD patients, but also a negative correlation between activity of MPFC and symptom severity in this disorder. This study is particularly important when taking into account studies regarding the role of the MPFC in fear extinction. Bremner (2005) have reported the decrease of MPFC and anterior cingulated cortex (ACC) activity in PTSD patients (compared to a control group) during the extinction of fear. Thus, recurrent flashbacks and intrusions and prolonging the chronicity of PTSD may be explained by an inability of the MPFC to extinguish fear and inhibit the exaggerated amygdala. However, it is worthwhile to note that some of the existing studies indicate a positive correlation between the MPFC and amygdala (e.g., Gilboa et al. 2004). Shin et al. (2006) have considered the duration of PTSD as a possible reason for that inconsistency. They have observed that an inverse relationship between the amygdala and the MPFC tends to appear in studies involving participants with longer duration of PTSD whereas the positive correlation tends to appear in studies regarding relatively more acute PTSD.

The ACC in many PTSD studies is perceived as a part of the MPFC. However, it seems to deserve special attention. It is related to error processing (Botvinick et al. 2004) and considered to be a part of system evaluating if there is a discrepancy between environmental demands and personal resources as well as between the used pattern of behavior and its outcome (Lewandowska et al. 2012). Many findings indicate structural changes in the ACC in PTSD patients, yet the character of these changes is not clear. Some researchers have reported ACC volume reduction (e.g., Woodward et al. 2006), other decrease of its gray matter density (Kasai et al. 2008), and other changes in ACC shape but not in its volume (Corbo et al. 2005). According to twin studies, it seems that structural changes in the ACC are related rather to a neurodegenerative process than a genetically determined susceptibility (Kasai et al. 2008) whereas functional changes may be considered as a familiar risk factor for the development of PTSD (Shin et al. 2011). In PTSD, functional changes in the ACC seem to vary depending on the part of the structure. Many findings indicate decreased activity in the ventral ACC (a part related to emotional processing) and increased activity in the dorsal ACC (a part involved in cognitive processing) during performing emotional tasks by PTSD patients (e.g., Fonzo et al. 2010). According to Bremner (2005), the ACC may be involved in inhibition of inadequate fear. Thus, decreased activation of the vACC in PTSD patients may lead to an increased feeling of fear when there is no real threat. On the contrary, increased dACC activity is positively correlated with hyper-arousal symptoms (Fonzo et al. 2010) and may predict symptom severity (Shin et al. 2011).

The hippocampus is another part of the brain strongly affected by PTSD. It is a structure related to memory processing, learning, fear conditioning, and context providing. Numerous studies indicate a smaller hippocampus volume in PTSD (e.g., Wang et al. 2010). Despite some findings considering it as a predictor of PTSD susceptibility (e.g., Gilbertson et al. 2002), results obtained in most of the studies suggest that such a volume decrease may be a result of a neurodegenerative process caused by prolonged stress. For instance, it has been reported that the hippocampus volume in PTSD patients neither decreases within the first six months after a traumatic event nor differs in comparison to healthy subjects (Bonne et al. 2001). It is also unchanged in children with PTSD but smaller in adults with PTSD who had experienced abuse during childhood (Woon and Hedges 2008). Thus, the disorder duration seems to be one of the most important variables although it may be also explained by the type of traumatic experience. The reduction of hippocampus volume seems to be related to the deficits in verbal declarative memory (see Heim and Nemeroff 2009). Because of its involvement in contextual information processing, hippocampus damage may be also related to dissociative symptoms in PTSD. PTSD patients display decreased hippocampal activation while performing unemotional memory tasks (e.g., Geuze et al. 2008) as well as during encoding of trauma-related stimuli (Hayes et al. 2011). Interestingly, hippocampal activity during memorizing of emotional faces appears to predict recovery from PTSD (Dickie et al. 2011). The degree of hippocampal activity in PTSD patients varies across sex, i.e., men have greater activity in this structure during fear processing (Felmingham et al. 2010). It may indicate that men have an enhanced capacity for contextualizing fear-related stimuli.

Apart from the changes in structures belonging to the neuro-circuitry system of PTSD, there are many other, mainly functional, changes in various parts of the brain, such as the parietal and posterior temporal cortex, posterior cingulated cortex, parahippocampus, insula, and inferior frontal gyrus, especially while serving trauma-related stimuli (e.g., Fonzo et al. 2010; Lanius et al. 2010). Moreover, some studies indicate changes in connectivity with and within some not mentioned neural structures and networks (such as the default mode network or insula) and these changes also seem to be a predictor of symptom severity (Lanius et al. 2010).

It is worth noting that the diversity of clinical symptoms among PTSD patients also may be explained at the brain level. There are some investigations differentiating between re-experiencing PTSD patients who are hyper-vigilant and tend to experience flashbacks and dissociative ones who show numbing in response to fearful stimuli. For instance, a study of Lanius et al. (2005) has revealed that dissociative patients, in comparison with re-experiencing ones, had stronger connections with the left inferior frontal gyrus (involved in representing bodily states) that may explain distorted body perceptions during some of the dissociative states. The study of Hopper et al. (2007) indicates specific difference in neural structure activation across three PTSD symptoms: re-experiencing, avoidance, and dissociation. For instance, the re-experiencing and avoidance states have been reported as negatively correlating with the vACC activity, and dissociation states seem to correlate positively. Moreover, only avoidance states correlate negatively with dACC activity. According to the results obtained in this study, it seems that some discrepancies between previously mentioned findings may be related to the differences in dominant symptoms among studied PTSD patients. Taking into account stages of degenerative changes in the brain caused by prolonged stress (see Roozendaal et al. 2009) and some suggestions that differences in the functioning of the neuro-circuitry system may be related to the PTSD duration (acute vs. chronic) (see Shin et al. 2006), it also may suggest that re-experiencing, dissociation, and avoidance are some steps in the disorder development. Extra studies are needed, but it seems that such findings might be a further step in PTSD therapy.

CONCLUSION

In the literature, we can find some conceptual frameworks describing the relationship between culture and trauma (e.g., van Rooyen and Nqweni 2012). This difficult issue is also slowly emerging in the field of neuroscience, for example, in the area of neuro-anthropology, which attempts to combine cultural formations, biocultural responses, and methods of investigating brain activity (Collura and Lende 2012). Studies concerned with biological mechanisms underlying self development in relation to culture and society (Kitayama and Park 2010) may also indirectly shed some light on the character of the link between trauma response and collective context. Furthermore, also a link between the psychological factors underlying trauma, the clinical manifestation of trauma-related symptoms, and their neural correlates is widely investigated. What is more, the newest scientific approach, i.e., real-time functional magnetic resonance imaging (rtfMRI) with neurofeedback, provides an opportunity to conduct psychotherapy inside the MR scanner. Providing real-time fMRI feedback during recalling autobiographical memories has been proven to make an individual more successful at changing the blood flow in the amygdala (Zotev et al. 2011), which potentially may be used in the treatment of anxiety and stress-related disorders, including PTSD. Thus, understanding of the aforementioned interdependence, likewise the improvement in neuroimaging methods, may contribute to the more effective therapy of trauma-related disorders. The holistic view of trauma may also increase the understanding of modern society problems both at the sociological and psychological level.

REFERENCES

Ansari, D. (2012). Culture and education: New frontiers in brain plasticity. *Trends in Cognitive Sciences* 16 (2), 93–95.

Antonovsky, A. (1987). *Unraveling the Mystery of Health: How People Manage Stress and Stay Well.* San Francisco: Jossey-Bass.

Armony, J.L., Corbo, V., Clement, M.H., Brunet, A. (2005). Amygdala response in patients with acute PTSD to masked and unmasked emotional facial expressions. *American Journal of Psychiatry* 162, 1961–1963.

Beck, U. (2006). *The Risk Society: Towards a new modernity.* Barcelona: Paidos.

Bisson, J., Andrew, M. (2009). Psychological treatment of post-traumatic stress disorder (PTSD), The Cochrane Library, 1.

Bloom S.L. (2006). Societal trauma: Danger and democracy. In N. Totton (Ed.), *The Politics of Psychotherapy* (pp. 17–29). London: New Perspectives, Open University Press.

Bonne, O., Brandes, D., Gilboa, A., Gomori, J.M., Shenton, M.E., Pitman, R.K., Shalev, A.Y. (2001). Longitudinal MRI study of hippocampal volume in trauma survivors with PTSD. *American Journal of Psychiatry* 158 (8), 1248–1255.

Botvinick, M., Braver, T., Yeung, N., Ullsprger, M., Carter, C., Cohen, J. (2004). Conflict monitoring: Computational and empirical studies. In: M. Posner (Ed.), *The Cognitive Neuroscience ofAttention* (pp. 91–104). New York: Guilford Press.

Bremner, J.D. (2005). Effects of traumatic stress on brain structure and function: Relevance to early responses to trauma. *J Trauma Dissociation* 6 (2), 51–68.

Bremner, J.D., Brett, E. (1997). Trauma-related dissociative states and long-term psychopathology in post-traumatic stress disorder. *Journal of Traumatic Stress* 37–49.

Carlson, E., Rosser-Hogan, R. (1994). Cross-cultural response to trauma: A study of traumatic experiencing and posttraumatic symptoms in Cambodian refugees. *Journal of Traumatic Stress* 7 (1), 43–58.

Collura, G.L., Lende, D.H. (2012). Post-traumatic stress disorder and neuroanthropology: Stopping PTSD before it begins. *Annals of Anthropological Practice* 36, 131–148.

Corbo, V., Clément, M.H., Armony, J.L., Pruessner, J.C., Brunet, A. (2005). Size versus shape differences: Contrasting voxel-based and volumetric analyses of the anterior cingulate cortex in individuals with acute posttraumatic stress disorder. *Biological Psychiatry* 58, 119–124.

Cozolino, L. (2002). *The Neuroscience of Psychotherapy: Building and Rebuilding the Human Brain*. W.W., New York, London: Norton and Company.

Curwen, B., Palmere, S., Rudell, P. (2000). *Brief cognitive Behavior Therapy*. London: Sage Publication, Ltd.

Dickie, E.W., Brunet, A., Akerib, V., Armony, J.L. (2011). Neural correlates of recovery from post-traumatic stress disorder: A longitudinal fMRI investigation of memory encoding. *Neuropsychologia* 49 (7), 1771–1778.

DiGrande, L., Perrin, M.A., Thorpe, L.E., Thalji, L., Murphy, J., Wu, D., Farfel, M., Brackbill, R.M. (2008). Posttraumatic stress symptoms, PTSD, and risk factors among lower Manhattan residents 2–3 years after the September 11, 2001 terrorist attacks. *Journal of Traumatic Stress* 21 (3), 264–273.

Dunmore, E., Clark, D.M., Ehlers, A. (2001). A prospective investigation of the role of cognitive factors in persistent posttraumatic stress disorder (PTSD) after physical or sexual assault. *Behaviour Research and Therapy* 39, 1063–1088.

Edward, R., Morris, D.B. (1984). Changes in object relations from psychosis to recovery. *Journal of Abnormal Psychology* 93, 209–215.

Ehlers, A., Clark, D.M. (2000). A cognitive model of posttraumatic stress disorder. *Behaviour Research and Therapy* 38, 319–334.

Enright, S.J., Beech, A.R. (1993). Reduced cognitive inhibition in obsessive-compulsive disorder. *Clinical Psychology* 32, 67–74.

Felmingham, K.L., Kemp, A.H., Peduto, A., Williams, L.M., Liddell, B., Falconer, E., Bryant, R. (2010). Neural responses to masked fear faces: Sex differences and trauma exposure in posttraumatic stress disorder. *Journal of Abnormal Psychology* 119 (1), 241–247.

Foa, E.B, Riggs, D.S. (1993). Post-traumatic stress disorder in rape victims. In: J. Oldham, M. B. Riba, and A. Tasman (Eds.) *Annual Review of Psychiatry*, Vol. 12 (pp. 273–303). Washington, DC: American Psychiatric Association.

Fonzo, G.A., Simmons, A.N., Thorp, S.R., Norman, S.B., Paulus, M.P., Stein, M.B. (2010). Exaggerated and disconnected insular-amygdalar blood oxygenation level-dependent response to threat-relates emotional faces in women with intimate-partner violence posttraumatic stress disorder. *Biological Psychiatry* 68 (5), 433–441.

Geuze, E., Vermetten, E., Ruf, M., deKloet, C.S., Westenberg, H.G. (2008). Neural correlates of associative learning and memory in veterans with posttraumatic stress disorder. *Journal of Psychiatric Research* 42 (8), 659–669.

Gilbertson, M.W., Shenton, M.E., Ciszewski, A., Kasai, K., Lasko, N.B., Orr, S.P., Pitman, R.K. (2002). Smaller hippocampal volume predicts pathologic vulnerability to psychological trauma. *Nature Neuroscience* 5, 1242–1247.

Gilboa, A., Shalev, A.Y., Laor, L., Lester, H., Louzoun, Y., Chisin, R., Bonne, O. (2004). Functional connectivity of the prefrontal cortex and the amygdala in posttraumatic stress disorder. *Biological Psychiatry* 55, 263–272.

Hayes, J.P., LaBar, K.S., McCarthy, G., Selgrade, E., Nasser, J., Dolcos, F., VISN 6 Mid-Atlantic MIRECC workgroup, Morey, R. A. (2011). Reduced hippocampal and amygdala activity predicts memory distortions for trauma reminders in combat-related PTSD. *Journal of Psychiatric Research* 45 (5), 660–669.

Heim, C., Nemeroff, C.B. (2009). Neurobiology of posttraumatic stress disorder. *CNS Spectr.* 14 (1) (Suppl 1), 13–24.

Hopper, J.W., Frewen, P.A., van der Kolk, B.A., Lanius, R.A. (2007). Neural correlates of reexperiencing, avoidance, and dissociation in PTSD: Symptom dimensions and emotion dysregulation in responses to script-driven trauma imagery. *Journal of Traumatic Stress* 20 (5), 713–725.

Horowitz, M.J. (1997). *Stress Response Syndromes: PTSD Grief and Adjustment Disorders*, Jason Aronson, Northvale.

Kasai, K., Yamasue, H., Gilbertson, M.W., Shenton, M.E., Rauch, S.L., Pitman, R.K. (2008). Evidence for acquired pregenual anterior cingulate gray matter loss from a twin study of combat-related posttraumatic stress disorder. *Biological Psychiatry* 63, 550–556.

Kilpatrick, W. (1975). *Identity and Intimacy*. New York: Harper and Row.

Kitayama, S., Park, J. (2010). Cultural neuroscience of the self: Understanding the social grounding of the brain, *SCAN* 5, 111–129.

Kelly, G.A. (1955). *The Psychology of Personal Constructs volume 1: A Theory of Personality volume 2: Clinical Diagnosis and Psychotherapy*. New York: Norton and Company.

Kolassa, I.-T., Elbert, T. (2007). Structural and functional neuroplasticity in relation to traumatic stress. *Current Directions in Psychological Science* 16 (6), 321–325.

Kuo, J.R., Kaloupek, D.G., Woodward, S.H. (2012). Amygdala volume in combat-exposed veterans with and without posttraumatic stress disorder: A cross-sectional study. *Archives of General Psychiatry* 69 (10), 1080–1086.

Lanius, R.A., Bluhm, R.L., Coupland, N.J., Hegadoren, K.M., Rowe, B., Théberge, J., Neufeld, R.W., Williamson, P.C., Brimson, M. (2010). Default mode network connectivity as a predictor of post-traumatic stress disorder symptom severity in actuely traumatized subjects. *Acta Psychiatrica Scandinavica* 121 (1), 33–40.

Lanius, R.A., Williamson, P.C., Bluhm, R.L., Densmore, M., Boksman, K., Neufeld, R.W., Gati, J.S., Menon, R.S. (2005). Functional connectivity of dissociative responses in posttraumatic stress disorder: A functional magnetic resonance imaging investigation. *Biological Psychiatry* 57 (8), 873–884.

Laposa, J.M, Rector, N.A. (2012). The prediction of intrusions following an analogue traumatic event: Peritraumatic cognitive processes and anxiety-focused rumination versus rumination in response to intrusions. *Journal of Behavior Therapy and Experimental Psychiatry* 43 (3), 877–883.

LeDoux, J.E., Schiller, D., Cain, C.K. (2009). Emotional reaction and action: From threat processing to goal-directed behavior. *The New Cognitive Neurosciences, 4th Edition.* Gazzaniga, M.S. ed. MIT Press.

Lewandowska, K., Wachowicz, B., Beldzik, E., Domagalik, A., Fafrowicz, M., Mojsa-Kaja, J., Oginska, H., Marek, T. (2012). A new neural framework for adaptive and maladaptive behavior in changeable and demanding environment. In *Neuroadaptive Systems: Theory and Applications,* eds. Fafrowicz, M., Marek, T., Karwowski, W., Schmorrow, D., Boca Raton, FL: Taylor & Francis.

Jobson, L., O'Kearney, R. (2008). Cultural differences in personal identity in post-traumatic stress disorder. *British Journal of Clinical Psychology* 47, 95–109.

Jobson, L., O'Kearney, R. (2011). Cultural differences in levels of autonomous orientation in autobiographical remembering in posttraumatic stress disorder. *Applied Cognitive Psychology* 25, 175–182.

Marcus, G. (2009). How does the mind work? insights from biology. *Cognitive Science* 1, 145–172.

Marsella, A.J. (2010). Ethnocultural aspects of PTSD: An overview of concepts, issues, and treatments. *Traumatology* 16 (4), 17–26.

Michael, T., Ehlers, A., Halligan, S.L., Clark, D.M. (2005). Unwanted memories of assault: What intrusion characteristics are associated with PTSD? *Behavior Research and Therapy* 43, 613–628.

Mitte, J. (2005). Meta-analysis of cognitive–behavioral treatments for generalized anxiety disorder: A comparison with pharmacotherapy. *Psychological Bulletin* 131, 785–795.

Oren, L., Possick, C. (2010). Is ideology a risk factor for PTSD symptom severity among Israeli political evacuees? *Journal of Traumatic Stress* 23 (4), 483–490.

Rogers, M. A., Yamasue, H., Abe, O., Yamada, H., Ohtani, T., Iwanami, A., Aoki, S., Kato, N., Kasai, K. (2009). Smaller amygdala volume and reduced anterior cingulate gray matter density associated with history of post-traumatic stress disorder. *Psychiatry Research: Neuroimaging* 174, 210–216.

Roozendaal, B., McEwen, B.S., Chattarij, S. (2009). Stress, memory and the amygdala. *Nature Reviews Neuroscience* 10, 423–433.

Shin, L.M., Bush, G., Milad, M.R., Lasko, N.B., Handwerger Brohawn, K., Hughes, K.C., Macklin, M.L., Gold, A.L., Karpf, R.D., Orr, S.P., Rauch, S.L., Pitman, R.K. (2011). Exaggerated activation of dorsal anterior cingulated cortex during cognitive interference: A monozygotic twin study of posttraumatic stress disorder. *American Journal of Psychiatry* 168, 979–985.

Shin, L.M., Kosslyn, S.M., McNally, R.J., Alpert, N.M., Thompson, W.L., Rauch, S.L., Macklin, M.L., Pitman, R.K. (1997). Visual imagery and perception in posttraumatic stress disorder: A positron emission tomographic investigation. *Archives of General Psychiatry* 54, 233–241.

Shin, L.M., Rauch, S.L., Pitman, R.K. (2006). Amygdala, medial prefrontal cortex, and hippocampal function in PTSD. *Annals New York Academy of Sciences* 1071, 67–79.

Shin, L.M., Wright, C.I., Cannistraro, P.A., Wedig, M.M., McMullin, K., Martis, B., Macklin, M.L., Lasko, N.B., Cavanagh, S.R., Krangel, T.S., Orr, S.P., Pitman, R.K., Whalen, P.J., Rauch, S.L. (2005) A functional magnetic resonance imaging study of amygdala and medial prefrontal cortex responses to overtly presented fearful faces in posttraumatic stress disorder. *Archives of General Psychiatry* 62, 273–281.

Silove, D., Brooks, R., Steel Bateman, C., Steel, Z., Fonseca, Z., Amaral, C., Rodger, J., Soosay, I. (2010). Social and trauma-related pathways leading to psychological distress and functional limitations four years after the humanitarian emergency in Timor-Leste. *Journal of Traumatic Stress* 23 (1), 151–160.

Sztompka, P. (2000). The ambivalence of social change: Triumph or trauma? Wissenschaftszentrum Berlin fur Sozialforschung, discussion paper, P00-001.

Terheggen, M.A., Stroebe, M.S., Kleber, R.J. (2001). Western conceptualizations and Eastern experience: A cross-cultural study of traumatic stress reactions among Tibetan refugees in India. *Journal of Traumatic Stress* 14 (2), 391–403.

van Rooyen, K., Nqweni, Z.C. (2012). Culture and posttraumatic stress disorder (PTSD): A proposed conceptual framework. *South African Journal of Psychology* 42 (1), 51–60.

Vyas, A., Mitra, R., Shankaranarayana Rao, B.S., Chattarji, S. (2002). Chronic stress induces contrasting patterns of dendritic remodeling in hippocampal and amygdaloid neurons. *The Journal of Neuroscience* 22 (15), 6810–6818.

Wang, Z., Neylan, T.C., Mueller, S.G., Lenoci, M., Truran, D., Marmar, C.R., Weiner, M.W., Schuff, N. (2010). Magnetic resonce imaging of hippocampal subfields in posttraumatic stress disorder. *Archives of General Psychiatry* 67 (3), 296–303.

Woodward, S.H., Kaloupek, D.G., Streeter, C.C., Martinez, C., Schaer, M., Eliez. S. (2006). Decreased anterior cingulate volume in combat-related PTSD. *Biological Psychiatry* 59, 582–587.

Woon, F.L., Hedges, D.W. (2008). Hippocampal and amygdala volumes in children and adults with childhood maltreatment-related posttraumatic stress disorder: A meta-analysis. *Hippocampus* 18, 729–736.

Young, M.B., Erickson, C.A. (1988). Cultural impediments to recovery: PTSD in contemporary America. *Journal of Traumatic Stress* 1 (4), 431–443.

Zotev, V., Krueger, F., Phillips, R., Alvarez, R.P., Simmons, W.K., Bellgowan, P., Drevets, W.C., Bodurka, J. (2011). Self-regulation of amygdala activation using real-time fMRI neurofeedback. *PLoS ONE* 6 (9), e24522.

Section III

Management

38 Specific Sources of Stress in the Teaching Profession

A Three-Factor Model—A Case Study

Zofia Lacala, Luiza Seklecka, and Tadeusz Marek

CONTENTS

INTRODUCTION

A review of the literature shows that although some authors have carried out studies of stress in specific professional groups (e.g., Liu et al. 2005, 2008; Taris et al. 2005), few have focused on the specificity of causes and effects of stress experienced by individuals employed in various professions.

A promising area of research appears to be the verification of the hypothesis that different models of stress may be appropriate for individuals employed in different fields, based not only on the content and type of work that they perform and its associated burdens, but also on the social status of the work (Menon et al. 1996) as well as the work environment resulting from gender domination of a particular field (Gonzales-Morales et al. 2010).

The results of the study reaffirm the validity of creating a general stress model and one that incorporates the dimensionality of the sources of stress, taking into account the specific character of a given profession (Boyle et al. 1995; Thompson et al. 2006).

A further justification for the creation of such a model seems to be the results of research on coping strategies in professions that are clearly dominated by one gender. Gonzales-Morales and her co-authors (2006) have confirmed in studies that in female-dominated professions the search for community support lowers stress among women and raises it in men. Regarding the application of active strategies for dealing with stress, some studies indicate a greater effectiveness for such strategies among men (Gonzales-Morales et al. 2006), and others (Kohler et al. 2006) do not detect gender-based differences in the influence of active strategies for stress perception.

Some of the characteristic sources of stress for teachers are high societal expectations, conflicts with parents and students, problems with the availability of resources, administrative restrictions (Gonzales-Morales et al. 2010, p. 33), work with underachievers, discipline issues during lessons, verbal and physical aggression on the part of the students, time pressures, work overload, assessment by others, relationships with coworkers in challenging and changing situations, management and administration, and poor working conditions (Kyriacou 2001).

It is worth pointing out here the results of a study of stress carried out by Chan and co-authors (Chan et al. 2010) because of the scale of the study, which involved 1710 teachers of primary and secondary school and was based on a self-administered questionnaire. Among the sources of stress most frequently mentioned by teachers were a heavy workload, time pressure, education reforms, external school review, pursuing further education, and managing students' behavior and learning. School management was also singled out as a source of stress by about 50% of those involved in the study. Interestingly, relatively few teachers mentioned getting along and maintaining working relationships with colleagues, salary cuts, or school violence as sources of stress.

A study carried out in Poland by Ogińska-Bulik (2006) names the following factors as the most stressful for those in the teaching profession: psychological stress, lack of reward, stressful contacts with the community and lack of control, conflict between one's own ideals and aims and the possibility of fulfilling them, the lack of professional preparation for new challenges, and the feeling of incompetence related to the loss of control and influence. Sources of stress among teachers were also studied in Poland by Grzegorzewska (2006). She used a questionnaire prepared for teachers (Travers and Cooper 1996). Based on a factor analysis, she isolated the following sources of stress: student aggression, management (including work organization and the direction of the school), physical working conditions (including class size, overcrowding, and noise), lack of input from teachers on the decision-making process, work overload, personnel management in the school, interpersonal relationships, relationships with management and other teachers, lack of support, lack of recognition and respect, poor possibilities for promotion, lack of professional competence (including fear of difficult situations in relationships with students and parents), lack of aptitude or pedagogic competence, job security, and lack of clarity about the teacher's role.

It is worth focusing on stress associated with the teaching process itself and with student-teacher relationships as well as that associated with aims resulting from the educational process. This type of interaction has been very concisely characterized in Polish literature by Sęk (2007, p. 149) as one which includes "great additional difficulties, resistance and contradictions" and in which "the effects of one's work are indirect, [and] gratification is uncertain and considerably delayed." In schools, incidents that highlight the discrepancy between the aims of education formulated by teachers and the attitudes of students to school and learning is one of the characteristic and particularly aggravating conditions leading to chronic stress in the teaching profession (Kyriacou 1987, 2001; Trendall 1989; Capel 1989; Leiter 1999). In large measure, the unpredictability of events in a school and the search for novel approaches as well as the constant feeling of risk and inability to predict all the consequences of one's actions create difficulties in the decision-making process for teachers.

Being a teacher demands that one fill various roles, which are often contradictory; on the one hand, teachers are expected to maintain friendly, sensitive, and supportive relationships with students, and on the other, the expectation that they are to instruct and, above all, assess students demands objectivity (Esteve 1989; Woods 1990). Finding a balance between these two approaches while simultaneously not losing the trust of the students is particularly difficult for the teacher.

In order to understand the emotional load borne by teachers resulting from interpersonal relationships, the authors propose an analysis of "emotional labor" defined as the following three strategies: surface acting, deep acting, and the expression of naturally felt emotion (Diefendorff et al. 2005; Judge et al. 2009; Rupp et al. 2009). In line with this research trend, Constanti and Gibbs (2004) believe that a teacher's effective work with students demands the management and control (aspects of emotional labor) of spontaneous feelings. In a study of teachers (Cheung et al. 2011), the relationship between occupational burnout, psychological capital, and the three emotional strategies mentioned above has been demonstrated. Naring et al. (2006) have shown that occupational burnout among teachers, especially depersonalization, is dependent on emotional labor, primarily concerning the application of the strategy of surface acting, or the control of displayed emotion.

Cherniss (1993, 1995) explains difficulties in meeting complex demands and challenges in the workplace with reference to deficiencies in professional competence and the feeling of one's personal efficacy. This problem is especially severe among workers with little professional experience. Some of the factors isolated by the author as sources of stress and potential causes of burnout (Cherniss 1980, pp. 158–180), however, seem to have a universal character independent of age or length of professional service. The possibility of seeking advice in problem situations for all employees, especially with superiors, can undoubtedly reduce risk and uncertainty when making difficult decisions in school. Generally speaking, feedback from superiors concerning the quality of performance on the part of employees significantly reduces the stress level known as role uncertainty. The maintenance of a balance between independence of action in the workplace and respect for management directives as well as the ability to limit the influence of management is essential for the sense of autonomy of the employee. In the social professions, the lack of autonomy is considered to be the main drawback and conditions job commitment (Fengler 2000). In turn, a situation in which an employee is left to deal with difficulties on his or her own may lead to a sense of alienation, both in the workplace and beyond.

In our times, it is a common belief that more and more is being expected of the employee, often disproportionately to his or her abilities. In the teaching profession, this does not only apply to the completion of tasks in the workplace, but also to the necessity of preparing materials for lessons and checking students' work, which normally results in "taking work home from the office." Additionally, the need to continually upgrade one's qualifications and keep up with systemic changes often results in a deterioration of family life. The job-home interface causes emotional exhaustion and chronic tiredness, leading to an inability to keep up with current tasks.

Work overload may also be caused by the willingness of the employee himself to take on extra assignments (Maslach and Leiter 2005).

The unequal distribution of additional assignments to only certain selected employees may result from unequal treatment of subordinates by the superior, which is usually understood as a kind of favoritism and discrimination. The delegation of assignments to chosen individuals, however, may be the result of the superior's assessment of the subordinate's competence, of their ability to organize work and use time wisely, and primarily as a guarantee of achieving a certain high level of execution of the tasks.

The possibility of comparing particular professions leads to the necessity of finding more general categories of the sources of stress. One proposition is the categorization model of Pines (1982), which isolates four dimensions of stress:

- The psychological aspect divided into cognition (e.g., range of autonomy, overload, variety of tasks) and the emotional (e.g., career development, recognition)
- Physical working conditions
- The social aspect (including the number of problems in relationships with others, support and relationships with coworkers, support and feedback from superiors)
- The workplace aspect (including bureaucracy, rules and regulations and the role of the individual in the organization, conflict and role uncertainty, and work overload)

Cooper and co-authors (Cooper et al. 1988) have isolated the following six stress areas in work:

- Factors intrinsic to the job
- The managerial role
- Relationship with other people
- Career and achievement
- Organizational structure and climate
- Home-work interface

POLISH CASE STUDY

Aim of the Research

The list of specific factors of stress in the teaching profession is long, and it seems natural to attempt to order them in broader categories, such as those suggested by Pines (1982) or Cooper et al. (1988). Such a solution could answer questions about the roles of particular dimensions of stress in occupational burnout syndrome. It would also permit comparison between various professions, the expediency of which in accurately describing the mechanisms of stress in different occupational groups is indicated by several studies. Therefore, the aim of this study is to identify the main and specific areas of stress in the teaching profession as well as their significance in the context of results.

Research Sample

Three hundred and twenty-three teachers in Polish primary and secondary schools took part in the study. The sample included 85.8% women and 14.2% men. The gender structure of the sample confirms the female domination of the profession. The subgroup of teachers 21 to 36 years of age represented 52.2% of the sample, from 37 to 55 years of age 45%, with only 2.5% above the age of 55.

Research Method

The "Sources of Pressure in Your Job" scale of the Occupational Stress Indicator (OSI) (Cooper et al. 1988; Robertson et al. 1990; Williams and Cooper 1998) was used in the study. The original scale includes 61 items that measure the following stress factors: factors intrinsic to the job, the managerial role, relationships with other people, career and achievement, organizational structure and climate, and the home-work interface. The participants registered their perception on a Likert scale ranging from 1 (it certainly is not a source of pressure) to 6 (it certainly is a source of pressure).

The justification for using such a scale and not a scale developed specifically for teachers is the possibility of making comparisons between various professions.

The Polish version of the OSI was developed in the 1990s in the Centre of Management Psychology and Ergonomics at the Jagiellonian University by C. Noworol and Z. Lacala. Work on the Polish version was carried out in accordance with the requirements of the cultural adaptation of tests.

Results of the Exploratory Factor Analysis

In order to extrapolate the structure of factors from the Polish version of particular scales of the OSI, an exploratory factor analysis (EFA) was used. Calculations were made using the SPSS 8.0 data analysis tool. A reduced method of principal components was used. Kaiser's Varimax rotation was applied to obtain a simple orthogonal structure.

In order to determine whether factor analysis is an appropriate model for working with the collected data, a preliminary analysis of the correlation matrix was carried out on a few test items.

The value of the Kaiser-Meyer-Olkin indicators (KMO = 0.930) can be interpreted as high. The value of this indicator allows us to accept the assumption that the test items are related and form strong correlations that can be interpreted using quality analysis. This assumption is also confirmed by the results of Bartlett's sphericity tests ($\chi^2 = 9915.878$, $p < 0.001$), which allow us to reject the hypothesis that the correlation matrix of the test items is an identity matrix, characterized by the absence of a common factor.

The main criterion and statistical estimation technique is the *sree* test.

Identification of the factor structure is based first on the high loading of test items on one of the factors and on the low loading on the remaining factors and, second, on the level of saturation of a given factor (with an accepted value of loading higher than 0.5).

Analysis of the relationships between factors allows us to assume a relatively high independence. Intercorrelation of particular factors is relatively low, indicating their orthagonality.

For the study group a three-factor solution was accepted (Tables 38.1 through 38.3). The order of factors identified is significant as is the associated size of variance exposed (Factor I 18.39%, Factor II 14.63%, Factor III 10.54%). In total, the isolated factors explain 43.56% of the variance of the test items.

Factor I comprises 20 items. Based on an analysis of their contents, the variable isolated concerns "content and type of work carried out." The sources of stress for teachers that make up this factor are the achievement of goals and standards in the face of low student motivation; dealing with "delicate situations," mainly resulting from the complexity and specific nature of student-teacher contact, which itself is often complicated by the transfer of responsibility for the upbringing of students to teachers and schools; making decisions in school situations often characterized by a high degree of uncertainty and indeterminacy as well as the difficulty of predicting the outcomes of these decisions (which is a characteristic difficulty for teachers); the necessity of adapting to constant change in the context of educational reforms as well as the need to continually upgrade one's qualifications and skills; lack of information regarding the work performed; the need to carry out extra duties and be flexible; and the desire to achieve one's own goals associated with career development.

Factor II comprises 13 items, concerning "school management and relations with superiors." It occurs that two particularly severe causes of stress are interactions with superiors caused by a lack of support or consultation in the solving of difficult problems and the feeling of discrimination and favoritism toward other employees, leading to a feeling of isolation.

TABLE 38.1
Factor I—Content and Type of Work Performed

Content of Test Items	Factor Loading
Achievement of self-defined levels of work	.730
Dealing with unclear or "delicate" situations	.647
Inappropriate use of time by others	.646
Being perceived as a boss or decision-maker	.643
Manner of self-development	.640
Increase in effects of minor tasks	.630
Characteristic features of the organizational model	.629
Changes in work requirements	.613
Necessity of risk-taking	.601
Important decision-making	.589
Lack of practical support from persons outside of the work environment	.581
Even division of work and responsibility	.578
Simply being visible and available	.557
Insufficient feedback on work	.539
Unclear job prospects	.537
Home life with a partner who also has a career	.536
Demands made by work on private life	.532
Values and atmosphere in the workplace	.518
Changes in career path	.516
Inability to delegate work	.500

TABLE 38.2
Factor II—School Management and Relationships with Superiors

Content of Test Items	Factor Loading
Lack of support from superiors	.804
Feeling of isolation	.797
Lack of recognition	.715
Lack of consultation and communication	.656
Inappropriate direction and support from management	.651
Role uncertainty	.645
Lack of job prospects	.638
Staff shortages and turnover	.597
Concealed discrimination and favouritism	.589
Threat of approaching dismissal or early retirement	.570
Contradictory tasks and demands in the workplace	.539
Lack of stability or family support	.526
Career stagnation or work below one's level of abilities	.500

TABLE 38.3
Factor III—Workload

Content of Test Items	Factor Loading
Taking work home	.689
Excessive amounts of work	.679
Promotion to positions beyond one's competence	.641
Inability to "switch off" at home	.616
Managing or supervising the work of others	.593
Coping with administration	.591
Necessity to work long hours	.504

Role uncertainty is associated with lack of knowledge regarding performance assessment, including feedback on whether performance is adequate and what, if any, changes or improvements are required from the teacher. Role conflict, in turn, is created by the lack of precise definition of demands from superiors, lack of clearly set realistic aims, both short-term and long-term, and lack of setting direct responsibility for situations that arise. In this context, it is important for the superior to enable clear communication among the interested parties and to achieve consensus among management, coworkers, and parents.

TABLE 38.4
Reliability Coefficient Cronbach's α, Means (*M*), Standard Deviation (*SD*), and Standard Error (*SE*)—"Source U" in the Group of Teachers

Factor	Cronbach's α	M	SD	SE
I	.930	4.21	.87	.054
II	.874	3.27	.95	.059
III	.815	3.70	.87	.054

Finally, Factor III comprises six items concerning "workload" caused by an excess of responsibilities and the necessity to "take work home from the office" as well as coping with administrative demands in terms of documentation and paperwork.

Table 38.4 presents the results of an analysis of reliability for three isolated scales in the teachers' study group as well as an estimation of means, standard deviations, and standard errors.

The values for Cronbach's α coefficient indicate a high level of reliability for the scales identified. Moreover, a comparison of the means indicates that the highest level of stress is associated with the specific character of work in the teaching profession.

CONCLUSION

Based on the literature, one can say that the sources of stress in the teaching profession have been precisely identified and described, but the current study additionally answers questions concerning the specific stress areas and their relative importance in stress models.

The analysis carried out indicates that the majority of specific causes of stress in the teaching profession relate to three main factors. The order and percentage of variation within individual factors allows us to conclude that the most important source of stress for teachers is the specific character of their work. A second factor identified as a source of stress for teachers is the manner in which their schools are managed, in particular the consequences of their interactions with superiors. The final area of stress concerns excessive workload, arising not only from an excess of responsibilities in the workplace, but also from "taking work home from the office."

In studies carried out by Boyle and his colleagues (1995), it occurs that two dimensions of stress, "workload" and "student misbehavior," accounted for most of the variance in the prediction of occupational stress among teachers. The considerable concurrence between the results presented in this article and those in the study by Boyle and co-authors should be noted.

Bearing in mind that the teaching profession is highly female-dominated, it may come as a surprise that interpersonal relationships do not constitute a separate area of stress in the school workplace. We must, however, differentiate here three categories of social interactions in schools: those with students, those with superiors and coworkers, and those outside of the workplace.

Stress caused by contact with students has been taken into account in the first factor. This seems obvious, as the specific nature of the teaching profession cannot be explained without reference to the number and quality of interactions of this type.

Relationships with superiors and especially the lack of support from them has been accounted for in the second factor, which gives us an image of how schools are managed. One can say with certainty that in the minds of teachers, the school management commits every type of error described in leadership literature. Leadership positions are filled by individuals who have experience as teachers, which should, in theory, make management of a school easier. School management, however, requires preparation and specific competences.

As far as relationships with other teachers as a source of stress are concerned, the results of the study are inconclusive. In studies carried out in Poland by Grzegorzewska and Tucholska, "interpersonal relations" and "inappropriate interpersonal relations in the teaching staff" were a substantial source of stress and emotional exhaustion. However, Chan and co-authors (Chan et al. 2010) achieved results from a large test group, which suggests that relationships with colleagues were considerably more rarely mentioned as a source of stress.

Alternatively, a path model of this category of sources of stress among teachers has been proposed by Boyle and co-authors (1995), in which "poor colleague relations" are a mediating variable.

Additional justification for research on the question of what role relationships with coworkers play in the stress model is given by a study carried out by Menon et al. (1996). According to the authors, the motivation for the maintenance of harmonious interpersonal relationships in a given profession is dependent on the level of status and autonomy in that profession. The authors believe that the level of dependence on others has a direct influence on the possibility of interpersonal conflicts arising. The converse of this can be seen among academic teachers, who enjoy a high degree of autonomy.

The methodological approach proposed here aims to identify the main and specific areas of stress in the teaching profession as well as their importance in the context of their irrefutable individual effects on occupational burnout, health issues, and lack of job satisfaction. These consequences of stress in turn are reflected in changes in the quality and quantity of individual contacts, and in the amount of attention paid to the needs of the students and the teaching process. For the functioning of the type of organization that a school is, consequences such as absenteeism and the tendency to leave the profession are not without significance.

Further studies will aim to check on the validity of the explanatory factor solution, which will require the use of confirmatory factor analysis for an independent measure and representative sample of teachers.

Part of the OSI as an instrument to measure sources of stress will also require improvement in order to ensure an adequate number of items that represent the main areas of stress and with the option to work out a model for individual professions that also has the potential to be used in comparative studies of different work environments and professions.

ACKNOWLEDGMENTS

The authors would like to express their gratitude to the master's degree students of professor Tadeusz Marek, Ph.D.: Bartłomiej Krysik (Jagiellonian University in Krakow, Poland), Renata Grabowska and Marzena Gruchała (Warsaw School of Social Sciences and Humanities in Warsaw, Poland) for collecting results and preparing them in electronic form.

REFERENCES

Boyle, G.J., Borg, M.G., Falzon, J.M., and Baglioni, A.J., Jr. (1995). A structural model of the dimensions of teacher stress. *British Journal of Educational Psychology, 65, 1*, pp. 49–67.

Capel, S. (1989). Stress and burnout in secondary school teachers: Some causal factors. In: *Teaching and Stress*. Cole, M. and Walker, S. (Eds.). Milton Keynes, UK: Open University Press.

Chan, A.H.S., Chen, K., and Chong, E.Y.L. (2010). Work stress of teachers from primary and secondary schools in Hong Kong. Proceedings of the International MultiConference of Engineers and Computer Scientists, Vol. III, IMCS, March 17–19, Hong Kong.

Cherniss, C. (1980). *Professional burnout in the human services organizations*. New York: Praeger.

Cherniss, C. (1993). Role of professional self-efficacy in the etiology and amelioration of burnout. In: W.B. Schaufeli, Ch. Maslach, and T. Marek (eds.). *Professional burnout: Recent developments in theory and research* (pp. 135–149). Washington, DC: Taylor & Francis.

Cherniss, C. (1995). *Beyond burnout: Helping teachers, nurses, therapists and lawyers recover from stress and disillusionment.* New York: Routledge.

Cheung, F., So-Kum Tang, C., and Tang, S. (2011). Psychological capital as a moderator between emotional labor, burnout, and job satisfaction among school teachers in China. *International Journal of Stress Management, 18, 4*, 348–371.

Constanti, P., and Gibbs, P. (2004). Higher education teachers and emotional labor. *International Journal of Educational Management, 18*, 243–249.

Cooper, C.L., Sloan, S.J., and Williams, S. (1988). *Occupational stress indicator.* Windsor, England: NFER-Nelson.

Diefendorff, J.M., Croyle, M.H., and Gosserand, R.H. (2005). The dimensionality and antecedents of emotional labor strategies, *Journal of Vocational Behavior, 66*, 339–359.

Esteve, J. (1989). Teacher burnout and teacher stress. In: *Teaching and stress.* Cole, M. and Walker S. (Eds.). Milton Keynes, UK: Open University Press.

Fengler (2000). *Helping is tiring: Burnout syndrome in professional work,* Polish ed. Pomaganie męczy. Wypalenie w pracy zawodowej. Gdańsk, GWP.

Gonzales-Morales, M.G., Peiro, J.M., Rodrigues, I., and Greenglass, E.R. (2006). Coping and distress in organizations: The role of gender in work stress. *International Journal of Stress Management, 13*, 228–248.

Gonzales-Morales, M.G., Rodrigues, I., and Peiro, J.M. (2010). A longitudinal study of coping and gender in a female-dominated occupation: Predicting teachers' burnout. *Journal Occupational Health Psychology, 15, 1*, 29–44.

Grzegorzewska, M.K. (2006). Stress in the Teaching Profession. Specifics, conditions, and effects. Wydawnictwo Uniwersytetu Jagiellońskiego (WUJ).

Judge, T.A., Woolf, E.F., and Hurst, C. (2009). Is emotional labor more difficult for some than for others? A multilevel, experience-sampling study. *Personnel Psychology, 62*, 57–88.

Kohler, J.M., Munz, D.C., and Grawitch, M.J. (2006). Test of a dynamic stress model for organizational stress: Do males and females require different models? *Applied Psychology: An International Review, 55*, 168–191.

Kyriacou, C. (1987). Teacher stress and burnout, an international review. *Educational Research, 19*, 146–152.

Kyriacou, C. (2001). Teacher stress: Directions for future research. *Educational Review, 53*, 27–35.

Leiter, M. (1999). Burnout among teachers as a crisis in psychological contracts. In: Understanding and preventing teacher burnout: A sourcebook of international research and practice. R. Vanderberghe, and A.M. Huberman (Eds.), New York, Cambridge University Press, pp. 202–210.

Liu, C., Spector, P.E., and Jex, S. (2005). The relation of job control with job strains: A comparison of multiple data sources. *Journal of Occupational and Organizational Psychology, 78*, 325–336.

Liu, C., Spector, P.E., and Shi, L. (2008). Use of both qualitative and quantitative approaches to study job stress in different gender and occupational groups. *Journal of Occupational Health Psychology, Vol. 13, 4*, 357–370.

Maslach, C., and Leiter, M.P. (2005). *Banishing burnout: Six strategies for improving your relationship with work*. San Francisco: John Wiley and Sons.

Menon, S., Narayanan, L., and Spector, P.E. (1996). Time urgency and its relation to occupational stressors and health care professionals. In: C.D. Spielberger and I.G. Sarason (Eds.) *Stress and emotion, Vol. 16*. Washington, DC: Taylor & Francis.

Naring, G., Briet, M., and Brouwers, A. (2006). Beyond demand-control: Emotional labor and symptoms of burnout in teachers. *Work and Stress, 20*, 303–315.

Ogińska-Bulik, N. (2006). Professional stress in the context of social services: Sources, consequences, prevention. Warsaw, Difin.

Pines, A. (1982). Changing organizations: Is a work environment without burnout an impossible goal? In: Paine, W.S. (ed.), *Job Stress and Burnout*, pp. 189–211, Sage Publications; Beverly Hills, CA.

Robertson, I.T., Cooper, C.L., and Williams, J. (1990). The validity of the occupational stress indicator. *Work and Stress, 4, 1*, 29–39.

Rupp, D.E., McCance, A.S., Spencer, S., and Sonntag, K. (2009). Customer (in)justice and emotional labor: The role of perspective taking, anger, and emotional regulation. *Journal of Management, 34*, 903–924.

Sęk, H. (2007). Professional burnout among teachers. Conditions and possibilities for prevention. In: *Professional burnout. Causes and prevention.* Sęk, H. (Ed.). Warsaw PWN.

Taris, T.W., Bakker, A.B., Schaufeli, W.B., Stoffelsen, J., and van Dierendonck, D. (2005). Job control and burnout across occupations. *Psychological Reports, 97*, 955–961.

Thompson, B.M., Kirk, A., and Brown, D. (2006). Sources of stress in policewomen: A three-factor model. *International Journal of Stress Management, 13, 3*, 309–328.

Travers, C.J., and Cooper, C.L. (1996). *Teachers under pressure: Stress in the teaching profession.* London: Routledge.

Trendall, C. (1989). Stress in teaching and teacher effectiveness: A study of teachers across mainstream and special education. *Educational Research, 31*, 52–58.

Williams, S., and Cooper, C.L. (1998). Measuring occupational stress: Development of the pressure management indicator. *Journal of Occupational Health Psychology, 3, 4*, 306–321.

Woods, P. (1990). Stress and the teacher's role. In: *Teaching and Stress.* Cole, M., and Walker S. (Eds.). Buckingham, UK: Open University Press.

39 Elements of a Successful Entrepreneurship/ Innovation Culture

David F. Barbe

CONTENTS

INTRODUCTION

Employing both strategic and opportunistic approaches, Mtech has developed a coherent set of programs and activities that has contributed to a robust campus entrepreneurship culture for faculty, graduate students, undergraduate students, high school students, and even middle school students.

It is instructive to first discuss the overall Mtech entrepreneurship-innovation ecosystem. An underlying principle is that student and faculty entrepreneurs and their entrepreneurial ventures need to be mentored and developed over a significant time period and through a diverse set of support activities if they are to be successful. Figure 39.1 shows the complete Mtech entrepreneurship-innovation ecosystem. Brief descriptions of each of the elements shown in the figure but not discussed in the body of the paper will be provided in Appendix 1; however, the complete ecosystem is shown at this point to illustrate the overall culture that this rich set of programs provides and how the programs serve students, faculty, and companies. The figure should impart the notion that the sources of innovation are research, concepts, and ideas produced by faculty and students and the desired outputs from the ecosystem are successful entrepreneurs, innovative companies, and innovative products.

Details of the programs listed in the figure that relate to the entrepreneurship and innovation culture for students and faculty are discussed below. Papers that give more thorough discussions of the complete ecosystem are provided in the bibliography.

EDUCATION PROGRAMS AND ACTIVITIES

The University of Maryland is a recognized leader in entrepreneurship and innovation education through the dynamic entrepreneurship courses and offerings of Mtech. At Mtech, we believe that a firm grasp of the entrepreneurial process and mindset benefits every person engaged in developing

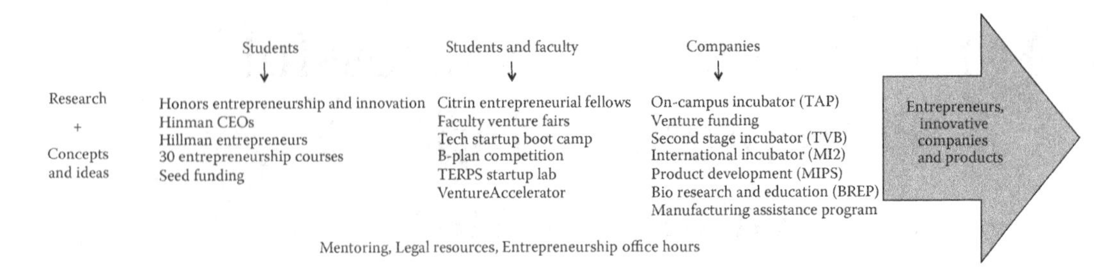

FIGURE 39.1 The Mtech entrepreneurship and innovation ecosystem.

technology. Our goal is to infuse technology-creating students, faculty members, and professionals with that knowledge and its accompanying skills. Armed with an entrepreneurial mindset, technology creators drive economic growth by launching successful ventures and bringing life-changing products and services to market (Barbe 2010).

Through courses, seminars, workshops, competitions, volunteerism, and their own companies, Mtech's students are part of a special experiential learning model. While providing entrepreneurship and innovation education and helping teams to start and operate ventures are important, a continuum of hands-on mentoring helps students not yet engaged in founding and managing start-ups to develop their entrepreneurial skills. Over the course of students' careers in Mtech programs, they can develop innovative ideas and write business plans. Students may also compete in the university's $75K Business Plan Competition, attend the Technology Start-up Boot Camp each fall, and volunteer through program-managed activities.

In contrast to many entrepreneurship programs across the country that are aimed at graduate students or upperclassmen, Mtech invests significant attention to undergraduate students as well. When entrepreneurship and innovation are introduced early, the knowledge gained significantly impacts direction toward more entrepreneurial and innovative careers. This knowledge combines with the community to create a dynamic hub for entrepreneurial skill building and innovative venturing.

CORNERSTONE ENTREPRENEURSHIP PROGRAMS

Mtech's anchor undergraduate entrepreneurship program is the *Hinman CEOs Program*, which was started in 2000 (Barbe 2012). Each fall, 45–50 incoming juniors are selected to enter the two-year program. The two main criteria are a demonstrable entrepreneurial spirit and a strong academic record. The application process occurs during the second semester of the sophomore year and culminates with personal interviews of those students still in the application pool. The acceptance rate is about 30%. Students of all majors are eligible for the program, and historically, about a third are engineering and science majors, a third are business majors, and the remaining third represent more than 20 other majors. The diversity of the students in the program is considered to be very important. These students take a course on entrepreneurship each of the four semesters that they are in the program.

The Hinman CEOs living-learning environment is housed in an exclusive, apartment-style residence hall designed to encourage the exchange of ideas. State-of-the-art technology in an incubator-like setting enables the free flow of ideas among the CEOs as well as easy interaction with experts outside of the university. Seminars by experts in company formation and by successful entrepreneurs are held to educate and inspire the CEOs. Mentoring is provided by onsite staff who have experience advising young entrepreneurs and by other volunteers from inside and outside the university who have expertise in business, legal, technology, and other fields important to entrepreneurs.

Essentially all of the students are involved in entrepreneurial endeavors, and about 20% of these endeavors generate revenues. In each of the past two years, student companies have generated more than $1 million in revenues while the students were still undergraduates. Although the program only launched in 2000, already two companies founded by Hinman students while they were in the program were cited in the 2009 *Inc.* 500 list of the fastest growing companies in the United States. Also, about 10% of the Hinman CEOs students work full-time in the companies that they start while students.

The *Entrepreneurship and Innovation Program* in the Honors College was modeled after the Hinman CEOs Program and was started in 2011. As part of the Honors College, the program combines small classes taught by exceptional faculty with the wide range of additional education opportunities offered by a large research institution. The Entrepreneurship and Innovation Program track starts with "Discovering New Ventures," a one-credit seminar focused on building the entrepreneurial mindset and introducing basic entrepreneurship principles and terminology. In the second semester, "Contemporary Issues in Entrepreneurship and Innovation" inspires innovation and creativity through interactive lectures, workshops, and case studies in contemporary issues to include energy, life sciences, health care, and technology. The third semester includes "International Entrepreneurship and Innovation," an introduction to the opportunities and challenges of entrepreneurship and innovation from an international perspective through lectures and guest speakers with international experiences. The fourth semester is the capstone course, with the "Social Entrepreneurship Practicum" enhancing strategic capabilities and leadership skills through the development of an innovative for-profit product or service concept with social benefits. The academic program also consists of Honors 100 and two Honors Seminars approved by the Entrepreneurship and Innovation Program. Honors seminars and courses are an excellent way to satisfy graduation requirements as many of these classes totaling 16 credits may count toward general education requirements and the student's major. As with the Hinman CEOs Program, these students are assisted in coming up with commercializable products and services and starting companies to bring them to society.

For the *Hillman Entrepreneurs Program*, selected entrepreneurial-minded students begin their education at a nearby community college and matriculate to the University of Maryland for their junior and senior years. Students selected for the program receive scholarships for all four years in the program. Without the scholarships, these students may not have the financial capacity to pursue a college degree or compete well for financial aid. The Hillman scholarships enable students to attend college full time and take a rich set of four specially designed entrepreneurship and leadership courses. The program includes out-of-the-classroom activities and mentoring to help them develop as entrepreneurs within a community of entrepreneurs. A full-time mentor is supported at each educational institution and each is charged with guiding student classroom education, ventures, and activities. In contrast to the Hinman CEOs Program, the Hillman Program is nonresidential. In addition to being nonresidential, the Hillman Program crosses two campuses. It faces the unique challenges of building a community for the students and a collaborative environment between the institutions. Also, the profile of the community college students admitted to the Hillman Entrepreneurs Program includes those who have faced economic and social challenges and many are nontraditional students. The community college campus serves people from a wide age range, including students with families, and the Hillman Program reflects this. It is indeed inspirational to see these students succeed in their academic and entrepreneurial endeavors.

The valuable academic experiences are complemented with the central offices of program staff. In this way, students can easily visit staff offices to discuss their latest new venture idea or to tackle a tough legal, financial, or ethical question. The program directors are seasoned in the entrepreneurial community, pairing practical experience with a top-tier education to coach teams and individually mentor students.

30 Entrepreneurship Courses: Including the course for the three cornerstone programs discussed above, Mtech teaches other courses to students from middle school to working professionals and for nontechnical students and technical students—including a course for life science students. Typically, these courses have enrollments of more than 1200 students per year. With such a wide range of student age, the course content is selected to correspond with the needs and knowledge of the students. For example, the curriculum for the course for middle school students is quite different from that for the course for graduate students, and the course for life science students is different from that for other students.

Seed Fund: A $50,000 seed fund is available each year for students of the Hinman CEOs Program, the Hillman Entrepreneurs Program, and the Entrepreneurship and Innovation Program. These funds are exclusively dedicated to new ventures making a positive social impact, typically in education, health care, environment, and related areas. Funds are provided as grants to students with no payback or equity requirement. Funds are to be used for R&D or working capital for student ventures.

Internships play an important role in the entrepreneurship and innovation experience. Students can learn hands-on with area start-ups, in the offices of venture capitalists, and in faculty laboratories. Applying their experience in these environments amplifies their learning, in turn, bringing energy and expertise back to their fellow Mtech students. On a competitive application basis, select students will be placed in these entrepreneurial internships each summer.

The University of Maryland $75K Business Plan Competition: Since the spring of 2001, Mtech has developed and managed a university-wide business plan competition. More than 360 entrepreneurs have vied for annual prizes of up to $75,000. The competition serves as a catalyst to encourage students and eligible alumni to strategize and present their best ideas and to write business plans.

The Technology Start-Up Boot Camp: Started in fall 2001, this annual one-day program serves more than 500 technical students and faculty annually from throughout the region and is devoted to exploring technology venture formation. Topics include evaluating business ideas, managing intellectual property, building a team, and obtaining financing. Boot Camp is the model for the National Collegiate Inventors and Innovators Alliance's Invention to Venture Workshops that have had wide national reach in impacting communities.

MTECH VENTURES

The primary programs that are aimed at company generation and growth will be discussed briefly. These programs are managed by a staff comprised of seasoned veterans of tech startups and venture capital firms.

TERP Startup Lab: The TERP Startup Lab is aimed at two types of faculty and student companies: ventures that are embryonic but have promise of getting to the high growth stage and "lifestyle" companies that are worthy of support but are not expected to become scalable. The well-equipped TERP Startup Lab provides a place in the on-campus incubator building where these companies can quickly develop their technology prototypes and get help in moving their ventures forward. The program strengthens the entrepreneurial community and provides shared networking/learning experiences with VentureAccelerator and on-campus incubator companies. Winners of the University of Maryland $75K Business Plan Competition and the Faculty Venture Fair are eligible for one free year at the TERP Startup Lab. Special accommodations are made for students in the Hinman CEOs and Hillman Entrepreneurs programs.

VentureAccelerator (VA): The Bayh-Dole Act provides for intellectual property (IP) resulting from federally sponsored research to be owned by the university and licensed for commercial purposes. Because the large majority of university research is federally funded, the university owns most of the IP research carried out by faculty and graduate students. If the university decides not to pursue protection of invention disclosures, it may assign its rights to the inventors. Additionally, in some cases, faculty, staff, and graduate students may develop IP through their own private activities for which the university does not own the rights. Furthermore, at many universities, IP generated by undergraduate students is owned by the students. Thus, IP can be either owned by the university or by faculty, staff, and students.

Most major universities have substantial research activities that result in invention disclosures and subsequent protected IP. Traditionally, universities seek to license this IP. VentureAccelerator provides faculty and students with opportunities to start companies based on IP with promising commercial potential. As discussed above, the IP can either belong to the faculty and students or the university, and if owned by the university, then the company will generally need to obtain a license to use the IP from the university.

The basic concept for VA is to select the most promising opportunities for commercialization, to form companies and develop the companies through intensive processes to the point that they can move forward on their own. As discussed above, in the case for which the university owns the IP on which the companies are based, the companies license the IP from the university. In cases for which the inventors and not the university own the IP, the inventors assign the IP to the companies.

There are several processes involved in VA. The first process is the selection of companies to enter VA. This is accomplished very carefully and involves four steps: identification of candidate opportunities, due diligence regarding IP, market analysis, and feasibility of assembling human resources. If these steps provide positive results, a panel of experts in the field of the company is assembled, and after a presentation by the inventor(s) and an extensive question-and-answer session, the panel advises VA regarding the viability of the venture. After the company is admitted to VA, the second process begins, involving extensive planning. The steps included in this process culminate in a well-developed, fundable business plan. The last processes include steps to obtain initial funding, to prepare to launch the product or service and to recruit the management team to take the company to the next level. At this point, the company is ready to "graduate" from VA.

On-Campus Incubator: The first company was admitted to the On-Campus Incubator in May 1985, and since then, more than 90 companies have graduated and populate the region. Of these, 65% are operating five years after graduation. This is much greater than the success rate for all startup companies and is due to two factors: the highly selective acceptance processes and the nurturing of those companies in the incubator. Two of the companies have been acquired for more than $1B each, and a third company developed the technology on which some hybrid-electric drives for automobiles are based.

Candidate companies are admitted into the incubator only after passing rigorous technical and business reviews. About one in six applicant companies are admitted. Once admitted, companies that are making reasonable progress can remain in the incubator for up to four years. All types of technology companies are acceptable. Biotech companies tend to stay the full four years, and software and information technology companies tend to stay for a much shorter time. Once admitted, companies enjoy increased credibility due to the rigorous admittance process and the dynamic atmosphere that creates an environment in which entrepreneurs flourish. The incubator is outfitted with flexible, furnished office and laboratory space, modern IT and biotechnology infrastructures, in-house business support, and convenient office facilities.

Companies pay a license fee that is based on the space occupied; plus they grant the University 1% equity per year of incubator occupancy. Table 39.1 provides a comparison of VA and the on-campus incubator. Note that VA is only for University of Maryland affiliated faculty, staff, and students whereas the incubator is for entrepreneurs in general.

TABLE 39.1
Comparison of VentureAccelerator and On-Campus Incubator

	VentureAccelerator	On-Campus Incubator
Eligibility	Current UMD affiliation	All entrepreneurs
Mtech role	Active: part-time interim CEO	Advisory, mentor
Location	Anywhere	On-campus incubator building
Duration	6–24 months	24–48 months
Terms	1%–3% equity per quarter Deferred service fees	1% equity per year Space fees below market rates Deferred service fees

Second Stage Incubator: Mtech operates a second-stage incubator located near campus. Companies that lease space in this incubator are graduates of VA or the on-campus incubator or other companies that want to be located close to the university. There is no special mentoring unless occupying companies request assistance. Space is leased at market rates, and the University does not receive equity and generally does not provide services other than those usually provided for leased space.

VENTURE SUPPORTING ACTIVITIES

Below is a discussion of activities that broadly support technology ventures.

The Maryland Intellectual Property Legal Resource Center (MIPLRC): The MIPLRC is a program of the University of Maryland's Law School, and its offices are located in Mtech's on-campus incubator building. It is led by a member of the law school faculty and involves law students specializing in intellectual property. It is both a training ground for the law students and provides free services to the entire university community. Under the supervision of the faculty member, the students provide free legal services for entrepreneurs and emerging technology companies, including patent applications, prior art searches, license agreements, nondisclosure agreements, and company formation. The law students provide considerable preliminary work for the startup companies associated with all of Mtech's programs. For example, they can develop drafts of patent applications, which then are finalized by professional patent attorneys, thus saving the companies substantial costs.

Entrepreneur Office Hours: Entrepreneur office hours are held the second Tuesday of each month for free mentoring and advisory sessions for students, faculty, and regional entrepreneurs. Advisors include individuals from Mtech, the business school's Dingman Center for Entrepreneurship, the licensing office, the IPLRC, experienced entrepreneurs, venture capitalists, state economic development and grant agencies, small business development center, and various outside entities. These sessions provide help and advice with how to build and finance a startup company, develop and protect intellectual property, navigate the technology transfer process, refine business strategies for rapid growth, and tap into other entrepreneurial resources. Typically, there are 15–20 appointments each month with 50% from campus and 50% from outside the University.

Faculty Venture Fairs: Each year, the University hosts two open houses: one in the fall with a life science focus and one in the spring with an information technology focus. Embedded in each of these open houses is a venture fair in which faculty present their inventions to a panel of regional investors and entrepreneurs. The winners generally form start-up companies, and this process identifies candidates for the TERP Lab or the VentureAccelerator.

BRIEF DESCRIPTIONS OF ADDITIONAL ELEMENTS OF THE MTECH'S ECOSYSTEM

Product Development Program (MIPS): This highly successful program provides incentives that draw University faculty and graduate students together with Maryland companies to develop new and improved products for the companies through on-campus projects carried out by faculty and graduate students.

Citrin Entrepreneurship Fellows: This is a special graduate student fellowship program to attract highly entrepreneurial engineering students to the University of Maryland for their graduate studies. Citrin Fellows are selected for their entrepreneurial spirit and business concepts, and while they are doing their graduate studies, an experienced mentor works with them to move their business concepts forward.

International Incubator (MI2): This incubator is located near the campus and provides space and services for international companies that desire to have close collaboration with the university and federal agencies located near the university and that desire to learn how to do business in the United States.

Bio Research and Education Program (BREP): A program that provides research, education, and development of biotech products and processes for companies through two bio-process scale-up facilities staffed with experts.

Manufacturing Assistance Program: This program includes a staff of manufacturing experts who provide critical solutions to help manufacturers grow and become more competitive.

Venture Funding: This is a program that is managed by Mtech for the Maryland Department of Natural Resources that supports startup companies that are developing innovative technologies that address environmental issues.

LESSONS LEARNED

Develop a comprehensive entrepreneurial culture, and successful entrepreneurs and entrepreneurial ventures will develop naturally. In order to develop an entrepreneurial culture, it is necessary to have both entrepreneurship courses and a variety of support activities. It is productive to begin teaching and involving children in entrepreneurial activities at an early age. When Mtech began teaching entrepreneurship courses for third- and fourth-year undergraduate students, some people suggested that it was too early; however, this has proven not to be the case. Mtech experienced demand for first- and second-year undergraduate students, high school students, and even eighth-grade students.

The most successful method to "teach" entrepreneurship is experiential, i.e., learning by doing, and, instead of relying on case studies, it is preferable to have students develop their own ideas for businesses and form teams to develop business plans around those ideas.

A solid understanding of the entrepreneurial processes: intellectual property, market analysis, business planning, financials, obtaining funding, managing a startup, etc. can and should be taught. While case studies may be used, they should not be relied on entirely. Entrepreneurship is not a "science," and it is much more meaningful to students to form student teams around their own ideas and to facilitate their development of plans for their *own* businesses. This experiential, "hands-on" method has proven to be very effective in providing a deeper understanding of entrepreneurship.

Most faculty know very little about the processes involved in forming a successful company and usually need a program that does it with and for them, and they need to be mentored through acceleration and incubation for an extended period of time in order to be successful.

Often faculty start companies for consulting purposes, and this doesn't require much business knowledge because they only involve getting paid for consulting hours worked through the company shell. On the other hand, high-tech, high-growth companies are another matter, and it is rare for faculty in engineering and science departments to have the knowledge to successfully start and manage such companies. Success is greatly aided by accelerators that provide experts to help, if not carry out, the work needed to move the companies forward. As the companies mature beyond acceleration, incubators that provide substantial mentoring and access to university resources are also important to the success of companies.

CONCLUSIONS

The mission of Mtech is four-fold: to provide entrepreneurship education for students, including middle school, high school, undergraduate, and graduate students and to working professionals; to create and assist in the development of new companies; to create mutually beneficial partnerships with existing companies; and to contribute to the economic development in the region. Especially within the past 10 years, Mtech has contributed significantly to the entrepreneurship and innovation culture at the university, regionally, and nationally. Some of its programs have been replicated widely across the United States, and many universities in the United States and abroad visit Mtech to learn about Mtech programs and processes. The activities of Mtech continue to evolve and are becoming more international. Taken together, the program offerings are very comprehensive and provide an entrepreneurship/innovation ecosystem for faculty, students, startup companies, and existing companies.

REFERENCES

Barbe, D.F., (2010) "A Model for Cross Disciplinary Education, Technology Transfer and Teaching Non-Technical Skills for Engineers," Proceedings of the IEEE Conference on Transforming Engineering Education, pp. 1–32.

Barbe, D.F., (2012) Proceedings of the International Conference on Engineering Education, pp. 64–71.

40 Ergonomics in a Sustainable Society
Well-Being and Social Inclusion

Giuseppe Di Bucchianico and Stefania Camplone

CONTENTS

TARGET OF ERGONOMICS: THE "COMFORT" OF THE INDIVIDUAL

Ergonomics aims to increase the "comfort" of the individual. This is an obvious statement, almost banal, confirmed, however, also by the main "official" definitions of the discipline itself. This makes wide reference to human well-being: sometimes in general reference to the design process, which must be based on psychological and physical needs of the individual [as reported in the International Ergonomics Association (IEA) website/homepage (www.iea.cc)]; and sometimes to emphasize the need to achieve a desired "optimization" between the well-being of individuals and the overall "performance" of the system that they use (IEA Council 2000).

Since 1949, therefore, when the psychologist K.H.F. Murrell founded the Human Research Group and suggested the word "ergonomics" to define a new way of studying and solving in an interdisciplinary way the relationship between people and contexts in which they operate, ergonomics pursues a "wellness" achieved through the design of "comfortable" tangible and intangible goods: objects, equipment, tasks, operational procedures, services, environments, and organizing systems of life and work. Using specific methods and advanced research instruments, developed within the areas of specialization and application of physical, cognitive, and organizational ergonomics, this multidisciplinary science is able to provide a constant contribution to innovation in different fields of design (from the best-known disciplines of industrial design, architecture, and urbanism to the design of organizational structures and processes in a broad sense). Such a research has been undertaken from more than half a century, with a constant adaptation and enlargement of its scope in relation to the evolution of human activities and to the changes requested by health and wellness.

In particular, the user-centered design (UCD) approach, which characterizes the latest development of ergonomics applied to the project, focuses on the centrality of its relation to the entire design process with an anthropocentric view that, although attentive to the relationship between the individual's needs and the context in which he or she works, it does not seem able, with its current theoretical apparatus and instruments, to face with some new issues raised by the contemporary society, which requires a dramatic change.

In the transition to a sustainable society (Brundtland 1987), in fact, the anthropocentric view of traditional ergonomics seems, sometimes, inadequate to the "systemic" vision demanded by new issues.

In other words, the individual and his or her "well-being" risk to become "isolated" from his or her "extended" context, which the theme of sustainability needs to confront with. For example, in some cases, the UCD approach, openly pursuing the comfort of products used "exclusively" by specific groups of users, does not seem to consider as central the theme of well-being deriving from a socially "shared" use of the same products.

Actually, some reflections on the theme of "sociopleasure" concerning the pleasantness of the products have been proposed some years ago as part of a classification of "pleasures" (Jordan 2000), even if it was related to the analysis of the "emotional" features of individuals, rather than to the current social and ethical dimension of design for sustainability (Manzini and Vezzoli 2007).

So the question is, can the attribute of "exclusivity" of a product, service, or environment be indifferent to its supposed "ergonomicity"?

"TRADITIONAL" IDEAS OF "EXCLUSIVE" WELL-BEING

In order to define the quality that refers to the concept of "exclusivity" of goods, it is needed to relate to the value that this concept takes within the society which it refers to.

The word "exclusive" has still a generally positive meaning. The prevailing meaning of "exclusive" is indeed "unique and unrepeatable" and, to a larger vision, that of "...reserved for a select few people on the basis of wealth or prestige" (Sabatini and Coletti 2004).

This meaning of the word is clearly tied to an idea of well-being that can be defined as "traditional," referring to the economic and cultural context of industrial societies of the last century, which identified the well-being with the "possession" of products (Manzini and Vezzoli 2007).

It is, in fact, a development model in which environmental resources seemed to be limitless, and where the joint development of science and technology extends the availability of unedited families of products, with falling prices and increasing quantities (Gershuny 1978). This idea of well-being, which in the last half century has been extended to a global scale, and that has been identified "*...nella possibilità di possedere, esibire e consumare individualmente i prodotti e, venendo verso tempi più recenti e società più ricche, nella possibilità di scegliere tra diverse opzioni e di ottenere dei set di prodotti e servizi personalizzati*" (translation: "...with the possibility to possess, produce and consume the products individually and, in recent times and wealthier societies, with the possibility to choose between different options and to obtain sets of customized products and services") (Manzini and Vezzoli 2007), first established the serious effects of environmental degradation known to everyone.

These problems have not been reduced even when we tried to reduce the "environmental load" (the ecological footprint) of many products by redesigning them, with the aim of improving their eco-efficiency. In fact, extending the field of observation from individual products to the overall system, it was soon realized that the situation related to the consumption of environmental resources was indeed tending to get worse (IPTS 1999), because "*...quando i prodotti diventano leggeri, piccoli, efficienti ed economici, tendono a cambiare di statuto e a proliferare, evolvendo verso forme di consumo più ampio e veloce, avvicinandosi ai cicli della moda (come nel caso degli orologi) oppure all'istantaneità dei prodotti usa-e-getta (come nel caso delle macchine fotografiche)*" [translation: "...when the products become lighter, smaller, more efficient and economic, they change their status and tend to proliferate, evolving into broader and faster forms of consumption, closer to the cycles of fashion (as in the case of watches) or to the immediacy of disposable products (as in the case of cameras)"] (Manzini 2007).

Inherently tying together the growth of welfare with the consumption of natural resources, therefore, the model of socioeconomic development of industrial societies has become inevitably not extendible to all inhabitants of the planet, and therefore, it had irreparably gone into crisis with the development of emerging countries and the expectations of economic and well-being growth by their citizens.

Meanwhile, in those societies with a more mature industrialization, a new idea of well-being has emerged and has spread, no longer linked to the possession and consumption of material goods. It is the so-called "access-based well-being" (Rifkin 2000; Manzini 2007), which places in relation the quality of life to the quantity and quality of services and experiences and intangible assets that can be accessed.

Also this form of well-being, however, is proved to be unsustainable in environmental and social terms.

In particular, in environmental terms, consumption of material resources has not been reduced (even increased): in fact, the increase in available information and in relationships and contacts, including the virtual ones, ultimately, tends to generate new opportunities for consumption (Manzini and Vezzoli 2007). On the other hand, in social terms, the access-based well-being has not encouraged the development of forms of social sharing, as it continued to refer to goods basically intended to be used "individually," often even with an exasperation of the concept of "exclusive and reserved access."

CONTRIBUTION OF ERGONOMICS TO "TRADITIONAL" IDEAS OF WELL-BEING

The atomistic and mechanistic culture, which has dominated the scientific and cultural scenario until the end of the 20th century, has involved all areas of contemporary knowledge, investing, as well as design research. In this context, ergonomics, according to the demand posed by the industrial and postindustrial society, has developed a conceptual and operational apparatus appropriate to support the design research of solutions related to the different "ideas of well-being" that has gradually established.

The rapidity of technological development, hypersophistication of means used in the different fields of design, complexity of operations, and the increasing level of precision and accuracy required for different scales of project led to a frantic hyperspecialization of skills.

The interdisciplinary approach to the project, promoted from ergonomics since its origin, if well coordinated, allows one of course to deal with any issue in a complete and competent way, actually using the different types of "technical" knowledge involved as they were many "lenses," returning a complete view of the problem and its solution from different angles.

In particular, the coordinated application of the ergonomic operating principles of "Globality of purposes," with which the design attention is directed to issues of general and broad-spectrum, and of "Interdisciplinarity of technicians," which requires involving and coordinating multiple disciplines within each design process development (Bandini Buti 1998), allows one to track extremely broad and yet detailed problem frameworks, able to define systems of needs, requirements, guidelines, and solutions, which are substantially complete with respect to the specific analyzed target. Ergonomics is successful, however, also in the essential aim to make the different involved disciplines "talk" to each other, thus avoiding the risk of an approach based just on the "sum" and on the "shoulder to shoulder" contributions of different disciplines, relating each other, if necessary, only after the design choices, so making their contributions sometime arbitrary and unilateral.

The contribution of ergonomics to the "traditional" ideas of well-being described above, therefore, was de facto materialized in the so-called UCD. This is a design approach that, in recent decades, has allowed the ergonomic design to pursue the more "traditional" kinds of comfort and well-being.

Placing at the center the design process, the knowledge of users and of their needs, and requirements concerning the use of artifacts in specific contexts, however, it has actually determined a user "targeting," creating products that tend to be used "exclusively" by groups of individuals. This is a direct consequence of the so-called "target segmentation," a concept developed at the end of the 21st century, within the discipline of marketing, as one of the possible corporate strategies to oppose to the pressures placed by the globalized society (Porter 1980).

FROM THE UNSUSTAINABILITY OF CURRENT MODELS OF DEVELOPMENT TO THE "NEW IDEAS" OF WELL-BEING

The real prospect of an imminent and irreversible social and environmental crisis, however, since some time ago requires the necessity of a new model of sustainable development for both mature industrial societies and those newly industrializing countries.

The theme of "unsustainability" of the present model of development of contemporary societies in fact has already entered the agendas of world policy and economy.

The concept of environmental degradation and of limited natural resources is also not always associated with the other "limits" of our planet, which equally affect the determination of well-being of present and future generations. We refer, for example, to the inevitable reduction in demand for some goods (due to markets saturation), to the inhomogeneous population growth among the different geographical areas and to the social issues arising from it (directly related to the problems of migration and racial and cultural intolerance), or to the limits set by the idea of an irresistible and inevitable economic "growth," still linked to patterns of development of industrial societies of the last century, but that now, at the scale of global markets, has proved to be incompatible with the emerging principles of social equity and environmental balance (Manzini 2004). These "limits," in fact, make ever more necessary and urgent a change of direction of the current dominant system of economic and social development, whose "unsustainability" makes them inevitably bounded to decline and to be replaced by other emerging models that are oriented to development sustainability, carrying innovative and promising ideas of well-being related to everyday local scenarios.

The prefiguration of new and hoped scenarios of sustainability, based on new ideas of "sustainable well-being," starts the debate both on the responsibility of design in relation to their social and ethical dimension (Maldonado 1970; Papanek 1973) and on the primary role of design disciplines, and among these disciplines on the role of ergonomics, which has the "well-being" even in its "statute," though closely related to human being.

It must first be clear, however, the meaning that is given to the concept of "sustainability" and, therefore, to determine which is the new idea of well-being to be associated with it.

The idea of "sustainable development," as intended in its first formulation in the "Burtland Report" (1987), does not refer to small improvements in everything concerning the environment generally, as often happens with some so-called "green washing" strategies, but relates to a process of transition to a sustainable society, which necessarily consists of the development of social learning, "*...grazie al quale, progressivamente, tra errori e contraddizioni, come sempre accade nei processi di apprendimento, la società umana imparerà a vivere meglio consumando (molto) meno e rigenerando la qualità del proprio habitat, cioè dell'ecosistema globale e dei contesti di vita locali, in cui gli esseri umani si trovano a vivere*" [translation: "...thanks to which, gradually, between errors and contradictions, as always happens in every learning process, human society will learn to live better by consuming (much) less and regenerating the quality of its habitat, that is of the global ecosystem and of local contexts of life, in which people live"] (Manzini 2004).

It is therefore important to identify a new form of well-being, so-called "context-based" (Manzini 2007), that allows one to recognize the value of diversity of physical and social contexts, in the idea that their valorization could carry to an overall improvement of quality of life.

WHICH ERGONOMICS FOR SUSTAINABILITY?

The question that arises at this point is what could be the contribution of the ergonomic design for a "sustainable well-being."

The inadequacy of the UCD approach to the complexity of the issues placed by sustainable development seems increasingly evident. Extending the argument to the same "interdisciplinary approach" to design that it offers, it is clear that it was possible and desirable until the "ideas of well-being" foreshadowed simple answers that were directly related to the "...analysis of human–system interaction and the design of the system in order to optimize human well-being and overall system performance" (IEA Council, San Diego 2000).

The complexity, the extent, and versatility of the problems that the transition toward sustainability places require an emblematic change of perspective in the theoretical definition and articulation of practical activities related to the ergonomic design, which go far beyond their simple "interdisciplinarity." On the other hand, the discipline continues to consider itself being officially divided into "areas of specialization" (IEA), apparently even separate and with centrifugal research paths.

Actually, however, many attempts have already been made to identify new paths of reflection in order to meet the new instances mentioned above.

Recently, a significant contribution toward a critical review of the operational principles of ergonomics has been offered by some reflections on "holistic ergonomics" (Bandini Buti 2008). This new perspective refers directly to the "holism" (From Greek ὂλος, meaning "the whole"), that is, a philosophical position based on the idea that the properties of a system cannot be explained solely through its components. The idea is that, if in the scientific–mechanistic approach, all the attention was focused on the detection of the constitutive "bricks" of the reality, with the holistic approach, also known as "systemic approach," the importance of studying is recognized, especially the "the connecting fabric." That is, no longer taking care of the "quantity of items" but of the "quality of relations" between them.

The more extended issue of ergonomic design for sustainability is currently approached in an international framework of "thinking tank." In fact, in 2009, the IEA has set up a Technical Committee on "Human Factors and Sustainable Development," organized into four subcommittees (STC). One of them is titled precisely "Ergonomics and Design for Sustainability."

Within this context, approximately a year ago, it was started up as a reflection that, starting from a terminological clarification on the meaning of the same title of the subcommittee, it has investigated the possibility and the need for ergonomic design to adapt itself to new models of sustainable development. In particular, the research unit on "Ergonomics and Design for Sustainability" at the University of Chieti-Pescara proposed a basis of argumentation for a first possible direction of research (Di Bucchianico and Marano 2010). Divided by points according to a logical structure, it expresses the need and opportunity for ergonomics to start an update of its current theoretical and procedural tools toward conceptual and operational apparatuses that are useful to look for new design solutions of "sustainable well-being," with possible variations in different applicative sectors.

This is based on the idea that ergonomics, being already oriented by definition to the "human" well-being, should be facilitated, and on a certain sense even prepared, to think and expand its own criteria of reasoning even to "new" parameters of sustainability. It lacks, however, the initiation of an evolutionary process of its theoretical and procedural apparatus, so that they can be foreshadowed equipment, tasks, environments, and systems, which are certainly "comfortable" (as this is a feature that should basically belong to all the artifacts and not just to those so-called "ergonomic"), but more and more system based on a new idea of "sustainable" well-being (Di Bucchianico and Marano 2010). Therefore, on the basis of these considerations, the above Research Unit at the University of Chieti-Pescara adopted some guidelines proposed by the Design for Sustainability for the generation of systems, scenarios, and ideas of well-being (Manzini 2004). These guidelines were developed by an ergonomic point of view (Marano 2010), in the idea that they could stimulate a reflection in the width of the specialized fields of research that confront with ergonomics. In particular, we refer to the following:

- *Contextualize the proposals and increase the value of what already exists*: optimize the ergonomic life of products, services, and workplaces
- *Share and socialize*: facilitate the way of sharing and socialize products, services, and activities of work and management
- *Increase the "systemic intelligence"*: choose "smart systems" for learning and development of technical skills and creative abilities
- *Create "oases of slowness"*: increase the value of contemplative time in actions with a specific purpose
- *Satisfy the desire for "savoir faire"*: regenerate the contexts of everyday life on its cultural, psychological, and functional (economic and productive) levels.

Starting from those five principles, later the Research Unit "Ergonomics and Design for Sustainability" at the University of Chieti-Pescara (Italy) identified also some specific "sustainable arguments" for reflection: the several disciplines that relate to ergonomics can "crosswise" compare with them to start the evolutionary process of their different theoretical and practical apparatuses (Di Bucchianico and Marano 2010).

But in order to operate in this new scenario, ergonomics is asked to aprioristically compare with two additional issues: one concerning with a kind of "approach" and the other dealing with a specific "concept."

From Knowledge Interdisciplinarity to Transdisciplinarity of Competences

The first matter dealing with the "approach" of ergonomic design concerns its relationship "WITH" the different design disciplines and, above all, the relationship "AMONG" the several areas of specialization that relate to it.

We have already highlighted the benefits that the interdisciplinary approach to "traditional" ergonomic design introduced to the development of "traditional" solutions of well-being.

However, considering the systemic complexity of the issues that the transition to sustainability poses, and the need for a holistic approach to their solution, it is necessary that also the ergonomic design changes its approach, moving from knowledge "interdisciplinary" to a kind of "transdisciplinarity" of competences: which is, to encourage the various "isms" that form it to identify a new theoretical and applicative "cross-setting," useful "*...per porre al centro dei suoi interessi l'uomo ed il suo contesto, nella loro totalità*" (translation: "...to put man and his environment, in their totality, at the center of its interests") (Bandini Buti 2008). This will probably allow one to recover the value of a unitary study on individual's needs (in this case, with a direct reference to the etymology of the word "individual," from Latin *individuus*, which makes it effectively "indivisible"), even in their connections with the "extended scene" of a society oriented toward sustainability, which are aspects that are apparently more intangible.

"Context" and the Principle of "Inclusiveness"

An equally desirable and necessary change in the transition toward sustainability is related to the common sense attached to the concept of "exclusivity" attached to products, environments, and organizational systems. In fact, in the traditional industrial societies, a product that is "exclusive" is generally considered unique, unrepeatable, elegant, precious, elitist, privileged, or reserved for a few (selected on the basis of wealth or prestige).

Actually, the root of the word "exclusive" (from Latin *ex claudere*: leave out) has a negative value, that is, leave out, exclude, reject, refuse, remove, isolate, segregate, or marginalize. In fact, the attribute of "exclusivity" of products tends to develop a social organization based on individualism, personalism, and, in a broad sense, selfishness.

Its opposite is the inclusion, that is, the act to enclose, contain, accept, accommodate, understand, incorporate, accept, emancipate, and, on the broad sense, empower. Literally "include" means "to understand or admit someone or something on a group context or framework, recognizing, respecting and preserving all the original characters, however." Actually, the term "inclusion" was originally used in geology, indicating the presence of foreign bodies within the minerals, a characteristic that often leads to specific lighting effects. The inclusion is something different from integration, which instead is "...entering organically to be part of a given context, social, political or cultural, accepting its system of rules and, therefore, giving up the original characters."

So changing the positive conception attributed to "exclusivity" of products and environments as the benefit of its opposite, that is, their "inclusive" dimension, appears a further step to the transition toward sustainability.

Actually, the five principles of design for sustainability above described already contemplate in them, more or less explicitly, the concept of inclusion and the qualitative aspects associated with it. In particular, it is possible to understand their innovative content, especially in relation to the social and cultural dimensions of human contexts, just interpreting and paraphrasing them with the filter of "social inclusion":

- *Contextualize the proposals and increase the value of what already exists*: appreciate the richness of natural and/or sociocultural resources that a wide variability of contexts and situations may be able to offer. This means giving value to the "differences," rather than demonize them, by considering them as a resource rather than a constraint.
- *Share and socialize*: update and strengthen the concept of "community," even by approaching individuals through a careful, shared, and collective use of goods. Self-enjoyment of goods by "everyone" becomes a point of departure, not of arrival, for social equality.
- *Increase the "systemic intelligence"*: create the most suitable conditions for the development of processes of social learning, including shared experiences. In fact, everyone can have some difficulties in performing a task, but they are always well balanced by his or her "residual abilities," which the social environment must invite to share with others.
- *Create "oases of slowness"*: promote the development of operative contexts having different temporalities, in order to encourage, next to high efficiency systems, the development of the so-called "contemplative time." The usability of products, measured by their efficiency and effectiveness, gives priority to new opportunities of socialization through a renewed physical and smart relationship with artifacts and systems.
- *Satisfy the desire for "savoir faire"*: to meet the "creative" needs related to the innate human desire to pursue a personal satisfaction through a good work, done with art, intelligence, manual wisdom, and knowledge. Comfort intended as the minimization of personal effort gives way to a comfort related with the satisfaction to develop a task with awareness and autonomy.

These guidelines underline the possibility of replacing the existing "dis-enabling" and "exclusive" products, environments, and services, with systems that represent new "enabling" and "inclusive" solutions. In other words, they underline the possibility for individuals and communities to improve their abilities to undertake, collaborate, and find their own ways to live better, through the development of the principle of inclusion.

In this direction, social inclusion, intended as an active, comfortable, and pleasant participation of "everybody" to everyday work, social, and recreational activities, expresses some unexpected potentials toward the definition of a new "wellness," which is both "human" and "sustainable."

ERGONOMICS, MATERIALS, AND SUSTAINABLE WELL-BEING

The field of research attributed to the theme "materials for design" is "crosswise" related also to the issues so far discussed. This is because of two factors: the apparent relationship between the materiality of designed artifacts and the matters dealt with in design for sustainability, and the relationship between materials characteristics and the possible value attributes that can be attached to them.

As regards the first aspect, the "material" dimension of objects and environments has always characterized every daily life context of humanity. To this, it has *de facto* been assigned the greatest responsibility for environmental degradation of the planet. The materials and their "concreteness" are, therefore, inevitably central to the interests of the discipline of design for sustainability. They have been in charge of it since the first definition of environmental requirements of products, declined through the principles of specific eco-strategies for life cycle design (LCD), which are minimizing resources, choosing resources and processes with low environmental impact, maximizing life of products, extending life of materials, and facilitating disassembly (Manzini and Vezzoli 2007).

Consequently, specific tools were defined dedicated to the choice of materials with low environmental impact, such as, for example, the IDEmat software developed by TU Delft University, Netherlands, which contains a database with information also on the environmental characteristics of materials, as well as on the physical, mechanical, and economic ones. The most recent discussions, however, are mostly focused on the social and ethical questions of the "material" dimension of products.

As regards to the second aspect, namely, the values delivered to the materials for design, the question takes on historical meanings. Without going too far back in time, looking for imitative episodes of the sensorial features of precious materials by cheaper ones, the imitative phenomenon manifested all its evidence with the spread of plastics in the last half century. Also called *materiali del possibile* (translation: "materials of possible") (Cecchini 2005), plastics were initially despised, especially on an aesthetic level. In fact, the changing nature of their appearance and the flexibility of production have often reduced plastics to "surrogates" of the most precious, upper, and, in one word, "exclusive" materials. Even when they gain their own language and express all the potential of composite plastic materials from being "specially designed" for specific uses ("materials not made for doing something but that do something!"; Manzini 1986), their exoticism and originality of expression were again associated with a new idea of "exclusivity." Thus, for example, the use of carbon, kevlar, titanium, etc. spread for uses not appropriate to their physical or mechanical properties, but only for the elitist character that the products take through their choice. It is clear, therefore, that the theme of the transition from exclusivity to inclusiveness as an aspect of determining the value of the artifacts, needed for the transition toward sustainability, directly involves also the world of "materials for design." Therefore, also here there is the need of critically reviewing the previous paradigms of choice of materials related to the well-being.

So the question is, which are the most appropriate materials for the "well-being based on the context"?

The need to refer to well-being "developed taking into account the whole scene in which people lives take place" (Manzini 2007) asks in the collective consciousness a new "identity" for the new materials, which in the meantime come to be a total of almost 100,000 different units, without considering their possible combinations. Again, therefore, the criteria for selection of materials for design can and should compare with the principles of design for sustainability previously described. Quickly scrolling the statements, it is immediately obvious that the references to a collective memory, to an enhancement of local identities and differences, and to an ethical approach to the project are all topics that have the potential to define new paradigms for selecting materials for design.

ACKNOWLEDGMENTS

This paper takes up and develops some reflections carried out under the coordination of the IEA Sub-Technical Committee "Ergonomics and Design for Sustainability," chaired by Giuseppe Di

Bucchianico and Antonio Marano. Some arguments were further developed from the authors on the occasion of a few conferences on the specific theme "Ergonomics, Design and Sustainability." In particular, this paper has been written by Stefania Camplone (paragraph "Ergonomics, Materials, and Sustainable Well-Being") and Giuseppe Di Bucchianico (the remaining paragraphs).

REFERENCES

Bandini Buti, L., *Ergonomia e Progetto. Dell'utile e del piacevole.* Maggioli, Santarcangelo di R., 1998.

Bandini Buti, L., *Ergonomia Olistica. Il progetto per la variabilità umana.* Franco Angeli, Milano, 2008.

Brundtland, G. H., *Our Common Future.* WCED (World Commission on Environment and Development), United Nations, 1987.

Cecchini, C., *Plastiche: i materiali del possibile.* Alinea, Firenze, 2005.

Di Bucchianico, G., Marano, A., Arguments of "ergonomics and design for sustainability." In: *Neue Arbeits und Lebenswelten gestalten, Proceedings of the 56th GfA International Congress (Gesellschaft fur Arbeitswissenschaft)*, Darmstadt, 24–26 Marzo 2010. GFA-Press, Dortmund, 2010.

Gershuny, J., *After Industrial Society? The Emerging Self-service Economy.* MacMillan Press, London, 1978.

IEA Council, *Definition of Ergonomics.* San Diego, Aug. 2000.

IPTS, Futures Project, Natural Resources and the Environment Panel Report, N° Series 05. IPTS, Seville, 1999.

Jordan, P. W., *Designing Pleasurable Products. An Introduction to the New Human Factors.* Taylor & Francis, London, 2000.

Maldonado, T., *La speranza progettuale.* Ambiente e società, Einaudi, Torino, 1970.

Manzini, E., Idee di benessere (ed idee sul benessere) (Tr.: Ideas of Wellbeing (and ideas on wellbeing)), speech at the Master's Course in Ergonomics, CRE-Firenze, April 2004.

Manzini, E., *La materia dell'invenzione.* Arcadia, Milano, 1986.

Manzini, E., Vezzoli, C., *Design per la sostenibilità ambientale.* Zanichelli, Bologna, 2007 (Engl. tr.: *Design for Environmental Sustainability.* Springer, London).

Marano, A., Ergonomics and design for sustainability: the benefits of "savoir-faire" in sustainable society. In: *Ergonomia: valore sociale e sostenibilità, Atti del IX Congresso Nazionale SIE (Società Italiana di Ergonomia)*, Roma, 27–29 Ottobre 2010, Centro Congressi CNR. Edizioni Nuova Cultura. Roma, 2010.

Papanek, V., *Progettare per il mondo reale. Il design come è e come potrebbe essere.* Mondadori, Milano, 1973.

Porter, M., *Competitive Strategy.* Free Press, New York, 1980.

Rifkin, J., *The Age of Access.* Punam, New York, 2000.

Sabatini, F., Coletti, V., *Dizionario della lingua Italiana.* Rizzoli Larousse, Milano, 2004.

41 Business Process Management in Polish Public Administration

Anthropological Aspects of Change in the Face of Technology

Roman Batko

CONTENTS

INTRODUCTION

Communism in Poland was ousted over 20 years ago. One might think that an interim of that length should be sufficient for the dubious practices and absurdities of procedure for which that regime was notorious to have been consigned to oblivion. Indeed, in the business world, they largely have been, but in public administration, change comes slowly. The foundations for the creation of a civil society were laid with the reform devolving power to local communities by establishing councils and other local government structures at the gmina (commune or borough), powiat (district), and voivodship (province) levels. These systemic transformations lacked sufficient reinforcement in the form of modern management methods and techniques, however, and the consequences of this are still emerging. Public administration in Poland is perceived as an overbureaucratized, inefficient golem that puts more effort into preserving the status quo and securing its own interests than into ensuring the well-being of the citizen. Residents are treated not as clients to be served (nor are they provided with assistance in need) but as unwelcome intrusions into the peace of the official's "work." This stereotypical opinion does not, of course, apply to all of the institutions in the country's public administration system, but, as stereotypes are wont to do, it holds considerable sway over the collective imagination. This is also reflected in colloquial language, where the term "administrative officer" has a distinctly pejorative tone that evokes neither sympathy nor trust. Small wonder, then, that one of the priorities for the Polish public administration system must be to modernize the way it is managed, with a focus on quality, effectiveness, rebuilding trust, and improving communication. The aim of this article is to profile a unique new method for implementing a wall-to-wall process management model in Poland's public administration. Its field of application is labor market institutions whose task it is to construct regional reemployment programs, provide support to both the jobless and potential employers, run training courses, and offer vocational counseling. I conducted my participation and observation (*action research*) throughout 2009 and the first 2 months of 2010. The research method I employ is founded on the premise that the interpreting researcher does not stand outside the organization attempting not to interfere in the processes taking place in it but, on the contrary, is actively involved in drafting and implementing change. This observation method is one method employed in organizational ethnography (Konecki 2000; Kostera 2003). Participating in the process of designing change on an equal footing with other members of the organization gives the researcher insight into something that Polanyi and Kostera (Kostera 2003) consider essential from the anthropological standpoint: *tacit knowledge*. This concept is what is expressed in the actions of a specialist who does his/her job well but does not stop to think about the principles that govern it. In 2009, the Malopolskie Voivodship Labor Office (WUP) embarked on a management modernization project based on process management, part funded by the European Union (EU) (The Malopolskie Voivodship Labor Office 2012). Involved in this project on the one hand as an external expert, but gradually, as I said, becoming increasingly immersed in the day-to-day life of the organization on the other hand, I was able to use qualitative methods to analyze the process of change initiated there.

BUILDING A MODEL PROCESS IN PUBLIC ADMINISTRATION

Polish labor market institutions have representation at regional (voivodship) level in the form of voivodship labor offices. The tasks of these offices include creating labor market strategy, conducting various types of studies and analyses designed to assist politicians in taking strategic decisions, and putting all sorts of programs and projects into action on the ground to encourage cooperation between public institutions and employers in order to reduce unemployment and get the jobless into work. In Poland, there are 16 such institutions, one in each voivodship (region). While each voivodship labor office has the same, or similar, objectives, the organizational structure and management methods and techniques of each one are different. The lack of shared standards or a common model for providing services hinders communication and, above all, the synergy that is generated by the exchange of experiences and by the shared aim of perfecting a single, shared management model. This situation is the legacy of the singular brand of antipathy with which any kind of centralization or imposition of operating systems is viewed in postcommunist Poland. Such initiatives are often perceived as attempts to curb freedom of operation or limit autonomy. Recently, however, an awareness of the need for cooperation and change has begun to emerge among the managers of these organizations—in response, of course, to the growing demands of stakeholders and clients. It is vital that everything possible be done to ensure that this will to change is not squandered on ineffective or pointless action. Change has to be wrought throughout the organization; it cannot be partial or superficial. Change management, in the words of Bourne (2008), involves implementing and monitoring new management principles and practices in organizations in order to improve their performance or effect strategic change. This is what is needed at the WUP, where the tasks with which this institution is charged (including management of the many priorities generated by the Operational Program Human Capital (PO KL) for the years 2007–2013, the biggest program in the history of the EU to have received funding [of over 11.5 billion EUR] from the European Social Fund) are creating the need for rapid and far-reaching changes to its organizational structure. Among the challenges it is facing are the doubling in size of its workforce in a very short period of time; the urgent need to initiate new, inexperienced employees in its work; the necessity for cooperation between large numbers of units and branches; and a call for project management capability. The WUP is also seeing rising numbers of clients (PO KL applicants) and stakeholders, and new challenges are also emerging, such as the need to manage huge budgets awarded in competitions for EU aid funds, to operate efficiently, to meet deadlines, and to cope with increasing risks. In this situation, new management solutions are essential for the organization to survive and develop. It is not only processes that will be subject to change but also the organizational culture—and this includes leadership models. For the purposes of this application project, a definition of the concept of "process" in public administration was established. Figure 41.1 shows a process model for a public sector organization, embedded in the context in which it always occurs. This context is, above all, a system of values common to both the clients of the process and its executors. In the name of these values, employees should not act unethically or inappropriately. This context also serves as a reminder of the overarching goal of public administration: the pursuit of the common good, to which every process should contribute a specific value added. Cultural codes also are of significance to the quality of public sector processes—hence, it is important that process executors be familiar with the local language (including, in some circumstances, dialects), customs, and history. The quality and effectiveness of these processes is also connected with the political situation, balance of political power, and political leanings of the clients involved. The political context of the processes is also determined to some extent by other levels of local and national government. The needs of clients that are the inputs to the process, are not only defined by current external legislation or internal regulations but also subject to regular monitoring in order to optimize the "product" that is the process to meet real needs. Clients' satisfaction with the service they receive (the process)

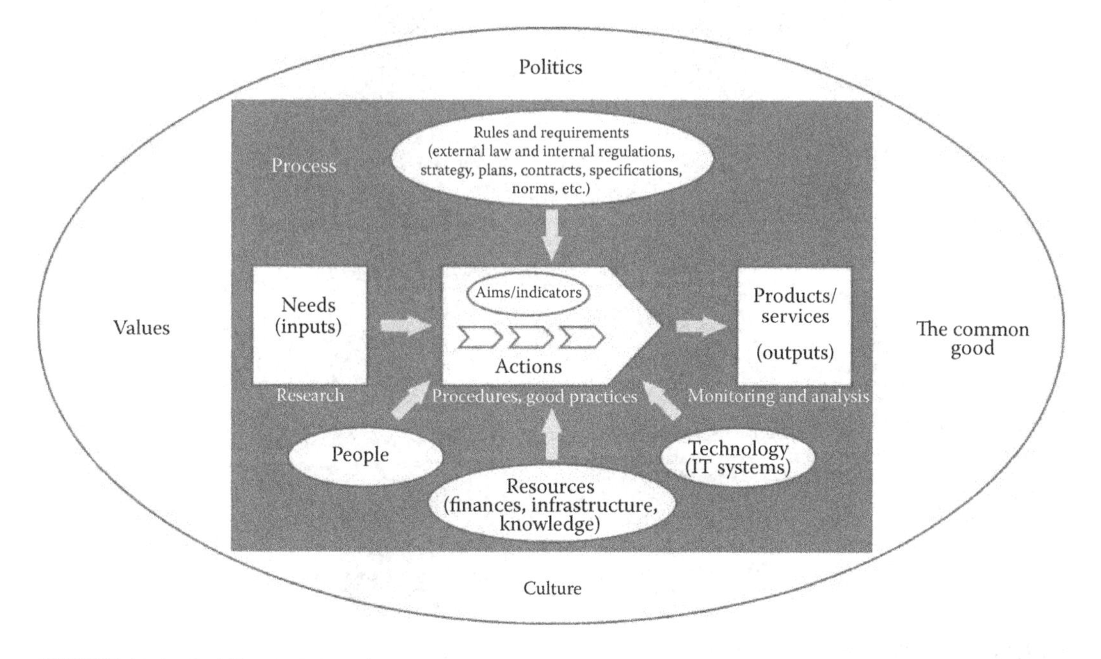

FIGURE 41.1 Public administration process model.

is monitored, as is the question of whether the outcome impacts subsequent efforts to improve the process, through feedback. The requirement to monitor the effectiveness of the process implicates the necessity for implementation of objective management tools. An objective is a projected future state or object that the organization aims to achieve; goals may be strategic—connected with the achievement of the organization's overall strategic plans—or operational (Burlton 2001; *Guide to the Business Process Management Common Body of Knowledge* 2010; Harmon 2007; Jeston and Nelis 2008; Rummler et al. 2010; Sharp and McDermott 2001; Smith and Fingar 2007).

STAGES IN THE IMPLEMENTATION OF PROCESS MANAGEMENT IN PUBLIC ADMINISTRATION (THE EXAMPLE OF THE VOIVODSHIP LABOR OFFICE IN KRAKOW)

The implementation of process management was based on the plan–do–check–act (PDCA) cycle familiar from implementation of quality management systems (Batko 2009a). It thus takes the form of feedback between the nine stages and their improvement in successive cycles by using knowledge of unsuccessful events in the previous cycle, as shown in Figure 41.2.

Each stage of the cycle contributes a value added and is a direct result of the previous stage.

PROCESS ANALYSIS AND IDENTIFICATION

Process identification involved leading workshops with the office's management staff, during which the needs of various client groups, identified through research, were examined in the light

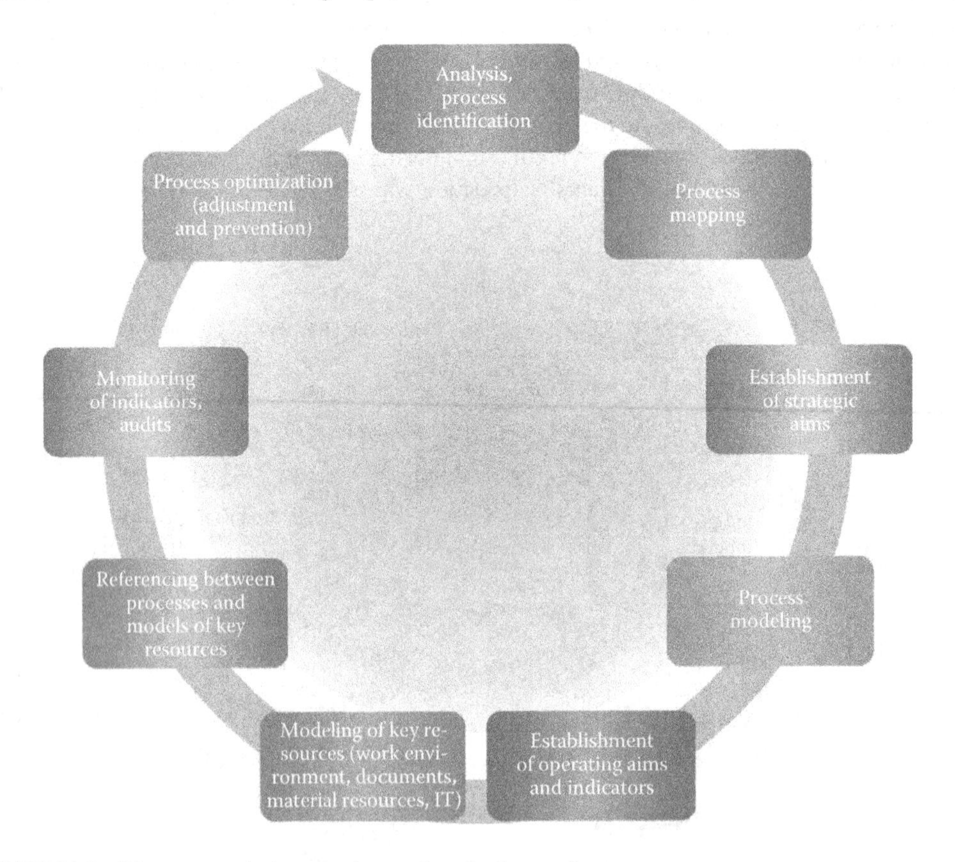

FIGURE 41.2 The process design, shaping, and perfecting cycle.

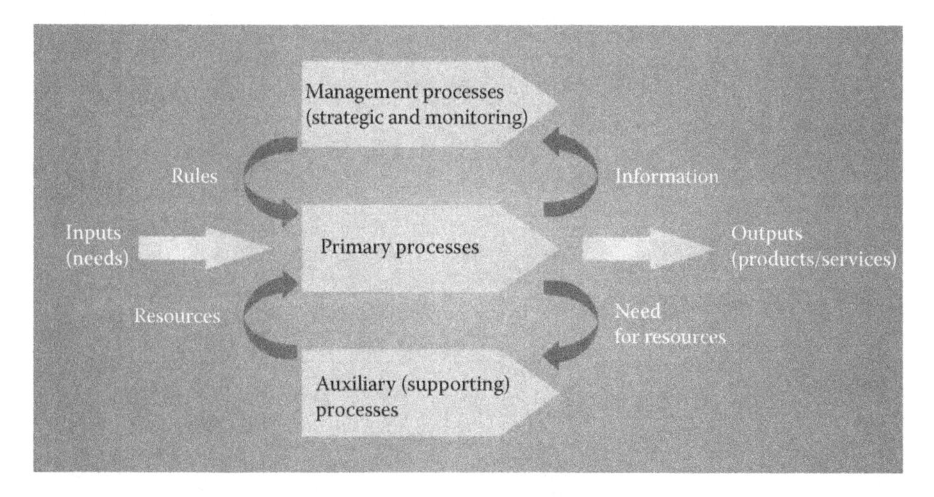

FIGURE 41.3 Interactions between the various types of processes.

of current legislation and internal strategic and planning documents in order to establish the "products" and services that are the process outputs and thus to identify the processes occurring in the office.

In accord with the method shown below, the processes were divided into three categories: management (strategy and monitoring), primary (operational), and auxiliary (supporting). Management processes in the office are those connected with strategic decisions affecting the development of the labor market, creation and monitoring of development plans, and optimization, using management controlling mechanisms. Primary processes are connected with meeting the needs of the office's external clients, while auxiliary processes meet the needs of the office's internal clients (employees). Without auxiliary processes, it would be impossible to execute the primary processes, as shown in Figure 41.3 (Chourabi et al. 2009).

Process Mapping

It was decided that the project management implementation process was to be supported by the ADONIS Business Process Management Toolkit created by BOC Information Technologies Consulting GmbH, which was established in 1995 in Vienna as a spin-off from the Business Process Management Systems (BPMS) Group from the Department of Knowledge and Business Engineering at the University of Vienna (BOC Information Technologies Consulting GmbH 2012). This program was used to create a process map and process models and to integrate various fields—human resources, documents, information technology, and risk management. The process map, as a graphic presentation at the most general level of all the processes identified in the office, divided into three groups—management, primary, and auxiliary—is shown in Figure 41.4.

The 19 processes foregrounded on the map are complex in structure, comprising several levels of subprocesses, down to reference models. An IT tool employing hyperlinks facilitates rapid exploration of the process structure and ease of access between connected processes. IT support is vastly significant, since a total of almost 300 processes and subprocesses have been identified and modeled in this institution. Without computer support, it would be far more difficult—in practice, impossible—to operate efficiently in such a complex information environment. One example of penetration of the complex structure of one process, "strategic planning," to a further level is presented in Figure 41.5.

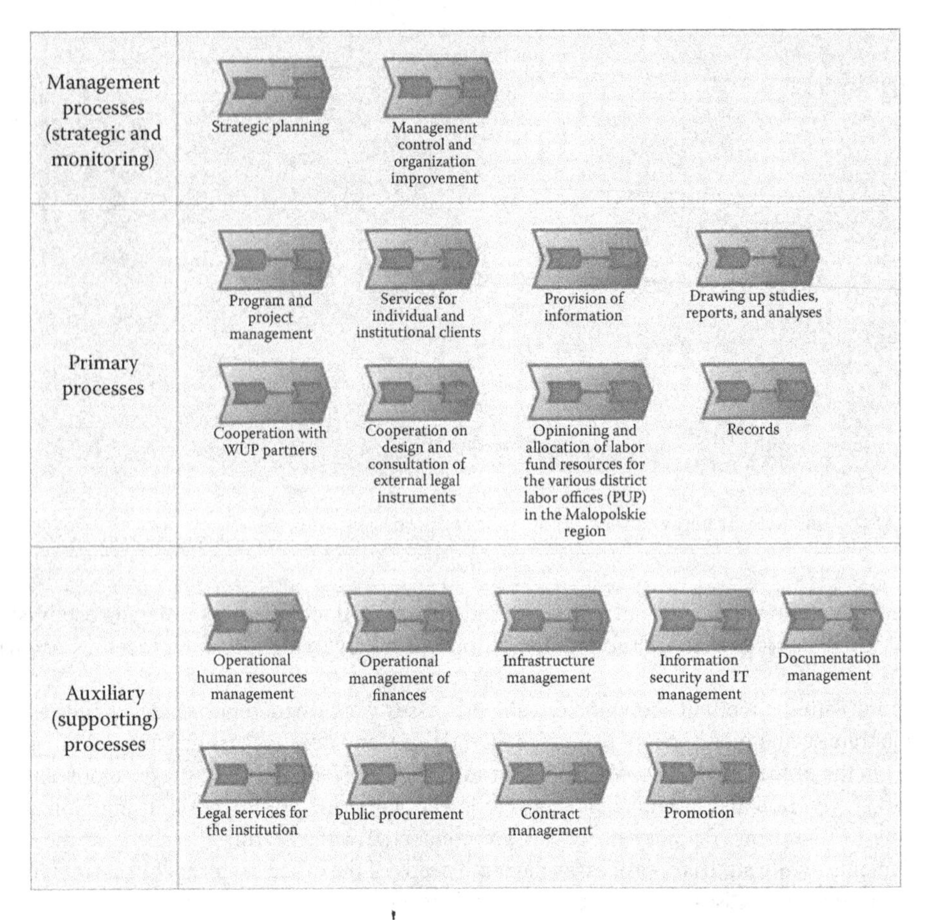

FIGURE 41.4 Process map for this regional labor office institution.

Establishing Strategic Goals

The WUP had already identified its strategic goals before embarking on the Business Process Management (BPM) implementation, so this stage involved revision and assessment of progress of the various strategic goals. An important systemic solution is the facility of using the IT support tool to reference-link the established strategic goals with the operational goals pursued in the operational processes and subprocesses.

Process Modeling

This stage involved gathering information on good practices in process execution and holding workshops with process owners to produce graphic presentations showing the ordered, focused, and effective sequencing of all the actions taken in the course of the process and identifying the inputs and outputs of the process. The notebook dialogue boxes dedicated to the objects, for example, "action," contain details of the various actions and decisions, as well as provisional estimates of the time they take to perform. Figure 41.6 shows an example of a process model created in this regional labor market institution that is already in use in the day-to-day provision of services by the institution.

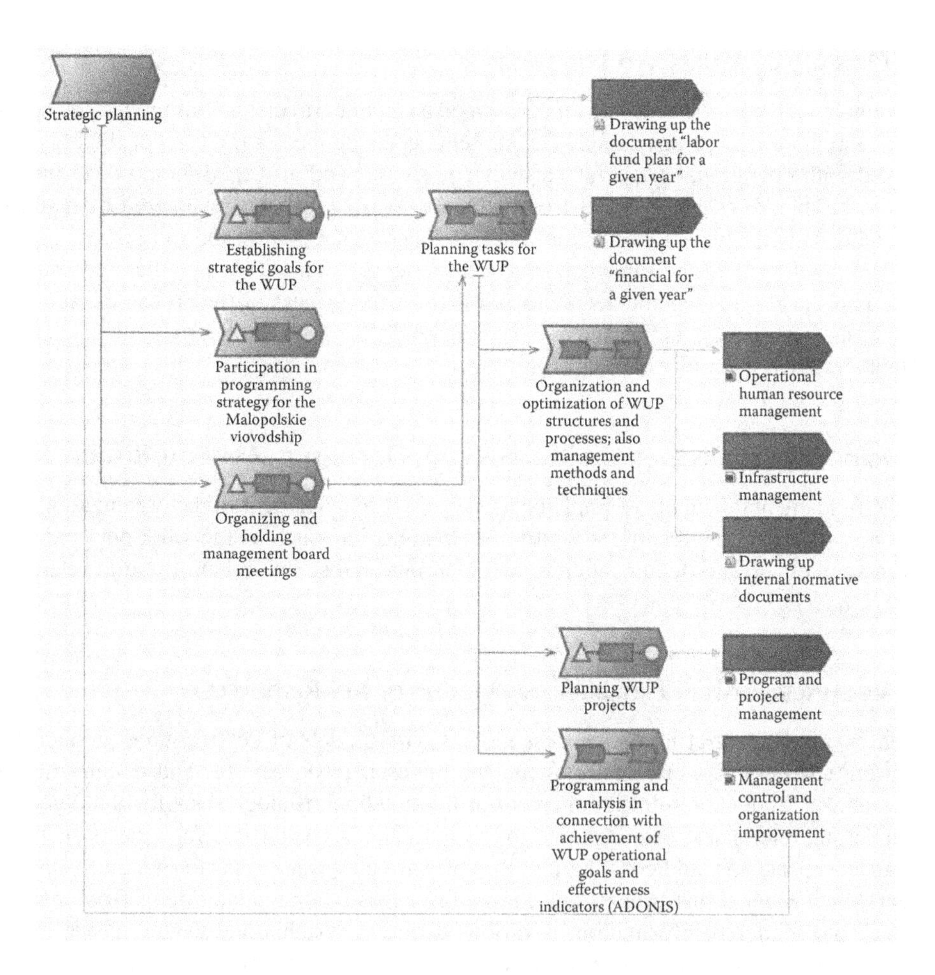

FIGURE 41.5 "Strategic planning" process map for this regional labor office institution.

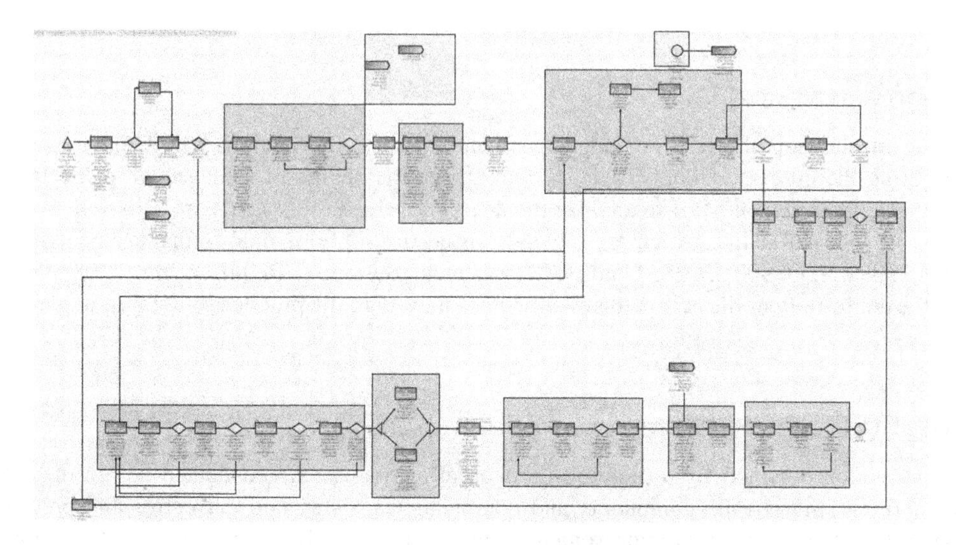

FIGURE 41.6 Example process model in the regional labor market institution (evaluation of the content of system applications and procedure for conducting negotiations).

Setting Operational Goals and Indicators

Management by objectives (MBO) was incorporated as an integral element of process management in the regional labor market institution. The strategic goals defined at the level of the organization were linked with operational goals for particular processes and allocated monitorable indicators. The data to be monitored are sourced from databases or spreadsheets or entered by hand. If the expected value is not achieved, the IT tool flags the operational goal in red to illustrate nonfulfillment of the intended plan. The same happens to the strategic goal with which that operational goal is linked. This enables management staff to assess the effectiveness of the processes on an almost real-time basis and to adjust and optimize the processes on the basis of data rather than simply intuition (Hubbard 2007; Rummler et al. 2010).

Modeling of Key Resources (Work Environment, Documents, Material Resources, IT)

One of the fundamental benefits of implementing process management in the organization was the integration of different key resources within the organization, such as the work environment, the material resources, IT and documentation, and identified risks. These help to build references in the process models.

Setting References between Processes and Models of Key Resources

This stage involved integrating the process models with models of key resources created for the organization, such as the work environment, the material resources, IT, and documentation, as described above. Under the relevant chapters in the notebook dialog boxes for the objects used in the modeling, references were included to the individuals responsible for aspects such as task performance, supervision, and consultation; the software necessary to perform a given action; and documents, such as legal instruments and forms. Prior to the implementation of BPM, employees had to seek out scattered information in various different systems and centers every time they needed it, which of course wasted time, compromised the credibility of the information (which might be incomplete, out of date, etc.), and was symptomatic of the flawed internal communication system.

Monitoring Indicators, Audits

With the introduction of process management, a number of authorizations were handed over, thus empowering the process leaders. Tools for monitoring the degree of achievement of operational goals enable management staff to analyze the degree to which these goals are achieved on a constant basis. Internal audits, which are tools also used in ISO 9001 quality management systems (Batko 2009a; ISO 9001:2008: Quality Management Systems—Requirements 2008), help to check the conformity of the accepted systemic principles with the practices employed in particular processes.

Process Improvement (Corrective and Preventive Actions)

The information obtained from monitoring, internal audits, checks conducted within the office, benchmarking, and clients' complaints and comments translates into corrective and preventive actions taken, that is, process improvement. This dynamic is vital in the present-day reality of constant change in which contemporary public administration functions. There is no saying that process models created today will still be current tomorrow. Clients' needs change rapidly, and the process should change with them.

Models constructed in this way produce a synergy effect and integrate the organization's scattered resources and management techniques. Quality management is also of key significance for understanding processes and implementing process management. It is the client (or, more precisely, his/her needs), from whose perspective the quality management system is built, who initiates the process, and it is the client who evaluates the process by evaluating the "product," or service, that the office has rendered. It is therefore vital to take care to research both clients' needs and their satisfaction and to use the conclusions reached on this basis to optimize the process. Implementing process management is a process in itself. It would seem rather irrational suddenly to abandon a functional structure based on specialized cells and tasks in favor of process management. Instead, the process should be an evolutionary one, in order not to create excessive organizational chaos and to avoid major tensions and conflicts in connection with changing competencies and loss of power and influences. A consistent, gradual process of moving from one model to another is likely to bring greater benefits. Employees become accustomed to participating in teams performing processes; start to understand the relationships between clients' needs, their own work, and the final product; and begin to perceive themselves as internal clients and demand quality in internal processes. The system of monitoring achievement of goals and internal audits helps management staff to overcome their fears surrounding employee empowerment and their anxiety that they will lose control of management of the office. An organization that is process oriented is flexible, fast acting, and client friendly (Kanter 1990). The famous saying by Brown and Grey (1995), "Processes don't do work, people do," encapsulates the truth that even the best organized management system will not bring the organization real benefits if it is not created by competent, engaged employees. Implementation of the solutions described here must therefore be supported by a training program and an incentive system. Knowledge management is a vital condition for effective process management, but paradoxically, it is implemented process management that facilitates better management of scattered knowledge and more rapid induction of new employees to task performance.

BENEFITS OF PROCESS STANDARDIZATION: ANTHROPOLOGICAL ASPECTS OF CHANGE IN THE PUBLIC SECTOR ORGANIZATION

Introduction of common standards across a given sector of public administration, as in the case discussed here, a labor market institution, brings significant benefits for the clients of its services and positive anthropological changes to public organizations. This study facilitated the identification of the following key benefits (Batko 2009b).

Design and Dissemination of the Most Effective Good Practice in a Given Area

There are two parallel issues here. One is connected with the "development" of standards. Standards should not be imposed, but a discussion forum should be created in which elements of standards will be agreed upon by equal partners. Only then is there a chance of discovering the optimum good practices while at the same time minimizing antipathy to the changes by means of cooperation and involvement. The other question is success in disseminating these standards. Even the best good practice will remain lifeless if it is not disseminated, that is, accepted and popularized, with training, publications, information portals, and modern IT tools to support the process.

Ease of Transposing the Standard to Other Institutions

A standard, as its definition suggests, assumes a certain universality of application. Thus, once devised, a standard may be implemented in other organizations of a similar nature and range of services. This transposition may of course be horizontal, for example, the standards devised and accepted by pilot offices may easily be implemented in similar offices, or it may be vertical—standards tried and tested in central administration may be used at various other levels of administration and also implemented in other public sector organizations, for example, government agencies and civil society organizations cooperating with these offices.

A Common Code, an Accepted Language for the Standard, to Enable Benchmarking

The basis for creating a standard is the acceptance of a common code. By this, I mean developing both a glossary of the definitions and basic terms used in the process descriptions and a shared modeling tool, to render standard processes comparable. This approach to building standards facilitates benchmarking of public organizations. As the way of performing processes covered by a standard code is very similar, it is possible to compare, for instance, the degree of fulfillment of assumed goals, both qualitative and quantitative. It is thus possible to compare the time taken to perform various tasks within the process, waiting times for particular services, numbers of errors made in the performance of the tasks, the level of employment in the process, and so forth. Significantly, this type of standardizing code need not be solely national in nature but may also be supranational, a basis for global benchmarking. Attempts to create such codes have been underway for over 20 years, to cite but the four versions of the international quality management standard ISO 9001, or the European self-assessment and qualitative reward method Common Assessment Framework (CAF) and European Foundation for Quality Management (EFQM) (Batko 2009a).

Cutting the Costs of Nonconformity

One of the sources of nonconformity is uncertainty, which will arise wherever the performer of a task has insufficient knowledge on how to perform a given task. Nonconformity always costs something. This is perhaps not always as clearly visible in the realm of public administration as in industry, where losses are often material, but they can be no less damaging. The cost is the time that employees have to spend correcting badly written letters, badly researched information, flawed decisions, and so forth. This cost tends not to be recorded anywhere and, in fact, may not even be treated as a cost. The costs of compensation paid out by the administrative body for a decision issued in contravention of the law and effectively appealed against by a citizen is, of course, a cost, however, and a very considerable one, as is the loss of society's confidence, or misguided strategic decisions, the costs of which may be visible for generations. Standardization of processes in administration can significantly cut the costs of nonconformity. Where there are designated procedures taking account of current laws and incorporating the essential information in terms of binding deadlines and forms, employees work in an environment where there is limited risk of error. More importantly still, optimization of standards means that the first time an instance of nonconformity arises, corrective action is taken on a system-wide scale, and if approached effectively, this eliminates the causes of the nonconformity and, hence, potential costs.

Improved Process Monitoring and Auditing

Calibrating processes helps to monitor their effectiveness. After establishment of the goals to be achieved through the process, control points may be set at intervals along the process model, where information is gathered on factors such as the timeliness or quality of task performance. Another valuable tool is the audit, which helps to monitor whether actions are being performed in conformity with the law and other standards, to identify weaknesses and risk areas, and to offer suggestions for optimization measures. Standardized processes are easier to audit because their progress is predictable, defined, and, as mentioned above, monitored in terms of quality and effectiveness.

Reduced New Employee Induction Time

Often, the weakest link in the public administration processes is employees who are ill prepared to perform them. This is why so many negative opinions regarding the quality of the work done by the public services focus on employees' competence, their personal manner, and their empathy. This applies to some degree to experienced employees who may be burned out professionally, but it also applies to new employees who have not been given sufficient training or support in the form of standards for action on processes. Often, they are aware of their lack of competence and attempt to cover up for it with an arrogant, unpleasant manner. A professionally constructed process model comprises not only procedure paths but also the communication relations that the performer of the actions should undertake (supervision, consultation, information); a base of easily accessible, up-to-date forms, legal documents, and so forth; and also information on deadlines, goals set, and process indicators. The time invested in designing models and standardizing processes will pay dividends, especially in situations where there is a high staff turnover and new employees frequently have to be inducted. Not only will it reduce the time it takes for a new employee to start performing tasks independently, but it will also reduce the number of nonconformities arising out of ignorance and uncertainty (and hence cut the costs of nonconformities, as shown above).

CONCLUSION

The work of this regional labor market institution (the Malopolskie WUP) to improve the management of its organization by implementing process management principles and elements of total quality management (TQM), such as continuous improvement, the internal audit, and standardization of services, is now fact. But it is worth looking not only at the benefits that the change process has brought but also at the obstacles that had to be overcome. Most of the threats to the process of change that were observed during the implementation work at the WUP were rooted primarily in the cultural resistance of the people involved, which was related to their fear of change. Change is a process that, by its very nature, impacts the *status quo*. The way in which this resistance to change was overcome was to plan a nine-stage management implementation process. Employees were involved in the conceptual work and the workshops, and they were aware of the goal and of the benefits this change could bring. They were also given appropriate training.

In my opinion, at this phase of project realization, the most important and clearly measurable benefit is the identification of the processes that go on at the WUP (and the numerous subprocesses—of which there are almost 300). This gave rise to a process map showing their interdependencies and setting out paths of communication and the need to take account within a given process of the performance and expectations of other process leaders. Modern administration is increasingly departing from the rigid procedures with which it had been invariably associated since the introduction of the Weberian model (Brodkin 2007), in favor of effective processes calibrated with measurable goals and managed in a way that harmonizes with the organization's strategy, while incorporating a large degree of delegated authority. Process description aids information management, even at the very rudimentary level of the basic knowledge essential to performing tasks at the workstation. Process management also helps in seeing the organization as a system of interconnected centers and resources and supplements the hierarchical model of functional management with horizontal structures focused on achieving process goals. Perhaps the most important value added of process management, however, is that it faces up to the challenge, diagnosed by Luhmann (1979) back in the late 1970s, of the increased complexity, uncertainty, and risk inherent in contemporary society. Employee and stakeholder confidence in modern management methods, including process management, will be built not on the false premise that "liquid modernity" (Bauman 2000) will be superseded once and for all by certainty and permanence but on the assurance that we can cope with it, be flexible and change, and keep abreast of it.

REFERENCES

Batko, R. (2009a). *Zarządzanie jakością w urzędzie gminy*. Kraków: Wydawnictwo Uniwersytetu Jagiellońskiego.

Batko, R. (2009b). Standaryzacja procesów jako podstawa doskonalenia administracji publicznej. In K. Lisiecka, and T. Papaj (Eds.), *Kierunki doskonalenia usług świadczonych przez administrację publiczną* (pp. 129–142). Katowice: Śląskie Centrum Społeczeństwa Informacyjnego and Akademia Ekonomiczna w Katowicach.

Bauman, Z. (2000). *Liquid Modernity*. Cambridge: Polity Press.

BOC Information Technologies Consulting GmbH. (2012). Retrieved from http://www.boc-group.com.

Bourne, D. (2008). Zarządzanie zmianą i badanie wartości organizacji z użyciem techniki siatki repertuaru. In M. Kostera (Ed.), *Nowe kierunki w zarządzaniu: Podręcznik akademicki* (pp. 495–510). Warszawa: Wydawnictwa Akademickie i Profesjonalne.

Brodkin, E. Z. (2007). Bureaucracy redux: Management reformism and the welfare state. *Journal of Public Administration Research and Theory*, *17*(1), 1–17.

Brown, J. S., and Gray, E. S. (1995). The people are the company. *Fast Company*, *Premie Issue*, *1*(1), 78–82.

Burlton, R. (2001). *Business Process Management: Profiting From Process*. Indianapolis: Sams Publishing.

Chourabi, H., Mellouli, S., and Bouslama, F. (2009). Modeling e-government business processes: New approaches to transparent and efficient performance. *Information Polity. An International Journal of Government and Democracy in the Information Age*, *14*(1–2), 91–109.

Guide to the Business Process Management Common Body of Knowledge. (2010). Chicago: Association of Business Process Management Professionals.

Harmon, P. (2007). *Business Process Change, Second Edition: A Guide for Business Managers and BPM and Six Sigma Professionals*. Burlington: Morgan Kaufmann Publishers.

Hubbard, D. W. (2007). *How to Measure Anything: Finding the Value of "Intangibles" in Business*. New Jersey: John Wiley and Sons.

ISO 9001:2008: Quality Management Systems—Requirements. (2008). Geneva: International Organization for Standardization.

Jeston, J., and Nelis, J. (2008). *Business Process Management: Practical Guidelines to Successful Implementations*. Oxford: Butterworth-Heinemann.

Kanter, R. M. (1990). *When Giants Learn to Dance*. New York: Touchstone.

Konecki, K. (2000). *Studia z metodologii badań jakościowych: Teoria ugruntowana*. Warszawa: Wydawnictwo Naukowe PWN.

Kostera, M. (2003). *Antropologia organizacji: Metodologia badań terenowych*. Warszawa: Wydawnictwo Naukowe PWN.

Luhmann, N. (1979). *Trust and Power*. New York: John Wiley and Sons.

The Malopolskie Voivodship Labor Office (WUP). (2012). Retrieved from http://wup-krakow.pl.

Rummler, G. A., Ramias, A., and Rummler, R. A. (2010). *White Space Revisited: Creating Value through Process*. San Francisco: Jossey-Bass.

Sharp, A., and McDermott, P. (2001). *Workflow Modeling: Tools for Process Improvement and Application Development*. Boston, London: Artech House.

Smith, H., and Fingar, P. (2007). *Business Process Management: The Third Wave*. Tampa: Meghan Kiffer Press.

42 Organizational Citizenship Behavior

In Search of a New Kind of Employee

Adela Barabasz and Elżbieta Chwalibog

CONTENTS

INTRODUCTION

Both managers and scientists often tell us that companies need to be flexible, adaptable, and agile. Faced with unpredictable competitors, they have to be in a constant state of a kind of alert, ready for changes, and developing new skills. In the face of a new economics, it makes sense for companies to pursue ever-greater levels of flexibility—flexibility and change are requirements. Companies usually reinvent themselves because competitors with more attractive products or efficient production systems are considered a threat for them. If they do not change, they will probably lose their customers, investors, and so on. It means they need a new kind of employees—involved, open to change, willing to learn, not afraid of challenges and eager to take them, and not afraid of failures. This last item is embedded by design in the organization's life, because with no failures and mistakes, it is difficult to imagine that we are actually building something truly new and original.

This article presents the basic assumptions of a concept that is not so well known in Poland yet, the concept of organizational citizenship behavior (OCB). According to the authors, these assumptions not only open a new perspective in the approach to so-called human potential management in the organization but also foster a look at the organization as a subject. This examination based on variables and psychological dimensions helps us to understand better the dynamism of inter-organizational forces. The current results of the research, also taken in Poland (Witkowski and Chwalibóg 2010), encourage us to refer to the concept of organizational personality (OP), comprehended as a dynamic instance integrating the functioning of an organization in the sphere of interpersonal and intergroup relationships, regulating the behavior of the individuals within the structure of the organization. Its objectives are developing behavior consistent with the organization's goals; reducing the level of fear and aggression through the usage of established defense strategies; and creating and transferring a system of common beliefs, judgments, evaluations, and visions onto the whole organization. A necessary condition for the forming of an organization's personality is frequency and intensity of relationships between the members of the organization. With a low frequency and intensity of relationships, the organization remains a set of loosely connected individuals. With a high frequency and intensity, a new "quality" emerges, with features resulting from the personalities of its most influential members (Barabasz 2008). Employees who believe that people outside the company think highly about their organization may develop OCB as part of being good citizens in a good organization. A strong affective commitment to an organization is likely to generate a high value of OCB (Carmeli 2005).

The article has a theoretical and conceptual character. It could be a base for the interpretation of the empirical research aimed toward finding the relationship between occurrence of the particular types of OCB and the dimensions of OP according to the W. Bridges typology. The results will be presented in the next planned scientific description.

Factors like inevitability of changes, meaning of innovativeness and knowledge, and processes of organizational learning in an organization all prompt us to acknowledge that employees inherent in human resources decide about the degree to which companies keep up with the demands of the internal and external environment.

According to Beugelsdijk (2008), human resources practices are strongly related to a firm's capability for generating some innovations. The study suggests that the five components of OCBs—altruism, courtesy, conscientiousness, sportsmanship, and civic virtue—have simultaneous and significant influences on knowledge sharing (Chieh-Peng 2008).

CONCEPT OF OCB

The concept of OCB was first introduced by Dennis Organ in 1983. He defined it as "individual behaviour that is discretionary, not directly or explicitly recognized by the formal reward system, and in the aggregate promotes the efficient and effective functioning of the organization. By discretionary, we mean that the behavior is not an enforceable requirement of the role or the job description, that is, the clearly specifiable terms of the person's employment contract with the organization; the behaviour is rather a matter of personal choice, such that its omission is not generally understood as punishable" (Podsakoff et al. 2000).

In the literature, we can find different forms of OCB. Meta-analysis done by Podsakoff et al. (2000) covers the concept more widely and distinguishes among the OCB:

1. **Helping behaviors**—are related to voluntary help to others in dealing with or/and preventing problems connected with work.
2. **Sportsmanship**—the desire\readiness\willingness to tolerate (without complaint) the inevitable inconvenience and enforcement of work and maintaining a positive attitude even when things go wrong; not being offended when someone does not stick to the comments; the readiness to sacrifice one's own business for the good of the team; the ability not to get affected by objection or denial.
3. **Organizational loyalty**—protecting the company and its goals, promoting the company, keeping up its appearance, protecting it from external threats, being loyal even in unfavorable conditions; identifying with the company and its leaders, taking care of the common good.
4. **Organizational compliance**—internalization and acceptance of rules and procedures that exist in the company, which are conscientiously and honestly abided by, even when there are no observers.
5. **Individual initiative**—voluntary going out of one's own role in the organization (containing behaviors connected with the task), going beyond minimal requirements or desired level of quality of the job done. These behaviors consider volunteer initiative, creativity, and innovation to increase the quality of the job for the company; enthusiasm when fulfilling their requirements; making an effort to complete the task; voluntarily taking additional

responsibilities; and encouraging coworkers to do the same. These are all the behaviors of going out of and beyond one's responsibilities. Organ thinks of these behaviors as being the most difficult to distinguish from behaviors connected with role or task.

6. **Civic virtue**—the interest and commitment to the organization as a whole. It can be observed in willingness to actively participate in organizational management (e.g., attendance at the meetings and at the debates concerning the company's policy, showing what strategy will be taken); monitoring the environment to find threats and opportunities for the company (such as watching the growth of the competition); and keeping an eye on the company's business (reporting risks, suspicious activities inside and outside the company) even when the personal costs of the one are high. These behaviors reflect one's feeling of being a part of the whole (the way a citizen feels to be a part of a country) and accepts the reasonability that comes with it.
7. **Self-development**—includes voluntary behavior of employees concerning widening of their own knowledge, skills, and abilities.

Concerning the direction of citizenship behavior, we can differentiate them after Williams and Anderson into the following (Podsakoff et al. 2000):

- OCB-I—Behaviors that immediately bring profit to specific individuals, and indirectly also to the organization (e.g., help after long absence or interest in other people's business). These behaviors fit in the category of helping behaviors.
- OCB-O—Behaviors that generally bring profits to the organization (e.g., informing them about not getting to work earlier, obeying the informal rules made to keep the order). These behaviors fit in the category of organizational obedience/subordination.

The concept of OCB has no strong theoretical background, but it is important that empirical studies have shown many important results.

RESEARCH FOCUSED ON OCB

Early research in the United States from the 1980s focused on finding a correlation between OCB and individual characteristics of workers, their attitudes—especially the morale of the underlying job satisfaction, organizational commitment, perception of fairness, and perceptions of leaders' support. Correlation between these characteristics and OCB ranged from 0.23 to 0.31 (Podsakoff et al. 2000) and, in other studies, even reached 0.44 (Organ et al. 2006).

Another area of research in this field is dispositions of employees, such as agreeableness, conscientiousness, and positive and negative emotionality. Studies have shown a positive, although weak, relationship between these dispositions and OCB. Correlations ranged from 0.04 to 0.30 (highest for conscientiousness). Predispositions to show OCB are empathy, need for achievements, affiliation need, proactive attitude, loyalty, and planning. The general trend of research shows that the relationship between personality and OCB is weak; perhaps this relationship is not direct, and attitude can intermediate (Organ et al. 2006). Polish studies have showed a clear positive correlation between the OCB and extraversion (0.35) and its components (activity, 0.42, and assertiveness, 0.30), as well as components of conscientiousness (striving for achievement motivation, 0.25, and self-discipline, 0.30), but negative correlations with a component of neuroticism (impulsiveness, –0.28) and a component of agreeableness (uprightness, –0.27). A clear correlation emerged between OCB and temperamental dispositions—liveliness correlated at 0.42 and activity at 0.38 (Witkowski and Chwalibóg 2010).

With regard to the perception of roles—the ambiguity or conflict of roles—the majority of studies showed a weak negative relationship with OCB (correlations from –0.02 to –0.16). These two phenomena are linked with job satisfaction, and satisfaction is correlated with the OCB, which is why it appears to be a mediator of that relationship. Demographic variables, like seniority in the organization and gender of the workers, are not correlated with OCB (Podsakoff et al. 2000).

The characteristics of the tasks show significant, but low, correlations with OCB. The subjects of the studies were primarily feedback, routinization of work, and satisfaction that comes with the job. Feedback (from 0.16 to 0.21) and job satisfaction (from 0.14 to 0.27) are positively correlated with OCB, whereas routinization is negatively correlated (from –0.10 to –0.30) (Podsakoff et al. 2000).

Research shows that formalization, inflexibility of the organization, support counselors, and distances in space are not correlated with OCB. Compounds that have been discovered relate to the consistency of the group (from 0.12 to 0.20), perceived support of the organization (0.31), and rewards beyond the control of the leader (from –0.03 to –0.17)—the first two proved to be positive and the third negative (Podsakoff et al. 2000).

Additional research was conducted in this area, and other relationships with OCB were discovered. Transformational leadership behaviors, developing a vision, providing an appropriate model, building acceptance for group goals, expectations of a high level of performance, and intellectual stimulation are positively correlated with OCB, oscillating at around 0.2. A positive correlation was also found with rewarding leader behaviors and negative one with punishment. Subsequent studies have shown that support leadership behavior and a clearly outlined role of a leader are also positively correlated with OCB. The general behavior for the leader-member exchange (LMX) of the organization correlates with OCB at 0.3 (Podsakoff et al. 2000). There is a suspicion that some of the leadership behaviors correlate with OCB through the influence of employee attitudes (e.g., sense of justice).

MECHANISMS OF OCB

Studies from United States have identified four mechanisms that can explain OCB. According to summary made by Blatt (2008), we can find the following:

1. Social exchange perspective and tit-for-tat mechanism show that workers who believe that they are preferably treated by the organization increase their involvement and often exhibit OCB. However, employees who perceive themselves as treated less favorably (e.g., due to work on short-term contracts, low sense of security, little chance for promotion) are less involved and less likely to exhibit OCB.
2. The mechanism of identification means that employees engage in OCB as soon as they join the organization in their social identity—thereby positively influencing OCB, which is perceived by them as beneficial to themselves as individuals. Identification with the organization leads to the internalization of its objectives by the employee and to the inclusion of OCB to behaviors connected with the organization role (in normal conditions, far beyond the role).
3. Impression management makes an employee want to improve his/her future in the organization by exhibiting OCB. Such strategic reasons cause the employee to try to impress both colleagues and superiors to increase the chance of promotion or positive evaluation.
4. Positive relationships may underlie OCB, primarily due to increase of empathy and responsiveness to the needs of others and inducing behavior associated with "being a good colleague."

These four mechanisms are based on social psychology. The most important question shows the connection between motivation and the definition of OCB—for example, is it still OCB if employee thinks about his impression management or maintains positive relations?

OCB IN THE CONCEPTUAL AGE

For years, organizational behaviors have been of interest for researchers and practitioners of management. It is valuable for every manager in the organization to discover the causes of the behavior of his/her subordinates. The multiplicity of factors that influence human behavior is so large that prediction of the next action is impossible. Practitioners of management should not believe that research on human behavior is, in this case, unnecessary. Each test, correlation, and experiment can capture certain relationships between variables (or a lack of them), which is very valuable each time. Finding a direct cause–effect relationships in this case is very difficult; however, obtained correlations of various factors of human behavior in organizations gives us a practical indication of the direction in which we can modify various management systems to make the occurrence of desired behaviors more probable. OCB is a specific type of organizational behavior. For each employer going beyond the formal system of rewards, organizational roles and responsibilities are valuable and can improve the effectiveness of the organization.

Looking through the nature of OCB, its mechanisms, and its types, we can see that every employer wants to have employees with a high level of OCB. It is low-cost benefit for the organization to have loyal, supporting, and involved workers. The very important question is still how to motivate employees who exhibit OCB. There can be a fine line between using that kind of employee and releasing their behavior. It is crucial to understand how basic the motivation for employees with high OCB is. We can suppose that without a positive reaction of the organization's management, the frequency of OCB can be reduced. And here is a place for nonmaterial motivation. The concept of OCB shows that person does not look forward to receiving a reward, but we should remember that creating new behaviors in the organization needs to show a new appropriate role model for other employees.

OCBs can be perceived as exceeding the requirements of the position or role in the organization, creative behaviors, or as triggering that kind of behavior. Because of that, they can be very useful for every employer searching for competitive advantages, especially in the conceptual age. Today's market situation brings new chances for companies implementing unconventional solutions. That is the reason for searching for new dimensions of "the ideal worker." Deep analysis of OCB's correlates can show new interesting ways of rearranging management systems.

The conceptual age shows new opportunities for organizations. The huge role of information, knowledge, and creativity, but also the concentration on cost management, can bring about important changes for every employer. The employee who shows OCB can share with others; he is loyal even if he has greater challenges or opportunities outside the company; he encourages new changes and projects; and therefore, he helps reach the goals, mission, and vision of the organization. It is crucial to understand that this kind of attitude can reduce employees' resistance to new ideas or fear of changes in the company, and it can also bring more a optimistic view on crisis inside or outside the organization.

NEW RESEARCH DIRECTIONS: CONCLUSION

The research on OCBs should focus on three perspectives: individual, group, and organizational. The areas of research explore individual variables (personality, temperament, character, attitudes, etc.); group variables (group roles, group leadership, etc.); and organization variables (organization structure, organization culture, climate, and personality, but also responsibilities and duties of the position, types of tasks, organization roles, etc.).

The results of research confirm the link between OCB and many of the variables mentioned above. The most important for Polish researchers is to review the results appearing in foreign studies on Polish ground. OCBs are a concept that still has to be defined, both theoretically and empirically. The importance of the issues the concept of OCB describes makes us want to look at them carefully and analyze them through further studies.

OCBs seem to be beneficial for the organization—the staff member who goes beyond his/her responsibilities both at the coworker and organization level contributes to better functioning in each of these areas. OCBs are, by nature, prosocial, and if we do not focus on the complex issues of motivation of such behavior, it appears that they usually manifest themselves when the employee and the organization are "well matched." The results of many studies may provide an important indication for employers. If it is relevant for the position to manifest the OCB, we should take into consideration the variables connected with the individual level. It is also important to find correlations between the group and organizational level, because conscious stimulation and change in these areas may affect the number of such desirable behaviors as OCB.

Previous studies focused on various aspects of OCB encourage us to take a chance to understand whether and how different forms of behaviors, described as "citizenship behaviors," influence, enhance, or attenuate each other; how can they be stimulated; whether they can be easy controlled; how far their existence in the organization is related and dependent on consciously introduced policy in terms of human resources management; and so on. What facilitates the idea of organization personality is the understanding of mechanisms of mutual relationships between personalities of organization members (individual level) and transference of particular individuals' characteristics into the group and organization level. The application of current studies in terms of OCB and the knowledge regarding mechanisms related to the personality of an organization, based on the psychoanalytical approach, allow mostly for the following:

- Diagnosis of human resources—which can be conducive to optimization of activities undertaken to adjust the company to the environment's conditions; indicate underlying causes of difficulties deeper than reasons consciously indicated by members of the organization or management representatives; and allow the formulation of effective ways of solving problems and foster problem prevention.
- Control of personal policy—indication of the type of company personality helps in the adequate choice of employees, both for key managerial positions, which are required at a particular stage of company development, as well as all other meaningful positions in terms of realized (or planned) strategy of operation; planning the trainings, identification of the required employee profile in particular field of company's operations.
- Organization improvement—with knowledge about the organization's weaknesses and strengths, it is much easier to indicate actions taken toward coping with difficulties and threats, which organizations with a particular personality face; it facilitates the matching of the strategy of solving conflicts, and it fosters the optimal utilization of the psychological potential of organization members.

Assuming that the perception of an organization determines the level of involvement of its members in order to realize the organization's goals, and that this involvement in turn influences the view of organization, the studies focused on the concept of OCB and the idea of OP mentioned above open new possibilities, especially for those representatives of management, who are aware of the meaning of a good match of employee personality with the organizational assumptions. It can be expected that employees will feel a strong relationship with the organization, which takes care of its good public image in the environment (Dutton et al. 1994), and this, in turn, enhance the process of identification with the organization. In accordance with the understanding of the mechanisms of OP, this should translate to better results in terms of tasks realized in the organization.

REFERENCES

Barabasz A. (2008). *Organizational Personality. Its Application in Management.* University of Economics Press, Wroclaw.

Beugelsdijk S. (2008). Strategic human resource practices and product innovation. *Organization Studies*, vol. 29(06), pp. 821–847, SAGE Publication, Los Angeles.

Blatt R. (2008). Organizational citizenship behavior of temporary knowledge employees. *Organization Studies*, vol. 29(06), pp. 849–866, SAGE Publication, Los Angeles.

Carmeli A. (2005). Perceived external prestige, affective commitment, and citizenship behavior. *Organization Studies*, vol. 26(3), pp. 443–464, SAGE Publication, Los Angeles.

Chieh-Peng L. (2008). Examination of gender differences in modeling OCBS and their antecedents in business organizations in Taiwan. *Journal of Business and Psychology*, vol. 22(3), pp. 261–273.

Dutton J.E., Dukerich J.M., Harquail C.V. (1994). Organizational images and member identification. *Administrative Science Quarterly*, vol. 39(2), pp. 239–263.

Organ, D.W., Podsakoff, P.M., MacKenzie, S.B. (2006). *Organizational Citizenship Behavior: Its Nature, Antecedents and Consequences*. SAGE Publication, Thousand Oaks, CA.

Podsakoff P.M., MacKenzie S.B., Paine J.B., Bachrach D.G. (2000). Organizational citizenship behaviors: A critical review of the theoretical and empirical literature and suggestions for future research. *Journal of Management*, vol. 26(3), pp. 513–563.

Witkowski S.A., Chwalibóg E. (2010). Organizational citizenship behavior—opportunities and a need for research, in: J. Stankiewicz (ed.), *The Social Dimensions of Business. Management. People. Organizational Culture. Social Responsibility*. University of Zielona Gora Press, Zielona Gora, pp. 558–579.

43 Cultural Determinants of Entrepreneurship Development in SME Sector

The Case of Poland

Dariusz Turek

CONTENTS

> If we learn anything from the history of economic development, it is that culture makes all the difference.
>
> **D.S. Landes (2008, p. 577)**

INTRODUCTION

Determinants of entrepreneurship development have in recent years become a crucial issue discussed by scientists focused on the field of economy. This interest stems from Schumpeter's (1934) belief that entrepreneurial activities not only make economic development more dynamic but also influence the richness and quality of individual lives, as well as whole local communities. The main reason, mentioned by numerous authors, is the increased significance of small- and medium-sized enterprises for the economic condition of economies (Noga 2001). It is estimated that in Europe itself, there are 23 million SMEs that generate 67% of employment in the private sector (European Commission SEC 2006). In Poland, according to the REGON Registry/Business Statistical Registry, in 2009, there were 3.75 million small companies registered, out of which 99.9% were micro-, small-, and medium-sized companies that jointly constitute almost 50% of the Polish national gross product and employ about 70% of all people employed in the economy (Mały Rocznik Statystyczny Polski 2010). Therefore, it is this sector, and to be more precise, the attitudes of individuals conducting business activity in this sector of enterprises (especially the owners), that determine today's competitiveness and the future and further development of the Polish economy.

Contemporary research concerning entrepreneurship, both in Poland and Western countries, is dominated by the economic perspective focusing on structural, regional, or economic and legal considerations. However, in recent years, scientists are more focused on another research concept—the social and cultural dimension of entrepreneurship (Berger 1994; Harrison and Huntington 2003; Glinka 2008; Freytag and Thurik 2010; Kochanowicz and Marody 2010).

In the frames of this approach, there are factors analyzed that determine the demonstration of entrepreneurial attitudes by individuals, expressed by, for example, establishing and developing companies (Drucker 1992). Such a microeconomic perspective of approaching entrepreneurship is a key factor to understanding the reasons or conditions that determine the fact that some individuals in certain countries or environments decide to establish their own companies while others prefer to work for the benefit of other people.

The thesis concerning cultural nature of deep-rooted entrepreneurial actions leads to conclusion that for the development of companies, especially the SMEs, legal regulations, direct investments, or natural resources are not sufficient. A certain cultural background that inspires people to work hard, seize opportunities, promote responsibility, have freedom of choice and possibilities, and observe law and be innovative is also necessary (Anderson et al. 2000).

CAN CULTURE INFLUENCE ENTREPRENEURSHIP AND ENTERPRISE DEVELOPMENT?

The question concerning the connection of national cultures with economic activity is not new. At the beginning of the 20th century, attempts to address this issue included Schumpeter's (1934) works in the field of economy, Weber's (1994) sociological research, and McClelland's (1961) work in the field of psychology. Although some of the conclusions of the above-mentioned authors have been revised by contemporary scientists (the criticism of Schumpeter's concept is presented by Kirzner [2001], Weber from Novak [1993], and McClelland from Beugelsdijk and Smeets [2008]), the fact remains that those authors started an important trend in research on reasons for development or stagnation of the world's economies.

Ignoring the concept of cultural determinants of economic development, various economic systems adapted in countries, or even business thinking and behavior, which are broadly discussed in reference books (Harrison and Huntington 2003; Kochanowicz et al. 2007; Landes 2008), it is worth focusing on the motives of individuals who consider establishing business activity, and then about its effective development.

The first publications concerning this issue date back to the 1960s to 1970s (Hoselitz 1957; Baumol 1968; Soltow 1968), although truly significant studies have only been published since the 1980s. Despite the fact that this issue has been discussed by scientists for over 30 years, there still appear opinions that not much is known in that field. Licht (2010, p. 27) indicates that the analysis of connection between entrepreneurship and culture at the national levels is at its initial stages, and there are answers lacking to the fundamental question concerning the way in which culture influences people's entrepreneurial behaviors. Such considerations result not from the poverty of theoretical analyses but from the available empirical research. Although there have been many studies concerning entrepreneurship in the intercultural or intracultural dimensions, most of them concentrated only on the entrepreneurs themselves (Hayton et al. 2002). Taking into consideration the findings of Baumol (1990), which state that apart from productive one, the entrepreneurial actions can have an unproductive and destructive character, the empirical data raise justified concerns about the actual condition of economies subject to analysis. Moreover, the establishment of a company is often a necessity (so-called forced entrepreneurship), resulting from the need to provide for oneself and family, from frustration with low salary, or simply from lack of satisfaction with employment history (Reynolds et al. 2004).

Another problem is the assumption of too much idealization and adaptation of *ceteris paribus* conditions (Wennekers et al. 2007). The functioning and development of companies depend on such a great number of variables, both endogenic and exogenic, that it is impossible to control all of them in the research procedure. Therefore, not all of the variables, such as the economic system or political determinants, can be clearly compared and included in research questionnaires.

An unquestionable limitation is also the fact—as indicated by Glinka (2008)—that most of the influential scientific concepts of entrepreneurship are established in the United States. Therefore, a question arises concerning the possibility of reliable transfer of scientific research results and generalization to other countries (p. 72). However, it seems that the main problem is the shortage of theoretical models on which the research models could be based. The dominant theory focuses on Hofstede's (2007) concept of cultural dimensions, which are commonly criticized (Myers and Tan 2002; McSweeney 2002). Apart from this theory, the authors of textbooks mainly refer to the concept of Schwartz (Siegel et al. 2008), studies of Inglehart (Inglehart and Welzel 2005), or the European Values Study (2008) and the World Value Survey (2005).

Therefore, being aware of those numerous limitations, it is worth describing the present state of studies on cultural determinants of entrepreneurship.

Hayton et al. (2002) made an attempt to revise and organize the knowledge about the ties between culture and entrepreneurship. When making the meta-analysis of the most important publications in that area, covering years of 1971 to 2002, there can be four dimensions of dependencies constructed. The first one pertains to the relation between national culture and individual features or dispositions of entrepreneurs. The second one relates to the intercultural roles between people conducting business activities and the ones who do not. The third one concerns cultural considerations of corporate entrepreneurship. The fourth one, which is rarely discussed by scientists, relates to the interdependency of culture (values, cultural dimensions) and entrepreneurship at the national and local level.

It is worth presenting, even if briefly, the most important of those findings, especially the ones relevant to this study, and concerning the first three dimensions.

1. *The connection between national culture and features of entrepreneurs.* It turns out that, depending on their cultural values, businessmen emphasize the difference in significance of motives for establishing business activities, and they define it in different ways. They also differ in intensification of features important for entrepreneurial activities such as innovativeness, location of control, risk taking, and activeness. In cultures characterized by a small distance of power (measured by Hofstede's questionnaire), the intensification of these features is higher, and in highly individualistic cultures with low levels of uncertainty avoidance, there occur higher rates of innovativeness and internal location of control. The distance of power and individualism also differentiates entrepreneurs with regard to cognitive schemata relating to knowledge, motivation, and ability to establish business entities. However, studies conducted by Del Junco and Brás-dos-Santosa (2009) on businessmen from Germany, Italy, and Spain, countries with different cultures, show that these entrepreneurs are similar in terms of their appreciated values, attitudes, and motivating factors (e.g., motivation to achieve).
2. *Intercultural differences between entrepreneurs and non-entrepreneurs.* The results from a great number of studies show that businessmen are characterized by a bigger need of affiliation (forming associations), autonomy, and the need of achievement than people who do not conduct business activities. Moreover, those conducting business activities demonstrate higher activity in terms of control and life planning. Other studies, not analyzed by Hayton et al. (2006) indicate that entrepreneurs are characterized by a lower level of aversion toward risks (Van Praag 1996), are more optimistic, do not avoid uncertainty so much (Gifford 2003), and also express internal location of control and more intensive belief in their own efficiency (Łaguna 2010).

3. *Dependencies between culture and entrepreneurship at the national and local level.* This level of analyses was the most neglected one, although recent reference books present an increasing number of studies relating to, among others, Romania (Iacobuţă et al. 2009), China (Kshetri 2007), Japan (Aoyama 2009), Taiwan, Vietnam and the United States (Nguyen et al. 2009), Cape Verde Islands (García-Cabrera and García-Soto 2008), or South Africa (Urban 2006). Generally, it can be said that individualistic culture with a small level of uncertainty avoidance and a high level of distance toward power positively determines the level of innovativeness of national companies, and their readiness to establish new economic initiatives. However, promoting the need for autonomy, achievements, and self-efficiency in the culture is linked to increased rates of entrepreneurship.

Hayton et al. (2006) having revised the most important studies available, constructed a model showing the most significant cultural determinants for entrepreneurship (Figure 43.1).

Since the publication of the article by Hayton et al. (2006) nine years have already passed. Since that time, new empirical data have appeared, which either revised their findings or, as in the case of cultural considerations of entrepreneurship at the national and local level, institutional considerations, or social and demographical ones, complemented them.

The new findings related, among others, to the connection between the dimension of "uncertainty avoidance" and entrepreneurship. In his first works, Hofstede indicated the relation between uncertainty avoidance and lower tendency toward starting economic activity. It is worth indicating that in the last edition of the book *Cultures and Organisations*, Hofstede revised his attitude, indicating that uncertainty avoidance can positively stimulate undertaking economic initiatives and self-employment (Hofstede and Hofstede 2007, p. 198). However, the studies of Wennekers et al. (2007) have revised this view. The analyses conducted in 21 member states of Organization for Economic Co-operation and Development (OECD) show that uncertainty avoidance as a dimension of culture positively correlates with undertaking individual business activity. It is the uncertainty that is a driving force for starting new economic initiatives. It is also conducive to the increase in self-employment (Wennekers et al. 2007).

Other important results were presented by Uhlaner and Thurik (2007). Their studies on post-materialistic values (related to quality of life) show that they have negative influence on undertaking

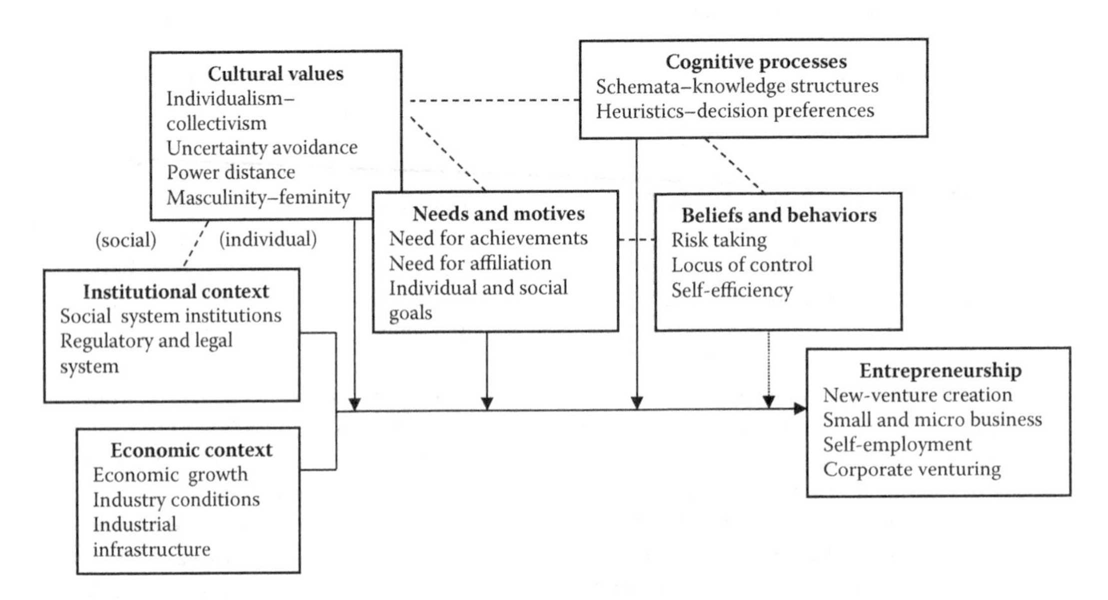

FIGURE 43.1 Cultural considerations for entrepreneurship according to Hayton et al. (From Hayton J.C. et al., *Entrepreneurship Theory and Practice*, 2002, Vol. 26, No. 4.)

economic activities. They established that the higher the level of satisfaction from life of an individual in a society, the lower the level of entrepreneurship among that society (Uhlaner and Thurik 2007).

In recent years, there have appeared more and more studies relating to *institutional* considerations of entrepreneurial actions and development.

Baumol (1990) indicated that the quality of institutions in a given country has a significant meaning for allocation of business activity, and it is in this regard that we can explain the higher level of economic activity among individual countries. However, Kirzner and Sautet (2006) stated that the quality of institutions allows the entrepreneurs to use more resources and to generate better opportunities for economic development. It is worth mentioning—as Grosse indicates—that different institutions in certain countries determine the management mechanisms in economy by lowering or increasing the efficiency of this system (Grosse 2010, p. 224).

The empirical studies in this regard show that entrepreneurship develops worse in those countries or regions where there exist "weak institutions" (Johnson et al. 2000). In addition, research conducted in 29 countries including Poland indicates that institutional considerations (education, tax levels, efficiency of government, law regulations) are strictly connected with undertaking and development of economic activities (Bjornskov and Foss 2008). These results are consistent with other empirical data suggesting a connection between economic freedom and quality of institutions, and the economic development (Berggren 2003).

When we mention institutions, we cannot forget about social capital and trust. Having revised the most important publications on that issue, Glinka (2008, pp. 75–76) indicates that trust increases economic efficiency and general economic activity; it is a factor that leads to reduction of transaction costs and risks, and is conducive to innovativeness, cooperation, and increase in resources.

The new area of research also focuses on *social and demographic variables* treated as cultural determinants of company and entrepreneurship development. It is worth mentioning though that most of the researchers think that this issue is of lesser importance and not worth further considerations due to ambiguity of the obtained results (Shane and Venkataraman 2000; Dorado 2006). However, if we ignore these opinions, it is worth presenting the most important conclusions concerning these variables.

The studies of Beugelsdijk and Noorderhaven (2004) postulate that it is not necessarily the national culture itself, or the beliefs or opinions, that is significant for entrepreneurship. The crucial factors are the values and attitudes of people inhabiting a given region. Their studies on 54 economic regions of Europe have proved that there exists a relation between attitudes shared by people inhabiting a certain area and their entrepreneurship activities and thus the development of this area (Beugelsdijk and Noorderhaven 2004).

The problematic variables for studies on entrepreneurship are age, sex, and education. The report of *Global Entrepreneurship Monitor* indicates that the highest level of prevalence rate of entrepreneurial activity is demonstrated in people aged 25–34 (Reynolds et al. 2004). However, other studies do not confirm that postulate (Wennekers et al. 2007). What we know about the age variable is the fact that younger people compared to their seniors are less thrifty, more likely to spend money, and characterized by lower intensification of fear in situations relating to business decision making (Goszczyńska 2010, p. 72). Similar concerns can be raised with regard to education and sex. Some of the researchers postulate a negative relation between higher education and entrepreneurial activity and self-employment (Uhlaner and Thurik 2007), while others either do not report any dependencies (Wennekers et al. 2007) or obtain positive correlations (Acs et al. 2004). The sex variable is even more confusing, as scrupulously described by Mueller (2004).

The analyses presented above indicate that cultural factors intensively stimulate the dispositions, features, or motives of businessmen themselves, as well as determine the establishment and development of business activities at both the national and local level. We can also find studies that do not confirm such findings (Mueller and Goić 2002). They are, however, rather limited in number and, as their authors indicate, in scope. Therefore, if culture can both stimulate or hinder development of companies, especially in the SME sector, the question arises concerning the type of culture present in Poland. Is it a proentrepreneurial culture or one that hinders business? It is worth trying to address this question.

ARE WE PRODEVELOPMENTAL? CULTURAL DETERMINANTS OF THE SME SECTOR IN POLAND

We have to agree with the remark of Glinka (2008, p. 77) that, in Poland, there exist a few studies relating *expressis verbis* to the relation between culture and entrepreneurship, and that there is a distinct shortage of quantitative research. Studies that were published in science magazines or books present a similar conclusion (Marody and Lewicki 2010, p. 125). In Poland—as the authors indicate—despite its transformation, as well as institutional and mentality changes, the culture still does not support entrepreneurial and innovative activities and the creation of public well-being.

Despite those explicitly negative remarks relating to the present state and perspectives of development for Polish companies, it is worth trying to take a holistic look at the Polish culture and conducting a deductive inference to check if those analyses are fully correct.

The cultural determinants of entrepreneurship development of the SME sector in Poland are difficult to describe as the culture and social values, as well as our economic imagination, which have been formed for generations, have been subject to crucial transformations in recent years. As a result, the description of those determinants can be inconsistent and full of paradoxes.

Values and Attitudes of Poles and Perspectives of Entrepreneurship Development of SME Sector

The relation between culture and the entrepreneurship in the SME sector, and to be more precise, the owners and employees of these enterprises, is limited. The only bases for expression of culture are values, attitudes, and behaviors. The evidence of the aforementioned dependencies is provided not only by the cited empirical studies but also by analyses conducted by scientists such as Hofstede, Schwartz, Triandis, Fukuyama, Putnam, Hampden-Turrner and Trompenaars, or Harrison and Huntington. For example, Grondona presents values that differentiate progress-prone cultures from behavioral ones. Among twenty features characteristic to the progress-prone culture, we can list a few beliefs:

- Belief in an individual and his/her creativity manifested by their innovativeness.
- Wealth is something that does not exist yet but can to be created.
- Success is a derivative of competition and competitiveness.
- Fair distribution with regard to interests of future generations, significance of saving, and investing.
- Work and entrepreneurship, which is highly respected.
- Intellectual pluralism.
- Education that does not impose universal rules but rather focuses on enabling one to find them.
- Focus on future.
- Future can be created.
- Optimism stemming from the control over the course of actions.
- Democratic governance (Grondona 2003, pp. 105–112).

Many scientists have tried to find out whether those values are also shared by the Poles. For example, research conducted by CBOS (Public Opinion Research Centre) in 2010 entitled *Co jest ważne, co można, a czego nie wolno—normy i wartości w życiu Polaków/What is important, what is allowed and what is not—norms and values in the life of the Poles* (CBOS, BS/99/2010) indicates that the dominant values are the following: health (97% participants think it is important, including 85% who think that it is very important) and family happiness (it is significant for 95% of respondents). Other important values are honesty (95%) and respect for other people (94%). After that, we can list tranquility (important for 91% of respondents), success of homeland (84%), friends (84%), career (80%), freedom of self-expression (79%), and education (78%). For two-thirds of research participants, religion is also important in everyday life (66%), a little fewer respondents think that participation in democratic social and political life is significant (62%), and a similar number of people appreciate contact with culture (61%). Half of research participants (51%) appreciate materialistic values such as welfare and wealth, and two-fifths (41%) indicate the significance of a rich, adventurous life.

It can be clearly seen that the Poles mainly focus on social values (family, respect for others, friends) and material ones (health, carrier, welfare, and wealth). The high frequency of these values was also confirmed by many other analyses (Skarżyńska 2005; CBOS BS/133/2005). For example, in "Diagnoza Społeczna 2009"/Social diagnosis 2009/, "health" was rated the highest (67%) among other researched values, and happy marriage (56%) and children (47%) were, respectively, at the second and third places; career and money held the fourth place (32%) (Diagnoza Społeczna 2009). Moreover, the research on values of inhabitants of Europe conducted every 9 years (European Value Survey 2008), as well as the *World Values Survey* from 2006, show that the Poles highly appreciate social and materialistic values, although in the recent years, their approach has been evolving toward postmaterialistic values (those relating to personal development and to quality of life) (Fijałkowska et al. 2005). Materialism can be observed in attitudes of the Poles toward work (Marody 1996). It turns out that the most important features of a good job are high income and job security, instead of for example, career opportunities, career path, or self-fulfillment (Skarżyńska and Chmielewski 2001). Moreover, 67% of research participants are convinced that work is predominantly a way to earn money and that they would not work if it were not for their subsistence costs (CBOS, BS/187/2006). From the analyses and studies of Doliński (1995), it can be concluded that the "ethics of productivity" is not an indicator of value order and behaviors of the Poles. According to the author, in Poland, there is no such thing as the cult or respect for work, which could lay foundation for generating welfare. Further studies of Wojciszke and Baryła (2002, p. 57) also show that "ethics of productivity" is the lowest appreciated value by the Poles.

A bit different results concerning the relation of Poles to work were obtained during research conducted by CBOS in 2006, according to which 92% of respondents believe that work gives a purpose in life and therefore it is worth working hard, and 92% think that professional duties should be carried out with full devotion, even if they are not so important. Eighty-nine percent believe that hard work is essential to achieving success in life, and 84% treat work as a moral obligation toward themselves and others. Furthermore, studies cited by Goszczyńska and conducted by Wesołowska et al. indicate that financial issues (*earning a bigger amount of money*) are not the main reason for undertaking economic activity—it is the freedom and the ability to make independent decisions (Goszczyńska 2010, p. 187–190).

Could it be that the attitude of the Poles evolves toward a more economy-friendly one? Maybe owing to their diligence, perseverance, and determination, the Poles are highly appreciated and well received by inhabitants of other countries?

In order to verify the opinion concerning the evolution of culture toward a proentrepreneurial one, it is worth to also analyze other spheres of the Polish "spiritual state"—predominantly the social values that are perceived as the most important by the Poles.

According to the majority of studies, the Poles form a collectivistic society, with a high dominance of social groups that provide care and security. According to Hofstede (2007), or Hampden-Turrner and Trompenaars (2006) though, this dominance is not so significant, and we are in fact closer to individualism, as the rate of individualism IDV in a percentile scale amounts to 60 (which rates Poland 22 out of 74 researched countries) (Hofstede 2007, p. 91). The analyses of Polish scientists confirm the hypothesis about the Polish collectivism. For example, Nasierowski and Mikuła (1998) using the tool for measuring cultural dimensions of Hofstede proved that collectivism of Polish managers is much higher as it amounts to 22 points in the IDV scale. Similar results were obtained by Sułkowski (2002) and Sitko-Lutek (2004), diagnosing Poland as more collectivistic than individualistic. From the studies of Skarżyńska (1991), it can be concluded that the most appreciated values are the ones connected with health, family life, and friends, and the studies of Schwartz (1999) show that the Poles manifest one of the lowest levels of recognition of egalitarian values and emotional autonomy, and a moderate level of recognition of intellectual autonomy. Similarly, according to Hryniewicz (2004), an important feature of the Polish society is the strong identification with small groups such as family and friends.

However, assuming strong collectivism in Polish employees is not fully justified taking into consideration the fact that the highly appreciated values presented in the studies also include freedom of self-expression (79%) or success and fame (27%) (CBOS, BS/99/2010).

Paradoxical nature of the presented results according to which the Poles are a collectivistic nation and simultaneously they are not can be only apparent. The recognition of a group as homogenous or uniform is only a methodological endeavor enabling generalization of the whole population. In the case of values and the national culture itself, this uniformity is not complete, which Budzyńska (2008) tries to highlight by presenting values that "divide" rather than unite the Polish society, or Stor (2009) by showing the existing dualisms in description of the state of the Polish culture. A similar interpretation is provided by Goszczyńska (2010), who claims that different outcomes of individual studies result from research methodology, in particular, from the theoretical assumptions concerning the issue of entrepreneurship, and construction of research tools.

However, if we assume that the Polish society is more collectivistic than individualistic, we can draw significant conclusions for entrepreneurship. As mentioned before, cultural individualism is better than collectivism for the development of entrepreneurship (Hayton et al. 2002). Therefore, Polish cultural basis is not conducive to development of companies from the SME sector. Of course, in individual Voivodeships, regions, or social groups—as proven by Beugelsdijk and Noorderhaven (2004)—values and attitudes of Poles can vary a lot. This hypothesis, however, requires further and deeper empirical analyses.

Religiousness and Morality of Poles and the Perspective of Entrepreneurship Development

The dimension of religiousness postulated by Max Weber can stimulate the development of entrepreneurship, and the studies of Jasińska-Kania (2007) show that the Poles declare higher level of moral rigorism than other nationalities. Moreover, the report of CBOS indicates that 31% of the respondents think that everyone should have distinct moral values and never abandon them (CBOS, BS/40/2009). Nevertheless, Mariański (2001, p. 11) claims that the state of Polish morality is ambiguous. In recent years, we have been able to observe a clear transformation of Polish morality toward secularization and permissiveness.

According to many scientists, the above-mentioned changes concerning the moral estimation of work by the Poles, which have occurred in recent years, are not random (Ziółkowski 2000; Morawski 2002). They result from natural transformations inspired by the process of modernization (e.g., economic, institutional, and technological development) and globalization (Giddens 2001). Both according to Jakubowska-Branicka (2008) and to the studies of CBOS, these processes cause changes in the awareness and identity of the Poles leading to the formation of increasingly liberal attitudes.

The results from a 2009 study conducted by CBOS (*Moralność Polaków po dwudziestu latach przemian/Morality of the Poles after twenty years of transformation/*) show that relativism as a phenomenon characteristic to moral transformations in the modern world is often also adapted by the Poles. However, it has to be mentioned that compared to results from 2005, the percentage of people convinced about the existence of clear rules determining which is good and which is wrong, which are to be observed by everyone regardless of circumstances (universalism), has increased (by 7 points, to 45%). Moreover, the report of CBOS from 2010 indicates that 90% of the Poles follow the Decalogue in their everyday life (CBOS, BS/99/2010). Nevertheless, the most common belief (47%) is that the notions of good and bad are not of objective nature. Whether an individual action is good or bad depends primarily on the circumstances (CBOS, BS/40/2009), which can lead to—as other authors indicate—corruptive behavior in the Poles.

According to the studies by CBOS conducted in 2010, 87% of respondents think that, in Poland, there exists corruption. This opinion dominates almost from the beginning of the transformation process, but nowadays, it is expressed more often than at the beginning of the 1990s (CBOS, BS/63/2010). It is one of the most significant problems of the Polish economy and a limitation to the dynamic development of the SME sector (Dylus 2006; Tymiński and Koryś 2006).

This dichotomy of moral rigorism of the Poles confirms the remarks of, among others, Skarżyńska (2005, p. 80) and Kiciński (2007, pp. 148–149). According to these scientists, in Poland, there have for decades existed significant discrepancies between the declared "lofty" values and the everyday practice in the sphere of morality. Moreover, as Glinka (2008, p. 105) indicates, the religiousness of Poles is closer to resentment rather than openness to success, and therefore, it does not constitute a strong basis for innovative and entrepreneurial behaviors. Lewicka (2005, p. 18) also confirms that rigorism in declarations of the Poles (ignoring the issue of their actual behavior) is a factor blocking the economic development, social welfare, and quality of life of individuals.

Self-Beliefs and the Image of Entrepreneurship in the Mass Media

According to psychologists specializing in entrepreneurship, there exist three basic beliefs strongly related to success in business: belief in self-efficacy, hope, and optimism (Łaguna 2010, pp. 89–102). Regardless of the obvious and homogenous dependencies of these beliefs on the structure of personality, it is worth indicating their cultural basis. Belief in self-efficacy, optimism, and temporal orientation, which is strongly connected with hope, are in a way rooted in cultural values (in cognitive patterns, to be more precise), but it can be externally stimulated by factors such as mass media coverage or opinions of authorities. Similarly to other beliefs, these ones are also subject to shaping, persuasion, or change. Therefore, two questions arise: Is the belief in own efficacy, hope, and optimism common among the Poles? Does the image of entrepreneurship presented by the mass media—the most powerful source of social influence—stimulate these beliefs? The first question can be answered by revising periodic studies conducted by CBOS, Eurobarometer reports, and Social Diagnosis (Diagnoza Społeczna 2009), as well as by revising the analyses of the Polish culture.

These analyses show that the Poles have recently become—especially according to Eurobarometer reports—more optimistic toward their future, although the rate of optimism in the studies of CBOS is average (CBOS, BS/106/2010). However, we do not know which sample group includes the entrepreneurs and thus cannot determine whether they are optimistic or not. When it comes to feeling in charge and self-efficacy, most of the studies indicate that we are, as a nation, capable of shaping and changing reality. As the authors of Diagnoza Społeczna (2009, p. 199) suggest, this could mean that the Poles are regaining the feeling of self-determination and associate their future with their own actions while experiencing less influence of actions of other people and external institutions.

The Poles are a forward-looking nation (it is worth indicating that at the beginning of the 1990s, the temporal orientation of the Poles was completely different [Tarkowska 1992]), accumulating resources and foreseeing future successes and victories. According to Sitko-Lutek (2004, p. 152), we are characterized by a long-term orientation, which translates into stubbornness, foresight, high adaptability, and readiness to pursue goals. These conclusions are not categorical though as the Poles try to be thrifty, but are far from efficient at it. From studies of Surmacz and reports of Pentor on saving, it can be concluded the Poles rely mainly on their intuition when it comes to investment decisions and allocate their savings in low-risk and ineffective saving systems (Goszczyńska 2010, p. 108). However, it is worth indicating that propensity to save is related to higher level of optimism and positive attitude toward the future, which increasingly characterize the Poles.

Glinka (2008) tried to analyze in her studies if those beliefs were stimulated by the mass media. She thoroughly analyzed pieces of mass media information concerning the image of businessmen and economic processes. According to empirical data she gathered, there were frequently published information pieces that repeated the following statements:

- Unethical behavior is common and pertains to almost all economic activities.
- The motive for entrepreneurial activities is profit, often at all cost.
- State policies in Poland have a destructive influence on economy.
- Entrepreneurs operate in a high-risk environment, and the government does not act in their interest.
- Competition is good, but not always honest.
- The owners of small- and medium-sized enterprises have it toughest in the market.
- Economic success results from illegal or dishonest activities.
- Employers exploit the workforce and infringe the Labor Code.
- Work is not an appreciated asset.
- Individual entrepreneurs have no influence on the economy, or their own economic condition (p. 156).

Similar statements are present in Polish cinema, television (TV series), and literature (Glinka 2008, p. 180). It is therefore not surprising, according to Glinka, that there are negative stereotypes concerning entrepreneurs and entrepreneurship in general among Poles.

What can be concluded from the above? It seems that despite frequent occurrence of articles in the mass media, which support negative beliefs and stereotypes about entrepreneurs and entrepreneurship, the Poles still believe in their potential, competencies, and power to create concerning entrepreneurial activities. They are also characterized—as described by Trzebiński and Drogosz (2005)—by proactive autonarrations, enabling them to overcome institutional limitations. As a common Polish saying goes, "Polak potrafi"/"The Pole can do it."

INSTEAD OF A SUMMARY, WHAT KIND OF CULTURE IS NEEDED FOR THE DEVELOPMENT OF ENTREPRENEURSHIP IN POLAND?

Cultural determinants of entrepreneurship have become such a significant issue that they were included in studies and recommendations of the European Union. The EU Economic and Social Committee has, in its opinion, related to cultural aspects and their importance with regard to development of entrepreneurship mindsets (Entrepreneurship mindset and the Lisbon Agenda 2008/C 44/20; Official Journal of the European Union; 16.2.2008). The opinion emphasized that the development of an entrepreneurial mindset in a given country relies not only on education (strategies promoting creativity and openness) but also on mass media (TV, radio, newspapers), which should stimulate such a mindset in individuals and help shape new cultural patterns.

Therefore, Poland faces a challenge with regard to the development and slow evolution of some of the dimensions of this "spiritual sphere," in particular, regarding the image of entrepreneurship and entrepreneurs. Thus, the obvious question arises: what else should we change, which direction should we head to, what is our goal? Should the values proposed by Grondona, discussed in this paper, become our values, or should we search for our own way, and if so, what way should it be? It seems that there is no perfect cultural model that stimulates entrepreneurial attitudes. Efforts to adapt and imitate values of others or their behavioral patterns are bound to fail. We have to focus on promoting individual entrepreneurial activities—accepting challenges, taking initiative, or risks—and we will transform the dominant forms of entrepreneurship in Poland.

REFERENCES

Acs Z., Arenius P., Hay M., Minniti M., *Global Entrepreneurship Monitor 2004: Executive Report*. Babson College, London Business School, 2004.

Anderson A.R., Drakopoulou-Dod S., Scott M.G., Religion as an environmental influence on enterprise culture—The case of Britain in the 1980s. *International Journal of Entrepreneurial Behaviour and Research*, 2000, Vol. 6, No. 1.

Aoyama Y., Entrepreneurship and regional culture: The case of Hamamatsu and Kyoto, Japan. *Regional Studies*, 2009, Vol. 43, No. 3.

Baumol W.J., Entrepreneurship in economic theory. *The American Economic Review*, 1968, Vol. 58.

Baumol W.J., Entrepreneurship: Productive, unproductive, and destructive. *The Journal of Political Economy*, 1990, Vol. 98, No. 5.

Berger B., eds., *Kultura przedsiębiorczości* (*Enterprise Culture*). Rój, Warszawa, 1994.

Berggren N., The benefits of economic freedom: A survey. *Independent Review*, 2003, Vol. 9.

Beugelsdijk S., Noorderhaven N., Entrepreneurial attitude and economic growth: A cross-section of 54 regions. *Annals of Regional Science*, 2004, Vol. 38.

Beugelsdijk S., Smeets R., Entrepreneurial culture and economic growth. Revisiting McClelland's thesis. *American Journal of Economics and Sociology*, 2008, Vol. 67, No. 5.

Bjornskov C., Foss N., Economic freedom and entrepreneurial activity: Some cross-country evidence. *Public Choice*, 2008, Vol. 134, No. 3–4.

Budzyńska E., Podzielane czy dzielące? Wartości społeczeństwa polskiego. In: *Wartości, postawy i więzi moralne w zmieniającym się społeczeństwie* (Divided or dividing? The values of the Polish society. In: *Values, Attitudes and Moral Bonds in a Changing Society*), eds. J. Mariański, L. Smyczek. WAM, Kraków, 2008.

Del Junco J.G., Brás-Dos-Santos J.M., How different are the entrepreneurs in the European Union internal market?—An exploratory cross-cultural analysis of German, Italian and Spanish entrepreneurs. *Journal of International Entrepreneurship*, 2009, Vol. 7.

Doliński D., *Etyka produktywności. Czy duch kapitalizmu krąży nad Polską?* (*The Ethics of Productivity. Does the Spirit of Capitalism Go Around Poland?*). "Kolokwia Psychologiczne" 1995, Nr 4.

Dorado S., Social entrepreneurial ventures: Different values so different process of creation, no? *Journal of Developmental Entrepreneurship*, 2006, Vol. 11, No. 4.

Drucker P., *Innowacja i przedsiębiorczość* (*Innovation and Enterprise*). PWE, Warszawa, 1992.

Dylus A., ed., *Korupcja. Oblicza, uwarunkowania, przeciwdziałanie* (*The Corruption. Nature, Determinants, Prevention*). Ossolineum, Wrocław-Warszaw-Kraków, 2006.

Fijałkowska M., Lewandowska J., Wiórka B., Wenzel M., System wartości materialnych i niematerialnych. In: *Polska, Europa, Świat. Opinia publiczna w okresie integracji* (The system of material and immaterial values. In: *Poland, Europe, the World. Public Opinion in the Age of Integration*), eds. K. Zagórski, M. Skrzeszewski. Scholar, Warszawa, 2005.

Freytag A., Thurik R., eds., *Entrepreneurship and Culture*. Springer, Heidelberg, 2010.

García-Cabrera A.M., García-Soto G., Cultural differences and entrepreneurial behaviour: An intra-country cross-cultural analysis in Cape Verde. *Entrepreneurship and Regional Development*, 2008, Vol. 20.

Giddens A., *Nowoczesność i tożsamość* (*Modernity and Identity*). PWN, Warszawa, 2001.

Gifford S., Risk and uncertainty. In: *Handbook of Entrepreneurship Research*, eds. Z.J. Acs, D.B. Audretsch. Kluwer, Boston, 2003.

Glinka B., *Kulturowe uwarunkowania przedsiębiorczości w Polsce* (*Cultural Determinants of Polish Enterprise*). PWE, Warszawa, 2008.

Goszczyńska M., *Transformacja ekonomiczna w umysłach i zachowaniach Polaków* (*Economic Transformation in the Minds and Actions of Poles*). Scholar, Warszawa, 2010.

Grondona M., Kulturowa typologia rozwoju gospodarczego. In: *Kultura ma znaczenie* (Cultural typology of economic growth. In: *Culture Matters*), eds. L.E. Harrison, S.P. Huntington. Zysk i Ska, Poznań, 2003.

Grosse T.G., Kulturowe podstawy zróżnicowań kapitalizmu w Europie. In: *Kultura i gospodarka* (Cultural basis for diversification of capitalism in Europe. In: *Culture and Economy*), eds. J. Kochanowicz, M. Marody. Scholar, Warszawa, 2010.

Hampden-Turrner Ch., Trompenaars A., *Siedem kultur kapitalizmu* (*Seven Cultures of Capitalism*). Oficyna Ekonomiczna, Kraków, 2006.

Harrison L.E., Huntington S.P., eds., *Kultura ma znaczenie* (*Culture Matters*). Zysk i Ska, Poznań, 2003.

Hayton J.C., George G., Zahra S.A., National culture and entrepreneurship: A review of behavioral research. *Entrepreneurship Theory and Practice*, 2002, Vol. 26, No. 4.

Hofstede G., Hofstede G.I., *Kultury i organizacje* (*Cultures and Organizations*). PWE, Warszawa, 2007.
Hoselitz B.F., Neoeconomic factors in economic development. *The American Economic Review*, 1957, Vol. 47, No. 2.
Hryniewicz J.T., *Polityczny i kulturowy kontekst rozwoju gospodarczego* (*Political and Cultural Context of Economic Growth*). Scholar, Warszawa, 2004.
Iacobuţă A.O., Baciu L., Asandului L., Institutional features and entrepreneurship development in Romania. A case study on the north-east region. *International Journal of Management Perspectives*, 2009, Vol. 4, No. 1.
Inglehart R., Welzel Ch., *Modernization, Cultural Change and Democracy*. Cambridge University Press, New York, 2005.
Jakubowska-Branicka I., Mentalność demokratyczna a dogmatyzm. Przemiany postaw społeczeństwa polskiego w procesie demokratyzacji. In: *Wartości, postawy i więzi moralne w zmieniającym się społeczeństwie* (Democratic mentality versus dogmatism. Changes in the attitudes of Polish society in the democratization process. In: *Values, Attitudes and Moral Bonds in a Changing Society*), eds. J. Mariański, L. Smyczek. WAM, Kraków, 2008.
Jasińska-Kania A., Przekształcenia moralności w Polsce i w Europie. In: *Wymiary życia społecznego. Polska na przełomie XX i XXI wieku* (Transformations of morality in Poland and in Europe. In: *The Dimensions of Social Life. Poland at the Turn of 21st Century*), eds. M. Marody. Scholar, Warszawa, 2007.
Johnson S., McMillan J., Woodruff C., Entrepreneurs and the ordering of institutional reform. *Economics of Transition*, 2000, Vol. 8, No. 1.
Kiciński K., Moralność prywatna a moralność publiczna. In: *Jedna Polska?. Dawne I nowe zróżnicowanioe społeczne* (Private morality versus public morality. In: *Old and New Social Diversity*), ed. A. Kojder. WAM, Kraków, 2007.
Kirzner I.M., *Competition and Entrepreneurship*. University of Chicago Press, Chicago, 2001.
Kirzner I.M., Sautet F., *The Nature and Role of Entrepreneurship in Markets: Implications For Policy*. Mercatus Policy Series, Mercatus Center, George Mason University Policy Primer No. 4, (June).
Kochanowicz J., Mandes S., Marody M, eds., *Kulturowe aspekty transformacji ekonomicznej* (*Cultural Aspects of Economic Transformation*). Instytut Spraw Publicznych, Warszawa, 2007.
Kochanowicz J., Marody M., eds., *Kultura i gospodarka* (*Culture and Economy*). Scholar, Warszawa, 2010.
Kshetri N., Institutional changes affecting entrepreneurship in China. *Journal of Developmental Entrepreneurship*, 2007, Vol. 12, No. 4.
Łaguna M., *Przekonania na własny temat i aktywność celowa. Badania nad przedsiębiorczością* (*Convictions About the Self and Intentional Activity. Research on Entrepreneurship*). GWP, Gdańsk, 2010.
Landes D.S., *Bogactwo i nędza narodów* (*The Wealth and Poverty of Nations*). Muza, Warszawa, 2008.
Lewicka M., "Polacy są wielkim i dumnym narodem", czyli nasz portret (wielce) zróżnicowany. In: *Jak Polacy przegrywają, jak Polacy wygrywają?* ("Poles are a great and proud nation", or our (greatly) diversified portrait. In: *How Poles Lose, How Poles Win?*), ed. M. Drogosz. GWP, Gdańsk, 2005.
Licht A.N., Entrepreneurial motivations, culture, and law. In: *Entrepreneurship and Culture*, eds. A. Freytag, R. Thurik. Springer, Heidelberg, 2010.
Mariański J., *Kryzys moralny czy transformacja wartości?* (*Moral Crisis or Value Transformation?*) KUL, Lublin, 2001.
Marody M., Psychologiczne nastawienia w zmieniającej się rzeczywistości. In: *Oswajanie rzeczywistości. Między realnym socjalizmem a realną demokracją* (Psychological Attitudes in a Changing Reality. In: *Between Real Socialism and Real Democracy*), eds. M. Marody. ISS UW, Warszawa, 1996.
Marody M., Lewicki M., Przemiany ideologii pracy. In: *Kultura i gospodarka* (Changes in the ideology of work. In: *Culture and Economy*), eds. J. Kochanowicz, M. Marody. Scholar, Warszawa, 2010.
McClelland D.C., *The Achieving Society*. Princeton University Press, Princeton, 1961.
McSweeney B., Hofstede's model of national cultural differences and their consequences: A triumph of faith—A failure of analysis. *Human Relations January*, 2002 Vol. 55.
Morawski W., Realizacja zasad sprawiedliwości społecznej w Polsce jako miara "powrotu do normalności". In: *Kondycja moralna społeczeństwa polskiego* (Implementation of the principles of social justice in Poland as a measure of "the return to normality". In: *Moral Condition of the Polish Society*), eds. J. Mariański. WAM, Kraków, 2002.
Mueller S.L., Gender gaps in potential for entrepreneurship across countries and cultures. *Journal of Developmental Entrepreneurship*, 2004, Vol. 9, No. 3.
Mueller S.L., Goić S., Entrepreneurial potential in transition economics: A view from tomorrow's leaders. *Journal of Developmental Entrepreneurship*, 2002, Vol. 7, No. 4.
Myers M.D., Tan F.B., Beyond models of national culture in information systems research. *Journal of Global Information Management*, 2002, Vol. 10, No. 2.

Nasierowski W., Mikuła B., Culture dimensions of Polish managers: Hofstede's indices. *Organization Studies*, 1998, Vol. 19, No. 3.

Nguyen T.V., Bryant S.E., Rose J., Tseng Ch., Kapasuwan S., Cultural values, market institutions, and entrepreneurship potential: A comparative study of the United States, Taiwan, and Vietnam. *Journal of Developmental Entrepreneurship*, 2009, Vol. 14, No. 1.

Noga A., *Makroekonomia a przedsiębiorczość* (*Macroeconomy and Enterprise*). WSUiB, Warszawa, 2001.

Novak M., *Liberalizm: sprzymierzeniec czy wróg Kościoła. Nauczanie społeczne kościoła a instytucje liberalne* (*Liberalism: The Church's Ally or Enemy. Social Teachings of the Church versus Liberal Institutions*). "W drodze", Poznań, 1993.

Reynolds P.D., Bygrave W., Autio E., *Global Entrepreneurship Monitor 2003: Executive Report*. Babson College, London Business School and Kauffman Foundation, 2004.

Schumpeter J.A., *Theory of Economic Development*. Transaction Publisher, New Brunswick-London, 1934.

Schwartz S.H., A theory of cultural values and some implications for work. *Applied Psychology: An International Review*, 1999, Vol. 48, Issue 1.

Shane S., Venkataraman S., The promise of entrepreneurship as a field of research. *Academy of Management Review*, 2000, Vol. 25.

Siegel J.I., Licht A.N., Schwartz S., Egalitarianism, cultural distance, and FDI: A new approach. *American Law and Economics Association Annual Meetings*, 2008, Vol. 133.

Sitko-Lutek A., *Kulturowe uwarunkowania doskonalenia menedżerów* (*Cultural Determinants in Management Training*). UMCS, Lublin, 2004.

Skarżyńska K., Czy jesteśmy prorozwojowi? Wartości i przekonania ludzi a dobrobyt i demokratyzacja kraju. In: *Jak Polacy przegrywają, jak Polacy wygrywają?* (Are we pro-development? Values and convictions of individuals versus the prosperity and democratization of the country. In: *How Poles Lose, How Poles Win?*), ed. M. Drogosz. GWP, Gdańsk, 2005.

Skarżyńska K., Położenie jednostki w strukturze społecznej a akceptowanie wartości. In: *Psychologiczny model efektywności pracy* (Individual's position in social structure versus accepted values. In: *Psychological Model of Labour Effectiveness*), ed. X. Gliszczyńska. PWN, Warszawa, 1991.

Skarżyńska K., Chmielewski K., *Praca w życiu Polaków: wewnętrzna potrzeba, konieczność bytowa?* (*Work in the Life of Poles: Internal Need, Living Necessity?*). "Studia Psychologiczne" 2001, Nr 1.

Soltow J.H., The entrepreneur in economic history. *American Economic Review*, 1968, Vol. 58.

Stor M., Dualizm polskiej kultury narodowej w praktyce organizacyjnej korporacji międzynarodowych w Polsce. In: *Kulturowe uwarunkowania zarządzania kapitałem ludzkim* (Dualism of the Polish national culture in organizational practice of multinational corporations in Poland. In: *Cultural Determinants of Human Capital Management*), ed. M. Juchnowicz. Wolters Kluwer, Kraków, 2009.

Sułkowski Ł., *Kulturowa zmienność organizacji* (*Cultural Changeability of Organizations*). PWE, Warszawa, 2002.

Tarkowska E., *Czas w życiu Polaków. Wyniki badań, hipotezy, impresje* (*Time in the Life of Poles. Research Results, Hypotheses, Impressions*). IFiS PAN, Warszawa, 1992.

Trzebiński J., Drogosz M., Historie, które kształtują nasze życie: o konsekwencjach proaktywnych i defensywnych autonarracji. In: *Jak Polacy przegrywają, jak Polacy wygrywają?* (The histories which shape our lives: On the consequences of proactive and defensive auto-narratives. In: *How Poles Lose, How Poles Win?*) ed. M. Drogosz. GWP, Gdańsk, 2005.

Tymiński M., Koryś P., eds., *Oblicza korupcji. Zjawisko, skutki i metody przeciwdziałania* (*The Faces of Corruption. Phenomenon, Effects and Prevention Methods*). Centrum Edukacji Obywatelskiej, Warszawa, 2006.

Uhlaner L., Thurik R., Postmaterialism influencing total entrepreneurial activity across nations. *Journal of Evolutionary Economics*, 2007, Vol. 17.

Urban B., Entrepreneurship in the rainbow nation: Effect of cultural values and ESE on intentions. *Journal of Developmental Entrepreneurship*, 2006, Vol. 11, No. 3.

Van Praag C.M., *Determinants of Successful Entrepreneurship*. Thesis Publisher, Amsterdam, 1996.

Weber M., *Etyka protestancka a duch kapitalizmu* (*Protestant Ethics and the Spirit of Capitalism*). Wydawnictwo "Test" Bernard Nowak, Lublin, 1994.

Wennekers S., Thurik R., von Stel A., Noorderhaven N., Uncertainty avoidance and the rate of business ownership across 21 OECD Countries, 1976–2004. *Journal of Evolutionary Economics*, 2007, Vol. 17, No. 2.

Wojciszke B., Baryła B., Potoczne rozumienie moralności. In: *Jednostka i społeczeństwo* (Popular understanding of morality. In: *Individual and Society*), eds. M. Lewicka, J. Grzelak. GWP, Gdańsk, 2002.

Ziółkowski M., *Przemiany interesów i wartości społeczeństwa polskiego* (*Transformations of Interests and Values of the Polish Society*). Fundacja Humaniora, Poznań, 2000.

INTERNET SOURCES

Co jest ważne, co można, a czego nie wolno—normy i wartości w życiu Polaków (What is Important, What is Allowed, What is Forbidden—Norms and Values in the Lives of Poles), CBOS, "Raport z Badań" BS/99/2010; http://www.cbos.pl/SPISKOM.POL/2010/K_099_10.PDF.

Diagnoza Społeczna 2009. Warunki i jakość życia Polaków (Social Diagnosis 2009. Conditions and Quality of Life of Poles), Rada Monitoringu Społecznego, Warszawa, 2009; http://www.diagnoza.com/pliki/raporty/Diagnoza_raport_2009.pdf.

European Values Study 2008, http://www.europeanvaluesstudy.eu.

GUS—Mały Rocznik Statystyczny Polski 2010 (Short Statistical Yearbook: Poland 2010); http://www.stat.gov.pl/gus/5840_737_PLK_HTML.htm.

Moralność Polaków po dwudziestu latach przemian (Poles' Morality after 20 Years of Transformation), CBOS, "Raport z Badań" BS/40/2009, http://www.cbos.pl/SPISKOM.POL/2009/K_040_09.PDF.

Nastroje społeczne w lipcu (Social Feelings in July), CBOS, "Raport z Badań" BS/106/2010; http://www.cbos.pl/SPISKOM.POL/2010/K_106_10.PDF.

Opinia Europejskiego Komitetu Ekonomiczno-Społecznego w sprawie: "Postawy przedsiębiorcze a strategia lizbońska" (The Opinion of the European Economic and Social Committee on the Case of: "Business Attitudes and Lisbon Strategy") (2008/C 44/20); Dziennik Urzędowy Unii Europejskiej 16.2.2008; http://eur-lex.europa.eu/LexUriServ/LexUriServ.do?uri=OJ:C:2008:044:0084:0084:PL:PDF.

Opinia publiczna o korupcji i lobbingu w Polsce (Public Opinion on Corruption and Lobbying in Poland), CBOS, "Raport z Badań" BS/63/2010; http://www.cbos.pl/SPISKOM.POL/2010/K_063_10.PDF.

Report on the Implementation of the Entrepreneurship Action Plan, European Commission SEC (2006) 1132, http://europedirect.esgs.pt/Grupos/Noticias/rapa.pdf.

Wartości i normy w życiu Polaków (Values and Norms in the Life of Poles), CBOS, "Raport z Badań" BS/133/2005, http://www.cbos.pl/SPISKOM.POL/2005/K_133_05.PDF.

World Values Survey 2005, http://www.worldvaluessurvey.org/wvs/articles/folder_published/survey_2005.

Znaczenie pracy w życiu Polaków (Meaning of Labour in the Life of Poles), CBOS, "Raport z Badań" BS/187/2006, http://www.cbos.pl/SPISKOM.POL/2006/K_187_06.PDF.

44 Professional Development in the Five Hundred Largest Polish Enterprises

Results of Own Research

Tadeusz Oleksyn, Justyna Bugaj, and Izabela Stańczyk

CONTENTS

INTRODUCTION

Professional development is one of the functions of human resource management (HRM) that is of particular importance for many reasons. First, an economy based on knowledge develops, and we aim at a knowledge society and, further in future—let us hope so—towards a society of wisdom. (As the ancient Greeks—including Socrates, Plato, and Aristotle—claimed, wisdom is more than knowledge as it is additionally combined with the skill of using this knowledge in one's life and activity. As they said, many people have knowledge, but there are definitely fewer wise people [Gadacz].) Second, the concepts of continuing education and learning organizations have been commonly accepted and respected for a few decades now. Third, knowledge and useful skills as well as related techniques and technologies get older quickly. In order to keep the ability to work and get promoted, one has to develop all the time; get rid of outdated knowledge, skills, and habits; and acquire new, more up-to-date and more effective ones.

Professional development can and, we think, should be connected with innovation; in many cases, it is supposed to lead to innovations. We think that the situation in which the relations between professional development and outlays on this development as well as its effects are not examined or analyzed is improper, except for the areas in which innovations are not anticipated or can even be a threat. On some positions, it is necessary to comply strictly with the mandatory procedures, while innovations are less important. However, in the majority of cases, innovations require support, especially in Poland, which is not perceived well in this respect when compared to the majority of highly developed countries. If there are no distinct relations between professional development and innovation, this may mean that development is misdirected and the pursuit of applying new knowledge and new skills is not strong enough. It can be assumed that the problem lies in the lack of proper registers and information used to examine such relations. When such information is not available, neither the effectiveness of professional development nor its influence on innovations can be tested.

This publication is aimed at presenting the results of research conducted among enterprises entered into the "500 List" ("*Lista 500*") in the period 2009–2011 within the research project NN 115 2600 36 financed by the Ministry of Science and Schools of Academic Rank. "500 List" includes the 500 largest companies in Poland (Oleksyn, Stańczyk, and Bugaj 2011). Ranking lists show not only the position occupied by the enterprise in a given year but also the industry it belongs to, its sales revenues, total revenues, gross profit, net profits, and employment. Lists of this kind have been published in Poland each year for more than a dozen years now. Our field of interest was not only professional development but also other issues, such as organization and functions of HRM, HRM subjects and distribution of their roles, human resource (HR) strategic management, control of fund flow and increasing the effectiveness of HRM, work assessment, employment development, relations with trade unions and perception of their role, collective labor agreements, salary systems, incentives, and culture and ethics of organizations within the HRM system. The effect of these works is a research report that is quite extensive (more than 200 pages) and is to be published by the end of 2011. The report has been based on a substantial research questionnaire (with 135 questions) and 10 case studies or interviews. Due to the extent of the problems tackled, we expected to have a low rate of returns, below 20%. But we treated the analysis of the cross section of various HRM areas and a broad image of HRM in these companies as more important that the number of businesses examined. If the questionnaire were less extensive, it would be easier to fill in, and we would obtain a higher rate of return, but then the survey would be more superficial. Eventually, we received questionnaires from 68 companies back, which is not a delightful effect. However, this is the reality and the price we have had to pay for our assumptions.

It would be impossible to present all results here, so we will only be restricted to one of more than a dozen research areas we have focused on, namely, professional development.

RESULTS OF QUESTIONNAIRES

Importance of Various Forms of Professional Development for Examined Companies

Various forms of an employee's professional development are used in the examined enterprises, and their usefulness is evaluated differently. An attempt was made to assess the importance of particular forms of professional development for the system of HRM and its effectiveness. The results are compared in Table 44.1.

The above data demonstrate that the following are *appreciated* the most as a way of employees' competence development in the hierarchy order specified below:

1. Internal training; the high position of this kind of training has been maintained for a few years, and it can be interpreted in such a way that this training is usually the most needed and best adapted to needs.
2. Working in project teams.
3. Working in problem and heuristic teams.

The following have been given a "strong" average evaluation:

1. Horizontal promotion, that is, changing the position without promoting to a (higher) managerial position (as an event that enforces the practical development of knowledge and skills)
2. Extension and change of the duties on the position, without changing the position

3. Internal training and self-education
4. Planned substitution
5. Institutional education
6. Job rotation
7. Development with a coach, mentor, supervisor, or coordinator

A relatively *low position* of the following forms of professional development attracts attention:

- Assistantship (developing managerial competencies by allocating young people with the right predispositions as assistants to experienced and talented managers in a given organization); it seems that this form has not been widely adopted in Poland yet. This can indirectly suggest reluctance to invest in the future but also disregard of the fact that outstanding managers are not too easy to find.
- National and foreign transfers/exchanges; this can be related to the fact that Poland is not sufficiently attractive as a labor migration destination and the low involvement of foreign investors in creating the potential for professional development of the Poles in foreign concern/holding entities.
- E-learning, Internet, extranet; this is probably a reaction to the previously overestimated expectations.

TABLE 44.1
Practical Significance of Particular Forms of Professional Development for the Execution of the Personnel Policy in Researched Companies—as Marked by Respondents

No.	Form of Development	None	Marginal	Low	Average	Extensive
1.	Institutional education (undergraduate studies, postgraduate studies, etc.)	0	21.6	8.9	37.8	32.4
2.	Internal training	2.7	5.5	5.5	25.9	67.6
3.	External training	0	0	13.5	45.9	40.5
4.	Self-education	2.7	5.5	13.5	45.9	29.7
5.	National and foreign transfers/exchanges	22.9	22.9	25.7	17.1	11.4
6.	Internship and professional practice in the country and abroad	20.0	25.7	31.4	17.1	5.7
	Assistantship (e.g., with the company's leading managers)	37.5	34.3	16.1	12.9	3.2
7.	Job rotation (not allowing one person to stay in one position)	20.0	14.3	25.8	34.3	5.7
8.	E-learning, intranet, extranet	8.6	28.6	22.9	20.0	20.0
9.	Planned substitution	5.7	8.6	40.0	37.1	0
10.	Working in project teams	5.9	11.8	17.6	29.4	35.3
11.	Working in problem teams and creative problem solving	5.4	10.8	21.6	16.2	29.7
12.	Extension and change of the duties on the position	5.3	10.5	17.6	47.0	17.6
13.	Vertical promotion	2.9	11.8	17.6	44.2	23.5
14.	Horizontal promotion—e.g., reaching the expert level	3.1	14.3	14.3	48.6	20.0
15.	Spiral promotion	24.1	28.6	61.9	14.3	4.8
16.	Development with a coach, mentor, supervisor, coordinator, etc.)	9.0	18.2	24.2	33.3	15.2

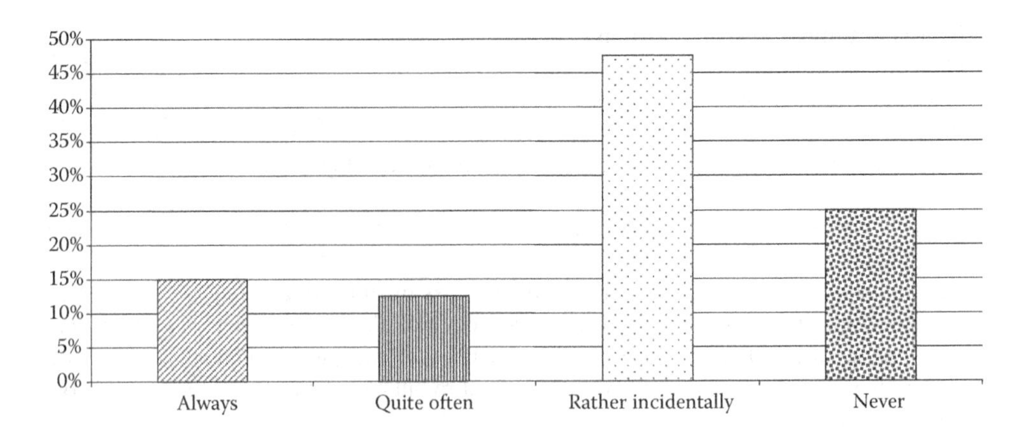

FIGURE 44.1 Scope of mentoring application.

The range of applying selected forms of professional development is presented below. It is not always closely correlated with the evaluation of the importance of these forms for the examined organization (Table 44.1 and Figure 44.1).

Mentoring

Mentoring is an organized process of transferring knowledge and skills from experienced people within the organization to young employees who are often only just taking their first steps in their professional career. The mentor not only helps develop professional competence but also assists in the social and professional development of a young employee in the new environment. In case of professional solutions, the mentor is usually the most experienced employee with certain pedagogical inclinations, a friendly and caring person, rather than a manager. This does not mean that the manager cannot get engaged in the process, but they usually do not have enough time to achieve meaningful results. Apart from that, especially in case of specialists, the best effects are obtained when one mentor takes care of one employee or the maximum of a few employees allocated to them. Obviously, this does not mean that this is the only thing the mentor does.

The scope of the mentoring application in the researched enterprises is presented in Table 44.2 and Figure 44.2.

As has been demonstrated, mentoring does not occur on a broader scale. In 75% of cases, it is never used or is used just incidentally. Additional statements have shown that it is relatively more common in the case of salespeople than in the case of other professional groups. What is more, it is also suggested that mentoring is rarely fully institutionalized (completely organized and formalized professionally). A looser form is more frequent, in which the mentor's tasks are not specified precisely and the final effects of mentoring are generally not organized or systematized in any way.

TABLE 44.2
Scope of Mentoring Application

No.	To What Extent Is Mentoring Used in Relation to Employees Starting Their First Job?	%% Structure of Answers Given (Percentage)	Accumulated Structure
1.	Always	14.7	14.7
2.	Quite often	13.2	27.9
3.	Rather incidentally	47.1	75.0
4.	Never	25.0	100.0

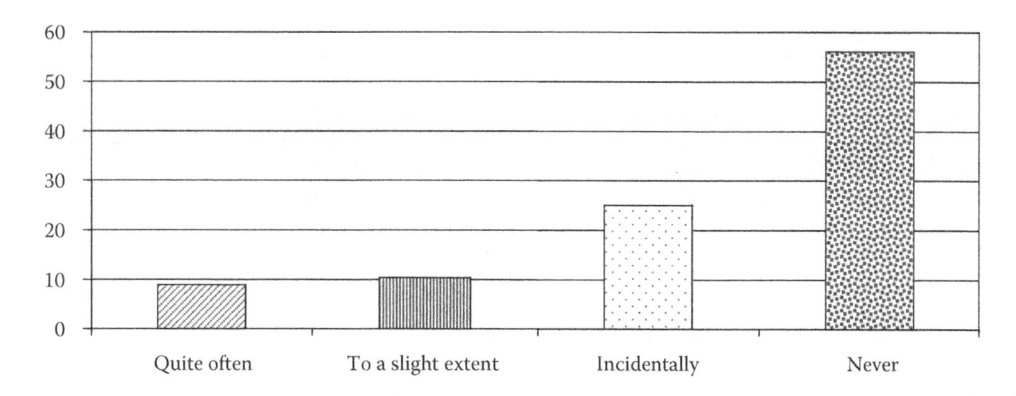

FIGURE 44.2 Frequency of coaching application in researched enterprises.

Coaching

Coaching refers to the relation between an experienced and creative employee and a less experienced person (but not a total beginner, as in case of mentoring). This relation is supposed to help accelerate professional development, increase work efficiency, and promote better work and organizational culture. An important objective of coaching is to increase creativity and innovation. Coaching is a relation characterized by greater partnership than in the case of mentoring. A coach cannot be bossy, and he/she should take into account the possibility that in some situations and in some issues, the partner can have better ideas and be intent on developing as well as applying them.

The solution where a manager is also a coach has its fans. In the case of small teams, when a manager has enough time and enthusiasm as well as the right professional and psychological predispositions, this solution can prove useful. However, we are of the opinion that generally, it is better when an experienced and creative employee with high qualifications cooperating with at least one partner (maximum of a few) becomes a coach. Then, there are chances of good cooperation, constant contact, mutual inspiration, quick development, and considerable effects. These remarks refer especially to professionals. When it comes to the development of employees who perform routine activities, for example, related to production, the above suggestions are less applicable. Their professional development can be entrusted to team leaders or development specialists functioning in specialist training centers. There are no solutions ideal for everyone.

The range of coaching application in the researched enterprises is presented in Table 44.3 and in Figure 44.3.

It is clear that coaching is generally used less often than mentoring. It is relatively more popular in the fields of design, technical service, marketing and logistics, production, and sales and trade. But generally, it is not too frequent, which is evidenced by the fact that in 81% of the examined enterprises, coaching is not present at all or happens only occasionally.

TABLE 44.3
Range of Coaching Application

No.	Range of Coaching Application in an Enterprise	%% Structure of Answers Given	Accumulated Percentage
1.	Quite often	8.8	8.8
2.	To a slight extent	10.3	19.1
3.	Incidentally	25.0	44.1
4.	Never	55.9	100.0

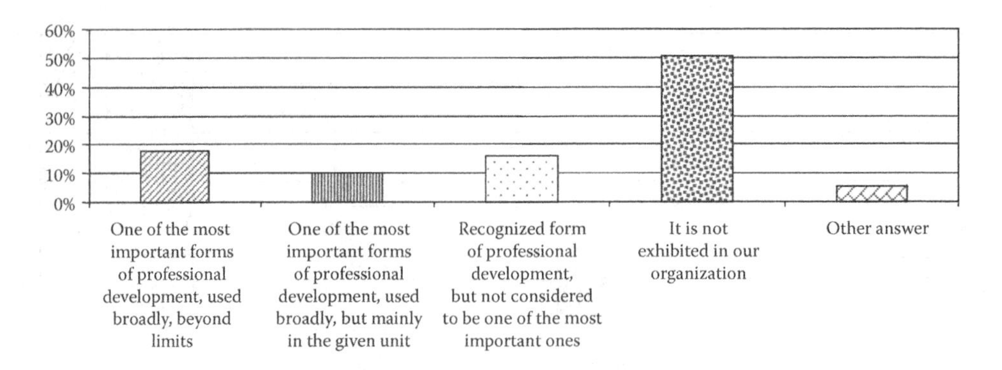

FIGURE 44.3 Role of job rotation in the execution of professional development policy.

We are aware that the descriptions "quite often," "to a slight extent," or "incidentally" are ambiguous and can be understood in various ways. However, we have found them necessary in the situation when enterprises generally do not have registers or statistics of the application of mentoring and coaching. There are also difficulties with the occurrence of different forms partially similar to mentoring and coaching (frequently not denominated this way). For this reason, the data contained in Tables 44.1 and 44.2 cannot be treated as strict data and are, rather, of demonstrative nature.

Job Rotation

Job rotation is conducive to professional development. Various research demonstrates that professional development as well as introduction of changes and improvements favorable to the organization are the fastest in the first few years of working on a given position. Later, the process of learning and introducing innovations gets considerably inhibited. There are many exceptions connected with professions requiring intensive learning (scientists, doctors, pharmacists, IT specialists, lawyers, auditors, etc.), but with reference to the majority of jobs, systemic conditions should be created for job rotation. This is also important from the point of view of developing employment flexibility and people's capacity to learn and adopt changes.

Opinions of the respondents with regard to the role of job rotation in the execution of the policy of professional development are presented in Table 44.4 and Figure 44.4.

There are different job rotations in various areas of the organization. Some of them are characterized by greater dynamics of changes in this field, which is motivated by the aim to increase

TABLE 44.4
Role of Job Rotation in the Execution of Professional Development Policy

No.	Job Rotation in Our Enterprise	%% Structure of Answers Given	Accumulated Percentage
1.	One of the most important forms of professional development; it is widely used, exceeding the limits of particular organizational units and fields	17.6	17.6
2.	One of the most important forms of professional development; it is widely used but mainly within a given organizational unit, seldom in a given field of activity	10.3	27.9
3.	Recognized form of professional development but not considered to be one of the most important ones	16.2	44.1
4.	It is not exhibited in our organization	50.5	94.9
5.	Other answer	5.1	100.0

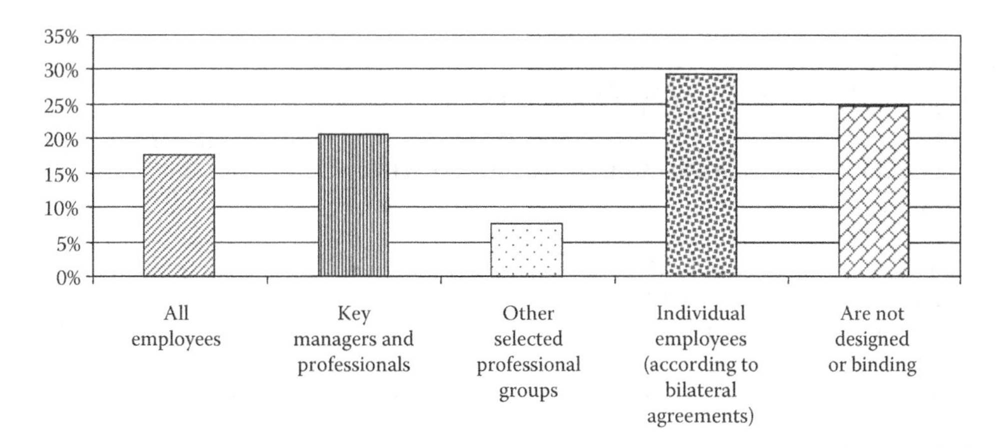

FIGURE 44.4 Developmental paths are valid for these.

employees' versatility and professional development. Another goal is to adjust employment better in time and space to the changing staff needs in particular areas. Greater dynamics of changes were indicated in such fields as marketing, sales and trade, and the technical division (designers, constructors, technologists). Long-term attachment to the same positions can be observed in financial and accounting divisions/departments, in the case of simple operational works (particularly uneducated employees), as well as many lower- and medium-level executives (rarely the highest level).

Paths of Professional Development/Career

Within the model of *personnel management* (1960s and 1970s in highly developed countries), it was more common to design development/career paths for the long term and for all or the majority of profession and qualification groups. This was especially true about large or very large organizations. Following the petroleum crisis and in the conditions of the predominance of the *HRM* model in the reality of the last three decades changing far more quickly, the development of career paths for all or most employees for a longer time was gradually abandoned. This was affected, for example, by the very critical opinions of E. Deming on the absurdity of these practices (high costs, inability to predict needs and opportunities connected with development that will emerge in a few or more years). Opinion of Peter F. Drucker was very critical too (Drucker 2004, pp. 285–286). The United States, European Union, Japan, Canada, and Australia began to give preference to the developmental paths for the period of a few years only for selected groups of positions, generally key specialists and some managers. Similar tendencies are now manifested in Poland. Nevertheless, the theme is approached in a flexible manner, and it is often said that the issue of professional development prospects can be raised by an employee and a candidate as well as that some arrangements in this respect may or even should be made by the employee and the employer's representative. Many organizations acknowledge the right of an employee to know what can be expected, at least in general and in a conditional manner. This, however, does not mean routine and common design and strict following of these developmental paths, especially that the resulting prospects for employees can turn out to be less attractive than the opportunities given by life.

The approach to *professional development paths* in the researched enterprises is demonstrated in Table 44.5.

The table shows that developmental paths are designed for all employees in every sixth enterprise of the researched ones. But this does not mean that all companies have fully developed procedures, suitable supporting IT systems, monitoring, and so forth.

TABLE 44.5
Developmental Paths in Researched Organizations

No.	Paths of Development/Career Are Designed and Executed For	%% Structure of Answers Given	Accumulated Percentage
1.	All employees	17.6	17.6
2.	Key managers and specialists	20.6	38.2
3.	Other selected professional groups (connected with sales and trade, other)	7.6	45.8
4.	Individual employees (pursuant to bilateral agreements)	29.4	75.2
5.	Are not designed or binding (a vacancy can be taken over by someone else if they are better)	24.8	100.0

It is more and more frequently thought that arrangements with employees concerning developmental paths and promotions should not be "rigid." If there is a better candidate, he/she should get promoted rather than the person to whom the developmental program refers. However, this is a controversial issue and one of the weaknesses of *professional development paths and promotions.*

Supporting Creativity and Innovations

In the contemporary reality, the development of creativity and innovativeness of the people related to the organization is of considerable importance. Respondents were asked to choose the methods used in their company in the examined area. The results are presented in Table 44.6.

TABLE 44.6
Use of the Techniques of Methodical Support of Creativity and Innovativeness in the Examined Enterprises

Applied Methods and Techniques		Percentage of Companies Using a Given Method/Technology
Reengineering		10.3
Kaizen		33.8
Rings		8.8
Heuristic techniques	Brainstorming	86.7
	Value analysis	25.0
	Objective tree	45.3
	Morphological method	23.5
	Six Hats/de Bono hats	3.2
	Synectics	3.2
	Devil's advocate	33.3
	Delphi method	5.4
Rationalization and inventiveness (methodical, organized)		100.0
Competitions		15.0
Evaluation systems that expose criteria related to creativity and innovation (especially with regard to engineers and technicians)		14.7
Motivation and remuneration system that offers appreciation of creative and innovative people		29.4
Promotion system taking into account creativity and innovativeness		16.2
None are applied		3.2

Note: Answers do not add up to 100%.

The credibility of the answers of HR managers to this question depends on how closely they cooperate with division heads and how well they recognize what others do. When it comes to the techniques described above, they are more often used in technical and development divisions than in HR departments. Obviously, the question did not concern HR issues only. It would be impossible for us to determine whether the techniques described above are used in a pure (original) form or in "mixed" versions. It is also unclear how widely and how often these methods are used. Nevertheless, the gathered material gives an overview. It is beyond any doubt that all these techniques should be broadly popularized.

Apart from the above, also, the following methods are used: cost deployment (CD), loss analysis, HERCA method, and such tools as 5S (Sort, Straighten, Sweep, Standardize and Sustain), 5G (5th Generation mobile network), 5W (Who, What, When, Where, Why are questions whose answers are considered basic information-gathering) + 1H, 4M (Manpower, Methods, Machinery, Materials), 5 Why, problem solving, ABL, Muri/Mura/Muda, NVAA analysis, Plan-Do-Check-Act (PDCA), MTS, and TIE 1.

Developmental Programs for Highly Skilled/Talented Employees

We have assumed that talent management is the process of attracting, keeping, motivating, and developing talented employees as well as achieving the best effects of their work, also as a result of inspiring and strengthening the "demonstration effect" and readiness of other employees to follow and imitate good examples.

Talent management programs are, in particular, based on the following assumptions (Armstrong 2007):

- Creating the organization that is an "employer of choice"—an attractive workplace for skilled and creative people
- Recruitment of highly qualified employees
- Designing positions and creating roles that give opportunities to use and develop the possessed skills, roles, and interesting positions that offer autonomy and are challenging
- Appreciating skilled employees by rewarding their professionalism, high quality of work, and achievements

This puts a demand on the organization to have a professional approach to the entire HRM process in the company and to make financial outlays, for example, on training courses (which are sometimes very expensive and completed with international certificates), promotions, bonuses, and rewards.

The attitude of the researched enterprises to this philosophy and this kind of program is demonstrated in Table 44.7.

In the majority (53%) of the examined organizations, no programs for talented employees are organized. Only 10% can boast considerable achievements, and every third has such programs implemented but without huge success.

TABLE 44.7
Attitude to Developmental Programs for Skilled Employees

No.	Do You Run Talent Development Programs?	%% Structure of Answers Given	Accumulated Percentage
1.	Yes; we attach great importance to them and have significant results	10.3	10.3
2.	Yes, but so far the effects have not been considerable	33.8	44.1
3.	No	52.9	97.0
4.	Other	3.0	100.0

PROMOTIONS IN THE ORGANIZATION

P.F. Drucker has noticed a characteristic change in people's motivation to develop professionally. While in the 20th century, the main motivation to develop was the desire to get promoted, starting from the turn of the 20th and 21st century, the most important motivation is aiming at keeping work, which also requires constant development (Drucker 2004, p. 289). The possibility of vertical promotion gets considerably reduced due to the widespread use of lean management since a range of indirect management levels is reduced and many organizations give up the positions of full-time substitutes. In this situation, most often, it is relatively easiest to become a project manager as one has the right competence.

It is mainly the development of the organization that gives chances of *vertical promotion*. In the developing organization, there are new teams requiring leaders and new organizational units requiring heads created. The development of the organization, in turn, depends on the quality and attractiveness of products for customers as well as the product quality and price ratio, so consequently, also on the level of sales dynamics. In other words, the prospects of promotion depend, to a considerable extent, on the market value and development of the organization itself. The organization coping with stagnation and regress offers few chances of promotion, especially vertical ones.

The researched enterprises exhibit the *democratization of promotion processes*, which is evident in

- Competitions for vacancies, which are open competitions more and more often.
- Promotion resulting from the previously determined procedures, but even here, candidates for promotion must compete against others; competencies and achievements at work are the decisive factors.
- Greater influence of employees on the choice of candidates for promotion, especially in case of first-line managers; however, the direct choice of leaders by employees still does not actually exist (unlike in the United States and Japan).

In the United States, this practice was initiated at the turn of the 1980s and 1990s in the Saturn vehicle assembly plant belonging to the General Motors concern, where the entire organization was based on the teams of 10–12 people managed by leaders chosen at the rank-and-file level, but employees' proposals were rejected very rarely.

Promotion to medium- and higher-level executives usually entails the need to have the successor prepared by the candidate for promotion, even if this condition is not formulated directly. Generally, it is implied as something obvious. Therefore, there is an informal rule applied according to which you do not get promoted further if you have not prepared a good successor on your own. In Japan, it is considered great dishonor if a candidate for a manager is to be found outside the organization. This is (partly) justified only when the company develops very quickly and enters new fields that have not been in the area of the company's key specializations so far. This approach is not common in Poland. However, internal promotions have been definitely predominant in the examined enterprises.

The *rate of promotion* to managerial positions and preferences connected with such promotions has been, to a certain extent, diversified in particular organizations:

- It has been most often claimed that age does not play a considerable role, but professional and leadership competencies are important (almost half of the answers).
- In about 15% of all cases (which is not a very high figure), it has been said that young and talented people are more eagerly qualified for promotion as soon as possible.
- The "American model" in which it is possible to "jump" one or even two levels up in promotion and where young, talented people quickly going through their career are openly welcome is even less popular (about 6% of responses).
- The "Japanese model" in which promotion by one level is only possible when one has proven to be a good employee and have good achievements at a lower managerial level is

more popular than the American one but still considered to be too rigorous (about 18% of responses).

- Young age can be an obstacle to the promotion to a managerial position, especially at a medium or higher level—even if the candidate meets all other expectations; this has been confirmed in the case of 13% of the examined companies.

According to our respondents, the *criteria of promotion* are the following:

- It has been most often said that only substantive criteria and the achieved successes are decisive of promotion (55% of respondents have chosen this option).
- Secondly, 34% of responses have indicated that even though substantive criteria and successes are decisive for promotions, there are also exceptions to this rule, which do not occur too frequently (no proper candidate, protection, etc.).
- Thirdly, it has also been mentioned in 8% of responses that even though mainly, the substantive criteria are of importance here, the number of undeserved promotions is quite considerable.
- For 6% of enterprises, the "glass ceiling" is a problem; competent people often cannot take the positions they are fit for as promotion depends more on informal arrangements—political, family, social, and other.

MASS TRAINING CONNECTED WITH CHANGING THE EMPLOYMENT STRUCTURE

Mass training for young people connected with the need to counteract the aging process of staff, deep restructuring, expanding the scale of activity, or all these factors in combination has been a specific challenge for many enterprises to face.

A considerable increase in the average age of the personnel resulted from the employment policy, which, in some cases, was the same since the beginning of the political transformation (1990) when considerable employment cuts were observed. These reductions were caused by liberating enterprises from "social employment," which was redundant from the perspective of their actual needs as well as the privatization and need to increase the competitive value. In the first years of the transformation, they also resulted from the "shock therapy" and the entailed temporary yet deep recession. Most enterprises faced with the pressure of long-term employees and trade unions (also representing mainly the interests of long-term employees) dismissed young employees who worked on a short-term basis and did not take new employees. There was a hope that the level of employment would be automatically reduced in consequence of the process of natural departure (retirement, pensions, etc.) and it would be possible to dismiss fewer people as this is always an unpleasant thing to do. This was comfortable but also had negative side effects as it generated a high unemployment rate among young people and, later, their high labor migration as well as considerable increase in the average age of employees, often even to the level of 50 years of age. Thus, it was necessary to revitalize enterprises by reversing these tendencies as a result of greater recruitment of young people, also due to the need to implement new technologies and related employment restructuring. At the same time, it was necessary to conduct mass training and catalyze the process of knowledge and skill transfer from experienced employees.

These processes are analyzed closer in the case study of Fiat Auto Poland.

CASE OF FIAT AUTO POLAND

In 2010, Fiat Auto Poland SA occupied the 7th position on the list of the 500 largest companies in Poland. Only the largest companies from the petroleum, gas, and energy industry were classified at higher positions (PKN Orlen SA on top). The factory was established in 1992 on the basis of the automobile factory Fabryka Samochodów Małolitrażowych (FSM) in Bielsko–Biała and in Tychy, after the takeover of 90% of the shares in FSM by the Fiat Group.

Initially, Fiat Poland continued the production of the Fiat 126, which was finalized as late as 2000, and the Cinquecento. (In 1993, the model took the second position in the competition "Car of the Year 1993," until 1998, when this model was substituted with Seicento—meaning "six hundred" in Italian.) In 1994, assembly was begun, and then the full range of the Uno model production was completed in 2002 (winning the title "Car of the Year 1994"). In the period 1995–2000, Punto was manufactured, Siena in the period 1997–2001, and Palio Weekend, Bravo, Brava, and Marea Weekend models in 1998–2004.

In 2000, the factory underwent more radical restructuring, which resulted in the transfer of the car manufacture process from Bielsko–Biała to Tychy, but the production of engines and gear engines remained in Bielsko–Biała. With time, the Fiat–GM Powertrain company that produces large quantities (700,000 a year) of 1.3 dm^3 "Multi Jet" diesel engines for over a dozen different models of passenger cars and vans for Fiat, General Motors, and Suzuki was established. The management board of the factory and some services remained in Bielsko–Biała.

In 2003, the manufacture of the Panda (the Fiat Panda was awarded the title of "Car of the Year 2004") began; since 2007, Fiat 500 Abarth, and since 2008, the new Ford Ka have been produced. *Fiat 500* received the title "Car of the Year 2008" and won the 12th edition (in 2011) of the oldest and most prestigious European competition of industrial design, "Compasso d'Oro ADI." Fiat Auto Poland got this award for the second time (for the first time in 2004 for the Panda).

At present, the factory in Tychy manufactures four models: Fiat Panda (the new model of the Fiat Panda is to be launched this year in Fiat's Italian factory in Campania, and Fiat Auto Poland will cease its production), Fiat 500, new Ford Ka, and Lancia Ypsilon.

The Tychy factory is the largest of Fiat's factories in Europe and the second in the world after the Brazilian one located in Betin. In terms of the quality, modernity, and work efficiency, it is considered to be the best Fiat plant worldwide (http://wikipedia.org/wiki/Fiat_Auto_Poland). In 2009, the volume of production in Tychy exceeded 600,000 vehicles. Fiat Auto Poland cars are exported to 68 countries. From 1992 to the end of 2010, a total of 6 million cars were produced here, with 4.6 million intended for exportation. The current level of employment in the factory is more than 6000, while all Polish companies within the group and three joint-venture companies employ almost 14,000 people.

The factory in Tychy is the leading plant in the entire Fiat Group in many respects. Its experience is propagated in other factories of the group in Italy and in other countries. In 1995, this factory was the first entity in the entire group to introduce the system of ISO 1991 and ISO 14001 norms (concerning environmental resource management). The factory also uses a total quality management (TQM) philosophy. As a result of using modern technologies (the factory has more than 500 industrial robots), the simultaneous use of four different coupled quality systems, and giving priority to quality, the number of defects detected in vehicles has dropped to a level comparable to the worldwide motor industry leader—the Japanese Toyota.

In 2004, the factory received the quality award *Polska Nagroda Jakości*, and in 2005, it was among the finalists of the European Foundation for Quality Management (EFQM) Excellence Award—as the first Polish company in history. Since 2007, the factory has been among the worldwide leaders using the World Class Manufacturing (WCM) system, that is, the best and most effective methods of manufacturing process management worldwide. (WCM is a complex and integrated system involving "10 management pillars" and "10 technical pillars," borrowed from Toyota.) Following the international audit, the factory was granted the Bronze Level certificate. In 2009, the group of international auditors supervised by Prof. Hajima Yamashina awarded the Silver Level certificate to the company.

A high evaluation in connection to WCM corresponds with a high quality of vehicles. In 2010, the Fiat Panda was classified in the first position in the class of city cars (segment A and B jointly) in the reliability ranking prepared by German Allgemeiner Deutscher Automobile Club (ADAC).

In 2010, the factory received the following titles and distinctions (http://www.pup.tychy.pl/firma/targi_15html):

- Good company
- Most rapidly developing company of the year
- Company with corporate social responsibility (leader in Silesia)
- Most valuable company in Poland

In the period of more than a dozen months from 2008 to 2009, the company radically changed the staffing structure, by employing and training more than 2000 young employees. This process was conducted by the plant on its own, without the help of outsourcers.

The professional development of employees is considered the company's key success factor. In 2010, the factory spent 1.053 million PLN on this goal, executing a range of various programs. A total of 6882 people took part in training courses. This is a number exceeding the level of employment in the factory, which means that some people participated in the programs more than once. Professional development related to both general knowledge and specialist, specific skills and expertise. Professional skills, communicativeness, innovativeness, entrepreneurial attitude, and leadership skills were developed. Emphasis was also laid on language courses: English and Italian.

The center of practical training located on a separate space and enabling manual workers as well as team leaders or foremen to learn quickly how to perform any assembly operation is worth special attention. They can be helped here by instructors, but also by colorful photos showing various parts and operations. The photos show about 4300 technological operations (all relatively more complicated processes) in a much clearer way than the technological process sheets. An employee having problems with any operation can come here (or be sent here) in order to learn the operation in a perfect way. They do not need to be trained in the assembly line, which would be more stressful and would temporarily entail lower quality.

Great importance is attached to ergonomics, the elimination of employees' excess efforts and movements thanks to, for example, placing assembly parts on suitable stands so that they can be collected and assembled fast and with maximum comfort. The work of people responsible for providing subassemblies and parts for car assembly in various equipment options has been facilitated by automatically highlighting one by one the elements that must be supplied to a given position to assemble a specific version of a vehicle in intermediate storage areas. Thanks to this, errors are eliminated, and the work of employees responsible for production logistics is much easier and faster.

Order and hygiene visible everywhere make a very good impression. They are the result of the full implementation and use of the 5S system, which is one of the elements of WCM.

The entire Fiat Group follows the matter-of-fact and highly professional *Code of Conduct* (2010). It is a compact document with 20 pages based on both international documents and the mission and standards mandatory in Fiat. The Universal Declaration of Human Rights, conventions of the International Labour Organisation, and the Guidelines of the Organisation for Economic Co-operation and Development (OECD) are intended for international organizations. "The Code is to be a guidebook and help for all board members, directors, managers and other employees of Fiat Group in the effective execution of the Group's mission. It is also the Group's Corporate Governance" (2010, p. 3). The code stipulates what follows:

- Selected rules of conduct in professional matters (concerning conflicts of interest, insider trading and the prohibition to use classified information, confidentiality obligations, counteracting corruption and illegal transfer of funds, preventing money laundering, protection of reputation and competition, respecting the embargo and export controls, data confidentiality)
- Selected issues connected with employees (freedom to join trade unions, equal treatment, harassment, work environment, salary and working time, recruitment and promotions,

internal control system, information on companies and accounting books, company's assets, external activities, obligations, executives, corporate officer)
- Selected issues connected with health care, security, environment (occupational health and safety, environmental protection in production processes, environmental impact and product safety)
- External relations (with customers, suppliers, public institutions, trade union organizations, political parties, local communities, company's public relations, relations with mass media)

The factory has a well-developed system of health care for employees, recreation and regeneration, as well as social benefits. Apart from obligatory medical tests, the factory ensures basic and specialist medical care in fully equipped health care centers in Tychy and Bielsko–Biała, located within the area of the factory. These centers perform, for example, specialist tests, rehabilitation, and dental treatment, including prosthetics. They also promote prevention (diet, healthy lifestyle); programs of heart protection; and combating hypertension, diabetes, obesity, and other social or civilization diseases. In 2010, the centers executed 8424 patient visits upon demand and 3582 obligatory visits involving both preliminary and periodic medical examination. The entire system is well organized and very comfortable for employees. In many locations within the production department, there are nice, separated rooms organized for getting and eating food during special breaks. It should also be pointed out that breaks from work are much longer than the minimum breaks specified in the Polish Labour Code: one half-hour break, one 15-min break, and one 10-min break.

Employees can also take advantage of the wide offer prepared by the Centre of Recreation and Culture of Fiat Auto Poland. The employees and their families can practice 28 various sports disciplines (football, cycling, tennis, bodybuilding, and others), sports and recreation (fitness, fishing, and other), and numerous cultural events in the center.

The tradition of Fiat Group (not only in Poland) is to reward the children of employees for good results at school. So far, Fiat Auto Poland has spent 9.5 million PLN on awards for 2527 kinds. Another tradition involves organizing family days (many times during the year, usually on Sundays). Family days are very popular among the families of employees as well as retired former employees. On this occasion, everyone can show their workplace and the entire factory or its part.

The factory cooperates with the Research and Academic Centre in Turin, the Warsaw University of Technology, the Silesian University of Technology, and the University of Bielsko–Biała (*Akademia Techniczno–Humanistyczna*). Master's thesis and doctoral dissertations concerning the automotive industry are rewarded, and diploma training as well as internships for students are organized.

ACKNOWLEDGMENTS

This paper is based on the information provided by the following persons (in alphabetical order):

- Jan Drapała, training department manager
- Włodzimierz Galas, industrial relations department manager
- Jerzy Kramarczyk, lidera Filaru WO
- Teresa Łukawiecka, development department manager
- Andrzej Piętka, board member, HR manager
- Zenon Raszyński, assembly manager

and the following materials about the factory:

- Corporate Social Responsibility (study of March 2011)
- Fiat Auto Poland (May 2011)

- Fiat around Us (June 2011)
- Code of Conduct 2010
- Polish Road of Fiat—90 Years of Fiat in Poland (December 2010).

REFERENCES

Armstrong M. Zarządzanie Zasobami Ludzkimi. Oficyna Wolters Kluwer Business, Krakow, 2007.
Corporate Governance FIAT Group. Turyn, 2010.
Gadacz T. Pochwała mądrości. Wykład profesora Tadeusza Gadacza. http://www.iumw.pl/o-madrosci/articles/wyklad-prof-gadacz.html.
Oleksyn T., Stańczyk I., Bugaj J. Diagnoza i kierunki zmian w zarządzaniu zasobami ludzkimi w przedsiębiorstwach z Listy 500. Oficyna Wydawnicza SGH, Warszawa, 2011.
Drucker P.F. Zawód menedżer. Wydawnictwo MT Biznes, Czarnów, 2004, pp. 285–296.

45 Medical Tourism as a Business Opportunity for Transforming Economies

Jacek Klich

CONTENTS

INTRODUCTION

Globalization processes cover, among others, an international exchange of services. Medical services constitute an important part of an international exchange in services. One may observe an increase in the share of health care services in world trade (Chanda 2002). This process is leveraged by changes in technology (telemedicine, e-commerce), ease of travel, and the continuous liberalization of markets. A separate category of tourism that is growing in importance is medical tourism (medical travel, health tourism, global health care). Although there is no single definition of health tourism, it could be broadly defined as people traveling from their place of residence for health reasons (Garciá-Altés 2005). Medical travelers are people whose primary and explicit purpose in traveling is to seek medical treatment in a foreign country (Ehrbeck et al. 2008).

Medical tourism possesses interesting spin-off economic and organizational effects including the establishment of various kind of intermediaries (for example, medical brokerage firms such as Planet Hospital or Global Choice Health Care, etc.); a plethora of medical travel facilitator websites (Cormany and Baloglu 2011); periodicals (*International Journal of Medical Tourism*); associations (like Medical Tourism Association); undertakings of various activities (World Conference on Medical Tourism, Global Health Congress, etc.); and ending up in educational spheres. An example is creating special courses (training) for medical tourism specialists. (For example, the University of Richmond and the Medical Tourism Association plan to offer a new Certificate in Medical Tourism Studies in 2011. This program will require six courses to complete the certificate. All participants will earn continuing education unit and a certificate from the University of Richmond.).

As it is acknowledged in the literature, the aggressive growth of non-Western health care institutions, complete with international accreditation, may have a significant impact on a physician's and hospital's income in Western countries, especially in the United States (Pafford 2009).

Medical tourism raises several questions, some of them still unanswered, especially in respect to legal issues, the quality of care, and business ethics. Receiving medical care abroad may subject medical tourists to unfamiliar legal issues, the limited nature of litigation in various countries, medical malpractice, and so forth. Hospitals and/or doctors in some countries may be unable to pay the financial damages awarded by a court to a patient who has sued them, owing to the fact that the hospital and/or the doctor does not possess appropriate indemnity coverage and/or medical indemnity.

Quality of care refers to measuring it, ranking it, and providing consumers with reliable and adequate information about the services available. As a practical matter, providers and customers commonly use informal channels of communication–connection–contract. In many cases, this tends to mean less regulatory or legal oversight to assure quality and less formal recourse for reimbursement or redress, if needed. There are a few organizations performing quality control tasks, like Joint Commission International in the United States, the Trent International Accreditation Scheme in the United Kingdom and Hong Kong, the Society for International Healthcare Accreditation (SOFIHA), HealthCare Tourism International, the United Kingdom Accreditation Forum (UKAF), and the International Medical Travel Association, (IMTA), based in Singapore. However, their actions are not coordinated nor standardized.

There can be major ethical issues around medical tourism. For example, the illegal purchase of organs (Budiani-Saberi and Delmonico 2008) and tissues for transplantation has been alleged in countries such as India and China. There is also a widening of the inequalities between foreign and domestic patients, which refer to preferences given to foreign patients in respect to the access to health care services. Medical tourism centered on new technologies, such as stem cell treatments, is often criticized on the grounds of fraud, a blatant lack of scientific rationale, and patient safety.

These and some other problems notwithstanding, one can maintain that the observed trend in medical tourism supported by econometric models shows that self-selected medical tourism is preferred over employer- or government-sponsored programs and over the status quo (Bies and Zacharia 2007). This indicates that medical tourism is a feasible business option mainly due to the rise in prices of medical treatment. This, coupled with the state of the economy, is of growing concern to consumers (Deloitte 2011b). Across the globe, many consumers are delaying care, altering household spending, and worrying about their ability to pay for future health care because of potential costs. In addition, consumers remain largely confused about their health care system; grade their system as underperforming relative to what they know of other systems; and believe that spending is wasteful in their country's health system.

All these lead toward the growing interest of health care consumers in receiving medical treatment abroad.

The economic and business consequences of medical tourism are positive for recipient countries (i.e., countries of medical travel destination) and result in considerable profits for both health care providers and national governments, but it can adversely affect the countries of a medical traveler's origin. It was estimated that in 2007 alone, some 500,000 Americans went abroad for health care (Roundtable Discussion 2008). Some estimate that medical tourism may represent $162 billion in lost spending in America by 2012 (Pafford 2009).

Although medical tourism is already well grounded, especially in Asian economies, there is still a considerable gap in the research on it. This refers to basic issues starting from a lack of the systematic collection of data on the quantity of trade in health services (Smith et al. 2009) and ending up with the effects of medical tourism (Youngman 2010). Consequently, there is a sufficient basis to maintain that what is known about medical tourism is mostly based on speculation (Youngman 2010), and discussions on medical tourism are influenced by ideology rather than evidence (Lunt and Carrera 2010). This refers entirely to the Central and Eastern European (CEE) countries as well, which are generally seen as attractive for foreign investors (Bremme et al. 2008).

This chapter is aimed at medical tourism in seven Central and Eastern European countries: the Czech Republic, Estonia, Hungary, Latvia, Lithuania, Poland, and Slovakia. The goal is to show the growing interest of CEE countries in the medical tourism market and to identify the main challenges these economies are facing in trying to strengthen their positions in this market. This chapter is organized as follows. First, basic trends in the tourism industry over the last couple of years are presented, and the share of the selected CEE countries in the European Union (EU) tourism market is shown. This is followed by four ways of supplying services as defined by the General Agreement on Trade in Services (GATS), and then medical tourism as an example of *consumption abroad* is presented. Then the competitive position of seven CEE countries in the European tourism market is briefly presented, accompanied by the pros and cons to obtaining health treatment in these countries, followed by the main challenges these countries are facing on their way to strengthen their position on the global and European tourism market.

TOURISM AND MEDICAL TOURISM

Over the last couple of years, the tourism industry has been among the most dynamic economic sectors, generating a wide range of benefits including a growing contribution to gross domestic product (GDP)—in some cases, over 10%—and substantial foreign exchange earnings (World Tourism Organization 2009). As with any other sector, the tourism industry was hit by the latest global financial crisis, and the impacts of the economic downturn on tourism in Asia and the Pacific, Europe, and North America have been researched (Papatheodorou et al. 2010). The negative trends have been observed worldwide (Figure 45.1) and in the EU countries (Figure 45.2).

In 2010, there were the first signs of recovery, and international tourism receipts were estimated at US $919 billion worldwide (EUR 693 billion), up from US $851 billion (EUR 610 billion) in 2009. In real terms (adjusted for exchange rate fluctuations and inflation), international tourism receipts increased by 5% as compared to an almost 7% growth in arrivals, showing the close correlation between both indicators and confirming that in recovery years, arrivals tend to pick up faster than receipts (World Tourism Organization 2009).

According to the forecast prepared by the United Nations World Tourism Organization, international tourist arrivals are projected to increase by some 4% to 5% in 2011 (World Tourism Organization 2011). One may conclude that the tourism sector offers considerable opportunities.

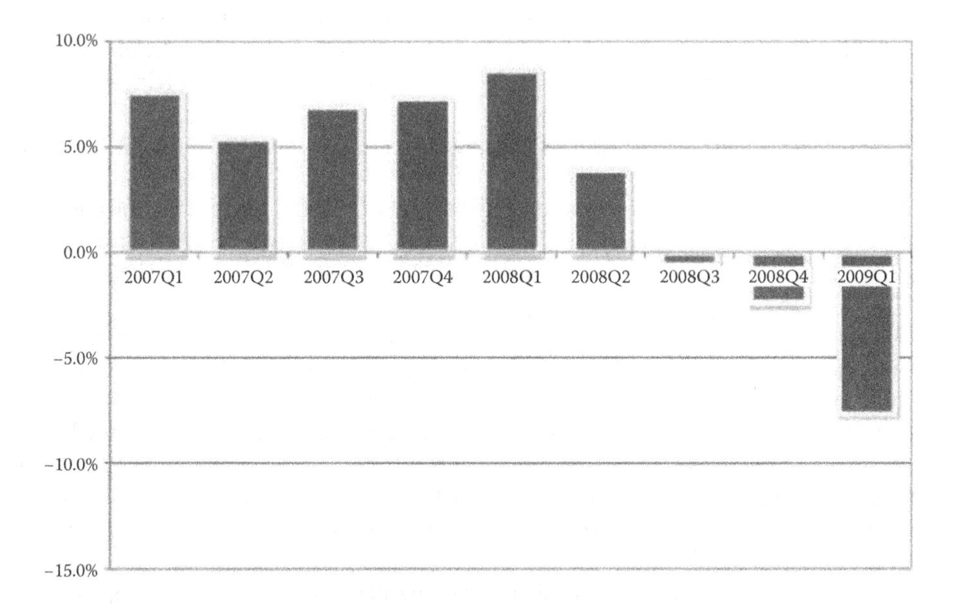

FIGURE 45.1 World international tourist arrivals. Note: Percentage change over the same period of the previous year. (From the *World Tourism Barometer*, United Nations World Tourism Organization, Madrid, 2009. After Papatheodorou, J. et al., *J. Travel Res.*, 49, 2010.)

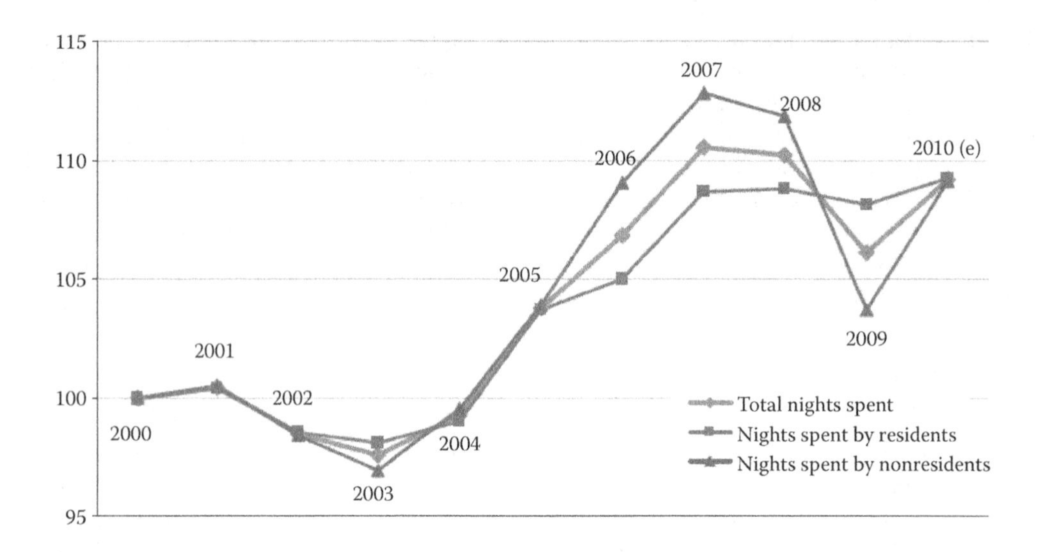

FIGURE 45.2 Number of nights spent in hotels and similar establishments, EU-27, 2000-2010 PNG. (From Eurostat, retrieved from: http://epp.eurostat.ec.europa.eu/statistics_explained/index.php?title=File:Number_of_nights_spent_in_hotels_and_similar_establishments,_EU-27,_2000-2010.PNG&filetimestamp=20110222145042.)

In the EU, tourism is an important and fast-evolving economic activity with social, cultural, and environmental implications. According to the World Tourism Organization, Europe is the most frequently visited region in the world. Five of the top 10 countries for visitors in the world are EU member states. The EU has a large number of small and medium-sized businesses, and their contributions to growth and employment varies widely from one region of the EU to another.

The economic importance of tourism for EU countries can be measured by looking at the ratio of international tourism receipts relative to GDP. In 2009, this ratio was highest in Malta (10.2%) and Cyprus (9.2%), confirming the importance of tourism to these island nations. In absolute terms, the highest international tourism receipts were recorded in Spain (EUR 38,125 million) and France (EUR 34,928 million), followed by Italy, Germany, and the United Kingdom (Tourism trends 2010).

One may conclude, then, that tourism can be perceived as an important component of the EU economy.

EU residents took more than 100 million holiday trips in 2009. Short trips (of one to three nights) accounted for slightly more than half (54%) of the trips made by EU residents (Figure 45.3); approximately three-quarters (76%) of the trips made were to domestic destinations, while 24% were abroad.

The CEE countries' share in the EU tourist market was low in respect to both nights spent in hotels and similar establishments and number of tourists. In 2009, the Czech Republic, Estonia, Latvia, Lithuania, Hungary, Poland, and Slovakia reported a total of 78,928 thousand nights spent in hotels and similar establishments. This number was far lower than in Spain alone, the EU leader in EU tourism, which reported 250,985 thousand nights in 2009 (*Eurostat regional yearbook 2010*).

In 2009, 24,499 thousand tourists visited the seven CEE countries, which was slightly more than half the number of tourists visiting Germany alone that year (44,715 thousand) (*Eurostat regional yearbook 2010*).

The low share of CEE countries in the EU tourism market (Figure 45.3) is accompanied by the relatively low share of tourism in GDP in the countries at hand (Table 45.1), especially in respect to Latvia and Poland, where the share of receipts from tourism in GDP in 2007 was 1.62% and 1.98% respectively.

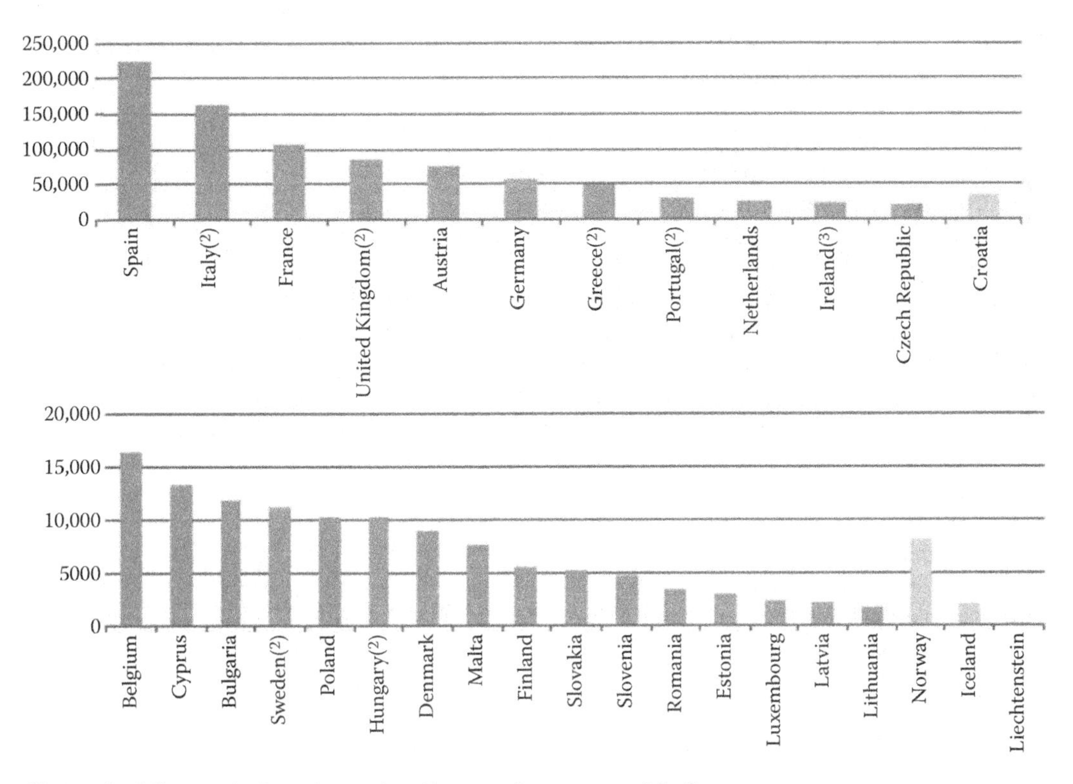

(1) Note the differences in the scales employed between the two parts of the figure.
(2) 2007.
(3) 2006.

FIGURE 45.3 Tourism destinations, 2008 (1000 nights spent in the country by nonresidents). (From Eurostat Methodologies and Working Papers. *Tourism Statistics in the European Statistical System: 2008 Data.* 2010 Edition. Publications Office of the European Union, Luxembourg, 2010, p. 47.)

Taking into account the considerable potential of the tourism market, one may maintain that actions aimed at increasing the share of CEE countries in it would be an attractive alternative, especially when taking into account the EU strategy toward the tourism sector.

The role of tourism in the EU economy was redefined in 2006, when the European Commission adopted a communication titled "A Renewed EU Tourism Policy: Towards a Stronger Partnership

TABLE 45.1
Tourism Receipts from Personal Travel in 2007: Selected CEE Countries

Country	Receipts in 2007 (EUR Million)	Relative to GDP, 2007 (%)
Czech Republic	3675	2.89
Estonia	572	3.74
Hungary	2792	2.75
Latvia	343	1.62
Lithuania	579	2.04
Poland	6160	1.98
Slovakia	1250	2.28

Source: Europe in figures, *Eurostat Yearbook 2010*, Publications Office of the European Union, Luxembourg 2010, p. 395.

for European Tourism." The document addressed a range of challenges to shape tourism in the future, including the following:

- Europe's ageing population
- Growing external competition
- Consumer demand for more specialized tourism
- Need to develop more sustainable and environmentally friendly tourism practices

Some limitations regarding obtaining data on medical tourism in the EU should be mentioned. Eurostat statistics on "tourism demand" refer to tourist participation as the number of people in the population who make at least one trip of at least four overnight stays during the reference period (quarter or year). They also look at the number of tourism trips made (and the number of nights spent on those trips), separated into tourism-related variables, such as the following:

- Destination country
- Departure month
- Length of stay
- Type of trip organization
- Transport mode
- Accommodation type
- Expenditure

The statistical data are also separated into sociodemographic explanatory variables, such as age and gender, but unfortunately, do not cover any trip goals other than "visiting friends."

This limitation contributes to a considerable gap in research on medical tourism in the EU acknowledged in the literature (Carrera and Lunt 2010).

As presented below, the concept of medical tourism responds to challenges in the tourism sector that the EU is facing. Thus, the EU and CEE countries may perceive medical tourism as a valuable business opportunity. In 2004, the value of the medical tourist market worldwide was US $40 billion and was estimated to grow to US $100 billion by 2012 (according to McKinsey and Company, the Confederation of India, and Grail Research, LLC).

Medical tourism has been rapidly developing in Asian and Latin American countries over the last several years. Regional and national governments in India, Thailand, Singapore, Malaysia, the Philippines, and Indonesia regard the "medical tourism" trade as an important resource for economic and social development (Kuan Yew 2006; Mudur 2003). In these countries, revenue generated from tourism is a significant part of the national economy.

The EU countries were not so active in fostering medical tourism but remain important players in medical markets too. The taxonomy of the EU countries in this respect is slightly blurred due to the fact that some countries play double roles as countries of origin of medical travelers and recipient countries (the United Kingdom, Germany), while others are rather recipient countries (CEE countries).

One may conclude that CEE countries are motivated to increase their share in the EU tourism sector and that the appropriate way to do this seems to be the development of medical tourism.

GATS AND MEDICAL TOURISM

Medical tourism can be perceived as a sign of globalization processes. They directly influence trade in health-related goods, services, and people—patients and health professionals. As mentioned above, there is a considerable lack of reliable data not only on trade of health care services but also on such basic phenomena as migrations of and among health professionals in the EU and globally (García-Pérez et al. 2007).

The general rules of the world trade in services are outlined in GATS. GATS defines the trade in services as "the supply of a service:

(a) from the territory of one Member into the territory of any other Member;
(b) in the territory of one Member to the service consumer of any other Member;
(c) by a service supplier of one Member, through commercial presence in the territory of another Member;
(d) by a service supplier of one Member, through presence of natural persons of a Member in the territory of any other Member" (GATS, Part 1, Article I, item 2) (GATS, p. 285–286).

In the health care sector, the equivalents of the above-mentioned four-way supply of services respectively could be

1. Telemedicine, which represents *cross-border supply*
2. Medical tourism illustrating *consumption abroad*
3. Foreign Direct Investments (FDI) in building hospital facilities, which shows *commercial presence*
4. Temporary movement of health service providers to other countries to work for a limited period (*migrations of health care professionals*) (Smith 2004)

In this chapter, only consumption abroad, that is, medical tourism, is addressed.

MEDICAL TOURISM CONCEPT

Although there are several definitions of medical tourism, a review of the literature and the Internet produce two sets of articles: travel for the purpose of delivering health care and travel for the purpose of seeking health care (Reed 2008). For the purpose of this chapter, the second orientation is chosen. Consequently, by "medical tourists," we mean patients trying to avoid treatment delays and obtain timely access to health care and traveling abroad for appropriate treatment. Medical travelers also include uninsured Americans and other individuals unable to afford health care in their home countries.

Medical services typically sought by travelers include elective procedures as well as complex specialized surgeries such as joint replacement (knee/hip), cardiac surgery, dental surgery, and cosmetic surgeries. However, virtually every type of health care, including psychiatry, alternative treatments, convalescent care, and even burial services, is available. A specialized subset of medical tourism is reproductive tourism and reproductive outsourcing, which is the practice of traveling abroad to undergo in vitro fertilization, surrogate pregnancy, and other assisted reproductive technology treatments, including freezing embryos for retroproduction.

Over 50 countries have identified medical tourism as a national industry.

On the demand side of the medical tourism market, the United States is probably the biggest market, taking into account roughly 40 million uninsured and millions of underinsured individuals. Consequently, the development of all activities connected to medical travel events in the United States should be carefully monitored. One should notice that the acceptance of medical tourism among American health care consumers is growing, reaching 40% of respondents willing to travel abroad for health care (Pafford 2009). As Deloitte's (2011b) Global Survey of Health Care Consumers of 2011 indicates, less than a quarter (22%) of US consumers grade their country's health care systems as "A" or "B." A supplementary study shows that rising health care costs and the recent economic downturn are prompting consumers to scale back, skip care, and consider nonconventional options (Deloitte 2011a). This, in turn, may suggest that there is considerable potential demand for medical travel.

Medical tourism was number 9 in the top 10 travel trends for 2010 in the United States (Pollard 2010).

TABLE 45.2
Costs of Selected Procedures in Selected Countries (in US $)

	Thailand	India	Singapore	United States	United Kingdom
Heart bypass graft surgery	7894	6000	10,417	23,938	19,700
Heart valve replacement	10,000	8000	12,500	200,000	90,000
Angioplasty	13,000	11,000	13,000	31,000–70,000	–
Hip replacement	12,000	9000	12,000	22,000–53,000	–
Hysterectomy	10,000	–	13,000	–	–
Bone marrow transplant	–	30,000	–	250,000–400,000	150,000
Liver transplant	–	40,000–69,000	–	300,000–500,000	200,000
Neurosurgery	–	800	–	29,000	–
Knee surgery	8000	2000–4500	–	16,000–20,000	12,000
Cosmetic surgery	3500	2000	–	20,000	10,000

Source: Smith, R.D. et al., *Lancet*, 373, 2009.

Consequently, there is sufficient ground to maintain that medical tourism is no longer a niche market centered on cosmetic surgery in exotic locations but is also growing quickly as a low-cost alternative to routine procedures in the United States, especially for the uninsured or underinsured. In addition, American insurers are supporting the expansion of medical tourism, and American consumers are willing to take advantage of financial incentives to travel for care. More and more companies are beginning to offer global health care options that will enable North American and European patients to access world health care at a fraction of the cost of domestic care.

The differences in prices seem to constitute one of the leading motivation factors for medical travelers (Table 45.2). There are numerous Web pages and medical brokerage firms offering information and advice about where patients can go to save money (Benzler 2011).

Companies that focus on medical travel typically provide nurse case managers to assist patients with pretravel and posttravel medical issues. They also help provide resources for follow-up care upon the patient's return.

TYPES OF MEDICAL TRAVELERS

Most of the people who travel to receive medical, dental, or cosmetic care fall into the following four categories:

Medical tourism for elective surgery: People seeking out elective procedures such as cosmetic, plastic, dental, and wellness treatments that are not covered by insurance plans.

Medical tourism for the underinsured: Because of cutbacks on coverage and considerable increases in insurance costs, more and more individuals find themselves "underinsured." Due to high deductibles, co-payments, out-of-pocket expenses, waiting lists, and limited physician choices, many patients are forced to seek of alternative treatments.

Medical tourism for the uninsured: This refers especially, if not exclusively, to Americans who usually are self-employed and resign from private medical insurance plans. In such a situation, medical illness may lead to bankruptcy; hence, medical tourism could be the only feasible option.

Medical tourism of others: This heterogeneous group is composed of those for whom the cost of treatment overseas is less than their out-of-pocket expense in the United States and other countries.

McKinsey's study (Ehrbeck et al. 2008) distinguishes between five categories of medical travelers based on their motivations to search for medical treatment abroad:

1. Most advanced technology, independent of cost
2. Better quality of medically necessary procedures
3. Quicker access to medically necessary procedures
4. Lower cost of medically necessary procedures
5. Lower cost of discretionary procedures

These taxonomies may illustrate that lower-cost motivations for medical travel notwithstanding, a considerable amount of attention is put on advanced technologies and quality of care, which medical travelers view as important.

GEOGRAPHICAL DISTRIBUTION OF INTERNATIONAL PATIENTS

Taking into account that there is a scarcity of research on flow patterns of foreign patients, one should be careful with conclusions originating from McKinsey's study (Ehrbeck et al. 2008) on the geographical distribution of medical travelers. This study shows that the majority of the North American international patients travel to Asia (45%), 26% to Latin America, and 2% to Europe, and 27% are receiving health care in other North American countries (Figure 45.4).

International patients from Europe travel for health services to Asia (39%), North America (33%), the Middle East (13%), Latin America (5%), and other European countries (10%).

As much as 95% of patients from Africa travel to Asia, while 4% travel to Europe, and only 1% travel to Latin America.

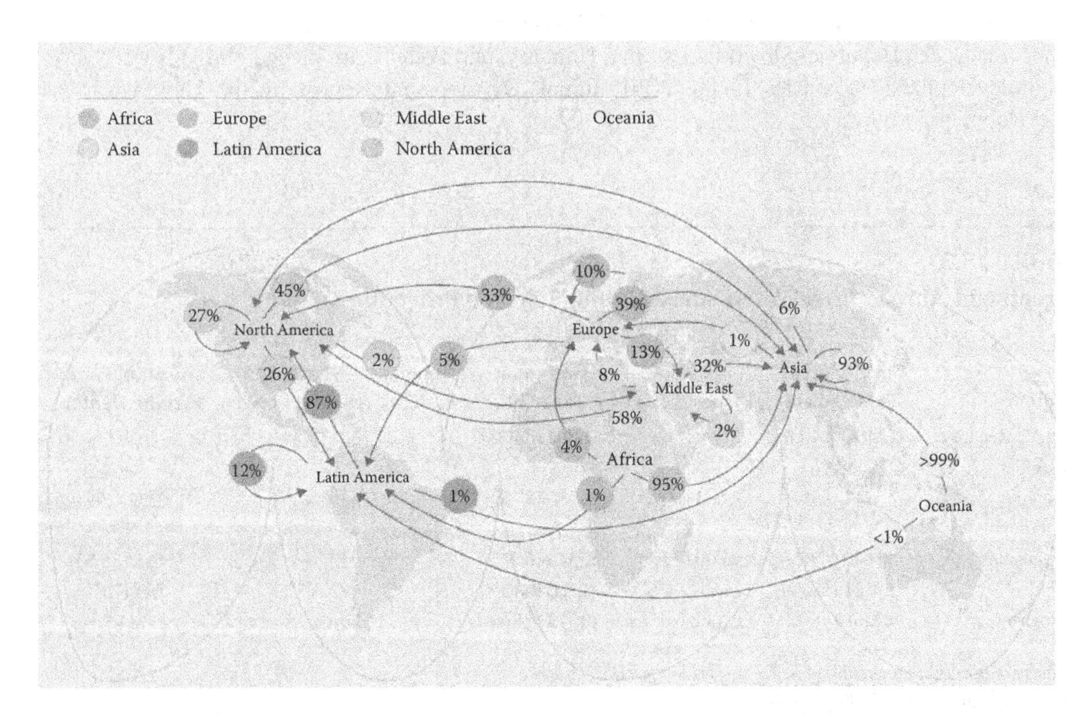

FIGURE 45.4 Medical travelers by point of origin. (From interviews with providers and patient-level data in the McKinsey analysis. From Ehrbeck, T. et al., *McKinsey Quarterly*, May 2008.)

Around 93% of patients from Asia travel to other Asian countries for health services, 6% to North America, and only 1% to Europe, with over 99% of patients from Oceania traveling to Asia for health services.

All this means that Asia is the most popular destination for patients from Oceania, Africa, North America, and Europe and is the second most popular place for patients from the Middle East. One may conclude that Asian countries possess a strong (and growing) position in the world market for medical tourism services.

MEDICAL TOURISM TO CEE COUNTRIES

New EU member states entered the EU medical travel market relatively late. There is some anecdotal evidence that the Czech Republic, Estonia, Hungary, Latvia, Lithuania, Poland, and Slovakia are trying to take advantage of their price competitiveness (Table 45.3) against old EU member states and the United States to attract patients from these (and other) countries. The medical tourism market in the CEE countries is hard to quantify due to fragmented and not fully reliable data, which is a common illness of the global medical tourism market.

According to some estimations, in 2006 in Poland, medical tourism patients accounted for more than 450,000 visitors to the country, which represented roughly 5% of the total foreign visitors to Poland that year. Unfortunately, Poland did not report on the number of medical tourists in 2009. Data available for the first half of 2010 show 120,000 medical tourists. The newly established (in 2010) Chamber of Medical Tourism (Izba Gospodarcza Turystyki Medycznej) estimated that the number of medical tourists to Poland on the whole for 2010 was approximately 250,000.

In the first half of 2010, the average bill of a medical traveler was EUR 1200.

As shown in Table 45.4 there are some differences between the national health care systems of analyzed CEE countries. This table also indicates that certainly, there is room for improvement in each of the seven national health care systems, especially when compared to Western ones, taking into account not only healthy life expectancy but also life expectancy at birth, infant mortality, and so forth. It should be noticed, however, that five out of the seven analyzed countries (the Czech Republic, Slovakia, Estonia, Hungary, and Poland) are placed on the list of "very-high-development" countries in the 2010 Human Development Report by the United Nations Development Program.

TABLE 45.3
Treatment Abroad PriceWatch Survey 2008 (Selected Procedures)

Country	Cost[a] of Breast Enlargement	Cost[a] of Liposuction	Cost[a] of a Face-Lift	Cost[a] of Dental Implants—Straumann System excluding Crown	Cost[a] of Partial Acrylic Dentures
Czech Republic	£2464 (43%)[b]	£1745 (42%)	£2304 (66%)	£778 (61%)	N/A
Estonia	N/A	N/A	N/A	N/A	N/A
Hungary	£2212 (49%)	£609 (80%)	£1955 (71%)	£783 (61%)	£255 (29%)
Latvia	N/A	N/A	N/A	N/A	N/A
Lithuania	£1842 (58%)	£800 (73%)	£2345 (65%)	£530 (75%)	£50 (86%)
Poland	£2121 (51%)	£1600 (47%)	£2190 (68%)	N/A	£94 (74%)
Slovakia	£2268 (48%)	£2068 (31%)	£1973 (71%)	N/A	£118 (67%)

Source: http://treatmentabroad.net/cost (various pages).

[a] UK sterling price including hospital and doctors' fees.

[b] Percentage saving on UK price (in parentheses).

TABLE 45.4
Main Features of National Health Systems of Selected CEE Countries

Country	Per Capita Total Expenditure on Health at Average Exchange Rate (US $)	Healthy Life Expectancy (Total Population)	Practicing Physicians per 1000 Population	Private Expenditure on Health as a Percentage of Total Expenditure on Health	Out-of-Pocket Expenditure as a Percentage of Private Expenditure on Health	General Government Expenditure on Health as a Percentage of Total Expenditure on Health
Czech Republic	1384	68	3.51	16.6%	90.1%	80.2%
Estonia	1004	63.1	3.35	20.9%	97.4%	75.5%
Hungary	938	64.1	2.48	27.8%	83.6%	69.6%
Latvia	750	62.2	3.1	39.5%	96.7%	60.5%
Lithuania	730	64.1	3.7	27.4%	97.9%	68.3%
Poland	804	66.2	2.18	25.4%	88.4%	68.2%
Slovakia	1373	66.6	3.03	28.2%	88.5%	67.3%

Source: WHO data for 2009.

When trying to compose a list of arguments for medical travelers to choose CEE countries as a point of destination, one could mention:

Competitive prices: Certainly, the CEE countries are not financially as attractive as Asian ones, but they may be considered as a feasible option.
Highly qualified medical personnel: Physicians and nurses, medical schools, and universities in the CEE countries are not broadly recognized and accredited, but the Bologna process and European standards are acknowledged by European citizens.
Good-quality medical facilities: It must be stressed here that there are mainly (if not exclusively) private health care providers offering treatment for foreign patients. Private premises have been constructed over the last couple of years and are well equipped.
Proximity: The facts that each of these seven countries can be reached by plane and that a journey will not take more than 3 h from any other European country make the CEE countries real competitors against Asian countries. This is particularly important in case of surgical treatments.
Comparatively cheap travel from Europe to the CEE countries is a logical consequence of the previous point.

As always, there are also cons to travel to the CEE countries for medical treatment.

Like in many other cases, medical centers offering treatment for foreign patients suffer from the lack of accreditation.
The number of options available in health care providers is limited. This is partly due to the very early stage of development of medical tourism in this part of Europe and partly due to the limited private resources involved in this sector.
A comparatively poor command of English among providers' staff.

This quick presentation of *pros* and *cons* to travel for medical treatment in the CEE countries can be supplemented by a list of barriers to medical tourism that these countries are facing. One may point at the following:

Restrictions on travel by governments: As *The EU Citizenship Report* (2010) shows, one may identify 25 obstacles that EU citizens face in exercising their rights across borders and suggest measures to tackle them. We should remember, however, that the new *EU Directive on Cross-Border Healthcare* may lead to considerable improvement in this respect.
Concerns about quality: A lack of commonly accepted standards and procedures as well as orchestrated actions does not serve to improve the whole situation. One may observe, however, the growing number of overseas hospitals certified for meeting US quality and safety standards. In 2010, there were 383 Joint Commission International accredited hospitals (compared to just 46 in 2004).
Regulations: Despite all of the efforts undertaken in the EU aimed at the harmonization of law, there still are considerable and, in some cases, striking differences among the EU member states that speak nothing about differences between the EU countries and third parties.
Litigation: This phenomenon arises directly from different legal regimes, mentioned above.

Looking at the current stage of development of medical tourism in and by the CEE countries, one may identify certain challenges:

An information gap: More empirical data have to be collected, in particular, data on the size of the medical tourism market, pricing of services, regional breakdown of markets, and the impact of the different forms of trade agreements. One should keep in mind that medical brokerages, destination facilities, and national governments emphasize the multiple benefits of international medical travel, while the possible shortcomings of health-related travel receive less consideration (Turner 2007).

Infrastructure and necessary resources for promotion and service delivery, along with effective partnerships with agencies around the world.

Marketing skills and training on global marketing (from Web marketing to enquiry conversion and service creation).

Getting accreditation/affiliation.

Appropriate insurance schemes.

CONCLUSION

This concise presentation allows for the following conclusions:

Medical tourism is growing in scale and importance, but is very poorly researched; thus, conclusions should be very carefully formulated.

Despite a lack of well-grounded examinations, there is a strong common belief that medical tourism possesses positive meanings for patients and health care systems.

Medical tourism to CEE countries is in its infancy, but is viewed as a way of fighting the recent downward trends in tourism profits and can be a tool leading to an increasing share of tourism in GDP.

CEE countries may offer competitive prices for medical treatment for European medical tourists, but geographical proximity should be more intensively used in marketing while competing against Asian countries.

CEE must build strategic alliances with recognized and respected Western medical brokerage firms.

REFERENCES

M. M. Álvarez, R. Chanda, R. D. Smith (2011). The potential for bi-lateral agreements in medical tourism: A qualitative study of stakeholder perspectives from the UK and India. *Globalization and Health*, Vol. 7, No.11. Retrieved from: http://www.globalizationandhealth.com/content/7/1/11.

D. Benzler (2011). Medical tourism map: Where patients go to save, from Medical Tourism City. *Global Healthcare Community*, June 9. Retrieved from: http://www.mint.com/blog/wp-content/uploads/2011/06/MedicalTourismInfographic.jpg.

W. Bies, L. Zacharia (2007). Medical tourism: Outsourcing surgery. *Mathematical and Computer Modelling*, Vol. 46, Nos. 7–8, pp. 1144–1159.

L. E. Bremme, R. De Vré, S. Goeller (2008). Pharma's generics opportunity in Central and Eastern Europe, *McKinsey Quarterly*, May 2008.

D. A. Budiani-Saberi, F. L. Delmonico (2008). Organ trafficking and transplant tourism: A commentary on the global realities. *American Journal of Transplantation*, Vol. 5, No. 5, pp. 925–929.

P. Carrera, N. Lunt (2010). A European perspective on medical tourism: The need for a knowledge base. *International Journal of Health Services*, Vol. 40, No. 3, pp. 469–484.

R. Chanda (2002). Trade in health services. *Bulletin of the World Health Organization*, Vol. 80, No. 2, pp. 158–163.

Comparing quality in medical tourism, *International Medical Travel Journal*, 11 March 2011. Retrieved from: http://www.imtj.com/articles/2011/blog-comparing-quality-in-medical-tourism-40165/.

D. Cormany, S. Baloglu (2011). Medical travel facilitator websites: An exploratory study of web page contents and services offered to the prospective medical tourist. *Tourism Management*, Vol. 32, No. 4, pp. 709–716.

Deloitte (2011a). Survey of Health Care Consumers in the United States. Key Findings, Strategic Implications, Deloitte Center for Health Solutions, Washington D.C., 2011.

Deloitte (2011b). Survey of Health Care Consumers Global Report. Key Findings, Strategic Implications, Deloitte Center for Health Solutions, Washington D.C., 2011.
T. Ehrbeck, C. Guevara, P. D. Mango (2008). Mapping the market for medical travel, *McKinsey Quarterly*, May 2008.
Europe in figures, *Eurostat Yearbook 2010*. Publications Office of the European Union, Luxembourg 2010.
European Commission (2010). EU Citizenship Report 2010. Dismantling the obstacles to EU citizens' rights, European Commission, Brussels, 27.10.2010.
Eurostat Methodologies and Working Papers (2010). *Tourism Statistics in the European Statistical System: 2008 Data*. 2010 Edition. Publications Office of the European Union, Luxembourg, p. 47.
A. Garciá-Altés (2005). The development of health tourism services. *Annals of Tourism Research*, Vol. 32, No. 1, pp. 262–266.
M. A. García-Pérez, C. Amaya, Á. Otero (2007). Physicians' migration in Europe: An overview of the current situation. *BMC Health Services Research*, Vol. 7, p. 201. Retrieved from: http://www.biomedcentral.com/1472-6963/7/201.
GATS. General Agreement of Trade in Services. Retrieved from: http://www.wto.org/english/docs_e/legal_e/26-gats.pdf.
L. Kuan Yew (2006). Excerpts from speech by Minister Mentor Mr Lee Kuan Yew at the SGH 185th anniversary dinner on 16 April 2006 at Ritz-Carlton Millennia. *Singapore Medical Association News*, Vol. 38, pp. 12–15.
N. Lunt, P. Carrera (2010). Medical tourism: Assessing the evidence on treatment abroad. *Maturitas*, Vol. 66, No. 1, pp. 27–32.
G. Mudur (2003). India plans to expand private sector in healthcare review. *British Medical Journal*, Vol. 326, p. 520.
B. Pafford (2009). The third wave-medical tourism in the 21st century. *Southern Medical Journal*, Vol. 102, No. 8, pp. 810–813.
A. Papatheodorou, J. Rosselló, H. Xiao (2010). Global economic crisis and tourism: Consequences and perspectives. *Journal of Travel Research*, Vol. 49, No. 1, pp. 39–45.
Poland: Recovery of medical tourism in Poland is encouraging (2010). *International Medical Travel Journal*, 17 September 2010. Retrieved from: http://www.imtj.com/news/?EntryId82=249214.
K. Pollard (2010). The outlook for medical tourism in 2010. *International Medical Travel Journal*. Retrieved from: http://treatmentabroad.blogspot.com/2010/01/outlook-for-medical-tourism-in-2010.html.
C. M. Reed (2008). Medical tourism. *Medical Clinics of North America*, Vol. 92, No. 6, pp. 1433–1446.
K. Ross (2001). Health tourism: An overview. HSMAI Marketing Review, Hospitality Net. Retrieved from: http://www.hospitalitynet.org/news/4010521.search?query=%22health+tourism%22ć.
Roundtable Discussion. Medical Tourism (2008). *Telemedicine and e-Health*, Vol. 14, No. 1, pp. 15–20.
R. D. Smith (2004). Foreign direct investment and trade in health services: A review of the literature. *Social Science and Medicine*, Vol. 59, No. 11, pp. 2313–2323.
R. D. Smith, R. Chanda, V. Tangcharoensathien (2009). Trade in health-related services. *The Lancet*, Vol. 373, No. 9663, pp. 593–601.
Tourism trends. Data from October 2010 Eurostat. Retrieved from: http://epp.eurostat.ec.europa.eu/statistics_explained/index.php/Tourism_trends#Top_destinations.
L. Turner (2007). "First world health care at third world prices": Globalization, bioethics and medical tourism. *BioSocieties*, Vol. 2, pp. 303–325.
World Tourism Organization (2009). Tourism: An engine for employment creation and economic stimulus. Retrieved from: http://www.unwto.org/media/news/en/press_det.php?id=3891.
World Tourism Organization (2011). International tourism: First results of 2011 confirm consolidation of growth. Retrieved from: http://media.unwto.org/en/press-release/2011-05-11/international-tourism-first-results-2011-confirm-consolidation-growth.
I. Youngman (2010). Do we actually know anything about medical tourism? *International Medical Travel Journal*. Retrieved from: http://www.imtj.com/articles/2010/what-we-know-about-medical-tourism-30083/.

46 Are All Managers Strategists? *Thinking and Behavioral Styles of Polish Managers*

Beata Bajcar

CONTENTS

STRATEGIC ASPECTS OF MANAGERIAL ACTIVITY

Management theories ascribe a significant role to the strategic activity of managers. Management, planning, and strategic thinking have been extensively developed. These areas of managerial activity come as a response to the growing challenges in the contemporary world. An individual must be capable of dealing with a high degree of complexity, dynamism, and insecurity in the personal and corporate environment. The radical technological advancement characteristic of the information age, information instantly becoming outdated, dynamism of social changes, and globalization processes constantly raise the degree of insecurity and unpredictability. Thus, in current times, there is a growing significance of strategic thinking in the context of economic, political, and managerial activity, or any other forms of purposive behavior. An increasing number of situations and problems require new conditions of functioning and more advanced skills. People must acquire and use new competencies such as considering and integrating various items of information, predicting consequences, and planning alternative scenarios. This means that people's competencies and knowledge are a key potential within the organization (Kaplan and Norton 2002).

Such difficult operating conditions in our civilization require strategic thinking from contemporary managers. The success of an organization largely depends on the manager's ability to deal with a complex and ever-changing world and, especially, on his/her strategic ability. This style of management entails investing in employees' qualifications and their management. Such an approach was largely responsible for corporate success in the age of information (Kaplan and Norton 2002). The 21st century gave rise to a *conceptual age*, in which expertise and results of strategic thinking are a necessary, albeit insufficient, prerequisite of managerial efficacy. The management process relies now, to a larger extent, on other behavioral factors, which foster the process of strategy creation and its successful implementation (Switzer 2008). A considerable role is ascribed to creativity, innovativeness, and flexibility of the manager, manifested in his/her readiness for change (Amabile 1997; Basadur and Hausdorf 1996). The development of an organization is believed to be further stimulated by empathy, ability to effectively regulate emotions, and interpersonal competencies (Goleman 2004). This requires the manager to utilize a variable range of core competencies and strategic competencies to solve increasingly complex problems in the organization (Bonn 2001; Mumford and Connelly 1991).

The psychological and organizational approaches to strategic behavior have a lot in common, but there are also distinct differences. In psychology, strategic competencies are perceived as general skills or cognitive mechanisms integrated with action. Theorists and researchers focus on determining how to use these cognitive mechanisms and actions effectively to transform ideas into results. Thus, strategic thinking can be seen as a form of controlling action in a complex and changeable environment and with unpredictable results of certain activities (Sternberg 1999). The main role of the strategist is to discover the structure of complexity and the sources of insecurity arising from a situation and to plan alternative courses of action depending on the possible scenarios (Kotarbiński 1965). Strategic thinking of the manager involves the following: (1) discovering new problems and presenting them in a solvable form; (2) indicating directions of development and activity; (3) suggesting innovative ways to deal with problems; (4) setting goals and planning long-term activities; (5) developing strategies for the achievement of these goals; (6) selecting strategies adequate for the requirements of a situation; as well as (7) controlling and evaluating the effectiveness of strategies or a change in the strategy (Liedtka 1998; Hussey 2001). Defined in such a way, strategic thinking relates to the self-regulation mechanism of purposive behaviors (Bandura 2009). The self-regulation mechanism is based on a sequence of cognitive operations and behavioral acts (from setting an objective to its achievement), interconnected based on a feedback system. In a full self-regulation cycle, goals, plans, and strategies constitute structures of the mind thanks to which a world of potential possibilities is transformed into a world of constructed creations and various solutions. Another important aspect of strategic thinking is the metacognitive components, that is, the ability to distance oneself from the situation, the ability to think within different time perspectives and to adjust strategies adequately to a problem, and the ability to use one's expertise and competencies at the right time and appropriately for the situation (Mumford and Connelly 1991; Zaccaro et al. 2004; Merkessini 1996).

Management theories recognize the role of a concept of strategic thinking and behavior (Switzer 2008). It constitutes an element of a more general system of strategic management in an organization. In this paradigm, the model of strategic thinking centers, rather, on future prospects of an organization's activity, which, in practice, means individual processes of constant assessment of the state of the organization, analysis of the situation, prediction of market trends, development of strategies, and planning the manner of their implementation (Guth and MacMillan 1986). A strategist often synthesizes expertise on how to successfully develop new ideas, how to transform them into results, and how to use thinking tools to solve problems within an organization (Loehle 1996). The strategist also integrates convergent thinking (critical and analytical thinking, predicting the future, ethical evaluation, and reflection) with processes of divergent thinking (creative, innovative, intuitive thinking, and openness) (Wootton and Horne 2010; Hussey 2001). Strategic thinking allows the manager to acquire a wider view of the organization and from a longer time perspective. Therefore, this is how an integrated system of managing of all the subsystems of the organization is created (Drucker 1992). This is not to say that manager's role should be limited to strategic activity. In the literature, there is a shift in emphasis from the executive to the conceptual work of managers. Thus, it is not so much a simple swap of roles but, rather, their proper coincidence. It should also be noted that the strategic aspect of managerial activity entails integration of the processes of analysis, evaluation, and creation with implementation (Oettingen and Gollwitzer 2010; Drucker 1992).

Psychological and organizational theories of strategic thinking and behavior offer different perspectives in the description and explanation of the essence of these notions. Since this multiplicity of notions and definitions is problematic, the issue calls for a systematization. Management approach refers to strategic activity from the individual (manager), organizational (management), and global level (position of the organization in the macroeconomic context) (Bonn 2005; Goldman 2008). Psychology sees strategic activity as a disposition (as strategic capabilities) or as a process (as regulation of purposive behaviors). The issues presented show that strategic activity is complex and multidimensional in character. It represents a dynamic model of managerial activity, combining cognition and pragmatism through the perspective of future events and action results (Gallén 2006).

In the literature, larger emphasis is given to strategic thinking and behavior as managerial competencies. These are especially useful in the diagnostic and conceptual work of top-level managers (Katz 1974). Faced with civilizational challenges, contemporary managers should be endowed with strategic competencies such as the ability to rightly judge the situation, to create the company's mission and directions of development, to predict market trends accurately, and to quickly respond to changes and use them as opportunities to start new activities. Managers should also be constantly ready to take risks and show persistence and perseverance in achievement of strategic goals as well as motivation for self-growth and development. Strategic competencies allow managers to focus not only on development of strategies for a shorter or longer time intervals but also on their successful implementation (Drucker 1992). Therefore, in the recruitment process, the focus is shifted from the intellectual potential of the manager towards success orientation. From the point of view of the organization's expectations as to the manager, there are two key issues. Firstly, it is important to ascertain whether a particular competency is a disposition and cannot be learned or whether it is an acquired skill and can be improved. Secondly, there is a need to diagnose strategic competencies of the manager in order to predict his/her efficacy in managing an organization. This has significant implications for the process of recruitment and supporting individual development of managers.

MEASUREMENT OF STRATEGIC THINKING AND BEHAVIOR DIMENSIONS

In this study, the *Strategic Thinking and Behavior Questionnaire (STBQ)* (Bajcar 2012), which measures 11 strategic dimensions, was used.

- The *Activity* scale diagnoses focusing on an activity, initiative, as well as inclination to take action in various situations.
- The *Flexibility* scale measures openness to novelty and changes as well as the ease of introducing changes in activities.
- The *Creativity* scale measures the ability to research, generate, and implement new activities as well as nonstereotypical and innovative solutions.
- The *Persistence* scale measures persistence as well as perseverance in achievement of goals set and activities undertaken.
- The *Risk Preference* scale measures the level of risk preference as well as the ease of coping with difficulties and poorly structured tasks as well as dealing with insecurity and unpredictability of the environment.
- The *Self-Efficacy* scale diagnoses the individual's attitude to efficacy and achievement of goals set.
- The *Analysis* scale measures the ability to analyze strong and week points of the problem as well as evaluating the chances and dangers in the process of problem solving. The scale also measures the ability to detect gaps and new aspects of the situation.
- The *Globality* scale diagnoses the ability to think in broad terms, the ease of assuming many different perspectives in evaluation of goals, tasks, and activities.
- The *Consequence Prediction* scale measures the ability to predict consequences of events from the personal and global perspective as well as the results of one's own and other people's activities.
- The *Long-Term Planning* scale measures the tendency to plan and structure activity over a long period of time.

- The *Strategic Evaluation* scale diagnoses an individual's tendency to analyze one's own activities and their results (successes and failures) with respect to exploitation of resources such as time, effort, and financial, emotional, and social costs.

As a result of exploratory factor analysis, two coherent functional constructs have been yielded. These are related to strategic dimensions of behavior. The first factor, labeled *strategic behavior style*, covers the first five scales of the STBQ and refers to the behavioral indicators of activity. The other factor, labeled *strategic thinking style*, is composed of the six remaining STBQ scales, referring to mental processes (Bajcar 2012). This duality in the structure of activity clearly manifests separate functional systems, which function as behavior regulators as well as regulators of mental processes (Sternberg 1999). Therefore, it is rational to refer to thinking and behavioral styles separately.

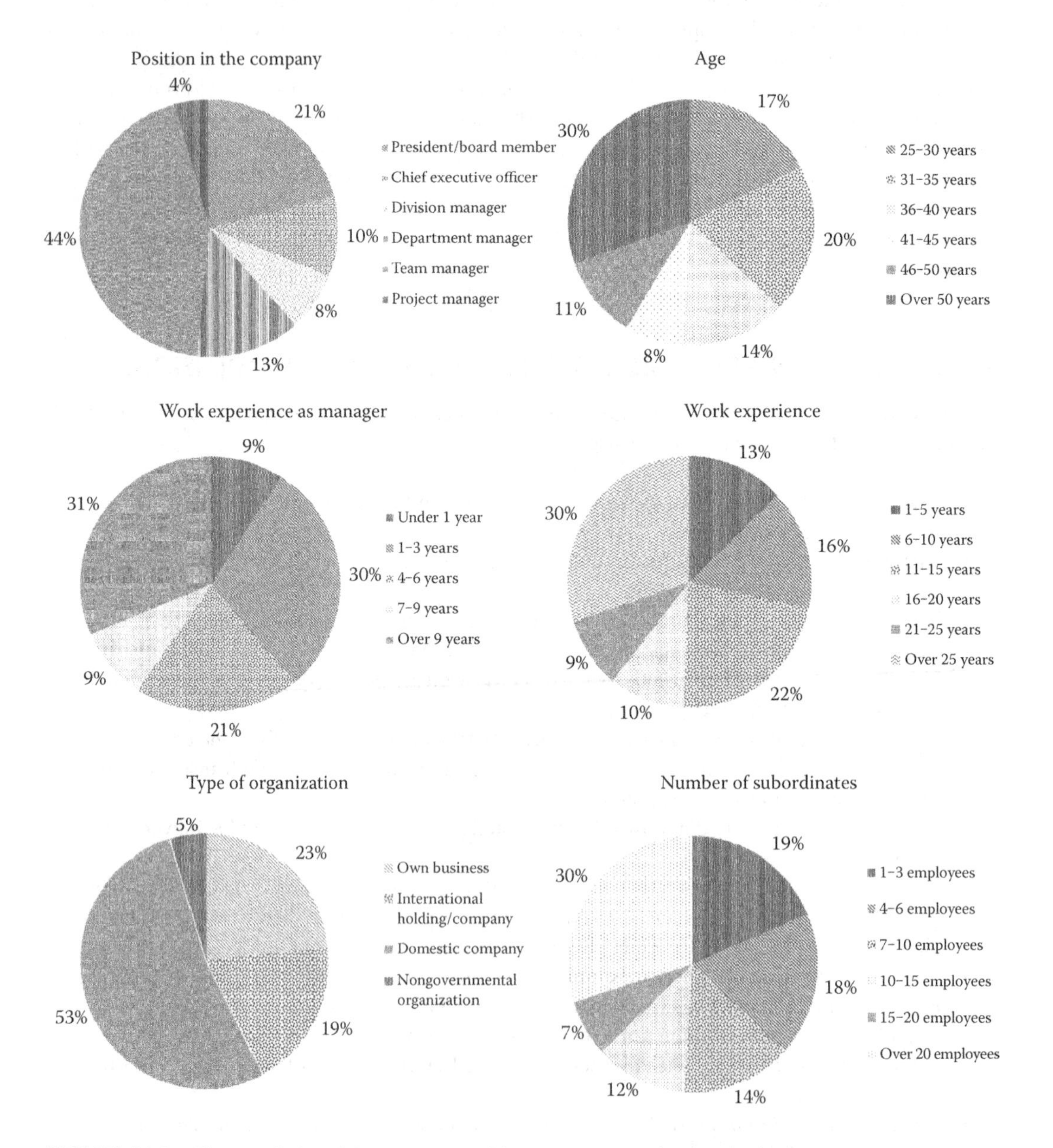

FIGURE 46.1 Characteristics of the test group with respect to organizational criteria.

STRATEGIC THINKING AND BEHAVIOR OF POLISH MANAGERS: RESEARCH RESULTS

The study was conducted on a group of 296 Polish managers occupying different positions and in different organizations, altogether, 140 women and 156 men aged 25–65. The sample included representatives from among top-level managers, directors of various department managers, and midlevel managers. Detailed characteristics of the sample are presented in Figure 46.1.

K-means cluster analysis (Romesburg 2004) was used in order to classify characteristic patterns of activity. As a result, four subgroups differing in thinking and behavioral style have been extracted: (1) activists, (2) thinkers, (3) passivists, and (4) strategists. Each group is characterized by a different level of characteristics describing strategic activity (Figures 46.2 and 46.3).

As can be seen in Figure 46.3, most managers are *strategists* (46% subjects), who are distinguished by the intensity of all dimensions of strategic thinking and behavior. As regards thinking style, they obtained a high level with respect to *analysis*, *consequence prediction*, *globality*, *long-term planning*, as well as *strategic evaluation*. At the same time, they obtained a high level of behavioral style dimensions, that is, *activity, flexibility*, *creativity*, *persistence*, and *risk preference*. *Strategists* easily integrate effects of analytical processes with practical implementation of

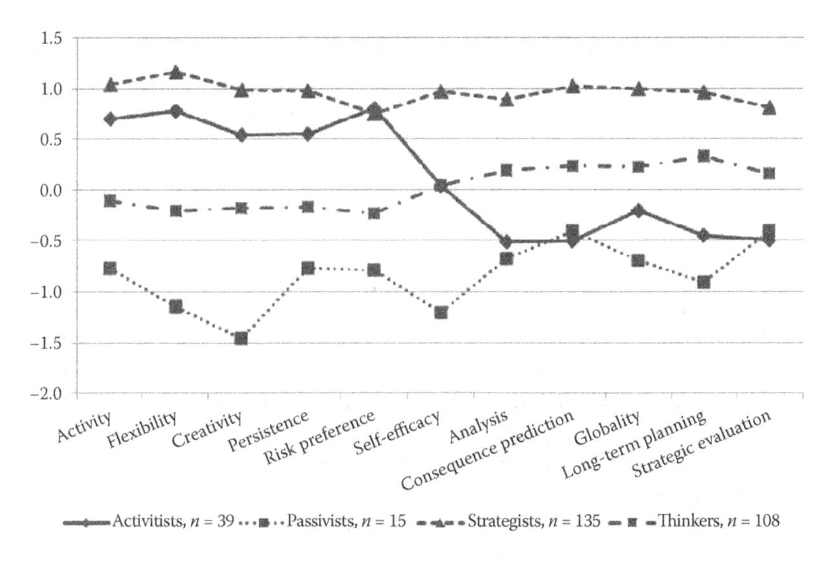

FIGURE 46.2 Strategic thinking and behavior styles of Polish managers.

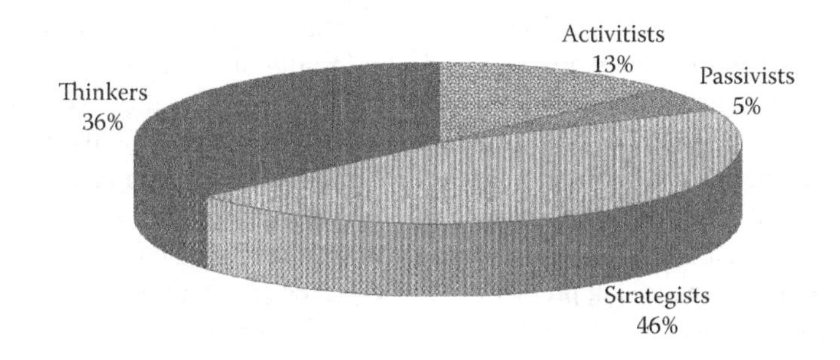

FIGURE 46.3 Frequency of strategic thinking and behavior of Polish managers.

concepts, ideas, and strategies. Therefore, they can more easily adjust to a particular situation. Thus, they deal with obstacles fast, effectively, and independently. Strategic style indicates a high intellectual potential in managers, which may lead to high efficacy in their work as well as organizational success. Therefore, such a large number of strategists among managers constitutes a very optimistic result for corporate development.

Over 36% of the managers tested are *thinkers*. They have high scores in dimensions referring to effective, deliberative thinking style (*analysis*, *strategic evaluation*, *globality*, *consequence prediction*, *long-term planning*, and *self-efficacy*) and low scores with respect to the dimensions of strategic behavior (*activity*, *creativity*, *flexibility*, *risk preference*, and *persistence*). This means that *thinkers* limit their activity to the sphere of thinking and conceptual activity, without any behavioral consequences. Such a result may signify that many managers focus on conceptual work and on decision making, delegating operational functions to their subordinates. This means that managers do not have to concentrate on the processes of implementation and achievement of goals and tasks in the organization. Nevertheless, it seems that the preparation of concepts and strategies requires integration of mental activity with the actions of the manager as well as the whole team. This tendency of managers to withdraw from action may be regarded as avoidance of difficulties connected with operationalization and implementation of their own ideas in the organization.

Thirteen percent of managers showed a style typical of *activists* (Figure 46.3). They have high scores in behavioral dimensions and low scores in cognitive activity dimensions. They are strongly activity oriented, are more creative and flexible, and also manifest a higher threshold of persistence and risk preference. This means that *activists* focus on implementation and maintenance of activity, at the same time decreasing the intensity of conceptual work. Such intense activity of managers without systematic in-depth analysis of the situation, prediction, and planning of the company's future does not increase efficacy at the individual or organizational level. This small representation of *activists* also confirms the key significance of cognitive competencies in managerial work.

The last subgroup of managers, *passivists*, accounted for a mere 5% of the sample. The style of their activity is characterized by low scores in cognitive activity and behavior, that is, low level of *activity*, *flexibility*, *creativity*, and *persistence*. Such persons are not very analytical and synthetic and do not think in broad categories. Also, they nor predict or plan the future. The passive style expresses the subject's tendency to preserve the status quo, to go with the flow, or to await what course events will take. Such a small percentage of *passivists* among sample managers may indicate that this style of activity is not appropriate for effective management of organizations. However, it may be worth taking a closer look at the work of passive managers.

The activity of Polish managers points to a highly strategic character of managerial work. Apart from the majority of managers being *strategists*, there are also other patterns of thinking and behaviors, varying in the degree of adaptive skills. The differences may derive from both the managers' individual characteristics and the various profiles of managerial functions and tasks at different levels of management. The role of managers with active and passive style of dealing with managerial problems and tasks should not be underestimated.

The next step was an analysis of the differences in the style of strategic thinking and behavior of managers in relation to gender and age. The analysis of variance (ANOVA) results do not reveal any significant differences between men and women in the profile of strategic thinking and behavior. This is an important argument in the program for fair chances and equal rights for both sexes with respect to managerial positions. The style of managerial activity differs slightly with respect to age. The analysis showed significant differences in the thinking style between managers under 40 years of age and those over 40 years of age (Figure 46.4).

The results showed that managers over 40 years of age are more active thinkers than younger managers. They are more analytical, predict the consequences more often, and do more long-term planning than younger managers. This means that the life experience fosters a more reflective, deliberative style of thinking. On the other hand, the strategic behavior dimensions of managers are not dependent on age.

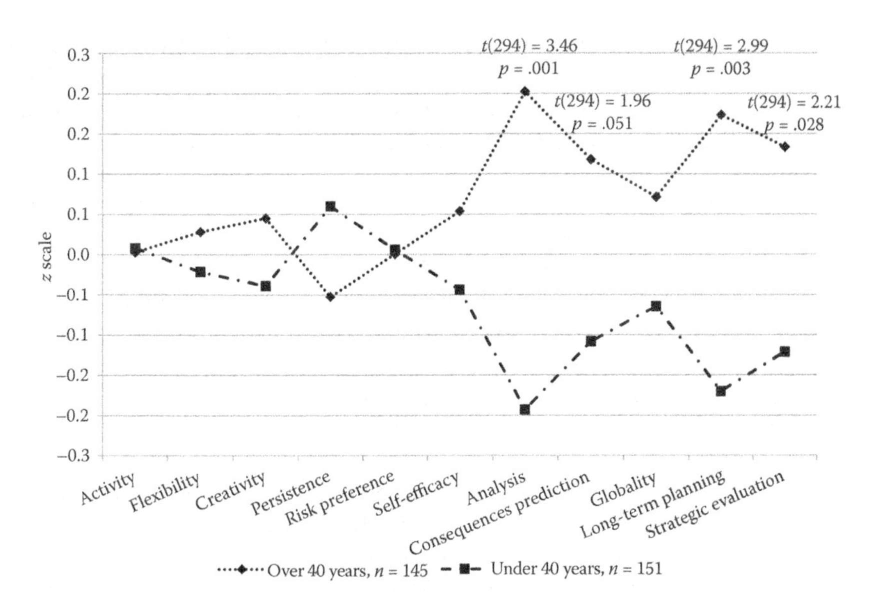

FIGURE 46.4 Strategic thinking and behavior of managers depending on age.

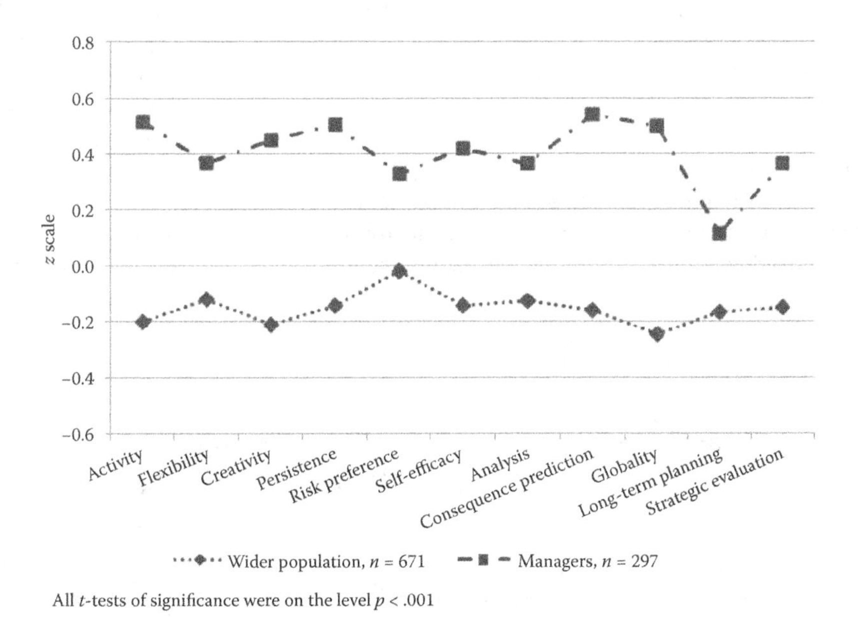

FIGURE 46.5 Strategic thinking and behavior of managers compared to wider population.

To gain a wider perspective on activity styles of Polish managers, their scores in dimensions of strategic thinking and strategic behaviors have been compared against the scores achieved by the wider population (Figure 46.5). The results of this analysis show that managers achieve significantly ($p < .001$) higher scores in all dimensions of cognitive and behavioral activity.

The differences in activity styles may indicate some important issues. Firstly, managers have high competencies with respect to conceptual, analytical, and strategic thinking and use them to manage the organization. Secondly, the character of their professional activity as well as economic and market requirements force managers to employ strategic thinking and behavior in problem solving. Thanks to frequent activation of various mental processes and undertaking creative, innovative, and long-term activities, they acquire the competencies essential in managerial work. It should also be noted that this considerably higher level of cognitive activity and behavior manifests a high intellectual and adaptive potential of Polish managers. This may lead to a high level of individual, organizational, and market efficacy.

To validate these assumptions, thinking and behavioral styles of managers and their subordinates have been compared. *t*-test analysis showed significant differences in all the dimensions of styles of activity between managers and employees (Figure 46.6).

Figure 46.6 shows that employees have a lower or average level of both thinking and behavior dimensions. Their scores in all dimensions of activity oscillate around the mean value. Compared to managers, employees obtain significantly lower scores in characteristics of mental activity; they are less analytical, predict less, and do not make long-term plans. Other employees are also less active, creative, innovative, daring, and persistent than managers.

The hypothesis concerning the possibility of managerial training in strategic thinking competencies may serve as a recommendation to introduce elements of strategic competency training in the education system of managers. Such practices are successfully used in organizations as part of manager training programs. In order to validate this assumption, the competency levels related to activity styles have been compared in managers and management students. The test was conducted on 201 students from public and private schools of higher education aged 19–25 years (123 women and 88 men). The analysis of the results indicates that, in all of the dimensions of thinking and behavioral styles, there are significant differences between current and future managers (Figure 46.7).

Regarding behavioral style, students are less active, creative, flexible, and persistent than managers, but they tend to take higher risk than managers. The nature of this excessively high risk preference, accompanied by an insufficient level of social maturity and a hedonistic approach to life, is that of a developmental effect. As regards thinking style, students display a significantly lower level of *analysis*, *self-efficacy*, *consequence prediction*, *planning*, and *strategic evaluation* of the situation. The biggest differences can be seen in *globality* of thinking. Compared to managers, students have problems with the global style of processing and thinking "in broad categories." The result can easily be explained with reference to thinking habits acquired during excessively convergent school education.

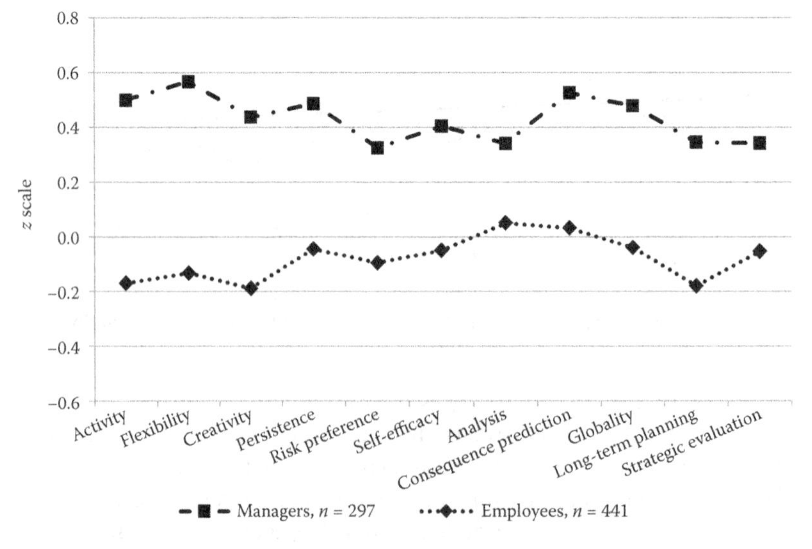

FIGURE 46.6 Differences in strategic thinking and behavior between managers and other employees.

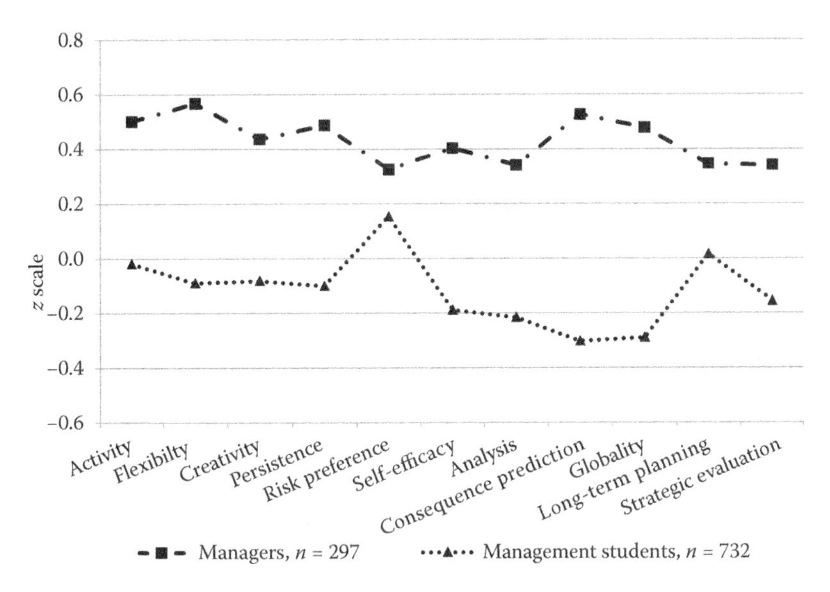

FIGURE 46.7 Differences in strategic thinking and behavior between managers and management students.

STRATEGIC THINKING AND BEHAVIOR OF POLISH MANAGERS IN RELATION TO ORGANIZATIONAL CRITERIA

To gain a wider view of the activity of Polish managers, the researcher conducted an analysis of strategic thinking and behavior dimensions in relation to professional variables such as work experience, work experience as a manager, the level of managerial position, and the type of organization. The results of the analysis will be presented respectively.

Management theorists and practitioners emphasize the growing significance of strategic competencies depending on the level of management (Katz 1974). In this analysis, managers have been divided into two groups: (1) top-level managers (presidents, chief executive officers, and department managers) and (2) lower-level managers (team and project managers). The research shows that there are differences in the characteristics of thinking and behavior of managers depending on their position in the organization (Figure 46.8).

Figure 46.8 shows that top-level managers are more active and flexible as well as more daring than lower-level managers. It seems that top-level managers respond more strongly and adapt more easily to changes in the environment. At the same time, they tend to take higher risks in order to achieve the company's strategic goals. Lower-level managers are slightly less pragmatic and daring, but this does not become an obstacle in the execution of their duties. It should be noted that the scores of both groups are high.

As regards strategic thinking style, top-level managers are significantly more predisposed to long-term planning than lower-level managers. In addition, regardless of their position, managers have a similar (high) level of *analysis*, *globality*, *consequence prediction*, as well as *flexibility*. The results confirm the role of cognitive competencies in top-level managers (Katz 1974). Figure 46.8 shows two competency profiles similar in shape but at a higher level in relation to top-level managers when compared with lower-level managers. Probably, there is a projection of the desired competency model in managers employed in different positions (Amitabh and Sahay 2008). Figure 46.9 illustrates significant trends in the frequency distribution of particular styles of activity depending on the managerial position, $\chi^2(15) = 31.346$, $p = .007$.

While analyzing sources of variability in thinking and behavioral styles of managers, the researcher conducted an analysis of differences in style dimensions depending on the length of

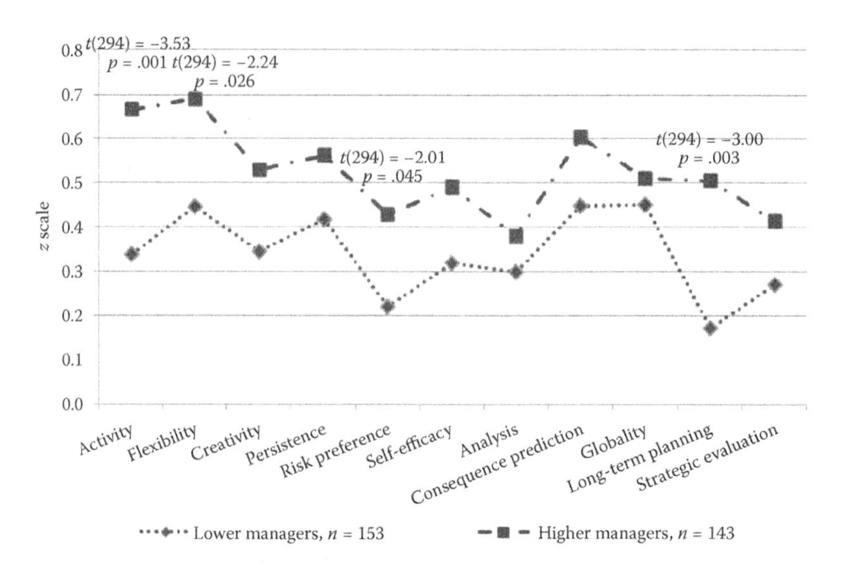

FIGURE 46.8 Strategic thinking and behavior of managers depending on level of management.

work experience and the length of employment in a managerial position. The length of work as manager varied considerably between managers tested, but the biggest differences in strategic thinking and behavior occurred in three groups of managers with different lengths of employment (Figure 46.10).

ANOVA indicates that the level of some strategic dimensions of activity depends on the length of employment. Figure 46.10 shows significant differences in thinking style dimensions, that is, *analysis, predicting consequences, long-term planning,* as well as *strategic evaluation.* This means that managers with longer work experience are more reflective, analytical, and strategically oriented than managers with less work experience. This may mean that the length of employment accounts for evolution of the style of activity. Less experienced managers tend to be *activists* highly motivated to work and involved in various initiatives and tasks. Average work experience fosters a more passive profile of managerial activity. Such a decrease in the level of activity and mental processes may be symptomatic of professional burnout. A long period of employment, however, allows managers to develop a strategic behavioral style.

No significant differences have been observed in the profile describing thinking and behavioral styles of managers depending on the period of employment in a managerial position. As may be noticed, only considerable work experience (not necessarily in a managerial position) facilitates effective management of the company. In addition, the style of activity of managers is universal and independent of the type and size of the company. However, this hypothesis requires further and more detailed research.

A further analysis of data showed, however, that there are differences in frequency of the four styles of activity of company owners and managers employed in Polish and foreign companies and in nongovernmental organizations, $\chi^2(15) = 23.817$, $p = .005$.

Figure 46.11 illustrates that 50% of managers employed in nongovernmental organizations are activists, while 35% are strategists. The results differ from the general trends observed during the research. In addition, almost half of company owners manifest strategic behavioral style, while 36% are *thinkers.* Such results may indicate that company owners often perform managerial, conceptual, and operational functions. The group of managers employed in domestic and international companies counts significantly more *strategists* and *thinkers* in comparison to the remaining styles of activity. A very small percentage of managers in each type of organization display the passive thinking and behavioral style.

The results of analysis of the strategic activity profile of managers in the context of their professional status presented in this paper do not constitute a full description of characteristic patterns of

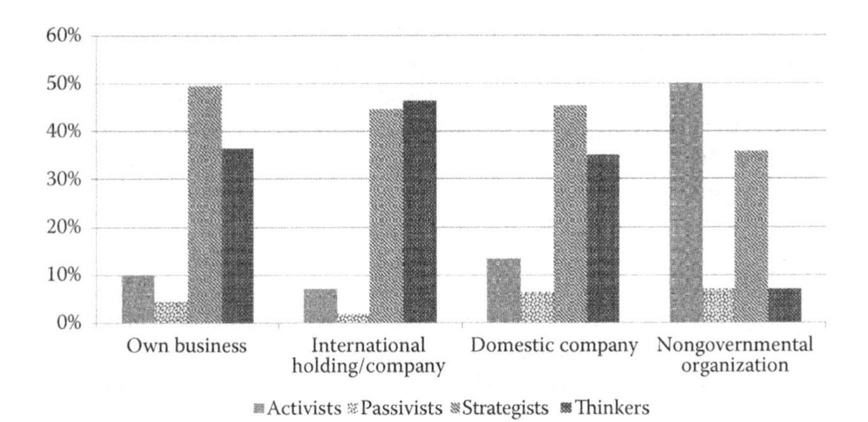

FIGURE 46.9 Frequency of strategic thinking and behavior depending on position in an organization.

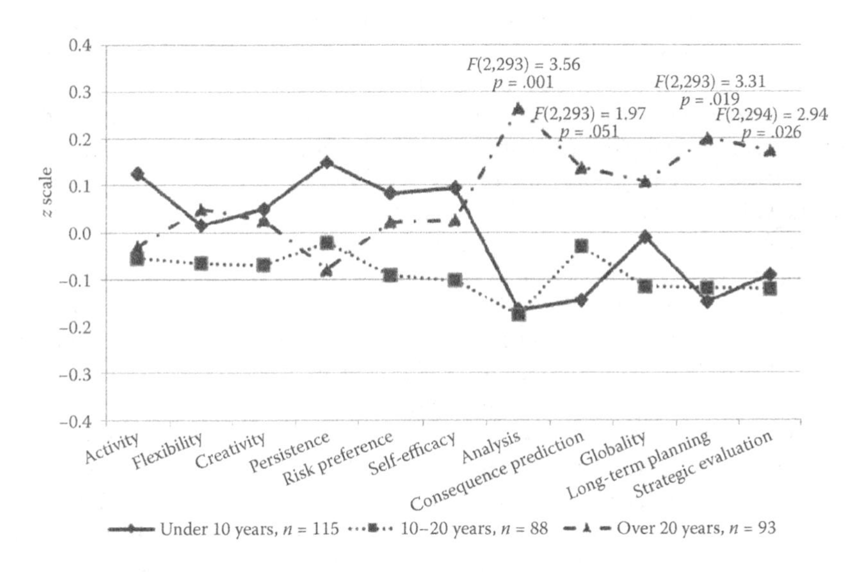

FIGURE 46.10 Strategic thinking and behavior of managers depending on work experience.

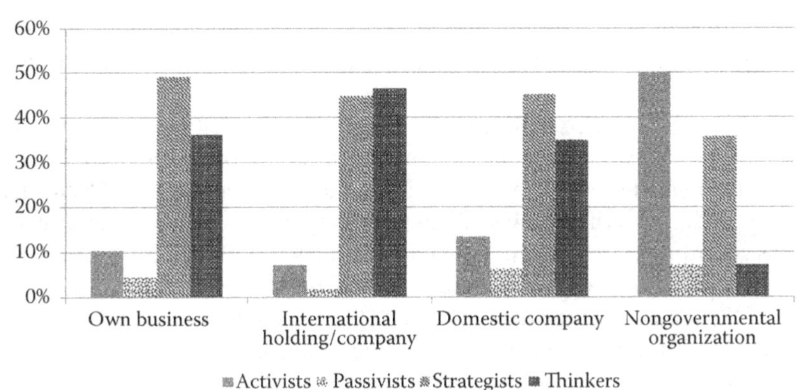

FIGURE 46.11 Frequency of strategic thinking and behavior depending on the type of organization.

managerial activity. It may be worth contrasting the characteristics of thinking and behavioral styles with other aspects of managerial activity, such as the management styles or styles of coping with stress. The research conducted so far demonstrates that the style of activity corresponds to the preferred style of leadership (Dragoni et al. 2011) and has strong personality and temperamental determinants (Gallén 1999; Bajcar 2012). It may thus be assumed that styles of strategic thinking and behavior of managers form a basic pattern of activities that regulate management processes.

CONCLUSION

Summarizing, Polish managers demonstrate different thinking and behavioral styles, which lend more or less strategic character to their activity. There is no definitive answer to the question of whether all managers are strategists. This study provides evidence, however, that a large number of Polish managers are strategists.

Among subjects, almost half of the managers displayed the strategic style of thinking and behavior. *Strategists* integrate the processes of convergent and divergent thinking, analysis and assessment of the situation, and development of strategies with implementation processes. The strategic style constitutes the most rational and innovative pattern of activity. It is highly desirable in managers. It increases managers' entrepreneurship and resourcefulness in dealing with organizational problems and economic challenges. That is why *strategists* probably, to a large extent, contribute to the increase in the development and success of the company. *Strategists* can be found most often among top-level managers employed in domestic and international companies, aged over 50 years, although also a large percentage of them are aged 25–35 years, with more than 20 years of work experience, and with different work experience in a managerial position.

Thirty-six percent of Polish managers in the sample were *thinkers*. They are characterized by a deliberative style with a predominance of mental activity, that is, analytical thinking, global thinking, predicting, and long-term planning. *Thinkers*' behavioral style is, however, more passive and stereotypical. At work, they instead focus on diagnosis, conceptual work, and decision making (Knootz and Weihrich 1990); they consider alternative courses of action but do not become involved in implementation of their ideas and designs. Generally, this style of activity seems to be inefficient, since a lasting preference for analysis and assessment of the situation does not contribute to achievement of goals. In management, the deliberative style may be treated as a set of competencies useful for the development of goals, strategies, and tactics within an organization. *Thinkers* delegate implementation of strategies to their subordinates. Notably, this style may manifest in excessive theorizing and procrastination in implementation of activity in the company. *Thinkers* are usually employed in domestic and international companies or are company owners. Most often, they are midlevel managers with varying lengths of employment in general and work experience in a managerial position. Is, therefore, the deliberative style most adequate for the midlevel management positions? Can team managers afford to delegate all executive functions to their employees? Are they excessively focused on conceptual work, thus neglecting execution and implementation? These questions require empirical verification.

There were merely 13% *activists* in the sample. Their style of activity is characterized by a strong representation of the processes of implementation of ideas and activity and a low level of strategic thinking. Such a predominance of behavioral style over thinking presents the Darwinian model of rational behavior. In this pattern, managers focus on testing the possibilities available and optimizing their actions according to the changing situation (Mintzberg 1994). Lack of the ability to predict opportunities and threats and lack of alternative strategies may result in *activists*' poor decisions and limitation of the direction of the development of the company to stereotypical, impulsive, and short-sighted actions. Thus, the active style has limited adaptive skills in the face of contemporary problems and challenges confronted by organizations. The most *activists* are among managers aged 30–40, with the shortest period of employment in a managerial position and with the shortest period of employment in general. They are most often team managers or presidents. It means that strong activity orientation is characteristic of lower-level managers. This seems to be rooted in the growing ratio of executive responsibilities in these positions.

A small percentage of managers are *passivists* (5% of the subjects). They display lower activity in both cognitive and executive processes. *Passivists* display a strong tendency to preserve the *status quo*, to go with the flow, or to await what course events will take. A low level of *analysis*, *consequence prediction*, and inventiveness indicates a high degree of unreflectiveness, superficiality, and stereotypy in thinking and behaviors. The passive style is highly dysfunctional in the context of organization management. Yet, its occurrence, especially among professionally active managers, is extremely interesting. However, the sample contained too few *passivists* to make far-reaching conclusions. From the psychological point of view, *Passivists* constitute especially intriguing research subjects. They can be encountered most often among men employed as lower-level managers with average work experience. A question arises as to how they deal with problems in their personal and professional life. A low level of involvement and motivation to work may lie at the root of such an approach to life.

In comparison with the wider population, the managers tested are better *strategists*. They display a higher level of activity of mental and behavioral processes. This puts managers in a good light. They cope with the dynamism and complexity of the environment considerably better. They actively and efficiently solve professional problems and tasks. As a rule, they are more analytical, better at predicting and long-term planning, and also more active and open to changes. In the light of the research results presented, differences in managerial activity style were observed with regard to age, work experience, and position in the company. The data obtained indicate a number of significant implications concerning the description and explanation of strategic activity, for example, the possibility to foster and develop strategic competencies of managers.

Interesting data were also obtained through comparison of activity styles in managers and management students (future managers). It appears that, compared to students, managers tend to display considerably more strategic thinking and behaviors, which fosters efficacy at work and development of the organization. This means that work experience activates key strategic competencies in managers. This effect has two important practical implications. Firstly, it confirms the need to develop cognitive and behavioral competencies in managers. Secondly, it indicates a need to review the existing programs of study for future managers so that students could acquire skills that would increase their efficiency in their future work. The need to improve the economics and management study programs is being widely discussed in academic circles (Kentworthy and Wong 2005; Payne and Whittaker 2005). Strategic competencies may be developed through confronting students with problems which stimulate their thinking, and not merely require identification and retrieval of the knowledge acquired.

Specialist publications list a large number of characteristics and skills of successful managers, with the profile of the efficient manager resembling "the Olympic model" (Simon 1983). The most rational approach in the development of the competency model of managers for the 21st century is to define more metacompetencies. These would allow the regulation of activity by adjusting basic competencies to solve problems within the organization and to achieve the strategic goals of the company. Managers would then know what to do, how to do it, and what skills can facilitate it. In the light of theoretical considerations and research results, it is necessary to integrate the set of managerial competencies and establish their hierarchy. It might be beneficial to develop an integrated model of strategic intelligence, which defines the high level of adaptive skills of managers in the global context of organizations, society, and economy.

REFERENCES

Amabile, T. M. (1997). Motivating creativity in organizations. *California Management Review*, *40*(1), 39–58.
Amitabh, M., and Sahay, A. (2008). Strategic thinking: Is leadership the missing link. Presented at the 11th Annual Convention of the Strategic Management Forum, Kanpur, India.

Bajcar, B. (2012). Kwestionariusz Styl myślenia i działania strategicznego—nowe narzędzie do pomiaru wskaźników myślenia strategicznego. [Strategic thinking and behavior style questionnaire—A new tool to the measuring of strategic thinking indicators]. *Psychological Studies*, *50*(2), 5–24.
Bandura, A. (2009). Cultivate self-efficacy for personal and organizational effectiveness. In E. A. Locke (Ed.), *Handbook of Principles of Organization Behavior* (2nd Ed., pp. 179–200). Oxford, UK: Blackwell.
Basadur, M. S., and Hausdorf, P. A. (1996). Measuring divergent thinking attitudes related to creative problem solving and innovation management. *Creativity Research Journal*, *9*, 21–32.
Bonn, I. (2001). Developing strategic thinking as a core competency. *Management Decision*, *39*(1), 63–71.
Bonn, I. (2005). Improving strategic thinking: A multilevel approach. *Leadership and Organization Development Journal*, *26*(5), 336–354.
Dragoni, L., Oh, I. S., Vankatwyk, P., and Tesluk, P. E. (2011). Developing executive leaders: The relative contribution of cognitive ability, personality, and the accumulation of work experience in predicting strategic thinking competency. *Personnel Psychology*, *64*(4), 829–864.
Drucker, P. F. (1992). *Managing for the Future*. New York: Harper Collins.
Gallén, T. (1999). The cognitive style and strategic thinking. Proceedings of the Leadership and Myers-Briggs Type Indicator (pp. 25–30). Washington DC, USA.
Gallén, T. (2006). Managers and strategic decisions: Does the cognitive style matter? *Journal of Management Development*, *25*(2), 118–133.
Goldman, E. (2008). The power of work experiences: Characteristics critical to developing expertise in strategic thinking. *Human Resource Development Quarterly*, *19*(3), 217–239
Goleman, D. (2004). Co czyni cię przywódcą? [What makes you a leader?]. *Harvard Business Review Poland*, *12*, 92–101.
Guth, W. D., and MacMillan, I. C. (1986). Strategy implementation versus middle management self-interest. *Strategic Management Journal*, *7*, 313–327.
Hussey, D. (2001). Creative strategic thinking and the analytical process: Critical factors for strategic success. *Strategic Change*, *10*, 201–213.
Kaplan, R. S., and Norton, D. P. (2002). *Strategiczna karta wyników [Strategic Scorecard]*. Warszawa: PWN.
Katz, R. L. (1974). Skills of an effective administrator. *Harvard Business Review*, *2*(4), 58–65.
Kentworthy, J., and Wong, A. (2005). Developing managerial effectiveness: Assessing and comparing the impact of development programmes using a management simulation or a management game. *Developments in Business Simulation and Experiential Learning*, *32*, 164–175.
Knootz, H., and Weihrich, H. (1990). *Essentials of Management*. New York: McGraw-Hill.
Kotarbiński, T. (1965). *Praxiology: An Introduction to the Sciences of Efficient Action*. New York: Pergamon Press.
Liedtka, J. M. (1998). Strategic thinking: Can it be taught? *Long Range Planning*, *31*(1), 120–129.
Loehle, C. (1996). *Thinking Strategically. Power Tools for Personal Advancement*. London: Cambridge University Press.
Merkessini, J. (1996). Executive leadership in a changing world order. Requisite cognitive skills. The first literature review. ARI Technical Report No. 96-05. Alexandria, VA: U.S. Army Research Institute for the Behavioral and Social Sciences.
Mintzberg, H. (1994). The rise and fall of strategic planning. *Harvard Business Review*, *72*(1), 107–114.
Mumford, M. D., and Connelly, M. S. (1991). Leaders as creators: Leader performance and problem solving in ill-defined domains. *Leadership Quarterly*, *2*(4), 289–315.
Oettingen, G., and Gollwitzer, P. M. (2010). Strategies of setting and implementing goals: Mental contrasting and implementation intentions. In J. E. Maddux, and J. P. Tangney (Eds.), *Social Psychological Foundations of Clinical Psychology* (pp. 114–135). New York: Guilford.
Payne, E., and Whittaker, L. (2005). Using experiential learning to integrate the business curriculum. *Developments in Business Simulation and Experiential Learning*, *32*, 245–254.
Romesburg, H. C. (2004). *Cluster Analysis for Researchers*. Morrisville, NC: Lulu Press, Inc.
Simon, H. (1983). *Reason in Human Affairs*. Stanford: Stanford University Press.
Sternberg, R. J. (1999). *Cognitive Psychology*. New York: Holt, Rinehart and Winston.
Switzer, M. (2008). Strategic thinking in fast growing organizations. *Journal of Strategic Leadership*, *1*(1), 31–38.
Wootton, S., and Horne, T. (2010). *Strategic Thinking. A Nine Step Approach to Strategy and Leadership for Managers and Marketers*. London: Kogan Page.
Zaccaro, S. J., Kemp, C., and Bader, P. (2004). Leaders traits and attributes. In J. Antonakis, A. T. Cianciolo, and R. J. Sternberg (Eds.), *The Nature of Leadership* (pp. 101–124). Thousand Oaks: Sage Publications, Inc.

47 Prodevelopmental Orientation as a Determinant of an Active Participation and Cooperation in a Reality Oriented toward a Global Change

Agnieszka Cybal-Michalska

CONTENTS

A humanistic discourse about the quality of a contemporary society should allow for the context of the social–cultural changes taking place in the globalizing world. The most characteristic feature of the times that came after the period of modernity is the rapid development of civilization. Civilization can be understood as an organized and highly pluralized form of life of a social community and as a peak or a twilight of a social–cultural development. The core of globalization, as a process of radical and permanent changes, is the cultural diversity of contemporary societies, which determines the axionormative chaos and emptiness, large character, disorientation, changeability, identity crisis, pluralism, and relativism of outlooks. According to Kwieciński (2002), "These problems mock borders." On the individual level, it means losing axionormative points of reference, the lack of roots, individualistic alienation, and the necessity to exist in an ambiguous and internally contradictory reality, which is not a monolith. (Cybal-Michalska 2006).

According to J.P. Hudzik (1997), the world around us today has lost the durability and stability, the visible unity and continuity, that were once attributed to it. There is also no common consent about any coherent set of models for a reasonable action, there is no one reliable prescription for a sensible and safe life in a sensible and safe world.

The conclusion that comes to mind from these considerations is that we live in times of crises. The essence of these times is the transition from the monoideological political correctness and the culture of collective axiological declamation to the pluralistic order in the political and cultural sense, to the culture of individual choices.

The spirit of the times implies yielding to individualistic tendencies, the disintegration of individualistic life projects; what we have instead is, as Z. Bauman (2000) said, "the kaleidoscope of independent episodes." But on the other hand, as G. Lipovetsky (1987) (Środa 2003) emphasizes, not having any aims or roots, any substance or horizons in the presence of which it could, interchangeably, define itself, individualism can now grow freely, sprouting in every possible direction. "Our fragile world," according to R. Kapuściński (2002), demands to look at globalization as "a human reality," whose task is to look, simultaneously, at "a man in the world" and at "a world in a man" (Wojnar 2003).

The phenomenon of the contemporary world is the cognitive globalization, which is distinguished by the extraterritoriality of knowledge that reduces the globe to the size of "the global village." The new status of knowledge, whose sources are everywhere and almost nowhere, as there are no privileged places that have the monopoly for trustworthy knowledge that is worth having, results from the explosion of information and the dependence of the quality of knowledge on the demands of the global market, which is, first and foremost, linked with productivity and the usage of information in action that aims at production, innovation, and management. We are stepping into the period that carries the shift from an author to a recipient, in the name of "the power over a text." Now, when we have new technologies at our disposal, we are going to create our own parcels of knowledge. The new quality of the information society that has become the reality, and not the intellectual abstraction, results from a rapid development of advanced information, computer, and telecommunication technologies. Those technologies, according to J. Naisbitt (1997), accelerate the speed of changes and overcome the inertia of information. The principle characteristic of the information society is the gathering, transformation, reconstruction and use of knowledge to actively cope with the situation of sociocultural change and the process of shaping the future. The global society has to be arranged according to specialized knowledge and according to human resources who have the knowledge and who are specialists. The life and functioning in the two cultures, "the intellectual" culture and "the executive person" culture, point at a dichotomous structure between "intellectualists" (who are interested in science as the most perfect way of cognition) and specialized managers (who are interested in the revolution of productivity and management). This dichotomous system demands reshaping into a new synthesis of organizing the variety of knowledge, and it constitutes the main challenge for an information society, that is, the society that is based on many various branches of knowledge and that is able to comprehend the multifaceted application of knowledge and its constructive use. According to A. Giddens (2001), the orientation towards permanent self-education and self-improvement, both in the individual and the wider, social sense, is an answer to general and prevailing tendencies to continual review of the image of reality and the development of abstract systems that favor the creation of modern forms of expertise and narrow specializations. Formal and abstract systems of analysis contribute to the devaluation of skills that concern almost all the levels of human existence. The expropiar effect of the abstract systems that contributes to the alienation and fragmentation of an individual identity reveals itself in creating new principles for an individual and common action and making changes in an objectified, symbolic, and mental world, as a result of internalization, recontextualization, and externalization of knowledge. The epistemological assumption concerning the learning object distinguishes the postmodern concept of innovative teaching. It has to do with the conviction that, in his/her research efforts, the subject should be able to use the knowledge that he/she has already acquired, remembering, at the same time, that this knowledge is not enough, as M. Mead (2000) said. The frames of articulation of different interpretations of the notion of "learning" as an immanent characteristic of a man correspond to the broad understanding of this category as a kind of such an attitude towards knowledge and life, in general, where you emphasize the meaning of human initiative to gain knowledge with the confidence that this knowledge is useful as an essential value to live in a world of permanent changes. The slogan "learn or die," according to J.W. Botkin et al. (1982), sounds like a challenge and a warning. It is the proclamation that calls for making an effort to clench "the human gap" (one can see bigger and bigger dissonance between humankind and the civilization it created) by the requirement of an innovative and anticipatory learning process that is not devoid of reflectiveness. Reflectiveness as an important component in the social consciousness means the ability of a given society to think critically about itself; to notice negative, pathological phenomena; to define future threats; and, on the basis of such an assessment, to take precautionary measures that will stop or turn away unfavorable trends, whose influence may be subjected to amendments (Sztompka 2002).

It seems a good moment to quote questions asked by Z. Kwieciński: Can a single man be prepared—through conscious organization of his actions and educational processes—to influence the effective solution of global problems, understood as a general threat to the survival and well-being of mankind, as well as general expectations about the establishment and growth of positive development tendencies of the contemporary civilization? These are elements of a bigger question about the quality of a global change society. I. Illich's (1994) deliberations give a partial answer to that question. According to the author, we can only live and breathe the change. You cannot plan, in advance, your road to humanity. Each of us and each circle of people that we live and work with must become an example for an epoch that we want to create. Z. Kwieciński adds that one thing is certain: The fuzziness of the perspective will be the permanent feature of the present and the future; we are, thus, doomed to make choices among various cultural offers. The cultural pluralism is already an irremovable feature of reality, and experiencing ambivalence is a life *sentence* or, even, a curse for a contemporary man. Since the diversity of cultures is a fact, it should be perceived as goodness and complexity of educational activities, as a challenge and a chance for education. In the context of the quoted remarks, it seems important to stress that the contemporary challenges that the society of a late modernity has to face and that are linked with the global cultural ecumenism provoke reflection that one of the very important factors that influence the condition of a contemporary man is global education, in a broad sense of this term. The aim of global education will be, on the one hand, to shape predevelopment orientation among the youth and, on the other hand, to educate teachers in a holistic, interdisciplinary, and developmental way—as it is, teachers who prepare the youth to live an active and conscious life and to cooperate in a local, national, European, and global society (Cybal-Michalska 2001). In the context of educational needs of a globalizing world, Kwieciński's statement has an important message: That the present of a ceaseless crisis demands a type of a man who can face up to difficult circumstances and tasks—a plenipotentiary man, with shaped competence to formulate and solve new, difficult tasks; a man who is wise, responsible, sympathetic, and able to loyal cooperation; a man who can critically choose and judge manifold and glimmering cultural offers. His criticism is based on universal values and rules. How a man is shaped, how he can deal with the complex challenges of the contemporary world, depends on the education itself.

In the world of changeable axiologies, a peculiar need for deepened "planetary consciousness" (Peccei 1987) or "globalized thinking" (Melosik 1989) appears. It results from the fact that people worldwide feel that they share a common lot. This need calls for intercontinental reconciliation. At the same time, the need for deepened and multiple considerations arises. This consideration concerns the quality of subjective existence in a heterogeneous world.

Freedom of an individual, which constitutes the highest goodness, may, at the same time, become a threat, and yet it does not mean "the end of a man" (Fukuyama 2004). It is, rather, a crisis of a certain particular role that was ascribed to him/her in the structure of a traditional society. According to L. Witkowski (1999), competences necessary to take part in a culture full of semiotic meanings have to differ from those that are part of a culture presented as a space that can be made clear, coherent, hierarchized, and organized with preserving harmony and a structure of unambiguous rules. Conscious participation in an ambiguous and ambivalent global reality demands the preparation of "a subject man," focused on self-education, open for the free self in the future, with full individual responsibility for one's actions. As R. Kapuściński (2002) stresses, there is nothing obvious in being a man. It does not happen just like that; you always have to aim at it, always urge oneself, always arouse the willingness to be a man.

A prodevelopment orientation plays an important role in the defense and development of "a truly humanistic civilization. It is also a remedy for a contemporary human being to be oneself again in this situation in world of careers "(Cybal-Michalska 2013). A context in which one uses the notion "orientation" means, in a broad sense, the ability to assess, discern, and interpret a situation, the knowledge of facts, and leaning towards something. According to T. Hejnicka-Bezwińska (1991), this term is sometimes defined as the ability of proper and quick establishment of external data and the data linked with oneself and with the direction of one's deliberate action. The link between orientation and physical activity of an individual and, especially, their attitude towards the surrounding world was S. Błachowski's subject of research. The author defined orientation as a primitive predisposition formed as a result of evolution. S. Szuman linked orientation with an action understood as a task, which has to be successfully solved to achieve a planned goal. A. Lewicki uses a different definition of orientation. He claims that orientation is a direct basis for an action (retrospection), which he attributes with an important meaning for forming experience as a basis for future actions. According to him, orientation has not only a descriptive component but also an evaluative one. K. Skarżyńska (1990) presented a definition of psychological orientation, while describing an egalitarian and nonegalitarian orientation. According to her, orientation is a more or less cohesive set of cognitive orientations, as well as motivating and moral ones, towards a given situation (determined by a kind of an interpersonal reaction), which serves as a guide for individual actions in a given situation. S. Gerstmann paid attention to the relation between orientation and knowledge. He described orientation as the third, next to scientific knowledge (theoretical, organized) and popular knowledge (practical, everyday), source of information about man and the surrounding reality. In this definition, orientation combines the above-mentioned types of knowledge with action. Orientation that takes into consideration action contains information about, on the one hand, the surroundings (orientation in the conditions of action) and, on the other hand, the subject of action (orientation in itself). With these assumptions in mind, the same author divided global orientation. He singled out a general orientation, typological–situational and situational. The first of these three forms itself during childhood, under the influence of upbringing, but it undergoes modifications during the whole life of an individual. Thanks to it, an individual formulates aims and ways of achieving them. The typological–situational orientation indicates the importance of aims, tasks, and the feeling of awareness of social importance. The situational orientation depends on the content of general and typological–situational orientation. It refers to facts and individual happenings. Taking into consideration the above assumptions, one may assume that orientation determines attitudes, and these attitudes lead directly to behavior. M. Ziółkowski's (1990) considerations on the subject of orientation correspond with the above conclusion. In this context, he defines orientation as generalized tendencies to perceive, judge (an individual has a system of values as well as categories of values), feel, and react to social reality (the author assumes that the subject can define, order, and react properly to the surroundings only when he/she can include it into a general category, not necessarily with full consciousness of the process). One may thus assume, that orientations consist of not only, lucid, clear, and conscious convictions, but they also have convictions that are only partially conscious, premonitions and feelings that are often on the verge of unconsciousness. Fundamental elements that make up the notion of orientation are values, popular knowledge, evaluation, and predisposition towards behaviors.

The nature of prodevelopmental actions, in the general perspective, constitutes actions (of an individual) concentrated on the realization of a certain vision of an ideal state. T. Zysk (1990), citing W. Reykowski and J. Kozielecki, emphasized that the most essential and characteristic feature of a human mind is the ability to formulate the vision of future state of things. The subject is convinced of the perfection of this state and takes action aiming at achieving it.

What are the psychological characteristics conditioning prodevelopmental behaviors? T. Zysk (1990) mentions the following: Inner motivation, attitude towards the future, and active coping with reality. He ascribed special meaning to the first two factors.

The inner motivation (defined as "growth" or aspiration) is determined by the vision of the ideal state and is most often characterized as an engagement of an individual in some kind of an action because it gives the subject pleasure or because it arouses the subject's interest. Motivations are obviously linked with revealing and fulfilling needs. Motivation also can be defined as a behavioral term. Deci and Ryan describe inner motivation in this way, defining it as behaviors that may occur and be sustained, even when there are no visible reinforcements of two fundamental factors: The need for competence and self-determination. The need for competence is linked with the need for rivalry, with a cognitive curiosity and with a tendency to show perfection in action. The self-determination may be identified with the feeling of being the author of things, the shaping of responsibility in an individual. The subject does not concentrate on the "fate" but on the independent creation of one's future, that is, on the "choice." It is accompanied by an antifatalistic attitude, as A. Sarapata (1993) indicates.

J. Kozielecki characterizes the inner motivation, which constitutes the basis for the prodevelopmental orientation, as a heterostatic motivation, which is determined by a set of inner reinforcements.

He assumes that it is based on keeping divergence between the present (real) state and a desired (ideal) state. Keeping this divergence is a source of satisfaction for the subject, and it results from the inner mechanisms of an individual.

Attitude towards the future is the second important element (next to motivation) in shaping prodevelopmental behaviors. J. Kozielecki thinks that the temporal orientation, that is, the ability of an individual mind to formulate visions of future states of things, is a basis for prodevelopmental behaviors. He emphasizes that the most fundamental and, at the same time, the most characteristic feature of a human mind is the ability to formulate visions of future states of things, which an individual is convinced are better and more perfect than the present ones. Moreover, an individual is ready to take action aiming at achieving them, despite the appearance of many objective obstacles. Actions that aim at accomplishing the vision, through changing and subduing reality, seem to constitute the nature of prodevelopmental actions. The author points out the transgressional nature of a human mind, the tendency to go beyond "what one is and what one has." It is assumed that the temporal orientation, with an assumption of a properly formulated state of perfection, guides an individual to activity, which favors developmental and creative actions. W. Łukaszewski (1984) pays attention to the fact that the temporal orientation has to have realistic and instrumental features in order to constitute a significant element of prodevelopmental actions of a subject. He distinguishes—in accordance with a classical understanding of time—three types of temporal orientation: Retrospectivism, presentism, and futurism. There may also occur a compilation of those three and then an individual turn to the whole horizon of time. Planning—as A. Sarapata (1993) emphasizes—characterizes active people, who control their own fate and are prudent and far-sighted, accomplishing their goals and tasks. In this category, one may also mention the individual's interest in the matters of their own country, continent, and the world. Active coping with the reality displays itself in an active reshaping of a present state aiming at the realization of a vision of ideal states. Diaz-Guerri, in his research, distinguished two styles of dealing with problems: Active (an individual makes changes in social surroundings, as a way of influencing reality) and passive (a subject adapts to the changes that take place in the surrounding reality). He stresses that active dealing with reality takes place not only in situations when an individual is responsible for what he does and what happens around him but, also, when the responsibility is subscribed to someone else. The essence of active coping with reality is, also, dealing with the results of someone else's actions, not only one's own. The subject handles the surrounding reality in an active way. He is convinced that there is a possibility of influencing the change of a situation; he has a great sense of power. An active individual is also characterized by the need for achievements, which manifests itself in a constant aiming at achieving the best results in conditions of competition; innovative predispositions; a constant will to take rational decisions, that is, for example, to foresee the results of risk; and a constant tendency to objective analysis and judgment of one's own actions.

The notion of prodevelopmental orientation is, thus, a broad term that includes different dimensions of the term "development." In this context, Inkeles, in his research, made an attempt to describe characteristic features of people take development actions. He distinguished the following three spheres of life: Political (the subject has a possibility to interfere in a direct or indirect way in the political life), economical (it determines the most prodevelopmental attitudes), and family (it displays itself through a tolerant and pluralistic attitude of a subject towards various values). Taking into account the above assumptions, T. Zysk (1990) made an attempt to characterize a prodevelopmental mind. Table 47.1 presents a set of features that make up an innovative personality.

The findings presented above probably do not eliminate all the misunderstandings linked with the usage of a notion of prodevelopmental orientation, but they definitely facilitate its usage. In the light of the presented definitions, the prodevelopment of an individual will manifest itself through the following features: Creativity, openness ("the demand for something new") or readiness ("the permission for something new") for new experiences, aiming at gaining knowledge, the acceptance of changes, innovation, tolerant thinking, the feeling of dignity and being the cause of things, confidence in other people, and an attitude towards future.

TABLE 47.1

Prodevelopmental Personality

	The Field of Political Life	The Field of Economical Life	The Field of Family Life
Cognitive flexibility: Understood as being open to new experiences, an ability to assimilate new experiences.	Tolerance for views of other people; the awareness of the existence of a variety of opinions, attitudes, or actions; a man who thinks in this way does not think that others have to think in this way too; he is convinced that dissimilarity of opinions supports individual and common interests; he is not afraid to accept different opinions from others, and he is convinced that the acceptance of views of other people does not have to necessarily lead to the destruction of his own views; he is a supporter of social changes, and he actively takes part in wide spheres of a political life, believing that through such an activity, he is able to influence this sphere of a political life; big social mobility.	Innovative actions, being open to new solutions, and aiming at involving new solutions in his actions.	Being open to new experiences; the awareness of a variety of values and tolerance for different cultures, habits, clothes, and behavior; religious tolerance; aspiring to include new solutions in the sphere of family life.
Cognitive curiosity: Understood as a high level of aspiration to gain knowledge; it manifests itself in a high assessment of usefulness of knowledge and a constant will to gain more knowledge.	Aspiration to actively gain knowledge about the mechanisms that rule the social world, with a conviction of the usefulness of such knowledge; readiness to express opinions about social phenomena in the situations that demand it; interest in general matters, not only in one's surroundings and profession; judging people not on the basis of their membership in a particular group or an institution but on the basis of their qualifications.	Aiming at improving his/her professional qualifications, at gaining knowledge about the mechanisms that rule economical life; the interpretation of economical phenomena on the basis of knowledge, not popular opinion; an ability to interpret economical phenomena in the categories of interests and general social purposes and not only one's own business; respect for technology and science; aspiration to understand technological processes.	Aiming at gaining knowledge about the mechanisms of a family life (e.g., sex life), needs and aspirations of other members of the family, and so forth; a conviction that family problems can be solved in a rational way; belief in a rational upbringing.
Temporal orientation: Understood as a sense of time, an attitude towards the future, and an ability to plan, an ability to change general aims and tasks into more specific actions.	An ability to select an action for a present situation; an attitude towards present, not past, happenings; the ability to change general aims and values linked with a political life (for example, democracy, freedom, etc.) into specific actions.	High sense of time, that is, the ability to realistically assess the course of action; punctuality, time rigorism; interest in the effects of one's own actions; an ability to plan and to organize a place of work.	An ability to formulate aims and actions for the future and to take actions to provide for the future (one's own, the future of one's children, etc.); being addicted to tradition; an ability to plan and organize a family life.

The feeling of being the cause of things: Understood as a sense of an ability to influence the course of action.	A feeling that you can do something, even when a country or circumstances impose "wrong" or unaccepted laws; a feeling that my actions mean something; a feeling of influencing political events (elections, expressing opinions etc.)	The feeling of being the cause of things and of being in control of what one does; a conviction that one may influence a place of work.	A feeling that family problems may be solved; a feeling of an ability to choose a shape of a family life.
Confidence in other people and in the world: Manifests itself mainly in the conviction that you can depend on other people, on their obligations, or on the effects of their work; a conviction that the social world that surrounds a man is comprehensible and friendly, that a man can have confidence in it.	A feeling that social institutions, delegates, and representatives of a society will act in the interests of general good; a conviction that political institutions function in the name of social good; a conviction that situations function in an objective and "honest" way, that they "do their job"; a conviction that a social life is not ruled by corruption or one's own interests.	Confidence in quality of work; an "objective" and "honest" access to goods; trusts in the rules of an economical life; trusts in the institutions of an economical life (bank, social security, etc.)	Confidence, not only in the closest people but also in others, who represent different habits or views or who believe in different values or ideologies; a conviction that people are "good and wise by nature" irrespective of race, religion, or political views; a general conviction that you may depend on people.
A feeling of dignity and mutual respect: Manifests itself mainly in a conviction about one's own self-esteem, a conviction that a man has rights to acknowledge some essential values; mutual respect for social life partners.	A conviction that all citizens, irrespective of position or convictions, have an indispensable right to realize the values they believe in; always respect partners or political opponents; mutual respect between the authority and a citizen.	A conviction that every subject of an economical life has a right to realize their own values; mutual respect between an employer and an employee, a clerk and a customer, a shop assistant and a client, and so forth.	A conviction that not only the closest people have a right to respect; respect for people who have different views, outlooks on the world, or behaviors; respect for cripples, for the sick, or for those who do not subscribe to the typical canon of normality; respect for fundamental values, including "past," traditional values, as they determine the continuity of an individual or group fate.

Source: Zysk, T., *Prodevelopmental orientation* [In:] Reykowski J., Skarżyńska K., Ziółkowski M. (ed.), *Social orientations as an element of mentality*, Poznań 1990.

Taking the road of "intellectual emancipation" (Jankowski 2003) means, as Bauman (1999) puts it, talking to people rather than fighting with them, understanding them rather than getting rid of them or destroying them as "mutants," and expanding one's own tradition thanks to the experiences of other people rather than cutting oneself off from the unknown ideological movements. The affirmation of the cultural variety, seeking harmony with the community, in a broader and broader sense, and the consciousness of starting a creative dialogue with "otherness" are extremely important for the global society of the future. A common statement that "we act locally thinking globally" clearly refers to the perspective according to which a man is a subject (has, according to M. Czerepaniak-Walczak (1994), a creative power, has a causative power, and is responsible for what he creates) who has a right to dignity, self-fulfillment, and development.

REFERENCES

Bauman Z., The fall of legislators. [In:] *Educational Forum* 1999, no. 1/2.
Bauman Z., *Postmodernity as a Source of Suffering*. Warszawa 2000.
Botkin J.W., Elmandijra M., Malitza M., *No Limits to Learning. Bridging the Human Gap*. Warszawa 1982.
Cybal-Michalska A., *Academic Youth and Career*. Kraków 2013.
Cybal-Michalska A., *Identity of Youth in the Perspective of the Global World*. Poznań 2006.
Cybal-Michalska A., *Proeuropean Orientations of Youth*. Poznań 2001.
Czerepaniak-Walczak M., *Between Adaptation and a Change. Emancipatory Elements of the Education Theory*. Szczecin 1994.
Fukuyama F., *Our Posthuman Future: Consequences of the Biotechnology Revolution*. Kraków 2004.
Giddens A., *Modernity and Self-Identity*. Warszawa 2001.
Hejnicka-Bezwińska T., *Life Orientations of Youth*. Bydgoszcz 1991.
Hudzik J., The head or the heart?—Towards post-modern deconstruction of the classic category of "phronesis." [In:] Hudzik J., Mizińska J., (eds.), *Memory—Place—Presence*. Lublin 1997.
Illich I., *Celebration of Consciousness*. Poznań 1994.
Jankowski D., Formal education versus self-education. [In:] Wojnar I. (ed.), *This World—A Man in this World. Spheres of Educational Contradictions*. Warszawa 2003.
Kapuściński R., Our fragile world. [In:] Domosławski A. (ed.), *Word Not for Sale. Talks Abort Globalization and Dissent*. Warszawa 2002.
Kwieciński Z., Education for global survival and development. [In:] Pacławska K. (ed.), *Tradition and Challenges*. Kraków 1996.
Kwieciński Z., Education in the face of differences and variability. An opening speech at IV all-Polish Pedagogical Convention in Olsztyn. [In:] Malewska E., Śliwerski B. (eds.), *Pedagogy and Education in the Face of New Communities and Differences in Uniting Europe*. Kraków 2002.
Kwieciński Z., Pluralism—A pedagogical chance or a burden? [In:] Kukołowicz T., Nowak M. (eds.), *General Pedagogy. Axiological Problems*. Lublin 1997.
Kwieciński Z., *Trails, Traces, Attempts*. Poznań-Olsztyn 2000.
Lipovetsky G., *L'empire de L'ephemere. La mode et son destin dans les societes modernes*, Paris 1987.
Łukaszewski W., *Chances of Personality Development*. Warszawa 1984.
Mead M., *Culture and Commitment*. Warszawa 2000.
Melosik Z., Education aimed at the world—Educational ideal of the 21st century. *Pedagogical Quarterly* 1989, no. 3/89.
Naisbitt J., *Megatrends*. Poznań 1997.
Nikitorowicz J., *Border—Identity—Cross-Cultural Education*. Białystok 1995.
Peccei A., *Future is in Our Hands*. PWN, Warszawa 1987.
Sarapata A., *Modernity of the Polish People*. Warszawa 1993.
Skarżyńska K., Egalitarian and non-egalitarian orientation. [In:] Reykowski J., Skarżyńska K., Ziółkowski M. (eds.), *Social Orientations as an Element of Mentality*. Poznań 1990.
Środa M., *Individualism and its Critics*. Warszawa 2003.
Sztompka P., *Sociology*. Kraków 2002.
Witkowski L., Meta-axiological premises of the educational reforms (the orientation triangle in culture). [In:] Pluta A. (ed.), *Sketches about Education and Culture*. Poznań 1999.
Wojnar I., Pictures of the world in the perspective of UNESCO. [In] Wojnar I., (ed.), *This World—A Man in this World. Areas of Educational Contradictions*. Warszawa 2003.

Ziółkowski M., Individual orientations and social systems. [In:] Reykowski J., Skarżyńska K., Ziółkowski M. (eds.), *Social Orientations as an Element of Mentality*. Poznań 1990.

Zysk T., Prodevelopmental orientation. [In:] Reykowski J., Skarżyńska K., Ziółkowski M. (eds.), *Social Orientations as an Element of Mentality*. Poznań 1990.

48 Information Overload and Human Information Needs

Marek Hetmański

CONTENTS

INTRODUCTION

Finding proper and valuable information and knowledge capable of controlling and managing our activities in everyday life, science, education, entertainment, or the public domain has become the focal point for practitioners (computer and information technology engineers, politicians, publishers, and educators) as well as theorists (cognitive and artificial intelligence scientists, psychologists, anthropologists, and philosophers). In the age of "the information turn," the problem of how to search, retrieve, process, and convey information in order to realize one's own practical interests and cognitive needs and, subsequently, satisfy institutional, sociocultural demands and standards is of particular importance. Immanuel Kant's (1998) three questions—"What can I know," "What must I do," and "What may I hope," stated in the century of complete, certain, and true knowledge ideally formulated by Newtonian physics—remain a challenge and demand new answers. Living in the decades of rapid scientific and technological progress—when ideas of complexity, nonlinear and dynamic chaos, as well as epistemological concepts of bounded rationality, unpredictability, and uncertainty, disclaim, or at least weaken, the previously accepted viewpoint—we have to focus our attention on the phenomena through which these questions are manifested.

Kant's last question—considered in the context of unpredictability and limited hopes to gain complete knowledge—does not seem to have an easy answer. Cognitive needs and wants like curiosity, searching for news, or more detailed research including searching for the truth are unquestionable. They are natural for all human beings. Owing to information–communications technologies, which amplify them instrumentally, these needs and wants are increasingly intense. The Internet, World Wide Web, net communicators, information forums, and e-mail discussion groups as well as educational institutions like long-distance education, permanent education, e-learning, or organizations like open universities—they all rely on and, consequently, radically modify human cognitive needs. Having such needs bodily and mentally rooted as well as possessing culturally inherited standards (norms) of such needs (a truth, trustworthiness, reliability, etc.), we find ourselves in a very specific and challenging, though epistemologically obscure, situation. Our cognitive needs are not ours; we tend to know, in fact, what is not needed; we take virtual pictures or simulations as real; being so deeply involved in technology, we seem to have lost our natural way of looking for really important information (Hetmański 2010). But, first of all, we are constantly forced into coping with *an increase in the volume of information*, and *a decrease in its meaning*, which is actually in line with the prophets of "the digital revolution," found neither among precipitous databases nor in global networked communication. Even in education, which is now almost completely shaped by technology, we feel lost and helpless. Pupils and students follow rules and instructions rather than their natural and spontaneous cognitive needs. Teachers, on the other hand, treat them as "natural born cyborgs" whose main task is just to process information given at the input. Even higher education has undergone similar changes.

EPISTEMOLOGY OF INFORMATION

Information gives humans an opportunity to acquire knowledge. However, this happens neither automatically nor always in the same way. It is only deliberately and selectively recognized signals that are absorbed and subsequently processed in the existing agent's cognitive structures (perceptual schemas, memory structures, intellectual frames as reasoning, conceptual thinking, or imagining) that actually constitute meaningful human knowledge. There are many bodily processes and mental operations on the one hand as well as different sociocultural circumstances on the other hand that determine when and why information becomes knowledge. Technological support, especially sophisticated computerized information technologies, is one of such factors. Nevertheless, they are not so crucial as it has been recently and repeatedly claimed in the theories of artificial intelligence and robotics (Hetmański 2005). Except instrumental information processing that takes place in the technologically advanced environments, real informational phenomena occur in human beings—their bodies, minds, and especially, the ways they interact with the environment being equipped with the instruments and tools. Information as such is a function of all these things and processes.

INFORMATION SEEKING AND INFORMATION NEEDS

The need for information is neither special nor the most important need of human beings. If "need" is an inner motivational state that brings about human thoughts and behaviors, the need for information (information need) is very closely connected with other cognitive mental states like believing, remembering, imagining, doubting, fearing, or expecting. It is an element in the structure of practical–cognitive attitudes that people assume in real-life situations. However, information need is a more complex psychosocial phenomenon than one could expect. As cognitive scientists have demonstrated, cognitive needs have to be separated from merely desires and wants, which, in fact, have no real impact on an agent's informational needs. Suggestive or intensive desire is not the same as a cognitive need; even an intentionally formulated plan or a strategy does not equal a cognitive need. One could want to, for instance, search all library catalogues looking for special data, whereas one's real information need would be just to browse or scan a concrete bulletin or journal. It implies that information needs depend on many objective as well as subjective circumstances. Generally speaking, information needs of an agent must be: (1) *instrumental* in reaching desired goals by using different means, tools, and instruments to accomplish these goals; (2) *necessary* in the sense of being important for the agent's vital (primary or secondary) life's needs, which may actually change one's course of conduct; and (3) *true* (not in the epistemological sense) with respect to the real (objective, not imagined) situations or positions of a person who seeks knowledge. Finally, another important criterion of a need, not a mere desire to be acquainted with more or less imagined or simply wanted things, is the agent's *behavior* in which the information needs are met. "Information seeking is the behavior that is the directly observable evidence of information needs and the only basis upon which to judge both the nature of the need and its satisfaction" (Allen 1996). Behavior such as acts, individual conduct, cooperation schemes, social attitudes, utterances, and so forth in which human needs are expressed as well as the instruments that effectively realize them are proper criteria that tell us when and why people have cognitive needs. A behavioral approach toward information needs offers an objective way of studying, measuring, and evaluating them. It also offers an opportunity, which is worth mentioning, to avoid false distinctions and misleading conceptions and theories.

There is another important distinction of needs versus *demands* that helps assess properly and justifiably real cognitive needs and avoids telling (suggesting to) people, which is not infrequent, what their "real" or "true" information desires or obligations are. To be more objective and critical, one should talk about the demands that people actually have, demands of which they are not fully aware, rather than rely on their declarations or apparently pseudo-cognitive attitudes. There are three strategies (approaches) one can find in the recent cognitive science literature that supply objective and verifiable methods on how to recognize cognitive needs and demands. The first strategy depends on distinguishing such features (states) of a particular realization of a need as unconscious, conscious, expressed in utterances, and finally, communicated. If the inquirer, after mental deliberations, puts in words or questions a cognitive need after receiving the answers from an information source/system, then he/she declares an actual informational demand; central to the entire process is the ability to communicate one's desires by negotiating questions and answers. The amount as well as the nature (the level of generality or incorrigibility) of the conveyed questions give information about the inquirer's information needs and real-life cognitive demands.

The second method, appealing to Shannon's concept of uncertainty (convergent with the thermodynamic definition of entropy), consists in juxtaposing the levels or states of information and knowledge in order to reach the final satisfactory state. People constantly compare current levels of knowledge against the goals they want to achieve and look for information that could reduce uncertainty involved in such situations. They are confronted with many states of anomalous, incomplete, or ambiguous knowledge and are forced or obliged to cope with them. Uncertainty is very closely connected with the feeling of anxiety, which is a strong and powerful cognitive motive. Conceived objectively (as a statistical measure of the states of information) as well as subjectively (as a experienced mental state), uncertainty in both cases is an originating state of any type of search manifesting in everyday life, mass communication, and entertainment, as well as more formalized research in science and domains of public affairs (systems of law).

If uncertain or incomplete data or the news brings about in the agent's mind a gap in previous or current knowledge, then a need to "make sense" occurs. Making sense, the third approach to information demands, is a type of compulsion experienced by the agent while confronting with the lack (gap) of sufficient information; unanswered, open questions; unsolved problems; and so forth. The search for sense (the search for meaning) constitutes what can be called precisely an information need. This need, however, arises in an individual's mind being at the same time strongly connected with or triggered by real-life situations. The emotional as well as cognitive side of "making sense" eventually shows that human information needs are complex and multilayered cognitive–practical undertakings.

The aforementioned approaches to the information needs and information demands bring to light a variety of strategies in which people reveal their real cognitive demands or wants and, by extension, a variety of research methods found in cognitive psychology. In the spectrum of these methods—from objective to subjective ones—one can recognize and analyze all human cognitive endeavors. As Donald Case (2007) holds:

> "The prototypical search for the Objective point of view is one in which there is a well-defined need to retrieve a specific fact to make a decision or solve a problem. From this perspective, information needs are thought to be relatively fixed. (...) In contrast, the Subjective pole represents the idealized view that many (an perhaps even the majority of) searches for information are prompted by a vague feeling of unease, a sense of having a gap in knowledge, or simply by *anxiety* about a current situation. This view does not deny that purposeful thought leads to information seeking, but rather emphasizes that humans are often driven to 'make sense' of an entire situation, not merely its component 'data', and that rational goals are often overstated. Under such a view, information needs are highly dynamic."

The results of research programs show additionally that people, no matter what the areas of their cognitive interests are or what the nature of processed information is, undertake specific, distinguishable activities that can be called information behavior. "I make a case for use of the term 'information behavior' as better suited to characterizing a broad range of relevant human behaviors dealing with information" (Case 2007). Information behavior is characterized thus by an agent's intentional, conscious, and effective interest in recognizing, processing, and conveying information gathered from the environment that helps to coordinate his/her activity in it.

INCOMPLETE INFORMATION AND UNCERTAIN KNOWLEDGE

Information conceived mainly as an objective, quantitative, and measured characteristic of events also has another, equally important aspect—subjective and qualitative, correlated directly with particular human agent needs and intentional demands, which decide what is information: information for an agent, information in respect to an agent's needs. The scientific perspective as well as the ordinary way of thinking both assume that information is a physical state or a natural commodity that is found in the real world, ready to be taken and easy to use. In short, they remain at the agent's disposal. However, this overall viewpoint obscures to some degree the real nature of information, especially its entanglement in human affairs. Information processing and managing imply vital questions, problems, and dilemmas worth discussing.

The most interesting and, therefore, practically important are the situations where information entails specific limitations and involvement with the particular agent's reactions toward the signals that carry (reveal) information. These reciprocal, objective–subjective interrelationships occurring in human experience are generally called knowledge but at the same time equivocally "incomplete" or "improper," "misleading" or "misinterpreted," as well as "overload" or "lacking" information. In fact, all these modifiers mean different things. According to the syntactic and semantic or pragmatic and communicational contexts of information, one can, in fact, distinguish between different meanings of information due to the nature of things (affairs) that generate it. All that happens, as Shannon and Weaver's (1964) mathematical theory of communication assumes, at the "source of information"—in the physical places and/or social situations where information is given, dispensed, and conveyed. These states of affairs take on, simply speaking, one of the two alternative statuses—they are or they are not, they occur or do not occur, they are conveyed or not, and so forth. In this sense, information as such exists or does not exist at all. Consequently, it is true or it is not true. The criterion of its existence is the agent's reaction toward it including selective recognition, measurement, processing, utilizing, and evaluating. Most of these processes and activities happen in communication undertakings. This ontological situation involves epistemological consequences in all areas of human knowledge and communication. As Fred Dretske (2008) asserts:

> "As the name suggests, information booths are supposed to dispense information. The ones in airports and train stations are supposed to provide answers to questions about when planes and trains arrive and depart. But not just *any* answers. *True* answers. They are not there to entertain patrons with meaningful sentences on the general topic of trains, planes, and time. Meaning is fine. You can't have truth without it. False statements, though, are as meaningful as true statements. They are not, however, what information booths have the function of providing. Their purpose is to dispense truths, and that is because information, unlike meaning, has to be true. If nothing you are told about the trains is true, you haven't been given information about the trains. At best, you have been given misinformation, and misinformation is not a kind of information anymore than decoy ducks are a kind of duck. If noting you are told is true, you may leave an information booth with a lot of false beliefs, but you won't leave with knowledge. You won't leave with knowledge because you haven't been given what you need to know: information."

In other words, humans cannot posses knowledge without previously acquiring proper, non-misleading information. The mental content of current beliefs or other mental states that the agent would have has nothing in common with the information occurring at "the source" and, finally, at "the destination." In other words, only information that is actually demanded with respect to a real agent's cognitive needs constitutes human knowledge and not merely elusive impressions or transient news. While human knowledge emerges from the process of transmitting the signals that constitute the messages conveyed through the communication channels, it is not reducible to that process. In fact, it appears in the receiver's reciprocal relations and responses to the sender's intentions. It is, generally speaking, an emergent product of complex, manifold, and multilayered information processing taking place between people and the world at large. This, however, does not appear automatically while conveying the signals, because information encoded digitally or in an analogue manner is merely a necessary, and not sufficient, element of human knowledge. Informational value of the external signals and signs that humans detect, process, and manage is deeply connected with the informational needs they have.

Really important cognitive situations are those in which human beings recognize, in a way that is not fully proper, albeit satisfactory, specific information that has meaning but is not epistemologically true. Such situations happen to people when they make decisions in a state of uncertainty on the basis of insufficient and incomplete knowledge. It takes place almost in all areas of human endeavors—not only in common everyday activity but also in business, management, and even, paradoxically, in scientific reasoning, estimating, or judging. Learning and education are not—which is worth mentioning and actually has far-reaching consequences—free of such ambiguous situations. Judging, evaluating, and behaving in a state of uncertainty are the subject of multidisciplinary research (psychological, anthropological, social, and political), which renders important and interesting results. This research analyzes different ways and methods people cope with not only the lack of proper and sufficient information but also redundancy and surplus of information as well. The methods used in such situations are strikingly rich and lead to more or less satisfactory results. This all proves that people do not need to posses full and complete knowledge in order to behave effectively and even rationally. Satisfactorily made decisions can be, at the same time, as valid as improper, accordingly to the rules of logical (rational) thinking.

Classic and standard psychosocial experiments done by Amos Tversky and Daniel Kahneman (1998) in the 1970s showed that people (laymen as well as scientists!) always rely on heuristics, which are not fully rational (algorithmic) inferences. Heuristics reduce complex cognitive tasks (of assessing possibilities and predicted values of perceived events) to simpler and useful strategies. They do not imply complete knowledge, relaying on opinions, incomplete information, and common sense or even intuition. Heuristics are a type of a biased cognition. Some examples include (1) insensitivity to prior probability of outcomes (due to additional, new information) that are not recognized in the stereotypes obscuring them; (2) insensitivity to predictability when people prefer forecasting in terms of favorableness of the description, ignoring prior, valuable, and sufficient information; and (3) illusion of the validity of the increasing redundancy among the correlated statistical data (which are the example of information overload), which, in fact, decreases accuracy of description and prediction. These heuristics are, as Kahneman and Tversky argue, unavoidable systematic errors that happen to people, especially when they are confronted with complex cognitive situations, mainly with information lack or overload. The researchers summarize their Cognitive Illusions Program with the conclusion that people rely on heuristics only because of their inevitable cognitive biases and illusions, which do not allow them to have full access to proper information. That is why more information and computation (if available) is always better for adequate and optimal estimations or predictions of complex situations. One should thus rely on heuristics only in routine decisions of little importance.

Gerd Gigerenzer is one of the cognitive scientists who conduct empirical research programs on how people cope with complex cognitive situations in the absence or overloading of information. However, he summaries his results in a different way than Kahneman and Tversky. People, he asserts, rely on heuristics due to the structure of the problem, not to their cognitive inclinations. In concrete situations, people choose, more or less reasonably, mostly intuitively, appropriable heuristic methods and strategies—"fast and frugal heuristics" that exist in the human mind as an "adaptive toolbox." Besides, relying on heuristics is not an error. It happens even in serious and important cognitive situations such as scientific research or medical statistical diagnoses. Good decision making or problem solving requires ignoring part of the available information and performing less complex estimations. Human beings have evolved, Gigerenzer (2008) admits, as "natural statisticians" who are rather good at simple, noncomplex tasks. Violations of logical rules are not cognitive illusions, but they are a manifestation of practical, bounded rationality.

> "The adaptive toolbox contains the *building blocks* for *fast and frugal heuristics*. A heuristic is fast if it can solve a problem in little time and frugal if it can solve it with little information. Unlike as-if optimization models, heuristics can find good solutions independent of whether an optimal solution exists. (…) Heuristics work in real-world environments of natural complexity, when an optimal strategy is often unknown or *computationally intractable*" (Gigerenzer 2008).

The main feature of such heuristics is looking for simple and discrete data and information because human beings have been biologically designed to cope with obvious, transparent, non-apparent situations. Even in complex, instrumentally mediated environments (not to mention the very recent networked, computerized systems functioning in the information society), people are still furnished with very simple conceptual tools useful in acquiring and producing even the most complex and sophisticated systems of knowledge. "Good decision making in a partly uncertain world requires ignoring part of the available information and, as a consequence, performing less complex estimations because of the robustness problem" (Gigerenzer 2008). In other words, even if confronted with the complex, obscure practical–cognitive situations, humans act surprisingly effectively relying on their natural faculties coping with uncertainty found in their environments. Neither lack nor surplus of information prevents them from fulfilling their real information needs. It appears, from the evolutionary perspective at least, that people are relatively well equipped with natural tools to manage such situations. However, they are frequently confronted with new intellectual situations. Looking at this from the civilization perspective, one can admit, nevertheless, that functioning in the technological environments, especially when people are confronted with the overload of signals, signs, the news, pictures, as well as models or simulations (simulacra), brings about new practical problems and epistemic dilemmas.

INFORMATION OVERLOAD

What, in fact, is "overload of information," as it is commonly and misleadingly used (and misused)? It is not a single phenomenon that could happen everywhere and anytime in the same way. Nevertheless, components of information technology, that is, hardware, software, networked systems, and so forth, are standardized and work (until they fail, which happens quite frequently) following an algorithm pattern. People use them differently for different purposes and in accordance with their computer literacy skills. As these skills differ significantly depending on age, training, gender, cultural standards, and so forth, computerized tools and systems do not guarantee the one and only effective means of access to information. It is one of the reasons for the occurrence of excess of information, and this is why people feel overwhelmed, if not threatened, with it. Interestingly enough, information overload not only is brought about through automatically incoming information, which is not welcome by users, but also, is caused by a lack of proper skills to manage the information systems. In short, this type of overload is due, not by accident, to the agents' inability to utilize the possibilities that modern technology offers and not to information technologies as such. As a matter of fact, this subjective factor is connected with the objective one—the automatic, exponential growth of digitally coded bytes; both are responsible for problems and dilemmas resulting from the phenomenon under analysis.

Information overload should be analyzed in three aspects: (1) its *reasons*, subjective as well as objective causes; (2) *mechanisms* that generate and govern it; and (3) immediate and direct *results* as well as secondary and indirect *consequences* caused by it. They, all together, enable us to explain the essence of the phenomenon in question.

Different factors cause certain agents (or systems) to struggle with the surplus of information. When signals and signs coming from many sources are not properly processed and managed (e.g., lack of effectiveness, inherent difficulty, low speed, high costs, etc.) by an agent and where information is redundant, the effect of overload appears. Overload is then a quantitative phenomenon. A similar situation occurs where signals do not match the information acquired or where prior expectations of the agent are not met. Consequently, problems with absorbing information come to the surface. However, results of information overload may *differ very much* because information absorbing, processing, or managing does not lead to the same consequences. There may be results of such information conceived as negative or positive, which is due to the role it plays in the structure of knowledge. If the general structure of knowledge—the sensual and conceptual schemas and frames, memory, imagination, and linguistic skills, which the agent is supplied with—is rich and functionally effective, then it is unlikely for information overload, considered at least from the subjective perspective, to happen. Nevertheless, any dysfunctional disturbance of this structure could disrupt smooth and easy absorption of information. Besides, the phenomenon in question largely depends on the place (library, school, company, service agency, communications operator, etc.) where transformation of information into valuable knowledge takes place.

It follows that one can distinguish a few different possible reactions and responses to overload, as defined above, namely, *omission*, *error*, and *escape*, which are, in fact, dysfunctional (they consist in failing to process some of the incoming and input information, including giving it up entirely). Others include *queuing*, *filtering*, and *approximating*, which are, this time, maladaptive. They consist in delaying (suspending) the inputs with the intention of catching incoming information later, processing information selectively, or lowering the criteria and standards of absorbed and processed signals and the news. The latter responses and attitudes are the errors that people make not as deliberately as they do in the former cases, where they less consciously but, nevertheless, still dysfunctionally (without any chance to succeed) try to cope with this difficulty.

Generally speaking, information overload does not always evoke the same subjective feelings in the agent. It may happen that he/she experiences, in such cases, not only anxiety (according to Abraham Maslow's (1954) statement that "we can seek knowledge in order to reduce anxiety and we can also *avoid* knowing in order to reduce anxiety") but also other, more positive and constructive mental states and feelings. As Case (2007) aptly puts it: "We often think of information as *reducing* anxiety, but such is not always the case." People do not always experience the mentioned effects negatively because they very often adapt in a smart way to having too much information and then take a more positive stance toward it. Living among a myriad of signals, images, the news, and other information that bombards people for a long time and with the same intensity makes them (especially the young) immune to them, changing the way they experience the world.

As psychological experiments, comparative cross-cultural studies, and theoretical analyses show, adequate and reasonable qualitative estimation of the situations in which people feel cognitive and intellectual discomfort, called "information overload," is relatively difficult. Apart from the technological circumstances, at the core of things lie an agent's attitudes and skills as well as his/her networks of connection with many types of informational environments (conceived as a cybernetic model of interrelations between inputs and outputs). In substance, only then can external circumstances codetermined with subjective factors constitute human cognition, learning, knowing, and communicating overload.

CONVERGENCE BETWEEN INFORMATION AND MINDS: TWO STUDIES—MANUEL CASTELLS AND JEROME BRUNER

Education is the place where the above-mentioned processes occur and undergo different and not-always-welcome changes. Considered functionally and not structurally (as a process and not a state), education is a distinctive phenomenon in which emergence of acquired knowledge accomplished throughout different types of information processing (perceiving, understanding, memorizing, imagining, speaking, etc.) includes all the human cognitive abilities and activities (training, learning, educating, specializing, etc.) and their objectified and externalized results (knowledge-how, knowledge-that, inference, language, formal knowledge, science, etc.). During the course of long-term education, humans are still confronted with many obstacles and challenges worth considering, apart from other interpretations, epistemologically. In order to be precise and concise, one can say that the education taking place on the all levels (elementary, secondary, higher, etc.) entangles its subjects and agents into long-lasting, multiple, repeated, and never-ending informational processes. Educational institutions like schools and universities provide a proper environment and material equipment, especially and lately, technological peripherals—ubiquitous computers that have dramatically changed our lives.

There is a widespread viewpoint, established by multimedia companies as well as psychologists, educators, and futurologists, that for the last two or three decades, we have been living in the age of *hypertext*. This effective and pervasive techno-cultural phenomenon—the complex text consisting of a combination of such elements like pictures, words, pictographs, sounds, and so forth—gives us unimaginable possibilities of experience, cognition, and knowledge. Hypertext and hypermedia are more and more perceived as an ideal model (symbol) of the essence of our civilization.

> "[T]he post-war challenge of managing information overload, a model of a mind as a web of trails and associations, and a concept of non-linear writing then extended to a freely accessible 'grand library' all of kinds of media, finally led us to the concept of hypermedia. This vision of the potential of the hypertext opens out to encompass an emancipatory configuration of human knowledge based in accessibility and manipulation through associative links" (Lister et al. 2003).

Not all practitioners and theorists of education concur with general image of modern civilization. Manuel Castells, a distinguished Internet researcher, doubts such an idea. It is too primitive a vision of culture and technology telling that the latter simply determines the former. A mutual convergence between human cognitive processes or ideas and information/communication technologies occurs in a much more complex and, thus, ambiguous way than the oversimplified viewpoint suggests.

> "Our minds—not our machines—process culture, on the basis of our existence. (...) Therefore, if our minds have the material capability to access the whole realm of cultural expressions—select them, recombine them—we do have a hypertext; the hypertext is inside us. Or, rather, it is our inner ability to recombine and make sense inside our minds of all the components of the hypertext that are distributed in many different realms of cultural expression" (Castells 2001).

Technology has consequences not only in the communication infrastructure but, more fundamentally, in the way people perceive, memorize, conceive, and imagine the world. This is why the main effects of information technologies happen in human minds, especially when people interchange the meanings of their experience in their social environments.

As Castells (2001) writes:

> "Our minds are not single, isolated worlds; they are wired to their social environment, so we process signals, and we look for meaning, according to what we perceive through the experience of everyday life. But in a social structure—the network society that induces structural individualism, and increasingly distinct social experiences, some of this shared meaning through practice is lost, so that areas of cognitive dissonance may grow proportionally to the extent of self-construction of meaning. The more we select our personal hypertext, under the conditions of the networked social structure and individualized cultural expressions, the greater the obstacles to finding a common language, thus common meaning."

In other words, relying only on subjective and individual impressions, no one could effectively process information even if technology would seem to supply him/her with unlimited possibilities.

Pure technological processing and managing of data, taking place without understanding, will confront people sooner or later with information overload.

To participate in the cultural hypertexts, as the author of *The Internet Galaxy* suggests, people must go up over individual experiences and images to the level of commonly shared and interchangeable universal meanings. It happens thanks to the mass communication based on, Castells (2001) says, the "existence of protocols of meaning." Culture and science very much help to achieve such goals, but learning and education contribute the most. But it is a pure possibility of the ideal—fruitful and effective communication based on the perfect commitment. There is also an alternative negative scenario—increasing misunderstanding and disagreement leading finally to the total social dispersion. "Lack of common meaning could open the way for widespread alienation among humans—everybody speaking a different language, built around his/her personalized hypertexts" (Castells 2001). But he does not believe such a course of happenings and makes a critical warning as regards the convergence between technology of information and human knowledge.

The same school of thought is presented by Jerome Bruner in his remarks on the "culture of education," which is determined, as he admits, more and more by the information technologies. His viewpoint reveals similarities with Castells' opinions about the consequences of improper use of technology in education. Relying too much on the technology may fail if we will not use previous, traditional "instruments" of education and learning like oral speech and writing, both going on in direct personal connections. Bruner confronts two models of education and, therefore, theories of the human mind—culturalism and computationalism. Both of them tend to explain how technology helps as well as disturbs us in acquiring information and building new knowledge. Bruner's sympathy is for culturalism because he doubts how people could develop properly their intellectual faculties relying only on the computerized technologies that supply them with enormously extended amounts of information, but do not give the tools to recognize where the meaning is (if any) among the information.

> "Like its computational cousin, culturalism seeks to bring together insights from psychology, anthropology, linguistics, and human sciences generally, in order to reformulate a model of mind. But the two do so for radically different purposes. Computationalism, to its great credit, is interested in any and all ways in which information is organized and used information in a well-formed and finite sense mentioned earlier, regardless of the guise in which information processing is realized. (...) Culturalism, on the other hand, concentrates exclusively on how human beings in cultural communities create and transform meanings" (Bruner 1996).

Education is a matter of long-term mutual interchanges of meaningful information taking place in institutions and between agents who have proper methods and procedures and, subsequently, agents who are open to be shaped and changed. The vital role in this process is played by the instruments, including technological ones. "How well the student does in mastering and using skills, knowledge, and ways of thinking will depend on how favoring or enabling a cultural 'toolkit' the teacher provides for the learner. Indeed, the culture's symbolic toolkit actualizes the learner's very capacities" (Bruner 1996).

REFERENCES

Allen, Bryce 1996. *Information Tasks: Toward a User-Centered Approach to Information Systems*. New York: Academic Press.

Bruner, Jerome 1996. *The Culture of Education*. Cambridge, Mass.: Harvard University Press.

Case, Donald 2007. *Looking for Information. A Survey of Research on Information Seeking, Needs, and Behavior*. Amsterdam: Academic Publications.

Castells, Manuel 2001. *The Internet Galaxy. Reflections on the Internet, Business, and Society*. Oxford: Oxford University Press.

Dretske, Fred 2008. Epistemology and information. In: P. Adriaans, J. van Benthem (eds.), *Philosophy of Information*. Amsterdam: Elsevier, pp. 29–47.

Gigerenzer, Gerd 2008. *Rationality for Mortals. How People Cope with Uncertainty*. Oxford: Oxford University Press.

Hetmański, Marek 2005. Artificial intelligence: The myth of the information science. In: M. Żydowo (ed.), *Ethical Problems in the Rapid Advancement of Science*. Warsaw: Polish Academy of Sciences, pp. 46–58.

Hetmański, Marek 2010. Technologized epistemology. In: M. Miłkowski, K. Talmont-Kamiński (eds.), *Beyond Description: Normativity in Naturalized Philosophy*. London: College Publications, pp. 112–132.

Kant, Immanuel 1998. *The Critique of Pure Reason*. Cambridge: Cambridge University Press.

Lister, Martin et al. 2003. *New Media: A Critical Introduction*. London: Routledge.

Maslow, Abraham 1954. *Motivation and Personality*. New York: Harper.

Shannon, Claude, Weaver, Warren 1964. *Mathematical Theory of Communication*. Urbana: The University of Illinois Press.

Tversky, Amos, Kahneman, Daniel 1998. Judgment under uncertainty: Heuristics and biases. In: idem (eds.), *Judgment under Uncertainty: Heuristics and Biases*. Cambridge: Cambridge University Press.

49 Development of Competition in the Polish Energy Sector

Sylwia Słupik

CONTENTS

INTRODUCTION

Energy sectors in many countries worldwide are currently undergoing dynamic and profound changes, related to gradual departure from monopolistic market structures, deregulation, and also privatization. All developed countries' governments, particularly in the European Union (EU), are restructuring this sector in order to decrease direct, state influence and introduce competition. Key objectives of the agenda on changes include increased economic efficiency in providing electricity and gas and obtaining lower prices for their consumers. Maintaining energy security due to timely and adequate investments in generating capacities also remains at the top of agenda. The very essence of changes in the energy sector is primarily based on functional decomposition of previously vertically integrated technical and technological structures: from generation through transmission to distribution and retailing, that is, unbundling, and on imposing a legal obligation onto owners of transmission or distribution grids to grant access to other market participants (third-party access [TPA] rule). Provided that energy generating and retailing are going to be under market regulation, while network collusion—constituting the real infrastructural natural monopoly—will remain in domain of specialized administrative regulation, the reform of the energy sector will take place from monopoly to market through regulation, being a certain substitute to a competitive marketplace. Energy is a fundamental commodity; therefore, its lack poses a threat to every country's security. Processes related to the energy sector are severely intrusive in terms of natural environment, and realization of the common market idea may only succeed when protection of competition legislation is secured. Hence, governments of member states exerting influence over energetics are transforming the radical regulation forms, being state ownership, into other forms complying with the notion of split supervision—into regulative and ownership functions.

The course and effects of liberalization processes in the energy sector in Poland, particularly in the first phase of those processes, are determined to a large extent by the regulative governmental activities, where the foreground objective is creation, protection, and development of competition (Szablewski 2011). However, introduction of the TPA rule to power networks in the EU and Poland stumbles across multiple hurdles and barriers along the way. The natural indicator of competition is the number of customers who have changed their electric energy and/or gas providers. It should be emphasized that among member states, the number of customers who swapped their suppliers has been constantly increasing. There are practically very small changes observed among entitled gas consumers and also in the group of small and medium enterprises and households. Among various barriers, the one responsible for lack of competition and the possibility of choosing an alternative provider would have to be the dominating position of entities—monopolistic energy enterprises and insufficient division of consolidated entities.

DETERMINANTS OF DEVELOPMENT OF ENERGY MARKET ENSUING FROM THE III ENERGY PACKAGE

European energy policy (An Energy Policy for Europe 2007), guaranteeing complete respect for member states' right to choose their own fossil fuel consumption structure in energetics and for their sovereignty regarding primary energy sources and in the spirit of solidarity between countries, seeks to realize the following three objectives:

- Improvement of competition in the energy market, owing to liberalization of gas and electric energy markets, and eventually, creation of homogeneous markets for those commodities.
- Guaranteeing energy security to the member states.
- Sustainable development of energetics, assuming rightful respect for natural environment, and counteracting climate change in particular.

Liberalization of the energy market in member states of the EU is simultaneously accompanied by the process of their integration. An endeavor aimed at building and developing a competitive energy market requires—both from European bodies as well as member states' bodies—a continuous search for and implementation of various legal regulations put in place to bring that objective to realization. The objective, which aspires to fulfill the supreme idea in the form of building a system, ensures undistorted competition throughout all the EU. Characteristics of the energy market being an economy's constituting part, whose functioning is based predominantly on infrastructural networks, in a significant manner determines the character of legal solutions within that sector. The EU has adopted so far three packages of directives set to reform domestic electric and gas energy markets and progress Europe forward in achieving the aforementioned objectives (Table 49.1).

Regulations following from the first and second market directives (Directive 1996/92/EC 1997 and Directive 2003/54/EC 2003) were not sufficient enough to achieve the common objective in the form of a liberalized, fully competitive, joint European electric energy market. Subsequent changes were brought about by the III liberalizing package, passed by the European Parliament and the Council of the European Union in 2009, comprising "new sector directives," which are the directive concerning the internal markets in electricity (Directive 2009/72/EC 2009) and natural gas (Directive 2009/73/EC 2009) and regulations on for example, cross-border exchange in electricity (Regulation No 714/2009 and Regulation No 715/2009) and establishing an Agency for the Cooperation of Energy Regulator (ACER) (Regulation No 713/2009.). Regulations previously put in place proved insufficient in the light of contemporary challenges posed by the necessity of diversification of energy sources and climate change. The main objective of the III package regulations is a competitive market offering to all consumers within the EU an undisrupted supply of electric energy, complying with ecological standards and with prices set in the market. The objective is set to be realized especially through

- Harmonization of improvements concerning national regulatory bodies and a guarantee of their independence
- Harmonization of tasks, obligations, and responsibilities resting on transmission system operators and a guarantee of their independence
- Establishing ACER and agencies for cooperation of operators (European Network of Transmission System Operators [ENTSO])

TABLE 49.1
Review of EU Directives Liberalizing the Energy Market

Directive 96/92/EC	Directive 2003/54/EC	Directive 2009/72/EC
• Unbundling of accounts • Market access under the TPA or single-buyer regulation • Seller selection right for consumers consuming >40 GWh/year within the EU • Obligation to designate transmission system operators (TSOs) • Very general requirements concerning the regulatory body, limited in essence to pointing to the necessity of creating such a body in member states	• Legal unbundling—obligation of organization isolation of DSO from integrated entities • Guarantee of TSO and DSO independence • Market access under the TPA regulation • Seller selection right for nonhousehold consumers since January 07, 2004 and since January 07, 2007 for all consumers • Appointment of supplier of last resort • Obligation to designate one or more bodies holding a function of regulator and quantitative enumeration of those bodies' competences	• Obligation to guarantee independency to regulatory bodies and to increase scope of their competence—i.e., in terms of responsibilities related to building a common energy market within the community • Tightening of the unbundling criteria • TSO–ownership unbundling or ISO or ITO (independent TSO with very restrictive criteria of its independence • Obligation of providing public service • Emphasizing the importance of consumer rights • Procedure of supplier change—maximum of 3 weeks • "Intelligent" metering systems at 80% of consumers' homes until 2020 • Establishment of ENTSO (European Network of Transmission System Operators) • Establishment of ACER

Source: TOE Report, Warsaw, Poland, 2010 p. 21.

- Common operating and network development principles, which are transparent and consulted on with market participants
- Introduction of additional regulatory tools, constraining anticompetitive practices of enterprises and increasing competition level
- Empowering energy consumers and others

Adopted solutions are featuring separation of network activity from generating and retailing energy, higher investment outlays on cross-border connections, collaboration and solidarity between member states, higher market transparency, and access to market information. Previous regulations required the energy transmission activity to be separated functionally and legally from the remaining activity of a vertically integrated enterprise. This requirement also concerns distribution from June 1, 2007, in the case of integrated enterprises serving an excess of 100,000 consumers (Nowak 2009). The new legislation assumes that a member state can choose between three methods of unbundling transmission and generation and delivery of energy: complete ownership unbundling, establishing an independent system operator (ISO), or establishing an independent transmission operator (ITO). In the first case, a vertically integrated enterprise is obliged to sell their transmission grids and create a separate entity, which will manage them. The second possibility—introduction of an ISO—would allow an enterprise to maintain the vertically oriented grid ownership rights, unless it is operated by an independent entity. And finally, the third option—an ITO—allows en enterprise to maintain integrity of generating, deliveries, and transmission of energy, but at the same time, it obligates them to observe precisely determined rules guaranteeing those two parts of the enterprise to act in practice independently.

Permitting three options of a transmission system operator's functioning instead of the option of complete ownership unbundling and exclusion of a distribution system operator (DSO) from this proposition at the current stage does not fulfill the objective of providing a level playing field in terms of access to network infrastructure, adequate investment, and security of supply under most economically effective conditions. Taking into account the lack of radical improvement in that area and the fact that the so-called third way emerged thanks to member states constituting the "blocking minority," in order to provide conditions for developing a common, competitive market on the European scale, one would have to reckon with the fact that the use of regulatory measures at disposal of the European Commission, especially the Directorate General for Competition, can be intensified.

The aftermath of implementing the III liberalization package is going to affect both the Polish energetics and energy consumers. Abolishment of internal boundaries within the EU is thought to be the consequence of the reform being introduced, aiming to enable free flow of electric energy and gas between community members—similarly as it happens in the case of commodities, services, or capital. It is synonymous with fiercer competition between energy companies in efforts to improve their efficiency and striving for customers. The client will be equipped with new instruments to protect his/her rights in order to render him/her an energy market participant with full rights. New provisions are intended to protect the most indigent energy consumers. First and foremost, member states are placed under an obligation to provide a high level of consumer protection through guaranteeing transparent terms and conditions of contracts being concluded, resolving disputes, and easy access to information about the energy supplied, especially through creating comprehensive information points. Moreover, the role and rank of the regulator would change, who should be authorized to make independent decisions, among others, by introduction of terms of office. Collaboration between a transmission system operator and its counterparts in other countries tightens, which should help to expand technically transmission networks.

The energy market in Poland was created on the grounds of an act from April 10, 1997—*Energy Law* (Polish Energy Law 1997)—guaranteeing equal rights to participants and the possibility of free access to the market, restricted only by necessity of satisfying technical and economic requirements. After the act was put into practice, energy companies have started to emerge, whose business was trading in electric energy, and consumers had their way paved to transmission services. Before the Energy Law act was put into practice, the Polish electric energy sector, already since 1990, had been restructured in terms of organization and ownership. The initial stage, aiming to demonopolize and transform the Polish sector, was elimination of the Energy and Black Coal Authority, and closely following was the formation of new organizational structure with a distinct divide in subsectors of generation, transmission, and distribution. Numerous amendments of the Energy Law act introduced legal grounds for operating and the scope of duties for transmission and distribution system operators. In accordance with that in the electric energy sector, separate subsidies formed performing functions of transmission system operators. In order to launch the segment of bilateral contracts, concluded between suppliers (producers and enterprises trading with electric energy) and consumers, a balancing market was created, which was managed by the operator of the transmission system. A transmission system operator (PSE Operator SA) constitutes a stand-alone legal person within the Capital Group Polskie Sieci Elektroenegetyczne Operator, which is also active in the generating subsector and is a party in long-term contracts concluded with certain power stations. The owner of PSE SA is the State Treasury. Moreover, the latest amendment stipulates that the end consumer has a right to choose from which sellers he/she would like to purchase gas or electric energy. Consumers will have the possibility to compare offerings from individual energy providers and will also be able to terminate an agreement, under which an energy company supplies gas and electric energy, without incurring additional costs. It also obligates energy sellers to draw up offers for sale of gas or electric energy, which should determine conditions of its sale, publish online and make available for public viewing at their headquarters information about offers for sale of gas or electric energy, and publish model contracts concluded with users of the system, especially model contracts concluded with end consumers. According to the amendment, the operator bears the compensatory responsibility for attributable disruptions in electricity supplies (Polish Energy Law 1997).

From the point of view of electric energy market, the most important seem to be the following, assumed by the legislator's amendments objectives:

- Change of gas or electric energy supplier procedure improvement through introduction of uniform rules for changing the supplier and imposing onto suppliers the obligation of publishing online information about sale prices and terms and conditions of use.
- So-called minimum trading quotes intended to prevent situations, where electric energy is sold by producers to a trading company operating within a single market group at a price significantly lower than the price, which is set beyond the group.
- Methane support system. Changes proposed in the act will allow those fuel gases to be covered by a support system as part of high-efficiency cogeneration.

Stipulated in the amendment changes are previous experiences related to functioning of the energetic sector. They are set to improve the security of energy supplies and more rational distribution of issue-related competences between various energy sector participants.

Propositions in the area of developing competitive fossil fuels and energy markets were enclosed in the document passed by the Council of Ministers in November 2009, "Polish Energy Policy until 2030," where future actions taken in favor of building a competitive energy market comprise:

- Implementation of the new electric energy market model, introducing for example, intraday market; markets for energy reserves, transmission rights, and generating capacities; as well as mechanism managing system services and reliability must-run generation.
- Facilitation of energy provider change, that is, by introducing nationwide (in Poland) standards regarding technical features, installation, and reading of electronic metering systems.
- Creation of conditions allowing shaping of reference prices of electric energy in the market.
- Optimization of operating environment for domestic energy-intensive consumers in order to prevent their products sold in worldwide markets from losing their competitive edge.
- Protection of the worst off consumer of electric energy from the results of increasing prices.
- Alternation of regulation mechanisms supporting competition in the gas market and introduction of market methods for shaping gas prices.

STATE AND STRUCTURE OF THE POLISH ENERGY MARKET

The present shape of the energy market is mostly influenced by demonopolization, liberalization, and privatization processes, which had commenced in Poland in 1997. Transformations taking place over years were heading toward implementation of market mechanism in this sector of the economy, aimed at obtaining an undisrupted functioning of the competitive marketplace, where competition between enterprises will have translated into a decrease in energy prices and improvement in customer service with the country's energy security being maintained. Despite multiple initiatives having been taken over, an excess of a decade mainly in the area of legislation, not all problems have found their resolutions. New challenges have also emerged, such as the way increasing demand for energy and supply are balanced, the latter being dependent on considerable investment outlays.

The situation in the energy market in Poland is determined by consequences of progressing consolidation in the electric energy industry and recentralization of trading in the gas industry, being detrimental for competition in the energy market. Despite the electric energy sector being diversified in regard to both business activity and ownership, the gas sector still remains fully monopolized. Such market structure, even with DSOs being isolated out of the equation, renders it impossible to have conditions favoring real competition coming into existence, and as a consequence, administrative regulation of a restrictive nature cannot be repealed in favor of competitive mechanisms.

The structure of the electric energy market is a consequence of the "Program dla elektroenergetyki" (program for energetics) passed by the Council of Ministers in 2006. Its realization fruited in the formulation of four vertically consolidated energy groups: that is, Tauron, Polska Grupa Energetyczna, ENERGA, and ENEA. Their joint market share in the energy-generating sector amounted to 62% in 2009. The three biggest producers had at their disposal over half of installed generating capacities and were responsible for 55% of electric energy produced. The largest share in generating subsectors belongs to Polska Grupa Energetyczna SA. The number of producers owning at least 5% market share remains invariable. Electric energy trading in the wholesale market is characterized with high concentration, especially inside vertically consolidated energy groups. In 2009, aggregate turnover among the aforementioned four groups reached 58.31% of the entire wholesale trading. High concentration level is confirmed primarily by all Herfindahl-Hirschman (HHI) indexes, whose value significantly exceeds the bottom limit of a high concentration level (1800). Although this indicator fell in value in 2009, it should be emphasized that the market remained highly concentrated, an indication of which is the retained 75% joint market share of the three largest entities (The President of the Energy Regulatory Office in Poland 2010) (Table 49.2).

Stock exchange transactions still remain fairly insignificant in Poland; hence, the dominating form of wholesale trading with electric energy is still constituted of bilateral contracts (over 90% of those sold in 2009 electric energy). The remainder of sales was completed in the balancing market and, to a lesser extent, in the *spot* markets. Polish Power Exchange trading, despite a 45% raise compared to 2008, amounted to 3.07 TWh. In contrast to the total energy consumption, in 2009, it constituted only 2.07%. Greater trading volume than the stock exchange characterizes the Platforma Obrotu Energią Elektryczną, where the trading volume was shaping in 2009 at the level of 4.36TWh, which constituted a 2.39% set against the domestic electric energy consumption (Energy Regulatory Office in Poland [ERO] data).

The structure of electric energy sales by producers was not subject to change, which would have allowed efficient functioning of the electric energy competitive market, despite termination of long-term contracts and realization of a public help program to cover the stranded costs.

The retail electric energy market, similarly to the wholesale market, did not record any considerable progress in implementation of competition. The chairman of ERO sustained the obligation of submitting tariffs for approval in respect of households (tariff group G), which does not entirely comply with the objective of liberalizing the electric energy market. Further, maintaining such a state of affairs not only has a negative impact on the development of competition in the Polish energy market but also can indirectly cause an increase in prices of services and consumer products.

TABLE 49.2
State of Concentration in the Polish Electric Energy Market

	HHI Index		
	Generating Subsector		
Year	Net Installed Capacity	Net Generation	Wholesale Market
2008	1592.6	1622.1	3632.6
2009	1617.6	1565.1	2850.6

Source: Urzad Regulacji Energetyki (URE; Energy Regulatory Office) based on data from Agencja Rynku Energii S.A. (ARE S.A.; The Energy Market Agency).

Similarly to previous years, participants of the retail market, besides end consumers (both in households and enterprises), include enterprises managing the distribution network (DSOs) and electric energy sellers (trading enterprises). The number of end consumers of electric energy remaining in demand amounts to around 16 m, of which, excess of 85%, is constituted by households. At the same time, the volume of energy sales to that group is not high and constitutes, in total, around 24% of total electric energy sales. In 2009, 28 enterprises were selling directly to end consumers, of which 18 are companies with domestic capital, whereas 20 sellers are associated with DSOs in terms of capital (The President of the Energy Regulatory Office in Poland 2010).

The highest share of electric energy sales is divided between 14 incumbent sellers, who, after the isolation of DSOs, remained rightful parties of comprehensive agreements (sales and distribution). They are holding the function of public sellers to households, which did not make a decision to switch energy providers. There are also approximately 20 active participants operating in the market, which do not originate from structures of former distribution companies. Approximately 200 other sellers are vertically integrated industrial power enterprises (of local range), providing distribution services apart from sales. The general number of entities holding a license to trade with electric energy is about 310.

The shape of the gas wholesale market was not subject to any significant changes in 2009. Close to monopolistic position of Polskie Górnictwo Naftowe and Gazownictwo SA (PGNiG SA)—an enterprise in majority owned by the state—results from the fact that they are primary gas importers into Poland, they have a 100% stake in gas extraction from domestic gas fields, and they are the sole owner of underground storage installations. The high concentration level in the gas wholesale market renders the share of active, independent market participants marginal, being approximately 2%. Those entities are purchasing gas predominantly from PGNiG SA. It is worth noting that the current structure of the gas wholesale market in Poland is increasingly deviating from the one functioning in other communities' countries, where the place for wholesale trading with natural gas is becoming stock exchanges or trade exchange hubs. It is difficult, then, to consider the gas market in Poland to be liquid, in a situation where trade transactions are exclusively made by bilateral contracts of long-term or indefinite nature. In consequence of the executed recentralization process of trading within GK PGNiG SA, small diversification of prices is notable in terms of wholesale and retail consumers. The characteristic of the wholesale market is completed by insufficient integration of the domestic transmission system with the systems of the European Community's countries and 100% reservation of transmission capacities by PGNiG SA at the outputs, resulting in inactivity in domestic intersystem exchange—independent of GK PGNiG SA—and foreign trading enterprises (The President of the Energy Regulatory Office in Poland 2010).

Similarly to the case of the wholesale market, it is difficult to put down year 2009 as a breakthrough year in retail trading in fuel gases. The retail market remains the market of one seller. The remainder of retail trading enterprises is predominantly focused on reselling natural gas purchased from PGNiG SA to end consumers. 2009 saw a 5% fall in sales volume caused above all by decreased demand for natural gas by the biggest industrial consumers, that is, nitrogen and petrochemical plants and other large consumers showing gas consumption above 25 million m3/year. The most prominent group of PGNiG SA's consumers are the households—97.5% of all consumers. Their contribution to the sales volume in 2009 amounted to 28%. The largest contribution to the natural gas sales volume by PGNiG SA was presented by industrial consumers—58.2%, and among those dominating were nitrogen plants, refineries, and petrochemical companies. Moreover, PGNiG SA sells gas to Gaz-System SA and PGNiG SA DSOs—for their own use and system balancing.

SITUATION OF END CONSUMERS IN THE ELECTRIC AND GAS ENERGY MARKET

As of July 1, 2007, domestic retail markets became completely open, from the legal point of view; hence, all European consumers can choose now their energy providers and take full advantage of existing competition. However, even though the growing number of consumers actively profit from the possibility of choosing the seller by either swapping providers or negotiating with a previous provider, a more favorable contract, still apart from several illustrious exemptions, the real activity levels in terms of changing providers in unimpeded electric and gas energy markets among small and medium users are very low (*Consumer on the Electricity Market* 2007).

Consumer activity in the electric energy markets in Poland manifested by using the right to change energy providers still remains very low. The share of large- and medium-size industrial entities, which have changed their providers, increased slightly, whereas the share of consumers in the G tariff group who have changed their providers is trace (0.007%). In total, by the end of 2009, there were 2599 consumers noted, including 1062 household consumers, who concluded a sale contract with a provider different from a trading company isolated from a vertically integrated enterprise operating within the DSO area, whose network those consumers are connected to (Table 49.3). The volume of electric energy purchased in that manner was higher in 2009 only by around 2% compared to that in 2008, which amounted to 12,920 GWh, so 11% of total supplies to end consumers. The current gas market structure in Poland results in zero rate of changed providers and, as a consequence, allows PGNiG SA to conclude so-called comprehensive agreements containing contractual provisions of sale contracts, providing transmission and distribution services, and also providing storing services (The President of the Energy Regulatory Office in Poland 2010).

Consumer passiveness can be attributed to a lack of attractive, competitive offers and the nature of those services. Nevertheless, it also reflects challenges with which regulation of competition is faced in unimpeded energy markets. Member states and national regulatory bodies have to provide transparent and straightforward procedures of changing their provider, in order to guarantee trustworthiness, which consumers need, and also carry out information and education activities.

The next important issue, for a regular consumer, is the impact that market liberalization has on electric and gas energy prices. Electric energy prices, offered to consumers who did not use their right to choose a seller, went up between the 4th quarter of 2008 and 4th quarter of 2009 by 31.7%. The biggest raise was recorded in prices for small industrial consumers, a raise of 39.3%, and the smallest for households, a raise of 24%. In case of consumers taking advantage of their right to choose electric energy providers, the price is determined in bilateral contracts (Table 49.4).

The highest raise in prices occurred in the group of small and medium enterprises showing low energy consumption. That group of industrial consumers was the least prepared for energy market opening and felt the highest raise in energy prices annually, over 30% In that group, especially among consumers belonging to the public finance sector, we have been observing the most intensive efforts aimed to change the previous provider. The experience shows that consumers stumble across

TABLE 49.3
Number of Consumers Who Change Their Supplier in the Polish Electric Energy Sector (State at the End of the Year)

Consumer Groups at Consumption Criterion (MWh)	Number of Consumers Who Changed Their Suppliers according to Energy Consumption			
	Big and Medium Industrial Entities and Small Enterprises		Households	
	2008 r.	2009 r.	2008 r.	2009 r.
>2000	56	232	–	–
50–2000	13	563	–	–
<50	16	742	905	1062
Total	85	1537	905	1062

Source: URE.

TABLE 49.4
Prices of Electric Energy in Poland, Applied to Consumers Having Comprehensive Agreements

Distinction		Electric Energy Prices		
		4th Quarter 2008 (zł/MWh)	4th Quarter 2009 (zł/MWh)	Change (%)
Consumers in General		202.53	266.83	31.7
Specifically	A tariff group consumers	189.64	245.42	29.4
	B tariff group consumers	204.61	273.05	33.4
	C tariff group consumers	215.36	299.91	39.3
	G tariff group consumers	195.37	242.2	24.0
	Including households	195.72	241.99	23.6

Source: ARE SA.

two fundamental barriers: lack of competitive offers issued by energy sellers and behaviors of DSOs noncompliant with independency rules. Taking into account the necessity of eliminating barriers constraining a consumer's ability to access competitive offers and also the fact that energy prices for households, purchasing power parity (PPP) adjusted, are among the highest in the EU; deregulation of prices in this segment should be preceded with procompetitive changes, showing distinctive and robust effects reaching both the retailers and wholesalers. In the gas sector energy, prices across the board remained regulated until the end of 2009. Compared to 2008, the average price of methane-rich natural gas delivery in 2009 fell by 3.3%, while for households—depending on the tariff group—the decrease ranged between 0.5% and 0.9%, for industrial consumers connected to the distribution network, 0.9% to 6.5%, and for industrial consumers connected to the transmission grid, 4.3% to 7.3% (The President of the Energy Regulatory Office in Poland 2010).

Energy prices can be problematic for some economically disadvantaged consumer groups and due to that fact in many countries, providers are obliged to offer to those consumer groups certain price discounts and limit the number of users being disconnected from the electric network. Implementation of market solutions and introduction of adequate regulatory procedures can lead to improvement of availability and standards of services provided both to lower-income consumers and remaining clients. However, it is going to be a long-term process, which has to be executed maintaining the reliability and quality of supplies.

The main barriers rendering it impossible for end consumers to benefit to a full extent from liberalization of energy markets are anticompetitive practices of incumbent enterprises, interested in maintaining the status quo, as well as lack of adequate market infrastructure, in a broad sense, comprising also the legal framework. Incumbent enterprises misinform and mislead consumers interested in changing the supplier or are piling up difficulties (e.g., set high technical requirements); as a consequence, the consumer would have to bear monumental outlays, often "consuming" an excess of expected benefits from changing the provider (*Consumer on the Electricity Market* 2007).

CONCLUSION

The underlying conclusion, which can be drawn from the conducted analysis in this chapter, boils down to the ascertainment that the course and effects of liberalization of network economic infrastructure sectors, especially in the preliminary phase of those processes, are determined to a large extent by the state's regulatory activity, whose prime objective is to create, protect, and develop competition.

Progressing hierarchical consolidation of the energy sector has led to the formation of a limited number of energy groups, showing great market strength. Freeing electric energy from the constraint of long-term contracts did not bring the expected outcomes in the form of increased competition, better market liquidity, and market transparency. Almost the entire volume of electric energy is being sold by means of bilateral contracts. Also, the competition in the retail market remains still very limited. Differences in prices offered by trading companies are not attractive enough to entice clients to swap their current providers. In relation to the gas sector, the assessment of its operating functionality is influenced by a lack of proved solutions, which would have systematically determined the target gas market model in Poland. A fundamental change is required in the way the Polish government approaches the gas industry, meaning abolishing the very strong—dictated by concern for guaranteeing an adequate security level—interventionist approach in favor of a strongly promarket approach.

An effective competition mechanism would have improved the productivity of enterprises' business activity, playing into the fuel gas consumers' hands, and also would have provided an appropriate—from the supply security point of view—level of investment outlays on network infrastructure development. Nonetheless, among positive changes in the electric energy market in Poland in 2009 and at the beginning of 2010, one can count the following:

- Introduced by the transmission system operator, changes in the balancing market (including the change in the price setting mechanism and enabled possibility of making transactions in the intraday market)
- Growing number of consumers being entrepreneurs, who benefited from the TPA rule—in the A, B and, C tariff group, an increase from 85 consumers at the end of 2008 to 1599 consumers by the end of 2009
- Development of the standard model of Generalnej Umowy Dystrybucyjnej (General Distribution Agreement)

The state of energy markets in the countries of the EU is showing multiple differences compared to the situation taking place in corresponding network infrastructure sectors in Poland. In terms of the implementation of the law in force and performed responsibilities and domestic regulators' activity in the sectors of electric and gas energy, Poland does not differ significantly from other EU member states. However, the actual state and scope of competition in those markets compared to other countries is not sufficiently satisfactory. The progress, however, in the implementation process of competition mechanisms in those areas of activity is not adequate. As far as the formulation of the energy market structure in Poland is concerned, two tendencies are shining through. The first are the efforts to introduce real competition to those sectors' segments of activity, where the introduction is feasible. The second is the tendency to create strength in terms of capital enterprises, the consequence of which is strengthening of position of companies already demonstrating certain supremacy. The European policy complying with passed directives definitely prefers the first model. Assuming that generation and commerce are functioning in line with market rules, transmission and distribution are supposed to function under natural monopoly conditions, supervised by the regulator. However, once multiple EU countries are subjected to scrutiny, a considerable concentration of capital is notable along with processes resulting in further strengthening of large entities in the market. Poland shapes its energy policy in terms of both electric and gas energy as a combination of those two tendencies. Currently, liberalization of the electric energy market is far more advanced than the gas market liberalization, where considerably higher entry barriers occur, topped with limited competition, but that state is predominantly caused by the monopolistic position of the Polskie Górnictwo Naftowe i Gazownictwo SA. Furthermore, a certain barrier standing in the way of creating competition in the electric energy market is still lack of consumers aware of the possibility of changing the energy provider. In the electric energy market, a legal separation of the network from trading activity has taken place in vertically integrated enterprises, which strengthened the transparency of the structure and market operating rules. Furthermore, eliminated in a systematic manner were previous development constraints of the wholesale market (e.g., long-term contracts, improved balancing market rules). The moment since energy markets were opened up on July 1, 2007, to all consumers, and purchasing rights were guaranteed to them was very important, as they constitute the necessary condition for the development of competition in the retail market. Also, the interest of the most disadvantaged consumer–household consumer is protected by the institution of public seller, who is obligated to conclude an agreement with the household consumer and provide to him/her energy supplies under the comprehensive agreement comprising sales and the service of distribution.

REFERENCES

Consumer on the Electricity Market. (2007). Warsaw: ERO Regulator Library.

Directive 1996/92/EC. (1997). The European Parliament and the Council of the European Union.

Directive 2003/54/EC. (2003). The European Parliament and the Council of the European Union.

Directive 2009/72/EC. (2009). The European Parliament and the Council of the European Union.

An Energy Policy for Europe. (2007). Brussels: Communication from the Commission to the European Council and European Parliament.

Nowak, B. (2009). Challenges of liberalisation—the case of polish electricity and gas sectors. *Yearbook of Antitrust and Regulatory Studies, 2*(2), 141–168.

Office of Competition and Consumer Protection. (2011). *Consumer on the Electricity Market*. Warsaw-Wroclaw: Office of Competition and Consumer Protection.

Polish Energy Law. The Act of 10 April 1997 The Energy Law, Journal of Laws of 2003 No 153, Item 1504, The Office of Sejm.

The President of the Energy Regulatory Office in Poland. (2010). *National Report to the European Commission*. Warsaw: Energy Regulatory Office.

Regulation No 713/2009. (2009). The European Parliament and the Council of the European Union.

Regulation No 714/2009. (2009). The European Parliament and the Council of the European Union.

Regulation No 715/2009. (2009). The European Parliament and the Council of the European Union.

Szablewski, A. T. (2011). The need for revaluation of the model structure for electricity liberalization. *Yearbook of Antitrust and Regulatory Studies, 4*(4), 201–223.

50 Benchmarking: The Best Practice for Implementation of PRME Principles

International Comparative Study

Agata Stachowicz-Stanusch and Anna Ptak

CONTENTS

INTRODUCTION

"We are responsible to employees, the men and women who work with us throughout the world. (...) We are responsible to communities in which we live and work and to the world community as well. (...) Our final responsibility is to our stakeholders. Business must make a sound profit." (Johnson 1943). The above citation is a part of the world-famous Credo written and published by General Robert Wood Johnson in 1943, the founder of the Johnson & Johnson Company. It shows both the presence and importance of business responsibility incorporated into the company's strategy, mission, and values, already in early 1950s. And this paper refers exactly to this strand of responsibility. Still, responsibility as a vocabulary expression is used by the authors in a wide sense, aiming to encompass diverse terms such as business ethics, sustainability, CSR, or responsible management. Nevertheless, it is rooted in the management context.

This paper aims to draw attention to responsibility in management education, as a mandatory condition for creation of a responsible future manager, who will have skills, knowledge, and particular features to conduct the business in a sustainable way. Of course, it is not a new topic in management, as presented in the very beginning. Actually, it goes back to the ancient ages. However, the nature and impact of responsibility on business faced a similar evolution as facilitation and performance of exchanges, which is central to business. The general idea of responsibility has never changed. But the approaches have changed. Nowadays, it is crucial for the success of the global business to gain young professionals, university graduates in management sciences, highly and appropriate educated, with well-developed managerial skills and a strong sense of business, economic, social, and green responsibility.

While the current approach lenses creation of a responsible future manager, this paper focuses on a relatively new, though popular idea of the PRME. This abbreviation stands for the principles for responsible management education and provides a helpful platform for all organizations in management education field worldwide. There are three key words in the paper's title that mirror the idea and content of the whole piece. It is benchmarking, the PRME, and study. Benchmarking was acknowledged as the best application for the PRME actions. Therefore, the next chapters will present various benchmarking approaches for implementation of the PRME principles. Eventually, the statement in the title referring to benchmarking as the best practice for implementation of the PRME principles will be presented and explained as applicable. One thing regarding benchmarking needs immediate clarification to avoid further justifications. Regarding the context of this paper, benchmarking will be considered as more in practice, tending to be more about sharing best practices than undertaking formal comparative measurements (Harvey 2004). Authors will look at benchmarking as a method of teaching an institution how to improve (UNESCO 1998). The last word that plays a main part here is study. It refers to the nature of the paper. Beginning with a short study on evolution of the responsible management education in a global aspect, authors focus on the examination of the state of implementation of the PRME practices among initiative's participants basing on reports on progress. This study entails a new research among selected universities from different countries, conducted for the purpose of this paper. Of course, its presentation is followed by detailed analysis of gained data to reveal results and finally to draw conclusions.

The structure of the paper remains consequent and transparent to the reader throughout the whole piece.

EVOLUTION OF THE RESPONSIBLE MANAGEMENT EDUCATION

The aspect of responsibility in management came into picture in the 1970s and 1980s in the United States. The country marked a changing attitude toward society and business already in the 1960s, while the economy developed. Not only national but also multinational organizations based in the United States were growing in size and importance. Ethical actions gained on importance while facing first big environmental damages, caused often by chemical industry. This led to a social response, also by business schools that incorporated social responsibility into their studies. In 1986, the Bedford Committee recommended to incorporate ethical standards in accounting education, and the American Assembly of Collegiate Schools of Business (AACSB) and the AICPA recommended ethics education for students, at the general, business administration level as well as in every accounting course (Haas 2005).

Numerous scandals in the 1970s and 1980s in the United States, racking at different government units, brought ethics into light. The term "business ethics" came into common use in the United States in the 1970s. In this sense, it is rooted in the academic writings, teaching, research, and publications (De George 2005). In a broader sense and more general usage, it referred to business scandals. International scandals like the Watergate or Iran Contra indicated the lack of ethics the executive level (Agacer et al. 1997). This tendency was also uncovered at lower organizational levels, including trading or fraud scandals. But it was not only the United States that faced such scandals. For instance, Japan witnessed a Recruit scandal with politicians and gangsters. The above happenings have only brought the problem of business ethics and corporate responsibility to regular citizens who responded strongly against such practices, and they have not been given responsible management education in the first place. Its origins reach back to ancient Greece, to actual origins of business, which means facilitation and performance of exchange. According to De George (2005), in a broad sense, ethics in business is the "application of everyday moral or ethical norms to business." Many of these norms, still currently in application, refer to religion and tradition like the Ten Commandments. Switching to philosophy, a similar long tradition can be observed. Already in ancient Greece, Plato was dealing in his discussions with justice and Aristotle with economic relations, commerce, and trade (Plato's *The Republic*).

Moving back to more contemporary times, in the past centuries, famous reformers like Luther or Calvin discussed trade and business, which finally entailed emergence of Protestant work ethic.

And almost 20 years ago, in 1993, the Parliament of the World's Religions made also a crucial step toward business ethics. It adopted a "Declaration of a Global Ethic." This declaration condemned abuses of the Earth's ecosystems, as well as poverty and hunger. Furthermore, it condemned abuses of the economic disparities that threaten many families (Parliament of the World's Religions, 1993).

Another aspect of business ethics is mirrored by introducing the citation of Johnson & Johnson Credo (Johnson 1943). It refers to incorporation of ethics in corporate culture and encompasses all activities toward compliance with ethical standards of a culture. Agacer et al. (1997) suggest correlation between ethics and increasing internationalization of business, giving as an example American managers who were stationing in a Third World country. They might have experienced cultural differences like significantly lower ethical standards than accepted at home country. That is why they could have been confronted with much lower ethical standards allowing, for instance, bribery or "under the table" amounts.

The beginning of the 21st century brought recent corporate accounting scandals back into limelight, which drew the attention on effective teaching. A study of McNair and Milam (1993) showed that just a little over three hours were spent on ethics in an accounting course. Ten years later, there were already new courses being offered on corporate governance, fraud detection, and professional responsibility. Of course, accounting is only one of the management fields that had to take a new route toward more responsible education. The idea of responsibility in education is an appropriate education and creation of a future manager, so that this person will act in a responsible way in a professional environment. This responsible way encompasses all aspects of social and business behavior that determines individual and organization activities and eventually their impact on global environment. This issue became much more complex than even 30 years ago because of globalization. Nowadays, the world functions as a huge network, and is often called as a "global village," because of a matrix of relations between its all members. In other words, every organizational activity impacts stakeholders or further environment in some way. Therefore, members of organizations must be aware of possible scenarios of their actions. Furthermore, they must be prepared (by universities) to take proper action. Last, but not least, they must be sensibilized, complying with global ethical standards and sustainability.

Nowadays, scale of global risks is at a level never witnessed before. They include environmental risks like climate change and the affiliated global warming, water scarcity, financial crisis, recession, and many more risks that have a direct impact on global economy and every-day life of an individual (Gitsham and Peters 2009). Gitsham and Peters (2009) see the necessity of transformational change. Management education and leadership development are the main actors in meeting this complex and global challenge. A former Global Board Member and Global Head of Human Resources at Unilever Andre van Heemstra said once that "leaders need to be able to introduce environmental and social criteria into strategic decision making from the start—not doing this is worse than stupid, it's reckless" (Gitsham). Both approaches address the striving toward a sustainable value. Cooperrider (2009) perceives sustainable value as the most important opportunity of the 21st century in terms of management education. Cooperrider finds it essential to enable young people to acknowledge their active participation in management history. Therefore, management education and development of a future responsible global leader play an important role in generating a sustainable value that lies behind the new global initiative, presented below.

Taking universal values and business into classrooms is the capacity of the Principles for Responsible Management Education (Ban Ki-Moon 2007). It is an initiative supported by the Organization of United Nations. The Principles are inspired by internationally accepted values such as the Principles of the United Nations Global Compact. The idea was officially introduced by the Global Compact Office in 2006, and it was launched one year later under the patronage of UN General Secretary Ban Ki-Moon. Academic institutions like AACSB, EFMD, Aspen Institute, EABIS, GRLI, and Net Impact co-convened the drafting process together with the UN Global

Compact. The PRME is a global call for business schools and universities worldwide to gradually adapt their curricula, research, as well as strategies and teaching methodologies to new business challenges and opportunities of the 21st century. Its goal is to promote corporate responsibility and sustainability in business education. "It aims to inspire and champion responsible management education" (PRME Web site). It provides a framework, which is a kind of a guide for participating schools to advance broader cause of CSR. It also strives to encourage academic institutions to incorporate universal values into curricula and research. The number of participating institutions is increasing every year.

The heart and the soul of the PRME are its principles. They were developed by the PRME task force consisted of sixty deans and scholars. Participating institutions regularly report on progress in introducing these principles into their strategies, values and cultures. There are six principles listed below, published at the PRME official website (www.unprme.org):

Purpose: We will develop the capabilities of students to be future generators of sustainable value for business and society at large and to work for an inclusive and sustainable global economy.

Values: We will incorporate into our academic activities and curricula the values of global social responsibility as portrayed in international initiatives such as the United Nations Global Compact.

Method: We will create educational frameworks, materials, processes and environments that enable effective learning experiences for responsible leadership.

Research: We will engage in conceptual and empirical research that advances our understanding about the role, dynamics, and impact of corporations in the creation of sustainable social, environmental and economic value.

Partnership: We will interact with managers of business corporations to extend our knowledge of their challenges in meeting social and environmental responsibilities and to explore jointly effective approaches to meeting these challenges.

Dialogue: We will facilitate and support dialog and debate among educators, business, government, consumers, media, civil society organizations and other interested groups and stakeholders on critical issues related to global social responsibility and sustainability.

Current participating institutions as well as joining institutions work on implementation of these principles at different levels. Some institutions start only with recommendations and first drafts of a plan for implementation. Others already put these plans into action, focusing on particular principles or sometimes all of them at once. Following chapters will address above presented principles in research studies, presenting different approaches of various participating institutions.

State of Implementation of the Principles

To assess the most up-to date state of implementation of PRME practices, two different reports have been examined. This resulted in creation of a new analysis based on the consolidation approach of the above-mentioned reports.

The first report that was analyzed is the one published by PRME and regards sharing information on progress (PRME 2010). It focuses on activities performed between 2008 and June 2010. It provides an overview for each principle, followed by detailed criteria groups that covered PRME members in their initiatives. The second report that was examined at this point is the Stachowicz-Stanusch's (2010) study of criteria for sharing information based on progress reports published by member organizations. It is also based on the reports on progress similarly to the first report, but with a slight difference in the analysis approach of initiatives.

The analysis of the above-described two reports resulted in a new consolidated comparison. However, the new comparison is not a 100% mix of them both. Before its creation, all similarities and differences in both approaches have been deeply examined. Therefore, it is rather an agreement than a supplementation. After a study of the latest reports on progress from 2011 and 2010 published on the official website of PRME, general criteria accompanied by detailed categories have been distinguished. The first principle is therefore described by criteria of faculty with establishment of PRME-related internal task force and development on responsible management, student initiatives like clubs and organizations, participation and/or organization of conferences, development of green and ethical campus, and development of skills through curriculum. Principle 2 focuses on current, new, as well as academic offering, which adds responsibility content. The third principle first goes to institutional solutions, like unit or person responsible for responsible management education, as well as procedures and norms. There are three key lenses of principle 4, namely, internal framework including research units, support for research teams, and research standards; research results with publications and seminars, conducted research, and grants and awards; and educational practices like e-learning and study tours. Main criteria for principle 5 include partnership with business, science, and other schools, government, students, and other organizations. The last principle is described by communication tools, best practices, and additionally collaboration with commercial, industrial, and agricultural networks, associates, student organizations, the media, and NGOs.

The examples gathered from the reports show that the most advanced initiatives focus on incorporation of values into curriculum as well as partnership and dialogue, very often handled as one group of common principles. Still, the state of implementation of PRME initiatives depends on the organization itself, including not only its budget, ideas, and plans but also practical features like the available infrastructure and climate conditions. For instance, relatively big campuses turned into more green, meaning environment-friendly and ethical. Another example is the case of Australian School of Business of University of New South Wales in Australia, which has planned water refill stations to be installed across campus last year. They would allow students, staff, and visitors to refill water bottles, at the same time reducing the need for a new plastic bottle. Royal Holloway of University of London in the United Kingdom has launched its Campus Community Garden, which encourages students to grow and eat their own fresh vegetables. On the other side of the globe, San Francisco State University in the United States went ethical by making a long-time commitment to social justice (Reports on Progress of the presented schools, 2010 and 2011: www.unprme.org).

As mentioned before, some members present only plans or even only general ideas for implementation of PRME initiatives without any concretes. Some schools follow all principles, while others choose particular principles to concentrate on from the start. Therefore, the implementation is very individual, and its performance is very much tailored to every school. However, it is observable that relatively big universities, meaning in high number of students, also abroad, with a big campus, infrastructure, and resources, of a rich culture in tradition, norms, and activities, and also internationally active via student exchanges or seminars, present more advanced implementation of PRME initiatives than other schools.

Of course, the new comparison is a compromise from the point of view of the authors. Further different consolidation approaches are not excluded, though not applicable for the purpose of this paper. The aim of the creation of this new comparison was to examine a current state of implementation of PRME initiatives as well as to gain a helpful tool for further research conducted among various schools worldwide, basing on the same pattern. Of course, the easiest way to get it was to choose the first report as a basis for further research. However, this approach could not reveal the complexity of implementation of PRME initiatives among different schools. Thanks to the analysis of both presentations as well as a deep study of the latest reports, a multidimensional image of the current state could have been delivered.

RESEARCH ON PRINCIPLES 5 AND 6

After a detailed analysis of the state of implementation of PRME activities among member organizations conducted in the previous chapter, this part focuses on the last two principles, namely, principles 5 and 6. These principles have been chosen for further research because of two main reasons. First of all, these two principles can be considered as one group of familiar activities. Principle 5 is about multidimensional interaction, resulting in various partnerships, whereas principle 6 focuses on dialog or, in other words, facilitation of communication between university and all its stakeholders. Secondly, both principles regard a very comprehensive set of related activities, targeting both immediate and further environment of a university or its department. As presented in the previous chapter, principle 5 was characterized by partnership with business, science, and other schools, government, and also locally; partnership with students in the form of cooperation with student organizations, support, and student exchanges; as well as partnership with other organizations in the form of joint projects, for instance. There is always a room for other related activities that are not included in the above classification to be implemented by universities.

Similarly, the last principle can be described by a multiple set of activities regarding communication tools like featured events or use of renewable energy and emission reduction. Further activities include best practices in form of contributions to society, as well as norms and related policies. Correspondingly to the former principle, also the aspect of collaboration with commercial, industrial, agricultural networks and associations, student organizations, media, and NGOs needs to be taken under consideration. Also here, there are organizations that can present other related activities that are not included in the above classification.

Three universities from different countries have been selected for this particular research. None of these universities is a current member of the PRME initiative. The purpose of setting these universities together was to examine how also nonmembers deal with responsible management, exploring similarities and differences in particular approaches. This university trio includes the University of Kassel from Germany (www.uni-kassel.de), the University of Otago from New Zealand (www.otago.ac.nz), and the Silesian University of Technology from Poland (www.polsl.pl). All universities or their departments have been examined due to criteria characteristic for both selected principles 5 and 6. Table 50.1 reflects PRME-related activities of three selected universities. Information included in this comparison was gained from universities' official Web sites. The combination of this set of analyzed universities was not random. On one hand, there are two schools, geographically and culturally close to each other, while one school is established thousands of kilometers from these two, in a totally different environment. On the other hand, each school is unique, in terms of tradition, educational focus, and internationalization.

Some of the cells were left blank in the table. This means that there was no information found regarding a related criterion. For some activities, specific examples could be found, whereas for other programs, only general associations were available. Therefore, this table must be analyzed globally regardless of these kinds of differences.

As presented in Table 50.1, appointing industry professionals in external boards is common both in Germany as well as in New Zealand, which is closely related to various business partnerships, including also further cooperation with a university's alumni. Partnership with students is more common among all selected schools. They include various forms like career offices, cooperation with student organizations, and popular exchange programs.

Table 50.2 shows interesting examples among the least described categories. This includes best practices, for instance. Among three selected universities, only one school presented a concrete program related to complying with norms and standards. In this case, it was about plagiarism policy, in the direct relation to responsible management. The issue of plagiarism may not be the most "trendy" but for sure it is "up to date."

Both tables mirror various activities that relate to different target groups of each school. They focus on the internal stakeholders of the school like students and staff, as well as on external stakeholders like business, other schools, government, or other organizations. The presented set of examples is a proof for implementation and incorporation of the "responsible" quantum into the strategy of a school that is not an active member of the PRME initiative. However, it is aware of the management responsible education and acts toward its development or enrichment at home. These results entail the question that must be put here: Could each of these schools be better off using international benchmarking for implementation of PRME-related activities?

TABLE 50.1
Principle 5: Comparison of PRME-Related Activities

		Example of a Program/Activity		
		University of Kassel (Germany)	**University of Otago (New Zealand)**	**Silesian University of Technology (Poland)**
1	Partnership with business			
	Responsible unit/person			Center for Innovation and Technology Transfer
	Appointing industry professionals	Viessmann Werke GmbH & Co. KG, SMA Solar Technology AG and others—10% shareholder—impact on strategic development of academic offerings	External Board of Advisors including business professionals (participation in the school's activities)	
	Collaboration with departments dealing with CSR	Through meetings/seminars, e.g., "CSR in einem Handelsunternehmen – ein Praxisbeitrag", Otto Group, Social Compliance Division Manager Sybille, Duncker	Through internships	
	Executive education	Study programs also for candidates with at least 3 years professional experience	Executive programs (e.g., MBAs)	Bookkeeping courses (involving, e.g., tax consultants)
2	Partnership with science and other schools	Technische Universitat Munchen, Germany, SAP University Alliance	Partnering for Innovation—joint initiative between the University of Otago, the University of Canterbury, and Lincoln University	Ministry of Science and Education in Poland (Research Department)
3	Partnership with (local) government	Stadt Kassel—IWU (International Winter Universitat) Stipendium	Dunedin City Council (financial support for the Centre of Entrepreneurship), focus on Maori Business (partnership with government)	President of Zabrze` (city of location of the Silesian University of Technology, Management Department)—Management Forum
4	Partnership with students			
	Cooperation with student organizations	UniKassel Trasfer (e.g., Bewerbungstraining)	Otago University Students' Association	Student career office
	Awards and other (financial) support for students	Studentenwerk (Service for students)	Career Development Centre	Student government
	Student exchange	Erasmus programs International Office—Go out! Study worldwide! (students inform other students)	Otago Global Student Exchange	Erasmus programs

(continued)

TABLE 50.1 (Continued)
Principle 5: Comparison of PRME-Related Activities

		Example of a Program/Activity		
		University of Kassel (Germany)	University of Otago (New Zealand)	Silesian University of Technology (Poland)
5	Partnerships with other organizations			
	Joint projects	Loewe—Landes Offensive zur Entwicklung Wissenschaftlich-okonomischer Exzellenz	Audacious—student support program (for University of Otago and Otago Politechnic students—help to start and run their own business)	Technopark Gliwice (High-Tech organizations)
	Awards and other (financial support)			
6	Other (not distinguished above)	EU programs (e.g., social funds)	Otago Chamber of Commerce (academic staff)	Academic Incubator of Entrepreneurship European Programs Office

Source: Information published on the universities' Web sites.

TABLE 50.2
Principle 6: Comparison of PRME-Related Activities

		Example of a Program/Activity		
		University of Kassel (Germany)	University of Otago (New Zealand)	Silesian University of Technology (Poland)
1	Communication tools			
	Featured events (e.g., symposia, conferences, websites)	Seminars with professionals dealing with CSR, e.g., "Ethische Grundsatze als Grundlagen von CSR"—Praxisbeispiel Korruptionsbekampfung with Daniel Kronen, Senior Manager Integrity, Corporate Legal and Compliance, Siemens AG	Public lectures (e.g., on occupational health), Annual PIM Conference 2010 (Partnership in International Management)	Management Forum, The Week of Entrepreneurship
	Use of renewable energy, emission reduction, etc.	Use of ecological products in canteen (Bio-Mensa)		Competitions: "My idea for environmental protection"
2	Best practices			
	Contributions to society	Learnsoftware "PT Cards" designed by students of University of Kassel for pupils to learn English words with help of cell phones, KinderUni (KidsUniversity)	Shave 11—raising money for The Leukemia and Blood Foundation (LBF) as a part of the Shave for a Cure appeal	Open Day to Girls

(*continued*)

TABLE 50.2 (Continued)
Principle 6: Comparison of PRME-Related Activities

		Example of a Program/Activity		
		University of Kassel (Germany)	**University of Otago (New Zealand)**	**Silesian University of Technology (Poland)**
	Others (e.g., norms, policies regarding misconduct)		Plagiarism policy (including consequences)	
3	Collaboration with			
	Commercial, industrial, agricultural networks, associates	AGIL Arbeitsförderungsgesellschaft des Landkreises Kassel	Enterprise South Islands (Partnering for Innovation)	Silesian Green Power Vehicle (built by students of Silesian University of Technology)
4	Student organizations	DGB Campus Office (job advisers)	OUSA—Fairtrade Easter Egg Hunt	Student Government, Student Science Clubs
5	Media	Kassel University Press GmbH	University of Otago on iTunes	YouTube, Facebook, newsletters
6	NGOs	KarolaPlassmannBahlStiftung (City Architecture)	Solar Action—the New Zealand Renewable Energy Society (Energy from Waste, Public Forum)	
7	Other (not included above)			EU structural programs

Source: Information published on the universities' Web sites.

BENCHMARKING: CONCEPT OF A NETWORK

The previous chapter presented examples of the PRME-related activities of selected schools as well as various factors, sometimes common and other times very unique, that impact these activities. The presentation of the comparison in Tables 50.1 and 50.2 entailed a question left without an answer. This question was: could each of these schools be better off using international benchmarking for implementation of PRME-related activities? The answer to this question is positive. Recalling the chosen definition of benchmarking, already some explicated examples are the proof to this answer. Looking back at activities related to partnership with students, all of the selected schools run programs of student exchange basing on some chosen international concepts that work. This is an example of a well-functioning network of relations that enables and encourages benchmarking. Becoming a part of this network, like Erasmus exchange program, that has clear structure and frame and that must be only accepted by a new member and complied with, could be considered as a type of benchmarking. Having a closer look at these three schools, there are differences in environments, especially locally. And still, various types of networking in the form of different partnership, for instance, enable this lifelong learning of a university, which then succeeds in more responsible education. A network is a kind of complex interdependence of its members. Its membership is a privilege with some constraints but first of all with opportunities. If these schools exchanged information about performed and planned activities related to responsible management education, they could profit from others' experiences, knowledge, and prognosis. And this could result in reduction of cost related to research, test programs, and possible failures or mistakes, which also means that time is needed for adaptation and implementation of new programs. Therefore, benchmarking does seem to be a good option for organizations that operate in a dynamic, local, regional, and international environment. This is exactly the environment in which all of the selected schools are situated.

PRME initiative by gathering and encouraging to join educational organizations from all over the world created a network that is based on sharing of information on progress. Of course, principles are its essence, but they are only directions for further development. The real power of this initiative is the ability and willingness to share information. Members not only express their engagement with responsible management education in terms of compliance with distinguished principles but also show their ideas, programs, and plans. This established and growing database accessible for everyone is an opportunity to learn from other participants' experiences.

CONCLUSION

Whitehead (1929) claims that education must have a certain rhythm. And it includes three phases, namely, romance, precision, and generalization. Romance refers to purpose, value, and the pure love of learning. It is when a student enjoys the freshness of a certain discipline. This involves feelings and emotions, both positive and negative. The middle stage, precision, concerns the development of a specialized knowledge. The last phase is generalization, or in other words, application of what has been learned and its creative modification into something new. It refers to fruition, meaning designing, co-creating, and anticipating. Education needs this rhythmic nature to achieve goals effectively (Riffert 2005). Cooperrider (2009) develops this idea into a new educational approach: management schools as design studios.

This approach corresponds very well with the idea of benchmarking as a helpful tool for implementation of PRME practices. First of all, benchmarking is about sharing best practices in order to grow as an organization. This emphasizes the necessary evolution of various benchmarking approaches depending on changing circumstances and preferences. And this involves creativity. Second of all, it pictures management schools as places that not only encourage creative application of acquired specialized knowledge but also foster all senses toward sustainable value. Benchmarking is not a copy–paste option. It entails a certain level of adaptation, and therefore both creativity and responsibility. And exactly these two values could make benchmarking work for every management university in the world to create a responsible manager.

REFERENCES

Agacer, G., Vehmanen, P., Valcarcel, L. (1997). Business ethics: Are accounting students aware? A Cross-cultural study of four countries. *Electronic Journal of Business Ethics and Organization Studies* Vol. 2, No. 1.

Alstete, J.W. (1995). Benchmarking in higher education: Adapting best practices to improve quality. ASHE–ERIC Higher Education Report No. 5.

Armitage, A. (2010). Learning to dialogue: Towards a critical pedagogy for public finance and accounting management practice. *Journal of Finance and Management in Public Services* Vol. 9, No. 2.

Ban Ki-Moon (UN Secretary General). Closing remarks at the 2007 Global Compact Leaders Summit on 6th July 2007.

Benchmarking in Higher Education—Findings of a two-year EU-funded project. (2008). ESMU (European Centre for Strategic Management of Universities).

Business Week, March 9th 1992, pp. 67–69.

Cameron, K., Dutton, J., Quinn, R. (2003). *Positive Organizational Scholarship. Foundations of a New Discipline*. Berrett-Koehler Publishers, Inc.

Cooperrider, D. (2009). Learning methodologies for the future of management pedagogy of design. PRME Official Web site.

Davis, K. (1960). Can business afford to ignore social responsibilities? *California Management Review* Vol. 2, pp. 70–76.

The Declaration of a Global Ethic. (1993). Given at the *Parliament of the World's Religions*. Chicago.

De George, R.T. (2005). A history of business ethics. Paper from the "Accountable Corporation" Conference, Markkula Center for Applied Ethics.

Escudero, M. (2007). Business schools to advance corporate citizenship through the principles for responsible management education. UN Global Compact, News & Events Archive.

Gitsham, M. (Lead author) and supporting authors. *Developing the Global Leader of Tomorrow*. Ashridge & EABIS. www.unprme.org.

Gitsham, M., Peters, K. (2009). Developing the global leaders of tomorrow. *EFMD Global Focus* Vol. 3, No. 1, pp. 58–61.

Haas, A. (2005). Now is the time for ethics in education. *The CPA Journal*.

Harvey, L. (2004). Analytic quality glossary. *Quality Research International*.

Humphries, M.T., Dyer, S. (2005). Transformation through critical management education. *Journal of Management Education* Vol. 29, No. 1. pp. 169–195.

Johnson, R.W. (1943). Our credo. Johnson & Johnson Company Website. www.jnj.com.

McNair, F., Milam, E. (1993). Ethics in accounting education: What is really being done. *Journal of Business Ethics*.

Milne, M.J., Tregidga, H., Walton, S. (2005). Playing with magic lanterns: The New Zealand Business Council for sustainable development and corporate triple bottom line reporting. New Papers on Higher Education. Studies and Research. Benchmarking in Higher Education. A study conducted by the Commonwealth Higher Education Management Service. UNESCO, Paris, 1998.

PRMESIP 1st Analysis Report. Sharing Information on Progress. (2010). www.unprme.org.

Riffert, F.G. (2005). *Alfred North Whitehead on Learning and Education: Theory and Application*. Cambridge Scholars Press, p. 35.

Stachowicz-Stanusch, A. (2010). Good practices for implementation of principles for responsible management education framework—Research results. *Organizacja i Zarządzanie* Vol. 4, No. 12. Wydawnictwo Politechniki Śląskiej, Gliwice.

Whitehead, A.N. (1929). *The Aims of Education*. The Free Press.

Working Paper. Department of Accountancy and Business Law, University of Otago, Dunedin, New Zealand.

Yasin, M.M. (2002). The theory and practice of benchmarking: Then and now. *Benchmarking. An International Journal* Vol. 9, No. 3, pp. 217–243.

51 Impact of Economic Crisis on HRM Practices
Evidence from Poland

Janusz Stružyna, Magdalena Majowska, and Tomasz Ingram

CONTENTS

INTRODUCTION

Economy goes its own way, and there are both upturns and downturns. While people expect upturns and economic growth, downturns usually surprise societies at large as well as companies in particular. The crisis that affected world economy in 2007–2009 was a surprise to nearly everyone, including politicians and managers. It challenged every part of societies, including governments, organizations, and families. Governments were forced to make unpopular decisions, and companies were focused on cutting costs because of the dramatic sales decrease; for people, it signified necessity to pay more for products and services and pay closer attention to budgets. Markets also suffered strongly during crisis. Financial and stock markets noticed considerable loss of indexes, and labor markets noticed increased unemployment. The strongest consequences were, however, related to people and their life. People all over the world suffered from unemployment; self-employed faced difficulties and frequently were forced to close businesses or at least suspend them.

The crisis, believed to begin with the burst of the so-called "housing bubble" taking its roots in the United States (2005–2006), affected also Poland (Polish Information and Foreign Investment Agency 2009). The year 2009 was challenging for Polish economy (Interlace Research 2010); however, comparing its breakdown to other European countries, the overall performance was relatively strong. Generally speaking, Polish economy has been among the region's best. Foreign investments driving Polish economy are believed to help Poland survive hard times and helped to become the country a "green island" in Europe (Esposito 2010). From 1991 onward, Poland became an important recipient of foreign development investments (FDI) in the region, originating initially from Western Europe, the United Kingdom, and the United States. Only during years 1997–2001 multinational corporations opened branches that gave work to nearly 650,000 employees. Also the FDI stocks by 2001 reached 20% of Poland's gross domestic product. All these helped Poland to become the only country in the region withstanding crisis and becoming the "green island" on the map of Europe.

Despite the overall condition of Polish economy, companies faced significant problems. Symptoms of crisis were similar to those of other countries in the region: decreased demand for products and services, financial problems, and cost reductions. These problems affected companies of every size and nearly every branch.

Theoretically, if employing the crisis management solutions presented in extant literature (Lagadec 1990, 1993; Mitroff et al. 1988; Pearson and Mitroff 1993; Perrow 1984; Roberts 1990; Weick 1988), companies were properly prepared to face the crisis. Both the knowledge about the crisis and its impact on management practices comprise a foundation upon which managers make decisions aimed at preventing crisis or reducing their negative consequences. Regardless of theoretical preparation to managing companies offered by the literature in most cases, companies were only able to limit the extent of losses. The simplest way to deal with crisis and decreasing demand was to cut all unnecessary (unrelated to core business) expenses (i.e., cutting training expenses, implement outsourcing), and if this did not solve organizational problems, dismiss employees and finally close the business.

According to several managers, crisis has also had other face in Poland. Some companies treated rumors surrounding bad condition of world economy as a good justification for solving employment problems they faced during previous years. It was perceived as a good period to deal with trade unions and dismiss unnecessary employees. In this light, crisis became a cover-up for deeper organizational changes and helped companies to optimize the way it works.

It does not matter if an organization was facing real crisis-related problems or just used the crisis to justify unpopular decisions; the approach to employee, employment, and work among managers changed, and crisis remained the trigger of these changes.

Consequently, the paper addresses several questions that all aim at discovering and describing how HR managers understand the crisis phenomenon and its impact on everyday organizations' life. In order to obtain this aim in the first part, a short literature review on three ways of organizing studies describing organizational change field (hierarchical, spatial, and temporal) is presented. In that part, we also revisit the concept of crisis from organizational perspective to see how it influences organization, particularly from the Human Resource Management (HRM) point of view. In the next part, results of grounded theory-based research carried on in five Polish organizations (carried on with the Tempo Group, the consulting company from Silesia) are presented. The paper finishes with short discussion, conclusions, and future research.

LITERATURE REVIEW

Crisis threatens people, relations, societies, countries, and finally the world. It also affects organizations (Hermann 1963). During the last decades of the twentieth century, crisis literature evolved to encompass psychology, sociology, disaster response, and theory of management (Booth 1993). While the background for current crisis management comprises the classic work of Perrow (1984), recent years of in-field research resulted in nearly incalculable number of crisis studies and works. There are at least three ways of organizing these studies discernibly: hierarchical, spatial, and temporal.

According to the hierarchical order, studies create a "pyramid" that describes different ranges or frequency (i.e., global, world economy, social crisis). On the bottom of the pyramid are located works presenting individual crisis or crises concerning relationships (e.g., matrimonial). On upper-level studies, crises affecting social groups (i.e., of specific age) are located (Lawrence 1980). On the third level, one can find studies describing organizational and organizational reputation crises (Laufer and Coombs 2006). On the fourth level, local and industrial crisis (Shrivastava et al. 1988) as well as public and market crisis (Abolafia and Kilduff 1988) are placed. Studies addressing global, world economy, social, or environmental crisis constitute the top of the pyramid. Such studies are relatively rare and largest in range.

In spatial order crises, studies are divided into types on the basis of fields of interest. Therefore, one can distinguish political, economic, organizational, information, decision, risk, learning from crises, and strategic oriented studies (Ausland 1966; Burnett 1988; Mitroff 1994; Rerup 2009). Thematic areas concerning crisis are virtually infinite.

Within the third, temporal order, the knowledge on phases before, during, and after crisis is being sought (Jaques 2009; Mitroff et al. 1987).

The presented short literature review implies that it is nearly impossible to carry on an exhaustive, complete review of crisis-related studies and typologies. Different approaches to crisis inflict difficulties in creating a coherent view from equipollent standpoints. All these standpoints contribute to better understanding of the crisis; however, regardless of numerous studies, crisis phenomenon remains poorly defined. For the purpose of this paper, it can be characterized as a situation of a person, group, or organization, in which they are unable to cope with by the use of "normal," routine procedures. Crisis also causes stress and forces sudden, unpredictable change. It poses a challenge for individuals, groups, and organizations. More specifically, crisis can be perceived as "a low-probability, high-impact event that threatens the viability of the organization and is characterized by ambiguity of cause, effect, and means of resolution, as well as by a belief that decisions must be made swiftly" (Pearson and Clair 1998). The uniqueness of the crisis causes difficulties in handling it properly, and the understanding of a crisis remains in relation to people's actions, because people think by acting (Weick 2001).

As mentioned above, crisis affects organizations as a whole, but also people in particular. People are first to suffer from the crisis in organization, while for years, the role of uniting people with organizations has been inclined (Chapman 1991). Studies on interrelations between the way organization treats employees (human resource management) and crisis were previously described (Sitalaksmi and Zhu 2010), but there are still questions of the function and the role of HRM in a sloping economy (Alen 2010). By contrast, the paper concentrates on the willingness to discover and describe how organizations in general, and HR managers in particular, understand the crisis; how they perceive the relation between the financial, economic crisis, and HRM practices in the organization; and how they tend to treat employees, their problems, and jobs in the times of crisis.

METHODOLOGY

In order to answer research questions, qualitative research methods—grounded theory in particular—were employed. In the first step of research procedure, empirical data from five Silesian organizations affected by the crisis were gathered. The data were collected by the use of semistructured, focused in-depth interviews. Interviews were focused on understanding how crisis influenced the functioning of the organization and the HRM, understanding actions taken in the HRM field, caused by the crisis, emotions, and feelings accompanying introduction of actions. We also tried to gather information on perception of actions undertaken in the HRM field and the willingness to maintain employed solutions when the crisis is over. The information was gathered in the late 2009 from the employees of HRM departments. All five participating companies were clients of the consulting company Grupa Tempo. Grupa Tempo is one of the biggest Polish recruitment consultancies, operating in most regions of Poland. Since then, it has specialized in providing professional HR management services for public and private enterprises. Operating since 2002, it has gained an extensive knowledge of the Polish job market and HR management techniques applicable in various organizations. Its tailor-made services include recruitment and selection, competency management and assessment, HR consulting, customized HR tools, business trainings, and coaching. Grupa Tempo closely cooperates with a range of HRM academics and practitioners in order to constantly update HRM trends and applicable solutions that strengthen and maximize the potential of workforce. Thus, the survey was carried out among a group of Grupa Tempo clients. The participants' survey consisted of managers and directors representing HRM departments whose impact on the HR decisions in their organizations was substantial. The reason behind the sample selection was the intention to obtain information on the influence of the crisis on the actions of HRM-related organizations.

After gathering the data, they were coded using open coding (following Charmaz 2006 guidelines). Sentences or their fragments were coded depending on the degree of statement content saturation. Coding was performed in a table consisting of three columns. The first contained line numbers, the second contained statement content, and the third contained code corresponding to the essence of the respondent's statement (code). In the second stage, the code contents were imported to the NVIVO 7 program and analyzed to identify most frequent words. On that basis, in the third stage, axial coding concentrating on connecting codes into nodes was performed. As a result of this stage, the set of nodes that would constitute elements of the research model was identified. In the next step, parts of the text were ascribed to nodes, and on that basis, relations between separate nodes were identified. Nodes and relations allowed model creation and consequent proposition building. Because of the space limitation, only an excerpt from the original study showing the most interesting results related to the topic is presented.

Findings

On the basis of analysis, two basic nodes in the empirical material, economic crisis and organization from the structural–subjective perspective, were identified. These main nodes, according to respondents, are interrelated. According to managers, economic crisis influences strongly the way organization functions, and it also affects the manner in which company treats work and human resources. The crisis node is hardly present in statements of respondents; therefore, it does not found a basis for creating a "tree node." On the other side, the node "organization from the structural–subjective perspective" is composed of three subnodes: "management and central office of the organization," "departments of the organization," and "employees from the psychological perspective." Figure 51.1 presents the model created during the research; nodes are described in the further part of the text.

Economic Crisis

The economic crisis was hardly present during the interviews—respondents just mentioned it happened, and they concentrated on organizational consequences rather than on economic crisis itself. Despite the lack of detailed information on the crisis, several interesting observations were made during interviews. Respondents basically understand crisis as a situation on the market reflecting organizational functioning. It also forces organizations to perform certain activities and introduce changes in the HRM field—affecting human resources of the company. Respondents during interviews used to associate crisis with the general decrease in orders or abandoning previously placed orders. It was also underlined that crisis began earlier than it was reported and presented by the media.

Managers also tend to perceive deteriorating situation of company business partners as a sign of crisis. This deficient situation was also perceived as causing problems to researched companies. It occurs when main product/service buyer encounters problems. Also the change in behaviors of customers influences organizational well-being. In respondents' opinion, economic crisis had an effect on top management team (TMT) decisions, and these decisions frequently forced HR departments to act.

Organization from Structural–Subjective Perspective

In crisis situation, HR managers tend to perceive organization as composed of several departments, with the head office clearly distinguished, emphasizing actions and emotions, and also are inclined to evaluate undertaken actions. Human resource specialists are also likely to focus on employees, their feelings, anxieties, and opinions. Therefore, the organization tree node of the model is described with the use of three nodes: management and central office of the organization, departments of the organization, and employees from the psychological perspective. Departments of the organization are further divided into different departments, HRM comprising one of them. The HRM department itself is described in terms of functional, emotional, and evaluation perspective Figure 51.1.

Management, Central Office of the Organization

According to interviewed HR specialists, it is the head office that is responsible for changing HR policy and practice in the company during crisis. While crisis is perceived as unnatural situation, organizations have a propensity to employ extraordinary actions/procedures, and HR departments do not have necessary power to introduce such actions. In a crisis situation, instructions "flow" from the top to the bottom of organizations, and HR departments are responsible for putting them into practice. The strategic nature of the HR departments in this case is replaced with operational or tactical issues. HR strategic aspects are subjected to the head office in the time of crisis. The management becomes the body giving instructions to the rest of the organization—other departments are only passive executors of orders given by the management in the critical times.

During the crisis period, the relation "we–they" is observable; on the one side are employees and on the other side are managers. Management or central office, in respondents' opinion, is responsible for making difficult, troublesome, or inconvenient decisions. These decisions are to be brought into practice by human resource management specialist, and they become the forehead of the management. This frequently leads to conflicts and misunderstandings within the organization. Instead

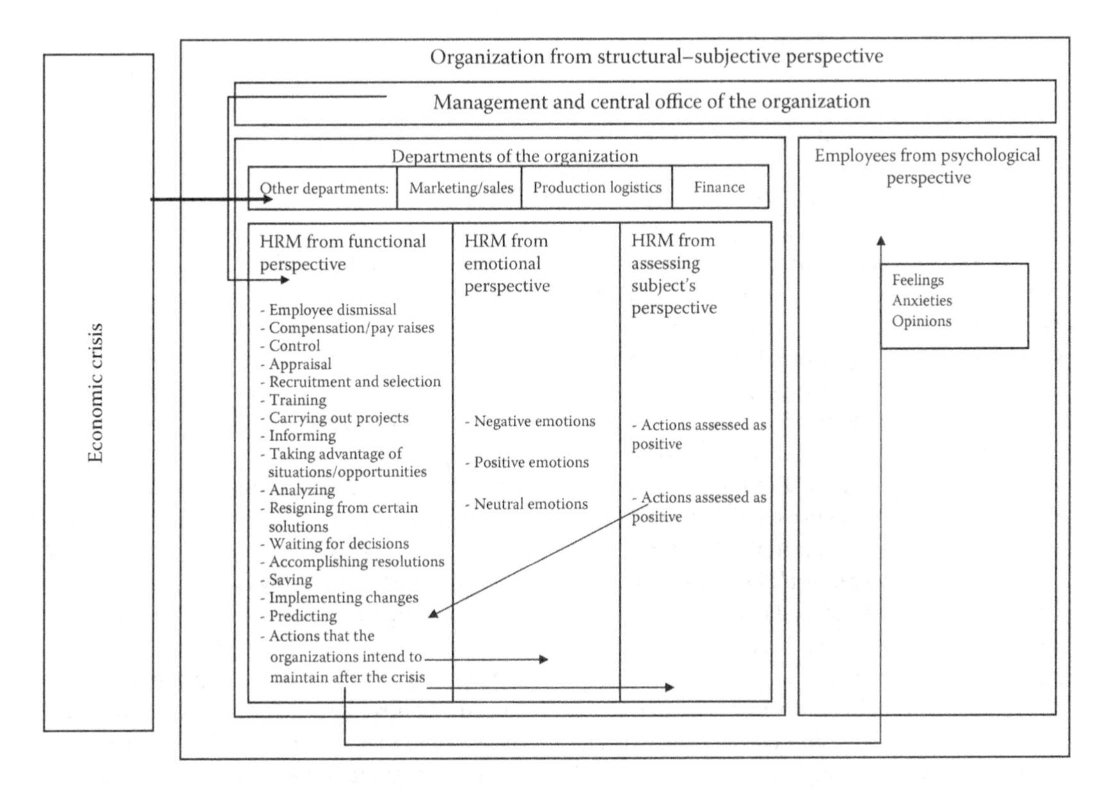

FIGURE 51.1 The HRM department described in terms of functional, emotional, and evaluation perspectives.

of a strategic role, human resource department tends to play intermediary role, mediating between the head office and employees.

Other Departments of the Organization

Among other organization departments (aside from HRM) mentioned by respondents, there are marketing, sales, logistics, production, and financial. During the crisis, these departments play a role of HRM partner in the decision-making process. They also facilitate implementation of decisions made by the central office. Members of these departments participate in problem-solving meetings, surpass information to the stuff, etc. Meetings are also supposed to lead to decisions concerning current and tactical functioning of the organization. It is also evident that in critical conditions, all the above-mentioned departments do not play a part as important as the management, being rather involved in completion of its instructions and recommendations.

Human Resource Management from Functional Perspective

Respondents, describing HRM department, usually underlined actions performed by the department, and they were characterizing currently performed duties and tasks. Research proves that the scope of HRM functions performed during crisis varies between typically performed actions. These functions can be both broader and narrower, depending on the situation. For instance, most important tasks performed by the HR department are those related to structural change with the purpose of staff reduction. During the staff reduction process, HR department plays a role of a mediator, representing social part of the organization—employee interests on one side and delivering decisions made by the central office on the other side. The major pressure during crisis is put consequently on (1) employee dismissals; (2) compensation issues; (3) performing control functions; (4) employee appraisals; (5) recruitment and selection; and (6) training and employee development. These main issues are enriched with such activities as carrying on projects, informing employees and head office, taking up opportunities, conducting analysis, predicting future market conditions, focusing on cost cutting, and resigning from unnecessary activities.

The most important question raised during crisis is concerned with staff reduction. On the basis of research, it is observable that organizations aim at avoiding dismissals and layoffs. Researched organizations' representatives (and human resource management employees) in each case declared that they made decisions to avoid layoffs; however, not in each case it was possible. Organizations are inclined to use specific solutions to help their employees during crisis, that is, changes concerning employee's working time; days off with an option to overtake in the next calendar year; two layoff days a week; classification of employees into three categories, A, B, and C, with the appropriate strategy regarding separate groups; dispatching employees to locations based in other countries; cooperation with other organizations that were not affected by the crisis; encouraging employees to put themselves through examination dealing with the degree of disability; proposing a three-fourths part-time work; and communication with temporary employment agencies. Organizations considered using the help of the State by means of PFRON—National Disabled Persons Rehabilitation Fund. One company was also willing to use anticrisis package.

The crisis forced organizations to cut costs much stronger than it used to during upturns. While compensation costs comprise a significant part of organizational costs, it is usually relatively simple to limit that large part. Surprisingly, research reveals that companies were inclined to freeze salary level rather than decrease the amount of money spent for compensation. However, in some cases, organization might be forced to change both financial and working conditions to guarantee employment for their employees. Frequently, organizations were inclined to change working conditions (e.g., working time, duties and tasks performed at work) than lay off some of its employees.

Organizations paid relatively high attention to performing control and assessment of employees. The basic aim of these procedures was to make the assessment and identify employees or groups of employees performing below standards. In the next step, organization would fire such employees having negative impact on organizational performance. One can also perceive it as a cover-up for dismissals while it certainly rationalized changes and tough decisions that were to be made. The huge amount of consideration was also given to employee appraisal issues. While there were no revolutionary changes in employee performance appraisal systems, nevertheless, some minor changes were implemented. In this case, organizations put greater emphasis on the level of goal accomplishment, and financial prizes were rigorously dependent upon them.

Respondents also mentioned some changes in recruitment and selection, training, and development of employees, but, as reported by researched companies, organizations were more willing to stop investments in employees than to further develop it.

It seems that the critical conditions did not change the scope of the performed actions, yet had a vital influence on the structure of interest dedicated to specific actions. Special concern was dedicated also to issues connected with informing employees, which is reflected in numerous meetings. HR specialists act as intermediaries between interests of management and interests of employees.

As for the scope of actions that the organizations intend to maintain after the crisis, the respondents unanimously declared themselves in favor of those actions they appraised as positive. Organizations declared the willingness to abandon services of external companies, to implement changes in bonus systems, to maintain balanced working time, to retain the compensation guarantees, to sustain communication, and to think of employees in key-employee categories (employees segmented into three groups). The necessity of maintaining employee achievement evaluation (at all ranks) and managing by objectives is also emphasized. Generally, the HRM departments would readily retain solutions that do not generate excessive costs.

HRM from Emotional Perspective

Regarding emotions perceived by HR managers, they felt discomfort caused by carrying out unaccepted recommendations of higher level managers (CEOs) or owners. Respondents reported stress, breakdown, irritation, sense of responsibility for decisions made (while managers were usually only executors of solutions), nervousness, lack of motivation, discouragement, lack of pleasure flowing from work, doubts, anxiety, and finally, gradual stabilization and extinction of negative emotions accompanying clarification and improvement of organization's situation. Respondents reported to feel uncertain and tired, "marked" for the rest of their life. They also sensed helplessness and lack of pleasure from work. However, there were also positive emotional aspects of crisis associated with signing and handling work return forms. Despite a few positive emotions, the whole bound of emotions surrounding critical situation can be called "negative."

HRM from Assessing Subject's Perspective

The HR specialists tended to make assessment of changes introduced within their companies. Assessments were either positive or negative, and no neutral evaluations were noticed. Positively assessed was the ability to work out a strategy model helping employees or the lack of layoffs. All changes that led to the greater trust in employees received positive judgment.

Organizations favorably assessed implementation of balanced working time and pay level guarantees, maintaining communication. Also the ability to adapt to new market conditions is composed of a source of positive opinions.

Generally, participants were disappointed with the lack of reception of signs of crisis. Managers felt unprepared to face necessity of changes. They reported lack of tools helping to manage staff in crisis conditions. They were also dissatisfied with the lack of confidence in weak signals. All necessary layoffs resulted from crisis were perceived negatively. However, respondents declared the will to maintain in future HR solutions and actions that were positively evaluated.

Employees from Psychological Perspective

Employees, according to HR managers, were suffering anxiety. Managers were experiencing negative feeling and attitudes from employees and from the tasks performed by the HR department. Crisis antagonized employees and the management team dividing them into two separate parts.

Anxieties of employees were—in respondents' opinion—related to the possibility of losing a job and uncertain future. Negative employee emotions were mostly associated with introduced changes. HR specialists reported perceived lack of confidence in decisions made by head office.

Relations between Nodes

The analysis revealed essential relations between above-described nodes. The first relation applies to the statement that the crisis had influenced the situation within the organization. In particular, an interrelation between critical situation and the level of orders emerged. HRM department functioning—in the light of respondents' statements—is not the answer to the functioning conditions of other departments, but it is directly dependent on the central office/owners/management. Actions undertaken by the HRM were reflected by the employees' feelings, yet considering the fact that the HRM department has a limited influence on the decisions made on higher levels, the possibility to moderate the feelings and anxieties of employees is also limited. Decisions taken by the HRM department raise specific emotions in the employees of this department and provoke evaluation of the performance (as well as the accuracy of decisions made on higher levels). Not every solution applied in the conditions of a crisis is assessed negatively. There is a relatively large set of actions assessed positively. The respondents declared the willingness to maintain these actions in the future.

DISCUSSION

Research results reveal several important implications dealing with working conditions and treatment of employees. That allows posing subsequent propositions described below. They may compose of future research questions/issues.

Crisis was associated with the decrease in orders and treated as external in relation to organization. As such, organization had no means to influence or to stop crisis. This is a highly simplified perception of economical phenomenon. The whole complexity and mutual interactions did not draw managerial attention; instead managers concentrated on relations with competitive environment and on finding solutions to existing problems. Although respondents declared that they were aware of a probability of economic crisis and arising problems to the company they were in, they did not perform preparation activities (i.e., through special training, preparing procedures, meeting with employees, performing audits, etc.; see Kash et al. 1998). Managers perceived a crisis as inevitable fate impossible to cope with. Based on the above-mentioned contention, we state the following proposition:

Proposition 51.1: Managers associate crisis with economic, external threats that are hard (or impossible) to avoid. ■

Crisis results also in creation of superficial barriers between parts of organization, even if such barriers did not exist before the critical situation. From the entirety (in times of economic growth), organization becomes a hierarchical, puzzled phenomenon in which the division "we–they" is evident on every level during crisis times. This separation results in antagonisms, anxieties, and conflicts. Therefore, we suggest the following proposition:

Proposition 51.2: During the crisis, organizations are perceived as divided into a collection of hierarchically organized parts and relations between parts become more formal than in economic growth times. ■

The crisis, perceived as external, independent from organizations' actions phenomenon, should be comprehended as complicated and sophisticated, including complex and dynamic relations.

During the analysis, we were not able to locate many connections between nodes, which suggest the following:

Proposition 51.3: Managers perceive crisis as a structural, hierarchical phenomenon and not as a dynamic, complex, interrelated one. ■

Crisis forces changes, but organizations, in order to prevent negative crisis effects, employ known and tried tools like the segmentation of employees, limited working time, help offered by the state, etc. The science brings more solutions (see Greiner 1972; Ramee 1987) to existing problems, but organizations either perceive them as invaluable or are just unaware of such solutions. Thus, we suppose the following:

Proposition 51.4: Managers considerably concentrate on well-known and tried tools and not on active seeking for sophisticated management solutions to crisis. ■

Crisis forces organizations to concentrate on critical issues, and organizational survival is one of them. Considering decreasing demand for goods, it can be obtained by cost reductions, and dismissals might be perceived as a good solution to high cost problem. Interviewed managers perceived two important issues: employees and cost reduction. Everything else was just treated as "surrounding" these key issues. Therefore, we suggest the following:

Proposition 51.5: Crisis forces managers to concentrate their activities on closest partners and resources. ■

Managers reported strong feelings related to crisis. In most cases, they were negative or highly negative, leading to dissatisfaction, lack of motivation, etc. These negative emotions influenced not only decisions made but also actions and behaviors. Therefore, we state the following proposition:

Proposition 51.6: Negative emotions associated with crisis strongly influence managerial actions and decision-making processes negatively influencing their performance. ■

Interviewed managers were inclined to treat crisis as any other difficult situation and were searching for solutions within well-known practices/activities. Traditional literature tools and practices, as presented in the textbook (i.e., Leopold et al. 2005), were used to prevent negative crisis results. They were neither modified nor redefined. Instead, managers used different sets of practices and emphasized different activities. Therefore, we suggest what follows:

Proposition 51.7: During crisis, HR managers prefer to concentrate on well-known practices but use them in diverse configurations comparing to economic growth times. ■

CONCLUSIONS

The 2007–2009 crisis had significant impact on the world. It is apparent not only in the state of the economy, prices of oil, rising inflation, or unemployment but also in everyone's life. It also changed the way organizations work. This study was aiming at understanding how managers perceive crisis and how it affected organizations. Research reveals that crisis images in managers' minds are far from literature patterns and concepts. Managers are inclined to perceive crisis as distant, external, and potentially harmful phenomenon; however, they do not allocate too much attention to understand it better. Their opinion about crisis surprises with simplicity, and their actions aim at restoring balance (the situation) they knew and experienced before the crisis occurred. Even frequent contacts with professionals (Tempo Group) who supplied respondents with news, information, and professional advice did not change the way interviewed managers perceived the crisis. At the same time, crisis changed many things—beginning with close-up of some companies, introducing new sets of known tools, relations between employees and managers within the companies that suffered it much, and finishing with the situation on the labor market.

The subject, respondents, and methods naturally put restrictions and determine limitations on obtained results. Polish managers constitute a very specific, unique group, and therefore, it would be interesting to carry on comparative studies—in different parts of the world—that would reveal more accurate and reliable patterns, ways of dealing with crisis in the HRM field. Considering the nature of the study (grounded theory), it would be interesting to include other research methods in future research on crisis impact on HR function within the company. It would help to expand the collection of data and increase the reliability of the crisis recognition.

Presented research results encourage further studies. It would be especially interesting to see how the locus of control affects the efficiency in coping with critical situations. How do the intuitive, natural ways of acting and informal relationships altogether affect the understanding of the crisis? What sets of HR tools are used and why during crisis? How do organizations in general and HR managers in particular learn from crisis? Which practices and solutions were kept after the crisis finished, which were not and why? There appears to be several issues worth understanding regarding the crisis and changes it made to the organization. In that light, further studies are clearly necessary, especially these helping to understand crisis before it happens again.

REFERENCES

Abolafia, M.Y., Kilduff, M. 1988. Enacting market crisis: The social construction of a speculative bubble. *Administrative Science Quarterly*, Vol. 33, No. 2, pp. 177–193.

Alen, B. 2010. Rethinking human resources in sloping economies: A strategic approach. *Advances in Management*, Vol. 3, No. 5, pp. 20–22.

Ausland, J.C. 1966. Crisis management: Berlin, Cyprus, Laos. *Foreign Affairs*, Vol. 44, No. 2, pp. 291–230.

Booth, S.A. 1993. *Crisis Management Strategy: Competition and Change in Modern Enterprises*. Routledge, London.

Burnett, J.J. 1988. A strategic approach to managing crises. *Public Relations Review*, Vol. 24, No. 4, pp. 475–488.

Chapman, J.A. 1991. Matching people and organizations: Selection and socialization in public accounting firms. *Administrative Science Quarterly*, Vol. 36, No. 3, pp. 459–484.

Charmaz, K. 2006. *Constructing Grounded Theory: A Practical Guide through Qualitative Analysis*. Sage Publications, London, Los Angeles, New Delphi, Singapore, Washington.

Esposito, F. 2010. Central, Eastern Europe still promise growth. *Plastics News*, Vol. 22, No. 3, pp. 12–14.

Greiner, L.E. 1972. Evolution and revolution as organizations grow. *Harvard Business Review*, Vol. 7–8, pp. 37–46.

Hermann, C.F. 1963. Some consequences of crisis which limit the viability of organizations. *Administrative Science Quarterly*, Vol. 8, No. 1, pp. 61–82.

Jaques, T. 2009. Issue management as a post-crisis discipline: Identifying and responding to issue impacts beyond the crisis. *Journal of Public Affairs*, Vol. 9, No. 1, pp. 35–44.

Kash, T.J., John, R., Darling, J.R. 1998. Crisis management: Prevention, diagnosis and intervention. *Leadership and Organization Development Journal*, Vol. 19, No. 4, pp. 179–186.

Lagadec, P. 1990. Communication strategies in crisis situations. *Industrial Crisis Quarterly*, Vol. 1, pp. 19–26.

Lagadec, P. 1993. *Preventing Chaos in a Crisis*. McGraw-Hill, London.

Laufer, D., Coombs, W.T. 2006. How should a company respond to a product harm crisis? The role of corporate reputation and consumer-based cues. *Business Horizons*, Vol. 49, No. 5, pp. 379–385.

Lawrence, B.S. 1980. The myth of the midlife crisis. *Sloan Management Review*, Vol. 21, No. 4, pp. 35–49.

Leopold, J., Harris, L., Watson, T. 2005. The strategic managing of human resources. Financial Times Prentice Hall, 2005.

Mitroff, I., Pauchant, T., Shrivastava, P. 1988. Conceptual and empirical issues in the development of a general theory of crisis management. *Technological Forecasting and Social Change*, Vol. 33, No. 2, pp. 83–107.

Mitroff, I.I. 1994. The role of computers and decision aids in crisis management: A developer's report. *Journal of Contingencies and Crisis Management*, Vol. 2, No. 2, p. 73–84.

Mitroff, I.I., Shriwastava, P., Udwadia, F.E. 1987. Effective crisis management. *Academy of Management Executive*, Vol. 1, No. 4, pp. 283–292.

Pearson, C.M., Clair, J.A. 1998. Reframing crisis management. *Academy of Management Review*, Vol. 23, No. 1, pp. 59–77.

Pearson, C.M., Mitroff, I.I. 1993. From crisis prone to crisis prepared: A framework for crisis management. *Academy of Management Executive*, Vol. 7, No. 1, pp. 48–59.

Perrow, C. 1984. *Normal Accidents*. Basic Books, New York.

Ramee, J. 1987. Corporate crisis: The aftermath. *Management Solutions*, Vol. 32, No. 3, pp. 18–22.

Rerup, C. 2009. Attentional triangulation: Learning from unexpected rare crises. *Organization Science*, Vol. 20, No. 5, p. 876–895.

Roberts, K. 1990. Some characteristics of high reliability organizations. *Organization Science*, Vol. 1, No. 2, pp. 160–176.

Shrivastava, P., Mitroff, I.I., Miller, D., Miglani, A. 1988. Understanding industrial crises. *Journal of Management Studies*, Vol. 25, No. 4, pp. 285–303.

Sitalaksmi, S., Zhu, Y. 2010. The transformation of human resource management in Indonesian state-owned enterprises since the Asian Crisis. *Asia Pacific Business Review*, Vol. 16, No. 1–2, pp. 37–57.

Weick, K.E. 1988. Enacted sensemaking in crisis situations. *Journal of Management*, Vol. 24, No. 4, pp. 305–317.

Weick, K.E. 2001. *Making Sense of the Organization*. Blackwell Publishing, Malden.

52 Information Technologies as a Pillar of a Knowledge-Based Economy
Some Remarks

Joanna Kalkowska, Hanna Wlodarkiewicz-Klimek, Edmund Pawlowski, and Stefan Trzcielinski

CONTENTS

INTRODUCTION

The Lisbon Strategy Importance

At present, one of the most important factors of socioeconomy development as well as improvement of competitiveness became an enterprise's transformation into requirements of knowledge-based economy. The enterprises are searching for competitive advantage concerning quality of products, manufacturing costs, time of launching products, and application of modern information technologies (IT). To achieve this, first of all, enterprises need to possess the ability of the knowledge potential proper usage because widely understood enterprise's development is connected with the permanent winning, transformation, and usage of knowledge and information. To fulfill that opinion, the European countries in 2000 accepted the common concept of knowledge-based economy. The development's postulate concerning the potential knowledge usage was presented in the Lisbon Strategy, which resolutions are still binding. The Lisbon Strategy focuses on four fundamental potentials:

- Human resources, that is, society of knowledge (which the part of knowledge is gathered in).
- Innovation system (with the entrepreneurship, more concentrated for operations of companies, but also on the cooperation with science); it creates new knowledge in result of discoveries and innovations.
- IT facilitating the exchange of knowledge, also with foreign countries.
- Institutional and legal environment, which creates conditions for development of presented domains; it constitutes various institutions, regulations, etc. (Kalkowska and Włodarkiewicz-Klimek 2009).

The Lisbon Strategy, adopted by union countries assuming creating the most competitive and dynamic knowledge-based economy in the world, in its conceptual and executive shape, is evolving up till today. On the level of individual member states, national conceptions of the development of the economy are being created taking into account the common Lisbon program. In Poland, the crucial document is the National Development Strategy 2007–2015, which is a superior document in a view of other strategies or programs. One of the most important elements of the vision of Poland till 2015 is the construction of the knowledge-based economy: "Poland have to develop its knowledge-based economy and apply a wide application of information and communication technologies in all areas…." The national strategy is realized through numerous strategic and operational documents (e.g., National Strategic Reference Framework 2007–2013 [NSRO] and related with operational program, Innovative Economy 2007–2013, etc.). Among other documents elaborating the Lisbon Strategy related to Poland and focused on the importance of IT in knowledge-based economy is a Strategy for Information Society Development 2007–2013.

Knowledge-Based Economy: Fundamentals

The widely understood knowledge became a fundamental driver of increased productivity and global competition. Knowledge is also the most valuable source of competitive advantage, and it is considered to be the prominent resource of enterprise in terms of its contribution to the value added and its strategic significance (Bolisani and Scarso 1999, p. 209). However, the dominance of the knowledge in the social and economic life resulted in the 1990s to introducing into the economic theory and practice the concept of "knowledge-based economy." Ambiguity in understanding the knowledge-based economy causes the fact that many sets of features with different degree of accuracy are being used in descriptions of this phenomenon. The classic definition presented in 1996 by the OECD (1996) shows a knowledge-based economy as an economy that directly is based on the production, distribution, and use of knowledge and information. World Bank is presenting another view point of the idea of knowledge based economy. Now it tends to claim that the economy is becoming "knowledge-based economy" when using and creating knowledge. That knowledge maintain the center of the economic development. A knowledge-based economy is an economy that uses knowledge as a motor of economic growth (World Bank 2006). Knowledge is treated here as a fundamental driving force of the economy, a factor stimulating progress.

The accepted conception of knowledge-based economy development supported by capital achieved from the European Union is a great background for enterprises' development. The direct and indirect opportunities that are created by knowledge development economy are a sign for adaptation of the enterprise development strategy to the new economy conditions as well as to proper selection and development of IT, which are one of the bases of adaptation to new reality. Taking into consideration the Lisbon Strategy, it can be stated that the enterprises that applied IT become a part of this strategy.

IT AND INFORMATION SOCIETY IN A KNOWLEDGE-BASED ECONOMY

IT and Management System

IT is a critical factor for the effective operation and prosperity of modern organizations; however, management and dissemination of information are central for the enterprise (Morgan et al. 2006, p. 980). Also, another result of IT application in hierarchical organizational structure in enterprise are communication barriers and mistakes in information flow as well as wasting human knowledge. Moreover, IT is considered one of the most important factors of modern enterprise development and competitiveness. IT usage influences also a number of indicators allowing for competitiveness estimation. These indicators are as follows: cost reduction, quality improvement, increasing production flexibility, product and technology innovativeness, extending product assortment and productivity, and introduction into international markets. Moreover, thanks to the information infrastructure and proper software, it is possible that a quite fast coordination of all tasks is carried out in an enterprise. Also e-business based on IT technology became an intensive dominant area. Besides, one of the key factors of the enterprise's management system in a context of knowledge-based economy is including IT into widely understood management processes. In that context, these technologies are treated as one of the determinants of enterprises' transformation into knowledge-based economy, enabling not only increasing but also creating new organizational knowledge. Furthermore, IT may dramatically enhance the coordination and control capacity of the enterprise, so this way it can be stimulated the increased use of the management system. IT removes the distance and time constraints in accessing required information flows and hence improves the coordination of activities within organizational boundaries. IT affects planning systems by improving organizational communication as well as by enhancing organizational flexibility (Spanos et al. 2002, p. 662); however, the application of advanced IT corresponds to an incremental process of organizational capability and strategic impact (Spanos et al. 2002, p. 661).

IT and Information Society in Poland against the Background of the European Union

Information and communications technology (ICT) is defined as the convergence of telecommunications and computing (Gibbs and Tanner 1997); simply it is called IT. In such general understanding, it can be a basis of further deliberations concerning information society.

The development of information society in Poland is understood as a common impulse and abilities of citizens, entrepreneurs, students, teachers, and office workers to use digital technologies in a daily social and economic life. Such understanding is a basis for efficient development of knowledge-based economy (Sartorius, p. 451). For better understanding of the idea of information society, the following factors are distinguished: existence of developed and commonly accessible information infrastructure, abilities of IT service and usage by majority of citizens, which is a basis for employment, as well as active collaboration with product and service sellers in the future, essential technology influence on civilization development, as well as cultural changes.

On the basis of the Europe's Digital Competitiveness Report-ICT Country Profiles from August 2009, it can be stated that the information society in Poland is still developing slowly, and in two-thirds of all benchmarking indicators, Poland is close to the bottom of the EU ranking. Poland is, however, taking active steps through a national ICT "Strategy for Information Society Development 2007–2013." This strategy is principally aimed at counteracting digital exclusion for low-income citizens and at widening access to the Internet for micro-, small, and medium-size entrepreneurs. The former plan aims at developing data communication systems used to carry out public tasks, while the latter focuses on the use of ICT for accelerating the growth of intellectual and social capital of citizens, for increasing productivity and competitiveness of Polish companies, and for improving the effectiveness of the public administration (Commission Staff Working Document 2009, p. 46).

Concerning broadband, the data included in the report shows that a broadband penetration (the number of total subscriptions to fixed broadband connections—households, enterprises, public sector by platform [DSL, all others] divided by the number of inhabitants) by population in Poland stood at 13.6% in 2008. It went up by almost 60% since 2010, but is still one of the lowest in EU27. Broadband is closely related with Internet usage. While the Internet usage in Poland has grown steadily over the past few years, rates of regular and frequent Internet are still relatively low compared to the EU average. Furthermore, about 44% of people never used the Internet as there are regular users. Concerning the availability of public services, Poland is far below the European average; take-up by citizens is low, while use by businesses has reached EU average levels. From the research included in the report, it results also that the investment in ICT R&D is very small, but progress in terms of ICT exports (in terms of ranking) confirms an important role for the Polish economy in the manufacturing of ICT goods. ICT take-up by businesses, reflected by e-commerce and e-business indicators, is still at a low level. The share of turnover generated through e-commerce is a quarter lower than on average in Europe, while the proportion of enterprises selling online is only half of the average figure. The automatic exchange of business documents, however, slightly exceeds the EU average and, together with e-government take-up, is a promise for progress in the further development of the information society (Commission Staff Working Document 2009, p. 46).

IT and Development of Information Society in Poland

According to the Strategy for the Development of the information society in Poland until 2013 elaborated by Ministry of Interior and Administration in December 2008:

"Poland's information society policy should respond to the specific needs of Polish society and be consistent with European policy, using its best experiences. The assessment of the status of information society development in Poland refers therefore, to, among other things, the fulfilment of the European policy priorities that are defined in the communication from the European Commission 'i2010-A European Information Society for growth and employment.' The European Commission has recommended the following three priorities in the information society policy:

- To create a single Information area in Europe that will support an open and competitive internal market in the scope of the information society and media;
- To accelerate innovation and investment in research in ICT in order to support growth and to create new and better jobs;
- To create an inclusive European information society that will support growth and the creation of new jobs in accordance with sustainable growth principles, while prioritising better quality of public services and life" (The Strategy for the Development of the Information Society in Poland till 2013, 2008, p. 3).

According to the above, the development of the information society in Poland should have the following permanent attributes:

- "Availability, security and confidence—access to reliable information or a secure service that is indispensable to citizen and businesses.
- Openness and diversity—no preferences in access to information; especially to public information.
- Universality and acceptability—efforts to ensure that participation in the information society is obvious and common to the maximum extent feasible, and that the information society products and services are as broad as possible.
- Communicativeness and interoperability—searching for and access to the desired information are secure, quick and simple" (The Strategy for the Development of the Information Society in Poland till 2013, 2008, p. 9).

In the context of information society, researches by the Central Statistical Office concerning the use of IT in Poland were carried out. The results of those researches were published in the Concise Statistical Yearbook of Poland in 2010. They include data collected from 2005 to 2009. The results are based on the information obtained from the persons taking part in the survey through face-to-face interviews. In 2009, the survey covered 8300 of households (i.e., 0.07% of total households), of which 11,200 of residents are of age 16–74 (i.e., 0.04% of the population). Data on enterprises concern economic entities employing more than 9 persons. In 2009, the survey covered 14,400 of enterprises (i.e., 14.9% of total enterprises).

First interesting research concerned the households equipped with computers and Internet access. Data concern households with at least one person aged 16–74 as well as access to the Internet by means of computers (desktop or laptop) as well as mobile phones, game consoles, or any other devices. The research results show that the amount of computers has grown from 2005 (40%) to

2009 (about 65%). Also the Internet access is better available, and it has grown as well from 2005 (about 30% and 15% with broadband access) to 2009 (about 58% and 51% with broadband access) (see Figure 52.1).

The results included also the place of use and the purpose of private Internet use. On this basis, it occurs that the Internet is used most often at home (38% in 2009) than at work (10% in 2009) and others (school, public places—12% in 2009). The respondents declared that they use Internet for sending and receiving emails (33% in 2009). About 29% in 2009 of people use Internet for participation of chat sites, online discussion forums, etc., and for finding information about goods or services (21% in 2009). Also quite high dynamic development can be seen according to purchase of goods and services (from 6% in 2005 to 12% in 2009), listening to Internet radio (from 6% in 2005 to 15% in 2009), and Internet banking (from 6% in 2005 to 12% in 2009). The detailed data comparing between 2005 and 2009 concerning the purpose of Internet use are presented in Figure 52.2.

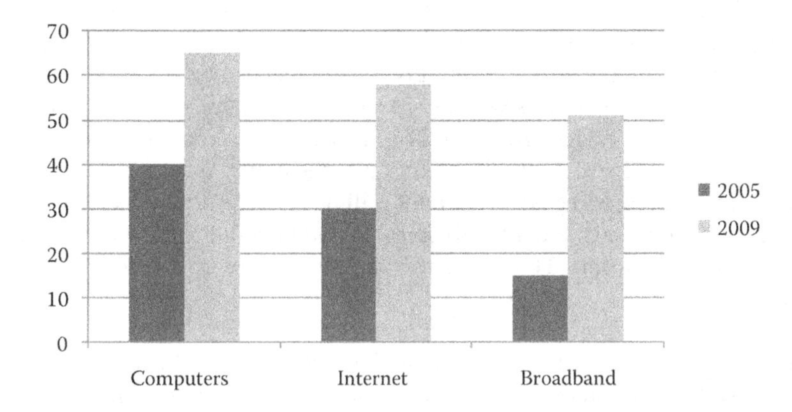

FIGURE 52.1 Households equipped with computers and Internet access. (From own study on the basis of data from Concise Statistical Yearbook of Poland, 2010, p. 302.)

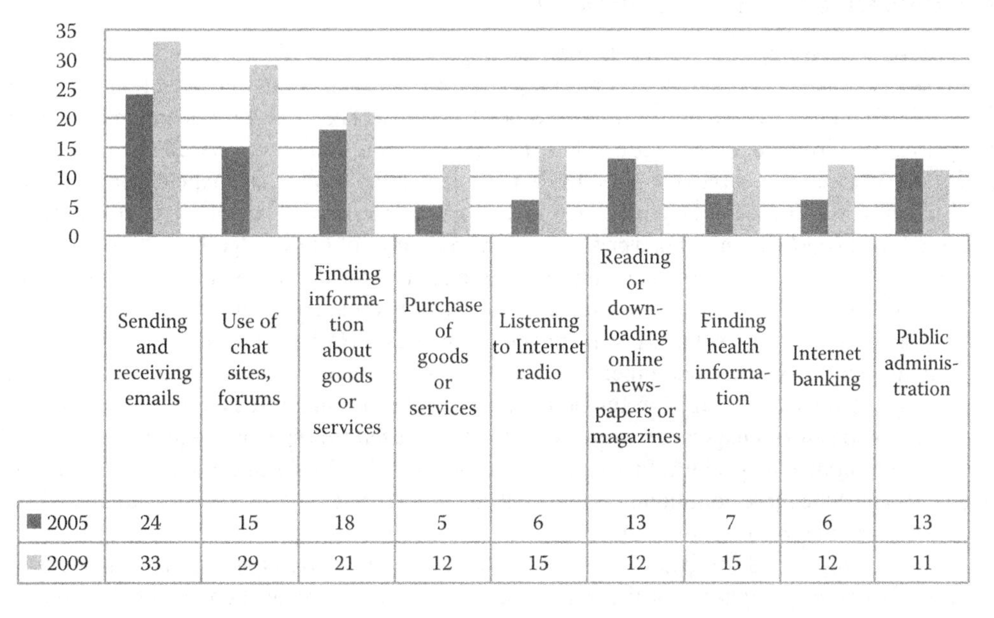

	Sending and receiving emails	Use of chat sites, forums	Finding information about goods or services	Purchase of goods or services	Listening to Internet radio	Reading or downloading online newspapers or magazines	Finding health information	Internet banking	Public administration
2005	24	15	18	5	6	13	7	6	13
2009	33	29	21	12	15	12	15	12	11

FIGURE 52.2 Detailed data comparing between 2005 and 2009 concerning the purpose of Internet use. (From own study on the basis of data from Concise Statistical Yearbook of Poland, 2010, p. 303.)

Second interesting researches in the context of information society concerned usage of selected IT by enterprises. There were taken into consideration following technologies: equipment in computers, Internet access, own webpage, local area network (LAN), Intranet, and Extranet, in particular branches like manufacturing, electricity, gas, steam and air conditioning supply, construction, trade, transportation, and professional, scientific, and technical activities. The data are presented on a very general level. It results that in 2009, about 93% of enterprises use computers and 90% has an Internet access. More than a half use LAN and have own webpage presenting basic information about the enterprise (Concise Statistical Yearbook of Poland 2010, p. 304).

On the basis on some facts presented above, the following relation can be visible: the better the Internet access both in households and enterprises, the bigger the percentage of Internet users, which directly influence their knowledge possession and development. Moreover, thanks to the Internet popularizing, it is a higher demand for digital services and e-economy products.

At present, a kind of growing trend of importance of IT and information society in Poland seems to be observed. The demand for IT products will also increase together with the economic growth. Such a situation can create a number of occasions that can be effectively used by capacity of knowledge-based economy enterprises. It also has been taken into consideration that the global or European development of IT affects the use of IT in Poland, which directly influences on the competitiveness of Polish enterprises. According to this, also the importance of different e-services like Internet banking, purchasing, and e-administration will increase. Skulimowski (2006) suggests that based on the economic forecasts, the infrastructural IT satiation in Poland will take place between 2012 and 2015. Also, at this time, IT literacy will replace poverty as a major digital divide factor (Skulimowski 2006, p. 19).

Nevertheless, in the development of information society, the problem of digital divide still has to be taken into consideration, which concerns not striving for social and economy development. To overcome this, investment in modern IT should be made, particularly in rural areas, and also the costs of Internet access should be systematically reduced.

IT IN SELECTED ENTERPRISES IN POLAND: SOME RESULTS OF PILOT RESEARCH

To improve the assumptions of the Lisbon Strategy concerning IT and information society, a pilot research was carried out in 20 enterprises in Poland, with a random selection that during 2007–2009 was introduced into the Warsaw Stock Exchange. The largest group among researched enterprises was production and trade enterprises (80%), then trade and service (10%), and trade (10%). The research was carried out in 2010. The subject of the researches included standard enterprise's documentation, secondary information, as well as interviews. The purpose of the research concerned the identification of enterprise tendency in a range of knowledge-based economy capability usage. The assessment of relatively well-developed information infrastructure was widely interpreted. First of all, it has been pointed out that all researched enterprises were equipped with information systems supporting management process. Half of enterprises had used integrated management systems, and others used especially dedicated systems in the following areas: accounting and finance, storage, production and procurement, human resource management, and customer relationship management. Moreover, about 20% used free specialist software available in the Internet or branch magazines. In a scope of Internet technologies usage, the influence of broadband access became more common. According to this, enterprises still use both wires and wireless Internet connection. Furthermore, the majority of researched enterprises (85%) use Intranet, and all of them has registered email account with own domain. On the basis of the conducted research, it can be also observed that the still essential equipment in enterprises constitutes desktop computers, and they are a majority compared to the laptops. These proportions are quite substantial. Mobile computers are used by management, managers of particular divisions, project managers,

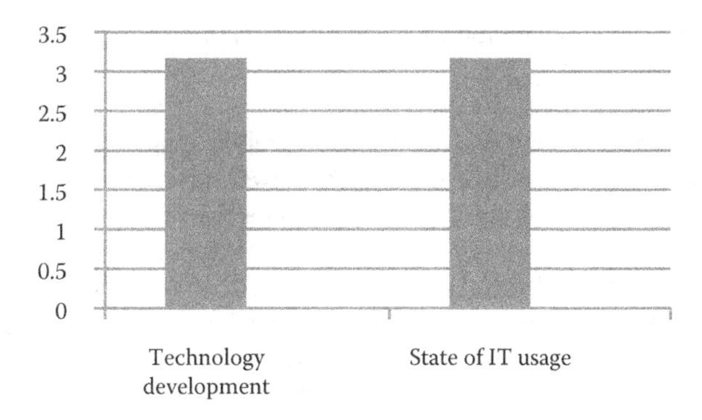

FIGURE 52.3 IT development and state of usage (average assessment). (From own study.)

as well as sales representatives. Other part of researches concerned the identification of advantages and barriers of IT usage. The barriers were identified as lack of financial resources and lack of modern technical infrastructure. By contrast, the advantages were identified as a speed of information exchange, communication efficiency, easy data access, and computer support of particular enterprise functional areas. The integral part of this research constitutes the level assessment of information system usage.

The research shows that companies applied advanced IT, but they did not use them effectively. Such a situation seems to be dictated both with the limitations concerning knowledge of this technologies application and infrastructure. Moreover, the respondents rather positively assess IT resources but comparing in relation to their modernity level, there is an observation that the IT is not fully used. It is also interesting that the arithmetical average of IT usage is exactly at the same level as the technology development in researched companies (Figure 52.3).

Analyzing the research results, it is important to remember that the companies were a different area of activity. Concerning IT infrastructure, the research included also questions concerning future investments and IT safety. Some respondents answered that they are going to invest in those technologies in the near future. Moreover, the companies are going to finance these investments partly with own capital, partly by loan bank, and if possible, with European Union support. The amount of this investment fluctuate from 100,000 to 400,000 Euro and will concern typical IT infrastructure and specialist software. Concerning safety, the main point concerned the protection of all IT resources and transfer of data (Kalkowska and Wlodarkiewicz-Klimek 2011).

CONCLUSION

IT provides a foundation for building up and applying knowledge in the private and public sectors. Countries with pervasive information infrastructures using innovative IT possess advantage for sustained economic growth and social development. In the turbulent environment, developing countries face opportunity costs if they delay greater access to use IT. IT is a key input for economic development and growth, offers opportunities for global integration while retaining the identity of traditional societies, can increase the economic and social well-being of poor people and empower individuals and communities, and enhances the effectiveness, efficiency and transparency of the public sector (World Bank Group Strategy 2002).

According to Korres, in OECD countries, access to telecommunications networks has increased in recent years by more than 10% a year especially in countries with lower penetration rates like Poland, Mexico, and Hungary. Wireless access has grown particularly fast. The Internet also continues to diffuse rapidly (Korres 2008, p. 200). Nevertheless, Poland still maintains a distance in comparison with European Union in an area that is described by knowledge-based economy indicators particularly concerning IT and inventions. On the other hand, on the basis of the researches conducted by the Department of Management Systems, Warsaw School of Economics, it results that Poland improved its knowledge-based economy index. According to this, Poland is placed in a half position of new EU members beyond Estonia, Czech Republic, Hungary, and Slovenia. Whereas in the development of information infrastructure, which allows one to follow world standards concerning information exchange, undertaking research collaboration by computing networks became a crucial element in creating a European civilization level in Poland.

The dynamic development of IT is not only a factor speeding up globalization processes but also a technology progress carrier influencing a number of transformations in enterprises. The particular significant in a process of transformation into knowledge-based economy is including IT to the widely understood management process. Moreover, the application of IT is one of the stimulant permanent adaptations to the environment changes. Furthermore, the availability of information systems in management has a significant importance in organizational knowledge increasing specially in a codified area. Also, the development of information systems is one of the most important aspects of internal structure adaptation to the knowledge-based economy. However, in that scope, some limitations have to be taken into consideration. A scarce interesting in modernization of widely understood technological infrastructure seems to be the biggest one. Such situation is caused not only by economic crisis but also with a lack of financial resources as well as lack abilities of using the European Union funds. On the basis of the conducted research, it can be observed that some enterprises indicated that they are going to invest in new technologies, as well as they determine the money amount of that investment.

Analyzing the availability and use of computers as well as IT in Poland, it seems to be necessary to improve conditions and information infrastructure including broadband Internet access. Such situation should positively influence both information society development as well as enterprise development into knowledge-based economy conditions.

REFERENCES

Bolisani E., Scarso E., Information technology management: A knowledge-based perspective. *Technovation* no. 19/1999, Elsevier Science Ltd.

Commission Staff Working Document, Europe's Digital Competitiveness Report-ICT Country Profiles, 2009, p. 46, (http://ec.europa.eu/information_society/eeurope/i2010/docs/annual_report/2009/sec_2009_1104.pdf).

Concise Statistical Yearbook of Poland, Central Statistical Office, Warszawa, 2010, (http://www.stat.gov.pl/cps/rde/xbcr/gus/PUBL_oz_maly_rocznik_statystyczny_2010.pdf).

Gibbs D., Tanner K., Information and communications technologies and local economic development policy: The British case. *Regional Studies* no. 31/8, 1997.

Kalkowska J., Wlodarkiewicz-Klimek H., Information technologies in concurrent engineering in selected knowledge based economy companies in Poland—some results of research. In: *Advances in Human Factors, Ergonomics and Safety in Manufacturing and Service Industries*, Karwowski W., Salvendy G. (Eds.). Boca Raton, Taylor & Francis Group, USA, 2011.

Kalkowska J., Wlodarkiewicz-Klimek H. (Eds.), *Managing Enterprises. Social Aspects*. Monograph, Publishing House of Poznan University of Technology, Poznań, 2009.

Korres G., *Technical Change and Economic Growth: Inside the Knowledge Based Economy*. Ashgate Publishing Group, Abingdon, Oxon, GBR, 2008.

Morgan A., Colebourne D., Brychan T., The development of ICT advisors for SME business: An innovative approach. *Technovation* no. 26/2006, 980–987.

OECD, *The Knowledge-Based Economy*. OECD, Paris, 1996.

Sartorius W., Information society and knowledge based economy in Eastern Poland [Społeczeństwo informacyjne i gospodarka oparta na wiedzy w Polsce wschodniej], http://www.mrr.gov.pl/rozwoj_regionalny/poziom_regionalny/strategia_rozwoju_polski_wschodniej_do_2020/dokumenty/Documents/3c06195d761142c9b489f66512d7a63cSartorius.pdf.

Skulimowski A.M.J., The Information Society in Poland: Recent developments and future perspectives, 2006, (http://www.scholze-simmel.at/starbus/r_d_ws1/poland.pdf).

Spanos Y.E., Prastacos G.P., Poulymenakou A., The relationship between information and communications technologies adoption and management. *Information and Management* no. 39/2002, 659–675.

The Strategy for the Development of the Information Society in Poland till 2013, Ministry of Interior and Administration, 2008, (http://www.mswia.gov.pl/portal/SZS/495/6271).

Wlodarkiewicz-Klimek H., Kalkowska J., Shaping the strategy of knowledge based economy of Polish enterprises. In: *Advances in Human Factors, Ergonomics and Safety in Manufacturing and Service Industries*, Karwowski W., Salvendy G. (Eds.). Boca Raton, Taylor and Francis Group, USA, 2011.

World Bank, *Korea as Knowledge Economy. Evolutionary Process and Lessons Learned.* World Bank, Overview, Washington, DC, 2006.

World Bank Group Strategy, *Information and Communication Technologies*. World Bank Publications, 2002.

53 Innovative Work Behavior (IWB) in View of the Analysis of Chosen Individual and Organizational Determinants

Agnieszka Wojtczuk-Turek

CONTENTS

INTRODUCTION

The increase of interest in employee innovative behavior stems from the search for factors that facilitate human capital–based competitiveness of firms, as it is more and more often stressed that, at present, economic value is created, to a significant extent, on the basis of intangible resources (Kianto et al. 2010).

Employee innovative behavior is related to the key aspects of organizational effectiveness: generation, promotion, and realization of new ideas that benefit performance (Sanders et al. 2010). Thus, both theoreticians and practitioners are concerned with determining individual and organizational predictors, which can later become a foundation for creative activities to be undertaken in a workplace. The recognition of such predictors is of a crucial relevance for the practice of human capital management in any organization—it allows us to stimulate the aforesaid behaviors by utilizing the competence potential of the employees and to create an appropriate organizational environment for such behaviors' development.

THEORETICAL CONSIDERATIONS AND HYPOTHESES

Innovative Work Behavior

The category of "innovative behavior" (West and Farr 1989; Scott and Bruce 1994; Janssen 2000; Kleysen and Street 2001; Yuan and Woodman 2010) is a construct related to an employee's individual characteristics within specifically undertaken forms of activity. It is defined as the sum of the individual's intentional actions that are aimed at generation, promotion, and realization of new ideas within a work role, group, or organization, in order to benefit role performance, the group, or the organization (Janssen 2000), at any level of organization (West and Farr 1989). It comprises, among others, the development of ideas related both to new products and technologies and to administrative procedures that serve to improve relations at a workplace and notably increase their effectiveness (e.g., search for new technologies, promotion of new means of goal achievement, application of new work methods).

The processuality of such behaviors is stressed (as it is possible to isolate certain stages in their course), and so is the fact that they encompass activities related strictly not only to the generation of ideas but also to taking actions that facilitate their promotion. From such a perspective, the innovative behavior reveals itself as a multiphase process, within which an individual recognizes a problem and, subsequently, generates new ideas, promotes them, builds support for their implementation, and finally, develops an appropriate model for using them to benefit the organization (Kleysen and Street 2001; Yuan and Woodman 2010).

Scott and Bruce (1994) point out the three types of behavioral tasks that constitute innovative behavior: generation, promotion, and realization of ideas. It has been suggested, however, to consider some other activities within innovative behavior (they could be treated as its dimensions): opportunity exploration, idea generation, championing, and application (De Jong and den Hartog 2010).

The analyses of various types of activities of creative nature conducted by Kleysen and Street (2001) allowed the isolation of 17 types of behaviors, which comprise 5 general dimensions of innovative behavior. They include *opportunity exploration*, treated as looking for and recognizing opportunities to innovate and gathering information about such possibilities; *generativity*, related to the activities that serve the good of others and bring forth development and beneficial changes in the organization (for the employees, but also in the area of products, services, and processes); *formative investigation*, concerned with analysis and support for ideas, solutions, and opinions (characteristic activities comprise formulating ideas and solutions and their testing and evaluation); *championing*, which comprises such behaviors as propagating and promoting, which are indispensable when carrying out potential ideas and solutions (mobilizing resources, gaining social influence, and negotiating); and *application*, which comprises implementation of solutions and their modification until they become a routine and are accepted in the organization as a standard and, as a result, the innovation loses its "novelty" status and turns into an "ex-innovation."

The above characteristics allow us to state that the competences necessary to undertake and efficiently realize innovative behavior exceed those which are usually associated with individual innovativeness; for example, creativity, since the key characteristic of such behaviors is the implementation of ideas (which does not have to be present in the case of creative behaviors). Nevertheless, the fact remains that implementation/realization of ideas has a creative character as well, since the process is often related to a need to solve all kinds of problems of organizational, technological, social nature, and so forth. From this point of view, it seems justified to search for predictors of innovative behavior within a wide range of individual and organizational (and also situational) variables and to determine the conditions under which they can be undertaken and successfully implemented.

The number of determinants that are included into the field of analysis related to this issue is constantly increasing and continues to stir scholars' interest. Both the individual and organizational

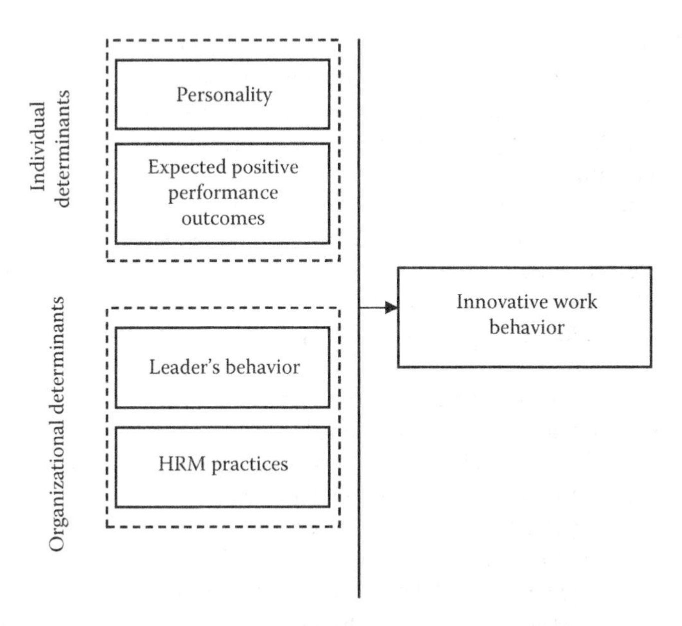

FIGURE 53.1 The hypothetical model of relationship between personality, expected performance outcomes, leader, HRM practices, and innovative work behavior.

determinants have been undergoing empirical verification. As far as *individual* determinants are concerned, the analyses included the relationship of innovative behavior with, among others, distributive and procedural justice (Janssen 2004; Ramamoorthy et al. 2005); perceived job challenge (De Jong and Kemp 2003); the perception of organizational climate (Scott and Bruce 1994); expected positive performance outcomes (Yuan and Woodman 2010); self-esteem and perceived insider status in organization (Chen and Aryee 2007); self-leadership competences (Houghton and Neck 2002; Carmeli et al. 2006); job satisfaction (Sanders et al. 2010); and commitment (Jafri 2010). In the case of *organizational* determinants, the following were examined, among others: degree of job autonomy (De Jong and Kemp 2003; Ramamoorthy et al. 2005); job demands (Janssen 2000); leadership (De Jong and den Hartog 2010); Human Resources Management (HRM) practices (Reuvers et al. 2008); and psychological contract (Ramamoorthy et al. 2005).

While attempting to specify the determinants of innovative behavior, it needs to be stressed that a model created to explain them should have an interactive character and consider both the individual and organizational determinants. In view of the analysis of literature on the subject of determinants, it is possible to suggest a hypothetical model for examination of the variables and their relationships (Figure 53.1).

Personality

In view of the conceptualization of the notion of "innovative behavior," one of the personality variables—creativity—gains a rather self-evident significance for the notion's explanation. Researchers stress the relationship between company innovativeness and employees' creativity, as the creation of values is subject to information and ideas (Alves et al. 2007). Creativity indicates generation of new and useful ideas related to the procedures and processes applicable in a workplace (Amabile 1988; Oldham and Cummings 1996). It forms a condition for new and original ideas to occur, although, as noticed by Amabile (1988), it differs from the innovative behavior in that creativity itself does not postulate the realization of the idea—innovative behavior is supposed to generate notable benefits for both the employee and the organization, resulting from the implementation of new solutions. Creativity can, however, form an important element of innovative work behavior (IWB), most visibly at the beginning of the innovation process. Within this context, the problem-solving style—organized, sequential, and logical as opposed to intuitive and involving the subconscious processes (Scott and Bruce 1994)—was mentioned as an individual attribute influencing innovative behavior. From the publications, there emerges a relationship, considerably well documented, between personality traits, conceptualized mostly in Costa and McCrae's (1992) Big Five model, and creativity (King et al. 1996; James and Asmus 2001; Wolfradt and Pretz 2001; George and Zhou 2001; Kelly 2006; Sung and Choi 2009), and determining the attempts at innovational activity. Several factors are mentioned as necessary for an individual to display a creative personality, such as tolerance for ambiguity, stimulus freedom, functional freedom, flexibility, risk taking, preference for disorder, delay of gratification, freedom from sex-role stereotyping, perseverance, and courage (Dacey and Lennon 1998).

It was also demonstrated that extraversion and openness to experience positively influence the undertaking of creative activities (George and Zhou 2001; Kelly 2006; Sung and Choi 2009). Both features are related to initiative, which is indispensable for innovation to occur (Talke et al. 2006). The recognition of opportunities/chances for implementation of innovation depends, to a large extent, on individual initiative and capabilities rather than on the routine practices and processes taking place in a company. In view of the initiative model based on individual competences and inspired by Amabile's (1988) concept, which refers to creativity and the theory of motivation, it can be stated that there are two basic stages relating to initiative emergence: the first one, during which generation of creative ideas occurs, activated by inner motivation (internal motivation, "operational knowledge," and creative cognitive style play an important role here), and the second stage, related to the selection of initiatives, which requires external (task-related and instrumental) motivation to be triggered and calls for strategic knowledge and an analytical cognitive style. Indeed, both the selection of ideas and the decision to undertake an initiative require a consideration of the enterprise's business goals, the costs of realization, the risk evaluation, and the assessment of resources and technological aspects. The choice of the most "promising" initiative has, therefore, a rational character (Talke et al. 2006).

Initiative is, in a natural way, related to proactiveness, which, in turn, is a product of personality predispositions and situational factors. A proactive personality is expressed in a tendency to "create" and influence the work environment. It has been shown that personality predispositions of this kind are related to professional success and that they explain employee effectiveness to a higher degree than, for example, conscientiousness or extraversion (Crant 2000). It has been empirically confirmed that a proactive personality is connected with employees' creativity (Kim et al. 2010) and their innovative behaviors, which are related to the development of new ideas and the demonstration of these in their work (Seibert et al. 2001).

While initiative constitutes a preliminary stage and a condition for the initiation of the innovative process, self-regulation of one's actions is indispensable for its continuation and successful completion. It has been established that there is a relationship between predispositions for self-leadership and innovativeness of work behaviors (Houghton and Neck 2002; Carmeli et al. 2006). Generation of creative activities consists of several levels of self-regulation: goal setting; action control and evaluation (cognitive, motivational); formulation of action plans and application of strategies (cognitive and motivational strategies); and the use of cognitive capabilities. As indicated by Houghton and Neck (2002), self-leadership strategies can be divided into three general categories: (1) behavior-focused strategies (the management of behaviors while taking into consideration the necessary but possibly unpleasant tasks); (2) natural reward strategies (stressing the pleasant aspects of a task); and (3) strategies based on constructive thought patterns (visualizing future activities as rewarded with success).

Despite the fact that the individual aspects that underlie the manifestation of innovative behaviors are numerable, in the publications on the subject, relatively little attention is paid to the examination of relationships of innovative behaviors, while, at the same time, taking into consideration the multiple personality traits/dimensions. The integration of the knowledge related to varied cognitive and motivational predispositions of an individual may provide a basis for establishing a fuller picture of the relationships of innovative behaviors at a workplace (IWBs) with the personality as a whole. Accordingly, hypothesis 1 was formulated.

Hypothesis 53.1: Personality is positively related to innovative behavior. ■

Expected Positive Performance Outcomes

From the point of view of undertaking and successfully completing IWBs, another individual dimension, defined in the literature as "expected positive performance outcomes," seems to gain significant prominence. It denominates the attitude that reflects the assumption within which an employee expects performance improvement or efficiency gains when convinced that his IWBs will result in improvement of productivity or effectiveness (as expressed, for example, in an increased ability to achieve goals, improved work quality, or decreased error rate) or in benefits for his work role or work unit (Yuan and Woodman 2010). Such an approach reflects the efficiency-oriented perspective, which dominates in organizations and which states that an employee's innovative behavior is shaped by his expectations as to the potential influence of such behavior on work efficiency (Yuan and Woodman 2010).

Efficiency dimensions are, naturally, varied, depending on an employee's job position. In relation to performance domains presented by Ostroff and Schmitt (1993), it may be stated that given organizational behaviors of employees (including innovative behaviors), which serve a goal and result in achievement, may be discussed in terms of input and output. Thus, in accordance with the exchange theory (Blau 1964), work involvement may be examined in the context of mutual exchange of resources between the employee (his time, involvement, loyalty) and the employer, who grants him economic, social, and psychological gains. Apart from the formal contract, the conditions of such an exchange are shaped by psychological contract, which influences behaviors initiated by the employee in the organization. This may explain the fact that the employees who undertake innovative behaviors expect financial rewards (Ramamoorthy et al. 2005). In this case, the important mechanism is how the employee perceives a degree to which his expectations are fulfilled by the employer, which, in turn, influences the employee's sense of obligation towards the employer and shapes the perception of procedural justice—which is, in the view of the research, related to innovativeness (Ramamoorthy et al. 2005). As it were, an employee feels obliged/bound to "be innovative," that is, to employ his/her unique knowledge in order to offer suggestions for improvement. In addition, this sense of obligation is influenced by support received by the employee in the organization while realizing professional goals (Aselage and Eisenberger 2003). Certainly, a sense of obligation does not constitute the sole factor explaining the undertaking of innovative behaviors, as, in view of the research on the relationship between such behavior type and organizational involvement, another factor turns out to be predictive, namely, emotional involvement, based on identification with goals and organizational values (Jafri 2010), and on the internal motivation (Amabile 1988).

Another important aspect of the belief in positive performance outcomes (especially with relation to idea generation and implementation) is formed by the conviction about one's *self-efficacy*, which forms a foundation for forecasting positive or negative outcomes of the undertaken activity, based on the subjective sense of the difficulty level in a given situation. The evaluation of the possibility for realizing the employee's own ideas is, naturally, accomplished not only on the basis of the employee's judgment of his/her competences, understood as his knowledge, skills, or attitudes, but also in the context of his entitlement to decide whether an idea—and which idea—will be chosen for realization, by the use of which methods, and with what access to resources. The conviction about one's self-efficacy, forming a component of the psychological capital (Luthans et al. 2007), exerts (apart from professional qualifications) a direct and indirect influence on a successful realization of professional goals (Stajkovitch and Luthans 1998). At the same time, for the achievement of professional outcomes, not only the psychological capital is important but also a positive supportive professional environment. It both creates a basis for maintaining a high level of work output and conditions the employee's involvement and sense of job satisfaction.

Conviction about one's self-efficacy may have a general character, but when it has a specific character, it applies to a particular range of activity, situations, or even tasks. In the context of innovative behaviors, it seems worthwhile to mention "the sense of creative self-efficacy," which reflects an individual's belief in possessing the ability to produce creative outcomes (Tierney and Farmer 2002, p. 1138). Creative activities, on the other hand, are facilitated by the possession of general skills, specific skills (for a given field), and task motivation (Amabile 1996). A high level of creative efficacy sense, together with a high level of self-expectations for creative behavior, is strongly related to creative work involvement (Carmeli and Schaubroeck 2007).

The above considerations allow for the adoption of the next research hypothesis:

Hypothesis 53.2: Expected positive performance outcomes are positively related to innovative behavior. ■

Leadership

Leadership is mentioned among other significant determinants of innovativeness in a workplace (Mumford et al. 2002; Mumford and Licuanan 2004)—especially, leadership of a transformational character (Reuvers et al. 2008; Rank et al. 2009; Dubey and Ghai 2010). Transformational leaders use their personality traits to create visions, to inspire and motivate employees. They build organizational culture and establish new, higher work standards, while taking into consideration the associates' individuality (Takala 2005).

Transformational leadership is shaped by four factors: (1) *idealized influence*—the leader is an exemplary role model, admired and respected; (2) *intellectual stimulation*—the leader encourages the employees to think for themselves, to challenge the established assumptions, and promotes their recognition of their own convictions and values; (3) *individualized consideration*—the leader devotes attention to employees, is aware of their professional needs, and creates opportunities for development and self-fulfillment; (4) *inspirational motivation*—the leader unfolds an optimist vision of the future and encourages the employees to increase their expectations and their efforts towards goal achievement (Bass 1990; Bass and Avolio 1994). The leader's activities mentioned above create favorable conditions for the employees to undertake increased efforts towards the realization of professional challenges, since such activities activate competences; become an opportunity for experiencing positive emotional states (curiosity, astonishment, fascination), provide a sense of efficacy and control, and induce cognitive conflict (divergence, inconsistency, deficit, indeterminacy, dissonance)—that is to say, situations well managed by creative employees.

From the point of view of motivating for implementation of generated ideas, it is an important behavior of a leader to signal his/her approval for an employee's innovative activity and to provide support during its realization (e.g., by providing resources or exercising his/her job position). The support for creative ideas forms an important component of pro-innovative organizational culture (West 2000; Tesluk et al. 1997; Ekvall and Ryhammar 1999). The results suggest that when the superiors are perceived as helpful in the realization of innovational activities, the associates feel encouraged to use their influence while employing innovative activities in a workplace (Janssen 2005).

As an employee perceives support on the part of a superior, his/her effectiveness increases, particularly when the employee displays low organizational-based self-esteem (Rank et al. 2009). Low self-esteem causes employees to doubt whether their efforts aimed at introducing new ideas are valuable for the organization, especially when encountering resistance. Thus, such employees may particularly benefit from the presence of a transformational leader, who instills optimism and inspires self-confidence (Rank et al. 2009).

In this context, it is also necessary to stress the delegation of functions, which forms a clear signal for the employee that his/her superior considers him/her to be competent, capable, important, and necessary for the organization (Chen and Aryee 2007). When an employee incorporates such positive information into his/her self-image, it causes an increase in his/her organization-based self-esteem.

Delegating is also based on acceptance of an employee as a member of the organization and on his/her perceived insider status. Within the theory of Leader–Member Exchange (LMX) (Dansereau et al. 1975), it is stressed that the quality of relations between the manager and the employees is varied and subject to change. The managers and the employees engage in a process of "role development," within which an understanding is reached as to the extent of decision-taking freedom and the influence and autonomy granted to the employees. In the course of time, the formal and impersonal relations between the leader and the employee (low quality of LMX) evolve into mature relations based on trust, mutual friendliness, and respect (high-quality LMX). The second type of relations provides the employees with larger autonomy and increased range of freedom in decision making. Delegating constitutes a clear signal for the employee that he/she is trusted, that he/she is granted insider status and made responsible for some undertaken activities. In this sense, such determinants perform the intermediary role in the influence of delegating on undertaking of innovative behaviors by employees (Chen and Aryee 2007), although their direct impact has also been pointed out (Scott and Bruce 1994).

Hypothesis 53.3: The leader's behavior and its perception by the employees are positively related to innovative behavior. ■

HRM Practices

Organizational practices within human capital management are regarded as another significant factor of individual innovativeness of employees (De Leede and Looise 2005; Dorenbosch et al. 2005; Shipton et al. 2006). They derive from the strategy accepted in a company and from its prevailing organizational culture, which expresses the most important organizational values. When innovativeness is one of these values, the management's activities will be directed at searching for, utilizing, and developing innovative potential related to human capital.

The scholars analyzing the issue of HRM in relation to innovativeness have indicated the following aspects: appropriate organizational structure, staffing of innovative organizations, key roles of individuals, individual development and careers, effective team working and leadership, extensive communication and participation, efficiency (together with its measures) and rewards, and creating a creative culture (De Leede and Looise 2005). According to the authors, traditionally realized HRM practices have to be "renewed" in line with organizational strategy and in accordance with the stages of innovative process.

The empirical research confirms the influence of HRM on such organizational mechanisms as development and use of intellectual capital (Wright et al. 2001), knowledge creation and development of new products (Collins and Smith 2006), or organizational learning (Snell et al. 1996). The latter aspect is particularly relevant in the case of knowledge-intensive firms, for example, service sector companies, where organizational learning processes and knowledge transfer form a basis for creation of innovative solutions. Such a basis allows for creating novel solutions within knowledge management and for the creation of the so-called "best practices manuals," or procedures for registration of problem solutions, which can be unrestrictedly used by the employees.

The practices that are related to innovativeness include trainings; employee induction in the organization (and providing him/her with the knowledge of goals, norms, and processes); teamwork; appraisal; and exploratory learning (Shipton et al. 2006). In particular, the last practice mentioned has a lot in common with innovative behaviors, since exploratory learning requires new ideas to be created by way of an active search for alternative viewpoints (McGrath 2001; Danneels 2002). It is accomplished both during knowledge exchange in an organization and in cooperation with individuals from outside. In the light of research on predictors of innovative behaviors, a relationship has been demonstrated between the factor of "generating ideas and implementing them" and the pro-innovative dimension of organizational culture: "positive interpersonal exchange and management support" (Wojtczuk-Turek 2010). This dimension is related to employees' participation, rewards gained for proposing and realizing ideas, promotion of cooperation and knowledge sharing, and open communication style.

In view of the analyses of organizational climate as a predictor of innovativeness, Scott and Bruce (1994) recognized the character of performed tasks as relevant for displaying such behaviors (together with an employee's decision making participation). In the case of realization of routine tasks and the low participation in decision making, the relation between the climate and innovative behavior is lower than when the task is of a nonstandard nature and the employee has a high level of autonomy in decision making. Naturally, it is the "psychological climate" that is of concern here—due to the fact that it expresses the individual's cognitive interpretation of the situation in the organization, as a place which favors, or is disposed against, the realization of one's own creative abilities. It creates a base for a sense of satisfaction and for forming positive judgments of the organization, including HRM practices. Satisfaction with Human Resources (HR) activities acts as an intermediary in the relation between LMX and innovative behaviors (Sanders et al. 2010).

One of the most important HRM practices, determining the occurrence of employee innovative behaviors, is an opportunity for competence improvement. Such a goal may be realized in the form of both trainings and job design operations, for example, a temporary position change or teamwork forming. What is aimed at here is the creation of an opportunity for knowledge transfer and for shaping a diversified job environment, thanks to which the employees will be able to fulfill the needs related to creative activities: the need for achievement, new experiences, varied stimulation, curiosity, self-fulfillment, success, self-efficacy and control, and realization of challenges. These factors draw on internal motivation, indicated as a basic mechanism for inducing and maintaining effort directed at realization of innovative goals. According to Amabile (2004), the creative influence of motivation results from the fact that "internally motivated" individuals present, during realization of job tasks, a higher involvement in activities of exploratory nature, which creates an opportunity for a more novel and increasingly flexible approach. However, the role of financial rewards is not without significance (Ramamoorthy et al. 2005).

The analysis of the relationship between HRM practices and innovativeness allows us to formulate the next hypothesis.

Hypothesis 53.4: HRM practices are positively related to innovative behavior. ■

METHOD

The goal of the research was to establish the chosen determinants of innovative behaviors, both of individual and organizational nature. The problem addressed in the research concerned the verification of the assumption that there is a relation between the occurrence of employees' innovative behaviors on one hand and their personality predispositions, expected positive performance outcomes, leader's behavior in a workplace and human resources management organizational practices on the other.

Examined Variables and Measures

The conducted survey took into consideration the following set of variables:

Dependent Variable

Innovative behavior—This is defined as generation, promotion, and realization of ideas. The variable was measured with a 14-item Innovative Behavior Questionnaire developed by Kleysen and Street (2001). The data were provided by choosing an answer to every statement from a 6-point scale, where 1 = "never" and 6 = "always." In the process of cultural adaptation of the instrument, statistic analyses were performed for the sake of secondary verification of the reliability. The coefficient of reliability α for the whole instrument amounted to 0.94. On the basis of the factor analysis (KMO = 0.929, $\chi^2 = 2016.359$, df = 91, $P < .001$) performed by the method of principal component analysis with varimax rotation (using Kaiser normalization), two factors were isolated: *recognizing problems and initiating activities* (factor 2, $\alpha = 0.83$) and *generating ideas and implementing them* (factor 1, $\alpha = 0.92$). Jointly, they account for 62% of variances.

Independent Variables

Personality—This is understood to refer to personality dimensions underlying human behavior and isolated on the basis of factor analysis. They were formulated as concepts on the basis of which an instrument for personality diagnosis was created—the Adjective Checklist of H. G. Gough and A. B. Heilbrun (1983). It provides a functional measure for defining the examined features, and the content characteristics was based on specific scales of the test. Thus, the personality traits were described in terms of Murray's system of needs, Berne's transactional analysis theory, and Welsh's conception. The variable was measured with a modified adjective check list (ACL) adjective test (Płużek 1978). Reliability coefficients (Cronbach's α) fluctuate in the range from 0.53 to 0.95. The test's factor analysis, performed with varimax method (using Kaiser normalization), with relation to the number of factors, allowed us to isolate six factors: *strength, self-confidence, socialization, self-control, kindness*, and individuality.

Expected positive performance outcomes—The variable expresses an attitude in accordance with which the expected performance outcomes are positive when the employee believes that innovative behaviors in a workplace will result in productivity or effectiveness improvement. In order to measure this variable, a questionnaire of attitudes, behaviors and organizational practices was employed, for which scale items were adapted from the instrument featured in the study of Yuan and Woodman (2010, p. 341), and which refer to a study by House and Dessler (1974). The instrument also examines leadership and HRM practices, which may favor IWBs. On the basis of explorative factor analysis by the method of main component analysis with varimax rotation (KMO = 0.895, $\chi^2 = 1716.222$, df = 136, $P < .001$), three factors were isolated: *leadership, HRM practices*, and *expected outcomes*. The coefficient of reliability (Cronbach's α) amounted to 0.88 for the whole instrument.

Leadership, HRM practices—They comprise a set of organizational factors that are related to a leader's behavior and his/her perception and to HRM practices that may support the occurrence of innovative behaviors: (1) *leader's behavior*—the superior's real support while employees' ideas are realized, possession of a considerable authority among employees and inspiring their trust, use of one's position to solve employees' problems, a sense of fair treatment by the superior and his/her understanding for employees' problems and professional needs; (2) *HRM practices*—among others, a temporary change of position in order to increase knowledge and competences, opportunity to gain knowledge that is relevant from the point of view of tasks performed at work, opportunity for improvement of competences that are not directly related to one's professional activity. In order to measure the variable of leader's behavior and HRM practices, a Questionnaire of Attitudes, Behaviors and Organizational Practices was employed, for which the scale items were adapted from the instrument featured in the study of Yuan and Woodman (2010, p. 342), and which refer to Scott and Bruce's study (1994).

The research methods also accounted for the controlled independent variables, relevant for the analyzed dependent variables, which comprise age, sex, education, job seniority, job position, company size, and line of business.

Sample and Procedures

The surveys covered 246 employees of firms diversified as to size and line of business. The majority of employees who participated in the survey represented large companies (68%) from the following sectors: *processing industry* (32%), *production* (27%), *financial agency services and banking* (12%), and *telecommunications* (10%). Among the respondents, the majority were employees within the age range of 26–35 (60%), with a university education (92%), holding managerial positions (69%), mostly with work experience of over 5 years (55%); among the respondents, 64% were female, 36% male. The survey was anonymous; the questionnaire was sent to the respondents via electronic mail. Five hundred thirty-eight questionnaires were distributed, and 246 of them were answered and returned.

Results

The first step in the research procedure was to search for a relationship between innovative behavior and measured variables, based on *Pearson's r* correlation analysis. Correlation coefficients for specific variables are presented in Table 53.1.

On the basis of the conducted correlation analysis, it was observed that there is a significant statistic relationship between the three personality factors, "self-confidence," "self-control," and "individuality," on one hand and two innovative behavior factors, "recognizing problems and initiating activities" and "generating ideas and implementing them," on the other. Thus, it appears that undertaking innovative behaviors is related to the personality dimensions that concern both occupational and social functioning. The strongest relationship of the two dimensions of innovative behavior was marked in relation to the personality trait of "self-confidence," whose psychological meaning is connected to such characteristics of undertaken social interactions, which feature drawing attention to oneself, establishing contacts easily, and ability to become a group leader, together with self-trust and lack of doubts about one's abilities. Such types of behaviors are extremely useful in the context of promoting one's ideas and gaining support for their realization. The analyses conducted by Robertson and Myers (1969) presented a distant but statistically relevant relation between innovative behavior and one of the personality traits, that is, the sense of self-esteem and self-acceptance.

The relationship between the competence factor of "leadership" and the dimension of "generating ideas and implementing them," revealed in another study concerned with competence predictors of innovative behaviors (Wojtczuk-Turek 2010), indicates the significance of the social context for the occurrence of innovational behaviors, especially during the stage of idea implementation. The said research showed a positive relation of innovational behavior dimensions with *the need to dominate* (Dom $r = .294$, $P < .01$ for "generating ideas and implementing them" and $r = .290$, $P < .01$ for "recognizing problems and initiating activities") and *the scale of self-trust* (S-Cf $r = .278$, $P < .01$ for "generating ideas and implementing them" and $r = .277$, $P < .01$ for "recognizing problem and initiating activities"), and a negative relation with *the need to abase oneself* (Aba $r = -.201$, $P < .01$ for "generating ideas and implementing them" and $r = -.208$, $P < .01$ for "recognizing problem and initiating activities"). High domination and self-trust together with a low level of the need to abase oneself signify determination, self-confidence in contacts with others, assertiveness, lack of discomfort in the situation of experiencing others' disapproval, and attention paid to making a "good impression." The achieved result confirms the legitimacy of assuming a "social–political" perspective while examining innovative behavior, which is created as an effect brought about by the manifestation of the employee's specific image in his/her organization (Yuan and Woodman 2010).

The second personality factor that is related in a significant manner to the innovative behaviors is "self-control." It is more strongly related to the dimension of innovative behavior defined as "recognizing problems and initiating activities." The dimension of "self-control" reflects the characteristic manner in which an individual functions in the task-related sphere, the said manner being

TABLE 53.1
Pearson *r* Correlations among Researched Variables

Variable	1	2	3	4	5	6	7	8	9	10	11	12	13	14	15	16
1. **Generating ideas and implementing them**	1															
2. **Recognizing problems and initiating activities**	$.788^{**}$	1														
3. Strength	.051	.095	1													
4. Self-confidence	**$.233^{**}$**	**$.254^{**}$**	−.029	1												
5. Self-control	**$.174^{**}$**	**$.204^{**}$**	$.355^{**}$	$.224^{**}$	1											
6. Kindness	−.066	−.080	$-.212^{**}$	−.019	$.223^{**}$	1										
7. Individuality	**$.136^{*}$**	**$.146^{*}$**	−.063	$.179^{**}$	$.645^{**}$	.054	1									
8. Leader's behavior	.083	.113	−.003	−.129	.129	.063	.100	1								
9. HRM practices	.116	.066	−.073	−.027	.076	−.030	.090	$.519^{**}$	1							
10. **Expected positive performance outcomes**	**$.145^{*}$**	**$.139^{*}$**	.025	−.025	.100	.008	.022	$.467^{**}$	$.318^{**}$	1						
11. Company size	.001	.015	−.067	−.050	−.057	−.034	−.089	$.133^{*}$	$.166^{*}$	$.204^{**}$	1					
12. Education	.012	−.031	−.022	.019	−.037	−.118	−.068	−.014	.073	−.054	.100	1				
13. Job position	**$-.231^{**}$**	**$-.203^{**}$**	.012	$-.161^{*}$	−.092	−.008	−.020	.007	.107	−.009	.058	$.153^{*}$	1			
14. Job seniority	.105	.114	.063	−.047	−.053	−.110	−.052	.116	.018	$.153^{*}$	$.298^{**}$	.053	$-.188^{**}$	1		
15. Sex	.119	.036	$-.219^{**}$	.016	.067	.082	.096	.000	.022	.064	.016	$.132^{*}$	−.069	−.078	1	
16. Age	.110	.095	.048	−.031	−.063	$-.150^{*}$	−.082	.040	.086	.055	$.314^{**}$	.056	$-.201^{**}$	$.502^{**}$	−.008	1

$^{*}P < .05$; $^{**}P < .01$.

characterized by the ability to undertake activities and to finalize them, conscientiousness, orderliness, planning, prudence, and fulfillment of obligations. It coincides with the data obtained in another research study, which indicated a relationship between innovative behavior and the ability of "self-leadership" (Houghton and Neck 2002; Carmeli et al. 2006), and especially with the dimension of "goal determination" (Wojtczuk-Turek 2010). The obtained result confirms the role of self-regulatory processes in the process of creating and implementing solutions.

In this context, it seems necessary to indicate the relationship revealed in the research between the innovative behavior and the *need for achievement* (Ach), included in the personality trait defined as "strength." Despite the fact that there were no correlations on the factor level, the analysis of separate scales of the ACL test revealed it (Ach $r = .216$, $P < .01$ for "generating ideas and implementing them" and $r = .244$, $P < .01$ for "recognizing problems and initiating activities"). Hence, it is possible to state that innovative behaviors are related to expectation to achieve long-term goals, high motivation for activity (resulting from aiming at fulfillment of high standards), and readiness to make an effort.

The factor of "self-control" is also related to an inquiring attitude towards the surrounding world, which forms an essential condition for recognizing problems and initiating activities. Openness for experience (George and Zhou 2001; Kelly 2006; Sung and Choi 2009) and proactivity form a significant factor of innovative behavior occurrence in a workplace (Crant 2000).

From the point of view of the innovative behaviors, the aspects expressed in the third personality trait of "individuality," which is more strongly related to the dimension of "generating ideas and implementing them," also are of marked significance. This factor covers cognitive abilities; originality of thinking and perception; logicality; appreciation for intellectual and cognitive issues; wideness of interests; esthetical sensitivity; drawing creative inspiration from what is complex, changeable, and disordered; speed and perceptivity when gaining insight into problematic situations; self-trust; courage; self-sufficiency; and strength of will. This description may be related to a cognitive style indicated by Scott and Bruce (1994), used for problem solving and forming an individual attribute determining innovative behavior. On the level of analysis of individual scales of the ACL test, the relationship with innovative behaviors was particularly noticeable with regard to the two scales that constitute the factor of "individuality": *creative personality scale* (Cps $r = .226$, $P < .01$ for "generating ideas and implementing them" and $r = .208$, $P < .01$ for "recognizing problems and initiating activities") and *the need for change* (Cha $r = .137$, $P < .05$ for "generating ideas and implementing them" and $r = .158$, $P < .05$ for "recognizing problems and initiating activities"). Moreover, though the relation of the need for change with innovative behavior is not close (while being statistically significant), it is necessary to point out the high significance of openness to new and complex situations, precisely in the context of initiating creative activities. The results obtained in the study confirm the importance of these personality predispositions that are related to creativity (King et al. 1996; James and Asmus 2001; Wolfradt and Pretz 2001; George and Zhou 2001; Kelly 2006; Sung and Choi 2009) in undertaking creative behaviors.

In the light of the obtained results, there are grounds for confirmation of hypothesis 1, which specifies the relationship between innovative behavior and personality traits.

Apart from personality variables represented in the ACL test, the presented research revealed the relation of innovative behavior with "expected positive performance outcomes." The condition for such expectations to arise is an employee's perception of the influence of undertaken innovative activity on the increase of his efficiency—expressed as, for example, increased capacity for goal achievement, job quality, or lower error rate (Yuan and Woodman 2010). A positive (distant, but statistically relevant) relation of "expected positive performance outcomes" was present with regard to both "generating ideas and implementing them" and "recognizing problems and initiating activities," which coincides with the results of the analyses of Yuan and Woodman (2010).

The achieved result appears consistent with the conclusions drawn from the analysis of the relation of personality traits with innovative behavior, where "self-confidence," including self-confidence in contacts with others, turned out to be crucial. Employee's attitude expressed as "expected positive performance outcomes" appears to be one of the key mechanisms regulating preservation of this activity, related to the implementation of ideas, which is accompanied by promoting the activity, gaining resources, and creating support for its realization. For initiating innovative activity, an employee's perception of the actual organizational financial support is significant, together with that element of pro-innovative culture, which is defined as "market orientation and realization of innovative ideas" (Wojtczuk-Turek 2010). The achieved result forms a basis for positive verification of hypothesis 2, which describes the relationship of innovative behaviors with this variable.

In this context, it is surprising that in the light of achieved research results, there was no marked relation of innovative behavior with either leader's behavior and leader's perception or HRM practices. Substantial research proves that these factors play an important role in creating IWBs (Reuvers et al. 2008). Thus, hypotheses 3 and 4 do not receive empirical support.

The analysis of secondary independent variables allows us to state that there is a relation of innovative behavior only with the job position (Table 53.1). The conducted comparison with the use of a *t* test between the persons in managerial positions and regular employees allows us to state that innovative behaviors are more often manifested by managers ($t = 3.555$, $P = .001$). The result is not surprising in the context of knowledge concerning the specifics of the occupational functioning of this group of employees, where the dominant mental activities cover problem solving, decision making, and strategic thinking—as a consequence of the increase of significance of operational, creative, and decisive abilities, caused by the increased complexity and uncertainty of the environment (Nosal 2001).

The last part of the analysis consisted of conducting an exploratory hierarchical regression analysis in order to establish the individual and organizational predictors of the occurrence of innovative behavior. Two independent models of regression were constructed with the use of a stepwise method (Table 53.2).

The first model, for the factor of "recognizing problems and initiating activities," is well fitting and reliable ($F = 8.592$, $P = .001$). It explains 9% of the variances of obtained results. The main predictors within this model are two personality traits, "self-confidence" and "self-control," together with "expected positive performance outcomes." However, the personality trait of "self-confidence" turned out to be the strongest predictor, which explains 5% of the variances of obtained results ($\beta = 0.266$). "Expected positive performance outcomes" explain 2% of the variances of obtained results. It allows us to state that the stronger "self-confidence," "self-control," and "expected positive performance outcomes," the more innovative behaviors are manifested by the employees of the organizations covered by the research.

The second model, for the factor of "generating ideas and implementing them," also appeared to be well fitting, and it explains 7% of the variances of the obtained results. In this case, the strongest predictor is one personal trait, "self-confidence," which explains 5% of the variances of obtained

TABLE 53.2

Results of Multiple Regression Analysis for Predictors of Innovative Behavior

Dimensions of Innovative Behavior	Predictors	B	SD	β	T	*P*	Model
Recognizing problems and initiating activities	Self-confidence	.049	.014	.226	3.460	.001	Adjusted $R^2 = .09$,
	Self-control	.024	.011	.149	2.149	.033	$F(222) = 8.592$,
	Expected positive performance outcomes	.223	.109	.130	2.039	.043	$P < .001$
Generating ideas and implementing them	Self-confidence	.116	.032	.237	3.683	.000	Adjusted $R^2 = .07$,
	Expected positive performance outcomes	.584	.249	.151	2.343	.020	$F(223) = 9.318$, $P < .001$

results. This model also covered "expected positive performance outcomes," which, as in the previous model, explain 2% of the variances of the results.

The obtained results allow for the statement that in the presented research, the key determinants of innovative behaviors are found on the side of individual factors (although their influence is not significant). The factors of organizational character, leader's behavior or HRM practices, were not present among the predictors, although the literature stresses the significance of the practices related to human capital management (De Leede and Looise 2005; Shipton et al. 2006), the person of the leader (Reuvers et al. 2008), or pro-innovative organizational culture (Ekvall and Ryhammar 1999; West 2000; Hunter et al. 2006).

Discussion

The conducted research focused on the search for relations of IWBs with individual factors (personality, "expected positive performance outcomes") and organizational factors (leader's behavior, HRM practices).

In the context of verification of research hypotheses, it was established that the manifestation of innovative behaviors related both to problem recognition and activity initiation and to idea generation and implementation is particularly strongly connected with these personality traits that are displayed in social situations, for example, self-confidence. It coincides with the data indicated in other research, in which individual innovativeness is associated with self-esteem and self-acceptance (Robertson and Myers 1969), experiencing a lower level of social anxiety (Hutchinson and Skinner 2007), or leadership competences (Wojtczuk-Turek 2010). The traits mentioned above are significant from the point of view of key types of innovative activity—presenting ideas, creating support for their realization, and implementation. One of the risks resulting from IWBs indicates, indeed, a destruction of the *status quo* (e.g., with regard to the established framework of task realization), which may lead to conflicts with associates (Shih and Susanto 2011). In this context, it seems justified to expand, in the further research, the analysis of the "social–political" perspective, which stresses the fact that innovative behavior is created as an outcome of an employee's manifestation of a certain image in the organization, translating, in turn, into gaining support and securing resources indispensable for realization of ideas (Yuan and Woodman 2010). It may well be that from this point of view, not only the "impression management" is of importance.

The obtained results also confirmed the significance, for the initiation and realization of innovative behaviors, of such personality predispositions as inquiring attitude towards the world, creativity, ability to undertake activities and to finalize them, conscientiousness, orderliness, planning, prudence, expectation to achieve long-term goals, and readiness to make an effort. The empirical analyses to date stress the relation of innovative behaviors to both "self-management" abilities (Houghton and Neck 2002; Carmeli et al. 2006) and, especially, the dimension of "goal setting" (Wojtczuk-Turek 2010), but also to personality predispositions that are related to creativity (King et al. 1996; James and Asmus 2001; Wolfradt and Pretz 2001; George and Zhou 2001; Kelly 2006; Sung and Choi 2009) or to taking initiative (Talke et al. 2006).

Another individual variable, crucial from the point of view of innovative behaviors, is the attitude defined as "expected positive performance outcomes," which coincides with the findings from the analyses of Yuan and Woodman (2010). This variable is related to such personality predispositions (self-confidence, self-control) that strengthen the employee's conviction of self-efficacy, especially in the aspect of undertaking innovative activities (Tierney and Farmer 2002). A strong sense of creativity efficacy together with a high level of self-expectations for creative behavior is strongly related to creative work involvement (Carmeli and Schaubroeck 2007).

Organizational factors—leader's behavior and HRM practices—were not found among the predictors of innovative behavior, although they are indicated as important determinants of employees' individual innovativeness (De Leede and Looise 2005; Reuvers et al. 2008). In the present research, innovative behaviors do not relate significantly either to the leader behaviors (actual support during realization of employees' ideas, using one's job position in order to solve subordinates' problems) or to the way the leader is perceived (possession of authority among employees and inducing trust, a sense of fair treatment by the superior, and his/her understanding for the problems and professional needs of the subordinates). In the literature, it is stressed that the perception of leaders as people who help to realize innovative activities encourages employees to make use of their influence (Janssen 2005), and when the employee perceives support on the part of the manager, his/her efficiency increases, particularly so when his self-esteem is low (Rank et al. 2009). In light of the obtained results, the respondents were characterized by a high level of self-confidence, and thus, in their case, taking advantage of managerial support is not essential. A leader's behaviors consisting of inducing optimism and creating self-confidence are valuable in the case of an employee's low self-esteem, since such an employee may doubt whether his/her efforts directed at introducing new ideas are valuable for the organization, especially when he/she encounters resistance (Rank et al. 2009). And, although in the research, "supportive attitude of the manager and associates during realization of ideas" ranked high among the factors of innovative involvement, the factors related in a significant way with the occurrence of innovative behavior are the following: opportunity for promotion, unrestricted activities, proposing and realizing ideas, teamwork membership, and risk-taking opportunity (Wojtczuk-Turek 2010).

Another direction of the interpretation of the obtained results may concern perception of the leader in terms of an authority figure, or the employee's expectations with regard to fair treatment by the leader. In the case when the employees perceive the manager as a person who cannot provide a source of inspiration and does not understand their problems, the strength of the leader's influence is much less perceptible. It may lead to the creation of a relation that is defined within the LMX theory (Dansereau et al. 1975) as a "low-quality leader–team member exchange," which definitely does not facilitate the creation of conditions for innovativeness. During the present research, the number of respondents who provided a positive answer to the statement "*My boss understands my job problems and needs*" amounted to 17%, "decidedly yes," and 33%, "yes, rather."

Moreover, if a subjective evaluation of leader status in a given organization is low, then the conviction that he/she could be helpful during realization of ideas may be weak. It is necessary to remember that in order to initiate innovative activity, it is important for the employee to perceive the actual organizational financial support and the dimension of pro-innovative culture, which is defined as "market orientation and realization of innovative ideas" (Wojtczuk-Turek 2010).

The unambiguous interpretation of obtained results is hindered by the fact that the research did not account for leader's behaviors related to transformational leadership and correlated with innovative behaviors: idealized influence, intellectual stimulation, individual consideration, and inspirational motivating (Bass 1990; Bass and Avolio 1994).

The obtained results of the research did not confirm the assumed relation with innovative behaviors of not only the leader's behavior but also the following HRM practices: the existence of a "best practices manual," a temporary change of job position in order to increase knowledge and competences, opportunity to gain knowledge that is relevant from the point of view of tasks performed at work, opportunity for improvement of competences that are not directly related to one's professional activity, and team forms of work. In the light of the research to date, the activities related to individual innovativeness are trainings, employee induction in the organization, teamwork, appraisal, and exploratory learning (Shipton et al. 2006). For that reason, it is particularly surprising that there is a lack of relation between innovative behaviors and opportunity for improvement of competences in the organization, as an answer to an employee's strongly expressed development needs. It is possible that it is precisely the innovative behaviors that provide an opportunity for accepting challenge, and thus, they form an area of improvement of competences within exploratory learning. The activities related with innovative behaviors—generating, promoting and implementing ideas—require the possession of a number of social and cognitive competences and, at the same time, provide a situational context for their natural training.

To sum up, it seems worthwhile to indicate the problems of the presented research and the future empirical perspectives. Firstly, from the point of view of continuation of analyses, it is indispensable to employ a larger number of tools for the measurement of organizational variables, which would allow us to consider a wider spectrum of a leader's behavior and HRM practices. In such a case, the examination of relationships between these variables would be more thorough. Secondly, the analyzed results were limited by the lack of control over the variable of "social desirability," which, in the case of scales based on self-report, may lead to a distortion of empirical material. It seems that the further research within the discussed field should also cover a larger group of variables of individual and organizational character, analyzed within the interactive model, since the expression of organizational behaviors (including the innovative behaviors) always occurs in a specific context, which may include factors of facilitatory nature and also factors of inhibitory nature.

REFERENCES

Alves, J., Marques, M.J., Saur, I., Marques, P. (2007) Creativity and innovation through multidisciplinary and multisectoral cooperation. *Creativity and Innovation Management*, Vol. 16, No. 1.

Amabile, T.M. (1996) *Creativity in Context*. Westview, Boulder, CO.

Amabile, T.M. (1988) A model of creativity and innovation in organizations. *Research and Organizational Behaviour*, Vol. 10.

Amabile, T.M. (2004) Stimulate creativity by fueling passion, in: *The Blackwell Handbook of Principles of Organizational Behaviour*, (Ed.) E.A. Locke. Blackwell Publishing, UK.

Aselage, J., Eisenberger, R. (2003) Perceived organizational support and psychological contracts: A theoretical integration. *Journal of Organizational Behaviour*, Vol. 24.

Bass, B.M. (1990) *Bass and Stogdill's Handbook of Leadership*. Free Press, New York.

Bass, B.M., Avolio, B.J. (Eds.) (1994) *Improving Organizational Effectiveness through Transformational Leadership*. Sage Publications, Thousand Oaks, CA.

Blau, P. (1964) *Exchange and Power in Social Life*. Wiley and Sons, New York.

Carmeli, A., Meitar, R., Weisberg J. (2006) Self-leadership and innovative behaviour at work. *International Journal of Manpower*, Vol. 27, No. 1.

Carmeli, A., Schaubroeck, J. (2007) The influence of leaders' and other referents' normative expectations on individual involvement in creative work. *The Leadership Quarterly*, Vol. 18, No. 1.

Chen, Z.X., Aryee, S. (2007) Delegation and employee work outcomes: An examination of the cultural context of mediating processes in China. *Academy of Management Journal*, Vol. 50, No. 1.

Collins, C.J., Smith, K.G. (2006) Knowledge exchange and combination: The role of human resource practices in the performance of high-technology firms. *Academy of Management Journal*, Vol. 49, No. 3.

Costa, P.T., McCrae, R.R. (1992) Four ways five factors are basic. *Personality and Individual Differences*, Vol. 13, No. 6.

Crant, J.M. (2000) Proactive behaviour in organizations. *Journal of Management*, Vol. 26, No. 3.

Dacey, J.S., Lennon K.H. (1998) Understanding creativity. *The Interplay of Biological, Psychological and Social Factors*. John Wiley and Sons, Inc. New York.

Danneels, E. (2002) The dynamics of product innovation and firm competences'. *Strategic Management Journal*, Vol. 23, No. 12.

Dansereau, F., Graen, G., Haga, W.J. (1975) A vertical dyad approach to leadership within formal organizations: A longitudinal investigation of the role making process. *Organizational Behaviour and Human Performance*, Vol. 13, No. 1.

De Jong, J., den Hartog, D. (2010) Measuring innovative work behaviour. *Creativity and Innovation Management*, Vol. 19, No. 1.

De Jong, J.P., Kemp, R. (2003) Determinants of co-workers' innovative behaviour: An investigation into knowledge intensive services. *International Journal of Innovation Management*, Vol. 7, No. 2.

De Leede, L., Looise, J.K. (2005) Innovation and HRM: Towards an integrated framework. *Creativity and Innovation Management*, Vol. 14, No. 2.

Dorenbosch, L., van Engen, M.L., Verhagen, M. (2005) On-the-job innovation: The impact of job design and human resource management through production ownership. *Creativity and Innovation Management*, Vol. 14, No. 2.

Dubey, R., Ghai, S. (2010) Innovative atmosphere relating to success factors of entrepreneurial managers in the organization agility—An empirical study. *International Journal of Innovation Science*, Vol. 2, No. 3.

Ekvall, G., Ryhammar, L. (1999) The creative climate: Its determinants and effects at a Swedish university. *Creativity Research Journal*, Vol. 12, No. 4.

George, J.M., Zhou, J. (2001) When openness to experience and conscientiousness are related to creative behaviour: An interactional approach. *Journal of Applied Psychology*, Vol. 86, No. 3.

Gough, H.G., Heilbrun, A.B. (1983) *The Adjective Check List. Manual.* Consulting Psychologists Press, Palo Alto.

Houghton, J.D., Neck, Ch.P. (2002) The revised self-leadership questionnaire. Testing a hierarchical factor structure for self-leadership. *Journal Managerial Psychology*, Vol. 17, No. 8.

House, R.J., Dessler, G. (1974) The path goal theory of leadership: Some post hoc and a priori tests, in: *Contingency Approaches to Leadership*, (Eds.) J. Hunt, L. Larson. Southern Illinois University Press, Carbondale.

Hunter, S.T., Bedell, K.E., Mumford, M.D. (2006) Climate for creativity: A quantitative review. *Creativity Research Journal*, Vol. 19.

Hutchinson, L., Skinner, N.F. (2007) Self-awareness and cognitive style: Relationship among adaptation-innovation, self-monitoring, and self-consciousness. *Social Behaviour and Personality*, Vol. 35, No. 4.

Jafri, M.H. (2010) Organizational commitment and employee's innovative behaviour. *Journal of Management Research*, Vol. 10, No. 1.

James, K., Asmus, Ch. (2001) Personality, cognitive skills and creativity in different life domains. *Creativity Research Journal*, Vol. 13, No. 2.

Janssen, O. (2000) Job demands, perceptions of effort–reward fairness, and innovative work behaviour. *Journal of Occupational and Organizational Psychology*, Vol. 73.

Janssen, O. (2004) How fairness perceptions make innovative behaviour more or less stressful. *Journal of Organizational Behaviour*, Vol. 25.

Janssen, O. (2005) The joint impact of perceived influence and supervisor supportiveness on employee innovative behaviour. *Journal of Occupational and Organizational Psychology*, Vol. 78.

Kelly, K.E. (2006) Relationship between the five-factor model of personality and the scale of creative attributes and behaviour: A validational study. *Individual Differences Research*, Vol. 4, No. 5.

Kianto, A., Hurmelinna-Laukkanen, P., Ritala, P. (2010) Intellectual capital in service- and product-oriented companies. *Journal of Intellectual Capital*, Vol. 11, No. 3.

Kim, T.Y., Hon, A.H., Lee, D.R. (2010) Proactive personality and employee creativity: The effects of job creativity requirement and supervisor support for creativity. *Creativity Research Journal*, Vol. 22, No. 1.

King, L., Walker, L., Broyles, S. (1996) Creativity and the five factor model. *Journal of Research in Personality*, Vol. 30.

Kleysen, R.F., Street, Ch.T. (2001) Toward a multi-dimensional measure of individual innovative behaviour. *Journal of Intellectual Capital*, Vol. 2, No. 3.

Luthans, F., Youssef, C.M., Avolio, B.J. (2007) *Psychological Capital: Developing the Human Competitive Edge.* Oxford University Press, Oxford.

McGrath, R.G. (2001) Exploratory learning, innovative capacity and managerial oversight. *Academy of Management Journal*, Vol. 44.

Mumford, M.D., Licuanan, B. (2004) Leading for innovation: Conclusions, issues, and directions. *Leadership Quarterly*, Vol. 15.

Mumford, M.D., Scott, G.M., Gaddis, B., Strange, J.M. (2002) Leading creative people: Orchestrating expertise and relationships. *Leadership Quarterly*, Vol. 13.

Nosal, Cz. (2001) *Manager's Psychology of Thinking and Acting*. Wydawnictwo AKADE, Wrocław.

Oldham, G.R., Cummings, A. (1996) Employee creativity: Personal and contextual factors at work. *Academy of Management Journal*, Vol. 39.

Ostroff, C., Schmitt, N. (1993) Configurations of organizational effectiveness and efficiency. *Academy of Management Journal*, Vol. 36.

Płużek, Z. (1978) Methods of personality examination, in: *Psychometric Problems in Peptic Ulcer Disease*, (Ed.) J. Łazowski. PZWL, Warszawa.

Ramamoorthy, N., Flood, P.C., Slattery, T., Sardessai, R. (2005) Determinants of innovative work behaviour: Development and test of an integrated model. *Creativity and Innovation Management*, Vol. 14, No. 2.

Rank, J., Nelson, N., Allen, T., Xu, X. (2009) Leadership predictors of innovation and task performance: Subordinates' self-esteem and self-presentation as moderators. *Journal of Occupational and Organizational Psychology*, Vol. 82.

Reuvers, M., van Engel, M.L., Vinkenburg, C.J., Wilson-Evered, E. (2008) Transformational leadership and innovative behaviour: Exploring the relevance of gender differences. *Creativity and Innovation Management*, Vol. 17, No. 2.

Robertson, T.S., Myers, J.H. (1969) Personality correlates of opinion leadership and innovative buying behaviour. *Journal of Marketing Research*, VI.
Sanders, K., Moorkamp, M., Torka, N., Groeneveld, S., Groeneveld, C. (2010) How to support innovative behaviour? The role of LMX and satisfaction with HR practices. *Technology and Investment*, Vol. 1.
Scott, S.G., Bruce, R.A. (1994) Determinants of innovative behaviour: A path model of individual innovation in the workplace. *Academy of Management Journal*, Vol. 37, No. 3.
Seibert, S.E., Kraimer, M.L., Crant, J.M. (2001) What do proactive people do? A longitudinal model linking proactive personality and career success. *Personnel Psychology*, Vol. 54.
Shih, H.A., Susanto (2011) Is innovative behaviour really good for the firm? Innovative work behaviour, conflict with co-workers and turnover intention: Moderating roles of perceived distributive fairness. *International Journal of Conflict Management*, Vol. 22, No. 2.
Shipton, H., West, W., Dawson, J., Birdi, K., Patterson, M. (2006) HRM as a predictor of innovation. *Human Resource Management Journal*, Vol. 16, No. 1.
Snell, S.A., Youndt, M.A., Wright, P.M. (1996) Establishing a framework for research in strategic human resource management: Merging resource theory and organizational learning. *Research in Personnel and Human Resources Management*, Vol. 14.
Stajkovitch, A.D., Luthans, F. (1998) Self-efficacy and work-related performance: A meta-analysis. *Psychological Bulletin*, Vol. 124.
Sung, S.Y., Choi, J.N. (2009) Do big five personality factors affect individual creativity? The moderating role of extrinsic motivation. *Social Behaviour and Personality*, Vol. 37, No. 7.
Takala, T. (2005) Charismatic leadership and power. *Problems and Perspectives in Management*, Vol. 3.
Talke, K., Salomo, S., Mensel, N. (2006) A competence-based model of initiatives for innovations. *Creativity and Innovation Management*, Vol. 15, No. 4.
Tesluk, P.E., Farr, J.L., Klein, S.A. (1997) Influences of organizational culture and climate on individual creativity. *Journal of Creative Behaviour*, Vol. 31, No. 1.
Tierney, P., Farmer, S.M. (2002) Creative self-efficacy: Its potential antecedents and relationship to creative performance. *Academy of Management Journal*, Vol. 45, No. 6.
West, M. (2000) *Creativity Development in Organization*. PWN, Warszawa.
West, M.A., Farr, J.L. (1989) Innovation at work: Psychological perspective. *Social Behaviour*, Vol. 2.
Wojtczuk-Turek, A. (2010) Organizational and competence predictors of innovative behaviours—Empirical Analyses. *Współczesne Zarządzanie*, Vol. 4.
Wolfradt, U., Pretz, J. (2001) Individual differences in creativity: Personality, story writing, and hobbies. *European Journal of Personality*, Vol. 15.
Wright, P.M., Dunford, B.B., Snell, S.A. (2001) Human resources and the resource based view of the firm. *Journal of Management*, Vol. 27, No. 6.
Yuan, F., Woodman, R.W. (2010) Innovative behaviour in the workplace: The role of performance and image outcome expectations. *Academy of Management Journal*, Vol. 53, No. 2.

54 External Conditions of Enterprise Development in a Knowledge-Based Economy

Hanna Wlodarkiewicz-Klimek, Joanna Kalkowska, Stefan Trzcielinski, and Edmund Pawlowski

CONTENTS

SOURCES OF ORGANIZATIONAL DEVELOPMENT IN A KNOWLEDGE-BASED ECONOMY

Dynamic changes in the environment and social evolution cause the necessity for adapting the organization internal system to the new situation and to the new requirements. While identifying organizations' development opportunities arising from the environment, the change of economy model resulting from the Lisbon Strategy must be considered. The Lisbon Strategy has assumed the construction of the most competitive knowledge-based economy in the world, which would be able to maintain a sustainable economic development and create more and better positions in jobs, as well as it would keep the social coherence.

The Lisbon Strategy focuses on four fundamental potentials, which can be distinguished as follows:

- Human resources, that is, society of knowledge (which the part of knowledge is gathered in).
- Innovation system (with the entrepreneurship more concentrated on operations of companies, but also on the cooperation with science), which creates new knowledge as a result of discoveries and innovations.

- Information technologies facilitating the exchange of knowledge, also with foreign countries.
- Institutional and legal environment, which creates conditions for development of presented domains; it constitutes various institutions, regulations, etc. (Włodarkiewicz-Klimek 2009).

Ten-year perspective of the strategy realization revealed a number of positive changes that took place in European countries thanks to actions directed in strategic prospects and to funds distributed through operational programs, even though changes were not as dynamic as it has been expected. The realization of the Lisbon Strategy has been seriously disturbed by the worldwide crisis that forces European countries to initiate actions for preventing negative results of the breaking economy. In 2010, the reformed concept of new Europe, included in the new strategy for Europe 2020—strategy for smart and sustainable and inclusive economy (Europe 2020, European Commission), has been accepted. The newly accepted strategy assumes three directions of actions determined as follows:

- Intelligent development—development of knowledge and innovation-based economy
- Sustainable development—support of economy that uses resources in a more efficient way and which is more friendly to the environment, as well as it is more competitive
- Inclusive development—support of the economy characterized with the high level of employment and economic coherence, which guarantees the economic, social, and territorial coherence

Operations characterized above represent the direct continuation of assumptions from the Lisbon Strategy. Such image results from the analysis of the strategic concept—Europe 2020 particularly from the point of view of enterprises, for which the sustainable development of knowledge-based economy gives significant opportunities. The knowledge-based economy, according to the World Bank (2006) definition, is interpreted as an economy in which the equable use and creation of knowledge are at the heart of its process of economic development. Knowledge-based economy uses knowledge as a motor of economic growth.

EVALUATION OF DEVELOPMENT AND KNOWLEDGE-BASED ECONOMY SHAPE IN THE CONTEXT OF CHANGES WITHIN MODERN ORGANIZATIONS

The knowledge-based economy development dynamics, as well as its influence on organization management, will be evaluated through identification of changes appearing in four principal potentials, that is, human capital, innovation systems, information and communications technology (ICT), and economic incentive and institutional regime. There are many methods of evaluation: starting from the analysis of individual macroeconomic indexes and ending on very complex measurement methodologies. Still, all examinations focus on the analysis of statistical data. They are very useful from the point of view of evaluation of the society development. Unfortunately, they do not represent a simple reflection of factors of the organization growth.

Therefore, searching the relation between the development of knowledge-based economy and the dynamics of the organization growth, we have to focus on interpreting opportunities that are possible to identify in the environment using statistical results. Authors have accepted results of evaluation of the knowledge-based economy obtained from the knowledge assessment methodology (KAM) (World Bank 2006). The KAM assumes an interactive diagnosis of the state of economy by using statistical quantitative and qualitative indexes, which enables identifying challenges and opportunities created in the transition into the knowledge-based economy. The KAM consists of 109 structural and qualitative variables for 146 countries to measure their performance on the four knowledge economy (KE) pillars: economic incentive and institutional regime, education, innovation, and ICTs.

Variables are normalized on a scale of 0 to 10 relative to other countries in the comparison group. The KAM also derives a country's overall KE index (KEI) and knowledge index (KI). The KAM KI measures a country's ability to generate, adopt, and diffuse knowledge. This is an indication of overall potential of knowledge development in a given country. The KEI takes into account whether the environment is conducive for knowledge to be used effectively for economic development. It is an aggregate index that represents the overall level of development of a country or region toward the KE. The KEI is calculated based on the average of the normalized performance scores of a country or region on all four pillars related to the knowledge economy (www.worldbank.org/kam).

In their research, authors have focused on the analysis of the knowledge-based economy on the basis of the KEI, which contains the following variables (Measuring Knowledge in the World's Economies, Knowledge Assessment Methodology, and Knowledge Economy Index).

1. The economic incentive and institutional regime:

- Tariff and nontariff barriers
- Regulatory quality
- Rule of law

2. Education and human resources:

- Adult literacy rate
- Secondary enrollment
- Tertiary enrollment

3. The innovation system:

- Royalty and license fees payments and receipts
- Patent applications granted by the US Patent and Trademark Office
- Scientific and technical journal articles

4. ICT:

- Telephones per 1000 people
- Computers per 1000 people
- Internet users per 10,000 people (www.worldbank.org/kam).

The presented variables are completed by two indicators that show the overall efficiency of the economy:

- Average annual GDP growth
- Human development index

Analysis of KAM variables enables assessing development and shape of knowledge-based economy in the context of sources of opportunities for development and changes within modern organizations.

ANALYSIS OF THE AREAS OF EXTERNAL CONDITIONS FOR DEVELOPMENT OF ENTERPRISES AND THEIR IMPACT ON ORGANIZATIONS' FUNCTIONING

Analyzing the potentials of different areas of knowledge-based economy and taking into account the statistical evaluation of the economy development, expressed in KAM indexes, we are able to characterize principle features affecting the changes in the behavior of the organizations. The following aspects are analyzed: innovation, changes in the structure of the human capital, technology development, as well as changes in the legal and institutional environment, which create conditions for the knowledge-based economy growth. Basing on the assessment of the knowledge-based economy development based on KAM indexes, a preliminary analysis concerning the external conditioning's influence on enterprise development will be carried out.

Analysis of the Knowledge-Based Economy Potential Innovation

Continuous development and stimulation of innovation are principal premises of the development in knowledge-based economy. Generally, innovations mean that the product or the service is different from all other similar products and services with its scale of novelty. In the context of the knowledge-based economy, innovations in its wider meaning must be examined. Then, innovations can be interpreted not only as changes concerning the product (introduced in the process of production and into the market of new or significantly improved products) and the methods of production (using new methods in production or real improvement), but also it can be understood as changes in work or production organization (new organizational solutions in their structural and process meaning, as well as significant improvements in existing processes), or changes of management methods that fulfill determined technical, economical, and social criteria, in order to obtain assumed social and economic benefits.

The climate of innovation created by the growing knowledge-based economy constitutes a specific element of external influences of modern organization development. These conditions can be considered in terms of the environment in the macroscale, in which they might be interpreted as the total of conditions created by the economy, with particular attention focused on activities of the European Union related to the realization of its strategy. Most important are as follows:

- Functioning of the European Research Area, which enables a fluent exchange of knowledge and experience between the state, science, and business
- Development of national innovation system, including the implementation of the state innovation policy
- Growth of national expenditure on research and development
- Introduction of legal regulations in order to facilitate widely the understood innovative operations, including simplification of the patent policy
- Creating conditions for development of the ICT
- Growth of the importance of proinnovative education in the society and diffusion of knowledge in society
- Redistribution of European funds in order to stimulate innovation of enterprises

In the microscale, external conditioning affecting the development of innovation focuses on direct relations created by the organization with its partners in the nearest environment—the sector or the region. Among them, it is possible to distinguish the following:

- Direct competitor innovation level
- Clients' expectation concerning innovative products
- Length of product life cycles in the sector
- Cooperation models between scientific and research centers in the sector (region)
- Regional innovation level

Human Resources

The development of knowledge-based economy causes a significant change in the approach and structure of human resources. People are treated as human capital, understood as a resource of knowledge, skills, health, and vital energy, included in every human and in the society as an entity that determines the work capacity to adapt to the changes in the environment and the possibility of creating new solutions. It is possible to analyze opportunities created by the development of human capital in two approaches:

- General approach to the entire society, in which the growth of human capital value is directly related to the creation of the national income
- Individual and personal approach, which focuses on the position of individual in the labor market, investment in the education and development, as well as with differentiation of individual incomes that results from it

The presented approach directly relates to the changes that take place in the new economy. It is necessary to enumerate the following factors in the general approach referring to the creation of the social life new model:

- Development of active policy of employment that aims to balance between flexibility and certitude of employment–flexicurity model
- Investment in human resources through realization of the concept of lifelong-learning
- Initiation of actions serving to forecasting and monitoring qualifications and skills needed in the future
- Creation of conditions for increasing the employees' mobility
- Expenditures for health protection, which affects the length of life, the strength, stamina, vitality, and vigor of people

Such social changes lead to changes of behavior and attitude on the level of individual person—both in private and professional sphere of life. The most important changes observed on the level of individual are the following:

- Changes of demographic structure
- Changes of the family model and family lifestyle
- Increase in the level of wealth resulting from the growth in knowledge and competence
- Growth of education level in families (investing in knowledge)
- Sensitivity of the labor market on individual worker characteristics such as intelligence, skills, entrepreneurship, commitment, health, and psychophysical condition
- Motivation for sharing information, knowledge, work in team, and focus on success

Information and Communications Technologies

According to the Lisbon Strategy, the motor of the KE are all activities that lead to increased competitiveness through research, innovation, ICT, and investment in human resources. In this context, ICTs become particularly important because they constitute one of the main drives fastening the process of globalization and technical progress, both in the aspect of forming the knowledge-based economy and in the context of individual enterprises. ICTs are one of the most important conditions of realization of the knowledge management concept in the company. A proper ICT structure enables increasing innovation and fast coordination of all activities in the enterprise. Moreover, those technologies constitute one of the key factors affecting competition of enterprises functioning in the knowledge-based economy. The application of ICT is one of the stimuli for continuous adaptation of the company to the changes occurring in the environment. Such changes can be transformed into potential areas of opportunities to be used or they might even be absorbed by enterprises. The following are factors that can be distinguished among external conditionings for the development of the organization, which relates to information technologies:

- Common adoption of ICT as a key factor of economic development
- Development of information society
- Possibilities of gaining European funds for development of ICT infrastructure
- Globalization of the market of ICT services
- Relatively well-developed information infrastructure

- Relatively high level of IT investment
- Better access to broadband technologies
- High level of education of the IT-specialized staff
- High level of business computerization
- Strong interest of e-business services
- Computerization of the public administration sector

Institutional and Legal Setting

The institutional and legal environment forms conditions for the development of determined areas. This sphere has particular impact on the development of business and entrepreneurship through its effort to control and adapt to European standard legal, economic, institutional, and capital systems. Within the frames of factors of the institutional and legal environment that support the development of organizations in the knowledge-based economy, we might enumerate the following ones (Włodarkiewicz-Klimek and Kałkowska 2010):

- Maintaining macroeconomic indicator regime
- Realization of structural reforms
- Simplification of legal regulations
- Creation of a dynamic business environment with less bureaucracy
- Creation of conditions supporting the functioning of enterprises in the EU single market

External Factors Affecting the Development of Polish Enterprises according to the KAM Assessment

Analyzing cumulative assessment of the development of knowledge-based economy in Poland, we can see, in general, a gradual, positive growth of indices (Table 54.1).

A particular growth in the area of the ICT structure (ICT index) can be observed, in which the most dynamic increase concerns the number of mobile phones, where the number of Internet users and people owning a computer gradually adheres. In the sphere of education and human resources (education index), the biggest growth refers to the education index for higher education, and among indicators of innovation, the royalty payments and receipts for research and development presented the strongest growth. Rates of the economical and institutional regime presented the weakest dynamics. The worldwide crisis explains this low level of dynamics.

Positive changes of indicators showing the systematic development of knowledge-based economy in Poland in connection with specific activities undertaken by the agenda of the European Union become a source of real opportunities for businesses; also, they are a major cause of internal organizational changes.

INFLUENCE OF THE KNOWLEDGE-BASED ECONOMY DEVELOPMENT ON CONTEMPORARY ENTERPRISES CHANGES

A company operating in the environment of the knowledge-based economy is subjected to its influence. The result of that effect is the adaptation of organizations' internal potentials enabling them to continuous development considering opportunities arising from the environment. Figure 54.1 presents a model that influences the knowledge-based economy on areas and potentials for internal adjustment of the organization.

Continuous development of four principal areas of knowledge-based economy (innovation, human capital, ICT, and the institutional and legal environment) has become the direct opportunity to implement changes in the organization. Organizations use the potential of the environment; however, the direct result of that activities are the changes appearing in the following areas:

TABLE 54.1
KAM Indicators of Development of the Knowledge-Based Economy

Variable	Poland (Most Recent) (Group: All Countries) Actual	Normalized	Poland 2000 (Group: All Countries) Actual	Normalized	Poland 1995 (Group: All Countries) Actual	Normalized
Annual GDP growth (%)	5.20	5.10	3.20	4.73	5.80	8.13
Human development index	0.870	7.55	0.852	7.61	0.822	7.33
Tariff and nontariff barriers	85.80	9.02	74.80	6.83	57.00	4.21
Regulatory quality	0.71	7.12	0.64	6.97	0.62	7.03
Rule of law	0.28	6.30	0.60	7.10	0.64	7.29
Royalty payments and receipts (US$/pop.)	43.82	7.14	15.32	6.89	1.24	4.52
S and E journal articles/mil. people	179.58	7.85	143.17	7.79	117.90	7.79
Patents granted by USPTO/mil. people	0.70	6.10	0.42	5.93	0.27	6.34
Adult literacy rate (% age 15 and above)	99.31	7.67	99.70	8.28	99.06	8.25
Gross secondary enrollment rate	99.55	7.99	100.39	8.72	96.32	8.26
Gross tertiary enrollment rate	66.95	8.41	49.70	8.34	34.97	7.76
Total telephones per 1000 people	1.360.00	7.19	460.00	6.99	150.00	6.07
Computers per 1000 people	170.00	6.69	70.00	6.64	30.00	6.67
Internet users per 1000 people	440.00	7.40	70.00	7.17	10.00	8.55

Source: http://info.worldbank.org/etools/kam2/KAM_page1.asp.

- Structure of the organization (interpreted as totality of internal relations in the organization management system).
  - Creating a new management system: resigning from bureaucratic solutions in favor of models of an intelligent and knowledge-based organization.
  - Forming structures of the organization to be more flexible basing on organizational processes, instead of hierarchy.
  - Creating participative management style and reliance of power on the authority resulting from knowledge and skills.
  - Shaping the organizational culture based on trust.
  - Introducing modern concepts and methods of management.
- Strategy (interpreted as a concept for a long-term development of the organization).
  - Change of competition model: from direct competition to engagement in cooperative activities.

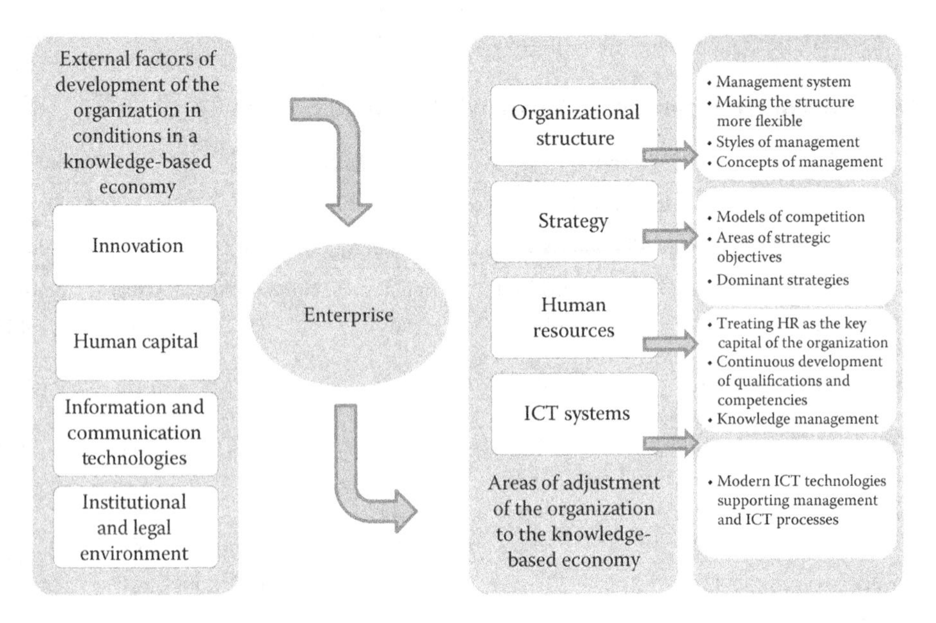

FIGURE 54.1 Areas and potentials for internal adjustment of the organization to the knowledge-based economy. (Own study.)

  - Focus of strategic objectives on obtaining prominent resources (through strengthening competences, knowledge, and skills), which form the bargaining power against suppliers and direct competitors.
  - Crucial importance of knowledge in creation of the organizational strategy.
- Human resource (understood as knowledge, skills, commitment, and possibilities of development for employed staff).
- Treating human resources as a key capital of the organization.
- Integration of the development of the organization with the knowledge development, qualifications, and competences of employed people.
- Motivating employees to create, develop, and share organizational knowledge.
- ICT systems (interpreted in the instrumental approach as creation of the information infrastructure; however, in the process approach, it is understood as creation of management support systems).
  - Increasing importance of ICTs as a basic tool for supporting management.
  - ICT as the basis of development of the operating systems management (e.g., in the area of logistics).
  - Key importance of ICT systems as instruments of knowledge and information management.

External conditions of adjustment of the company to the knowledge-based economy force also redefine management, particularly because of the change of the organization resources system and the necessity of focusing management functions on immaterial resources. Therefore, taking into consideration resource systems in the knowledge-based organization, the management will become a system of activities leading on carrying out the management functions on material and immaterial resources of the organization (including personalized, codified, and well-grounded knowledge). It should enable efficient (effective, beneficent, and economic) way of achieving goals of the organization.

REFERENCES

Europe (2020) A strategy for smart, sustainable and inclusive growth European Commission, Brussels, March 3, 2010.

Measuring Knowledge in the World's Economies, Knowledge Assessment Methodology, and Knowledge Economy Index. Available at www.worldbank.org/wbi/k4d.

Włodarkiewicz-Klimek H. (2009) Shaping the personal function of contemporary enterprises in the knowledge based economy conditions. In: *Management Enterprises Social Aspects*, Poznań: Poznan University of Technology, pp. 59–68, 2009.

Włodarkiewicz-Klimek H., Kałkowska J. (2010) Shaping the strategy of knowledge based economy of Polish enterprises. In: *Advances in Human Factors, Ergonomics and Safety in Manufacturing and Service Industries*, Karwowski W., Salvendy G. (eds.). Boca Raton: Taylor and Francis Group, Advances in Human Factors and Ergonomics Series, pp. 42–51.

World Bank (2006) *Korea as Knowledge Economy. Evolutionary Process and Lessons Learned.* Overview, Washington: World Bank.

55 Endogenous and Exogenous Determinants of the Development of Modern Towns

Andrzej Olak

CONTENTS

INTRODUCTION

Globalization processes have not decreased the interest in local development perceived as a category of socioeconomic growth. This development will be constantly identified with creating the best conditions of life in the local environment as possible by striving to achieve effective organization and structure as well as correct functioning of the local social system, in other words, a certain territorial unit, that is, a city, a commune, a region. The crucial problem has become to recognize and appropriately choose the development factors.

The goal of this article is to focus on the fact that local development is always three-dimensional. That is why the networking of socioeconomic space and larger influence of exogenous factors should be taken into consideration. The exogenous factors become a specific reflection of the given goals of local development. They embrace the given set of data illustrating the utility values, both the ones created by nature and by people, human capital, institutions that decide about the development of the given country, culture and economic traditions of the given territory, and competitiveness of the places to stay.

It seems to be equally important to focus on primary theoretical concepts of local development, which, when used at a certain time and place, lead to positive changes in this development. Particular attention has been paid to the concepts connected with the current, that is, taking place at this moment, discourse of economic circles about imitation (convergence) and polarization of development. The review and the systematization of the main theoretical concepts concerning determinants of local development may serve to determine the most frequently used conditions in the process of urban development.

THEORETICAL ASPECTS OF LOCAL DEVELOPMENT

The term *local development* comprises a process of positive changes that refer to a quantitative growth and a qualitative progress, which appear at the given area including the needs, preferences, and the value hierarchies appropriate for it. It has also become a subject of many studies and analyses, which has been reflected in the heterogeneity concept and scientific theories. It is also connected with regional development, because regional changes result in local changes (economic development occurs at a certain place—space).

The term itself has not been defined in a straightforward manner in the professional literature. One of the reasons for the lack of a definition of this process is its interdisciplinary meaning and because experts in many fields of science and economic life (economy, economic geography, space politics, economic history, mezoeconomy, sociology, ecology) are willing to research this process. The process category itself is directly related to economic terminology, entailing the fact of connecting it with the local economy (cities, communes, districts). The differences between the local development and the territorial one are fluid. That is why the factors, determinants, and characteristics determining the development of the transformation process become vague (Alińska 2008, p. 49).

The local development category puts an emphasis on the fact that it refers to the socioeconomic activity scale and claims that the local development is based on endogenous factors of development, understood as their local needs, resources, organizations, society, and economic entities. It becomes a specific supplement to the regional development, which proceeds in relation to the exogenous factors. Such understanding of the development notion cancels the applicable restrictions, both material and specific ones, and therefore

- It embraces a wide range of economic activities aimed at raising the level of material and immaterial sphere of the living place.
- It can be identified with the creation of desirable conditions of life in the local environment (effectiveness of an organization, structure, activity of the local social system; Noworól 2008, p. 15).

The prerequisite of local development is the local government taking responsibility for regional development with reference to the needs and expectations of its society. The crucial factor and the basis of local development will be a marketing approach to economy, because it favors entrepreneurial development.

Referring to the definition of local development, Noworól quotes French theoreticians. J. L. Guigou claims that "local development expresses the local solidarity which creates new social relations and demonstrates the microregion inhabitants' will as far as valorisation of local wealth is concerned, which creates economic development altogether" (Noworól 2008, p. 16). O. Godard defines "local development as a process of diversification and enrichment of economic and social activity of the given territory which has its source in immobilisation and coordination of its resources and energy. It is the outcome of the inhabitants' effort, it is connected with the existence of a development project which integrates the economic, social and cultural elements, it creates the space of active solidarity from neighbouring spaces." For B. Gesner, local development is "the process through which the population obtains larger autonomy in defining and satisfying their needs, solving their problems, and defining their future. It is the effect of shaping collective dynamics in which the economic, social, cultural and territorial dimensions overlap and which combines local initiatives with supralocal activities and resources" (Alińska 2008, p. 16).

The theoretical elaborations comprising noneconomic ways of territorial development define it as a long-term process of changes taking place in stages, being clearly differentiated. The basic determinants necessary to define territorial development are

- The object which is being changed
- The change understood as the next stage of the object (Kudłacz 1999, pp. 16–17)

At this point, territorial development can be seen as an operational activity in which comparable stages (several of them) can be distinguished and differences between them determined the object that changes; the change is perceived as successive transformations of the object (Kudłacz 1999, p. 18).

In this respect, the object will be understood as

- A physical and geographical space; the material and immaterial capital (experience and knowledge).
- A system of social and economic relations having a market or nonmarket character.
- Rate of learning, where the main role is to be played by formal and informal institutions and the manner of information circulation (extension of closeness and reduction of transaction costs). The potential consequences may involve positive effects.
- Participation in local management.

It obviously does not entail the mere perception of a territory as the object. The immense number of theoretical views of the problem to be found in the literature of the subject may induce thinking about territorial phenomena and dependences from a systemic perspective. In this respect, territories are seen as systems that open to the environment. Their existence will be determined by the correct relation with the surrounding (Noworól 2008, p. 19), which calls for an example of living systems "for which the following continuous cycle is characteristic: taking outlays, their internal transformation (complete), making the product and feedback (with the help of which one element of the experiment influences the next). [...] The surrounding and the system should then be perceived as interacting and mutually dependent objects" (Noworól 2008, p. 19).

Local development is often equated with economic development, which is perceived as acting to the benefit of the local population as far as the material sphere is concerned as well as the goods and services. It is often forgotten that it is a very important activity as far as the development of immaterial services is concerned, for example, education and culture. Therefore, it evolves into economic development, that is, material production, and social development, where the availability and quality of services are the main criteria (Domański 2001, pp. 127–134).

Efficient and more importantly effective functioning of a local economy require planning and investments performed by the local government and businessmen. It gives an impulse to further development and supports innovation and prosocial attitudes of local leaders who undertake actions to the benefit of local inhabitants. Their actions often contribute to arrival of the first major investors, which attract successive ones.

The theory of local innovative environments assumes that technical factors play the main role. In a very essential manner, it is related to the influence of innovativeness of places on the local development as well as the learning abilities of local inhabitants. However, what is worth stressing is that the connection between the above-mentioned innovation and the local sociocultural context, especially from the perspective of its historic context, still seems to be insufficient. Concentration of small- and medium-sized companies dealing with computer programming and modern products is commonly observed in urban agglomerations. They are becoming an environment in which accumulation of knowledge and skills favors the development of innovative economic activities on a larger scale and in a longer period of time (Kukliński and Pawłowska 1998).

Local/regional development theories may be found in studies conducted in numerous scientific fields. The most common ones, deserving the greatest attention, are as follows (Alińska 2008, p. 58):

- *The neoclassic regional growth theory*, which refers to sustainable economic growth of a region as a result of full utilization of production factors. This theory is based on the assumption that "emerging deficiencies of one of the production factors in the region" are compensated by their translocation among regions.
- *The post-Keynessian theory* stressing investments as the most important factor of the development, taking the multipliers into consideration.

- *The endogenous concept of growth* noticing that stimulation of the internal potential is the basis of the region's development; "the use of competitive advantage and the development determinant the given region is in disposal of enables achieving a higher level of socio-economic development."
- *The concept of growth poles* stressing innovations as activities of prodevelopmental nature.
- *The concept of ecodevelopment*, which assumes that the ecodevelopment is the result of safe socioeconomic development, both today and in the future; "according to this concept, special attention should be paid to management resources so that they improve people's living conditions and the natural environment status."
- *The concept of development based on innovation processes*, which refers to the use of new technologies as well as modern organizational solutions and knowledge for the sake of the regional development, because it is becoming a warranty of substantial return on investments.
- *Territorial production systems* understood as a marriage of "one or several types of production" assuming such a form of activity that allows for creating a local network of experience and knowledge sharing, which reduces the operating costs involved.

The above concepts stress different aspects of local/regional development. Some of them speak of developmental factors, others focus on the matters of organization and management, and yet others mainly address the economic and social surrounding as well as ecology or the human factor. As far as local development is concerned, the above theories are complemented by an analysis of the theoretical concepts of local development by T.G. Gross, which "stresses the theories of the basic product, the new theory of trade, the geographical centres of growth, the core and periphery model, the production cycle, the network structure, groups (clusters) and social capital" (Alińska 2008, p. 59). K. Olejniczak focuses on the local development theories referring to networks (Alińska 2008, p. 59), clusters, and regional innovative systems. The last concept is particularly worth mentioning as it refers to both endogenous and exogenous factors of local development.

The theories and concepts presented prove that defining the notion of local/regional development is an open sphere for further studies and analyses. The functioning theories indicate the potential tools and conditions to be reached for that may allow for successful accomplishment of the goal assumed, which is becoming the quantitative and qualitative progress observed at the given territory, entailing the needs, preferences, and hierarchies of the values suitable for it.

CITY AS SPATIAL AND SOCIOECONOMIC ARRANGEMENT

According to the legal definition of a city, it is a settlement unit "dominated by compact housing development and non-agricultural functions, vested municipal rights or a municipal status granted according to the applicable regulations. Under the administrative system, cities have the status of independent communes (urban communes) or rural-urban communes" (Majer 2010, p. 121).

Processes connected with cities include the following:

- Urbanization characterized by rapid population growth of the central city areas compared to the areas outside the main city. The number of people inhabiting an agglomeration increases as a result of favorable population growth and migration. The influx of inhabitants to the city is large and results from the growth of employment, mainly in the sphere of industry.
- Suburbanization—more rapid population growth in suburban areas surrounding the city than in the city center. The birth rate in the outskirts is higher, and some central areas display negative birth rates, which consequently leads to the development of an agglomeration. Migrations decide about population changes and employment growths, in general, but in certain central districts, one may observe them to decrease. The sphere of services is widely developed.

- Deurbanization—it is characterized by a decreasing number of people both in the city area and in the outskirts. The decline causes a decrease in the agglomeration as a whole, and the rate of migration from the city areas to suburbs and to smaller towns exceeds the growth of the agglomeration outskirts. In this phase, the functions of city centers are subject to intense erosion. The residential function disappears, the service activity focused on satisfaction of the inhabitants' needs also declines, the administrative function and the employment increase, and the infrastructure intended for transport purposes occupies larger and larger areas, which clashes with the residential function. The social and cultural functions played by city centers are at a decline.
- Reurbanization—it takes place when the share of the city population in the overall number of inhabitants initially increases due to a dropping rate of decline, followed by a population increase in this area. In the process of the city's revival, centripetal forces dominate, contrary to the deurbanization phase, when centrifugal forces are the predominant ones (Węgleński 1986, p. 12).

Cities do not constitute monolithic chains. Diversity is their distinguishing feature as regards the potential socioeconomic scale, the functions performed in the urban network structure, the geographical location, and the historic tradition. Their fundamental quality is that they do not function as separate elements of the space but remain in a complex settlement structure.

Bearing in mind the historical course of development of towns into an urban system, one should agree with the thesis that "one of the most important factors of their development is the location in the geographical network" (Ziolo 1999, pp. 120–132). It caused that, in the most conveniently situated cities, predominant social, cultural, and economic potentials were gathered, and it contributed to the enterprise development. This factor is particularly crucial when determining the city's position in the development of the urban network.

What is also important for the city development is the influence of the course of transport routes. The cities situated along the existing or planned motorways will be more competitive.

Equally important are the demographic and economic potentials, administrative functions, and the geographic location. The fraction of people working in the given city exceeding the number of inhabitants implies a higher degree of competitiveness. Similarly, the fact that local self-governmental authorities (administrative functions) reside in the given city strengthens its functions.

Modern cities, depending on their rank, are connected with the regional, national, and international environment. This causes that, in the process of functioning, links of functional nature are established (Szymla 2001, p. 35). Positive influence on the developmental process is also exerted by a location next to the border with the European Union countries; however, a negative impact results from being close to the eastern (currently external) borders of the EU. A preestablished function of the city will influence the assessment of potential advantages from the perspective of the capital invested in every city (according to the maximum profit criterion). Cities will strive to raise their market capacities compared to the surrounding in order to succeed.

Transformations of the functional structures of contemporary cities, and particularly the deindustrialization and the related decline of the exogenous sector's share in general employment, on the concentration of exogenous functions, have caused changes to the strategic management. Favorable living and development conditions have been measured by the level of urbanization and industrialization. At present, these conditions are becoming a burden as regards the development and the quality of life. What is crucial, however, is the diversity of the economic structure, a well-developed telecommunication, road and transport infrastructure, the presence of R&D institutions, and human capital, which, as J. Heckman writes, "is so much more than the mere intelligence quotient measured in tests." The formation of new enterprises, the development of the service sector, and the concentration of foreign investments are becoming the actual measures of contemporary development, both with regard to the local and the regional sphere (Majer 2010, p. 333).

The noticeable concentration of investments and modern types of economic activity in large cities and their functional regions causes selective migration to these locations. This phenomenon is increasing their human capital resources at the expense of smaller (peripheral) areas and conditions larger opportunities, both economic and developmental ones, of large/larger cities. Cities are becoming attractive places for new inhabitants who are the sources of new ideas and solutions to the existing or the emerging issues and problems (Gorzelak and Smętowski 2005, pp. 58–63).

Continuous elimination of free trade limitations, the freedom of capital flow, privatization of companies and sectors, and liquidation or limitation of the state's functions (deregulation) all create favorable conditions for economic growth. The basic driving force in the economic sphere is gradually becoming competition. It is very strongly connected with the notions of subjectivity and objectivity of the market, both of which are characterized by demand restrictions and concentration of entrepreneurs. A company, in order to survive and thrive, must read the signals received from the surrounding area, interpret the facts and events, and build useful knowledge for itself based on the data and information acquired from external sources (Morawski 2009, p. 17). The competitive edge will mainly depend on the conditions created at the given market where activity is conducted as well as on its nature and attractiveness. This circumstance is perfectly noticeable in relation to space and specific territories where each of them creates a complex and unique product. It is becoming the source of values for the partners to the exchange and the abundance of market entities (both internal and external ones, numerousness of entities influencing target markets, creating partnership networks; Rupik 2005, p. 23).

Cities must seek the means allowing them to overcome and solve the existing problems concerning local development, related to transformations of the natural environment and its condition. At present, this problem mainly concerns

- Changes observed in the spatial environment that threaten the ecological and climate balance (mainly through contamination of soil, water, air, and climate changes)
- Urbanization and suburbanization processes (the increase in peripheral territories of cities assumed to be developed)
- Globalization of the cultural environment, which manifests itself in making the civilization products similar (dominating infrastructural objects like motorways, airports, large chain stores)

In many documents pertaining to sustainable development, three domains of our reality are mentioned, namely, the economic growth, the social progress, and the changes to the natural environment, as correlative phenomena. "The awareness of the need for joined solving of those three problems of the world has resulted from the failures of the former, sector-based approach to those tasks, and at the same time, it has confirmed the breakthrough made in the evaluation of the past forms of human economic activity," Zabłocki (2002, p. 42) claims. This diagnosis results from the fact that "[…] the results of such an activity, both the achieved and the predicted ones, perceived from the global perspective, appeared to be less positive than it had been thought of earlier" (Zabłocki 2002, p. 42). For specific relationships have been noticed that, in the social, environmental, and energy related domain, "[…] assumed the nature of <<ceiling effects>>" (Zabłocki 2002, p. 42). "Many cooperative situations" were critically judged because […] material stability cannot be provided when, in the sphere of developmental policy, no attention is paid to such conditions as changes in the access to resources and changes in the distribution of costs and benefits brought by the development" (Zabłocki 2002, p. 42). It should also be remembered that "[…] the combined analysis of the economic, environmental and social processes meant that during the evaluation of the present actions and planning of the future ones, a wider circle of phenomena has been taken into consideration. The acceptance of these additional, justified criteria (for example, ecological damage) changed the result of the general evaluation and stopped numerous kinds of activities, which has been considered beneficial" (Zabłocki 2002, p. 43). The activities considered positive were those that had positive effects in the economic development sphere as well as in the remaining domains, that is, the social development and the environmental protection (Zabłocki 2002, p. 43).

In the opinion, "[…] about the correlation and the equivalence of economic growth, the condition of the natural environment and the social development" (Zabłocki 2002, p. 43), the World Commission on Environment and Development concentrates on the economic and ecological threats of the contemporary world, focusing on the fact that "even a narrow-scale perception of material stability means one should care about social equality between generations, which, logically understanding, must be extended on the equality of every generation" (Zabłocki 2002, p. 43). As a result of such a perspective, the relevant scheme changes, namely, one "[…] in which the former human–environment relationship has been substituted with a configuration of three components: economy, society and nature" (Zabłocki 2002, p. 43). Contrasting of the threats that appear in the contemporary world assumes two goals: (1) procuring changes in the economic development causing an increase in expenditures spent to improve the condition of natural environment and growing social welfare; and (2) a change in people's attitude toward one another expressed through "[…] abandoning deleterious forms of the nature exploitation and reducing those economic processes that introduce profound differences in the global social structure" (Zabłocki 2002, p. 44).

In the economic nomenclature, sustainable development means an economic growth leading to an increase in social cohesion, including the reduction of social stratification, proving equal opportunities, counteracting marginalization and social discrimination, and raising the quality of natural environment among other aspects, by limiting the harmful impacts of production and consumption on the condition of the environment and protecting natural resources.

This process is responsible for low unemployment, efficiency of institutions and companies, high level of social trust, and proper functioning of the civil society. The very essence of sustainable development is the perception of social, economic, and ecological arguments as equally valid.

The main factors decisive about the competitiveness also include spatial accessibility understood as investment opportunities. It comprises convenient conditions of communication, road transport, rail transport, access to modern telecommunication services, fiber-optic lines, telephony, well-developed water supply and sewage disposal systems, reserve capacities in the power engineering industry, telecommunication and water intakes, existence of a ready-to-use documentation for infrastructural investments, free areas and structures for development, comprehensive financial service, advantageous tax systems, engagement of local self-governments, and inhabitants in the performance of investments. It seems accurate to claim that "[…] it is hard to imagine efforts undertaken by local authorities to the benefit of creation and stimulation of economic growth in isolation from the magnitude and properties of the territory where it takes place or will take place in the future" (Zalewski 2005, p. 94).

It may be assumed that the competitive advantage of a city/territory will be defined by two elements:

- Capacities of subsystems
- "The ability to reach target groups which may be interested in this potential and migrate to the territory or invest their assets on it, at the same time, contributing to the change of the entire system" (Noworól 2008, p. 38)

What seems crucial in this context is the way to reach individual groups of decision makers, or in other words, territorial marketing defined as "the entirety of coordinated activities of local, regional or nationwide entities aimed at creating the processes of exchange and influence by recognition, creation and satisfaction of the needs and desires of inhabitants" (Szromnik 1997, p. 38). The constantly increasing competition and the fast socioeconomic development imply problems cities must face forced to search for new ways of efficient management. Marketing is then becoming "an attempt to face the challenges resulting from the globalisation processes and the resulting need for competition between territorial units" (Noworól 2008, p. 38). Noworól (2008, p. 39) mentions a group of entities and persons that territorial marketing will address:

1. Foreign investors and those in disposal of resources
   Traditional industries (e.g., heavy industry)
   "Clean" economic activity (advanced technologies, services)
   Investors representing different commercial real estate sectors
   Investors in the residential sector
   Entities operating in the public sphere allocating funds (for e.g. state budget, aid schemes)
2. Export markets
   Other territorial units operating in the domestic market
   Foreign markets
3. Visitors
   Tourists focused on learning, aesthetic experience, and what is referred to as sociospatial expansion
   Visiting entrepreneurs (participants to congresses and conferences as well as businessmen interested in purchasing local assets, e.g., real estates)
4. Inhabitants
   Professionals (scientists, freelancers)
   People with a high financial status
   Local businessmen
   Employees (blue collar workers), both qualified and unqualified
   People who study, especially higher education students

With regard to the above issues, territorial competitiveness may be perceived as the ability to undertake actions leading to "an increase in the attractiveness of one's own potential" in economic, environmental, and human terms compared to other territories and target groups (visitors, export markets, and inhabitants).

Therefore, a modern city should be perceived as a certain socioeconomic and spatial arrangement, which consists of different parts performing specific functions in its structures: main (international, domestic), complimentary (to the benefit of leading segments), standard (to the benefit of the quality of elements having regional functions), and local (satisfying the city's internal needs and those of the surrounding area). However, one must not disregard the fact that "territorial development is a truly complicated process which cannot be brought down to mere economic and social aspects in their narrow sense, and neither should it be controlled by the politics of predominantly economic nature [...]. Spatial development is focused on the balancing of areas and regions entailing the spatial aspects and factors: from the economic ones to the cultural ones, from the natural ones to the social ones" (Noworól 2008, p. 51).

SUBURBANIZATION PROCESSES ON CITIES

The structural changes taking place in the economy and the globalization processes intensify the development of the largest cities, turning them into centers of economic growth by increasing the role of services, including the market ones, entrepreneurship, innovation (technological and product related), influx of foreign capital, better access to the sources of information about innovations, as well as qualified employees. At the same time, deconcentration of population could be observed in the metropolitan scale being a phenomenon related to increasingly stronger processes of suburbanization in the surroundings of larger cities. It also applied to enterprises because foreign and domestic investors chose their localizations near large cities more often. This allowed them to use the urban outlet market and labor market as well as the available technical and social infrastructure (Noworól 2008, p. 72).

The expansion of urban infrastructure toward the city surrounding (forest areas and farmlands adjacent to the compact urban space) is a phenomenon that may be observed all around world. It is particularly intense in areas surrounding large urban agglomerations, regardless of the given country's developmental level. More and more people who earn more than just the average as well as those receiving smaller wages (often from rural areas or small towns) decide to move to the suburbs of urban centers. In such cases, the urbanized area is becoming larger and larger at the expense of meadows, forests, waters, and arable lands. This process is referred to in the literature of the subject as "urban sprawl" (of both the city and the suburbs), and most frequently, it is connected with the development of residential areas (Czerny 2007, p. 8).

More and more areas within urban or rural communes are intended for residential development as well as for erection of supermarkets, shopping malls, business parks, warehouses, new industrial institutions, roads, car parks, sports and recreation centers, and other forms that do not have agricultural character. This spatial expansion of the city worries the ecologists who perceive negative consequences of similar actions. They believe that the area of arable lands surrounding towns is clearly shrinking. Local authorities represent an opposite standpoint. This phenomenon, in their opinion, exerts positive influence on the local and regional economy, and it has a more beneficial social dimension (attracting investors, government subsidies, and development schemes for motorways and mass transit routes, expansion of the road network, transport development, higher standards of education in municipal schools, competence of local authorities, better job offers on the local market). Therefore, they believe that this process should be supported. The decisions made by authorities on the local and regional levels favor urban augmentation leading to increasing intensity of the building sector far away from the city. This involves construction of roads meeting specific transport standards, development of the public transport, and the grid of access roads. The business sphere, which used to develop rather chaotically and rapidly at the beginning, more and more often settles in a permanent form of urbanized zones in suburban areas, situated along main roads. Shopping centers, warehouses, wholesale depots, business parks, and special economic zones are established there, all of that to serve the purpose of attracting new investors. Also new housing estates, schools, hospitals, and other service centers are gradually being erected.

INTERNAL AND EXTERNAL PREMISES OF MODERN CITY DEVELOPMENT

Local development is perceived as a complex, multilayered phenomenon. It is conditioned by many factors as well as internal and external conditions. The notion of a local development factor can be understood as a set of "driving forces, conditions, decisions and situations which are essential to local development" (Alińska 2008, p. 65).

A review of empirical literature dedicated to this problem allows for making a distinction between two basic factors:

1. Endogenous, which comprises the widely understood internal and external spheres and include human capital
2. Exogenous, which includes a widely understood external sphere related to the surrounding and contains nationwide conditions, regulations, politics, and competition

This breakdown is characterized by a considerable interdependence and, in terms of its range, mutual interpenetration. The exogenous division is based on external markets as well as domestic and international ones, which causes the influx of external development factors to cities/regions, namely, those assuming the form of investments, knowledge, experience, modern technologies, and innovations. The endogenous division is based on internal resources, namely,

- Local society
- Local culture
- Internal development conditions

and as such, which seems to be the most important from the perspective of local development, it is based on the human and social capital, individual and local resourcefulness, entrepreneurship and acceptance of changes, and social activeness consisting in "self organisation aimed at creating interest groups and social movements, and striving to achieve collective support for their goals" (Alińska 2008, p. 66).

The aforementioned factors of local development mutually penetrate and remain dependent on one another, such as, for example, distances from main roads or agglomerations, distances from the country borders, tourist attractiveness, and spatial possibilities. If one was to attribute higher efficiency to one of the groups, "the relative importance of external (mainly connected with the location) and internal (endogenous), specific and individual factors for every region should be investigated" (Alińska 2008, p. 67). If perceived in such a manner, the factors in question may be divided into the following categories:

1. *External factors*, which the local community is not able to influence. They are, however, relevant for local development. Location, financial support of the country, the EU funds, credit and tax systems, and accessibility of the production means are some of them (Alińska 2008, p. 67). Efficient management of these factors performed by local authorities and economic entities may have an effect on local development as well as that of economic entities.
2. *Internal factors*, which depend on the attitude of local authorities and the local community to a considerable extent. Some of them are broadly understood human capital, social involvement, infrastructure, and business support organizations (regional development agencies, loan and guarantee funds, marketing and promotional activities, tax preferences). The individual differentiations of these factors may be perceived as variables intensifying local development, which has been shown in Table 55.1 (Alińska 2008, p. 67).

Yet another classification of the local development factors has been proposed by S. L. Bagdziński (Geise 2009, p. 48) (Table 55.2).

In the first category of factors, the greatest relevance is attributable to political factors that establish the framework of local activities and seem to be the most beneficial ones as regards the democratic nature of authority, since they involve participation of the local community in making decisions on the crucial matters for its development. Equally important seem to be the competences of local authorities, that is, their prerogatives, as far as the development trend is concerned. They function under specific statutory and constitutional framework, and the only, yet crucial factor of restrictions is a too limited financial independence expressed in the small number of available means to achieve one's own developmental goals or initiatives and innovations undertaken. The factors of authority, that is, the authority–society relationship and the level of society's acceptance of the authority, remain interdependent. They are determined by mutual relations such as partnership or trust, and the determinant can be the level of social involvement in local actions and ventures.

The social factor seems to be a lot more complicated bearing its specific nature in mind, since the needs and values condition the other ones. The crucial characteristic feature of this factor is an earlier level of development, which influences aspirations, human relations, entrepreneurship, and the attitude toward the broadly understood development and changes. Geise (2009, p. 50) stresses the

TABLE 55.1
Relevance of Internal Factors of Local Development

	Size of Local Administrative Units		
Categories of Explanatory Variables	**Large Cities**	**Medium-Size Cites**	**Small Towns**
Human capital	****	*****	*****
Civil society	****	***	***
Infrastructure	***	***	*****
Business support institutions	****	*	*
Leadership	**	**	0
Cooperation	0	**	**
Marketing/promotion	0	*	****
Reduction of local costs	0	0	0

Source: P. Swianiewicz and J. Łukomska, *Władze samorządowe wobec lokalnego rozwoju gospodarczego. Które polityki są skuteczne? (Local self-governmental authorities versus local economic development. Which policies are more efficient?), Samorząd Terytorialny*, 204, no. 6, in: A. Alińska, *Instytucje mikrofinansowe w lokalnym rozwoju społeczno-gospodarczym (Microfinancial institutions in local socio-economic development)*, Warsaw School of Economics, Warsaw, 2008, p. 67.

Note: Six-grade evaluation scale where 0 means a lack of statistic influence and **** mean very significant influence.

fact that "social capital, which is the outcome of networking relations between autonomous partners involved in local development, favours long-term cooperation, helps deepen economic relationships and contributes to the trust building. The ability to participate in cooperation and partnership is referred to as a social treasure."

Major influence on the local market effectiveness is exerted by the level of development of entrepreneurship in the economic sphere. The denser the network, the larger the possibility of external

TABLE 55.2
Local Development Factors

Political	**Social**	**Economic**
1. Nature of the authority—political system	1. Needs and values	1. Natural resources
2. Range of competences—prerogatives of different level authorities	2. Aspirations of inhabitants and representative bodies	2. Natural environment qualities
3. Governance type—authority–society relationships	3. Interpersonal relationships	3. Workforce resources and their qualifications
4. Level of the authority acceptance	4. Attitude toward reforms, innovations, and technical progress	4. Economic potential and its structure including "hard" and "soft" infrastructure
	5. Private enterprise	5. Capital investments
	6. Attitude toward local authorities and entrepreneurship	

Source: S. L. Bagdziński, *Lokalna polityka gospodarcza (Local economic policy)*, Toruń, 1994, p. 18, in: M. Geise, *Wpływ wybranych problemów gospodarki globalnej na rozwój lokalny (Influence of selected problems of global economy on local development)*, University Publishing House of the Higher School of Economics in Bydgoszcz, Bydgoszcz, 2009, p. 48.

capital influx and structuring of the economic network, which leads to the increase in competition within a given territory (Geise 2009, p. 51).

Another group of local development factors is the economic factors that the author considers to be among the extensive factors stimulating economic growth. It features the resources of knowledge and experience, that is, the human capital. High qualifications of employees and specialists are expected to increase the developmental potential of a city, commune, or region. It is the personnel qualifications that trigger the given corporation's interest in the given location.

The factors mentioned by S. L. Bagdziński have external and internal provenance. The first ones, that is, endogenous, are related to the local space. The exogenous (external) ones are thought to be "regional, domestic, global conditions." There are dependences and relations between them that lead to the development if only they have been appropriately channeled.

In the modern market dependences, important factors of local development are the networks (network society) defined as "formal and informal organisations and platforms, which enable the exchange of information and technology, and strengthen the cooperation between the network participants. Depending on the network type, its participants may include regions, companies, scientific centres or individuals" (Alińska 2008, p. 69).

The endogenous and exogenous factors exert the greatest influence on local development. Some of them are particularly characteristic owing to their individualized nature, being a consequence of emergence in a different territory and a coincidence of numerous phenomena. However, it is always their goal to stimulate the social and economic activity for the sake of development.

CONCLUSION

Due to the spatial dimension of local development, in order to analyze it, one should entail its networking socioeconomic sphere and the larger influence of exogenous and endogenous factors, which become a specific reflection of the commonly envisaged goals of local/urban development. The endogenous factors comprise a certain set of data that indicate the utility values as well as those created by both nature and men, the human capital, institutions deciding about the development of the given territory, culture and economic traditions of the given area, and competitiveness of the places where one resides. The exogenous factors, that is, ones that the local population cannot affect by any means whatsoever, are crucial to local development. They are based on external markets as well as the domestic ones and international ones, which triggers the influx of external development factors, such as investments, knowledge, experience, modern technologies, and innovations, into cities/regions. Efficient management of these factors performed by local authorities and economic entities may have an effect on the level of local development as well as that of economic entities.

Spatial dimensions are significantly translated into the conditions, resources, and opportunities of economic and social entities. They will always be identified with creation of the best available conditions of living in the local environment through the pursuit of effective organization and structure as well as appropriate functioning of the local social system.

REFERENCES

Alińska A., *Instytucje mikrofinansowe w lokalnym rozwoju społeczno-gospodarczym, (Microfinancial institutions in local socio-economic development)*, Warsaw School of Economics, Warsaw, 2008.

Czerny M., Łuczak R., Makowski J., *Globalistyka: Procesy globalne i ich lokalne konsekwencje (Globalisation: Global processes and their local consequences)*, PWN, Warsaw, 2007.

Domański B., Czynniki społeczne w lokalnym rozwoju gospodarczym we współczesnej Polsce (Social factors in the local economic development of the contemporary Poland). In: I. Sagan, M. Czepczyński (eds.), *Wybrane problemy badawcze geografii społecznej w Polsce (Selected research problems of social geography of Poland)*, Institute of Economic Geography of the University of Gdańsk, Gdynia, 2001.

Geise M., *Wpływ wybranych problemów gospodarki globalnej na rozwój lokalny (The influence of selected problems of the global economy on local development)*, Publishing House of the Higher School of Economics in Bydgoszcz, Bydgoszcz, 2009.

Gorzelak G., Smętowski M., *Metropolia i jej region w gospodarce informacyjnej (The metropolis and its region in the information economy)*, the Centre for European Regional Studies of the University of Warsaw, Wydawnictwo Naukowe "SCHOLAR", Warsaw, 2005.

Kudłacz T., *Programowanie rozwoju regionalnego (Regional development programming)*, Wydawnictwo Naukowe PWN, Warsaw, 1999.

Kukliński, A., Pawłowska, K. (eds.), *Innowacja, edukacja - rozwój regionalny*, Higher School of Business—National Louis University, Nowy Sącz, 1998.

Majer A., *Socjologia i przestrzeń miejska (Sociology and urban space)*, Wydawnictwo Naukowe PWN, Warsaw, 2010.

Morawski M., *Zarządzanie profesjonalistami (Managing professionals)*, Polskie Wydawnictwo Ekonomiczne, Warsaw, 2009.

Noworól A., *Planowanie rozwoju terytorialnego w skali regionalnej i lokalnej (Planning of territorial development in the regional and local scale)*, Publishing House of the Jagiellonian University, Krakow, 2008.

Rupik K., Rozwój koncepcji marketingu terytorialnego. Analiza historyczna (Development of the concept of territorial marketing. A historical analysis). In: H. Szalce, M. Florek (eds.), *Marketing Terytorialny (Territorial Marketing)*, Publishing House of the Poznań University of Economics, Poznań, 2005.

Szromnik A., Marketing terytorialny—geneza, rynki docelowe i podmioty oddziaływania (Territorial marketing—origins, target markets and subjects of influence). In: T. Domański (ed.), *Marketing terytorialny (Territorial Marketing)*, University of Łódź, Łódź, 1997.

Szymla Z. (ed.), *Konkurencyjność miast i regionów (Competitiveness of cities and regions)*, Krakow University of Economics, Krakow, 2001.

Węgleński J., *Urbanizacja. Kontrowersje wokół pojęcia (Urbanization. Controversies around the notion)*, PWN, Warsaw, 1986.

Zabłocki G., *Rozwój zrównoważony—idee, efekty, kontrowersje (Sustainable development—ideas, effects, controversies)*, the Nicolaus Copernicus University in Toruń, Toruń, 2002.

Zalewski A. (ed.), *Nowe zarządzanie publiczne w polskim samorządzie terytorialnym (New public management in the Polish local self-government)*, Warsaw School of Economics, Warsaw, 2005.

Zioło Z., Model funkcjonowania przestrzeni geograficznej jako próba integracji badań geograficznych (Functioning model of a geographical space as an attempt to integrate geographical studies). In: *Geografia na przełomie wieków—jedność w różnorodności (Geography at the turn of the centuries—unity in diversity)*, Faculty of Geography and Regional Studies of the University of Warsaw, Warsaw, 1999.

Section IV

Higher Education

56 Believing in the University in a Conceptual Age

Ronald Barnett

CONTENTS

INTRODUCTION

The idea of the present age as a "conceptual age" is a provocative idea, in the best sense of being "provocative." Far from this being a conceptual age, it could be said that the present age is marked precisely by a diminution of concepts. Certainly, words and phrases and even neologisms abound; there is a cascade of new terms, not least in a digital age. But concepts, it may be felt, are thinning and even dissolving, for concepts imply a public able to share ideas and to engage in debate such that those ideas are deepened and strengthened into collectively understood concepts. In a liquid age, as Bauman (2000) has termed it, that very idea of an educated and reflexive public that shares fundamental ideas has to be in doubt.

But there *are* concepts; they are, though, increasingly the concepts of the dominant powers in society. The market, the individual as responsible for his or her life trajectory, competition, regulation, audit, the knowledge economy, and skills: in relation to higher education, these surely are indicative of the dominant concepts of our age.

The University is implicated in all of this. The very idea of the University itself thins and even threatens to dissolve, as each "university" is challenged to identify its own mission. Any connection between the ideas of "University" and "universality" is also dissolving as the university becomes excessively parochial, attending to its own impact on its region and its engagement with the potential employers of its graduates.

Against these reflections, the questions in front of us are clear but profound: To what extent can the University be associated with the largest possible ideas—of truth, of knowledge, of dialogue, of human being, and of societal well-being? To what extent is there still space for the University to articulate creative concepts that are critical of the dominant concepts of the age? Is any kind of universality still available to the University? What, indeed, has become of the idea of the University itself?

MOVING CURRENTS

There are many currents that beset the modern age, but two are especially relevant to our purposes here. On the one hand, there is a burgeoning of new ideas. The intercommunication—or to put it more graphically, the colliding—between different disciplines and perspectives is one cause of this creativity. And that phenomenon, of an interdisciplinarity in which discipline speaks unto discipline, is partly a recognition of problems, either theoretical or practical, that require a complex of intellectual resources if they are going to be seriously addressed. Engineering, mathematics, genetics, creative arts, humanities, medicine, biochemistry, informatics, computer science, anthropology, philosophy, and sociology—all these and many other areas are to be found abutting and even transgressing each other, as they are turned upon complex problems, whether in the personal or societal or physical environments. Not surprisingly, then, there is a profusion of new ideas, of new terms, in the world of inquiry.

On the other hand, questions can be legitimately posed as to the status—the epistemological depth—of the new ideas that are emerging. A liquid age surely bequeaths liquid concepts. Words and metaphors swim or are transported across networks and are taken up in different settings. In the process, the meanings of those words and metaphors transmogrify. And the networks themselves are often ephemeral, set up, or simply emerging to address an issue or problem, which perhaps spans the policy, academic, political, and practitioner domains (concerned, say, with the building of a dam in an area in which human or natural habitats are implicated). What we are seeing, under such conditions, is the formation of ethno-epistemic assemblages (Irwin and Michael 2003), that is to say, fuzzy and dynamic conglomerations of knowing activities. Knowledge here, it will be noticed, gives way to "knowing," attempts to hazard some understanding that is at once practical, academic, and policy-imbued and in which multiple networks participate. Characteristically, too, such knowledge activities occupy contested territory, as the different networks secrete their own perspectives and aspirations in the formation of collective representations.

A term such as "uncertainties" can, in the United Kingdom, lend itself to a multiagency research program spanning different disciplines* precisely because it can be approached through many perspectives. Relevant perspectives not only are within the academy but go wider to embrace political and public interests and those of the corporate world. In the contemporary age, many groupings will not just have cognitive perspectives on "uncertainties" but will have practical stances towards "uncertainties." Can it be said, therefore, that there is a concept of "uncertainties" that is held in contemporary society? Or should we say, rather, that there is a medley of concepts of "uncertainties"? Or, rather instead, that there is a cluster of sentiments, at once variegated with and even antipathetic to each other? Surely, it is this last option that deserves our support: What we are faced with in the present age is not so much a welter of concepts but a liquid mess of sentiments, complex currents moving in different, and even conflicting directions. Rather than living in a conceptual age, we live—or so it may be suggested—in a Sentimental Age, an age, that is to say, of multiple sentiments.

The idea of the University is itself implicated in these reflections. Especially, as higher education expands into a mass system (in very many countries) and as research is increased but also is expected to engage with the wider society in a multitude of ways, multiple ideas of the university emerge. Dimensions of the idea of the university connect the university variously with ideas of the market, entrepreneurialism, the knowledge economy, the learning economy, citizenship, digital learning, and professionalism, to identify but a few. In this milieu, universities are conjoined to find their own niche in the national and global panoply of tertiary institutions and to develop their own "missions."

Against this canvas, the concept of the university can be understood in sharply differing ways. On the one hand, it may be felt that the concept of the university is here expanding and is being enriched, as it is stretched this way and that. There may be no essence to the concept of the university now, just—in true Wittgensteinian mode (1978)—an interplay of overlapping ideas that cluster together but that have no single unifying feature. Private and public, multifaculty and specialist,

* http://www.globaluncertainties.org.uk/.

mainly undergraduate or almost entirely postgraduate, local and global, research-intensive or teaching-intensive, focused on the natural sciences or focused on the creative arts, favoring face-to-face teaching or e-learning at a distance, characterized by "pure" inquiry or more applied forms of learning and knowing (or even "work-based learning"): all these and yet other options are open to institutions that take the name of "university." Any common thread that runs through all these "universities" is difficult to discern.

In this milieu, ideas of the university proliferate, even ahead of forms of the university. Just a few of the many ideas of the university already being voiced (or implicit in currently circulating ideas) are the following:

- The borderless university
- The capitalist university
- The civic university
- The chrestomathic university
- The collaborative university
- The commercialized university
- The cosmopolitan university
- The disciplinary university
- The ecological university
- The edgeless university
- The fully functioning university
- The enquiring university
- The enterprise university
- The entrepreneurial university
- The university as fool
- The global university
- The injured university
- The learning university
- The liquid university
- The networked university
- The posthistorical university
- The pragmatic university
- The schizophrenic university
- The sentimental university
- The service university
- The socially engaged university
- The supercomplex university
- The theatrical university
- The virtual university
- The university of wisdom
- The world university

This is a supercomplex world (Barnett 1990), in which ideas—here of the university—proliferate in a bewildering way, crisscrossing each other and setting off and contending against each other.

On the other hand, another interpretation presents itself in this conceptual landscape. This is that the concept of the university has thinned close to its point of dissolution. After all, now, it seems, all connotational bets are off: there appears to be neither any condition that an institution has to fulfil in order to warrant the title of 'university' nor any limits to an institution that might be termed 'university' or, if there are such limits, where they might lie. It is almost as if, conceptually, anything goes so far as the idea of the university is concerned. As such, it can fairly be said that the idea of the university has disappeared.

THE END OF THE IDEA OF THE UNIVERSITY?

There is—as it might be termed—a deeper ontological substrate that is implicated in these reflections. At one time, the idea of the university was associated with a certain kind of value structure (cf. Taylor 1969), which included the following kinds of belief. A university offered a social space for disinterested inquiry, for a collective search for truth (however that might be understood and successively redefined), and for an insistence that those inquiries be conducted according to publicly attested standards, and a sense that such a collective inquiry was personally and socially edifying, assisting personal development and the growth of public reason and understanding in society. Such a set of assumptions was accompanied with a belief in the value of both institutional autonomy and academic freedom, as concomitants of and as a safeguard of this idea of the university.

With the repositioning of the university such that it is much more an institution *of* society, and with its proliferating and diverse forms, the academic culture which that idea of the university reflected is in jeopardy. Indeed, an attack on the somewhat self-concerned stance of the university which that earlier pure idea of the university reflected was precisely part of the purpose of the repositioning of the academy, namely, to effect an assault on the academy's inwardness, marked out by that earlier idea of the university.

At least a thousand flowers bloom in this new academic world. Very heaven must it be to be in this open and free academic age, when a university can be more or less anything it wants, unconstrained by any particular idea of the university. In this milieu, those who lead universities, who plan national systems, and who regulate transnational systems of quality control need have no consideration for any idea of the university. Now, there is conceptual anarchy, so far as the idea of the university is concerned. Is this not just what a marketized system requires: a free market not only in knowledge production and knowledge services (a kind of "knowledge capitalism" (Murphy 2008)) but a free market also in the interpretation and reshaping of the idea of the university?

As remarked, this bewildering array of forms adopted by contemporary "universities" presents us with two conceptual possibilities: that the concept of the university is now rich and varied and open to creative interpretation and that the concept has thinned to the point where it is about to dissolve. There is, however, a third option available to us, an option that is already implicit in these reflections. This third option stems from a recognition that the forms and accompanying ideas of the university are but reflections of the empirical forms and ideas of the contemporary university. At a deeper level, in a globalized academic world, worldwide currents of marketization, entrepreneurialism, performativity, and commodification, linked, in turn, to the world's knowledge economy, are affecting the academic world on a global scale. And so we may say that we are seeing the birth of *the marketized university*, and here, tacitly, is a new idea of the university, the university as a promoter of marketized knowledges (cf. Molesworth et al. 2011). Far from the idea of the university dissolving, the idea of the university is flourishing, even if—for the many who oppose the marketization of higher education—somewhat unhappily. Nevertheless, the continuing presence of the idea of the university, albeit in new form, is testimony to its continued longevity.

A CONCEPTUAL MATRIX

Concepts, we may say, therefore, have their presences at different levels. There is the empirical level, the level of experiences. And in an age that is at once globalized and networked, with its speeding up of the dynamics of life and increasingly complex interconnectivity, we can assume that understandings of terms will multiply as they are taken up and given somewhat varying interpretations across the world. Whether one perceives this phenomenon as a proliferation of concepts—and the dawn of a conceptual age—or as a dissolution or a hollowing-out of concepts is, I think, a moot point. At any rate, as noted, this phenomenon is present in the academic world.

But given that this globalized world is dominated by large interests of markets, flows of information for economic gain, the growth of the knowledge economy, competition, and an agenda of "skills," these interests (endorsed by the World Bank and other transnational agencies) are bequeathing tacit concepts that are, in turn, reshaping the university. Accordingly, the proliferation of forms and ideas of the university that are empirically evident are but epiphenomena to which, at a deeper level, the underlying ideas of the markets and competition are giving rise. Far from there being a conceptual free-for-all, the diverse forms and ideas of the university that are emerging are, to a large part, in the service of the marketized university. This particular conceptual playing field is far from even, but is weighted towards certain dominant concepts.

There is, therefore, as we may term it, a conceptual axis of *depth*. To what extent does an idea of the university extend into the deep structure of the university, to take account of the underlying forces at work? To what extent does the concept have security partly in virtue of its recognizing that the university is an institution that has its presence in society through its being called upon to serve many masters, including those of the global knowledge economy and of national advancement? Or, to the contrary, is the concept superficially attractive but even utopian and yet without purchase in the deep structure of the real world and so is illusory and flimsy?

Crossing this axis of depth is another axis, that of *criticality*. At one end of the axis lies a position of *endorsement*, in which an idea of the university in effect acts to sustain the current power structures. At the other end of this axis lies a position of criticality, in which would lie ideas that pointed the university in the direction of being a center of critical consciousness in society.

These two axes, placed against each other, generate four quadrants: (1) superficial/endorsement (quadrant A), (2) superficial/critical (quadrant B), (3) deep/endorsement (quadrant C), and (4) deep/critical (quadrant D) (Figure 56.1).

In quadrant A would lie, for example, ideas of the university built around the ideas of quality, audit, and excellence that treat such matters as unproblematic. Examples include those of "the world-class university" and "the university of excellence." Such ideas provide superficial endorsement of contemporary policies in higher education. In quadrant B would lie an idea of the university such as "the edgeless university" (Bradwell 2009), "the virtual university" (Robins and Webster 2002), or

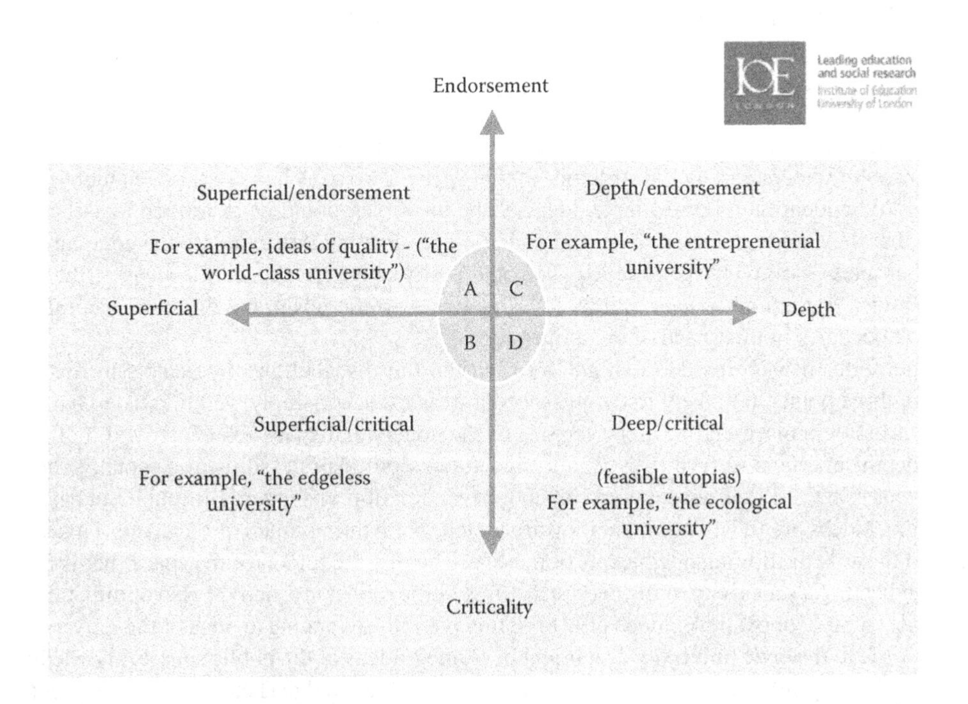

FIGURE 56.1 Ideas of the university: a conceptual space.

"the borderless university" (Middlehurst 2001; Erdinc 2002). Such ideas either are explicitly critical of the contemporary character of universities ("the edgeless university") or are tacitly critical of universities ("the virtual university" and "the borderless university"). All are urging the university to embrace modern technologies fully and become much more fluid in the forms of their knowledge generation and communicative processes. But such ideas are prone to a kind of superficiality, being concerned with the manifest form of the university.

All of these ideas that focus on the superficial forms of the university, whether endorsing or are offered in a more critical vein, are ideological, serving in effect to sustain contemporary policies or to urge the development of universities such that they are liable to work in the interests of the large power structures in society. This can also be said for ideas that sit in quadrant C. However, quadrant C ideas are not merely ideological but are particular pernicious ideologies in that they are of the moment and take on the sense that there could be no other. The most prominent such ideology at present is that of "the entrepreneurial university" (Clark 1998; Shattock 2009). This is projected vigorously as if it is inevitable and right at the same time. The world of higher education is going in the direction of the entrepreneurial university, *and* this direction of travel is right and proper. It is an endorsing idea of the university, therefore, but it is also attending to its deep structure. For it comes armed with a sense of the deep structure of higher education, of the underlying presence of the global economy, and is seized with a sense of the value of markets and competition. And so it not only has inevitability attaching to it; it even has a kind of *gravitas*. It seems to have done its homework and claims to be an insightful response to the nature of the current age.

The fourth quadrant is that of ideas of the university that are both deep and critical. These are ideas of the university that also take account of the deep structure of higher education in the current age but, far from endorsing the emerging entrepreneurial university, contend against it. Such ideas of the university would play up possibilities for the university in bringing about a more equitable society or improvements in social and personal well-being or in helping to develop the public sphere. Ideas such as these constitute a set of—as it were—*feasible utopias* (Barnett 2011). These are utopian ideas in that they are reaching out to large visions of the university but, being grounded in an insight into the deep structures of the university, are feasible. These ideas have their feet on the ground; they are in the real world, in the deep structures of the world.

THE CONCEPTUAL SPACE OF THE UNIVERSITY

There is, then, in principle, a conceptual space opening for ideas of the university. And there are various ways of understanding the structure of that space. Two axes have so far been identified, but others could be identified. For example, ideas of the university could be examined to ascertain the position that they hold on an *optimistic–pessimistic axis*. Putting this axis into the company of our two earlier axes—so giving us three axes—would really open up a space for ideas of the university. Each idea of the university could be interrogated as to the position—or, more accurately, the region—it occupies in this space of the university.

The question then is this: How might we describe the distribution of concepts in this space? There are three points to make. Firstly, the space is perhaps surprisingly well populated, with ideas of the university being found in many regions of the conceptual space (characterized by the three axes of depth/superficiality, criticality/endorsement, and optimism/pessimism). Secondly, the space is somewhat tilted, in that concepts in the endorsement camp are favored in public debate; other more critical ideas are to be found, but they are muted, not gaining much of an airing. This space is weighted towards the dominant concepts of our age. Thirdly, the region of the space that is (1) critical, (2) indicating a sensitivity to the deep structures of the university, and (3) also optimistic is thin, though not empty. For example, much effort is being put into advancing an idea of the university that gives new life to the civic university as it might be termed: Ideas of the public sphere, knowledge as a public good (Marginson 2007; Holmwood 2011), socialist knowledge (Peters and Besley 2006), the public benefits of learning, civic engagement, and academic openness are all being called into action.

This cluster of ideas of the university, intent on orienting the university towards the wider society rather than the economy per se, are present in the literature but are not much part of public debate.

Lastly, we can observe that there is another feature that attaches to the dominant concepts of the university, now being built around such mantra ideas of "diversity," "competition," "particularity" (with each university identifying and projecting its own distinctiveness), and the proclaimed value of universities attending to the needs (largely economic needs) of their own region. Through all of this, and in their responses, universities are becoming *parochial* institutions in several ways. Geographically, universities are being oriented more to their local region (whereas from their inception in medieval times, they were always global in their outlook); epistemologically, they are coming to concern themselves with the problems and concerns of their region, and those problems are characteristically practical problems, whereas originally, universities had been concerned with truth in a transcendent sense. Even as universities are encouraged to understand themselves in global terms, still, they are coming to have their being in their local communities.

RECOVERING THE UNIVERSAL

We can reflect on our explorations so far in this essay. In a marketized age, the conceptual space of the university is—at least potentially—opening up. The universities have their being in a liquid age, so it is inevitable that ideas of the university will expand. At the same time, however, some ideas emerge as dominant, so the conceptual space of the idea of the university is tilted—towards ideas of competition, global "excellence," and contribution to the global economy and of the market itself. This is a Darwinian situation of the survival of the fittest, and concerns with global league tables and institutional solvency support such a perspective.

It follows that it is not true to say that the idea of the university has hollowed out; rather it has *thinned* out. It has thinned out in that most of the ideas of the university currently available are somewhat superficial, insufficiently taking account of the deep structure of the being of the university in the world in the 21st century; it has thinned out in that weight is being put behind ideas of the university—such as the entrepreneurial university—that are losing contact with the historic depth of the idea of the university.

This may seem to contradict our earlier observations in which it was suggested that the idea of the entrepreneurial could at least come to reflect a deep reading of the university, attending to its structural position in the global economy, the marketization of knowledge, and the developing demand for higher education. However, there is a thinness attaching to the idea in that it has abandoned the deep value structure of the Western university, built around a valuing of critical reason, academic freedom, and knowledge as "its own end" (Newman 1976). Now, activities of the university are governed by considerations of market positioning and institutional projection.

There is, here, a further sense of the emerging parochialism of the university, additional to those just noted. In a liquid age, in an age in which institutions such as the university are caught in vortices of knowledge transmission and generations that transgress societal boundaries, the university is much less in thrall to Truth as a project. For 700–800 years, the Western University was associated with a sense of truth that stood independently of projects and persons. A sharp distinction came to be made—by Karl Popper and Ernest Gellner among others—between knowledge in the context of discovery and knowledge in the context of justification. Now, those two domains are colliding. This collision of two worlds of knowledge lies at the heart of Lyotard's (1984) "postmodern" concept of performativity: Now, what counts for knowledge is precisely how it performs in the world. Here, surely, we have a drawing in of the world of the university; now, knowledge has to be put to work in furthering its own projects and positioning. No longer is it serving the interests of truth as such. (Is it coincidence that, currently, we are seeing scandals of data misuse, plagiarism, and so forth within the scientific community?) This amounts to a new parochialism, an ontological parochialism in which the university is content with pursuing its own projects on this Earth, rather than allying itself to the somewhat larger project of understanding man's place in the metaphysical universe.

Another way of putting these observations is to suggest that we are seeing here the rise of particularism and the loss of universalism. Formerly, it was precisely a sense of its being associated with sacred realms of knowing and being that supplied legitimacy to the university. (Popes and then kings almost fell over themselves to provide privileged sanctuaries to their universities so that, freed from the profane and mundane world, those universities could contemplate and even live in a transcendent world.) Neither the knowledge of individuals (not even of the pope or king) nor its own knowledge, the university was allied to a project of knowledge *per se*. The idea of objective knowledge was, in retrospect, merely a means through which the university could safeguard and fulfill its interest in the universal order of things rather than the particular. Now, a university will blithely abandon a department—chemistry, modern languages, philosophy—simply because it is in its own interests. The University of Today heeds no universal calling.

CHALLENGES AND PROSPECTS

A conceptual age, if that is what we have, has conflicting implications for the university. On the one hand, a space opens for the idea of the university to become multitudinous idea*s* of the university. And this has happened, to some extent. There are many ideas of the university that are emerging. On the other hand, in this liquid and fast-moving age, there is a conceptual thinness to many ideas. Many ideas either are not rooted in or are sensitive to the deep (societal and global) structures in which the university has come to have its being. Many ideas of the university, in turn, are ideological, supporting the dominant interests in capital accumulation (personal or corporate), and are losing sight of any larger concerns with the well-being of the world. The historic connection between the ideas of "University" and "universality" is dissolving as the university becomes excessively parochial, as it turns its attention to projects that have an immediate return to the university. The university is no longer content (or able) to live in a metaphysical world, with a sense of furthering humanity's relationship with the universe; now, the university's gaze has shrunk to a concern with its fortunes in this immediate world.

And so our opening questions return: To what extent can the University be associated with the largest possible ideas—of truth, of knowledge, of dialogue, of human being, and of societal well-being? To what extent is there still space for the University to articulate creative concepts that are critical of the dominant concepts of the age? Is any kind of universality still available to the University? What, indeed, has become of the idea of the University itself?

Another way of putting this set of questions is to observe that we do not so much need more concepts of the university but that we need *better* concepts, concepts that are at once (1) sensitive to the deep structure of the university (in its societal and global base), (2) utopian, (3) feasible, (4) optimistic and (5) universalistic. The cluster of concepts of the university that are now emerging around the idea of the (new) civic university—connected with ideas of democracy, the public good, civic engagement, public dialogue and openness—are promising to come close to satisfying these five conditions, demanding as they are. These ideas are sensitive to the deep structure of the university; they are utopian (in reaching out to a better world that may not be realized); they are feasible (or, at least, they are keen to demonstrate their credentials as to their feasibility); they surely spring from an optimism that things really can go better for the university; and they are universalistic in their orientation, being concerned for the global good.

We might place all of these more socially oriented concepts of the university under the umbrella concept, so to speak, of *the ecological university* (Barnett 2011). The ecological university has a care or concern (to draw on Heidegger) for its total environment, it is aware that it itself is held in all manner of networks (social, economic, cultural, personal, and at once local and global) and is therefore sensitive to its interconnectedness and strives to do all it can to maintain the sustainability of that total environment. As a corollary, it will eschew activities that would have the effect of diminishing its interconnectiveness through, for instance, excluding individuals and communities from its resources. Indeed, it will be proactive in engaging with the wider society in putting its resources into the service of society, indeed the world. The continuing development of the well-being of the various ecologies in which it is implicated is its main purpose.

Such a concept of the university would need to be worked out, and that work lies ahead. There are major issues, not least over the sustainability of that very idea, the ecological university. In particular, might resources be forthcoming for such a concept of the university to be developed and translated into practical projects? As remarked earlier, it is understandable if a pessimistic outlook presents itself, given the weight of the forces propelling the university in other directions—of competition, of inwardness, and even, in some ways, of exploitation. But there are both theoretical and practical grounds for optimism, too. Precisely because the world is faced with major social and environmental problems, it is likely to enlist its universities in attending to such mega-problems.

CONCLUSION

In a liquid world, many ideas of the university abound, swimming in the conceptual currents of the age. Some ideas of the university, however, are swimming with the tide and are given greater impetus by the surrounding forces. "The entrepreneurial university" grows in strength and size as a result. Others swim against the tide and struggle somewhat: The "socially engaged university" and the "ecological university" have to swim vigorously to make headway. But currents may change. Tributaries can merge into the mainstream. New concepts can emerge and grow in strength. Most such concepts, admittedly, lack strength, fail to make any impact, and quickly die.

Much, indeed, is needed if a fledgling concept is to succeed against the odds; practical, ideological, and financial resources are needed, certainly. There has to be a real prospect that such assistance will be forthcoming, and help may come from unlikely quarters. The corporate world may just find that it has some need of an ecological university; democratic societies may also just become alert to the potential that an ecological university may offer, a university that concerns itself with universal interests and the well-being of the world. It still makes sense to believe in the university in a conceptual age.

REFERENCES

Barnett, R (1990) *Realizing the University in an Age of Supercomplexity*. Buckingham: Open University Press.

Barnett, R (2011) *Being a University*. Abingdon: Routledge.

Bauman, Z (2000) *Liquid Modernity*. Cambridge: Polity.

Bradwell, P (2009) *The Edgeless University: Why Higher Education Must Embrace Technology*. London: Demos.

Clark, B R (1998) *Creating Entrepreneurial Universities: Organizational Pathways of Transformation*. Oxford: Pergamon/IAU.

Erdinc, Z (2002) 'Australia online; Borderless university', *Turkish Online Journal of Distance Education* 3 (4).

Holmwood, J (ed) (2011) *A Manifesto for the Public University*. London: Bloomsbury.

Irwin, A and Michael, M (2003) *Science, Social Theory and Public Knowledge*. Maidenhead: Open University Press.

Lyotard, J-F (1984) *The Postmodern Condition: A Report on Knowledge*. Manchester: Manchester University.

Marginson, S (ed) (2007) *Prospects of Higher Educations: Globalization, Market Competition, Public Goods and the Future of the University*. Rotterdam: Sense.

Middlehurst, R (2001) 'University challenges: borderless higher education, today and tomorrow', *Minerva* 39, 3–26.

Molesworth, M, Scullion, R and Nixon, E (eds) (2011) *The Marketisation of Higher Education and the Student as Consumer*. Abingdon: Routledge.

Murphy, P (2008) 'Defining knowledge capitalism', in M A Peters, S Marginson and P Murphy (eds) *Creativity and the Global Knowledge Economy*. New York: Peter Lang.

Newman, J H (1976) in I T Ker (ed) *The Idea of a University*. Oxford: Clarendon Press.

Peters, M and Besley, A C (2006) *Building Knowledge Cultures: Education and Development in the Age of Knowledge Capitalism*. Maryland: Rowman and Littlefield.

Robins, K and Webster, F (eds) (2002) *The Virtual University? Knowledge, Markets, and Management.* Oxford: Oxford University Press.
Shattock, M (ed) (2009) *Entrepreneurialism in Universities and the Knowledge Economy.* Maidenhead: McGraw-Hill.
Taylor, C (1969) 'Neutrality in political science', in P Laslett and W G Runciman (eds) *Philosophy, Politics and Society*, 3rd series. Blackwell: Oxford.
Wittgenstein, L (1978) *Philosophical Investigations.* Oxford: Blackwell.

57 The Function of Imagination in University Science

David Brian Hay

CONTENTS

I have never let my schooling interfere with my education.

Mark Twain (1892)

INTRODUCTION

This paper concerns the function of the imagination (cf. Bakhtin 1981, 1986) in scientific research and in science education. Michael Bakhtin's dialogic theory—sometimes referred to as a theory of the literary or social imagination (see Holquist 2005)—is a robust account of intertextuality (Kristeva 1986a), whereby new insight and new imaginative understanding are seen to be produced in the inter-animation of one and "an other" text (Wegerif 2007, 2008). In this paper, using dialogic theory to explore the results of several case studies, I show how the imagination is indispensible in science practice and also fundamentally inseparable from *understanding* scientific data. While teaching creativity is a central theme for the educational practices of schools (e.g., Alexander 2008a; Wegerif 2010), imaginative process is a much neglected issue in university science (Hay 2010). Throughout this paper, I set out to show how the development and exercise of imagination is a means by which students and researchers gain entreaty to an *understanding* of science (Northedge 2003). The analysis is developed in order to explore how the quality of teaching university science may be enhanced by more deliberate focus on the imaginative function. Related to this, I discuss the importance of researchers' making their *informal* texts available to students. I conclude that learning and practicing science both depend on imaginative process, but the imaginative function is often a hidden virtue in formal science.

THE PROBLEM

It is tempting to suggest that only conspicuous scientific breakthroughs are the products of remarkable imagination and insight. This would be wrong. As Kuhn (1977) has shown, "scientific revolutions" tend to accrue *within* existing scientific paradigms, and the work of Nobel Laureates is *not* essentially *ex situ*. Without detracting from the notable achievements of Prize winning science, the point is that *if* imagination has a place in science, then it is probably indispensible throughout *all* the practices of research and teaching: the routine, day-to-day, as well as the outstanding. If we want to know more about the function of imagination in science, we must turn toward the "daily practices" of scientists for analysis. Here, through what is sometimes called "Science Studies," there is a small but growing literature attesting to the ways in which imagination is vital for research science (Latour 2005; Myers 2008, for example). But while the imagination is conspicuous in the texts that constitute the social sciences and the arts and humanities (see Bruner 1986; Kristeva 1986a, b for analysis), in most science, the imagination is a hidden virtue and sometimes treated as a problem. This is a consequence of failing to acknowledge that while "imaginative" can be used to label fanciful conjecture or even pathology, Bakhtin's (1981, 1986) dialogic theory is a robust account of human understanding in which imagination has a central role and is defined transparently. In order to go further, we need to be specific about what the imagination *is* and thus also where and how we look for it, but first we should explore what understanding science means since this will be the context for considering imagination as a process: after all, "knowing" textbook information does not need imagination, just the exercise of memory.

UNDERSTANDING SCIENCE

Most of us agree (I take it) that to practice science means much more than verbatim recall of information. In order to be "a scientist," one must understand the ways that scientific data are constituted and how the texts of science work (i.e., how scientific texts function and how they can be used/extended). Just knowing (recalling) how a particular concept is *given* (as a formula, a model, a theorem, or an image) is not enough, and a student of science must begin (at least) to recognize the work that science-texts achieve (see Box 57.1). Thus, Perkins (2007) distinguishes between scientific "knowledge" (as the assimilation of inert and static information) and understanding how science is essentially "performative" (i.e., where knowledge is working, functioning, and available for use). As Latour (1987) puts it, there is a conspicuous difference between "readymade science" and "science in the making."

For Perkins, thinking outward from the "givens" of scientific knowledge is the hallmark of scientific understanding (see Perkins 1981). This performance can be visualized in text (communication), and to *show* an understanding, it is not always necessary for an individual to design and carry out a laboratory experiment or fieldwork. In fact, scientific performance resides in all our acts of science—writing, speaking, drawing. etc., and broadly the academic literacy of science (Lea and Street 1998; Street 2004) can be treated as synonymous with the ways scientists show their science practice (Kress 2003, 2010). In other words, researchers' scientific practice is made visible in their texts of science, and arguably, their science work is constituted there (in text) rather more than it is in laboratory work. The issue is that most scientific discoveries are not merely serendipity but consequences of dialogue with science (i.e., dialogue with "other" scientists or dialogue with "otherness" in general [Wegerif 2007]). This occurs in text (*is* text) including the "texts" of private thoughts, and it is the imaginative propensity of text *itself* that generates new insights, new text, and new potential outcomes. This is not to say that experimental work is only a process of verifying what has been projected in a text beforehand. New experimental data are often surprising, but then, a response to that surprise depends on further use of other texts, so that the published work and thought of others are included too. Box 57.1 exemplifies these issues through case study of the intertextual work of one applied chemist.

BOX 57.1 SCIENCE TEXTS THAT WORK

In some of my recent research, my colleagues and I have used a novel approach to concept map analysis (Novak 1998) to document the ways that one of us (Stuart) uses text to carry out his science work. Stuart is an applied chemist and we explored how his academic texts (about pharmaceutical design) start *before* his work in the laboratory, carrying through and after his "lab-time" in his reading, conversation, and the imaginative projection of how new drugs *might* be known and made. When Stuart made concept maps about the subjects that he taught his students (i.e., his work with liquids [aqueous] and solids [insoluble] drug compounds), his maps were shown to carry out the *work* of Pharmaceutical Science. Each zone of the mapping text was composed of pairs of concept labels denoting (a) what is known about the material properties of a particular "thing" and (b) the theories or methods by which these properties are identified and potentially manipulated. These dyads (of theory/method and material/property) also occurred as sets or series of interactions (Figure 57.1), so that looking *through* each one was the "image shape" of his practice work (as *either* work with solids or with liquids).

At each intersection point and throughout each series as a whole, many different drugs were implicated (implied as theoretical outcomes). Effectively the working outcome of the text was to *make* (in text) a series of different pharmaceutical products, some of which exist already (like Calpol [Figure 57.2]) and others that are new potentials and are currently the subjects of Stuart's new patent applications.

Our conclusions were that Stuart's text production is an important locus for his work in science: or said another way, his science-work is often situated (practiced) in his texts that work as productive science. These texts work (i.e., carry semiotic productivity) because they constitute creative (making) processes even if they also must be used like recipes to show material outcomes in the laboratory. Perhaps this is an obvious point. Except for the most serendipitous mistakes of protocol, most work in the laboratory is carried out to plan, and this implies a schedule and *projected* outcomes first. All too easily, however, we speak of

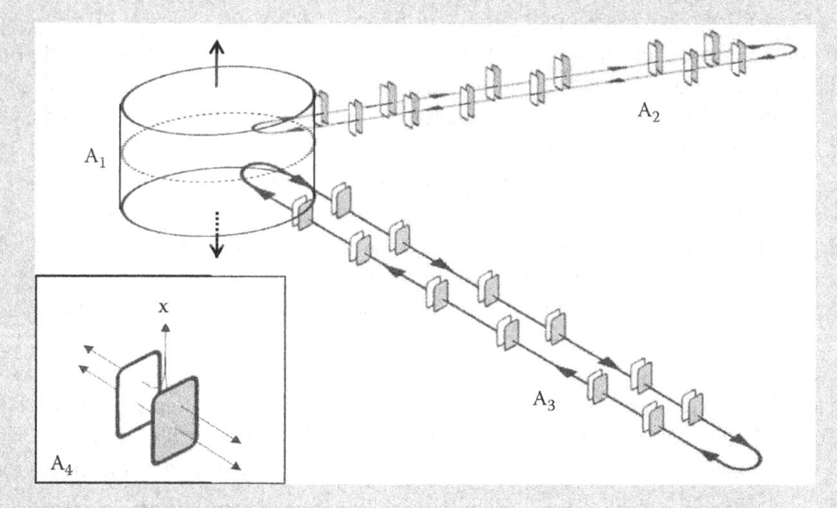

FIGURE 57.1 Stuart's concept mapping texts comprised a central zone of pharmacy that was *not* his practice but his practice context (A_1) and two branches of work with either pharmaceutical compounds that are water soluble (A_2) or solids that are not (A_3). Both of these branches comprised successive dyads of labels denoting theory/method and material/property (A_4). In these dyads (as well as in the whole of each series), different results or outcomes (X) were projected as consequences of the text itself.

(*continued*)

BOX 57.1 (Continued) SCIENCE TEXTS THAT WORK

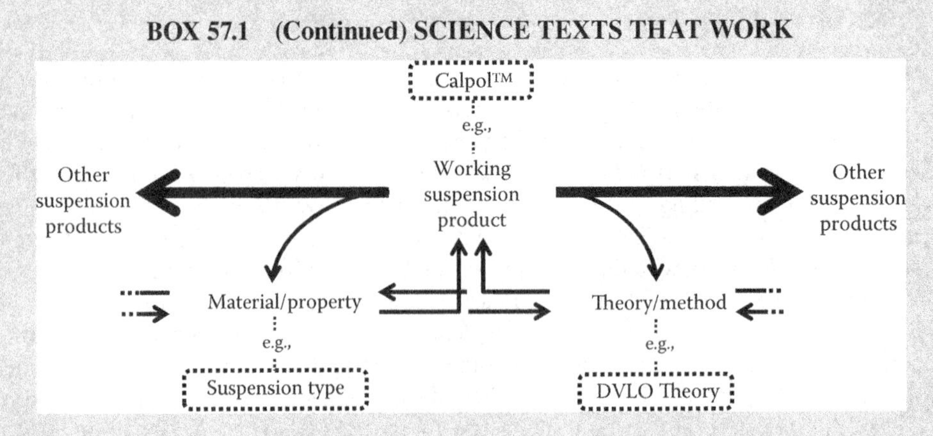

FIGURE 57.2 Specific products were implicated (like Calpol) as a consequence of showing how the properties of suspensions can be known and manipulated using the theories and methods of physicochemistry (i.e., DVLO theory that predicts the behaviors of two different solid particles in a single carrier). Thus, the text "made" Calpol (in text), and this product then became available as another means of knowing and working with other suspension type properties and theories.

"writing up" experiments or fieldwork only after these are done. In fact, Stuart's "writing" (or better, dialogue) carries through these processes including much more of his *practice* as a scientist than just using instruments correctly or making careful observations, etc. There is a fundamental (if subtle) issue at stake here. It is not only that Stuart's texts are plans, then tests, and then explanations or reports: it is that *in* the process of producing text, the *text* itself throws up productive insights that Stuart could not otherwise achieve. In Stuart's hands (or mind) the *text* does work! By writing or drawing or speaking, etc. (and/or in thought), Stuart draws many texts together into contact "producing" new outcomes (or reproducing existing ones like Calpol) as particular combinations of theory/method and material/property project these outcomes. These "products" arise from using what Stuart knows already (including his knowledge of the research texts of others), but some of the results may be new to him (a new personal understanding) or even new to science (a new discovery or product). Then what is new enters into his (and other scientists') "image-shape" of practice acquiring a new role in the text series through which Stuart (and others) knows and shape their drug design (Figure 57.1). The process is imaginative because *text* has imaginative propensity: projecting a potential outcome that is otherwise intangible. This is a robust account of scientific imagination that does not depend on either supernatural cause, or on serendipity, and it also implies that Stuart *is* a scientist largely because of how he operates *within* the scientific texts he also uses/makes.

Now we are able to be more precise about what it means to practice applied chemistry in text production (like Stuart). Rather than focusing upon the "correctness" of an individual's knowledge of scientific information (as we tend to do in science teaching), we have effectively shifted our attention toward the ways that scientists use and constitute their scientific dialogue. But if we acknowledge that scientific practice resides in imaginative use of text, then we must also ask some fundamental questions about higher education that teaches students only to "know" the worked-out already.

KNOWLEDGE VERSUS PRACTICE

In the last two decades of research of science education, analysis of students' scientific misconceptions has received more attention than any other issue (Chang et al. 2010). This coincides with a general tendency to "teach" science *not* as researchers practice their discipline or craft (creatively, imaginatively and in the reciprocally participative function of text), but *as if* learning to be a scientist could depend on knowing science "correctly." This is a subtle point, since of course participation (as a scientist) implies an understanding of the work of others, but the problem may be illustrated by thinking about the different roles usually accorded researchers *versus* students. As I have shown in Box 57.1, the researcher *uses* and participates in creating new potential; but more often than not, the student's position is the inverse of this; "learners" being asked to work backward from the "worked-out already" to produce a logical justification from the readymade and "organizing" principle. This *might* also result in potential for projection *forward* (from the known toward unknown), but often it does not and rote or verbatim learning is commonplace (see Novak 1998; Hay 2007). The particular problem in science is that the *formal* system of scientific knowledge does not include the *difference* of perspective that is necessary to engender imagination (understanding). This issue can be explored by thinking about how science operates as "language."

TWO POLES OF LANGUAGE

Bakhtin distinguishes between two poles of all human language (including the specific language genres that the sciences have made its own). Pole 1 is the formal language *system*, changing over time (at a cultural level), but essentially nonnegotiable and predetermined for the "speaker." This system supplies a meaning *a priori*, distinguishing words from the cries of a baby or requiring that in a science context, a label such a "Neo-Darwinism" is used specifically. The opposite end of the language continuum is the *events* pole (pole 2). It is here that the results of language *use* arise (as understanding), and this is always unpredictable, depending on the unprecedented and spontaneous events occurring between people who are connected in dialogue (Bakhtin 1986, pp. 159–172). The potential for this understanding arises because of *difference* (between different people and their different viewpoints; Box 57.2).

BOX 57.2 BAKHTIN'S DIALOGIC THEORY

Bakhtin's theory is a general account of human understanding as this depends on the social imagination and relations to others (people) or to "otherness" in general (Wegerif 2007). For Bakhtin, understanding arises in text (including all communication: writing, speech, music, art, gesture, etc.) and is a consequence of bringing the *difference* of different texts together in contact. In his words, "The text lives only by coming into contact with another text (with context). Only at the point of this contact between texts does a light flash... joining a given text to a dialogue" (1986, p. 162). In this "conversation," insights arise including the "other" (or "otherness")—and for Bakhtin, even our private thoughts are conversational (inclusive). Different from Vygotsky's (1986) view of dialectic process, language is *not* a tool (a means to ends) but "an end itself" (Wegerif 2007, p. 28). As Wegerif (2008) describes it, dialogic theory depicts the consequence of *"telling" something twice* or *seeing things from two perspectives at once* (Wegerif 2010): first as one already perceives it (through ones precondition) and second, imaginatively, as *if* one were another person (like a teacher or another scientist with a different vantage point). So long as the different texts (or views) of these "two" do not or cancel one another entirely, each viewpoint starts to inter-animate the *difference* in between them, and in this "dialogic space," a new (or "third") position arises ("vertically"—in relation to the horizontal plane of "voices" joined in conversation). As this "third" develops and is shared (communicated), so new labels (new words, pictures, gestures, etc.) also start to figure in the text, extending and developing it as a new and more inclusive understanding that arises from *within* the process (see Box 57.5).

The particular problem for the sciences, however, is that the knowledge of science is synonymous with the *system* pole (pole 1) of scientific language—and here there is none of the difference between texts (perspectives) necessary for the work of imagination. This is because science is about *things* (not people), and our formal scientific system makes room for just one causal explanation (paradigm) at a time. For nonscience disciplines, the two poles of language are contiguous because the *systems* of language in the arts and humanities and the social sciences are developed for *thinking about the thoughts of others*. But in science, connectivity is severed between the language system and the events pole. This is the problem of the *worked out already*: in order to become accepted as a part of science, knowledge really is *worked-out* until there is no *difference* left within it. Thus, *formal* scientific publications do *not* comprise our *understanding* of science, and it is only in (less formal) "talk" *about* these science texts that understanding may arise. Thereby, scientific practice is starkly divided between the formal work of knowledge (system) and the *practice* work that occurs when different viewpoints are brought together informally. But even while this informal practice is the means by which new knowledge is created, the *interpersonal* difference that is necessary for "the new" can never find its way back into formal science—that would not be science and would create a different social-subject system.

None of this is difficult for the initiated (i.e., the practicing researcher), who can navigate back and forth between science as a formal system and informal language use (maybe without even being aware that they are doing so), but it is intensely problematic for students if their point of contact with the discipline is formal (i.e., with text books, reviews, academic papers, and even lectures that recapitulate the formal "known already"). Then the student does not have access to the resources that make science knowledge available as understanding—since these resources are really interpersonal and relational, including the perspectives of the *people* that practice science. It is these differences that make each researcher participant/constituent of a different scientific relationship toward their objects of inquiry, and it is only in the relations of these (informal) practices that understanding is available (Box 57.3).

It is exactly these issues that Alberts (2005) addresses when he advocates teaching about "intelligent design" alongside the scientific theories of evolution. This is not because Alberts thinks that creationism is a valid way of knowing scientifically, but because it is against this very different (nonscience) background that the critical features of evolution (as a scientific theory) become apparent. Without a difference (or variation), our canonical science texts (like evolution by natural selection) are just static givens, being in and of themselves nonresponsive "things" and admitting to only verbatim learning. For just the same reason, there are many advocates of teaching science history (see Matthews 1994; Monk and Osborn 1997, for example). But we can also find examples of this principle without turning to other disciplines (like history) or even to nonscience (like "intelligent design").

THE CELL AS SUBJECT

Alongside Darwin's theory of evolution by natural selection and the three-dimensional structure of DNA, the cell concept is one of the "biggest" ideas of modern bioscience. As a *formal* text, the principles of cell biology are simple to state and without imagination they are relatively easy to memorize:

- All living organisms are composed of cells.
- Cells develop from other preexisting cells.
- The cell is the source of the vital functions of an organism.
- Cells contain the genetic information necessary to regulate cell function and also for transmitting information to the next generation of cells.

But "the cell" also has a powerful *informal* function that is not readily accessible except from *within* the field of research work in cell biology. Instead of being just a set of definitions, the cell biologists' paradigm is a lived experience—a means of inhabiting the imaginative space (or "otherness")

BOX 57.3 EXPERIENCE OF THE "OTHERNESS" OF NEURONE CELLS

In the last few years, I have been working with Richard Wingate and Darren Williams (both researcher/teachers in the MRC Centre for Neuroscience at King's College London) in order to explore the ways the researcher's imagination figures in inquiry into "things." Darren and Richard are trying to understand the ways that neurone cells locate themselves and function in relation to each other in the developing embryo. Shown in Figure 57.3 is Richard's drawing of a "neurone cell" (A). This is an imaginative (and informal) image; *not* a picture Richard might ever exhibit in a paper, book, or journal, but it is how he sees (or senses) "the neuron" in the eye of his mind. Next to his drawing (B) is a formal textbook image of a neurone—redrawn from Gray's Anatomical Textbook (Warwick and Williams 1973). Below these both are also six student's neurone drawings. These were chosen at random from among the 174 drawings made by third year students of the "Neuroscience" module that Richard and Darren teach together. Like Richard's image, these were drawn from the mind (without prompts).

The most obvious feature of these collated drawings is that all the students' images are (more or less) copies of the text book image, and while I have just shown six from 174 drawings, *any* would have done to demonstrate this reproductive stance. But Richard's drawing is different from the rest. Just as an image, the impetus of Richard's "neuron" is downward (from the lighter area included by the axon body and its branches (at the top) toward the reaching tips of the branching dendrite (that

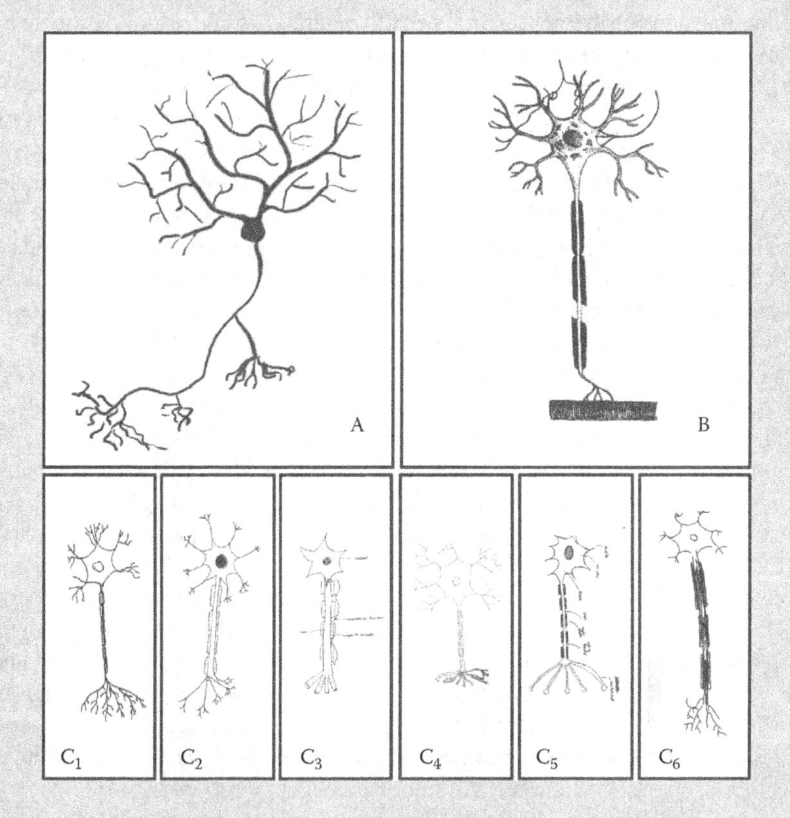

FIGURE 57.3 Drawings of the neuron cell identity: (A) the drawing of one researcher/teacher (Richard). (B) The textbook image. (Redrawn from Grays Anatomy.) (C_1–C_6) The responses of third year undergraduates to the prompt "please draw a neuron."

(*continued*)

BOX 57.3 (Continued) EXPERIENCE OF THE "OTHERNESS" OF NEURONE CELLS

are "weighted" [below] by their darkness [see Klee 1973: Figure 76]). Opposite to this, the text book image and all the student drawings "move" in the other direction (upward), standing upright (as it were) from a black origin that is white-in-white (below), toward a black end (above). Thus, it is only Richard's drawing that *embodies* the known direction of neuronal signals (downward as all these drawings are orientated). This is an important issue; in fact, the textbook image often includes an arrow pointing downward, labeled to show the direction of the neuron signal. This is to segregate the function and structure of the formal image (i.e., to separate information about neuron shape from information about its signal). In Richard's drawing, function and structure are one exhibiting the important bioscience axiom that structure follows function.

To go further, however, we also need to know about the different purposes of Richard's *informal* image and the text book drawing (that the students have copied). The text book image is derived from another image drawn by Lewellys Barker *circa* 1890 (see Daston and Gallison 2007; Wingate and Kwint 2006). Barker was a bioscience illustrator: *not a neuroscience researcher* (Daston and Gallison 2007; Wingate and Kwint 2006) and "Barker's neurone" is not intended to be an image of a neurone cell. Rather it is reference text—designating the archetype (or class) of (neurone) cells and giving the researcher *method* for addressing questions like: "does the cell I'm looking at fit the category of neurone"; or if it does in parts, "do departures from the standard indicate pathology." These are the functions of the images in a truth-to-nature scientific atlas (Daston and Gallison 2007). Thus, it is also important that while Richard's image is *not* a copy of the "Barker neurone," nevertheless, Barker's image enters into Richard's text (as part of its constitution). Richard's image of a "neurone" includes "Barker's neurone" since Barker's image is inseparable from how researchers practice neuroscience. But lots of other lived experience also enters into Richard's drawing. First, the angles of branching, fluidity of form, and balance in Richard's drawing are all consequences of Richard's work with neurones and his analysis of the images and video produced by others. Thus, his drawing conveys a familiarity with the neuron structure that would be lacking in the work of a draftsman working from a written description or attempting to extrapolate a general form from a single photograph. Second, the image is a working model: it expresses potential areas of new inquiry as well as gathering together what is known. It is therefore strictly informal, acting as a dialogic text because it is enactment of a conversation with other researchers. What is not depicted, but is nevertheless implicit, is the complex of interaction within the cell's environment. While growing, the neuron reaches out and contacts other cells and its shape is molded by environmental forces. In the adult brain, dendrites are studded with synaptic contacts, each of which represents an interface with another, equally complex branching neurons. Even though the neuron sketch looks complete in itself, this hidden other world is present. For example, in this instance, the axon structure infers three distinct targets for information distribution that would label this cell as a "projection neuron" (type) in the central nervous system. To the expert eye, this can be "read" in the three distinct branches that Richard shows in the bottom portion of his image.

Above all however, Richard's informal image is imaginative (in ways that the text book drawing and the students' "copies" are not). Richards' image is an act (performance) and it is indicative of his *outward* looking "gaze." Situated in the field of neuroscience research, Richard senses (even feels) what is plausible for neurones and what is sensible to ask about them. This sensing combines all his experience, not just as knowledge or information, but *constituting a research project*: and his image is shaped as *if* he experiences what it might mean to *be* a neurone cell. This is imagining an experience of "otherness" and this imaginative potential is what gives Richard his "position" *in* research (his "signature"). The question we must now explore is how some of this imaginative propensity might be made available to students (Box 57.5), particularly if the tendency to copy formal images is so strong and problematic (Box 57.4).

of the cell itself (where the "cell identity is subject")—so that each researcher can project what is plausible or even sensible to *ask* about a cell (about its function and its structure, etc.). This projection is embodied (see Myers 2008), constituted in the craft work of research and also in all the different "conversations" that researchers must enjoin in research practice. Thus, the concept of the "cell as subject" is also inseparable from the people/practices that are party to the ways "the cell" is known (i.e., researcher judgment and method, visual culture etc., all of which must enter into *what "the cell" is known to be* and how *"it" might be rendered*). This becomes clear in researchers' *practice* (texts), but it is generally hidden (or implicit) in the formal "cells" of textbooks or academic papers. Thus, the researcher has a *subject–practice* image that includes knowledge of how they and other researchers show and talk about the cells they work with that is often unavailable to nonresearchers. This is not a contest with "objective science" (see Latour 1999); it is merely acknowledgement that in practice, method, culture, judgment, etc., are all part of *what is known* by experts (Daston and Gallison 2007).

We have already explored the creativity of the texts of applied chemistry, seeing how one researcher's text predicts new drug products, but now we can also look for the function of imagination in the practices of researchers who are looking into "things" (i.e., hidden [unknown] phenomenal identities like neuron cell function/regulation). This is because we can now acknowledge that the "thing" (a cell for example) also becomes a correspondent in the *process* of inquiry (Box 57.3).

TEACHING IMPLICATIONS

Thus far, I have been trying to illustrate three important and related issues:

1. *That the practice of research is essentially imaginative*—since the "perfomativity" of research work occurs in text where dialogue *lends* the researcher the potential of another viewpoint—the experience of an "other" (person) or of an "otherness" *per se.*
2. *That location in the practice field is inseparable from the imagination of the researcher*—people (like Richard and Stuart) being party to the constitution of the academic subject (as text), whether their goal is applied (making products like Stuart) or natural science (knowing neurones like Richard).
3. *That the agency and difference of the subject do not enter into formal Science*—being available only in practice where experience of the subject's constitution is unavoidable.

Now we need to explore the implications of these issues for our science teaching. Briefly, my point is that university teaching cannot prepare students to *be scientists* unless it also "teaches" them to use imagination (in the same ways that researchers exercise this function). Several consequences immediately arise. First, and somewhat controversially perhaps, we might suggest that it is only practicing researchers that ought to be teaching university students. Nonresearcher *teachers* can only draw upon the formal texts of science (Box 57.4), and also they may not critique a student's potential projections forward or outward *from* the known. Second, even practicing researchers will fail to engender students' imaginative work if they do not show *themselves* in teaching. This is to say that researchers ought to teach informally, exhibiting the craft work and imagination that includes them as people and participants in science. In one way or another, it is the constitution of the researchers' practice image (like "Richard's neurone" in Box 57.3) that needs to be made available to students. Nevertheless, however science is taught, we must acknowledge that some of our students are still able to wrestle out the hidden dimension of "the subject" and to use this in order to practice thought in the ways researchers do (see Hay 2010; Box 57.5). After all, some of our students will go on to become new scientific leaders and the scientific leaders of today were also students for a time. The issue is that if scientists practice in text, then hidden scientific practice *may* still be conveyed in the lecture to the degree that it is possible to infer the informal (social) stance, which organizes a more formal exposition. But first we should explore why often, in the teaching-only context, this becomes prohibited (Box 57.4).

BOX 57.4 THE PROBLEMS FACING NONRESEARCHER TEACHERS

I have already used analysis of one researcher's neuron image to illustrate the ways that this has a different (informal) potential from the use of formal texts (like "Barker's neuron"). But we can also extend this commentary to explore the problems that accrue when teaching is led by those that *do not* practice in the corresponding field. Like Darren, Richard, and I, Sisika Ranaweera and Lisa Montplaisir have been interested in bioscience student's drawings. The work of Ranaweera and Montplaisir (2010) includes analysis of neuron images, but different from us, these authors score the quality of their student's drawings using Barker's image as the yardstick, giving the highest marks to images that copy it exactly. From a researcher's practice perspective, this is mistaken. First it means that the students' productivity is capped by the "correctness" of the authorized image: a student's drawing cannot be *more right* than Barker's drawing. Thus, here there is no potential for going beyond the given (Bruner 1960). Second, as we have already seen (Box 57.3), Barker's image is *not* a drawing of "a neurone": it is a standard and it is intended to be used as *method*. Taking Barker's drawing as an image of a neurone is therefore intensely problematic. Using Barker's image in this way is to substitute the craft work of neuroscience for its "object" (i.e., the "thing" that is the purpose of inquiry). But what else can nonresearchers do?

This is not to say that researchers always make good teachers: many might not (see Kinchin and Hay 2007). But at least researchers have the *potential* to teach in the same ways that research is practiced—imaginatively.

IMAGINATIVE STUDENTS

Most researchers tend to teach in formal ways (and as we have seen, nonresearchers do not have an alternative). First, the university science curriculum tends to be composed of formal content. Second, formal science has an authority (and even perhaps a perceived respectability) that is "safer." While the informal conversations that take place at conferences and in other social settings might be vital in shaping how research is practiced, they are seldom modeled in the lecture theater (or even the tutorial). Nevertheless, a single module is often taught by many different "experts," and the *difference* between them (as the different speakers of respective academic subjects) is sometimes sufficient for students to win-out the *subject*. Box 57.5 is an explicit illustration of this, showing how one student (Lisa) uses her imagination to depict the ways a supposedly common curriculum-target might be told as *if* it were spoken by several different researchers. The data that I use to show this process are already published (see Hay 2010), and there is a website that provides examples of this student's work as well as their commentary on learning (see dialogueonthepage.com). Briefly in a paraphrase of Lisa's words, "When I listen to a lecture or when I read a scientific paper, it is not important to me how I might understand these texts. I am only a student and my grasp of the subject is limited—so that I will only introduce my own mistakes. Instead I listen (or read) as if I were one of the other researchers who teach me. So when I listen to Dr. Williams give a lecture, I write my notes as if I were Dr. Wingate: imagining how he would respond and re-pattern Dr. Williams' explanation. Afterwards I throw my notes away and I repeat the process from another perspective (in another lecture or while reading another paper). Eventually, by doing this, the subject just appears for me—as mine—as a consequence of bringing all of these (people) together."

BOX 57.5 TAKING THE PERSPECTIVES OF "ANOTHER"

Lisa (a third year university student) shared her study-work with me during a year-long study of her understanding of neuroscience. We met every second week and Lisa would show me the sketches (notes, drawings, and concept maps) that she made as well as talking

BOX 57.5 (Continued) TAKING THE PERSPECTIVES OF "ANOTHER"

about them (and her experiences of learning). At every stage, Lisa's learning work was imaginative as she went about repatterning one researcher's text as if she were standing in the shoes of another researcher. Thus, for example, the dialogueonthepage.com website shows a reenactment of how Lisa reads a paper: first, to grasp the gist; second, to rewrite it as she imagines one or more of her lecturers might have read it; and third, to gather new labels and identify the identities (that arise as new insight). Lisa did this every time she read and she used the same process in her lectures, treating these as "readings" too. Often she would talk about the relationship that she was developing with her teachers and other scientists by doing this, and in the end, she described

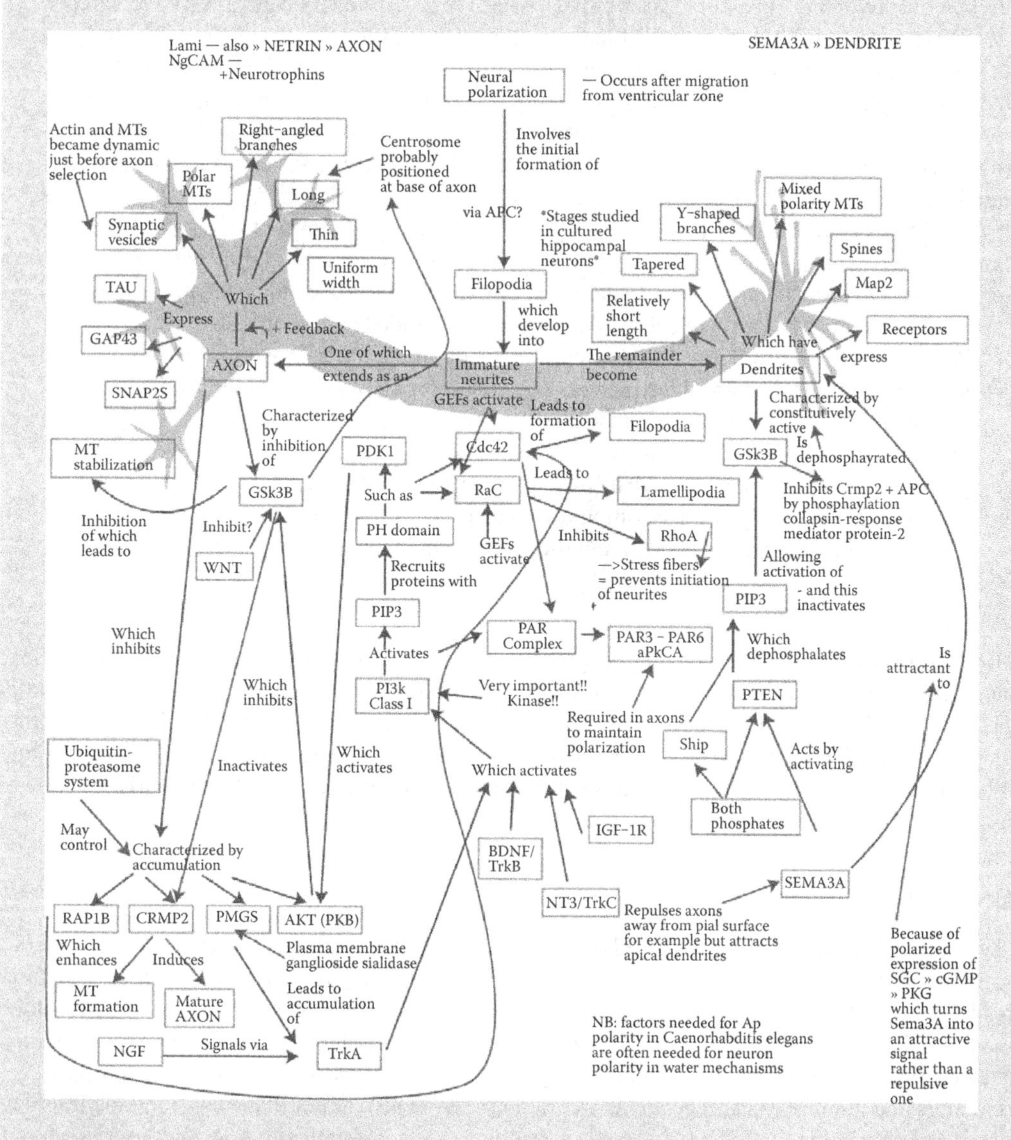

FIGURE 57.4 One student's (Lisa) revision concept map of third year neuroscience illustrating the neuron cell identity she implicates (and labels) through her learning work.

(*continued*)

BOX 57.5 (Continued) TAKING THE PERSPECTIVES OF "ANOTHER"

how her own relationship developed *within* the academic field—so that *she* was also part of it. Figure 57.4 is one of twenty "maps" that Lisa made two weeks before her final examinations. It was drawn without prompts. I (not Lisa) drew the blue graphic neuron image underlying the text and boxes in the map. This is explained below.

When Lisa spoke to me about her map (Figure 57.4), she burst into spontaneous laughter—as she traced the outline of "the neuron" (shown) with her: "Of course you can see what I've done—I have drawn the neuron structure as I imagine it. That's amazing! I didn't know that I was doing that but of course I feel the neuron—as a pattern—and then all the things I know about the neuron are a consequence of this." In order to verify what Lisa had done, I asked Richard (as Lisa's researcher/teacher) to inspect her map. All the distinctive features of the expert "gaze" are there. Her image is a moving one: it is predictive and suggesting potential inquiry as well as showing what is known. Also it comprises questions, effectively projecting how a neuron might develop and chasing issues of neuron-structure back toward the development processes that are (and might be) drivers of its form and function. In summary, Lisa had done what Richard showed us in his drawing of a neuron (Box 57.3). In her intensely interpersonal study work, Lisa has acquired a perspective from *within* the neuron cell identity (where the "cell is subject") and projecting outward from this all her knowledge is available—including new potentials of neuroscience inquiry. In this map (and others), Lisa is working forward from research (like a researcher), but she has acquired this potential as a student, winning it from the people whose texts she collided in new insight.

When we looked at all her other maps, this same imagination was evident and also the issue of relationship became increasingly conspicuous. These were relationships extended toward the neuron cell phenomenon that Lisa studied (shaped), but also toward the people that taught her. For example, Lisa's image of "dendrogenesis" was a tree-like shape, organizing and predicting the process of " dendrite development" as "roots" and "leaves" and "branches" and including Lisa's affective disposition toward these "images" as well as toward the people that were part of them (see Hay 2010 and the website dialogueonthepage.com).

Box 57.6 shows a model of the social process that Lisa used to constitute her understanding from within the subject field. It is a general model of imagination, explaining how the academic subject (that includes its speaker) can arise spontaneously in dialogue. The point is that through the dialogic process, Lisa has become a researcher, not because she learned the formal texts of science but because she finds the neuron subject shaped between the texts of those that already embody a "signature" image of the hidden neuron cell identity.

CONCLUDING COMMENTS

In presenting my analysis, I acknowledge that I have taken considerable license. All my juxtapositions (of "understanding" versus "knowing" or "formal" versus "informal" science, etc.) are intended as vignettes, and in fact, these distinctions are rather more blurred than I have tried to show them here. My approach is a means of drawing attention to the issues that interest and concern *me*, and I trust my exercise is not a fundamental sleight of hand. Likewise, the distinction I draw between researchers and nonresearcher teachers is a caricature, and I also admit that a considerable fraction of the science teaching that occurs in research intensive settings is already informal: journal clubs, fieldwork, laboratory exercises and lectures, tutorials, etc. Also (as I illustrate) the processes of students' "reading" are sometimes already an opening toward the interpersonal contact that makes dialogue happen (Hay 2010). In the last six years, however, I have watched more than 500 hours of science lectures, and on balance, the majority of teaching still remains a formal "telling." A major

BOX 57.6 THE INTER-ANIMATION OF TWO "SPEAKERS" TEXTS

When two (or more) texts are brought together, there are three potential outcomes: (1) synthesis; (2) one writing out the other; or (3) the inter-animation of the two. The first two of these outcomes are (Hegellian) dialectic products: the third is dialogic (cf. Bakhtin 1981, 1986) and essentially imaginative. In this model (Figure 57.5), I project the inter-animation of two different (but related) texts. These are shown as the texts of two lecturers (lecturers A and B: or in Lisa's case, the texts of Darren Williams and of Richard Wingate). Now the student (Lisa) may simply try to learn each text separately (as two different given "things"): but she does not do this; instead she brings them together because of trying to imagine how one would "read/rewrite" the other in their tensions. Since Richard's and Darren's texts are actually constituent of different identities (different people; different viewpoints—projecting different potential neuron cell identities), they never fit exactly together, and only a new insight produces a way of seeing both their texts simultaneously. This insight is imaginative (and creative) occurring in the "space" of tension in between the *difference* of the two, and as this insights start to be labeled, it becomes a new text (a new image of the academic subject: shaping Lisa's identity in the field of neuroscience as well as potential identity of "the neuron," which she labels). This is new development of the subject *per se*, and it includes Lisa's development as a new speaker of her object of inquiry.

This is the process by which new personal understandings arise, but it is also the process by which scientists constitute insights that are new to science. To acknowledge the function of imagination in science is to orientate students and researchers together in the same dialogic direction of scientific practice (communicative and therefore inter-personal) rather than to juxtapose two titles: one of "novice" and one of "expert."

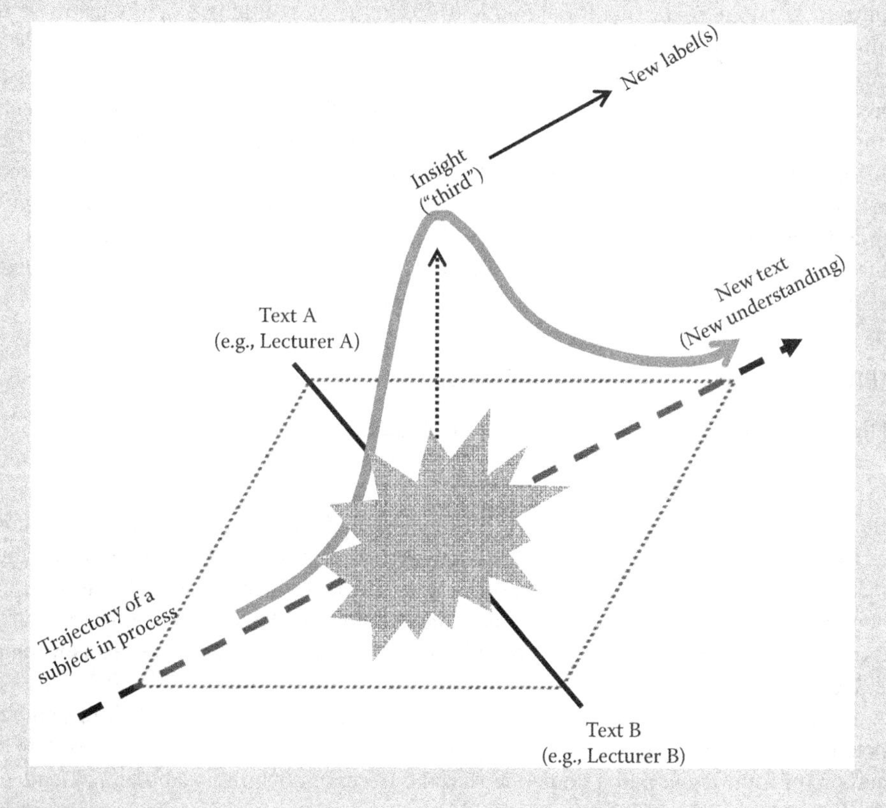

FIGURE 57.5 A graphic representation of the arising of new insight and the "label" of new identity.

culture change is needed since this telling cannot "place" the student as *participant* in the scientific process (Kress 2003). In many ways, what I have been trying to show throughout this paper corresponds with Gunther Kress' description of a shift away from "how we tell the world towards the way we show it" (paraphrase, Kress 2003, p. 140). This is part of the semiotic project (Kristeva 1986b). Writing in the *Higher Education Research Network Journal*, Alison Harvey also gives the issue voice:

> The dichotomies of teaching/learning and research/teaching are pervasive in the academy. In conceptualising these dichotomies primacy is given to the person (teacher/learner/researcher). Knowledge is conceived as product. The academic, in their role as researcher, produces knowledge, which they come to possess. This knowledge is then transmitted from the academic in their role as teacher to the student—or acquired by the learner, depending on whether the prevailing fashion prioritises the teacher or the student. [But] Shifting the emphasis to knowing as a process challenges these dichotomies, suggesting that knowledge is practiced not possessed [so that] ways of knowing and ways of being a knower are intimately linked [and] knowing is a matter of seeing ....
>
> **(Harvey 2011, p. 34)**

This is the viewpoint that Kristeva (1998) encapsulates in her notion of "a subject in development," and here the function of imagination is vital—indispensible. Bakhtin's dialogic theory is one of the most important sources of modern semiotics (instigating semiotic theory and included by it), and Bakhtin's work is to show how all our human texts (including texts of science) are intrinsically imaginative (creative or recreative, because of the dialogic propensity of text *per se*), even when that imagination is hidden in the finished account. But there is one last issue that is also vital to consider, since fundamental to the function of imagination, Bakhtin also shows relationships as the precondition of dialogue. The quality of relationships, or at least anticipation of what these relationships might be (become), is what facilitates a dialogue (Hobson 2002; Wegerif 2010). As Wegerif (2007) explains, in order to see the subject as another sees it, first it is necessary to put aside one's fixed preclusive view. This is against "the self" in favor of "the other." Now any distinction between "teaching" and "research" pales beside the issue of responsibility toward the quality of dialogic relationship *per se* (see Alexander 2008b). Relational commitment is therefore foregrounded, not because it is an *ex situ* moral virtue but because it *is* the subject in development: synonymous with the future development of the people that "speak" for themselves, their objects of inquiry, and of course their "teachers"—in these simultaneous regards. For entirely rational reasons, relationship becomes the most precious attribute of understanding science.

SPECIFIC TEACHING RECOMMENDATIONS

To end this paper, I offer several explicit recommendations for researchers who teach in higher education science. These are as follows:

1. Be clear about where and how the practice of your field is mediated (i.e., in the context of this paper, *either* in the manipulation of signs denoting material property and theory/method *or* in the embodied imagination of the scientist where relationship toward a hidden phenomenal identity is vital)—since understanding these different epistemic cultures (Knorr Cetina 1999) is the key to very different "signature pedagogies" (see Gurung et al. 2010, eds.).
2. Acknowledge that researchers who teach are more than role models for their research practice; they are part of the fabric of research. As Knorr Cetina puts it, "... as well as being authors of knowledge and a component of the [research] setting, [laboratory] leaders are the integrating element [of research]" (Knorr Cetina 1999, p. 217): they are the elements (or principals) who constitute both proto-knowledge and new knowledge making culture.

> Thus, in lectures, rather than telling "this slide shows ... this or that," the actor/agents of research have the potential to exhibit "science in the making" (Latour 1987) exhibiting how the claims of any "data" already resides within its form of label but the label also functions scientifically toward a new contention. This is important because when leading scientists offer labels for invisible identities, they also label how they hope to make a research contribution—and in doing so they offer means organizing a particular laboratory or field: organizing teaching—and thus also they are furnishing new labels for their own identity in teaching and research.

This second point leads to some important contentions that are somewhat different from the formal science-teaching model, specifically because of implicating the importance of the researcher *in the flesh*. Thus, I suggest that the potential distinctiveness of researchers' teaching resides in

1. A willingness/ability to tell the stories of laboratory life and culture—since these (informal) narratives can figure a hidden practice identity in ways that formal science texts cannot (see Knorr Cetina 1999, p. 220; Bruner 2002).
2. Exhibiting mistakes and u-turns in research—since the polished (finished) academic text rarely shows its means of constitution except to those who are experts in the practice-field already (see Weller 2010).
3. Display of the imaginative (informal) images that exhibit hedges, bets, and the image of unique research potentials (i.e., the contributions that researchers hope to make)—these can make the hidden purpose/object/method of inquiry more available to students (see Bruner 2002) as a combinatory identity.
4. Teaching in teams where the differences between leading researchers are deliberately played upon (used to generate a "tension")—since this tension is potentially the "material" with which students might begin to imagine, label, and thereby, participate in new knowledge-making culture (see Wegerif 2007).
5. Experiments with ways of "placing" students *in* the practice field (e.g., where "the cell is subject"), using choreography (dance and drama and other means of engendering embodiment).

Any claim that the teaching of science ought to change must contend with the truism that the science leaders of today have acquired their capacity for leading science through experience of preexisting science-teaching culture. I reply that perhaps we have the potential to develop more and better scientific leaders than we do right now, and while recent higher education policy has been toward widening participation, much of this has also been directed toward "employability" and development of scientific "workers." This is vital; but also scientific research is a cutting-edge for new employment and new generation of an economic/social good. It is equally important to develop the next generation of creative scientists.

ACKNOWLEDGMENTS

I am particularly grateful to Darren Williams, Richard Wingate, Stuart Jones, and Lisa for their insights and materials. The case studies reported in this paper were developed with funding from the Society for Educational Studies and the Higher Education Academy UK.

REFERENCES

Alberts, B. (2005). Commentary: A wakeup call for science faculty. *Cell* 123, 739–741.
Alexander, R.J. (2008a). *Essays on Pedagogy*. London: Routledge.
Alexander, R.J. (2008b). *Towards Dialogic Teaching: Rethinking Classroom Talk* (4th edition). York: Dialogos.

Bakhtin, M. (1981). *The Dialogic Imagination: Four Essays by M. Bakhtin*. In: M. Holquist (Ed.), trans., C. Emerson and M. Holquist. Austin, Texas: University of Texas Press.
Bakhtin, M. (1986). *Speech Genres and Other Late Essays*. In: C. Emerson and M. Holquist (Eds.), trans., V.W. McGee and C. Emerson. Austin, Texas: University of Texas Press.
Bruner, J. (1960). *The Process of Education*. Cambridge, Massachusetts: Harvard University Press.
Bruner, J. (1986). *Actual Minds, Possible Worlds*. Cambridge, Massachusetts: Harvard University Press.
Bruner, J. (2002). *Making Stories: Law, Literature, Life*. Cambridge, Massachusetts: Harvard University Press.
Chang, Y.-H., Chang, C.-Y. and Tseng, Y.-H. (2010). Trends of scientific education research: An automated content analysis. *Journal of Science Education and Technology* 14 (4), 315–331.
Daston, L. and Galison, P. (2007). *Objectivity*. New York: Zone.
Gurung, R.A.R., Chick, N.L., Haynie, A. and Ciconne, A.A. (Eds.) (2010). *Exploring Signature Pedagogies: Approaches to Teaching Disciplinary Habits of Mind*. Sterling, Virginia: Stylus.
Harvey, A. (2011). Disciplines and knowing: Exploring the implications of disciplinary context for pedagogic practice. *Higher Education Research Network Journal* 3, 39–48.
Hay, D.B. (2007). Using concept-mapping to measure deep, surface and non-learning outcomes. *Studies in Higher Education* 32 (1), 39–57.
Hay, D.B. (2010). The function of imagination in learning: Theory and case study data from third year undergraduate neuroscience. *Psychology* 17 (3), 259–288.
Hobson, P.R. (2002). *The Cradle of Thought: Exploring the Origins of Thinking*. London: Macmillan.
Holquist, M. (2005). *Dialogism*. London and New York: Routledge.
Kinchin, I.M. and Hay, D.B. (2007). The myth of the research-led teacher. *Teachers and Teaching: Theory and Practice* 33 (1), 43–61.
Klee, P. (1973). *Pedagogic Sketchbook*. London: Faber and Faber.
Knorr Cetina, K. (1999). *Epistemic Cultures: How the Sciences Make Knowledge*. Cambridge, Massachusetts: Harvard University Press.
Kress, G. (2003). *Literacy in the New Media Age*. London: Routledge.
Kress, G. (2010). *Multimodality: A Semiotic Approach to Contemporary Communication*. London: Routledge.
Kristeva, J. (1986a). Word, dialogue and novel. In: *The Kristeva Reader*, T. Moi (Ed.). Oxford: Blackwell, pp. 34–61.
Kristeva, J. (1986b). Semiotics: A critical science and/or a critique of science. In: *The Kristeva Reader*, T. Moi (Ed.). Oxford: Blackwell, pp. 74–88.
Kristeva, J. (1998). The subject in process. In: *The Tel Quel Reader*, P. Ffrench and R.-F. Lack (Eds.). London and New York: Routledge, pp. 133–178.
Kuhn, T.S. (1977). *The Essential Tension: Selected Studies in Scientific Tradition and Change*. Chicago and London: University of Chicago Press.
Latour, B. (1987). *Science in Action: How to Follow Scientists and Engineers through Society*. Cambridge, Massachusetts: Harvard University Press.
Latour, B. (1999). *Pandora's Hope: Essays on the Reality of Science Studies*. Cambridge: University of Cambridge Press.
Latour, B. (2005). *Reassembling the Social: An Introduction to Actor-Network-Theory*. Oxford: Oxford University Press.
Lea, M. and Street, B.V. (1998). Student writing and staff feedback in higher education: An academic literacies approach. *Studies in Higher Education* 23 (2), 157–172.
Matthews, M. (1994). *Science Teaching: The role of History and Philosophy of Science*. New York: Routledge.
Monk, M. and Osborn, J. (1997). Placing the history and philosophy of science on the curriculum: A model for the development of pedagogy. *Science Education* 81, 405–424.
Myers, N. (2008). Molecular embodiments and the body-work of modelling in protein crystallography. *Social Studies of Science* 38 (2), 163–199.
Northedge, A. (2003). Enabling participation in academic discourse. *Teaching in Higher Education* 8, 168–180.
Novak, J.D. (1998). *Learning, Creating and Using Knowledge: Concept Maps as Facilitative Tools in Schools and Corporations*. Hillsdale, New Jersey: Lawrence Erlbaum.
Perkins, D. (1981). *The Mind's Best Work*. Harvard: Harvard University Press.
Perkins, D. (2007). Theories of difficulty. In: *Student Learning and University Teaching*, N. Entwistle and P. Tomlinson (Eds.). Leicester: The British Psychological Society, pp. 31–48.
Ranaweera, S.P.N. and Montplaisir, L.M. (2010). Students' illustrations of the human nervous system as a formative assessment tool. *Anatomical Sciences Education* 3 (5), 227–233.
Street, B.V. (2004). Academic literacies and the 'new orders': Implications for research and practice in student writing in HE'. *Learning and Teaching in the Social Sciences* 1 (1), 9–32.
Vygotsky, L. (1986). *Thought and Language*. Cambridge, Massachusetts: MIT Press.

Warwick, R. and Williams, P.L. (Eds.) (1973). *Anatomia del Gray*, Vol. 2. Bologna: Zanichellli.
Wegerif, R. (2007). *Dialogic, Education and Technology: Expanding the Space of Learning*. New York: Springer.
Wegerif, R. (2008). Dialogic or dialectic? The significance of ontological assumptions in research on educational dialogue. *British Educational Research Journal* 34 (3), 347–361.
Wegerif, R. (2010). *Mind Expanding: Teaching for Thinking and Creativity in Primary Education*. Maidenhead: Open University Press/McGraw-Hill.
Weller, S. (2010). Comparing lecturer and student accounts of reading in the humanities. *Arts and Humanities in Higher Education* 9 (1), 85–104.
Wingate, R. and Kwint, M. (2006). Imagining the brain cell: The neuron in visual culture. *Nature Reviews Neuroscience* 7 (9), 745–752.

58 World, University, Class, and Identity

Radosław Rybkowski

CONTENTS

In 2009, the World Bank published a book by Jamil Salmi, *The Challenge of Establishing World-Class Universities*. The author is a prominent Moroccan education economist, who graduated from the French Grande École ESSEC (*École Supérieure des Sciences Économiques et Commerciales*), and holds a master's degree in Public and International Affairs from the University of Pittsburgh and a Ph.D. degree in Development Studies from the University of Sussex (UK). After working for some time at the National Institute of Education Planning in Morocco, in 1986, he has joined the World Bank where he is a Tertiary Coordinator; he is also an expert collaborating with Organization for Economic Co-operation and Development (OECD). Jamil Salmi (2009, 2010) is an excellent example himself of the globalization of higher education at work.

According to the vision presented in his earlier publications, including *Constructing Knowledge Societies: New Challenges for Tertiary Education*, he names four main features of the knowledge-based economy: "an appropriate economic and institutional regime, a strong human capital base, a dynamic information infrastructure, and an efficient national innovation system." Each of these four pillars is based upon solid tertiary education, which turns out to be an essential driving force for national economic prosperity in the times of global competition. Therefore, the future of democratic societies and sustainable economies depends on new visions of higher education and of the role it plays in modern world (Salmi 2002, 2009).

Many authors have published books and articles on the nature of higher education nowadays and the changes it faces: growing interdependencies and desirable wider outreach of academic programs. Salmi suggests that a university can no longer be a top-class institution if it did not leave an international intellectual footprint. This assumption leads us to the question whether universities aiming at educating global citizens can protect national identity (Hayes 2002; Green 2006; Currie 1998).

The paper begins with the short overview of the globalization and internationalization of higher education, main trends and forces that shape the world of colleges and universities. The changing environment, as well as new approach to higher education policies, resulted in redefining the core of higher education: purposes, outcomes, and even the clientele of academic educational services. New definition of the university functions must result in a new concept of a classy institution, which is the main focus of the third part of the paper. The last question discussed is identity: do universities still play an important role in creating elites that shape national identities? Or maybe they are just top-class training centers that supply best educated and mostly needed workers? More than 900 years after the first European university was founded, policy-makers, university administration, and the public must once again answer the question, what the higher education is for?

HIGHER EDUCATION AND THE WORLD

Philip G. Altbach, the Director of the Center for International Higher Education, recommends careful distinction among three different processes influencing higher education institutions (HEIs) worldwide: *globalization*, *internationalization*, and *transnational education*. Although in many aspects similar, they employ different forces and lead to different ends. The most general is the process of *globalization* that includes "the broad, largely inevitable economic, technological, political, cultural, and scientific trends that directly affect higher education" (Altbach 2005). In case of the globalization it is impossible to name one source or reason of the changes, because it is a complex result of many interactions. The growing international trade exchange (understood traditionally as the exchange of goods, but also with a rapidly expanding share of exchange of services) makes public policies vulnerable to actions or crises in distant regions. The reduction of trade barriers, culminating in the creation of the World Trade Organization (WTO), influences so many aspects of everyday life that even as old institution as the university cannot hide in the ivory tower any longer. One of the obvious results of the globalization is the use of English as a common language of knowledge (Altbach 2009).

According to Altbach (2005), *internationalization* describes "policies and programs adopted by governments, and by academic systems and subdivisions to cope with or exploit globalization." Internationalization does not exclude independent state or institutional initiatives. A government can encourage and support educational and research collaboration with institutions in the other state. An institution can sign a mutual agreement with the school on the other side of the globe. Both state and institutional actions can be controlled (e.g., forbidden, as is the common case of the totalitarian regimes). The example of internationalization is Bologna Process in Europe and the creation of the European Higher Education Area.

Transnational education (or *multinationalization*) "means offering one country's academic program in other countries." Very often, transnational education is offered by a good quality university opening a branch campus in the other country or launching a double degree program with a partner on-site. Transnational education does not have to be supervised or initiated by multilateral agreements on the governmental level. Branch campuses booming in the Persian Gulf are specific joint ventures in which usually a government invites a particular school to open and run various programs of studies in the Gulf. Such transnational programs expand very rapidly as a response to unmet educational demand (Altbach 2005; McBurnie and Verger 2006).

The neoliberal vision of the economic and social growth calls for more openness in international trade. Thanks to the constant flow of goods and services, the less developed states should fulfill their potentials faster, while the developed countries of the First World should profit from exporting their material and nonmaterial commodities. Neoliberal governments opt for open market, free trade, and substantial reduction of the public sector. Within the realm of higher education, in many regions of the world traditionally supported and supervised by national governments (including many European states), the neoliberal approach forced HEIs to treat their educational services as any other commodity that can be offered in the free market and bought by the customers. Moreover, because of the process of economic globalization, universities can offer their services abroad (Robertson, Bonal, and Dale 2006; Olssen 2006; Bagalkot 2003).

The World Bank, WTO, OECD, and even some agencies of the United Nations (e.g., UNESCO) promote the neoliberal vision of globalization in which higher education is just another, although to some extent special, kind of negotiable services. The General Agreement on Trade in Services (GATS), still being negotiated under auspices of WTO, is the example of the ultimate neoliberal globalization in higher education. According to some proposals, mostly favored by developed states with top quality HEIs (United States, Australia, New Zealand, and United Kingdom are among strongest supporters), an educational services provider from any country should be able to deliver its offering to any other country in the world. Thus, the foreign university should be treated as equal to home institutions and should face no special obstacles in launching its activity. Carlos Alberto Torres (2009, p. 16) argues that the agenda of OECD and World Bank's sponsored globalization "includes a drive toward privatization and decentralization of public education, a movement toward educational standards, and the testing of academic achievement to determine the quality of education at the level of students, schools, and teachers. Accountability is another key tenet of the model."

Top class universities, offering education of good quality, and decently endowed can easily profit from this kind of globalization. The negotiators representing the United States in GATS favor such perception: greater openness, education without barriers means larger market, more high school graduate applying for college, and, eventually, larger profit. Educational activity in underdeveloped regions as well as the successful operation in the Gulf of Southeast Asia can add some extra value to the prestige of the school; sometimes it is as important as material profit (Rumbley 2007).

The opponents of globalization point out, however, that such openness could be threatening to less stable, less developed higher education systems. Newly established universities in such countries as Malaysia or Vietnam could not compete with branch campuses of prestigious and rich institutions that easily import their programs of studies along with professors. In this case, free market, decentralization, and even the quest for higher educational standards might slow down the growth and deeply hurt national systems of higher education (McBurnie and Ziguras 2006; Sirat 2002).

The critics do not share the optimism of international organizations; they warn that there will be winners and losers. Not every university will be able to attract foreign students or rich enough to pay foreign professors. Not every institution will be able to organize branch campuses across the world. But there is no doubt that due to the information technology development, no university can resist global changes.

FUNCTIONS OF UNIVERSITY

The interdependencies of affecting HEIs in every region of the world have challenged the very core of the higher education philosophy. The accountability and social responsibility of universities call for the answer what higher education is nowadays and what it is for. The problem of balancing higher education as a *public* versus *private good* is as difficult as it has been never before. The functions of the university must be redefined; otherwise, it would not be possible to sketch any vision of the university for the decades to come. The analysis of public and private outcomes of education is not just another popular academic pseudo-dilemma. It leads to very practical governmental decisions: how much money spent on higher education.

Howard R. Bowen, in his seminal book *Investment in Learning. The Individual and Social Value of American Higher Education*, analyzed the nature of higher education. Although the book was published more than 40 years ago, it has not lost its importance, including Bowens description of university's functions:

> "higher education engages in three principal functions: education, research, and public service. Education as here defined includes both the curricular and extracurricular influences on students. Its purpose is to change students in both the cognitive and affective aspects of their personalities and to prepare them for practical affairs. Research, broadly defined, includes the scholarly, scientific, philosophical, and critical activities of colleges and universities, as well as their creative contributions to the arts. The purpose of research is to preserve, acquire, disseminate, interpret, and apply knowledge, and to cultivate creative frontiers in arts and sciences. The clientele includes students, professional peers, various groups (such as government, business, farmers, labor unions, professional practitioners), and the general public. The public service activities include health care, consulting, off-campus lectures and courses, work performed by interns, artistic performances and exhibits, spectator sports, and so on" (Bowen 1977, pp. 7–8).

All of these functions are influenced by globalization and might be further reshaped by internationalization and transnational activities. In the 21st century, students can easily access information, not only scientific, from every place of the world. They can, legally or less legally, watch films, listen to music, or be an active part of the global society. The higher education will never be the same. The university administration and professors must admit that the student body has changed and course delivery has to be adjusted to this new situation (Starke-Meyerring 2003, pp. 151–172; Low 1997). The best examples of the changing nature of the educational practices are branch campuses and offshore HEIs. The institutions established in Southeast Asia and the Gulf provide the space for the mutual interactions of people, both students and faculty members, from various countries and cultures. These interactions are as important as the teaching–learning process itself (Stasz, Eide, and Martorell 2007; Otter 2007).

The second function of the HEI, the research, has a visible international character in the 21st century. As Salmi states in *The Challenge of Establishing World-Class Universities*, modern HEIs build their prestige mostly on the research outcomes—the international rankings focus on the measurable results of their achievements: publications in prestigious journals and important awards. Natural science experiments as well as social science investigations could be so costly and so complicated that no longer one scientist or even a group from just one school can produce valuable knowledge. In 2007, *Annual Review of Biochemistry*, the journal with the impact factor of 33.45, published articles mostly with two or more authors. Volume 70 of *Physical Review D* in 2004 published an article titled "Search for B-meson Decays to Two-Body States with a_0(980) Mesons" authored by 608 persons, representing 78 institutions from 10 countries (Aubert et al. 2004).

The quest for the ablest researchers leads to the *brain gain–brain drain* practices. The institution looking for better position in international ranking could hire scientists from other countries. The process can be further supported by the state migration policy. Faculty mobility is, on the one hand, desirable, because it supports personal development. On the other hand, it can lead to the weakening of already less developed institutions. The results of such process can be observed in the US HEIs: the percentage of Ph.D. students of foreign origins grew significantly from 1980 to 2004. In medicine, it grew from 17.6% to 30% and in physics, from 21.6% to 42.3% (*Education Trends* 2005).

The impact of globalization and internationalization in research along with new information and communication technology has changed the nature of knowledge distribution. Online databases provide access to books, articles in scholarly journals, and other academic resources. Today, one can have access to the latest publications from any place in the world. The access, however, is not always for free. Therefore, there could appear a 21st century digital divide between states that can and cannot afford paying for commercial databases with the most important results of worldly research (Odin 2004).

Salmi's (2009, p. 7) report points out that there are three key factors leading to the creation of the world-class university: "(a) a high concentration of talent (faculty and students), (b) abundant resources to offer a rich learning environment and to conduct advanced research, and (c) favorable governance features that encourage strategic vision, innovation, and flexibility and that enable institutions to make decisions and to manage resources without being encumbered by bureaucracy." Concentration of talents and learning environment for advanced research cannot be achieved against the globalization and internationalization processes. The role of government is to assist HEIs in their autonomous decisions on the way to envisioning the dreams of the best, most open universities.

NEW MEANING OF CLASS?

The definition of *class* is never easy in case of the institutions of higher education. The book by Mitchell L. Stevens, *Creating a Class*, tries to grasp the core of the process of creating and protecting "a class." Although he focuses on US institutions, some basic findings could be applied anywhere. Stevens argues that "class" does not refer to the quality of institution only. The most important problem is that higher education in the 21st century is the means of creating social stratification. He comes to the conclusion that "higher education has not been the great American equalizer. To be sure, there are proportionally more college graduates in this country than in any previous era, but, with only a few exceptions the overall distribution of educational attainment remains stubbornly correlated with socioeconomic background." The most efficient way of establishing "a class" is by turning away less able applicants: thus, selectivity, not the number of Nobel prized professors or top quality publications, is the real indicator of the quality of US colleges and universities (Stevens 2007, p. 14; Golden 2007).

Salmi in his *The Challenge of Establishing World-Class Universities* actually rephrase Steven's recipe: the most important sources of international institutional success are "high concentration of talent (faculty and students)" along with "abundant resources." Salmi in his report argues, however, that sufficient financial support (that stays for the "abundant resources") usually is not sufficient to raise overall quality of the institution. The most important task of the university officers together with the government officials is to facilitate "the concentration of talents." Without the ablest professors, working with able students, even best laboratories could do nothing for the prestige of a school. Therefore, the real secret of world-class university are the people working and studying there.

One of the functions described by Howard Bowen is not directly reflected in Salmi's approach—that is, public service. A careful reader of the World Bank report would notice that "off-campus lectures and courses, work performed by interns, artistic performances and exhibits, spectator sports" do not count for the world excellence of HEIs. On the other hand, Michell Stevens considers social services to be fundamental for two theories explaining the role of higher education: *reproduction* and *transformation*. *Reproductions* thesis is built upon Marx's critique how groups of power create desirable social and cultural systems to legitimate their class advantages. Higher education, in this vision, is a means of reproducing the social class. Such reproduction is not achieved by mere teaching–learning process but rather by a kind of class initiation. *Transformation* thesis emphasizes the role of education in a constant social progress toward modernity. The new generation should not only reproduce the existing order, cultural, and political regime. It should help to achieve status and power, and therefore, the university education is so important because it confers advantages rather independently of students' social background (Stevens 2007).

The social value of higher education is more difficult to define even if one assumes that social services of HEIs are as important as education and research. In the context of international and global interactions of institutions and people involved in higher education, the social services change their nature: similarly to the other two functions of colleges and universities. Twenty-first century graduates of a good university must be responsible citizens of their home countries but also responsible citizens of the world. In the times of global challenges (e.g., global warming, climate changes, water scarcity, regions with persistent famine), these graduates are the people that can find the solutions to the existing problems (Wagner 2004).

The interactions with students and faculty members from foreign countries should help in understanding and appreciating cultural diversity. This feature of modern graduates gains importance in a globalized world: many companies have multinational workers; they operate in different countries. In such instances, global awareness should not be just a fancy of rich elite; this is a necessary condition for economic growth. The ability to work in almost any place in the world is also a great asset of a modern college graduate. Understanding cultural background of the contractors and trade partners can save time and enlarge profits (Singh 2005; Rizvi 2004, 2007).

This new approach to international dimension of higher education has already changed the US higher education and the attitude of American students. Previously they were known for their reluctance for going abroad as a part of the academic program. Nowadays, no university could aspire to high status without offering study abroad programs. The time spent abroad is important not only as a part of regular teaching–learning process but also as a necessary element of "global socialization" that can produce a mature citizen of modern world (Wasserman 2009; Mills 2008; Al Karam and Ashencaen 2006; Eisenstadt 2000).

WHO NEEDS IDENTITY?

The education of "global citizens" reveals one problem of modern higher education. No matter how open the university is, it is always located in a particular state. No matter how diverse the student body is, there is always one dominant group. In connection with the public versus private good dilemma, both institutional and governmental policy should approach the issue of identity. Should students be citizens of the world first of all, or should they be rather the citizens of the country of graduation? In other words, in the times of globalization, is identity an issue at all? Maybe infusing the graduates with up-to-date knowledge is enough to educate mature citizens (Takayama 2007; Schofer and Meyer 2005).

The complexity of the special social character of higher education has been revealed in case of transnational activities. Rich HEIs from the developed countries can transfer their educational offer to the other state, including teachers, textbooks, programs of studies, examination procedures, etc. Majority of institutions that are expanding this way use English as the language of instructions, which is undoubtedly the most common language of knowledge in the 21st century. In many regions, the possibility to study in English and to interact with native speakers is an additional bonus, and students prefer such education because it better suits their future careers.

Grant McBurnie and Christopher Ziguras in their book, *Transnational Education: Issues and Trends in Offshore Higher Education*, analyze the delicate issue of creating national identity through higher education as it has been managed by the Malaysian government. Because of the unmet demand for higher education, the government decided to open their educational market and invite foreign universities to organize transnational education. Australia, due to the strong economic ties and relative proximity, was deeply interested in transferring the educational offer to Malaysia. Public Australian universities in collaboration with partners on-site opened several institutions that offered a regular Australian curriculum. In Malaysia, there are many different ethnic groups that speak different languages, and the government soon noticed that the graduates do not have sufficient language skill to be able to work at public administration. Therefore, the graduates of Australian–Malaysian transnational institutions worked either at private business or migrated abroad looking for a better job. This trend was completely against the original concept of education as the means for faster economic development. After some deliberation, the government made education in Malaysian language mandatory. It was not only the language itself that was important—the language stood for the national identity as well (McBurnie and Ziguras 2006, pp. 63–65).

The graduates of universities should not only be efficient and creative workers but also be aware of the global issues and problems. Howard Bowen observed in the 1970s that higher level of education leads to greater involvement in community buildings. Although the process of globalization calls for redefinition of territorial community, there is still great importance of locality. Universities and colleges are places of constant social interactions that help in finding the core of the community development. The universities play also an important role as mediators between researchers and society in more general sense. Even the use of new technology could be beneficial, since "universities can play an important role in reducing social exclusion in communities. Where service universities are operating effectively, academic communities and local communities have developed and become equal partners, as seen in the creation of partnership networks between universities and communities, the development of joint projects and the like" (Leliûgiené and Barsauskiené 2007).

Identity is not only the problem of students; it can affect faculty members, too. Oili-Helena Ylijoki, in her article, "A Clash of Academic Cultures: The Case of Dr. X," describes the experience of one of Finnish scholars that had rather wanted to remain anonymous. The changing environment, both local and global, led Dr. X to lose the confidence in his work as a teacher and researcher—"Dr. X might quite easily end up with burnout and totally lose his motivation for and enjoyment in his work" (Ylijoki 2008, p. 84). The case analyzed by Ylijoki proves that identity is still an important element of personal and social development. Both professors and graduates who are losing their identities, not confident in moral and political values, could neither be active members of local community nor citizens of the state and of the world. Despite the development of new technologies and despite the student and faculty mobility, the main task of the university will remain to prepare a person to be active in his/her professional, political, and private life.

The study of American and Australian students proved that contacts with other nationalities is always a test for one's personal identity. "Australian students, with a rather amorphously defined sense of national identity in terms of prior upbringing and socialization, rather readily adapt to differences in foreign settings and enjoy the experience of broadening horizons. Americans, in contrast, with a much more intense understanding of and commitment to national identity, more easily feel lost outside the home environment." The young Americans that visit Australia had to renegotiate their own national identity by defending the US foreign policies. The survey conducted by the National Survey of Student Engagement emphasized the correlation between studying overseas and reflective learning. Thus, the national identity can be constructed and negotiated not only at home institution, while interacting with compatriots. Going abroad can be a better test for national identity. This proves, however, that social function of the university still should be one of top priorities of higher education (Steams 2009, p. 85).

CONCLUSION

The world-class university, as it is envisioned by the international organization such as World Bank or OECD, is the result of the concentration of talent, surrounded by abounded resources and facilitated by a liberal policy of government. The combination of these factors is a necessary condition for any attempt to achieve world status. Salmi explains that the international dimension has become an integral part of top-class universities; therefore, there could be no quality without advertising the institution's offer worldwide and without inviting people from abroad (students and faculty members alike).

The very same organizations that promote various forms of internationalization among HEIs because of the need for open-minded graduates understand the role of higher education for local development. The national and regional identity should still remain an important part of the educational process. Even Salmi, who definitely support the idea of world universities, has no doubt that institutions should not forget about the closest surrounding. In his article, "Nine Common Errors When Building a New World-Class University," he describes one of the errors of aspiring institutions:

> "Hiring foreign academics is common practice to accelerate the launch of a new university. Indeed, it makes good sense to bring experienced teachers and researchers to help; it can also be a very effective capacity-building strategy when an important part of the mission of the foreign academics is to train younger, less experienced academics in the host country. On the other hand, it can be a counterproductive strategy in the absence of systematic efforts to attract and retain qualified national academics" (Salmi 2010).

When a person is leaving Harvard University through one of the gates, there is an inscription saying "Depart to serve better thy country and thy kind." This is still the advice to all institutions aspiring to greater status and prestige. The vision of the university should not be based on the demands of the world and needs of international corporations. The source of inspiration should be local problems and challenges, and local need and aspirations. The OECD report of 2007, *Higher Education and Regions: Globally Competitive, Locally Engaged*, emphasizes the local engagement—"to interact with society and promote the social impact of their scientific and cultural activities" is still the key mission of higher education (*Higher Education and Regions* 2007, p. 83). Thus, the world-class university should remain local institution. Knowing the importance of international interactions, the national governments and university officials must always have in mind the most important stakeholder – the society. The only reason for the existence of universities.

REFERENCES

Abdulla Al Karam, Andromeda Ashencaen, Creating international learning clusters in Dubai. *International Educator*, vol. 15, no. 2 (2006), p. 12–15.

Philip G. Altbach, Globalization and the university: Myths and realities in an unequal world. *The NEA 2005 Almanac of Higher Education*, p. 63–74.

Philip G. Altbach, Higher education: An emerging field of research and policy. In: Roberta M. Bassett, Alma Maldonado-Maldonado (eds.), *International Organizations and Higher Education Policy*. New York: Routledge, 2009, p. 9–25.

B. Aubert et al., Search for B-meson decays to two-body states with $a_0(980)$ mesons. *Physical Review D*, vol. 70, no. 11, 111102, 2004.

L.H. Bagalkot, Implications of globalisation for university reforms: Concern for quality and equality. In: Josephine Yazali (ed.), *Globalisation and Challenges for Education. Focus on Equity and Equality*. Delhi: Shipra Publications, 2003, p. 667–681.

Roberta M. Bassett, *The WTO and the University. Globalization, GATS, and American Higher Education*. New York: Taylor & Francis Group, 2006.

Howard R. Bowen, *Investment in Learning. The Individual and Social Value of American Higher Education*. San Francisco: Jossey-Bass, 1977.

Janice K. Currie, Janice Newson, *Universities and Globalization: Critical Perspective*. Thousand Oaks: Sage Publications, 1998.

Education Trends in Perspective; Analysis of the World Education Indicators, 2005 Edition. Paris: UNESCO Institute for Statistics, OECD World Education Indicators Programme, 2005.

Shmuel N. Eisenstadt, Multiple modernities. *Daedalus*, vol. 129, no. 1 (Winter 2000), p. 1–29.

Daniel Golden, *The Price of Admission*. New York: Three Rivers Press, 2007.

Andy Green, Education, globalization, and the nation state. In: Hugh Lauder, Phillip Brown, Jo-Anne Dillabough, and A.H. Halsey (eds.), *Education, Globalization and Social Change*. Oxford: Oxford University Press, 2006, p. 192–198.

Dennis Hayes, Robin Wynyard (eds.), *The McDonaldization of Higher Education*. Westport: Sage Publications, 2002.

Higher Education and Regions: Globally Competitive, Locally Engaged. Paris: OECD, 2007.

Clark Kerr, At large: Howard R. Bowen (1908–1989): Fiat Lux et Justitio Omnibus. *Change*, vol. 22, no. 2 (1990), p. 78–79.

Irena Leliûgiené, Viktorija Barsauskiené, The role of the university in community development: Responding to the challenges of globalization. In: David Bridges, Palmira Juceviciené, Roberta Jucevicius, Terence McLaughlin, Jolanta Stankeviciûtè (eds.), *Higher Education and National Development. Universities and Societies in Transition*. New York: Routledge, 2007, p. 227–236.

Morris Low, Japan: From technology to science policy. In: Henry Etzkowitz, Loet Leydesdorff (eds.), *Universities and the Global Knowledge Economy, A Triple Helix of University-Industry-Government Relations*. London: Pinter, 1997, p. 132–140.

Grant McBurnie, Christopher Ziguras, *Transnational Education: Issues and Trends in Offshore Higher Education*. Florence: Routledge, 2006.

Andrew Mills, Emirates look to the west for prestige. *The Chronicle of Higher Education*, vol. 55, no. 5 (2008), p. A1, A23–A24.

Jaisbree K. Odin, New technologies and the reconstruction of the university. In: Jaisbree K. Odin, Peter T. Manicac (eds.), *Globalization and Higher Education*. Honolulu: University of Hawai'i Press, 2004, p. 149–162.

Mark Olssen, Neoliberalism, globalization, democracy: Challenges for education. In: Hugh Lauder, Phillip Brown, Jo-Anne Dillabough, and A.H. Halsey (eds.), *Education, Globalization and Social Change*. Oxford: Oxford University Press, 2006, p. 261–287.

Darron Otter, Globalisation and sustainability: Global perspectives and education for sustainable development in higher education. In: Elspeth Jones, Sally Brown (eds.), *Internationalising Higher Education*. New York: Routledge, 2007, p. 42–53.

Fazal Rizvi, International student mobility and its limits: The Australian case. *International Higher Education*, vol. 37 (Fall 2004), p. 7–9.

Fazal Rizvi, Postcolonialism and globalization in education. *Cultural Studies < = > Critical Methodologies*, vol. 7, no. 1 (2007), p. 256–263.

Susan L. Robertson, Xavier Bonal, Roger Dale, GATS and the education service industry: The politics of scale and global reterritorialization. In: Hugh Lauder, Phillip Brown, Jo-Anne Dillabough, and A.H. Halsey (eds.), *Education, Globalization and Social Change*. Oxford: Oxford University Press, 2006, p. 228–246.

Laura E. Rumbley, Philip G. Altbach, *Observatory on Borderless Higher Education, International Branch Campuses Issues*. London: Observatory on Borderless Higher Education, 2007.
Jamil Salmi, *The Challenge of Establishing World-Class Universities*. Washington: The International Bank for Reconstruction and Development/The World Bank, 2009.
Jamil Salmi, Nine common errors when building a new world-class university, Inside Higher Ed, The World View, August 22, 2010, http://www.insidehighered.com/blogs/the_world_view/nine_common_errors_when_building_a_new_world_class_university (accessed: April 2, 2011).
Jamil Salmi (ed.), *Constructing Knowledge Societies: New Challenges for Tertiary Education*. Washington: The International Bank for Reconstruction and Development/The World Bank, 2002.
Evan Schofer, John W. Meyer, The worldwide expansion of higher education in the twentieth century. *American Sociological Review*, vol. 70. no. 6 (2005), p. 898–920.
Michael Singh, Responsive education: Enabling transformative engagements with transitions in global/national imperatives. In: Michael W. Apple, Jane Kenway, Michael Singh (eds.), *Globalizing Education: Policies, Pedagogies, and Politics*. New York: Peter Lang Verlag, 2005, p. 113–134.
Morshidi Sirat, Managing the interface with the region: The case of Universiti Sains Malaysia, Pulau Pinang, Malaysia. In: Jean L. Pyle, Robert Forrant (eds.), *Globalization, Universities and Issues of Sustainable Human Development*. Cheltenham: Edward Elgar Publishing, 2002, p. 194–211.
Doreen Starke-Meyerring, Re-visioning higher education on the internet: A cross-cultural rhetorical study of higher education policy in Germany and in the United States. PhD dissertation presented at the University of Minnesota, 2003.
Cathleen Stasz, Eric R. Eide, Francisco Martorell, *Post-Secondary Education in Qatar*. Santa Monica: RAND Corporation, 2007.
Peter N. Steams, *Educating Global Citizens in Colleges and Universities. Challenges and Opportunities*. New York: Routledge, 2009.
Mitchell L. Stevens, *Creating a Class. College Admissions and the Education of Elites*. Cambridge: Harvard University Press, 2007.
Keita Takayama, A nation at risk crosses the pacific: Transnational borrowing of the U.S. crisis discourse in the debate on education. *Comparative Education Review*, vol. 51, no. 4 (2007), p. 423–446.
Carlos A. Torres, *Education and Neoliberal Globalization*. New York: Routledge, 2009.
Peter Wagner, Higher education in an era of globalization: What is at stake? In: Jaishree K. Odin and Peter T. Manicas (eds.), *Globalization and Higher Education*. Honolulu: University of Hawaii Press, 2004, p. 7–23.
Gary Wasserman, Bridging cultures in Doha. *The Chronicle of Higher Education*, vol. 55, no. 55 (2009), p. B9–B11.
Oili-Helena Ylijoki, A clash of academic cultures: The case of Dr. X. In: Jussi Välimaa, Oili-Helena Ylijoki (eds.), *Cultural Perspectives on Higher Education*. Dordrecht: Springer, 2008, p. 75–89.

59 Subjective Perspective on Experiences in Project Management

The Implications for Improvements in Higher Education

Vaiva Zuzevičiūtė

CONTENTS

INTRODUCTION

Some of the theorists note that some of the European states have worked toward the welfare state. However, even in 21st century, there are still questions, especially in the field of socially sensitive spheres, that remain to be addressed (Pestieau 2006; 1). On the other hand, a fact that most of the European Union states aim to meet the needs of those who are unable to take care of their needs should be considered the main achievement of contemporary societies. Even if today we enjoy the highest standard of living the humankind ever enjoyed, there are groups of people and individuals' whose basic needs are neglected. In education, there are many groups whose access to recourses is limited due to background or historical circumstances. Older adults, migrants, and learners with special needs not always have access to the educational (or other) necessary services. Sometimes it is more expensive to provide these services; sometimes there is a lack of competence within community of educationalists to provide services; or both. Project activities in the field of education provide opportunities to consolidate recourses, human, financial, and organizational, for the provision of services that otherwise would be delayed or not accessible (Zuzevičiūtė and Žvinienė 2007).

In this context, however, it is crucial to note that a lot of nonmanagement professionals (social workers, artists, teachers, and representatives of other professions) undertake project activities and face the need to implement project management. Project management, therefore, becomes a rather generic competence than a competence that merely professionals in management possess.

This fact shapes a *problem question* that is being addressed in the study presented: What are subjective perspectives of nonmanagement professionals (educationalists; in this case: teachers and teachers of adults), developing a project proposal/application with the focus on how respondents interpret their experiences about the two dimensions of project management (technical and sociocultural).

It is worth following this study theme because, even if there is a long and rich tradition of studies in the field of management, there are not, however, many studies on the way individuals interpret their experiences in the process. The issues of what is easy and what additional support do individuals need remain to be further addressed. Moreover, there are not many studies dedicated to the experiences of nonmanagement professionals in the field.

Therefore, *scientific innovation* of the study stems from at least two aspects. Firstly, it focuses on subjective perspectives on those involved into project management. Secondly, a study focuses on perspectives of the nonmanagement professionals who were faced with the need to undertake project management in addition to their regular activities. In this case, representatives of the teaching profession participated in the study.

Methods of critical reference analysis, also an empiric study, based on qualitative research methodology in social sciences, were used in the process of designing the study and for developing this paper.

PROJECTS IN EDUCATIONAL SPHERE

Even if today we enjoy the highest standard of living the humankind ever enjoyed, there are groups of people and individuals whose basic needs are neglected. In education, there are many groups whose access to recourses is limited due to background or historical circumstances. Older adults, migrants, and learners with special needs not always have access to the educational (or other) necessary services. Sometimes it is more expensive to provide these services, sometimes there is a lack of competence within community of educationalists to provide services, or both.

Project activities in the field of education provide opportunities to consolidate recourses, human, financial, and organizational, for the provision of services that otherwise would be delayed or not accessible (Baca and Jansen 2003; Corrigan 1997; Zuzevičiūtė and Žvinienė 2007). Therefore, project activity in education is a necessity and also, in a way, a demonstration of citizenship (Gasiūnaitė 2008; Zuzevičiūtė and Žvinienė 2007). Many teachers, adult educators, and researchers decide to use opportunities for additional financial recourses because they believe in the necessity and duty to implement some initiative as early as possible, and for as many recipients as possible. In 2008–2009, world economy suffered from a recession, and the processes analyzed above became even more evident. It is important to think and act for ourselves and those that need support.

However, project activity is an activity within domain of management. Therefore, those teachers who decide to engage into project activities inevitably face the need to act as managers. Engagement in projects means expanding field of expertise and experience. Project requires teachers to learn quickly and to develop their competence in a completely new field for them: management. These new requirements may create additional tensions and fears for teachers; however, there are not so many studies in Lithuania and globally on the subjective perspectives of teachers in the field (Gasiūnaitė 2008).

CHALLENGES OF A CONTEMPORARY WORLD

An equal distribution of social well-being covers economic well-being. Total well-being, also accessibility to and quality of education, health care, as well as smooth functioning of the legal system, is a complicated task for all states. Even welfare states (Finland, Sweden, and Norway) tackle these issues (The World Bank 2002; Pestieau 2006). In comparison with Lithuania, where, according to the Department of Statistics, in the year 2009, nearly 17% of low-income workers made up 15% of the population, the rates are rather low. For example, in Finland (approximately 4%), however, the problems remain (www.std.lt, accessed on April 17, 2010). In any case, the distribution of economic well-being is not equal. Activity of state, nongovernmental organizations, and various funds allowing both groups and individuals to learn, get employed, and seek self-realization is one of the processes of creating a society we and our children deserve and want to live in. To contribute to the social integration process through an education improvement project is to contribute to the personal integration process in order for everyone to have the same opportunities to accumulate social capital. Different authors define the concept of *social capital* differently. For example, Coleman (1990) defines social capital as social structures (with learning opportunities as an integral part) that facilitate individual action. Bourdieu (here from Fowler 2000) sees it as the aggregate of actual or potential resources based on institutionalized relationships. According to Putnam (1993), social capital refers to social networks and the norms of reciprocity and trustworthiness that arise from them. Every person who takes part in such networks, that is, who has an opportunity to use existing institutions, understands and adheres to the agreed norms of activity and common values, has someone to share his/her achievements and frustration, is much stronger and ready to meet challenges than the one who has become isolated early, or later in a lifetime, and has not got an opportunity to make use of the common capital. The amount of income may reveal different level of social capital rather than different skills (competencies of medical doctors in Lithuania, Ireland, or Serbia are similar; however, the salaries might differ dramatically).

Additional and targeted recourses can help to aid those who do not have an immediate and direct access to social capital. Projects are one of the channels to pursue the goal; therefore, their importance increases and increases the need for nonmanagement professionals to undertake the activity.

Since early times, which historiography traces back, reallocation of recourses left after meeting primary needs, to establish a fund to finance other activities, was commonplace. One way of reallocation is when a state accumulates residual finances; second is when it is made by private persons; and third is when it is made by religious or political organizations (Kučinskienė and Kučinskas 2005; Neverauskas et al. 2005).

There are several concepts and definitions of a project that have some differences but that share more similarities, on the other hand. Project is being defined as prearranged and preplanned creation or change of an object (simple or complex), having a well-thought-out, clear, and public goal (why and what is going to be created or changed); and also, having clear, well-defined, reasonable, and targeted resources: finances, manpower, time, information, etc. (Charvat 2003).

Any project has an environment in which the project is being implemented. Thus, project execution is always related to, for example, organizational culture (Zakarevičius 2003). Project has results. Also, a project may be understood as complex, harmonized, one-time efforts undertaken for a certain defined period of time, with an established budget and assigned resources, and oriented toward satisfaction of end users (Gray and Larson 2000).

In most of the definitions, we find that a project is a one-time work having a clear goal with parameters of quality, cost, and schedule and a temporary organizational structure. From the systemic point of view, a project is an organized change of an initial situation to achieve a desired result. Projects are associated with various funders (organizations, regional, national, EU or world programs, private persons, strategic attitudes of various agencies, and invitations to contribute with projects) and their different goals (Zuzevičiūtė and Žvinienė 2007).

PHASES OF PROJECT MANAGEMENT: IDEA AND PROPOSAL DEVELOPMENT

There are several phases in project management: conceptual (generating idea), proposal development, selection (developers do not participate in this phase), implementation and internal evaluation, and external assessment (Kučinskienė and Kučinskas 2005; Zakarevičius 2003).

Project idea covers several elements. On the one hand, it is necessary to know and clearly define what problem needs to be solved; on the other hand, an original, creative, and realizable solution of a problem needs to be suggested. A problem in social field is always related to people. A human being is a complex creature who takes part in many interactions. Every person is a member of some social institution, for example, has a family, works, and participates in social activities. Some of them face personal challenges; others experience employment difficulties. Thus, a project idea necessarily deals with a target group, because it is impossible to introduce a universal project that could help to overcome difficulties of the whole society.

Radical changes, for instance, fundamental change of the established social order, are defined as a destabilization factor. History proves that destabilization brings more pain and loss than before making radical decisions. For example, French Revolution or events of 1917 in Tsarist, Russia, brought more sufferings than an order that these changes were opposing (Holman 1978).

Three main elements of a project idea are identified (Kučinskienė and Kučinskas 2005; Neverauskas et al. 2005; Zuzevičiūtė and Žvinienė 2007):

1. Problem (we try to clearly identify it, anticipate its solutions, and solve it)
2. A target group, which has a certain problem

These first two elements are closely interrelated.

3. An original, creative, and feasible suggestion how to define and solve problems of a certain target group

As it has already been mentioned, in social reality, every human being is in constant interaction with others. An initiator, organizer is always associated with others. "Others" may be called "stakeholders," and it is the fourth element of a project idea. There may be many stakeholders. One more element to be considered in the project idea stage is a team, that is, whether a group of people competent to identify and solve a problem can be forced.

It is evident that a project idea is not only the beginning of a huge work but also the product of an immense preparatory work.

This phase and the subjective experiences within the phase were at focus of the part of the study that is presented in this paper. As Neverauskas et al. (2005) emphasized, project management has at least two dimensions: technical and sociocultural. This separation was also taken into an account while organizing a study, and specifically, formulating items for written reflections. Any project has several phases:

1. Preinvestment phase distinguishes the following steps (further on project proposal/application development):
   a. Identification—defining the need
   b. Initiation—informal and semiformal conversations with potential team members
   c. Concept development—identification of project purpose and goals
2. Investment phase:
   a. Detailed project design activity
   b. Project marketing (for services, production of a product) activity
3. Closing phase

Project review, evaluation, and approval (D'Orr 2007); however, just the first phase (project proposal/application development) is at focus in this particular chapter, and the subjective experiences within the phase are analyzed further.

METHODOLOGY OF AN EMPIRIC STUDY

Though ample literature is available on project management, not many studies were dedicated to subjective experiences of those involved into project management, especially those whose major field is other than management (Gasiūnaitė 2008). This empiric study was focused on teachers. In a contemporary world, as it was emphasized above, teachers find themselves in situations where they are faced with necessity to implement project activities. Projects allow teachers to implement innovations in their educational practices and organizations, and also to share best practices beyond immediate users. The aim of the empiric study was to collect perspectives in order to identify which of the activities and aspects of project management (phase-proposal development) are easy, which create additional tensions, and what additional help is needed for teachers. An empiric longitudinal study of three years with two separate groups of respondents was being implemented since 2007 (2007, 2008, and 2009) in order to identify in what way do teachers interpret their experiences in project management. One of the groups of respondents was asked to share their subjective perspectives on one of the phases of project management, and specifically, development of a proposal for a project activity. Another group of respondents was followed with an aim to collect their subjective perspectives during the process of both proposal development and implementation of the successful project proposal/application.

In this paper, subjective perspectives of the first group respondents are presented. Due to ample data, only findings from 2009 are presented. All of the respondents received consultations on the proposal writing in the framework of either in-service training seminars or individual consultations based on the in-formal networks.

PROFILE OR RESPONDENTS

Respondents of the first group are teachers or adult educators who have just started their careers and found themselves in situations where their supervisors, employers, or colleagues expected them to develop educational project proposals. The expectations are grounded on the fact that a young person is flexible and knows how to use Internet and other recourses well. This fact, and the fact that young people in most cases are competent users of foreign language, serves as impetus for expectations and also pressures that young teachers experience.

The respondents are of age 24 to 31 years old; their total number is 65 (Table 59.1). In most cases, a team of two or three respondents worked on one proposal/application; however, they shared their perspectives at an individual basis in written reflections.

As it is evident, the respondents are young people just starting their careers. On the one hand, they just had had an access to higher education and its ample recourses. On the other hand, the start of the career might serve as an additional and powerful source for stress in itself.

TABLE 59.1
Profile of Respondents (Group 1)

Year	No. of Respondents	Profile of Respondents and Data
2007	21	Teachers, teachers of adults. Ages from 24 to 29; 19 women, 2 men (168 written reflections; 2 respondents worked on the second proposal, 19 on the first proposal)
2008	23	Teachers, teachers of adults. Ages from 24 to 31; 21 women, 2 men (207 written reflections; 3 respondents worked on the second proposal, 20, on the first proposal)
2009 (due to ample data, some findings of this stage are presented in this paper)	21	Teachers, teachers of adults. Ages from 24 to 30; 18 women, 3 men (251 written reflections; 4 people worked on the second or third proposal, 17 worked on the first proposal)
Total no. of respondents	65	

Therefore, even if the data from an empiric study are ample, and respondents were approached in a period of three years, the *limitations* of the study are evident. Limitations stem from the factors mentioned above, and specifically, that the subjective perspectives might be influenced by the stress and tensions a young person experiences at this point of his/her professional life. On the other hand, the influences are counterbalanced by at least two factors. Firstly, the number of respondents is sufficient to have a balanced picture of the situation. Secondly, starting to get involved into the new activity, such as project activities, poses stress for teacher (or any other nonprofessional in management) at any point of his/her career. Another source of limitation comes from the fact that reflections are written in Lithuanian language, and both the interpretation and the translation may result in some discrepancies. This risk, however, is inevitable in qualitative studies.

PROCEDURE OF A STUDY

Respondents were asked to share their personal experiences during the development of their first or second project proposal in the form of written reflections. Respondents were asked to provide information on their experiences about two dimensions of project management (Neverauskas et al. 2005), and specifically, technical and sociocultural dimensions of project management. Respondents were asked to share their experiences, to identify difficult issues they come across in the process, and also to identify sources for support and issues they found rewarding in the process. Each respondent was given a number, and a year code. For example, code R308 means that a respondent, according to the list completed according to maiden surname list that year, comes third, and this particular respondent shared his opinions in 2008.

Teachers worked both on national (Lithuania, proposals under the framework of Structural Funds, or under the framework of municipality or calls by the Ministry of Education and Science or some other ministry) and international proposals (such as LLL program Comenius, or Leonardo, or transversal programs). Therefore, the scope of the proposals was quite different; however, on average, respondents reported to have spent from 2 to 3 months for the work. Respondents were asked to share their experiences on a weekly basis. Therefore, it resulted in 7 to 12 reflections for each respondent. This resulted in an ample pool of information (Table 59.1). Some of the reflections were sent by e-mail each week to a researchers' team; some of the respondents chose to share with researchers with the full set of reflections by the end of the period. Length of each reflection varied from several lines to two pages in some cases. Respondents were asked to identify their reflections by their mother's maiden surname in order to ensure anonymity and respect for our respondents.

In data analysis, phase texts were analyzed applying software for qualitative data analysis: Weft QDA. One person‘s contribution in a separate reflection was used as a unit; therefore, in this case (for year 2009), the number of possible contributions does not exceed 251.

FINDINGS OF SUBJECTIVE PERSPECTIVES ON PROJECT PROPOSAL/APPLICATION DEVELOPMENT

As it was already mentioned, the phase of development of proposal/application development was at focus during the study. The analysis revealed that a majority of contributions were predominantly focused on one of the dimensions of project management, and specifically, the sociocultural dimension. Less than a fourth of contributions were directly related to the technical dimension. That is, respondents shared their impressions and emotions that the process of developing a project proposal/application triggered.

Analysis of the data reveals that the phase of development of a proposal generates a lot of stress for young teachers, when it is the first or the second or third proposal they are working on. In the project development phase in majority of cases, respondents shared having experienced a stress with a varied intensity during the whole period.

According to respondents (149 contributions), development of project proposal could be divided into six steps, with each having its own emotional contents (Table 59.2).

TABLE 59.2
Steps of Project Proposal Development Phase ($N = 149$)

Step	Emotional Contents	No. of Contributions and Examples
1 Generating an idea of decoding an idea	Stressful (difficult, challenging, responsible): both generating and decoding an idea of employers or colleagues for which they ask respondents to develop a project proposal for additional or full financing	$N = 78$ "I don't know what is more difficult. To find out that I am expected to work on the proposal or to understand at least to some extent what my boss means. Especially when I did not dare taking too much time. I think she <boss> told me her expectations in three sentences or so. I am so afraid that I did not get it right what the school wanted to do....there are people who help me out, but it is so difficult, I don't want to take their time..." (R1409)
2 Searching for a suitable framework/ funder	Stressful (a lot of information, no one to consult, different regulations)	$N = 41$ "We learn project management at university, but- when you start looking—so many funders, so many programmes. And when you think that already found the one that suits you, it turns out the call is closed, and you have to start again. Sometimes it is so stressful and disappointing." (R309)
3 Choosing a framework/funder to address	Extremely stressful (almost random choice at some point)	$N = 71$ "I just decided that I have to choose any of the three more or less suitable. I had enough. The time is running short, I spent three weeks on that, one of the tenders closes up in three weeks, so I just picked up the one that seemed the easiest, or I will not be doing it anymore." (R1909)
4 Starting to get acquainted with a proposal form/ application	Extremely stressful (unexpectedly complicated, software needs applications that are not yet available at school/learning center, more information that is possible for a person to collect, information is privileged; information that only other people can provide, e.g., organization balance sheets, CVs)	$N = 111$ "Last week I thought I'd better go and tell her I am not doing this. First the application form did not open up, it asked for the 8 version of Acrobat reader, so I asked our computer lab person to help me, then it turned out I have to provide short CV's of all people involved in the project. Names, degrees, major fields are needed. When I asked two or three people to provide me with info I either received their full CVs—so additional work, to read through several pages, and to extract what I need, or they just tell me. I am not a recorder, so I wrote what I remember, but do I remember well? And when I started working on WPs, it is a good thing we learnt about them and Gantt at university, and that we still receive consultations. Otherwise I really should better stop this." (R1409)
5 Completion of a proposal form/ application	Stressful (some of the parts of application are unclear, some sections repeat themselves, unclear requirements, difficult to use codes, difficulties in communicating with colleagues and future partners)	$N = 93$ "I am just tired. I want this to be over, I work day and night, time is running out, partners don't answer my messages, I don't know if they agree to take up some activities. Just decided they do in some cases, and that they don't in other. It is difficult to understand those indicators of target groups, what is the code of each." (R209)

(continued)

TABLE 59.2 (Continued)
Steps of Project Proposal Development Phase (N = 149)

Step	Emotional Contents	No. of Contributions and Examples
6 Calculating budget	Extremely stressful (unexpectedly difficult, unclear regulations, too many regulations, no one to consult, unclear situation in market, difficulties of harmonizing national and international regulations)	N = 133 "I remember now when my professor told not to postpone budget till the results and activities are completely clear. She told us that at least a full week is needed for the task, but I did not believe this was so difficult. It was so clear and easy to watch professor calculate budget on whiteboard. But now I would really let everything go if not for the immense work already invested. I think I never felt so frustrated, scared and lost. With 400,000 EUR in front of me I feel paralysed. If not for the consultations, this is just too much to imagine how and when to spend. At school no one imagines this kind of sum too." (R1109)

Approximate level of stress (*this estimation has no numeric value; it is just an illustration of the nature of contributions within the category*)

Extremely stressful			√	√		√
Stressful	√	√			√	
Neutral						
Steps of development of a proposal/application (*according to respondents*)	1	2	3	4	5	6

FIGURE 59.1 Approximate illustration of stress levels respondents shared having experienced during the steps of project proposal/application development steps.

The categories and emotional contents within each of the steps enable illustration that approximates the perspective of proposal developers (Figure 59.1).

As it is approximated, it is evident how difficult it was for teams of nonmanagement professionals to act in the process. No numeric value should be attributed to the estimation, because it is merely a visualization of respondents' contributions on how they felt on the dimensions of project management (technical and sociocultural).

DISCUSSIONS

These findings are of extreme importance because they show that even technical dimension of project management has a profound impact on developers of proposals/applications. The findings did reveal several contributions about rewarding experiences. Several of them were on the unexpected support they received from their colleagues, people who had participated in projects for a decade or so, and who chose to consult, provide information, explain, or even help with some of the tasks (in 13 reflections, but several times it was the same person to share; therefore 8 respondents). Several of them were on the feeling of accomplishment at the moment of submitting a proposal/application (14 contributions, 11 respondents).

Majority of contributions reveal that for nonprofessionals in management (in this case—teachers and teachers of adults), starting the initiative for project, developing a proposal/application is a stressful situation. They feel a lack of support, lack of information, even from the agencies and professionals who should be providing needed information, and lack of encouragement. Respondents' perspective identifies that the phase of proposal/application development has six steps: generating/decoding of an idea; searching for a suitable framework/funder; choosing of a framework/funder;

starting to get acquainted with a proposal form/application; completion of a proposal form/application; and calculating budget. All these steps generate stress of different intensity. Respondents share that the most stressful is choosing a funder, starting to get acquainted with the proposal/application form, and calculating budget.

The findings of an empiric study are important because of at least three reasons:

Firstly, this study adds to findings of Gasiūnaitė (2008) about the lack of support from the agencies of funders, about the complicated procedures and obscure requirements, and also technical difficulties that make the process even more difficult. When it is not here argued that funders should change the requirements, it is argued that if there is a plan to attract creative thinking, new people into the initiatives, the agencies, support, and information systems should further be developed in order to provide more user-friendly services.

Secondly, study reveals that nonmanagement professionals need a lot of additional support. It seemed that the ones with an initial background in the field, for example, who studied project management at university, felt more confident and at least knew the possible domains of difficulties and possible sources for support. In a contemporary world, where many of the nonprofessionals in management undertake project activities due to various reasons, it is evident that initial professional education (for sure, of educationalists), at least in higher education, has to take into account this aspect of education (project management).

Thirdly, it is also evident that the subjective perspectives of respondents are negative and reveal a lot of stress in the process, and therefore, further studies are needed on what ways respondents themselves identify for remediating situation.

CONCLUSION

1. The study revealed that from respondents' perspective, the phase, which in project management theory is identified as a single activity, and namely, of proposal/application development, is not perceived as such by respondents.

The study revealed that respondents identify six steps of a single activity in project management (proposal/application development):

a. Generating/decoding of an idea
b. Searching for a suitable framework/funder
c. Choosing of a framework/funder
d. Starting to get acquainted with a proposal form/application
e. Completion of a proposal form/application
f. Calculating budget.

All these steps generate stress of different intensity. Respondents share that the most stressful is choosing a funder, starting to get acquainted with the proposal/application form and calculating budget.

2. Study reveals that respondents still lack support from the agencies of funders, about the complicated procedures and obscure requirements, and also technical difficulties that make the process even more difficult. When it is not here argued that funders should change the requirements, it is argued that if there is a plan to attract creative thinking, new people into the initiatives, the agencies, support, and information systems should further be developed in order to provide more user-friendly services.
3. Study reveals that nonmanagement professionals need a lot of additional support. Subjective perspectives of respondents are predominantly negative and reveal a lot of stress in the process. It seemed that the ones with an initial background in the field, for example, who studied project management at a university, felt more confident and at least knew the possible

domains of difficulties and possible sources for support. In a contemporary world, where many of the nonprofessionals in management undertake project activities due to various reasons, it is evident that initial professional education (for sure, of educationalists), at least in higher education, has to take into account this aspect of education (project management).

REFERENCES

Baca C., Jansen P. (2003). *Project Management Professional Workbook.* San Francisco and London: Sybex.

Bourdieu P. (2000). Reading Bourdieu on society and culture. Edited by Bridget Fowler, *Sociological Review Monograph Series.* Oxford: Blackwell: Sociological review.

Charvat J. (2003). *Project Management Methodologies.* New York: John Wiley and Sons.

Coleman J. (1990). *Social Capital. Foundations of Social Theory.* Cambridge, MA and London: Belknap Press and Harvard University Press.

Corrigan P. (1997). *The Sociology of Consumption.* London and Thousand Oaks: SAGE.

D'Orr A. (2007). *Advanced Project Management. A Complete Guide to the Processes, Models and Techniques.* London: Kogan Page.

Gasiūnaitė G. (2008). *Andragoginės veiklos tobulinimas projektine veikla.* (Bakalauro baigiamasis darbas, rankraštis). Kaunas: VDU.

Gray C. F., Larson E. W. (2002). *Project Management. The Complete Guide for Every Manager.* New York: McGraw-Hill, Hill International. 2002, ISBN 0071376011.

Holman R. (1978). *Poverty: Explanation of Social Deprivation.* Bungay, Suffolk: Martin Robertson, Chanter Press.

Kučinskienė R., Kučinskas V. (2005). *Socialinių projektų rengimas ir valdymas.* Klaipėda: Klaipėdos Universiteto Leidykla.

Neverauskas B., Stankevičius V., Viliūnas V., Černiūtė I. (2005). *Projektų valdymas.* Kaunas: Technologija.

Pestieau P. (2006). *The Welfare State in the European Union.* New York: Oxford University Press, ISBN 0-19-926101-6.

Putnam R. (1993). *Making Democracy Work. Civic Traditions in Modern Italy.* Princeton, NJ: Princeton University Press.

The World Bank. (2002). *Constructing Knowledge Societies: New Challenges for Tertiary Education.* Washington, DC: The International Bank of Reconstruction and Development/The World Bank.

Zakarevičius P. (2003). *Pokyčiai organizacijoje.* Kaunas: Pačiolis, VDU.

Zuzevičiūtė V., Žvinienė V. (2007). *Projektų rengimas ir valdymas.* Kaunas: VDU.

60 The Idea of Social Responsibility of a University

Marcin Geryk

CONTENTS

INTRODUCTION

Universities are important elements of the public space. They are the source of the intellectual elite; they determine the level of society's education. Their influence and the potential they accrue make them play a special role. That role goes beyond what is traditionally understood as universities' tasks: education and research. According to ever-increasing social needs, universities should be involved in many more actions, all of them being socially responsible.

TASKS OF ORGANIZATIONS AND BACKGROUND OF SOCIAL RESPONSIBILITY

Organizations play two important social roles—they generate products and services, and then they deliver them to the market expecting a return on their investment. They also promote public responsibility, in its various forms. On the other hand, social entrepreneurship releases innovation and facilitates mobilization to act for society. It is a promising path that leads to social changes, also through the use of the power of dialogue, thanks to the system of partnerships and networks. The platform of dialogue allows organizations to interpret social needs correctly (Hyunbae Cho 2010).

Globalization processes and the development of the "knowledge society" have generated far-reaching implications for higher education through the way in which they have transformed the nature of business and social life. It requires every organization to implement new strategies and focus more on social needs (McCaffery 2005).

The idea of social responsibility has a rich historical background, starting from the ethical issues in Aristotle's works, through the ethical considerations in all religions, and in the dynamic transformations resulting from the industrial development (Geryk 2009a).

The development of philanthropy led to the creation of very close contacts between business representatives and social care, education, and cultural institution workers. It was up to the corporation managers what direction or preference they followed, but these also depended on regional conditions and the companies' current circumstances. The support they provided has had a huge impact on the development of many organizations, including educational, cultural, and civic ones. Thanks to the good relations, the sharp divisions between public and private organizations, paid and voluntary jobs, as well as between governments, universities, or corporations started to disappear.

As a result of those actions, both the economic and social systems developed in new dimensions (Heald 2005).

Thus, the idea of social responsibility has its roots in many historical breakthroughs. It is connected with the development of societies but first of all with civilizational progress. The dynamic industrial development only enhanced those events and accentuated their significance. The maturation of the awareness of social needs was slow, and it developed on many different levels. It seems, however, that these days, organizations and stakeholders are aware of their objectives and needs to such an extent that it may be easily assumed that further intense development of this idea will follow (Geryk 2009b).

The awareness of both society and corporation management has currently reached such a level that the problem of whether to implement the issues of social responsibility does not seem to exist. The debate has been dominated by the question of how to do that (Zerk 2006). At the same time, a growing number of business, social, government, and multilateral organizations as well as companies show a great interest in the ideas of social responsibility. The institutionalization of this process by setting up departments dealing with CSR or with codes of conduct has increased this interest. Universities play an important part in the popularization of CSR ideas. They have introduced those issues to their curricula, thus raising the awareness of the future social and economic leaders. The constantly growing number of consultants, legal regulations, and international actions promoting socially responsible business and the practice of social reporting also play a significant role (Utting 2003).

The basic idea is that businesses, and the executives who manage them, actually do and should create value for customers, suppliers, employees, communities, and financiers. The crucial things are the quality of the management of those relationships and the value created for stakeholders (Freeman 2008).

Pursuant to the social policy sketched by the European Commission in Lisbon in 2000, the issue of social responsibility gained a new momentum. In Poland, besides other actions, it has been pointed that socially responsible actions of organizations should be made public. It was suggested, however, that such a publication should take the form of an announcement or information put up on an Internet site addressed to stakeholders. It is a pity, however, that it has not been accepted that such reports should be assessed and be given opinion by an independent institution. Their contents, then, would gain the value of impartiality and greater credibility (European Commission 2007).

It testifies to organizations' understanding of the needs put forward by society. Naturally, the problem of the implementation of those rules as well as the effectiveness of actions and the achieved social effects will be different. However, the mere fact that such actions are consciously undertaken shows that both organizations and their stakeholders mature together.

The transparency of actions combined with honest information play an important role. Information ought to reach stakeholders in the form of reports, which should be, preferably, prepared by independent institutions. Then their credibility is much higher (Henriques 2010).

Naturally, the right to know concerns those actions that may have an influence on the social environment. Therefore, stakeholders expect explanations, rationale for the undertaken actions, which have an economic, social, organizational, or environmental influence, after all. They also want to know about the risk or potential hazards that those actions may entail. What is interesting is that stakeholders expect the right to participate in the development, monitoring, as well as evaluation of those decisions that have, or may have, an influence on their lives and their environments (Dubigeon 2006).

THE UNIVERSITY'S RESPONSIBILITY BEFORE SOCIETY

Having a university in its neighborhood is an ennobling situation for the local community. The high opinion of a university as a place where knowledge is accumulated and the high opinion of the people who pass on this knowledge lead to the situation where educational institutions face increasingly high expectations. Similarly to other organizations, universities are expected to be able to pay close attention to the needs of their stakeholders, who, in this case, are an expanded collective of social groups. One may even assume that in the modern society, it is difficult to discern who is not part of the stakeholders' circle. All participants of public, social, and economic life are or, probably, will be interested in the correct operation of educational institutions. However, the universities' situation is one where they face growing social expectations while financial resources for their statutory activity are decreasing (Clark 2007).

All human activity is, after all, based on access to knowledge. Its scope and availability are part of universities' domain (Woźnicki 2007). The quality of their activity influences many other organizations and offices—in short, all organizations that use the human resources or scientific achievements of universities.

One of the ways universities can became more socially responsible is by engaging in intellectual entrepreneurship. According to research done by K. Ramanathan (2005), it could be the development of technological start-up entrepreneurs, the commercialization of nanotechnology, and the promotion of technology commercialization and of the newly established venture.

The scope of social responsibility is still being discussed. Can a university, as a higher education institution, undertake socially responsible actions, and if so, in what scope? Naturally, stakeholders will have different expectations towards a university as compared from the expectations towards a company. It seems that the discussion over the scope of university activity is important. However, it is difficult to find opponents of universities' better relations with their immediate environments (Geryk 2009b).

While undertaking many socially responsible actions, universities also, incidentally in some measure, develop their own image, thus creating a strong brand. However, those actions should be addressed to stakeholders in order to activate certain social roles among them and build stronger relations. It is only in such a form that image-oriented actions will be socially responsible (Geryk 2009b).

There are also many examples of how educational institutions treat their students as stakeholders. The body of students creates the school culture. Another way of linking the school management and students is to involve them in the study of best practices and programs. They can also research the same topics and compare findings and recommendations. Students can also provide valuable insights into new plans prepared by the school (Holcomb 2007).

Proper information on pro-social actions is also one of the expectations. It requires an understanding and appreciation of the value of proper communication with stakeholders. Its significance is highlighted by the rule of transparent operation of universities, where they provide credible information on their on-going and planned actions. It is worth pointing out that effective presentation of appropriate actions is met by the environment with a positive response (Geryk 2010). Those changes are representative of society's growing involvement, whose need to participate in deciding on the shape of the public space seems to be gaining more and more importance. It is so because it leads to the expansion of civic society (Evers and Heinze 2010).

Bearing social expectations in mind, organizations undertake pro-social actions and make them a key aspect of their activity. Being responsible for its environment is a result of the growing social conviction that the benefits that an organization reaps from its operation in a given environment serve as a basis for stakeholders' claims to some of those benefits. It is becoming more of a demand rather than a recommendation. Therefore, the development of those social expectations should be examined with close attention (Hess et al. 2008).

It should be borne in mind that a university's social responsibility has to be an integral part of its policy (Geryk 2012). It is the most important connection between an organization and the market, in its economic, financial, and technological meaning, and between the organization and the nonmarket areas of social life, politics, and culture. The interactions between universities and stakeholders are so close that the distinctions between the market and nonmarket areas are practically nonexistent (Perrini et al. 2006).

Being aware of the difficulties in transforming an organization into one that is managed according to the ideas of social responsibility makes it possible to judge the size of the whole enterprise properly. Its multidimensional character and the strong influence on the organization's operation entail significant organizational changes. However, it is only a strategic attitude that will lead to a synergy in action and bring benefits in the form of positive feedback leading to significantly better social relations (Dymowski 2011).

Transforming a university into a socially responsible organization may be difficult. It requires significant changes in the way it operates and in the way it is managed. However, such a transformation deals mainly with the changes in the organization's values, its vision, and the practices and procedures that are closely connected with those values (Andersen 2004).

Universities should play a more active role as innovators. There is a need to focus on, and develop, the efficiency and effectiveness of the different forms of technology transfer. These actions should be undertaken to bridge the gap between university research, technological development activities, and the commercial market. The main restriction is the traditional nature of universities that may limit the development of relationship building skills, which are so important for universities to become socially responsible (Jones-Evans 2003).

The worldwide financial crisis provided another impulse for organizations to increase their interest in their environments (Geryk 2010). Also universities became a subject of expectations and pressure on account of the inefficiencies in business ethics and social responsibility education. Appropriate education of future managers was supposed to protect the economy from the recurrence of crises caused by irresponsible investment (Ghoshal 2005). One way or the other, ethical education, if it were to be compliant with the market, requires a re-organization of curricula, which makes it possible to analyze not only the traditional but also the alternative markets (Khurana 2007).

However, what can be done about the suspicion and acrimony, which, according to Abrams (2008), "exist among people and seem to be growing stronger"? As he puts it, "(...) management must apply to its relationships with the rest of the community the same type of searching analysis that that it would make of its more usual business problems" (Abrams 2008).

The RESPONSE program makes it possible to change the hitherto teaching methods into an educational path that increases students' awareness of the need for pro-social activity. It contributes to the possibility that such actions will be undertaken in the future and that their spontaneous diffusion will take place within organizations (Zollo et al. 2006).

Universities will have to redefine themselves to operate successfully at the forefront of change—to ensure flexibility while still maintaining loyalty to traditional values. Managers of the educational institutions and the body of professors have a shared responsibility in ensuring that universities choose the proper way to serve their stakeholders (McCaffery 2005).

RESEARCH ON SOCIAL RESPONSIBILITY OF UNIVERSITIES

In the years 2007–2010, a research project, "Social Responsibility of the University as Perceived by Its Stakeholders," was conducted. The aim of the project was to examine the position and role of social responsibility in the process of university management. The core of the research project was the characteristics of social responsibility in the sector of higher education and an attempt to assess the influence of pro-social actions on the improvement of management efficiency both in short-term and long-term perspectives. It was also researched whether the way universities are managed is in line with stakeholders' expectations and whether universities acknowledge the necessity to analyze their opinions. The important question is to what extent the issue of social responsibility may be referred to higher institution educations (Geryk 2010).

In the qualitative research, which took the form of in-depth interviews, respondents claimed that the notion of social responsibility was understandable. At the same time, they pointed out that although such actions are perceived as positive, they are not fully appreciated by university managers. When talking directly about the social responsibility of universities, the majority of the respondents focus on the fundamental university activity, that is, education and formation (Geryk 2010). This role corresponds with the expectation that universities will prepare appropriately educated graduates (Geryk 2012). They also assume that the implementation of those tasks, included in a university's mission, is equivalent to being socially responsible.

An important observation is the fact that universities readily engage their students and employees in pro-social actions. This may be a result of a premeditated marketing action. Pro-social activity employed for the purposes of the promotion of a university and the development of its brand is not perceived as contradictory with the ideas of social responsibility. It is so because it turns out that a strong brand and the reputation that goes with it may be an important asset (Geryk 2010).

Another stage of the qualitative research was addressed to university stakeholders, defined as students, members of business groups and associations, journalists, and local authorities. The understanding of social responsibility used to be narrowly understood as a relation only with students. Among many examples of pro-social actions, the following were frequently mentioned: society education; voicing opinions on socially important issues; scientific research; close cooperation with employers; actions for children and the disabled; actions for health care, sport, culture, and the arts; actions for the natural environment; and charity. However, one of the most significant expressions of universities' social responsibility should be actions for students and university employees.

High-quality education was among the respondents' expectations, yet it should be combined with development opportunities outside universities. Stakeholders also hope to increase their chances on the job market through the development of better relations with employers. This would find its expression in a comprehensive system of work experience, apprenticeship, or other forms of cooperation with the business sector. On the other hand, the employers who took part in the research stress the importance of adjusting the curricula as well as university profiles to the dynamic changes on the market. The local authority representatives who took part in the research stress the need to educate experts, their expectation being that those experts are open to changes and their attitude is formed as one full of humanism and moral sensitivity. Yet on the other hand, journalists stress competence, that is, good preparation for future jobs. They also voice a critical opinion that universities only produce diplomas. This group of respondents also hopes for a greater openness of universities to cooperation with not just the media but also local governments, businesses, and institutions.

The respondents expect that universities will be more attentive to society's voice. They also point to a group—students—who might participate in university management. Anyhow, it is expected that this group, widely involved in education institution management, should include representatives of other stakeholder groups.

In the subsequent research stages, a cycle of quantitative studies was carried out. They were conducted on representative samples: the first one on 300 adult inhabitants of Gdańsk, Gdynia, and Sopot, and the following one on a 500-strong representative sample of the inhabitants of the whole country. The former of the studies served as an inspiration to further, deeper studies over the issue of the social responsibility of universities. The social expectation concerning universities' greater involvement was confirmed. Also, more than 90% of the tri-city inhabitants declared their need for the local community to be informed on universities' pro-social actions. Thereby, they expect information on how pro-social initiatives are supported by universities. They also expect reports on those actions. The respondents also pointed to the difficulty they had in defining the notion of social responsibility; every third of them did not have an opinion on that matter.

It is worth pointing out that as few as 24% of the inhabitants of the country (in 2008) had come across the notion of social responsibility. Within this respondent group, the social responsibility of universities is associated with attention to good-quality education. On the other hand, among the actions for society, it was taking care of universities' own employees that took the first position—more than 98% of the respondents considered this factor as important or very important. At the same time, as many as 93% admit that the main incentive for those actions is the need to improve a university's own image. This should lead to increased competitiveness, as is claimed by 91% of the respondents. When giving examples of expected actions, as many as 94% of the respondents claim that a university's role is to support children and youth education.

The aim of the third research stage, which included quantitative research done on a sample of 81 Polish university managers, 41 from private universities and 40 from state ones, was an evaluation of pro-social activity of the universities. Also, an attempt was made at an analysis of the influence of those actions on the universities' management.

Nearly 93% of the private university managers and more than 87% of the state university managers had come across the notion of social responsibility of universities. More than 90% of the respondents provided a lot of examples. They concern such actions as actions for the disabled, information society development, support for talented students, and environment protection.

At this stage, the collected results were divided into those from the representatives of state and those from the representatives of private universities. As many as 73% of the representatives of the former type of university see significant differences in pro-social activity between commercial and state universities. Another difference that is statistically significant is the evaluation of the importance of support for secondary schools. The state university respondents considered this action as important twice as frequently.

The more commercial character of private universities and the different social expectations towards schools of this type were pointed out as the reason for undertaking such actions. What is interesting is that the respondents from the private universities stressed the need to increase the efforts to maintain good relations with their environments. It was so because state universities are commonly perceived as better in this respect.

Similarly to the previous stages of the research, it was confirmed that the most important reasons why universities undertake pro-social actions are the need to improve their position on the education market and the development of their brands as well as—to a lesser degree—the sense of moral duty and the need to improve their relations with local communities.

The last stage of the research on the social responsibility of universities was its foreign part. It was carried out among university managers from 46 countries, which made it possible to look at the national results from a much wider perspective. Its aim was to study the notion of the social responsibility of universities and to determine the main factors creating a university's image, the reasons why universities undertake pro-social actions, and the forms of information on the actions.

It appears that the respondents were familiar with the issue of the social responsibility of universities; it is actually something quite natural. Such a university should offer better education; prepare its graduates for modern life; educate ethical, responsible managers; and influence other organizations, such as cultural institutions or secondary schools.

The majority of the foreign university managers did not observe significant differences in pro-social activity between state and private universities. Among the reasons for those actions, 92% pointed to the need to improve their relations with their environments, and 89% pointed to the need to improve their brands. At the same time, 73% point out that an evaluation of universities' curricula by independent accreditation institutions is important for their image. What is interesting is that 73% of the respondents declared environment protection projects as an example of actions undertaken by the universities in question.

As K. Obłój (2010) points out, the gradual disappearance of differences between trades or the deregulation and intensification of globalization processes lead to such big transformations of the environment that the way it functions becomes incomprehensible for managers and it is difficult for them to see a predictable logic in it.

The globalization processes, the increasing migration of students, the exchange of academic staff and, above all, the growing competition from many dynamically developing research centers and many foreign universities are a great challenge for universities. It pertains to a stronger connection with local communities and a closer attention to the needs of local environments. This is the only way to be able to compete over the influence on the future shape of societies. And this is going to be a manifestation of the social responsibility of the university in its full sense—an ethical institution, responsive to the needs of its environment, taking care of the natural environment, and promoting innovativeness and creativity.

CONCLUSION

The analysis conducted on the basis of the multistage research allowed for a formulation of recommendations for university managers. Universities should integrate their actions on the strategic level with the ideas of social responsibility. Besides, they should aim to improve the quality of their relations with their stakeholders, who should be identified as all institutions and all people. It is a result of the wide scope in which universities influence their environments. While undertaking prosocial actions, care should be taken to inform stakeholders about them. What is the most important, however, is that university managers should be aware of their responsibility for their actions; they have to be in line with stakeholders' expectations. All of them should lead to mutual benefits, both for the universities and their environments, and the high ethical, moral, and qualitative requirements should find their reflection in the mission and policy of the institution of higher education.

REFERENCES

Abrams F.W., In: *Management's Responsibility in a Complex World, Stakeholder Theory*, eds. A.J. Zakhem, D.E. Palmer, M.L. Stoll. Prometheus Books, New York, 2008, pp. 26–31.

Andersen B., *Bringing Business Ethics to Life. Achieving Corporate Social Responsibility*. ASQ Quality Press, Milwaukee, 2004.

Clark B.R., *Creating Entrepreneurial Universities: Organizational Pathways of Transformation*. Emerald, Bingley, 2007.

European Commission, *Corporate Social Responsibility. National Public Policies in the European Union*. Office for Official Publications of the European Communities, Luxembourg, 2007.

Dubigeon O., Legal obligations and local practices in corporate social responsibility. In: *Corporate Social Responsibility. Volume 1: Concepts, Accountability and Reporting*. Palgrave Macmillan, New York, 2006, pp. 254–283.

Dymowski J., The social responsibility and business. In: *Social Services of Responsible Business*, eds. M. Bonikowska, M. Grewiński. The Publishing House of the Higher School of Pedagogy of the Society of Public Knowledge, Warsaw, 2011, pp. 79–105.

Evers A., Heinze R.G., The social policy: Risks of economization and opportunities for integration. In: *The German Social Policy: Economization and Crossing the Barriers*, eds. A. Evers, R.G. Heinze. The Publishing House of the Higher School of Pedagogy of the Society of Public Knowledge, Warsaw, 2010, pp. 15–34.

Freeman R.E., In: *Managing for Stakeholders, Stakeholder Theory*, eds. A.J. Zakhem, D.E. Palmer, M.L. Stoll. Prometheus Books, New York, 2008, pp. 71–88.

Geryk M., A historical view of corporate social responsibility. *International Journal of International Institute for Advanced Studies in Systems Research and Cybernetics,* Vol. 9, No. 1, 2009a, pp. 13–18.

Geryk M., The issue of social responsibility of higher education institutions. Theoretical considerations and research. In: *Advances in Education*, Vol. IX, eds. G.E. Lasker. The International Institute for Advanced Studies in System Research and Cybernetics, Tecumseh, 2009b, pp. 43–51.

Geryk M., *Social Responsibility of the University*. The Publishing House of the Warsaw School of Economics, Warsaw, 2012.

Geryk M., *Social Responsibility of the University as Perceived by Its Stakeholders. Research Report*. The Publishing House of the Warsaw School of Economics, Warsaw, 2010.

Ghoshal S., Bad management theories are destroying good management practices. *Academy of Management Learning and Education,* Vol. 4, No. 1, 2005, pp. 75–91.

Heald M., *The Social Responsibilities of Business. Company and Community 1900–1960*. Transaction Publishers, New Brunswick, 2005, pp. 265–269.

Henriques A., In: *Corporate Social Responsibility, Sustainable Development and Triple the Guiding Line*, eds. A. Henriques, J. Richardson. The Publishing House of the Higher School of Pedagogy of the Society of Public Knowledge, Warsaw, 2010.

Hess D., Rogovsky N., Dunfee T., The next wave of corporate community involvement: Corporate social initiatives. In: *Dorporate Social Responsibility. Readings and Cases in a Global Context*, eds. A. Crane, D. Matten, L.J. Spence. Routledge, London—New York, 2008, pp. 270–285.
Holcomb E.L., *Students are Stakeholders, Too!* Corwin Press, Thousand Oaks, CA, 2007, pp. 22–26.
Hyunbae Cho A., Politics, values and social entrepreneurship—A critical analysis. In: *The Social Entrepreneurship*. The Publishing House of the Higher School of Pedagogy of the Society of Public Knowledge, Warsaw, 2010, pp. 47–69.
Jones-Evans D., In: *Developing Entrepreneurial Universities. Cases of Good Practice in Academic-Industry Linkages, Intellectual Entrepreneurship through Higher Education*, eds. S. Kwiatkowski, J. Sadlak. Publishing House L. Kozminski Academy of Entrepreneurship and Management, Warsaw, 2003, pp. 175–202.
Khurana R., *From Higher Aims to Hired Hands. The Social Transformation of American Business Schools and the Unfulfilled Promise of Management as a Profession*. Princeton University Press, Princeton–Oxford, 2007.
McCaffery P., *The Higher Education Manager's Handbook*. Routledge Falmer, London and New York, 2005.
Obłój K., *The Passion and Discipline of Strategies. How to Build a Successful Business with Dreams and the Decision*. The Publishing House of Poltext, Warsaw, 2010.
Perrini F., Porutz S., Tencati A., *Developing Corporate Social Responsibility. A European Perspective*. Edward Elgar Publishing, Cheltenham–Northampton, 2006.
Ramanathan K., In: *Universities as Facilitators of Courageous Acts of Intellectual Entrepreneurship, Intellectual Entrepreneurship and Courage to Act*, eds. S. Kwiatkowski, N.M. Sharif. Publishing House of L. Kozminsky Academy of Entrepreneurship and Management, Warsaw, 2005, pp. 155–185.
Utting P., CSR and development: Is a new agenda needed? In: *Corporate Social Responsibility and Development: Towards a New Agenda?* United Nations Research Institute for Social Development (UNRISD), Geneva, 2003, pp. 6–9.
Woźnicki J., The *Academic Schools as the Institutions of Public Life*. Polish Rectors Foundation, Warsaw, 2007.
Zerk J.A., *Multinationals and Corporate Social Responsibility: Limitations and Opportunities in International Law*. Cambridge University Press, Cambridge, 2006.
Zollo M. et al., Understanding and Responding to Societal Demands on Corporate Responsibility (RESPONSE). Final Report, Brussels, 2006.

61 Technology-Enhanced Learning in Higher Education

A Proposal for Developing a Quality Evaluation Framework

Diogo Casanova, Nilza Costa, and António Moreira

CONTENTS

INTRODUCTION

The higher education (HE) space has been suffering profound changes, most of which related to the potential of the use of technologies to support teaching and learning (T and L) practices (Allen and Seaman 2003). New forms of learning mediated by technology are emerging, promoting flexibility and removing the remaining obstacles to the free mobility of the learning process. This mobility mediated by technology can lead to the promotion of internationalization and the exchange of values, knowledge, and cultures, promoting new competences and fostering the emergence of new virtual environments (Daukšienė et al. 2010). These changes, however, are not running at the same pace as the *boom* of new technologies, brought in by Web 2.0, would suggest. Social networking, social media, social bookmarking, collaborative tools, and instant messaging rapidly changed the habits of Internet users, promoting new trends, practices, and motivations (McGreal and Elliott 2004). The use of the Internet by society rapidly increased with the so-called *democratization of the Web* (Reilly 2007), but HE practices did not follow this sudden change. The teaching staff have been struggling with enhancing the students' learning experience using technology mainly because the design of the course, the learning strategies, the learning materials, the monitoring, and the feedback process are not aligned with the demands of learning environments mediated by technology (Oliver 2006; Riley 2007). Moreover, technology in HE is mainly used in indirect activities that support the T and L process such as communication tools, repositories, and dissemination tools (Stensaker et al. 2007).

In compliance with this deficiency it is important to develop frameworks about quality related with the use of technology in T and L that, on one hand, allow its use as quality standards to lead to good practices and, on the other hand, allow teachers and universities to monitor and evaluate the impact of the use of technology in T and L practices. These actors can then understand if technology is, in fact, enhancing students' learning.

Taking into consideration the previous argument, this paper focuses on the development of a quality evaluation framework for the use of technology-enhanced learning (TEL) in HE. We will present the first stage of this study, which refers to the data collection process that set the scaffolding for the design of criteria and performance indicators to evaluate the quality of TEL.

THEORETICAL BACKGROUND

Much has been said about the definitions of e-learning, online learning, and distance learning (Guri-Rosenblit 2005; Moore et al. 2010). These definitions drive us to the process of learning also as a result of the use of one medium. The use of e-learning, or any other definition of learning supported by technology, does not constitute *per se* an enhancement of the quality of T and L or that the learning process in fact exists. It is, however, considered to have the potential to transform the learning process, enabling such enhancement.

Foundations of TEL

Technology can be used in different contexts, with different objectives and forms. The teacher can use technology to promote face-to-face (f2f) strategies, to develop autonomous learning, to extend virtually f2f sessions, or to develop distance learning. The use of effective technology in pedagogical strategies must assume TEL as being directly linked with a student-centered orientation as opposed to a teacher-centered orientation. Student-centered learning foundations reflect a more user-centered view about the nature of knowledge and the role of the learner (Wang and Hannafin 2005) that has to be more active in pursuing the acquisition of knowledge. Thus, this more active role of the student in the learning process engages with constructivism paradigms that sustain that the learner determines what, when, and how the learning process will occur (Wang and Hannafin 2005). Also, the course has to be designed aiming to promote forms of active learning given that the more active a student is in the learning process, the more student-centered the learning process also becomes (Tigelaar et al. 2004). Therefore, in TEL, the course design and the learning environment must profess meaningful scenarios in the form of a problem or a specific goal that has to be connected to the learning strategies and activities. Hence, it should promote decision-making, problem solving, manipulating, interpreting, hypothesizing, and experimenting (Kim and Hannafin 2011). These strands will demand new roles and competences from both the learner and the teacher. Learners have to develop individual learning plans that require skills to judge their individual needs, while the teacher has to guide this process in the form of tools, resources, and, if needed, direct instruction:

> "In re-designing more student-centred curricula, institutions must foresee that students will need more guidance and counselling to find their individual academic pathways in a more flexible learning environment" (Reichert and Tauch 2005, p. 18).

The teacher has to be constantly present in the formal learning environment, but s/he has also to have an informal presence that would point towards the adoption of a more consulting and advising role (Goodyear et al. 2001) than an instructing one. In order to develop TEL, the teacher has to be more aware of the use of technology. Mainly, s/he must be updated with new trends and pedagogical uses of ICT and capable of adapting to new educational contexts to provide learners with more and better educational materials and better learning environments. The teacher has to be more present, even when s/he is not in the same room as the student, and should be capable of promoting different learning paths personalizing each learning experience (Siemens 2006).

This change of educational practices will have consequences in (1) educational policies, (2) teachers' training, (3) understanding the best ways to integrate ICT in T and L practices, and (4) the design of information systems and technological applications (Lea et al. 2001). Teachers, learners, and institutions have, therefore, to adapt to new demands in order to use technology in a manner that enhances the learning experience.

What Is Quality in E-Learning?

The definitions of quality vary and commonly reflect the different perspectives of the individual and of society. Pawlowski (2007) defines quality as *"'appropriately meeting the stakeholders' objectives and needs, which are the result of a transparent, participatory negotiation process within an organization"* (p. 4). The same author refers that, in e-learning, quality can be related to all the processes, products, and services supported by ICT. Thus, the concept of e-learning quality must comprise all factors in the learning process. Inglis (2005) describes quality as related with comparative judgments, whereas quality assurance makes a comparison with a predetermined standard (minimum standard), and quality improvement (enhancement) is related with the relation between the current standard, the benchmark, and the pathway to achieve this benchmark. Mellar and Jara (2009) agree and sustain that "quality assurance and quality enhancement can be seen as parts of a larger process of quality management: assurance being concerned with determining that objectives and aims have been achieved, while enhancement is concerned with making improvements" (p. 20). Harvey (2006), however, prefers to address the quality assurance process to the practice of ensuring that there are mechanisms and procedures to assure that the standard quality is delivered. These processes of quality measurement involve the judgment of performance against a set of criteria. We agree with the definition offered by Sadler (1987) that says that a criterion is

> "a property or characteristic by which the quality of something may be judged. Specifying criteria nominates qualities of interest and utility but does not have anything to offer, or make any assumptions about, actual quality" (p. 194).

Why Evaluate?

Evaluating T and L quality is a complex process that, based on a referential and on systematic data collection, analysis, and interpretation, conducts a judgment that should promote a better understanding of quality of T and L and the enhancement of such quality. This process of evaluation must be structured and consolidated, respecting all stakeholders' objectives and needs (Pawlowski 2007), and must comprise a set of benchmarks that allow practitioners and evaluators to conduct their judgment on a sustained basis. Thus, during this process, it is necessary to understand (1) the expectations of different stakeholders (Prosser et al. 2003), (2) what criteria and performance indicators emerge from the literature, and (3) a measure of the extent to which these expectations and referential are considered and achievable (Kirkwood and Price 2005). The stakeholders must be engaged in this process reflecting and giving their feedback, fostering a participated evaluation process. In this research, we consider the evaluation process as a valuing process, a strategy, or a product based on criteria and performance indicators. This valuing process will trigger a set of contextualized information that will allow the various stakeholders to ascertain that the quality exists and, if needed, to ensure that practitioners have the necessary tools to enhance the process, the strategies, and the products based on a set of standards (Scheerens et al. 2003).

There are several reasons to promote the evaluation of T and L in HE. Scheerens et al. (2003) highlights three reasons: (1) to formally regulate the desired level of quality of educational results, (2) to render the practitioners accountable for their T and L practices, and (3) to promote the enhancement of its quality.

RESEARCH OVERVIEW

As mentioned in the introductory part, there has been a loophole in the research related with the evaluation of the use of ICT in HE T and L practices that has to do with the definition of quality standards in TEL and with the need for holistic models for evaluating the impact of ICT use, which at the same time can be used by practitioners and researchers to enhance the learning experience (Jara and Mellar 2010). This apparent loophole justifies the study we are conducting, which is confined to the following general objectives:

1. To design a TEL in the HE quality reference framework
2. To develop and validate an evaluation model that will allow practitioners and researchers to evaluate and monitor TEL practices in HE

So as to fulfill these goals, our study is divided in two different phases (Figure 61.1): (1) to design a *Reference Framework* (stages 1 and 2) and, based on this framework, (2) to develop an *Evaluation Model* (stages 3 and 4) for evaluating TEL in HE.

The difference between both of these phases is that the *Evaluation Model* will be directed towards a specific context and reality. It is not our intention to enclose our *Reference Framework* in a specific context; we want to allow other researchers and practitioners in HE in general, and in Portugal in particular, to use their own evaluation model based on the *Reference Framework* we are developing.

In this paper, we will focus on the first phase of this study, the design of a *Reference Framework*, which comprises two stages: the data collection process and the selection of criteria.

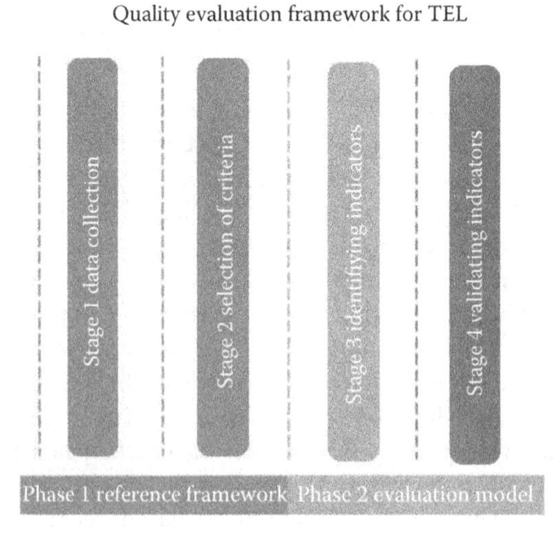

FIGURE 61.1 Research design overview.

DESIGN OF THE REFERENCE FRAMEWORK (STAGES 1 AND 2)

As mentioned above, the first phase of our research is the development of a *Reference Framework*. For the *Reference Framework*, we use the definition proposed by Figari (1994), which refers to the development of a framework of references (internal and external) of one object or reality from which two diagnostic outcomes can be derived: evaluation processes and training programs. The object underlined in this study is the quality in TEL in HE.

The *Reference Framework* design process is an effective practice that allows contextualization, transparency, data triangulation, and knowledge production, while involving all educational stakeholders in the process (Reis and Alves 2009). Using this practice, we distance ourselves from the role of the evaluator by focusing on the process of developing sustained and contextualized knowledge in a dialectic process where all actors contribute with their input (Figari 1994). According to Alves (2001), the design process of the *Reference Framework* allows us to find and/or develop references, diagnose, define evaluation dimensions, and justify the chosen criteria.

Our *Reference Framework* is designed based on stage 1, comprising three periods of the data collection process: (1) literature and legislation and regulations review (white circles), (2) interviews (gray circles), and (3) meta-analysis of competences needed for TEL (black circle) (Figure 61.2).

Period 1 is divided in two different parts. The first part is related with the research conducted worldwide on the use of technology in T and L practices and the evaluation of this use. Furthermore, we will also take into account the research conducted in the University of Aveiro related with the evaluation of online activities (Dias 2010) and with the use of e-learning. A second part is related with the review of legislation, regulations, and reports from the institution and from national and international organizations. This literature and regulations review period not only enabled the collection of several data related with the research topic (white circles), it also enabled identifying relevant research related to the competences needed for TEL practices (period 3).

In period 2, we aimed to understand the conceptions of representatives of the various stakeholders, namely, the teaching staff, the administrative staff, students, and the management bodies (gray circles phase).

Period 3 is related with a meta-analysis study aiming to collect key competences for TEL practices both for students and teachers (black circle phase). This period is linked with period one since the relevant research that allowed identifying the competences was selected within the literature and regulations review period.

With the data retrieved from the data collection stage, we can then start stage 2 (Figure 61.3), which includes three periods: the selection of data (white circles), the selection of statements (gray circles), and the transformation of statements into criteria (black circles).

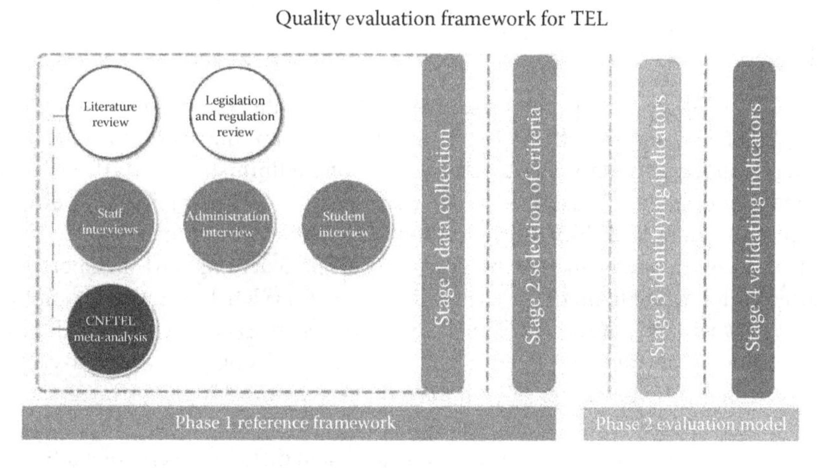

FIGURE 61.2 Research design—focus on stage 1.

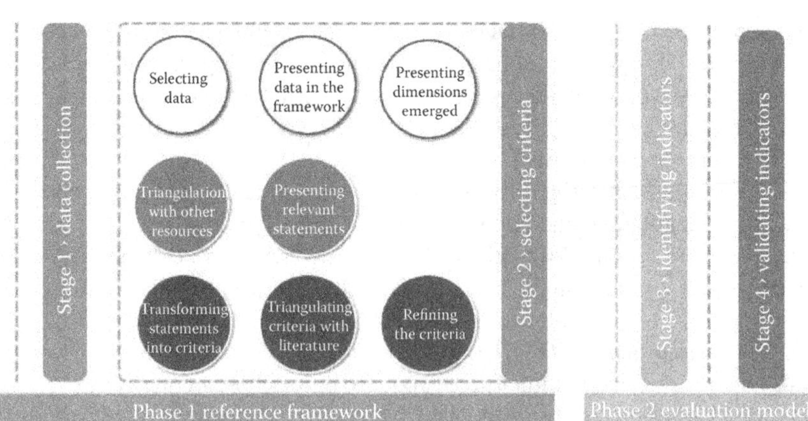

FIGURE 61.3 Research design—focus on stage 2.

Period 1 is divided in three parts: (1) selection of data, (2) systematization of the data into a framework relating it to a dimension, and (3) presenting the dimensions emerged.

In period 2, we triangulate the data collected from one source with data collected from other sources so that we can identify resemblances. We can then present relevant statements and findings from our data collection stage.

In period 3, we transform the relevant statements into intelligible criteria. We can then triangulate each criterion with the literature in order to refine our *Reference Framework* with more adequate and comprehensible statements.

As referred to above, our research design for this paper comprises the two stages of the development of our *Reference Framework*, the *data collection stage* and the *selecting criteria stage.*

METHODOLOGY

The *data collection stage* comprises three periods: (1) literature and regulations review, (2) stakeholder interviews, and (3) meta-analysis of competences needed for TEL. We will now present the methodology for periods 2 and 3 of this stage.

INTERVIEW PERIOD

The stakeholders' interview period comprises two different data collection techniques: individual semistructured interviews to members of staff and focus group interviews to students. The staff interviews were conducted with seven teaching staff, one administrative staff, and one element from the management body. (These interviews were held at the University of Aveiro during the months of April and June of 2010.) All the respondents had at least a master's degree and represented different scientific areas and teaching positions. The teaching staff members were chosen by the e-learning supporting team of the University of Aveiro (UOEL)—the criterion for choosing the teaching staff was to be a user of ICT in T and L practices. Each interview had the maximum duration of one hour. The students' focus group interview was conducted with six students. The design of a mini–focus group was considered to be the best methodological strategy since we were looking for reflected and varied contributions (Morgan 1997). The students were selected bearing in mind that they had to be from different scientific areas and different cycles of study and teaching delivery methods.

Three researchers, one of whom is an expert in qualitative analysis, validated the interview scripts, which were sent to the respondents before the interviews took place. It was considered relevant for this study that the respondents could reflect on the questions before the interview. At the end of the process, the transcriptions were sent to each respondent for validation. Data analysis was conducted using the Computer Aided Qualitative Data Analysis Software NVIVO9.

Six dimensions arose from the analysis of the interviews: (1) the context of the study, (2) the practices conducted, (3) existing support, (4) the tools used, (5) the competences needed, and (6) the quality related with the use of ICT in teaching practices. From a global data analysis of the interviews, 168 nodes emerged and, among them, 20 categories and 148 subcategories. We coded 376 citations in 1110 references within the 9 interviews.

CNETEL Meta-Analysis Phase

The second data collection period was the meta-analysis of needed competences for TEL. For the definition of competence, we considered all the personal characteristics, behaviors and attitudes, skills, and knowledge required (Lucia and Lepsinger 1999) to develop and/or conduct a TEL practice. From the literature review (LR) process, we selected eight academic papers/reports that explore the competences that teaching staff need to develop ICT mediated T and L strategies. We considered the use of ICT in f2f and distance-based courses and therefore did not make any distinction between these delivery methods of T and L.

The first step was to find and select all text references in the papers/documents that could be referred to as competences. In a first reading, we selected 320 references. For consolidation of the data retrieved, we did a second and a third reading, getting to a final result of 323 references. This validation process allowed us to start the categorization of the references into competences through the aggregation of one or more references into one competence class. With this process, we created 189 competence classes, the most often mentioned being the following: "know how to give feedback" (7 references), "being open to collaboration and group work" (6), "having basic technological skills" (6), "knowing how to access useful technology" (6), "knowing how to evaluate" (5), and "actively promote the participation of students" (5).

Although we identified the most important competences, we still had an excessive number of categories. So we tried to develop a process of refinement, creating clusters that comprised the characteristics, behaviors and attitudes, skills, and knowledge in just one competence; that is, if one teacher has, at the same time, "the willingness to promote collaborative learning" and the "skills to develop strategies for promoting collaborative learning," we would consider just one competence cluster: "promoting strategies of group work and collaborative learning." This clustering process allowed us to reduce the number of categories to 34. We then decided to consider only those categories that were mentioned in more than 50% of the documents analyzed, so we only selected 18 final competences categories.

RESULTS AND DISCUSSION

The interviews and CNETEL periods led to a set of information that can be linked to criteria to determine the quality of TEL. These data are, however, still not considered criteria since we have to triangulate them with the literature and regulations review period to identify similarities.

Relevant data were retrieved from the literature and regulations review that are mainly related with the evaluation of the use of technology to support T and L. From the several models that emerge from the LR, there are some reflections that arise about the perception that staff and students have about the use of ICT in T and L. Sun et al. (2008) reflect about the criteria that promote the effectiveness of TEL, having concluded that aspects such as students and staff satisfaction, flexibility, usefulness, and usability from learning environments are key factors that have to be taken into account. Mahdizadeh et al. (2008) aim to understand what ICT features staff are using and for what purpose. From this research, it is inferred that administrative features (calendar, schedules, and alerts); information repositories (articles and slide presentations); and communication features (email, mailing-lists and forums) are the features most valued and used by staff in the T and L process. This study points towards the conclusion that previous staff perceptions and previous experience working with a specific feature will influence the future use of this feature as opposed to using other features or tools.

Other studies point towards a more relevant role given to the student and his/her satisfaction towards the quality of instruction (Stewart et al. 2004). Palmer and Holt (2010) choose to evaluate what features of the Learning Management Systems (LMS) students find more relevant and those that are considered in fact capable of enhancing learning quality. From this research, it is concluded that students are satisfied with having access to the course presentation, the notes given by the lecturer in the lab or in f2f sessions, and forum discussions, considering these important features. There are, however, other features that are considered of greater importance and that students do not feel to be fully met, such as having teachers' feedback about the assignments and access to their grades and to their individual learning progress in the course (Palmer and Holt 2010). Several authors also evaluate the quality of educational resources. Leacock and Nesbit (2007) suggest aspects such as the content quality and its alignment to the learning objective, the existence of feedback mechanisms, the presentation, the usability, and reusability of educational resources. Shee and Wang (2008) research points out that the interface quality is one relevant aspect in the students' experience of interacting with the learning environment.

The evaluation of TEL, in HE, has been confined, in most cases, to one dimension of the T and L process rather than sustained in a more holistic evaluation as the framework proposed by McGorry (2003). This author triangulates different dimensions of the T and L process such as flexibility, feedback, interaction, technical support, students' learning, and their level of satisfaction. Just like McGorry, one of the first results that emerged from our data collection process was the need to organize each aspect in their respective dimension. This deduction that emerged from the respondents' feedback and from the LR suggests evaluating more than one dimension for understanding all the processes of T and L. This allows the promotion of a more holistic view about quality evaluation. We agree with this view mainly because it allows a more flexible and personalized *Reference Framework* that can lead to an *Evaluation Model* with just one or more dimensions, depending on the need and on the context. Thus, we suggest five dimensions to evaluate the quality of TEL:

1. Competences: all criteria related with the competences needed by the actors in the process of T and L—teaching staff and students. As mentioned above, we considered, for the definition of competences, all characteristic behaviors and attitudes, skills, and knowledge that one actor has to possess to take part of TEL practices.
2. Logistics and support: all the criteria related with the logistics and equipment needed for a TEL practice and the support given by the university in the form of tools, helpdesk, and training.
3. Learning environment and learning resources: all the criteria related with the quality of the learning environment designed by the practitioner, the learning resources proposed, and the context they are proposed for.
4. T and L strategies and practices: all criteria related with the strategies that both teachers and students develop when facing TEL practices.
5. Expectations and perceptions from stakeholders: all the criteria related with the expectations and the perceptions of stakeholders when facing TEL practices and whether these expectations are fulfilled.

The development of such dimensions allowed us to group, in an easier manner, the statements that emerged in the data collection stage, as presented in Table 61.1. For each emerging statement, we suggest the respective dimension(s) and the period(s) that led to the emergence of such statement.

TABLE 61.1
Statements That Emerged from the Data Collection Phase

Statement	Period(s)
T and L Strategies and Practices Dimension	
Different actors must collaborate with each other	CNETEL
The use of ICT makes the communication process with students easier	Interviews
The teacher should develop suitable T and L strategies	CNETEL
Students' online participation must be assessed	Interviews
Those involved must make provision for different students' profiles	Interviews, CNETEL
The teacher has to promote active learning	CNETEL
To monitor and track students' learning process	Interviews, CNETEL
There must be feedback on assignments	LR (Palmer and Holt 2010)
Competences Dimension	
The students must respect teaching staff availability	Interviews
Those involved must have skills to communicate online	Interviews, CNETEL
The teacher must be available to listen to the student	CNETEL
The teacher must have the necessary skills to manage discussions	CNETEL
The teacher has to possess knowledge for the evaluation process	CNETEL
The teacher has to understand existing needs and limitations of each student	CNETEL, Interviews
Teaching staff must know how to integrate technology into pedagogy and have pedagogical knowledge	Interviews, CNETEL
Teaching staff must have technological knowledge	Interviews, CNETEL
The students must possess study methods	Interviews
The teacher must be updated and must contribute to change for the best	CNETEL
The teacher must have deep knowledge of scientific content	CNETEL
The teacher must perform as an interface with existing regulation and policies	CNETEL
Expectations and Perceptions Dimension	
Expectations before the course	LR (Sun et al. 2008)
Satisfaction from the stakeholders	LR (Wasilik and Bolliger 2009; Shee and Wang 2008)
Impact of the use of technology	LR (Liaw et al. 2007)
Learning Environment and Learning Resources Dimension	
It is expected that teaching staff make available all relevant documents in the learning environment	Interviews, LR (Palmer and Holt 2010)
The teacher must have the skills to develop and manage the learning environment	CNETEL
Access to grades and individual learning process in the course	LR (Palmer and Holt 2010)
The quality of the interface	LR (Shee and Wang 2008)
The quality of the content	LR (Leacock and Nesbit 2007)
The alignment with objectives and learning outcomes	LR (Leacock and Nesbit 2007)
Content usability	LR (Leacock and Nesbit 2007)
Content reuse	LR (Leacock and Nesbit 2007)
Logistics and Support Dimension	
To monitor and track students' learning process	Interviews, CNETEL
The existence of interaction services	Interviews
Technical support must always be present	Interviews
There is pedagogical support to both teachers and students	Interviews
The university is evaluating the impact of ICT use in T and L	Interviews
Teacher's effort is recognized	Interviews
There is training support to teachers	Interviews

As mentioned above, for our *Reference Framework*, each statement must be validated using the criteria of being referred to in more than one period of the data collection stage. So if a statement is referred in the Interviews and in the LR, then it will be considered validated. The triangulation of the data that emerged from the different data collection periods led us to a preliminary version of our *Reference Framework* with six statements for evaluating the quality of TEL (Table 61.2). To finalize the process, we drew a correspondent criterion for each validated statement.

After collecting the data from the focus group interview and finishing the LR period, we will conclude phase one of our study, which is confined to our *Reference Framework*. We will have a framework of referenced criteria that can be used to determine if quality in TEL practices exists. This *Reference Framework*, and the criteria within, can be used for diagnosis, evaluation, or action (i.e., developing a training program), and it must have a characteristic of openness that comprises being capable to adapt to different contexts, in different forms, or by different actors.

We are aware that the number of validated statements ($n = 6$) suggests a deeper review on instruments and models for evaluating the quality of TEL. We reached a point of the research that suggests one of the following steps: (1) change the validation method overcoming the triangulation process or (2) increase the LR phase widening the search criteria filters. In any case, we come to the conclusion that the empirical data retrieved from the interviews and CNETEL are relevant data that can be important to sustained quality statements for the use of TEL.

Within the quality statements, we reached the conclusion that there is a strong need from both students and teaching staff to monitor the learning process and that there should exist appropriate strategies and logistics to do so. Another point worth mentioning is the needed competences. This field means that teaching staff should have the necessary competences about how they should communicate online, to integrate the use of ICT in the T and L strategies, and that they should have technological competences. The integration is probably the most challenging task from the curriculum design point of view since it is expected that ICT barriers are not felt during the learning process by the student. However, data collected from the students' interview point out that it would come as a surprise if the teacher asks students to search for documents on a learning topic on the Web instead of expecting to find the document in the LMS. Students interviewed expect that important information is given directly through the LMS instead of being integrated with ICT resources.

TABLE 61.2
Criteria Emerged from Triangulation

Dimension	Statement	Period(s)	Criteria
T and L strategies and practices	To monitor and track students' learning process	Interviews, CNETEL	Teaching staff monitor and track students' learning process
Competences	Those involved must have skills to communicate online	Interviews, CNETEL	Communication between actors is conducted effectively
Competences	Teaching staff must know how to integrate technology into pedagogy and have pedagogical knowledge	Interviews, CNETEL	Teaching staff possess the required pedagogical competences to integrate technology
Competences	Teaching staff must have technological knowledge	Interviews, CNETEL	Teaching staff possess the required technological competences
Learning environment and learning resources	It is expected that teaching staff make available all relevant documents in the learning environment	Interviews, literature review (Palmer and Holt 2010)	Documents, presentations, and notes are available in the learning environment
Logistics and support	To monitor and track students' learning process	Interviews, CNETEL	There is logistics for monitoring and tracking students' learning process

Another important competence from the students' point a view is the communication competence. Communicating online has different characteristics from to communicating f2f; however, students expect the same availability from the teacher, which means that they expect the teacher to be available every time, and this expectation must be managed before the course.

FINAL CONSIDERATIONS

This paper stems from the development of a Quality Evaluation Framework for TEL. Specifically, we aimed to present the first phase of developing a framework that relates to the development of a *Reference Framework* and its design process. We presented several statements and linked criteria that can be used as standards for TEL quality and, therefore, capable of being used to evaluate TEL practices. In the development of our framework, we took the position that it would be relevant to design an open framework that could be adapted to different contexts and objectives. With this in mind, it would be possible to add new dimensions or criteria to our *Reference Framework* to adapt it to different contexts or objectives. This characteristic of openness is an asset of our framework, allowing other researchers and practitioners to use the framework and design an evaluation that is tailored to their own specific contexts.

ACKNOWLEDGMENTS

This research was funded by the Portuguese Foundation for Science and Technology (reference SFRH/BD/60295/2009). The paper was supported by the LLL Erasmus project Tea-Camp (502102-LLP-1-2009-1-LT-ERASMUS-EVC).

REFERENCES

Allen, I. E., and Seaman, J. (2003). *Sizing the Opportunity: The Quality and Extent of Online Education in the United States, 2002 and 2003*. Needham, Massachusetts: The Sloan Consortium.

Alves, M. P. (2001). *O papel do desempenho do professor nas suas práticas de avaliação*. Braga: Universidade do Minho.

Daukšienė, E., Teresevičienė, M., and Volungevičienė, A. (2010). Virtual mobility creates opportunities. *Informacinių technologijų taikymas švietimo sistemoje, 2*(December), 30–35.

Dias, A. B. (2010). Proposta de um Modelo de Avaliação das Actividades de Ensino Online. Unpublished PhD, Universidade de Aveiro, Aveiro.

Figari, G. (1994). *Évaluer: quel référentiel*. Bruxelles: De Boeck-Wesmael s.a.

Goodyear, P., Salmon, G., Spector, J. M., Steeples, C., and Tickner, S. (2001). Competences for online teaching: A special report. *Educational Technology Research and Development, 49*, 65–72.

Guri-Rosenblit, S. (2005). 'Distance education' and 'e-learning': Not the same thing. *Higher Education, 49*, 467–493.

Harvey, L. (2006). Impact of quality assurance: Overview of a discussion between representatives of external quality assurance agencies. *Quality in Higher Education, 12*(3), 287–290.

Inglis, A. (2005). Quality improvement, quality assurance, and benchmarking: Comparing two frameworks for managing quality processes in open and distance learning. *International Review of Research in Open and Distance Learning, 6*(1), 1–13. Retrieved from www.irrodl.org/index.php/irrodl/article/view/221/304.

Jara, M., and Mellar, H. (2010). Quality enhancement for e-learning courses: The role of student feedback. *Computers and Education, 54*(3), 709–714.

Kim, M. C., and Hannafin, M. J. (2011). Scaffolding problem solving in technology-enhanced learning environments (TELEs): Bridging research and theory with practice. *Computers and Education, 56*(2), 403–417.

Kirkwood, A., and Price, L. (2005). Learners and learning in the twenty-first century: What do we know about students' attitudes towards and experiences of information and communication technologies that will help us design courses? *Studies in Higher Education, 30*(3), 257–274.

Lea, L., Clayton, M., Draude, B., Manager, A. S., and Barlow, S. (2001). Revisiting the impact of technology on teaching and learning at Middle Tennessee State University: A comparative case study. Paper presented at the TN Higher Education IT Symposium 2001.

Leacock, T. L., and Nesbit, J. C. (2007). A framework for evaluating the quality of multimedia learning resources. *Journal of Educational Technology and Society, 10*(2).

Liaw, S.-S., Huang, H.-M., and Chen, G.-D. (2007). Surveying instructor and learner attitudes toward e-Learning. *Computers and Education, 49*(4), 1066–1080.

Lucia, A. D., and Lepsinger, R. (1999). *The Art and Science of Competency Models: Pinpointing Critical Success Factors in Organizations.* San Francisco: Jossey-Bass.

Mahdizadeh, H., Biemans, H., and Mulder, M. (2008). Determining factors of the use of e-learning environments by university teachers. *Computers and Education, 51*, 142–154.

McGorry, S. (2003). Measuring quality in online programs. *The Internet and Higher Education, 6*, 159–177.

McGreal, R., and Elliott, M. (2004). Technologies of online learning (e-learning). In T. Anderson and F. Elloumi (Eds.), *Theory and Practice of Online Learning* (pp. 115–135). Athabasca, Canada: Athabasca University.

Mellar, H., and Jara, M. (2009). Quality assurance, enhancement and e-learning. In T. Mayes, D. Morrison, H. Mellar, P. Bullen and M. Oliver (Eds.), *Transforming Higher Education Through Technology-Enhanced Learning* (pp. 19–31). Heslington: The Higher Education Academy.

Moore, J. L., Dickson-Deane, C., and Galyen, K. (2010). E-learning, online learning, and distance learning environments: Are they the same? *Internet and Higher Education, 2*(2), 1–7.

Morgan, D. L. (1997). *Planning Focus Group*. Thousand Oaks, CA: Sage.

Nesbit, J., Leacock, T., Xin, C., and Richards, G. (2004). Learning Object Evaluation and Convergent Participation: Tools for Professional Development. *Proceedings from the Seventh IASTED International Conference*, Hawaii.

Oliver, M. (2006). Editorial: New pedagogies for e-learning? *ALT-J, 14*(2), 133–134.

Palmer, S., and Holt, D. (2010). Students' perceptions of the value of the elements of an online learning environment: Looking back in moving forward. *Interactive Learning Environments, 18*, 135–151.

Pawlowski, J. M. (2007). The quality adaptation model: Adaptation and adoption of the quality standard ISO/IEC 19796-1 for learning, education, and training. *Educational Technology and Society, 10*, 3–16.

Prosser, M., Ramsden, P., Trigwell, K., and Martin, E. (2003). Dissonance in experience of teaching and its relation to the quality of student learning. *Studies in Higher Education, 28*(1), 37–48.

Reichert, S., and Tauch, C. (2005). *Trends IV: European Universities Implementing Bologna.* Retrieved from http://www.eua.be/eua/jsp/en/upload/TrendsIV_FINAL.1117012084971.pdf.

Reilly, T. O. (2007). What is web 2.0: Design patterns and business models for the next generation of software. *Communications and Strategies, 1*, 17–37.

Reis, P., and Alves, M. P. (2009). Observação de aulas em contexto de ADD: Um projecto de Escola à procura dos referentes. Paper presented at the X Congresso Internacional Galego-Português de Psicopedagogia, Braga, Universidade do minho.

Riley, D. (2007). Educational technology and practice: Types and timescales of change. *Educational Technology and Society, 10*, 85–93.

Sadler, D. R. (1987). Specifying and promulgating achievement standards. *Oxford Review of Education, 13*(2), 191–209.

Scheerens, J., Glas, C., and Thomas, S. M. (2003). *Educational Evaluation, Assessment, and Monitoring.* Lisse: Swets and Zeitlinger B.V.

Shee, D., and Wang, Y. (2008). Multi-criteria evaluation of the web-based e-learning system: A methodology based on learner satisfaction and its applications. *Computers and Education, 50*, 894–905.

Siemens, G. (2006). *Knowing Knowledge*. Lulu.com, ISBN: 978-1-4303-0230-8.

Stensaker, B., Maassen, P., Borgan, M., Oftebro, M., and Karseth, B. (2007). Use, updating and integration of ICT in higher education: Linking purpose, people and pedagogy. *Higher Education, 54*, 417–433.

Stewart, I., Hong, E., and Strudler, N. (2004). Development and validation of an instrument for student evaluation of the quality of web-based instruction. *American Journal of Distance Education, 18*(3), 131–150.

Sun, P., Tsai, R., Finger, G., Chen, Y., and Yeh, D. (2008). What drives a successful e-Learning? An empirical investigation of the critical factors influencing learner satisfaction. *Computers and Education, 50*, 1183–1202.

Tigelaar, D. E. H., Dolmans, D. H. J. M., Wolfhagen, I. H. A. P., and van Der Vleuten, C. P. M. (2004). The development and validation of a framework for teaching competencies in higher education. *Higher Education, 48*, 253–268.

Wang, F., and Hannafin, M. (2005). Design-based research and technology-enhanced learning environments. *Educational Technology Research and Development, 53*(4), 5–23.

62 Design of an E-Learning System Based on the Socialization, Externalization, Combination, and Internalization Model

Dong Yun Lee and Jussi Kantola

CONTENTS

INTRODUCTION

The Internet has changed the way people learn in their daily lives. People use the Internet to search for valuable information for their learning or to collect materials for their research. Organizations such as companies and schools are actively utilizing the technological advance by introducing electronic learning (e-learning). As Hiltz and Wellman (1997) said, e-learning has several advantages, such as time and location flexibility, cost and time savings, self-paced and just-for-me learning, collaborative learning environment, and unlimited access and use of learning materials. In addition, e-learning offers improved methods of instructional delivery with the assistance of information and communication technologies and multimedia technologies, which enables high quality of learning.

With these benefits, the majority of current e-learning systems focus on how to manage learning content and learners; how to deliver the materials effectively; or how to design personalized learning environments according to learners' learning style, interest, and background. However, the important issue related to the fundamental goal of learning should not be neglected. That is, knowledge has the best value and the most powerful asset in the knowledge society as the created knowledge can have great influence on the survival and competitiveness of organizations (Nonaka 1994). In this regard, the essence of an e-learning system, which is the top priority of the system, is knowledge creation and management. An e-learning system should be expected to support knowledge creation of organizations, which goes beyond just delivery and understanding of learning contents.

To design an e-learning system that supports knowledge creation of organizations, this paper adopted the knowledge management (KM) approach. The KM field describes an effective way of sharing individual knowledge with other people in the level of an organization and the mechanism of creation of organizational knowledge. In detail, (1) the mechanism of knowledge creation is adapted from the Socialization, Externalization, Combination, and Internalization (SECI) model created by Nonaka (1994), which serves as the conceptual model of this paper. (2) The learning process model is built from cases of 3M Company that were successful in knowledge creation and innovation. (3) Based on the related works, an e-learning system model is presented, and it will be illustrated by the empirical data from the interview with the staff of 3M Company.

The primary objective of this paper is to present an e-learning system model that supports the learning process for innovation from the perspective of KM. To build the learning process model, a case study about 3M Company's innovation was conducted. Then, an e-learning system model was proposed along with the learning process model based on the related works. Finally, scenario analysis through the interview with the staff of 3M Company was conducted to show now the e-learning system model can strengthen the knowledge activities of the learning process. This paper aims at answering the following questions to meet the objective:

- Which process in learning does an individual or team typically go through to achieve an innovation?
- How effectively are e-learning methods including KM techniques designed and applied to enhance the learning process?

RELATED WORKS

In the literature review of KM by Liao (2003), there are seven categories in KM technologies: KM framework, knowledge-based systems, data mining, information and communication technology, artificial intelligence/expert systems, database technology, and modeling. Among them, the KM framework is the structure of all an individual's and organization's activities or performances related to knowledge. In general, the KM framework describes the KM lifecycle, considering such activities as knowledge creation, storing, measuring, transferring, and reusing.

Nonaka (1994) suggested a dynamic theory of organizational knowledge creation through continuous interaction between tacit knowledge and explicit knowledge. Szulanski (1996) analyzed the process model of knowledge transfer and developed the measure of stickiness of transferring at each stage. Alavi and Leidner (1999) studied how to make organizations act effectively with the help of KM and analyzed the necessitated nature of KM systems through three perspectives: information based, technology based, and culture based. Wiig et al. (1997) suggested that effective KM is influenced by the selection of methods, techniques, and tools for four sequential activities, such as reviewing, conceptualizing, reflecting, and acting. Wilkins et al. (1997) defined the knowledge assets and proposed a method of measuring them.

With the development of KM, Nonaka's theory, which is called SECI model (Figure 62.1), has been proven in that organizations that have successfully applied his research have proliferated (Ichijo and Nonaka 2007; cf. Nonaka et al. 2008). Especially, his view on the tacit–explicit dichotomy in knowledge categorization led to a universal knowledge life cycle. His dynamic spiral model of knowledge conversion portrays the process through which personal knowledge becomes more widespread in an organization and evokes new knowledge creation through interactions with other people.

According to the theory of organizational knowledge creation introduced by Nonaka and Takeuchi (1995), knowledge is created in a spiral process where tacit and explicit knowledge interact. This SECI process is based on four different modes of knowledge conversion: socialization, externalization, combination, and internalization. Socialization (from tacit knowledge to tacit knowledge) is a process of sharing experiences; externalization (from tacit knowledge to explicit knowledge) is a process where tacit knowledge is articulated to explicit concepts; combination (from explicit

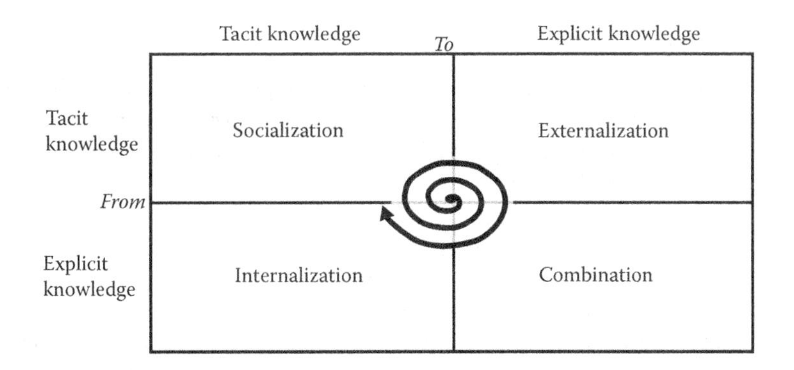

FIGURE 62.1 SECI model. (From Nonaka, I., *Organization Science* 5 (1), pp. 14–37, 1994.)

knowledge to explicit knowledge) is a process where concepts are systematized into a knowledge system; internalization is a process where explicit knowledge is embodied in tacit knowledge and is closely related to learning by doing (Nonaka and Takeuchi 1995). Organizational knowledge creation starts at the individual level and then moves up through communities of interaction crossing sectional, departmental, divisional, and organizational boundaries (Nonaka and Takeuchi 1995). This spiral process of organizational knowledge creation presents a systemic view on how organizations create new knowledge.

METHODOLOGY

As a case study, 3M (Minnesota Mining and Manufacturing) Company was examined following the principles of qualitative research (Creswell 2009). Data on how the organization performed successful knowledge creation in new product development or new technologies were collected.

There are some reasons why 3M Company was selected for case study. Firstly, 3M Company is one of the very few organizations that are successful in innovation, although many organizations try to find necessary factors for innovation and achieve it. 3M Company developed 50,000 products in 7 main businesses, like the office business, medical business, display and graphic business, and so on. In addition, it is still displaying more than 500 new products every year. Secondly, other companies in various fields are trying to go deep into the study of the innovation of 3M Company and make progress by benchmarking it. Finally, 3M Company is acknowledged by many researchers in organizational learning, business consultants, and authorities. For example, Collins and Porras (1994), the authors of *Built to Last: Successful Habits of Visionary Companies*, said that they would choose 3M Company among the companies introduced in their books if they had to entrust only one company that has continuous adaptability with their life in the next 50 or 100 years. It means that many experts believe in the sustainable development of 3M Company, and even competitors voice strong support for its innovation.

For this study, a document analysis of the literature by Gundling (2000), *The 3M Way to Innovation: Balancing People and Profit*, was performed; 3M Company's successful development of two new products was investigated. From the two cases, we proposed a learning process model of innovation. Then, we established an e-learning system model based on related works. For scenario analysis of the model, a semi-constructed interview via e-mail is on going with the staff in 3M Company. The interviewees have experienced various projects of new product/technology development. In the first half of the interview, we tried to identify how an individual or team created knowledge that was needed for its goal, such as new development/technology development. Then, we obtained their opinions about the behaviors in utilizing KM tools, such as groupware, knowledge portal, and so on. The collected data were analyzed for case study, which resulted in the learning process model and e-learning system model.

Case 1: Post-it Notes (Gundling 2000)

In 1970s, Post-it Notes, one of the five most popular office supply products in the United States, was developed by Art Fly and Spence Silver, who were scientists at 3M Company. In fact, the original concept of the product was bookmarking. Art Fly had trouble in marking his hymnbook whenever he had choral practice. Though he put a small piece of paper into the book as a temporary solution, the paper always fell, and he forgot the page that he should pay attention to. The scientist started to think about the solution for how he could easily bookmark.

In the course of creation of the bookmarking product, Art Fly had help from his coworker, Spence Silver, who had invented temporary adhesives. At that time, the temporary adhesives failed in commercialization. However, the two scientists believed in the success of their research about paper that is easily attached to a book and easily detached from it. Furthermore, they enlarged the concept of the product from just bookmarking paper to scratch paper. They handed out free samples to other members in 3M Company, and Commercial Office Supply Division finally permitted the commercialization of the product in 1980. Post-it Notes was given an award for outstanding new product in 3M Company. In this case, the procedure of developing Post-it Notes is described in the following steps.

1. Recognition of an inconvenience
2. Setting a goal of developing a new product and understanding available resources
3. Research on a temporary adhesive and its application as a product
4. Development of a bookmarking product and its negative evaluation
5. Resetting a goal of developing a memo pad
6. Experiments on a memo pad with an adhesive property
7. Development of Post-it Notes and its positive evaluation and commercialization

In the first step, the innovator found some remarkable stimuli during choral practice. The innovator typically experienced all kinds of stimuli that came from inconvenience, and they were occurring everywhere in daily life beyond the boundary of the innovator's workplace. Based on the interest and values of the innovator, however, some remarkable stimuli were filtered among all stimuli. That is, they are the product of the innovator's intentional choice of consciousness from numerous activities. In this case, dissatisfaction in bookmarking was the remarkable stimulus and regarded as the main problem to be solved. In the second step, the innovator set the goal to be the invention of a bookmarking product. At the same time, he grasped the constraint that limited his usage in resources of 3M Company. In this case, the innovator remembered that he had attended the presentation about a temporary adhesive by his coworker. In the third step, the innovator easily acquired and utilized the knowledge related to the adhesive property through cooperation with the coworker. In the fourth step, the two innovators developed a bookmarking product and faced difficulties in persuading the Commercial Office Supply Division of 3M Company. After the evaluation of the first invention, they decided to magnify the application from a bookmarking product to a memo pad. As a result, they went back to the second step to modify the first goal and finally developed Post-it Notes.

Case 2: Hydrofluoroethers Development Program (Gundling 2000)

In the 1990s, the US government banned the use of chlorofluorocarbon (CFC), which was employed as a refrigerant, a blowing agent, and a solvent, as people started to understand how the material could damage the ozone layer. The environmental regulation led 3M Company, which had sold perfluorinated chemical products in air conditioning and manufacturing processes, to develop a new replacement, which is called the hydrofluoroethers (HFE) development program. For the project, 3M Company put together a team that was comprised of experts in areas like chemistry, engineering, patents, environmental policy, and project management. In the early part, the team interviewed with government officers and clients to seize the environmental index and expected functions of a future replacement. Then, the team created HFE that is less toxic and subsists in the atmosphere for a brief space of time, after it analyzed hundreds of compounds to create a new molecular structure through various experiments measuring chemical characteristics. 3M Company also studied the program for purifying and recycling HFE and the application of the substitute in domains such as aerospace engineering, computers, and electronics, based on its market analysis. Finally, HFE was presented in a technology exhibition in Anaheim, California, in 1996 and got approval from the Environmental Protection Agency.

In this case, the following steps can illustrate the procedure of creating a new material called HFE, technologies of purifying or recycling the material and product as a new refrigerant.

1. Prohibition of using CFC by the US government
2. Setting the goal of creation of a new material and gathering experts in various fields
3. Research on the expected properties of the new material
4. Development of HFE
5. Renewing the goal of developing a purification and recycling program of HFE
6. Simulation of a toxicology prediction model
7. Generation of research on the program
8. Renewing the goal of researching the application of HFE for products
9. Development of products, like a refrigerant, and their positive evaluation

RESULTS

Proposed Learning Process Model

The analysis of the cases in the previous section indicates that the four steps are a series of learning processes and that innovation is the outcome of the learning process. It means that learning includes *problem recognition*, *goal creation and constraints understanding*, *knowledge activities from acquisition to utilization*, *result creation*, and *evaluation*, as it entails a newly changed inner space, knowledge, and behaviors.

Besides the procedure of innovation, several factors for innovation are derived from the above case study. The first one is curiosity, which is a driving power of creativity. Curiosity made the innovator continuously ask some questions to himself, such as "Why do I feel uncomfortable in a specific situation?" or "How can I solve the dissatisfaction?" These kinds of questions increase the desire for learning and knowledge, getting an innovator to look for necessary books or to go and see people who have expertise. The second factor is inspiration, which enables an innovator to stumble over great ideas. Creative ideas are the starting point of innovation and essential for invention. Until the development of new products or technologies, the 3M Company members in the cases made mistakes while conceptualizing for new products or doing experiments. They even recycled prior ideas and technologies that had been discarded because of low commercial values. However, inspiration from the lesson of failure was helpful in achieving innovation. The last factor is spontaneous and intentional immersion of innovators. By thinking about ideas and solutions repeatedly, the above innovators experienced immersion, which elevated their concentration, tenacity, and problem solving. Immersion is promoted by 3M Company policy, which does not care for mistakes or failure a bit. Through the case studies, a learning process model is generated (Figure 62.2).

In this proposed learning process model, an innovator or team recognizes a problem in the first step. There are a lot of stimuli around an innovator beyond the boundary of his or her workplace. Stimuli have a variety of forms such as dissatisfaction, inconvenience, failed products or technologies, and so forth. Having curiosity about phenomena that people usually experience unconsciously,

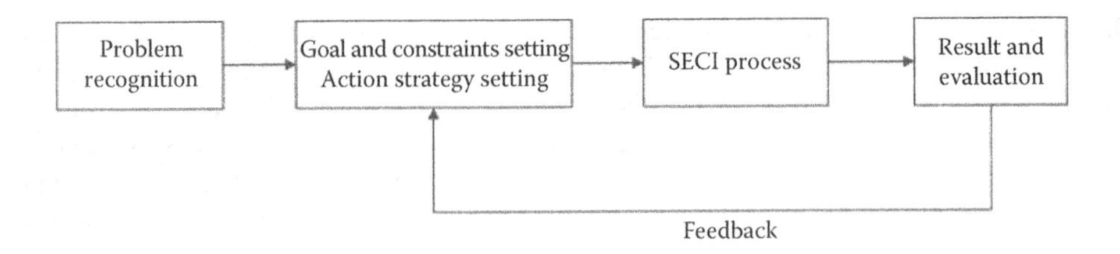

FIGURE 62.2 Proposed learning process model.

an innovator constantly observes all kinds of stimuli and finds what makes him or her interested. During the intentional observation, some remarkable stimuli can be filtered as mysteries to be solved. The mystery, or main problem, can be developed into a personal project and be the subject that an innovator is immersed in.

In the second step, an innovator or team sets a goal, constraints, and action strategy. This does not mean that the three elements should be always specified at the same time. An innovator longs for change by solving the mystery. Developing a new product or technology and implementing a new idea can be examples of a goal. They are the ultimate outcome, which should be innovative. In 3M Company, innovation generally includes the outcome of creativity and practicality making great profits. Constraints indicate conditions that limit an innovator or team in achieving a goal. For example, an innovator understands his or her prior knowledge related to the main problem and how much an intellectual or financial resource can be utilized in an organization. Furthermore, an innovator can form a team whose members have different expertise from one another. Action strategy is connected with decisions on how an innovator or team would obtain knowledge that is necessary to solve the main problem. It also covers an issue about how to exploit available resources.

In the third step, an innovator or team is close to the goal through SECI process that is mentioned in the related works. In this proposed learning process model, an innovator or team creates new knowledge in "Ba." For example, an innovator can catch customers' needs by observing them in the field as the innovator's goal is to develop a new product. The innovator can have conversation about his or her ideas with other coworkers, like a manager, via groupware. The innovator can search a database system or knowledge portal to get advice from experts in a certain field. The innovator also reads some material from knowledge assets in his or her company. The innovator can carry out an experiment to test the performance of product that he or she developed.

In the fourth step, the result is generated and evaluated with regard to the perspective of marketability and profitability. If the result does not have a good rating, an innovator or team should go back to the second step and reset the goal, constraints, and action strategy. In this process, feedback with single-loop learning or double-loop learning is provided. The loop will be iterative until the result is approved as innovation.

The proposed model has some differences from the existing popular learning models of an individual and organization (Figure 62.3)—single-loop learning (Are we doing things right?) and double-loop learning (Are we doing right things?)—created by Argyris (1977). First of all, while learning in the existing model is just acquiring new knowledge and behavioral changes or the changes of fundamental theory or goals, the proposed model includes innovation as the outcome of learning. In this regard, there is an assumption that changes in behavior and fundamental goals do not guarantee innovation. The proposed model emphasizes that the goal, constraints, and action strategies can be renewed until the best result, innovation, is produced. Innovation is ultimately achieved when feedback does not happen anymore. Secondly, the starting point of learning is problem recognition, unlike Argyris' model, which starts from goal setting. Problems that are ahead of goal setting can be motivation of innovation. It has various forms of stimuli such as interest, dissatisfaction, inconvenience, and so on in everyday life. Innovators should intentionally observe and recognize core problems among the stimuli. Moreover, elements like curiosity, inspiration, and immersion were essential to innovation in all the steps. Thirdly, goal setting and action strategy are

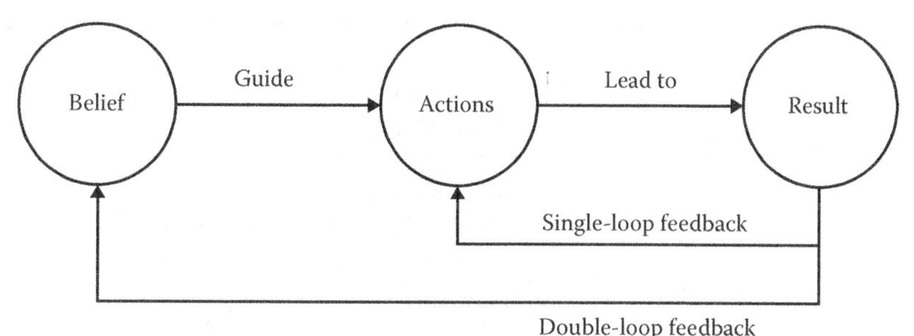

FIGURE 62.3 Single-loop learning and double-loop learning. (From Argyris, C., *Harvard Business Review* 55 (September–October), pp. 115–125, 1977.)

in the same stage, while Argyris set them in the linear relationship. This does not mean that a goal should precede an action strategy. In the absence of a goal, an action strategy can impact on goal establishment or vice versa. Therefore, feedback includes single loop and double loop simultaneously. As shown in Figure 62.2, however, both of them should be specified before the SECI process, the third step. Without a concrete goal and action strategy, tacit knowledge and explicit knowledge gained through the SECI process would be vague and scattered.

Proposed E-Learning System Model

For innovation, each step on the proposed learning process model should be performed effectively. Through the literature review on KM, we suggest that some required functions can support the steps (see Table 62.1).

Based on the functions, the proposed e-learning system model with modules is shown in Figure 62.4. In problem recognition, an innovator can have an interface to see and experience many stimuli, like others' ideas or information on a new product or technology. Further, an innovator can search some data, information, and knowledge. An intelligent agent sometimes provides personalized stimuli according to the interest or background of an innovator. In the second step, an e-learning system should help an innovator to set a goal and to begin his or her project. The innovator can be given the information on the available resources, such as financial assistance or an expert who has domain-specific knowledge. KM tools are useful in scheduling and deciding roles of departments in cooperation as well as manage the process and analyze workflow. In the third step, an e-learning system has to support knowledge creation ability to achieve a goal. With collaboration tools and

TABLE 62.1
Learning Process and Required KM Functions

Learning Process	Required KM Functions
Problem recognition	Knowledge portal/knowledge analysis/search engine
Goal setting	Project management/search engine
Constraint setting	Project management/search engine
Action strategy	Project management/workflow analysis
Socialization	Collaboration/knowledge capture/knowledge storage
Externalization	Collaboration/knowledge representation
Combination	Collaboration/knowledge capture/knowledge analysis/ knowledge sharing and transferring/search engine
Internalization	Knowledge utilization/search engine
Result and evaluation	Evaluation/project management

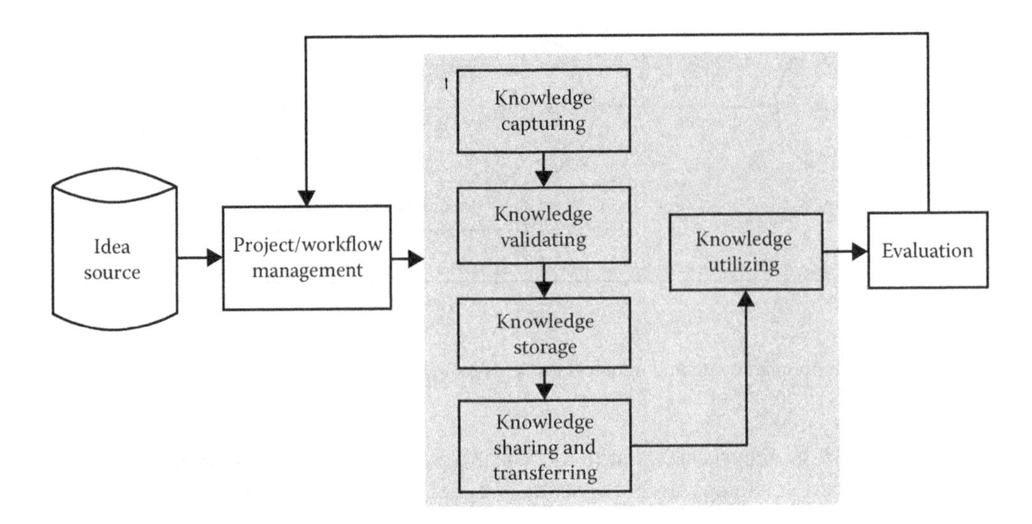

FIGURE 62.4 Proposed e-learning system model.

a knowledge repository, an innovator acquires the needed knowledge, expresses it as a concept explicitly, shares it with his or her team members, and practices the knowledge. This third step of the model is in an analogue to the SECI process (Nonaka and Takeuchi 1995); it is based on the SECI process. In Figure 62.4, this step is represented on a shaded background. In the fourth step, an e-learning system can provide an analysis tool for the evaluation of generated results. Unless the evaluation is positive, the system informs an innovator or team of lessons learned or gives some recommendations. Then, a project management tool enables the project to be renewed.

Most organizations have introduced up-to-date e-learning technologies with an indefinite expectation that they would be useful in overcoming time and place restrictions. However, a technology push strategy can be ineffective as technologies cannot be balanced with an organization's business model or the existing system. In this section, however, the proposed e-learning system framework provides an outline of how e-learning technologies can be properly deployed by matching the technologies to the learning process model. In other words, the proposed e-learning system model can be implemented as a consortium formed by e-learning technologies, most of which are from KM tools, to enhance knowledge-related activities in the learning process model. As a result, organizations will be able not only to adopt suitable e-learning technologies but also to establish an e-learning system as a general or specific solution for their innovation.

CONCLUSION

In this paper, we have proposed a learning process model of an innovation and an e-learning system model from the perspective of KM. In detail, we have described a specific learning procedure for an innovation through the case studies of 3M Company. We also built an e-learning system model that fits the proposed learning process based on the literature of review of KM tools by considering the relationship between knowledge activities and KM functions that supports the activities.

For the further study, we are currently conducting scenario analysis after the interview with the staff in 3M Company to illustrate how the proposed e-learning system model can practically work in business.

Practical implications of this research may be organizational e-learning tools/solutions for organizations and companies where people at work can officially participate in innovation and new knowledge creation. Potentially, these kinds of proposed organizational e-learning tools may increase the feeling of participation among people at work and may support faster innovation processes in organizations and companies. But of course, more research is needed to validate these potential implications. We intend to follow these exiting lines of investigation in the future and continue our research and application in different kinds of practical settings.

REFERENCES

Alavi, M. and Leidner, D.E., "Knowledge management systems: issues, challenges, and benefits," *Communications of the AIS* 1 (7), pp. 1–37, 1999.

Argyris, C., "Double loop learning in organizations," *Harvard Business Review* 55 (September–October), pp. 115–125, 1977.

Collins, J.C. and Porras, J.L., *Built to Last: Successful Habits of Visionary Companies*. Harper-Business, New York, 1994.

Creswell, J.W., *Research Design: Qualitative, Quantitative, and Mixed Methods Approaches*. SAGE, Thousand Oaks, California, 2009.

Gundling, E., *The 3M Way to Innovation: Balancing People and Profit*. Kodansha International, New York, 2000.

Hiltz, S.R. and Wellman, B., "Asynchronous learning networks as a virtual classroom," *Communications of the ACM* 40 (9), pp. 44–49, 1997.

Ichijo, K. and Nonaka, I., *Knowledge Creation and Management: New Challenges for Managers*. Oxford University Press, New York, p. 323, 2007. ISBN 0195159624.

Liao, S., "Knowledge management technologies and applications: Literature review from 1995 to 2002," *Expert Systems with Applications* 25 (2), pp. 155–164, 2003.

Nonaka, I., "A dynamic theory of organizational knowledge creation," *Organization Science* 5 (1), pp. 14–37, 1994.

Nonaka, I. and Takeuchi, H., *The Knowledge-Creating Company: How Japanese Companies Create the Dynamics of Innovation*. Oxford University Press, New York, 1995.

Nonaka, I., Toyama, R. and Hirata, T., *Managing Flow: A Process Theory of the Knowledge-Based Firm*. Palgrave Macmillan, New York, 2008.

Nonaka, I., Toyama, R. and Konno, N., "SECI, Ba and leadership: A unified model of dynamic knowledge creation," *Long Range Planning* 33 (1), pp. 5–34, 2000.

Szulanski, G., "The process of knowledge transfer: A diachronic analysis of stickiness," *Organizational Behavior and Human Decision Processes* 82 (1), pp. 9–27, 1996.

Wiig, K.M., Hoog, R. and Spex, R., "Supporting knowledge management: A selection of methods and techniques," *Expert Systems with Applications* 13 (1), pp. 15–27, 1997.

Wilkins, J., Wegen, B. and Hoog, R., "Understanding and valuing knowledge assets: Overview and method," *Expert Systems with Applications* 13 (1), pp. 55–72, 1997.

63 Non-Online and Online Business Instruction

Comparative Outcomes

Thaddeus W. Usowicz

CONTENTS

INTRODUCTION

Budget austerity and the associated teaching load increases over the past half-decade have critically affected the local state-funded college and university system. At the College of Business, the number of enrolled declared students has increased by 16%, while the average student-to-faculty ratio has doubled from about 20 to 40. The increased load has led to larger section sizes. In order to maintain manageable sizes for the core and concentration courses, significant increases in size have been made to prerequisite course sections, with the resulting deterioration of traditional face-to-face interaction between the students and the instructor. Fortunately, over the same period, the text publisher for this course had been developing and enhancing a text-oriented Web site for students to learn online through solving generated problems. The site eventually provided feedback showing correct solutions upon request, but did not provide instruction in conceptual principles. The course thus evolved into a *hybrid* or *blended* learning course. The former term is falling out of use, and the meaning of the latter term is controversial as being imprecise. Blended learning is used here to indicate that a mix of media is employed. The blended combination of media for teaching classroom environments considered here is traditional face-to-face lecturing and instruction, text, and the interactive web-based online text-oriented site provided by the publisher.

The degree of impact that interaction in problem solving with an instructor has on the outcome performance of students is of special interest. Is the instructor critical for effective teaching of the material within the context of a blended learning course? The face-to-face interaction is useful for motivating students who normally might not be interested in the material. Providing examples of solving realistic problems appears to be useful but is time consuming. Many students procrastinate, and some demonstrate weakness in learning the material through practice assignments prior to tests. Periodic testing also enhances motivation. In the course for introductory business mathematics studied here, the final examination accounts for 33% to 41% of the total points gained in test and examination outcomes. Short tests ("quizzes"), longer midterm exams, the final examination, and a short project are the primary determinants of the student's final grade. Graded assignments provide practice problem-solving experience for the students.

The topics are prerequisite course knowledge for the core courses leading to a degree in business. The core courses must result in a minimum satisfactory grade; prerequisite courses need to receive credit and do not have the same urgency for a letter grade for the student; obtaining credit that does not impact on the student's grade point average is sufficient.

The topics covered are quadratic, exponential, and log functions as a review of algebra; the mathematics of finance; and an introduction to differential calculus. Advanced placement courses at the secondary school level can cover much of this material, except for the mathematics of finance, prior to entrance into the university. Students may also challenge their level of required knowledge for this course by taking a preliminary examination.

An unfortunate interesting side issue that continues from traditional class environments is that controls are needed to ensure that the student receiving credit for acquiring the knowledge is the same person taking the tests. Specifically, tests must be monitored and the student's identification card produced. There has been a history of very few incidents when students attempted to have substitutes take the tests, especially the final examination. A lab with online workstations was not available for courses, outside of those focusing on information technology application skills; however, several monitored online test-taking labs, for example, for taking tests posted on the university iLearn system, have been made available on campus for courses' online exams. Regardless, in various courses, a few students have attempted to take the online tests from an unmonitored location, despite being warned that grades would not be accepted if the student's location could not be verified. For the samples we consider here, Web site–generated tests are printed, thereby introducing extra steps, which did not take advantage of the online environment fully. In traditional teaching environments, the class section size was often small enough for the instructor to know each student by name, making control easier. The issue of control and university credit excludes consideration of pure distance self-regulated e-learning courses.

REVIEW

The Web-based text companion site was a Moodle iLearn site from the text publisher. As the functionality of the site improves, answers could be entered and checked using a formula editor, or they could be drawn graphically. A multiple-choice question may also be generated. The site provides feedback of correct solutions upon request.

Blended learning involves the combination of face-to-face and Internet-provided instruction. Other diverse media may be included (Osguthorpe and Graham 2003; Woods et al. 2004; Barnard et al. 2009; Bonk and Graham 2006). Among the many perspectives covered by the articles in the book by Bonk and Graham (2006), one by Graham and Moore points out that blended learning is playing a significant role in tough economic times. Moore, however, identifies the face-to-face teacher as an expensive communications technology. This perspective neglects to recognize that the face-to-face instructor is adaptive. The instructor has a brain that is a true learning machine, wired very differently than a programmed computer that may win chess games or a quiz like Jeopardy. The instructor can, therefore, adapt lectures, give feedback on problem solving, and help in response to changing requirements and student needs. The character of this course is to develop students' conceptual knowledge and analytical mathematical skills. It would be expected that students will learn to apply diverse cognitive processes for these learning objectives. An instructor would recognize the learning process for supporting different individuals. Paechter and Maier (2010) show that students prefer face-to-face learning when conceptual knowledge is the subject matter or when skills in the application of one's knowledge are to be acquired.

Problems faced by faculty in using a Web-based learning site for instruction, other than for Information Systems computer tool skills courses, are similar to those identified by Ocak (2011). The problems most often encountered are specifically related to the availability of instructional laboratories with computer workstation. Students at the prerequisite level do not have universal access to personal laptops to take advantage of wireless networking in the classroom. The advantages of using online learning sites to support the instructor are similar to those observed for older students by Ke (2010). The advantages at the university level are offset by student behavior and habits such as procrastination (Michinov et al. 2011; Odaci 2011). Hershkovitz and Nachmias (2011) also showed that over half the students tended to stop, slow down, or delay online interaction until the end of the semester when using a course Web site.

METHODOLOGY

Sources for Data

The data available to the author were based on three different classroom environments over a four-year period during which the Web-based site appeared and evolved while maintaining a focus on the text. The text did change editions to accommodate integration with the Web site, but the chapter-by-chapter content used for the course was the same, although it may have been reorganized slightly. The three environments are identified below as (1) traditional face-to-face, (2) transitional blended, and (3) Web-based blended. The transitional environment was also Web-based but was not relied upon as heavily by students as in the last example. The first two environments used more time and effort on feedback in the classroom.

All classes meeting with the instructor were in regular classrooms with no computer terminals universally available to the students. Online assignments could be done in generally available computer labs. Quizzes addressed the most current topical material, while midterm and final examinations were cumulative. A short project associated with the mathematics of finance, which focused on discount rates, was also assigned in each class to develop decision-making problem-solving skills beyond the text-based assignments. The instructor assigned the project grade prior to the final examination; the analysis does not include these data in the strictly quantifiable graded assignment data. Teaching assistants were no longer available to support the instructors, but several graduate students staffed a part-time math and statistics lab to help students do assignments.

How assignments and tests were compiled and reviewed varied by the classroom environment sample and is described below.

The traditional face-to-face sample had a small class size: starting with 15 and ending with 11 students registered at the end of the semester. A large proportion also signed up for credit/no credit grading, which would have affected the level of effort they would give for a top scoring outcome. This was an evening section, yet attendance was high at about 80% to 90%. Graded assignments on problems selected by the instructor from the ends of chapters were augmented by recommended non-graded practice problems. Solutions for assignments were covered in class upon a student's request for a specific problem; students often asked that solutions to problems be covered. For tests (quizzes, midterm examination, final examination) the instructor often created an evolving scheme for answering the problems in steps. The answers to the steps were scored consistently. The intent of the scheme was to have the student recognize how skills learned earlier were used in the construction of more complex solutions. The final examination provided 33% (60/180) of the outcome scores studied. Solutions for all test problems, except for the final exam, were covered in class with a randomly chosen student for each problem. This feedback procedure was used in all three class samples and provided the primary corrective mechanism for face-to-face instruction. The instructor's perception was that a lower proportion of students participated in the feedback sessions in the third sample.

The transitional blended sample had a large class size with 37 students by the end of the semester. The estimated attendance was lower at about 50%. Assignments were based on selections from the Web site for grading. They were supplemented by ungraded recommended selections from the end of the text's chapters. The graded assignments were done online outside of the classroom, and the Web site provided feedback and graphics capability. In the classroom, the instructor had network access available through a laptop to go over similar problems at the request of students. The instructor rarely employed network access since the problem solution could be done more dynamically and quickly with regular whiteboard work.

Test questions were generated from the online system solely. Questions were multiple-choice. The instructor modified some of the questions. Finally, the tests were printed on paper for control purposes. The final examination provided 41% (90/220) of the outcome scores studied.

The Web-based blended sample had two lecture-size classes sized at over 50 for each section by the end of the semester. Attendance was low on nontest meetings at about 20% to 30%, indicating that most of the students chose the Web-based site for learning. Attendance increased in the final weeks of the semester, but reliable data were not captured. The chapter-by-chapter topical practice depended on online problems with an assignment graded on completing a given percentage successfully. The site captured data on the actual number of problems answered successfully. The student chose the problems to solve with the recommendation that the student attempt a full spectrum of the problem types. All generated questions were either formulas or multiple-choice; however, animated solutions were available for certain problems, in addition to feedback and graphical solutions. The system generated all test questions. Multiple-choice questions were chosen and were copied to a printed-paper test sheet. The tests were answered and graded using Scantron sheets. Again, the instructor employed this method for control purposes. The instructor would also have some face-to-face interaction with the students before the test. The final examination also provided 41% (90/220) of the outcome scores studied. Test questions generated by this site are very similar to the assigned practice problems; therefore, they do not challenge stretching the student's problem solving skills beyond those directly covered.

Statistical Tools

Regression analysis captured the underlying analyses of variance (Neter and Wasserman 1974). Quantitative totals of quiz and examination outcomes were related as a function of graded assignment measures. The traditional assignment scores were for problems or portions of problems uniformly graded by the instructor. For the online Web-based generated assignments, this measure was the number of problems successfully answered over the semester.

Hypotheses

The null hypothesis is, effectively, that there is no impact on the relationship between graded practice assignments and outcome performance, that is, the regression slope is not significantly different from zero. In other words, the instructor can interact with the students to decouple the outcome from performance on practice problems. If this hypothesis is true, then the instructor's face-to-face interaction with the students is very important.

The alternative hypothesis is that the student attempts to derive sufficient skills from performance of graded practice assignments to obtain outcomes as tested. The goal, of course, is to obtain good outcomes, but this hypothesis effectively states that outcomes are related to graded Web-based assignments, that is, the effort and usage the student applies to the online site. This is not to imply that the instructor cannot impact on outcomes of individual students who seek out face-to-face interaction. It also makes no statement as to reasons why a student would not attempt to use the Web site for practice through the assignments.

RESULTS AND DISCUSSION

The statistical analysis results for test outcomes related to assignment scores are given in Table 63.1. Data distributions and the regression lines for the samples are plotted in Figures 63.1 through 63.3. The basis of the outcome measures, that is, how the tests and the assignments were constructed,

TABLE 63.1
ANOVA and Regression Analysis of Test Outcomes as a Function of Graded Scores for Assigned and Online Generated Problems

	Sample Model		
	Traditional Face to Face	**Transitional Blended**	**Web-Based Blended**
Multiple *R*:	0.031786	0.481925	0.708605
R Square:	0.001010	0.232252	0.502121
Adjusted *R* Square:	–0.109988	0.210316	0.497191
Standard Error:	19.491376	17.126392	14.571807
Observations:	11	37	103
ANOVA			
Regression			
df	1	1	1
SS	3.458213	3105.561753	21628.80718
Residual			
df	9	35	101
SS	3419.223	10265.96527	21446.09352
Total			
df	10	36	102
SS	3422.681	13371.527	43074.900
F	0.00910	10.58786	101.86048
Significance *F*	0.92608	0.00252	5.610E-17
Regression Line			
Intercept:	132.48	55.53	41.44
X Variable Coefficient (Slope):	–0.149	0.051	0.091

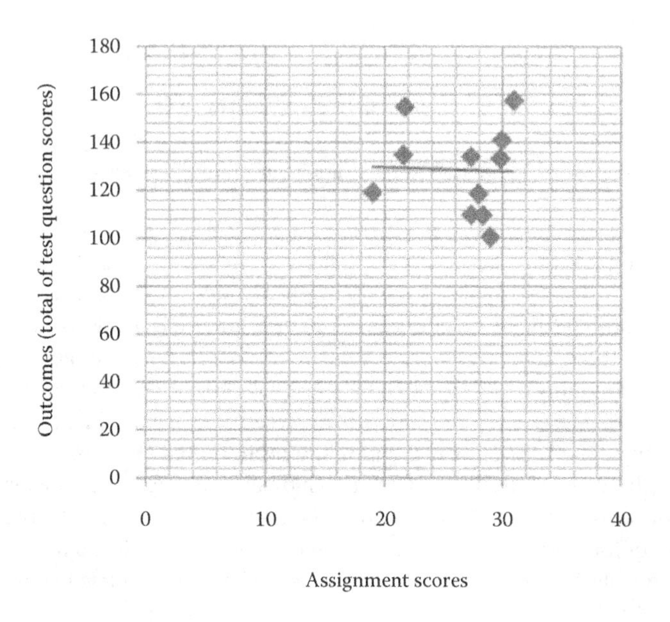

FIGURE 63.1 Outcomes as a function of graded assignment scores for the traditional face-to-face sample.

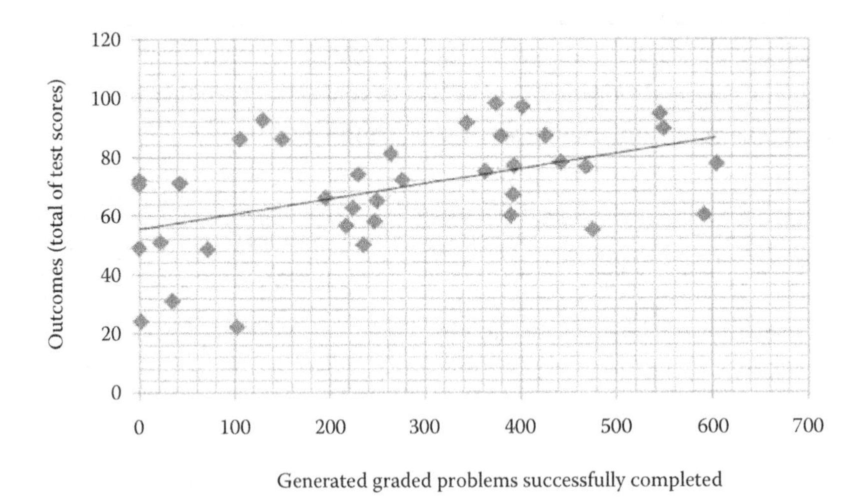

FIGURE 63.2 Outcomes as a function of graded assigned online problems for the transitional blended sample.

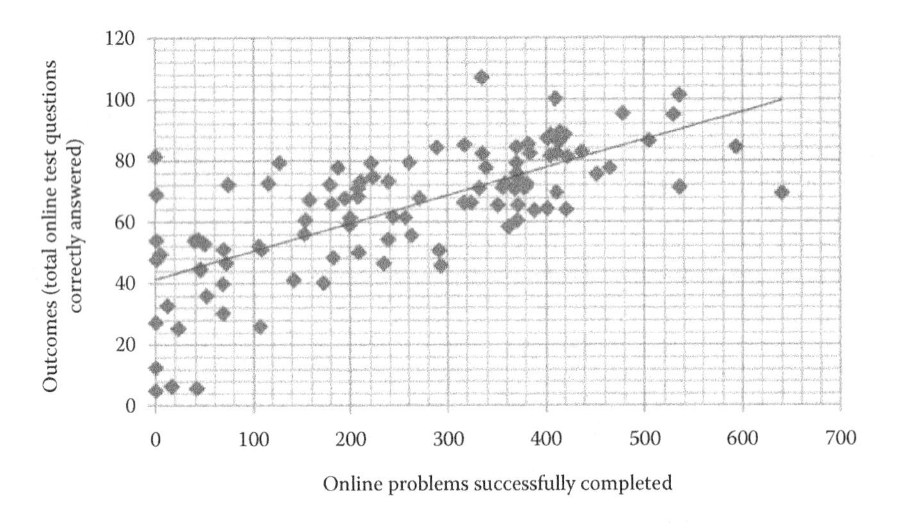

FIGURE 63.3 Outcomes as a function of graded assigned online problems for the Web-based blended sample.

varied among the samples; they would not be strictly comparable to each other. The final grades were curved since the instructor wanted to challenge the students to greater effort even though they would be aware of the likelihood of curving. The final grades were derived from the measured outcomes, so the quantified outcomes were better for comparison within the samples. The average percentages of the maximum possible outcome scores for the three samples were 71.4%, 62.9%, and 55.9%, respectively. The decline in scores, as the test questions became simpler, may indicate a decrease in the motivation of the students who relied more on the Web-based site. Alternatively, it may be a failure in the Web site design. It may also indicate a change in character of the students as they come from the secondary school system into the university; anecdotally, instructors who have been at the university for over two decades have voiced this concern. It is a matter requiring further study before any conclusions on student behavior in a mathematics course can be reached.

The null hypothesis holds for the traditional face-to-face interactive instructional environment, supporting the view that the instructor can significantly influence the students' performance and outcomes. The alternative hypotheses hold for the two blended samples ($P < .005$). The majority of the students chose to go the path of using the Web-based assignments as the primary instruction tool for themselves. Data tying individual student attendance and face-to-face interaction to outcomes had not been collected; the instructor's perception was that these students received the better outcomes. Another study could clarify these relationships; note that the original data were collected by actual observation of behavior and were not collected with this study design in mind.

CONCLUSION

Can the instructor have a significant impact on student outcomes in a mathematics course through face-to-face interaction rather than through text-oriented Web-based sites? The data indicate a positive response. The role of the instructor is very important, since the instructor can adapt to the diverse styles of learning by individual students. However, a more dynamic site design and online environment with mobile access, podcasts, games and simulations, and competitive collaborative systems may affect the future role of the instructor. Better feedback and help provided on the site would certainly improve its appeal; however, the face-to-face encounter for explaining concepts and providing adaptive feedback is crucial. It may take a couple of decades before mathematical site design catches up to these requirements for capturing and keeping students involved. Techniques addressing student attitudes, procrastination, motivation, and persistence to increase reliance on the Web-based site need to be developed. Many of them would be applied during the face-to-face sessions. In the meantime, the face-to-face interaction in higher education for teaching material that is required knowledge, but not of great appeal for the student, appears to be irreplaceable.

REFERENCES

Barnard, L., W. Y. Lab, Y. M. To, V. O. Paton and S. Lai (2009), Measuring self-regulation in online and blended learning environments. *Internet and Higher Education*, 12, 1–6.

Bonk, C. J. and C. R Graham (Eds.) (2006), *The Handbook of Blended Learning: Global Perspectives, Local Designs*. San Francisco, Wiley.

Hershkovitz, A. and R. Nachmias (2011), Online persistence in higher education web-supported courses. *Internet and Higher Education*, 14, 98–106.

Ke, F. (2010), Examining online teaching, cognitive, and social presence for adult students. *Computers and Education*, 55, 808–820. Available at www.elsevier.com/locate/compedu.

Michinov, N., S. Brunot, O. Le Bohee, J. Juhel and M. Delaval (2011), Procrastination, participation, and performance in online learning environments. *Computers and Education*, 56, 243–252. Available at www.elsevier.com/locate/compedu.

Neter, J. and W. Wasserman (1974), *Applied Linear Statistical Models: Regression, Analysis of Variance, and Experiment Designs*. Homewood, IL: Richard D. Irwin, Inc., 1974.

Ocak, M. A. (2011), Why are faculty members not teaching blended courses? Insights from faculty members. *Computers and Education*, 56, 689–699. Available at www.elsevier.com/locate/compedu.

Odaci, H. (2011), Academic self-efficacy and academic procrastination as predictors of problematic internet use in university students. *Computers and Education*, 57, 1109–1113. Available at www.elsevier.com/locate/compedu.

Osguthorpe, R. T. and C. R. Graham (2003), Blended learning environments: Definitions and directions. *The Quarterly Review of Distance Education*, 4(3), 227–233.

Paechter, M. and B. Maier (2010), Online or face-to-face? Students' experiences and preferences in e-learning. *Internet and Higher Education*, 13, 292–297.

Woods, R., J. D. Baker and D. Hopper (2004), Hybrid structures: Faculty use and perception of web-based courseware as a supplement to face-to-face instruction. *Internet and Higher Education*, 7, 281–297.

64 Do We Really Equip Our Students with Inquiry Skills?

Małgorzata Krzeczkowska and Iwona Maciejowska

CONTENTS

INTRODUCTION

Much discussion is being held on whether the objective of higher education is to prepare graduates for carrying out their professions or for living in the contemporary society, or rather, it is a pursuit of the Truth within the community of masters and students, and so forth (Baumann 2010). Regardless of the model selected, one of the desired outcomes of each level of education is to develop higher-order cognitive skills (HOCSs). Many publications have been devoted to definitions, the change in the paradigm (Zoller and Scholz 2004), teaching methods (Zoller 1993; Zoller and Pushkin 2007), assessment methods (Kulm 1990; Zoller 2001; Palmer and Devitt 2007), teacher training (Leou et al. 2006), and so forth.

A major purpose of contemporary science education is the development of students' reasoning and ability to think critically in the context of both the specific content of traditionally disciplines as well as the processes and interrelationships concerning societal, economical, scientific, technological, and culturally bound issues within real-world complex systems, locally and internationally. HOCSs are valued because they prepare students not only for advanced academic work but also for the challenges of adult work and daily life, it means, for rational and responsible active participation in the democratic decision-making process (Zoller 1993; Zoller et al. 1998). "Higher order cognitive skills include the abilities to: make meaning, by interpreting information, forming and applying concepts and principles, critical analysis, synthesis into coherent wholes, generate ideas, using innovative thought, creativity, take decisions, using procedures, algorithms, strategies, heuristics and judgments about applicability, reflect on own purposes and processes, including justifications for judgments and decisions, possibilities of transferability" (The TELRI approaches).

It may seem that, thanks to many university activities—seminars, mini projects, diploma theses (bachelor, master, doctoral)—development of HOCSs (including inquiry skills) should not pose any difficulty for the students. While working on the ESTABLISH project (http://www.establish-fp7.eu/) the research team of the Department of Chemistry Education at Jagiellonian University had an opportunity to check how far removed from reality the statement above may be.

METHODS

Implementing inquiry-based science education (IBSE) at schools constitutes one of the European Union's (EU's) priorities, thanks to which it is one of the leading issues that Framework 7 Programme has been working on. The ESTABLISH project, realized within this program, involves IBSE workshops for students and future teachers. The course was attended by 77 students (62 women and 15 men) who were mostly third year (bachelor's degree) and some of them first year (master's degree) students of chemistry and environmental protection (academic year 2010/2011). Apart from their basic study subject, these students are also attending courses preparing them to become teachers (for more details, see *The System of Education in Poland*. EURYDICE). It is the so-called concurrent system of preservice teacher training. Groups consisted of 6–10 students, which during class would be divided into smaller groups of 2 or 3, depending on the task and the number of present students.

The training, which constituted a part of the project, consisted of two stages: theoretical and practical. At the theoretical stage, the training uses a method of role-playing. Students were to imagine that they work for a research and development (R&D) institution and plan stages of their research activities (in groups). Preparation of an algorithm of future actions and decisions was offered for students as a useful tool. The schedules of research work were presented and discussed with the whole group. Next, the students were to formulate an interesting question for pupils, relating to everyday life, whose objective would be to engage them in IBSE, for example, "Why does a cookie becomes soft and a baguette hard if left uncovered overnight?" In the second part of training, students should plan and make a simple IBSE experiment to be carried out at schools.

In the experiments, students were to investigate, among others, the dependence of solubility of sugar in water on temperature and absorbency of a disposable diaper and to compare solid and foamed polystyrene using corresponding cups. Students had access to daily-used objects only, that is, glasses, cups, spoons, hot and cold water, and so forth. Academic teachers supervised the work of students. The transparencies that students used for the presentations of their results were collected along with comments and remarks. At the end of the 3-hr class, students were asked to complete an evaluation survey.

The evaluation questionnaire consisted of 13 open-ended questions and 10 close-ended questions such as the following: "Asses if during that class you had opportunities to develop … skills"; "What was: interesting–boring, new–known, easy–difficult, …"

The research described in this paper applied participant observation conducted by two teachers simultaneously and document analysis: research plans, results of experiments, and evaluation questionnaires (Pilch and Baumann 2001). Participant observations were based on an adapted Pathways to Inquiry (project of Louisiana State University, retrieved June 1, 2011, from http://pti.lsu.edu/Default.aspx).

1. *Identify Questions for Scientific Investigations*
   - 1.1 Identify testable questions
   - 1.2 Refine/refocus ill-defined questions
   - 1.3 Formulate hypotheses
2. *Design Scientific Investigations*
   - 2.1 Design investigations to test a hypothesis
   - 2.2 Identify independent variables, dependent variables, and variables that need to be controlled
   - 2.4 Identify flaws in investigative design
   - 2.6 Conduct multiple trials
3. *Use Tools and Techniques to Gather Data*
   - 3.1 Gather data by using appropriate tools and techniques
   - 3.2 Measure using standardized units of measure
   - 3.3 Compare, group, and/or order objects by characteristics
   - 3.5 Use consistency and precision in data collection
4. *Analyze and Describe Data*
   - 4.1 Differentiate explanation from description

4.2 Construct and use graphical representations
4.3 Identify patterns and relationships of variables in data
4.4 Use mathematic skills to analyze and/or interpret data
5. *Explain Results and Draw Conclusions*
5.1 Differentiate observation from inference
5.2 Propose an explanation based on observation
5.4 Form a logical explanation about the cause-and-effect relationships in data from an experiment
7. *Communicate Scientific Procedures and Explanations*
7.1 Communicate experimental and/or research methods and procedures
7.2 Use evidence and observations to explain and communicate results
7.3 Communicate knowledge gained from an investigation orally and through written reports, incorporating drawings, diagrams, or graphs where appropriate

The following steps of research (based on the available literature, such as *Concepts, Principles, and Key Terms in the History, Philosophy, and Practice of Science*, from Loma Linda University, http://resweb.llu.edu/rford/courses/spol624/science_concepts.html; *Your Research Project How and Where To Start?*, from the University of Leeds, http://www.rdinfo.org.uk/flowchart/flowchart.html; and Trochim's *Research Methods Knowledge Base*, from http://www.socialresearchmethods.net/kb/index.php, all retrieved June 1, 2011) were developed by authors for teaching purposes:

1. Formation of the topic, analysis of the phenomenon
2. Identifying the research problem
3. Searching the existing literature and critical appraisal of the literature
4. Developing the questions and/or construction of hypothesis
5. Planning an experiment
6. Carrying out an experiment:
   a. Sampling and data collection
   b. Qualitative and/or quantitative data analysis
   c. Replication and verification
7. Interpretation of results, answering the research question or revising of hypothesis, iteration if necessary
8. Drawing conclusions and recommendations
9. Dissemination (report, publication, présentation, etc.)

Order of steps may vary depending on the subject matter. Iteration can occur at three levels (Figure 64.1).

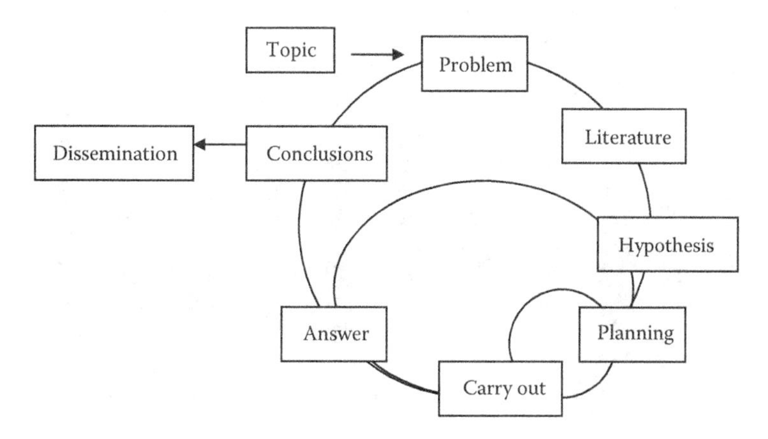

FIGURE 64.1 The scientific research process.

RESULTS

Below, exemplary results of the research described above are presented.

Stage 1

Twenty-seven "research plans" were collected and compared with the schedule provided above. Three of them were very far from the subject of that task. Twenty-four were taken into consideration. Figure 64.2 presents numbers of indications of each stage/step of the research process done by students' groups.

The most popular steps, mentioned by three-fourths of students, were review of the existing literature, planning an experiment, carrying out an experiment, and dissemination. Slightly more than half of the students marked "developing the research question or hypothesis" and, later on, "interpretation of results" and "answering/revising."

It should be added that less than half the groups specified some "substeps":

1. Sampling and data collection—13%
2. Qualitative and/or quantitative data analysis—50%
3. Replication and verification—42%

Eleven groups marked "iteration" (Figures 64.3 and 64.4), that is, repeating the experiment in different conditions or using another thesis, and so forth.

Stage 2

The observation of the planning process as well as IBSE experiments themselves revealed that students encounter most problems at the following stages/competencies:

- 1.1 Identify testable questions
- 2.1 Design investigations to test a hypothesis
- 2.2 Identify independent variables, dependent variables, and variables that need to be controlled
- 2.4 Identify flaws in investigative design
- 2.6 Conduct multiple trials
- 3.1 Gather data by using appropriate tools and techniques
- 3.5 Use consistency and precision in data collection

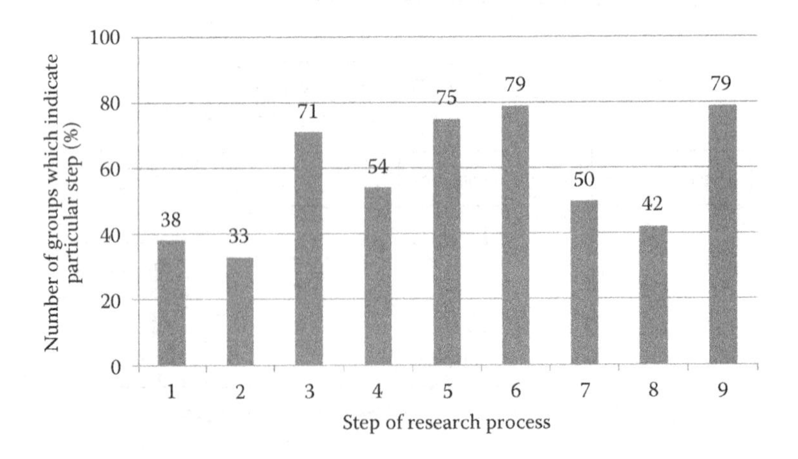

FIGURE 64.2 Numbers of indications of each step of the research process.

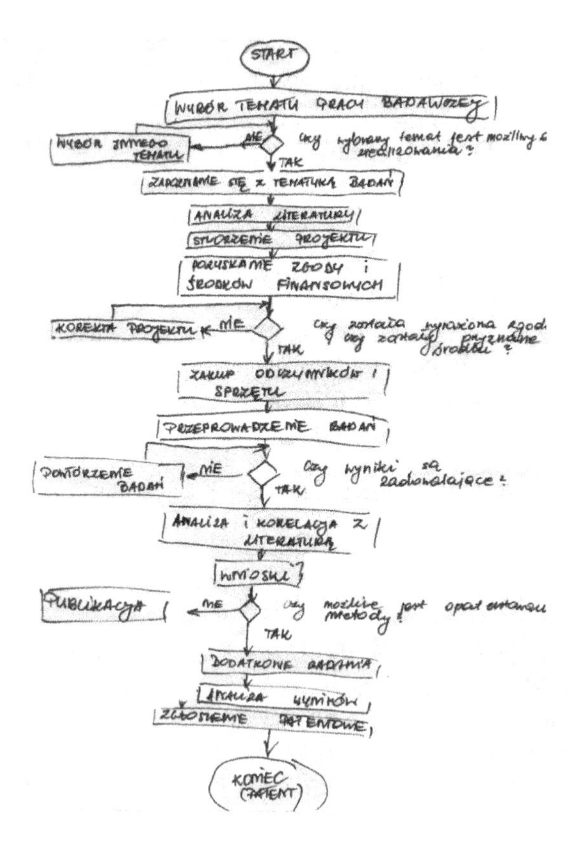

FIGURE 64.3 Student's work A (iterations included).

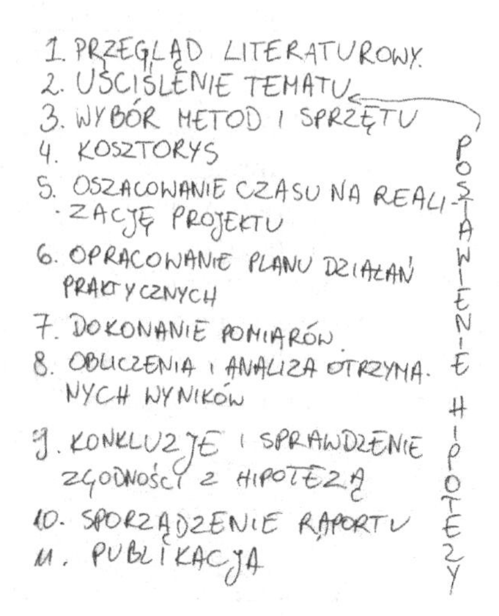

FIGURE 64.4 Student's work B (without iteration).

With a certain task to do, though given time to plan the experiment, the students took to work spontaneously without any plan. It was observed that in doing their work the students did not at all realize the importance of repeating the experiments to ascertain the reproducibility of the results. None of the groups decided to do that. When students had the possibility to select experimental tools on their own, about 60% of the groups chose the best ones available, that is, tools that enable the most consistent, repeatable results. If they do not use sophisticated equipment typical for university laboratories, they quite automatically do not care about consistency and precision in data collection.

Stage 3

The analysis of evaluation questionnaires showed the following:

According to students, the most difficult task (one-third of the answers) was to formulate preliminary questions that were to engage secondary school students in the investigation. The same number of students found the questions interesting. One-sixth of the respondents said that they found it hard to plan even a simple experiment. Preparing a plan of research (RPG) was a completely new task to 10% of students. The biggest group, 44% of participants, considered preparing and conducting an experiment related to everyday life interesting.

Some of the reflections shared by the students are worth quoting, such as "I learn about complicated issues at the university and yet I find easy tasks difficult to deal with" and "There are many interesting questions that I would like to know answers to."

CONCLUSION AND RECOMMENDATIONS

The results of that preliminary research seem to be interesting and allow us indicate areas that demand improvement. Without doubt, one of them is identifying the research problem and planning a given experiment, which appears to be a weak point of the focus group. It should be noted that even though since the last reform of the Polish education system in 1999, the ability to "plan an experiment" is mentioned in the core curriculum for secondary schools and is verified in the final chemistry exams, it still has not been improved. It might be because tasks included in the final exam, such as "plan an experiment showing…," used to concern only experiments described in textbooks and done/discussed in class. As we can see, the truth is that this task may only reveal whether a secondary school student remembers and is able to recall particular previously acquired information instead of whether they are able to plan a new experiment—the lowest category of objectives by Bloom et al. (1956).

It is very likely that the improper choice of experimental details, such as the methods, conditions and the error analysis, derive from the fact that throughout the course of studies up to the point of this research, our students are often, if not always provided with detailed manuals that they are expected to conscientiously follow. In school conditions, it is still rare to apply a fully developed version of IBSE as far as chemistry education is concerned (Bernard et al. 2012).

Drawing conclusions (among others, curve extrapolation was observed) is not students' strong point either, which, according to the authors, is a result of the most popular form of laboratory classes, the so-called "cookbook chemistry lab." During that kind of class, the students are not supposed to "discover" certain dependence but, rather, to verify and correctly apply a widely known algorithm.

The third weak point of our students is that they showed that they do not see the importance of confronting their experimental results with those collected by other students or reported in the literature. It was proved both at the theoretical stage and by the observations of the research carried out by the students. It may result from arrogance (Obuchowska 1995), which is typical of young people, and allows them to claim that each thought is revealing and each activity is unique and correct. Unfortunately, if this type of attitude is not modified during the education process, it might have a negative impact on the students' future adult life, especially the decision-making process.

The difficulty in formulating engaging questions may relate to little curiosity about the world. Nobody asks more questions than small children do. Also, creativity understood as an ability to propose many, new, unusual, interesting ideas is inversely proportional to a pupil's age. "Our decline in creativity does not start when we are 40 or 50. It starts around about the age when we enter school," some experts tell us (e.g., *Age and Creativity*, retrieved June 1, 2011, from CreatingMinds.org, http://creatingminds.org/articles/age.htm). A suggestion that creativity may change rather than decline with age is provided by some others (Abra 1989). On the other hand, the students' interest in experiments related to everyday life supports the thesis concerning teaching in context not only in primary and secondary schools but also in higher education institutions (Overton et al. 2009).

It may be claimed that, as it takes place at lower levels of education, university students encounter problems employing knowledge acquired at particular levels of education in other courses, faculties, or nonacademic circumstances, and this especially concerns basic inquiry skills, such as determining the objective of the research; developing a research hypothesis; standardization of research conditions; establishing an independent, controlled variable; and selecting a research method. This may result from a "hand-in-hand" guidance method applied in the process of preparing bachelor diploma theses.

In the next phase of the study, authors are going to ask students about their opinions related to the following statement: "A common misunderstanding is that by this method a hypothesis could be proven or tested. Generally a hypothesis is used to make predictions that can be tested by observing the outcome of an experiment. If the outcome is inconsistent with the hypothesis, then the hypothesis is rejected. However, if the outcome is consistent with the hypothesis, the experiment is said to support the hypothesis but alternative hypotheses may also be consistent with the observations" (*Research*, retrieved Jun. 1, 2011, from http://en.wikipedia.org/wiki/Research).

REFERENCES

Abra, J., Changes in creativity with age: Date, explanations, and further predictions. *International Journal of Aging and Human Development*, (1989), 28, 2, pp. 105–126.

Baumann, B., *Making the Magna Charta Values Operational. Theory and Practice. Proceedings of the Conference of the Magna Charta Observatory*, 16–17 September 2010, Bolonia University Press.

Bernard, P., Maciejowska, I., Odrowąż, E., Dudek, K., Geoghegan, R., *Introduction of Inquiry Based Science Education into Polish Science Curriculum—General Findings of Teachers' Attitude*, (2013), 17(1–2), pp. 49–59.

Bloom, B.S., Engelhart, M.D., Furst, E.J., Hill, W.H., Krathwohl, D.R., *Taxonomy of Educational Objectives: The Classification of Educational Goals; Handbook I: Cognitive Domain*. New York, Longmans, Green, (1956).

Kulm, G. (ed.), *Assessing Higher-Order Thinking in Mathematics*. Washington, DC, American Association for the Advancement of Science, (1990).

Leou, M., Abder, P., Riordan, M., Zoller, U., Using "HOCS-centered learning" as a pathway to promote science teachers' metacognitive development. *Research in Science Education*, (2006), 36, 1–2, pp. 69–84.

Obuchowska, I., *Drogi dorastania. Psychologia rozwojowa okresu dorastania dla rodziców i wychowawców [Roads of Adolescence. Developmental Psychology of Adolescence for Parents and Educators]*. Warszawa, WSIP, (1996).

Overton, T., Bers, B., Seery, M.K., Context- and problem based learning in higher level chemistry education. In: Eilks, I., Byers, B. (eds.), *Innovative Methods of Teaching and Learning Chemistry in Higher Education*. London, RSC Publishing, (2009), pp. 43–60.

Palmer, E.J., Devitt, P.G., Assessment of higher order cognitive skills in undergraduate education: Modified essay or multiple choice questions? *BMC Medical Education*, (2007), 28, pp. 7–49.

Pilch, T., Baumann, T., *Zasady badań pedagogicznych [Principles of Educational Research]*. Wydawnictwo Akademickie "Żak", Warszawa, (2001).

The System of Education in Poland. EURYDICE. http://www.eurydice.org.pl/sites/eurydice.org.pl/files/the_system.pdf, Retrieved June 1, 2011.

The TELRI approaches. http://www.warwick.ac.uk/ETS/TELRI/About_TELRI/Approaches/index.html, Retrieved June 1, 2011, from the University of Warwick.

Zoller, U., Alternative assessment as (critical) means of facilitating *HOCS*-promoting teaching and learning in chemistry education. *Chemistry Education: Research and Practice*, (2001), 2, 1, pp. 9–17.

Zoller, U., Lecture and learning: Are they compatible? Maybe for LOCS; unlikely for HOCS. *Journal of a Chemical Education*, (1993), 7, pp. 195–197.

Zoller, U., Fastow, M., Lubezky, A., Tsaparlis, G., College students self-assessment in chemistry examinations requiring higher and lower order cognitive skills (HOCS and LOCS). An action-oriented research. *Journal of Chemical Education*, (1998), 76, pp. 112–113.

Zoller, U., Pushkin, D., Matching higher order cognitive skills (HOCS)–promoting goal with problem-based laboratory practice in a freshman organic chemistry course. *Chemical Education Research and Practice*, (2007), 8, 2, pp. 153–171.

Zoller, U., Scholz, R.W., The HOCS paradigm shift from disciplinary knowledge (LOCS) to interdisciplinary evaluative system thinking (HOCS): What should it take in science-technology-environment-society-oriented courses, curricula and assessment? *Water Science and Technology*, (2004), 49, 8, pp. 27–36.

65 Rethinking Teaching in Higher Education

Analyzing Interaction in Blended Learning Environments

Danny Barrantes Acuña

CONTENTS

A NETWORK OF KNOWLEDGE

Among all different active discussions, the advent of a digital revolution (Raschke 2003) challenges society and, consequently, formative practices. The vehement clamor emphasizes reconsidering digital literacies and pedagogical principles regarding new conceptions of time and space.

Narrative structures are one of those elements necessary to understand after our current context. Landow (2004) lights up this current debate when he alludes, "...teachers, educational technologists, Web masters, and software developers—base our ideas about the nature of reading, the purpose of documents, and their relation to individuals and communities on the mistaken assumption that electronic documents are essentially the same as books. They are not."

Many could argue that a screen, the greatest graphical terminal, represents nothing more than a shared arena to mediate our everyday interaction with machines and their sets of programmed applications, that is, until a new update, a new machine, or a new technology arises. This affirmation is especially intriguing for most of us who remember different interfaces that ceased to exist five, eight, or 15 years ago, leaving behind just a vague nostalgic "old school" memory. However, when we think back on these previous electronic environments, i.e., ALGOL 60, LaTeX, DOS, and WordPerfect among many others, we can find that they had a particular level of "abstractness," different from interfaces nowadays on their constant quest for optimization and human language simulation. Computing machines became massive trading objects; consequently, they had to emerge beyond incomprehensible codes meant for programmers and technicians. It needed to be promoted as an informational management tool that allows private citizens around the world to administrate information in electronic formats. Not long ago, concepts such as *user-centered design* (Norman 1988) were conceived to study and develop computer languages and applications that could emulate human natural semiotics to help diminish the gap between humans, computers, and their interactions. This helped digital devices become "fashionable" among social sectors worldwide, strengthened with additional elements—such as the World Wide Web—that nest and flourished into crucial variables that today impact both technological and social realms.

There are, however, side effects that remain either possibilities or challenges when it comes to education and social promotion. This can be perfectly framed when we look into the word *user.* According to RAE (http://buscon.rae.es/draeI/SrvltConsulta?TIPO_BUS=3&LEMA=usuario), a user is one that uses something, someone who has the rights to use something that doesn't belong to him or her and has certain limitations. Going back into its Latin root, the word *usuarĭus* is an adjective that describes something "that may be used by one other than the owner but not for profit; (object/slave)." A *user,* unlike an *editor,* an *administrator,* a *RID,* a *super administrator,* or a *developer* remains passive and diminished, being able to participate and offer feedback when allowed by someone who is a creator or administrator, an old subject of historical remembrance that comes inherent with human relationships: power struggle and control. Is this so with our students or any learner? Are they still *users*? This can have remarkable results that are important to reconsider when trying to build educational and pedagogical scenarios based on digital mediums. One of contemporary education's roles is to reframe these kinds of traditional conceptions for contextual and pertinent understanding within formative processes. As Jenkins et al. (2009) point out, "We must push further by talking about how meanings emerge collectively and collaboratively in the new media environment and how creativity operates differently in an open-source culture based on sampling, appropriation, transformation, and repurposing." Such actions can't be executed by passive *users* but by a collective of decision makers, critical thinkers, researchers, and craftsmen who can elaborate integral answers. This is nothing new, and history has shown us valuable examples of reconfigurations that weave "hard" and "soft" dimensions of knowledge. Ehn (1998) reminds us of the case of the Enlightenment project and how it has reached its original purpose developing "hard" sciences to its expectations, proof of "the digital revolution, the exponentially growing information and communication technology. In contrast, however, the more 'soft' expectations of the Enlightenment project concerning values, art, aesthetic ideals, ethics and politics have in no way been met during the last centuries."

An endless amount of electronic applications, devices, and technologies have been developed, (with many more to come), which feed the public's hunger for more. However, contemporary universities should plead for critical positions, promoting balanced conjectures based on theoretical principles and clear practical procedures. Our role is to avoid any form of techno-centrism but, instead, to face a new challenge that cares for the validity of content creation in the middle of an informational sea that moves slowly, thick and senseless. Just like Harper (2010) reminds us, this is "the age when technologically mediated communication has reached its greatest volume and it might be encouraging commentators to forget the thing that is communicated (intentions) and the creature that creates and expresses these intentions [the human] and to focus instead on the sociological and organizational marvels that the technology enables." What was intended with the following educational settings deepens this understanding of integral elements, thus rescuing important flexible variables such as emotional processes, critical thinking, and hypertext theories.

Blended educational spaces host playful scripts and confront every element comprehended within information ecologies (see Nardi and O'Day 1999). It has to do with a reconsideration of the medium, but as well, it must include communication codes, our understanding of time and space, networks, and collaborative cognitive construction.

PEDAGOGICAL DIMENSIONS AND INSTITUTIONAL CONSIDERATIONS FOR BLENDED LEARNING ENVIRONMENTS

Any formative process comprehends a complex collection of elements that work in an interrelated manner with each other. Traditional paradigms—as a first approach—have normally stressed the elements of a basic relationship between one professor and his pupil. Next to it, content and its relevance become the trading material, carried through a power struggle interaction. As Raschke (2003) expresses, "The learning spaces that have dominated in each of the historical archetypes… so far have reflected a conventional learning culture centered on control of the classroom by the *instructor*"—this as a common behavior among different types of information transactions. However, the advent of the digital revolution permits a new generation of possibilities that open across this typical and intimate student-teacher relationship and make way for networked learning communities. This emphasis can be better explained if we realize "schools currently are still training autonomous problem solvers, while as students enter the workplace they are increasingly being asked to work in teams, drawing different sets of expertise, and collaborating to solve problems" (Jenkins et al. 2009).

Landow (2004) explains that "the blend of deconstruction and hypertext theory, or hypertext illuminated by a deconstructive approach, provides a wonderfully appropriate way of designing curricula," which enacts a completely different understanding of pedagogy on its familiar didactical sense. Cognitive processes and knowledge construction reach a new level that surpasses traditional paradigms in which "skill at copying quickly, accurately, and legibly counted far more than did the student's ability to abstract and synthesize."

What is thrilling about such strategies is the possibility to connect and develop study objects that merge and link to different cultural and social hyperlink nodes. Content behaves therefore in an endless "work in process" state that is constantly undergoing revision to the degree that texts are disseminated and distributed, and that content is "deconstructed" through infinite revisability; this way the university itself becomes de-localized (Raschke 2003).

As a necessary starting point, we understand a didactical strategy as a set of actions that are designed and put in place in an orderly structure to achieve a particular purpose in learning spaces. This configuration takes into account seven different elements:

1. *Educative goal:* comprehends the previous analysis to declare the pedagogical terms and content direction; a sum of a teacher's elements on which content is relevant at that particular moment. What are the student needs and expectations and the resources available?
2. *Context:* to understand the formative atmosphere in which the strategy is contained, (what university, what discipline, etc.), a particular historical moment, implementation possibilities, and interactional formats.
3. *Student's role:* comprehended by all pedagogical features that defines its role: learning styles, active participation, and communicative lines to optimize processes of knowledge construction.
4. *Teacher's role:* involves the action of acknowledging pedagogical tendencies and styles in order to develop communication and emphatical skill levels. When this element is highlighted, one has to analyze the new actions and responsibilities expected that are normally declared as institutional profiles. Contemporary regulations demand teachers to be active learners and be capable of designing formative scenarios pertinent to their context.
5. *Content nature:* due to a disciplinary object. It is necessary actions such as to delimitate an informational body that will be mediated and constructed. Narrative traces and interrelated routes among different contents are included in this element.
6. *Learning scenarios:* understood as the didactical proceedings after pedagogical principles. These scenarios are conformed by techniques and resources.
7. *Evaluation, monitoring, and feedback activities:* all activities held during formative processes that demonstrate the achievements of the different goals projected during the *educative goal* stage. Activities and actions can take part throughout the whole process, and they are not necessarily meant for quantitative-conditional assessment.

It is clear that for all teaching experiences, there has to be a previous consideration of all elements, thus pertinent design and practical implementations. There is, however, a common ground to any formative space design that has to do with philosophical dimensions, pedagogical tendencies, and developed skills. The teaching experiences to be portrayed in the next pages were based after a critical-constructivist tendency that, according to Hernández, Montenegro, Francis, and Gonzaga (2009), are conceived after dialectic and constructivist's epistemological bases, and they aim for development of critical and reflexive thinking to encourage the student into a process of deliberation followed up by purposeful actions. There is also a deep interest in fostering mental structure development through social interactions and through the study of object relations. The student should, at the end, be able to understand social reality challenges in front of them and configure integral answers.

For the purpose of this paper, blended learning environments (BLEs) have been selected to be analyzed. However, deciding to use BLE strategies calls for previous phases and considerations; therefore, this section highlights the importance of previous pedagogical elements, conditions, and principles that must be fulfilled for a later implementation of these kinds of scenario formats. There has to be, as well, a profound study of all different elements to implement digital spaces close to institutional regulations, course goals, content nature, student's learning styles, and professor's didactical skills. Based on previous experiences, pedagogical principles must be configured for any teaching strategy foundation (Marín et al. 2010):

1. Digital learning spaces must always prioritize the educative scope beyond the strategy to be developed and its technical implementations.
2. The effectiveness and relevance of media used in educational processes depends on the integration and mediation to be carried out within the formative space.
3. Technological applications should always reflect the pedagogical approach for any educational process. This will allow an active construction of knowledge out of the learning stage.
4. All formative environments comprehend a communicative nature; therefore, digital media should promote a process of teaching and learning based on active, fluid communication between the different participants.
5. When designing learning strategies, one must take into account the multiplicity of existing digital mediums, their theoretical and practical implications, and all pertinent content displaying formats.
6. The implementation of digital applications should be selected after taking into consideration the student's previous experiences, content to be developed, suggested activities, and information structure.
7. All information portrayed must generate expectations and interest among students, thus helping every learner achieve new findings, developments, and integral answers during their learning process.

In addition to the pedagogical foundations, any BLE should have a formal institutional body supporting its actions. Eventually, a project that began after individual attempts can escalate to be part of local online communities that move around in collaborative ways. The University of Costa Rica (UCR), for instance, formally settles institutional interest toward teaching, research, and social practices based on information and communication technologies (ICT). As part of its organic statute, a "profile of generic competencies for teachers of UCR (Acta de la Sesión No. 493. Consejo Universitario) defines a criteria list to describe expected skills when it comes to teaching: to be able to make use of ICT, develop abilities to use equipment, instruments and high-tech tools and the quest for flexibility and be adaptable to changes, new perspectives and innovations in knowledge."

When we take a glance at the national and regional picture, there are different in-house instances, international organizations (CSUCA [www.csuca.org], DAAD [http://www.conare.ac.cr/daad/spanish/retrato/ci_centroamerica.htm], HRK [http://www.hrk.de/de/hrk_international/staaten_und_regionen_1132.php#Zentralamerik], GUCAL [http://www.gucal.org/], and OEI [http://www.oei.es/index.php], among others), and national programs during the last decade, centering their actions on ICT and technologies development throughout the Central American region. Naturally, each country will have strategic agendas to promote joint actions. Education, economy, social development, ministries, and other sectors must work side by side in order to make a successful impact that, until now, as far as technologies and digital uses are concerned, are trying to bridge the digital gap and improve the possibilities of citizens in terms of connectivity. According to the Planning Office for Higher Education (OPES, http://www.conare.ac.cr/), the Costa Rican entity that coordinates technical and advisory processes for all public universities, there is a current program that is taking care of national access and uses and appropriation of ICTs. Three main goals are described below (Chacón 2009):

- Strengthen technological innovation for the development of the academy
- Ensure implementation and access and improve all national and regional networks in priority areas of national development, in conjunction with other institutions
- Strengthen the uses and adaptation of new ICTs at university activities

As shown above, most of the goals emphasize a technical-instrumental dimension. However, no further events strive for development of pedagogical thinking and digital literacies or electracies as Morrison (2001) suggests. The challenge remains in finding an integral bridge between theory and practical implementations.

TRAVELING WITHOUT MOVING: INTERNATIONAL PEER REVIEWING

Still, as an art student, one of our ultimate satisfactions after finishing any work is being able to find important outlets that reach as many people as possible. It represents public spaces where our social action—concerning any formal artistic extensions—mainly happens: a shared arena that allows us face-to-face public opinion with our given statements on society, philosophy, aesthetics, or inner memories out of our thought process. Like any other interactional process, feedback would eventually be received after public reaction, suggestions, and critics. However, the art and design guild in Costa Rica behave somewhat in a closed fashion. Most of the time, feedback comes from known faces, family, and friends and, if lucky, someone who makes decisions and will invite you to their next gallery show.

Digital Media 1 (taught during I Semester 2010) is a regular course offered at Visual Arts School, Faculty of Fine Arts at the University of Costa Rica. It is intended for graphic design students to take during the first semester of their undergraduate sophomore year. Most of them are very young—18 to 20 years old—and are up for new experiences, alternative learning spaces that let them propose and innovate while discovering; they behave in an enthusiastic way always in a quest to define personal and professional action lines.

While the title of the course can be understood in its broad sense, its contents and goals are true to traditional *design software development* practical workshops. Its main goal is to "get to know the theoretical and practical bases of the electronic edition of images composed by pixels and vectors," which provide a very basic and instrumental objective. The rest of the programmatic elements—description of the course, secondary goals, and suggested activities—eventually completed a clear picture in which teachers normally, in the past, have instructed students how to use Adobe Illustrator and Adobe Photoshop optimized for graphic design typical tasks (Table 65.1).

After our first meeting, a short *learning style* questionnaire, followed by discussion, revealed that students had further questions and expectations for all contents and thematic areas to be developed within this course, especially because 10 of them already had basic Adobe knowledge. There were three main challenges to be solved: (1) collaborative work in terms of connectivity at the physical laboratory (there was no Internet available, and it is strongly recommended to have access every time you initiate new software explorations), (2) amount of computers available for student use (one computer per two students proportion) and, finally, (3) lack of contextual-thematic areas to promote

TABLE 65.1
Digital Media 1, Course Overview

Course	Digital Media 1
Number of students	25 (13 male, 12 women)
Student ages	18–20 years
Hours per week at class	4 hours (Mondays 8–12am)
Physical classroom	Mac Laboratory (12 computers, no internet connection)
Hours per week at home	2 hours
Internet access at home	100%
Digital classroom	Wordpress installation (www.dannybarrantes.com/curso) WP-Polls, PDF files, .png Images, email account, online community services (picasa.com)

critical thinking and allow significant learning throughout. Students, at this point, had no strong backgrounds as far as establishing project development structures or administrative and contextual criteria for practical use on software were concerned.

After reviewing formal conditions, there was a common agreement to create a complementary online environment to allow extra interaction after class (http://dannybarrantes.com/curso/), such as debates and information sharing (Figure 65.1a and b). A special request from the students was to try something different other than the official UCR's Learning Management System Platform (http://mediacionvirtual.ucr.ac.cr/- Moodle LMS) because they thought of it as a "plain, unfriendly, boring, and inflexible" environment. Another feature they didn't like was that a new username and password ID was necessary in order to see all of the course content, and the customized environment created afterwards didn't require a password at all.

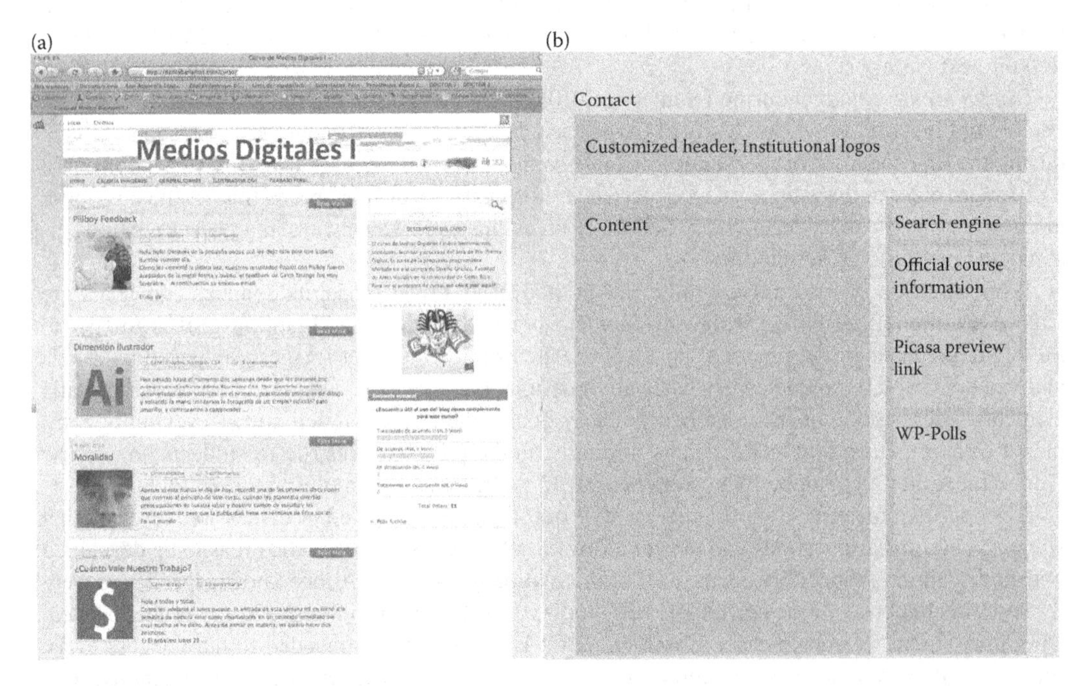

FIGURE 65.1 (a) Wordpress installation preview. (b) Wordpress interaction layout.

As shown, the Digital Media Online Classroom is designed to hold parallel activities that accompany each week's physical sessions. Alternative open source software tutorials, project galleries, productive work environments with relevant links and images, hardware discussion, PDF work guides, case studies, and budget-related issues were typical content featured. Anyone participating in this space was free to do so because there was no official pronouncement according to the course's evaluation. All students were invited to be part of discussions and practices offered, and their role over weeks was varied: Sometimes they would come, ask, and behave in rather active ways, and at other times they felt comfortable just working individually and keeping their website visits strictly an informational source.

Concerning interaction levels, participants were able to write comments and comment on their classmates' graphical processes after posting their final projects each week (https://picasaweb.google.com/diseno.grafico.ucr/) (Figure 65.2). One of the online classroom sections was dedicated to formative evaluation purposes, within which quick questions were asked concerning future content to be shown online, suggestions, and new learning achieved for each of the students' cases. For a more discreet environment, all of the evaluations didn't require any password; in this way, students were able to give their opinion anonymously, thus raising their motivation to communicate. Erickson (1997) claims that the Internet offers, like reality, the possibility to create social networks with the difference that it allows the person to stay anonymous, a possibility that surely has consequences for the individual. On the Internet—contrary to real life—a person is no longer part of a small group but a global network, thus representing a portrait means representing an unknown audience; therefore, allowing any possible coded message without personal repercussions over the physical individual.

As part of the evaluative strategies, Digital Media Online Classroom was constantly monitored by Google Analytics (http://www.google.com/intl/en_uk/analytics/), a service that allows Web analysis in terms of traffic data for particular Web spaces (Figure 65.3). It showed a rather positive reaction stating that website visitors spend at least 03:51 minutes each time they entered the online space. There was also a tendency to have regular visits from people during specific days throughout

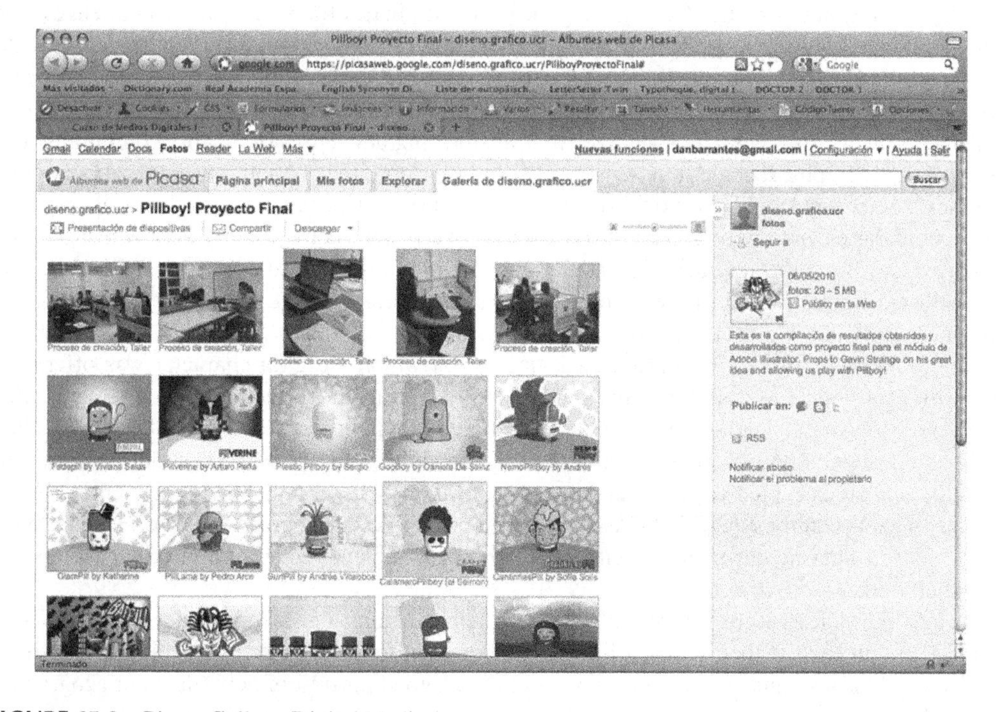

FIGURE 65.2 Picasa Gallery, Digital Media 1 course.

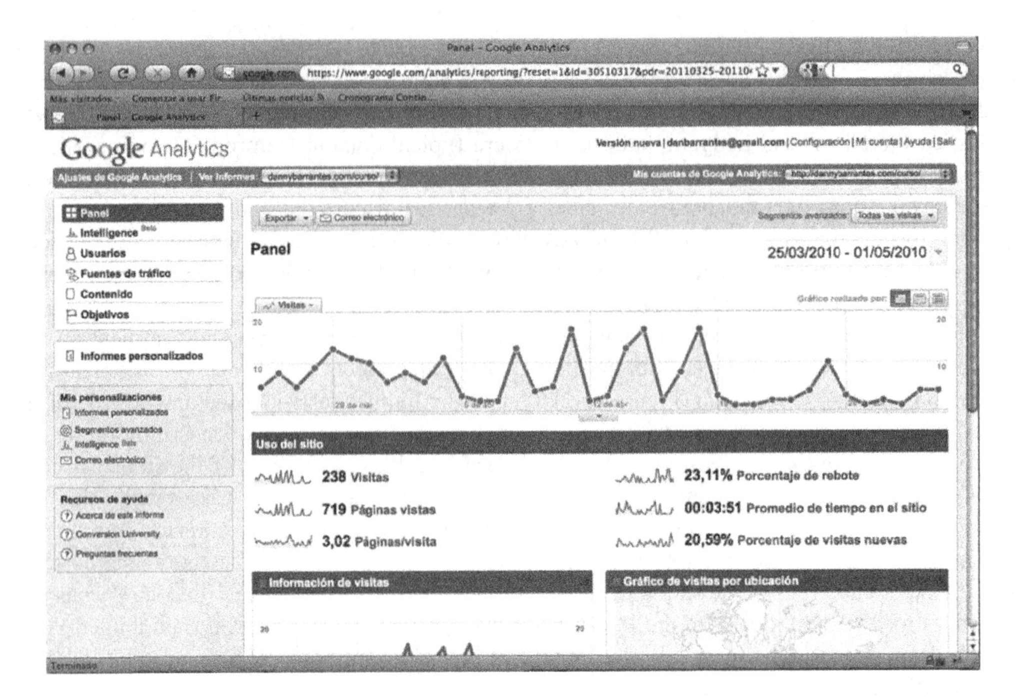

FIGURE 65.3 Online monitoring resources.

the week, proof of acceptance and steady interest in actions taking place there. In addition, an evaluation poll showed that 55% of the digital students found the online classroom a useful complement to the regular course, and 45% agreed on it as quite good.

Retaking some of the ideas explained at the beginning, important venues at which to show art works are always welcome for feedback and discussion. Good peer reviews and experienced interlocutors become essential for maturing ideas and aesthetics lines. One of the challenges with this course was directly related to space and connectivity issues; therefore, the presence of the Digital Media Online Classroom became a new-networked "venue" that enacted *death of geography*, explained by Berker et al. (2006), "a commonplace that the new networks of electronic communication, in and through which we live, are transforming our senses of locality/community and, on a wider geographical scale, our senses of 'belonging' to either national or international communities." After such concepts, a final Illustrator project was proposed to the students, opening our course boundaries and inviting Gavin Strange (http://news.jam-factory.com/), a known successful designer from Bristol to interact with us. Some of Strange's work on the web invites everyone to play with his creations, and this was the original idea for our course's final project (http://dannybarrantes.com/curso/?p=81). Pillboy (http://www.flickr.com/photos/jamfactory/sets/1494047/) (Figure 65.4) was a project he started back in 2005 in which a digital character was offered as vector files for people to customize and send back to be part of a collection made by many other designers. Students in Digital Media 1 took Pillboy as their base design to continue with their work, and an invitation (via email) was sent to Gavin so he could pick the best designs after his original idea. He accepted the offer and, as part of his review, he published a new entry (http://news.jam-factory.com/post/636770244/costa-rica-is-a-place-i-never-imagined-my-work) (Figure 65.5) on his website, sharing our material with his visitors: a complete venue for a visual artist to show his or her work.

Student's responses were full of enthusiasm. Their creations and technical developments were accurate, and in each of their cases, a final image was sent for final evaluation. Out of 25 of them, 100% passed the test, and level achieved was optimal: a good portrait of how "death of geography" concept looks.

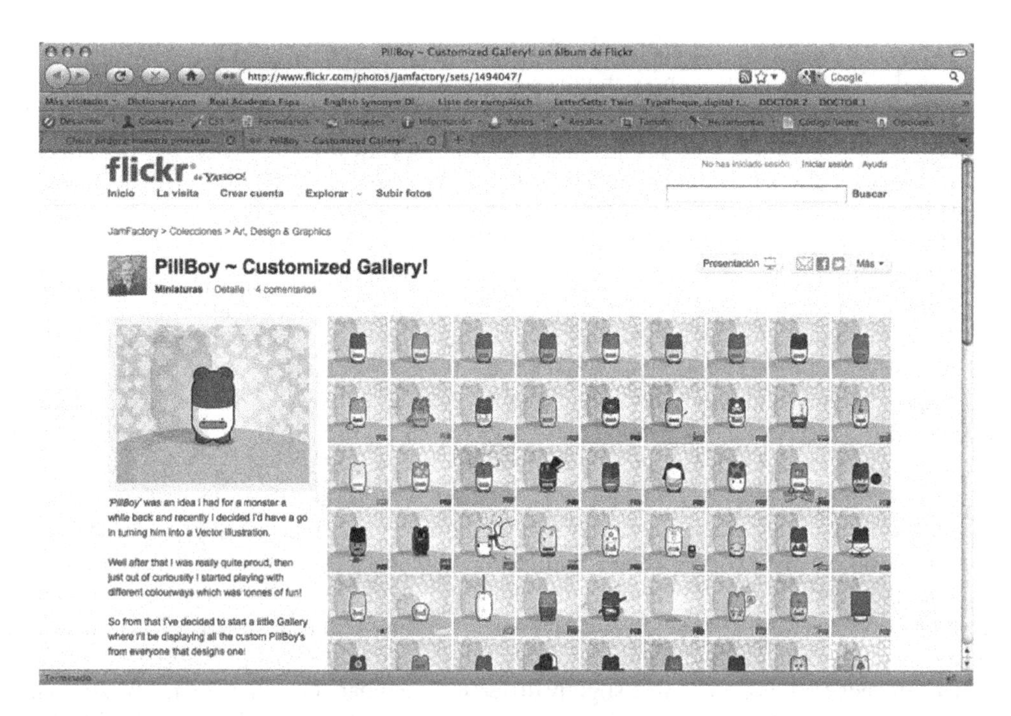

FIGURE 65.4 Pillboy by Gavin Strange.

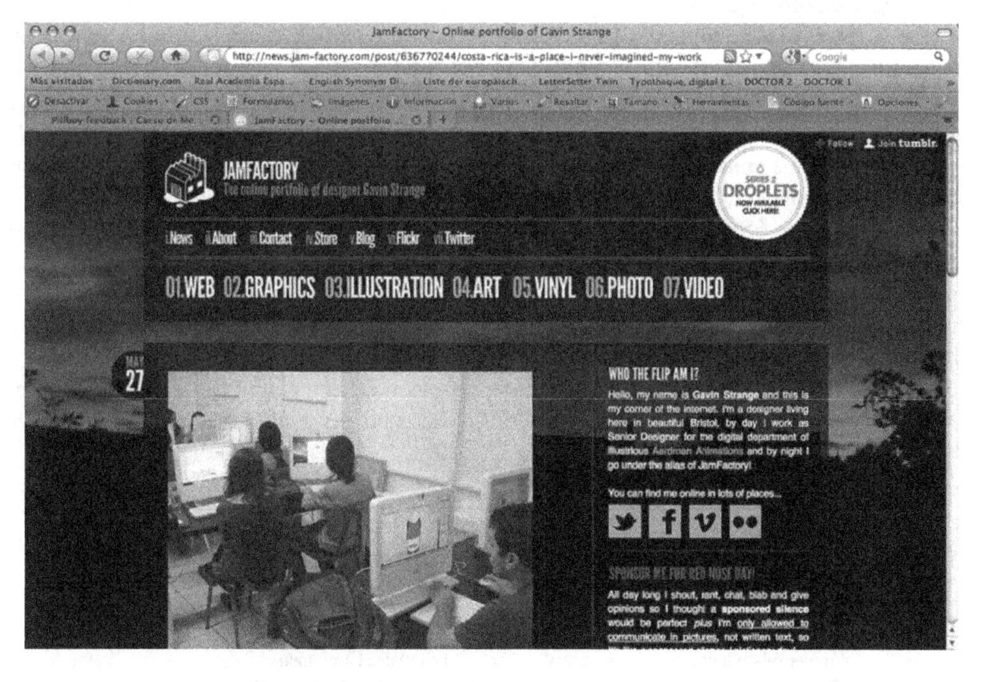

FIGURE 65.5 Gavin Strange's website entry.

REDISCOVERING NARRATIVES BEHIND WEB APPLICATIONS

Italian Futurism founder Tommaso Marinetti (Schmidt-Bergmann 1993) once affirmed, "Engines, are really mysterious.... They have their moods, unexpected bugs. It seems that they have personality, soul, will. It is necessary to stroke them and behave with respect to them." It is a special relationship, human-machine, always in the midst of its struggles and passions. Still today, both sides remain apparently unconnected, and until new concepts emerged, such as *digital domestication* (Berker et al. 2006), could we witness an intrinsic bond: machines, codes, and humans.

Web Applications (taught during November 2010) is a regular course offered at the Graphic Design Master's Program at Don Bosco University in El Salvador's capital city. It is intended for graphic designers, journalists, and computer scientists who want to specialize in Web design and multimedia uses. This course is one of their last, just one month before they present a final thesis. One of the features presented is that all students of this class (20 of them) maintain a day job, and at night, they come to class and study. Because of the level of intensity, sessions tend to be tiring for most of them, and attention levels sometimes drop after couple of hours (Table 65.2).

At this point in the program, all students involved have developed a number of technical skills, and they can easily confront basic challenges concerning web development; however, not all of them have the same instrumental level because their backgrounds are different. This course's main goal is "*understanding web concept and its implications and contemporary theories on usability*" which, again, suggests a tool-driven course.

During our first session, we discussed our main interests. It turned out that they were really interested in having some theoretical class that could help them understand how narratives and philosophical lines behaved under web application environments—something they said they lacked throughout their previous classes. In order to achieve this, three main challenges were presented: (1) develop theoretical classes that could connect to their previous technical knowledge, (2) exhausting/intensive class schedules, and (3) distance learning for two weeks (full online work). It was clear as well that demands for this course were slightly different in comparison to the *Digital Media 1* class, especially because all of the master program students had already achieved technical skills within their previous formative spaces.

Time and space was a challenge for this class, and as explained above, online environments are potential mediums that serve as cognitive bridges between human's processes for knowledge construction. Once they were asked for suggestions, students agreed on an online classroom (http://dannybarrantes.com/udb/) (Figure 65.6a), which was designed in a customized way to fulfill their needs. Immediacy was an advantage for such a proposal, and this allowed them have information, hints, and instructions with just one "click"; no password or username was necessary.

As shown in Figure 65.6b, a general layout for the actual classroom, the online space described this time had fewer interaction (blue zones) areas when compared with *Digital Media 1*. Hyperlinks to other relevant websites, work guides, articles, and a few examples were typical content here featured. It can be said that the system worked more like a non-located informational repository, in which everyone could meet and, if desired, participate.

During their distance sessions, the online classroom worked as a communication bridge. A customized forum engine (Figure 65.7) was installed so that students could give their opinions. Their

TABLE 65.2
Web Applications, Course Overview

Course	Web Applications
Number of students	20 (8 male, 12 women)
Student ages	23–40 years
Hours per week at class	2 weeks (Monday to Friday, 6pm–9pm, Sat. 8am–5pm
Physical classroom	Traditional classroom (each student has a laptop)
Hours per week at home	2 weeks, individual online work
Internet access at home	100%
Digital classroom	Wordpress installation (http://dannybarrantes.com/udb/) WP-Polls, PDF Files, HTML, email account.

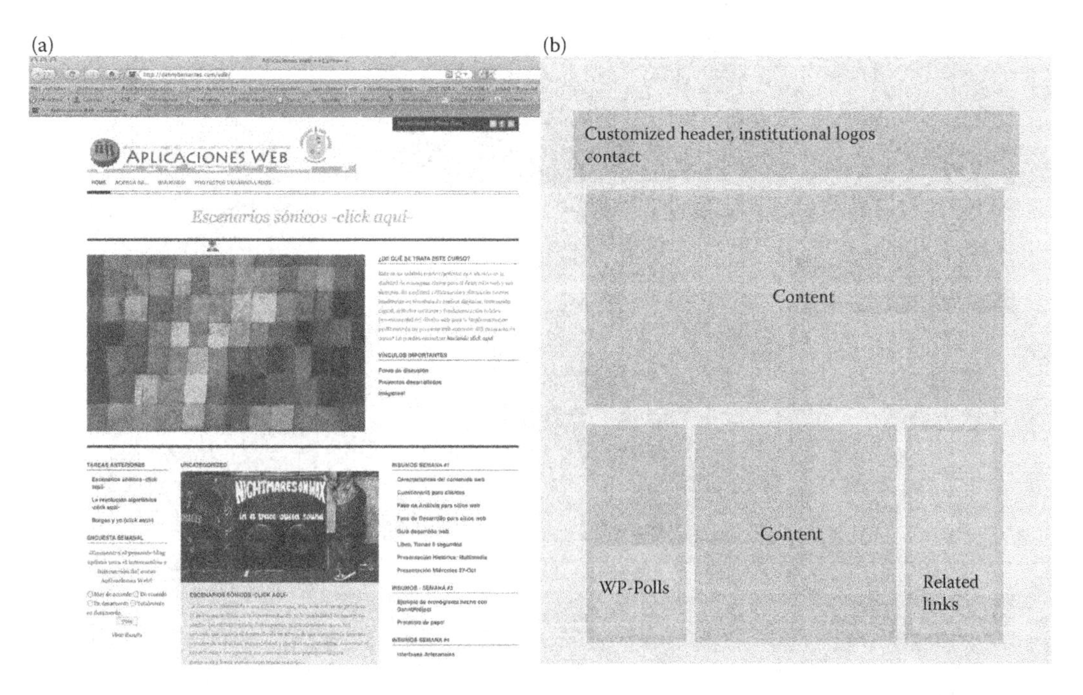

FIGURE 65.6 (a) Wordpress installation preview. (b) Wordpress interaction layout.

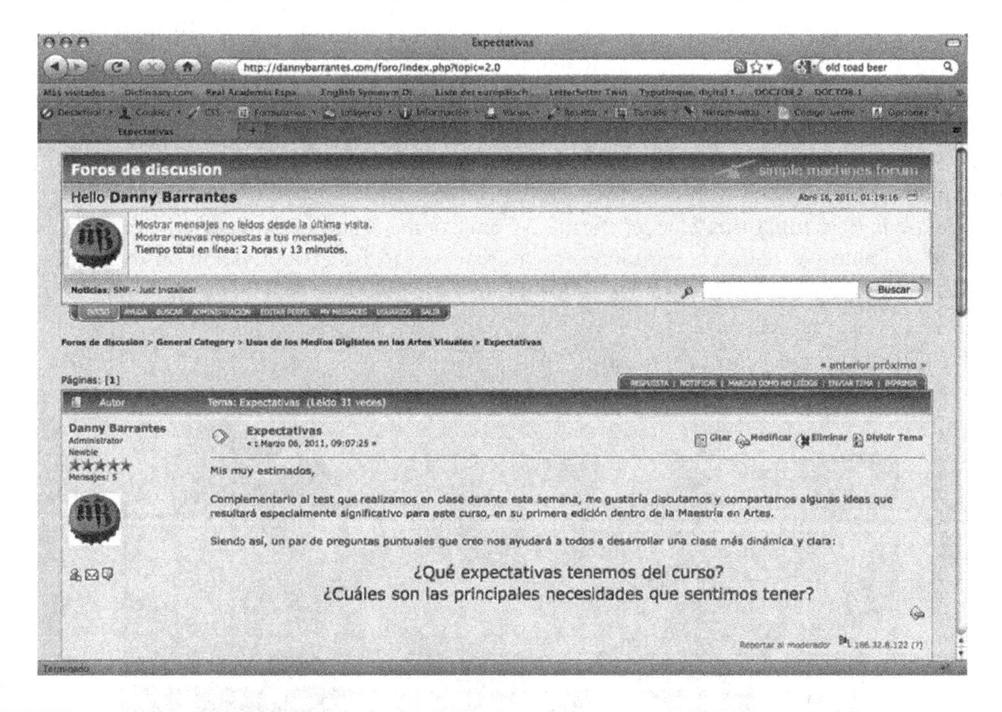

FIGURE 65.7 Customized forum engine, Web Applications course.

contextual reality, however, didn't allow them to be as active as desired, and it seemed that they weren't comfortable enough to work individually, away from any physical contact. The strategy that followed was to offer them a series of relevant readings and analysis as background for physically located discussions. This was especially important because one of the original challenges was to create pathways between their practical and theoretical sides (Figure 65.8).

FIGURE 65.8 Evaluation, class environment.

Exercises held and discussions as well as the nature of the content itself became their motivation because of a non-computer–centered approach. During Exercise One for example, they were asked to read poetry, and deconstruction theory should help them as background for remixology (Amerika 2008) exercises, so they could build their own version. It was hyperlink theory from its bases as in paper prototype relations for further discussion and new ways of discovering narrative. Their creations and scenes, however, were kept in our online classroom.

Evaluations stated that this blended formative environment had been received well. Out of all voters, 88% found the online classroom to be a great tool to foster motivation on their quest for new ideas and discussions. The strategy followed was effective for them—as expressed—because

FIGURE 65.9 Saturday discussions at the beach, Web Applications course.

of differentiation of environments, a good balance between physical and digital interfaces. The machine this time was just a bridge to enhance better physical interactions along a formative environment (Figure 65.9).

DIGITAL LEARNING ENVIRONMENTS FOR WORLD WIDE WEB OUTSIDERS

Our attention among these three different teaching experiences has moved around the relationship between humans and computers within formative environments, a balance that varies depending on different elements but surely makes special emphasis pedagogy as the integral object that oversees goals, mediums, message, and participants in a context. As Sherry Turkle (2008) mentions, "Neither physical nor digital objects can be taken out of the equation; nor should either fetishized. Over the past decades, we have seen an ongoing temptation to turn to computers to try solve our educational crisis. It is natural, in a time of crisis, to avidly pursue the next new thing, but we need to not lose sight of the things that have already worked."

University Didactics Online (Taught from 2006 until December 2010) is a regular course offered at the University Teaching Department from the UCR, and it is intended for active professors among all the different disciplines in the university. This course is taught each semester, and since 2006, an online version of the regular course has been adapted for all regions different than the central (Table 65.3).

Among all the courses previously portrayed, participants here are the ones who distance each other because of their completely different backgrounds, ages, and technical skills as they come freely to attend this course in search for pedagogical bases to strengthen teaching actions.

Previously mentioned, this course has been running for years already, and it was not until 2006 that a new online version of it was structured. Traditionally, this is a blended space in which 50% of its development is online, and 50% is physically located. However, conditions for the online version vary (see Table 65.1) because of geographical issues. Most of the professors would have to spend long hours on traveling and keeping a regular meeting during the whole semester while their offices are located circa up to 200 km away from Costa Rica's central region, where the university teaching department is. This strategy allows different regions and UCR professors studying worldwide to collide in the same space and share experiences while learning pedagogies to improve teaching in each of their domains (Figures 65.10 and 65.11).

Most of an assistant's first approach to this formative space is enthusiastic; however, in many of their cases, this will be the first time that they experience an online-based interaction environment. This can produce fear and frustration levels that can become major challenges for this course development.

Experience and evaluation made during the last four years has shown that 78% of them feel much more secure every time there is a physical meeting, and discussions are held while interacting in a common place. Because of this, this course maintains three physical sessions during the semester,

TABLE 65.3
University Didactics, Course Overview

Course	University Didactics
Number of students	20
Student ages	26–60 years
Hours per week at class	13 weeks online, 3 physical meeting
Physical classroom	Traditional classroom (each student has a laptop)
Internet access at home	90%
Digital classroom	Wordpress installation (http://mediacionvirtual.ucr.ac.cr) PDF files, Sound files, FLV Videos, Software Tutorials

FIGURE 65.10 Physical and digital-synchronic meeting.

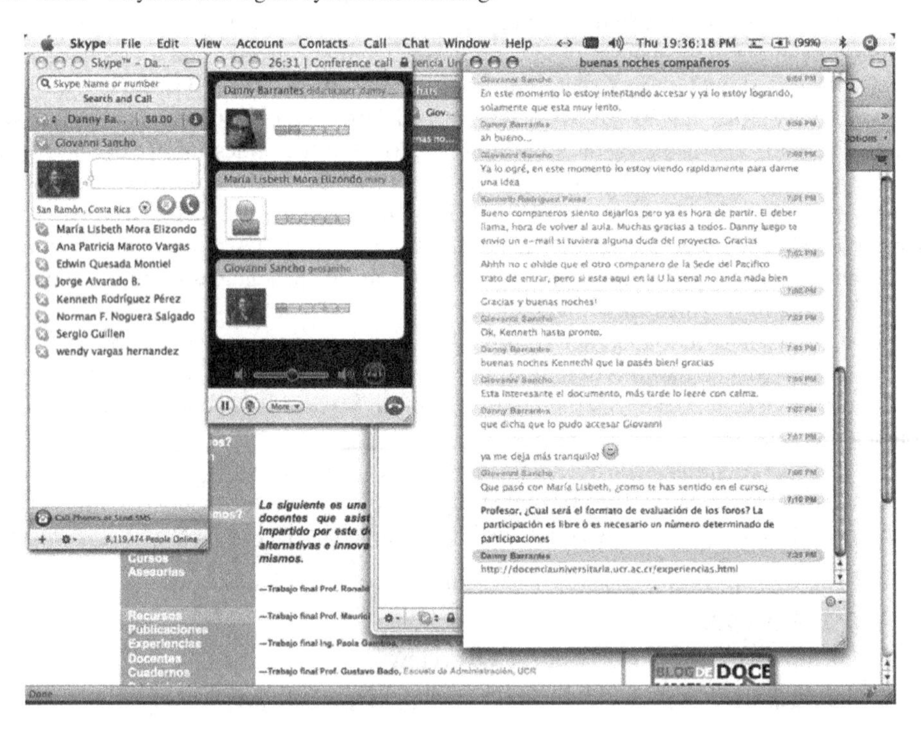

FIGURE 65.11 Online synchronic session.

and the first session becomes the most important of them all, mainly because it helps people to understand the logics, content, and technical demands that they will challenge. The other two sessions are normally held during week eight and the last session for their final project presentation (consists of a didactical experience that must be implemented and documented during the semester in one of the courses they teach) in week 16.

During their first session, we share four hours together. This time is devoted to round table action in which everyone debates about the needs and teaching contextualization in different regions of Costa Rica, expectations, and challenges to confront. As shown in Figure 65.10, not all of the assistants are physically present: Some of them are finishing postgraduate studies outside the country and using a Skype account, so they can connect during the whole session and participate equally (Figure 65.11).

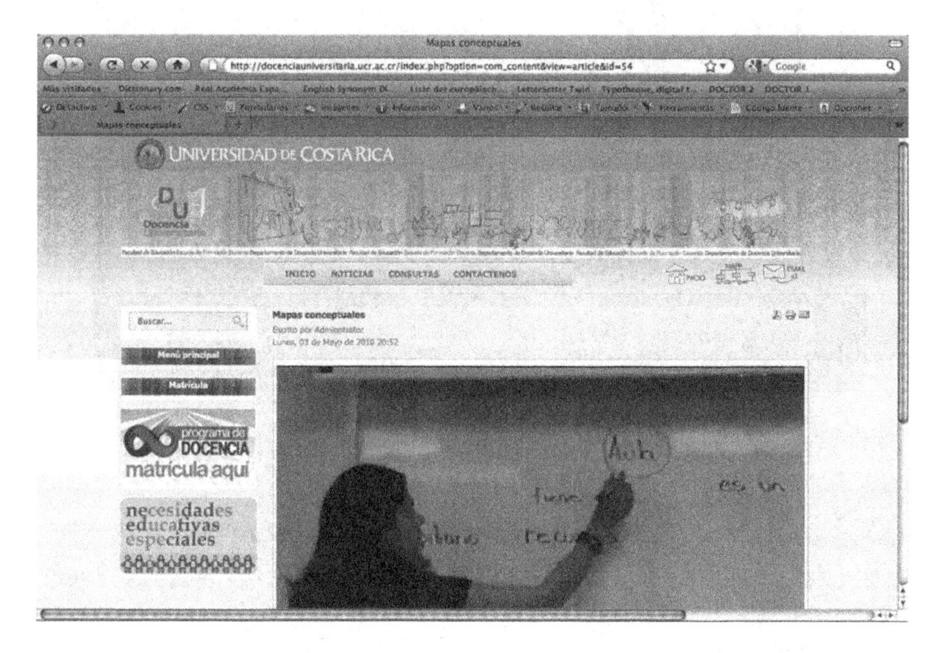

FIGURE 65.12 "Mind maps" online video tutorial.

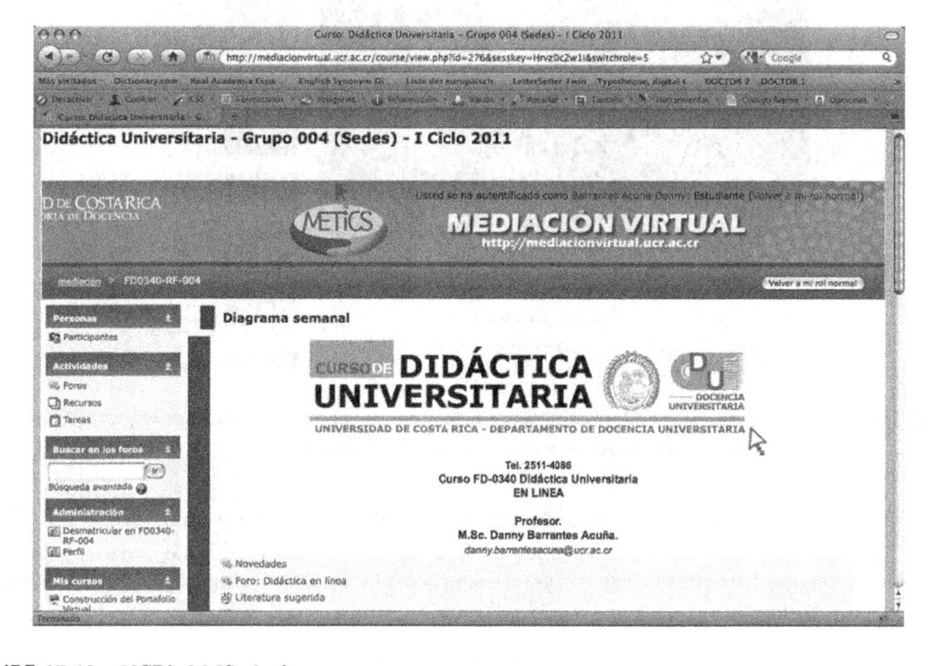

FIGURE 65.13 UCR's LMS platform.

FIGURE 65.14 Physical debate session.

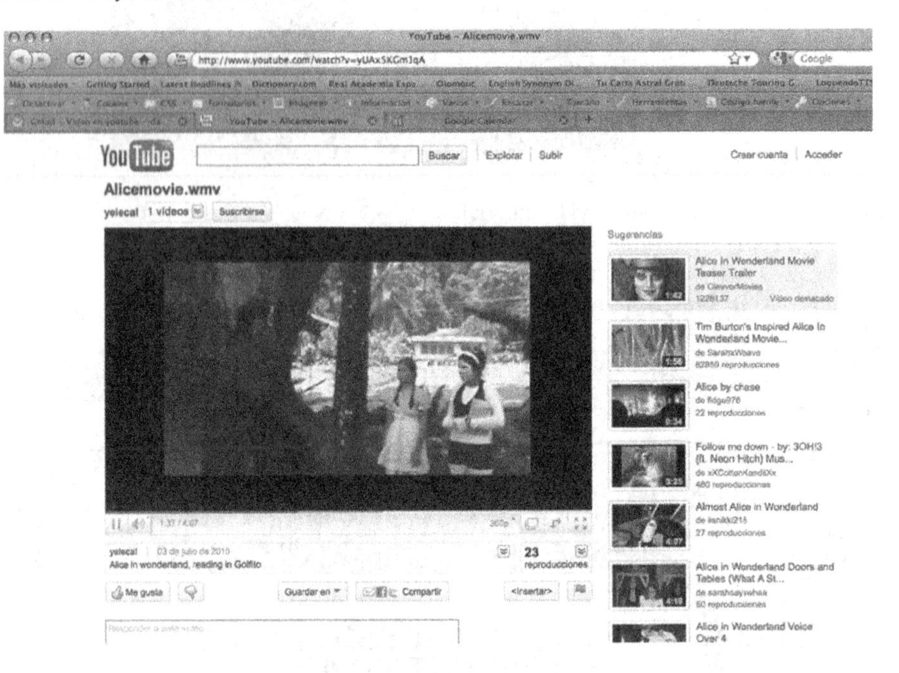

FIGURE 65.15 Alice in Wonderland, final project (http://www.youtube.com/watch?v=yUAx5KGm1qA, by Prof. Francisco Rodríguez. Golfito Area).

As shown in Figures 65.12 and 65.13, this course's online contents are managed at the official UCR's Learning Management System Platform,* a Moodle Learning Management System that is technically handled by the METICs Department—a global ICT promoter instance for educational practices at UCR—a clear and basic structure that helps online learners feel comfortable

* http://mediacionvirtual.ucr.ac.cr/ - Moodle LMS.

with information access and collects different formats and tasks (forum, documents uploads, wikis, video, hyperlink, audio, among others) within modules. As professors begin to work in such environments, they begin to develop regular actions, and it works as a "safe zone," in which everything is placed and actions take place.

Special requirements for these kinds of environments are mainly meant for whoever is in charge of promoting interactions and designing didactical materials online. In such cases, technologies, formats, software, plug-ins, or anything technically related has to be solved beforehand; this way, students care exclusively with content away from instrumental struggling (Figures 65.14 and 65.15).

REFERENCES

Acta de la Sesión No. 493. Consejo Universitario. Perfil de competencias genéricas para el profesorado de la Universidad de Costa Rica. Retrieved from: http://www.cea.ucr.ac.cr/cuestionarios/Perfil_docente_univ.pdf (April 23, 2010).

Amerika, M. 2008. Retrieved from: http://www.vjtheory.net/web_texts/text_amerika.htm (January 10, 2011).

Berker, T.; Hartmann, M.; Punie, Y.; Ward, K. 2006. *Domestication of Media and Technology.* Glasgow: Bell and Bain Ltd.

Chacón, S. 2009. La promoción de la virtualidad como recurso de apoyo a la docencia en la educación superior [The promotion of virtuality as teaching resource to support higher education]. In Badilla, E.; Cabrero, J.; Chacon, S.; Francis, S.; Mora, A. I.; Perez, L.; Revuelta, F. I.; Chambers, I.; Salas, F. E.; Silva, J. E. *La Docencia Universitaria en los Espacios Virtuales [University Teaching in Virtual Spaces].* AECI, UCR. Sección de Imprenta del SIEDIN [SIEDIN Printing Section].

Ehn, P. 1998. Manifesto for a digital Bauhaus. *Digital Creativity*, 9(4), 207.

Erickson, T. 1997. The World Wide Web as Social Hypertext. Retrieved from: http://www.pliant.org/personal/Tom_Erickson/SocialHypertext.html (January 10, 2011).

Harper, R. 2010. *Texture: Human Expression in the Age of Communications Overload.* Kindle Edition. Cambridge, MA: MIT Press.

Hernández, A. C.; Montenegro, M. L.; Francis, S.; Gonzaga, W. 2009. *Estrategias Didácticas en la Formación de Docentes.* San José – Costa Rica. Editorial Universidad de Costa Rica.

Jenkins, H.; Purushotma, R.; Weigel, M.; Clinton, K.; Robison, A. 2009. *Confronting the Challenges of Participatory Culture, Media Education for the 21st Century.* The John D. and Catherine T. MacArthur Foundation reports on digital media and learning. Cambridge, MA: MIT Press.

Landow, G. 2004. The paradigm is more important than the purchase. In Liestøl, G.; Morrison, A.; Rasmussen, T. *Digital Media Revisited: Theoretical and Conceptual Innovations in Digital Domains.* Cambridge, MA: MIT Press.

Marín, P.; Barrantes, D. 2010. La bimodalidad en los cursos universitarios: Los espacios formativos en el DEDUN [Bimodality at university courses: The training spaces in DEDUN]. In *Nuevos formatos para la función docente universitaria [New Formats for University Teaching Function].* Sección de Impresión del SIEDIN [Print SIEDIN Section], University of Costa Rica.

Morrison, A. 2001. Electracies: Investigating Transitions in Digital Discourses and Multimedia Pedagogies in Higher Education. Retrieved from: http://www.media.uio.no/personer/morrison.electracies/selchs.html (January 10, 2011).

Nardi, B.; O'Day, V. 1999. *Information Ecologies: Using Technology with Heart.* 6th Edition. Cambridge, MA: MIT Press.

Norman, D. 1988. *The Design of Everyday Things.* Reprint edition September 9, 1998. England, UK: MIT Press.

Raschke, C. 2003. *The Digital Revolution and the Coming of the Postmodern University.* Routledge.

Schmidt-Bergmann, H. 1993. *Futurismus—Geschichte, Ästhetik, Dokumente [Futurism—History, Aesthetics, Documents].* 2nd Edition. Germany: Verlag [Print] Rororo.

Turkle, S. 2008. A passion for objects: How science is fueled by an attachment to things. *The Chronicle of Higher Education*, 54.

World Economic Forum. 2011. The Global Information Technology Report 2010–2011. Geneva, Switzerland. SRO-Kundig.

66 Bridging the Gap between Industry, Science, and Education

Industry Experience to Students, Computer Science Methods to Administration

Bert Van Vreckem, Dmitriy Borodin, Wim De Bruyn, Victor Gorelik, Sergey Zhdanov, and Alexander Rodyukov

CONTENTS

INTRODUCTION

Aligning Education with Industry

Aligning higher education with industry needs is an ongoing concern for colleges and universities, especially in fields that evolve rapidly (Masurel and Nijkamp 2009), e.g., information and communication technologies (ICT). This "gap" between education and industry can be characterized from different points of view, and we illustrate this with a few examples and points for further discussion in this paper. First, we see a significant discrepancy between competencies demanded by the labor market on the one hand and those acquired by college or university graduates on the other (Sajid Sheikh et al. 2009; Froeschle 2010). Additionally, some of the existing practically useful scientific research results rarely or even not all are applied in real life (Dosi et al. 2006). In this paper, we share our experience with regard to bridging the gap between science, higher education, and industry. We present a case of planning an industry-friendly study program in the field of ICT and propose how to use a computer science–based approach to automate this planning process.

Development of a study program that reflects the industry needs with regard to graduates' skills and competencies has been widely studied and discussed (Liu and Jiang 2001; Daniels et al. 2007; Reif 2007), and different solutions were suggested. Some companies started their own initiatives intended to educate specialists according to industry needs. Well-known examples are certification programs such as Microsoft Certified Systems Engineer (MCSE) or Cisco Certified Network Associate (CCNA). However, many industry professionals and educators alike question the quality of brand-aligned certifications, and these programs do not train for more general competencies required from employees (e.g., teamwork, communication, etc.).

Responding to Changing Needs

Teaching staff in rapidly evolving disciplines have to be on the lookout for relevant changes continuously while keeping a delicate balance between following the latest trends and sticking to stable technologies that have proven their value. Methods, technologies, or paradigms that are state of the art now may be obsolete within a few years by the time a student graduates. However, today's hype will not necessarily ascend to tomorrow's mainstream. In any case, regular updates to the curriculum are necessary in order to remain relevant.

Another issue in planning study programs is global changes in systems of higher education, for example, the consequences of the Bologna declaration (Borodina et al. 2011). This means that every single study program in every institution of higher education has to be audited and—if necessary—adapted in order to get accredited under the new system.

Planning and organizing courses is tedious work and rapidly becomes intractable for larger institutions or departments. People responsible for this task do a terrific job with the limited resources available to them. More often than not, planning is largely a manual process, often based on implicit knowledge or "intuition." This motivated extensive scientific research on automating the planning process in education on different levels. Two levels can be distinguished within this planning: global and detailed (Zhdanov et al. 2009; Borodin and Tokarev 2008). The detailed level is represented by course scheduling, in which five different subproblems were classified by Carter and Laporte (1997): (1) course scheduling, (2) class-teacher scheduling, (3) student scheduling, (4) teacher assignment, and (5) classroom assignment.

The first subproblem—course timetabling (or course scheduling)—is solved for one semester or term taking the course assignment to semesters as an input. At the global level, the course assignment to semesters is provided as an output of the course planning problem. Course scheduling has been widely studied (for a good overview see, e.g., Rudová et al. 2011) while the course assignment to semesters is treated by administrations of educational institutions as problems of planning study programs (e.g., bachelor, master) or individual student curricula. The current paper formalizes this problem as a mathematical model, thus contributing with an automated course planning problem. The model is applied on a case study of a bachelor program of a Belgian university college. Courses were algorithmically assigned to semesters based on course prerequisite information. We then compared the result of the automatically acquired results with the actual study program, which looked quite promising.

The rest of the paper is structured as follows. Section 1 describes the case of converting a study program to fit the Bologna declaration while incorporating current industry requirements. In Section 2, the mathematical model for planning a study program is presented; the practical contribution and particular features of the model and of the case are discussed. The last section concludes the paper.

CONVERSION OF THE STUDY PROGRAM "GRADUAAT TOEGEPASTE INFORMATICA" INTO "PRBACH APPLIED COMPUTER SCIENCE"

The Faculty of Business Information and ICT of University College Ghent (HoGent) organizes two Bachelor-grade study programs, viz. Applied Computer Science and Office Management. A few years ago, in light of the Bologna process to standardize higher education degrees, the old program, called "Graduaat Toegepaste Informatica" (Graduate Applied Computer Science), had to be thoroughly reviewed and audited in order to have it accredited under the new system. This resulted in the program "Professional Bachelor Applied Computer Science" with a revised and updated curriculum and detailed course descriptions compliant with the European Credit Transfer and Accumulation System (ECTS). In this section, we discuss the methodology used, focusing on how to systematically take industry needs into account in the curriculum.

In order to allow comparison of study programs between institutions, a common reference level framework for competencies and qualifications is needed, expressed as learning outcomes in terms of knowledge, skills, and competencies (Winterton et al. 2005). To prepare the new curriculum, we first looked at which competencies an IT professional needs. We distinguish between the following types:

- *General competencies,* e.g., "being able to reflect critically," "being able to acquire and process data," "being able to communicate effectively," "an attitude toward life-long learning," "creativity," etc. In Flanders, these competencies are goals set by the government in legislation concerning higher education (Flemish Government 2003).
- *General professional competencies,* e.g., "being able to work in a team," "being solution-oriented," etc. These are also specified in Flemish legislation.
- *Specific professional competences,* e.g., "development of computer programs," "maintaining database systems," "system and network administration," etc. These competencies were not set by law but were compiled by the department and teaching staff.

In Flanders, several governmental and professional organizations have been gathering information on necessary competencies for specific vocations. We used the following sources to select the professional competencies for our study program:

- The "Competence and Occupations Directory for the Labour Market" (COBRA) of the Flemish Employment Service (VDAB): http://vdab.be/cobra/info.shtml, retrieved May 2, 2012.
- Professional competence profiles compiled by the Social and Economic Council of Flanders (SERV): http://production.competent.be/competent-nl/, retrieved May 2, 2012.
- Professional profiles from Agoria ICT, a section of the federation of the technological industry in Belgium.
- The social network of teaching staff, specifically alumni, contacts with companies that accommodate internships for our students, and the "resonance committee" (an advisory board in the department consisting of ICT professionals that provides continuous feedback on the curriculum and industry needs).

All competencies gathered in this way were compiled in a list that was then refined and expanded by small task teams consisting of teaching staff, according to their expertise. For every selected competence, we specified a number of indicators that show how the acquisition of the competence can be assessed (see Table 66.1). This makes the assessment process verifiable and focused. Finally, all competencies were assigned to courses, about three to four per course. In some cases, this resulted in redirecting focus of courses to more topical subjects and even the introduction of new courses or removing courses from the program.

AUTOMATED PLANNING OF A STUDY PROGRAM

After incorporating some of the industry needs into the course list, we consider the problem of planning all the courses over the study period. A course has a number of credits that indicates its importance within the study program. Typically, a full time enrollment is about 60 credits, and a full study program for a bachelors degree takes 180 credits. Also, courses are interdependent, that is, in order to enroll for some courses, students must have acquired credits for other courses. For

TABLE 66.1
Competencies and Indicators for the Course *Network and System Administration*

Core competency 1. Installation, configuration, and security of networks and operating systems
Indicators:
- Knowing the basics of network and system administration
- Knowing the basics of network and system security and being able to apply these
- Being able to track down and solve security risks

Core competency 2. Configuration, administration, and testing of complex network, telecom, and server systems
Indicators:
- Being able to map out, analyze, configure, and test a network infrastructure in a mixed environment (Windows, Linux, Cisco)
- Being able to automate system administration tasks using scripts
- Being able to document configurations and procedures

General competency. Executing assignments with perseverance, responsibility, and stress resistance
Indicators:
- Being able to systematically analyze and solve network problems (troubleshooting)
- Being able to set up a complex network infrastructure quickly and in teams

Note: Some competencies are evaluated through a classic written examination (e.g., basic knowledge), others through lab exercises (e.g., network server installation and documenting the process). Also, a few times per semester, students get a group assignment in which they have to build a complex network (including services such as e-mail) in teams in the course of a single day.

example, in order to follow a course named "Advanced Java Programming," a student would first have to succeed in the course "Introductory Java Programming."

The goal is to automate the creation of a study program from scratch, taking into account all available course data. In a more strict way, the general study program planning problem can be formulated as follows.

All the courses must be assigned to all the semesters in the way that all the requirements and limitations are satisfied while the value of some target indicator is optimized. Note that this model does not take financial quantitative indicators into account as proposed in McNamara (1971).

Depending on the system of education, the following requirements and limitations can be applied:

- A specified number of credits per semester and per academic year
- A specified number of hours per semester and per academic year
- Precedence relationships between courses

Examples of target indicators include the following:

- Sum of all distances between interdependent courses should be minimized. Number of credits per semester should be distributed as evenly as possible.

To model this assignment problem, we make use of an integer programming (IP) formulation (Williams 2009). IP is a mathematical method to determine the best outcome (e.g., maximum profit or minimum cost) of a certain linear objective function, given a number of constraints expressed as linear relationships. For reasonably sized problem instances, efficient algorithms exist that find an optimal solution, i.e., an assignment of values to all decision variables so all constraints are met and the chosen target indicator has the best possible value in reasonable time.

Now, we provide a mathematical model of the problem.

We use the following notation in the model

- $P_{i,k} = \begin{cases} 1, \text{if course } k \text{ is a soft prerequisite for course } i, \\ 0, \text{otherwise} \end{cases}$
  - *Course k is a **soft** prerequisite for course i* if and only if course k must be studied either in an earlier or in the same semester as course i;
- $H_{i,k} = \begin{cases} 1, \text{if course } k \text{ is a hard prerequisite for course } i, \\ 0, \text{otherwise} \end{cases}$
  - *Course k is a **hard** prerequisite for course i* if and only if course k must be studied in an earlier semester than course i;
- c_i denotes the number of credits for course i;
- C is a constant that denotes the number of credits per academic year y. (We assume an academic year consists of two semesters);
- S is a constant that denotes overall number of semesters in the program;
- N is a constant that denotes overall number of courses in the program.

Decision variables

$$a_{i,j} = \begin{cases} 1, \text{if course } i \text{ is assigned to semester } j, \\ 0, \text{otherwise} \end{cases}$$

where $i = 1..N$ and $j = 1..S$.

And the objective function

$$Minimize \sum_{j=1}^{S} \left| \frac{C}{2} - \sum_{i=1}^{N} a_{i,j} c_i \right| \tag{66.1}$$

Subject to

$$\forall i = 1..N \quad \sum_{j=1}^{S} a_{i,j} = 1 \tag{66.2}$$

$$\forall i,k = 1..N \quad P_{i,k} \left(\sum_{j=1}^{S} j a_{i,j} - \sum_{j=1}^{S} j a_{k,j} \right) \geq 0 \tag{66.3}$$

$$\forall i,k = 1..N \quad \left| \sum_{j=1}^{S} j a_{i,j} - \sum_{j=1}^{S} j a_{k,j} \right| \geq H_{i,k} \tag{66.4}$$

$$\forall y = 1..3 \quad \sum_{i=1}^{N} a_{i,2y+1} c_i - \sum_{i=1}^{N} a_{i,2y+2} c_i = C \tag{66.5}$$

The objective function (Equation 66.1) assures that credit assignments are balanced between semesters.

The constraint (Equation 66.2) ensures that each course is assigned to exactly one semester. A course cannot be given before its prerequisite courses (soft prerequisites), which is expressed in the constraint (Equation 66.3). In a similar way, constraint (Equation 66.4) encodes hard prerequisites. Finally, constraint (Equation 66.5) fixes the number of credits per academic year.

CASE STUDY

In order to validate the proposed model, we applied it to the curriculum of the study program "Professional Bachelor Applied Computer Science" mentioned above. The program consists of 34 courses taught over three academic years. The total number of credits per academic year is 60. An academic year consists of two semesters, which we will refer to as the fall (September–February) and

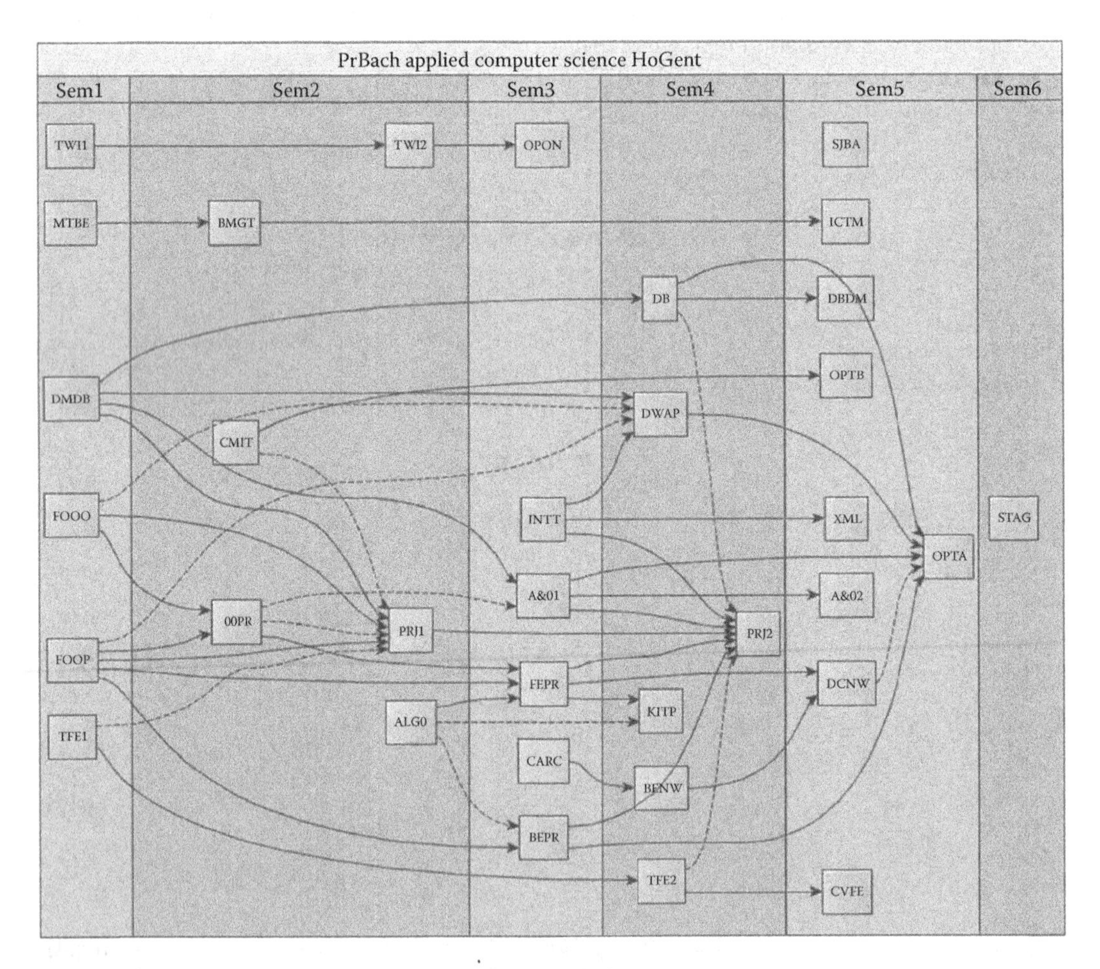

FIGURE 66.1 Study program proposed by the algorithm. Edges denote prerequisite relations between courses as specified in the ECTS sheets. Edges with dashed lines denote courses that can be assigned to the same semester even though one is a prerequisite of the other (*soft* prerequisite). Columns denote the six semesters in the program. Some edges are not shown for clarity. The course "STAG" (representing internships, "Stages" in Dutch) has *all* other courses as a (hard) prerequisite. Also, some superfluous prerequisites were removed from the graph by transitive reduction. For example, prerequisite TWI1 → OPON is already expressed by prerequisites TWI1 → TWI2 and TWI2 → OPON.

spring semester (February–July), respectively. The academic year starts the last week of September, so the first, third, and fifth semesters are fall semesters, the second, fourth, and sixth are spring semesters.

The model was implemented in CPLEX, a well-known commercial solver for mixed integer linear programming (MILP—IP is a special case of MILP) problems developed by IBM (http://www.ibm.com/software/websphere/ilog/, retrieved May 2, 2012). We took the prerequisites for each course that were specified in the ECTS sheets (http://ects.hogent.be/index.cfm?event=oplinfo&p=67, retrieved May 2, 2012 [in Dutch]) and encoded them in the model. It typically takes less than 30 s to find an optimal solution. To place this into perspective: A straightforward approach to solve this type of assignment problem is to let a computer evaluate and compare all possible course assignments. In our case, the number of possible assignments of 34 courses over six semesters is six to the power of 34. Even if the computer could process one million candidate solutions per second on average (which is rather optimistic), it would take about 660 times the estimated age of the universe to finish.

TABLE 66.2
Current Course Assignments

Semester 1	c_i	Semester 2	c_i	Semester 3	c_i	Semester 4	c_i	Semester 5	c_i	Semester 6	c_i
CARC	6	ALGO	5	A&O1	6	BENW	5	A&O2	4	STAG	27
DMDB	5	CMIT	4	BEPR	5	DB	5	CVFE	3		
FOOO	5	INTT	4	BMGT	4	DWAP	5	DBDM	4		
FOOP	6	MTBE	5	FEPR	5	OPON	4	DCNW	3		
TFE1	5	OOPR	4	TWI2	6	PRJ2	7	ICTM	3		
TWI1	4	PRJ1	7	XML	3	TFE2	5	KITP	3		
								OPTA	7		
								OPTB	3		
								SJBA	3		
Totals:	31		29		29		31		33		27

Note: This table shows the courses as they are currently assigned in reality, including credits per course and totals per semester.

TABLE 66.3
Proposed Course Assignments

Semester 1	c_i	Semester 2	c_i	Semester 3	c_i	Semester 4	c_i	Semester 5	c_i	Semester 6	c_i
DMDB	5	ALGO	5	A&O1	6	BENW	5	A&O2	4	STAG	27
FOOO	5	BMGT	4	BEPR	5	DB	5	CVFE	3		
FOOP	6	CMIT	4	CARC	6	DWAP	5	DBDM	4		
MTBE	5	OOPR	4	FEPR	5	KITP	3	DCNW	3		
TFE1	5	PRJ1	7	INTT	4	PRJ2	7	ICTM	3		
TWI1	4	TWI2	6	OPON	4	TFE2	5	OPTA	7		
								OPTB	3		
								SJBA	3		
								XML	3		
Totals:	30		30		30		30		33		27

Note: This table shows the course assignments as proposed by the algorithm, including credits per course and totals per semester. Differences are *emphasized*.

The study program proposed by the algorithm can be found in Figure 66.1 and Table 66.3. The four first semesters contain courses each totaling 30 credits. In only the last two semesters, it was not possible to assign 30 credits to each. During the final semester, students do not have regular classes anymore but instead do an internship of three months in a company. The course corresponding to internships (denoted STAG in the figure, short for "stages," i.e., internships in Dutch) has 27 credits, and all other courses are hard prerequisites. Consequently, it must be assigned to a semester by itself, and the fifth semester will have three credits extra. In this proposed solution, 26 of the 34 courses were assigned to the same semesters as currently in reality. The changes are fairly reasonable. In fact, credits are spread out more evenly as in reality (see Tables 66.2 and 66.3).

CONCLUSION

This paper covered two related topics in education: (1) alignment of study programs to be compliant with industry needs and (2) automated planning of courses in such study programs. Additionally, we presented a case study that spans both topics.

Regarding further research, it would be useful to link semester assignment with the following step in course planning, viz. scheduling. Furthermore, taking financial quantitative indicators into account should provide a considerable improvement over current results. Indeed, assigning teaching staff to courses each academic year is an important consideration for administrators. Courses are added or removed as dictated by industry needs. Organizational changes within the institution induce changes in the program. The number of students enrolled in each course fluctuates each year, either because of new enrollments, drop outs, or students retaking courses they failed. Some courses require more intensive guidance from teaching staff than others, so more staff should be assigned per capita. Teaching staffs tend to specialize in specific domains. Also, large differences in load between the fall and spring semester should obviously be avoided at all cost. Knowing how many teachers must be assigned to each course depends on the expected number of enrolled students and the "intensity" of the course (which can be expressed as a scalar value). For example, you need more teaching staff for coaching teams of students doing a programming project than for a theoretical course that can be taught in a big auditorium. The proposed mathematical model for course assignment can be aligned with the models for course scheduling and strategic university planning. However, mathematically calculated curriculum plans do have their limits. For example, a reassignment of courses could cause some teachers to lose their favorite course(s). This is a sensitive matter that would result in tension among the staff members. Consequently, we see algorithms for course planning as a decision *support* system rather than a decision system. The algorithm is useful to quickly calculate a feasible course plan, a prototype, as it were. This can then be further refined by discussion and negotiation, also taking human factors into account.

REFERENCES

Borodin, D., and Tokarev, A. (2008) "Mathematical models for educational planning," *Quality Journal*. 8, 8, pp. 5–14. 10 p. (*in Russian*).

Borodina, G., Zhdanov, S., Gorelik, V., Borodin, D., and Rodyukov, A. (2011) "Doctoral education: Comparing Bologna and Russian issues." Proceedings of the 4th International Conference: Innovations, ICT and their Application in Education, 8 p.

Carter, M. V., and Laporte, G. (1997) "Recent developments in practical course timetabling," *Practice and Theory of Automated Timetabling II*, Eds., Burke, E. and Carter, M., Springer-Verlag, Lecture Notes in Computer Science 1408, Berlin, pp. 3–19.

Daniels, C., Lynch Hale, N., and Feather-Gannon, S. (2007) "Bridging the Gap Between Higher Education and IT Industry Expectations: The First-Year Experience in Service Learning Projects," OSRA Conference 2007, San Diego, CA.

Dosi, G., Lerena, P., and Labini, M. (2006) "The relationships between science, technologies and their industrial exploitation: An illustration through the myths and realities of the so-called 'European Paradox,'" *Research Policy* 35, no. 10 (December 2006), pp. 1450–1464. http://linkinghub.elsevier.com/retrieve/pii/S0048733306001533.

Flemish Government. (2003) "Decreet betreffende de herstructurering van het hoger onderwijs in Vlaanderen (Decree concerning the restructuring of higher education in Flanders)." Available at: http://www.ond.vlaanderen.be/edulex/database/document/document.asp?docid = 13425. Retrieved 2008-04-28.

Froeschle, R. (2010) "Labor Supply/Demand Analysis: Approaches and Concerns. *TWC Labor Market, and Career Information (LMCI),*" 14 p., available at http://socrates.cdr.state.tx.us/iSocrates/Files/SupplyAndDemandAnalysis.pdf.

Liu, H., and Jiang, Y. (2001) "Technology transfer from higher education institutions to industry in China: Nature and implications." *Science* 21, pp. 175–188.

Masurel, E., and Nijkamp, P. (2009) "Bridging the gap between institutions of higher education and small and medium-size enterprises," Serie Research Memoranda 0037, VU University Amsterdam, Faculty of Economics, Business Administration and Econometrics: Amsterdam, The Netherlands.

McNamara, J. (1971) "Mathematical Programming Models in Educational Planning," *Review of Educational Research*, 41, pp. 419–446.

Reif, G. (2007) "Higher education's missing link: Examining the gap between academic and student affairs and implications for the student experience." *Higher Education* 28, pp. 90–99.

Rudová, H., Müller, T., and Murray, K. (2011) "Complex university course timetabling," *Journal of Scheduling,* 14, no. 2, pp. 187–297, doi 10.1007/s10951-010-0171-3.

Sajid Sheikh, M., Aurangzeb, M., and Tarique, I. (2009) "Bridging the gap between higher education and the Telecommunications Engineering Sector." *Education* 6, no. 8, pp. 62–67.

University College Ghent. (2008) "Zelfevaluatierapport van de opleiding professionele bachelor in de toegepaste informatica (Self-Assessment Report of the Professional Bachelor in Applied Computer Science)," Department of Business Information and ICT and Department of Business Administration, University College Ghent: Ghent, Belgium.

Williams, H. P. (2009) Logic and Integer Programming. Springer.

Winterton, J., Delamare-Le Deist, F., and Stringfellow, E. (2005) "Typology of knowledge, skills and competences: Clarification of the concept and prototype." European Centre for the Development of Vocational Training (Cedefop): Luxembourg.

Zhdanov, S., Gorelik, V., and Borodin, D. (2009) "Educational planning optimization: Review of mathematical models and methods." Mathematics, Informatics, Physics and Their Teaching: Dedicated to the 75th Anniversary of the Mathematical Analysis Chair, Moscow State Pedagogical University, pp. 152–160 (in Russian).

67 Education for Sustainable Development as a Reaction to Modern Challenges

Victor I. Karamushka

CONTENTS

INTRODUCTION

In a general sense, education is the process of transferring knowledge, skills, values, and experiences from one generation to another. This process is realized by the educational system consisting of the object (recipient of knowledge), subject (provider of knowledge), and materialized contents of education (educational programs and materials) as well as the educational environment (needs, requirements, regulations, technical support, etc.). Key objectives of an educational system are (i) to transfer or share knowledge, (ii) to train skills on how to use knowledge, and (iii) to train skills on how to get knowledge needed. Eventually, an educational system has an apparent impact on the mind, character, and physical capability on individual and social levels, making people prepared to function in different areas of human activity. Due to this, an education is a powerful factor of indirect transformation of the society and the environment. However, individuals, social groups, and society as a whole, meeting individual and common needs in the process of social activity, generate positive outputs and outcomes as well as cause multiple insignificant and essential negative changes, which may become later problems. With the lapse of time, a critical mass of such changes are transforming into challenges, which threaten the development of society and even human civilization.

In this regard, one may ask a question about how education, as one of the basic areas of human activity, should react to the changes caused by human activity? Has the educational system a potential to forecast and prevent negative consequences of human activity? This paper is an attempt to get answer to this question.

RESULTS OF ANALYSIS

The author has analyzed social processes taking place in Ukrainian society, mainly related to the educational sector during last two decades (in fact, since gaining independence by Ukraine in 1991). Sources of data and information include but are not restricted to scientific and educational publications (monographs and periodicals), national statistical data, national and international legal documents, informational resources of international organizations, etc.

Results of theoretical exercises are summarized in the following. Figure 67.1 represents a generalized scheme of knowledge transformation and formation of educational needs in society. On one hand, independent on the educational level of the society, human activity is aimed first of all at the meeting basic human needs. For this purpose, people (individuals, communities, and society) utilize different resources – in a larger sense, physical substances, energy, and information. All these resources are extracted from the natural, social, and anthropogenic environment. Utilization of the resources is accompanied with the generation of waste and emission of pollution, which are disseminated and/or deposited in the natural environment. All these processes are mediated by the economy and inevitably lead to transformation (changes) in the environment (first of all, the natural one) and relationships in a society as well as changes of the human mind, character, attitude, etc. In this respect, we consider a society as a part of the eco-social system.

Apparently, the actors (both public leaders and ordinary people) with new competencies are required by the society to manage changes in the eco-social system and meet new challenges. Therefore, evident or invisible messages or requirements are directed to the system, stipulating the development

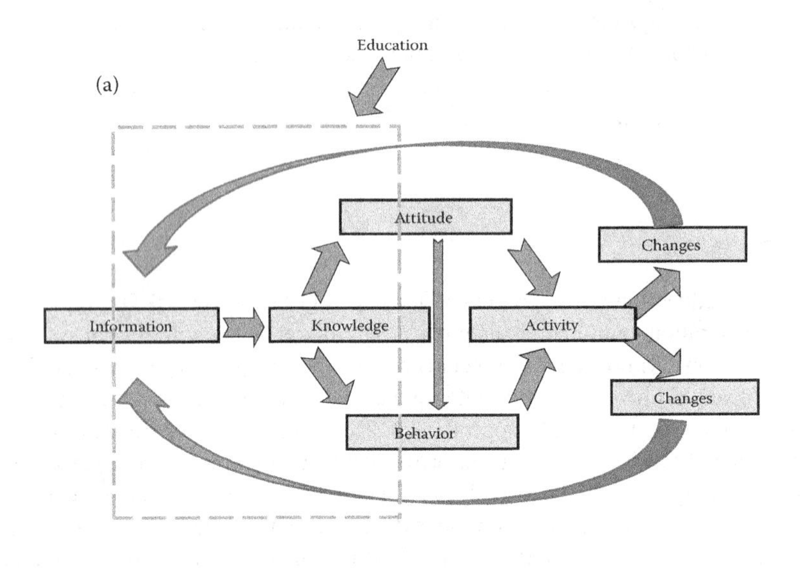

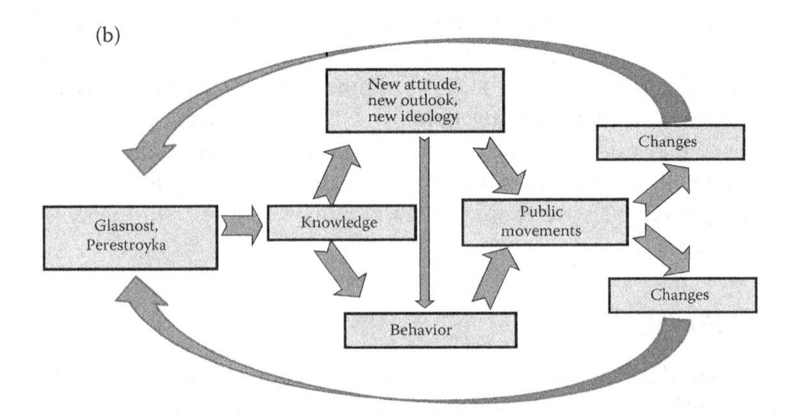

FIGURE 67.1 General scheme reflecting the role of education in a society (a) and the scheme of interrelation between educational inputs and societal changes in the former USSR at the end of Soviet era (b).

of such competencies. This system is dealing with an education in the broad sense of this term. During the educational process, inputs (information) are transferring into knowledge, skills, abilities, and other characteristics, creating background for development of attitude and influencing human behavior. Speaking about attitude and behavior, we keep in mind both individuals and community groups. Knowledge, attitude, values, etc. become apparent in behavior and predetermine forms and content of human activity aimed in the end on the improvement of the quality of human life.

The scheme depicted on the Figure 67.1a demonstrates the crucial role of the educational system in the cycle of transferring knowledge in the society. The scheme demonstrates that, on one hand, changes in the society generate new requirements for the educational systems, and, on the other hand, innovations in the educational systems fundamentally determine development of the society. Figure 67.1b illustrates the interconnection and interdependence between educational inputs and societal changes having a place in the former USSR in the last decades of the 20th century (Ukrainian Society 2005). In the middle of 1980, Ukraine, being a part of the USSR, entered into the phase of crucial contradictions between productive forces and dominating relationships in the society. Accumulations of the society's contradictions put into the agenda new requirements for ensuring human rights and freedoms. Proclaimed by the totalitarian regime policy of access to information (well known as "glasnost") immediately impacted the system of education and public information through filling in with new information, new interpretation, and a new outlook. Putting aside details of this process (well known as "perestroika"), we deliberately concentrate focus on the triggering effect of this aspect of societal transformations. Access to new information and new knowledge stimulated public activity, which was institutionalized in new public organizations and public movements. Finally, it ended up with a collapse of the socialist system and its transition to the market economy on the whole post-Soviet area, including Ukraine. Adjusting to the contemporary requirements, the educational system as well entered into a long phase of transformations. This phase is still ongoing.

FEATURES OF SOCIETAL CHANGES

To clarify the role of the educational system in managing changes in society and the environment, it is necessary to discuss the character of changes, which are consequences of the activity of educated members of the society. First of all, it is important to emphasize that in a larger sense, human activity is aimed at increasing the quality of life and ensuring its safety. Therefore, it is expected that positive changes are the main outcome of human activity. However, in reality, positive changes are inseparably connected with negative ones. For example, the manufacturing of customer goods is inevitably accompanied with waste generation. It is the first feature of changes resulting from human activity.

Second, with the lapse of time, positive changes may be transformed into negative ones or give rise to negative consequences. To give a simple example, let's be reminded that DDT (the synthetic pesticide dichlorodiphenyltrichloroethane) – a remarkable output of creative activity – first was synthesized in 1874 and since 1939 had been broadly used as an effective pesticide in agricultural and medical practice. The insecticide properties of DDT were discovered in 1939 by Swiss chemist Paul Muller, and in 1948, he was awarded the Nobel Prize in Physiology and Medicine for his discovery of the high efficiency of DDT as a contact poison against arthropods (mammals, birds, insects transmitting deceases). In 1940–1960, using DDT saved millions of lives from malaria in Asia, Africa, and Latin America. However, currently DDT is banned for use in 173 countries due to its persistence in the environment and its cumulative and toxic effect on the majority of living organisms (DDT 2012). As we can see, euphoria with the positive properties of DDT was transformed into frustration with its toxic effect and finally ended up in a ban on its production and use.

Another aspect of the issue is that positive changes require proper management and control to sustain their positive effects and avoid or prevent risks and threats to human safety. Creation of technologies of energy generation by nuclear power plants used to be considered as a progressive step in the technological development of society; however, weakening or loss of control over this sector may be followed up with tragic consequences. Do we need to be reminded of what Chernobyl and Fukushima mean for the contemporary world?

Finally, accumulation of positive changes in the eco-social system leads, in general, to negative effects. Moreover, accumulation of negative effects over critical mass becomes a challenge for societies, states, and entire human civilization. This thesis finds confirmation in the dynamics of the Earth's population. Statistical data convince us that population growth has become unprecedently high since 1940 (World Population Prospects 2006). This phenomenon has many reasons; however, many researchers consider the introduction of usual practice antibiotics as a key factor in triggering demographic processes (Catton 1980). Undoubtedly, antibiotics promoted to saving and extension of human life. As a result, the current annual increase of Earth's population is exceeding 70 million. Scientific forecast confirms that this tendency will be observed many decades in the 21st century and lead to overcrowding of the planet.

REACTION OF EDUCATIONAL SYSTEM ON SOCIETAL CHANGES

Evolution of the educational system in Ukraine during last two decades is a remarkable example of an educational response of socio-economical changes and demand of the society. Table 67.1 represents quantitative indicators of educational dynamics (data provided by the State Statistics Service 2012). There are at least two general conclusions resulting from the overview of these data.

First of all, the number of secondary schools and vocational educational institutions has been gradually decreasing starting from the middle of 1990. Accordingly, the number of students in secondary schools and vocational institutions as well as the number of specialists graduating from vocational establishments are decreasing considerably. In spite of this, the system of higher education demonstrates opposite tendencies: Since 1990, the number of universities in Ukraine has increased by 2.35 times, and the number of university students has increased accordingly (by 2.4 times). It means that the transition of Ukrainian society to a market economy provoked a demand for higher education on the one hand and, on another hand, created opportunities and facilitated access to higher education through commercialization of educational services. Undoubtedly, the quality of educational services provided by most of the newly established universities is problematic; however, this issue is a topic of other studies. In the light of the focus of this paper, it is important to consider the tendencies in the content of education and type of graduators from educational institutions.

As far as changes in society have resulted in specific demands for educational services, the educational system rapidly adapted to the pragmatic requirements of transforming society: in the near-term period (operational level), the educational system met the demand for teaching foreign languages (the country became open to the world), preparing accountants (demand resulted from increasing entrepreneurship and small business activity), training new drivers and vehicle service providers (parking for vehicles is growing rapidly), etc. The educational system slowly met the demand of the society in medium-term priorities (tactical level): preparation of specialists for growth (fundamental sciences, engineers, constructors, etc.) by classic universities sufficiently decreased. At the same time, preparation of specialists for management and regulation of current activities (business administrators, finance and economy managers, IT specialists, lawyers, psychologists, etc.) increased enormously. For example, the former Kyiv Institute of Engineers of Civilian Aviation was training specialists exclusively for the needs of the aviation and aerospace sector. In 1990, the Institute was reorganized as the National Aviation University. Currently, most students of the University are trained in specialties, which have very indirect relationships to aviation and space technologies. In 2009, the University got governmental financing for preparation of 499 masters of sciences degrees, and only 172 of them will be prepared for basic sector (Ministry of Education, Science, Youth and Sport of Ukraine 2009). Currently, many other specialties look quite exotic for this university (such as psychology, interpretation, international law, eco-biotechnology, social service, etc.).

It means that the university is functioning under pressure from the demand originated by changes in the society. As it follows from Table 67.1, the number of newly established universities in Ukraine

TABLE 67.1
Educational Institutions in Ukraine in 1990–2011 (Data Represent State of the Art at the Beginning of the Teaching Year)

	Secondary Education		Higher Education (Number of Institutions)		Number of Students (Thousands)		Graduated Specialists (Thousands)	
	Number of Secondary Schools (Thousands)	Number of Pupils (Thousands)	Vocational Institutions, Colleges	Universities	Vocational Institutions, Colleges	Universities	Vocational Institutions, Colleges	Universities
1990/91	21.8	7132	742	149	757	881.3	228.7	136.9
1991/92	21.9	7102	754	156	739.2	876.2	223	137
1992/93	22	7088	753	158	718.8	855.9	199.8	144.1
1993/94	22.1	7096	754	159	680.7	829.2	198	153.5
1994/95	22.3	7125	778	232	645	888.5	204.3	149
1995/96	22.3	7143	782	255	617.7	922.8	191.2	147.9
1996/97	22.2	7134	790	274	595	976.9	185.8	155.7
1997/98	22.1	7078	660	280	526.4	1110	162.2	186.7
1998/99	22.1	6987	653	298	503.7	1210.3	156.9	214.3
1999/00	22.2	6857	658	313	503.7	1285.4	156	240.3
2000/01	22.2	6764	664	315	528	1402.9	148.6	273.6
2001/02	22.2	6601	665	318	561.3	1548	147.5	312.8
2002/03	22.1	6350	667	330	582.9	1686.9	155.5	356.7
2003/04	21.9	6044	670	339	592.9	1843.8	162.8	416.6
2004/05	21.7	5731	619	347	548.5	2026.7	148.2	316.2
2005/06	21.6	5399	606	345	505.3	2203.8	142.7	372.4
2006/07	21.4	5120	570	350	468	2318.6	137.9	413.6
2007/08	21.2	4857	553	351	441.3	2372.5	134.3	468.4
2008/09	21	4617	528	353	399.3	2364.5	118.1	505.2
2009/10	20.6	4495	511	350	354.2	2245.2	114.8	527.3
2010/11	20.3	4299	505	349	361.5	2129.8	111	543.7
2011/12	19.9	4293	501	345	356.8	1954.8	96.7	529.8

during the last two decades has increased enormously. Most of them have private status and, of course, are oriented exclusively toward market demand. Thus, speaking on the whole educational system in Ukraine, one may conclude that it is evolving according to the short- and medium-term priorities of the society and complying with current demand.

Meanwhile, there are many contradictions, which are important to consider. The educational system is training people for future life and activity; however, educational programs are based on the current demand of the society. In other words, it used to be that educational programs did not consider potential future changes in the social and natural environment, which have to be managed by those who are graduating from universities at present. A simple case may illustrate this situation. Transport facilities during last two decades in Ukraine increased enormously. In the city of Kyiv (population 2.7 million), the capital of Ukraine, one million vehicles were registered in August 2008. Every day, at least 1.5 million cars are running in the city, and their number is still increasing. Similar tendencies are observed in other cities of Ukraine.

This means that (i) formal and informal educational institutions fully met increased demand in educational services at the operational level (training car drivers, mechanics, etc.); (ii) educational institutions are meeting demand in educational services at the tactical level by means of preparing professionals to manage increased traffic issues, transport logistic facilities, regulate sales and services, etc. However, there is no evidence that educational policy is based on determination of long-term priorities of sustainable development of the society. Meanwhile, it's clear now that increasing the number of cars apparently results in increasing fuel consumption and pollution of the environment. It leads to depletion of nonrenewable natural resources (oil and gas), climate change, and becoming one of the key challenges of human development. In general, this strategic aspect is still out of practical consideration both at the policy and educational levels.

This case demonstrates that the logic of changes in the society (both in economic and social sectors) inevitably leads to recognizing the concept of sustainability and principles of sustainable development (SD). Consequently, it sets up specific requirements to the contemporary educational system.

EDUCATION FOR SUSTAINABLE DEVELOPMENT IN UKRAINE

The paradigm of sustainable development got a theoretical background in the second part of 1900th (see, for example, Meadows et al. 1972; Toward 1973; Daly 1996), introduced by the UN Commission on Environment and Development (Brundtland Commission) in 1987 (Our Common Future 1987) and becoming well known since the UN Conference on Environment and Development (Rio 1992) (Agenda 21 1992). During the last two decades, many countries have developed, adopted, and are dealing with the implementation of national strategies of sustainable development. Several attempts to develop a national SD strategy were undertaken in Ukraine in the period 1995–2012; however, not one of them was successful. Taking into account that education is exclusively an important factor for human society's transition to sustainability (Agenda 21 1992), European countries under the leadership of the UN Economic Commission for Europe have adopted the Strategy for Education for Sustainable Development (UNECE Strategy 2005). In spite of the fact that Ukraine has participated in the development and approval of this policy act, still there is no national strategy or any other policy document determining goals, objectives, priorities, action plans, etc. in the field of education for sustainable development (ESD).

Meanwhile, there are some positive intentions declared at the governmental level. The Law of Ukraine "On the Basics of National Environmental Policy of Ukraine for the Period until 2020" (approved on December 21, 2010) envisages the elaboration and implementation of the "Strategy of Environmental Education for Sustainable Development of the Ukrainian Society and Economy of Ukraine" by 2015. As of May 2012, still there is no output of this process; however, there are a lot of progressive initiatives aimed at incorporation of SD principles into educational programs of secondary, vocational, higher, and post-graduate educational institutions. At the time when governmental bodies, in fact, ignore the ESD issue, these initiatives demonstrate the effectiveness of a bottom-up approach.

ESD initiatives in the Ukraine have been realized by means of (i) elaboration of manuals (see, for example, Lessons 2007; Karamushka et al. 2011) and teaching the lessons for SD in selected secondary schools, the number of which is gradually increasing (Education 2012); (ii) elaboration of the programs of specific disciplines, devoted to SD, and incorporation of them into curricula of more than 20 universities; (iii) incorporating SD aspects into a state educational standard of specialty "Ecology." Apart from this, a master's program, "Public administration and sustainable development," was opened at National University "Kyiv Polytechnic Institute," and the first 20 specialists graduated in 2011. The educational process has been supported with recently elaborated and published qualitative teaching materials (manuals, textbooks, etc.), promoting understanding of fundamental (e.g., Krysachenko and Khylko 2001; Melnyk 2005; Social-Economic Potential of Sustainable Development 2007; Sustainable Development: Theory, Methodology, Practice 2009) and practical (e.g., Karamushka 2009) SD aspects.

The basic objectives of education are transferring knowledge and formation of competencies for life activity. These objectives are supplemented with SD objectives, which are (i) formation of knowledge and competencies for forecasting of the consequences of life activity, evaluation, and prevention of its negative tendencies; (ii) formation of needs to change life objectives, which dominate under current market conditions; (iii) formation of the specific features of social consciousness, which correspond to the requirements of sustainable development of the society. As we can see, these objectives are corresponding to the requirements raised by evolving modern society (see the previous part of the paper).

CONCLUSION

The Ukrainian society is still searching for its own way of development. Governmental institutions, fighting with multiple economic, social, and other current problems, in general, ignore or pay little attention to the determination of the strategic long-term objectives and priorities of national development. The educational sector as well concentrates efforts mainly on meeting current demand of the society; however, logic of societal changes convinces us of the necessity of forecasting future societal needs and creating facilities to comply with them. Therefore, Ukraine, in spite of requirements of the UNECE Strategy for Education for Sustainable Development (UNECE Strategy 2005), still has not approved either a national strategy or a national action plan in this area. Such tendencies are typical for many other countries, in particularly, for former Soviet Union republics, e.g., Belarussia, Russian Federation, the Republic of Moldova.

In fact, bottom-up ESD movement in the educational sector is a protective reaction of the social system on challenging changes in the society. As we can see, the educational system initiatives are focused, among others, on the sustainable development as a fundamental approach of meeting social and other needs of society by means of its economic activity exclusively within the framework of carrying capacity of the natural environment (Our common future 1987; Agenda 21 1992). Moreover, according to the dominating opinion of the science and educational leaders, this approach is the only productive alternative. Therefore, one of the key objectives of the national educational system at present is to transfer knowledge on sustainability and form basic skills to apply sustainable development principles in practical life activity of the general public, public leaders, and decision makers. As it follows from previous parts, the educational system is contaminated with many progressive SD initiatives at the levels of secondary, higher, and post-graduate education institutions. However all of these initiatives still are looking like point interventions rather than having the character of planned and systematic actions. Having in mind that the importance of these SD educational initiatives is increasing, we may expect sufficient changes in the human capacity for understanding and applying SD principles in the near future.

The point is that ESD requires more understanding and support from governing institutions. In this regard, legally approved governmental decisions to develop and implement a national Strategy of Education for SD gives hope for educational leaders and professionals in Ukraine that integration of sustainability principles in educational programs will be supported by national government and new educational opportunities that will help people to meet and tackle modern civilization challenges.

The two following years will be very important for strengthening the ESD process in Ukraine. Failure of this process may result in the rising demand for the educational sector mainly in preparation of crisis managers.

REFERENCES

Agenda 21 (1992): http://www.un.org/esa/dsd/agenda21/res_agenda21_00.shtml.

Catton, W. R., Jr. Overshoot. *The ecological basis of revolutionary change.* Urbana and Chicago: The University of Illinois Press, 1980.

Daly, H. E. *Beyond Growth. The Economics of Sustainable Development.* Boston: Beacon Press, 1996.

DDT (2012): http://en.wikipedia.org/wiki/DDT.

Education for Sustainable Development in Actions: Familiarizing Ukrainian teachers with experience, methodology of teaching students skills and abilities to build their own life, 2012: http://esd.org.ua/En.

Karamushka, V. I. Introduction to the practice of sustainable development: Manual. Kyiv Lviv: Kray, 2009. 240 p. (in Ukrainian).

Karamushka, V. I., Pometun, O. I., Pylypchatina, L. M., Sushchenko, I. M. Lessons for Sustainable Development: Manual for 9(10) Grade Pupils. Kyiv: Publishing House Osvita, 2011. 144 p.

Krysachenko, V. S., Khylko, M. I. Ecology. Culture. Politics. Kyiv: "Znannia", 2001. 598 p. (in Ukrainian).

Lessons for Sustainable Development: Manual for 7–8 Grades Pupils. Kyiv: Litera Ltd, 2007. 96 p.

Meadows, D. H., Meadows, D. L., Randers, J., and Behrens, III, W. W. *The Limits to growth: A report for the Club of Rome's project on the predicament of mankind.* Universe Books, 1972.

Melnyk, L. G. (ed.) *Basics of Sustainable Development: Manual.* Sumy: University book, 2005. 654 p. (in Ukranian).

Ministry of Education, Science, Youth and Sport of Ukraine (2009): http://www.mon.gov.ua/newstmp/2009_1/28_07_1/1/

Our Common Future. Report of the World Commission on Environment and Development. Oxford, New York: Oxford University Press, 1987.

Social-Economic Potential of Sustainable Development: Manual. Ed. by L. G. Melnyk and L. Hens. Sumy: *University book*, 2007. 1120 p. (in Ukrainian).

State Statistics Service of Ukraine (2012): http://www.ukrstat.gov.ua/

Sustainable Development: Theory, Methodology, Practice. Manual. Ed. by L. G. Melnyk. Sumy: *University book*, 2009. 1216 p. (in Ukrainian).

Toward Global Equilibrium. Ed. by Meadows, D. L., and Meadows, D. H. Cambridge, MA: Wright-Allen Press, 1973.

Ukrainian Society. Ed. V. S. Krysachenko. Kyiv: "Znannia Ukrainy", 2005. 792 p. (in Ukrainian).

UNECE Strategy for Education for Sustainable Development (2005): http://www.unece.org/fileadmin/DAM/env/documents/2005/cep/ac.13/cep.ac.13.2005.3.rev.1.e.pdf.

World Population Prospects: The 2004 Revision Analytical Report. UN, New York, 2006.

68 Overseer or Mentor? *Future Roles of Supervisors at Universities*

Justyna Bugaj and Aleksandra Fedaczynska

CONTENTS

Changes connected with the execution of the assumptions of the Bologna process and introduction of the National Qualifications Framework are superimposed on different expectations of young people toward the effects of education. However, traditional education still remains predominant in Polish universities. The discussion concerning what the final diploma examination should look like (which has so far been most often connected with the defense of the previously written thesis) has not been finalized yet.

The requirement set for candidates for a degree (participants of a seminar) and the theses they write are varied and often depend on the attitude and role (Unsworth et al. 2010) assumed by the supervisor. Mentors help students choose the topic of the thesis, according to their interests. They also let them make choices connected not only with the fulfillment of seminar requirements but also with their further life and career and assist them in these choices. Seminars take place on the dates usually determined individually. Students who choose as a supervisor an overseer type select the topic of the thesis that reflects the supervisor's interests. The seminar, which can be either obligatory or optional, is held a few times a semester for the entire seminar group. If necessary, extra meetings with students and individual discussions of their theses are arranged.

Competition in the market of higher education institutions is so high that students begin to pay more attention to the market value represented by the diploma rather than to the diploma itself. This value can be recalculated as the indicator of the rate of finding new work or degree of its maintenance (indicator of employer's satisfaction with the newly hired employee). This, in turn, is connected with the field of study and specialization, i.e., the topic of the diploma thesis. Therefore, more and more often, students seek the assistance of such supervisors with whom the process of writing a diploma thesis has a practical dimension or is connected with spending time in the presence of an authority. Such theses can be projects executed upon the order of employers (business, industry, research labs).

In this study, the two most frequent cases of the role played by the supervisor (an overseer and a mentor) are presented. Differences in their behavior result from the way they establish relationships with students. Will any of these relationships be predominant in the future? Which one will be preferred by students? Due to the length of the text, the variants of these two cases will not be analyzed from the perspective of the degree of experience (number and quality of supervised diploma theses) and higher or lower requirements set for students. The most frequent kinds of supervisors: fatherly type, laissez-faire type, master type, and partner type will be discussed in further parts of this paper.

TRADITIONAL ATTITUDES OF STUDENTS AND TEACHERS (SUPERVISORS)

Traditional education that takes advantage of delivery methods gives the initiative of all activities to the academic teacher who has a direct influence on the students' learning process. The teacher is a source of knowledge, which is most often transferred in the form of a lecture. Students must be present at the lecture to remember the presented information. In this case, students spend the majority of their time in class with the academic teacher playing the role of a lecturer, regardless of the type of classes (lectures, classes, seminars).

In class, the academic teacher can take different roles (http://www.tlumaczenia-angielski.info/metodyka/teacherrole.htm [20.08.2010]), including the role of a facilitator (organizer) who creates the conditions of work, a controller who verifies the progress in thesis writing (e.g., classes with automatic repetition of content, reading aloud), disciplines, instructs, gives information (which task is to be executed, at what time, in what group), engages students in planning and conducting classes and informs them what they are supposed to do and why, evaluates or presents opinions on learning progress (feedback), rewards progress and corrects mistakes, and informs about requirements and assessment as well as remains fair and sensitive to student reactions. Other popular roles: actor standing in front of the whole class, gardener sowing the seeds and observing their development, guide leading students to discover and get to know the world. Denek also mentions another role of a teacher, namely the guide in the world of knowledge, the animator of learning, the organizer of work, the co-participant and co-partner (Macfarlene 2011).

Kember (1997) and later Harden and Crosby (2000) have elaborated two basic teaching models that are most frequent in higher education institutions: the *teacher-centered model* in which expert knowledge is transferred to novices (students) and the *student-centered model* in which knowledge is gained with the active role of the student and the teacher functioning as an organizer (facilitator). The comparison of the two concepts is presented in Table 68.1. In practice, there are two cases characterized by the predominance of the elements of one concept.

In practice, the teacher is often referred to as the advisor. Then, the teacher is described as a friend offering advice, as an understanding helper who uses the ethos of the academic teacher or ethos of the teacher service. Table 68.2 presents the roles of the academic teacher based on Arden's model (Harden and Crosby 2000).

The role of students also changes, and they become the participants of the learning process equally responsible for this process as the academic teacher. In the case of communication with the

TABLE 68.1
Comparison of Model Concepts of Teacher-Centered and Student-Centered Learning

TCL	SCL
Low level of student choice	High level of student choice
Student passive	Student active
Power is primarily with the teacher	Power is primarily with the students

Source: O'Neill, G., and McMahon, T., *Student-centred learning: What does it mean for students and lecturers?* University College Dublin http://www.aishe.org/readings/2005-1/oneill-mcmahon-Tues_19th_Oct_SCL.html [20.08.2010].

Note: SCL, student-centered learning; TCL, teacher-centered learning.

TABLE 68.2
Roles of the Academic Teacher

Role	Aspect
Provider of information	Lecturer
	Exemplar of practical skills
Behavior pattern	As a representative of a specific field
	As an active, conscious citizen of an information society
Supervisor (facilitator)	Master – guide
	Mentor – advisor
Creator of teaching materials	Textbooks, teaching aids
	Leaflets and brochures concerning studies
Organizer	Of the teaching process
	Of individual classes
Evaluator	Of students
	Of school or university curricula

Source: Own elaboration based on Harden, R. M., and Crosby, M. J., *AMEE Education Guide* No. 20, *Medical Teacher* (2000) 22, 4, 334–347; Kędzierska, B., Frankowicz, M., Mirecka, J., *Nowa rola nauczyciela a technologie informacyjne*, in: Kędzierska, B. and Migdałek, J. (eds.), (2004) *Informatyczne przygotowanie nauczycieli: Internet w procesie kształcenia,* Kraków Rabid, pp. 53–58 (English title: *The new role of teachers and information technology*).

supervisor–mentor/facilitator, the role of the student is to develop their skills, abilities, and predispositions with the use of the possibilities offered during seminars. The student is supposed to be independent, creative, and responsible (to themselves for their own progress and development to the academic teacher for learning results). They are supposed to find an independent point of view, have motivation to search and gain knowledge on their own. Finally, they are supposed to make decisions about their career path, ways of achieving goals, which is directly connected with their desires, aspirations, needs, and functioning in the knowledge society. The changes in the academic methodology and the role of students in the learning process are described more broadly by Krajewska (2003).

Depending on the motives they are driven by, students will tend to choose different promoters. Individuals who are engaged, who believe that work done while studying will be important for their future career, and who are ambitious will rather choose the mentor. Under the guidance of an overseer, they may feel suppressed, and if they want to develop themselves, they encounter obstacles. The command and instruction approach inspires their rebellion and the feeling of being deprived of their right to develop their outlooks on learning and working independently. The mentor stimulates their natural motivation and involvement. The support and praise they get for independent work and intellectual efforts are the best reward for them (Pausch and Zaslow 2008).

Students who choose the overseer are motivated by other factors. Those who make this choice usually think that the goal of studying is to complete studies and get a diploma that will increase their value in the labor market, both as a candidate for work and in the eyes of the current employer. The fact that the objectives are restricted to finishing studies is connected with the lack of broad academic interests. When the topic, literature, and requirements are strictly determined by the supervisor, it is a guarantee that the way to achieving the goal (finishing studies) is simple and not too difficult with minimum work outlays (Hoskins and Newstead 2009).

Figure 68.1 shows the process of communication between an academic teacher and a student. It will be hereinafter abbreviated as the NSN model. In the light of this model, the teacher is not a medium who acts in one direction, but he or she is concentrated on the current interaction with the

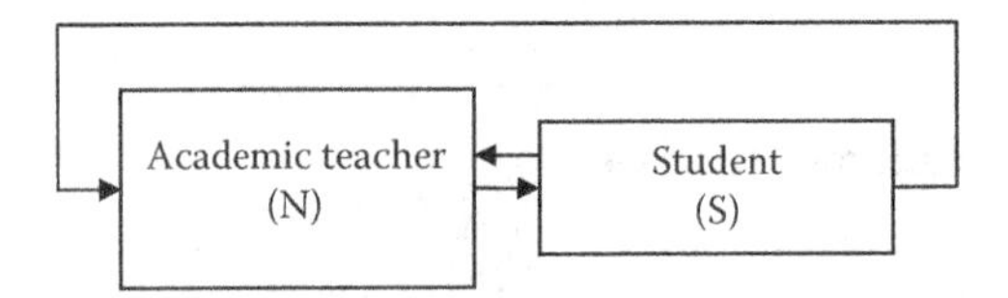

FIGURE 68.1 NSN model. (From Bugaj, J., *SIS—koncepcja prowadzenia dydaktyki akademickiej—metody i techniki szczegółowe,* in: Drzewowski M., Maliszewski W.J. (eds.), *Komunikacja społeczna a zarządzanie we współczesnej szkole,* Wydawnictwo Adam Marszałek, Toruń 2008, pp. 247–257 (English title: *SIS—The Concept of Conducting Academic Teaching—Detailed Methods and Techniques.*)

student and on taking adequate stimulating activities. The teacher is ready to cooperate, exchange information, and learn together. Here, the academic teacher stimulates the process as a mentor or facilitator, depending on the possibilities and abilities of a seminar group, the degree of its involvement, and preparation for the tasks the group faces.

There are two loops (a small one and a large one) shown in Figure 68.1. The large loop means stimulation to modify work with students (curriculum, presented examples, range, and topics of projects for student's own work, etc.) planned for a successive semester. The small loop means operational contacts (during the semester, during the seminar) and the resulting changes at short intervals (preparing extra tasks or problems for outstanding students to solve, reacting to press information, especially influencing the topics of supervised diploma theses, reacting to inventions or technological facilities, extra interpretation of the already discussed issues, etc.). The discussed NSN model also refers to the possible roles the academic teacher can assume, depending on his or her approach to the profession.

SUPERVISOR–OVERSEER

By adopting the traditional role, the overseer focuses on determining requirements toward seminar participants and enforcing them. This refers to typically formal requirements, such as thesis volume, bibliographic description, structure of theses written under his or her supervision, graphic layout, and contents. This kind of supervisor determines the number of chapters and the range of content in each of them. The student is also required to read certain literature and describe the issues they have become familiar with.

The overseer largely interferes with the student's thesis, but he or she also has high responsibility. In the traditional role, the supervisor is responsible for the thesis submitted by the student, its academic level, and progress made by the student in the writing process. Therefore, such a supervisor makes key decisions to determine the directions of the student's development (indicates literature, specifies the way the problems are tackled and defines the problems that must be discussed in the paper). The overseer also specifies the range of knowledge that must be included in successive parts of the thesis. Regardless of the fact of whether the requirements are selected individually for students or specified for the entire seminar group, it is the overseer who determines the way the topic is to be presented and verifies to what extent the student has fulfilled the given task.

The students must meet the requirements set by the overseer, taking into account the criteria their supervisor has determined for completing the seminar and finishing studies. The traditional approach to supervising the thesis largely limits the possibility of students' independent creation of work but, at the same time, gives them the sense of security. The main responsibility for the thesis lies on the part of the supervisor who is a creative author while the student is left with the role of an executor.

The main method that can be used by the overseer during the seminar is a lecture. During the lecture, he or she can clearly and efficiently present expectations toward the thesis and the way it should be written as well as the schedule for handing of individual parts of the thesis. It is a very effective form if we assume that each student understands and accepts the rules mandatory when writing the thesis and all requirements set by the overseeing supervisor.

Another method that proves useful during the seminar conducted by the overseer is the method of controlling student progress with a test. Short tests administered during regular meetings make it possible to verify whether the student can correctly determine successive parts of the thesis. This mainly refers to the empirical part of the thesis. In this way, the overseer can check, for example, the skill of formulating research questions as well as the knowledge and level of understanding of the methods that have been recommended for the seminar.

SUPERVISOR–MENTOR

The second role that can be played toward students is that of a mentor. The mentor is a wise and trusted counselor or an influential supporter (http://dictionary.reference.com/browse/mentor, 01.09. 2011). Mentors are usually senior individuals who can offer their own experience, knowledge, and wisdom (Parsloe 2003). Unlike the supervisor, the mentor leaves the entire responsibility for and burden of the diploma thesis to the student. "Among the people we meet, some remain in our memory forever and even in specific conditions they can be transformed into active elements of our personality" (Wołoszyn-Spirka 2001). The adoption of the role of the mentor has already been described in the business environment as the technique of development support called mentoring.

The concept of mentoring has been weakly defined, and in principle, it refers to coaching. But the majority of definitions take advantage of the definition created by Megginson and Clutterbuck (2005), according to which "mentoring relates primarily to the identification and nursing of potential for the whole person. It can be a long-term relationship, where the goals may change but are always set by the learner. The learner owns both the goals and the process."

In our further consideration, we will use the definition stating that "mentors are people who help others use their own potential through their own work and activity, so the goal of mentoring is the long-term grasp and practical use of new skills by way of counselling and consultation" (Parsloe 1992).

The stages of the cooperation process between the master and the pupil are described differently, but they are most frequently composed of four stages (Parsloe 2003):

- Approval of a personal development plan (understanding and acceptance)
- Inspiring independent management of one's learning process (motivation to act)
- Giving support in the execution of a personal development plan (supporting plan execution)
- Assistance in evaluating results (evaluation and maintenance of dynamics)

The supervisor-mentor emphasizes the change in the student's role from a passive one to an active one in which the student must work harder than the teacher (Edgerton 2004). They can also cause a shift of the center of gravity in relations between the teacher and the student (toward cooperation, negotiations). It requires the different behavior of students, their participation in classes on equal terms, which can cause their resistance (e.g., fear of the unknown, earlier bad experience, wrongly planned classes). By predicting this behavior, the teacher can get prepared constructively. During the first classes, the mentor can present the advantages of the applied methods to students as well as explain why these and no other methods have been chosen and why they may be advantageous for both parties. The supervisor playing this role should react to all doubts and explain them on a current basis. Additionally, teachers should create activities so as to encourage students to active participation in classes. They should also explain which teaching method can be helpful in achieving individual educational effects and react to students' ideas.

Direct interaction and responsibility for the achieved goals as well as reflection on the undertaken actions through discussion teach students to express their own thoughts and justify judgments.

This enables the exchange of differing opinions and allows for working out a common standpoint. It also arouses intellectual curiosity and interest as well as motivates students to do their independent searches. These activities create the atmosphere of cooperation in the educational process based on mutual respect, trust, and kindness. It is further strengthened by mutual relationships and the responsibility of students for fulfilling their role, but it also increases the importance of their subjectivity, activity, and independence in the process.

Unlike in the traditional seminar, work with the mentor requires more independence of students and the skill of not only searching for information but also finding the right development paths. This refers not only to the diploma thesis but also to professional or academic choices (Gruszczyńska-Malec and Rutkowska 2005). The role of the mentor is mainly to provide feedback to support the learning process and help shape the "searching" learning style (Carless et al. 2011, pp. 396– 398). An inexperienced student may fail to see the end of the path they have chosen. An interview with the mentor leads to more conscious choices that can be supported with different mentoring techniques.

The techniques of stimulating creative thinking belong to more pleasant and useful techniques in the mentoring process. They help find a way to the goal and "produce" many solutions from which one can choose real and achievable ones. On the other hand, they cause the lack of discipline and help break the deadlock. Many of them can be used at the beginning of work with the student when they choose their own way of academic interests. These methods prove useful when the student has problems specifying what they want to write about or what they would like to study.

One of the methods of stimulating creativity is the exercise in which the promoter asks the student to close his or her eyes and move spiritually into another place. "Moving" to the subordinate's desk at work is an excellent opportunity to discover the student's interests and professional goals as well as obstacles to overcoming them. We request the student to feel, see, and experience the moment when he or she does what he or she likes at work and then when he or she does what he or she does not like. The student must think that he or she tells his or her best friend about it. What the student becomes aware of during such a journey can be more valuable than the conclusions drawn so far (Silberman and Auerbach 2006).

While writing a thesis, the student can get lost in a dead end. Then, the supervisor may use the mentoring technique of a metaphor. If properly used, the metaphor helps understand the situation. Three steps are necessary for applying the metaphor used to introduce an effective change. The first step involves the choice of the metaphor. It is good to let the student choose the metaphor to develop from among a few proposed by the mentor. It is important for the mentor to be able to place himself in the reality he or she will cope with when writing the diploma thesis. Each of them should put the student in one specific situation with its background and the role played.

The second task is to transfer the metaphor to the reality. At this stage, the student should be able to seek similar situations in his or her seminar paper and find it among the current experiences. Obviously, there is a probability that the metaphor quoted by the supervisor only partly coincides with the reality. The discussion at this stage should lead to describing these changes. The third stage that summarizes this exercise is the lesson the student has taken from the metaphor. Has the awareness of the situation made him or her more optimistic or more pessimistic? Or has it perhaps mobilized him or her to act or bored him or her? It is possible that the metaphor will not be satisfactory for the coached individual due to not too strict adjustment to his or her own situation. Then it is good to look for convergent points and draw conclusions for the student and the decision made (Megginson and Clutterbuck 2005).

The mentor accompanies the student and supports him or her in the learning process as well as helping make decisions concerning the diploma thesis in the short-term perspective and concerning the student's life and career in a broader context. The most important thing that can be given to the student is the tendency and ability to reflect upon oneself (Pausch and Zaslow 2008). The overseer gives support and is the oasis of safety; he or she takes responsibility for the paper and its creation. The student is supposed to write the thesis but does not need to stretch his or her wings. The basic differences between the mentor and the overseer are shown in Table 68.3.

TABLE 68.3
Basic Differences between the Mentor and the Overseer

	Mentor	Overseer
Basic objective	Choice of the thesis topic and range that reflect the student's interests, writing the thesis, development of the student in the selected range and supporting academic as well as professional aspirations	Writing the diploma thesis by the student, passing the diploma examination and completing studies
Initiative	On the part of the student, the student proposes the topic, scope of the thesis, their ideas are discussed; the speed of work and its direction depend on the student's choice	On the part of the supervisor, thesis topic, perspective of matter description, speed of work depend on the supervisor's requirements
Voluntariness of meetings	High level, apart from introductory classes, other classes depend on the degree of the student's involvement and speed of work	Low level, students are obliged to attend regular meetings with the supervisor and submit assignments regularly
Emphasis laid on	Development of the student, their interests, choice of an academic and/or career path	Writing the diploma thesis according to the formal requirements
Relationship between parties	Partnership but unbalanced	The supervisor is the student's superior

Source: Own elaboration.

Both as an overseer and as a mentor, the supervisor can exhibit different styles of work. Therefore, from the point of view of the student, promoters can be divided into four types (own elaboration, based on the consultation with Marek Frankowicz, PhD):

- Fatherly type
- Laissez-faire type
- Master type
- Partner type

Supervisors of each type have high demands toward future participants of the seminar, but each of the types is driven by other motives.

The first type is characterized by a caring attitude toward the students. They care for them, defend them if necessary and strive for projects for them. They organize work, tasks, and particular activities—at the same time requiring total obedience. Responsibility lies directly on them. None of the decisions can be made without their knowledge, even the most insignificant decisions. The atmosphere in their team is friendly and one can feel as if it's a family, so it is easy to work in the team. The seminar group is stable and hermetic, and everyone has a strictly determined role that is rarely changed.

The laissez-faire type is extraordinarily demanding and authoritarian. Students are required to act obediently and to execute their tasks on time. When working with the promoter who takes a laissez-faire attitude, one acts independently as the promoter only allocates tasks and does not intervene in their execution. But the supervisor reacts immediately in case the deadline has been exceeded. In this case, the atmosphere of work is built by the students themselves, and they decide about the speed of work and the level of involvement. Students are eager to join the seminar conducted by this promoter, appreciating their outstanding position as a mentor who gives them much freedom in executing individual works.

The master type is also extremely demanding even though they follow the rule that it is the greatest reason to be proud if the pupil is better than them. They are always ready to help and discuss, stimulating the development of all seminar participants. Work under the supervision of the promoter is very inspiring and based on clearly determined objectives that are individually adjusted to each student.

The last type of supervisor, the partner, focuses on cooperation in the team. Their main task is to arouse the interest and creativity of seminar group members. They lay emphasis on independence, the ability to make difficult decisions and react in critical situations. They also help choose the development path but do not control it. Students supervised by the partner type must exhibit independence in searching knowledge and should come to seminar meetings with their own ideas supported with literature. Only then will work with the supervisor-partner be really fruitful.

The discussed types of supervisors differ in terms of the indicator of involvement in relationships between the teacher and the student, the indicator of requirements (that can be raised or decreased during the cooperation) and the partnership indicator (that manifests the strength of the relationships, trust, responsibility, etc.). Each of these types can assume one of the roles described above: either an overseer or a mentor. Even though none of the types has a role ascribed, the roles of the overseer and mentor are to a certain extent closer to one type of the supervisor than to others.

In this context, the relationship built between the supervisor and the student is of particular importance. For this reason, the teacher's involvement in the relationship with the student is more frequent in case of the partner type or the master type. The two types are considerably focused on building relationships. They are governed by a high level of partnership between the student and the supervisor yet with the inequality of relationships resulting from the university environment preserved (Rodziński 2004).

The laissez-faire type and the fatherly type establish less intimate relationships with students or even minimize them. Supervisors who conduct seminars in this way are more bossy and less prone to accept independent solutions proposed by students.

CONCLUSION

When taking care of students as a supervisor of their diploma thesis, an academic teacher undertakes a task that involves supporting them at the stage of choosing the range, topic, and execution of the thesis. They are also obliged to control progress in thesis writing throughout the year or semester and to review it as well as allow for its defense (diploma examination). The involvement of the supervisors and their responsibility for the theses written under their supervision do not form the entirety of the role they play in the process of educating a young person just beginning his or her adult life (Franke and Barbro 2011).

When choosing the supervisor, the young person takes into account the potential benefits it can bring. They try to choose the person under whose guidance they will be able to develop their academic and career interests. Another criteria of choice is the way the supervisor works, requirements toward the diploma thesis, the level of control over it, and the student's progress. Changes to the role of academic teachers and supervisors are constantly adapted to the conditions and requirements of the higher education market.

The range of the student's interests, the degree of specialization in the process of writing the diploma thesis, and finally, the choice of the topic and the perspective from which the student describes this topic are apparently the choices necessary for finishing this stage of studies. In fact, it is life choices that are decisive for the student's further career and, as a result, the student's life. The support, indication of ways and possible consequences of choices made by the students at this stage is of key importance for their future development. Therefore, the supervisors and the role they assume for candidates for a degree as well as the relationships established with them (Kek and Huijser 2011, pp. 188–189) are important elements of education.

REFERENCES

Bugaj, J. (2008), SIS—koncepcja prowadzenia dydaktyki akademickiej—metody i techniki szczegółowe, in: Drzewowski, M., Maliszewski, W. J. (eds.) Komunikacja społeczna a zarządzanie we współczesnej szkole, Wydawnictwo Adam Marszałek, Toruń (SIS—The Concept of Conducting Academic Teaching—Detailed Methods and Techniques, in: Drzewowski, M., Maliszewski, W. J. [eds.], *Social communication and management in contemporary school*).

Carless, D., Salter, D., Yang, M., and Lam, J. (2011), Developing sustainable feedback practices, *Studies in Higher Education*, Vol. 36, No. 4, SRHE Routledge. Taylor & Francis Group, New York.

Edgerton, R. (2004), Getting from here to there... or the strategy for change, in: Gil, V.M.S., Alarcao, I., Hooghott, H. (eds.) *Challenges in teaching and learning in higher education*, Aveiro: University of Aveiro.

Franke, A., Barbro, A. (2011), Research supervisors' different ways of experiencing supervision of doctoral students, *Studies in Higher Education*, Vol. 36, No. 1, SRHE Routledge.

Gruszczyńska-Malec, G., Rutkowska, M. (2005), *Mistrzostwo osobiste a wybór kariery zawodowej*, Zarządzanie Zasobami Ludzkimi 2/2005; (*Personal mastery and career choice*).

Harden, R. M., and Crosby, J. (2000), AMEE Guide No 20: The good teacher is more than a lecturer-the twelve roles of the teacher. *Medical Teacher,* Vol. 22, No. 4.

Hoskins, S. L., Newstead S. E. (2009), Encouraging student motivation, in: Heather, F., Ketteridge, S., and Marshall, S., *A Handbook for teaching and learning in higher education. Enhancing academic practice.* London, New York: Routledge.

Kędzierska, B., Frankowicz, M., Mirecka, J. (2004), *Nowa rola nauczyciela a technologie informacyjne*, in: Kędzierska, B., Migdałka, J. (eds) *Informatyczne przygotowanie nauczycieli: Internet w procesie kształcenia*, Krakow: Rabiol (The new role of teachers and information technology).

Kek, M., and Henk, H. (2011), Exploring the combined relationships of student and teacher factors on learning approaches and self-directed learning readiness at Malaysian university, *Studies in Higher Education*, Vol. 36, No. 2, SRHE Routledge. Taylor & Francis Group, New York.

Kember, D. (1997), A reconceptualisation of the research into university academics conceptions of teaching. *Learning and Instruction,* Vol. 7, No. 3.

Krajewska, A. (2003), *Rola studentów w nowocześnie pojmowanym procesie kształcenia*, Uniwersytet w Białymstoku, http://gazeta-it.pl/200305225097/Rola-studentow-w-nowoczesnie-pojmowanym-procesie-ksztalcenia.html [20.08.2010]; (*The role of students in the modern educational process*).

Laissez-faire, http://en.wikipedia.org/wiki/Laissez-faire (2009.05.15).

Macfarlene, B. (2011), Professors as intellectual leaders: Formation, identity and role, *Studies in Higher Education*, Vol. 36, No. 1, SRHE Routledge. Taylor & Francis Group, New York.

Megginson, D., and Clutterbuck, D., (2005), *Techniques for coaching and mentoring*, Oxford: Elsevier.

O'Neill, G., and McMahon, T. (2005), *Student-centred learning: What does it mean for students and lecturers?* University College Dublin http://www.aishe.org/readings/2005-1/oneill-mcmahon-Tues_19th_Oct_SCL.html [20.08.2010].

Parsloe, E. (1992), *Coaching, Mentoring and Assessing*, London: Kogan Page.

Parsloe, E., and Wray, M. (2003), Trener i mentor. Udział coachingu i mentoringu w doskonaleniu procesu uczenia się. Oficyna Ekonomiczna, Kraków (Coaching and mentoring: Practical methods to improve learning).

Pausch, R., and Zaslow, J. (2008), *Ostatni wykład* Warszawa: Nowa Proza (*Last lecture).*

Rodziński, S. (2004), Relacja mistrz—uczeń. Anachronizm czy nowa szansa, in: Skulicz, D. (ed.), W poszukiwaniu modeli dydaktyki akademickiej, Wydawnictwo Uniwersytetu Jagiellońskiego, Kraków. (*Master—student relation. An anachronism or a new opportunity*, in: Skulicz, D. (ed.), *In Search of Models of Academic Teaching*).

Silberman, M., Auerbach, C. (2006), ACTIVE TRAINING: A handbook of techniques, designs, case examples, and tips, Jossey-Bass Pteiter: San Francisco.

Unsworth, K. L., Turner, N., Wiliams, H. M., and Piccin-Houle, S. (2010), Giving thanks: The rational context of gratitude in postgraduate supervision, *Studies in Higher Education*, Vol. 35, No. 8, SRHE Routledge. Taylor & Francis Group, New York.

Wołoszyn-Spirka, W. (2001), W poszukiwaniu realistycznych podstaw moralnego postępowania nauczyciela, Wydawnictwo Uczelniane im K. Wielkiego, Bydgoszcz (*In search of realistic grounds for teachers moral behavior*).

69 Quality of Virtual Mobility

Juris Dzelme and Ivars Linde

CONTENTS

VIRTUAL MOBILITY, DISTANCE EDUCATION, AND E-LEARNING

The development of information technologies and the increasing need for in-service training creates the demand for the wider use of e-learning, distance education, and different forms of mobility in formal, nonformal, and informal education. The most interesting and less investigated in the context of lifelong learning (LLL) is virtual mobility.

We shall investigate virtual mobility in the framework of formal education. There are different approaches to distance learning or e-learning, distance education, blended learning, virtual and physical mobility (Byrne et al. 2007; Cuadrado and Ruiz 2009; Dzelme 2003; Kristoffersen et al. 1998; Sloka 2007; Sullivan 2008). We propose to use three levels of analyses for the investigation of the learning process and virtual mobility (VM): (1) technical level, tools; (2) organizational level, content of education and management; (3) social level, expectations and aims of different involved parties. All three levels interact and usually overlap; therefore, definitions could differ significantly and often are unclear. To introduce regulations for VM and similar new activities, to receive public support and public financing, the common understanding and *transparency* should be achieved.

At the technical level, the attention is paid to the tools, the means of interaction of the participants of different learning and teaching processes. The technical side is emphasized, the process of learning to use information and communications technology (ICT) (information technology [IT] supported learning), and the use of video and audio equipment are analyzed. Online and off-line forms of distance learning (DL) or e-learning (e-L) and virtual learning (VL) could be used for formal, nonformal, and informal education. (Formal education means fixed and confirmed learning outcomes [LO]; nonformal education has aims to achieve some LO, not clearly fixed before the beginning and/or not evaluated and confirmed LO; informal education is without clear aim and structure, without planned LO, usually some useful experience.) In the case of distance learning, students much of the time are physically distant from the core premises of the provider; e-learning means the use of electronic (digital) techniques. Usually physical contact of students and teachers also are included in the DL, but there are not agreed borders between face-to-face learning, DL, and blended learning. The reality is that e-learning is usually used in conjunction with face-to-face techniques—technology-enhanced learning or blended learning. Some border values should be suggested, but they should be flexible and overlapping. For example, learning could be recognized as face-to-face if the direct contact between the student and teacher is more than 15% of the average time spent for the studies by the student. Learning could be recognized as DL if the direct contact between the student and teacher is less than 25% of the whole time spent for the studies.

At the organizational level, the attention is paid to the LO. In the case of distance education (DE), feedback must exist; the results of the learning activities, learning outcomes, must be controlled. DE (formal and nonformal) is DL with at least some feedback. Evaluation must be included as a part of the DE process with a clear end (outcomes) and assessment of the process and results. LO are fixed for the DE. According to the Latvia Law on HEI: *Distance education is an extramural method for acquiring education, which is characterized by specially structured educational materials, individual speed of learning, specially organized evaluation of educational achievement, as well as utilization of various technical and electronic means of communication.*

The third, social level, includes expected social and cultural aims of the education and different social activities linked with education. VM (Erasmus) is DE (only formal HE) with social interaction. (VM does not replace physical mobility but gives a new step for the development of mobility. VM is not only between different cultures [not only intercultural].) The aims to achieve personal development, to develop democratic procedures, and to support research and dissemination of knowledge are included in VM (but not always in DE).

PRINCIPLES OF EVALUATION OF VM

Multilevel (hierarchical) structure of the evaluation, which mainly coincides with the structure of the system (or activity) under investigation, should be used (Dzelme 2003; Kristoffersen et al. 1998; Sloka 2007). Evaluation could integrate several actually existing levels of the system because evaluation is oriented to integration. In higher education, two main levels are the study program (granting degree; also diploma or other academic award) and the higher education institution as a whole. Both these levels should include the evaluation of the arrangements for the DE and VM.

A combination of *internal and external* evaluation should be used. At each level, responsibility for the activities and results is mainly within this level. The main part of decisions (and, consequently, responsibility) about activities at a certain level must be taken at this level (not as orders of "higher power"). This is the main idea of democracy. The autonomy and responsibility at lower levels must grow together with the growing amount of information and complexity of tasks.

At each level, at least two different evaluations should be combined.

1. Internal evaluation created at this level (self-evaluation) means critical thinking and inclusion of all activities in more relationships, contexts, and a wider environment. This internal evaluation goes beyond the requirements directly necessary only to fulfill the task. Evaluation activities (criticism, feedback) must go parallel continuously (but with periodically changing intensity) to the direct activities (main task). All decision makers should be involved in critical review and continuous search of the improvements of their decisions.
2. External evaluation of decisions (and of their results), of future possibilities, and strategic development are carried out by persons or organizations that are not involved in the decision making under investigation (at least at the same level). This should be a view from outside without constraint (interest, responsibility) to continue the way it started (i.e., without conflict of interests) and with a partly other, preferably wider, context.

The same evaluation can be described (and normally should be) at the same time as internal and as external but for different levels inside a multilevel system. For example, the evaluation of the study program by the Senate (or its representatives) of the HEI is external for the study program but internal (part of system of internal evaluation activities) for the HEI. In the case of DE and VM, there are more different interactions between levels, and the definitions of levels change. Special attention should be paid to the establishment of clear and transferable rules, but the involvement of different levels should be recommended.

The tools used in the evaluation must be oriented to synthesis and integration. DE and VM allow putting together very different information. This advantage should be used. All facts and situations must be investigated using context. In education, as in the main part of other important fields of people's activities, the most complicated and most important part is the interaction of the involved personnel. The tools used for evaluation should be oriented to the investigation of communication between people. Interviews, focus groups, brain-storming, meta-modeling, and similar methods are more interesting (and preferable) than tests, questionnaires, and different statistical data, but both kinds of tools (data, activities) are necessary for the evaluation. Special attention should be paid to the style of interactions in the course of the evaluation. Communication must be constructive, directed to the improvement and to the finding of new possibilities in the framework of existing systems and/or switching to other, better solutions and finding new ways. Emotional intelligence (attitudes, competence) and its place in the system (especially in the case of education) must be investigated and evaluated first of all, leaving knowledge and skills in the second position. Consequently, in the course of evaluation, the communication between people and creative interaction are more important than testing, the collection of statistical data, factual findings, and evidence. DE and VM can create more problems for the interaction of people. Special measures to guarantee good personal communication in virtual space should exist and must be evaluated. Social and cultural problems of the DE and VM must be investigated using physical contacts and IT. The tools used for DE must have rather high capacity to allow different social and cultural activities.

CRITERIA FOR THE EVALUATION OF STUDY PROGRAMS WITH VM

Evaluation of study programs should be based on the use of LO. The basic principles of the use of LO are described in the European qualification framework (EQF). VM should be included in the study process, and its evaluation as an additional possibility is one more way, which leads to almost the same LO (Sloka 2007). VM should not exclude physical mobility but improve it. At the same time, additional possibilities, given by VM, must be supported by appropriate additional parts in the whole study process, including internal and external evaluation (also self evaluation). The criteria used for the evaluation without VM should be expanded. Searching for answers for all the questions about technical support, communication, feedback, etc. must be added as specific parts about virtual space and possibilities to achieve the aims of studies using virtual space (e-L, DE, VL). Critical analysis should show new opportunities and should give recommendations as to how to avoid difficulties. In general, the same main approach and the same criteria could be used, but in the self-evaluation and internal and external evaluation, the knowledge and understanding of new technologies and linked with them communication and other new opportunities must be added.

The activities of staff members (academic personnel) in education and research outside their own HEI could be investigated by using the approach accepted for diploma recognition. Especially the supplement to the Lisbon Convention about transnational education should be used. The following main questions from this supplement are important for the evaluation of VM.

1. Compliance of the activities of the partners (their teachers) with the national legislation of the respective partner's country.

   Availability of the documentation concerning the teaching activities in virtual space (outside coordinators HEI):

a. The policy and the mission statement for the courses and the whole module
b. Management structure of the module
c. The goals, objectives, and content of the module, set of six courses of study, and educational services

Existence and transparency of legally binding agreements or contracts setting out the rights and obligations of teachers in participating higher education institutions (coordinator and partners). Control and monitoring by the coordinator of the information made public by partners.

2. Comparability of the academic quality and standards of the courses of the module are necessary:
   a. To those of the higher education institution of the coordinator
   b. To those of the modules (similar study programs) and higher education institutions in the partner's country

   The partner provides transparency of the procedures and decisions concerning the quality of educational services.
3. Proficiency of the staff members teaching on the courses established in the partner's HEI in terms of qualifications, teaching, research, and other professional experience. Effectiveness of the measures organized by the coordinator (its higher education institution) to review the proficiency of the staff, delivering programs in the partner's HEI.
4. Responsibility of the coordinator for issuing the certificate (qualification) resulting from its module in the partner's HEI. The transparency of the information on the certificates (qualifications), in particular through the use of the Certificate Supplement (similar to the Diploma Supplement), facilitating the assessment of the qualifications by competent recognition bodies, the higher education institutions, employers and others. The inclusion of the information about the nature, duration, workload, location, and language(s) of the courses (parts of the module), lead to the certificate (qualification) in the Certificate Supplement.
5. Equivalence of the requirements of the admission of students for a course of study, the teaching/learning activities, the examination and assessment, provided by the partners (in their HEI), to those of the same or comparable courses (of similar programs) delivered by the coordinator's higher education institution. Existence of clear indications of any difference in this respect between the partner's and the coordinator's higher education institutions.
6. Existence of clear indications of any difference in the academic workload, duration of studies or otherwise between the partner's and the coordinator's higher education institutions and a clear statement on the rationale of the differences.

The evaluation of study programs with VM should be carried out with the use of similar additional criteria. Special attention should be paid to the combination of *virtual and physical mobility*. The evaluation should not be divided for virtual and physical mobility, but the combined use of virtual and physical mobility creates specific problems in each part of the evaluation methodology, the organization of meetings, and the use of questionnaires.

Better accessibility of higher education, development of lifelong learning, and in-service training are possible using the combination of *virtual and physical space, including virtual and physical mobility*. We have investigated new, developing forms of higher education, in which mistakes are more dangerous, but advice and transfer of the examples of good practice and experience of others are more valuable and effective.

Combination of virtual and physical mobility is a new and effective way to improve higher education and its accessibility. The new form, virtual mobility, is a new object for the investigation and quality assurance (QA). The new form needs new arrangements for the improvement and control—the two main parts of QA.

Quality assurance of virtual mobility must include self-evaluation and external evaluation of the newest parts of higher education system: *virtual space and virtual mobility*. We have analyzed different tools used in virtual mobility. The same tools are valid for quality assurance, but additional efforts, including national and global regulations, are necessary. Special attention must be paid to the compliance with different regulations, different traditions, and the social and cultural environment.

To achieve the common learning outcomes, different ways must be investigated during self-evaluation and external evaluation of study programs and higher education institutions. Quality assurance is a very good tool to harmonize the global system and to introduce improvements in it. The right use of external experts with good qualifications and good authority is the best and most effective way to implement and improve the quality of such a new form as the virtual mobility.

TOOLS FOR THE EVALUATION OF STUDY PROGRAMS WITH VM

The tools used in the evaluation must be oriented to synthesis and integration. All facts and situations must be investigated using context. In education, as in the main part of other important fields of people's activities, the most complicated and most important part is the interaction of involved personnel. The tools used for evaluation should be oriented toward the investigation of *communication* among people. Interviews, focus groups, brain-storming, meta-modeling, and similar methods are more interesting (and preferable) than tests, questionnaires, and different statistical data, but both kinds of tools (data, activities) are necessary for the evaluation. Special attention should be paid to the style of interactions in the course of the evaluation. Communication must be constructive, directed to the improvement and to the finding of new possibilities in the framework of existing systems and/or switching to other, better solutions and finding new ways. Emotional intelligence (attitudes, competence) and its place in the system (especially in the case of education) must be investigated and evaluated first of all, leaving knowledge and skills in the second position. Consequently, in the course of evaluation, communication between people and creative interaction is more important than testing, collection of statistical data, factual findings, and evidence.

A significant part of the QA of study programs with VM is comparison and analysis of the existing practice of internal QA for Erasmus virtual mobility in partner institutions. The main questions, which could be included in questionnaires, interviews, or discussions at the special meetings at the partner institutions, are the following. (We use the definition "*Quality is fitness for purpose*" and abbreviations: DE – distance education, HE – higher education, HEI – higher education institution, QA – quality assurance, SP – study program, VM – virtual mobility.)

MAIN QUESTIONS ABOUT THE VIRTUAL MOBILITY

Mission (Aims, Means)

- Are the DE and VM included in the mission of HEI?
- How?
- Are the aims for VM different from the aims of physical mobility (in the mission of HEI)?

Stakeholders (Target Group)

- Are the students involved in DE different from a full-time (face-to-face) group?
- Are employers interested in DE?
- How important it is in service training?

Mentors (New Approach, Paradigm)

- Is there a tendency to shift the acquisition of skills from the HEI to the work place?
- Are there any needs to organize the interaction of teachers (mentors) with graduates?

- Is the VM (of teachers and/or students) used to continue learning after leaving the HEI?
- Are there any needs to prepare mentors (using VM)?

Internal QA of Teachers

- What are the main tasks of teachers (different levels from assistant to professor)?
- What are the main criteria for the election in the academic position?
- In what cases is there a real competition for the academic position?
- Are the VM and/or DE included in the criteria for QA of teachers?
- What are the possibilities to use the achievements in DE and/or VM to confirm the high level of the teacher's skills?
- How many levels (chair, department, university, etc.) of QA are used inside the HEI?
- How effective (useful) are different levels of QA used inside the HEI?

Criteria

- How important are social criteria (communication skills, emotional intelligence, etc.) for QA of teachers?
- How important are research activities for QA of teachers?
- How important is mobility (international cooperation) for QA of teachers?
- Are there different approaches to the evaluation of social skills and research activities for face-to-face education and DE?
- Are there different approaches to the evaluation of physical mobility and VM?

QA of Study Programs (SP)

- Are there internal regulations for the QA of SP?
- Who is responsible for QA of SP?
- Who participates in the QA of SP?
- Are there different regulations for DE SP?
- How are the e-learning and/or DE evaluated?
- Are there different approaches to face-to-face education, DE, blended learning?

QA of Modules, Courses

- Are there any specific rules for QA of modules and courses?
- Are there different (face-to-face education, DE, blended learning) modules in one SP?
- Are there different approaches to face-to-face education, DE, blended learning in one module?

Ranking

- Is there ranking for the HEI, SP, and modules in use?
- Are there any specific criteria for DE in ranking?
- Is ranking used in QA?

CONCLUSIONS

Schematic relations between e-L, DE, and VM in Erasmus projects are the following:

1. DL and e-L means the use of ICT and the Internet and the use of electronic digital techniques for learning.

2. DE is DL (e-L, VL) with feedback. The evaluation and assessment of the process and results must be included. LO are fixed. The physical mobility is desirable.
3. VM (Erasmus) is DE (formal HE) with social and cultural interaction. The aims to achieve personal development, to foster democratic procedures, and to support research are included.

The combination of internal and external evaluation should be used in the framework of multilevel (hierarchical) structure of the evaluation. At least two different evaluations should be combined: internal evaluation (self-evaluation), which includes critical thinking, contexts, and a wider environment, and the independent external evaluation of decisions (and of their results) for future possibilities and strategic development. For the DE and VM, there are more different interactions between levels, and the definitions of levels change; therefore, special attention should be paid to the establishment of clear and transferable rules.

VM should be included in the study process and its evaluation as an additional possibility, one more way, which leads to almost the same LO (Sloka 2007). VM should not exclude physical mobility but improve it. At the same time, additional possibilities, given by VM, must be supported by appropriate additional parts in the whole study process, including internal and external evaluation (also self evaluation). The criteria used for the evaluation without VM should be expanded.

The combination of virtual and physical mobility is an effective way to improve higher education and its accessibility. Virtual mobility is a new object for the investigation and quality assurance (QA) and needs new arrangements for improvement and control—the two main parts of QA. Different tools, used in virtual mobility, are valid also for quality assurance, but the changes of the existing national, regional, and global regulations and traditions are necessary. Attention must be paid to the social and cultural environment. The new form needs new arrangements for the improvement and control—the two main parts of QA. QA must include self-evaluation and external evaluation of the e-learning and virtual mobility.

ACKNOWLEDGMENTS

The investigation is supported in the framework of the LLL Erasmus project Tea-Camp (*LLL Erasmus program Multilateral Virtual Campuses project TeaCamp - Teacher Virtual Campus: Research, Practice, Apply [Project ID 502102-LLP-1-2009-1-LT-ERASMUS-EVC]*).

REFERENCES

Byrne N., Coca M., Cuadrado M. (2007). Interdisciplinary university e-learning through weblab systems. A two-country project. Valencia: INTED Conference.

Cuadrado M., Ruiz M. E. (2009). University students' satisfaction on virtual platforms: An international e-learning program based on Moodle, e-Learning (EL 2009) - IADIS Multi Conference on Computer Science and Information Systems (MCCSIS 2009) (Algarve, June 17–20).

Dzelme J. (2003). Structure of higher education quality assurance system in Latvia, in *Systems of education quality assurance in CIA and Baltics: Moscow* – Yoshkar-Ola, pp. 39–52.

Kristoffersen D., Sursock A., Westerheijen D. (1998). *Manual of Quality Assurance: Procedures and Practices.* United Kingdom: Phare, European Training Foundation.

Sloka B. (2007). Study of labour market "Compliance of Professional and Higher Education Programmes with the Requirements of Labour Market." LU: Rīga, 2007, XII + 247 pp.

Sullivan P. (2008). HEFCE's Crystal Ball? Students as scholars, employees, leaders and managers? *In Practice (17):* Leadership Foundation for Higher Education.

70 Virtual Mobility Phenomenon
Educational Perspective

Margarita Teresevičienė, Airina Volungevičienė, and Estela Daukšienė

CONTENTS

THEORETICAL BACKGROUND OF VIRTUAL MOBILITY CONCEPT

Virtual mobility (further VM) is rather a new phenomenon and has been influenced by the development of ICT very much. Therefore, we start our research from the theoretical background analysis of the concept in order to identify its characteristic features and to define the virtual mobility concept from an educational perspective.

The "virtual mobility" concept is usually referred to as either mobility or technology-enhanced (sometimes referred to as virtual) learning. Although there have been very different approaches to the concept discussed, here we briefly describe these approaches in the chronological order, however, focusing only on the educational approach of the phenomenon.

VIRTUAL MOBILITY CONCEPT

The first notions on virtual mobility are indicated in the last decades of the 20th century and the beginning of the 21st in some research papers (Bunt-Kokhuis 2001) and project result reports (*Humanities project report* 1995), (Spot+ project team 2001). S. Van de Bunt-Kokhuis (2001) creates a rather interesting although specific definition of VM in which it is described as "the collaborative communication between a faculty member and his/her counterpart(s) mediated by a computer. More often, these meetings will be interactive and take place across national borders and across time zones."

In the Humanities project (1995), the concept of virtual mobility is considered to be "effective networking" whereas the "Spot plus" project team (2001) widens the understanding of this concept and brings in two forms of mobility: physical and virtual. Virtual mobility is defined in this project outcome as a situation within a university that implies a "possibility to attend classes, seminars, and other events held in a place located anywhere in the world; the possibility to access reference materials and contents at a distance by using ICT–based solutions; the possibility to communicate with other people located anywhere" (Spot+ project team 2001, p. 10). The idea of virtual mobility as a "hybrid model" in which a distance-learning module is introduced into normal curricula can be recognized here. The training material also indicates that "virtual mobility includes all forms that are communication intensive and run at international level" (Spot+ project team 2001, p. 12). "Virtual Erasmus student" training material, referring to the Humanities project, specifies that virtual mobility is constituted of the following elements:

- Transnational lectures and/or learning materials
- Cross-border recruitment of students
- Intensity of communication flows
- International accreditation of achievements
- Multilingualism
- Complementarity between virtual mobility activities, traditional lectures, and physical mobility
- International recognition and accreditation of study achievements (Spot+ project team 2001)

To sum up, the above descriptions of virtual mobility elements suggested that in 1994–2001 virtual mobility concepts covered formal and informal education activities with the main focus on communication and collaboration, using different resources that were not location dependent.

A different kind of approach to virtual mobility was introduced by J. Silvio (2003), who distinguished three types of space—geographic, social, and virtual—and described virtual mobility as a new phenomenon. He indicated virtual mobility as a movement "from one place to another in a new space called virtual space … enabled by computer-mediated communication" (Silvio 2003). Supporting the French philosopher Pierre Levy's ideas, Silvio (2003) suggested that "virtual is not opposed to the real." Explaining the basis of the virtual reality term and ideas that a language, online course, or the whole university campus can be represented by digital numbers, the author used M. Dertouzos' words to define the main principles of the virtualization process and computer-mediated communication. Thus using the mentioned concepts and taking Dertouzos' "pillars" into account, J. Silvio defined virtual mobility as a representation of physical mobility taking place in a virtual space, implying no movement of persons in a geographic space, in which "information and the objects represented by them 'move' electronically from the computer center of one university or enterprise to another located in different places in the geographic space" and drew a conclusion that "virtual mobility is mobility of 'bits' instead of 'atoms' " (Silvio 2003). Similarly to Silvio (2003), Vilhelmson and Thulin (2005) defined virtual mobility as "physical transportation and face-to-face contacts, replaced, complemented, or even generated by virtual ones."

As virtual mobility started to be included in the agenda of more and more political documents, project activities, or education-related conferences, more of the different virtual mobility aspects and characteristics appeared in the concept definitions. Schreurs et al. (2006) stressed that "virtual student exchange allows collaboration with foreign students and teachers that are no longer location dependent. The exchange might range from a single course to a full academic year. Through Virtual Mobility a university can also offer international experience for students and staff through an international discussion group, an international seminar, an international learning community with regard to a theme of a course or a cluster of courses."

A more full-scale interpretation, which included an intercultural aspect and the reference to education, was provided in the *Best-Practice Manual on Blended Mobility* (Op de Beeck et al. 2008) in which the authors consider virtual mobility to be "a form of learning which consists of virtual components through an ICT supported learning environment that includes cross-border collaboration with people from different backgrounds and cultures working and studying together, having, as its main purpose, the enhancement of intercultural understanding and the exchange of knowledge."

A very diverse approach of VM, in which its reference to physical mobility was presented, can be found in the Glossary of the Lifelong Learning Programme 2007–2013. Virtual mobility is characterized here as "a complement; or as a substitute to physical mobility (Erasmus or similar) in addition to a type of independent mobility which builds on the specific potentials of on-line learning and network communication. It may prepare and extend physical mobility, and/or offer new opportunities for students/academic staff who are unwilling or unable to take advantage of physical mobility. It involves the development of virtual mobility for academic staff. It means that full academic recognition is given to the students for studies and courses based on agreements for the evaluation, validation and recognition of the acquired competences via virtual mobility. In this context, cooperation agreements are the key to ensuring sustainable mobility schemes" (EC Glossary on the LLP 2007–2013, 2007).

In late autumn 2009, the Teacher Virtual Campus: Research, Practice, Apply (TeaCamp 2009–2011) project launched its analysis on the existing virtual mobility practices in the six consortium countries. One of the problems with the identification of the existing virtual mobility practices indicated in the analysis results was a "diversified definition of virtual mobility" (Dauksiene et al. 2010). Thus, as the processes of the European youth and academic mobility policy and higher education modernization facilitated virtual mobility, it became emphasized, used, and discussed in various research papers; however, common agreement of the virtual mobility concept had not been crystallized out.

To sum up, the following characteristics of virtual mobility can be distinguished on the basis of the above mentioned publications: cooperation of universities as well as students and teachers and international study experience with the stress on cultural aspects, recognition possibilities, and different kinds of activities that may lead to physical mobility or exist separately as virtual exchange or virtual mobility. As the concept of virtual mobility is only being implied and changing, it will be referred to and described here indicating the characteristics and activities it may include.

Characteristics, Components, and Activities of Virtual Mobility

Different research studies and papers pointed out a lot of different *characteristics* of virtual mobility. In 2010, the TeaCamp project consortium indicated virtual mobility in higher education concept as an activity or a form of learning, research, and communication and collaboration, based on the following characteristics:

- Cooperation of at least two higher education institutions
- Virtual components through an ICT–supported learning environment
- Collaboration of people from different backgrounds and cultures working and studying together, creating a virtual community
- Having a clear goal and clearly defined learning outcomes
- Having, as its main purpose, the exchange of knowledge and improvement of intercultural competences
- As a result of which the participants may obtain ECTS credits and/or its academic recognition will be assumed by the home university
- Providing visibility of university in higher education area, capitalization of educational process
- Leading to the integration of ICT into their mainstreaming in academic and business processes (Tereseviciene et al. 2011)

About at the same time, in September 2010, the Movinter project white paper "In Praise of Virtual Mobility: How ICT Can Support Institutional Cooperation and Internationalization of Curricula in Higher Education" (Dondi and Salandin 2010) stressed the need to reformulate the existing virtual mobility concept while assessing the possibilities created by virtual mobility. They also mentioned that the virtual mobility concept is often misconceived with distance learning or education and e-learning; however, "e-learning and distance education do not necessarily imply internationalization of learning/knowledge, intercultural dialogue and cross-border academic cooperation … which are, in turn, the major VM drivers" (Dondi and Salandin 2010). Discussing the benefits and possibilities created by virtual mobility they also identified 10 major virtual mobility descriptive elements and called them VM *components* (here from Montes et al. 2012):

1. International student groups
2. Interactivity and communication between students of different countries through ICT
3. International teaching groups
4. Multicultural exchange
5. Use of appropriate technological solutions
6. Joint choice of the subject to be studied through VM
7. Joint curricula design
8. Joint production of learning resources
9. Joint titles
10. Mutual confidence relationships

The indicated VM components are rather broad; notwithstanding that they may exist in different backgrounds via various models and include from some to all of the mentioned 10 components. However, "regardless of the learning format and the involved VM components, three key elements are essential to design an internationalization strategy" for a successful virtual mobility. These key elements are interculturality, all partners' participation, and strong communication aptitude (Montes et al. 2012).

To be more precise and to better represent the possible virtual mobility organization ideas as well as the understanding of the phenomenon in higher education, the main activities of virtual mobility will be briefly discussed here. Referring to virtual mobility types indicated by Bijnens and Op de Beeck (2006) or the categorization of VM activities presented by different authors (such as Silvio 2003; Pigliapolo and Bogliolo 2007; or others), the broadest categorization of virtual mobility activities was presented in the "European Cooperation in Education through Virtual Mobility" (Bijnens et al. 2006), in which VM activities are characterized

- By their degree of virtualization
- By the technologies used for the activities
- By the teaching and/or learning scenarios
- Based on the circumstances in which the virtual mobility activity takes place

This categorization provides possibilities for different authors to specify VM from different perspectives. As the paper analyzes VM phenomenon from the educational perspective, virtual mobility activities discussed here refer to the ones based on the circumstances in which virtual mobility activity takes place. The typology of virtual mobility *activities*, following the "European Cooperation in Education through Virtual Mobility" (Bijnens et al. 2006), is described in the following way:

1. A virtual course (as part of a program) or seminar (series)
2. A virtual study program
3. Virtual student placements
4. Virtual support activities to physical exchange

The first three activities of virtual mobility may be taken as a complement or as a substitute to physical mobility, and the last type of activities are set as a complement to physical mobility. The above presented categorization lacks one more factor or characteristic of virtual mobility that can be noted here; a virtual study course(s) or a program can be designed and/or provided for students by more than one host university, which is not usually possible in a physical mobility case. This way, the students from different institutions can attend the course(s)/seminar/program studying and working collaboratively in a more diverse international group, gaining experience in dealing with cultural differences and opinions from more than one cultural background. This kind of virtual mobility was organized and implemented in the TeaCamp project and will be explored and analyzed in more detail in discussing the participants' attitudes, skills, and competencies acquired during virtual mobility as well as virtual mobility recognition possibilities.

VIRTUAL MOBILITY CASE AND ITS ORGANIZATION PECULIARITIES

The LLP Erasmus program project TeaCamp (Teacher Virtual Campus: Research, Practice, Apply) team, aiming to increase virtual mobility among academic staff, prepared an inter-university study module, realized via virtual mobility activities, and performed research in order to identify competencies acquired by students during virtual mobility studies, attitudes, and recognition possibilities of the participating academic staff. The virtual mobility case represented and analyzed in this paper is based on the preparation, delivery, and organization of the virtual mobility module "Virtual Learning in Higher Education." The research was implemented in three diagnostic surveys carried out at the beginning and the end of the module.

The module was prepared by an inter-university team of 13 teachers and implemented as a study module in the 2010 autumn semester at six partner institutions. The aim of the Virtual Learning in Higher Education (VLHE) module was to enable students to plan and experience virtual mobility sessions by practicing video lecture participation, performing group and individual online activities, and using and sharing virtual resources in a multicultural virtual learning environment.

The module preparation was organized online as well as in international group discussions at face-to-face and virtual meetings in April–September, 2010. Twelve learning outcomes were developed for students to be acquired during this course, and six sub-modules, during which these learning outcomes were to be reached and virtual mobility competencies were to be improved, were elaborated. The course curriculum was developed using the Moodle virtual learning environment. Each sub-module contained compulsory readings, group or individual assignments, recommended recourses, sub-module guidelines, and discussion forums.

The module delivery was organized in September–December, 2010. There were 29 learning participants (further on *the students*) and 13 teachers attending the module. They all were from the following higher education institutions: University of Aveiro (Portugal), Jyväskylä University (Finland), UNIOVI – University of Oviedo (Spain), Jagiellonian University (Poland), BETI (represented by the students from Kaunas University of Technology, Lithuania), and Vytautas Magnus University (Lithuania) and interacting within preset international student groups (Figure 70.1).

The module delivery was organized in synchronous and asynchronous ways. There were 13 virtual synchronous meetings (video conference sessions) organized at the mentioned institutions each Friday from September to December, 2010. So the students staying at their home university participated in the virtual lectures, delivered by professors from various universities (depending on the sub-module), during which sub-module assignments were described, various unclear issues discussed and clarified, and feedback on the performed assignments provided. During the week, student learning was organized in VLE Moodle, in which all study materials and assignments were uploaded and discussion forums and consultations were organized. All video conference sessions were recorded and uploaded in VLE for the students who could not participate in the video conference to be revised.

The research was implemented in August, 2010, through February, 2011, in three stages: preparation of research instruments, implementation of the research, and analysis of the research data.

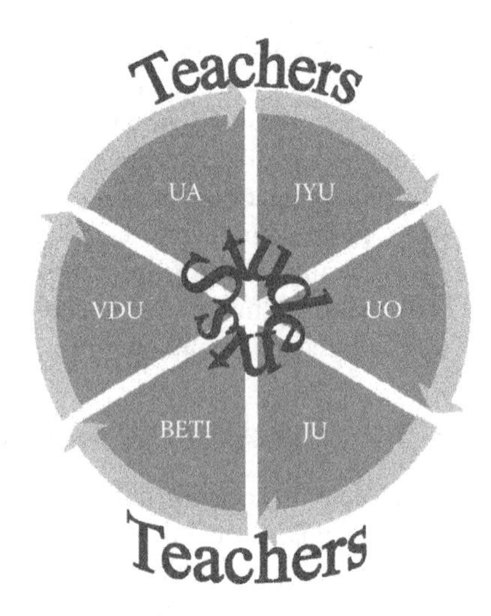

FIGURE 70.1 Virtual mobility case module delivery design.

There were three types of questionnaires prepared and presented for the module "Virtual Learning in Higher Education" participants. The first online questionnaire, "A pre-session diagnostic survey," was aimed at the identification of students' achievements, referring to the module learning outcomes, and targeted to be improved at the module and had 29 respondents—module students. The second questionnaire, "A post-session diagnostic survey of students," was performed in order to identify students' achievements, competencies, attitudes, and recognition possibilities and was responded to by 23 respondents—module students. The third questionnaire, "A post-session diagnostic survey of teachers," was performed to support the VM organization peculiarities and compare teacher and students' VM competencies, attitudes, and indicated recognition possibilities. It had 13 respondents—module teachers.

VIRTUAL MOBILITY COMPETENCIES IMPROVED DURING VLHE MODULE

The virtual mobility module "Virtual Learning in Higher Education" was prepared and realized in order to improve certain competencies of the participating students—referred to as module learning outcomes—and virtual mobility competencies that were specified as the English language competency, ICT competency, intercultural communication, and personal and social competencies.

The data analysis of the pre-session diagnostic survey shows that before starting attending the module students felt most comfortable with *information analysis and evaluation* ($N = 29$, only one student indicated minimal level achievements). The learning outcome of *knowing different technological resources for collaborative online learning* was also self-assessed by the students to be at least at a minimal level. The students felt themselves to be most professional in *understanding the skills needed to facilitate and manage collaborative online learning* (11 students (out of the total number of 29) and *comparing learning styles and learning strategies* (nine students, $N = 29$, stated that they had achieved this learning outcome).

After the module (see Figure 70.2), analyzing the achievement of the learning outcomes self-assessed by the module students ($N = 23$), it was evident that only one student indicated that he or she had not achieved at least a minimal level of one learning outcome. All the other learning outcomes had been achieved by all the participants at least at a minimal level. Comparing the learning outcome achievement of students before ($N = 29$) and after ($N = 23$) the module, improvements in all indicated course learning outcomes were found.

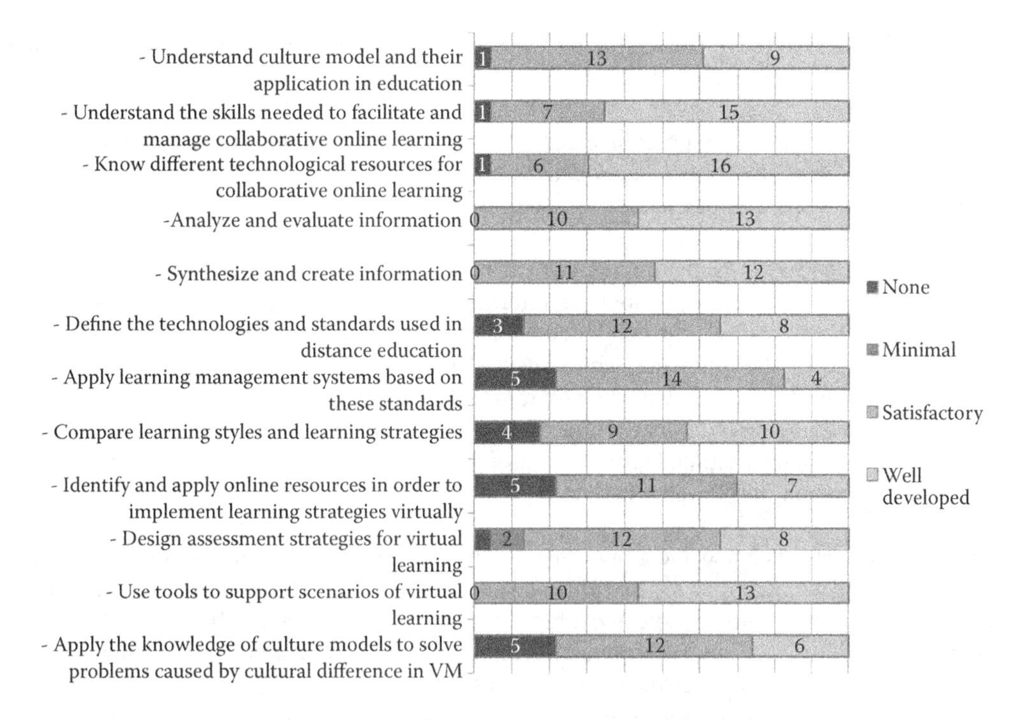

FIGURE 70.2 The students' competency (referred to as learning outcomes) improvement (in numbers of students) after the module.

Virtual mobility competencies, improved by the students after the module, were indicated during the second stage of the research in "A post-session diagnostic survey." Virtual mobility competence was suggested to be formed of intercultural communication, ICT, foreign language (English being the case in the module), and personal and social (such as being structured and self-organized, keeping time and meeting deadlines, respect for others, working in groups, etc.) competencies. All the respondents ($N = 23$) admitted that their competence improved at least minimally as almost all ($N = 20–22$) assessed their competence improvement at a satisfactory level. Virtual mobility competence improvements in detail are shown in Figure 70.3.

"A post-session diagnostic survey" also asked the students to add any additional competencies or skills that had been improved during virtual mobility studies. There were very different outcomes indicated, such as "profession-related competence" or "data gathering," "working late nights and

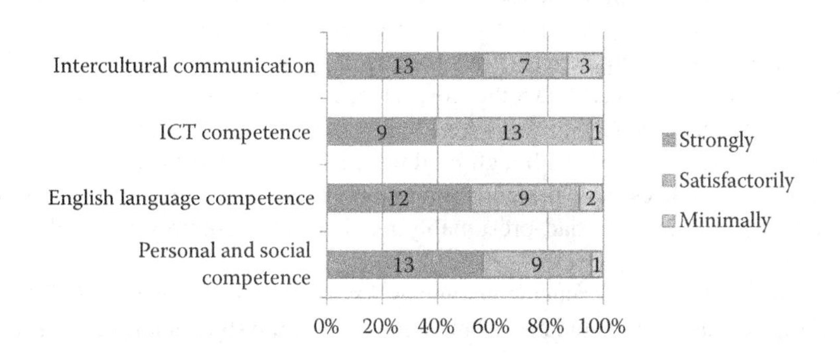

FIGURE 70.3 The students' improvement of virtual mobility competencies (in student numbers).

weekends," etc. Some of them stressed more specific skills or competencies that might be related to the above-mentioned ones and which comprise a general virtual mobility competence:

- Intercultural communication competence: "international communication," "communication skills with people from other countries, cultures..."
- Personal and social competence: "tolerance," "patience," "I can organize my time schedule of studies better than earlier"
- ICT competence: "information literacy competence," "virtual communication skills," or "skills related to the evaluation of information (information literacy)" or a broader approach was suggested—the combination of ICT and learning (individual and in group) was indicated by the following achievements: "learning to work cooperatively in a virtual environment," "one or two new programs," "use different tools for group work." The "international group work" and "collaborative work competence" as well as "experience in working in an international group" or "group working and encouraging members of my group to common work" were also stressed.

Other respondents specified the achievements related to module learning outcomes, such as "more defined understanding of my learning strategies" or "competence related with ... learning strategies and methods." Others indicated a broader approach with "some difference between cultures and learning styles," "new viewpoints of cultural differences in studying," or, in general, "a skill to analyze learning outcomes/recourses was improved."

Most of the students had a positive approach and valued their participation: "We all wanted to reach the same goals; we could feel the spirit and support from the national group" or "I am happy that I participated although it was not always easy. I have many positive feelings about the course. I think it was a valuable experience; I better understand virtual mobility and appreciate this form of learning. I see many possibilities of creating such courses in my field of studies, on a national or international level..." One of the students noted his or her personal discovery while participating in the module: "I realized that communication face to face is very important because when humans are communicating online, it is difficult sometimes."

The teachers' virtual mobility competencies were indicated in a "post-session diagnostic survey of teachers." During the survey, teachers were asked to assess if planning, delivering, and organizing this course improved their virtual mobility competences. All of them ($N = 13$) marked that intercultural communication and personal and social competence were at least minimally improved. However, one of the teachers admitted having not improved his or her e-competence at all, and two of them had not improved their English language competence at all. The assumption for the lack of improvement might be attributed to a short time of partner institution responsibility for sub-module delivery, which took only two weeks. That is rather a short time, especially having in mind that more than one teacher were responsible for five out of six sub-modules. For a more detailed illustration of teachers' virtual mobility competence improvement see Figure 70.4.

Among any additional competencies that had been improved by the teachers "curriculum development and course design," "teaching competence concerning evaluation of e-learning or VM courses, e-assessment, or ICT tools usage," and "planning and promoting of learning in virtual environment" were indicated. Although module teachers had too little time to improve their virtual mobility competencies, still the collaborative preparation for the virtual mobility course, course organization and delivery had presumably resulted in the improvement of some of their competencies.

Comparing the student and teacher competency improvement in virtual mobility competencies, higher progress was seen in students' competencies. Module student and teacher improvement in virtual mobility competencies and their comparison are shown in Figure 70.4, in which some teachers admitted having not improved some of the competencies, and the students upgraded their competencies at least at a minimal. The same assumption with regard to failure to improve their

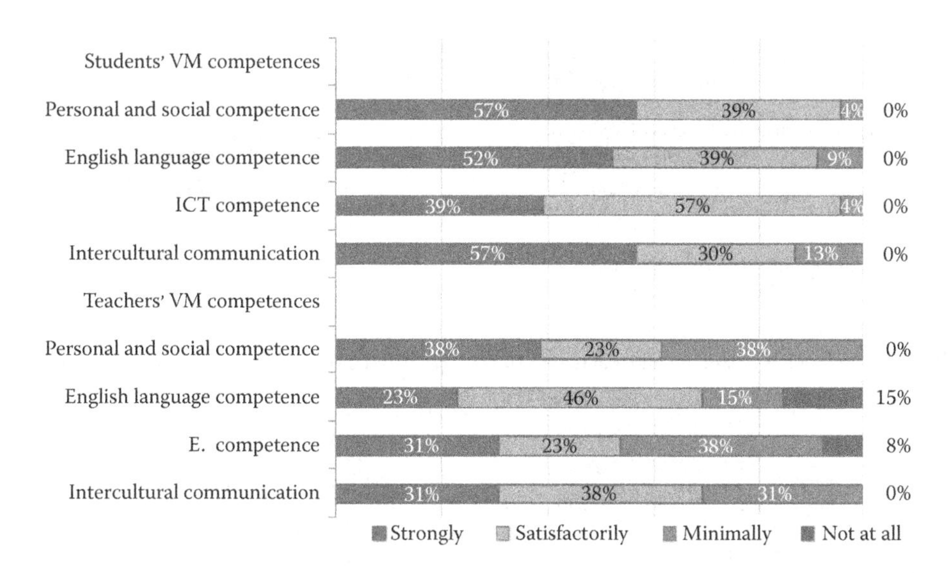

FIGURE 70.4 Improvement of teacher and student virtual mobility competency (improved categories in percentages).

competencies due to too little time for teacher competency improvement might be considered here. It could also be noted that there were some comments from the teachers that they would like to try delivering more than two weeks of the virtual mobility course.

To sum up, before starting attending the module, the students felt most comfortable with information analysis and evaluation or with knowing different technological resources for collaborative online learning. The students felt most professional in understanding the skills needed to facilitate and manage collaborative online learning and comparing learning styles and learning strategies. After the module, the students noted their improvement in all learning outcome–related competencies as well as a strong improvement in most of the virtual mobility competences: intercultural communication, English language, personal and social competencies, and a bit smaller but adequate improvement of ICT competency together with some additional skills or competencies. After the module delivery, half of the teachers improved their virtual mobility competencies at least at a satisfactory level. As students spent more time learning than teachers did delivering, students' virtual mobility competency improvements were higher than that of the teachers.

PARTICIPANTS' ATTITUDE AND VIRTUAL MOBILITY RECOGNITION

The last chapter of the paper addresses the participants' attitude toward virtual mobility and its recognition possibilities. The issues were addressed in post-session diagnostic surveys of students ($N = 23$) and teachers ($N = 13$).

Attitude toward Virtual Mobility

The participation in a virtual mobility course not only helped most of the students and teachers to better understand the virtual mobility concept and its realization in practice. The experience also allowed most of the participants to see more of the positive aspects of virtual mobility (indicated by 72% the participants) although one fourth (25%) of the participants had not changed their attitude toward VM; still it remained positive (see Figure 70.5). Most of the students ($N = 17$) also noted that motivation to choose the module changed during virtual mobility sessions. Only one teacher saw more negative aspects of virtual mobility after the virtual mobility course.

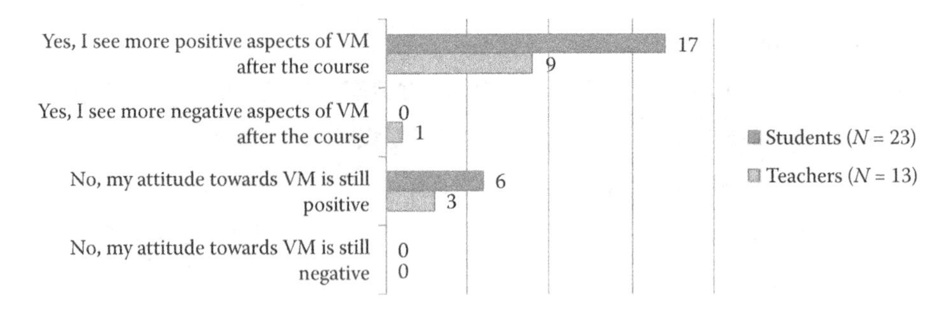

FIGURE 70.5 Changes in student and teacher attitude after virtual mobility course (numbers of respondents).

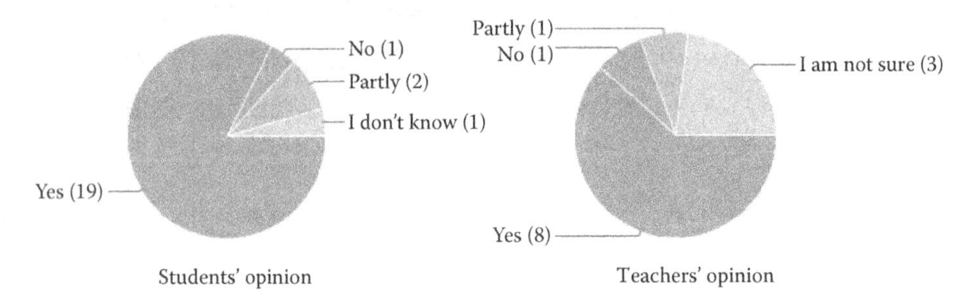

FIGURE 70.6 Respondents' opinions if virtual mobility sessions should be treated or recognized as a supplement to physical Erasmus mobility (in numbers).

Concerning the provision of a similar virtual mobility course in the future, 19 students ($N = 23$) and 11 ($N = 13$) teachers indicated their interest in participating again, given the possibility. The diagnostic surveys of the students and teachers also asked for respondents' opinion regarding physical and virtual mobility relations—if virtual mobility sessions should be treated or recognized as a supplement to physical Erasmus mobility (see Figure 70.6).

Seventy-five percent of the respondents on average (83% of the students and 62% of the teachers) agreed that virtual mobility sessions should be treated or recognized as a supplement to physical Erasmus mobility. One of the students, who had participated in physical mobility before the module, opposed that virtual mobility sessions should be treated or recognized as a supplement to physical Erasmus mobility "because virtual mobility is not equal at all to physical Erasmus mobility"; however, the other three students who had some physical mobility experience, and 16 who had not ($N = 19$) as well as eight teachers (who had not indicated their physical mobility experience) agreed that virtual mobility sessions should be treated or recognized as a supplement to physical Erasmus mobility. Two more students (who both had not participated in physical mobility programs before) and one teacher indicated that virtual mobility sessions could be partly treated as physical Erasmus. One of them explained: "Virtual mobility sessions should be treated as a supplement to physical Erasmus mobility just partly because nothing can replace study in a different country and getting to know other cultures during learning."

Virtual Mobility Recognition

The last section of the surveys intended to identify virtual mobility recognition possibilities at participating institutions. At the beginning of the TeaCamp project, while preparing a "State of the Art" report, systemic analysis of documents, legal acts, and scientific literature was implemented in project partnership institutions, and a pilot survey was organized. The aim of the survey was to identify the existing virtual mobility practices at TeaCamp partnership institutions and countries. The survey provided only some existing practices that should be discussed and analyzed in order to be recognized as existing virtual mobility. The survey respondents, being institution experts in the virtual learning area, were asked to indicate if there were any legal restrictions for VM of teachers and students to occur. The respondents of the pilot survey indicated no legal restrictions for VM of teachers and students at any of the participating institutions.

Thus VLHE module teachers were addressed with a similar question: If there were any legal possibilities at their institutions (that they were aware of) for the recognition of this course. Six teachers from five HEIs admitted being aware (the other seven were not) of the legal possibilities for the course recognition. Module students and teachers were also asked to indicate if the module would be recognized at their HEI institution. Seventeen of the students (N = 23) and five of the teachers (N = 13) from Vytautas Magnus University (Lithuania), University of Aveiro (Portugal), University of Jyvaskyla (Finland), and Jagiellonian University (Poland) indicated that the VLHE module would be recognized at their HEI; the other respondents were not sure if the module would be recognized at their HEIs. One of the module teachers also suggested that VM mobility sessions were to be "arranged as a part of a regular study program." This was noted to be an easier way to recognize the module, which was also one of the important issues addressed by the teachers to be improved. Module institutional recognition was also indicated by the respondents as leading to a higher commitment of participants.

CONCLUSIONS

1. Although virtual mobility is at the core of European education policy, modernization trends of higher education institutions, and European project activities, its benefits are discussed and stressed constantly; the concept itself is still changing and sometimes misconceived with the distance education and e-learning. The recent projects' initiatives have tended to harmonize the perceptions of its definition; however, some time for the emergence of common agreement regarding virtual mobility awareness is still needed.
2. Virtual mobility at the TeaCamp project was organized as the delivery and organization of a joint study module "Virtual Learning in Higher Education," which involved 13 teachers and 29 students from five countries and six higher education institutions in participation in synchronous (virtual sessions with discussions and lecture delivery, group assignments) and asynchronous (virtual learning organization in VLE Moodle with virtual sessions' records, group and individual assignments of students, and other learning organization methods used) ways. After the module, which was the virtual mobility case analyzed in this paper, students and teachers identified a strong or satisfactory level of improvement in virtual mobility competencies that can be divided into ICT, English language, intercultural communication, personal and social competencies, learning outcome–related competency, and some additional skills, such as time management, learning to work cooperatively in a virtual environment, and better understanding of virtual mobility as a form of learning.
3. The analysis of student and teacher attitudes toward virtual mobility showed that the experience of virtual mobility course participants changed their attitude and more positive aspects of virtual mobility were identified after completion of the course. The peculiarities of virtual mobility studies were discussed from the teacher and learner perspectives, and a lot of improvement or different organization scenarios were drawn; virtual mobility experience was valued from a positive perspective. Most of the students and teachers who have already experienced virtual mobility supported the idea that virtual mobility sessions should be treated or recognized as a supplement to physical Erasmus mobility. This suggests that not only virtual mobility activities should be organized, but virtual mobility could also enrich physical mobility.
4. After the virtual mobility experience in the VLHE module, the majority of the students and half of the teachers indicated that the module would be recognized at their HEIs; the others were not sure or aware of the module recognition possibilities. The uncertainty of

students and teachers confirmed that the phenomenon is rather new at traditional universities, and a lot of procedures have to be implemented in order for students and teachers to benefit from the possibilities and advantages created by virtual mobility.

REFERENCES

Bijnens, H.; Boussemaere, M.; Rajagopal, K.; Op de Beeck, I.; Van Petegem, W. (eds.). (2006). *Best practice manual "European Cooperation in Education through Virtual Mobility."* Retrieved from http://www.europace.org/articles%20and%20reports/Being%20Mobile%20Manual%20-%20Internet%20version.pdf.

Bijnens, H.; Op de Beeck, I. (2006). *Elearningeurope.info.* Retrieved October 22, 2009, from The Integration of Virtual Mobility in Europe: http://www.elearningeuropa.info/directory/index.php?page = doc&doc_id = 7245&doclng = 6.

Bunt-Kokhuis, S. G. (2001). Academic Pilgrims: Faculty Mobility in the Virtual World. *On the Horizon, 9* (1), 1–6.

Dauksiene, E.; Tereseviciene, M.; Volungeviciene, A. (2010). Virtual Mobility Creates Opportunities. *Application of ICT in Education 2010: Experience, issues and perspectives of e-studies.* Conference Proceedings, November 18, 2010, Kaunas, Lithuania. ISSN 1822-7244, pp. 30–35.

Dondi, C., Salandin T. (2010). The Movinter White Paper: *"In praise of VM. How ICT can support institutional cooperation and internationalisation of curricula in higher education."* Retrieved August 21, 2011, from http://www.vertebralcue.org/images/stories/movinter_white_paper.pdf.

European Commission Glossary on the Lifelong Learning Programme 2007–2013. (2007). Retrieved May 2, 2010, from Virtual Mobility: http://ec.europa.eu/education/programmes/llp/guide/glossary_en.html#117

Humanities project report. (1995). Retrieved March 3, 2011, from http://tecfa.unige.ch/tecfa/research/humanities/humanities-report.html.

Montes, R.; Gea, M.; Dondi, C.; Salandin, T. (2012). *Virtual mobility: the value of inter-cultural exchange.* eLearning Papers, Special edition 2012. Opening learning horizons, 2012. Retrieved on April 26, 2012, from www.elearningpapers.eu.

Op de Beeck, I.; Bijnens, K.; Van Petegem, W. (2008). Home and Away. Coaching exchange students from a distance. A best-practice manual on blended mobility. Heverlee Belgium: EuroPACE ivzw.

Pigliapoco, E.; Bogliolo, A. (2007). *Accessible virtual mobility.* Retrieved January 21, 2010, from EADTU's 20th Anniversary Conference "International courses and services online: Virtual Erasmus and a new generation of Open Educational Resources for a European and global outreach": http://www.eadtu.nl/conference-2007/files/I6.pdf.

Schreurs, B.; Verjans, S.; Van Petegem, W. (2006). Towards Sustainable Virtual Mobility in Higher Education Institutions. *EADTU Annual Conference 2006 Proceedings,* Tallinn (Estonia), November 22–23, 2006.

Silvio, J. (2003). *Global Learning and Virtual Mobility.* (T. Varis, T. Utsumi, and W. R. Klemm, Eds.) Retrieved November 20, 2010, from http://www.friends-partners.org/glosas/Global_University/Global%20University%20System/UNESCO_Chair_Book/Manuscripts/Part_IV_Global_Collaboration/Silvio,%20Jose/Silvio_web/SilvioD9.htm.

Spot+ project team. (2001). *Training Module 2: A Virtual Erasmus Student.* Retrieved November 30, 2010, from http://www.spotplus.odl.org/downloads/Training_module_2.pdf.

TeaCamp project. (2009–2011). Retrieved from www.teacamp.eu on January 15, 2011.

Tereseviciene, M.; Volungeviciene A.; Dauksiene, E. (2011). *Virtual Mobility for Teachers and Students in Higher Education.* Comparative research study on virtual mobility. Vytauto Didziojo Universitetas, Kaunas, Lithuania.

Vilhelmson, B.; Thulin, E. (2005). Virtual Mobility of Urban Youth: ICT-based Communication in Sweden. *Tijdschrift voor Economische en Sociale Geografie (Journal of Economic and Social Geography), 96* (5), pp. 477–487.

71 Continuous Business and Educational Process Improvement in Higher Education Institutions through Implementation of New Technologies

Svetlana M. Stanišić, Verka Jovanović, and Nemanja M. Stanišić

CONTENTS

BUSINESS PROCESSES REENGINEERING IMPLEMENTATION IN HIGHER EDUCATION INSTITUTIONS

According to the contemporary requests for higher education institutions' workflow improvement, deployment of the appropriate strategies is found to be very important (Jacobs and Van Der Ploeg 2006). Therefore, business processes reengineering (BPR) within higher education institutions represents a paradigm shift that mainly refers to segments such as teaching, student service affairs, scientific research, and logistics (Srikanthan and Dalrymple 2003). In historical terms, the implementation of new ways of business system organization and operation became possible after information technologies had made a significant development. The previous theories of business development were established upon guiding principles such as

- Formation of separated organizational subunits charged with specialized tasks
- Assignation of the tasks to single-discipline specialized operators
- Generation of the conflicting work goals between organizational subunits

With time, mentioned business guidelines became limiting factors for organizational growth and sustainability, since the old-established business development model had to deal with internal competition between organizational subunits instead of being focused on profit. On the contrary, according to the contemporary business development model, a holistic approach to business processes is implied. In contrast to the older model, the BPR is focused on continual business process integration, which leads to business efficiency enhancement (MacIntosh 2003). Thus, a successful workflow followed by the growth of the education institutions is ensured. Thereby, each organizational subunit is perceived as a part of an overall system. and the business process is performed by teamwork. The success of BPR implementation is assessed through business efficiency improvement, beneficiaries' satisfaction increase, and preservation of top management and employees' interests therewithal, and it does not refer to simple automation of the process.

The purpose of this paper is to discuss the need for the BPR implementation in higher education institutions and to present the successful implementation of an organizational and educational model through a case study of Singidunum University, the largest private university in Serbia. At present, Singidunum University has 10,000 students, with 1500 new students enrolled each year. The issues investigated are mostly relevant for large educational institutions, and the discussion proposed can be useful for strategic, tactical, and operational-level managers involved in development and design of higher education institution systems. Among the main triggers for the BPR implementation are the growth of the institution with the subsequent lack of process organization and control; information technologies revolution; and the increasing level of speed, density, and diversity of business processes.

The redesign of a business process can be achieved with respect to both decision-making and operational levels. On the strategic and decision-making level of top management, corresponding targets are set, and total resources of the institution are defined. Further, most of the resources are provided, information is exchanged, and tasks are defined on the tactical level of the institution, while on the operational level, assigned task performance is ensured. Respecting the fact that reengineering is mentioned only in connection with business processes, one should always bear in mind the following question: What are the expectations and demands of our beneficiaries? The redesign in order to achieve integration and streamlining of the business processes starts with the need to define which of the previous business rules will be further used and which of the new ones are going to be introduced. This can be done only after efforts are made in order to understand and identify every step of the workflow and to further analyze and update the way the work steps are performed. The BPR methodology can be framed in four different phases, as described further in the text.

1. Organization guidelines should be defined according to the vision of the top management team. With respect to the legal status of the higher education institution, the integrated university should be seen as a complex business system wherein rights, obligations, and responsibilities are defined by the university statute. The integrated university has its own organizational units, as follows: chancellor's office, faculties or/and departments, university institute, university library, and so forth. The implementation of BPR involves changes in information flow and decision-making concepts, which are suggested in order to by-pass the differences in two information processing strategies, top-down and reverse bottom-up. From the standpoint of top management, the top-down strategy is used for defining business process goals and resources, and such an approach demands wide and global insight into subsystem organizations without specifying details. On the contrary, the bottom-up strategy involves precise actions conducted for specification of individual base elements within the system and their function, in order to achieve business process optimization. The BPR applies both strategies to determine the way subsystems are regrouped, in order to accomplish time targeted objectives according to the given priorities.
2. Creation of the information model, that is, IDEFIX and IDEFO (the Integrated Definition Language Extended) diagrams, used as a representation of concepts and relationships between the entities and functions of the system should be undertaken. The BPR is relying on the information system alteration and IT experts and top management cooperation to a great extent. A complex organizational structure of a university system can be presented by a simple system diagram that shows the interfaces between investigated entities (Figure 71.1).

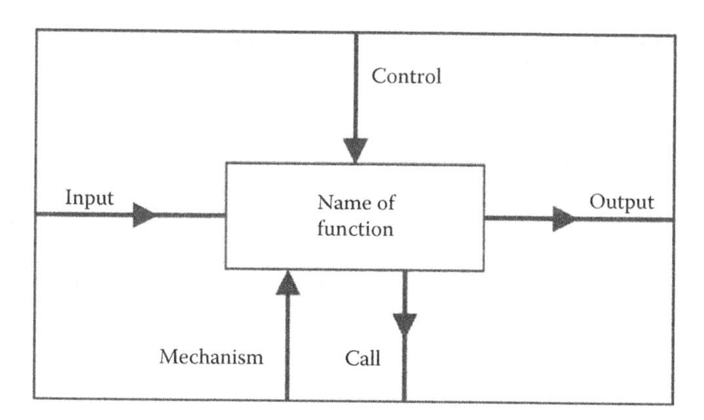

FIGURE 71.1 The basic element diagram.

The decomposition of business processes was undertaken by accomplishing subordinate tasks, such as defining boundaries of the business model, that is, system limits, and defining the way business activities are connected. Thereby, the horizontal linkages between the entities or functions indicate data and information flow direction. The vertical linkages within the system indicate the type of interactions between the entities, depending on the arrow direction. The arrow that enters the entity from the upper side indicates the controlling process related to the external control sources, laws, and regulations. On the contrary, the arrow that enters the entity from the down side indicates the internal control mechanism, related to the top management or supervisors of the higher education institution. The structure activity diagram of university business processes (Figure 71.2) is designed in order to represent the number of levels and entities in the hierarchy. This is the vertical aspect of business process organizing and refers to the pattern of relations between different system elements. Also, a diagram of a system as a whole and its inputs and outputs from/to external factors shows effective communication, coordination, and integration of the business processes. These operations depend on the horizontal aspect of organizing that contributes to the information flow and refers to the pattern of interactions among system elements. As shown in Figure 71.2, business system activities within an integrated university consist of administration and management, educational, student service, scientific research and publishing, human resource, and logistic activities. The diagram of the business processes starts with liabilities of the administration and management sector. Further, each diagram element can be decomposed into more detailed diagrams, as shown in Figure 71.3. For instance, administration activities comprise the following processes: planning and decision making, operations in the field of quality management system, information system operations, economy and financial operations, legal affairs operations, and marketing and public relations (Figure 71.4).

3. Relational database application development software CASE (Computer Aided Software Engineering) tools, such as BPwin and ERwin, represent a modeling tool fully compliant with the IDEFO business process modeling standard aimed at analyzing, documenting, and planning changes in complex business processes (Serova 2008). At Singidunum University, Serbia, research was conducted in order to provide a comprehensive overview of business processes, and thereby, the CASE ERwin tool was used for process modeling, according to ISO 9001:2000 quality requirements, while the CASE BPwin tool was used for data modeling and for generating a database scheme of selected systems in database control. The results have shown that about 446 business processes are taking place at the same time at the University (Jovanović and Veljović 2010).

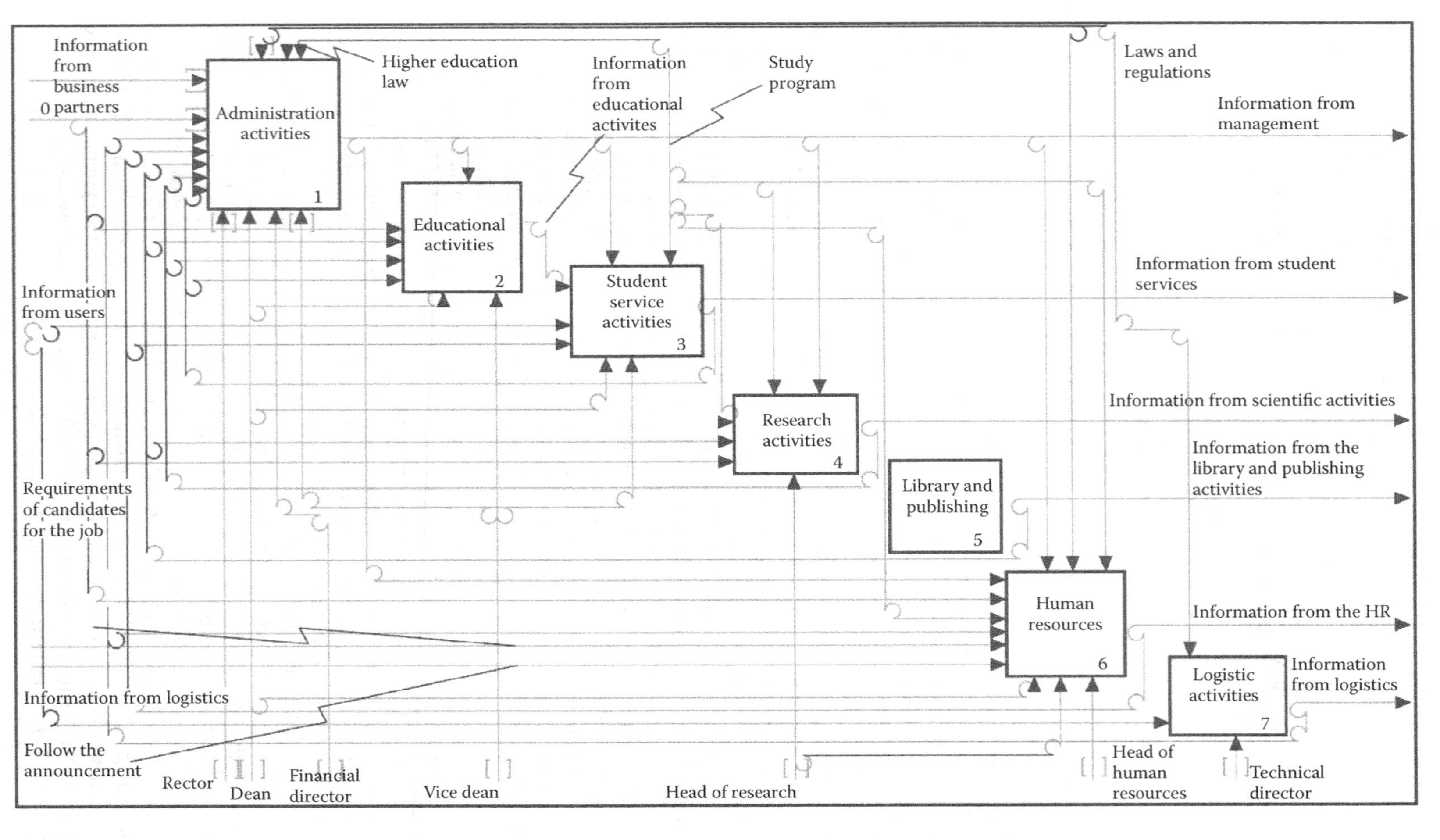

FIGURE 71.2 Analyzing system processes: the entity relationship diagram.

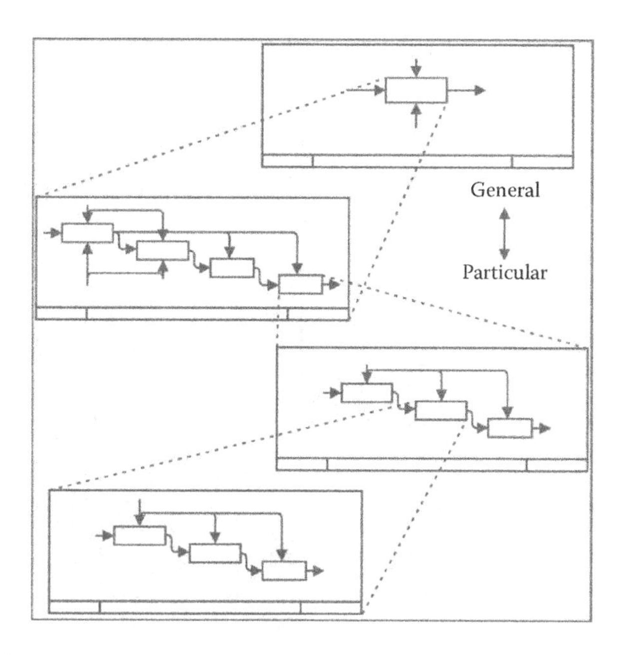

FIGURE 71.3 The diagram provides more information (to particular) or less (to general) on each entity.

4. The large institutions often struggle with the implementation of a new system organizational model that goes beyond changes related to single departments at the institution. The proposed business process model is implemented at Singidunum University, and the results indicate that reduction in costs of development, improvement of the system elements communication, reduction in time taken for non-productive activities, and reduction in education activities organization failure are achieved. As we concluded, the support of the top management and a change of employees' mindsets towards accepting that a lot of improving has yet to be done at all levels are needed throughout the implementation process.

The BPR is being positively received within the business world and could potentially be applied to fields less related to business techniques, such as university management (Adenso-Diaz and Canteli 2001). There are no strict rules or legislatives related to the implementation of the BPR as a strategic organizational process, so one must find its own solutions from case to case, while the experience of others may be utilized only as a pattern. The continuous quality, communication enhancement, reduced development costs, and decrease of charges are proven to be achieved. Relating to this, the establishment of technical conditions, innovative changes in human resource management, and development of awareness on an institution's strategic goals and founding values are required for introducing the BPR model in higher education institutions.

The BPR within higher education institutions is shown to be efficient in terms of solving organizational issues and developing leadership and change management. However, the emphasis on information technologies introduction in the BPR refers to the human resources change, which is obligatory for significant gains in efficiency (Laguna and Marklund 2005). Regarding this, failing to change people and inherited organization culture has been a major barrier to success. Further, a few factors more were found to be critical to BPR implementation success, such as teamwork performance, quality management system and satisfactory awards, effective change management, adoption of new information technologies, effective project management, and adequate financial resources (Ahmad et al. 2007).

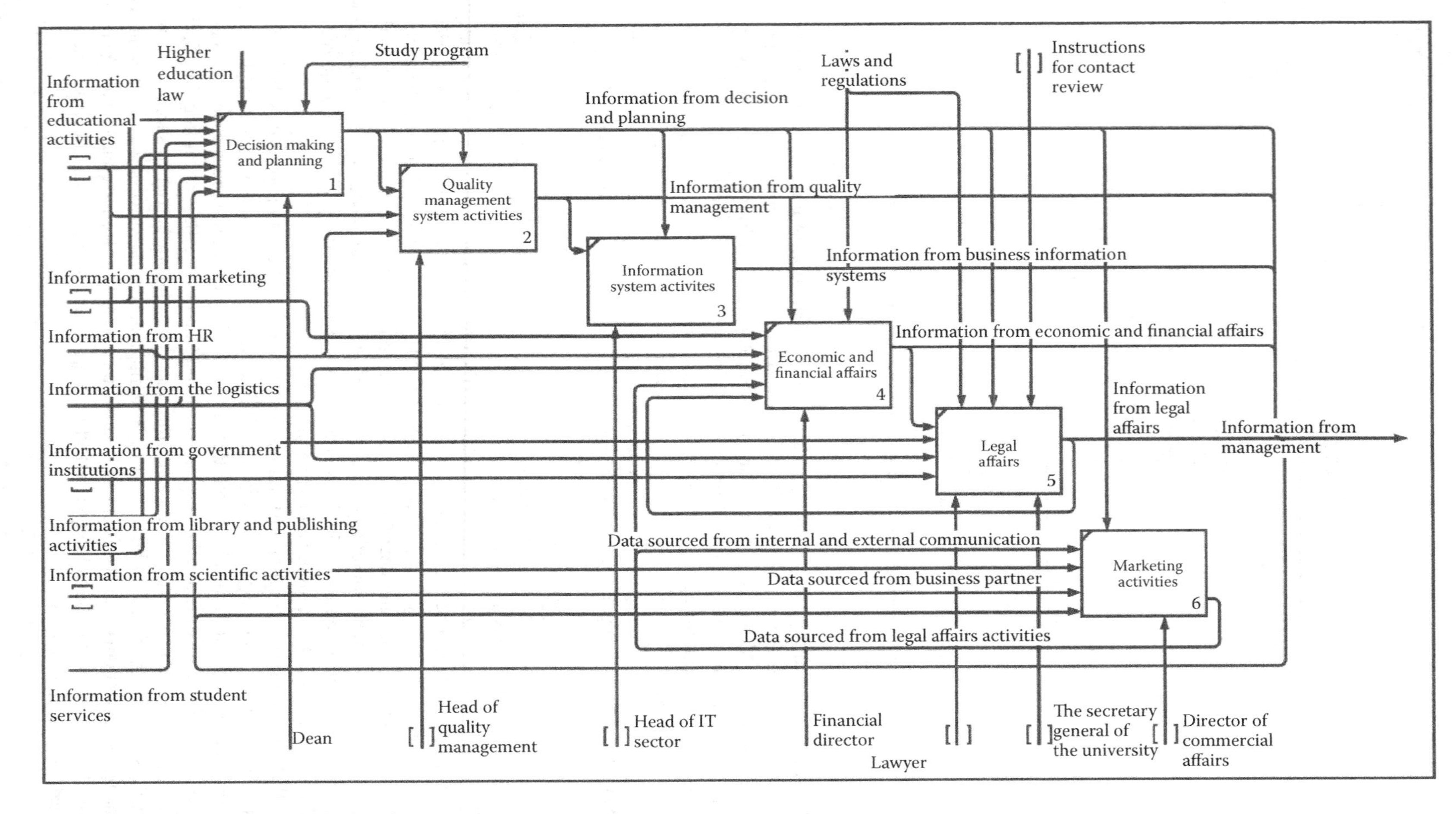

FIGURE 71.4 Administration activity diagram.

DISTANCE EDUCATION MODEL AS PART OF THE BPR IN HIGHER EDUCATION INSTITUTIONS

In facing the challenges of globalization, information technology, and knowledge-driven economy, the education goals are changing towards student-centered education and lifelong learning (Cheng 2005). The lower cost of computer hardware, software, and telecommunications services; subsequent better access to the Internet in the general population; and the fact that the young generation is computer skilled have influenced the rapid development of the distance education model in recent years to a great extent (Walker and Fraser 2005). Thereby, the new technologies have become a structurally integrated element of the learning environment, and an appropriate foundation for improving the quality of the learning process has been created. The term e-learning refers to a segment of distance education relying on a computer-based delivery system and, in a broader context, includes not only online learning but also the use of information and communication systems for learning process implementation in the traditional classroom environment (Khan 2005).

Economic pressures make it difficult for individuals to take several years off from work to attend university on a full-time basis (Bartolic-Zlomislic and Bates 1999). Online learning programs have seen a rapid increase in student enrollments in recent years mainly due to the following advantages:

1. Learning content delivery can be synchronous and asynchronous, depending on the type of media and students' choice.
2. The learning process can be self-paced, be lifelong, and occur independently of time and place.
3. The student is allowed to attend more learning programs at the same time, without change of residence.
4. Learning method boundaries are extended to include different interaction levels.
5. The use of information technologies in the learning process provides more skilled students and enables them to reach the level of competence they will need for their future job.

Singidunum University provides an opportunity for studying through a distance learning system (DLS), which can reach those disadvantaged by limited time, distance, or physical disability. Presently, the number of DLS students exceeds 1500, with a tendency to increase every year, and includes students living in three cities: Belgrade, the University center location; Subotica; and Niš. The distance learning process is adopted by the use of the MOODLE e-learning software platform, an asynchronous distance education tool with optional conventional class attendance for the students with residence in Belgrade. About 160 learning courses are generated for MOODLE users. For the students residing outside the university center location, besides the MOODLE e-learning software platform, additional videoconferencing systems are adopted as a synchronous education tool for two way-live video lessons. Besides MOODLE, all DLS students are able to access the designed Web site, which contains forums and video-recorded lessons and provides relevant information on the study programs. Throughout the development of Singidunum University, distance learners were not the primary target group, but as it turned out, distance learning has huge potential, and development-related costs were worth funding. Total costs of the DLS implementation and maintenance include the following:

1. Costs for the purchase of equipment—These costs were related to the purchase of the video-conferencing system, since MOODLE is a free and open-source software platform. The rest of the technical resources are to be found in the university as standard equipment.
2. Costs that occur on an ongoing basis—These costs are associated with technical equipment maintenance.
3. Costs associated with the development of learning content—These costs are related to the teachers' salaries and depend on the number of learning courses. Most of the teachers are already employed by the university in conventional learning programs.

4. Costs associated with delivery—These costs are related to site administrators, who have exclusive access to the system and have an important function in managing the MOODLE site and video conferencing system.

For the students outside of the university center location, 75% of the total lessons are video conference mediated, and the rest are held in the traditional classroom setting. From the standpoint of top management, DLS implementation has several advantages:

1. High availability to an unlimited number of beneficiaries
2. Reduction in total costs of ownership, development costs worth funding
3. Scalability, in terms of DLS ability to handle an increased number of beneficiaries with the same resources without loss of performance quality
4. Interoperability, the ability to be integrated with any existing software service
5. Ability to be customized and modified to fit beneficiaries' learning preferences
6. Allowing reports generation and providing data on the effectiveness of teaching method
7. Variety of capabilities and tools
8. User-friendliness, in terms of being structured in an easy way to understand and navigate
9. Ability for administrators to adjust the security level and to retrace all steps off logged-in students
10. Adequate security level, not lower compared to the other information system components

The effectiveness of DLS is influenced by numerous factors, such as media and course delivery mode selection, technology adoption, change implementation, strategies to increase interactivity, policy and operational issues, as well as a high degree of trust between distance learning students and the project developers (Sherry 1995). According to some exploratory studies (Motiwalla and Tello 2001) conducted in order to determine the effectiveness of the e-learning system, students have a positive experience with the online learning environment. Besides the forenamed, when creating courses for Internet distribution, the university and teachers benefit from the identification of a developmental model taking into account both learning and well-structured design principles (Passerini and Granger 2000). The challenge connected to e-learning project development refers to the developers' ambition to add functionality and to control time and resources while maintaining quality (McVay and Roecker 2007). However, when compared to traditional teaching, distance education methodology is perceived by students as having advantages that are not necessarily knowledge related.

Students' perception of personal participation and overall interaction are strong predictors of satisfaction (Fulford and Zhang 1993). The most significant challenge in distance learning is related to the student–student and student–teacher interaction issue. Although the e-learning environment can pose a critical barrier for discussion, encouraging students' interaction in conventional classrooms is not an obligatory practice.

Besides the fact that an effectively designed e-learning system offers the possibility for communication and interactive learning, students can choose to attend classes and meet teachers over the length of the course. The education providers, both administrators and teachers, have the obligation to make efforts to initiate students' interaction by use of group projects or discussion sessions. At Singidunum University, quality evaluation of DLS was conducted in order to determine the students' satisfaction. While most of the students had a positive experience with the video conference mediated courses, the most common complaints related to the MOODLE were about the disengagement among old-generation teachers to provide the learning materials and contents that meet the highest standards of quality and student usability. This problem was solved by having a training course with mandatory attendance and participation of all teachers, organized by the DLS administrators.

When it comes to the online presence and activity, our study results showed that about 18% of the distance learning students have logged in to MOODLE courses once per week or less. Before we can draw the conclusion that the percentage of the distance learning students who do not graduate is too high, we should compare it to the results related to the conventional learning students.

Therefore, the e-learning concept requires adaptability to new technologies, strong motivation, and self-discipline to enhance learning outcome. Also, there are different reasons mentioned in the literature for the failure of e-learning projects, such as failure to understand e-learning courses as dynamic entities that require maintenance, failure to manage risk and dedicate long-term support to the e-learning concept, and failure to implement different types of multimedia effectively. Some distance education projects result in quite positive outcomes, while others do not. But the same variability appears in conventional learning. The quality of the educational service depends on educational content, education providers, and the type of methodology applied in the teaching process (Šimić and Ivančić 2008). Top management team members' perceptions of identity and image as well as their ambition to create a proactive institution are the key elements in the strategic governance of a university.

CONCLUSION

The application of information technologies enables data control in higher education institutions and subsequent introduction of the BPR in the most efficient way. Computer models of complex management systems are used for several advantages, such as to reduce the number of instances of and to accelerate decision-making process, to enhance organizational culture, to ascribe responsibility, and to reduce administrative costs and help large university systems gain a competitive advantage. However, the BPR in higher education institutions is not limited to changes in administrative service but also includes attempts to redesign teaching and learning processes, and it remains a considerable challenge for universities. DLS implementation with the use of a video conference system and the MOODLE e-learning software platform provides higher education institutions the dual benefit of extending their outreach and keeping the cost economical. Online learning can open up new markets and give to students and older, mature, lifelong learners access to learning content, even if they are separated by time and space.

REFERENCES

Adenso-Diaz B., Canteli A.F., "Business process reengineering and university organisation: A normative approach from the Spanish case," *Journal of Higher Education Policy and Management*, Vol. 23, Iss. 1, 2001.

Ahmad H., Francis A., Zairi M., "Business process reengineering: Critical success factors in higher education," *Business Process Management Journal*, Vol. 13, Iss. 3, 2007.

Bartolic-Zlomislic S., Bates A.W., "Investing in on-line learning: Potential benefits and limitations," *Canadian Journal of Communication*, Vol. 24, Iss. 3, 1999.

Cheng Y.C., *New Paradigm for Re-Engineering Education: Globalization, Localization and Individualization*. Springer, Netherlands, 2005.

Fulford C.P., Zhang S., "Perceptions of interaction: The critical predictor in distance learning," *American Journal of Distance Education*, Vol. 7, Iss. 3, 1993.

Jacobs B., Van Der Ploeg F., "Guide to reform of higher education: A European perspective," *Economic Policy*, Vol. 21, Iss. 47, 2006.

Jovanović V., Veljović A., "Business process reengineering in high educational institutions by using IDEFO standards," *Singidunum Review*, Vol. 6, Iss. 2, 2010.

Khan B.H., *Managing E-Learning: Design, Delivery Implementation and Evaluation*. Hershey, PA; Information Science Publishing, 2005.

Laguna M., Marklund J., *Business Process Modelling, Simulation and Design*. Pearson Prentice Hall, 2005.

MacIntosh R., "BPR: Alive and well in the public sector," *International Journal of Operations and Production Management*, Vol. 23, Iss. 1, 2003.

McVay Lynch M., Roecker J., *Project Managing E-Learning*. Routledge, Taylor and Francis Group, 2007.

Motiwalla L., Tello S., "Distance learning on the internet: An exploratory study," *Internet and Higher Education*, Vol. 2, Iss. 3, 2001.

Passerini K., Granger M.J., "A developmental model for distance learning using the internet," *Computers and Education*, Vol. 34, Iss. 1, 2000.

Serova E., "Modern approaches for computer modelling for business-tasks decisions," Proceedings of the Ninth All-Russian Symposium, Strategic Planning and Evolution of Enterprises, CEMI RAS, Moscow, Russia, 2008.

Sherry L., "Issues in distance learning," *International Journal of Educational Telecommunications*, Vol. 1, Iss. 4, 1995.

Šimić M.L., Ivančić I., "Marketing concept implementation as an innovative approach to education system improvement in Croatia," Proceedings of the International Innovation Conference for Co-operation Development, InCoDe, Pecs, Hungary, 2008.

Srikanthan G., Darlymple J., "Developing alternative perspectives for quality in higher education," *International Journal of Educational Management*, Vol. 17, Iss. 3, 2003.

Walker S.L., Fraser B.J., "Development and validation of an instrument for assessing distance education learning environments in higher education: DELES," *Learning Environments Research*, Vol. 8, Iss. 3, 2005.

72 Changing the Subject
Funding Higher Education in Scotland

James Moir

CONTENTS

INTRODUCTION: THE FUNDING CONTEXT

The publication of the Browne Report in October 2010 (http://hereview.independent.gov.uk/hereview/report/) on higher education funding in England in the United Kingdom led to much debate and political activism. Its broad acceptance by the UK Conservative/Liberal Democrat Coalition Government, coupled with a period of substantial spending cuts in terms of block teaching grants, has effectively ended state support for higher education, except for some limited funding for science, technology engineering, and mathematical (STEM) subjects. However, the report points out that only the top 40% of earners will pay back all charges paid out by the government upfront; that the 20% lowest earners will pay less that the current system; and that the return for graduates will be, on average, 400%. Therefore, in cost–benefit terms, it would appear that in the words of the report, study will be a "risk-free activity."

The Government legislated to allow higher education institutions (HEIs) in England to raise their tuition fees from the current maximum level of £3290 to a potential £9000 in order to offset the reduction in teaching grants. However, it was thought that HEIs would be deterred from charging the maximum due to a levy over and above £6000, which is based on a sliding scale up to the maximum fee cap. Therefore, institutions are currently being faced with some difficult decisions as it is generally recognized that the £6000 level of fees will barely meet, or come in at just under, actual teaching costs. What now seems to be the case is that the majority have set their fees for next session at, or close to, the maximum level. Although, perhaps politically embarrassing, and calling into question the up-front costs of this funding model, it is evident that institutions are engaged into trying to "play the market," so to speak. In this context, there is a concern that some less overtly vocational subjects may be more vulnerable to a market-like situation where employability and earning potential are the drivers of subject choice. These are not easy times for the arts, humanities, and social sciences.

In Scotland, by contrast, higher education is currently free, and there is a cross-party consensus among politicians that direct fees will not be introduced. Instead, it has been argued that a "Scottish solution" should be found to address the funding crisis in Scottish HEIs brought about by the reduction in the overall block grant from the Westminster Government. The extent of that funding gap, although a source of dispute between the Scottish universities and the Scottish Government, is officially put at between £155m and £202m by 2014/2015 (http://www.scotland.gov.uk/Topics/Education/UniversitiesColleges/16640/stakeholdergroups/FinalReport). This is based on modeling the average English fee at £7500, the first figure without being indexed to inflation and the second with inflation built into the calculation.

Although Scotland does have its own parliament and some limited tax-varying powers, it is widely accepted that it will face its share of budget cuts in the national priority to reduce the United Kingdom's deficit. The minority Scottish National Party Government's Green Paper released in December 2010 (http://www.scotland.gov.uk/Publications/2010/12/15125728/0) is a consultation document that lays out some funding options for consideration. However, given cross-party election commitments to free higher education, rationalization and regional collaboration would appear to be the only way forward. Moreover, it is also the case that the potential problem of English "fee refugees" applying to Scottish HEIs in large numbers may require a higher fee levy for English students studying at Scottish universities. If this were to come into operation, then the government could raise £62m. With the addition of efficiency savings of £26m added to the calculation, it is argued that the funding gap would, in effect, be around £93m. However, whichever way the figures are calculated, it is clear that Scotland's universities face considerable shortfalls in income and, as a result, will face challenging times ahead in terms of resource decisions, as well as what they can offer in terms of "market value" and "brand."

Throughout all of this discussion about the future of higher education, there is one question that has dominated the agenda: What contribution do graduates make to national competitiveness and, more broadly, to a globalized knowledge-based economy and society? What is their *value*? This has become acute in the face of the current economic context and resulting need for competitive advantage. Graduates are required to be adaptable, multiskilled and entrepreneurial, and able to meet the needs of a rapidly changing world. However, there has been a tension between the focus on the economic costs and benefits of higher education versus the notion of social capital investment in relation to the knowledge society.

A critical evaluation of, and position on, this issue is advanced in four stages within the paper through considering the nature of Scottish higher education in the current global context and what it can offer as a "brand." The first stage considers that context in terms of a brief overview of notions the knowledge society and knowledge economy, which are treated as complementary rather than identical or mutually exclusive. The next stage in advancing a position turns to address the current "Graduates for the 21st Century Theme" in terms of the need to engender attributes that enable graduates to operate within a rapidly developing knowledge society and economy. The third stage situates the process of developing graduate attributes (GAs) within the context of global capitalism and the project of self-realization. The fourth stage in the development of the position taken offers an alternative way of conceptualizing the current theme through reconnecting with the tradition of the "democratic intellect" in Scottish higher education. Finally, the paper concludes by calling for a "bran" of Scottish higher education that builds upon that tradition rather following the current unidimensional focus on competitiveness with respect to employability.

THE KNOWLEDGE SOCIETY AND THE KNOWLEDGE ECONOMY

There has been something of a debate around differentiating the terms "knowledge society" and "knowledge economy" and the relevance of this for higher education (Sörlin and Vessuri 2007). The two terms tend to be used interchangeably, but it is evident that although they are interrelated, there are differences between them. The term *knowledge society* arose from the earlier term "information society" in order to move beyond the notion of technological change and to include of social, cultural, economic, political, and institutional transformation. Some writers, notably Castells (2000), refer to the informational society, making the comparison between industry and industrial. In this regard, Castells draws attention to the ways in which his use of the term *informational* is indicative of a form of social organization in which information is fundamentally linked to productivity and power. Castells goes on to consider how informationalism reinforces control over the labor process and extends capitalism through a "networked" *modus operandi* around the world.

However, while his position has much to say about the reach of informationalism within the knowledge society in which knowledge is aligned with production, he has less to say about knowledge as a potentially negotiable, conflicting, and perhaps liberatory aspect of human activities and social relations. In other words, there is a conflation between the knowledge society and the knowledge economy that reduces the latter to the former. However, the knowledge society may also be viewed as distinct from the knowledge economy in the sense that it is more than a commodity or something to be managed. It can also be regarded as something that is a participative and interactive process that is for the public good.

For the purpose of this paper, my position is not to become ensnared in debating the appropriateness of one term or another but, rather, to consider how higher education can both support a transformative role as well as deliver graduates who are able to operate within the knowledge economy and drive it forward. My focus is, therefore, on how students develop both in relation to the knowledge society and also with respect to their potential employment as part of the knowledge economy. As Probst et al. (2000) point out, knowledge is based on information but is always bound to persons and constructed by them. This is a key issue in considering the next stage in developing my position.

GAs: GRADUATES FOR THE 21ST CENTURY

This twin focus on the knowledge society and the knowledge economy is a feature of the "Graduates for the 21st Century Enhancement Theme" within Scottish higher education. A major focus of this work is the development of GAs in terms of the qualities that students acquire during the course of their learning. These qualities are therefore concerned with the ability to adapt to changing circumstances, to work across knowledge boundaries, and to become active and engaged citizens.

Scotland has looked to the Australian system of higher education for inspiration, and in particular, Simon Barrie's (2004, 2006, 2007) work has had a significant impact on thinking about the nature of generic GAs. It is clear from this work that participation and engagement are considered as being crucial to the development of GAs. This has gained expression through curricular reforms that encourage active learning and personal development planning. Therefore, the challenge is to ensure that these activities manifestly demonstrate the development of GAs.

This challenge is not unique to Scotland, but the strong sector-wide focus on GAs arguably throws it into greater relief. Take the increased diversity of the student population that enters the first year, which is a result of the widening of participation. How can we ensure that this diverse population acquires those GAs that we say are crucial to the purpose of higher education? And how do the varying personal, cultural, and economic circumstances of students impact upon the development of these attributes? How do students identify with their place in higher education as *students*, rather than as, for example, consumers in the light of increased fee payments or contributions? In this regard, there are parallels to be drawn with the critiques of the consumerist culture in schooling (see Sandlin and McLaren 2009).

The widening of participation, and the concomitant increased diversity of student backgrounds, has forced us to rethink how we encourage the development of GAs across such a diverse body of students. There are, therefore, questions about the sociological impact on higher education and in what ways the focus on the development of certain attributes can be squared with such diversity. In a wide-ranging review, David (2007) notes:

> In the early twenty-first century, there are clearly rich and diverse studies about and on higher education within a sociological methodological framework. While many of the studies point to the malign effects

> of globalization and neo-liberalism on the processes of managerialism and bureaucracy, masquerading as quality assurance, within higher education they also celebrate the ways in which the new forms of 'academic capitalism' allow for a diverse and potentially inclusive form of higher education.... It remains an open question about what the future of higher education may hold for subsequent generations into the twenty-first century: equity or diversity or both? (p. 687).

Given the impact of the current global economic situation and the imperative that has been placed on higher education to "deliver" on employability, David's point above becomes all the more acute in terms of the drive to develop the social capital required for the knowledge economy and society. However, there is something of a danger here in trying to differentiate between GAs within the Scottish system and the rest of the United Kingdom, or rest of the world for that matter. The trap here is one of trying to develop a set of "tartan attributes" that mark out Scottish higher education. This is clearly a difficult, if not untenable, position, given that many attributes are precisely the ones that are required in common to be able to operate in, and develop within, the knowledge economy and society.

However, there is also another aspect to the development of GAs, perhaps a more manipulative one in terms of the engineering of self-identity. This is related to the issue of social capital and the need to steer and drive oneself through the complexities of life in the knowledge economy and society. It is this issue that I now turn to in the third stage in the development of a position on the future of Scottish higher education.

GAs, SOCIAL CAPITAL, AND SELF-REALIZATION

If GAs are considered as a means of developing social capital, then a question that follows is how are they related to notions of self-development? It is at this point that I wish to draw upon the contemporary writing of Petersen (2011), who draws attention to the way in which authentic self-realization has become *the* guiding normative demand in modern society. His focus is very different from mine in that he is concerned with how this is related to the pathology of depression when people fail to maintain a sense of self-development. However, I want to focus on the flip side of this in terms of the drive to develop GAs as a form of self-realization. Therefore, although I do not share his focus on the sociological roots of pathology, I am nonetheless interested in the manifestation of authentic self-realization through the vehicle of GAs.

Petersen begins by aligning himself with Boltanski and Chiapello's (2005) insight in *The New Spirit of Capitalism*, with regard to the emergence of a new ideological commitment to capitalism based on self-realization. They argue that this commitment is premised on a different normative foundation than previously adhered to, one based on people's engagement with the capitalist system in terms of the project of self-realization. Boltanski and Chiapello argue that capitalism has thrived on critique in order to mutate and adapt to changing socio-historical circumstances. That critique of capitalism has taken two broad forms: social and artistic. The social critique has involved exposing the exploitative nature of capitalism in terms of such as aspects as poverty, social injustice, and rampant individualism. The artistic critique has focused on the ways in which capitalism reduces people to being cogs in an economic machinery and thereby delivering a technocratic and dehumanized society. In other words, they point out that the artistic critique is concerned with the oppression of individual creativity, autonomy, spontaneity, and authenticity through the relentless pursuit of standardization (Boltanski and Chiapello 2005, p. 37). However, this latter form of critique has taken hold to such an extent that capitalism has had to move with the times and has, in large measure, absorbed it into its own ideological basis by stressing the need for individuals to seek self-realization through the various interconnecting spheres of an individual's life (in work, through education, at home, etc.).

This new form of ideology requires that the individual considers himself or herself as an ongoing "project" of authentic self-realization. This is attained through activities that allow this aspect of self to be developed in the workplace, private life, and leisure. Given this logic, and the rapidly changing world of the knowledge economy and society, individuals must learn to develop and deploy a range of attributes that allow them to be flexible, mobile, enterprising, creative, adaptable, and malleable. As Petersen notes in direct reference to the work of Chiapello and Fairclough:

> What is relevant is to be always pursuing some sort of activity, never to be without a project, without ideas, to be always looking forward to, and preparing for, something along with other persons, who are brought together by the same drive for activity? (Chiapello and Fairclough 2002, p. 192).

It is but a short step from this characterization of modern subjectivity to consider the common stock of GAs such as "flexible collaborators," "active enquirers," and "creative handlers of complexity" as nothing less than an attempt to specify the normative content of self-realization. And yet, there is a clear tension here for some between what they regard as the academic nature of personal development leading to personal growth and the concomitant contribution to an educated citizenry, and the underlying national imperative that requires knowledge linked to economic wealth creation. However, in an era of mass higher education, it is often the latter that is a priority for governments, and no more so than in the current economic climate. The graduate contribution to society is considered in terms of both individual and wider societal wealth creation. This is considered as the return on the financial and personal investment in higher education.

It is therefore the case that with the concept of GAs, the purpose and meaning of higher education qualifications now extend to that of individual behavior. An individual's personal and social patterns of behavior have become normalized as part of his or her portfolio of GAs related to "employability skills." This new rhetoric represents fundamental change in how higher education is legitimated, one in which it is less in terms of subject-specific qualifications and more towards the possession of attributes that equip graduates to respond to the changing nature of the labor market. In this sense, the personal is made public and, in effect, codifies desired individual behavior resulting from the educational process. The whole notion of GAs is one that is bound up with an ideology of self-realization in relation to changing circumstances. This discourse taps into zeitgeist of the times, an age of insecurity and risk (Beck 1992); of individualism set in relation to appeals to market-like structures and globalization where these are considered as a normative ethic for guiding action (Sennett 2006); of constant self-reinvention capable of producing greater freedom but also anxiety (Elliot and Lemert 2006). The requirement for graduates to be adaptable and entrepreneurial has, therefore, never been greater set within a world of instability and uncertainty. Indeed, this is still further compounded by the increasing market-like rhetoric surrounding student choice and the selection of subjects and courses on the basis of perceived employability advantages. "Choice" is one of those universally accepted benefits of a market, and it is argued that students are choosing, and will increasingly choose, on the basis of employment prospects and earning potential. Of course, there are those who may argue that it depends who is doing the choosing: students or parents.

In the fourth stage of my position, I want to turn to offer another vision of Scottish higher education. It is rooted in tradition and yet is not out of place. Indeed, my argument is that it fits well with the changing times: a place for tradition in the modern age, if you will.

Changing the Subject: GAs and the Democratic Intellect

It would be fair to say that higher education is in a state of transformation across the world as it responds to the growth of the knowledge economy. There has also been a realization that the process of globalization requires undergraduates to be exposed to an education that will develop self-realization that is aligned to citizenship. The 2009 synthesis report from the Global University Network for Innovation (GUNI) entitled *Higher Education at a Time of Transformation: New Dynamics for Social Responsibility* draws attention to the many challenges confronting the sector that stem from those of wider society: beyond the "ivory tower" or "market-oriented university" towards one that innovatively adds value to the process of social transformation. The report argues that the creation and distribution of socially relevant knowledge is something that needs to be at the core of university activity, thereby strengthening their social responsibility. As the GUNI report puts so well, this calls for us to rethink the purpose of higher education, a purpose that is one of transformation rather than transmission:

> The central educative purpose of HEIs ought to be the explicit facilitation of progressive, reflexive, critical, transformative learning that leads to much improved understanding of the need for, and expression of, responsible paradigms for living and for 'being' and 'becoming', both as individuals alone and collectively as communities (GUNI 2009, p. 7).

This notion of higher education as educating citizens with a sense of civic awareness chimes with another Scottish higher education tradition: that of the democratic intellect (Davie 1961). Davie examined the decline of a type of higher education offered in Scottish universities after the Enlightenment of the 18th century that encouraged breadth of study and a commitment to public engagement through the study of philosophy and a broader concern with theoretical and conceptual issues. Even today, the notion of a broad higher education, at least to begin with in the early part of a program of study, is still with us in many of Scotland's four-year degree programs.

Davie advanced the argument that the democracy of the democratic intellect lay in the way in which the generalism of the Scottish philosophical tradition acted as a barrier to an individualistic notion of learning and, in so doing, bridged the gap between the expert few and the lay majority. It was argued that this created a sort of intellectual bridge between all classes in which the Scottish intelligentsia remained in touch with its popular roots, retaining a strong sense of social responsibility. Davie argued that a "common sense" developed, in which the expert knowledge of individuals was enhanced by, and held accountable to, the understanding of the wider public. This was "democratic" in as much as there was a social distribution of intellectual knowledge. This "democratic intellect," therefore, runs contrary to the notion of intellectual elites and rule by experts.

The democratic intellect concept clearly resonates with the "Graduates for the 21st Century" Theme and its focus on a more holistic approach to higher education. However, a note of caution needs to be sounded in that it is set within the context of ever-increasing costs for those entering higher education and, as previously noted, a legitimating rhetoric of "employability." There is little room here for the notion of citizenship and the democratization of knowledge that involves not simply the development of expertise but also the importance of bringing in knowledge from "outside" in terms of forging a real connection with lived experience. To do otherwise might risk opening up new spaces for critical debate and alternative ideas and practices. As Lyotard (1984, p. 4) put it in *The Postmodern Condition*, we are left with an "exteriorization of knowledge with respect to the 'knower,' at whatever point he or she may occupy in the knowledge process."

This issue is at the very heart of what GAs are being developed for and the nature of university education. It has long been recognized that learner autonomy is a crucial aspect of these attributes in terms of such aspects as self-regulation of learning and time management, the ability to generate one's own sense of enquiry, and the development of academic skills in making connections between conceptual knowledge and practice. However, given the points made above about the rapidly changing nature of knowledge society and economy, learner autonomy also involves the ability to adapt to change and to be able to evaluate different kinds of knowledge. In addressing this aspect, it is worth considering the words of Ron Barnett (2011), who writes:

> "In a world of liquid knowledge, where knowledge has become knowledges jostling and even competing with each another, there are knowledge spaces: universities can approach knowledge—and are doing so—in radically different ways" (p. 32).

It is therefore arguable that HEIs need to consider GAs related to the notion of self-realization as something that needs to be developed for, and beyond, the knowledge society in an effort to preserve the standing of higher education as an institutional public good. Otherwise, a focus on the corporate ends of the knowledge economy and the, at times, excessive mantra of preparing the "knowledge workers" of the future will dominate. HEIs have contributed more than their share in preparing graduates for the economy, but they have perhaps been less successful in developing the attributes of citizens who are able to advance the general democratic quality of their society and workplace. The concluding section considers Scottish higher education in terms of offering historical continuity within an era in which, although knowledge is everywhere, the age-old means of capitalizing on it and improving our lives lies in exchanging and sharing it.

CONCLUSION

So where does Scottish higher education position itself among all of this discussion of the *value* of graduates? For one thing, the current theme's focus is squarely on the sharing and exchange of knowledge within the context of employability and citizenship. The inclusion of GAs into programs of study has focused attention on the requirement for a broad education that engenders a wide range of attributes and skills. This has ensured that HEIs in Scotland tend to avoid narrow specialization in most of their programs of study. But more than this, the focus on GAs encourages students to engage with personalization through both practicing and reflecting upon it. This approach encourages students to consider the mutual relationship between self and discipline as part of an ongoing communicative activity, rather than as simply an instrumental document-driven process. There is more than a little of the democratic intellect tradition here than has perhaps been acknowledged. In this regard, perhaps Scottish higher education should position itself as offering a continuity of tradition that can thrive in a global context. Given the need to focus on knowledge application and transformation as related to employability and citizenship, the current Theme is worth pursuing as an integral democratizing aspect of higher education and not as a kind of latest trend initiative to be dispensed when some new one is required as part of a rhetoric meeting the challenges of a rapidly developing knowledge economy and society.

The nature of that society is something that is, of course, contested itself and how one acts on the basis of knowledge. Whether the knowledge economy and society is regarded as a projection for future development, a here-and-now reality, or even, as some would have it, a possible blight on our lives, requires higher education to address the educational and social issues that it raises. Whatever attributes are developed, and whatever economic or social investment is made in higher education, it would be desirable to obtain some balance between a concern with knowledge not only as a form of work but as something that is connected with a desire to understand and learn from others; not principally as a means of individual competitiveness in the knowledge economy but as something that is considered as relevant to the widening participation agenda and to the development of educational opportunities; and not simply as a way of operating in the knowledge society but as a responsibility to develop a broader, more cosmopolitan understanding of different forms of knowledge and how these can lead to a more democratic sense of community.

REFERENCES

Barnett, R. (2011) *Being a University*. Abingdon and New York: Routledge.

Barrie, S.C. (2004) A research-based approach to generic graduate attributes policy. *Higher Education Research and Development*, 23(3): 261–275.

Barrie, S.C. (2006) Understanding what we mean by *generic* attributes of graduates. *Higher Education*, 51(2): 215–241.

Barrie, S.C. (2007) A conceptual framework for the teaching and learning of generic graduate attributes. *Studies in Higher Education*, 32(4): 439–458.

Beck, U. (1992) *Risk Society: Towards a New Modernity*. London: Sage.

Boltanski, L. and Chiapello, E. (2005) *The New Spirit of Capitalism*. London: Verso.

Castells, E. (2000) *The Rise of the Network Society*. Oxford: Blackwell.

Chiapello, E. and Fairclough, N. (2002) Understanding the new management ideology: A transdisciplinary contribution from critical discourse analysis and new sociology of capitalism. *Discourse and Society* 13(2): 185–208.

David, M. (2007) Equity and diversity: towards a sociology of higher education for the twenty-first century? *British Journal of Sociology of Education*, 28(5): 675–690.

Davie, G. E. (1961) *The Democratic Intellect.* Edinburgh: Edinburgh University Press.

Elliot, A. and Lemert, C. (2006) *The New Individualism: The Emotional Costs of Globalization.* London: Routledge.

Global Network for University Innovation (GUNI) (2009) *Higher Education at a Time of Transformation: New Dynamics for Social Responsibility.* Basingstoke: Palgrave Macmillan.

Lyotard, J. F. (1984) *The Postmodern Condition: A Report on Knowledge.* Manchester: Manchester University Press.

Petersen, A. (2011) Authentic self-realization and depression. *International Sociology*, 26(1): 5–24.

Probst, G., Raub, S. P. and Romhardt, K. (2000) *Managing Knowledge: Building Blocks for Success.* Chichester: Wiley.

Sandlin, J. A. and McLaren, P. (Eds.) (2009) *Critical Pedagogies of Consumption.* London: Taylor & Francis.

Sennett, R. (2006) *The Culture of New Capitalism.* New Haven, CT: Yale University Press.

Sörlin, S. and Vessuri, H. (Eds.) (2007) *Knowledge Society vs. Knowledge Economy: Knowledge, Power, and Politics.* UNESCO Forum on Higher Education, Research and Knowledge, International Association of Universities, New York: Palgrave Macmillan.

73 Engineering Ethics Education through Project-Based Learning

Masakatsu Matsuishi, Kazuya Takemata, Shigeo Matsumoto, Taketo Yamakawa, and Toshiaki Isozaki

CONTENTS

INTRODUCTION

Technologies can bring great benefits but can also bring harm to the well-being of individuals, society, and the environment. The impact of technology has been expanding and accelerating because of widespread economic activities and rapid progress in distribution and communication. Under such circumstances, engineers should do their best to employ their engineering knowledge to improve the well-being of the public and to maintain the global environment.

The Code of Ethics for the National Society of Professional Engineers, USA, states that engineers in the fulfillment of their professional duties shall (The National Society of Professional Engineers 2010)

1. Hold paramount the safety, health, and welfare of the public
2. Perform services only in areas of their competence
3. Issue public statements only in an objective and truthful manner
4. Act for each employer or client as faithful agents or trustees
5. Avoid deceptive acts
6. Conduct themselves honorably, responsibly, ethically, and lawfully so as to enhance the honor, reputation, and usefulness of the profession

The engineering accreditation agency, the Accreditation Board for Engineering and Technology (ABET), regards engineering ethics as an important component of engineering education. ABET's Criterion 3(f) states that "an Engineering Program must demonstrate that their students attain an understanding of professional and ethical responsibility" (The Accreditation Board for Engineering and Technology, Inc. 2010). Thus, engineering ethics has become a required component of engineering education at institutions of higher education.

Engineering ethics is (1) the study of moral issues and decisions confronting individuals and organizations involved in engineering and (2) the study of related questions about the moral ideals, character, policies, and relationships of people and corporations involved in technological activity (Martin et al. 1997). Teaching engineering ethics can achieve at least four desirable outcomes: (1) increased ethical sensitivity, (2) increased knowledge of relevant standards of conduct, (3) improved ethical judgment, and (4) improved ethical willpower (Davis 1999).

The Kanazawa Institute of Technology (KIT) implemented engineering ethics education in nine technical courses in order to achieve our goal of "Ethics Across the Curriculum (EAC)" (Nishimura 2006). KIT decided to employ "the Micro-Insertion Technique" to introduce engineering ethics and related topics into the courses. The Micro-Insertion Technique introduces ethics and related topics into technical courses in small enough units not to push out technical material (Davis 2006).

In response to the engineering ethics education at KIT, the authors implemented engineering ethics in two introductory project-based learning (PBL) courses: Engineering Design I (ED I) for freshmen and Engineering Design II (ED II) for sophomores. These introductory PBL courses are mandatory for all freshmen and sophomores (Matsuishi et al. 2004). Every year, approximately 1700 freshmen and 1700 sophomores take ED I and II, respectively. Students choose a PBL project of their interest, conducting a needs assessment and analyzing the results, determining the design specifications, and generating viable design solutions.

The following ethics-related learning activities are present in the PBL courses:

1. The instructor delivers a lecture on engineering ethics and the ethical roles and responsibilities of engineers.
2. Students are required to confirm that their design solutions do not harm the global environment, do not endanger human life or health, contribute to human welfare and happiness, and do not infringe on intellectual property rights. Students are also required to demonstrate that their design solutions are ethical during their final oral presentation.
3. The instructor gives students advice on engineering ethics during an office hour meeting at the students' request.
4. Students evaluate their performance and progress with regards to engineering ethics at the beginning, middle, and end of the course. Students are able to recognize their current levels of achievement and try to attain higher levels of ethical awareness.

Assessment of objectives and data is discussed in this paper.

ENGINEERING ETHICS EDUCATION AT KIT

Ethics across the Engineering Curriculum

KIT implemented engineering ethics education in nine technical courses to achieve our goal of "Ethics across the Curriculum (EAC)." The nine technical courses are Introduction to Engineering I, II, and III; Science, Technology, and Society; Humanity and Nature III; Science and Engineering Ethics; and ED I, II, and III. As shown in Figure 73.1, the courses are offered across all four years of undergraduate education. Primary ethical outcomes to be achieved in the nine engineering ethics courses are shown in Figure 73.2.

KIT decided to employ "the Micro-Insertion Technique" to introduce engineering ethics and related topics into the nine technical courses. The Micro-Insertion Technique introduces ethics and related topics into technical courses in small enough units so as not to push our technical material. Micro-insertion requires neither new courses nor radical changes in existing courses (Davis 2006). Thus, engineering ethics education can be achieved over a large portion of the engineering curriculum.

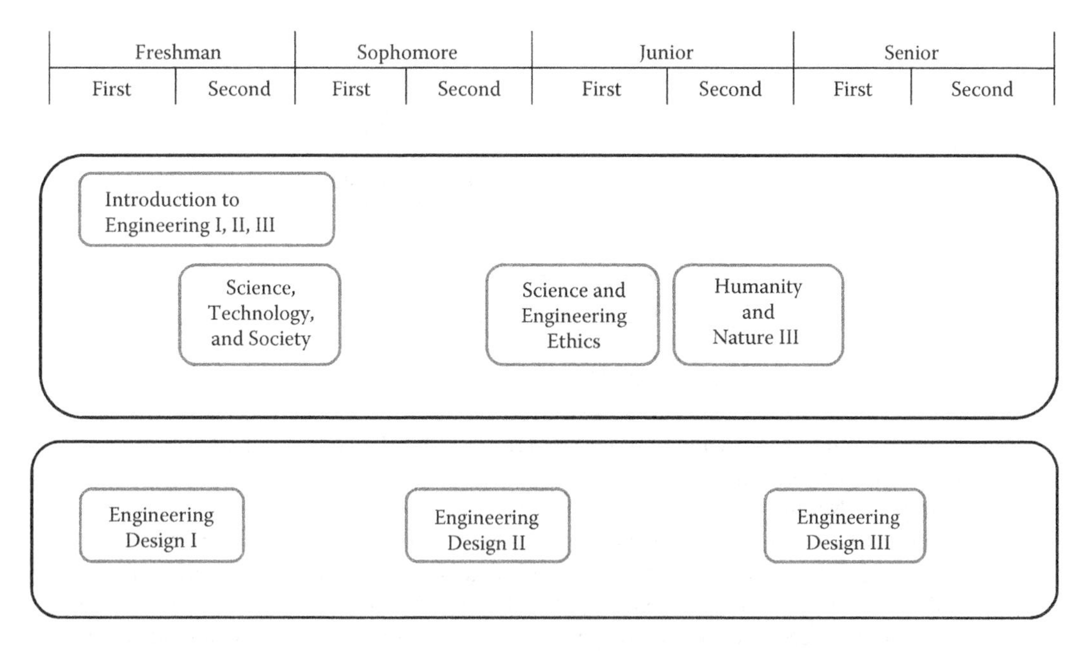

FIGURE 73.1 Engineering ethics courses.

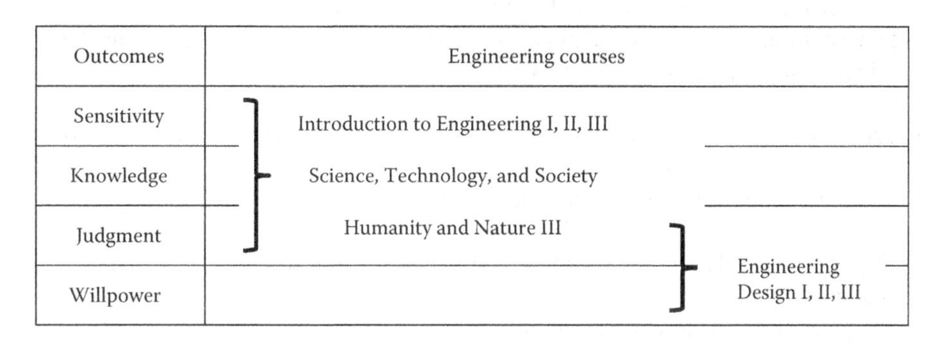

FIGURE 73.2 Primary ethical outcomes achieved by each technical course.

Outline of Engineering Ethics Taught in a Variety of Engineering Courses

1. Introduction to Engineering I, II, and III
   Introduction to Engineering I, II, and III are mandatory for freshmen. Students learn skills and knowledge essential for learning independently and taking responsibility for their own learning. Students learn the professional obligations of engineers and the social impact of technology. Students conduct case studies of past accidents, for example, the Space Shuttle *Challenger* disaster (Roger et al. 1986), in order to understand and appreciate the professional obligations of engineers. Their learning objectives are to develop ethical sensitivity, knowledge, and judgment.
2. Science, Technology, and Society
   Science, Technology, and Society is mandatory for freshmen. Students learn the relation of Science, Technology, and Society and recognize the complexity of technical challenges and professional responsibility. A case study of historical accidents and disasters like the Chernobyl nuclear power plant accident (The International Atomic Energy Agency 1992) allows students to go through iterative steps of reflecting, investigating, analyzing, and

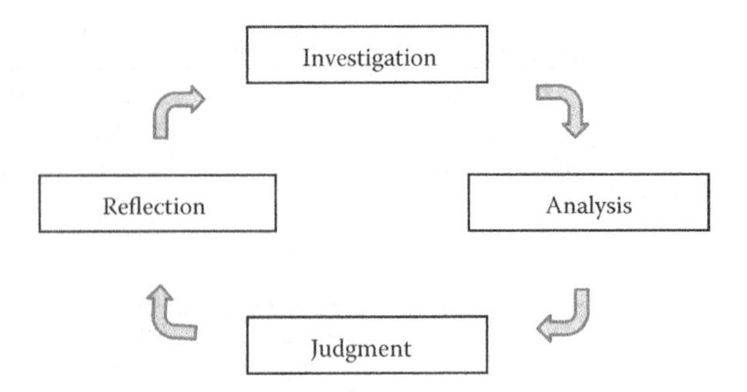

FIGURE 73.3 Iterative process of engineering ethics study.

making judgments, as shown in Figure 73.3. The learning objectives are to develop ethical sensitivity, knowledge, and judgment.

3. Humanity and Nature III
   Humanity and Nature III is mandatory for juniors. During this course, students stay at KIT's Anamizu Seminar House for three consecutive days, located in an area of great natural beauty and contact with nature. They learn skills and knowledge essential to be professional engineers. An instructor gives a keynote speech on the professional obligations of engineers. Students conduct group discussions based upon the keynote speech by the instructor. They understand and appreciate the professional obligations of engineers and the environmental and social impact of science and technology. Its learning objectives are to develop ethical sensitivity, knowledge, and judgment.
4. Science and Engineering Ethics
   Science and Engineering Ethics is optional for juniors and a specialized course for engineering ethics. Its learning objectives are
   - To acquire knowledge of engineering ethics
   - To understand that science and technology have a profound impact on society
   - To understand that engineers shall hold paramount the safety, health, and welfare of the public
   - To judge from an ethical view point
5. ED I, II, and III
   ED I, II, and III are mandatory for freshmen, sophomores, and seniors, respectively. The outlines of these courses are shown in Table 73.1.

The objectives of these design courses are not only to provide students with superior technical capabilities, but also to enable them to identify and tackle ill-structured and open-ended problems, to generate a set of distinct and creative design concepts, to develop the engineering design expertise and abilities for tackling problems independently and in a team, and to acquire important basic skills such as communication and leadership. Students are required to generate viable design

TABLE 73.1
Engineering Design Courses

Course Name	Students	Period (Credits)	Type
Engineering Design I	Freshman	One semester (2)	Project-based learning
Engineering Design II	Sophomore	One semester (2)	Project-based learning
Engineering Design III	Senior	One year (8)	Research project capstone design

solutions that meet ethical requirements. Their ethics learning objectives are to develop ethical judgment and willpower.

ENGINEERING ETHICS COURSE CONTENT IN INTRODUCTORY PBL COURSES

KIT developed two introductory PBL courses: ED I and ED II. ED I and ED II are required for all freshman and sophomore students, respectively.

Procedures covered in the PBL courses are

- To identify design opportunities
- To characterize the design project
- To generate design concepts
- To evaluate the design concepts and to select the most promising one
- To design in detail
- To presents results

Learning objectives of the courses are

- To be able to perform market research and address needs of clients
- To be able to determine design specifications
- To be able to generate viable design solutions and choose the best one
- To be able to design in detail
- To be able to work in a team and to communicate effectively

The authors integrated engineering ethics into the two introductory PBL courses. The following ethics-related learning activities have been added to the courses.

- The instructor delivers a lecture on engineering ethics and the ethical roles and responsibilities of engineers. Figure 73.4 shows some of the slides used for a lecture on engineering ethics.
- Students follow the engineering design process to develop a design solution that will be able to contribute to the welfare and happiness of the human race. An instructor gives advice to students on their consideration, judgment, and decision-making from an ethical view point at an office-hour meeting.
- Students are required to confirm that their design solution does not harm the global environment, does not endanger human life or health, contributes to human welfare and happiness, and does not infringe on intellectual property rights. Students are also required to demonstrate that their design solutions are ethical during their final oral presentation.
- Students are required to evaluate their performance and progress with regards to engineering ethics at the beginning, middle, and end of the course. Students are able to recognize their current levels of achievement and try to attain higher levels of ethical awareness.

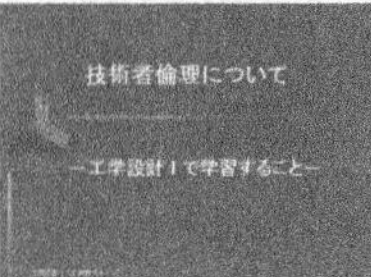

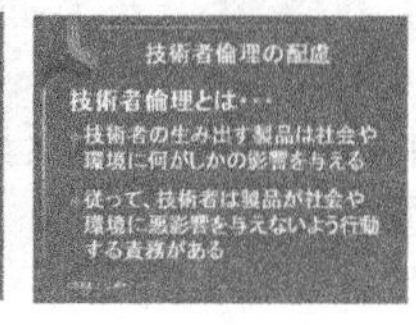

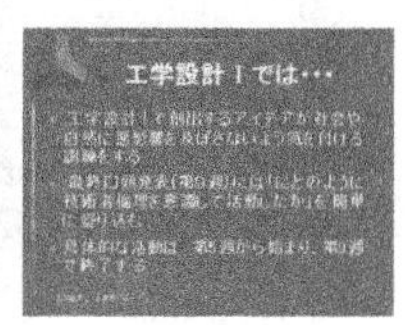

FIGURE 73.4 An example of a lecture on ethics.

RESULTS OF ENGINEERING ETHICS EDUCATION

1. Ethical design solutions
   Students generate ethical design solutions according to the requirements that their design solutions do not harm the global environment, do not endanger human life or health, and contribute to human welfare and happiness. At the end of the term, design teams gave final oral presentations in which they demonstrated that their design solutions were ethical.
   As an example of ED II projects, design activities of a design team are discussed in this section, whose project theme was "Designing an eco-friendly indoor lamp." Their goal was to develop an indoor lamp that can lessen consumption of electricity and therefore decrease CO_2 discharge. They succeeded in developing a design solution to lessen energy consumption and working to prevent global warming.
2. Assessment of performance and progress with regards to engineering ethics
   At the end of the term, a questionnaire was given to all students in the PBL courses. The authors wanted to find out how well the students understood the importance of engineering ethics from the lectures given in those classes. Figure 73.5 shows the results of the survey. Approximately 87% of students understood the importance of engineering ethics from the lecture.

Students evaluated their ability with regards to engineering ethics at the beginning, middle, and end of the course using the following questions:

Q1: Can you review and improve design specifications and design solutions to benefit society?
Q2: Can you make judgments and behave based upon the principles of engineering ethics?
Grade scale: 1, Very Poor; 2, Poor; 3, Average; 4, Good; and 5, Very Good

Students were able to recognize the current level of their achievements and tried to attain the higher levels. This process is referred to as the PDCA Cycle of Self-Learning. Figure 73.6 shows students' assessment of their performance and progress. It was found that students' understanding and behavior with regards to engineering ethics advanced steadily during the PBL courses, as shown in Figure 73.6.

A questionnaire was given to all students in the PBL courses in order to find out when the students gave consideration to and made sensible judgments about engineering ethics. Figure 73.7

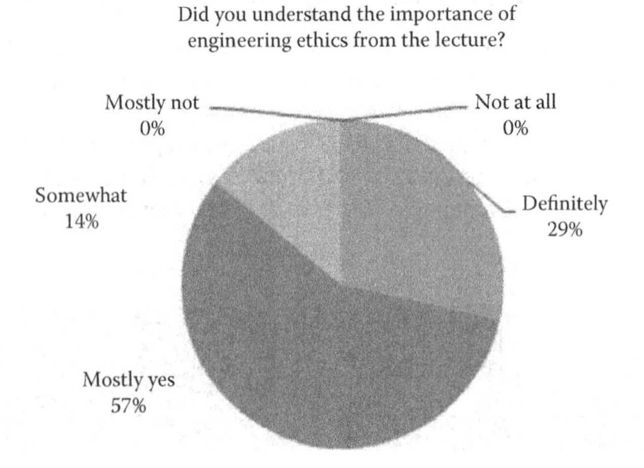

FIGURE 73.5 Understanding of the importance of engineering ethics.

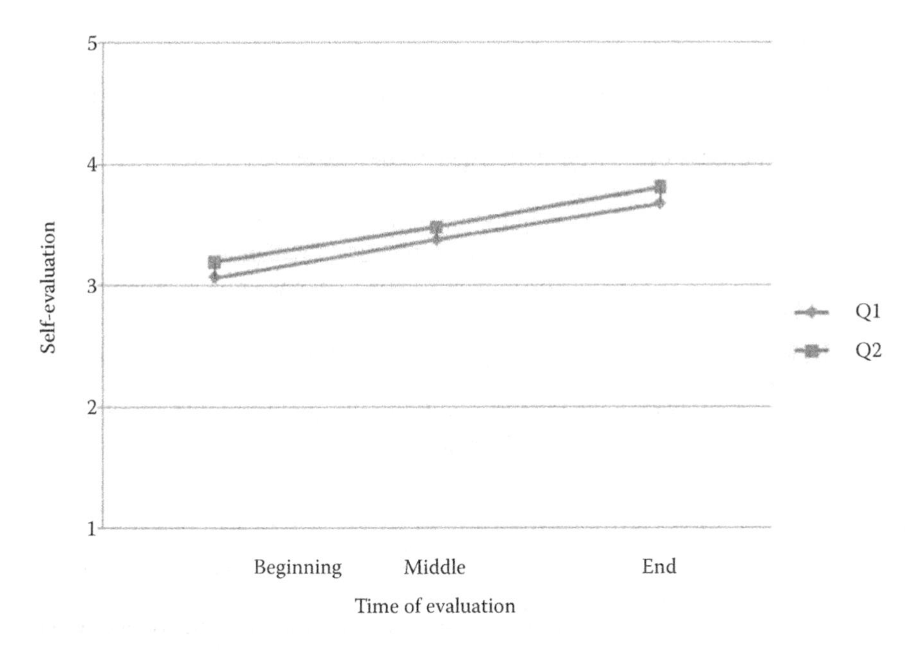

FIGURE 73.6 Assessment of performance and progress with regards to engineering ethics.

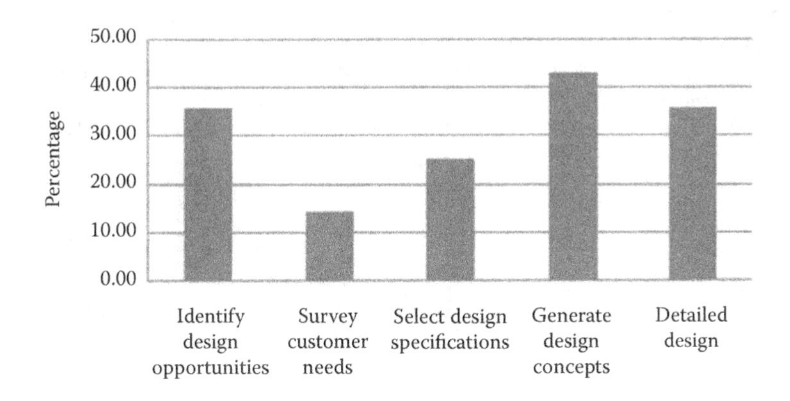

FIGURE 73.7 Percentage when students consider and judge about engineering ethics.

shows that students gave consideration to and made sensible judgments about engineering education mainly at the following stages:

1. Identifying design opportunities
2. Generating design concepts
3. Carrying out detailed design

The authors surveyed how students understood the role of engineering ethics in their future career. It was found that most students understood that the application of engineering ethics would be one of the great industrial, social, and moral issues of their future career.

CONCLUDING REMARKS

1. KIT implemented engineering ethics education in nine technical courses in order to achieve our goal of "EAC."
2. KIT employs the "Micro-Insertion Technique" to introduce engineering ethics and related topics into the nine technical courses.
3. The authors integrated engineering ethics into project-based learning (PBL) courses.
4. Students generated ethical design solutions according to the requirements that their design solutions did not harm the global environment, did not endanger human life or health, and contributed to human welfare and happiness.
5. Students' understanding and behavior with regards to engineering ethics advanced steadily during the PBL course.
6. Students understood that the application of engineering ethics would be one of the great industrial, social, and moral issues of their future career.

REFERENCES

The Accreditation Board for Engineering and Technology, Inc. (2010). Criteria for accrediting engineering programs. Effective for Evaluations during the 2010–2011 Accreditation Cycle.

Davis M. (1999). Teaching ethics across the engineering curriculum. The OEC International. Conference on Ethics in Engineering and Computer Science.

Davis M. (2006). *Integrating Ethics into Technical Courses: Micro-Insertion, Science and Engineering Ethics.* Springer, Vol. 12, Issue 4, pp. 717–730.

The International Atomic Energy Agency. (1992). The Chernobyl Accident: Updating of INSAG-1.

Martin M. et al. (1997). Ethics in Engineering, Third Edition. McGraw-Hill.

Matsuishi M. et al. (2004). Evolution of engineering design education at KIT. Proceedings of the 2004 Annual Conference and Exposition, American Society for Engineering Education.

The National Society of Professional Engineers. (2010). Code of Ethics for Engineers.

Nishimura H. (2006). Approach of Kanazawa Institute of Technology for the achievement of ethics across the curriculum. *Journal of JSEE*, Vol. 54, No. 4, p. 631–634 (in Japanese).

Roger W. et al. (1986). Report of the Presidential Commission on the Space Shuttle Challenger Accident.

74 Essential Roles of Language Mastery for Conceptual Understanding and Implications for Science Education Policies

Liliana Mammino

CONTENTS

INTRODUCTION

Science teaching and learning and, consequently, the extent and quality of students' acquisition of scientific knowledge, rely on communication: teacher–learner and learner–teacher communication as well as the communication realized through books and other study materials. Language is the most fundamental communication tool. It is also the essential instrument for the development of thought (Bruner 1975) and for any process of knowledge acquisition (Chomsky 1975). Thus, language is the essential tool for a student to be able to reflect on the themes that are objects of attention within a course, to analyze them and consider all their individual aspects, the connections between individual aspects, and the ensuing implications. As a result, language is the most essential tool for a student's engagement in his/her learning process, leading to the gradual and continuous acquisition of knowledge that is the objective of education. This implies the need to ensure the attainment of adequate language-mastery as a prerequisite largely conditioning the efficiency of science education.

This work analyzes and discusses the main issues regarding the role of language in science education: the importance of language mastery to enable science understanding; the importance of adequate language-mastery sophistication to match the demands of the continuous increase in science conceptual sophistication; the issue of the language of instruction and the importance of mother tongue instruction for science understanding; the inferences prompted by already existing massive observations; and the consequent indications for educational policies. The *leit motiv* throughout the analysis and discussion responds to focused attention on the objectives of science education and on the importance of ensuring that the students' contact with science is as broad and as comfortable as possible, so that it can lead to effective familiarization with a conceptual world whose relevance in modern society is continuously increasing.

Conclusions and inferences are largely (but not solely) based on the results of a systematic investigation—throughout the last 15 years—of language-related difficulties encountered by tertiary-level chemistry students. The sources of information for the investigation comprised the analysis of students' written works, the information stemming from classroom interactions, and occasionally, also personal interviews meant to ascertain the reasonability or correctness of interpretations (Mammino 1998a, 2005a, 2006a, 2007, 2009, 2010, 2011). Overall, the collected documentation amounts to several thousand concrete examples. Their analysis utilized a method that Brooks (2006) most aptly defines as "using linguistic tools borrowed from cognitive linguistics and systemic functional grammar" to investigate the relationships between language-usage and science understanding in students' answers. It is believed that, by integrating the language perspective and the science perspective, such an approach can provide an optimal basis for inferences on the role of language mastery in science learning and on the corresponding implications for educational policies.

LANGUAGE MASTERY AND SCIENCE LEARNING

Language mastery conditions science learning both in a direct way and because of being instrumental to the acquisition of other skills that play important roles in the learning process. These two modes are analyzed in the next subsections.

Language Mastery and Conceptual Understanding in the Sciences

Language and science are intimately connected because of the nature of language as an instrument of thought and communication. New developments in the sciences imply the generation of new ways of expressing the new concepts (Sutton 2003). The conceptual investigation of the relationships between language and science expands through philosophical issues as various as the nature, reliability, and validity extent of scientific knowledge; the incidence of uncertainty; the degree of fuzziness inherent in language (Russell 1923) but also in various science concepts, including several from chemistry (Rouvray 1997); and the search for ways of introducing language components into mathematics (one of the potentialities of fuzzy logic). A deeper review of these themes would go beyond the scope of the current work, whose major focus is on education. However, it is crucial to bear in mind that the importance of the relationships between language and science constitutes the foundation of the roles of language mastery in science education. Language mastery is an essential key to conceptual understanding in the sciences because science is built and communicated through language and because the teaching–learning process is a communication-depending activity. The educational implications are obvious: language mastery conditions science understanding and determines the extent of science knowledge acquisition.

The role of language in the teaching and learning of science and the question of how to utilize language to facilitate understanding have been the object of intensive educational research for decades, stimulating an enormous number of studies and implementation of projects in many countries and contexts. Even a fast review would require much more space than an article, to be sufficiently representative; therefore, only some works whose conceptual approaches relate more closely to the focus of the current discussion are briefly recalled here. "Classical" works pose the issue of the role of language at the core of investigation and reflection (Munby 1976; Carré 1981;

Lahore 1993; Wellington and Osborne 2001). Since all the aspects of the learning process are conditioned by language mastery, the analysis of specific educational issues or learning options often incorporates or reveals language connections. The difficulties that many students experience on reading and understanding science books (Mallinson et al. 1952; Davies and Greene 1984; Fang 2006) are largely determined by language mastery inadequacies; therefore, the design of "teaching strategies to help students cope with these challenges" (Fang 2006) needs to incorporate language-related components. Typically, language-related activities like reading and writing—the cores of literacy—may constitute important tools for science learning. Attending to literacy "in its fundamental sense" when teaching science can enhance science literacy (Norris and Phyllis 2003). Reading a science text means active engagement in understanding sentences and the concepts they convey (Norris and Phyllis 2008, Phyllis and Norris 2009). Writing is apt to promote conceptual understanding because expressing things through a written text requires adequate clarity in the way concepts are understood and, therefore, the effort to find ways to construct sentences expressing the wanted concepts involves intensive reflection on those concepts (Beall 1991, 1993, 1994; Beall and Trimbur 1993; Cooper 1993; Halliday and Martin 1993; Kovac 1999; Bressette and Breton 2001; Mammino 2013). The use of language by the teacher or by teaching materials has crucial roles for learners' approach to science. Attending to the discourse utilized on teaching provides a key to convey fundamental aspects of the scientific method (Newton and Newton 2000); all the details of explanation-wording are important, because each detail conveys specific direct information and specific implications (Mammino 1998b, 2000). The concept of "teaching science as a language" may span from the pedagogical use of historical perspectives (how terms developed together with concepts) (Sutton 1992, 2003) to the connections with everyday language (Brown and Ryoo 2008).

Inadequate levels of language mastery unavoidably correspond to inadequate acquisition of science knowledge. This correspondence is highlighted by the difficulties experienced by many students, consequent to the current language-mastery deterioration in various contexts—a phenomenon that is raising deep concerns among educators. For instance, the language-mastery levels of students entering universities are often inadequate for the conceptual demands of modern science, or language mastery–dependent abilities, like the ability to organize statements and information rationally, are often inadequate for thesis-writing, sometimes up to the point of "producing the impression that those students have not yet learnt to reason" (*Il Tirreno* 2009). Diagnoses of this type confirm the tight dependence between language-mastery and reasoning abilities and, therefore, the tight dependence between language mastery and knowledge acquisition. The analysis of the difficulties experienced by students who, because of historical reasons, are forced to approach sciences through a language different from their mother tongue provides information on the correspondence between inadequate language mastery and inadequate acquisition of science knowledge, which can be considered systematic and complete because of the abundance of diagnosed features and details, enabling pattern identification. Such information is presented in more detail in a separate section, in view of its role as massive experimental documentation.

Language Mastery and the Mastery of Other Skills Crucial for Science Learning

Language mastery is an essential key also for the development of other communication skills important in the sciences, like visual literacy (the ability to read images and to communicate through images), the use of complex drawings (like those utilized in engineering), and the use of symbols and systems of symbols. Moreover, it is also an essential key for the development of abilities crucial to thinking processes, like logical thinking or abstract thinking abilities. The mechanisms through which language mastery determines the acquisition of these other abilities are analyzed and documented in Mammino (2010), and therefore, only a brief review is given here.

Visualization has a variety of roles in science, from general-type ones, like the expression of trends through diagrams (whether mathematics-related diagrams or diagrams based on experimental data), to roles specific to individual fields, like the visualization of different features of molecules in computational chemistry. In science education, visualization is fundamental to attract and focus students' attention on specific aspects of the description of the objects of interest, to clarify concepts, or to familiarize students with entities that are not part of direct experience, like the invisible world of atoms and molecules in chemistry. The ability to read images, that is, to detect and understand the information they convey, develops through *ad hoc* education, which is realized through language (i.e., through explanations highlighting the meaning of images). Thus, language mastery inadequacies hamper the development of visual literacy, and poor language mastery is often associated with poor visual literacy.

Symbols or systems of symbols are communication tools with essential roles in some sciences. In chemistry, systems of symbols compress information. In order to identify and read the compressed information, it is necessary to have firstly understood the "uncompressed" information, and this understanding is attained through language; for instance, in order to understand chemical formulas, the learner needs to have understood the relevant information about atoms and molecules and about elements and compounds and how all this information is compressed into formulas. In mathematics, symbols represent abstract entities; in order to be able to handle abstract entities, the learner needs basic familiarization with abstract reasoning, which can only be acquired through language.

Logical thinking enables the identification of connections between different pieces of information. It has fundamental roles in many aspects of science learning. It is necessary for conceptual understanding, because descriptions and models contain various pieces of information that connect to each other to build a meaning. It is necessary in practical/laboratory work, for the interpretation of experimental data. Its level is closely depending on the language mastery level. As the content of science courses becomes more advanced, the demands on logical thinking abilities increase sharply, requiring the ability to handle complex sentences and complex discourses, that is, requiring increasingly high sophistication in the learners' language mastery.

LANGUAGE OF INSTRUCTION AND CONCEPTUAL UNDERSTANDING IN THE SCIENCES

Until recently, the issue of the language of instruction for science teaching–learning concerned only countries which, for historical reasons mostly related to colonial heritage, continue using the language of the former colonial power as a medium of instruction. This comprises most of Sub-Saharan Africa, whose situation thus provides extensive documentation about the negative impact of second language instruction on the acquisition of science knowledge and on development in general. In recent years, the current role of English as *lingua franca* in science communication has prompted initiatives (in some contexts), for which selected science courses are taught through English at a preuniversity level; some impacts are already diagnosable and appear to deserve more detailed investigation.

Evidence from the Sub-Saharan Africa Experience

The documentation from the Sub-Saharan Africa experience offers the most complete and systematic information on the impacts of second-language instruction because of the enormous number of learners affected and because of the long number of years through which it has been practiced. The use of a second language is so generalized for science subjects that many learners have never approached a science through their mother tongue, even in their youngest years. This has serious impacts on science learning because of the fundamental roles of language in the expression and understanding of science. It also impacts on the general perceptions about science (Mammino 2006b). The fact that instruction utilizes a language different from the learners' mother tongue, and the heavy impact it has on the acquisition of knowledge and expertise—above all science and technology expertise—is recognized as a major obstacle to development (Prah 1993, 1995, 1998, 2002; Owino 2002; Brock-Utne and Hopson 2005; Prah and Brock-Utne 2009).

As already mentioned, a 15-year investigation of language-related difficulties experienced by students taking tertiary-level chemistry courses, and of how these difficulties decrease the effectiveness of the teaching/learning process, has been carried out at the University of Venda, a "Historically Black University" in South Africa mostly serving a poor rural community and suffering from historical and contextual disadvantages, described in more detail in Mammino (2008). A disadvantaged context provides a more complete picture of possible difficulties, because underprivileged students are affected more extensively by difficulties-generating factors; it thus constitutes a more informative investigation ground and, simultaneously, it offers a scenario of the possible ultimate consequences of certain options. Illustrative examples of the main categories of identified difficulties are included in separate works (Mammino 1998a, 2005a, 2006a, 2007, 2009, 2010, 2011); they are not included here for obvious space reasons, as a selection of examples would need to comprise more than a few cases in order to be at least minimally representative. Overall, the investigation provided an ensemble of diagnoses, comprehensively outlined in Mammino (2010) and which can be summarized as follows:

- A number of errors concerning both the selection of individual words and the combination of two or more words to build sentences are clearly to be ascribed to inadequate mastery of the language that is the medium of instruction. Many of these errors would never occur in the mother tongue, because of the deep internalization of the word-concept correspondence within the mother tongue. The confusions associated with these errors decrease students' possibility to attain a clear understanding of the science concepts concerned.
- The identification of relationships among different pieces of information is often vague, incorrect, or absent. Some fundamental relationships, like the condition–consequence relationship (if…then…), remain beyond the reach of many learners, which seriously affects the understanding of models and of model generation. The difficulties increase sharply as the complexity of the discourse (the complexity of the concepts and, therefore, of their expression) increases.
- The ability to utilize other communication tools, including visualization, is often inadequate or poor, because its development has been hampered by inadequate language mastery.
- The awareness of their language-mastery inadequacies generates diffuse shyness, preventing learners' active participation and interaction in the classroom. The potential benefits of interactive teaching and active learning options remain largely unexplorable or unattainable.
- Most learners experience serious difficulties in reading textbooks because language-mastery inadequacies hamper understanding. The most frequent (nearly universal) response in the face of these difficulties is resorting to passive memorization, that is, memorization without understanding. This defeats the objectives of science education, as both the conceptual content and the way of proceeding of the given science remain unfamiliar (even alien) to most learners. It does not help even in terms of basic performance for assessment purposes, as students often assemble memorized clauses or sentences unrelated to each other (and to the question concerned), with partial or total loss of meaning; in addition, language-mastery inadequacies prevent the ability to analyze sentences after writing them and to detect the actual meaning conveyed, or the absence of a detectable meaning.
- The fairness and reliability of assessment decrease sharply as language-related factors become "confounding variables" (Clerk and Rutherford 1998) shading the learner's potentialities and abilities for science learning: the extent of learning is decreased by language-mastery inadequacies, and these same inadequacies limit the expression of the knowledge that the learner may have acquired.
- At the teaching level, the quality of the presentation of material is restricted by the need to utilize only grammatically and logically simple sentences, as most students would not be able to follow even only slightly complex logical constructions. This prevents expansions

> that would be vital for the development of science-related creative-thinking abilities, like expansions into the consideration of the validity of models; into historical features; into the challenges of current crucial research questions; or into interfaces with other branches of the given discipline or with other sciences. In this way, most aspects of the meaning of science, or of doing science, remain beyond the domain of what students come in contact with. This hampers the potentialities of the most gifted students, by preventing explorations that could be appealing to them. It is particularly sad when students become aware that language-related difficulties prevent them from exploring interesting domains, as it happened (on more than one occasion) with students taking a quantum chemistry course, on attempts to outline more complete views of the theoretical foundations than those sketched in the course handout (Mammino 2005b). The damage to innovative research potentialities is enormous.

The Sub-Saharan Africa situation provides experimental evidence also of the importance of a good theoretical knowledge of the mother tongue for the development of language-mastery abilities, including good mastery of the second language: "the growth in a second language depends on the level of the first language" (Qorro 2011). The intimate relationships between language and thought, and the deep internalization of all the aspects of the mother tongue (from sound-concept correspondence to the meaning of individual words and of the possible word combinations to express information) make the mother tongue the ideal ground for the development of the awareness of the meaning of logical relationships and of the ways in which they are expressed. Once this awareness is acquired, it remains as part of a person's acquired knowledge and is easily transferred to any other language that that person learns and utilizes. If this awareness is not acquired, the learner remains deprived of the possibility of fully expressing his/her views and conceptions in any language. An example is the situation of many students taking science courses at the University of Venda, who are not in a position to adequately express things through English and, on the other hand, although speaking their mother tongue fluently, are not in a position to use it to express things in writing, because of not having learnt its theory adequately. Such a situation, which can at best be described as the presence of "two underdeveloped languages" (Qorro 2011), implies huge limitations in the development and realization of a learner's potentialities both during his/her studies and as a future professional.

Experiences in Traditionally Mother-Tongue-Instruction Contexts

In recent years, experiences with the teaching of selected science subjects through a language different from the mother tongue in contexts in which the mother tongue is generally utilized as a medium of instruction are highlighting features that recall some of the observations from the Sub-Saharan Africa experience. Ongoing investigations are identifying striking similarities, which confirms the dominant impact of language-related difficulties with respect to other factors: learners from average or wealthy families in a developed European country and learners from an underprivileged African context may end up experiencing analogous difficulties, once they are made to approach science through a language different from their mother tongue (Brock-Utne 2011).

Another aspect that should raise major concern for science education is the unavoidable oversimplification that appears to be inherent in teaching through a second language, whichever the context (whether developing or developed). The teacher needs to utilize simple, often one-clause sentences to enable learners to understand the meaning conveyed by the sentence. In this way, the relationships among different pieces of information are not expressed, the complexity of the scientific discourse gets lost, the richness of a science as an investigation process disappears, and explorations aimed at providing glimpses of how much more there is, beyond what is part of the course, become impossible. The learner's familiarization with the given science suffers. It would be important to investigate the extent to which it suffers, both through typically science-education research tools (verification of learners' acquired knowledge) and through statistical investigations like, for instance, a comparison of the proportion of students selecting a science that was taught through a second language with respect to those who were taught that same science in the mother tongue, or the proportion of students who, at university admission exams, fail the exam in a science that was taught through the second language, while passing all the other exams well. This would provide important information on the extent to which learners are disadvantaged by being taught a certain science through a second language when they first encounter that science, and on the impacts on the learner's further educational career and the selection of the professional career.

Detailed documentation from the experience of countries that have tried teaching science through English for a certain period, and reinstated mother-tongue instruction after the negative impact became evident, would bring important contributions to the overall information.

MAJOR PARADIGM SHIFTS IN EDUCATIONAL POLICIES

The observations outlined in the previous sections prompt a variety of reflections on the best approaches to science education in view of the importance of generalized science literacy and of an adequate presence of science professionals in a knowledge-based society. The fundamental role of language mastery for science learning calls for a paradigm shift in which this role is fully acknowledged. The shift needs to take into consideration three major aspects, which will be analyzed in detail in separate subsections:

- The importance of ensuring adequate language-mastery sophistication for students to be able to understand the science discourse
- The importance of mother tongue instruction to ensure adequate familiarization and conceptual understanding
- The importance of the acquisition of adequate communication skills in the current *lingua franca* (English), to ensure the possibility of communicating across countries and cultures

Importance of Ensuring Adequate Levels of Language Mastery

The importance of ensuring adequate levels of language mastery becomes an educational first priority, as a pre-requisite to the acquisition of science knowledge and of the ability to creatively think in the sciences. This requires the generation of suitable interfaces between language education and science education, above all (but not only) at a preuniversity level, to set firm foundations. In this context, the "language education" concept refers to the mother tongue as the natural medium for setting the bases of language mastery, that is, developing those abilities and knowledge that enable full use of the expression power of a language. Learning the theory (grammar, syntax) of the mother tongue implies learning about logical relationships (space and time relationships; cause–effect; condition/hypothesis–consequence and its various possibilities [reality, possibility, unreality]; etc.) and the ways in which they are expressed. These relationships are fundamental not only for understanding the science discourse, but also for thought-development and communication in any field. The reflection on their nature and roles, inherent in the study of how they are expressed, builds a background of thinking abilities that is precious for science learning.

Building extensive interfaces between language education and science education requires the design of innovative approaches—a challenge to educational research in both areas. Interfaces need to be carefully balanced: language education and language mastery need to be presented as a value on its own (a human value, providing thinking and communication tools), carefully avoiding any risk that it might be perceived only as an ancillary to science education. On the other hand, the role of language mastery in the sciences needs to be given specific attention within the presentation of science content. This responds to the suggestion of "teaching sciences as language," with the science teacher being simultaneously a language teacher in order to facilitate understanding and stimulate creative thinking (Sutton 2003). In addition to highlighting how each new step in the sciences implied the generation of a suitable way to express it (Sutton 2003), "teaching sciences as language" may involve a variety of practical components: stressing aspects like the importance of rigor in the expression of science, and showing how the requirement of *being rigorous* is realized through accurate selection of individual words and of the ways of combining them to form sentences (Mammino 1995); turning language rigor into a pedagogical tool (Mammino 2000); and including language-related exercises into science textbooks, above all for preuniversity education (Mammino 2003) and, correspondingly, including examples from the sciences into the ensembles of examples from different knowledge areas utilized in language textbooks. The major challenges for developing and implementing these innovations stem from the cross-disciplinary nature of the *language-of-science* (Mammino 2005c), requiring simultaneous knowledge of language aspects and science aspects; on the other hand, it is a field where investigation can lead to interesting new outlooks, expanding its value to many areas of educational research and to the exploration of interfaces with other disciplines, including philosophy.

Ensuring the development of adequate language mastery and creating adequate interfaces between science education and language education is functional to prevent the problems observed in several contexts in the last decades, when students entering university science courses do not have adequate language-mastery sophistication to meet the conceptual demands of those courses. The problem has prompted immediate intervention measures like the organization of a *Written Italian Service* for freshmen, including a 30-hour language course, at the University of Venice (Ca' Foscari) in 2005, after 44% of entering students failed the Italian language test and 25% of the answers to questions simultaneously testing language and logical abilities were incorrect. The stated objective of the course, "enabling students to learn the bases of Italian language up to attaining expression and writing abilities including complex discourses," clearly highlights the prerequisite character of adequate language mastery for pursuing science or engineering studies and, simultaneously, highlights the course remedial role for abilities that—following normal expectations—should have been developed and firmly established in earlier instruction stages. Measures of this type, although necessary to address a diagnosed problem, cannot substitute accurate pre-university education, as they cannot reach the level of internalization and thinking-capacity generation that can be fostered by gradual and focused training across several years of pre-university instruction. Ensuring such training is the only durable option to enhance science literacy and the understanding of science, thus also becoming a prerequisite to stimulate science vocations.

Importance of Mother Tongue Instruction

The issue of the language of instruction requires deep reflections in a variety of directions, considering information from history; the evidence provided by contexts with generalized second language instruction, in which all sciences are approached only through a second language; and the evidence from mother tongue instruction contexts, in which some subjects have been taught through a foreign language at a pre-university level in recent years.

The history of European science, and of technological and industrial development in Europe, shows that development accelerated sharply after science started being written in the national languages, abandoning the former *lingua franca* (Latin). This shift, pioneered by Leonardo and Galileo, enabled faster and easier access to science content, making it accessible to many more persons and expanding the human resources engaged in the generation of science. Today's reality also leads to the same conclusion: countries that have been developed for long, or that have experienced fast development in recent decades, use mother tongue instruction, while the use of a second language constitutes one of the major causes of the slower development rate of Sub-Saharan Africa (Prah 1993, 1995, 1998).

Educational considerations also lead to the same conclusions about the importance of mother tongue instruction for science courses. The core objective of a science course is the familiarization with the given science. All educational efforts within a science course should be aimed at pursuing this objective and, therefore, teaching/learning to communicate through another language should never prevail over the core objective of a science course. Learning and understanding science is not easy, and many learners encounter a variety of difficulties. Science education research aims at designing options to facilitate understanding, to make the learning process easier and more comfortable. The way in which science is communicated is the major aspect determining the level of clarity and comfort with which a student can approach it. The comfort offered by the mother tongue is unmatchable. A language different from the mother tongue poses a barrier that enormously increases learning and understanding difficulties, that is, it actually opposes the objectives of science education and science education research. The scenario depicted by Rubanza (2002):

> When teachers and learners cannot use language to make logical connections, to integrate and explain the relationships between isolated pieces of information, what is taught cannot be understood—and important concepts cannot be mastered.

This is not consistent with any science education objective: it depicts a situation in which teachers and learners are forced to use a language that is not their own, and which they do not master to the full extent with which one masters the mother tongue, with the result that the communication of science fails, that is, science education is defeated.

Increasing the awareness (also at the level of public perception) of the role of language in science, and of the importance of mother tongue instruction for science learning, requires the clarification of some misconceptions concerning the nature of what can be called the *language-of-science*, the difference between *language-of-science* and *lingua franca*, and the importance of language mastery in the sciences. Some of these misconceptions are present in several contexts, often to an extent that turns them into public perceptions, while others are more related to the language and instruction policies of specific contexts. They are briefly discussed in the next paragraphs.

By its nature, the term *language-of-science* refers to the mode of expression typical of the sciences, whose requirements (Mammino 1995) are determined by the need for rigor (consistency with the characteristics of the system or phenomenon concerned, as we know them) and clarity (communication efficiency and avoidance of the risk of ambiguities). On the other hand, the *language-of-science* concept is often taken as a synonym of *terminology*, whereas it is totally different. Terminology is an ensemble of names—the names denoting the objects of interest in a given science. Being an ensemble of names, terminology is not, by itself, a language, as it is not capable of communicating information. The communication of information relies on the common-language components linking the technical terms and building a message (Mammino 2006b). For instance, the sentence "*In an ideal gas, there are no intermolecular interactions*" conveys information because the technical terms (*ideal gas*, *intermolecular interactions*) are linked by the common-language words *in*, *an*, *there*, *are*, and *no*, which build the message. This role of common-language components is at the basis of the importance of language mastery in science learning and science practice, and the rigorous use of common words in a scientific discourse is the key both to correct expression and to the pedagogical contribution stemming from the way of wording.

The identification of English as the *language-of-science* is a preconception more diffused in contexts where instruction is performed through English as a second language, generating the perception that science can only be expressed through English (Mammino 2006b); it is currently expanding to other contexts. The preconception involves confusion between the *lingua franca* concept (the language utilized for communication across different linguistic groups) and the *language-of-science* concept (the mode of expression typical of the sciences, as outlined in the previous paragraph). Clarifications are vital to engender the awareness that any language can be utilized for expressing science in accordance with the requirements of the *language-of-science*.

The importance of language mastery in the sciences in not always sufficiently recognized. The opinion that language is not-so-important in the sciences, although based on obsolete views about a complete dichotomy between science and humanities, is still extant in some practices, for example, the practice of lowering the language-mastery requirements for students entering science faculties in some tertiary institutions. All that has been discussed in the previous sections shows the paramount importance of language mastery for science learning as well as for the development of scientific thought, that is, for the further progress of science. Once language mastery is acknowledged as a major key to science understanding, it is easy to infer that the language that a student masters best and more completely (i.e., the mother tongue) provides the best tool for science learning, as it provides the best medium for understanding: "students understanding of what they are taught is the basis of all learning" (Qorro 2011).

The nature of the objectives of different courses may also require clarifications in terms of information to the public. The increasing importance of a good knowledge of English for finding jobs is generating the perception that "learning English" is the dominant objective of instruction. This is observed among parents and pupils in Sub-Saharan African contexts (Brock-Utne 2011; Ziegler 2011), but also in developed contexts that have so far utilized mother tongue instruction. It happens to hear parents in some European countries who are ready to consider it as positive if their child learns chemistry (or physics, or mathematics) through English, because "in this way, he will learn English better"; what those parents overlook is the question of whether, in this way, their child will learn chemistry (or physics, or mathematics) to a reasonable literacy level. Distinguishing the educational objectives of different courses becomes of paramount importance: the objective of a foreign language course (e.g., an English course) is to enable learners to learn and master the foreign language, but the objective of a science course is to enable learners to learn the given science. Within science courses, all efforts are to be put into ensuring better quality of science learning. Mother tongue instruction is one of the conditions creating the grounds to pursue it.

Importance of Good Communication Skills through English

Mastering communication through the *lingua franca* is important, more so as the fast globalization makes contacts across language groups increasingly common. The importance of mother tongue instruction to ensure science understanding does not exclude the presence of options to adequately train learners in the use of English for science communication. As previously mentioned, this should not be done at the expenses of science learning and, therefore, it needs to have independent schedules and time-framing with respect to the science courses. Then, the training in the use of English (or other languages different from the mother tongue) would respond to the separation of the understanding component (which is easier, more effective, and more complete through the mother tongue) and the expression component, which can be realized through more than one language: through the mother tongue within the science course (to verify and reinforce conceptual understanding) and through another language, either within the foreign language course or within separate time-slots devoted to interfacing science learning and the expression of the acquired knowledge through the foreign language. This option guarantees higher-quality expression of the science content through the foreign language, as learners would express material that they have understood and whose correct expression they have already analyzed within the science course. Moreover, expert guidance within the interface slots may turn the activity into something more complete than purely expression exercises, by encouraging an analysis of the content in language terms (analysis of the reasons for the selection of specific words, analysis of the relationships among different pieces of information) with reference to the way the other language utilizes words and expresses relationships; this would then be beneficial both to a clearer understanding of the content and to language learning. Interactive options for the analysis are apt to enhance learners' active reflection on the science content as well as on how to express it. In this way, the training to express science through the second language would

- Become beneficial to science understanding through the analysis, from the points of view of different ways of expressing them, of science concepts that the learner has already acquired firmly.
- Highlight the cross-language character of the requirements of the *language-of-science* (Mammino 2006b) by showing, through concrete practice, how the same requirements (being rigorous, being clear) are pursued through different languages on the basis of the same references (the science concepts, i.e., the characteristics of the system or phenomenon that is the object of interest and the way in which we know them). This enables a more holistic perception of the nature of the global scientific endeavor.

The challenges facing educators and educational research are analogous to those described for the integration of mother-tongue language education and science education, that is, the cross-discipline character of the activities. On the other hand, the educational value of the expected benefits constitutes important motivations for active search aimed at designing effective approaches, capable of maximizing the benefits to science education, to the mastery of the foreign language to express science, and to the stimulation of the perception of the cross-culture nature of science and its language.

DISCUSSION AND CONCLUSION

Generalized science literacy and the presence of an adequate number of science vocations are fundamental requirements of a knowledge-driven society. The quality of science education determines the level of science literacy acquisition as well as the possibility of preparing youths who can engage in science careers and advance its frontiers. This quality largely depends on learners' language mastery.

The recognition of the importance of language mastery for science understanding leads to a consistent set of recommendations to enhance attention to this role in the design of educational policies and approaches. They can be summarized as follows:

- Giving adequate space to ensure the development of language mastery up to the sophistication level that is required for science understanding. This includes the mastery of logical relationships and complex discourse, up to acquiring a good working familiarity with them. It is more effectively pursued within the mother tongue and, once acquired, can be easily expanded to other languages.
- Maintaining consistency with the general objectives of science education: making the acquisition of science concepts as complete as possible (for the given level of instruction) and as comfortable as possible and ensuring learners' familiarization with the sciences in the sense of learning (coming to know) as much as possible about their questions, their approaches and modes of proceeding, and their history.
- Acknowledging the essential role of mother tongue instruction for science understanding.
- After science concepts are clearly understood through the mother tongue, utilizing the training to express them through other languages as an additional tool to enhance the clarity of science understanding and to better familiarize learners with the requirements of the mode of expression in the science and their cross-language nature. This will also increase the sophistication level of the mastery of the other language/s concerned.

The practical realization of these goals requires careful investigation and innovative design, so as to utilize all the potentialities enabled by the incorporation of language aspects into science education, with the objective of facilitating and enhancing science learning. In this way, specific attention to the roles of language components in science education becomes the foundation not only to ensure the understanding of individual science concepts, but also to enable science education to go beyond the provision of individual pieces of information, or of a fragmented set of practical skills, and expand into real familiarization with the conceptual frameworks and the reasoning and investigation approaches of a given science (as analyzed, for example, by Russo [1998] for the case of mathematics education). Although more detailed discussion of these aspects of science education would go beyond the scope of the current work, it is important to recall that a knowledge-driven society requires generalized abilities for creative thinking, which can develop only from the consideration of relationships among individual pieces of information, and among self-consistent ensembles of pieces of information, up to entire logical and conceptual frameworks. Therefore, ensuring learners' acquisition of adequate language mastery becomes *conditio sine qua non* not only for effective science education, but also for the objectives and needs of a knowledge-driven society and for the further development of science.

REFERENCES

Beall, H. 1991. In-class writing in general chemistry: A tool for increasing comprehension and communication. *Journal of Chemical Education* 68(1): 148–149.

Beall, H. 1993. Literature reading and out-of-class essay writing in general chemistry. *Journal of Chemical Education* 70(1): 10–11.

Beall, H. 1994. Probing student misconceptions in thermodynamics with in-class writing. *Journal of Chemical Education* 71(12): 1056–1958.

Beall, H. and J. Trimbur. 1993. Writing as a tool for teaching chemistry. Report on the WPI conference. *Journal of Chemical Education* 70(1): 478–479.

Bressette, A.R. and G.W. Breton. 2001. Using writing to enhance the undergraduate research experience. *Journal of Chemical Education* 78(12): 1626–1627.

Brock-Utne, B. 2011. Teaching science for development—In whose language? Paper presented at the 8th LOITASA Workshop, Cape Town.

Brock-Utne, B. and R.K. Hopson, eds. 2005. *Languages of Instruction for African Emancipation—Focus on Postcolonial Contexts and Consideration.* Cape Town: CASAS.

Brooks, D.T. 2006. The role of language in learning physics. PhD diss., State University of New Jersey.

Brown, B.A. and K. Ryoo. 2008. Teaching science as a language: A 'content-first' approach to science teaching. *Journal of Research in Science Teaching* 45(5): 529–553.

Bruner, J. 1975. Language as an instrument of thought. In *Problems of Language and Learning*, ed. A. Davies, 61–88. London: Heinemann.

Carré, C. 1981. *Language Teaching and Learning: Science*. London: Ward Lock.

Chomsky, N. 1975. Reflections on language. In *On Language*, ed. N. Chomsky. New York: The New Press.

Clerk, D. and M. Rutherford. 1998. Language as a confounding variable in diagnosing misconceptions. *Proceedings of the Sixth Annual Meeting of the SAARMSE, UNISA*, 126–131.

Cooper, M.M. 1993. Writing—An approach for large enrolment chemistry courses. *Journal of Chemical Education* 70(6): 476–477.

Davies, F. and T. Greene. 1984. *Reading for Learning in the Sciences*. Edinburgh: Oliver and Boyd.

Fang, Z. 2006. The language demands of science reading in middle school. *International Journal of Science Education* 28(5): 491–520.

Halliday, M.A.K. and J.R. Martin. 1993. *Writing Science: Literacy and Discursive Power*. London: Falmer Press.

Il Tirreno. 2009. E anche le tesi di laurea sono incomprensibili. Dec. 12, p. 25.

Kovac, S. 1999. Writing in chemistry: An important learning tool. *Journal of Chemical Education* 76(10): 1399–1402.

Lahore, A.A. 1993. Lenguaje literal y conotado en la enseñanza de las ciencias. *Enseñanza de las Ciencias* 11(1): 59–62.

Mallinson, G.G., H.E. Sturm and L.M. Mallinson. 1952. The reading difficulty of textbooks for high school chemistry. *Journal of Chemical Education* 29: 629–630.

Mammino, L. 1995. *Il linguaggio e la scienza*. Turin: Società Editrice Internazionale.

Mammino, L. 1998a. Science students and the language problem: Suggestions for a systematic approach. *Zimbabwe Journal of Educational Research* 10(3): 189–209.

Mammino, L. 1998b. Precision in wording: A tool to facilitate the understanding of chemistry. *CHEMEDA, the Australian Journal of Chemical Education* 48–50: 30–38.

Mammino, L. 2000. Rigour as a pedagogical tool. In *The Language of Science*, eds. S. Seepe and D. Dowling, 52–71. Johannesburg: Vyvlia Publishers.

Mammino, L. 2003. *Chimica aperta*. Florence: G. D'Anna.

Mammino, L. 2005a. Language-related difficulties in science learning. I. Motivations and approaches for a systematic study. *Journal of Educational Studies* 4(1): 36–41.

Mammino, L. 2005b. Method-related aspects in an introductory theoretical chemistry course. *Journal of Molecular Structure (Theochem)* 729: 39–45.

Mammino, L. 2005c. Some general-character reflections on the relationships between humanities and sciences. *Ometeca* IX: 156–178.

Mammino, L. 2006a. Language-related difficulties in science learning. II. The sound-concept correspondence in a second language. *Journal of Educational Studies* 5(2): 189–213.

Mammino, L. 2006b. *Terminology in Science and Technology*. Thohoyandou: Ditlou.

Mammino, L. 2007. Language-related difficulties in science learning. III. Selection and combination of individual words. *Journal of Educational Studies* 6(2): 199–214.

Mammino, L. 2008. Teaching chemistry with and without external representations in professional environments with limited resources. In *Visualization: Theory and Practice in Science Education*, eds. J.K. Gilbert, M. Reiner and M. Nakhlekh, 155–185. Dordrecht: Springer.

Mammino, L. 2009. Language-related difficulties in science learning. IV. The use of prepositions and the expression of related functions. *Journal of Educational Studies* 8(4): 142–157.

Mammino, L. 2010. The mother tongue as a fundamental key to the mastering of chemistry language. In *Chemistry as a Second Language: Chemical Education in a Globalized Society*, eds. C. Flener and P. Kelter, 7–42. Washington DC: American Chemical Society.

Mammino, L. 2011. Importance of language mastering and mother tongue instruction in chemistry learning. Paper presented at the 8th LOITASA Workshop, Cape Town.

Mammino, L. 2013. Teacher-students interactions: The roles of in-class written questions. In *Chemical Education and Sustainability in the Global Age. Proceedings of 12st ICCE*, ed. M.-H. Chiu, 34–48, Dordrecht: Springer.

Munby, A.H. 1976. Some implications of language in science education. *Science Education* 60(1): 115–124.

Newton, D.P. and L.D. Newton. 2000. Do teachers support causal understanding through their discourse when teaching primary science? *British Educational Research Journal* 26: 599–613.

Norris, S.P. and L.M. Phillips. 2003. How literacy in its fundamental sense is central to scientific literacy. *Science Education* 87: 224–240.

Norris, S.P. and L.M. Phillips. 2008. Reading as inquiry. In *Teaching Scientific Inquiry: Recommendations for Research and Implementation*, eds. R.A. Duschl and R.E. Grandy, 233–262. Rotterdam: Sense.

Owino, F.R. ed. 2002. *Speaking African—African Languages for Education and Development*. Cape Town: CASAS.

Phillips, L.M. and S.P. Norris. 2009. Bridging the gap between the language of science and the language of school science through the use of adapted primary literature. *Research in Science Education* 39: 313–319.

Prah, K.K. 1993. *Mother Tongue for Scientific and Technological Development in Africa*. Bonn: German Foundation for International Development.

Prah, K.K. 1995. *African Languages for the Mass Education of Africans*. Bonn: German Foundation for International Development.

Prah, K.K. 1998. The missing link in African education and development. In *Between Distinction and Extinction*, ed. K.K. Prah, 1–16. Johannesburg: Witwatersrand University Press.

Prah, K.K. 2002. The rehabilitation of African languages. In *Rehabilitating African Languages*, ed. K.K. Prah, 1–6. Cape Town: CASAS.

Prah, K.K. and B. Brock-Utne, eds. 2009. *Multilingualism—An African Advantage. A Paradigm Shift in African Languages of Instruction Policies*. Cape Town: CASAS.

Qorro, M. 2011. Language planning and policy in Tanzania: When practice does not make perfect. Paper presented at the 8th LOITASA Workshop, Cape Town.

Rouvray, D.H. ed. 1997. *Fuzzy Logic in Chemistry*. New York: Academic Press.

Rubanza, Y.I. 2002. Competition through English: The failure of Tanzania's language policy. In *Rehabilitating African Languages*, ed. K.K. Prah, 39–51. Cape Town: CASAS.

Russell, B. 1923. Vagueness. *Australasian Journal of Psychology and Philosophy* 1: 84–92.

Russo, L. 1998. *Segmenti e bastoncini*. Milan: Feltrinelli.

Sutton, C.R. 1992. *Words, Science and Learning*. Buckingham: Open University Press.

Sutton, C. 2003. Los profesores de ciencias como profesores de lenguaje. *Enseñanza de las Ciencias* 21(1): 21–25.

Wellington, J. and J. Osborne. 2001. *Language and Literacy in Science Education*. Buckingham and Philadelphia: Open University Press.

Ziegler, R. 2011. Because I want to know English. Paper presented at the 8th LOITASA Workshop, Cape Town.

75 The Estimation of the Possibility of Introducing New Patterns into Office Space for Academic Staff at Polish Universities

Elżbieta Niezabitowska and Barbara Urbanowicz

CONTENTS

INTRODUCTION

The turn of the 20th and 21st centuries is a time of various changes. These changes are also observable at universities and in educational approach. The reasons for this are numerous. The most important of these are the growing number of students; introduction of increasingly sophisticated information technology; transformation of a society into a knowledgeable society; and finally, changing requirements of employees, who are currently required to be creative, flexible, and innovative. Consequently, such a change forces a modification in teaching methods and, as a result, shapes educational space. In the United Kingdom, the previously declared reasons have become the grounds for developing fresh ideas regarding the arrangement of academic office space.

The aim of the article is to present these ideas, to discuss the attitude of Polish academics to proposed solutions, and to analyze the possibility of introducing innovative solutions in the existing architectural structure, which will be further presented. The described conclusions are based upon research conducted at the faculties of the Silesian University of Technology in Gliwice.

WHY ARE NEW SPATIAL SOLUTIONS INTRODUCED INTO HIGHER EDUCATION?

Some universities, especially in Britain, during modernization or expansion occasions, apply new solutions to the office space for academics. The reasons for this are numerous, but most of them result from exploration and introduction of new spatial arrangements (the so-called *innovative office*) in commercial office buildings. Listed below are objectives set by the business environment expected in a modern office, which are also factors influencing the development of new concepts (van der Voordt 2003):

- Increased productivity achieved through cost reduction associated with an object (such as rent, property tax, HVAC, equipment depreciation, the annual cost of managing the office space, and issues related to the protection and safety) (Duffy 1997)
- Greater efficiency in the use of space and other resources, understood as the use of space in a way that improves the quality of work performed in it, for example, by providing an employee workplace depending upon the tasks to be performed
- Greater flexibility in shaping the office space (less time-consuming and less costly changes in layout), for example, in case of the need to reduce or increase the number of employees
- Better communication and cooperation, a greater degree of interaction
- Desire to create a professional image (to attract and retain potential employees, customers, business partners, etc.)
- Employee satisfaction with work (as well as with a workplace)

Therefore, the question arises—What is an innovative office?

WHAT IS AN INNOVATIVE OFFICE?

Innovation of offices refers to different scales and different areas. We can divide innovation into three fields—location, spatial arrangement, and the use of the workplace. The first of these aspects is location (van der Voordt 2003):

- *Central office*—tasks performed with colleagues under daily supervision of the management team, a permanent place of work, often in the city center, the headquarters of the company
- *Telework office*—work performed outside the headquarters
  - *Satellite office*—offices created away from the headquarters for members of the same organization, for example, to shorten travel time
  - *Business center, hotel office*—in the business center and hotel office areas (some organizations may rent office space with equipment)
  - *Guest office*—in the client's office
  - *Home office*
  - *Instant office*—work performed in other locations, for example, train, airplane, hotel

Another aspect is the arrangement of the space (types of offices) where we distinguish (van der Voordt 2003)

- *Private office* (*cellular office*)—with 1–3 workplaces
- *Group office*—with 4–12 workplaces
- *Openplan office*—with over 13 workplaces
- *Combi office* (*cocoon office*)—individual workplaces arranged along the outer walls (everyone works in his/her mini cell [cellular office]), and the center is a place of group work, gathering resources (which can be shared), and both formal and informal interactions
- *Cloister office*—an office where employees work in the open space or in the group offices, with the possibility to move to cellular offices in order to perform work that requires concentration

The last level of innovation is the introduction of new time–space strategies, allowing a better exploitation of the workplace and reducing the demand for space (van der Voordt 2003):

- *Personal office*—an office with one user assigned (1:1)
- *Shared office*—an office shared at different times (different times of the day or weekdays) by a few permanent users (1:x)
- *Nonterritorial office*—a flexible nonterritorial office, with changing workplace, designed for any potential user, usually poorly equipped (y:x), for example, *hot desking, drop in, hotelling* (Figures 75.1 and 75.2)

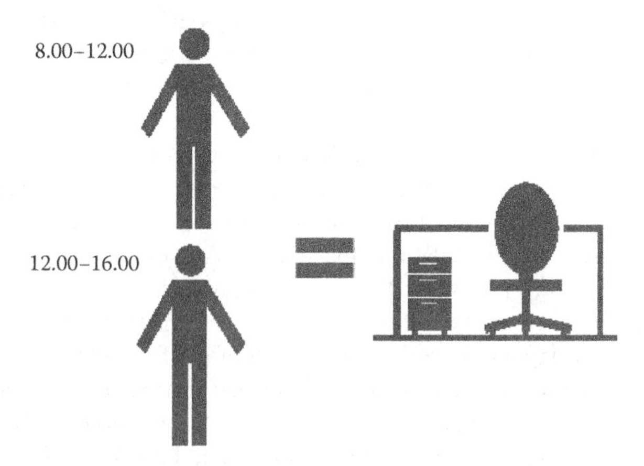

FIGURE 75.1 Shared office. (From Urbanowicz, B., *Wpływ światowych trendów w przestrzeni biurowej na stanowiska pracowników naukowo-dydaktycznych uczelni wyższych* [*Effect of Trends in Office Space on Academic Staff of Universities*]. The Silesian University of Technology's Scientific Papers, post-conference publication [KDWA4 Conference—Szczyrk 2010 r.], The Silesian University of Technology, Gliwice, in printing.)

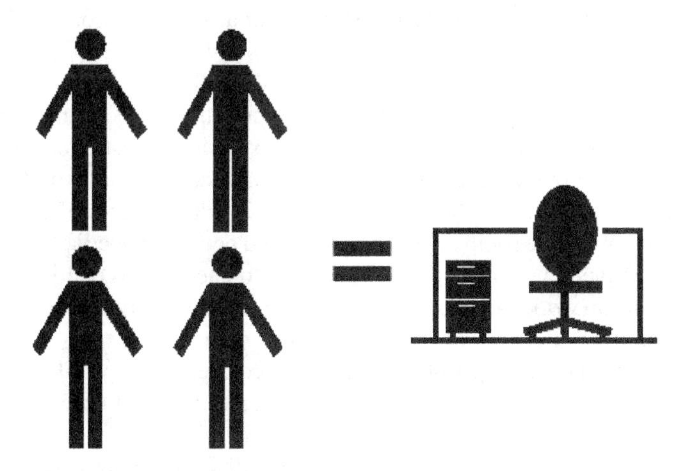

FIGURE 75.2 Nonterritorial office. (From Urbanowicz, B., *Wpływ światowych trendów w przestrzeni biurowej na stanowiska pracowników naukowo-dydaktycznych uczelni wyższych* [*Effect of Trends in Office Space on Academic Staff of Universities*]. The Silesian University of Technology's Scientific Papers, post-conference publication [KDWA4 Conference—Szczyrk 2010 r.], The Silesian University of Technology, Gliwice, in printing.)

Often, the described solutions are accompanied by supporting facilities, such as (van der Voordt 2003) work spaces that require concentration (*cockpits, hubs*); rooms for team work (*team rooms*); workplaces for short-term work, for example, for just checking e-mail (*standing workplaces/ touch-down workplaces*); conference rooms; meeting rooms (*meeting and conference rooms*); places to have coffee (*coffee corners*); places to read (*reading tables*); and reception/admission areas.

MODELS FROM DEGW

The discussed solutions come from a document written by Andrew Harrison and Antonia Cairns (2008), "The Changing Academic Workplace." The authors present in the article concepts of new solutions that were created based on case studies performed in the newly established or modernized buildings of universities in the United Kingdom (Queen Mother Building, University of Dundee, Dundee; Edinburgh's Telford College, Edinburgh; Queen Margaret University, Edinburgh; Law School [Lord Hope Building], University of Strathclyde, Glasgow; Department of Civil and Building Engineering, Loughborough University, Leicestershire; Faculty of Health and Wellbeing, Sheffield Hallam University, Sheffield; White Space, University of Abertay, Dundee). These models are ideograms, aiming to serve as a basis for discussion, not as determinants of spatial solutions. They are characterized by a different degree of meeting the needs of the worker (e.g., privacy, informal interactions, team working, etc.), which is the result of a different way of spatial arrangement. The schemes presented below include previously described features of innovative offices. Each of them contains a central area, which consists of the kitchen, print corner, and a place for informal meetings, and most of them presume work to be performed in the described system of *shared office* or *nonterritorial office*. Some of the initial data assisting in the creation of the presented proposals were the results of studies conducted in the Department of Civil and Building Engineering, Loughborough University (Harrison and Cairns 2008; Austin et al. 2006), displaying that an academic spends an average of 30% of work in his own office (PhD student, 50%; a researcher 70%). To compare, the rate in corporations does not exceed 35%–40%.

The following are the models of arranging the office space for academics, proposed by Harrison and Cairns (2008):

- *Studies* (*combi office*)— The suggestion considers small private offices arranged around a common central area. In comparison to other models, this solution most effectively meets the needs of privacy and provides needed space for required concentration and isolation; however, it impedes teamwork. An additional drawback is a higher cost of construction (a lot of walls) and supplying the rooms with equipment and necessary media (Figure 75.3).
- *Quarters*—rooms occupied by 4–7 people. The space is to encourage good interpersonal relations rather than group work. Next to the rooms, there are mini work rooms (to do work that requires concentration) and a short- rest area (Figure 75.4).
- *Clusters*—rooms occupied by 6–12 people; larger and more open. The spatial arrangement is to encourage teamwork and interaction. However, it creates a risk of excess interaction (Figure 75.5).
- *Hub*—the most "open" suggestion. A high degree of visibility contributes to cooperation. Rooms for over 12 people. In this solution, however, employees may disturb each other, and it may be necessary to introduce elements of "humanizing" such a multiscale office work environment (Figure 75.6).
- *Club*—the most diverse model, which provides a variety of workplaces. Allows an employee to select an optimal workplace depending on the type of work and frequency of staying in the school building (the appropriate place for "regulars" as well as "nomads"). Such space requires continuous management (some workplaces are shared) (Figure 75.7).

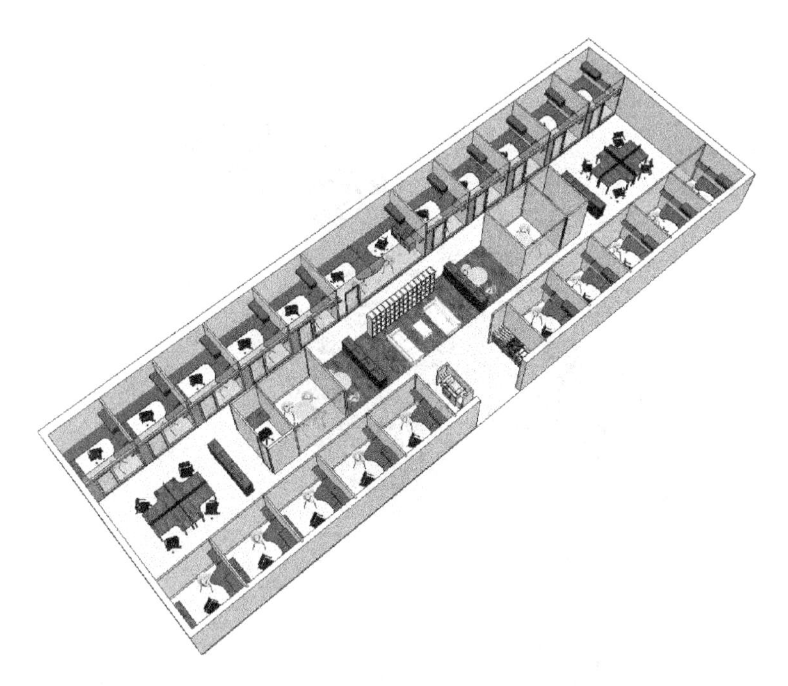

FIGURE 75.3 Studies. (From authors' own work on the basis of Harrison, A., Cairns, A., The Changing Academic Workplace. DEGW UK Ltd, 2008, retrieved from http://www.degw.com/knowledge_studies.aspx.)

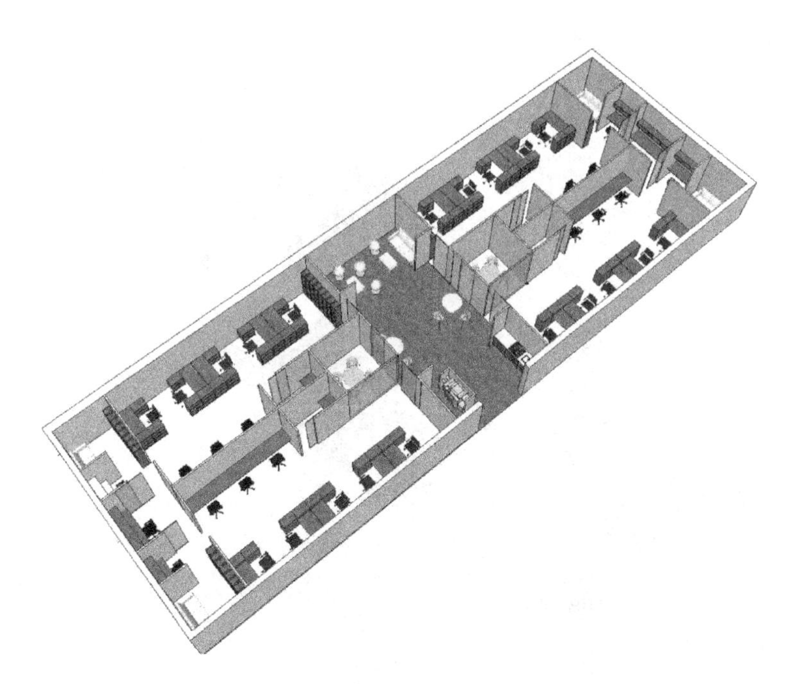

FIGURE 75.4 Quarters. (From Klaudia Szymańska and Agnieszka Szymiczek on the basis of Harrison, A., Cairns, A., The Changing Academic Workplace. DEGW UK Ltd, 2008, retrieved from http://www.degw.com/knowledge_studies.aspx.)

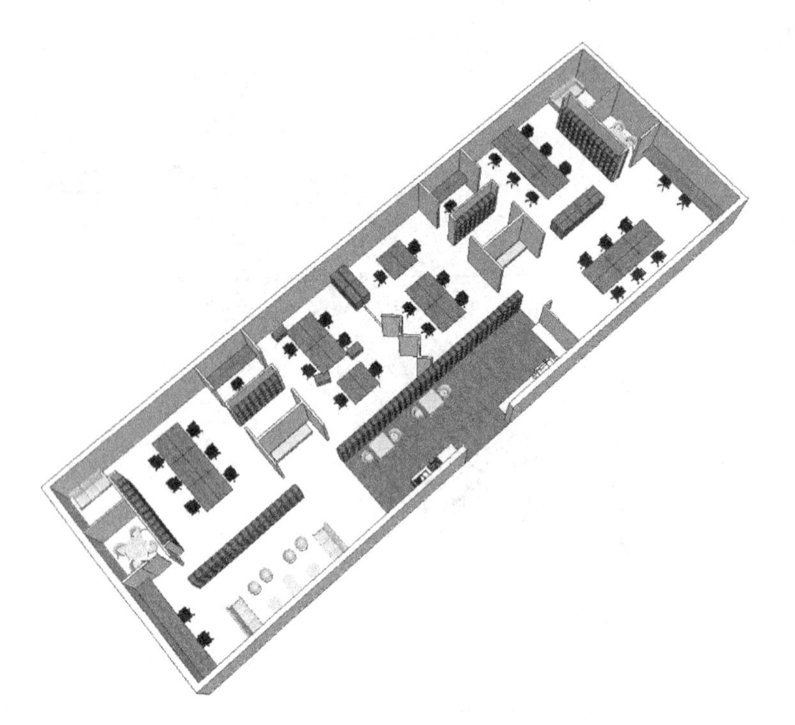

FIGURE 75.5 Clusters. (From authors' own work on the basis of Harrison, A., Cairns, A., The Changing Academic Workplace. DEGW UK Ltd, 2008, retrieved from http://www.degw.com/knowledge_studies.aspx.)

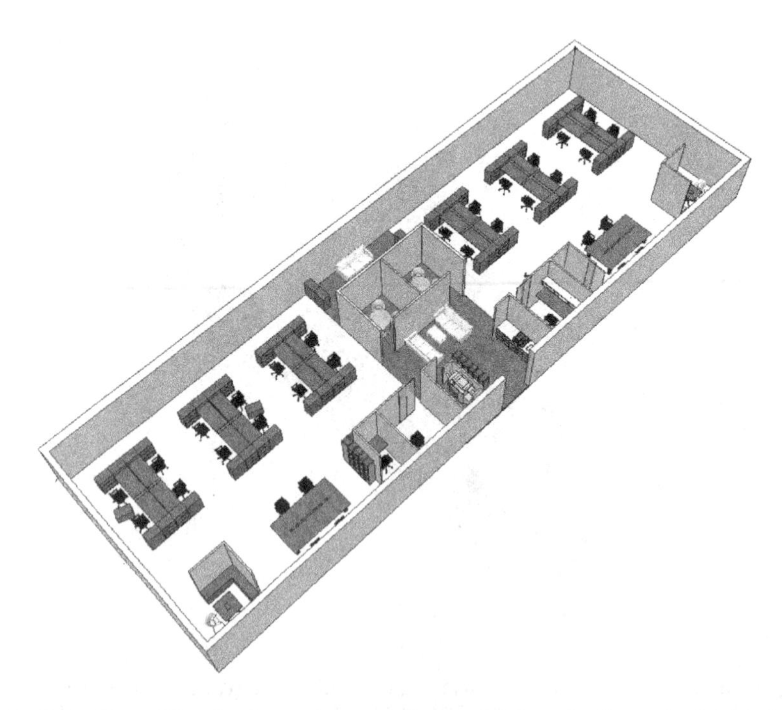

FIGURE 75.6 Hub. (From Klaudia Szymańska and Agnieszka Szymiczek on the basis of Harrison, A., Cairns, A., The Changing Academic Workplace. DEGW UK Ltd, 2008, retrieved from http://www.degw.com/knowledge_studies.aspx.)

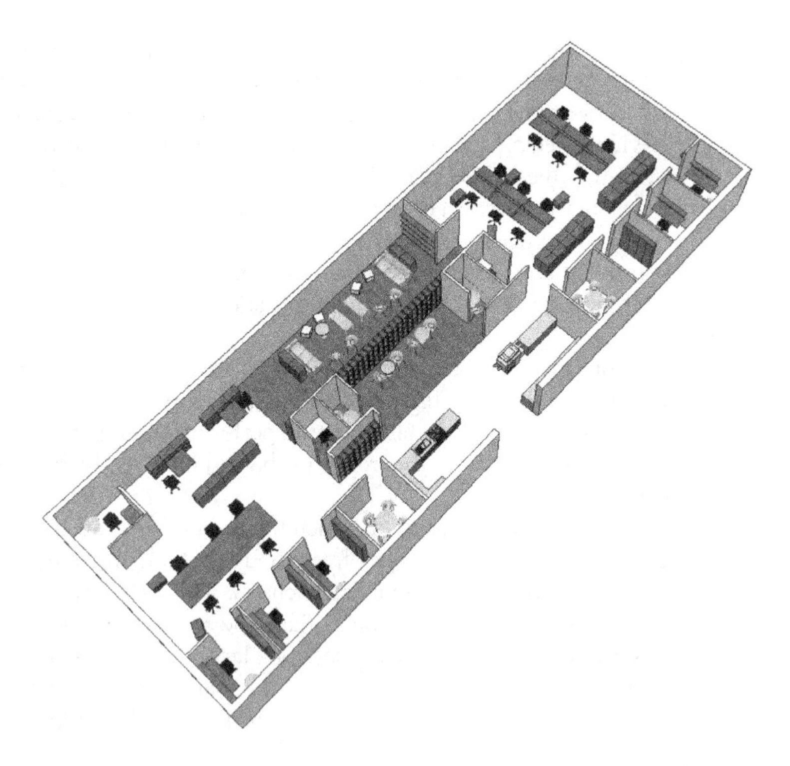

FIGURE 75.7 Club. (From authors' own work on the basis of Harrison, A., Cairns, A., The Changing Academic Workplace. DEGW UK Ltd, 2008, retrieved from http://www.degw.com/knowledge_studies.aspx.)

ACADEMICS IN POLAND AND THE UNITED KINGDOM

As previously explained, spatial solutions were proposed after a study of cases (*case study*) of newly created or modernized buildings of universities in the United Kingdom. According to the authors, these models allow the fulfillment of the diverse needs of the university employees. The researched objects in most cases included workplaces for academics (including researchers), administrative staff, and other support staff (e.g., technicians). In their research, the authors of this article also included university buildings with workplaces for the same groups of office workers; however they did not meet the researchers in person. (A similar situation occurs in the majority of Polish universities.) According to Dziennik (2005, No. 164, art. 1365), academic staff is teaching staff, academic teachers, researchers, qualified librarians, and records and information science staff. This fact is emphasized because of its significant impact on shaping space for academics, which will be discussed later in the article.

ATTITUDE OF POLISH ACADEMIC STAFF TOWARD THE NEW SPATIAL SOLUTIONS

The studies showed that open-space solutions are almost unacceptable—the most critically evaluated were *quarters, clusters, and hubs*. The most positive evaluation was given to *studies* and *clubs*, although not without criticism. Reasons for rejecting the more "open" solutions are numerous. Employees of Polish universities are mostly scientific and teaching staff. Besides their research work, they provide consultations for students. Consultations have always been held in the employee's room. Typically, they were conducted for one or two students at the same time. In cases where the room was occupied by more than one person, there was a distraction to others present in the room. Additionally, in an open-space environment, just walking around (e.g., a student who needs a consultation) causes distraction, whereas scientific work requires great concentration (one of the respondents claimed that this solution might be good at foreign universities, in which part of the staff [scientists] is not visited by students). This fact perhaps was the reason to criticize the solution of *studies*—despite having a personal single room, the possibility of being observed by others through the glass wall separating the shared part (disrupting privacy) and people walking across the shared section in the center were not acceptable. Another argument against the multipeople solutions was a way to establish workplaces—the respondents did not approve of others walking by behind their backs or possibility looking at their computer screens; they preferred to sit facing passing people. Another element mentioned as disturbing concentration was sitting in front of another employee (especially too closely).

The advantage of rooms for one to two persons is related to the character of work being performed—usually individual (scientific work), in many cases connected with team work, for example, the grant, but the latter is based on assigning tasks that are performed individually with periodic inspection and advisory meetings. Therefore, according to the respondents, there is no point in creating workplaces for team work.

In each proposed solution, a large table was used, which the respondents saw as a comfortable place for consultations (possibility to display designs freely and to review the effects of students' work), and an isolated room also equipped with a table and chairs, which also gave a possibility to hold consultations and receive visitors without disturbing others and conducting confidential discussions (currently, visitors usually are received at either a desk or a larger table in the room or near the room of the head of the faculty). The conclusion is that it is necessary to create areas of differing levels of privacy, starting with semi-private (e.g., a place for consultation, the reception of visitors) and finishing with a private area (for individual work).

Another solution, which in most cases was unacceptable, was a policy of *desk sharing, nonterritorial office,* and so forth. The respondents claimed that a computer is a very personal tool, as is a desk, which allows, for example, leaving unfinished work for the next day. In the case of the proposed solutions, a desk has to be cleaned up (a policy of "clean desking").

A shared kitchen proposed in each model was approved in most cases. (Any possible criticism in this case was caused by the concern of preserving cleanliness; the reason stems from unpleasant past experience—"borrowing" clean cups, throwing used tea bags into clean cups, etc.). According to the respondents, a shared kitchen would allow the consumption of a meal without disturbances (e.g., by incoming students), or without disturbances of different smells (a closed room); it could also function as an integration place (where co-workers could meet in one location).

The respondents agreed that recreation facilities often used in modern offices (such as a room with comfortable seating or entertaining games such as the so-called "foosball") are unnecessary for them because, quoting one of the statements, "we come to the school building to work, and after work, we rest at home."

CHARACTERISTICS OF EXISTING SOLUTIONS IN ACADEMIC OFFICE SPACE AT UNIVERSITIES

Another very important factor influencing the use of the *innovative office* in Polish universities is the present structure of the buildings and their capacity for transformation. The study indicated some characteristics of university buildings and the limitations in their transformations. These conditions are as follows:

- Usually, buildings are arranged in a corridor layout, triple row or one of its variants—from a linear arrangement to a complex multitrunked arrangement (Figure 75.8).
- The depth of the course of rooms is from 5 to 6.4 m; the depth of the corridor is from 2.7 to 3.3 m; total width of the building approximately 15 m (Figure 75.9).

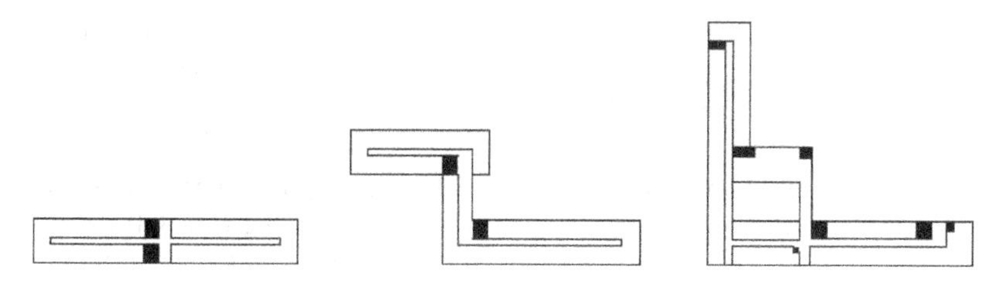

FIGURE 75.8 Variants of the corridor layout buildings.

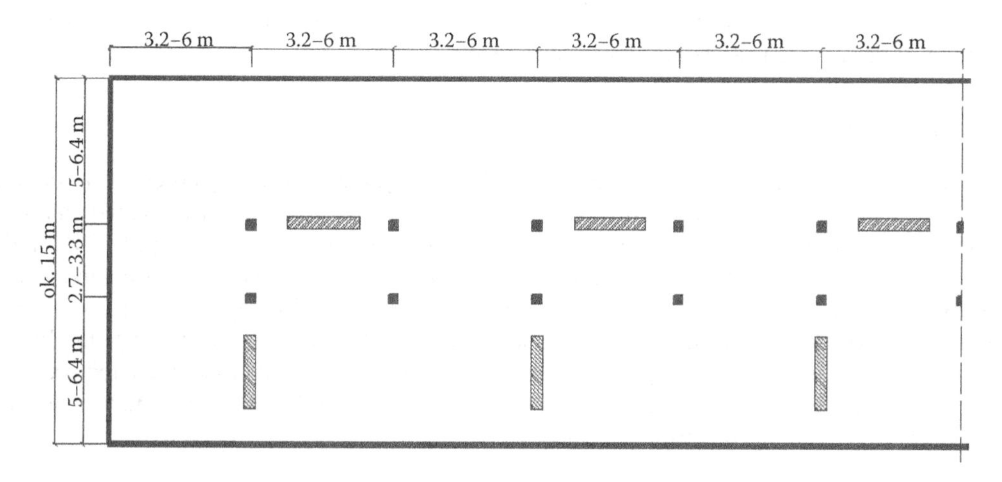

FIGURE 75.9 Construction and localization of the chimneys and installation in the buildings (black: construction, hatch: chimneys and installations).

- Nearly always, it is impossible to demolish the walls forming a corridor due to a dense arrangement of structural or installation building elements (chimneys, installation). The study encountered shafts along the supporting structure, which was running through the center of the building (along the corridor) and crosswise, at irregular intervals (Figure 75.9).
- Usually, there was neither a raised floor nor a lowered ceiling.
- The layout design hampered the distribution of media—whole beams (not openwork or prefabricated) running along the building. When attaching the installation along the corridor, it is very difficult to distribute it to the sides of the rooms.
- The ceiling height oscillates between 3 and 3.5 m. According to Polish law, as stated in Dziennik No. 75, work rooms designated for a permanent or temporary presence of not more than four persons must be of a minimum height of 2.5 m in the light; for a higher number of persons, the minimum height is 3 m. In addition, as stated by Masły (2009), according to the decree of the Minister of Infrastructure (Dziennik 2002, No. 75, art. 690), it is possible to reduce the height of 3 m to a minimum of 2.5 m when mechanical supply-exhaust ventilation or air conditioning is installed and the Provincial Sanitary Inspector issues his/her approval. Accordingly, modernization would allow, in many cases, putting the installation (HVAC, telecommunications, plumbing, alarm, fire, etc.) in the raised floor (from 20 cm) or the dropped ceiling (to 60 cm).
- Window divisions—the spacing of elevation elements (a so-called module of interior design) in the analyzed cases ranged from 3 to 3.2 m (in one case, 1.65 m). This value determines the size of the smallest room. According to D. Masły (2009), the combi office is the smallest of the currently available systems of spatial arrangement, and the width of

the rooms is generally 2.4 or 2.7 m, and the module of interior arrangement used for such width is, respectively, 1.2 or 0.9 m. Due to that fact, it is possible to apply different space solutions (width of the room as multiples of the module).

Due to the described conditions, it would be very difficult to apply solutions other than a cellular office to the existing structure. To compare, referring to Daniels, the text will describe a flexible solution in the form of the so-called reversing office (Masły 2009). Such a building is double row with structural axis spacing of 6–9 m (15 m in total). Reinforced concrete floor slabs are suspended and supported by a row of columns (one row of pillars within the course, the other structural components in the exterior walls), which gives great possibilities in arrangement. Openwork suspension allows for easy running installation along the corridors to work areas.

CONCLUSION

The study involved several faculties at the Silesian University, but preliminary observations on this small scale allow us to conclude that there are certain common guidelines for the design (regardless of the faculty) and a similar approach to innovative systems of arranging office space. The study demonstrates that a part of the proposed innovative solutions can be accepted by Polish universities. These proposals would simplify the management of space and would allow savings. However, the general approach to the proposed solutions is doubted probably due to the character of the Polish higher education and cultural background. (Europeans prefer more closed offices than, for example, Americans.) Another difficulty is the present architectural structure (buildings) in which it could be possible to apply a new spatial arrangement.

REFERENCES

Austin, S., Lansdale, M., Parkins J., *Research Environments for Higher Education*. Department of Civil and Building Engineering and Human Sciences Loughborough University, January 2006 (retrieved from http://www.academicworkspace.com/content/view/35/48/).

Duffy, F., *The New Office*. Conran Octopus, London, 1997.

Dziennik, U., 2002, no. 75, art. 690.

Dziennik, U., 2005, r. no. 164, art. 1365.

Harrison, A., Cairns, A., *The Changing Academic Workplace*. DEGW UK Ltd, 2008 (retrieved from http://www.degw.com/knowledge_studies.aspx).

Masły, D., *Jakość budynków biurowych w świetle najnowszych metod oceny jakości środowiska zbudowanego (Quality of Office Buildings in the Light of the Latest Methods to Assess the Quality of the Built Environment)*. The Silesian University of Technology, Gliwice, 2009.

Urbanowicz, B., *Wpływ światowych trendów w przestrzeni biurowej na stanowiska pracowników naukowo-dydaktycznych uczelni wyższych (Effect of Trends in Office Space on Academic Staff of Universities)*. The Silesian University of Technology's Scientific Papers, post-conference publication (KDWA4 Conference—Szczyrk 2010 r.), The Silesian University of Technology, Gliwice, (in printing).

Van der Voordt, D.J.M., *Costs and Benefits of Innovative Workplace Design*. Center for People and Buildings, Delft, 2003.

76 Development of the Education Quality Assessment System for Programs of Study in Economics at Polish Institutions of Higher Education in the Years 2002–2011

Elżbieta Izabela Szczepankiewicz

CONTENTS

INTRODUCTION

In Poland, ensuring good education quality at public universities and non-public colleges of economics is a very complex process. This process necessitates a number of actions on the part of both the state and the institutions of higher education themselves. Such actions should contribute to increasing the higher education institutions' competitiveness as regards the education services provided. The quality of education in Europe and in Polish institutions of higher education is tied to the existence of accreditation systems.

There are a number of higher education accreditation institutions in Europe. A state accreditation institution that represents the state authorities and the state accreditation system is established in each country that joins the Bologna system. Moreover, there are many community institutions established by the academic community that award accreditations in particular fields of knowledge. The most famous institutions that award accreditations to programs of study in economics and management include the *Association to Advance Collegiate Schools of Business* (*AACSB*) and the *European Foundation for Management Development* (*EFMD*), which offers the *European Quality Improvement System* (*EQUIS*).

At present, two methods of external education quality assurance are implemented in Poland. The first one is achieved through the state accreditation system, which, periodically and free-of-charge assesses, all the public and nonpublic institutions of higher education in Poland. The second external education quality assurance method is the community accreditation system, which was initiated by various academic communities. This type of accreditation is voluntary and paid. Accreditation should constitute a reliable proof of the fact that an institution of higher education holding an accreditation certificate provides its students with an adequate level of education.

The purpose of the present work is to discuss the process of formation and development of the state and community accreditation systems in Poland in the years 2002–2011. Moreover, it points to the European initiatives and standards that influenced the development of legal regulations governing the said systems. The role, the aims, and the results of both state and community accreditation in ensuring good education quality of programs of study in economic sciences were also discussed.

EUROPEAN EDUCATION QUALITY ASSURANCE STANDARDS IN HIGHER EDUCATION

In 2003, in Berlin, the Ministers of Education from the European countries that declared their will to implement the Bologna Process established the so-called Berlin Communiqué. It was addressed to the *European Network for Quality Assurance in Higher Education* (*ENQA*). In their Berlin Communiqué, the European Ministers pointed to the need to promptly develop a set of uniform standards and guidelines for assuring quality in higher education in Europe. Moreover, the Berlin Communiqué indicated the need to establish the procedures for supervising the implementation of the said standards and create a uniform European accreditation system.

As a response to the Berlin Communiqué, in 2005, in Helsinki, ENQA announced a report entitled "Standards and Guidelines for Quality Assurance in the European Higher Education Area" (2005). The said document was the result of collaboration between ENQA and a number of European institutions interested in this topic, such as the European University Association (EUA), the European Association of Institutions of Higher Education (EURASHE), the National Unions of Students in Europe (ESIB), the European Consortium for Accreditation (ECA), the Central and Eastern European Network of *Quality Assurance* Agencies in Higher Education (CEE), and the European Commission.

The Report was addressed to the Ministers of Higher Education of the European countries committed to implementing the Bologna Process as well as to accreditation institutions and authorities of European universities and colleges. The report comprises the following sections (Standards and Guidelines for Quality Assurance in Higher Education Area 2005, pp. 9–33):

1. Introductory section, which discusses the conditions, aims, and general principles regarding the quality of education in the European higher education system
2. European standards and guidelines for internal quality assurance within higher education institutions
3. European standards and guidelines for the external quality assurance of higher education
4. European standards and guidelines for external quality assurance agencies (e.g., accreditation committees)
5. Peer review system for quality assurance of the accreditation process and accreditation agencies or bodies
6. Future perspectives and challenges in education quality assurance

The purpose of introducing the foregoing standards, guidelines, and principles in higher education institutions and accreditation institutions all over Europe was to create the possibility of using common criteria for assessing education quality assurance in European higher education.

The standards constitute the most important guidelines for developing internal quality assurance systems in higher education institutions operating in the countries that joined the Bologna system and contributed to unification of the principles of education quality assessment and accreditation in those countries.

In the opinion of the authors of the present work, several internal quality assurance standards within higher education institutions, which exerted an important influence on the subsequent Polish regulations concerning the principles of implementation of internal education quality assurance systems in Polish institutions of higher education, deserve to be mentioned.

European standards for internal quality assurance of higher education, which exerted an important influence on the subsequent Polish regulations in Polish institutions of higher education, are as follows:

1. Policy and procedures for quality assurance
2. Approval, monitoring, and periodic review of programs and awards
3. Assessment of students
4. Quality assurance of teaching staff
5. Learning resources and student support
6. Information systems
7. Public information

European standards for external quality assurance of higher education, which exerted an important influence on the subsequent Polish regulations governing the operation of the state and community accreditation system and the regulations in Polish institutions of higher educations, are as follows:

1. Development of external quality assurance process
2. Criteria for decisions
3. Processes fit for purpose
4. Reporting
5. Periodic reviews
6. System-wide analyses

The said standards are presented in Table 76.1, whereas Table 76.2 presents the list and the essence of selected external education quality assurance standards that significantly influenced the subsequent Polish regulations governing the operation of the state and community accreditation system.

METHODS OF EDUCATION QUALITY ASSURANCE IN POLISH HIGHER EDUCATION INSTITUTIONS

Currently, there are two methods of assuring education quality in the Polish higher education institutions, namely, the external education quality assurance system of higher education and the internal education quality assurance systems, developed individually by particular higher education institutions. *The external quality assurance system* of higher education is achieved through the supervision of the Minister of Science and Higher Education over higher education institutions and the periodical assessment carried out by the State Accreditation Committee (Szczepankiewicz and Kiedrowska 2011, p. 92). Currently two forms of accreditation are used in Poland:

- State accreditation
- Community accreditation

TABLE 76.1
European Standards for Internal Quality Assurance of Higher Education

Standard	The Gist of the Standard
1.1. Policy and procedures for quality assurance	Higher education institutions should • Have a policy and associated procedures for the assurance of the quality and standards of their programs and awards • Commit themselves explicitly to the development of a culture that recognizes the importance of quality, and quality assurance, in their work • Develop and implement a strategy for the continuous enhancement of quality (the strategy, policy, and procedures should have a formal status and be publicly available)—include a role for students and other stakeholders in the quality management system
1.2. Approval, monitoring, and periodic review of programs and awards	Institutions of higher education should have official mechanisms for • Approval • Cyclical reviews • Monitoring of the programs and awards offered
1.3. Assessment of students	The students should be assessed in accordance with the published and consistently applied criteria, regulations, and procedures.
1.4. Quality assurance of teaching staff	Institutions should have ways of satisfying themselves that staff involved with the teaching of students are qualified and competent to do so. They should be available to those undertaking external reviews and be commented upon in reports.
1.5. Learning resources and student support	Institutions should ensure that the resources available for the support of student learning are adequate and appropriate for each program offered.
1.6. Information systems	Institutions should ensure that they collect, analyze, and use relevant information for the effective management of their programs of study and other activities.
1.7. Public information	Institutions should regularly publish up-to-date, impartial, and objective information, both quantitative and qualitative, about the programs and awards they are offering.

Source: Author's own based on the *Standards and Guidelines for Quality Assurance in the European Higher Education Area, 2005, The European Association for Quality Assurance in Higher Education (ENQA).* Helsinki. Finland, p. 6.

State accreditation is the basic method of external quality assurance in higher education. This system is governed by the Higher Education Act. The main task of the State Accreditation Committee is the cyclical assessment of the quality of education offered by all institutions of higher education in all programs of study. The accreditation carried out by the State Accreditation Committee is *obligatory* for all higher education institutions operating in Poland, both state-owned and non-state-owned. The accreditation awarded by the State Accreditation Committee is temporary, which means that it is awarded for a specific period, upon the lapse of which the higher education institution is subject to yet another accreditation assessment. Accreditation is awarded free of charge—the higher education institutions are not required to pay for it.

Community accreditation is the second method of external education quality assurance. Community accreditation is *voluntary*, as the decision to undergo accreditation is made by higher education institutions that are interested in the confirmation, assessment, and improvement of the quality of education offered. Such accreditation is *periodic*, as it is awarded for a specific period, for example, 3 or 5 years. Upon the lapse of this period, the higher education institution may apply for yet another accreditation. The higher education institutions that decide to undergo accreditation must cover 100% of its costs, as it is *not free of charge*.

Internal education quality assurance systems are designed and implemented individually by each higher education institution. The structure of internal education quality assurance systems should be based on the European standards for education quality assurance as well as on the Polish regulations governing the same. The quality assurance processes are implemented directly by the authorities of a given higher education institution and, in particular, by the rector and the unit (team)

TABLE 76.2
European Standards for External Quality Assurance of Higher Education

Standard	The Gist of the Standard
2.1. Development of external quality assurance process	The aims and objectives of quality assurance processes should be determined before the processes themselves are developed, by the Ministry, accreditation institutions, and higher education institutions, and should be published with a description of the procedures to be used
2.2. Criteria for decisions	Any formal decisions made as a result of an external quality assurance activity should be based on explicit published criteria that are applied consistently.
2.3. Processes fit for purpose	All external quality assurance processes should be designed specifically to ensure their fitness to achieve the aims and objectives set for them.
2.4. Reporting	Reports on the education quality assurance should be published and should be written in a style that is clear and readily accessible to their intended readership. Any decisions, commendations, or recommendations contained in reports should be easy for a reader to find.
2.5. Periodic reviews	External quality assurance of institutions and/or programs should be undertaken on a cyclical basis. The length of the cycle and the review procedures to be used should be clearly defined and published in advance.
2.6. System–wide analyses	Quality assurance agencies should produce from time to time summary reports describing and analyzing the general findings of their reviews, evaluations, assessments, and so forth.

Source: Author's own based on the *Standards and Guidelines for Quality Assurance in the European Higher Education Area, 2005, The European Association for Quality Assurance in Higher Education (ENQA).* Helsinki. Finland, p. 7.

headed by the coordinator for education quality assurance. In many higher education institutions, these actions are supported by a periodic assessment carried out by an internal auditor or an external auditor from a specialist company (Szczepankiewicz and Kiedrowska 2011, pp. 92–93).

Scheme 76.1 presents the methods of education quality in Poland.

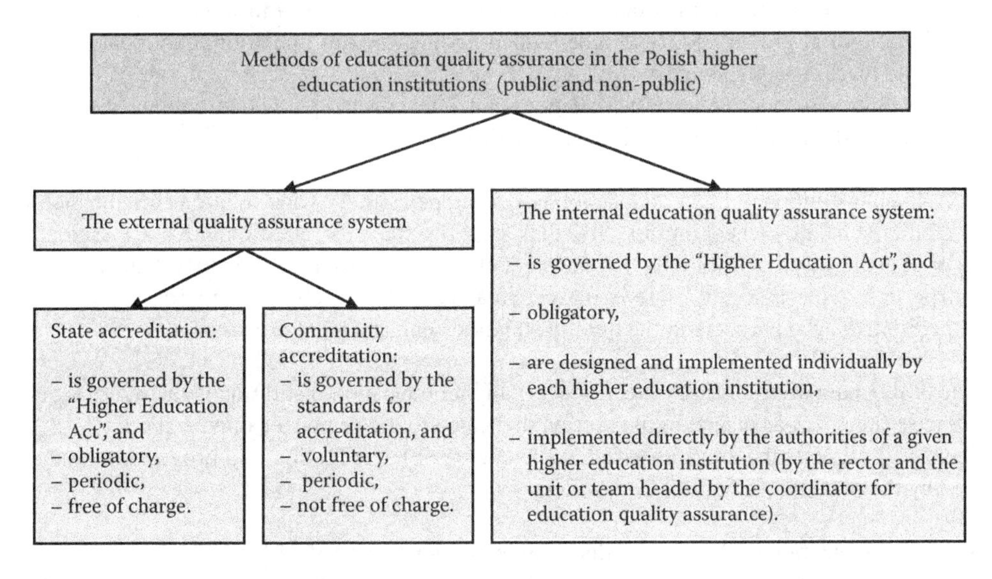

SCHEME 76.1 Methods of education quality in Poland.

FUNCTIONING OF THE STATE ACCREDITATION SYSTEM FOR PROGRAMS IN ECONOMIC SCIENCES

The state accreditation system is the basic method for external quality assurance of higher education in programs in economic sciences in Poland, while the voluntary community accreditation is an additional method for achieving the same goal.

Officially, the state accreditation system in Poland became operational in early 2002, before Poland joined the Bologna system. It is subject to continuous improvement. In order to ensure the minimum acceptable level of education in both state-owned and non-state-owned higher education institutions, the "Law on Higher Education" was passed, and the *State Accreditation Committee* (*PKA*) was established.

The State Accreditation Committee cooperates with various domestic and international organizations whose main activity is the standardization of accreditation principles and education quality assessment and accreditation, including community accreditation organizations. In 2009, when the process of adapting the regulations to European standards was completed, the State Accreditation Committee became a rightful member of ENQA. Since 2010, the Consultative Board operates at the State Accreditation Committee, which gives opinions on and provides counseling regarding the Committee's strategic goals and areas of activity, its development, as well as the standards and procedures it applies. The Consultative Board of the State Accreditation Committee is composed of

- The President of the Polish Confederation of Non-public Employers "Lewiatan"
- The President of the Conference of Rectors of Academic Schools in Poland
- The President of the Conference of Rectors of Vocational Schools in Poland
- The President of the Confederation of Polish Employers
- The President of the Polish Business Roundtable
- Independent international experts in the field of education quality assurance

Pursuant to Article 49 of the Higher Education Act, the State Accreditation Committee submits the opinions and proposals concerning the establishment of a given higher education institution or establishment of a faculty at an existing higher education institution to the Minister. It also gives opinions on the possibility to authorize institutions of higher education to administer particular programs of study at a given level.

In the course of assessment and awarding accreditation to a given program of study offered by a higher education institution, the State Accreditation Committee examines many areas, including the legal and formal grounds for the higher education institution's operation, qualifications of the research and teaching staff, scientific and research activity, quality of education, internal education quality assurance procedures, learning resources, and interuniversity and international cooperation. The State Accreditation Committee assesses whether

- A given institution of higher education has developed system-wide solutions for enhancing quality standards (checking, in particular, how the students' questionnaires are used and what is the scope of questions that the students answer when assessing their teachers)
- The unit being assessed reviews on a regular basis the teaching standards applied and determines who partakes in the quality standards enhancement process.

One of the main areas of assessment of every higher education institution is the internal quality assurance system. It was placed in the section dedicated to the principles of education.

Table 76.3 presents the areas assessed in the course of accreditation of a program of study performed by the State Accreditation Committee:

1. Brief presentation of the higher education institution
2. Presentation of the administrative unit

TABLE 76.3
Areas Assessed by the State Accreditation Committee during Accreditation Process

Assessed Areas	Detailed Requirements
1. Brief presentation of the higher education institution	• Legal conditions, including mission, strategic goals, external relations, higher education institution's structure, proportions of the recruitment process in the past three years, the number of graduates in the past three years, the overall budget structure • The number of lecturers employed by the higher education institution • The number of undergraduate, graduate, and PhD students
2. Presentation of the administrative unit	• Information on the administrative unit, including its staff and their development, as well as the research and learning resources • The number of lecturers employed in the unit • The number of academic degrees and titles awarded to the unit's staff in the past 5 years • The number of the unit's undergraduate, graduate, and PhD students
3. The staff of the inspected program of study	• A list of lecturers suggested for the minimum staff of the program of study being assessed, along with a list of their publications • A list of the remaining lecturers teaching in the program of study being assessed • The number of technical and engineering staff and administrators • Staff policy and its implementation
4. Education	• Internal education quality assurance system • Recruitment principles • The number of students and graduates of the program of study assessed • Graduate's qualifications structure • Schedules of study and curricula • Teaching methods and education organization applied • The aims of training programs, forms of their administration, control systems, credit award • Dropping out of students at particular levels of education • Diploma award principles
5. Research activity of a given administrative unit	• The category awarded to the administrative unit by the State Committee for Scientific Research (KBN) • Scientific achievements, including description of the research carried out, the grants obtained, the system for teaching staff development, and a list of publications of the administrative unit's staff • The link between the research carried out and the teaching process, including publications co-authored by students • Organization (co-organization) of scientific conferences associated with the program of study assessed
6. Teaching resources	Description of teaching resources, lecture halls, workshops, and laboratories and their equipment, computers and Internet access, own library resources, and access to other libraries
7. Students' affairs	Student organizations; student associations; scholarships; tuition fees; students' social affairs (clubs, dormitories, material assistance); sports and leisure facilities
8. Interuniversity and international cooperation	• Student and research and teaching staff exchange • Subjects of research and teaching works created in collaboration with domestic and foreign institutions
9. The most important achievements of the faculty (unit)	• Nationally or internationally recognized staff members • The most important scientific achievements (external awards) • Significant enrichment of research and teaching resources
10. Strengths and weaknesses	Description of the strongest and weakest points of the administrative unit
11. Future plans	Description of plans for the near future

Source: Author's own study based on the Resolution No. 873/2007 of the Presidium of the State Accreditation Committee in guidelines for self-evaluation report, http://www.pka.edu.pl/www_en/index.php?page = dokumenty_pl.

3. The staff of the inspected program of study
4. Education
5. Research activity of a given administrative unit
6. Teaching resources
7. Students' affairs
8. Interuniversity and international cooperation
9. The most important achievements of the Faculty (Unit)
10. Strengths and weaknesses
11. Future plans

Currently, 250 higher education institutions, including 73 state-owned ones, are administering programs of study in economic sciences. What follows from the State Accreditation Committee's report is that in 2009 and 2010, as in previous years, the State Accreditation Committee assessed 6 programs in economic sciences: management, economics, finance and accounting, logistic, it, and econometrics and spatial development.

In 2009, the assessed programs included the following: economics, 23 grades; finance and accounting, 16 grades; spatial development, 4 grades; IT and econometrics, 5 grades; and management, 46 grades. In the year under consideration (2009), the program of study in logistics was not assessed. As a result of the assessment, the State Accreditation Committee formulated 79 grades for education quality. As for 12 units, the State Accreditation Committee withdrew from the assessment, while in the case of 3 other units, the assessment was suspended for one year.

In 2010, the assessed programs included the following: management, 99 grades; economics, 30 grades; finance and accounting, 30 grades; IT and econometrics, 5 grades were assessed for the second time. As a result of the assessment in 2010, the State Accreditation Committee formulated 164 grades for education quality. In the year under consideration, the program of study in logistics and spatial development was not assessed.

Table 76.4 presents the results of accreditation performed in 2009 in juxtaposition with the average results from previous years.

It should be emphasized that the participation of negative and conditional grades in the overall amount of grades awarded by the State Accreditation Committee has grown alarmingly from 19.1% in the last two terms of office of the State Accreditation Committee to 29.1% in 2009. What is particularly disturbing is the nearly 300% increase in the number of negative grades in comparison with the average grades for the preceding 7 years. The State Accreditation Committee's report indicates that all the conditional and negative grades pertained to non–state-owned institutions of higher education.

TABLE 76.4
Results of the State Accreditation Committee's Assessments in 2009 and the Preceding Years

PKA Grade	Grade Average in the Past 7 Years	Grade in 2009	Dynamics (2009/Mean)
Positive	80.9%	70.9%	−12%
Conditional	15.4%	16.5%	+7%
Negative	3.7%	12.6%	+241%
Total:	100%	100%	X

Source: Author's own study.

In the 2009 accreditation of public institutions of higher education, the State Accreditation Committee awarded only positive grades for education quality. In 94% of cases, the accreditations were awarded for the second time, upon the expiry of the previous ones. This leads to the conclusion that state-owned institutions of higher education tend to maintain a good quality of education. In 2009, 63.5% of all non–state-owned institutions of higher education received positive grades for the quality of education offered. One in five non–state-owned institutions of higher education received a conditional grade, while one in six received a negative grade.

The State Accreditation Committee may withdraw from or suspend the assessment if a higher education institution included in the schedule of accreditations for a given year has closed down a particular program of study, for example, if no students have been recruited or if it has applied with the Ministry of Science and Higher Education for the right to supplement the minimum academic staff within the next 12 months. The State Accreditation Committee's report indicates that in 2009, 12 units of higher education institutions decided to close down certain programs of study. The main reason for liquidating the programs was the lack of prospective students. For the first time, three units closed down, the finance and accounting program, while one institution of higher education liquidated the program in management. In 2009, 86.7% of decisions to withdraw from or suspend the education quality assessment were made with reference to non–state-owned institutions of higher education.

Table 76.5 presents the summary of positive and negative grades awarded as a result of the inspections held in 2010 in public higher education institutions of economics.

Table 76.6 presents the summary of positive and negative grades awarded as a result of the inspections held in 2010 in nonpublic higher education institutions of economics.

In 2010, 76% of all institutions of higher education received positive grades for the quality of education offered. One in 12% institutions of higher education received a conditional grade, while 4% received a negative grade.

The conditional grades awarded in 2009 and 2010 for the quality of education were awarded for three main reasons:

- Failure to meet the requirements for minimum academic staff
- Poor quality of diploma works
- Incomplete implementation of the internal education quality assurance system (including, for example, failure to implement system-wide solutions for the stage process of verification of students' knowledge, inadequate organization of end-of-term examinations, the lack of criteria for the assessment of the students' knowledge and skills)

TABLE 76.5

Types and Number of Grades Awarded for the Quality of Education in Public Higher Education Institutions of Economics Accredited by the State Accreditation Committee in 2010

	Faculty of Study				
PKA Grade	**Economics**	**Management**	**IT and Econometrics**	**Finance and Accounting**	**Total**
Positive	6	38	2	18	64
Conditional	1	5			6
Negative		1			1
Withdrawal		2	1		3
Total	7	46	3	18	74

Source: Author's own study.

TABLE 76.6
Types and Number of Grades Awarded for the Quality of Education in Nonpublic Higher Education Institutions of Economics Accredited by the State Accreditation Committee in 2010

	Faculty of Study				
PKA Grade	**Economics**	**Management**	**IT and Econometrics**	**Finance and Accounting**	**Total**
Positive	12	40	2	6	60
Conditional	6	6		1	13
Negative	2	2		1	5
Withdrawal	3	5		4	12
Total	23	53	2	12	90

Source: Author's own study.

In the 2009 and 2010, negative grades were awarded as a result of the following:

- Failure to comply with the requirements for the minimum academic staff
- Lack of an internal education quality assurance system
- Inconsistency of the topics of diploma works with the subject matter of the assessed programs of study and their poor quality
- Inadequate teaching concept and quality of programs offered
- Lack of appropriate library resources
- Inadequate appointment of teaching staff
- Inadequate use of teaching methods and techniques for the purpose of the distance learning process
- Lack of scientific research
- Deficiencies in the organization of students' training programs
- Administering interdisciplinary studies or maintaining off-campus faculties without a due authorization of the Ministry of Science and Higher Education

An institution of higher education that has been awarded accreditation with *at least a positive grade* meets the education quality requirements for a particular level of studies. These conditions were indicated as criteria for assessment in particular resolutions of the State Accreditation Committee. Therefore, any institution of higher education must meet the requirements regarding the following:

- Formal and legal aspects of education (Resolution of the Presidium of the State Accreditation Committee No. 217/2008)
- Minimum academic staff (Resolution of the Presidium of the State Accreditation Committee No. 828/2008)
- System for verification of education results (Resolution of the Presidium of the State Accreditation Committee No. 219/2008)
- Schedules and curricula of the programs of study (Resolution of the Presidium of the State Accreditation Committee No. 501/2008)
- Scientific research in a discipline or an area associated with a particular program of study (Resolution of the Presidium of the State Accreditation Committee No. 94/2007)
- Students' affairs (Resolution of the Presidium of the State Accreditation Committee No. 218/2008)

The results of the assessment discussed in the 2009 and 2010 Reports of the State Accreditation Committee indicate that higher education institutions do an increasingly better job in defining the qualifications structure of their graduates, establishing the schedules and curricula of the programs of study, and specifying the education process results. When institutions of higher education have developed and implement internal education quality assurance systems, they expand their learning resources and handle students' affairs more effectively. The strengths of public institutions of higher education include their research and teaching staff as well as their scientific research. In many nonpublic higher education institutions, the documentation regarding the progress of study and the personal matters of research and teaching staff requires continuous improvement. External relations in the context of the students' vocational training, on the other hand, show an increasingly more efficient cooperation with the business sector. Moreover, the number of eminent practitioners partaking in the implementation of the educational process is on the increase, which results in postgraduate studies and research work. However, all the institutions of higher education still show a low level of internationalization of the teaching process.

FUNCTIONING OF THE COMMUNITY ACCREDITATION SYSTEM FOR PROGRAMS IN ECONOMIC SCIENCES

The community accreditation system is also very important for external education quality assurance of economics programs. The most important institution in the voluntary community accreditation system of the programs of study in economic sciences is the Foundation for the Promotion and Accreditation of Economic Education EPOQS (FPAKE).

FPAKE was established by the Conference of Rectors of Higher Education Institutions of Economics. It was created on the initiative of five state-owned higher education institutions of economics, including the Warsaw School of Economics (SGH), as well as four universities of economics in Katowice, Cracow, Poznań, and Wrocław. Today, FPAKE also cooperates with other universities and schools of economics in Poland. FPAKE's goals include performing and awarding accreditations to programs of study in economics and management sciences. FPAKE cooperates with various organizations in developing accreditation procedures and standards and ensures comparability of the adopted standards with both European standards and the standards applied by other Polish accreditation and education quality assessment institutions (Szczepankiewicz and Skoczek-Spychała 2011, pp. 254–255).

FPAKE's accreditation is addressed to all institutions of higher education (both state-owned and non-state-owned, economic and noneconomic, vocational and academic) that administer undergraduate (bachelor's) and graduate (master's) studies in finance and accounting, management, IT and econometrics, economics, international relations, commodity science, and spatial development. The FPAKE accreditation process is centered on the key areas of assessment, which are indicative of the quality of education offered by a given higher education institution. Key education quality areas in accordance with the FPAKE standards are mission and strategy, students, lecturers, scientific research, teaching resources, teaching processes, and impact on the community.

In 2003–2011, the state institutions of higher education that founded FPAKE were the first ones to undergo the FPAKE accreditation: the Warsaw School of Economics (SGH) as well as four universities of economics in Katowice, Cracow, Poznań, and Wrocław.

Due to the fact that this type of accreditation is voluntary, few higher education institutions decide to undergo it, and they undergo it infrequently. Some nonpublic schools of economics, too, decided to undergo accreditation. Table 76.7 presents certificates awarded by FPAKE for the quality of education in nonpublic schools of higher education in 2004–2011. Table 76.8 presents certificates awarded by FPAKE for the quality of education in public schools of higher education in 2003–2011.

TABLE 76.7
Certificates Awarded by FPAKE for the Quality of Education in Nonpublic Schools of Higher Education in 2004–2011

Name of the Higher Education Institution	Faculty of Study	Course of Study	Year
Warsaw School of Economics (Warszawska WSE)	Faculty of Economic Relations and Management	Management and marketing	2004
Warsaw School of Banking	Faculty of Finance and Management	Finance and banking	2005
		Finance and accounting	2009
	Faculty of Finance and Management	Management and marketing	2006
		management	2009
Białystok School of Economics	X	Economics	2005
Dąbrowa Górnicza School of Business	Faculty of Management, Computer Science and Social Sciences	Management and marketing	2006
Toruń School of Banking	Faculty of Finance and Management	Finance and accounting	2007
			2011
Wrocław School of Banking	Faculty of Finance and Management	Finance and accounting	2009
		Management	2009
Poznań School of Banking	(Chorzów Branch)	Finance and accounting	2009
Kielce School of Economics and Law	Faculty of Economics	Economics	2009

Source: Author's own study based on the Accreditation and certificates, http://www.fundacja.edu.pl/modul/1/47/50/certyfikaty_akredytacyjne.html (accessed: 2011-04-04).

TABLE 76.8
Certificates Awarded by FPAKE for the Quality of Education in Public Schools of Higher Education in 2003–2011

Name of the Higher Education Institution	Faculty of Study	Course of Study	Year
Warsaw School of Economics (SGH)	X	Finance and banking	2003
		Finance and accounting	2008
	X	Economics	2003
	X	Quantity methods and Information systems	2003
			2008
	X	Management and marketing	2003
		Management	2008
	X	International relations	2004
			2008
Wrocław University of Economics	Faculty of National Economy	Finance and banking	2003
	Faculty of National Economy	Management and marketing	2003
	Faculty of Management and Computer Science	IT and econometrics	2003
	Faculty of Management and Computer Science	Management and marketing	2003
	Faculty of Regional Economy and Tourism in Jelenia Góra	Economics	2003
			2007
	Faculty of Management and Computer Science	Finance and banking	2004
	Faculty of National Economy	International relations	2006
	Faculty of Economic Sciences	Management	2007
	Faculty of Economic Sciences	Finance and accounting	2007

TABLE 76.8 (Continued)
Certificates Awarded by FPAKE for the Quality of Education in Public Schools of Higher Education in 2003–2011

Katowice University of Economics	Faculty of Management	Management and marketing	2003
	Faculty of Management	Management	2010
	Faculty of Management	IT and econometrics	2004
	Faculty of Economics	Economics	2005
	Faculty of Finance and Insurance	Finance and banking	2005
Poznań University of Economics	Faculty of Economics	IT and econometrics	2003
	Faculty of Management	Management and marketing	2003
	Faculty of Management	Management	2011
	Faculty of Management	International relations	2003
	Faculty of Economics	Finance and banking	2003
		Finance and accounting	2008
	Faculty of Economics	Economics	2004
			2009
	Faculty of Commodity Science	Commodity studies	2007
Cracow University of Economics	Faculty of Commodity Science	Commodity studies	2006

Source: Author's own study based on the Accreditation and certificates, http://www.fundacja.edu.pl/modul/1/47/50/certyfikaty_akredytacyjne.html (accessed: 2011-04-04).

CONCLUSION

The European standards influenced the development of the Polish laws and regulations governing the education quality assurance and assessment in higher education institutions in Poland. Present-day institutions of higher education have sufficient formal and legal powers to ensure a high quality of education. It should be firmly emphasized that it is not just the founders and authorities of higher education institutions who determine the quality of education but also the teaching staff as well as the administration staff. Moreover, the state and community accreditation systems play an important role in assuring the quality of higher education in Poland. Currently, state and community accreditation systems are quite effective in ensuring the minimum quality of education for programs of study in economic sciences. The legislative body, too, supervises the quality of education. The systems should contribute to enhancing the qualifications of Polish economic staff and to raising the quality of education of the entire society. Moreover, the accreditation market itself should, in time, eliminate those higher education institutions that are believed to offer poor education quality and inadequate programs of study.

REFERENCES

Accreditation and certificates, http://www.fundacja.edu.pl/modul/1/47/50/certyfikaty_akredytacyjne.html (accessed: 2011-04-04).

The Act of 27 July 2005 Law on Higher Education, DzU 2005, No. 164, item 1365, with amendments.

Resolution No. 873/2007 of the Presidium of the State Accreditation Committee in guidelines for self-evaluation report, http://www.pka.edu.pl/www_en/index.php?page=dokumenty_en (accessed: 2011-04-04).

Resolution of the Presidium of the State Accreditation Committee No. 76/2009 on the determination of general criteria for the quality assessment of education, http://www.pka.edu.pl/www_en/index.php?page=dokumenty_en (accessed: 2011-04-04).

Resolution of the Presidium of the State Accreditation Committee No. 94/2007 on the criteria to assess compliance with the requirements for conducting research in a discipline or area related to a given field of study, http://www.pka.edu.pl/www_en/index.php?page=dokumenty_en (accessed: 2011-04-04).

Resolution of the Presidium of the State Accreditation Committee No. 217/2008 on the criteria to assess formal and legal aspects of education, http://www.pka.edu.pl/www_en/index.php?page=dokumenty_en (accessed: 2011-04-04).

Resolution of the Presidium of the State Accreditation Committee No. 218/2008 on the criteria to assess fulfillment of student matters requirements, http://www.pka.edu.pl/www_en/index.php?page=dokumenty_en (accessed: 2011-04-04).

Resolution of the Presidium of the State Accreditation Committee No. 219/2008 on the assessment criteria concerning the learning outcomes verification system, http://www.pka.edu.pl/www_en/index.php?page=dokumenty_en (accessed: 2011-04-04).

Resolution of the Presidium of the State Accreditation Committee No. 501/2008 on the criteria to assess study programmes and curricula, http://www.pka.edu.pl/www_en/index.php?page=dokumenty_en (accessed: 2011-04-04).

Resolution of the Presidium of the State Accreditation Committee No. 828/2008 on the criteria to assess compliance with the minimum staff requirements, http://www.pka.edu.pl/www_en/index.php?page=dokumenty_en (accessed: 2011-04-04).

Standards and Guidelines for Quality Assurance in the European Higher Education Area, 2005. The European Association for Quality Assurance in Higher Education (ENQA). Helsinki, Finland.

Szczepankiewicz E.I, and Kiedrowska M., 2011. Internal education quality assurance system at the universities and schools of economics in the light of European Standards and Polish regulations. Folia Pomeranae Universitatis Technologiae Stetinensis 287, Oeconomica 63, West Pomeranian University of Technology, Stetin.

Szczepankiewicz E.I., and Skoczek-Spychała M., 2011. The role and aims of state and community accreditation as an element of the education quality assurance system of programs of study in economic sciences in Poland. Folia Pomeranae Universitatis Technologiae Stetinensis 287, Oeconomica 63, West Pomeranian University of Technology, Stetin.

77 Changes in the System of Higher Education in Poland

A Historical Perspective and Present Challenges

Małgorzata Cieciora

CONTENTS

INTRODUCTION

Throughout the last 20 years after the end of the so-called centrally planned economy in Poland, a series of significant changes in the Polish system of higher education could be noticed. Just at the very beginning of the economic transformation (often referred to as "Balcerowicz Plan" or "shock therapy"), one could observe the tremendous impact that the new, market economy had on the academic sector. First, nonpublic schools were opened as early as in 1990, and state-owned universities started to offer commercial, paid courses as well. The number of schools increased from 112 in 1989 to 461 in 2010, and the number of students rose almost 5 times.

However, similarly to reforms conducted in other branches of the economy (Balcerowicz 2009; Kowalik 2009), for example, the health care or social insurance systems, the changes in the higher education system evoked mixed reactions, and the number of their advocates equaled the number of their fierce opponents. On the one hand, there was a striking need to change the inefficient, old system; on the other hand, though, the way of conducting the reform was heavily criticized for lack of professionalism; possible misappropriation of funds; and unsatisfactory, partial results.

Furthermore, throughout the years, new challenges appeared, mainly connected with tightening ties with other European countries. In 1994, Poland became a member of the European Union (EU), which entailed various socioeconomic and legal consequences, also for the sector of higher education. A need to harmonize some important parts of the systems in a group of countries in collaboration (e.g., the division of the structure of studies) appeared. This, in turn, had an impact on the activity of the domestic schools.

Another factor of utmost importance for Polish universities are the demographic changes that are taking place, above all the noticeable aging of the population.

The aim of this article is to discuss the most important of the changes in higher education in Poland and the possible future directions. Statistical data provided by the Polish Central Statistical Office (Główny Urząd Statystyczny) were analyzed as well as bills on higher education and findings of reports commissioned by the government and public institutions.

The Higher Education Sector after 1989

As has already been mentioned, the number of students increased dramatically (from around 400,000 in 1990 to 1,900,000 in 2010) (GUS 1998, 2010). New higher schools were founded—the number of schools increased from 112 in 1990 to 461 in 2010 (GUS 1998, 2010); almost all of them are nonpublic institutions. The dynamics of this growth was very strong in the 1990s; during the last decade, the trend noticeably slowed down. It is worth noticing that the number of foreign students more than tripled after the transition of 1989 (GUS 1998, 2010). Unfortunately, not all of the changes can be classified as positive. For example, the number of students from farmers' families decreased—whereas in the 1980s, 14% of children from farmers' families undertook studies, in the 1990s, the statistics dropped to 1.4% (MEN 2000). Moreover, most students from poorer backgrounds lose the race of free, day studies in prestigious public schools and have to undertake paid studies.

It is worth noticing that, despite the dynamic development of private schools, most students (over 60%) (GUS 2010) still study in public institutions. The best and most popular universities are very selective with regard to the enrollment for nonpaid, day studies. It should be stressed that public schools extended their educational offer; for example, the introduced interdisciplinary studies and courses conducted in English. It should also be stressed that public schools—as opposed to most of the newly founded, private ones—not only are didactic activity oriented but also conduct scientific research. It is worth mentioning, however, that an increasing number of courses is offered in commercial, nonregular (evening and weekend) modes of study. Consequently, nowadays, only about 40% of students are students of the traditional, day, regular studies. The situation is quite controversial as the right to free education at every level is guaranteed to every citizen by the Polish Constitution. When, however, in the 1990s, state budget subsidies to public universities fluctuated between 0.7% and 0.8% of the GDP and were not sufficient to cover the expenses of the didactic activities of schools, school senates decided to introduce this controversial, halfway solution. Since then, only a defined number of candidates who performed best on the entrance examination have been privileged with the right to study for free throughout the whole course of studies. The other ones could only study on a paid basis—in nonregular modes of studies in public schools or in nonpublic schools.

The first nonpublic school after the collapse of the previous economic system was registered in 1991. Then, as has already been mentioned, there was an avalanche-like development of nonpublic schools. To a great extent, that was the reason for the tremendous increase in schooling ratios in Poland. Another positive aspect connected with their activity is the increased attractiveness of the profession of university teacher, as the rates that they offer are quite high (though they offer very few full-time jobs). Moreover, they are not a burden to the state budget, and their average costs of education are lower than in public schools. However, there is also the flip side of the coin. First of all, education in Polish nonpublic schools is paid. Second, a lot of these institutions are quick profit oriented, which results in a "massive" production of low-quality graduates. Contrary to the best public schools, they are not selective at the enrollment phase. Most of the teachers are employed only on part-time contracts, which do not guarantee professional or social security. Moreover, they are not eager to engage in scientific research, and they focus only on teaching, limited to a narrow number of low-cost "market-oriented" fields of study. They do not educate a lot of new researchers and lecturers, either. One could say that they just "consume" the human capital and scientific achievements of public universities. What is more, the ratio of nonregular (evening and extramural) to regular students is higher than in public schools, as is the ratio of students per one teacher.

Finally, unlike in the United States, in Poland, paid (sometimes very expensive) studies at nonpublic schools are still considered to be less valuable and less prestigious than nonpaid, day studies at old, public universities.

Economic, Demographic, and Legal Environment of the Polish Higher Education System

Every organization (and its internal organizational culture) is affected by the external environment. This also applies to universities. The main factors that can influence the activity of Polish higher schools are the economic/demographic conditions and legal conditions.

As far as economic conditions are concerned, it is worth mentioning that Poland belongs to a group of middle-income countries. It has been a member of the EU since 2004. In terms of GDP, Poland is the 8th biggest economy in the EU and the 21st biggest economy in the world (MF 2011). During the recent crisis, Poland was the sole European country that recorded positive economic growth (1.8%) (MF 2011). It has to struggle, however, with the problem of rising unemployment (13.1% in March 2011) (MF 2011), increasing inflation (4.3%) (Trading Economics 2011), and a budget deficit (48.3 bn PLN) (PAP 2011). As far as the structure of the Polish economy is concerned, for a long period of time, it was based on agriculture and the heavy industry. Nowadays, Polish agriculture accounts for about 5% of GDP—more than twice than in the United States or Germany (Polska 2030 2011). It should be noted, however, that the situation is still changing. A growing share of GDP is produced in services and less and less in manufacturing and agriculture. It seems that the main obstacles to a more dynamic GDP growth include the following factors (Polska 2030 2011):

- The problem of the reallocation of the labor force to the other sectors of the economy
- Slow structural changes in the rural areas
- Shortage of well-qualified human capital
- Obsolete production technologies

Therefore, there seems to be a need to increase the innovation, knowledge-based economy, especially since countries with GDP per capita far below the EU27 (which is the case for Poland) are eager for innovations as they give them a real chance for rapid and sustained growth; thanks to innovations, less developed regions can not only accomplish the process of modernization of their economies but also significantly improve living standard of their inhabitants (Wildowicz-Giegiel 2011). Unfortunately, the innovative capacity of Polish R&D (measured by the number of patents) is currently quite low. In 2006, there were about 230 Polish patents filed. According to the Eurostat, the Polish ratio (3.4 patents per 1 million inhabitants) is twice as low as that in Slovakia and 70 times lower than in Finland (Ernst and Young, Instytut Badań nad Gospodarką Rynkową 2009).

Another important socioeconomic challenge, which the whole of Europe is going to face in the coming decades, is the rapid aging of the population. Current statistics show that average life expectancy has increased by eight to ten years since the 1950s (although it should be stressed that there are still significant variations between old and new EU member states) (Winkelmann-Gleed 2011). The problem concerns Poland as well—the proportions of people under 20 and over 64 years of age in the whole population changed from 33%–10% in 1990 to 21%–13.5% in 2008; moreover, according to the Eurostat forecasts, the proportions will reach 17%–23% in 2030 (Bukowski 2005). Older people (whose economic activity requires an update of knowledge and skills acquired at school) will, in the near future, account for a significantly greater proportion of the workforce than today. This will exacerbate problems associated with adaptation to labor market needs of a large number of people over 50. The development and popularization of lifelong learning for adults is, therefore, becoming an urgent need in Poland and in Europe (Bukowski 2005).

It is worth mentioning that, in comparison to other European countries, Poland is characterized by a relatively high intensity of formal education for persons under 25 years of age. According to Organization for Economic Co-operation and Development (OECD) (2009), investment in higher education in Poland results in one of the highest premiums among the OECD countries. It not only stems from significant differences in expected wages of people with and without higher education but is also connected with low direct costs associated with undertaking studies and the low opportunity cost of studies (the income that students would have earned if they had decided to work instead). The higher the rate of return on such investments, the greater the income inequality in society. In OECD countries, the least educated people are more disadvantaged in the labor market, both in terms of unemployment and the average salary. The higher the qualifications one has, the higher his/her wages and the more stable his/her employment (Bukowski 2005, 2010). However, in Poland, currently, a distressing phenomenon of increasing unemployment among workers with higher education can be observed. Whereas in 1997, university graduates accounted for only 1.4% of all the unemployed in Poland, the number soared to 10.5% in 2010 (ComPress SA 2011). The university diploma is no longer an effective shield against unemployment. Over the next few years, the number of unemployed graduates in Poland could rise by up to 3 million (Gazeta Prawna 2010). It seems that this is due to the fact that the educational offer of Polish schools does not match the labor market demand. Universities still "produce" too many specialists in humanities (languages, history, sociology, psychology) or marketing and management, though the market for them has been saturated for a long time. Instead, there is a growing need for the following specialists: computer scientists, construction engineers, mechatronics engineers, financial analysts, biotechnologists, and telecommunications engineers (Dudzik 2011). Unfortunately, the most sought-after graduates in sciences and engineering account for only around 10% of the total number of graduates (Dudzik 2011). The majority of the schools, however, do not monitor the careers of their graduates or pay attention to the market trends. Therefore, the programs they offer are often too theoretical, out-of-date, and, above all, not labor market oriented.

As has been already mentioned, active participation in the socioeconomic life in a rapidly changing modern world requires people to acquire new knowledge and skills continuously, not only during their formal schooling days but also after graduation. This process is called continuing education or lifelong learning. Currently, in OECD countries, more than 20% of adult citizens update their knowledge systematically. Poland occupies a position at the end of the scale, with the percentage of adult participation in continuing education fluctuating around 5% (Bukowski 2005, 2010). There are a few important aspects concerning the situation on the Polish labor market (especially of workers who are over 50 years old) and the concept of lifelong learning.

For example, one of the characteristics of the Polish labor market is a very low percentage of staff over 55 years old participating in any form of training. In 2005, only 13% of them participated in training organized by the employer (the EU27 average was 24%). This small percentage of seniors upgrading their qualifications is due, in large part, to the early retirement age in Poland. In Poland, very few people over 55 years old are still professionally active. Employers are not willing to pay the costs of increasing the qualifications of workers who are not likely to use them at work for a reasonably long period of time (Bukowski 2005, 2010).

Another serious barrier to investing in human resources by Polish entrepreneurs is the information barrier. There are a lot of educational institutions, many of them offering poor-quality services, and one may find it difficult to evaluate their offers properly. In this situation, entrepreneurs, especially SMEs, may resign from the organization of training, despite the public subsidies that they might receive. Sending workers to a low-quality training institution is regarded as more harmful than the resignation of any form of training. And the situation seems to be distressing. In 2008, the Supreme Court Council (Najwyższa Izba Kontroli) conducted an audit of the system of continuing education in Poland in terms of its adaptation to the needs of the labor market. The audit covered the activities of 40 public educational institutions in 2005–2008. The results of the audit showed a very low rate of class attendance. A lot of participants decided also to withdraw from the training programs; there were even cases in which all of the trainees resigned after the first semester (Bukowski 2010).

The last important conclusion concerning the demographic factor is that there seems to be an important correlation between the present demographic situation and the higher education sector. In the years 2008–2020, the number of people in the age group 18–24 will fall by 1.5 million (Ernst and

Young, Instytut Badań nad Gospodarką Rynkową 2009). This means that by 2020, the number of students may have decreased by as many as 600,000 to 800,000 (Ernst and Young, Instytut Badań nad Gospodarką Rynkową 2009). Consequently, the demand for teachers will diminish (by around 25% in 2020). It is likely that the average age of teachers will increase, as the number of habilitation dissertations remains stable and the dynamics of the growth of the number of PhD dissertations has dropped since 2006. Currently, over 30% of full-title professors are over 70 years of age (Ernst and Young, Instytut Badań nad Gospodarką Rynkową 2009).

As has already been mentioned, another important factor that has a significant impact on the functioning of higher schools in Poland is legislation. The most important legal document in Poland is the Constitution of the Republic of Poland, passed on April 2, 1997. Article 70 specifies the citizens' rights concerning education (Sejm 2011):

The Constitution of the Republic of Poland

Article 70

1. Everyone shall have the right to education. Education to 18 years of age shall be compulsory. The manner of fulfilment of schooling obligations shall be specified by statute.
2. Education in public schools shall be without payment. Statutes may allow for payments for certain services provided by public institutions of higher education.
3. Parents shall have the right to choose schools other than public for their children. Citizens and institutions shall have the right to establish primary and secondary schools and institutions of higher education and educational development institutions. The conditions for establishing and operating non-public schools, the participation of public authorities in their financing, as well as the principles of educational supervision of such schools and educational development institutions, shall be specified by statute.
4. Public authorities shall ensure universal and equal access to education for citizens. To this end, they shall establish and support systems for individual financial and organizational assistance to pupils and students. The conditions for providing of such assistance shall be specified by statute.
5. The autonomy of the institutions of higher education shall be ensured in accordance with principles specified by statute.

As far as other legislation on the higher education system is concerned, it should be mentioned that in Poland, in the two decades after 1989, there were bills passed and agreements signed that allowed for the following: (1) foundation of nonpublic schools; (2) foundation of higher vocational schools; and (3) adaptation of the Polish educational system to Western European standards (e.g., by the introduction of a two-level system of Master studies or the introduction of the ECTS grading system).

The results of the foundation of nonpublic schools were already discussed in a previous subchapter. An impactful change that requires additional comment is Poland's joining the Bologna Process.

In 2000, in Lisbon, the European Council set out a plan for the development of the EU economy. Its aim was to make the EU "the most competitive and dynamic knowledge-based economy in the world capable of sustainable economic growth with more and better jobs and greater social cohesion" (Euractiv 2011). At that time, the EU economy was lagging behind the other two economies of the triad, that is, the United States and Japan.

The overall goals of the strategy were defined as follows:

- Preparing the transition to a knowledge-based economy and society by better policies for the information society and R&D, as well as by stepping up the process of structural reform for competitiveness and innovation and by completing the internal market

- Modernizing the European social model, investing in people, and combating social exclusion
- Sustaining the healthy economic outlook and favorable growth prospects by applying an appropriate macroeconomic policy mix

One of the most important subtargets of the strategy was the improvement of employability. Therefore, it was decided to promote lifelong learning and mobility within the EU (both among students and workers). As far as building the knowledge-based economy and promoting mobility in the EU are concerned, the Lisbon Strategy was supposed to be supported by a series of reforms in higher education, which were called the Bologna Process. The Bologna Process officially started on June 19, 1999, when education ministers from 29 countries (including Poland) signed the Bologna Declaration. Its goal was to create the right conditions for increasing the mobility (and, consequently, cooperation) of citizens; adapting the educational system to the needs of the labor market; and increasing the attractiveness and strengthening the competitive position of the educational system in Europe (Ciucanu 2011). During the next eight years, the number of countries that signed the Declaration increased to 47. The main aims of the Bologna Declaration, elaborated during the next meetings of Education Ministers in Prague (2001), Berlin (2003), Bergen (2005), London (2007), Louvain-la-Neuve and Leuven (2009), and Budapest and Vienna (2010), are as follows (Dunkel 2011):

- Creation of a system of easily understandable and comparable degrees, including through the introduction of the Diploma Supplement
- Creation of a two-tier system of degrees (consecutive study programs, undergraduate/graduate)
- Introduction of a credit system, the European Credit Transfer System (ECTS), as well as modularization
- Promotion of mobility by the removal of obstacles to mobility, referring not only to geographical mobility but also to cultural competences, mobility between higher education institutions, and training programs
- Qualitative development of higher education through faculty development, study program accreditation, and promotion of European cooperation on quality development
- Promotion of the European dimension in higher education
- Lifelong and/or life-wide learning
- Student participation (participation in all decisions and initiatives at all levels)
- Promotion of the attractiveness of the European Higher Education Area (EHEA)
- Dovetailing the EHEA with the European Research Area (ERA), in particular, by incorporating doctoral studies in the Bologna process

The implementation of the Bologna Process is realized on international, country, and school levels. As far as the international level is concerned, Education Ministers meet every 2 years to take strategic decisions concerning the Bologna Process and to evaluate the current results. At the country level, the Bologna Process is supervised in Poland by

- The Ministry of Science and Higher Education, which is directly responsible for the adaptation of Polish law on higher education to the Bologna Process requirements
- The Conference of Rectors of Academic Schools in Poland (CRASP)
- The Conference of Rectors of Non-University Higher Education Institutions in Poland
- The General Council for Higher Education, the State Accreditation Committee
- The Students' Parliament of the Republic of Poland
- The Bureau for Academic Recognition and International Exchange
- The Socrates-Erasmus National Agency

As far as the university level is concerned, each institution is expected to

- Offer the recommended three (or at least two) level studies
- Provide the graduates with an opportunity to continue their education at a higher level
- Build and implement the internal quality assurance system
- Use the ECTS system
- Supply their graduates with Diploma Supplements (containing detailed information about the qualifications obtained)
- Promote the mobility of students and university staff
- Promote the European dimension in higher education (e.g., by starting cooperation with other European universities and offering classes in foreign languages)
- Promote and implement *lifelong learning*

The Lisbon Strategy for making Europe the most competitive and dynamic knowledge-based economy in the world is not considered to have been successful and needs to be redefined. In June 2010, the Heads of State and Government adopted the successor to the Lisbon Strategy, "Europe 2020: A strategy for smart, sustainable and inclusive growth," which, though it preserves the original goals of the Strategy, "no longer adopt[s] an incantatory tone as in 2000" (Pépin 2011). The Bologna Process, meant to support the Lisbon Strategy, also evokes mixed feelings. The main criticism concerns, among others, the following points (Szapiro 2011):

- Loss of diversity in the programs of study as a result of unification and standardization of the curricula
- Artificial and arbitrary assignment of credits to courses
- Lower level of two-level studies (a shortened version of the previous five-year system)

Apart from the highest level legislation, such as international regulations or the Constitution, the activity of the higher education sector in Poland is regulated by laws elaborated by the Ministry of Higher Education. Recently, a number of new reforms were proposed, and on April 5, 2011, Bronisław Komorowski signed the amendment to the Act on Higher Education (MNiSW 2011). According to the new law (whose main aims are to extend the autonomy of Polish universities and introduce the European Qualifications Framework),

- Universities that have full doctorate and habilitation granting rights will have more freedom in choosing curricula and creating new faculties.
- Public universities will receive financial help from the state budget in accordance to the level of education they offer.
- Universities will adopt the European Qualifications Framework, which will make Polish diplomas comparable to qualifications obtained elsewhere in Europe.
- The best faculties (of both public and nonpublic institutions) will be eligible for the status of the country's leading research centers (Krajowy Naukowy Ośrodek Wiodący [KNOW]), valid for 5 years. During this period, they will receive significantly increased funds. The number of KNOW in one area of knowledge and education may not exceed 3.
- Students of public schools who have decided to study at two faculties will have to pay for the second one (though the best students will be exempt from the payments).
- The so-called multiemployment of academic teachers will be limited; a public university lecturer will be allowed to find employment only "at one additional employer who conducts educational or scientific research activity" (at the consent of the rector).
- The habilitation procedure will be simplified—for example, the habilitation colloquium will be gotten rid of.
- Assistant professors and associate professors (with master's and PhD degrees, respectively) will have only eight years to obtain the higher titles; otherwise, they will lose their jobs.

The law will come into force at the beginning of the next academic year. As one might expect, it has already been met with both positive and negative comments (Sitek 2010; Perspektywy 2011; Dolinabiolotechnologiczna 2011).

It is generally appreciated that universities are going to enjoy more autonomy in developing curricula or that the scientific career paths are to be simplified. Imposing upon schools the duty to monitor the careers of their graduates also seems to be a positive change. What raises questions is, for example, giving the rectors the right to make decisions (on an unspecified basis) concerning the extra work of lecturers or favoring public universities as far as public financing of higher education is concerned. There are also some concerns raised concerning the (insufficient) protection of the social sciences and humanities.

CONCLUSION

Human capital is nowadays considered to be one of the key factors influencing economic growth and social development in the long run. In a societal perspective, the average level of education and competence of the population (the average level of human capital) affect the quality of labor resources and, thus, the productivity and the level of manufactured product. In fact, the presence of high-quality human capital in a given society results in external benefits, it influences growth and entrepreneurship, strengthens social cohesion, has a positive impact on the well-being of all citizens (not just those with the highest competencies), and increases trust among people. From the point of view of an individual, educational investments result in measurable returns in the form of an increased salary, greater opportunities for career progression, and a better position in the labor market (e.g., there is a lower risk of losing one's job). In other words, the higher the worker's qualifications, the better the wages and chances of stable employment (Bukowski 2010).

Therefore, a well-organized system of higher education is an essential element of a sound state economy as it provides it with highly qualified human resources and scientific-technological innovations. They are necessary for the proper development of modern enterprises. Consequently, the proper functioning of this sector should be considered as one of the pillars of pro-growth policy, especially at the beginning of the 21st century, often referred to as the "Information Age," the times of domination of the "Knowledge Class," whose success is based not on traditional, material capital but on knowledge. Poland is no exception in this regard.

Unfortunately, the present system of higher education in Poland is far from ideal. After 1989, the schooling ratios increased, and new schools were founded, but the prestige of the high school diploma decreased. Although, theoretically, every citizen has a constitutional right to free education—also on the higher level—in reality, most students have to pay for their studies. As a result, one can observe a dramatic decrease in the number of students from less affluent strata of society. Moreover, paid studies are often a lower-quality "service" as the "service suppliers"—organizers of evening or extramural studies at public universities or owners of private schools—are more concerned about getting access to students' fees than equipping them with valuable knowledge. Despite the start of activity of the State Accreditation Commission in 2002, a fully operational and efficient system of quality assurance has not been built yet. Another negative phenomenon is a noticeable mismatch between the educational offer of Polish universities and the requirements of the labor market, resulting in an increase of the unemployment rate of graduates. The situation of the "academic human resources" is also distressing—the present system of payment and promotion does not encourage talented graduates to choose a university career. In the academic environment, there is still a fierce, ongoing discussion concerning the academic career model. According to some academicians, the reforms conducted since 1989 went too far (they, among others, "devaluated the Professor title" by allowing university senates the right to grant the "university Professor" title to their employees). In the opinion of others, they were insufficient—for example, the habilitation was not abolished (though the last amendment got rid of the highly unpopular habilitation colloquium), and the "feudal" internal university structure was preserved. It seems that further changes are a necessity.

In order to conduct further reforms, now, one also has to take into consideration the following factors:

- Universities are facing demographic decline.
- The Polish society is aging.
- The Polish system of higher education is a part of the Bologna Process; therefore, there is a need to take into account its requirements, regardless of the perceived disadvantages of the Process.
- There is a need to continue the transformation of the Polish economy into a knowledge-based economy.

In this situation, the new challenges that the sector of Polish higher education will have to face include the following:

- Offering a properly composed mixture of various kinds of knowledge (containing the well-calculated proportions of theoretical knowledge on the one hand and the skills needed to ensure the success of graduates in the labor market on the other).
- Developing a high-quality training offer in line with the strategy of lifelong learning, directed largely at the more mature workers, with a view to updating their skills and strengthening their position in the labor market; it should be good enough to encourage employers to invest in this kind of training for their employees.
- Supplying the economy with R&D, which may be a valuable contribution to the knowledge-based economy.

The tasks are challenging, but if they are completed successfully, Poland will use its chance to become a modern, dynamically growing, affluent knowledge-based economy.

REFERENCES

Balcerowicz L. (2009) 800 dni Krótka historia wielkiej zmiany 1989–1991 (800 days A short history of a great change 1981–1991). Polityka, 2009.

Bukowski M. (ed.) (2005) Zatrudnienie w Polsce 2005. Praca w cyklu życia (Employment in Poland 2005. Job in the lifecycle). Warszawa, 2005. http://www.mg.gov.pl/NR/rdonlyres/80DEE1CA-2D83-4B2C-8FD6-5B27B4EA1153/14572/zatrudnienie2005rok.pdf [20 April 2011].

Bukowski M. (ed.) (2010) Zatrudnienie w Polsce 2008. Praca w cyklu życia (Employment in Poland 2008. Job in the lifecycle). Warszawa, 2010. http://ibs.org.pl/site/upload/publikacje/ZWP2008/Zatrudnienie_w_Polsce_2008.pdf [20 April 2011].

Ciucanu I. (2011) Higher education area: Redefining European education borders. Eurolimes [serial online]. Academic Search Complete, Ipswich, MA. [24 April 2012].

ComPress SA (2011) Jak na studiach dostosować się do rynku pracy (How to adapt to the job market during studies)? http://praca.gratka.pl/tresc/art/jak-na-studiach-dostosowac-sie-do-rynku-pracy-11044.html [20 March 2012].

Dolinabiolotechnologiczna (2011) Nowa Ustawa o Szkolnictwie Wyższym uderza w interesy nauczycieli akademickich (The New Act on Higher Education strikes against the well-being of teachers). http://dolinabiotechnologiczna.pl/polecamy/nowa-ustawa-o-szkolnictwie-wyzszym-uderza-w-interesy-nauczycieli-akademickich/ [20 April 2011].

Dudzik I. (2011) Polskie uczelnie nie wiedzą, dlaczego i po co kształcą (Polish higher schools do not know why and what for they teach). http://forsal.pl/artykuly/488347,polskie_uczelnie_nie_wiedza_dlaczego_i_po_co_ksztalca.html [20 April 2011].

Dunkel T. (2011) The Bologna process between structural convergence and institutional diversity. http://www.cedefop.europa.eu/etv/Upload/Information_resources/Bookshop/570/46_en_Dunkel.pdf [20 April 2011].

Ernst and Young, Instytut Badań nad Gospodarką Rynkową (2009) Strategia rozwoju szkolnictwa wyższego w Polsce do 2020 roku (A strategy of development of Polish higher education in Poland up to 2020). http://www.uczelnie2020.pl [03 May 2010].

Euractiv (2011) Lisbon agenda. http://www.euractiv.com/en/future-eu/lisbon-agenda/article-117510 [20 April 2011].

Gazeta Prawna (2010) Młodzi bezrobotni: nawet 3 miliony w 2011 roku (The young unemployed: Even 3 million in 2011). http://praca.gazetaprawna.pl/artykuly/447959,mlodzi_bezrobotni_nawet_3_miliony_w_2011_roku.html [20 April 2011].

GUS (1998) Szkoły wyższe w roku szkolnym 1996/97 (Higher schools in the school year 1996/97). GUS, Warszawa, 1998.

GUS (2010) Szkoły wyższe i ich finanse w 2009 r (Higher schools and their finances in 2009). http://www.stat.gov.pl/gus/5840_1177_PLK_HTML.htm [20 April 2011].

Kowalik T. (2009) www.polskatransformacja.pl (www.polishtransformation.pl). Warszawskie Wydawnictwo Literackie MUZA, Warszawa, 2009.

MEN (2000) Informacja na temat dostępności wyższego wykształcenia, dokument wewnętrzny MEN, opracowany na potrzeby Komisji ds. Pożyczek i Kredytów Studenckich na podstawie raportu przygotowanego przez ekpertówany UW pt. "Dostępność wyższego wykształcenia—materialne i społeczne uwarunkowania" (Information concerning the Commission on Student Loans and Credits, Based on reports by Warsaw University experts titled "Accessibility of higher education—material and social background").

MF (2011) Macroeconomic analysis of Polish economy. http://www.msp.gov.pl [20 April 2011].

MNiSW (2011) Tekst ujednolicony—Prawo o szkolnictwie wyższym (Consolidated text—Law on higher education). http://www.nauka.gov.pl/szkolnictwo-wyzsze/reforma-szkolnictwa-wyzszego/tekst-ujednolicony-prawo-o-szkolnictwie-wyzszym [20 April 2011].

OECD (2009) Education at a glance 2009—OECD indicators. http://www.oecd.org/dataoecd/1/28/43654482.pdf [20 April 2011].

PAP (2011) Deficyt budżetowy w 2010 r. niższy niż planowane 48,3 mld zł - źródło w MF (Budget deficit in 2010 lower than the planned 48.3 bn - source in MF). http://biznes.interia.pl/news/deficyt-budzetowy-w-2010-r-nizszy-niz-planowane-483-mld-zl,1580888 [20 April 2011].

Pépin L. (2011) Education in the Lisbon strategy: Assessment and prospects. *European Journal of Education* [serial online]. Academic Search Complete, Ipswich, MA. [24 April 2012].

Perspektywy (2011) Prezydent podpisał! Rewolucja na uczelniach od października (The President has signed! A revolution in higher schools from October). http://www.perspektywy.pl/index.php?option=com_content&task=view&id=3521&Itemid=0 [20 April 2011].

Polska 2030 (2011) Wyzwania rozwojowe (Development challenges). http://www.polska2030.pl [20 April 2011].

Sejm (2011) The Constitution of the Republic of Poland of 2nd April, 1997. http://www.sejm.gov.pl/prawo/konst/angielski/kon1.htm [20 April 2011].

Sitek B. (2010) Evaluation of the proposed reform of higher education in Poland against the background of sustainable development principles. *Human Resources: The Main Factor of Regional Development* [serial online]. Business Source Complete, Ipswich, MA. [24 April 2012].

Szapiro T. (2011) Proces Boloński: Nowe szanse, czy nieznane zagrożenia (The Bologna Process: New chances or unknown threats). http://akson.sgh.waw.pl/~tszapiro/zwiad/SzapiroBologna.pdf [20 April 2011].

Trading Economics (2011) Poland inflation rate. http://www.tradingeconomics.com/Economics/Inflation-CPI.aspx?Symbol=PLN [20 April 2011].

Wildowicz-Giegiel A. (2011) The relationship between intellectual capital and innovativeness of Polish economy. *Journal of US-China Public Administration* [serial online]. Academic Search Complete, Ipswich, MA. [24 April 2012].

Winkelmann-Gleed A. (2011) Demographic change and implications for workforce ageing in Europe. *Contemporary Readings in Law and Social Justice* [serial online]. Academic Search Complete, Ipswich, MA. [24 April 2012].

78 The Value System and Professional Aspirations of Polish University Graduates in Historical Perspective

Piotr Górski

CONTENTS

Throughout the last two years in Poland, and especially in the academic environment, there have been lively discussions on the condition of higher education (Strategia Reformy 2010). The Ministry of Science and Higher Education has prepared the project of new legislation concerning institutions of higher education. The analysis and propositions of solutions that were put forward indicated the significance of the following conditioning and civilizational changes:

- Globalization, demography, and socio-economical changes in Poland
- Law, public finances, cultural factors, and the preparation of graduates to face labor market challenges

While the strategy of higher education in Poland was being prepared, its connection with national and foreign strategic documents was being underscored. Legal and financial mechanisms were being created in order to lead to changes not only on the level of management of individual institutions of higher education, but also on the level of activities undertaken by candidates willing to pursue higher education (for instance concerning their preferences in faculty choice) as well as by students. These issues have been appropriately documented within the numerous reports ordered by the Ministry of Science and Higher Education that were uploaded to its official Web site (Reforma szkolnictwa wyższego).

In the following article, I propose to look at the issue of higher education from a slightly different angle, namely the angle of the youth who are currently studying. This aspect is usually considered in the context of the number of studying youngsters, higher education availability, and the relationship between current faculties and economic demand. In the approach that I am putting forward, I assume the thesis that meeting the economic requirement and student expectations is not the sole aim of an institution of higher education. It should also shape these expectations. Not only should it transfer knowledge and enable students to develop their skills, but also it should provide conditions enabling versatile development and prepare students to make responsible decisions. Apart from that, this perspective underscores the meaning of other figures participating in the process of preparing students to play their future professional specialist roles and the significance of employers and professional associations. It means that such categories as professional competencies and patterns and representations concerning professional roles will be useful in our analysis. I suggest looking at this issue from a historical perspective, which calls attention to the following conditions. First, higher education is connected with other components of social reality and with social structure due to the fact that during the last century it became the cradle of Polish intelligentsia. It is also connected with national culture and science (as it was influencing the creation and popularization of such values through research activities and education) and with economy (as despite the humanistic orientation of traditional Polish intelligentsia, the development of research activities and education were becoming more and more oriented toward the economic growth of the society). Second, the influence of institutions of higher education on the shaping of their graduates' social personality contributed to the formation of such types described by sociologists as "a well-bred person" or "a specialist-technocrat." I would like to highlight the factors that influence the content of these values and aspirations and indicate the tasks for the higher institutions resulting from this diagnosis. This is possible due to the analysis of the results of research on the value systems and professional aspirations of graduates. In this research, the alterations pertaining to the system of higher education and Polish society that occurred in this decade have been considered. In my deliberations, I will be referring to the research conducted among the students finishing their studies at the faculty of management at the Jagiellonian University (the research from 2007) (Górski 2009) and at the University of Science and Technology (the research from 2010) (Górski and Michniak 2011).

SOCIAL AND CULTURAL CONDITIONING OF THE ALTERATIONS OF HIGHER EDUCATION IN POLAND

When Poland regained its independence in 1918, there were six institutions of higher education functioning, including two (Polytechnic and University of Warsaw) that had been created in 1915 during the First World War. In 1937, there were 32 registered institutions of higher education. Their development was dominated by the schools specializing mainly in the area of humanities, jurisprudence, and natural sciences. The number of polytechnics, trade schools, and schools educating in terms of economy totaled to only nine. On the wane of the Second Polish Republic, there were 50,000 students being educated in all the types of high schools present in Poland, which amounted to 0.14% of the population. This evidence is proof of the fact that higher education was elitist. However, it did not differ much from the gross enrollment ratio on the same level in other European countries. According to contemporaries, the Polish educational system was characterized by the insufficient development of vocational and specialist schools. Not only did it result from the university traditions based on the idea of the university by Humboldt and Polish traditions of educating the intelligentsia, but also from the lack of financial resources. Educating lawyers and philologists was cheaper than educating physicians or engineers, and the labor market had many more employment opportunities for graduates at the time (Niezgoda 1993).

The model of "an intellectual" that Florian Znaniecki rightly characterized as that of "a well bred person" (Znaniecki 1974) was predominant in the environment of the educated in Polish culture of this period. The university degree, just as the secondary school certificate a bit earlier, was the prerequisite of becoming a member of the intelligentsia. It was the interwar secondary school that had a major influence on the shaping of social personality in the case of an intellectual. Intellectuals had to have general knowledge and be active in the area of culture. Professional aspirations were reflected by the willingness to perform the role appropriate to the social position of an intellectual; thus it was especially popular to be employed in a role connected with working in state institutions, freelancing, or journalism.

Despite their high social rank, the interwar intelligentsia experienced substantial deterioration of their financial condition, which was the outcome of the recession of the 1930s (Górski 2007).

The university professors were of the opinion that the quality of graduates completing their studies had worsened. Aleksander Hertz, a sociologist, underscored the fact that the intelligentsia lost their ability to perform leadership functions due to the growing popularity of bourgeois patterns among the youth studying at the institutions of higher education and due to the increasing antidemocratic tendencies.

The Second World War caused great population losses, which especially affected the intelligentsia. The political and economic alterations introduced by the pro-Moscow communists resulted in new concepts in the area of higher education. As a result, institutions of higher education developed, especially those educating engineers, economists and—in the area of humanities—teachers. Education was now financed by the state and became available to the youth of peasant and laboring origin that had been disfavored earlier. In the mid-1970s, when the generation of the post World War II baby boom was being educated at Polish institutions of higher education, the number of universities amounted to 89, and there were 468,129 students attending them (*Rocznik Statystyczny Szkolnictwa* 1976).

Despite the political and social changes, the image of the interwar intelligentsia was still common, and very often it was idealized by the youth and their parents. It was particularly evident when the differences between the level of general knowledge, manners, and cultural activity of the people educated in the interwar Poland and those educated in the People's Republic of Poland were being underscored. However, the socio-economic changes and the development of education at technical universities brought about the creation of a new model of an intellectual, the intellectual-specialist. This model was created as a result of reconciliation of certain elements of the socialist ideal (such as a sociocentric orientation and social relationship democratization) with the politics of the contemporary authorities that promoted accelerated economic growth, in which engineers and economists (viewed as technocrats) were supposed to achieve economic goals imposed by the authorities (Górski 1990).

The political and economic transformations at the beginning of the 1990s were followed by substantial changes in the sphere of higher education. The indicators of the net enrollment ratio increased from 9.8% to 35.3% in the years 1990 to 2004 whereas the gross indicators rose from 12.9% to 46.4%. In "Strategy for higher education development in Poland until 2010," which was written at the turn of this century, it was estimated that in 2010 the gross enrollment ratio would amount to 65% (*Strategia Rozwoju szkolnictwa wyższego w Polsce do* 2010). Although this indicator is lower than in the most developed countries of the world, it resulted in the substantial increase of the number of employees with higher education who were active in terms of their professional activities. It was especially characteristic of the employees who entered the labor market in the last decade. Another change concerned education profile. Not only do the researchers highlight the development of non-public schools, but they also turn our attention to the growth of the number of people studying at humanistic, social, and management faculties. These are the faculties that did not necessitate large expenditures, especially as the development of higher education was made possible due to the substantial use of non-budget resources. This educational structure is assessed negatively when confronted with the economic challenges. It is predicted that there will be a lack of specialists in the area of natural and technical sciences; moreover, students graduating from humanistic, social, and management faculties will face difficulties on the labor market (*Oferta szkolnictwa wyższego a wymagania rynku pracy*).

In 1990–1991, the number of students in Poland amounted to approximately 400,000 whereas in 2005 it amounted to approximately two million. This tremendous boost was connected with the increase in the number of the institutions of higher education—in 2005, there were more than 400 of them. This development consisted of the substantial growth of non-public schools (in terms of their number) and in the emergence of the institutions of higher education in the new urban areas that had not had these types of schools previously. The growth of number of the youth pursuing their studies was not only caused by the increase in terms of educational aspirations and in terms of higher education availability. Apart from that, it was the result of the conviction that the university diploma supports a young person in their efforts on the labor market and gives them a chance of getting better remuneration and fulfilling their professional aspirations (especially in the circumstances of the unemployment rate boosting). However, recently the number of studying young people has started to decrease, which is connected with the attainment of high value in terms of the gross enrollment index in the circumstances of a demographic low.

However, this colossal increase in the number of the institutions of higher education and the studying youth has brought about great differentiation in terms of diploma value. It has been reflected by the professional activity of their holders (their professional path and package altogether with the unemployment rate, which has grown for the university graduates this year). The number of the unemployed university graduates (even though it is still a small proportion of the unemployed) has recently risen almost 30%. In March 2011, merely 0.7% of all the employment offers ware made to the university graduates, and almost one third of these offers were, in fact, internship proposals (*Absolwenci zaczynają od bezrobocia* 2010). This deteriorating situation of the graduates is not only the result of the incongruity between education and labor market needs. It also resulted from the systematic knowledge level decrease and the lack of preparation of the people who pursue higher education. The vast majority of the institutions of higher education ceased to conduct any selection of candidates and started to accept everybody who would like to be educated in fear of the demographic low. Being more flexible in terms of requirements accompanied by the increase in the number of students, the insufficient number of research and teaching staff, and fewer resources spent on higher education led to the marked differentiation of education quality in the institutions of higher education.

A human being is an innovative factor in the economy. Thus, his or her efficiency in the labor market and feeling of satisfaction depends, first of all, on his or her level of competencies, attitudes, life orientation, and ability to make relevant and responsible decisions. In the next part of the article, I will present the value system and professional aspirations of university graduates.

THE VALUE SYSTEM AND PROFESSIONAL ASPIRATIONS OF UNIVERSITY GRADUATES

According to the research conducted by the Centre for Public Opinion Research, work was ranked as the second most important value (while family was the first one) in the hierarchy of values of Poles. The majority expect work that might provide them with a sense of security, and far fewer people are oriented toward a professional career, understood in terms of power and high salary. The research confirmed that the Poles ascribe great significance to education. Professional and educational aspirations were the strongest among the young respondents (18–24 years old) (*Aspiracje Polaków w latach 1998 i 2008* 2008).

The research on the educational aspirations of the Poles (*Czy warto się uczyć?* 2007) has shown that the chance to get a higher salary is seen as the reason why it is beneficial to decide to pursue education by the most numerous group of respondents (63%). Forty percent are attracted by an opportunity to have an interesting profession whereas 31% is of an opinion that the educated have “a better life.” Only 23% of the researched Poles claimed that pursuing education serves as a means of personal development and self-improvement. Every fifth respondent believed that being educated might help them to avoid unemployment (this belief was especially strong in the previous decade). However, the proportion of respondents sharing this motivation was two times smaller than in the research from 2004. A similar number of respondents (19%) claimed that the decision to pursue education might enable them to work abroad whereas the conviction that being educated might make it possible for them to set up their own business or work independently was shared by a slightly smaller number (14%). For 20% of the researched, striving to achieve recognition and the respect of others was the greatest motivation to be educated. This research has revealed that the common approval for the question *Is it beneficial to study?* (93%) is accompanied by quite different justifications. The level of education is an important variable differentiating the attitudes of Poles.

People with higher education most often point to an opportunity to perform an interesting role as their primary motivation (61% in this category of respondents), then to self-improvement and intellectual development (49%) and only then to high salary (45%).

The motive of "personal development" as a motivation to pursue higher education was most often chosen by the students I was researching (71.4% of respondents). The second one, highlighting the relationship between higher education and the fulfillment of professional aspirations, "achieving employment in the preferred profession," was mentioned by 59.6% of the researched. A little more than one third of the researched students finishing their studies at the Faculty of Management (AGH University of Science and Technology) associated their decision to pursue studies with the conviction that "in the present circumstances of competition on the labor market, studies enable acquiring skills facilitating success in rivalry with others" (36.2%). This motive was equally important with regards to pursuing education for the students of management in tourism that were researched in 2007. For about 25% of respondents from both groupings, the decision to choose a university faculty and specialization is connected with the realization of plans related with the willingness to run their own company sometime in the future or to be in charge of leadership positions. It seems to be quite understandable in the case of management students. Thus, on the one hand, the motivations pertaining to the decisions connected with university studies indicated by the research provides evidence for the strong internalization of higher education value in intrinsic categories with a strong orientation toward personal development. On the other hand, they bear witness to the necessity of acknowledging labor market challenges and the inevitability of rivalry with others in their professional strategy. What follows from this is that university studies may *enable* future professional achievements but not *guarantee* them as many years earlier when graduation itself was the source of social position and prestige.

It is possible to infer certain conclusions concerning the influence of institutions of higher education on the shaping of representations on performing future professional roles on the basis of the assessment of the usefulness of knowledge gained during university studies in future professional work. The vast majority of students from both universities expressed the conviction that this knowledge is useful with regards to their future professional career. It especially pertained to the expertise in the area of management and specialist professional competencies. Recently, the significance ascribed to language and computer skills has not been that substantial, which should not be interpreted in the categories of their decreasing importance. It rather means that it is common to have this skill set among university graduates. Its lack or inadequate level has become an obstacle hindering the fulfillment of professional plans unlike some time ago when having it was sufficient to get an attractive professional role. The importance of knowing foreign languages might be corroborated by the fact that every third respondent attended language courses, independently of participating in language classes at university.

The influence of higher education institutions might also be noticed in the representations students have concerning the professionalism of management (Górski 2008). In the analysis of the utterances of the students of tourism management, this influence consisted of pointing to the importance of knowledge in the context of competencies that a manager has and in a recognizing of the tasks they face and methods of their realization using the categories extracted from manuals on management and marketing. In these representations, awareness of competitiveness with regards to the environment in which a touristic enterprise functions and with regards to customer expectations is fairly distinct.

Treating university studies as a kind of activity enabling the building up of qualifications and competencies helpful in fulfilling professional plans is quite explicit when it comes to underscoring the significance of professional experience perceived as something that is most expected of employees by employers. As much as 62.7% of those researched asked to indicate what employee traits are particularly valued by an employer chose professional experience. Slightly less than half opted for communication skills and the ability to cooperate with others (42%). Then the choices included ethical features connected with an attitude toward professional duties, such as responsibility (38.2%) and integrity that was chosen by 30.7% of the respondents.

Apart from that, 77% of those researched indicated professional experience as a highly important factor that is crucial in the fulfillment of professional aspirations. These convictions were accompanied by undertaken activities as the vast majority of the researched students (91.3%) used to work during their studies, and 64.9% of them worked on a full-time basis. The declarations made by 40% of those researched show that taking up professional work was connected with the willingness to gain professional experience.

The researched students were aware of the significance of social competencies connected with communication skills, the ability to work in a team, or leadership. What is interesting with regards to these competencies, the difference between assessing their importance in the labor market and assessing the role of university in shaping them, is substantial. It clearly points to the necessity of considering creating such forms of student activity that would facilitate the development of these competencies. Similar conclusions concerning the research on assessing competencies of future managers from the point of view of management students and entrepreneurs results from Ryszard Walkowiak's analysis (Walkowiak 2007). I presume that the development of various competencies is not as influenced by the university curriculum as it is by student activity in associations of different kinds during their studies and by seizing the opportunity given by performing roles useful in future professional life. Of those researched, 10.6% declared that they were active in student organizations or other associations, which shows that student activity in this area is much less visible than their professional activity. The latter enables them to earn money and familiarize themselves with employee expectations whereas it does not give them an opportunity to play the role of leader and perform in front of greater audiences. In turn, this opportunity is offered by participating in various associations and student scientific movement.

Whereas in the declarations of students concerning their motivations for pursuing higher education, the desire for personal development is most prominent; in their answers to the question concerning expectations with regards to their future professional work, it is the desire for satisfactory remuneration and stable employment guarantee that dominates. Then there is the desire for professional development and desire for work enabling to show one's independence and providing respect. These expectations are correspondent with the expectations directed toward employers. The expectation of good relationships between employees is most often mentioned, and employer integrity has been indicated by 97.7% of those researched. The second group involves the expectations connected with employment guarantee, adequate remuneration, and appropriate work-life balance. Guaranteeing appropriate workload, possibility of professional development, independence, and opportunity of professional progression were mentioned less often.

If we notice that such employee traits as independence, enterprise, or innovativeness were less often enumerated as these expected by employers, then we get the image of graduates who interpret the information from the labor market as not so conducive to their development. The declared aspirations connected with personal development and the choice of university faculty enabling work in one's dream opportunity give way to the expectations of a sense of stability, and good interpersonal relationships are more valued than challenges enabling career growth. Assigning great significance to employer integrity means that people are convinced that one may be cheated by their employer. This is further corroborated by the frequent records concerning failing to provide timely salary and evading signing employee contracts, confirmed by employee complaints submitted to the National Labour Inspectorate (Państwowa Inspekcja Pracy). The majority of these complaints pertain to violation of rules regulating remuneration and employee documentation, for instance, failing to issue employment certificates (Błaszczak 2010).

The expectations of the respondents concerning remuneration turned out to be moderate. The minimum net wage acceptable for 40% of those researched amounts to 2000 PLN and to more than 3000 PLN for only 20% of them. A similar image of the pecuniary aspirations of university graduates may be inferred from the survey of "Gazeta Wyborcza." Every fifth respondent would accept any job offer even if it would not meet his expectations. Many would undertake work with a salary lower than the national average salary (*Niech młody nie znaczy zbędny* 2011). The analysis of professional aspirations of the researched students has shown that they are willing to face expectations that employers have. In turn, they expect a moderate salary and a sense of security in the working environment (meaning stability guaranteed by an employee contract and integrity of future employers), which is rather the testimony of their fears and lack of trust toward employers found to be quite common in the environment of young people. Work-life balance is also crucial for them. However, they declare their flexibility in terms of working hours, readiness to face various challenges posed by an employer, and willingness to work in a role not related to their university faculty. Their expectations are much less connected with social functions of a workplace, so characteristic of the generation of their parents.

The way in which young people interpret the stimuli coming from the labor market is not only expressed by their perception of employer expectations, but also by the perception of circumstances in which they will have to fulfill their professional goals and by assessing the chances of their realization. Almost half of the respondents chose the answer "*admittedly I will find some job, but it is hard to get appropriate employment*" when they were asked to assess these chances, which means that they are aware of the hardships in fulfilling the professional goals they are dreaming of. In order to point to the factors conditioning the success in realizing one's professional aspirations, 93.6% of them referred to being consistent in acting in such a way as to achieve one's goals and to the conviction that they expect being supported by people who are close to them (90.4%). Slightly less than a half (42.6%) were of an opinion that the success in realizing their professional goals will depend on acquaintances and connections. Every third person claims that they are dependent on the economic situation in the country whereas every fifth person considers the kind of university they graduated from to be advantageous. Thus, we may observe that work experience gained during university studies altogether with knowledge, consistency of acting, social competencies at one's disposal, and support from people close to one constitute the capital that, according to the respondents, would be a decisive factor when it comes to their future professional success.

Getting to know the professional aspirations of the respondents altogether with assessing the chances of their fulfillment helps us to highlight an increasing role of information coming from the labor market. It is this information that is being interpreted by the students and graduates and influences their behaviors in the labor market, both on the level of goals and of the conditions of their realization. From this perspective, they are trying to assess the input of a given institution of higher education to the development of their professional competencies. Even though the role of a university in shaping and developing social competencies has been described as not that significant, the majority of the respondents have expressed their satisfaction with regards to a university faculty they graduated from (86.0%) and declared that they would recommend it to their younger colleagues (82.0%). It proves, by all means, that the function of higher education has changed. It is now perceived as a precondition of realizing professional aspirations by the young Poles and their employers. However, lack of success in achieving these goals may lead to sense of frustration among graduates, and it proves the existence of the crisis of the basic assumption (so crucial for democratic ideologies of the 20th century), according to which social promotion was possible due to education.

CONCLUSION

Thus, what kinds of students are educated by Polish universities? In order to answer this question, journalists often reply that they are educating the future unemployed, frustrated people belonging to the lost generation. Apart from letters from disappointed graduates, press releases contain specialist advice pointing to the necessity of choosing another faculty or to starting professional work before making a decision on choosing a university faculty. Looking at these problems from a historical perspective, we may observe that, above all, the function of higher education in Poland has changed. It should be emphasized that not only has the number of universities (altogether with the number of people studying there) increased, but also some faculty changes have occurred, and above all, the relationships between higher education and the key subjects from its environment have altered. This situation has contributed to the change in graduate position both when it comes to social position and when it comes to the expectations connected with professional role realization. In the interwar period, graduating from an institution of higher education was equal to entering the social layer of intelligentsia whereas preparing oneself to perform a professional role was taking place in particular organizations. Being an intellectual was not only connected with one's status, but also with the realization of a particular cultural pattern. It was still attractive in the postwar period as despite the different political and social circumstances the model of an intellectual was still a point of reference for people graduating from university, and for many of them, it meant social promotion. In the system of planned economy, university faculties and the number of studying youth were meant to be adjusted to economic needs. The effect of this planning was often divergent from the assumptions, notwithstanding. It was one of the traits of this socio-economic political system. As a result, the number of people working in roles inconsistent with their education was increasing. This incongruity appeared due to the divergence between status elements: The social position of the intelligentsia often was not accompanied by appropriate financial situation.

The economic and social changes that followed after 1989 in Poland contributed to the emergence of a labor market with differentiated employee positions, depending not only on occupational prestige, salary, and work autonomy, but also on promotion chances and dangers of losing a job. This position was largely determined by the employee's level of education. On the one hand, the development of education was directed at education areas that were not addressed in the postwar period (especially those pertaining to the requirements of market economy), but on the other hand, it was connected with the development of private schools. These changes also concerned the revaluation of education—treating it as the capital consolidating the position of graduates in the labor market. This strategy is gradually becoming more and more fallible as first of all the education of large numbers of students is being conducted in faculties that are inadequate taking into consideration labor market demand, and second, education quality has deteriorated.

In effect, not only graduates are dissatisfied, but also employers and institutions of higher education are. The former are critical of what they receive in the teaching process and highlight especially the low practicality of education. Despite this, they are convinced that failure to graduate from university is a hindrance for their professional career, so it becomes the prerequisite of getting a satisfactory job. Employers are most often dissatisfied with the inadequacy of education with regards to their company needs. Very often, it is a result of Polish economic changes: the development of the services sector and emergence of vacancies in which higher education is not necessary. Employers complain about the lack of employees whereas employees—university graduates—complain that they did not study to work in the roles not necessitating higher education. In turn, institutions of higher education are dissatisfied with the worsening level of preparation in the case of applying candidates and with the necessity to educate large numbers of students, which is the prerequisite of being granted adequate subsidies by the Ministry of Science and Higher Education. The Ministry is trying to deal with this issue, in turn, by promoting the faculties that are ordered due to being adequate to the economic needs and by obliging institutions of higher education to monitor the fate of their graduates.

Such administrative solutions seem not to be sufficient, though. It is necessary to prepare and competently implement a strategy for individual institutions of higher education as well as a regional development strategy. There has to be some kind of rapprochement between schools, employers, and professional associations. I think that employers should be proactively engaged with the education of specialists by universities. First of all, it should pertain to curricula preparation, apprenticeship realization, and linking diploma subjects with the issues of an organization in which future specialists will find employment. These activities should be supported either by the ministry responsible for science and higher education or social policy altogether with associations, including economic associations. The educational profile, for instance, in the case of management faculty, should also be adjusted to trends and national or regional economic development strategy. It was Ryszard Rutka who highlighted this issue already in 2007 and pointed to the special needs of micro and small enterprises in the area of management. Their specificity meant that education in the area of management should have a complex character and should not purely focus on such specializations as finances, quality management, or marketing as it often happens. The second area that he referred to, equally crucial for the economy, is the one of executive staff in large companies. Their significance, especially in the conditions of the flattening of organizational structure, is increasing, which means that the demand for the development of their knowledge and competencies connected with organization and management is increasing as well (Rutka 2007).

However, we should bear in mind that the ultimate goal of education is the development of the young, who more and more often enter universities with vague expectations and convictions concerning the value that knowledge may have for them. Thus, it is the duty of the university and of professors and PhDs employed there to influence the attitude of students toward knowledge and to create appropriate conditions for their development. It means that emphasis needs to be put on the necessary activity of the young and on facing challenges that are posed by higher education. Such activities will not only be conducive to the development of professional competencies in the case of prospective specialists, but also to their responsibility and cognitive independence. They might enable them not only to meet labor market demands, but also to contribute with their creativity to the company where they will find employment.

REFERENCES

Górski P., *Postawy studentów wobec pracy i ich aspiracje życiowe a tradycja inteligencka*, (*Student's Attitude towards Work and their Life Aspirations in Intelligentsia Tradition*) "Zeszyty Naukowe AGH, Zagadnienia Społeczno-Filozoficzne" 1990, z. 33.

Górski P., *Problematyka inteligencji w pracach polskich socjologów okresu międzywojennego*, (*The topic of the intelligentsia in the Works of Polish sociologists in the Inter – war period*) "Kwartalnik Historii Nauki i Techniki" 2007 nr 3/4.

Górski P., *Profesjonalizm zarządzania. Wyobrażenia studentów i menedżerów*. (*Professionalism of management. The social representation of students and managers*). [w] Edukacja w gospodarce opartej na wiedzy. (*Education in the knowledge-based economy*) Pod redakcją Marii Fic. Uniwersytet Zielonogórski. Zielona Góra 2008.

Górski P., *Studenci zarządzania w turystyce o swoich studiach i aspiracjach zawodowych* (The Students of tourism management, their studies and professional aspirations). "Zarządzanie Publiczne" 2009 nr 1.

Górski P., Michniak J., *Absolwenci Wydziału Zarządzania AGH o swoich studiach i aspiracjach zawodowych*, (*Graduates of Management Faculty of the University of Science and Technology about their studies and professional aspirations*) "Zarządzanie Publiczne" 2010 nr 3.

Niezgoda M., *Oświata i procesy rozwoju społecznego: Przypadek Polski. Studium socjologiczne* (*Education and Processes of Social Growth: The Polish Case*), Wydawnictwo UJ Kraków 1993.

Rocznik Statystyczny szkolnictwa (*Statistical Yearbook of Education*), Główny Urząd Statystyczny, Warszawa 1976.

Rutka R., *Kształcenie kadr w szkołach wyższych dla potrzeb zarządzania. Kogo? Czego? Jak uczyć?* (*Universities Educating Future Managers. Who? What? How?*) "Współczesne Zarządzanie" 2007 nr 1.

Walkowiak R., *Kształcenie menedżerów w opinii studentów i przedsiębiorców*, (*Educating Managers opinions of students and entrepreneurs*). "Współczesne Zarządzanie" 2007 nr 2.

Znaniecki F., *Ludzie teraźniejsi a cywilizacja przyszłości*, (*Contemporary People and Civilization of Future*) Wydawnictwo Naukowe PWN Warszawa 1974.

INTERNET SOURCES

Absolwenci zaczynają od bezrobocia (*Graduates start from unemployment*) "Gazeta Prawna" z 28. 04. 2010 http://praca.gazetaprawna.pl/artykuly/416976,absolwenci_zaczynaja_od_bezrobocia.html [27 April 2011].

Boguszewski R., *Aspiracje Polaków w latach 1998 i 2008*, (2008), (*Aspirations of Poles in the years 1998 – 2008*). Komunikat z badań, Warszawa, http://www.cbos.pl/SPISKOM.POL/2008/K_155_08.PDF [27 April 2011].

Błaszczak A., *Przybywa skarg pracowników* (*Complaints of employees grow*) "Rzeczpospolita" z 17. 11. 2010 http://www.rp.pl/artykul/564994.html [27 April 2011].

Szczepańska J., *Czy warto się uczyć?*, (Is it worth to study?) (2007), Komunikat z badań, Warszawa, http://www.cbos.pl/SPISKOM.POL/2007/K_072_07 [27 April 2011].

Niech młody nie znaczy zbędny, (*Let the young doesn't mean not needed*) "Gazeta Wyborcza" z 24. 03. 2011 http://wyborcza.pl/1,86116,9304603,Niech_mlody_nie_znaczy_zbedny.html [27 April 2011].

Oferta szkolnictwa wyższego a wymagania rynku pracy, (*The offer of higher education and the requirements of the labor market*) http://www.frg.org.pl/page=zakończone_projekty_isw&projekt=34 [27 April 2011].

Reforma szkolnictwa wyższego, (*The reform of higher education*) http://www.nauka.gov.pl/szkolnictwo-wyzsze/reforma-szkolnictwa-wyzszego/ [27 April 2011].

Strategia Reformy (Strategy of Reform) "Forum Akademickie" 2010 http://forumakademickie.pl/strategia-reformy/ [27 April 2011].

Strategia Rozwoju szkolnictwa wyższego w Polsce do 2010, (*Development Strategy* for the *Higher Education* System in *Poland* up to the Year 2010) http://www.mimuw.edu.pl/~sjack/usw/strat_pliki/strategia.htm [27 April 2011].

79 Experimenting with Teaching Contexts

Maria L. Ekiel-Jeżewska

CONTENTS

INTRODUCTION

This story begins like many others: "Once upon a time, there were students and their teacher." Tell me what happened later; did they all learn happily ever after? In this paper, we are discussing examples of the processes that could help to make such a happy ending probable.

The literacy surveys, which focus on young people's ability to use their knowledge and skills to meet real-life challenges (PISA 2009), indicate that a deep educational change is necessary. New teaching methods at schools to raise students' interest in science and increase their skills, together with actions to improve the quality of teaching and support teacher's motivation, are called for (Rocard 2007), such as the inquiry-based science education and problem-based learning (La main à la pâte).

Searching for a sustainable improvement, it is good to remember that active educational methods, with hands on experience and learning by inquiry, have been investigated, promoted, and implemented in numerous contexts for more than a hundred years. Examples could be the heuristic educational method, described and applied a long time ago, see e.g., Dewey (1899), Armstrong (1903), and Smoluchowski (1917), and contemporary research on science education, e.g., by McDermott et al. (1996). In contrast to their quality, the learning processes described in these examples have not disseminated. Smoluchowski's scientific papers have thousands of citations and are widely known while his guide for self-study in science has not been translated and is hardly available even in Polish. There are many missing links in reforming education (Sarason 1990), other than inquiry-based or problem-based learning methods. One of them is the following.

In science, we intensively and continuously compare with peers the results of our research by discussing, reading, and writing articles and watching and preparing conference and Internet presentations. We benefit significantly from such activities; they help us to construct and redesign scientific models. But we practically do not exchange ideas about how we teach and learn, how we measure the productivity of this process, and how we estimate a sustainable increase of knowledge or development of lifelong skills. However, sharing, discussing, and evaluating our teaching experience is a prerequisite of a successful educational redesign.

In this paper, an example of a teaching context is discussed: a course and training programs on micro-hydrodynamics for undergraduate and graduate students.

TEACHING MICRO-HYDRODYNAMICS

Students, Framework, and Context

In both semesters of the academic year 2009–2010, I was guiding a course on micro-hydrodynamics, focused on the motion of micro-particles in fluid flows. It was a part of the educational program organized by the Institute of Fundamental Technological Research for its Ph.D. students and available also for graduate and undergraduate students from other institutions. The participants were Ph.D. students from various institutes of the Polish Academy of Sciences and undergraduate students at the Faculty of Physics, University of Warsaw. The latter asked me to teach them, and this is how the course took place. The teacher and the students participated in the course as the result of their own choice with no curriculum specified at the beginning and enough freedom to adjust the taught subject and level to the interests and skills of the learners. This, however, was difficult because their mathematical background, basic knowledge of hydrodynamics, interests, and activities differed significantly from person to person. Table 79.1 provides statistical data about the course.

The tablet notes, written and displayed during the lectures, are available at the Web site (Micro-hydrodynamics). To receive credits for the course, the students were asked to prepare "something" useful to guide the next generations of students who will learn micro-hydrodynamics. They were asked to choose the form and method of their projects in agreement with their own feeling of how the teaching should look like. Some examples were provided, such as the existing multimedia on the fluid dynamics, e.g., Taylor (1967) and Homsy (2008), in which basic concepts and links between them are explained with the use of pictures, graphs, and movies.

From the beginning of the course, the participants accepted the idea that the results of their individualized projects and cooperative task-oriented work would be shown on a special Web site and that they would be involved in designing this Web site. The presentations prepared by the students were collected and displayed in section "Subject." The other section, "Learning," shows the educational context of the course, and it serves the less obvious goal: sharing our teaching and learning experience with others and passing to the students of the course the message that discussion about such an experience is meaningful.

TABLE 79.1
Basic Information about the Course on Micro-Hydrodynamics, 2009–2010

			Number of Participants				
	Number of Hours	Number of Lectures	Present at More than One Lecture	Present at More than Five Lectures	Average per Lecture	Guiding Science Picnic	Completed with Credits
First semester	30	9	17	12	12	N/A	11
Second semester	45	9	7	5	5	11	12

There were two types of outcomes from the students' projects: reports and experimental modules, and both will be discussed later in separate sections. The experimental modules included the whole day of teaching hundreds of children at the Science Picnic in Warsaw in 2010, helping them to make experiments on their own and to discover properties of viscous fluids. It also meant hours of designing the sequence of experiments and completing the equipment. The course in 2009–2010 was followed by training programs and other Science Picnic activities in 2011. During two years, 35 projects were completed by 24 undergraduate and graduate students. Eleven reports were prepared during the first semester of the course, one in the second semester and five during training programs. Eleven students guided children at the Science Picnic 2010 and seven during the Science Picnic 2011.

Learning Principles

The motivation for this project came from a few basic learning principles. Students learn productively by teaching (Wilson and Davis 1994). The art of storytelling is significant; a good teacher helps students to make their stories attractive. We saw an example of such story told by Maria Siemionow at the conference Science, Technology, Higher Education, and Society in the Conceptual Age (STHESCA), 2011.

It is important to create an image of the concepts learned, i.e., visually think (Arnheim 1969). This image is needed to self-correct mistakes; to develop a better understanding; and in this way, to help how to progress (Dewey 1899; Schaefer-Simmern 1961). An example of productive interaction between a teacher and a seven-year-old child was illustrated by Dewey (1899) who showed the child's progress in drawing; see Figure 79.1. The teacher encouraged the child to look closely at trees and compare those seen with the one drawn and to draw trees from observation and, in this way, involved him in a new process, essential for learning. A detailed analysis of how to teach art and measure progress in learning was performed, described, and shown by Henry Schaefer-Simmern (1961).

The essential aspects of productive learning are also student-centered teaching (Dewey 1899; Sarason 1995; Głazek and Sarason 2007), the individual autonomy and the group self-governance (Jobert 1941; Korczak 1967; Piaget 1948; Drucker 1966), and productive thinking (Wertheimer 1959). Recognition of outcomes and skills is essential (Werquin 2010).

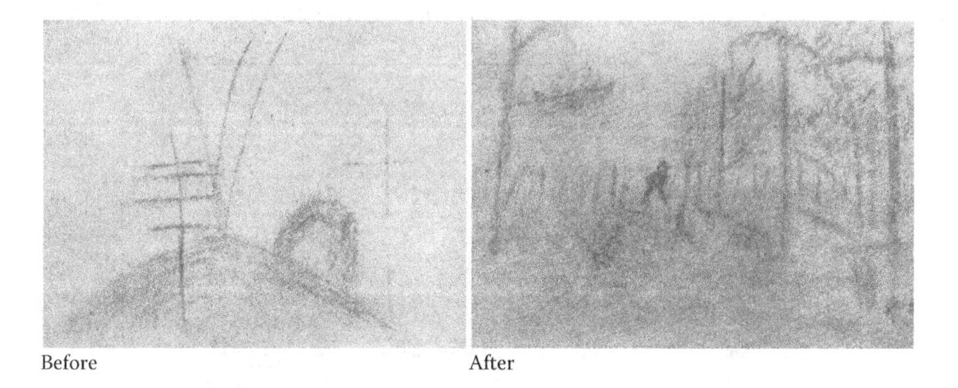

FIGURE 79.1 Influence of interaction with teacher on the progress in drawing of a seven-year-old child. Left: Child's drawing of a cave and trees. Right: Child's drawing of a forest. (J. Dewey, *The School and the Life of the Child*, Chapter 2 in *The School and Society*, University of Chicago Press, Illinois, 1899.)

RESULTS

Reports

For graduation at the end of the first semester, the students were asked to prepare a short report, which could be used by others to learn basic principles and features of micro-hydrodynamics, discussed during the lectures. The chosen form and topic were supposed to reflect the way and the content the students would be willing to learn themselves.

The students designed learning modules of the following types (the corresponding number of reports is indicated in parentheses):

i. Typing by a single person the lecture content from the teacher's tablet hand-written notes (2)
ii. Report prepared by a single student on a chosen subject, related to the lecture content, based also on other resources (3)
iii. Short article about own research, prepared by a single student or a pair of them (3)
iv. Presentation of important concepts from the lecture in a concise and visually attractive form, prepared by a single student in cooperation with the teacher (4)

The reports can be found on our Web site (Micro-hydrodynamics). Each of them is on a different subject. They provide interesting information about the students' initial point of view on a typical educational pattern; for most of them, personal relationships seemed to be insignificant in learning.

A characteristic feature of the types i–ii is that the students prepared their reports separately from each other without a personal contact with the teacher and in absence of any learners, who would read the report and give feedback. An example of type ii is shown in Figure 79.2.

The Ph.D. students involved in the team research projects prepared two reports (type iii) cooperating in pairs. Their discussion with the teacher was limited to the last stage of writing, when they were asked to include in the report a missing connection between their work and the basic principles discussed during the lecture.

The reports i–iii were finished in the first semester. The authors of the modules i and iii did not continue in the second semester, and it seems to correlate with the reports, which display their mental separation from the course. In contrast, the students from group ii involved in the second part of the course in a similar way as they had involved in active problem solving in their reports in a close connection with (and interest in) a subject discussed during the course.

The reports of type iv were created at the end of the second semester or even much later as the result of a very intensive, demanding, and rewarding cooperation between a student and the teacher.

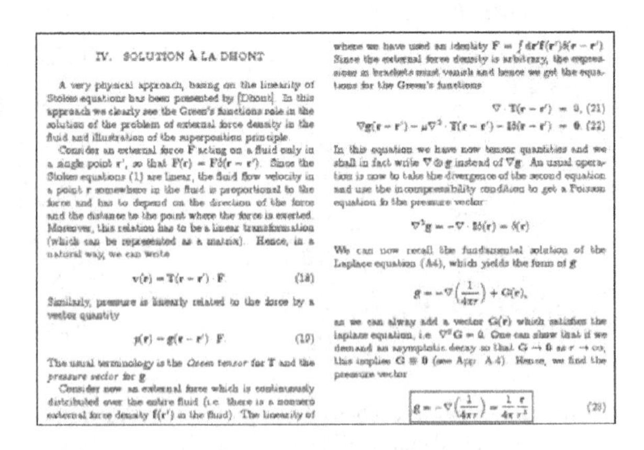

IV. SOLUTION À LA DHONT

A very physical approach, basing on the linearity of Stokes equations has been presented by [Dhont]. In this approach we clearly see the Green's functions role in the solution of the problem of external force density in the fluid and illustration of the superposition principle.

Consider an external force $\mathbf{F}$ acting on a fluid only in a single point $\mathbf{r}'$, so that $\mathbf{F}(\mathbf{r}) = \mathbf{F}\delta(\mathbf{r}-\mathbf{r}')$. Since the Stokes equations (1) are linear, the fluid flow velocity in a point $\mathbf{r}$ somewhere in the fluid is proportional to the force and has to depend on the direction of the force and the distance to the point where the force is exerted. Moreover, this relation has to be a linear transformation (which can be represented as a matrix). Hence, in a natural way, we can write

$$\mathbf{v}(\mathbf{r}) = \mathbb{T}(\mathbf{r}-\mathbf{r}') \cdot \mathbf{F} \qquad (18)$$

Similarly, pressure is linearly related to the force by a vector quantity

$$p(\mathbf{r}) = \mathbf{g}(\mathbf{r}-\mathbf{r}') \cdot \mathbf{F} \qquad (19)$$

The usual terminology is the *Oseen tensor* for $\mathbb{T}$ and the *pressure vector* for $\mathbf{g}$.

Consider now an external force which is continuously distributed over the entire fluid (i.e. there is a nonzero external force density $\mathbf{f}(\mathbf{r}')$ in the fluid). The linearity of

where we have used an identity $\mathbf{F} = \int d\mathbf{r}'\mathbf{f}(\mathbf{r}')\delta(\mathbf{r}-\mathbf{r}')$. Since the external force density is arbitrary, the expressions in brackets must vanish and hence we get the equations for the Green's functions

$$\nabla \cdot \mathbb{T}(\mathbf{r}-\mathbf{r}') = 0, \quad (21)$$

$$\nabla \mathbf{g}(\mathbf{r}-\mathbf{r}') - \mu\nabla^2 \cdot \mathbb{T}(\mathbf{r}-\mathbf{r}') - \mathbb{I}\delta(\mathbf{r}-\mathbf{r}') = \mathbf{0}. \quad (22)$$

In this equation we have now tensor quantities and we shall in fact write $\nabla \otimes \mathbf{g}$ instead of $\nabla \mathbf{g}$. An usual operation is now to take the divergence of the second equation and use the incompressibility condition to get a Poisson equation fo the pressure vector

$$\nabla^2 \mathbf{g} = -\nabla \cdot \mathbb{I}\delta(\mathbf{r}) = \delta(\mathbf{r})$$

We can now recall the fundamental solution of the Laplace equation (A4), which yields the form of $\mathbf{g}$

$$\mathbf{g} = -\nabla\left(\frac{1}{4\pi r}\right) + \mathbf{G}(\mathbf{r}),$$

as we can alway add a vector $\mathbf{G}(\mathbf{r})$ which satisfies the laplace equation, i.e. $\nabla^2\mathbf{G} = 0$. One can show that if we demand an asymptotic decay so that $\mathbf{G} \to 0$ as $r \to \infty$, this implies $\mathbf{G} \equiv \mathbf{0}$ (see App. A.4). Hence, we find the pressure vector

$$\boxed{\mathbf{g} = -\nabla\left(\frac{1}{4\pi r}\right) = \frac{1}{4\pi}\frac{\mathbf{r}}{r^3}} \qquad (23)$$

FIGURE 79.2 An excerpt from a report (type ii) prepared by a student at the beginning of the course. (From Micro-hydrodynamics, http://hydro.ippt.pan.pl.)

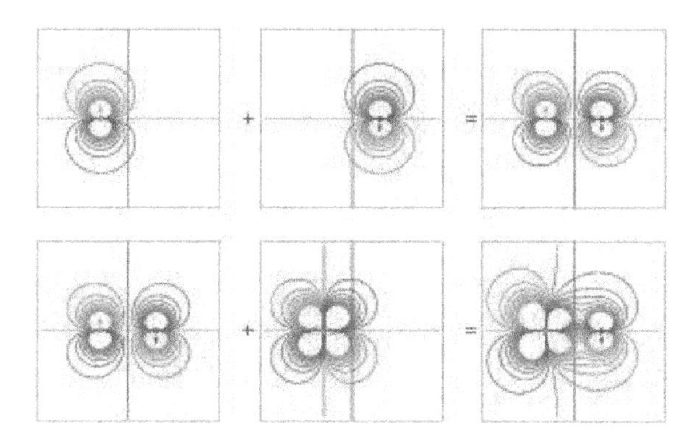

FIGURE 79.3 A page from another report (type iv) prepared by the same student, in cooperation with the teacher, at the end of the course. (From Micro-hydrodynamics, http://hydro.ippt.pan.pl.)

The students were involved in a productive thinking process (Wertheimer 1959), which included understanding solutions of complex equations by constructing them from basic elements and "translating" them into a graphical representation (Arnheim 1969). According to Arnheim, "The clarification of visual forms and their organization in integrated patterns as well as the attribution of such forms to suitable objects is one of the most effective training grounds of the young mind." An example of the outcome is plotted in Figure 79.3, which shows the fluid pressure field generated by a point particle settling under gravity close to a vertical wall.

A comparison of Figures 79.2 and 79.3 illustrates the student's progress in constructing a more complex and more realistic image of teaching and learning in a similar way as one can observe progress in constructing a more complex and more realistic image of an object in a process of unfolding the artistic activity, compared with pairs of Figures 79.1 and 79.2 in Dewey (1899), reproduced as Figure 79.1 in this paper and figures 41 and 51 in (Schaefer-Simmern 1961).

A similar progress was observed when the Web site was constructed. The pictures and plots created by the students were, at the beginning, put on a Web page at random; then they slowly got organized in patterns and focused to pass the main message as clearly as possible. Compare, for example, the first and the present versions of the home page in Figures 79.4 and 79.5. The present version (Micro-hydrodynamics) is more clear and better fitted to present the main goal

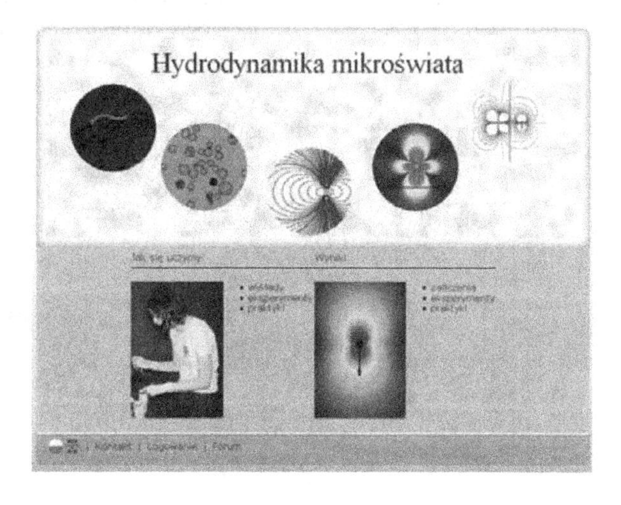

FIGURE 79.4 First version of our home page. (From Micro-hydrodynamics, http://hydro.ippt.pan.pl.)

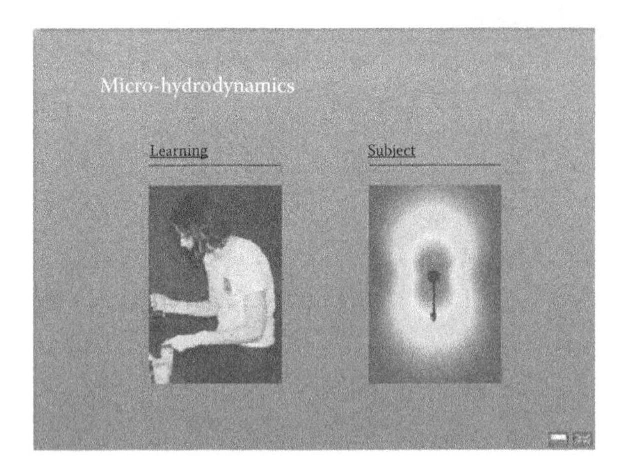

FIGURE 79.5 Present version of our home page. (From Micro-hydrodynamics, http://hydro.ippt.pan.pl.)

of the Web site: showing the learning process in addition to showing concepts and principles of micro-hydrodynamics.

Experimental Modules

Readers of the reports on the Web site usually learn in isolation. In the second semester of the course, we focused on creating a context of learning with personal contacts teachers-teachers, teachers-learners, and learners-learners. In connection with the lectures, we designed and prepared experimental inquiry-based modules for children and non-expert adults and applied them at the 14th Science Picnic in Warsaw (Science Picnic). A year later, a similar activity was repeated.

In a sequence of interrelated experiments, we guided the Science Picnic participants to understand the concept of viscosity, discover the differences between the micro and macro motion in fluids, and learn how is it possible to see the laws of micro-hydrodynamics by a naked eye (Micro-hydrodynamics). In Figure 79.6, we illustrate how students learned by teaching children, and in Figure 79.7, we show the model of "a swimming bacteria" (Taylor 1967) to give an idea of what can be found at the "subject" part of our Web site (Micro-hydrodynamics).

Behavior of children at the Science Picnic is illustrated in Figures 79.8 through 79.14. At the beginning, they stayed at a distance (Figure 79.8). Then, they get fascinated by the story told and shown by the students (Figure 79.9). They heard, "Do touch everything!" and started to experiment

FIGURE 79.6 Learning by teaching children.

FIGURE 79.7 The model of a swimming bacteria.

(Figures 79.10 and 79.11). Personal contact between children and students mattered to all of them (Figures 79.12 and 79.13).

In the project, the focus was on the learning process of the students, not the children. Photographs illustrate what the students did see and how they committed to teach.

The goal of designing and guiding the experimental modules was to involve graduate and undergraduate students in learning by teaching. Their learning process had two different meanings. While explaining the essential concepts and laws of micro-hydrodynamics, which I taught during the course, the students could fill in the gaps in their own understanding of physics. They did it by asking questions, reading, or their own thinking. They could also discriminate between productive thinking and superficial opinions and overcome false intuition by performing experiments.

FIGURE 79.8 Staying far away.

FIGURE 79.9 Getting fascinated.

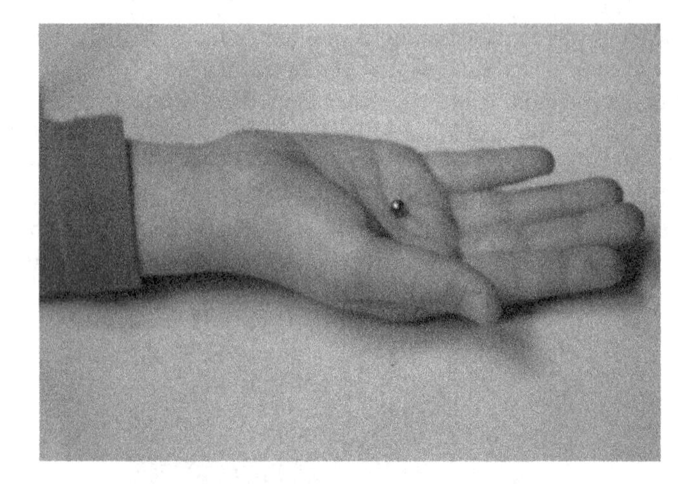

FIGURE 79.10 Involving.

FIGURE 79.11 Experimenting.

FIGURE 79.12 Being helped while experimenting.

FIGURE 79.13 Interacting with teachers.

Moreover, they had a chance to apply and develop the new teaching methods and principles used during the course. They had a personal contact with their children students and could sense what does it mean to teach students in contrast to teaching subject matter (Dewey 1899; Sarason 1995; Głazek and Sarason 2007). In addition, they involved in cooperation and the teamwork, improved their management and communication skills, and developed internal motivation to teach and improve teaching in future.

Evaluation

Evaluation of the learning efficiency was based on individual observations and discussions during the lectures and the accompanied activities, watching the design and the guidance of the experimental modules, and the content of the students' reports. Learning by preparing a report on microhydrodynamics has been analyzed in an earlier section. The team activities at the Science Picnic have been self-evaluated by the instructors in a sequence of reports, which are also available at the Web site (Micro-hydrodynamics). A more efficient process to self-correct misconceptions before passing them to children should be developed. In future activities, a quantitative observation survey (Clay 1993) of the learning process would be valuable.

FIGURE 79.14 Pleasure of teaching.

Could students be involved in redesigning school education on a large scale? Students participating in the Science Picnic 2011 were asked if they would be willing to volunteer in teaching children again in a similar way and why. They answered yes and pointed out their pleasure in helping children to understand some aspects of science. Moreover, they were convinced that such a learning pattern, typically absent in schools, is very important for children and for the whole society. This motivation, based on values, indicates that undergraduate and graduate students could significantly contribute to creating productive learning contexts in schools if such activities were included in their standard education (La main à la pâte) and were guided by faculty experienced in developing efficient methods of self-correcting teaching habits.

CONCLUSION

To involve each student in lifelong productive learning is a challenge for the modern society. The scientific community could contribute to this goal by, among other activities, experimenting with productive teaching contexts at the level of higher education and sharing the outcomes. Collecting, exchanging, discussing, and evaluating different teaching and learning methods is important for their global improvement in the modern society. Such a process is absent in the current educational system. We point out that it can be initiated from small-scale activities at the level of higher education. Scientific institutions that provide a framework to experiment freely with teaching and to disseminate the results, together with researchers who perform such experiments and evaluate, analyze, and exchange their outcomes, can play a significant role in supporting educational innovation. They can help to create contexts of productive learning at all levels if there is a significant number of them and if they actively cooperate or compete and use a scientific approach to their teaching and learning experiments.

A learning process of graduate and undergraduate students during a course and training programs of micro-hydrodynamics has been outlined in this article and shown on an accompanied Web site. The participants were involved in learning by teaching. Some of them were supposed to write a learning module for the Web site on a subject related to the lecture content. There were no other constraints on the form and topic. Others had a chance to prepare and guide inquiry-based modules for children and non-expert adults who wanted to learn the basic concepts of micro-hydrodynamics by making experiments; these students were deeply involved in the teaching process and declared their commitment to volunteer again in a similar activity. While interacting with their pupils, they judged that the learning process is productive and of a great value for children and for the society.

There are two main conclusions following from this project. First, it seems that undergraduate and graduate students consider redesigning education in a productive way as a very important goal, and therefore they could contribute significantly to such changes. Second, the results of the project illustrate that individual learning and teaching processes could be described, collected, displayed, discussed, and compared with other examples and analyzed and redesigned. The contexts of lifelong productive learning hinge on telling and listening to stories about how we teach and learn and on evaluation of the outcomes.

ACKNOWLEDGMENTS

I thank A. Myłyk for cooperation in constructing the Web site and the participants of the course and of the later training programs, M. Lisicki, A. Myłyk, R. Boniecki, M. Borkowska, K. Churski, K. Dudyński, J. Jaruszewicz, M. Karpiński, D. Lamparska, D. Milej, P. Kocieniewski, M. Oszmaniec, M. Pyzalska, B. Szewczak, M. Sikora, A. Słowicka, A. Strzałkowski, R. Szczypiński, K. Wędołowski, and N. Wójtowicz, for their contribution to the content of the Web site and numerous discussions. Figure credits: M. Lisicki (Figures 79.2 and 79.3), I. Z. Ekiel (Figure 79.6) and myself (Figures 79.7 through 79.14).

REFERENCES

Armstrong, H. E. *Teaching of Scientific Method and Others Papers on Education*, London, Macmillan, 1903, p. 471.

Arnheim, R. *Visual Thinking*, University of California Press, Berkeley, 1969.

Clay, M. M. *An Observation Survey of Early Literacy Achievement*, Heinemann, Auckland, 1993.

Dewey, J. *The School and the Life of the Child*, Chapter 2 in *The School and Society*, University of Chicago Press, Illinois, 1899 (not in copyright), digitized at http://www.archive.org/details/schoolsociety00deweiala.

Drucker, P. F. *The Effective Executive*, Harper & Row, New York, 1966.

Głazek, S. D. and Sarason, S. B. *Productive Learning: Science, Art, and Einstein's Relativity in Educational Reform*, Corwin Press, Thousand Oaks, 2007.

Homsy, G. M. (ed.), *Multimedia Fluid Mechanics*, DVD-ROM Second Edition, Cambridge University Press, Cambridge, 2008.

Jobert, A. *La Commission d'éducation nationale en Pologne: 1773–1794, son oeuvre d'instruction civique*, (Commission of the National Education in Poland: 1773–1794, Its Work of Civic Instruction), Les Belles lettres, Paris, 1941, in French.

Korczak, J. The children's home, in: *Selected Works of Janusz Korczak*, published for the National Science Foundation, Warsaw, 1967, http://www.januszkorczak.ca/legacy.

La main à la pâte, (Hands On), http://lamap.inrp.fr/.

McDermott, L. C. et al., *Physics by Inquiry*, Wiley, 1996.

Micro-hydrodynamics, http://hydro.ippt.pan.pl.

The OECD Programme for International Student Assessment – PISA 2009 Results: Executive Summary, OECD 2010, http://www.oecd.org/edu/pisa/2009.

Piaget, J. *To understand is to invent: The future of education*, Grossman, 1973 (English translation of *Où va l'éducation?* UNESCO, Denoël/Gonthier, Paris, 1948).

Rocard, M. *A Renewed Pedagogy for the Future of Europe*, 2007, http://ec.europa.eu/research/science-society/document_library/pdf_06/report-rocard-on-science-education_en.pdf.

Sarason, S. B. *The predictable failure of educational reform*, Jossey-Bass Publishers, San Francisco, 1990.
Sarason, S. B. *School Change: The Personal Development of a Point of View*, Teachers College Press, New York, 1995.
Schaefer-Simmern, H. *The Unfolding of Artistic Activity*, University of California Press, Berkeley, CA, 1961.
Science Picnic in Warsaw, http://www.pikniknaukowy.pl/standard/en/.
Smoluchowski, M. *Poradnik dla samouków*, Wydawnictwo A. Heflicha i St. Michalskiego, Warszawa, 1917, in Polish (a guide for self-study in science, including a resource on physics education at all levels).
Taylor, G. I. *Low Reynolds number flows*, Encyclopaedia Britannica Educational Corp., Chicago, 1985, video N 21617 (originally a colour sound film, Educational Services Inc., 1967), http://web.mit.edu/hml/ncfmf.html.
Werquin, P. *Recognising Non-Formal and Informal Learning: Outcomes, Policies and Practices,* OECD, Paris, 2010.
Wertheimer, M. *Productive Thinking*, Enlarged Edition, Harper & Row, New York, 1959.
Wilson, K. G. and Daviss, B. *Redesigning Education*, Teachers College Press, New York, 1994.

80 New ISO 29990:2010 as Value Added to Non-Formal Education Organizations in the Future

Jürgen Heene and Lilianna Jodkowska

CONTENTS

INTRODUCTION

The education system is an important part of social and economic policy in that it leads to professional workers being trained for the labor market, ideally so that the demand for work and the availability of work are offset against each other. For this reason, continued education and/or further training should be made available to all age groups.

Expanding and supporting of education throughout life is the declared goal of the International Labor Organization, which sees in the acquisition of knowledge in the course of education and training the possibility of acquiring new skills ("promote access to education, training and lifelong learning for people with nationally identified special needs") (*The new ILO Recommendation* 2006).

Attempts at creating a uniform international quality management system for educational organizations have been undertaken for a number of years. In the meantime, there have existed in Germany many various national quality management systems (QMS), created for educational organizations. They were and still are prevalent in varying degrees in a given country and offered by different certification organizations.

In the past years, among others, the European Community and other international organizations paid more attention and made endeavors to assure quality in education. Here may be counted among other things the Copenhagen Process, The European Network on Quality Assurance in Vocational Education and Training (ENQA-VET), a program for assuring quality in vocational education and training (EQARF), educational conventions aimed at leading to integration and inclusion, and ISCED UNESCO (types of schools are classified and characterized by the International Standard Classification of Education of 1997), PISA research (program of international assessment in learning) Organization for Economic Co-operation and Development (OECD); specialist standards, e-learning, ISO standards. The World Trade Organization (WTO) in particular was preoccupied with this problem within the framework of General Agreement on Trade in Services (GATS) round and pointed out the necessity of establishing uniform standards. In order to standardize systems of education quality certification, in 2007 a new international standard was prepared in the forefront there being the creation of a generic quality model for occupational practice, rendering services and creating common references for organizations rendering teaching services and their customers for planning, working out, and conducting education and further education (cf. Chapter "Introduction" DIN ISO 29990, 2010). Organizations providing non-formal education have consistently criticized the existing system of ISO 9001 in its 2001 version and subsequent one of 2008 as not adapted to the requirements of this specific product, which is learning, and not including the processes of learning service realization sufficiently.

The aim of the study is to present selected quality management systems (QMS) in organizations offering various forms of training and learning services in Germany having been used so far. The first part of the study presents selected QMSs currently subject to certification in Germany; the second part describes the process of preparing the new standard ISO 29990 for learning services. The third part of the study concentrates on the requirements of the standard itself; expected advantages for the actors of learning services market have been gathered in the final remarks.

EDUCATION CERTIFICATION IN GERMANY

The certification landscape in Germany developed over years. There are a lot of organizations offering various certificates or quality marks. This variety corresponds to various gravity points and to diverse trends of certifying organizations deeming certain criteria more important than others; at the same time, this makes certifications incomparable due to different requirements. One of the typical German problems is the fact that the policy of education, and so the rules concerning it, are left to the laws of individual states and the more difficult it is to introduce a uniform quality management system throughout the country.

Table 80.1 presents selected QMSs. They are distributed in various ways over the country, and also, on an international scale, other QMSs are functioning. The number (about 20 in Germany) and the variety of systems reflect the market's needs, and they are not a problem by themselves; a problem is posed by the comparability of the systems, due to their specificity and different requirements. Problems are created by the variety of the systems when a decision is being made by the potential customer about choosing an organization that offers education. In this context, a new international system provides a stable basis for the comparability of education services. (The number of organizations that have opted for QMS according to ISO 9001 standard is growing continuously from approximately 40,000 organizations in 1993 in 30 European countries to about half a million in 2009 in 50 countries, followed by the number of certificates granted.) Unfortunately, these statistics do not take into consideration the kind of an organization's activity; these are accumulated data. In spite of all, even in the past years of the financial and economic crisis, the annual growth in percentage equaled 4%, 6%, and 10%, which are analogous data for the years 2007 to 2009 (cf. ISO 9001 – Quality management systems – Requirements, http://www.iso.org/iso/catalogue_detail?-csnumber=46486).

It is worthwhile noting a few peculiar characteristic features of the systems mentioned in Table 80.1. System LQW 3 was prepared for the needs of the organization with the process of educating the person of the learner in the center of attention; it puts strong stress on the processes taking place during learning and teaching, their quality (as the central value), and on the emergence of the "learning organization" (Jodkowska, Szczecin 2008). This system is most widely distributed in the land of Lower Saxony (Niedersachsen) (in the phase of initiation, it was supported by means from the target fund). In Germany, 476 organizations possess this certificate; in Austria, there are 82 (data from May 2011) (http://www.artset-lqw.de/cms/index.php?id=organisationen). In 2008, attempts were made to initiate LQW in Poland and some other new EU member states. In Poland, the attempts have not been successful so far, and it may be assumed that the new ISO standard will considerably restrict any other initiatives.

A specific German QMS is the AZWV (Anerkennungs – und Zulassungsverordnung – Weiterbildung – Regulation on Recognition and Admission – Further Education AZWV). The

TABLE 80.1
QMS Examples

System	World	Europe	Germany	Regional German
TQM		EFQM (general)		
MMS	ISO 9001 (general) ISO 29990			
ISO derivatives 9001			DVWO QM BQM QVB	
ISO 9001 realization support	IWA 2 AQW		AWQ	
Quality systems	ISO/IEC 19796-1 IACET (USA)	EQUALS Q-For	AZWV LQW QESplus	The Bremen Model, the Hamburg Quality Mark, GAB-Quality Mark, Siegel Quality-proof Further Education et al.

Source: Henke H., Presentation of 28.1.2011 during training of certificating auditors "DIN ISO 29990:2010-12," Bonn, p. 5.

regulation concerns recognition and admission of further education courses, which, in its approach to the teaching process, is somewhat similar to the LQW system yet directed to different receivers. The AZWV aims at assuring quality in those educational programs, which are realized from means of the Federal Labor Agency (Bundesagentur für Arbeit) and was called by the Federal Ministry for Economy and Labor in 2004. In the scope of the AZWV system the whole organization offering education can be certified, and next, the particular courses offered by it (Trägerzulassung [carrier admission] § 84 and Maßnahmezulassung [admission of means] § 85 SGB III [Third Book of the Social Code]). Certifying particular courses assures the organizations of receiving donations from the Labor Agency. Due to its specificity, the system can be a pattern for those countries that carry out labor market programs, Poland included. It particularly stresses watching trends in the labor market and devising such courses that raise the chance for reintegration.

The rationale behind the emergence of the AZWV system is included in the document published by the Ministry of Economy and Labor under the same title (Begründung zur AZWV) with a recommendation to introduce the QMS (it can be, e.g., the ISO 9001, now also the ISO 29990, or a system prepared according to one's own criteria). This means that the organizations certifying according to this system have free reign when initiating the QMS. The AZWV standard requires among other things control of the processes of managing the organization, information flow, determining the organization's methods of checking their attainment, description of the QMS applied, way of treating the employees (how new employees are acquired, teachers in particular, possibilities of training for employees engaged in the learning process), description and availability of infrastructure of the organization used in the teaching process, etc. A QMS description is included analogically to other QMSs in the quality manual, in which each employee can obtain information concerning the functioning of the organization, information flow, the flow and accessibility of documents and regulations concerning to the most important processes described (division into system processes [of management], critical processes [main and basic], and auxiliary ones [subsidiary, supporting/*Division of processes* according to Sikora], and it need not be adhered to so rigorously). A system process is the process of acquiring participators, training the staff, watching the labor market, and conceiving new innovative curricula. As in other systems, the organizations certified according to the AZWV system are obliged to have a quality policy and to make it available to potential customers.

Since 2009, the number of organizations certified according to AZWV has increased from 4434 through 5204 in 2010 to 5417 and 2011 and testifies to interest in this federal system. It should be mentioned here, however, that the certificate qualifies to acquire means from the Labor Agency, which is unquestionably an additional motivation to acquire it.

Since April 2012, the system AZWV has been raised from the rank of regulation to the level of legal rules collected in the Third Book of the Social Code (SGB III/SGB III) and entered into force after the amendments as AZAV (Akkreditierungs – und Zulassungsverordnung Arbeitsförderung) with the reform of labor market instruments.

Especially for educational organizations but also as a supplement to the AZWV, the Quality Standard for Support of Employment Promotion (Qualitätsstandard für Träger der Arbeitsförderung [BQM]) has been prepared (QM in education). BQM is an initiative of the educational branch and was set up explicitly as a quality union, whose members pledge themselves to adhere to Federation of Institutions of Vocational/Professional Education (Bundesverband der Träger beruflicher Bildung [BBB]) quality rules (Federal Union of Vocational Training Units). The standard is tailored to the needs of the educational branch.

The PAS 1037:2004 is another system worth mentioning, which is available in Poland, too, through the DQS-Group Poland. This system, similarly to the AZWV system, addresses organizations offering educational curricula in the scope of CPD (continuous professional development). In the future, in the whole EU, only organizations having a quality certificate will be able to apply for financial support from the Labor Office (in consequence of the Copenhagen Declaration, the number of certificated institutions is one of the 10 indexes of education quality in the EU). It is thus an offer valid for the needs of training the unemployed and such people who are interested in in-service vocational training in Poland. The PAS system is offered in three levels: basic, standard, and excellence (Kompendium 2004). Basic level requirements are the smallest and pertain to the processes' documented quality and organizational capacity. On the standard level, the QMS functions have to be broadened by preparing and implementing a customer-related QMS. The excellence level necessitates the introduction and implementation of requirements according to European comparative criteria. The PAS system distinguishes main processes of added value and supporting ones. The introduction and maintenance of this system requires a comprehensive way of action (development and assurance of quality pertain not only to particular training and education scopes, but form a complex quality management system) but also customer orientation and communication (Figure 80.1).

The selected QMSs described are subject to voluntary certification; apart from them and others, quality marks are functioning, assigned according to established criteria. Due to the variety of

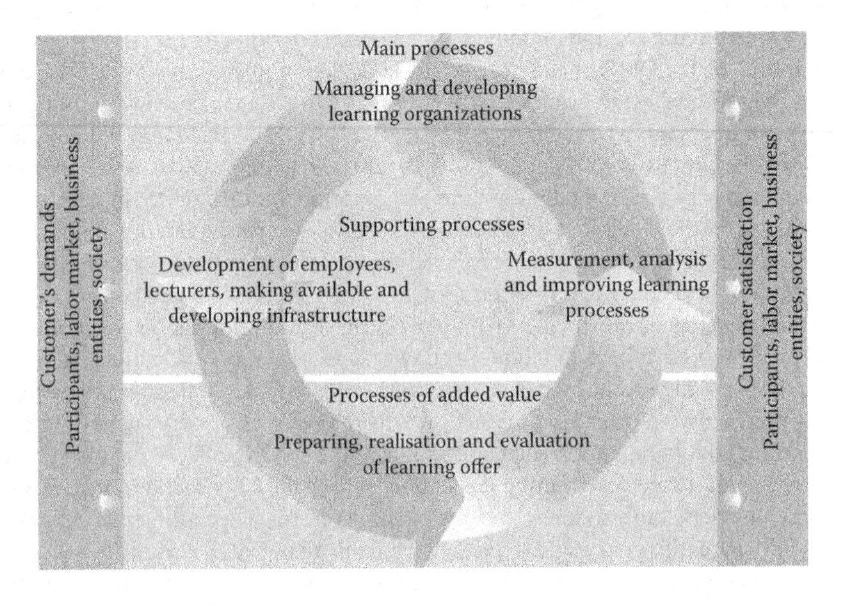

FIGURE 80.1 Model PAS 1037 – process approach.

available systems, without an expert's support, it is difficult to make a strategic decision of obtaining a certificate according to a given system.

Analogously to the variety of German systems on various teaching levels, the attention to education quality, school education included, appears in Poland, too, in various forms. It is possible to obtain a quality certificate according to ISO 9001 standard, as owned by an adult secondary school in Łódź "AP Edukacja sp. z o.o.," a public school of the "Lifelong Learning Center" from Nowa Huta, In-service Training Team from Warsaw, and many others. In Germany, state and vocational schools are still subject to teaching quality evaluation requirements described by respective supervisory units on the level of "Land."

The system of advisory and certificatory organizations in Poland is complicated; organizations have been active so far that specialize in QMS certification in education that were branches of German organizations on Polish territory, e.g., DQS Polska. It is also possible to obtain a certificate from organizations seated in Germany like DeuZert from Wildau near Berlin or Certqua from Bonn. A certificate of this organization is owned *inter alia* by the Enterprise Academy of Starachowice.

Apart from the above-mentioned certificates, there are also available quality certificates initiated as quality marks, in particular, districts and subject to the Boards of Education, like the Pomeranian Quality Certificates, Certificates of Highest Education Quality in Wielkopolska (Great Poland), and the "School of Enterprise" quality certificate. These certificates are granted after fulfilling particular requirements and are restricted in time.

The variety of certification forms in Poland, too, testifies to such need and striving to reach learners' satisfaction, employees' development, and school management. Similar to Germany, regional initiatives emerged, among other things, not one common system. Introducing the new ISO standard into organizations offering non-formal education may preclude the further development of diverse forms of certification and quality marks and will be a warranty of fulfilling uniform requirements.

The subject matter of attention to higher education quality has also been recognized and is realized in an analogous way to a certification method of study fields accreditation by accreditations commissions (in Germany by the Foundation for Accrediting Courses of Study [Stiftung zur Akkreditierung von Studiengängen] and in Poland by the State Accreditation Commission). The new ISO 29990 standard does not pertain, however, to the field of accreditation and certification last mentioned here.

ISO 29990:2010 PREPARATION FOR ORGANIZATIONS THAT OFFER EDUCATION

The process of creating each new international standard for quality (ISO) takes several years, during which the relevant bodies set requirements for future standards. The procedure for the development of the ISO 29990 standard was launched in February 2006 with the filing by the German Industry Standards (Deutsche Industrie Normung/DIN) a proposal in the International Standardization Organization/ISO concerning a new area of standardizing learning services. An inauguration session took place in Berlin in March 2007 of the ISO TC 232 standardization committee under the name "Learning Services for Non-formal Education and Training." The PAS 1037 system requirements were suggested as a basis for outlining the new standard's requirements. In April 2008, the main structure of the future standard was presented, and in March 2009, the first official outline (Draft International Standard) was presented to the world as ISO/DIS 29990:2009.

The Final Draft International Standard was made by the TC 232 Standardization Committee in London in January 2010. The preparation of a standard on this new, particular area of learning services in only three years, testifies to great global interest in the subject matter.

In September 2010, there was published ISO 29990:2010 "Learning Services for Non-formal Education and Training – Basic Requirements for Service Providers" (the final voting took place without any opposition on the part of ISO). In this way, it was possible for the first time to create a worldly consensus pertaining to assuring service quality for LSPs (learning service providers).

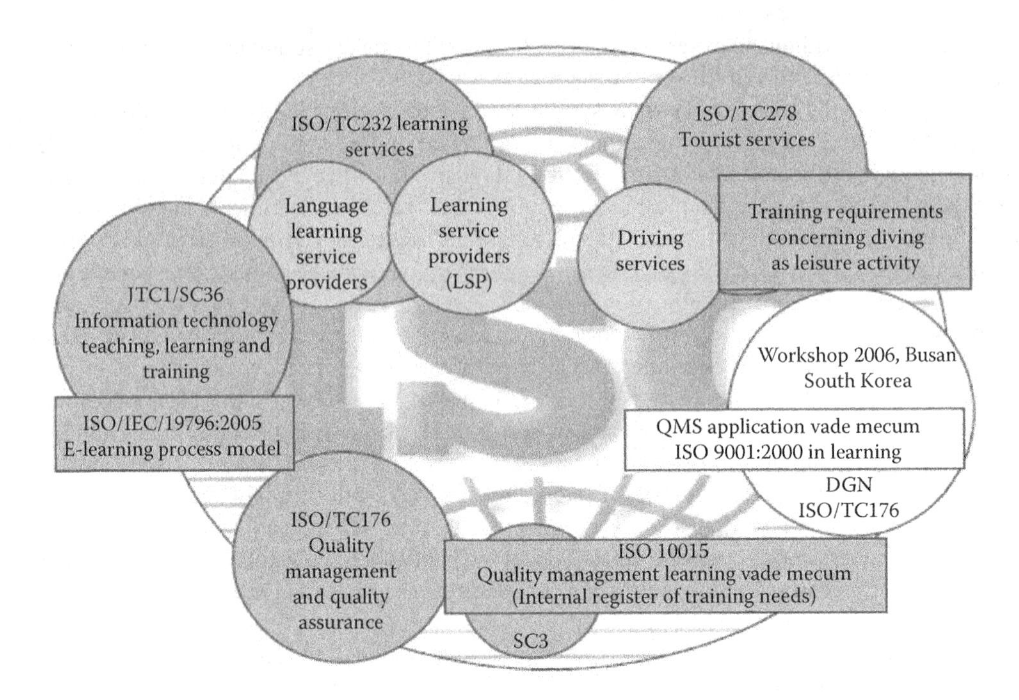

FIGURE 80.2 Plan for combining standards.

Besides Germany's representatives, many other countries, by their representatives, contributed to the successful standard preparation, including the United States, Canada, Great Britain, France, Ireland, Poland, Australia, Japan, Russia, Kenya, Austria, *et alias*.

When creating the new standard, stress was laid on creating a standard for institutions offering education as service in the free market. At the same time, the standard is fit for educational offers in the economic sector, like training in companies. There should be assured the quality and effectiveness of education and training and increased durability of education as well as transparency and comparability of services provided. The working out of a uniform standard was immeasurably important due to the fact that the number of LSPs (learning service providers) is growing as is the number of the services themselves. At the same time, attempts were made to combine the initiatives made so far (Figure 80.2).

Essence of ISO 29990:2010

ISO 29990:2010 was conceived as an "umbrella standard" to provide the basis for further standards for particular educational branches. In this scope, work has already commenced under the leadership of China concerning a possible ISO 29991 "Learning services – special requirements for language providers."

ISO 29990:2010 determines its application scope by itself: defining basic requirements for learning services and service providers in education and training. The concepts of education and training are made precise as "organized educational activities not counted among recognized formal areas of elementary education or regular school or university education" (cf. Chapter 2.15 ISO 29990:2010). The background is provided by the internationally recognized UNESCO definition of 1997 describing what "non-formal education" is any organized and sustained educational activities that do not correspond exactly to the definition of formal education and the derivative premise that education scopes cannot be the object of international standardization insofar as they are under the direct influence of the state or take place in the framework of exerting state authority

TABLE 80.2
Structure of ISO 29990

No.	Preface	Description
	Introduction	Aims of the standard: • Creating a generic quality model for the learning service provider • Creating a common reference for learning service providers and their customers
1.	Scope of application	"This international standard defines the basic requirements for learning services and service providers in education and training"
2.	Concepts	18 special concepts are introduced like competency, competence, curriculum, learning transfer
3.	Providing learning service	One of two (capable of being certified) main chapters with five subchapters along the process chain of a learning service
4.	Management by the learning service provider	The other of the two (capable of being certified) main chapters with 10 subchapters that guarantee a system able of being certified according to ISO criteria
	Annex A (informational)	Contents of economic (business) plan
	Annex B (informational)	Information for evaluating management system
	Annex C (informational)	Preventive and corrective measures
	Annex D (informational)	Examples of essential (indigenous) competencies of educational service providers
	Annex E (informational)	References between ISO 29990:2010 and ISO 9001:2008

Source: DIN EN ISO 29990, Learning services for education and training – basic requirements for service providers ISO/TC 232, Secretariat: DIN, Geneva 2010.

(*Guidebook for Planning Education in Emergencies and Reconstruction*. Chapter 12, Non-formal Education, Unesco 2006).

The scope of application of the new standard is nevertheless large. In Germany only, it is assumed in professional circles that there are more than 10,000 education and service providers in the concept of the new international standard of "learning service provider." They are currently most frequently certified according to the AZWV system insofar as they function in the promoted scope of education and training. As it has no application outside Germany's borders, it can be assumed that due to progressing globalization in the educational market more than 10% of educational and training providers from Germany will be active internationally. ISO 29990:2010 will be thus the basic certification standard, fulfilling the major role showing the international market quality ability of the learning service provider, which will enhance transparency and comparability outside of the country's boundaries. In Asia, e.g., in Japan and China, this is already perceived much more intensely than in Germany.

The concept of "providing learning services" is consistently used in the standard instead of training in order to focus attention on the learners and the results. The competencies of the learning service provider are in the center of attention. The attention of the new ISO 29990:2010 is support when selecting the suitable learning service provider. The structure of ISO 29990 standard has been presented in Table 80.2.

In the main chapter "Concepts," 18 special concepts are introduced that are important for understanding the new standard; basic examples have been gathered in Table 80.3.

The requirements of this international standard are given in the vernacular of the learning service provider. The user will also recognize it in the main chapter "Providing learning service" in which requirements are defined in the process chain of a learning service, the following being the most important:

TABLE 80.3
Selected Special Concepts of ISO 29990

Competence (Chapter 2.4 DIN ISO 29990:2010)	Knowledge, understanding, abilities, or attitudes as observable or measurable features or both, and that are applied and mastered in a particular work situation and also in professional and/or personal development
Curriculum (Chapter 2.6 DIN ISO 29990:2010)	Curriculum prepared by learning service provider describing goals, content, results, methods of teaching and learning, evaluation process etc.
Transfer of learning (Chapter 2.18 DIN ISO 29990:2010)	Application of learning results in other situations

Source: DIN EN ISO 29990, Learning services for education and training – basic requirements for service providers ISO/TC 232, Secretariat: DIN, Geneva 2010.

1. Establishing an educational need
2. Shaping learning services
3. Preparing learning services
4. Monitoring learning services
5. Evaluation by LSPs

These processes have been graphically presented in Figure 80.3.

The main chapter "Management by the service provider" gives in 10 subchapters the requirements for a QMS, which go beyond the "pure quality management system," e.g.,

- Strategic management of the organization
- Financial management
- Risk management
- Staff management

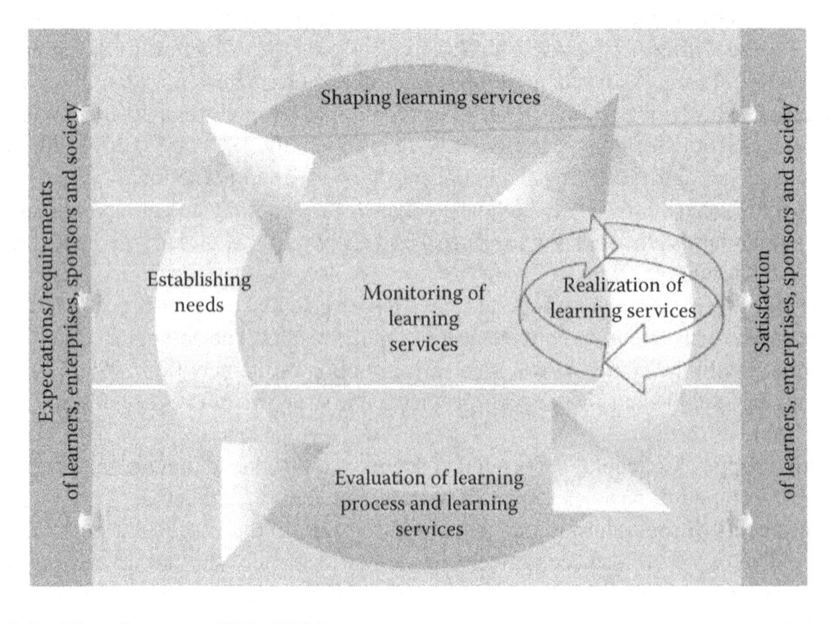

FIGURE 80.3 Flow diagram of ISO 29990.

- Communication management
- Handling complaints

The strategic management of the organization becomes particularly clear in the requirement of submitting a business plan (cf. Chapter 4.2 and Annex A DIN ISO 29990:2010). Other interesting aspects of this international standard are the requirements concerning documentation; more stress is laid here on verifiable effectiveness, modern suitable requirements concerning evaluation or guaranteed compatibility with ISO 9001:2008 (cf. Annex E [DIN] ISO 29990:2010).

The requirements of the standard are rigorous yet adapted, at the same time, to the specificity of organizations active in this branch.

FINAL REMARKS

Work on the new standard is completed, yet the process of bringing it into effect is not. The first ISO 29990 publication took place on September 1, 2010. The publication as a DIN standard appeared in January 2010. In the first quarter of 2011, a motion was put forward for the recognition by EN and CEN.

Advantages are expected from the new standard by all actors of the learning service market. Table 80.4 compares certain expected advantages for selected groups.

Internationally approved certification, in particular, brings added value to each international cooperation, a common basis for quality through uniform standards and common understanding of quality.

TABLE 80.4
Possible Advantages of Certification according to ISO 29990 for Selected Groups

Advantages for Providers

- Possibility of entering the market in various countries
- Meeting national certification requirements
- Optimization of the process (P-D-C-A)
- Clearly defined and documented processes
- Development of internal quality improvement abilities
- The new standard may become the basis for branch solutions (e.g., language learning, distance learning)
- Motivation of employees and their active participation in introducing changes in the organization

Advantages for Learners

- Transparency and comparability of offers
- Transparency and comparability of offers in various countries
- Reliability of offers
- Information pertaining to learning evaluation criteria
- Optimization of processes
- Improvement of learning environment
- Assuring suitable resources, incl. teachers (trainers) with suitable qualifications

Advantages for Enterprises

- Improved comparability of LSPs for the needs of internal and external training
- New training structures in the companies are not permanent and are not developed without engaging the LSP
- Standards facilitate interaction between educational institutions and enterprises and/or learners
- Education is increasingly perceived as service and has to be competitive

Advantages for Customers (Offices and Institutions), e.g., Labor Offices, International Organizations

- Defined uniform requirements
- Comparability
- Possibly conformity to law

Source: Own study.

At present, a certificate of this standard can already be acquired in Germany at the certification body DeuZert Deutsche Zertifizierung in Bildung und Wirtschaft GmbH at Wildau in Brandenburg, which is the first certification organization in the world to have conducted a certification procedure according to DIN ISO 29990:2010. The first to take up the challenge was an internationally renowned LSP, the Health Academy of Berlin Charity clinic (Gesundheitsakademie der Berliner Charité). The certification was successful and encouraged other organizations to obtain a certificate according to this standard. As quality certificates are valid throughout the world, Polish organizations can already obtain a certificate in Germany. It can be expected that certification according to this standard will in the future be provided by, e.g., DQS Polska.

As the first international body, the EADL (European Association for Distance Learning) has prepared a quality guide for applying the ISO 29990:2010 standard, which facilitates the understanding and interpretation of the standard's requirements.

Further development of the ISO 29990:2010 standard provides for its specialization, i.e., preparing a version for, e.g., language course providers. It is nevertheless difficult to predict in the first quarter of 2011 how the standard will be received on the LSP market. It can, however, be recognized that its preparation by an international body is an assurance of its success. At the same time, the ISO certificate is without doubt a recognizable category for all customer groups, influencing the process of decision making.

To sum up, it should be stated that, after years of cooperation and efforts of national and international bodies, a new ISO standard has been successfully created, tailored to the needs of educational organizations and also the comparability of the educational services they provide. So far, the problem always raised has been the lack of specificity of this industry in the requirements of ISO 9001 standard. The new ISO 29990 standard is addressed only to the organization that offers informal forms of education.

REFERENCES

Akkreditierungs – und Zulassungsverordnung Arbeitsförderung, Gesetz zur Verbesserung der Eingliederungschancen am Arbeitsmarkt (Accreditation and Licensing Regulation Labor demand, Law to improve the chances integration in the labor market), 20.12.2011, Bundesgesetzblatt Jahrgang 2011 (Federal Law Gazette) Teil I Nr. 69 (Part I No. 69), ausgegeben zu Bonn am 27 (Bonn 27). Dezember 2011 (December 2011).

DIN EN ISO 9001, Qualitätsmanagementsysteme – Grundlagen und Begriffe (Quality Management – Fundamentals and vocabulary) (ISO 9001:2008), Dreisprachige Fassung EN ISO 9001:2008, DIN Deutsches Institut für Normung e. V. (German Institute for Standardization), Beuth Verlag, Berlin 2008.

DIN EN ISO 29990, Lerndienstleistungen für die Aus- und Weiterbildung – Grundlegende Anforderungen an Dienstleistende (Learning Services for non-formal Education and Training – Basic requirements for service providers), ISO/TC 232, Secretariat: DIN, Geneva 2010.

Guidebook for Planning Education in Emergencies and Reconstruction. Chapter 12, Non-formal Education, UNESCO 2006.

Henke H., Presentation of 28.1.2011 during training of certificating auditors "DIN ISO 29990:2010-12," Bonn.

Kompendium. Przewodnik i poradnik do wdrożenia i utrzymania PAS 1037:2004 (Compendium. Guide and tutorial to implement and maintain PAS 1037:2004), RKW Rationalisierungs- und Innovationszentrum der Deutschen Wirtschaft (Productivity and Innovation Centre of the German economy), Brandenburg, Potsdam 2004.

Quality Guide ISO 29990. Leitfaden zur Anwendung und Hilfen zur Umsetzung, EADL and Forum Distance E-Learning, Grootebroek, Hamburg 2010.

Sikora T. (ed.) Zarządzanie jakością według norm ISO serii 9000:2000 (Quality Management according to ISO 9000:2000 series), Wyd. AE w Krakowie (Academy in Krakow), Krakow 2005.

INTERNET RESOURCES

Artset GmbH, http://www.artset-lqw.de/

ENQA-VET, http://www.eqavet.eu/gns/home.aspx

EQARF, http://www.ecvet.de/c.php/ecvetde/eqarf/instrumente/referenzindikatoren.rsys
ISO, http://www.iso.org/iso/catalogue_detail?csnumber=4648
Parlament Europejski (The European Parliament), http://ec.europa.eu/education/lifelong-learning-policy/doc1134_de.htm
UNESCO, http://www.unesco.org/new/en/unesco/

81 Polish Universities
Toward a Real Science–Business Cooperation

Krystian Gurba

CONTENTS

INTRODUCTION

Knowledge plays such an important role in a modern economy like electricity in the times of industrial economy. It is a core factor of production (Nelles and Vorley 2010). In the 1960s, the term "the knowledge industry" (by F. Machlup) and later "the knowledge economy" (by P. Drucker) began to become popular. The crucial role of knowledge was best emphasized by Drucker: "knowledge has become the central factor of production in an advanced developed society." In simple words, he stated that "the key to producing more work was to work smarter (instead of harder); the key to productivity was knowledge (instead of sweat)" (Beerkens 2008).

Knowledge created at universities has various tasks in the economy. First of all, it can be a subject of technology transfer, which enables a flow of innovations to the private sector and can provide universities with an additional income. It can also be—in a form of intellectual property rights—a contribution to companies. The efficient knowledge management can also be an additional source of awarding scientists for their creative work.

A change in the rationale for the higher education and science policies can be observed in a modern society. "From a safeguard of national culture and opportunity for social mobility and development it is becoming the basis for knowledge society and the driver of the knowledge-based economy" (Beerkens 2008). Universities fulfill this task by providing qualified human capital—knowledge workers—but also providing knowledge embedded in new technologies.

This new task for universities is often referred to as "a third mission" or creation of "entrepreneurial universities" (more in Nelles and Vorley 2010). Studies have defined it as "blurring the boundaries between the public and private science and technology" (Etzkowitz, Leydesdorff) or "a phenomenon, articulated in policy, in which universities are encouraged to realize their broader socioeconomic potential through knowledge exchange and partnership." It is assumed to focus on a commercial engagement, especially licensing and spin-off activities, having to fulfill both economic and societal roles (Nelles and Vorley 2010).

This article describes a phenomenon of entrepreneurial universities (especially factors of successful cooperation between science and business) both in the theory of economy and management and in the practice of selected Polish higher education institutions. Analysis of the literature, own desk research, as well as interviews and a case study of Jagiellonian University (JU) in Krakow based on survey designed for this purpose will enable to verify the hypothesis that Polish universities are becoming more business-friendly. Collected data will be also used to illustrate the importance of creation of technology transfer offices and implementation of internal commercialization policy.

FORMS OF PARTNERSHIP BETWEEN SCIENCE AND INDUSTRY

The World Intellectual Property Organization (WIPO) recognizes that the partnership of science and industry includes a wide range of relations from informal contacts (e.g., publication of articles, the mobility of employees, and direct, personal relations) to more formal ones, which are a basis of long-term cooperation. This second type usually involves a legal relationship between the institutions (contract) and includes issues such as commercialization of research results (licensing, spin-off companies), joint research projects, or contract research (more in Nezu; Polt). This article focuses on formal cooperation.

Studies suggest that licensing is a base for commercialization activities (Thursby et al. after Conti et al. 2009). On the other hand, most of the respondents in the survey conducted among European technology transfer centers (TTO) answer that they spent more time negotiating and managing contract research than license contracts (Conti and Gaule 2008).

The spin-off company concept is an alternative form of a commercialization since it requires a long-term commitment of a university into a business/risky project. Studies have repeatedly found that spin-off companies have a heterogeneous nature. It is related to different legal backgrounds in different countries as well as structure and strategy of universities that create spin-off companies (Bathelt et al. 2010). Furthermore, previous success in commercialization is crucial for further chances in the creation of spin-off companies (O'Shea et al., after Siegel et al. 2007).

Analysis of results of benchmarking in different European countries shows that different forms of industry and science relations are thriving in different countries. A contract and collaborative research is the most popular in the United Kingdom, Finland, Belgium, and Ireland, while institutions in Germany, Sweden, and also in the United Kingdom tend to use patents and licensing as a mean of commercialization. The third tool, spin-off companies, is often utilized in Germany, Belgium, and at some universities in United Kingdom (Polt et al. 2001).

CONDITIONS OF A SUCCESSFUL INDUSTRY AND SCIENCE COOPERATION

Studies have adopted a concept of entrepreneurial architecture (Burns, after Nelles and Vorley 2010). It defines five key analytical categories that interact to optimize an organizational innovation. These five dimensions are structures, strategies, system, culture, and leadership. While examining the innovation as a change within university to become more entrepreneurial, it is important to take into consideration all these five aspects. Despite that, most of the studies of the subject focus on just one dimension (Nelles and Vorley 2010).

As Lache sums up, major changes occurring in a university, which takes an entrepreneurial approach, are as follows:

- An increase in the management efficiency and strategic planning
- Expanding activity into adjacent areas (the establishment of specialized units, new research centers in science-based technologies sectors with close links to industry)

- Diversification of funding sources (private sources, technology transfer)
- Growth of an entrepreneurial culture (Lache, after Marszałek 2010; Beerkens 2008)

Polt suggests other classification of conditions:

- Legislation: positive or negative legal regulations
- Efficient or not efficient promotion programs
- Existing or not existing competences of intermediary structures
- Institutional settings, including decentralization or centralization of TTO actions (Polt et al. 2001)

In this paper, modified classification of factors is proposed: legislation—as an important framework for all actions of university; institutional settings—as relations between actors taking part in the process of enhancing cooperation between science and business; and structures—a role and efficiency of technology transfer offices and other intermediary units within university, and promotion and culture of cooperation.

Legislation

The evidence that a legal background has significant influence on results of science–business cooperation has been given by the United States. Passing the Bayh–Dole Act in 1980 has been a breakthrough that enabled universities to manage intellectual property resulting from government-financed research (Ponomariov 2008). Those changes led to increased labor mobility, development of venture capital institutions, and, as a result, a more rapid technology-based growth (Bathelt et al. 2010). Results of a legislative change were observed in the United States a few years later. The number of patents granted annually to US universities increased from less than 300 in 1980 to 3278 in 2005. There has been a rise in the number of licenses (four times from 1991 to 2005) as well as in license revenue (almost ten times) (Siegel et al. 2007).

Such a spectacular legal revolution was not needed in Europe. A model that is common in most European countries provides institutional ownership of IP rights (more on the other model used in Sweden and Italy—professor's privilege—in Conti and Gaule 2009). The most important advantage of an institution ownership model is that the research institution can be more effective in negotiations with business partners. So it is more probable that the invention would be successfully commercialized. This model is also represented in Polish law.

Innovation policy in Poland started in the early 1990s as a result of political and economic transition from communism toward a free market economy (Onak-Szczepanik 2007). Polish legislators understood that universities should be supported and that they should be encouraged to collaborate with a private sector. That is why the Higher Education Act declares that "universities cooperate with industrial environment, especially by sharing results of research with business and promoting entrepreneurship in the academic environment" (Higher Education Act, *Journal of Laws* of 2005).

Unfortunately, the underfinancing of Polish science has been one of the factors behind the poor performance of academic institutions in terms of cooperation with business. Apart from the low spending on R&D in general, Poland has an unfavorable structure of expenditures described as a share of funding for basic and for applicative research. In the innovative economies, part of funding for basic research is at around 20% of all R&D spending. In Poland in 2005, it amounted to 37.4% (Kasperkiewicz 2008).

Also internal regulations in each institution are important to create a framework for cooperation with business. Personal desk research conducted in 2011 shows that most of the academic universities in Poland have already implemented an internal intellectual property and commercialization policy (eight universities, half of them in the last two years) or are at the stage of projecting and implementing such regulations (three universities). Less than half of specialized universities (medical, agricultural, or economic) introduced above-mentioned regulations. Surprisingly, many technical universities above-mentioned still have not been put in place. On the other hand, most of them are planning to amend this situation.

The process of implementing a commercialization policy should accelerate after entering into force of amendment of Higher Education Act in October 2011, since it imposes a duty to introduce such internal regulations on each university (Higher Education Act, Art. 86c).

Institutional Settings

When relations between science and business are considered, the Triple Helix model is often used to describe them. It consists of three elements: academia, industry, and government and interactions between them (Leydesdorff and Meyer 2006). Several theoretical papers found that the Triple Helix model is not constant. As Etzkowitz writes, "Each sphere plays the role of the other in this new system. University becomes a company, industry becomes an educator, through corporate universities. A government becomes an important venture capitalist through programs focused on the applied research" (Etzkowitz, after Neto et al. 2009). The university takes over models of the behavior of enterprises in certain areas: the activity in raising capital and using the created and owned intellectual resources for its benefit (Marszałek 2010).

Selected studies recommend that universities have a leading opinion on how the science–business cooperation system of each of them is composed (Nelles and Vorley 2010). This decentralization enables for better flexibility and adjustment. The role of public policy at the beginning of changing universities to a more entrepreneurial organism is crucial. Legal changes are leading them to become more transparent, efficient, and economically accountable (Edquist, after Nelles and Vorley 2010). After initial actions to stimulate proinnovative changes, the state should act more like a coordinator, regulator, and facilitator with an important but limited role.

A dangerous phenomenon is described in literature as "decoupling" between a model of entrepreneurial architecture implemented at the university and practice. Sometimes formal changes are only nominal and fail to achieve their goals (Beerkens 2008).

Universities such as Harvard, Cambridge, Stanford, and Oxford are subject to copying of models of entrepreneurial architecture (Beerkens 2008). Universities tend to adapt to these models and this—in a global context—leads to convergence, isomorphism, homogenization, and standardization. The alternative approach to copying standard models is their "reinvention" and modulation to adopt them to a regional and local environment. Some authors, however, observe that these implemented models are more of symbolic modifications than significant adaptations (Beerkens 2008; Nelles and Vorley 2010).

Structures

Centers for innovation and entrepreneurship or technology transfer offices are names used for institutions that act as intermediaries between science and business (Santarek 2008). Literature describes their task as

- Promotion of entrepreneurship and self-employment
- Creation and development of small and medium enterprises
- Transfer of new technologies
- Stimulation and monitoring of cooperation of all actors on the local or regional level
- Human capital quality improvement (trainings, education, and consulting) (Santarek 2008)

TTOs are careful in new transactions since they do not want to be accused of "giving away" university-owned technologies (Siegel et al. 2007). It is especially a problem for Polish TTOs.

On the other hand, some authors draw attention to the negative effects arising from the fact that TTOs are assessed by university authorities. Therefore, they must present themselves as active, which can lead to overvaluation of inventions (Clarysse et al., after Siegel et al. 2007).

In the survey organized by Swiss researchers, TTO representatives were asked to rank by importance the effects of TTO actions. Majority of respondents answered that extremely or very important are industry-sponsored contract research income as well as the number of established spin-off companies. The following factors were also highlighted as important: the number of patents awarded and license income. Finally, TTO professionals recognized a number of license agreements executed only as somewhat important (Conti and Gaule 2008). A comparison with a similar research conducted in the United States shows that American TTOs focus more on generating revenue from licensing (for 71%, it is an extremely important effect) and the number of license deals (extremely important for 49% respondents) (Thursby, Jensen, and Thursby, after Conti et al. 2007).

Although the first TTOs in Poland were established at the beginning of the 1990s, their growth started in early 2000s with access to various support programs and European financial instruments for universities. Although the Higher Education Act of 2005 allows one to create TTO as a separate entity (company or foundation), most of Polish universities form TTO as a part of central administration.

PROMOTION AND CULTURE OF COOPERATION

Another important matter when analyzing the issue of entrepreneurial universities is teaching entrepreneurship. Also that function of universities has its origins in the United States. Anderseck (2004) writes that the practical approach to teaching entrepreneurship by sharing own experiences is the most successful and desired one.

A subject of entrepreneurship education seems to be especially important, as it is apparent from the survey conducted for the purpose of this article that there are almost as many students that plan, after graduation, to open their own business as those that want to have a full-time employment. The majority of interviewed students stated that during the studies, they had a contact with the subject of entrepreneurship, innovation, and cooperation between science and business, while half of them acknowledge that they would not know how to start a business. They also admit that although they are aware that the university provides support to young entrepreneurs, they have no information about specific forms of assistance (survey on a group of 50 master students at the Faculty of Management and Social Communication at JU).

Measuring the Cooperation between Science and Business

An important question that arises when analyzing the performance of universities in cooperation with a business is how to measure the effects of such cooperation. A number of patents cannot be a reliable factor since patent procedure takes up to five to six years; therefore, the result received will be outdated. The number of patent applications can be seen as more accurate. On the other hand, it must be underlined that not all inventions disclosed to the institution are patentable. Some of the innovations will therefore be "lost" while using this factor. Furthermore, the number of patents and patent applications does not directly result in the increased collaboration with a business, although obviously it increases the attractiveness level of the university.

The best outputs are connected with those incentives that are seen as the most important by TTOs and are most closely associated with the market: the number of licenses, revenue from licensing, the number of spin-off companies, and the amount of industry-sponsored research (Conti and Gaule 2008).

Jagiellonian University in Krakow: A Case Study

A survey designed to evaluate changes at Polish universities for the purpose of this article and for the purpose of future research consists of 16 questions regarding various aspects of changes enforced at universities to facilitate science–business cooperation. The survey has been designed for managers of technology transfer at each university. Questions in the survey have been divided into four chapters, each describing the most important elements of a system of technology transfer from academia to industry. Questions are described in Table 81.1.

Answers from JU will be used in this article as a part of a case study showing changes of the university to become more entrepreneurial. In addition, in the subsequent parts of the article, selected data from interviews with managers from other Polish universities will be presented. Survey can be used in future research to provide more in-depth analysis of performance of Polish universities.

At JU, the following units or institutions involved in science and business cooperation can be identified: technology transfer office—CITTRU (established in 2003), Academic Entrepreneurship Incubator (since 2006), Life Science Technology Park managed by the company JCI Ltd. (the park opened in 2009), and the Office of Career Services (established in 2001). No separate Department of Cooperation with Industry can be identified, but its role is executed by CITTRU and for the Medical School by Office for the Scientific Research.

According to the responses of CITTRU, all units involved in working with a business employ between 20 and 30 persons. It indicates the wide scope of JU actions and formalization of cooperation with business process at JU.

TABLE 81.1
Survey Designed to Describe Various Aspects of Changes Enforced at Universities to Facilitate Science–Business Cooperation

Chapter	Questions
Organization of cooperation between science and business	1. Which of the following units engaged in science and business cooperation and development of academic entrepreneurship exist in the institution?
	2. How many persons employed in the institution are directly involved in the organization of cooperation with business?
	3. What changes aimed at promoting cooperation between science and business have been introduced in the institution in the past 15 years?
	4. What has stimulated the changes designed to promote cooperation between science and business?
	5. What is the role of university faculties in efforts to enhance science and business cooperation?
Identification and management of innovations created in the university	1. What are the figures on the protection of intellectual property in recent years?
	2. Who covers the costs of legal protection of innovations created in the institution?
	3. How usually are identified innovations created in the institution?
	4. What are the main problems associated with the identification and management of innovations in the institution?
Promotion	1. What actions aimed at promoting cooperation between science and business accompanied changes introduced in the institution?
	2. Did the institution hold courses for students in intellectual property or entrepreneurship?
	3. Have the institution issued or implemented written internal regulations on the intellectual property rights or academic entrepreneurship?
Effects	1. What are the effects of activities related to academic entrepreneurship and intellectual properties (number of contract research agreements, number of license agreements, number of spin-off companies)?
	2. How do you assess in general the effects of changes in the organization of science and business cooperation, introduced in the institution (main positive and negative effects)?
	3. Among what units the revenue from university activities related to cooperation with business are distributed?
	4. Which are the most attractive areas of science in the institution in terms of cooperation with business?

JU is a pioneer in adapting the university to cooperate with a business. Over the past 15 years, all of the changes aimed at facilitating cooperation mentioned in the questionnaire were introduced, including creation of units aimed at organizing collaboration with a business; introduction of internal rules on the intellectual property, rules regarding the contract research, and rules governing the creation of companies for the commercialization of research results; and conclusion of a long-term agreement regarding cooperation with companies and participation in cluster initiatives.

On the other hand, there is a lack of involvement in international organizations facilitating science–business cooperation, which indicates the underestimation of the internationalization of science and business cooperation. It is a common attitude for Polish universities.

As incentives for changes at the university aimed at promoting cooperation between science and business, CITTRU representative states the following: the growing interest of business in cooperation with the university, the growing interest of university staff in cooperation with business, the desire to reduce the so-called "gray area" (i.e., the research collaboration of individual scientists with the business excluding the involvement of the university as an institution), as well as the observation of good practices from other foreign or domestic universities. A similar motivation is mentioned by the director of TTO at Lodz University, who sees changes at the university, especially implementation of intellectual property policy as a tool for TTO that should act as a guard and as a broker of technology.

According to JU, the aim of obtaining additional income was not an important motive for changes. It is consistent with observations of Conti and Gaule (2009) that there is nothing in economic theory that suggests that TTO's role is to maximize license revenue. It is hard to argue with their statement that "social welfare might be better served by TTO's facilitating local economic development or helping with the translation of academic research into products" (Conti and Gaule 2009).

Faculties at JU show a modest activity in the field of interaction with the TTO. According to CITTRU, they cooperate only in selected activities, and not all departments are interested in opening toward a private sector. The attitude is different at Warsaw University, which is a rare example of the decentralized approach: the costs of IP protection is to be covered in large part by faculties, and therefore, units receive a more significant portion of future income than in other universities. At Lodz University, enforcement of the innovation policy was accompanied by a meeting with an appointment of proxies in all faculties. They are contact points for TTO working normally as scientists. This attitude, which is close to the concept of network organization, is based on the success of TTOs in the United Kingdom. An example that is often cited is the Cambridge Enterprise. Its structure can be described as a widely networked organization. The task of parent institution—Cambridge University—is mainly providing exploitable results, competencies for an opportunity recognition, as well as networking and management of knowledge through a long-standing reputation (Franzoni 2004).

According to CITTRU managers, the most important positive effects of proentrepreneurial changes at the university are as follows (in the descending order of importance): an increased interest of researchers in cooperation with business, an increase in the number of joint research projects of universities and enterprises, and an increase in awareness of university researchers in areas related to cooperation with business (intellectual property, academic entrepreneurship). Similar conclusions arise from interviews with managers from University of Lodz and University of Technology and Life Sciences in Bydgoszcz. As representatives of the latter university say, "it is hard to see effects connected with changes done at the university already"; but as they admit, it enabled the scientists to see how important property protection can be for the research intellectual. Moreover, the number of patent applications has increased rapidly. In 2008, the university filed 25 patent and utility patent applications, and in 2009, 15 applications were filed. In 2010, the year when the innovation policy has been introduced, 44 patent applications, utility patent applications, and designs have been filed. In the case of Lodz University, a similar improvement can be seen already. The TTO director highlights that scientists come to TTO carrying printed regulations in their hands. In 2009, the university had 5 patent applications, in 2010 it was 6, and, in the first two months of 2011, the university had already filed 14 patent applications.

Like in most model TTOs, revenues from the collaboration are shared at JU between the central budget, faculty, and inventors. JU does not provide direct participation of TTO in profits. Some authors highlight that the "royalty distribution formula" has a significant influence on the involvement of single inventors and faculties or schools in the university (Link and Siegel, after Siegel, p. 648).

According to CITTRU representatives, formal regulations have been made concerning only certain aspects of commercialization, intellectual property, and academic entrepreneurship. On the other hand, the university is advanced in adapting the curriculum to the needs of education in these areas: students of all faculties have as obligatory course the foundations of intellectual property, and the postgraduate course on the intellectual property management will be carried out.

The costs of patent applications at JU is covered mostly from public funds obtained for this purpose (such as EU programs and Higher Education and Science Ministry programs), but funds from the university budget are also used, in particular, where there is no possibility of obtaining external funding. The possibility of funding from its own sources is a major advantage, particularly in the context of the high cost of international protection. Many Polish public research institutions withdrew from international protection of their inventions because of lack of own funds. In 2010, costs of patent protection of inventions from JU was estimated at 304 000 PLN. Most actions to sell technology (meetings with companies, trips to conferences, and trade fairs) are also financed from the external funds (CITTRU Annual Report 2010).

Representatives of CITTRU list the following sources of information about new technologies (inventions) created within the university:

- The most important: individual meetings with scientists initiated by TTO employees and direct knowledge of TTO staff of the research fields of university units and of scientists working in the them
- Important: the use of database of research projects at the university and individual meetings with scientists initiated by them
- Of average importance: trainings, workshops, or seminars for researchers
- Less important: an analysis of publications prepared by the scientists at the university
- Least important: the information obtained from faculties and the information obtained from outside the university

As the main problems associated with the identification and management of innovations disclosed in the JU, CITTRU considered the following:

- Limited financial resources of units involved in cooperation between science and business
- Passivity of academics or negative attitude to cooperation from business
- Insufficient interest from the business
- Low awareness in intellectual property and academic entrepreneurship issues
- Legislation impeding efforts toward cooperation of science and business

In comparison to the statistics from ProTon Europe (association of technology transfer offices) report, presenting the average of European TTO, the performance indicators of CITTRU show that the technology transfer office of JU is of an average size (when considering employees involved directly in business cooperation issues). The number of disclosed inventions per year is also similar to the European average. CITTRU files significantly more patent applications, which should result in the increase in the number of received patents in the upcoming few years. A creation of wide portfolio of patents should also allow JU to achieve better result in licensing. However, in 2010, a major difference between this oldest Polish university and European average could still be observed (Table 81.2).

As to the most important TTO outputs, the following positive effects at JU can be observed since 2006:

- An increase in the number of disclosed inventions: 0 in 2006, 12 in 2007, 15 in 2008, 16 in 2009, and 21 in 2010
- An increase in the number of patent applications: 1 in 2006, 5 in 2007, 23 in 2008, 36 in 2009, and 26 in 2010
- An increase in the number of contract research (since the procedure is centrally monitored and coordinated)
- An increase in the value of contract research agreements, from 770,000 PLN in 2009 to 1,500,000 PLN in 2010

The data presented above show that the number of patent applications as well as innovation disclosures increases after a creation of a specialized unit for technology transfer, passing internal regulations in the scope of technology transfer and IP management and accompanying educational actions. On the other hand, the number of contracts with companies and income received from technology transfer is not growing rapidly yet. It must be taken into consideration that there is a significant "time lag" between an implementation of changes and their results (Polt et al. 2001). Furthermore, there is a significant time needed for a single scientific research result to go through the pipeline of commercialization, from disclosure to its appearance on the market.

JU was one of the first nonpurely technical universities in Poland to implement an internal policy in the scope of innovations and technology transfer. It consists of two coherent parts: rules for the intellectual property and rules for the spin-off company creation. The purpose of the introduction of JU policy innovation was not so much to improve the statistics on the number of patent applications, but more so to increase the number of commercialized inventions. The intention was to change the common practice where the patented technologies remain in the bottom drawers of researchers' desks.

In 2009, JU also enforced a procedure for a contract research. Offers from the demand side (usually business companies) are directed to the technology transfer center (CITTRU). The contract research is most commonly conducted by scientists from chemistry, biology and earth sciences, biochemistry, biophysics, and biotechnology, as well as representatives of humanities—the Faculty

TABLE 81.2
Performance of Average European TTO and CITTRU (TTO of JU)

Indicator	Average TTO (in 2008)	CITTRU (JU) in 2010
Age of TTO	12.4 years	8 years
Employment of full time staff	9.7 persons	20 persons (10 directly involved in organization of cooperation with business)
Number of invention disclosures	19.9 inventions	21 inventions
Number of patent applications filed	10 patent applications	26 patent applications (including international procedure)
Number of patents granted	3.4 patents	0 patents
Number of license agreements	12.4 agreements	1 agreement

Sources: The ProTon Europe Fifth Annual Survey on Knowledge Transfer Activities in European Universities; CITTRU Annual Report 2010.

of Philosophy (Center for Evaluation and Analysis of Public Policies). It is important to outline that one third of European universities admit that contract research amount is not recorded in their institutions and no bidding procedure for scientists exists (Conti and Gaule 2008).

Various other initiatives of JU have one common purpose: to create an atmosphere of entrepreneurship. Academic Incubator of Entrepreneurship at JU has been opened in May 2006. It is a place for development of 20 companies. Since December 2010, Incubator is managed by CITTRU. Technology Park Life Science, which is being built at the III Campus of the university by the JCI—a company fully owned by JU—is an example of action to create the infrastructure for science–business cooperation (Marszałek 2010). The proximity of academic centers has a significant positive influence on the performance of the high-tech companies. They can benefit from the research conducted in the university and by using highly skilled human capital "produced" by universities (Bathelt et al. 2010).

Based on the comprehensive data from JU, direct and indirect effects of changes leading toward business-opened university can be shown. First of all, a significant increase in the number of patent applications can be observed. It was visible especially in the first year after the intellectual management policy has been enacted. A significant change of the quality of applications can also be observed (first joint-patent applications with foreign renown institutions such as Centre National de la Recherche Scientifique [CNRS] (The French National Centre for Scientific Research) from France illustrate this improvement). In 2010, the invention entitled "Construction and application of Salmonella-based anti-cancer vaccine" received Polish Product of the Future prize in the category of technologies in the preapplicative phase. That same year, the first license income has been received as a result of the agreement signed with a global biotech company.

The experience of JU shows that the creation of clear rules must be the basis for promotion of academic entrepreneurship, which will be a motivating factor for scientists. The university should become aware that the intellectual property protection is an investment. Lastly, it ought to be emphasized that the commercialization process requires cooperation with scientists at every stage and creation of a relationship based on trust.

CONCLUSION

The role of universities as a source of knowledge can be fulfilled only when they will not function as an isolated island but instead cooperate with private sectors.

The system of science–business cooperation is based on five key factors: legal framework, relations between institutions of the innovation policy (academia, industry, and government), intermediary structures (TTOs), and culture of cooperation.

The analysis of the Polish innovation policy as well as instruments of support offered to universities shows lack of aid in the creation of model procedures and codes of good practice in science–business cooperation such as Lambert Model Agreements in the United Kingdom. This results in necessity for universities to develop their own way of becoming more business-oriented. Furthermore, additional government funding is needed to address shortage of public and private investments in cooperation between science and business.

Currently only reliable indicators of development of technology transfer in Polish universities appear to be the number of disclosed innovations and filed patent applications. It is too early to observe the increase in the most important output: the number of contracts and revenue from technology transfer. An increase in new innovation disclosures, first license agreements, and revenues from the commercialization shows that universities that introduced changes in structure, procedures, and practice of research management, like JU, are becoming more entrepreneurial. However, as illustrated by this article, when measuring the effects of technology transfer as a number of license agreements, leading Polish universities and their TTOs are still performing significantly worse than average European institutions. Lastly, due to the recent substantial amendments in the Higher Education Act, it is expected that all universities take more active approach toward entrepreneurship, including creation of spin-off companies.

REFERENCES

Anderseck K., Institutional and academic entrepreneurship: Implications for university governance and management. *Higher Education in Europe*, XXIX(2), 193–200, 2004.

Bathelt H., Kogler D.F., Munro A.K., A knowledge-based typology of university spin-offs in the context of regional economic development. *Technovation*, 30, 519–532, 2010.

Beerkens E., University Policies for the knowledge society: Global standardization, local reinvention. *Perspectives on Global Development and Technology*, 7(1), 15–31, 2008.

Chu S., Ritter W., Hawamdeh S. (ed.), Managing Knowledge for Global and Collaborative Innovations, Series on Innovation and Knowledge Management—Vol. 8. World Scientific Publishing, Singapore, 2010.

CITTRU Annual Report 2010, CITTRU, Kraków, 2011.

Conti A., Gaule P., Are United States outperforming Europe in university technology licensing. Copenhagen Business School Summer Conference, June 17–19, 2009.

Conti A., Gaule P., *The CEMI Report of University Technology Transfer Office in Europe*. CEMI, Lausanne, 2008.

Conti A., Gaule P., Foray D., *Academic Licensing: A European Study*. CEMI Working paper, CEMI, Lausanne, 2007.

Franzoni Ch., Organizing the office for technology transfer. Ceris-Cnr, Working paper no. 15/2004, Turin, Italy.

Higher Education Act, Journal of Laws of 2005, No. 164, Item 1365 (consolidated text), http://www.nauka.gov.pl/fileadmin/user_upload/szkolnictwo/Reforma/20110523_USTAWA_z_dnia_27_lipca_2005.pdf [30.04.2011].

Interview with Dr. Dariusz Trzmielak, Director of TTO at the Lodz University [2.03.2011].

Interview with Katarzyna Piecuch, University of Technology and Life Sciences in Bydgoszcz [1.03.2011].

Kasperkiewicz W., Innovative strategy of the Polish economy. In: Okoń-Horodyńska, E., Zachorowska-Mazurkiewicz A. (ed.), *Trends in the Innovative Development of Polish Companies*. The Knowledge and Innovation Institute, Warszawa, 2008.

Leydesdorff L., Meyer M., Triple Helix indicators of knowledge-based innovation systems: Introduction to the special issue. *Research Policy*, 35, 1441–1449, 2006.

Marszałek A., Rola uczelni w regionie, Difin, Warszawa, 2010.

Nelles J., Vorley T., Entrepreneurial architecture: A blueprint for entrepreneurial universities. *Canadian Journal of Administrative Science*, 2010.

Neto B.H., De Souza J.M., De Oliveira J., Technological and knowledge diffusion through innovation network. In: Chu S., Ritter W., Hawamdeh S. (ed.), *Managing Knowledge for Global and Collaborative Innovations*, Series on Innovation and Knowledge Management—Vol. 8. World Scientific Publishing, Singapore, 2010.

Nezu R. (ed.), Technology transfer, intellectual property and effective university-industry partnerships. The experience of China, India, Japan, Philippines, the Republic of Korea, Singapore and Thailand, WIPO, 2007.

Onak-Szczepanik B., Innovativeness of Polish economy and state innovation policy. In: Woźniak M.G. (ed.), *Social Inequality and Economic Growth*, Knowledge-based economy, Issue 11. The Publishing Office of the University of Rzeszów, Rzeszów, 2007.

Polt W., Rammer Ch., Schartinger D., Gassler H., Schibany A., Benchmarking industry—Science relations in Europe—The role of framework conditions. Report to the Federal Ministry of Economy and Labour, Austria and to the European Commission (Enterprise DG), Vienna/Mannheim, 2001.

Ponomariov B.L., Effects of university characteristics on scientists' interactions with the private sector: An exploratory assessment. *Journal of Technology Transfer*, 33, 485–503, 2008.

The ProTon Europe Fifth Annual Survey on Knowledge Transfer Activities in European Universities, http://195.88.100.72/news/ProTonEurope2009SurveyReportsSignificantIncreasesinKnowledgeTransferActivitiesinEuropeanUniversities-262.aspx [30.04.2011].

Santarek K. (ed.), *Technology Transfer from Universities to Business. Creating Mechanisms of Technology Transfer*. Polish Agency for Entrepreneurship Development, Warszawa, 2008.

Siegel D.S., Veugelers R., Wright M., Technology transfer office and commercialization of university intellectual property: Performance and policy implications. *Oxford Review of Economic Policy*, 23(4), 640–660, 2007.

82 Positive and Negative Aspects of Educational Organizations' Development in Ukraine

Ludmyla M. Karamushka

CONTENTS

INTRODUCTION

Life today is characterized by change. Changes take place in society, organizations, or individuals. The change is a part of human nature and therefore continuous. But today's change has become faster, deeper, and more complex, bringing with it a new phenomenon in technology, society, politics, and economy, adaptation to which becomes a fundamental condition of survival for individuals and organizations (Wamwangi 2003).

One of the change management strategies is *Organizational Development*, which has been in operation for the last 40 years. The Basics of Organizational Development were created by Lewin (1958). Organizational Development is a process by which behavioral science knowledge and practices are used to help organizations to achieve greater effectiveness, including improved quality of work, life, and increased productivity (Cummings and Huse 1989; Cummings and Worley 1997).

Organizational development may be described as a methodology or technique used to make changes in an organization or section of an organization with a view to improving the organization's effectiveness. Organizational development has the following attributes: it is a planned process of change; it applies behavioral science knowledge; it aims at the change of organization culture; it aims at reinforcement of organizational strategies, structures, and processes for improving an organization's effectiveness and health; it applies to an entire system of an organization, department, or group as opposed to an aspect of a system; and it targets long-term institutionalization of new activities, such as operation of self-managed or autonomous work teams and other problem-solving capabilities (Wamwangi 2003). Underlying organizational development are *humanistic values*. Margulies (1972) articulated the humanistic values of organizational development as follows: providing opportunities for people to function as human beings rather than resources in the productive process; providing opportunities for each organization member as well as for the organization itself to develop to his or her full potential; seeking to increase the effectiveness of the organization in terms of all of its goals; attempting to create an environment in which it is possible to find exciting and challenging work; providing opportunities for people in organizations to influence the way in which they relate to work, the organization, and the environment; treating each human being as a person with a complex set of needs, all of which are important in his or her work and in his or her life (Margulies 1972). The goal of organizational development is to increase the long-term health and performance of the organization while enriching the lives of its members (Organizational development in education - colleges, universities, and other schools 2001).

The ideas of organizational development are currently under development in many ways, which include comparative infrastructure development of organizational development (Organizational development in education - colleges, universities, and other schools 2001), social areas, for example, in finance (Ortego 2009), long-lasting effects of organizational development, in particular, for senior employees (Roux 2008), analysis of the conflict of values being an obstacle for the organizational development (Zakharova 2008).

One of the important areas to realize the ideas of organizational development is *education*, as, in our opinion, it must promptly respond to changes in the society and is focused on training, development, and education of the growing individuals who have quickly to adapt to the society. Moreover, it is the educational organization that needs the application of humanistic ideas the most. Analysis of the relevant literature shows that the idea of organizational development in education has been widely discussed in western countries, especially in the United States. It is done through the analysis of the nature of organizational development, the development of information-based change tools (feedback and action planning, surveys, interviews, direct observations, etc.) and process tools (process consulting, role and responsibility charting, etc.) (Organizational development in education - colleges, universities, and other schools 2001), setting up counseling centers for Organizational Development and Leadership (ODL), which provide consulting, program development, and research services for executive leaders and academic departments in the areas of leadership development, planning, and assessment within higher education nationally and internationally (see, Center for Organizational Development and Leadership 2009). It is also done through staff organizational development training (see, Doctor of Philosophy in Educational Leadership and Organizational Development 2012), creation of special public organizations (see, UNNATI Organization for Development Education 1990) and networks (see, Professional and Organizational Development Network in Higher Education 2007) specializing in organizational development as well as special organizational development departments in universities (see, the The Organizational Development Division of the North Carolina Administrative Office of the Courts 2011).

As for Ukraine, in spite of the fact that various aspects of change management are widely represented in educational organizations (Karamushka 2008), the problem of educational organizations' development has not yet attracted much attention of researchers and practitioners. It can be accounted for, primarily, by the country's current social and economic difficulties and educational organization managers' and staff's poor knowledge of the subject.

The organizational development method uses *the planned research-based change* to increase motivation, remove obstacles, and make change easier (Organizational development in education 2001). However, the preliminary analysis of the work of Ukrainian educational organizations has found that the problems of organizational development are usually resolved spontaneously and without clearly defined research methods and regard to psychological factors and conditions that eventually lower educational organizations' performance. It should be noted that according to the public opinion there are very significant differences between educational institutions of the traditional type (conventional state-owned secondary schools) and the innovative type (private schools, school complexes, gymnasiums, lyceums, etc.).

Hence, the relevance and scientific importance of the problem in question have determined the following *research objectives*:

1. To find out how different aspects of organizational development are reflected in the activities of educational organizations in Ukraine.
2. To analyze the main initiators and agents of educational organizations' development.
3. To determine conditions for educational organizations' development.
4. To identify, according to the predetermined indicators of organizational development, the differences, if any, between the educational organizations of traditional and innovative types.

RESEARCH METHODS AND SAMPLE

For the survey, we applied a comprehensive *tool* "Organizational development of the organization" (Karamushka 2009), which included six sets of questions about various aspects of educational organizations' development. One of the sets included questions about the state of organizational development (presence of the ideas of organizational development and its main initiators and agents as well as the conditions for educational organizations' development).

The investigation was conducted in eight secondary comprehensive educational institutions of traditional and innovative types (four schools of each type) in the city of Kiev and the Kiev region (Ukraine) in 2011. The sample included 402 school employees (248 from the innovative organizations and 154 from the traditional ones). The data were processed using Statistical Package for the Social Sciences (SPSS) (v. 13) (methods of cross-tables and chi-square coefficient).

MAIN RESULTS

Analysis of the answers to the question about organizational development *in the respondents' educational organizations'* found that most respondents (60.2%) believed that their organizations paid much attention to organizational development (Table 82.1). Of the respondents, 23.1% said that their organizations gave attention to organizational development rather than not. A small part of the respondents had negative answers to that question: 1.5% of them said that their organizations didn't give attention to organizational development rather than did, and 0.3% of the respondents said that their organizations didn't pay any attention to organizational development. Of the respondents, 14.9% found the question difficult to answer.

On the whole, the answers could be considered as an evidence that, organizational development was adequately represented in educational organizations.

The investigation found statistically *significant differences regarding this problem between the educational organizations of the traditional and innovative types* ($\rho < 0.001$) (Table 82.2).

Educational organizations of the innovative type considerably prevailed ($\rho < 0.001$) over those of the traditional type regarding different manifestations of organizational development. Thus, 75.8% of the respondents from the innovative type of organizations and only 35.1% of those from the traditional type of organizations believed that much attention was paid to organizational development in their organizations. It's noteworthy that in the traditional type organizations almost every third respondent left the question unanswered whereas in the innovative organizations it was done only by almost every 30th respondent (33.8% against 3.2%, respectively).

On the whole, it can be concluded that different aspects of organizational development were more present in the innovative organizations than in the traditional ones.

As to the *main initiators and agents of* organizational development, it was found that top managers (school principals) were the most active agents (57.2%), followed by the organizations' collectives (46.1%) and individual employees (37.6%) (Table 82.3). Far less important roles in solving

TABLE 82.1
Presence of the Problem of Organizational Development in Educational Organizations (% the Respondents of the Total Sample)

Presence of the Problem of Organizational Development in Educational Organizations	%
My organization pays much attention to organizational development	60.2
My organization pays attention to organizational development rather than not	23.1
I find the question difficult to answer	14.9
My organization doesn't pay attention to organizational development rather than does	1.5
My organization doesn't pay any attention to organizational development	0.3

TABLE 82.2
Presence of the Problem of Organizational Development in Educational Organizations of Traditional and Innovative Types (% the Respondents of the Total Sample)

Presence of the Problem of Organizational Development in Educational Organizations	Educational Organizations	
	Traditional Type	Innovative Type
My organization pays much attention to organizational development	35.1***	75.8***
My organization pays attention to organizational development rather than not	27.3***	20.6***
I find the question difficult to answer	33.8***	3.2***
My organization doesn't pay attention to organizational development rather than does	3.2***	0.4***
My organization doesn't pay any attention to organizational development	0.6***	0.0***

***–ρ < 0.001.

TABLE 82.3
Initiators of Organizational Development in Educational Organizations (% the Respondents of the Total Sample)

Initiators of Organizational Development in Educational Organizations	%
Top managers (heads of educational organizations)	57.2
Collectives	46.1
Individual employees	37.6
Midlevel managers (heads of chairs, heads of methodological associations)	21.9
Superior organizations	21.9
Supervising organizations	18.2
Sponsors	15.2
Line managers (creative group leaders, task group leaders)	13.4
Recipients of educational services (parents, students)	10.7

the problems of organizational development were played by midlevel managers (heads of departments)—21.9%, superior organizations (district departments of education)—21.9%, supervising bodies—18.2%, sponsors—15.2%, line managers (heads of creative and professional groups)—13.4%, and clients of educational services (students and their parents)—10.7% (Table 82.3).

To our mind, the obtained data show certain positive tendencies in solving the problem of educational organizations' development: the problems of organizational development are solved within organizations rather than under the pressure from outside, which suggests that educational organizations have some autonomy and independence. Besides, it is positive that an important role in organizational development is played by the organizations' collectives.

However the recipients of educational services are surprisingly the most passive in solving organizational development–relevant problems. Low activity of the midlevel and line managers also slows down the realization of many organizational development ideas.

Comparative analyses of traditional and innovative educational organizations in relation to the main initiators of organizational development found the following tendencies (Table 82.4). The traditional and innovative educational organizations did not statistically significantly differ in organizational development initiators represented by higher organizations, supervising bodies, and sponsors.

However, the traditional and innovative educational organizations statistically significantly differed regarding the following groups' role in organizational development initiation: top managers

TABLE 82.4
Initiators of Organizational Development in Educational Organizations of Traditional and Innovative Types (% the Respondents of the Total Sample)

Initiators of Organizational Development in Educational Organizations	Educational Organizations	
	Traditional Type	Innovative Type
Top managers (heads of educational organizations)	35.1***	71.0***
Collective	37.7**	51.4**
Individual employees	48.7***	30.6***
Midlevel managers (heads of chairs, heads of methodological associations)	13.0***	27.4***
Superior organizations	22.1	21.8
Supervising organizations	15.6	19.8
Sponsors	13.0	16.5
Line managers (creative group leaders, task group leaders)	0.6**	21.4**
Recipients of educational services (parents, students)	6.5*	13.3*

***–$\rho < 0.001$; **–$\rho < 0.01$; *–$\rho < 0.05$.

(heads of educational organizations) ($\rho < 0.001$); collectives ($\rho < 0.01$); midlevel managers (heads of chairs, heads of methodological associations) ($\rho < 0.001$); line managers (creative group leaders, task group leaders) ($\rho < 0.01$), and recipients of educational services (parents, students) ($\rho < 0.05$).

The educational organizations of the innovative type were ahead of the traditional organizations regarding the participation of the above-mentioned categories of people in organizational development initiation: top managers (heads of educational organizations)—71.0% and 35.1%, respectively; collectives—51.4% and 37.7%, respectively; midlevel managers (heads of chairs, heads of methodological associations)—27.4% and 13.0%, respectively; line managers (creative group leaders, task group leaders)—21.4% and 0.6%, respectively; recipients of educational services (parents, students)—13.3% and 6.5%, respectively.

The traditional and innovative educational organizations statistically significantly differed in the role of individual employees in organizational development initiation ($\rho < 0.001$): In 48.7% of the traditional organizations and 30.6% of the innovative organizations, organizational development was initiated by individual employees.

To our mind, it suggests that the traditional and innovative educational organizations do not differ regarding the solution of the organizational development–relevant problems at the level of external factors. However, the traditional and innovative educational organizations considerably differ at the level of internal factors: The innovative organizations are ahead of the traditional organizations regarding the role of almost all the key participants in the educational process in solving the organizational development problems.

The investigation found *three groups of conditions for educational organizations' development*: (a) informational and self-educating, (b) professional interactional and professional relational, and (c) psychological (Table 82.5).

The analysis of the data found the following hierarchy of conditions:

The most developed were shown to be the *informational and self-educating* conditions, which included educational organization employees' self-education (59.2%) and informational support (51.2%). As the name suggests, the leading characteristic of this group of conditions is reception of information and improvement of knowledge through self-education.

Next came the conditions of *professional interactions and professional relationships*, which included educational organization employees' training (51.5%), formation of groups of professionals to share experiences (47.5%), contacts with partners abroad (31.6%), and contacts with Ukrainian

TABLE 82.5
Conditions of Educational Organizations' Development (% of the Respondents of the Total Sample)

Conditions of Organizational Development	%
Self-education and self-analysis	59.2
Training	51.5
Informational support	51.2
Professional experience sharing groups	47.5
Contacts with foreign partners	31.6
Contacts with Ukrainian organizations	30.8
Individual psychological counseling	16.7
Group psychological counseling	12.9

educational organizations (30.8%). It is noteworthy that being less numerous compared to the first group, this group of conditions dominated in terms of the number of conditions and dealt with the professional interactions and establishment of good relationships in the context of organizational development.

The least presented were the *psychological conditions*, which comprised individual and group counseling (16.7% and 12.9%, respectively). This group, which was three to four times less developed than the first two, related to direct rendering of organizational development–relevant psychological help to employees in educational organizations.

The investigation found statistically significant *differences between traditional and innovative educational organizations* in relation to the determined conditions of organizational development (Table 82.6).

As to the informational and self-educating conditions, the innovative educational organizations were ahead of the traditional organizations both by the level of employees' informational readiness ($\rho < 0.001$) and by the level of employees' self-education ($\rho < 0.001$).

The same situation was observed regarding the psychological conditions: The innovative educational organizations led both in individual ($\rho < 0.01$) and group counseling ($\rho < 0.001$).

Regarding the conditions of professional interactions and professional relationships, the investigation didn't find statistically significant differences between educational organizations of

TABLE 82.6
Comparative Analyses of Conditions of Organizational Development of Educational Organizations of Traditional and Innovative Types (% of the Respondents of the Total Sample)

Conditions of Organizational Development	Traditional Type	Innovative Type
Self-education and self-analysis	46.1^{***}	67.1^{***}
Training	51.3	51.6
Informational support	29.9^{***}	64.5^{***}
Professional experience sharing groups	42.2	50.8
Contacts with foreign partners	27.3	34.3
Contacts with Ukrainian organizations	37.7^{*}	26.6^{*}
Individual psychological counseling	10.4^{***}	20.6^{***}
Group psychological counseling	5.2^{***}	17.7^{***}

***–$\rho < 0.001$; **–$\rho < 0.01$; *–$\rho < 0.05$.

innovative and traditional types in most conditions (organization employees' training, formation of groups of professionals to share experiences, and contacts with partners abroad). The traditional type of organizations outperformed the innovative organizations only in contacts with Ukrainian educational organizations.

Thus, it can be stated that informational and self-educating and psychological proper conditions were most important for the educational organizations of the innovative type whereas the professional interactions and professional relationships were significant for the organizations of the both types; that is, they were the common conditions of organizational development.

CONCLUSION

1. The investigation found a number of positive tendencies in educational organizations' development, which included (a) the presence of the ideas of organizational development in most educational organizations; (b) initiation and realization of the ideas of organizational development from within rather than outside the organizations as the result of organizations' independence and activity; (c) both the managers' and collectives' participation in putting forward the ideas of organizational development; and (d) professional interactions and professional relationships were significant for the organizations of the both types; that is, they were the common conditions for organizational development.
2. However, a number of negative tendencies in educational organizations' development were found, too: (a) The recipients of educational services were passive in initiating and realizing the ideas of organizational development in both types of educational organizations; (b) both types of educational organizations had a gap between top managers and midlevel and line managers regarding their participation in initiation and realization of the ideas of organizational development, which might have negative effects on organizations' development; (c) informational and self-educating and psychological proper conditions were most important for the educational organizations of the innovative type; and (d) on the whole, the traditional type of educational organizations lagged behind those of the innovative type regarding the presence and realization of the ideas of organizational development.
3. The obtained findings may be helpful for educational organization managers and employees in the process of organizational development as well as for educationists' organizational development training at the institutions of postgraduate education and retraining courses.

REFERENCES

Center for Organizational Development and Leadership (ODL), (2009) http://www.odl.rutgers.edu/aboutodl.htm.

Cummings, T. G., and Huse, E. (1989). *Organization Development and Change.* St Paul, MN: West Publishing Company.

Cummings, T. G., and Worley, C. G. (1997). *Organization Development and Change*, 6th Edition, ITP, USA.

Doctor of Philosophy in Educational Leadership and Organizational Development, (2012) http://louisville.edu/graduatecatalog/programs/degree-programs/academic/ge/elodphd/.

Professional and Organizational Development Network in Higher Education, (2007) http://www.podnetwork.org.

Karamushka, L. M. (2009). The design and diagnostic tools of the investigation of the distinctive psychological features of organizational development. *Organizational Psychology. Economic Psychology. Social Psychology*, 2009, #24, pp. 196–208. (in Ukrainian).

Karamushka, L. M. (Ed.) (2008). *Technology of training of personnel of organizations to work at the conditions of social-economic changes*: Textbook. 2nd ed./Scientific editing: L. M. Karamushka. – K.: Naukovy Svit, 2008, 230 p. (in Ukrainian).

Lewin, K. (1958). *Group Decision and Social Change.* New York: Holt, Rinehart and Winston.

Margulies, N. (1972). *Organizational Development: Values, Process, and Technology.* New York: McGraw-Hill Book Co.

The Organizational Development Division of the North Carolina Administrative Office of the Courts (NCAOC), (2011) http://www.nccourts.org/citizens/jdata/documents/organizational_development_facts.pdf.

Organizational development in education - colleges, universities, and other schools, (2001) http://www.toolpack.com/education.html.

Ortego, J. (2009). Organizational Development: Impact of a coaching process for managerial skills development in a financial services cooperative, J. Ortego, A. Mata, M. Etxebarria, Abstracts of 14th European Congress of Work and Organizational Psychology: Developing people in 21st century organizations: Global and local perspectives. – Santiago de Compostela, Spain, May 13–16. (Abstracts, CD-ROM).

Professional and Organizational Development Network in Higher Education, (2007) http://www.podnetwork.org.

Roux, P. (2008). Long-term effects of organizational development in residential care for the elderly focused on the interests of both residents and employees, P. Roux//XXIX International Congress of Psychology (July 21–25, 2008. – Berlin, Germany) (Abstract, CD-ROM).

UNNATI Organization for Development Education, (1990) http://www.unnati.org/.

Wamwangi, K. (2003). 5th Urban and City management course for Africa: Face – face and distance learning version. *Module II. Organizational development as a framework for creating anti-poverty strategies and action including gender mainstreaming*. Tanzania.

Zakharova, L. (2008). Unconscious conflicts of values as socio-psychological barriers to organizational development, L. Zakharova, XXIX International Congress of Psychology (July 21–25, 2008. Berlin, Germany) (Abstract, CD-ROM).

83 IT Solutions in Assessment

Zoltán Kovács, Dóra Tasner, and Péter Volf

CONTENTS

INTRODUCTION

There are many challenges in today's higher education. One of them is the large number of students. "How is an instructor expected to track and to give credit to 200 or more individuals per class...?" asked Ray (2004). In the case of this or higher number of students, even the layout of seats requires careful design.

Computer-aided assessment has a relatively long history compared to general computer applications. Steele et al. (1980) presented a computer-assisted examination resource already in 1980. An early application called MICHELE can be seen as a forerunner of the more sophisticated systems we use today.

Researchers at National Cheng Kung University, Taiwan, describe a design issue aimed at the development and implementation of an adaptive testing system. The system can support several assessment functions and different devices (Huang et al. 2000).

WebCT was the first system that we tried in 1997. Training materials, including texts, quizzes, and glossaries were developed. Students have registered, but after a short operational period, universities stopped the WebCT subscription, so all invested work was lost.

Learning from the first attempt, the next period was passed working out such solutions that can be moved into electronic systems with the least amount of time and that are affordable. There were two directions of developments:

- E-conform paper-based tests
- Spreadsheet models for classroom works and generating quantitative problem for exams

Next we introduce the related solutions for these two. Both solutions use the import function of Moodle. The first decision that had to be made was the format. Fortunately Moodle accepts several import formats in the case of text input. Viewing over the opportunities, we found GIFT format as an optimal solution.

Although it cannot contain all the information related to questions, after the import is completed, some further settings are needed; due to its simple structure, the text file production can be kept simple. Text file coding must be UTF-8.

COMPUTER-AIDED EXAM LAYOUT DESIGN

Exam seating plan can be critical in the case of the following:

- Large number of participants
- Different (variation of) tests
- Low room capacity
- Few labor for monitoring exams
- Risk of cheating

The idea of making an exam seating plan came when it took more than an hour to start an exam (Figure 83.1).

The software—like the others—is spreadsheet based.

Below are the components of the system (each is a worksheet):

- Check-ins come from the central registration system.
- Preassignment (courses–groups; participants in the same groups sit normally in columns).
- Assignment (participants–groups).
- Test material production can be paper or computer based. In the case of paper, we print the most important information on the headline.
- Notification: via central registration system, e-mail, or paper-based announcement at the gate.

Sor\oszl	1	2	3	4	5	6	7	8	9	10
1	1	2	3	4	5	6	7	8	9	10
	Lami ad	Géczi Ā		Kiss Ni	Palkó Z		Kovács			
2	11	12	13	14	15	16	17	18	19	20
	Toldi K	Kazaku		Antal K	Kanyó Z		Horváth			Szabó V
3	21	22	23	24	25	26	27	28	29	30
	Pinizsi	Réti Sz		Jánosi	Katona		Ács Ist			Mezeiné
4	31	32	33	34	35	36	37	38	39	40
5	41	42	43	44	45	46	47	48	49	50
	Di Pol	Jánosi		Lisztma	László		Baji Év			Pozsgay
6	51	52	53	54	55	56	57	58	59	60
	Bognār	Gergely		Kiprich			Barta L			Kele-Ma
7	61	62	63	64	65	66	67	68	69	70
	Gyurkō	Letenye		Kuthy P	Kalapos		Málits			
8	71	72	73	74	75	76	77	78	79	80
9	81	82	83	84	85	86	87	88	89	90
	Birton	Merena		Czafit	Tóth Gy		Kanyó Z			
10	91	92	93	94	95	96	97	98	99	100
	Pintēr	Novák F		Fazekas			Barta L			Mojzer
11	101	102	103	104	105	106	107	108	109	110
	Brenner	Nagy Ge		Csur Jó	Somogy		Letenye			
12	111	112	113	114	115	116	117	118	119	120
13	121	122	123	124	125	126	127	128	129	130
	Szendre	Tamné		Gyozsár	Horvát		Merena			
14	131	132	133	134	135	136	137	138	139	140
	Mosō Er	Szilágy		Halász			Farkas			Németh
15	141	142	143	144	145	146	147	148	149	150
	Moroz K	Jeges E		Jeges E			Szepesi			

FIGURE 83.1 Exam seating plan.

CONVERSION OF PAPER-BASED TESTS INTO AN ELECTRONIC SYSTEM

There are many organizations that move from paper-based testing toward computer-based testing (CBT) (Prometric 2011; Examsoft 2011). Researchers also studied extensively the characteristics of CBT (Thelwall 2000; Clariana and Wallace 2002; Noyes et al. 2004; Macedo-Roueta 2009).

The authors of this paper have a sizeable collection of paper-based tests. In the last 10 years, the tests were prepared in a way that sooner or later they would be moved into an electronic system. For this reason, not only questions but also the right answers were hidden in some way in the text. An example is shown in Figure 83.2.

Virtually all the information is available for the transfer into an electronic system, in which a specific system was not known at the time when this solution was introduced. Years later—in 2008—decision was made that the electronic system would be the well-known, popular open-source Moodle. Since tests consist of mostly multiple-choice questions, there is a specific algorithm for evaluation that is slightly different from what we used during manual correction.

Below are the steps of conversion.

1. Fine setting of the .doc file. Layouts of each question were optimized for paper appearance. Saving space in order to save paper was one main objective. Questions and answers need a line in electronic version. We do this trimming manually because other incorrectness also might require human intervention.
2. Text import into spreadsheet. Like in the next cases, we use a spreadsheet model to generate GIFT text.
3. Checking and saving GIFT format file. Errors mostly occur in the case of not "well-conditioned" text file: missing one or two many new line characters, delimiter setting, or extra space. Steps 2 and 3 can be seen in Figure 83.3.

1. What is rather the characteristic of (physical) production? ≤≤<<•E••>>≥≥
   A. Front Office
   B. It can be done only by internal department
   C. Direct connection between the production and consumption process
   D. It is easier to define quality/conformity requirements••
2. You made tea for you breakfast. What kind of process was that? ••≤≤<<••D•>>≥≥••
   A. Project, **B.** Jobbing, **D.** Batch, **E.** Line, **F.** Continuous-flow

FIGURE 83.2 Paper-based tests contain the correct answers.

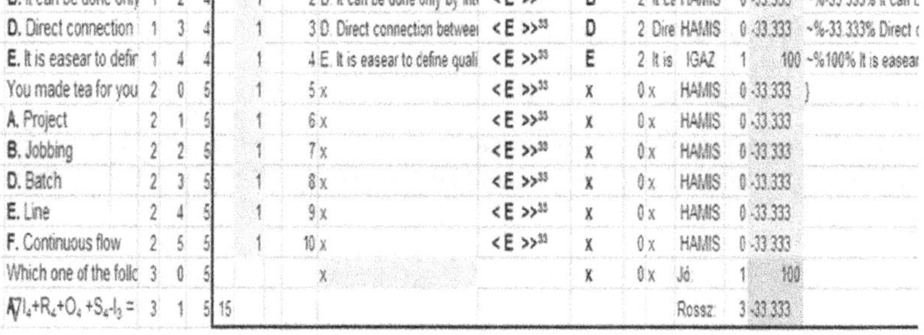

What is rather the ch	1	0	4	1	1	0	What is rather the characteri	££<< E >>[30]	<E >>[1]	61	What is rather the	What is r			:What is rather the ch
A. Front Office	1	1	4	4	1	1	A. Front Office	<E >>[33]	A	2	Fror	HAMIS	0	-33.333	~%-33.333% Front C
B. It can be done only	1	2	4		1	2	B. It can be done only by int	<E >>[33]	B	2	It ca	HAMIS	0	-33.333	~%-33.333% It can b
D. Direct connection	1	3	4		1	3	D. Direct connection betweer	<E >>[33]	D	2	Dire	HAMIS	0	-33.333	~%-33.333% Direct c
E. It is easear to defir	1	4	4		1	4	E. It is easear to define quali	<E >>[35]	E	2	It is	IGAZ	1	100	~%100% It is easear
You made tea for you	2	0	5		1	5	x	<E >>[33]	x	0	x	HAMIS	0	-33.333	}
A. Project	2	1	5		1	6	x	<E >>[35]	x	0	x	HAMIS	0	-33.333	
B. Jobbing	2	2	5		1	7	x	<E >>[30]	x	0	x	HAMIS	0	-33.333	
D. Batch	2	3	5		1	8	x	<E >>[33]	x	0	x	HAMIS	0	-33.333	
E. Line	2	4	5		1	9	x	<E >>[35]	x	0	x	HAMIS	0	-33.333	
F. Continuous flow	2	5	5		1	10	x	<E >>[33]	x	0	x	HAMIS	0	-33.333	
Which one of the follc	3	0	5				x		x	0	x	Jó:	1	100	
$A\sqrt{7}I_4+R_4+O_4+S_4-I_3=$	3	1	5	15								Rossz:	3	-33.333	

FIGURE 83.3 Converting test questions into GIFT text.

4. Import GIFT format file. This step is controlled by Moodle. We have to give the category in which the question will be the format (GIFT) and the file name (Figure 83.4).
5. Change setting. After import we change some default parameters that cannot be set by the GIFT format file: total pint, penalty points.

Spreadsheet allows one to generate/convert a maximum of 20 questions, but text files can be merged. Fractional points for correct and incorrect choice are calculated automatically. Imported question among other questions in the question bank can be seen in Figure 83.5.

```
::What is rather the characteristic of (physical) production?
A1::What is rather the characteristic of (physical) production? {
~%-3.333% Front Office
~%-33.333% It can be done only by internal department
~%-33.333% Direct connection between the production and
consumption process
~%100% It is easier to define quality/conformity requirements
}
::You made tea for breakfast. What kind of process was that?
A1::You made tea for you breakfast. What kind of process was
that? {
~%-25% Project
~%-25% Jobbing
~%100% Batch
~%-25% Line
~%-25% Continuous flow
```

FIGURE 83.4 GIFT text.

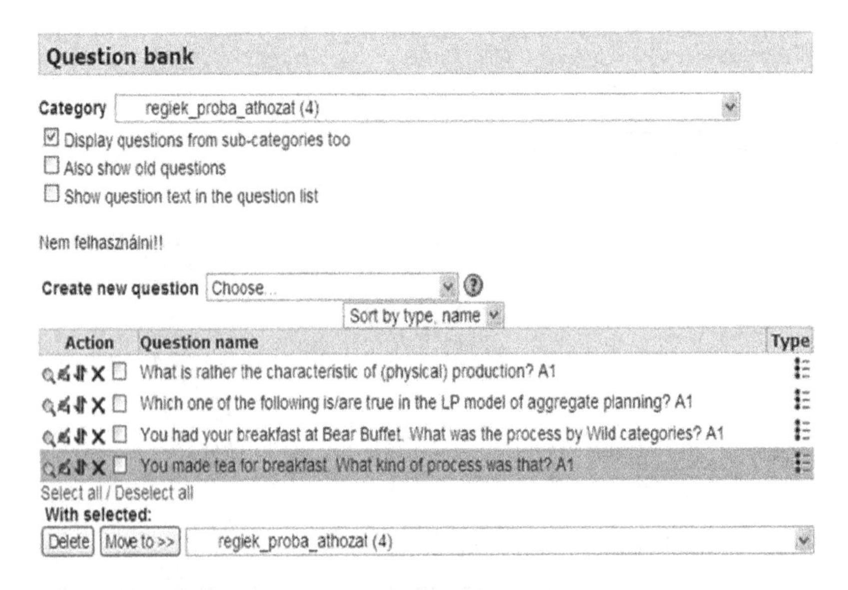

FIGURE 83.5 Questions in question bank.

QUANTITATIVE PROBLEM GENERATION

Problem solving is a special class of tests. Baker and O'Neil (2002) give a review of the status of technology and educational reform, present the traditional approaches to problem-solving measurement common today in technological settings, and propose how problem-solving assessment may change, in particular, the role that authoring strategies may play in measuring problem-solving performance successfully in computer environments. They suggest four areas that must be considered in which computers would increase the fidelity and validity of measures of complex problem solving: (1) the intentions and skills of assessment designers; (2) the range of performance that counts as problem solving; (3) the ways in which validity evidence can be sought; and (4) the degree to which the measurement produces results that generalize across tasks and contexts.

Problems we generate are formally multiple-choice tests, but in reality they are not. Typically answers belong to more results (subquestions), and each subquestion has only one correct answer. There are two types of quantitative problem generator developed.

Problem-Specific Model

This uses almost the same calculation model as what is used during the classroom work. We ask the values of certain cells of the model. Questions are formulated in whole sentences.

This kind of a generator consists of the following parts (worksheets):

Model Page(s)

This page contains the formulation of the problem to be solved. The user also gives here the questions. Depending on the settings, the software later multiplies them. Figure 83.6 shows the well-known economic order quantity model.

The model page also contains the text of questions and the correct answers (Figure 83.7). Questions can be in a different language.

Question Page

A user gives here a reference to the questions. Depending on the settings, the software later manifolds them (Figure 83.8).

Parameters for multiple choices can be set here: the number of choices per question and which one should be the correct one. For example, "2" as a correct answer number means that the correct answer will be "B."

	A	B	C	D	E	F	G
1	r	1000	kg/day	Using formula	T	28,28427	day
2	c1	0,001	euro/(kg·day)	EOQ:	q	28284,27	kg
3	c3	400	euro		Inv. Hold.	14,14214	euro/day
4					Supply	14,14214	euro/day
5	Money	euro	€	9,01561146	Total	28,28427	euro/day
6	Quantity	kg	2000000	2000			
7	Time	day					
8	T	q	Inv. Holding	Supply	Total		Delta:
9	13,52829	13528,29	6,764146504	29,5676624	36,331809		
10	15,03144	15031,44	7,515718338	26,6108961	34,126614		0
11	16,7016	16701,6	8,350798154	23,9498065	32,300605		1
12	18,55733	18557,33	9,278664615	21,5548259	30,83349		2
13	20,61925	20619,25	10,30962735	19,3993433	29,708971		3

FIGURE 83.6 An example for model page: EOQ.

Consumption of a part is constant in a system: 1000 kg/day.
Ordered parts arrive without delay. Backlog can not occur, machines are reliable ones. Inventory holding cost: 0.001 euro/(kg•day), cost of one supply: 400 euro.

a. What time should be between two orders in order the minimum total cost in a time unit? (28.2842712474619 day)
b. What quantity to order? (28,284.2712474619 kg)
c. How much will be the inventory holding cost during a time unit in optimal case? (14.142135623731 euro/day)
d. How much will be supply cost during a time unit in optimal case? (14.142135623731 euro/day)
e. How much will be the total operating cost (invetory, holding + supply) during a time unit in optimal case? (28.2842712474619 euro/day)

FIGURE 83.7 Question and solutions in the model.

EOQGIFT012
Problem
Consumption of a part is constant in a system :1000 kg/day . Ordered parts arrive without delay. Backlog can not occur, machine

Question	Név	Unit	No. of choices	Correct answ.
a. What time should be between two orders in order the minimum total cost in a time unit?	T	day	3	2
b. What quantity to order?	q	kg	3	1
c. How much will be the inventory holding cost during a time unit in optimal case?	Inv. hold.	euro/c	3	3
d. How much will be supply cost during a time unit in optimal case?	Supply	euro/c	3	1
e. How much will be the total operating cost (invetory holding + supply) during a time unit in	Total	euro/c	3	2

FIGURE 83.8 Question settings.

Interval Generation Pages

These pages serve to calculate the borders of multiple-choice intervals (Figure 83.9).

The multiple-choice generator, located on question page, compiles the GIFT text, including the evaluation of correct and wrong answers (Figure 83.10).

Figure 83.11 shows how a question looks in the Moodle system.

A	B	C	D	E	F	G	H	
		Choice_gen		0 - 90%	Min. 0			
	Alapértelmezések		12,00%	40,00%	0	1		
	Solution	Answer	Interval	Asszimme	Gap	Round	Error	Result
1	28284,27	B						A: 23532,5 ≤ X < 26926,6, B: 26926,6 ≤ X
2	14,14214	A				2		A: 13,47 ≤ X < 15,17, B: 15,17 ≤ X < 16,87
3	14,14214	C						A: 10,1 ≤ X < 11,8, B: 11,8 ≤ X < 13,5, C:
4	28,28427	A						A: 26,9 ≤ X < 30,3, B: 30,3 ≤ X < 33,7, C:
5	0	X						A: 0 ≤ X < 0, B: 0 ≤ X < 0, C: 0 ≤ X < 0, D

FIGURE 83.9 Generation intervals for multiple choices.

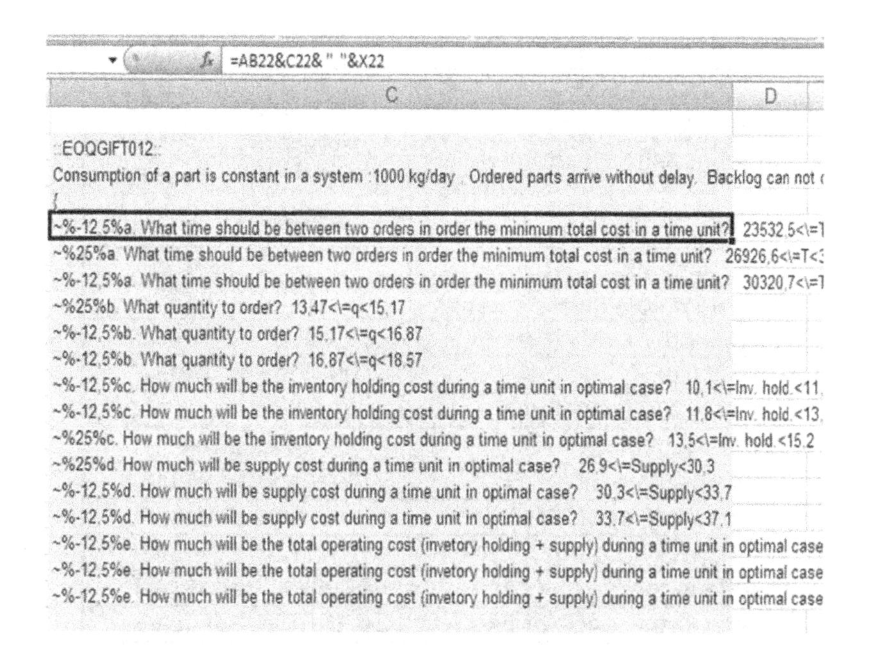

FIGURE 83.10 Generation of multiple choices.

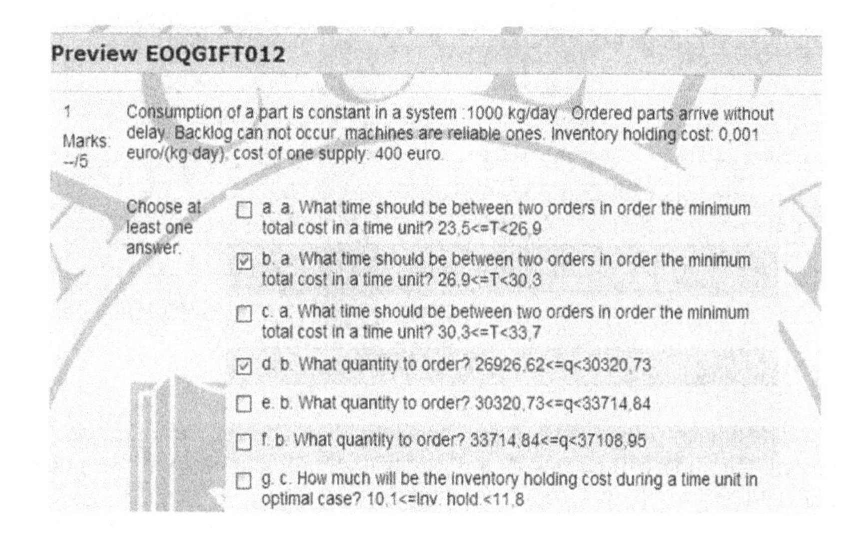

FIGURE 83.11 Imported test question in the Moodle system.

A Problem-Independent General Framework

The general framework provides an area where the model of the quantitative problem can be built up. It uses a kind of "mark and hide" technique. It consists of basically the same pages that were introduced above, but the philosophy is different.

This consists of the following pages.

Model Page(s)

This page also contains the formulation of the problem to be solved, including a *problem table* is the original (mathematical) model. It must be structured and headlined in the way that all the

Rate of consumpti	r	500	kg/week
Inventory holding c	c1	4	€/ (kg week)
Supply cost	c3	313,6	€
Quantity to be ord	q	750	kg
Order interval	T	1,5	week
Total inventory hol	Kh	1500	€/ week
Total supply costs	Ks	209,06667	€/ week
Total costs	Kt	1709,0667	€/ week
Quantity to be ord	qo	280	kg
Order interval in t	To	0,56	week
Total inventory hol	Kho	560	€/ week
Total supply costs	Kso	560	€/ week
Total costs in the	Kto	1120	€/ week
Saving in the case	Ksav	589,066667	€/ week
Saving in the case of EOQ		34,467156	%

FIGURE 83.12 Problem model in well-structured table.

GIFTv003h_sulypB_001b_10p_004

2	3	4	5	6	7
te of consumption (demand)		500,00	kg/week		
Inventory holding cost		4,00	€/ (kg week)		
Supply cost		313,60	€		
Quantity to be ordered		a:	kg		
Order interval		1,50	week		
Total inventory holding costs		b:	€/ week		
Total supply costs		c:	€/ week		
Total costs		d:	€/ week		
the case of optimum (EOQ)		e:	kg		
the case of optimum (EOQ)		f:	week		
the case of optimum (EOQ)		g:	€/ week		
the case of optimum (EOQ)		h:	€/ week		
the case of optimum (EOQ)		i:	€/ week		
Saving in the case of EOQ		j:	€/ week		

x	x					
	2	3	4	5	6	7
		x			x	x
x	x	x	x	x	x	x
		x			x	x
		x			x	x
x	x	x	x	x	x	x
		x	a:		x	x
		x			x	x
		x	b:		x	x
		x	c:		x	x
		x	d:		x	x
		x	e:		x	x
		x	f:		x	x
		x	g:		x	x
		x	h:		x	x
		x	i:		x	x
		x	j:		x	x
x	x	x	x	x	x	x
x	x	x	x	x	x	x

FIGURE 83.13 Mark and hide.

contents—text and numbers—have to be unambiguous for the reader, because explicit question will not be formulated. Question will be raised in a way that certain numbers will be hidden (such as sudoku) (Figure 83.12).

In a *mark and hide session*, one can mark the cells, rows, and columns. Letters "a:" and "b:" mean that values in these cells will be questions "a" and "b." Unmarked cells will be visible (Figure 83.13).

Question Page

Due to another way of questioning, this page is simpler. Questions are marked with letters such as "a," "b," "c," etc. (Figure 83.14).

Interval Generation Pages

Interval generation pages are same than it was presented in Figure 83.9. *GIFT generator* is rather complicated because it contains html tags for structuring table and reference for formula collection. (In the case of previous version—Figure 83.10—they were manually inserted.) Importable GIFT text can be seen in Figure 83.15.

Questions	Name	Menny	Number of questions	N. of correct	ctrl	1	Correct ans'
1. What is value a ?	a		3	3	3	4	750
2. What is value b ?	b		3	3	3	7	1500
3. What is value c ?	c		3	1	1	10	209.07
4. What is value d ?	d		3	2	2	13	1709.07
5. What is value e ?	e		3	1	1	16	280
6. What is value f ?	f		3	3	3	19	0.56
7. What is value g ?	g		3	2	2	22	560
8. What is value h ?	h		3	2	2	25	560
9. What is value i ?	i		3	1	1	28	1120
10. What is value j ?	j		3	1	1	31	589.07

FIGURE 83.14 Question table.

```
::GIFTv003k_eoq1A_001e_10p_004::  [html]::GIFTv003k_eoq1A_001e_10p_004::
Calculate certain missing values!
<table cellspacing\="0"cellpadding\="0" border\="0" style\="width\:780px;
height\: 200px;"><tbody>
<col width\="600" style\="width\: 600pt;"></col>
<col width\="60" style\="width\: 60pt;"></col>
<col width\="120" style\="width\: 120pt;"></col>
<col width\="0" style\="width\: 0pt;"></col>
<col width\="0" style\="width\: 0pt;"></col>
<tbody>
<tr><td>Rate of consumption
(demand)</td><td>500,00</td><td>kg/week</td><td></td><td></td></tr>
<tr><td>Inventory holding cost </td><td>4,00</td><td>€/ (kg
week)</td><td></td><td></td></tr>
<tr><td>Supply cost</td><td>313,60</td><td>€</td><td></td><td></td></tr>
<tr><td>Quantity to be
ordered</td><td>a:</td><td>kg</td><td></td><td></td></tr>
<tr><td>Order interval</td><td>1,50</td><td>week</td><td></td><td></td></tr>
<tr><td>Total inventory holding costs</td><td>b:</td><td>€/
week</td><td></td><td></td></tr>
<tr><td>Total supply costs</td><td>c:</td><td>€/
week</td><td></td><td></td></tr>
<tr><td>Total costs</td><td>d:</td><td>€/ week</td><td></td><td></td></tr>
<tr><td>Quantity to be ordered in the case of optimum (EOQ)
</td><td>e:</td><td>kg</td><td></td><td></td></tr>
<tr><td>Order interval  in the case of optimum (EOQ)
</td><td>f:</td><td>week</td><td></td><td></td></tr>
<tr><td>Total inventory holding costs  in the case of optimum (EOQ)
</td><td>g:</td><td>€/ week</td><td></td><td></td></tr>
<tr><td>Total supply costs  in the case of optimum (EOQ)
</td><td>h:</td><td>€/ week</td><td></td><td></td></tr>
<tr><td>Total costs  in the case of optimum (EOQ) </td><td>i:</td><td>€/
week</td><td></td><td></td></tr>
<tr><td>Saving in the case of EOQ</td><td>j:</td><td>€/
week</td><td></td><td></td></tr>
</tbody>
</table>
<br /><img hspace\="0" height\="400" border\="0" width\="800" vspace\="0"
src\="http\://moodle.gtk.uni-pannon.hu/file.php/1455/kepletek.gif"
alt\="Kérje a képletgyűjteményt!" title\="Ask for formulas!" /><br /><br />
{
~%-5%1. What is value a   ?  555<\=a<630
~%-5%1. What is value a   ?  630<\=a<705
~%10%1. What is value a   ?  705<\=a<780
~%-5%2. What is value b   ?  1110<\=b<1260
~%-5%2. What is value b   ?  1260<\=b<1410
...
~%-5%9. What is value i   ?  1276,8<\=i<1388,8
...
...
~%10%10. What is value j   ?  553,7258<\=j<612,6328
~%-5%10. What is value j   ?  612,6328<\=j<671,5398
~%-5%10. What is value j   ?  671,5398<\=j<730,4468
}
```

FIGURE 83.15 GIFT standard text.

Below are the advantages of this general framework:

- General questions. In the case of a new model, the only model that has to be created, questions can be given by marking the missing and asked values.
- Easy polyglot support.
- Orients students to organize their calculations in a table/spreadsheet way.
- Easy to enforce the progressive (forward) and retrograde (backward) calculation path. It is only the question of marking: which are the known, missing, and asked variables.
- One-step import into the Moodle system.

EXPERIENCES AND CONCLUSION

Using these IT solutions, exams became more efficient and effective. Seating plan accelerated dramatically the start of exams. Generally, it takes not more than 10 min to start the work. Since participants receive the most important information—name, course title, room, seat coordinates, and layout plan—they are already sitting on their place when the teaching staff arrives. Optimizing the distance of the same subject writers, cheating and other disturbing communication can be minimized.

Converting from paper-based tests permits us to generate (to recycle) hundreds of proven test questions into Moodle test bank rapidly. Quantitative problem generators allow us to generate test problems in a short time with a low error rate.

Student feedback is good. They prefer Moodle against paper. An additional gift for them was the evaluation algorithm. What Moodle uses is a little milder than what we used before during paper-based exams.

REFERENCES

Baker, E.L. and O'Neil, H.F. Jr. (2002): Perspectives on computer-based assessment of problem solving. *Computers in Human Behavior*, 18(6), 605–607.

Clariana, R. and Wallace, P. (2002): Paper-based versus computer-based assessment: Key factors associated with the test mode effect. *British Journal of Educational Technology*, 33(5), 593–602.

Examsoft (2011): http://www.examsoftsolutions.com/go/examsoft-transition-from-paper-and-pencil-testing (30 April, 2011).

Huang, Y.M., Lin, Y.T. and Cheng, S.C. (2000): An adaptive testing system for supporting versatile educational assessment. *Computers and Education*, 34(1), 37–49.

Macedo-Roueta, M. (2009): Students' performance and satisfaction with web vs. paper-based practice quizzes and lecture notes. *Computers and Education*, 53(2), 375–384.

Noyes, J., Garland, K. and Robbins, L. (2004): Paper-based versus computer-based assessment: Is workload another test mode effect? *British Journal of Educational Technology*, 35(1), 111–113.

Prometric (2011): Converting from paper based to computer based testing, https://www.prometric.com/reference/pbt_to_cbt.htm (30 April, 2011).

Ray, R. (2004): *Adaptive Computerized Educational Systems: A Case Study Evidence-Based Educational Methods*. 143–170. Elsevier, Amsterdam.

Steele, A., McCumber, M.J. and Davis, P.J. (1980): A computer-assisted examination resource. *Computer Programs in Biomedicine*, 11(3), 238–248.

Thelwall, M. (2000): Computer-based assessment: A versatile educational tool. *Computers and Education*, 34(1), 37–49.

84 Engineering Education in India
Challenges for the Twenty-First Century

N. C. Shivaprakash

CONTENTS

INTRODUCTION

Higher education has been in the spotlight recently in several countries for its current and potential role in economic and social development. In the case of India, the Knowledge Commission has articulated the concerns and outlined the strategies that can, among other things, help to build excellence in educational system to meet the knowledge challenges of the 21st century and increase India's competitive advantage in the field of knowledge. Education is the cumulated heaps of discoveries. The primary objective of the higher education is to prepare students for employment; the secondary objective is to promote equality and social justice to develop humanism and scientific outlook (Berlia 1998; Report by MHRD Government of India 1990). This paper deals with the higher education in India in general and technical education in particular. The scenario of the education system in India has completely changed in the 21st century. Higher education encompasses science, engineering, and medicine. National institutions in the country deal mainly with research; on the other hand, universities concentrate only on teaching. A good institution should have both components of research and teaching in the right proportion.

The interdisciplinary aspect is important in education in the present context (Froyd 2005). There has been a paradigm shift in research and development for bringing out new technologies, materials, and processes. The innovations emerging out of such different R&D are no longer solely monodiscipline based but are predominately interdisciplinary in nature. The concepts, principles, and procedures drawn from various disciplines are being used in that case either to develop new products and services or improve existing products and services to meet changing needs of society and economy. Hence, there is a need to adopt interdisciplinarity and interdisciplinary education and research in overall education system as the environment is ever changing and further becoming more and more multidimensionally complex. The concept of curriculum revision, getting academic autonomy, and becoming deemed to be universities is the order of the day. In this process, aspiring for accreditation of institutions plays a major role in the country. The details of all these aspects are discussed in this paper.

CURRICULUM DEVELOPMENT

The quality of education completely depends on the quality of the curriculum. Undergraduate teaching is always challenging. The technical education curriculum revolves around a certain universal set of concepts, principles, and practical skills and understanding, which both industry and society would agree are important for students to learn. The current crisis faced by the universities in the country is more intellectual (Yehuda 2011). Due to this, curricula developed in the undergraduate program mainly deviate from research, and only information of static nature rather than dynamic situation is provided to the students. The development of social, intellectual, and technological change within and outside universities is increasing out of proportion. Most of the universities are going for structural and radical transformation and thereby giving less importance to the curricula. The curricula have to be the core domain of the university along with other issues involved. The redesign of the curricula need not be universal, and at least regional importance has to be given. Twenty-six thousand intuitions should not have curricula of their own; as far as possible, core curricula should be the same for all institutions. The following are some suggestions to arrive at reasonably good curricula, keeping in mind that curricula should not be region specific:

1. Teach disciplines rigorously in introductory courses.
2. Do not teach just factual subject matter; highlight the challenges.
3. Create awareness of the great problems of humanity, for example, climate and health.
4. Demonstrate and rigorously practice interdisciplinarity.
5. Do not treat knowledge as static.
6. Provide fundamental understanding of the basics of natural and social sciences.
7. Emphasize broad and inconclusive evolutionary mode of thinking in all areas of the curriculum.
8. Familiarize students with nonlinear phenomena in all areas of knowledge.
9. Apply knowledge to real-world problems.
10. Implicate modern communication and information technologies for education.

The curricular changes of this magnitude and significance demand structural changes and institutional profiles of the universities. It demands radical changes in leadership, finance, rewards, assessment, and incentives. A dedicated group of scholars and educators should be part of curriculum changes.

ACADEMIC ASSESSMENT AND ACCREDITATION

Academic assessment and accreditation play an important role in parting good quality education. Setting up standards in education through various statutory bodies shall enhance the quality in education but may not be adequate to provide the total quality education (TQE). The system imparts TQE, which helps provide flexible learning methods, improved infrastructures, model development, need- and industry-based curricula, networked education, collaborative learning programs, industrial involvement in education, affordable educational policies, practical orientation, and progressive performance evaluation and development system. The desired results are directly dependent on two factors: (1) the involvement of all stakeholders and (2) action-oriented implementation of the policies. The quality is achieved by strategic plans like bringing in quality students, system of evaluation of students, and examination system. In a developing country like India, reforms in university education have the potential to achieve sustainable and far-reaching improvement in the lives of millions of Indians.

India is witnessing tremendous growth in terms of the number of engineering and technical colleges (3000 colleges) with an intake of above 1 million every year. Indian universities and engineering colleges mostly follow a syllabi system. Though this helps provide the subject knowledge, it is ineffective in developing manpower with abilities for this profession. The new educational policy in 1986 had recognized the need for the statutory body at the national level responsible for overseeing the growth and quality of technical education in the country. Accordingly, All Indian Council for Technical Education (AICTE) was established by an Act of the Parliament in 1987. As part of this, a National Board of Accreditation (NBA) was established in 1994 to access the quality of the technical education system. The board undertakes evaluation for the purpose of accreditation of programs and gives feedback to the institutions. The following eight criteria are used in the process (NBA Manual 2007):

1. Organization and governance
2. Financial resources
3. Physical resources (central facilities)
4. Human resources (faculty and staff)
5. Human resources (students)
6. Teaching learning processes
7. Supplementary processes (cocurricular and extracurricular activities)
8. Research and development

If we look toward the accreditation of other countries like the United States, United Kingdom, and Japan, we find that, for example, in the United States, the Accreditation Board for Engineering and Technology (ABET 2000) is trying to maintain the accreditation in their system. The various criteria for accreditation of programs of ABET are

1. Students
2. Program education objectives
3. Program outcomes
4. Continuous improvement
5. Curriculum
6. Faculty
7. Facilities
8. Support
9. Program criteria

If we compare the Indian accreditation system with that of the United States, we find that criteria 1, 2, and 3 are missing. Similarly, Engineering Professors Council (2002) London has formulated engineering graduates' output standard in which seven generic abilities have been identified. The Japan Board of Accreditation for Engineering (2002) has clearly laid down criteria on the basis of certain knowledge and abilities. Based on these trends in other countries, India needs to reconsider engineering and technical education and its accreditation system. Each college should have close linkages with selected industries, which will be helpful to decide the abilities; otherwise the output of the students will be poor and will not be useful for engineering and economics activities. India is being an associated member of Washington accord for technical education accreditation, making efforts to reform the present accreditation process by incorporating the ABET policies.

GAPS BETWEEN INDIAN INSTITUTIONS AND COUNTERPARTS ABROAD

The future of the youth depends on knowledge rather than resources. The status of India in the world, leadership, income level, and well-being of people depends on scientific and technological capability. India, to cope with challenges, will have to enlarge the scope of institutions over research, innovation, creativity, technology creation, patent, and intellectual property rights (IPRs). To achieve this, the following needs to be addressed:

1. Creativity, discoveries, and innovativeness
2. Institute–industry interaction
3. Commercial exploitation of research—patent
4. Production of more postgraduates and doctorates
5. Optimization of size of the institutions
6. More budgetary allocation
7. Academic freedom
8. Active and productive research
9. More technical universities

India has to identify these gaps and implement them by institutions to cope up with the Western world. Lots of collaborative work and interaction with external world are required to achieve the goal. Opening of more technical universities is the order of the day. India has 18–20 technical universities as compared to 84 in Germany and more than 100 technical (in US). The university education system when revitalized with strategic components adopting the changing needs and technology would address to the continuous improvement of the students in the nation and fulfill the emerging gap between the academia and the industrial world. It is essential that an appropriate mechanism be built in universities to encourage academic entrepreneurship as well as facilitate easy and effective technology transfer to the industry.

CONCLUSION

An attempt has been made to review the engineering and technical education in India. The 21st century poses challenges for good education. The issues involved in developing a reasonably good curriculum are discussed. A detailed account of assessment and accreditation process is described here. Finally, issues regarding gaps between Indian institutions and their counterpart abroad are presented.

REFERENCES

Accreditation Board for Engineering and Technology—ABET, Criteria for accreditation of engineering programs, 2000: www.abet.org/images/criteria/17.04.pdf.

Berlia S., Theme presentation on higher education summit road map for the future. UNESCO, World Conference on Higher Education in the Twenty–First Century, Paris, 5–9 Oct 1998.

Froyd, Integrated engineering curricula. *Journal of Engineering Education, American Society for Engineering Education*, 94, 147, 2005.

NBA Manual, Criteria and weightages, 2007: www.nba.aicte.ernet.in/parameter.doc.

Report on Education and National Development. Department of Education, Ministry of Human Resources and Development, Government of India, 1990.

Yehuda E., Lecture delivered at Indian Institute of Science, Bangalore, India, January 2011.

Section V

Education in Modern Society

85 Results of the Polish Policy in the Field of Education and Development of Young Scientists in the Years 1944 to 2012

Zofia Godzwon

CONTENTS

INTRODUCTION

This paper presents the results of the Polish policy in education and development of young scientists from World War II until 2012. The article discusses the development of higher education, the goals and tasks of the university under the law, and quantitative development of academic staff in the years 1944 to 2012. This paper shows the main causes of social and political changes affecting higher education. Laws of 1944 to 1989 and their implementation until the meeting of the "Round Table" show that those were seemingly the days of policies that build on the work carried out by the authorities as "happiness and human development." In 1976, politicians introduced into the Constitution that "the Polish United Workers' Party (PZPR) is the guiding force of the Nation"; in fact, the current investigation was "who is not with the Party, is against it."

Optimistic conclusion resulting from the material gathered is achieved by the Polish higher education, principles of education, and development of young researchers, comparable with the leading countries of the European Higher Education Area (EHEA). Barrier to the development of Polish higher education is the value of gross domestic product spent on research and higher education. Although it is a % of GDP, comparable with other countries in the EHEA, the quota value is several times lower than in these countries.

This article is based on statistics, laws, and source materials of institutions that have been important to the ongoing changes in higher education.

DEVELOPMENT OF HIGHER EDUCATION IN POLAND

The Second World War brought huge losses in teachers. Out of 4452 academic teachers, who worked in 1937/1938, 942 people involved in science and higher education were killed or died in the years 1939–1945. The scientific community in the Polish capital lost 416 university researchers over the war years, that is, 44.2% of the total losses. As many as 49% of Polish war losses of the scientific community were in group professors and associate professors. At the time of the war, losses in the academic environment amounted to 27.4% of young scientists (up to 35-year-olds). In contrast, before World War II, 88,150 teachers were employed in the educational system (schools and universities). During the war, 9662 of teachers of schools were killed. The biggest losses were among teachers of elementary schools; 7389 teachers of elementary schools, who constituted 76.3% of the total loss of teachers, were killed or died in warfare.

In June 1946, in Poland, there were 16 public higher education schools and universities, 4 nonpublic university colleges, 5 public higher vocational education schools, and 3 nonpublic school of higher vocational education. In Poland after World War II, the "Decree on the organization of science and higher education in 1947" upheld the distinction between academic institutions, with the right to confer degrees of MA and PhD, assistant professor, and academic title of professor, and division colleges, suitable for bachelor degrees and professional engineer. The Act of 1951 allowed for the existence of nonstate schools, which operated on the basis of regulations issued by the Ministry of Higher Education. In 1958, the law was written to distinguish a polytechnic, an academy, a school principal, and a university. Higher education institutions were granted the right to confer professional degrees: master's degree—an engineer and a doctor. The right to confer the degree PhD, associate professor, and professor has been subject to the staffing of departments that provide training. The list of authorized departments is determined by the Minister of Higher Education.

In 1965, the Law on Higher Education re-split into higher education establishment and higher vocational education school, established by the Council of Ministers. In institutions of higher education, full time, evening, and weekend learning can be conducted. Higher education institution graduates may obtain a master's degree also as an engineer or a doctor. However, higher vocational education school graduates receive diplomas with professional title, referred to in the Ministerial Regulation. In December 1968, higher vocational education schools were transformed into higher education institution (universities). Since 1969, institutions of higher education can lead vocational training, master's degree, doctorate, postgraduate, and professional training courses and other special studies and courses. In the 1960s, "higher education institution was actively involved in the construction of socialism in Poland," (Law of 11.05.1958 years, of higher education, paragraph 1) performing tasks related to education in the university. At that time, the number of higher education institutions amounted to 75 state universities. In the early 1970s, there were a total of 85 schools. In the years 1973 to 1977, there were 89 higher education institutions (Statistical Yearbook of the Republic of Poland 1974 (XXXV) to 1978 (XXXVIII). The following year, the number increased by three universities and the total remained until 1988. However, the academic year 1989/1990 was already inaugurated in 97 universities (Statistical Yearbook of the Republic of Poland 1991 [LI]).

The year 1990 began with major changes in higher education in Poland as a new Law on Higher Education was introduced. This act amended the tasks of the ministry responsible for higher education by deleting the existing record "implementation of state policy in the field of education, teaching and education of children and youth in need of a socialist society and the state." Higher education institution autonomy has been ensured. At the same time, this law also introduced the ability to create nonstate universities. Bednarska (Social Policy No. 3/2007, p. 7) indicates that "due to the transformation of the Polish economy and changes in the occupational structure also a different attitude to education gave notice. Educational aspirations of Polish society were stimulated, because education is treated as an investment in human capital, in the future, giving a profit reimbursed by sufficiently high wages and reducing the risk of unemployment." In the 1990s, many changes occurred due to the new legal basis regulating the functioning of the institutions of higher education in Poland. This created and developed nonstate universities. An act diversified master system studies by reintroducing a two-tier education system. In addition, third-degree studies created were run by nonuniversity research institutions (including the Polish Academy of Sciences). Already at that time, some schools were introduced in the framework of the existing studies in regular classes such as foreign language instructions. Students also had the opportunity to choose an individual course of study. Since 1998/1999, under the new law, higher vocational education schools were created. These schools were designed to prepare students to perform specific jobs and gave bachelor or engineering degrees.

In 2005, a new document, Law on Higher Education, came into force. It was an act that fulfills the conditions to adapt the higher education system to European standards. It was also a step in the direction of arrangement of higher education in Poland through a combination of the Higher Education Act of 1990 and the Law on Higher Vocational Schools issued in 1997. The number of schools in Poland in the years 1950 to 1989 was stable, varying between 75 and 86 universities. The year 1990 for Polish higher education proved to be a turning point. The Higher Education Act of 2005 introduced the possibility of creating nonstate universities. At the same time, the Act authorized the payment of certain educational services. These changes led to the development of the system of paid and part-time evening courses and postgraduate studies in public universities. The effect of the provisions of the Higher Education Act is presented in Figure 85.1.

Between 1960 and 1961, there were 75 higher education institutions, and in 2008/2009, there were 455 universities. There was a sixfold (606%) increase in the number of higher education institutions in Poland. The breakthrough was the year 1990/1991, when nonpublic schools began to emerge faster. In the academic year 1996/1997, they equaled the number of state schools. In the academic year 2008/2009, there were 324 private higher education institutions, while there were 131 state universities. In the same year, 131 state universities, which accounted for 28.8% of all higher education institutions, constituted up to 82.9% of working teacher positions; 324 nonstate higher education institutions, representing 71.2% of those schools, gave work to only 17.1% of teachers. The nonpublic schools constituted 33.32% of the total number of students.

In Poland in November 2011, there were 462 institutions of higher education. These were public schools (127), churches (7), and private schools (328). Among them, there were 111 academic institutions (i.e., with at least one power to confer doctoral degrees) and 351 vocational schools (leading only studies I and/or II level and/or cycle programs). Among the academic institutions, there were 91 public universities, 5 church universities, and 15 private universities. This statistic does not include seminaries run by churches.

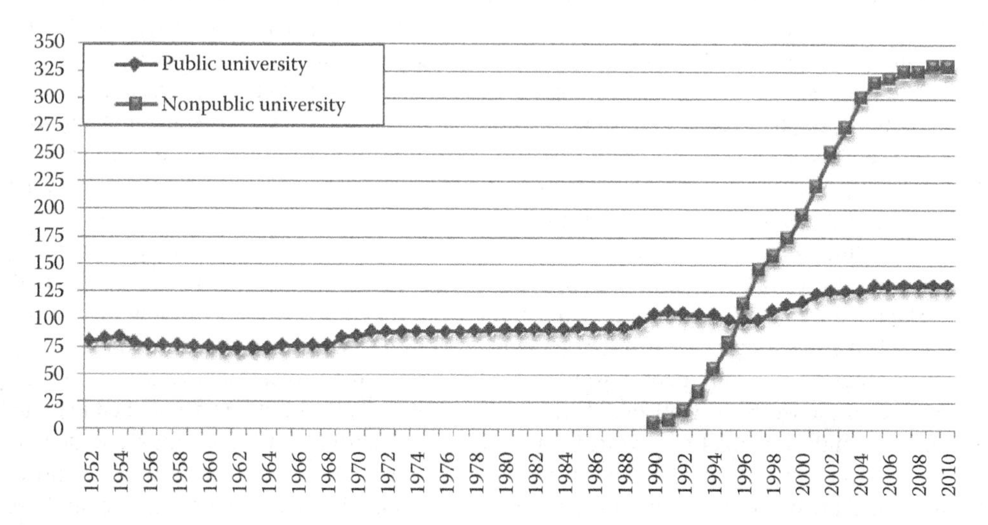

FIGURE 85.1 Number of public and private schools in the years 1952–2010. (From own study based on *Statistical Yearbook of Polish*, Central Statistical Office, Warsaw, year XXIV [1955] to year LXXI [2011].)

OBJECTIVES AND TASKS OF HIGHER EDUCATION INSTITUTIONS

In 1944, the Polish Committee of National Liberation announced, as the most urgent tasks in the areas, to liberate and rebuild schools and provide free education at all levels. Strict enforcement of universal education and immediate reconstruction of schools were announced. A special environment of the Polish intelligentsia was decimated by the Germans, especially men of science and art (Manifesto PCNL 1944). Immediately after the cessation of hostilities, teachers started to be trained not only in higher education but also in high schools and in an annual educational course for teachers for those with maturity exam (after high school).

The decree of October 1947 stated that research in Poland as free and creative is under special care. This purpose is served for academic higher education and higher vocational schools. While the task of higher vocational schools was to educate candidates for the highest level of practical profession, the academic mission of higher education institutions was to organize and conduct research, training and prepare theoretical and practical candidates for the scientific work. At the same time, the role of universities was to promote cooperation in the knowledge society.

The law on higher education and scientific workers of 1951 stated that universities, educating and raising people's intelligence in the spirit of selfless service of homeland, the struggle for peace, and socialism, prepare employees with the highest level of professional qualifications by making them capable of independent scientific development issues in the performance of their profession; and train scientists and prepare them for teaching and scientific research, and organize and conduct scientific research and participate in the promotion of science, technology, and the spread of a scientific view of the world. Half a year later, in the Constitution of July 22, 1952, the Polish People's Republic was proclaimed and approved the Stalinist system.

According to the Law on Higher Schools of 1958, universities are actively involved in building and strengthening socialism by conducting creative research, education, professional training, and preparation for their research and teaching, training, and education of highly qualified staff of professional intelligence, prepared to perform jobs requiring mastery of the art and scientific knowledge and to develop and nurture a national culture and technological progress (*Journal of Laws of 1958*, no. 68, item 336, art. 1).

Changing the Higher Education Act of December 1968 stated that the basic principle in the work of the university is the unity of science, teaching, and education. To this end, plans and curricula of university level should be determined by the requirements of science, economy, and culture of socialist society, and innovated with the development of science and technology progress. This law states that higher education institutions should actively participate in the construction of socialism in communist Poland through training and education of professional intelligence capable of active participation in the development of the economy and national culture, and in the socialist reconstruction of social relations. The purpose of the university was education and training of new scientific staff, able to ensure sustained progress of Polish science and its relationship with the social and economic practice, scientific research, in close connection with the needs of life, and the prospects for development of the country. The next task was to nurture and develop the national culture and cooperation in the development of technical progress and popularization of achievements of science, and their practical application in the national economy. Change in the Higher Education Act of 1968 enumerated the main factors of educational institutions. These were as follows: (1) the teaching process harmoniously combining theory and practice; (2) raising the educational progress of young people waiting for their civic and professional tasks; (3) enabling students to scientific work; (4) working ideologically—developed in cooperation with educational, political, and youth organizations, with pupils aiming to instill scientific outlook and socialist morality; (5) developing student's self-governance, shaping their social activity and teamwork skills and sense of social responsibility; and (6) cooperation with the social environment, based on the scientific output of universities and revealed in its popularization and practical applications, in collaboration with learned societies and professional and academic student activities (*Journal of Laws of 1968*, no. 46, item 334, art. 1). One of the duties of a rector in 1968 was managing the process of teaching and education of young people and ensuring that this process took place in socialist, patriotic, and internationalist way. Introducing the principle in matters of political and ideological activities, educational institutions cooperate with the authorities of institutional organizations of the Polish United Workers' Party. The youth and student associations should cooperate with the university in the ideological education of the young, helping to prepare for active participation in the life of a socialist society. These organizations should also develop and deepen young people's interest in the field of arts and culture, sports, and tourism (*Journal of Laws of 1968*, no. 46, item 334, art. 67).

Change in the Constitution of the Polish People's Republic (*Journal of Laws of 1976*, no. 7, item 36) introduced a provision that the leading political force in the construction of socialist society is the Polish United Workers' Party. At the same time, it was added (in art. 6 point 2) that Poland in its policy refers to the noble tradition of solidarity with the forces of freedom and progress, strengthening friendship, and cooperation with the Union of Soviet Socialist Republics and other socialist states. Economic and social life actually rule the party leadership at the head of the Secretariat of the Central Committee of the Polish United Workers' Party, with the dominance of the First Secretary.

During the period of martial law, the content of the Polish Constitution was again changed (*Journal of Laws of 1982*, no. 11, item 83). Two months later, the new Law on Higher Education came into force. Higher education was considered state entities established to conduct research and training and provide socialist education to students, according to the Constitution of the Polish People's Republic. They were self-governing communities, academics, students, and other staff. Higher education institutions have participated in the work of the comprehensive development of science, culture, and the national economy, and educated students in accordance with the ideals of humanism and social justice, to make them aware of their obligations and rights of the citizens of the socialist state. Higher education institutions are guided in their activities by Polish People's Republic and the principles of freedom of science and art. Record of Law pointed to the development of science and diversity while respecting the autonomy of artistic worldview (*Journal of Laws of 1982*, no. 14, item 113, art. 2).

The Higher Education Act of 1990 indicated that higher education institutions should be organized and operate on the principle of academic freedom, freedom of artistic expression, and freedom of teaching, because they are part of the Polish science and the national education system. The main tasks of higher education institutions were as follows: (1) training students in a given branch of knowledge and to prepare them to pursue professions; (2) conducting research or creative work of art; and (3) learning to make knowledge and expertise of people who have titles in professional practice and perform jobs. At the same time, according to the law, the task of higher education institutions was to develop and disseminate national culture and technological progress, cooperation in spreading knowledge in society, and care for the health and physical development of students. The role of higher education institutions was also education of students in the spirit of respect for human rights, patriotism, democracy, and responsibility for the future of society and the state, and making an effort to reign in the academic cult of truth and hard work, and an atmosphere of mutual goodwill (*Journal of Laws of 1990*, no. 65, item 385, arts. 2 and 3). The great change introduced by the Higher Education Act of 1990 was to allow for the creation of nonstate universities. At the same time, the act authorized the fees for some educational services, which allowed the system of paid evening courses and part-time and postgraduate studies in public universities. It changed a systematic approach to higher education. Meeting the costs of education has been transferred from the state budget only cost to the wider area. This act was a breakthrough in the financing of education. Additionally, universities could charge fees for educational services other than studies (*Journal of Laws of 1990*, no. 65, item 385).

In 1997, the Polish Parliament passed a new constitution of the Republic of Poland, stating that "recovered, in 1989, the possibility of a sovereign and democratic determination... we set the Polish

Constitution as the basic law for the state based on respect for freedom and justice, government cooperation, social dialogue and the principle of subsidiarity, strengthening the powers of citizens and their communities" (*Journal of Laws of 1997*, no. 78, item 483). Poland is a sovereign Polish United Workers' Party disbanded.

In the same year, the Law on Higher Vocational Schools states that among the basic tasks of professional education (state or nonstate) of students in the field and profession are to prepare them for the profession, to teach them to make specialized knowledge and skills, or to retrain them in special occupation; another task is educating students in the spirit of respect for human rights, patriotism, democracy, and responsibility for the welfare of society, the state, and own work. Vocational school graduates have the right to broadcast professional degree: bachelor's degree or engineering (*Journal of Laws of 1997*, no. 96, item 590).

The new Act of 2005: Law on Higher Education stated that universities are autonomous in all areas of its operations under the terms of the act. Government bodies and bodies of local governments can make decisions about college only in the cases provided for by law. Universities are guided by the principles of freedom of teaching, research, and artistic creation, with its mission of discovery and communication of truth carried out through research and education of students. In this way, higher education institutions are an integral part of the national system of education and science. Universities collaborate with the business environment, in particular, the sale or free transfer of results of research and development enterprises, and foster entrepreneurship in academia in the form of economic activity, which is organizationally and financially separate from the core university (*Journal of Laws of 2005*, no. 164, item 1365).

Introduced in 2011, the amendments to the Law on Higher Education in 2005 state that universities are cooperating with economic and social environment, in particular, in the field of research and development for businesses and distinct forms of activity, including through a special purpose company. Representatives of employers take part in the work of the university, the development of training programs, and the teaching process. The provisions of the Law on Higher Education, amended in 2011, determine what the primary mission of any institution of higher education is. Generally, it is described as the discovery and communication of truth. They also point the way in which the university should strive for its fulfillment. This should be done by taking action in the two basic functions of the university, namely, the conduct of its research and the education of students. In carrying out such a mission, the school should have regard to certain guiding principles, including those arising also from the Constitution, guaranteeing the correctness of its actions. It should take into account the principle of freedom of education, freedom of research, and freedom of artistic creativity. The law also requires that school in its operations be guided by the principle of respect for the rights of intellectual property protection, which indicates the importance of the legislature by law to have the operation and development of the university. The Law on Higher Education forms the basis for the functioning of the university, which (1) provides a high-quality education (educates students on a relatively high level in comparison to the global level) and employs staff who ensure the achievement of this objective; (2) grows research and development, in addition to teaching business; (3) opens to the environment, and (4) cooperates with the social and economic environment in defining and carrying out its duties. To ensure the effective implementation of this vision, the statutory requirements for plans of study and training programs take into account the effects of training in accordance with the National Qualifications Framework for Higher Education. So far, centrally defined "educational standards" have been considered for different fields of study and levels of education. The Law on Higher Education provides for cooperation between universities and the socioeconomic environment, particularly in research and development for businesses and distinct forms of activity, including through the creation of a special purpose company, as well as the participation of representatives of employers in the development of training programs and in the teaching process. Law on Higher Education shows a special mission university in cooperation with the Business (*Journal of Laws of 2011*, no. 84, item 455).

Broadcasting Degrees and Titles

In 1950, 78 schools had 125,096 students. In 1960, 165,687 students were trained at 75 universities. The number of students increased steadily, together with the baby boom in the second half of the 1970s—these students achieved a high level such as in 1977, amounting to 491,400 students. In 1980, in the Polish higher education there has been a sudden drop in the number of students and young scientists. In the years 1979–1985, this fact did not change the number of institutions of higher education (91). However, the number of trained students decreased to 340,700 at universities in 1985. It is a time of heightened political and social movements—the creation of the Independent Self-Governing Trade Union Solidarity, martial law, and the replacement of the socialist planned economy: the free market economy. In the year 1990/1991, around 403,800 students were educated at universities, 2.23% of whom were nonpublic school students. Changes in laws in 1990 and 1997 resulted in the increase in the number of universities and increased the number of students, reaching 1,953,882 students at Polish universities in 2005 (the peak of the second postwar baby boomers). With the demographic decline in the age group 19–24 years currently being projected by 2030, there were 1,841,251 students in 2010. The gender ratio of students strongly changed in the last half-century. In 1950, 65.48% of the studied were men, despite the fact that from World War II to 1962, in the age group 20–24 years, there was a predominance of women over men. The political situation in ideology created opportunities for major social groups of the 1960s and 1970s—this resulted in equal numbers of male and female students, which took place in the late 1970s. The year 1991 was the last year similar to the gender balance of students. Then the number of women surpassed that of men. In 2010, in Poland, 41.21% of men to 58.79% of women studied at universities.

In the year 2007/2008, 1,276,937 (65.9%) students studied in state colleges and universities and 660,467 (34.1%) in nonpublic institutions of higher education. In public universities, in the academic year 2007/2008, there were 803,473 (62.92%) full-time educated students and 473,464 (37.08%) part-time students. In private colleges, the situation was opposite because there were 136,731 (20.70%) full-time educated students and 523,736 (79.30%) part-time students. Analysis of full-time education shows that 85.45% of students are educated in public schools and 14.55% in private schools. For part-time programs in public universities, there were 47.48% of students in transient mode and 52.52% of the students in private colleges and universities. However, in 2008/2009, the number of students decreased, but a further increase in the number of nonpublic schools was noticed.

Out of the 131 state universities in the academic year 2009/2010, there were 96 institutions with university academic authority. In the same year, 1,164,461 people studied at academic universities. The three largest universities—University of Warsaw, Adam Mickiewicz University in Poznan, and the Jagiellonian University in Krakow—educated 150,587 students, including full-time and part-time (no postgraduate) students, which accounted for 12.93% of the total number of students of academic public schools. In a total of 17 public universities, 589,880 students were educated—half of the students in the academic institutions of higher education. The other half (574,581 students) pursue their studies in the remaining 79 schools of this type.

The dynamics of the promotion of graduates with a master's degree was adequate for the number of students. In Poland in 1960, there were 19,433 people who graduated from universities with a master's degree. The increase occurred gradually until 1976, in which 44,385 masters were promoted. In the following year, 1977, the number jumped rapidly to 66,424 masters, which made a change in education, introduced by the Act of 1968, and the baby boom in the age group 20–24 years. The annual number of promoted masters remained at a similar level to the beginning of the 1980s, when the number began to decline among both educated students, doctoral students and graduates.

In 1989, the decline in the number of graduates was so significant that in Poland, only 46,200 masters were produced, which is comparable with the year 1975. Amendment to the Act in 1990 led in 1995 to the promotion of 70.029 masters. At the peak of the baby boom in 2007, the higher education institutions produced 218,445 masters. Despite a decline in the number of students, the number of masters promoted continues to increase, reaching 224,476 in 2010. In 2010, the number of promoted bachelor degree graduates (53.13%) for the first time exceeded the number of promoted masters (46.87%). At the same time, more female masters (65.61%) than males (34.39%) were promoted in the academic year 2009/2010. Also, more females than males were promoted to the professional title of BA (65.26%).

In Poland, doctoral dissertation, prepared under the supervision of a promoter, should provide a genuine solution of the scientific or artistic problem and have general knowledge of the candidate's theoretical scientific or artistic discipline and the ability to independently conduct scientific or artistic work. Such a thesis can be a work of design, construction, technological, or artistic, or an independent and separate part of a collective work, if it has a unique contribution to the candidate. PhD thesis writing is a representation of the solutions posed on the basis of their own scientific research problems (nontrivial, unresolved, and practical) through novel (original) approach to the problem and properly selected methods and tools adopted to prove the hypothesis (*Journal of Laws of 2003*, no. 65, item 595, art. 13).

Regulations for obtaining a doctoral degree in Poland due to the Law on Scientific Degrees and Scientific Title and Degrees and Title in the Scope Art are uniform throughout the country. A doctoral degree is awarded at the request of the applicant for the award of the degree. It is awarded to a person who has fulfilled all the following three conditions:

1. A master's degree—an engineer, a doctor, or an equivalent.
2. Passed doctoral examinations in the range specified by the board of the organization. Doctoral examinations are conducted in the following areas: basic discipline corresponding to the theme of the doctoral dissertation, the discipline of extra set of rules, and a modern foreign language.
3. Presenting and defending a doctoral dissertation.

Doctoral programs in full-time or part-time studies are preparing for the transition process granting doctoral degrees. However, subject to the doctoral degree is a positive move of doctorate-granting procedures. You can begin the study, but also approach it outside studies. Therefore, in Poland, the doctorate-granting procedures have assistants who work at universities and other persons that meet the statutory requirements, working under the care of an independent researcher. These people are PhD students from the moment of embarking on a doctorate degree, by the faculty council, which has the power to confer the degree of doctor of science in the discipline chosen by the candidate.

The Higher Education Act of 1968 introduced a doctoral program in which a graduate student receives a scholarship. It became possible to achieve it in the second half of the 1970s—it included about 7000 students for doctoral studies. The 1970s for doctoral studies were a very favorable period. Doctoral students could not be employed or perform work under civil law contracts, except for doctorate consent of the head. The agreement had to be then linked to the objectives of the doctoral studies. In the 1980s in Poland, there was a huge drop in the number of trained graduate students due to the political and economic situation (dramatic drop in the value of a scholarship due to inflation) and demographic decline, which brought a decline in the number of graduates.

In the 1970s, a three-year study must have been performed very well by the nonworking graduate students to receive scholarship. The students received the PhD scholarship on two conditions: they could not take up a job while studying and they agreed to work for three years after graduation in an academic institution designated by their university. Thus, in the 1970s, there was high efficiency of training of young researchers. The beginning of the 1980s brought in Poland socioeconomic benefits that substantially reflect also higher education. Supply problems quickly led to the progressive inflation and increasing emigration. The demographic age group 20–24 years old came to this difficult situation in the 1980s.

The 1980s gradually lacked candidates for doctoral studies, as there was a statutory requirement that doctoral students ought not to be employed. However, despite the numerous regulations amending the amount of the scholarship and payment of compensation for the increase in prices of basket of basic social goods, PhD students with a scholarship could not be maintained, at least in part, in the 1970s. In the deepest crisis in doctoral studies in 1986, a similar regulation for creating and conducting doctoral studies was issued, under the Act of 1965, which was still in force.

In 1992, the Ministry of Education introduced a new way of allocating grants to state universities teaching activities. The grant was calculated using relatively simple mathematical formula. This algorithm has been subjected to change over the years. Indicators and its valid version were published annually in the Regulation of the Minister of Science and Higher Education. After the first divisions of the algorithm, the university saw a real opportunity to influence the growth of subsidies by increasing the number of students, graduate students, and teachers with degrees and academic title. Budget subsidy allocation algorithm specified the percentage of funding transferred from the previous year, while about one third of the subsidy was calculated by the number of students, graduate students, and teachers with a degree and academic title, multiplied by the appropriate factors and indicators of the costs for the field of study. The algorithm for the conversion value established for five PhD students at the student's daily value is equal to 1 point and 0.4 point for extramural students. This decision contributed to the increase in the number of graduate students admitted to full-time PhD. In 1994, for the first time after the break, there were more participants to PhD studies than the average in the second half of the 1970s. At the same time, many universities for doctoral studies began moving assistants to doctoral studies. In the following years, the university has not hired assistants – masters, because the budget subsidy which is calculated on the basis of the algorithm does not provide their share of scoring in this algorithm. The increase in doctoral students in 1995 was also due to the fact that this was the first year study on the basis of the Act of 1990 resulting in the increase in numbers of graduates with a master's degree.

As a result of these changes, between 1995 and 1998, there were 100% of the doctoral students. The peak year in terms of the number of doctoral students was 2004, in which more than 33,000 graduate students were educated. In 2005, the rules for the distribution of grants to institutions of higher education were changed. Budget subsidy allocation algorithm for teaching included the conversion rate of teachers with a master's degree. After more than 10 years, universities have started to receive a grant again taking into account budgetary assistants working with a master's degree, working full-time teaching and making a research, and at the same time preparing a doctoral thesis (Figure 85.2).

In 2004, there were 17,093 men and 15,253 women with a PhD title. The year 2005 and the following years brought gradual but steady decline in the number of doctoral students. The Higher

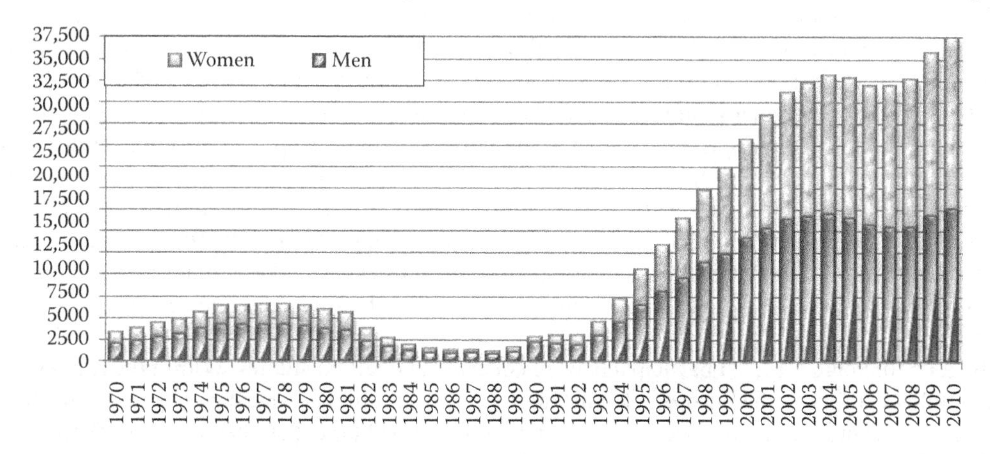

FIGURE 85.2 Number of doctoral students, broken down by gender. (From own study based on *Statistical Yearbook of Polish*, Central Statistical Office, Warsaw twentieth year [1960] to year sixty-seven [2007], Central Statistical Office, Colleges and their finances 2008, 2009, 2010. Information and statistical papers, Warsaw, 2009, 2010, 2011.)

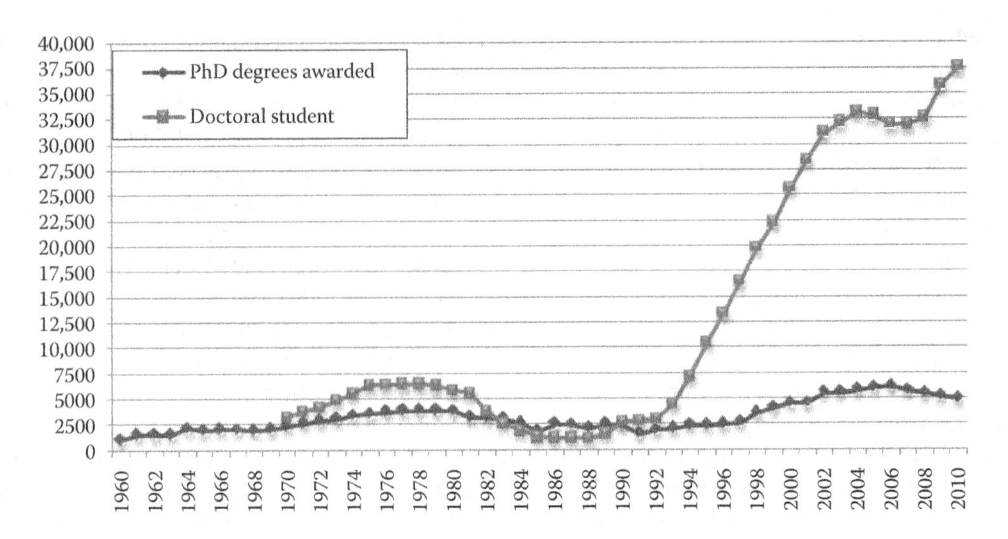

FIGURE 85.3 Number of doctoral degrees granted and the number of doctoral students in the years 1960–2010. (From own study based on *Statistical Yearbook of Polish*, Central Statistical Office, Warsaw twentieth year [1960] to year sixty-seven [2007], Central Statistical Office, Colleges and their finances 2008, 2009, 2010. Information and statistical papers, Warsaw, 2009, 2010, 2011.)

Education Act of 2005 changed them in stationary and nonstationary third cycles. Therefore, it is impossible to use these data and give a single naming. However, the peak of the baby boomers were in 2005 and should not reflect the decreasing number of doctoral students. Despite that, there was a process of reducing the number of men involved in doctoral studies. In 2008, 14,855 men studied at the PhD level—2228 fewer than in 2004. At the same time, the number of doctoral studies has increased by 1627 when it comes to women. In 2009, the number of participants in the third cycle amounted to 18,693 women and 16,978 men (Figure 85.3).

RESULTS OF TRAINING AND DEVELOPING ACADEMIC STAFF

In 1970 to 1979, about 7000 doctoral students were educated annually, each of whom, after signing a declaration that does not work, and will work, as indicated by the university, after obtaining a doctoral degree, received a scholarship for the first year salary of an assistant and the next few years older wage assistant. For example, in 1978, there were 6454 participants at PhD studies, while the number of promoted students amounted to 3824. These days, one must write doctorate, in addition to studies, and be employed as an assistant at the university and scientific research institutes. The social and economic crisis of the 1980s brought huge losses for Polish science. As a result of rising unemployment galloping inflation, and a large emigration in Poland, there were only 1123 participants at doctoral studies, and 2380 doctoral degrees were given. The introduction of paid evening courses, part-time and graduate in state universities, led to the rebirth of doctoral studies. PhD students still do not have the right to work, being participants of the study, but in place of scholarships receive only the possibility of getting receipts.

In 2005, in Poland, 32,725 participants were educated at doctoral studies. Within the framework of the Bologna Process, the European Ministers responsible for higher education began to talk about the third-level studies. In Poland, the peak period of operation of these studies was after World War II. In 2006, 6072 doctoral degrees were given. Now, after a slight decline since 2008, steadily the number of participants in studies III level is increasing, totaling in the academic year 2010/2011 to 37,492 participants. Unfortunately, since 2007, there has been a steady decline, with the number of doctoral degrees granted totaling to 4815 in 2010.

In 2008, there was a turning point when two women surpassed the symbolic number of doctoral degrees obtained over men. Already in 2010 was predominance of 285 women more than men. This means that there are 53.06% of women in the group of persons who completed PhD. Between 1994 and 1995, 2300 graduate students per year were promoted, which is comparable with the year 1970 (2290 promoted). Since the mid-1990s, the number of doctorates granted is steadily growing. In 1999, for the first time after 19 years of decline, the number of doctoral degrees granted, which started in the 1980s, exceeded 4000 doctorates, being compared to the second half of the 1970s, when around 3800 doctors were promoted a year.

The analysis of the long-term chart shows a tendency to achieve a doctoral degree by the greater number of men than women. However, this situation changed in 2006, as the number of doctoral degrees received by women and men caught up. Since then, we can observe a slight advantage of women who receive a doctorate degree. In 2010, women received 53% of the doctoral degrees (Figure 85.4).

In the 1970s, there was a twofold increase in doctorates granted in technical sciences. In Poland, it was a period of strong economic growth based on debt, but the average citizen saw the development of industry, economy, and infrastructure of society. Lots of building works, not only of buildings but of factories with complete equipment, brought a huge demand for engineers and highly skilled specialists in the field of engineering. The crisis in technical sciences started in 1980 and lasted until 1997. In recent years, giving a doctoral degree in technical sciences has dropped to the level of the early part of the 1960s in Poland. Only since 1998, the number of doctorates in technical sciences began to increase. It currently reaches about 80% of the doctorates granted in the late 1970s. In the 1970s, medical science, humanities, and sciences are areas in which the number of doctoral degrees conferred was the second largest group. However, each of these disciplines reached only about half of the doctorates granted in comparison to the technical sciences. In the subsequent years that followed, there has been a steady increase in doctoral degrees conferred in the fields of medical sciences and humanities. Currently, the most number of doctoral degrees in Poland are conferred in humanities. Arts and humanities achieved over the 50 years (1960–2010) reached a fivefold increase in doctoral degrees conferred. From 1988 to present, not only humanities but also medical doctoral degrees conferred annually gain a greater number. Medical professions in Poland

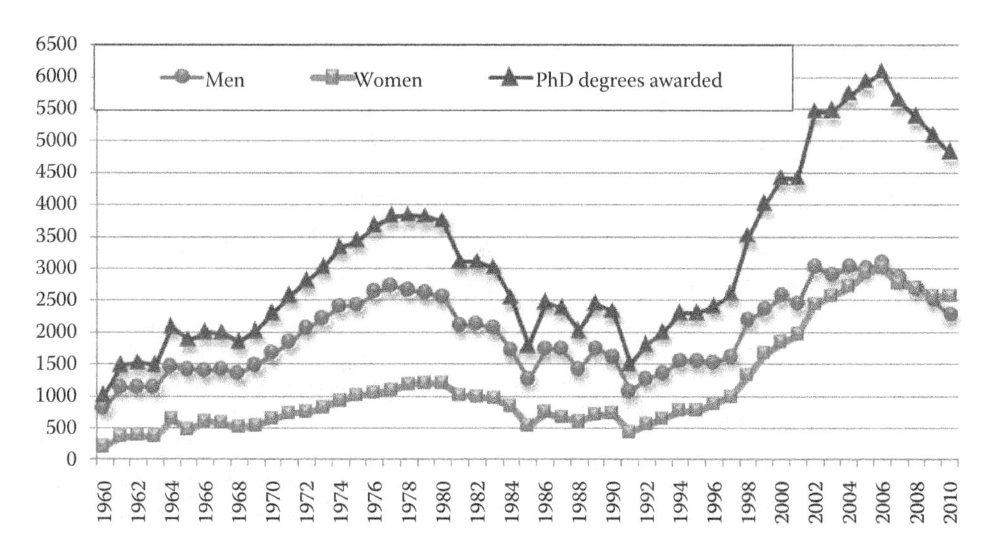

FIGURE 85.4 Number of doctoral degrees granted in the years 1960–2010 by gender. (From own study based on *Statistical Yearbook of Polish*, Central Statistical Office, Warsaw twentieth year [1960] to year sixty-seven [2007], Central Statistical Office, Colleges and their finances 2008, 2009, 2010. Information and statistical papers, Warsaw 2009, 2010, 2011.)

have always had a great prestige, and a doctoral degree in this profession significantly facilitated the rise. However, the humanities during the crisis were good ground. With a much smaller financial outlay than in the sciences or natural sciences, the humanities could develop and promote a large number of doctors.

A significant change to the Law on Scientific Degrees and Scientific Titles was introduced in 1968. The degree of docent replaced the postdoctoral degree, as an independent researcher who has the right of taking care of doctoral research. In Poland, for nearly 30 years—since the 1960s to 1988—the level of postdoctoral degrees conferred was at a stable level, ranging from 500 to 600 degrees a year. The reform of higher education, introduced with the new Law on Higher Education in 1990, a year ahead, highlighted the scientific community to prepare for the upcoming staffing needs. Years 1990 and 1992 were exceptional years in postwar Poland, which have contributed annually to the highest number of postdoctoral degrees. These were the years during which the state university opened a lot of evening courses and part-time and postgraduate studies. Also the demand for independent, academic, teacher giving permission to carry out the directions in the newly created private universities increased. There has been a steady increase of on average 200 postdoctoral degrees per year, which continues to the present. Concerning the number of postdoctoral degrees conferred, the increase in the degrees earned by women is noticeable. While in the case of men for nearly 15 years, a steady trend has been maintained in the mean constant, with systematic variations in the number of fixed degrees conferred with an annual amplitude, for women, the trend is of soft, steady growth of the number of postdoctoral degrees. The size of the increase in the number of employees cannot be looked at independently of the number of students trained per year in Polish universities. In 1990, 403,800 students were studying in 112 schools, while in 2005, there were 1,953,882 students in 130 public and 315 private institutions of higher education. Numerically, there was more than a fivefold increase in the number of students. In this situation, academics, especially independent workers who were a minimum of staff, with an increasing number of universities began working in several jobs (Figure 85.5).

In the years after World War II, the title of professor was given by the Polish president according to a prewar in this area of the law school academics. Constitution of the Polish People's Republic, July 22, 1952, abolished the office of the president. He took over the role of the Council of State,

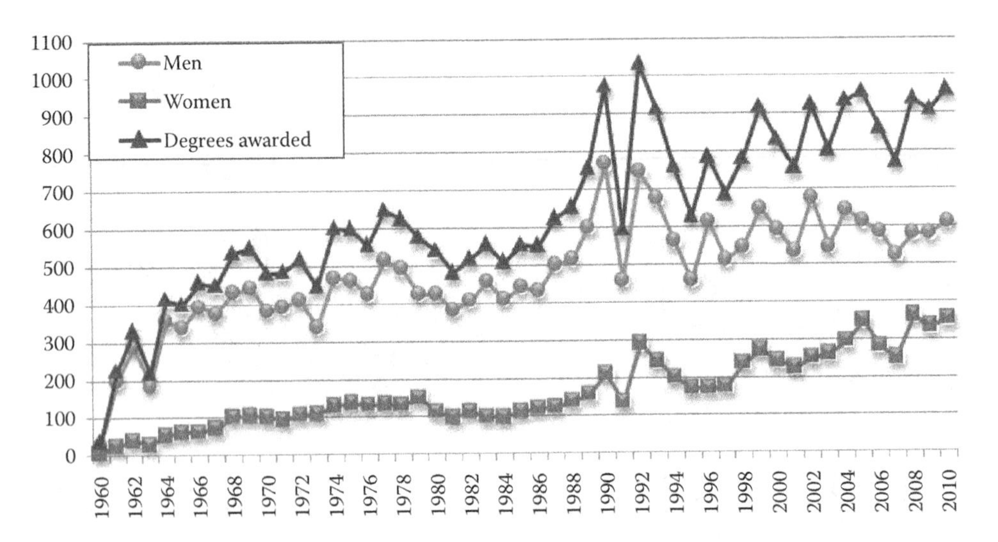

FIGURE 85.5 Number of granted, postdoctoral degrees in the years 1960–2010 by gender. (From own study based on *Statistical Yearbook of Polish*, Central Statistical Office, Warsaw twentieth year [1960] to year sixty-seven [2007], Central Statistical Office, Colleges and their finances 2008, 2009, 2010. Information and statistical papers, Warsaw 2009, 2010, 2011.)

which already existed under the Small Constitution of 1947, being collegial constitutional head of state, headed by the President of the State Council. Between 1952 and 1989, full professor and associate were given by the State Council at the request of the Prime Minister. In Poland, the number of titles professor per year broadcast does not change, but lately having a slightly increasing trend. In 2010, 25% of among those who have received this prestigious title were women.

Among those who were authorized to doctorates in the academic year 2010/2011, there were 30,343 people with the title of professor or postdoctoral degree who are employed as full-time researchers in Poland. However, if you analyze the employment of this group for the first time, then it drops to 18,490 people, that is, 60.94% of all employees in the group. Available statistics indicate only the sex of total employment (at all positions)—among independent researchers, 25.39% are female and 74.61% male as potential mentors and promoters. Growth trend is visible: women with PhD is more (from 2008) than men with PhD. And, in 2010 was 52.57% women and 47.43% of the men in doctoral studies. There are no available statistics on the number of preparing for the defense of the doctoral dissertation assistants who are not participating in the third cycle. With the current system of funding for these studies, nothing will prevent employer assistants from putting themselves on their doctoral studies at the university (Figure 85.6).

At the same time, the relationship between academic positions strongly changed. In 1965, there were 7.5% with the position of professors, 8.5% with the position of associate professors, and 26% of assistant professors, collectively accounting for 42% of the teachers. In 2010, this group reached 67.7% of the academic staff, and the positions of professors reached 23.9%. Over the 45 years, employment of an assistant decreased from 39.6% to 12.2% of all employed teachers. A group of other teachers in relation to these positions did not change, amounting in 2010 to 20.1% of all academic positions. Consequently, the position of professor of Polish universities is almost two times more likely than an assistant. About 3.5 times more are employed in positions of assistant professors than positions of an assistant. Therefore, the number of graduate students studying third degree fully in school is important, because they are more than three times of the assistants to be legally assigned to practice in the conduct of classes.

Currently, it is required in higher education institutions to implement the university system to maintain and improve the quality of education, which includes the supervision of the monitoring of the development of academic and verifies this development through the evaluation of performance.

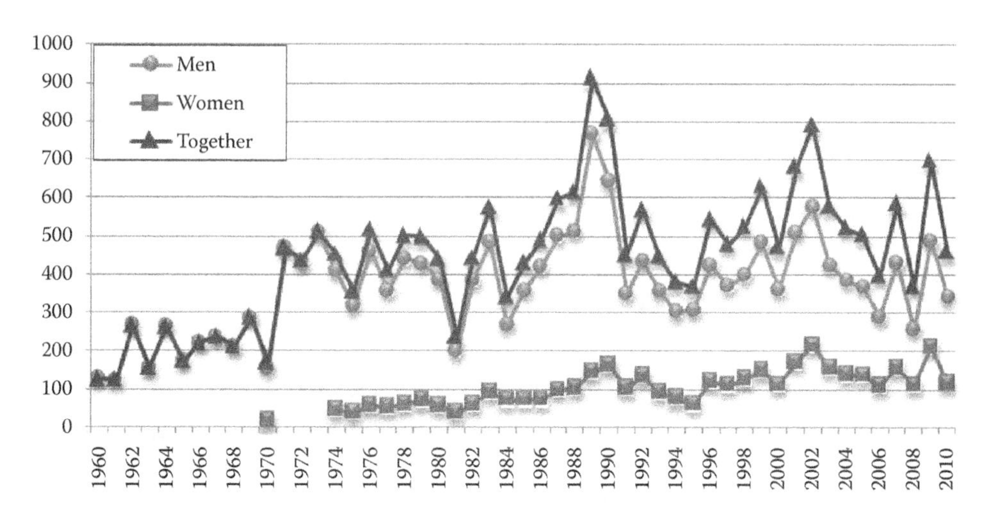

FIGURE 85.6 Number of titles of professor, granted in the years 1960–2010, by gender. (From own study based on *Statistical Yearbook of Polish*, Central Statistical Office, Warsaw twentieth year [1960] to year sixty-seven [2007], Central Statistical Office, Colleges and their finances 2008, 2009, 2010. Information and statistical papers, Warsaw 2009, 2010, 2011.)

REFERENCES

Bednarska H., Conditions of development, non-state higher education transformation period in Poland. *Monthly Social Policy* No. 3/2007.

Central Statistical Office, Colleges and their finances 2008, 2009, 2010. Information and statistical papers, Warsaw 2009, 2010, 2011.

Journal of Laws of 1933, no. 1, item 6, Act of 15.03.1933 on the academic schools.

Journal of Laws of 1947, no. 66, item 415, Decree of 28.10.1947 on the organization of science and higher education.

Journal of Laws of 1951, no. 61, item 38, Act of 15.12.1951 on the higher education and science workers.

Journal of Laws of 1958, no. 68, item 336, Act of 11.05.1958 on higher education.

Journal of Laws of 1965, no. 14, item 101, Act of 31.03.1965 on the degrees and academic titles.

Journal of Laws of 1968, no. 46, item 334, Act of 20.12.1968 amending the Law on Higher Education.

Journal of Laws of 1969, no. 4, item 31, Notice of the Minister of Education and Higher Education of 17 January 1969 on the publication of the consolidated text of the Act of 5.11.1958 on Higher Education.

Journal of Laws of 1969, no. 4, item 32, Act of 31.03.1965 on the degrees and academic titles.

Journal of Laws of 1976, no. 7, item 36 Notice of the President of the State Council of 16 February 1976 on the publication of the consolidated text of the Constitution of the Polish People's Republic passed by the Legislative Sejm (Parliament) on 22 July 1952.

Journal of Laws of 1982, no. 14, item 113, Act of 05.04.1982 on higher education.

Journal of Laws of 1987, no. 33, item 178, Act of 28.10.1987 on the establishment of the Ministry of National Education.

Journal of Laws of 1990, no. 65, item 385, Act of 12.09.1990 on higher education.

Journal of Laws of 1991, no. 58, item 249, Regulation of the Minister of National Education of 10.06.1991 on doctoral studies and scholarships.

Journal of Laws of 1997, no. 96, item 590, Act of 26.06.1997 on higher vocational schools.

Journal of Laws of 2003, no. 65, item 595, Act of 14.03.2003 on Academic Degrees and Title and Degrees and Title in the field of art.

Journal of Laws of 2005, no. 164, item 1365, Act of 27.07.2005 on the Law on Higher Education.

Journal of Laws of 2011, no. 84, item 455, Act of 18.03.2011 on the amendment of the Law on Higher Education.

Official Journal of the Ministry of Education PKWN, Nos. 1–4, Head of the Ministry of Education, Decree of 01/09/1944 on the organization of teacher training in the school year 1944/1945.

Official Journal of the Ministry of Education PKWN, Nos. 1–4, Call. For the Polish Teachers, Head of the Ministry of Education, Stanislaw Skrzeszewski PhD, Lublin 01.08.1944.

86 Creative School
Renewing Leadership for Creativity

Vilmos Vass

CONTENT

> It is quite strange how little effect school—even high school—seems to have had on the lives of creative people.
>
> **Mihaly Csíkszentmihályi**

When we are looking for some adjectives regarding schools—let us put our hands frankly on our hearts—the word "creative" does not come first. Sometimes (positively) we find the word "good," which is quite a general expression and does not help in digging deeper inside the school as an organization. In his famous powerful book *Creativity*, Mihaly Csíkszentmihályi stated that "creativity does not happen inside people's head, but in the interaction between a person's thoughts and a social cultural context" (Csíkszentmihályi 1996, p. 23). Based on this famous quotation, there are several reasons why I would like to analyze creative schools in connection with renewing leadership and not focusing on personal creativity, which is obviously an exciting topic itself. But the paradigm change of reshaping and redesigning school has taken the first place in the system level, namely, the organizational culture.

Creativity has always been a hot topic for researchers and practitioners. But on the ideal pulpit of different fields of creativity, I would admit that education is not in the first position. The "gold medal" is given to arts. When we think of the golden age of different periods of the history of arts, it is quite easy to emphasize some very high dominant milestones. Think about humanism and renaissance or impressionism. If we recall the creativity of famous painters, architects, and sculptors, in this sense, we are in the middle of personal creativity. In its narrow sense, the focus is only the personal skills and talents, but using analytical assumption, the "sociocultural context" is a highly important issue. The "silver medal" is given to economics. The creative industry has a high added value to the economic and social growth as well not because of the evidently strong coherency between creativity and innovation. "Creative industries represent highly innovative companies with a great economic potential and are one of the Europe's most dynamic sectors, contributing around 2.6% to the EU GDP, with a high growth potential, and providing quality jobs to around 5 million people across EU-27" (Green Paper 2010, pp. 2–3). It is worth to stress some economic-based concepts behind the progression of creative industry. Nevertheless, effectiveness, efficiency, and quality come into prominence in the economics and education sector as well. From this point, it is only a small jump into a school, but before it, I need to mention that the "bronze medal" concerning creativity is given to sports. If we read the reports of different sporting events, especially regarding teamwork, we regularly face the following expression: creative teamwork and creative offensive football. It is a positive fact the personal sudden solutions and smart steps would guarantee success, but if t not coupled with creative teamwork, the victory is not secured.

Finally—just dropping behind the ideal pulpit—we have arrived at the education sector. From a scientific point of view, creatology has become an almost new discipline in the last decades and its research findings are remarkable. The starting point was the early 1950s, precisely when the

President of the American Psychological Association, J.P. Guilford, took his remarkable address and gave the attion of the importance of creativity (Guilford 1950). His research has focused on the connection with creativity, intelligence, and problem solving, and that is why the first "pioneers" concentrated on individual creativity. There is no surprise that the mainstream of the creativity in education is highly dominated by the individuum. Nevertheless, creativity is a complex concept itself, and there are a lot of meanings and much more misconceptions about it. As Csíkszentmihályi pointed out, there are a lot of problems using creativity with some different meanings. The first usage, which is really widespread mainly in the public, is brilliant. This meaning of creativity is based on unusual thoughts. The second usage is "personally creative," which is focused on new, important discoveries. We can find its origin from Guilford's research, and Csíkszentmihályi's critic is "the subjective nature of this form of creativity; it is difficult to deal with it no matter how important it is for those who experience it" (Csíkszentmihályi 1996, p. 25). The third meaning of creativity, according to Csíkszentmihályi, is based on the public's critics and assessment. This is the most important concept in order to move from the individuum to the creative organization. But according to my assumption, in order to grow the creative capacity, it is important to focus on the change of the school organizational and educational culture in parallel with renewing leadership in the 21st century. I do not want to ignore the efficiency of the personalized creative processes, but in order to reach systematic and sustainable change on the school, it is necessary to "dig deeper" on the topic.

There are three approaches regarding the connection between creativity and renewing leadership for creativity. The first approach is under the umbrella of the knowledge-based society and the lifelong learning paradigm of the new meaning of learning. Creative schools have focused on the ancient principle of "learning by doing," namely, the pupils' activities, the instrictic motivation, the participation, the personal involvement, and the importance of the learning environment. Note that this is not only the cognitive-based meaning of learning, but highly emphasized the attitudes and the affective elements of the learning process as well. Never forget that the old meaning of learning has stressed on the short-term memory and high-level concentration. As Csíkszentmihályi stated in the above-mentioned powerful book:

> "To keep up the interest in a subject, a teenager has to enjoy working in it. If the teacher makes the task of learning excessively difficult, the student will feel too frustrated and anxious to really get into it and enjoy it for its own sake. If the teacher makes learning too easy, the student will get bored and lose interest. The teacher has the difficult task of finding the right balance between the challenges he or she gives and the students' skills, so that enjoyment and the desire to learn more result." (Csíkszentmihályi 1996, p. 175)

Summarizing the above thought concerning the new meaning of learning, on the one hand, the key elements are interests, enjoyment, and desire. From these key elements, there is only the small step to reach the creative learning and teaching process in the school level. On the other hand, every individuum has "the Element." Ken Robinson stated, "The Element is the meeting point between natural aptitude and personal passion" (Robinson and Aronica 2009, p. 21). From this point of view, every Element has "two features": aptitude and passion. Robinson has divided into two parts these features in order to analyze the conditions of creativity, namely, attitude and opportunity. Focusing on the opportunity leads us to the second approach.

The second approach is the new meaning of school organization. Creative school is based on the new meaning of learning. But the other side of the coin is collaboration. "One swallow does not make summer." This is an ancient proverb, but is really true in order to understand the essence of the new meaning of school organization. It would be an oversimplified schema: creative school = creative pupils + creative teachers and creative leadership. Creativity is a more complex and systematic process and is based on high-level professionalism of every actor inside an organization. The old-fashioned schools are isolated, which implies "ad hoc," unconsciousness creativity. This isolation is based on the medieval principle, namely, "My house is my castle." In practice, in this school, every teacher concentrates on his or her own subjects and pupils. There is no sharing in vision, decision making, and last but not least, learning. Without any exaggeration, one of the most decisive books in order to change this isolated school world is written by Peter Senge (*The Fifth Discipline*). In his book, he took the definition of the learning organization, where "people continually expand their capacity to create the results they truly desire, where new and expansive patterns of thinking are nurtured, where collective aspiration is set free, and where people are continually learning how to learn together." (Senge 1990, p. 3). Comparing with the key elements by Csíkszentmihályi, the similarities are obvious: desire and aspiration, but the key is "learning together." This is the basic phenomenon of the professional learning community (PLC), which is the most frequent concept of the theory of organization. The PLC is strongly connected to the creative school on the basis of several features.

The third approach is the new meaning of leadership. The well-known Seven C's can help one to understand the connection among renewing learning, PLC, and the leadership. Hord and Sommers (2008) analyzed the Seven C's connected with "encouraging, enhancing, and sustaining" professional learning communities. The Seven C's are as follows:

1. Communication
2. Collaboration
3. Coaching
4. Change
5. Conflict
6. Creativity
7. Courage

As we know, school leadership matters and improving and redesigning the leadership roles are a new priority all over the world. The leadership roles have changed rapidly and significantly. "The job used to be bells, buildings, budgets, and buses; now the pendulum has swung to instructional leadership" (Barber et al. 2010, p. 6). There are some challenging roles behind the instructional leadership such as supporting and evaluating pupils' learning and teachers' quality, setting and sharing school vision and goals, strategic and systematic thinking allocated to personal and financial resources, collaborating with the communities, and inspiring creative and innovative organizational culture. In order to maintain the sustainable change in the school in Hall and Hord's work, there are three styles of a principal:

1. Responder
2. Manager
3. Initiator

The main criticism of the first style is "little to move a school toward collegial learning and collaboration." This style has full reactions (slow or fast does not matter), but it cannot handle the change systematically. The manager style is full of proactive reactions, but systematic mind is weak. The initiator means that "the principal is consistently talking about students and programs to benefit students, initiating change, leading changes, and participating in monitoring the effects of the change." Because the last style is based on the individual capacity of the principal, the bricks of the building are not stable. That is why Hall and Hord (2006) emphasized the fourth style: collaborator that is based on sharing, effective cooperation, and sustainability.

Focusing on point 6 of Seven C's, creativity, in the narrow sense, Hord and Sommers (2008) emphasized the strength of challenges on creativity especially because of the organizational and implementation process, for example, finding the time and space for collaboration. As the above-mentioned approach has focused on the connection among creativity, leadership and school, mainly from the organizational issue, the next part it is important to analyze the implementation processes.

The next part it is obvious analyzing the implementation processes. Where are the organizational elements in the implementation? There are some levels of the implementation process, namely,

1. **Supra**: international, comparative
2. **Macro**: system, society, nation, state
3. **Meso**: school, institution, program
4. **Micro**: classroom, group, lesson
5. **Nano**: individual, personal (Letschert 2005)

As we would see, the individual creativity is only the starting point in order to understand the main phenomena of creative school; for that reason, from the bottom up, the nano level's implementation is excellent, but the meso level (school, institution, program) has become increasingly important over the last decades first of all due to the coherency between redesigning school and renewing the leadership. Behind the levels of the implementation, there are some horizontal points of view as well. One of them is the competency-based system, which is a high-level trend globally. Turning back to the connection between the new meaning of learning and lifelong learning paradigm, it is not a huge surprise that the competencies are "ante portas"; with the more optimistic view, these are inside the school. The European Union (European Parliament and the Council 2006) declared, defined, and structuralized eight key competencies. The recommendation contains the eight key competencies (see below) on the basis of the following structures: knowledge, skills, and attitudes. These key competencies are

1. *Communication in the mother tongue*, which is the ability to express and interpret concepts, thoughts, feelings, facts, and opinions in both oral and written form (listening, speaking, reading, and writing) and to interact linguistically in an appropriate and creative way in a full range of societal and cultural contexts.
2. *Communication in foreign languages*, which involves, in addition to the main skill dimensions of communication in the mother tongue, mediation and intercultural understanding. The level of proficiency depends on several factors and the capacity for listening, speaking, reading, and writing.
3. *Mathematical competence and basic competences in science and technology.* Mathematical competence is the ability to develop and apply mathematical thinking in order to solve a range of problems in everyday situations, with the emphasis being placed on process, activity, and knowledge. Basic competences in science and technology refer to the mastery, use, and application of knowledge and methodologies that explain the natural world. These involve an understanding of the changes caused by human activity and the responsibility of each individual as a citizen.
4. *Digital competence* involves the confident and critical use of information society technology (IST) and thus basic skills in information and communication technology (ICT).
5. *Learning to learn* is related to learning, the ability to pursue and organize one's own learning, either individually or in groups, in accordance with one's own needs, and awareness of methods and opportunities.
6. *Social and civic competences.* Social competence refers to personal, interpersonal, and intercultural competence and all forms of behavior that equips individuals to participate in an effective and constructive way in social and working life. It is linked to personal and social well-being. An understanding of codes of conduct and customs in the different environments in which individuals operate is essential. Civic competence, and particularly knowledge of social and political concepts and structures (democracy, justice, equality, citizenship, and civil rights), equips individuals to engage in active and democratic participation.

7. *Sense of initiative and entrepreneurship is the ability to turn ideas into action.* It involves creativity, innovation, and risk-taking, as well as the ability to plan and manage projects in order to achieve objectives. The individual is aware of the context of his/her work and is able to seize opportunities that arise. It is the foundation for acquiring more specific skills and knowledge needed by those establishing or contributing to social or commercial activity. This should include awareness of ethical values and promote good governance.
8. *Cultural awareness and expression,* which involves appreciation of the importance of the creative expression of ideas, experiences, and emotions in a range of media—music, performing arts, literature, and the visual arts (European Parliament and the Council 2006).

Regarding the connection between creative school and the meso level implementation of the key competencies, there are some multidisciplinary fields among these key competencies such as critical thinking, creativity, initiative, problem solving, risk assessment, decision taking, and constructive management. Focusing on the effective implementation of the key competencies, it is obvious that the first phenomenon of the creative school is structuralization. Guilford structuralized the intelligence as a 120-factor model, which could differentiate convergent and divergent thinking. The structure of creativity is based on some phenomena, such as fluency, flexibility, elaboration, sensitivity, synthesis, analysis, transformation, etc.

The above-mentioned phenomenon of the structure on creativity is a remarkable result in order to analyze the organizational issue as well. What are the main characteristics of the creative school that makes it become a professional learning community? Firstly, the creative school has an *innovative learning environment with full of stimulus.* This school has opened spaces, the creative architecture, where fantasy, inquiry, and curiosity have come into prominence. Regarding the classroom management, creative school can promote learning via curriculum planning, learning-based competency development, and formative assessment. In basic words, the creative school has focused on cooperative and inquiry-based learning, motivation-centered methodology, and differentiation. The innovative learning environment is connected to the personalized processes, for instance, diagnostic assessment where there are lots of self-evaluation, self-reflection, and personal involvement. Secondly, the creative school has *effective collaboration*, which is an important part of the abovementioned Seven C's. Nevertheless, there are some conditions of the effective collaboration such as trust, creative ethos, courage, and shared vision.

Summarizing the main characteristics of creative school on renewing leadership for creativity, there are three models: facilitative, constructivist, and distributed leadership (Roberts and Pruitt 2009). The facilitative leadership is not based on a dominant leader; this is not a hierarchical organization. This model is focused on collaboration and promotion. Promoting students' learning and staff's development means redesigning the structure of the school in order to handle the growth and change. This is the entrance of the PLC because of a lot of discussion, common decision-making processes, and cooperative workshops. The constructivist leadership is based on prior knowledge, whole school approach, and a lot of self-reflections, brainstormings, mindmapping, and any other collaborative learning that can promote the mutual notions, concepts, and understandings in the school community. This is the ongoing learning and knowledge management where all teachers are the leaders as well. This is the second important step toward PLC. The distributed leadership is a practice-oriented model that can promote formal and informal learning among the school community and is based on frequent interactions and not simply on leaders' knowledge and skills. This is the third important step to PLC. But where are the creative schools? How many steps are there in order to enter the creative school or at least knock the creative school's door? You can find the potential piece of answers in this article, but taking the overall picture is a creative task. These questions have been opened and finding the answers is an inspiring and ambitious activity. As creative school and redesigning leadership for creativity.

REFERENCES

Barber, M., Whelan, F. and Clark, M. (2010): *Capturing the Leadership Premium: How the World's Top School Systems are Building Leadership Capacity for the Future*. McKinsey, London.

Csíkszentmihályi, M. (1996): *Creativity*. Harper Collins Publishers, New York.

European Parliament and the Council (2006): *Recommendation of the European Parliament and of the Council of 18 December 2006 on Key Competences for Lifelong Learning* (Official Journal L 394 of 30.12.2006). Available at: http://eurlex.europa.eu/LexUriServ/LexUriServ.do?uri=OJ:L:2006:394:0010:0018:EN:PDF (Accessed: 15 May 2012).

Green Paper (2010): *Unlocking the Potential of Cultural and Creative Industries*. European Commission, Brussels. Available at: http://ec.europa.eu/culture/ourpolicydevelopment/doc/GreenPaper_creative_industries_en.pdf (Accessed: 15 May 2012).

Guilford, J. P. (1950): Creativity. *American Psychologist*, 5, 444–454.

Hall, G. E. and Hord, S. M. (2006): *Implementing change: Patterns, Principles, and Potholes*. Allyn and Bacon, Boston.

Hord, S. M. and Sommers, W. A. (2008): *Leading Professional Learning Communities. Voices from Research and Practice*. Corwin Press, California.

Letschert, J. (ed.) (2005): *Curriculum Development Re-Invented*. SLO, Netherlands Institute for Curriculum Development, Leiden.

Roberts, S. M. and Pruitt, E. Z. (2009): *Schools as Professional Learning Communities. Collaborative Activities and Strategies for Professional Development*. Corwin Press, California.

Robinson, K. and Aronica, L. (2009): *The Element*. Allen Lane an imprint of Penguin Books, London.

Senge, P. (1990): *The Fifth Discipline. The Art and Practice of the Learning Organisation*. Doubleday, New York.

87 New Ways of Defining Teacher Competencies

Demand-Driven Competency Definition and Development

Tibor Baráth

CONTENTS

INTRODUCTION

Focusing on a Hungarian development, this present study will describe the manner and method of defining the competencies teachers must attain during their training. The project was carried out by the University of Szeged, Juhász Gyula Teacher Training Faculty in 2010–2011. It aimed at establishing a regional research and service center for contributing to the renewal of teacher education. As a part of the project, six teacher vocations were selected to (re)define the competencies that are important to carry out the profession at a high level. The development is based on an approach, according to which during the training and preparation for various teaching vocations, the basis for developing the necessary competencies can comprise the mutually formed expectations of the users of the expertise (the employers of the teachers), trainers (university teachers), and the actual practitioners of the vocation (the teachers of specific areas). The structure of defining competencies is in line with EU expectations (European Commission–Directorate–General for Education and Culture 2004). In its focus—based on behavior science—lies the identification of activities and behavior forms that make teachers successful in their profession in the 21st century. For the technical validation of this method, we used a competency model (RDA, role diagrammatic approach) that enables a multidimensional understanding and description of human behavior. At the same time, the model is also capable of handling individual features together with the characteristics of training, i.e., which competencies can be expected to develop in the given training, in the same structure. Thus an optimal training structure, the actual existing training, and the competency profile of the individual can be depicted in the same structure. Comparing the profiles can assist both the training programs and the development of individuals.

Next, via describing an example of teacher training, we are going to give details of the aim and the procedure of the development, and we are also going to list some results. When we give an account of the results, we shall not provide the reader with the description of the entire competency expectation system; rather we will focus on the method and application possibilities.

CONTEXT OF DEFINING TEACHER COMPETENCIES

As a result of global changes, the formation of knowledge-based societies has made lifelong learning an integral part of everyday life. The speed at which knowledge can become outdated with new knowledge appearing, the rapid formation of new professions together with the necessary knowledge and the amount and rate of technological, social, and organizational changes have all made it unavoidable and indispensable to learn new things and to adapt to changes.

Speed of Change and Knowledge

If by "knowledge" we mean knowing about facts and correlations (generally, this is what we mean by factual knowledge), we can conclude that the amount of information available to us is growing at a dramatic rate. Around 15 to 20 years ago, we saw that the knowledge available for mankind doubles every seven years, and this rate dropped to every two years a few years ago (in around 2006). This means that if 15 to 20 years ago we said that a teacher's career lasted 35 to 36 years—assuming a constant rate of information growth speed—the amount of information grew 32-fold during the span of one career. By doubling the amount of information every second year, the information load has now increased to 266,144-fold. If we compare these two figures, it would be similar to having the spine of a medium-thick encyclopedia (approximately 4 cm) compared to the height of the Eiffel Tower (approximately 326 m). Talking about knowledge, we could go on by saying that it takes approximately two years for our knowledge to become outdated. Therefore, we need to live together with the paradox that while no knowledge exists without facts and information, the content and validity of these are changing rapidly. Probably, there is a good reason why Encyclopaedia Britannica, Inc. decided not to have its publication issued in a printed version in the future, but only its online version (See http://www.guardian.co.uk/books/2012/mar/13/encyclopedia-britannica-halts-print-publication) will be available. It would be an intriguing competition with Wikipedia.

School and Learning Science

What has been said above automatically leads us to a new field of science, that is, the science of learning. Learning science draws our attention to the necessary transformation of schooling, which has a fundamental impact on the competencies necessary for successful teachers.

Before, we knew little about the construction process of knowledge. Mental patterns in our brain influence elements of new impulses (new knowledge and experience), how they are incorporated into the existing system, and how they transform it. The change in the process of learning is manifested in the fact that besides individual learning, the importance of group learning is also growing radically. This phenomenon is closely related to the concept of lifelong learning, which, besides formal learning, now comprises nonformal and informal learning, as well. Knowledge management can be linked to learning within groups or organizations, which is termed knowledge transfer. Therefore, it can be of fundamental importance whether the above issues are present or not in the schools and the teacher-driven process of learning and, if they are present, how they are manifested.

Now let us provide an overview of some of the milestones of learning science with the help of a study written by Sawyer (2008). The present form of schooling was shaped in the times of the industrial society and became publicly available at the beginning of the 20th century. Education as one of society's subsystems also became one of the biggest and most heavily burdened systems by red tape. At the time of its formation, there was no deep and soundly based knowledge available about human learning that school education could have been built upon. This, in many respects, was based on experiences gained so far, common sense, and pre-suppositions; however, these experiences had not been scientifically examined, and therefore their validation had not been done either. According to this model of schooling, which Sawyer calls standard,

- The basis of schooling is learning facts, rules, and procedures.
- The level of schooling is demonstrated by the amount of the above.
- Knowledge is owned by the teacher whose task is to pass this on to the students.
- Learning goes from simple to complex, but what is called simple or complex is not determined by the manner of the child's learning but education specialists.
- The success of schooling can be measured by the level of attaining the facts and procedures (Sawyer 2008, pp. 45–47).

The transformation of an industrial society into a knowledge society—accordingly, the transformation of an industrial economy into a knowledge economy—fundamentally affected the world of knowledge and part of it, the school. In this "new world," the economy is built more on the production and allocation of information than on the production and allocation of tools and assets (Drucker 1993, p. 182). If we link this to the much-debated but also very influential concept of Florida (2002), i.e., the role of talent, technology, and tolerance in economy and competitiveness, then we can see another reassurance for how necessary it is to change the concept of learning. We can see that an open and receptive culture is linked to the economy and wealth even if no causal relationship can be offered between the two. In a value study conducted by Inglehart and Baker, it was found that the values present in an economy and society built on knowledge and creativity, rationality and tolerance or trust go hand in hand (Ingelhart and Baker 2000).

Wisely selected facts, rules, and procedures are not enough for adapting in the world of work and for a successful career because the complexity of work itself has grown radically, basically in all vocations and fields of work. A good example for this is computer use, which 20 years ago was only the privilege of a few people, mainly in positions requiring high qualifications, and by today, it has become a general issue.

World of Work and Learning

The changing link between work and learning is demonstrated by the fact that the labor market defines more and more precisely and accurately what competencies are expected from individuals in various jobs and positions, and at the same time, the role, the task, and the responsibility of the individual is also increased in their own self-development process. As a result, what also changes is the role of the sector that enables the preparation of the individual and supports learning, thus creating a bridge between the world of the individual and that of work.

In developed countries, we can increasingly see that the various players of society, i.e., the world of work, trainers, and researchers; the state; and the civil sector, establish clearly defined standards for workers in various sectors and people fulfilling various positions. Defining these standards is aimed at targeted training, further training, the coordination of individual and organizational aspirations—all in all, the achievement of a successful operation.

Defining competencies at a system level can be a pioneer example, and it has been realized in Great Britain. An organization that supports lifelong learning—Lifelong Learning UK (LLUK)—is one of the 25 sector skill councils that defines competency expectations for a given sector. Of the 25 sector skill councils, Lifelong Learning has an outstanding role, and it encompasses five areas: community learning; further training programs (nonuniversity programs for over 16s); higher education; library, archive, and information service; and work-based learning. These play a vital role in training specialists in other sectors. This is the exact reason why special attention was given to defining necessary competencies for workers operating in the fields of lifelong learning. (For the description of the standards for the sector, see http://www.excellencegateway.org.uk/node/57. For materials related to teacher training reform in the sector of further training, see http://www.excellencegateway.org.uk/node/64. For archived literature on LLUK, see http://webarchive.nationalarchives.gov.uk/20110414152025/http://www.lluk.org/).

International Examples for Defining Teacher Competencies

The European Union regards teachers as key players in training future generations and adults alike; therefore education systems receive an elevated importance in increasing the competitiveness of the European economy and the establishment of knowledge economies of outstanding performance. As a response to the challenges articulated in the Education and Training 2010 Programme, the European Council and the Commission provided recommendations to decision-makers for teachers' lifelong learning, professional development, and creating opportunities for quality learning. They also made a recommendation for what competencies must be developed by national systems in the course of training teachers-to-be. The recommendation defines common European principles that aim to stimulate the creation of high-level education (a high quality of professional operation, positioning the profession in the context of lifelong learning, mobility, and cooperation), and it defines related competency categories that national systems must create: work with knowledge, technology, and information; work with others; work with and in society (*The Recommendation of the European Parliament and the Council* 2006).

We can see that several countries have their competency expectations defined for teachers. In the UK, the competencies to be achieved and attained by teachers have been defined. If a teacher wishes to work in state schools, a prerequisite is to fulfill accreditation expectations, i.e., to reach a Qualified Teacher status. (For information about QTS, see http://www.education.gov.uk/get-into-teaching/faqs/becoming-a-teacher/qualified-teacher-status.aspx.) Reaching this status is supported by training programs and clearly defined competency standards. (For the description of standards, see http://www.education.gov.uk/consultations/downloadableDocs/98_1.pdf; the revised version can be found here: http://www2.warwick.ac.uk/fac/soc/wie/partnership/tda_standards.pdf.)

In the Netherlands, a development program was launched in 1993 as a result of which a law was adopted in 2004 on what competencies teachers are expected to have and how schools must see to the assessment, evaluation, and development of these competencies every six years. (For information about the system, see http://www.onderwijscooperatie.nl/.)

DEVELOPMENT METHODS AND THE MODEL

For defining the competency expectations for various teacher professions, we used the British approach. Defining the required competencies for trainers and teachers who work for the lifelong learning sector is based on the results of qualitative and quantitative research. The latter was supported by already existing databases as well as new data. The procedure covered the following:

- The analysis of a wide range of professional literature
- The definition and analysis of existing standards and sector-specific data sources

- Talks and interviews with the key stakeholders of the sector
- A broad-scope questionnaire study with service providers and employers working in the sector
- Future planning workshops (LLUK 2006, pp. 16–26)

We adapted the LLUK method to the Hungarian circumstances and the project requirements. Within the framework of the project, we developed competency profiles for six teaching professions: nursery school teacher, junior school teacher, medical rehabilitation teacher, school teacher, technical vocational instructor, and mentor. Development methods and results are introduced by this study based on the final report of the project (Baráth et al. 2010). We carried out the development in phases described below.

Preparing a Professional Literature Study

The development process started with the analysis of a wide range of professional literature, the aim of which was to provide the specialists involved in the development process with sufficient background information to create a common understanding of the concept of competency and to form a common professional ground for the development process. The analysis covered the following:

- The correlation between teaching efficiency and the economy
- The correlations between teaching efficiency and the quality of the teacher
- The selection and training of teachers necessary for quality learning and teaching, and related to this are the following:
  - The description and analysis of OECD and EU recommendations
  - The description of the teacher competency system of various countries together with their applications
  - Describing the LLUK competency standards in the field of teaching and learning
  - Analyzing Hungarian experiences in the field of defining competency standards in vocational and higher education with special regard to the training and output requirement of the education government

Analyzing these correlations together by taking a look at the international and national practices laid the foundation for those professional workshops where the actual definition of the standards for the six teacher professions took place.

Competency Expectation Defining Professional Workshops

Typically, 15 to 20 participants took part in the workshops from three target groups: employers (those who employ the teachers with the given qualifications), trainers (university teachers who prepare students for the teaching profession), and teachers (who actually work in the given professional field: junior school teacher, medical rehabilitation teacher, school teacher, etc.). Occasionally, besides the above participant groups, parents as well as leaders working in local education management were also invited. After the participants had learned about the project objectives and methods, they worked in mixed groups all throughout the workshops. The aim was to create structured and thematic debates about the given teaching profession and define the following: (a) professional expectations, articulated in a structured form; (b) define and identify the activities, actions, and forms of behavior that characterize the teacher who is capable of achieving high performance in the given profession in the 21st century and who can manage the students' work efficiently. The process of learning about each other's way of thinking and expectations together with a continuous reflective process made the three key stakeholder groups (employers, trainers, and the profession) reach an agreement on realistic expectations. Besides a professional focus, an intensive and deep

TABLE 87.1
The Composition of the Competence Defining Workshop Participants

Function	Junior School Teacher	School Teacher	Medical Rehabilitation Teacher	Nursery School Teacher	Mentor	Vocational Instructor	Total	%
Employer	4	8	4	5	6	7	34	31.78
Teacher	6	4	5	2	6	0	23	21.50
Maintainer	1	3	1	1	1	3	10	9.35
Instructor	7	5	5	6	8	4	35	32.71
Parent	0	1	2	1	0	1	5	4.67
Total	18	21	17	15	21	15	107	100.00

debate and interpretation process enhanced the participants' dedication toward the development itself, which was one key factor in assuring quality.

As an end product, the workshop participants produced a detailed list of sentences ranked in order of priority, which described the professional activity, behavior, and attitude of a newly graduated teacher who is efficient in his or her profession (with the exception of a mentor, who is a teacher with already existing experience and whose task is to support and promote the professional inclusion and development of new colleagues). The composition of the workshop participants is described in Table 87.1.

During the time of development, the workshop participants had access to the description of the competency profiles so that they could give their opinions on them and make suggestions and recommendations for corrections or alterations.

Questionnaire Study

Based on the list of sentences articulated in the workshops and describing the expected activities, forms of behavior, and attitudes, a questionnaire was created for each profile, and a questionnaire study was conducted mainly among the employers and the teachers. Fourteen hundred feedback comments were received for the six profiles (a minimum of 111, a maximum of 422). In the questionnaire study, the elements of the sentence list had to be ranked in the order of importance. At the same time, from an identical list for each profile, those characteristics had to be selected that were deemed necessary for the given profile. As described above, the questionnaire study served the purpose of fine-tuning the profiles articulated in the workshops. As we experienced, the proportion of correction remained quite low, and the content of the profiles developed during the workshops was fundamentally supported by the empirical study. (The limitations of this study do not allow for a more detailed description of the differences.)

Finalizing the Competency Profiles

The final version of the competency profiles was based on the expectations articulated at the workshops and the findings of the questionnaire study. The completed profile contains training and output requirements described by the education government. (For the training and output requirements—only available in Hungarian—see http://www.nefmi.gov.hu/kkk.) Technically, this was feasible by comparing the sentence lists articulated at the workshops and fine-tuned during the empirical study with the training and output requirements of the training knowledge, skills, and attitudes; then we united these. We had to pay special attention to defining the concepts of knowledge. This was either directly specified by the training and output requirement description or we used the forms of behavior and attitudes identified during the workshops when we wanted to define the knowledge content, and then we deducted the underlying necessary knowledge (e.g., if a teacher can motivate his or her students, then he or she must be aware of the concept of motivation and motivation theories).

The integrity of the development process was ensured by having a group of leaders of competency profile developers for teaching professions, representatives of employers and maintainers, as well as international experts harmonizing and finalizing the professional materials. In the course of this process, this group examined the following:

- The adequacy of the profile descriptions (applying the aspects of the target group)
- The correction of the profiles developed at the workshops (this was built on a broad-scope questionnaire study)
- The horizontal examination of the profiles (the presence of justified identical elements and differences in the competency expectations of the teaching professions)

The group with the above composition brought the final decision at the competency profile finalizing workshop. The final content of the competency expectations, e.g., required at least 60% consent of the expert group. The experts finalized the profiles and the descriptions alongside the following aspects:

- A review of the descriptions
- The clarity and straightforwardness of the descriptions for the target groups
- The interpretation of the descriptions for teaching activities
- Verifiability of identical elements and differences between the various professional fields
- Taking into consideration the training and output requirements specified by the education government

In all cases, the description of competencies took place in two coordinated structures. One description was developed from a competency model enabling measurement and comparison and containing characteristics of behavior alongside several dimensions (dynamic-stable, content orientedness–relation-orientedness, leader–operational elements). The other description contained an elaboration of the knowledge, skills, and abilities, the attitude and motivation necessary for operating in the profession, all based on the key competencies of the EUs recommendation.

Assessment and Evaluation of Competencies

A key requirement when developing competency profiles was to come up with a system that is capable of an accurate, quantitative, and structured description of the given professional competencies and which is also capable of defining the differences between the actual state of an individual (the collection of competencies possessed) and the target profile. For this purpose, we used the role diagrammatic approach (RDA) model, which examines the competencies in a system holistically and in the correlation of the roles fulfilled by the individual. It also gives assistance in identifying development directions; therefore, it can be used for both individual and group development plans.

RDA is a model capable of describing human behavior, and it helps identify what a person shows to the world, to his or her environment. In fact, it is like a map that contains the characteristics of a person in a system of approximately 30,000 words and expressions. We placed the contents of the prioritized sentence lists developed at the workshops onto this map and identified the key words of the sentences, which were then related to the axes of the RDA system. This is how we got the axis values that enabled a visual representation (see Figures 87.1 and 87.2).

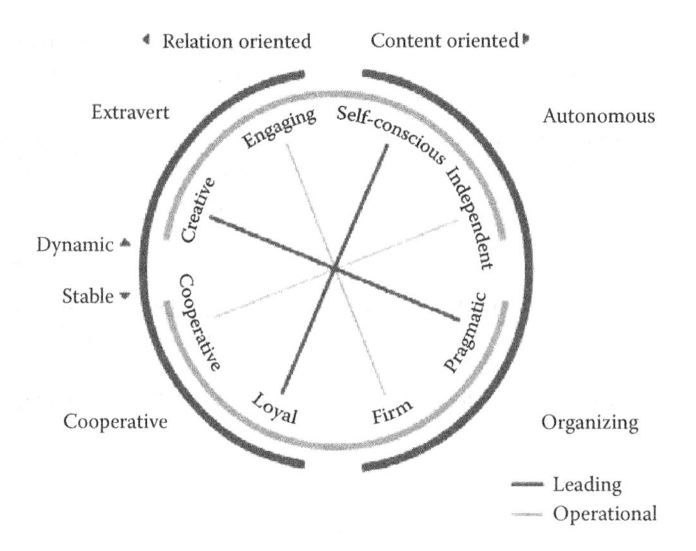

FIGURE 87.1 RDA axes and dimensions.

Figure 87.1 serves the purpose of straightforward understanding and a visual representation, in which the dimensions and axes of the RDA model can be seen. A person's behavior and system of values can be described along these dimensions and axes. For a detailed and comprehensive description, see the study written by Harten and Wolbers (2005) and an overview by Baráth (2010, pp. 37–40).

RESULT OF DEVELOPING COMPETENCY PROFILES

As a result of applying the development model introduced above, the planned six profiles were completed. Now, we shall provide the reader with an example for the form of the model and its use in training development, via the profile of a junior school teacher. We will describe the competency profile in the RDA system; then we are going to highlight some elements in light of the EU's key competency structure. Figure 87.2 shows the profile of a junior school teacher, in which the profile was described as detailed above.

The expectations from a junior school teacher can be pooled together in a competency matrix in which one dimension comprises the EU recommendations, and the other comprises—according to the widely accepted interpretation of the concept of competency—knowledge, skills, and abilities, attitude, and motivation. Figure 87.2 shows how many (elementary) expectations are present alongside the competency descriptors of the EU key competencies. Next, we will give an example for identifying competencies. We can see that in several instances, there is a proven close triple link among knowledge, skills, and attitude (Table 87.2).

Work with Knowledge, Technology, and Information

Knowledge

- Knows about the learning organization procedures, the aims, and criteria of their application as well as their methods of realization.

Skills and Abilities

- Consciously plans, organizes, and manages classroom processes adequately using various learning organization procedures, adaptive in selecting methods, adopted to individual learning routes. In his or her work, he or she focuses on skill development via activities and provides plenty of opportunities for enjoying the act of creation.

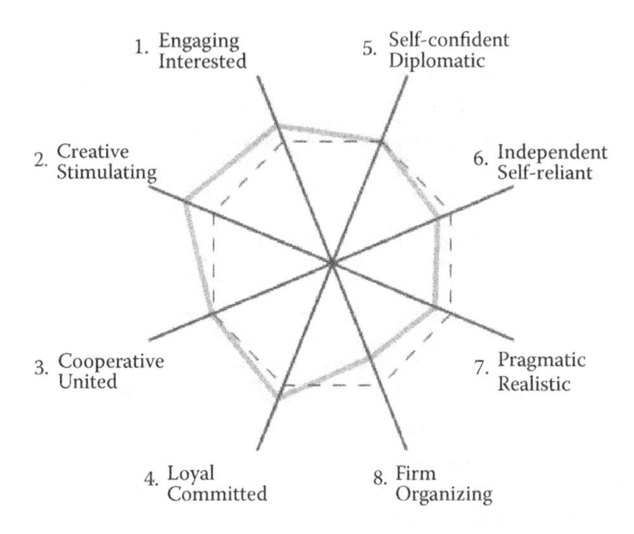

FIGURE 87.2 Profile of a junior school teacher.

TABLE 87.2
Competence Matrix of a Junior School Teacher

Key Competences/Competence Categories	Knowledge	Skills and Abilities	Attitude and Motivation
Work with knowledge, technology, and information	13	17	14
Work with others	4	8	9
Work with and in society	3	3	3
All	20	28	26

Note: In the rubrics you can see the number of expectations related to the given field.

Attitudes and Motivation

- Creates a positive and warm, supportive atmosphere in the lesson.

Work with Others

Knowledge

- Has the necessary socio-psychological knowledge to practice his or her profession.

Skills and Abilities

- Shares his or her experience and knowledge with his or her colleagues.

Attitudes and Motivation

- Interested in the opinions and suggestions of the children. He or she accepts and encourages student initiatives. Stimulates and supports alternative solutions different from his or her own and gives the opportunity to see this work in practice.

Work with and in Society

Knowledge

- Is prepared for the integrated teaching and education of children of ethnic and other nationality origin or those from different socio-cultures.

Skills and Abilities

- Creates such an acceptive and inclusive educational atmosphere in which children can feel valuable and accepted, in which they can learn to respect the peers from ethnic and other cultural origins. Educates the children to notice and accept common values.

Attitudes and Motivation

- Respects and accepts cultural differences.

TRAINING DEVELOPMENT

Measuring the competencies is of vital importance regarding the development of training programs aimed at developing competencies. The RDA model briefly described above is capable of examining the training programs from the aspect of to what extent they are suitable for the accepted competency profile. This means that knowing the content and the operation of a training program, the detailed profile of a training program can be compiled in the same manner as we have seen in the case of the competency profiles of various teaching professions. The profile of a given teaching profession can then be compared with the training program profile in operation. Differences or gaps can be analyzed on the basis of which suggestions and recommendations can be made for the development of the training program. The procedure itself will again be described via junior school teacher training. Here again, our aim is not to give a detailed analysis of all training elements but to provide an overview of the procedure.

Developing the Profile of Teacher Training

Defining the competency profile of a training program is based on two sources. The first one is the content analysis of the written materials of the training program. This analysis serves the purpose of judging how the compulsory core courses of the training program support the development of the competencies described in the profile. In the case of an eight-semester program, this analysis encompassed 19 courses of four semesters and the conceptional examination of the compulsory practical training. We selected those courses for analysis that are compulsory for all students and form the backbone of the training program. Divided into seven semesters, the practical training contained 366 lessons of individual teaching practice together with other fieldwork and a group teaching practice of 18 lessons per week for one semester.

The other source for the analysis came from a focus group interview with the teachers and third- and fourth-grade students participating in the teacher training program. In the course of this interview, we asked the subjects to tell us how the training program takes place in practice and to what extent it is in harmony with what the written material of the training program contains and also where and what kind of differences can be identified. This led to the formulation of the competency profile of the existing training program.

Comparing the Competency Profiles of an Accepted (Ideal) and the Existing Training Program of Junior School Training

Finally, the two competency profiles—that of the planned (ideal) and the existing one—were compared with the help of the RDA model used for the unified evaluation. Figure 87.3 shows the two profiles in the RDA system. From the differences, we can deduct how the basic principles defined during the formation of the ideal profile are present, where and what kind of changes are necessary to approximate the competencies acquired by the students throughout and as a result of the existing training to the ones described in the expected competency profile.

If we compare the profiles, we can see that the training program focuses on content, i.e., teaching knowledge and respecting administrative discipline. At the same time, it pays special attention

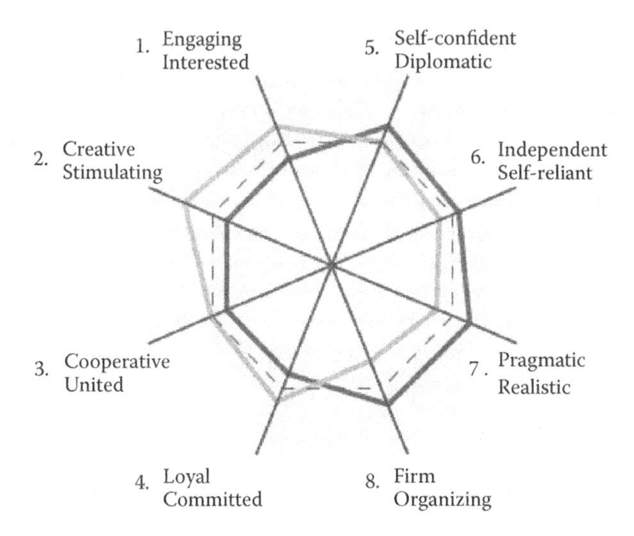

FIGURE 87.3 The planned (ideal) profile (grey) and the existing training profile of junior school teacher training (black).

to the formation of a credible professional approach as well as conscious effort making. (Compare Figure 87.3 and Figure 87.1 showing the RDA model). We can see that in Figure 87.3, the red profile has high values on the self-confident, independent, and practical, firm axes, and the green profile of the ideal training program expects exactly the opposite, i.e., a rather flexible, cooperative, engaging, and stimulating form of behavior.

The course descriptions typically focused on the fact that the students should be able to apply their knowledge in practice. The descriptions featured self-awareness, self-development, personality development, cooperation, and the preservation of health, but their accumulated weight even proved to be too small against skills such as analysis, independence, proliferation of knowledge, and being self-organized.

We could go on with the examples, but probably this is enough for us to see that formulating a competency profile for junior school teachers inevitably can lead to the development possibilities of the training program and the measurability of the competencies.

CONCLUSION

In this study, we described a model and a procedure with the help of which preparatory training programs for various professions can be linked with the experiences of those working in the profession. Thus the competencies to be acquired throughout the training programs can be built on the experiences and opinions of those who act as key players in the actual production and the application of knowledge. The definition of competencies built on facts, the debate, and consensus of representatives of the groups involved (employers, trainers, actors of the profession) can ensure future-orientedness and practicality at the same time. Making competencies measurable enables us to compare the ideal profile, i.e., the competencies necessary for the profession with the competencies actually developed during the existing training program, i.e., the profile of the existing training program. As we have seen in the example above, the differences can lead us to the correction of the training content and the applied methods, together with the identification and preservation of the proper emphases. As measurability enables us to compare the students' individual profiles to the ideal profile, we can find out with what targeted content and methods we can develop the students' competencies both individually and in groups.

The approach described here can clearly boost the sense of responsibility for quality learning in all players involved:

1. In the students, who presumably do not only wish to get a degree qualification but to gain applicable knowledge; therefore, they have to assume responsibility for the management of their own learning process and enhance the awareness about this process.
2. In the teachers, who can assess how the learning process managed by them can fit in the entire training process, what can be done and how to optimize students' learning in light of the output requirements.
3. Finally, the training institutions themselves, which continuously have to develop the training program and see to its applicability with the labor market needs and the continuous training of trainers.

The focus on demand-driven competency identification, training, and development built on facts can probably significantly change the methodology of planning university training programs, learning taking place in higher education together with its planning and assessment. We can see that all this results in the worlds of work and learning coming closer to each other, the importance of nonformal and informal learning gaining impetus, similarly to the concept of knowledge transfer, the everyday realization of which together with its methods and tools can become part of higher education.

REFERENCES

Baráth, T. (2010): Methodology applied in the project. In: Abari-Ibolya-Baráth (eds.): *Improving school leadership in Central Europe.* Tempus Public Foundation, Budapest. pp. 33–40.

Baráth, T.; Cseh, G.; Kézy, Á.; Kígyós, T.; Simon, M.; Tóth, L. (2010): *Final report – Competence based development of organization, content and methodology of teacher education on regional level.* Juhász Gyula Teacher Training Faculty, University of Szeged, Szeged. (manuscript).

Drucker, P. F. (1993): *Post-capitalist society*, Harper Business, New York.

European Commission – Directorate-General for Education and Culture (2004): *Common European Principles for Teacher Competences and Qualifications.* 21/03/2012 sighting http://ec.europa.eu/education/policies/2010/doc/principles_en.pdf

Florida, R. L. (2002): *The rise of the creative class: And how it's transforming work, leisure, community and everyday life.* Basic Books, New York.

Harten, H. and Wolbers, H. O. (2005): *Syllabus; RDA basics certification course.*

Inglehart, R. and Baker, W. E. (2000): Modernization, cultural change and the persistence of traditional values. In: *American Sociological Review*, Vol. 65, 2000. pp. 19–51.

Lifelong Learning United Kingdom or Lifelong Learning UK (LLUK) (2006): *A sector skills agreement for the life long learning sector.* Stage 1 – skills needs assessment. London.

Sawyer, K. (2008): Optimising learning: Implications of learning sciences research by R. Keith Sawyer. In: OECD (2008): *Innovating to learn, learning to innovate.* OECD Publishing, Paris.

88 Teachers as Citizens
Professional Preparation and Development

Grzegorz Mazurkiewicz

CONTENTS

INTRODUCTION

The most important result of a teachers' work is a student who learns. This is an obvious statement, which becomes complex when one starts to clarify what leaning means. There are numerous understandings of the concept of learning as well as numerous understandings of its aims. We know that the context of human life and also the context of education have changed dramatically during first decade of the 21st century. It is not necessary to repeat here the diagnosis of a contemporary globalized world hardly influenced by the development of technology and hurt by economic, environmental, and social crisis, but it is necessary to be aware of it. That awareness helps to comprehend the challenges arising in front of existing education systems and its main workers: teachers. We will also not discuss the process of learning itself and the aims that should be taken under consideration in the modern education, but we will rather focus on the general conditions of learning and on the way of thinking about teachers, their role, profile, and preparation.

Is it enough to draw new aims, directions, and tasks for teachers in order to change contemporary school? Is it sufficient, for the purpose of school improvement, to describe needed competencies to support teachers' development? The typical, quite narrow approach to reality is determined by a framework in which the only thing we need is to understand current conditions, discover future challenges, and create appropriate solutions. That kind of approach is our heritage from industrial and modernization times and is difficult to transform, but it is simplistic and naive. Yes, it would be very easy to improve our lives if it worked in such a predictable and organized way. Unfortunately, we often fail while using that approach. For example, decision makers very often truly believe that in order to improve the situation and to provide the labor market with skillful employees, we simply need to prepare an adequate educational path (institutions and teachers) and to recruit appropriate students. They also believe that in a few years the problem will be solved. Probably the popularity of this strategy is one of our problems. Difficult tasks never have simple and direct solutions. It is time to state it loudly and time to stop looking at school as an easy to use device.

Together with the development of widespread technology, the employees' permanent competencies have lost their meaning. It is no longer critical to be prepared before starting a new job; now it is more important to be able to learn needed competencies rapidly when a particular situation appears. A new ideal employee continuously acquires new skills. The way of functioning for the majority of contemporary institutions is determined by short-term, differentiated tasks. It is not what we can do but what we need to do that shapes organizations today. Casualization of jobs leads to short contracts often renewed, but always adjusted to new aims and projects. Flexibility, decreased hierarchy, and no linear organization of work change the structure and organizational climate, but in general, they influence the basic rules of cooperation (Sennett 2010). Unclear rules and unclear expectations create situations demanding people's independence in order to bring quick results. Only responsible and brave action might lead to success when it (success) is not predefined. That situation creates a high level of stress and anxiety especially because we are not prepared to understand the reality and not ready to accept it.

Unfortunately, one of the main outcomes of this new reality is growing inequality, which is the most dangerous phenomenon for social integration and trust (Wilkinson and Pickett 2009). Therefore, the main function of school should be creating opportunities for everyone to overcome "natural" obstacles connected with their place of birth (in a geographical and sociological sense). Democratic society requires democratic schools, and the latter require democratic structures and processes. Education is a group investment, not a personal agreement, between the individual and the institution of the acquisition of a right to have a good job or obtain a degree, which facilitates receiving a higher wage (Mazurkiewicz 2012).

I will stress Aronowitz's and Giroux's classic opinion that the crisis of the school occurs everywhere that the crisis of the citizenship idea is noticeable, where the social and historical awareness is missing, where inequality and injustice are ignored, and where the fundamental democratic values are compromised (1991). Where is the place for teachers in schools fighting with a crisis of democracy? What kind of people should they be in order to be able to help schools in the task of creating opportunities? I believe that we need to find a balance in a discourse about teachers' professionalism that would allow the inclusion of the concept of a teacher as a citizen, a member of civil society who is using the professional competencies not for proving mastery in what he or she does or for answering the bureaucratic (and not only) expectations, but for an independent search of the meaning and the construction of reality with responsibility and solidarity with all students.

The concept of teachers' citizenship allows more fully than the concept of teachers' professionalism an understanding and, in a certain way, the management of the complexity of the teachers' role. The discourse around teachers' professionalism should be broadened beyond the current stage and include thinking of them as citizens who are responsible for development of democratic societies. The discussion, research, and training should include three broad domains. One of them is the teachers' understanding of the world, the intellectual ability to work in a certain social and political context, influencing theirs and their students' lives. Second is the approach to reality and readiness for interaction with the social world that would not allow them to stay outside of the important processes in their communities and again in their students' lives. Third is the whole universe of teaching in which the term professionalism is natural and obvious and in which it means constant reflection, dialogue, and development, which leads to strengthening the self-regulating profession.

TEACHERS AS TRENDSETTERS

Teachers (and future teachers) too often face professional reality difficulty to such an extent that they feel helpless. Increasing expectations, lack of clarity about responsibilities, and deficits in professional preparation create situations in which teachers feel lonely and not ready for the work. Without support but with elaborate bureaucracy, suffering because of stress, competition, and "theatrical" gestures, teachers globally are afraid of being not able to fulfill plans and needs (Polak 2012). This leads to a question about teachers' roles and also opens a discussion about teachers' preparation.

Nevertheless, the picture of education would be too dark if we focused only on negative aspects of challenging conditions in schools and whole educational systems. John MacBeath (2012) points our attention to the fact that in our time schools have become better places for children because teachers and other experts have developed an understanding of the relationship between sanctions and incentives; motivation and demotivation; the school and classroom environment, which can both promote and inhibit learning and effective teaching; the impact of home and peer groups; learning disabilities; and also the damaging effects of discrimination. It is also a better place for teachers, which is a result of rising standards of teacher qualification and professionalism. Moreover, in the most privileged countries, they teach in classrooms that are better resourced and with smaller class sizes. They also enjoy more opportunities for continued learning and professional development (but, at the same time, at the end of the first decade of the 21st century, many countries are facing a recruitment and retention crisis) (MacBeath 2012, pp. 8–9).

In Poland, it is necessary to look back at the time after the political transformation of 1989 because it was a time of critical change but also a time of widening teachers' autonomy and increasing energy for work and development. Unfortunately, the enthusiasm of early change slowly disappeared and was substituted by the tyranny of political, social, and economical order. The speed of the change was really high, and very often changes of consciousness were much slower. Together with the loss of the ownership, the feeling of helplessness has appeared (Polak 2012).

Preservation or change—this is a dilemma, solutions to which determine the educational reality. What do societies really want from schools—protecting unfair status quo, supporting existing order, "improving" reality as it is or rather creating conditions for radical transformation, questioning the situation, and showing the direction of action that would lead toward equality and solidarity? Until now, education mostly served as a tool for explaining existing social stratification; for rationalization of economic structures; and for adjusting to, as it was believed, objective reality. But, according to Bauman's visions, in the contemporary liquid reality, time of chaos, when we cannot pretend any longer that it is possible to put all human affairs in order (because everything is intertwined and interdependent), we need to reject the delusion of utopia of order and faith in one-directional progress, objective knowledge, ability to control, and also obsession of structured reality. We cannot count on clarity over chaos as we believed in the past. Every initiative was driven by hope that one more effort will allow us to find a universal solution or will lead us to the state of happiness. It will not happen. This hope was a force behind the progress of the modernization era, but today it is over (Bauman 2011). School however, dependent on interpretation, was not able to fulfill the expectations of modern times or just participated in manipulation and preservation of the unfair social order. Today, it (school) should become an important and independent player in the game of interpretation of reality. Today, teachers as well need to understand that the whole burden put on their backs by societies is a result of the wrong priorities that will never come to reality (for example, demand for a quick fix on the labor market). If teachers are those to blame, it is not about "bad" teaching but about passive acceptance of those priorities.

Our lives were not given to us as an objective role to play, and education is not a process of passing a list of precise rules of conduct. It is difficult to live in the time of uncertainty, so teachers and whole societies have a right to feel helpless and anxious. It is easy to give advice to accept that state of affairs, but certainly this is not a solution. The only way to survive is to be active—in decision making, planning, and in the implementation of committed decisions. Of course, we will never have the certitude of the appropriateness of those decisions, but it is natural in a state of freedom without supervision of universal truths, solid social structures, "authentic" and "clean" national identities, etc. We need to be heroic, says Bauman, when liquid reality demands from us a lonely struggle with the consequences of our choices and deeds. How to be a teacher in this situation, which is lacking stable points of orientation and is characterized by increasing expectations from the outside world (authorities, employers, parents, and others) influenced by unrealistic aims and outdated beliefs (for example, about the economy or school–labor market relations. Teachers need to be citizens of the world who share responsibility for it with all co-citizens, not experts knowing all answers, citizens who, commonly with others, interpret the experience of life in order to build a safe environment for everyone and who do it in the name of solidarity and democracy, not competition and income.

The concept of citizenship seems old fashioned lately and, for sure, not so popular in the 20th century (at least in Poland during the 1990s). An entrepreneur may be an employee or sometimes a flaneur more visible in a public discourse—people rather individualistic with a loose (flexible) connection to the community in which they stopped for a moment. Citizenship publicly exists rather as a narrow concept of belonging to a specific group defined by state borders and is more important for those who are excluded and make an effort to join the particular community. Those who gained citizenship through birth value it very rarely. I would postulate that teachers might be trendsetters for revitalizing the idea of a citizen, responsible, interested in what is going around, active, involved, critical but not negative, creative but with an understanding of the context and abilities.

Teachers in search of professional perfection need to stop and reflect on the idea of citizenship in school. What does it mean, and how could it be used? First at all, it is worthwhile to remember that key factors forcing teachers to leave the profession are associated with a lack of trust and support, workload, little contact with other adults, lack of availability of information, not being in control of their work, inappropriate expectations of what schools and teachers could achieve, and intensified pressure (MacBeath 2012). Those factors are definitely tied to the area of citizenship that should be built, maintained, and promoted in education.

EDUCATION BUILT BY CITIZENS

It is difficult to talk about "the reform." We are bored to death by waves of the reforms focusing on change in schools but only on a surface, impacting only structures, never this, which is an essence of the education process—beliefs, attitudes, convictions. Teachers, citizens of civil society, do not have to suffer the agony of constant bureaucratic reforms because being a citizen means being actively involved in improving the current situation. Change is embedded in action and everyday life because change is an integral part of responsible self-organization of civil society. Teacher-citizens are not sense seekers operating on social order but sense constructers operating because of their inner conviction. So we do not need another democratic school reform, we need democracy, which will never appear without citizens. Through the years, a lot was done to stop teachers from being citizens and from acting as citizens, mainly through taking autonomy and independence away from them (Hargreaves and Shirley 2009).

As it happens quite often, some mistakes are caused by good intentions. After all, education constantly remains one of the major points of interest for governments all over the world because it is still perceived as the best investment for the future. In most countries, it is a significant aspect of public spending and still remains the warranty of success. The belief of usefulness of education for individual and social success is confirmed by various statistical data (OECD 2010). Unfortunately, the research stemming from that belief and focusing on school effectiveness brings blurred and difficult-to-use results. Schools differ so much in relevant aspects, such as the causes underlying their specific performance, capacity for change, contextual characteristics, etc. These differences are stressed when considering the practice of importing school effectiveness models from one country to another. That is to say that a one-size-fits-all solution cannot be used to improve school performance; instead school improvement efforts should carefully consider the power of site or place. Everything that goes on inside a school is tied to local culture, attitudes, values, tradition, and beliefs. Schools cannot be understood as detached from the world beyond their walls, but the more complex measures, correlations, effect sizes, and similar disputes are ruling the educational world, the more teachers rely on experts instead of on their compass and professionalism (MacBeath 2012, pp. 41–43). The previous sentences should not lead anyone to reject results of scientific research on teaching and learning from decision-making and strategy planning but to a more inclusive approach to the process of decision-making. Civil society needs a variety of stakeholders, and teachers should never be excluded from it.

We know from research that learning is an active socially constructed process situated in a broad socio-economic and historical context and is mediated by local cultural practices and perspectives, taking place not only in school but also in multiple places such as their homes and communities and that support of a variety of institutions is needed (MacBeath 2012). This is work for citizens who are able to create a network that would be a support system and a place for learning. It demands negotiating and communicating skills and also understanding from the side of learning professionals that learning is tightly connected with emotions, motivation, and individual differences among learners, including their prior knowledge and identity.

Teacher-citizens are open to deliberation about everything and are committed to change because it is not an event in their lives but the typical situation. Their fundamental task should be to help the student feel like an autonomous person and support him or her in perceiving himself or herself as a partner for others. Every human being, in order to feel safe, has to have a sense of control over what happens to him or her and to know that others notice him or her, respect his or her opinions, and take them into account. The school community should be a place of development and inspiration and not a place of forced work. Therefore, the school must build a culture of cooperation and appreciation instead of creating a culture of competition. Fear is the enemy of learning; it is also the enemy of all relationships. When we are afraid of our own safety, we rarely establish genuine and open relationships with others. If fear accompanies the educational process in the school, it leads to disastrous consequences. However, teachers must face the culture of fear (Palmer 1998) although that involves the necessity to question the traditional approach to interpersonal relationships in the professional context. The culture of fear manifests itself in reflexes, priorities, and practices concentrated mainly on guaranteeing one's own safety by proving one's own usefulness and infallibility. It is closely connected with what happens outside school. We will not escape the serious problem in the crisis of interpersonal relationships. Democracy does not depend only on the arrangements in the political system, the manner of voting, control, human rights, etc. Democracy is based, above all, on trust in others and on the belief that the persons who make decisions on our behalf do that on the basis of sensible opinions and judgments (Meier 2000). When that trust is missing, the democratic system starts to fail. The same applies to the school. It is absolutely essential to build trust and the sense of safety already in schools.

The first task for all who are struggling with the current educational challenges is to design and prepare a learning environment that would enable learning. This environment should take into account conditions that allow learning to appear and which are a stable identity, safety, and motivation. Learners participate fully in the educational process when they understand who they are and what they need to grow and develop. They learn when they feel safe and operate in a climate of mutual trust in which they know that they have the right to make mistakes and suffer failure because it will support their development in the long term. Students also need to really want to learn. What is important is the belief that it is worth learning and that it is possible. Three quite simple matters—identity, safety, and motivation—are starting points in thinking and designing schools (or, better, about education) and decide on the quality of learning.

What the teachers should do in the first place to increase students' motivation to learn is to encourage them to think about their own school experience. They have to answer themselves also the "identity" question: What do they go to school for? Building the habit of critical thinking reinforces their motivation to learn. School education is not only powerful machinery established, kept, and managed with great financial and organizational efforts by people hired by the system, but it is also a social experience of millions of young people who come to school with their own plans and dreams. They sometimes agree to cooperate; they sometimes choose a form of resistance (Shor 1992). If we fail to convince them that it is worth making an effort for their own development, instead of genuine learning, we will observe tricks of slaves trying to prove that they do everything that is expected from them.

However, the dialogue on the importance of the school cannot take place when teachers do not know the aim of the school teaching and are not able to show it to their students. In order to learn to inspire students, teachers have to first learn to reflect on their own actions. The teacher has to be ready to ask difficult questions and be open to inconvenient answers. It is the ability to reflect as well as the preceding curiosity and willingness to understand what happened that are the conditions for a high-quality educational process. Willingness to learn is better seen with the people who firmly believe in what they do and are convinced that they have knowledge, skills, and attitudes appropriate to do that. Self-confidence enables teachers to support the autonomy and freedom of students. Only suitable competence gives such a sense of confidence, which makes it possible for "the power to be shared." Therefore, only those who trust themselves and are able to prove their own qualifications for their profession can motivate their students to make efforts. You cannot teach without that. In a way, that means that the teachers who "have not done their homework" and have not gained that competence do not have a moral right to teach. Lack of professional competence completely destroys teacher's authority (Freire 2001, p. 85). The teachers who do not treat their development seriously cannot serve as leaders of the young.

A safe learning process and good motivation are the prerequisites for a genuine process of educating. Effectiveness of that process is higher when students understand their context and situation, when they are aware of their own identity. The educational process too often proceeds as if all its participants were the same; individual needs of students, their history and experiences, are not taken into account, and their future plans are not considered. The school must make it possible for students to reflect on who they are and what they expect from themselves, from the school, and from the world. That moment of reflection enables students to understand the specificity of their own situation and, because of that, owing to full awareness of who they learn for and what they learn, to intensify the process of learning.

Vaclav Havel believes that the only hope for our world is us ourselves and that we can make a change only by using our ability to reflect and our sense of responsibility. Such attitude is a worldwide revolution in people's minds. Nothing will change for the better without it, and nothing will save us from the inevitable catastrophe, which we are approaching (Palmer 1998, p. 20). If we manage to properly lead the process of change, there is a chance that students will not only be participants of the educational process who carry out orders automatically, but that students will become people aware of direct responsibility for their community and the whole world. Students of the civil school will be able to learn responsibility, responsibility for themselves; for what they do; how they study; and for other people, friends, neighbors and fellow citizens; for the community we live in. We have to remember that every valuable action starts the moment we realize its necessity and our individual responsibility for all its consequences. The responsibility for oneself and the world is the effect of education but also a prerequisite for education. We will teach well when our students are becoming responsible and are willing to shape the reality that surrounds them according to what they have been taught in a responsible school. How to focus teachers' attention (and their development) on issues fundamentally impacting learning process?

TEACHERS' EDUCATION AS CIVIC EDUCATION

All processes of teachers' development should lead to the connection of teachers' experience with professional development that serves to strengthen their social and human capital, which will support teachers' significant participation in sustainable development of societies (through students' learning). For now, our educational systems are rather outdated and not sufficient for our dreams. We should start thinking about radical changes, but do not have a vision and resources. The easiest way would be to assign teachers themselves to prepare the scaffoldings for that radical change, but of course, it is not possible. Every successful initiative needs at least two sources of motivation and action: top down and bottom up. It is a good point how to reconcile those two engines in the case of teachers' civic education.

First of all, we need to take care of the professional aspect of teaching. Outside the area of education, a professional is somebody who belongs to an occupational group that has a well-developed technical culture, high skill levels, special knowledge, acceptable standards, professional ethics, a long period of training, and a high degree of autonomy (Sachs 2004). It is also important for members of this group to be able to build their professional knowledge on their own, conduct research, publish articles, and hold discussions because it is them who know best and see most in this area. They need help understand what they observe and enable them to share their observations with others. This is the moment when a need and a chance for autonomy comes up. In uncertain situations, professionals make use of their independent judgment rather than routine habits or regulations (Mazurkiewicz 2012). It is important to help teachers to become professionals as a group because their professionalism would create the opportunity for bottom-up initiatives and allow the establishment of a strong support system for improvement.

The second pillar of the teaching vocation should be citizenship. The teacher-citizen develops the ability to be open to the world and to other people and works together with other teachers to understand their context and the complexity of the social processes and difficult role of the school. A professional teacher can prove his or her experience and expert knowledge, shows altruism toward those whom they serve, and presents mature autonomy that allows seeing inclusion not as a threat or artificial expectation. The following three domains, mentioned already in the introduction, are necessary to be taken under consideration while deliberating on the teacher's role:

1. A deep awareness of one's own attitudes, theories, or even limitations that determine the way we and other people function in the world and at school and a willingness to serve others in the process of growing up and development. I propose to call it intellectual sensitivity.
2. Activity for a social change understood as a main goal of pedagogical work through involving into projects inside and outside the school and inspiring others to make the effort for public good. Such attitude might be defined as willingness to be an educational activist.
3. A scientific approach to the process of teaching and learning that enables constant revision of the knowledge possessed, mastering the skills, and active and independent (autonomous) construction of own profession through research, reflection, dialogue, and cooperation with others. Generally, it should be defined as a professional approach to education (Mazurkiewicz 2012).

We need to help teachers to increase self-awareness; they need to know themselves and, at the same time, be convinced that they can do something, that they are effective. A responsible teacher-citizen understands that the reality, society, and school are all products of contradictory forces, but they are also the effect of people's actions, and as such, they are still unfinished and thus subject to change. Teachers cannot accept the conditions under which they live and work as something that exists objectively. They should try to decode how the authority, culture, or history determines certain choices that influence the school and conditions of students' life. Discussion on pedagogy and establishing common theoretical ground through sharing conviction on teaching and learning will help to understand the complex system of ideologies, mental models, and theoretical approaches owned and shared by the human network functioning within one school (Mazurkiewicz 2012).

Teachers, with all their capabilities, need to actively participate in social life. When we take an active part in shaping our own environment, we show that we care about it. We also show respect to the environment and the people who live in it. Such respect means a belief that each of us is an autonomous individual. If we respect one another, by giving ourselves the right to actively participate in life and transformations in the school, we will respect our students in a natural way. These two kinds of respect are inseparable.

In order to be able to fulfill his or her role, the teacher must acquire some intellectual skills or methodology that enables him or her to constantly monitor and master his or her own practice and actions. They need to secure a professional dialogue and, through this, to combat the traditional isolation of teachers. One of the possible ways to do it is common discussion on accessible data that allows them to show that they are professionals who can creatively solve problems and understand global and local conditions (Stringer 1999).

The presented three domains necessary in thinking about the teachers' role in designing and leading schools that would focus on the aspect of civil society, which would allow a push for a community of citizens, as a learning community is just an invitation to a discussion about both the concept of citizenship in education itself and necessary issues included in that concept. What is important and needed is a public discourse on the issues raised above.

I will repeat after Henry Giroux that we need new language that might reinvigorate the relationship among democracy, ethics, and political agency by expanding the meaning of the pedagogical as political practice. It is important to recognize that pedagogy has less to do with the language of technique and methodology than it does with issues of politics and power. Educators should raise question such as what is a relationship between social justice and the distribution of public resources and goods? (Giroux 2011, p. 72).

Teacher-citizens carry a huge responsibility, but it is not imposed on them but taken by them (in a free and thought-out way). Today, we may take a risk and say that there is no chance for improvement of the school's quality if the paradigm followed by this institution is not significantly changed.

REFERENCES

Aronowitz, S. and Giroux, H. A. (1991) *Postmodern Education. Politics, Culture and Social Criticism*, University of Minnesota Press, Minneapolis, London.

Bauman, Z. (2011) *Culture in the Fluid Modernity*, Polish Ed. *Kultura w płynnej rzeczywistości*, Narodowy Instytut Audiowizualny, Warszawa.

Freire, P. (2001) *Pedagogy of Freedom. Ethics, Democracy, and Civic Courage*, Rowman and Littlefield Publishers, Inc., Boulder, New York, Oxford.

Giroux, H. A. (2011) *The Promise of Critical Pedagogy*, in: *On Critical Pedagogy*, The Continuum International Publishing Group, New York.

Hargreaves, A. and Shirley, D. (2009) *The Fourth Way: The Inspiring Future for Educational Change*. Corwin, Thousand Oaks, CA.

MacBeath, J. (2012) *Future of Teaching Profession*, Leadership for Learning, The Cambridge Network, Cambridge.

Mazurkiewicz, G. (2012) *Edukacja i przywództwo. Modele mentalne jako bariery rozwoju*. Education and Leadership. Mental Models as Obstacles in Development. Kraków, Wydawnictwo Uniwersytetu Jagiellońskiego.

Meier, D. (2000) *The Crisis of Relationship. W: A Simple Justice. The Challenge of Small Schools.* Ayers, W., Klonsky, M. and Lyon, G. (ed.), Teachers College Press, New York and London.

OECD. (2010) *Education at a Glance*. OECD Publishing, Paris.

Palmer, P. J. (1998) *The Courage to Teach. Exploring the Inner Landscape of a Teachers' Life*, Jossey-Bass Publishers, San Francisco.

Polak, K. (2012) *Helplessness of a Teacher*, Polish Ed. *Bezradność nauczyciela*, WUJ, Kraków.

Sachs, J. (2004) *The Activist Teaching Profession*, Berkshire, New York: Open University Press.

Sennett, R. (2010) *The Culture of New Capitalism*, Polish Ed., *Kultura nowego kapitalizmu*, Warszawskie Wydawnictwo Literackie MUZA SA, Warszawa.

Shor, I. (1992) *Empowering Education. Critical Teaching for Social Change*, The University of Chicago Press, Chicago.

Stringer, E. T. (1999) *Action Research. Second Edition*, Sage Publications, Thousand Oaks, London, New Delhi.

Wilkinson, R. and Pickett, K. (2009) *Spirit Level: Why More Equal Societies Almost Always Do Better*, Allen Lane, London.

89 Social Learning–Socialization

Understanding Aims of Social Education against a Backdrop of Sociological Discourse on Socialization

Barbara Behrnd-Wenzel and Hartmut Wenzel

CONTENTS

INTRODUCTION

In contemporary education policy discussion, new demands are being made, especially on school personnel as a consequence in many countries of growing school problems of various kinds, such as violence, radicalism, vandalism, and also truancy, classroom disruption, etc. as well as unsatisfactory behavior of children and young people in and out of school.

For almost any social problem, whether prevention of violence or drug abuse, health education, combating radicalism and vandalism, and much more besides, the school is expected to develop a program to overcome the problem. The school is to impress its influence on pupils, deal with problems in an appropriate manner, and "teach" socially desirable behavior for society and the workplace.

In this context, there is constant reference to "socialization," by which is meant in particular the establishment of desirable social behavior. It is all very well to formulate these demands to make school a significant authority in solving social problems, but to translate them into sustainable school practice is fraught with problems and difficulties. For, unlike the development of structured knowledge, social behavior is not derived from a cognitive school learning process but, above all, from lifelong biographical processes of adaptation to a society with its various settings, norms, values, and structures and, at the same time, from individuals actively coming to terms with a particular social and material environment. In this, school is itself an institution, which does and should represent values and norms. These values and norms often differ considerably from those of the school's clients, namely pupils and parents. So conflicts are already built in, arising from discrepancies between the educational outlooks and aims of school and family, and these are well worth addressing. Furthermore, there is an awareness that the organizational rules of the school itself have effects on the development of personality, this being to some extent a "hidden curriculum" of the school.

In the last few decades, the social sciences have developed a differentiated understanding of socialization in the complex processes of interaction of the individual and the environment in personality development. This understanding goes beyond intended educational impact and includes undesired side effects. Theories of socialization have emerged in which a variety of traditional theories are embedded. These include, in particular, ingredients of sociology, anthropology, psychology, biology, cultural anthropology, psychotherapy, pedagogy, and also theories of learning and cognition concerned with the complex process of personality development in children and young people.

The first part of this chapter will pursue the understanding of socialization. This will be followed by a section to consider the school as an instrument of socialization with samples of two significant extensions of socialization theory. Thereafter, I will look at some results of school studies based on socialization theory.

UNDERSTANDING SOCIALIZATION

It is a fundamental anthropological fact that human beings, as a biological species, are born, grow up, age, and then, at some point, inevitably die. If, despite this biological "fact of development," a society with the stage of culture it has reached wishes to continue to exist and develop, beyond the individual lifespan of its members, then the responsible generation of adults has an important task. It has so to influence and support children and young persons as they grow up that they are later in a position to take over responsibility for the direction of their individual lives and, at the same time, for the development of society, including transmission of its culture. That means making it possible for the next generation to develop its own personality together with assimilation into existing society in such a way that an increasingly autonomous lifestyle becomes possible in the social context. So alongside individual development runs a process of social reproduction, which relates to required qualifications as well as social structures and socio-cultural values and norms. The theoretical understanding of these related aspects of personality development has been emphasized in recent decades, first in the social sciences and then, increasingly, in pedagogy, in the concept of socialization, and a new scientific direction has emerged for theory and research, the theory of socialization.

By socialization is understood the "process by which personality emerges and develops in reciprocal dependence with the socially transmitted human and material environment." Its predominant theme is "how the human being develops into a socially competent subject" (Tillmann 1993, p. 10). In the context of this understanding of socialization, education plays its part, namely in the intended influence of the adult generation on the future one. Central to this definition, as Tillmann indicates, is that all social as well as material elements can have an influence on personality development, that they are all "socially transmitted," i.e., that they can be experienced by an individual in a situation bearing the imprint of society and with meaning bearing the stamp of that society.

When we speak of personality, we mean here the specific make-up of features, qualities, attitudes, and capabilities, which characterize an individual. This make-up—the psychological structures and features of a person—has been built on the biological foundation of human life through the experiences encountered in the course of his or her life history. This understanding of personality is not, therefore, confined to externally observable modes of behavior but includes inner psychological processes and conditions; thus socialization is essentially about the formation of "individuality" and at the same time "social character." Socialization theory must therefore be concerned with the relationship of the development of aspects of inner reality and the conditions of external reality. The genesis of personality in the social process leads at the same time to socialization and individualization. Socialization theory has to do, therefore, with basic statements about the relationship of the individual and the social environment and postulates that this relationship is to be understood taking into account the active influence of the individual on that environment.

Considered in this light, human beings are not just the victims of their socialization, but they also have an effect on themselves and on their social environment and develop, in this interactive mode, into competent beings, to being a subject that is a socially competent being capable of initiating independent action. Recent socialization theory with this understanding is distinct from all social-determinist concepts in which socialization is misunderstood as a one-sided imprint. Socialization is not simply the transmission—whether voluntary or enforced—of society's expectations into psychological structures but is a process of active adoption of social conditions by human beings. There is a tension between the principle of scope for a human being to relate with a social environment actively, individually, and according to a particular situation and society's demands, which are often directed toward accepting norms. For this tension to be understood conceptually, socialization theories are not to be developed merely from the individual as a passive recipient object but must take into account the capability to shape actively as a subject.

Recent socialization theory is distinguished therefore from biologically based conceptions in which genetic dispositions determine and specify later behavior; such genetic determinism is seen as unproven speculation although clearly the biological basis has to be considered. Recent socialization theories are equally distanced from idealistic individualism and emphasize the reciprocal effects of subject and environment.

Anyone using the notion of socialization in the sense described above will refuse to reduce the analysis of personality development to planned educational actions. Rather, he or she will emphasize that the totality of life circumstances is important for the development of a subject. The influence of educational transactions is not denied but is included systematically. Education as a conscious and planned influence by adults on those growing to adulthood is seen as a part, or subcategory, of the process of socialization and respected accordingly.

SCHOOL AS AN INSTRUMENT OF SOCIALIZATION

In the course of history, social reproduction processes, that is, the younger generation growing into an existing society, have become more difficult and complex. Over a long period, it was possible for an overwhelming majority of society to grow into a community just through living in it and so, in a more or less natural growth following the path of their parents, to find their own place in society and contribute to its perpetuation. But in the course of time, with the growing complexity of societies and the progressive differentiation of work situations, more extensive measures have become necessary to bring about the reproduction and development of society. Ethnologists have identified specific educational praxis and mechanisms for socialization for the various structures of society, such as initiation rites and other special upbringing practices, which serve to reproduce society with its norms, values, and structures.

In modern societies, school, alongside the family, has a leading part in social reproduction. As well as the task of promoting individual learning and education, it also has the function of contributing to the reproduction of a society with its norms and values and also its social structures. To this extent, it is a social instrument, which, in recent decades, has grown enormously in importance for the upbringing of children and young persons. Childhood and youth, after all, are today largely periods of schooling.

If school is considered as a social instrument, then the development of pupils' personality becomes central, and the question becomes in what ways does the overall institution of "school" influence young people's development as an active person (subject)? And what is the relationship between in-school socialization processes and other social influences? Then the traditional pedagogical perspectives and analysis, limited to intended learning transactions, and the interaction of pupils and teachers are inadequate, and the additional influence of the school as an institution and sphere of living, with its impact on personality, has to be included systematically. In this way, the unplanned effects of school as an institution come under scrutiny alongside its officially intended goals. At the same time, an exclusively in-school view of the learning process is abandoned, and a systematic attempt is made to understand personality development in school as part of the reproduction process of society. What is transacted in school is, in this respect, always embedded in social connections and influenced by both positive and problematic developments in the family, the community, and likewise, the world of work.

TWO EXAMPLES OF EXTENSIONS OF SOCIALIZATION THEORY

Internationally, two theory extensions are of particular importance in describing these connections between individual development and social reproduction. These examples will now be considered. They both concern themselves with school socialization but from different perspectives. First, there is the structural-functionalist extension of Talcott Parsons (1902–1979)—more a macro-sociological one—and then the more micro-sociological extension of symbolic interaction developed from the work of Georg Herbert Mead (1863–1931) and Erving Goffman.

SCHOOL SOCIALIZATION IN A STRUCTURAL-FUNCTIONALIST PERSPECTIVE

It was Talcott Parsons in particular whose structural-functionalist perspective of a theory of schooling probed into the connection between educational measures in the classroom and processes of social reproduction (Parsons 1968). For him, school as distinct from family is a formalized system of roles, in which universalistic values are at play. The school classroom is a new world, to which the school beginner must adapt and accommodate. The enacted socialization process of internalizing roles is linked to a relatively systematic and continuous evaluation of school achievement, thus embarking on an increasing differentiation, which is later adopted as the basis of allotting social status. In this respect, in addition to its educational function, school becomes in equal measure, for individual and society, an instrument of selection.

It is to Parsons' credit that he developed this general interweaving of the school's socialization function with the selection that accompanies its qualification function. From the perspective of an observer, he points out how the school undertakes significant social functions. But this brings out, at the same time, a problem area, which is of considerable significance today. For, if outside the school, conditions of employment change and school-leaving qualifications can no longer guarantee direct entry to the world of work—when school awards and vocational positions no longer match—then efforts to achieve in school lose their sense for an increasing number of pupils. At the same time, there is, in the school, added pressure to conform and competition for high marks. School objectives, such as the development of autonomy and independence, come into conflict with socially desirable aims, for example, conformity and socially acceptable behavior. Such contradictions can be a burden on the atmosphere or climate of a school. Dissident behavior in school can then lead to more severe sanctions beyond the school. That, of course, has consequences for personality development and for the ways individuals manage the increased pressure to conform. And here, the school-family relationship plays an important part.

Meanwhile, extensive studies have been carried out in the critical engagement with the structural-functionalist approach, and these underpin the great importance of the school system's extended role in measuring and assessing achievement for individual development in the interplay of school and family.

SCHOOL SOCIALIZATION IN AN INTERACTIONIST PERSPECTIVE

A different perspective on processes of personality development in school is adopted on the basis of symbolic interactionism (Mead 1969; Goffman 1973). At the heart of the theory of symbolic interactionism is the description of the process of communication between subjects, as a social "process from which identity develops" (Mead 1969, p. 207). The subject is actively engaged in this communication process, interpreting situations and demands, and responding to them with self-conceived actions. In each action, there is always an enactment of individual identity. This is to be understood as a balance between personal and social identity. Habermas describes it as follows:

> Personal identity is expressed in an unmistakable biography, the social identity possessed by one and the same person in frequently incompatible sets of relationship. While personal identity guarantees something like the continuity of the Self in the sequence of changing circumstances of a life-history, social identity preserves a unity in the diversity of various role-systems, which have to be managed simultaneously. Both identities can be conceived as the result of a synthesis which extends across a sequence of conditions in the dimension of social time (life-history), i.e., across a variety of simultaneous expectations in the dimension of social space (roles). The identity of the Self can then be conceived as the balance achieved in sustaining both identities, the personal and social (Habermas 1973, S. 131).

Self-identity is thus described as a balance, which must be achieved in each interaction. On the level of social identity, the individual is required to adjust to norms of expected behavior: The enactor has to be like all the others. On the level of personal identity, the individual is required to present as unmistakable and unique, to be unlike anyone else.

In the theory of socialization, it is very important, as Tillmann (1993) points out, that the process of personality development is conceived as a unity of societal norming and individualization: By reflectively adopting the language symbols, values, and norms of the social environment, the individual becomes a competent member of society and, at the same time, an unmistakably unique individual. Because self-identity thus understood has constantly to be re-transacted in the interaction processes, the biographical dimension comes into play: Early childhood, parents, family, school, university, marriage, etc. present a sequence of experiences, in which individual identity is reinterpreted and further developed (cf. Tillmann 1993).

When applied to schooling, this theory offers a micro-analysis, which clarifies what "logic" is being adhered to by the actors within the institution and how identities are represented, upheld, or damaged in the process. This theory develops an altogether critical perspective of institutions. It points out that role structures anchored institutionally can have a repressive effect and that this may be the origin of damaged identities and lifelong failure. Children and young people do not have the same opportunities as teachers to bring their out-of-school identities into school processes. This makes it difficult for them to find their identity balance. Thus comes about a differentiated analysis of the relationship between "normality" and "deviation," which is of special importance in the school context. Such a concept has considerable strengths in the analysis of successful and unsuccessful socialization processes at the micro-level in describing the relationship between subject and the environment with which they communicate.

Pupils who obviously do not match the school's expectations for achievement and behavior are threatened not only by material sanctions from the school or from teachers; it is not only the individual reported action that is officially defined as *deviant*, but very quickly the person behind the action is typified as *deviant* or *conspicuous*. Especially when *deviance* is serious or frequent, it is progressively attached to the personality of the pupils concerned. Such processes of labeling deviant behavior can be the starting point of negative school careers.

It is teachers who determine whether or not an action is to be categorized as deviant, according to their individual conceptions of normality. Linked to their categorization, as can be seen from the perspective of Parsons, are rankings of status. For example, a pupil who has struck another is rarely seen as someone whose behavior is often quite different, but has just this once hit out in a particular situation and is all too quickly seen as someone who hits others or as physically aggressive. Such categorizations can be adopted by those with whom they interact and have an influence on their social position. One can speak of labeling or stigmatizing when pupils are defined as being deviant and when this categorization becomes known to others with whom they interact (e.g., other teachers and fellow pupils).

SOME RESULTS OF CRITICAL STUDY OF SCHOOLING

If one starts with the understanding that personality development is a process of active engagement of the individual with the social and material environment and so is also a process of developing identity, then a range of specific problem areas appear in the institutional framework of the school, exposed by recent critical study of schooling based on socialization theory (cf. Holtappels 1996):

1. The contents of school study programs are usually too little related to the everyday life praxis of pupils and thus are felt largely to lack relevance, reality, and practicality. Often, what is learned in school has little to do with the everyday questions and problems that engross children and young persons in particular phases of development. The complexity and contradictory nature of present-day and future life demands are largely blotted out. If school learning remains predominantly focused on grades and qualifications, it loses importance in proportion to the decline of the trading value of school-leaving qualifications in the market of training and employment. Today, the rationale for teaching and learning is becoming ever more fragile in the context of extensive unemployment.
2. In the organization of school learning, four problem areas are particularly exposed: First, school learning processes are usually pre-planned and ordered in defined units of time and task. This inflexible rhythm of the timetable narrowly restricts complex processes of learning. Moreover, single-function school rooms and restrictions of learning spaces dominate in school learning, i.e., the opportunities for experiential learning are usually confined to the classroom with its standardized conditions and materials. Related to this are formal role expectations in the learning and social behavior of pupils (classroom discipline). School aims are mainly cognitive learning processes rather than life and communication space; time and space for manual crafts, creative arts, and sporting activities are very limited. Furthermore, there is a predominant fragmentation into single subjects of complex life relationships and areas of knowledge instead of a stronger ingredient of cross-subject interdisciplinary learning. Last, forms of open and differentiated learning are seldom applied.
3. School learning processes are targeted predominantly on formal evidence of achievement (grades, reports). All expressions and learning transactions of pupils can be subordinated to school judgment and assessment. Not infrequently, school processes of recording achievement in the official documentation of results are associated with ceremonial downgrading and negative labeling. After all, the recording of results determines the status of a pupil in the class and therefore in the social order. Pupils are made to compete with each other as part of a hidden curriculum. The success of one group of learners determines the failure of another and vice versa.
4. School life and school climate and the corresponding forms of interaction and relationships are determined in large measure particularly by the daily time rhythms, the formal achievement-related character of classroom processes, and the hierarchical role structure. Linked to this is the fact that, contrary to education law and school regulations, pupils' opportunities for sharing decision-making and representing interests are confined to quite narrow channels. The asymmetric distribution of roles and responsibilities is shown by the fact that it is the teachers who organize the teaching program and who are in charge of time, materials, and activities; they are equipped with incomparably stronger powers to define and to sanction and thereby with incomparably greater opportunity to establish norms and exercise social control. Forms of social interaction and relationships between pupils and teachers and among pupils themselves are all too often determined solely, and therefore one-sidedly, by experiences of expectations of achievement and behavior and the formalized learning and interaction processes of the classroom teaching without any communicative development, reflection, or revision of routines of interaction or forms of

relationship. School, then, has the appearance of an educational institution divorced from its local socio-cultural environment as long as it neglects a reshaping of school life open to its environment. In their school learning, pupils seldom encounter serious situations based on real life, practical social experiences, or formative activities. And so the school is not viewed by its members as having any real social and cultural significance beyond the acquisition of knowledge as determined by the school and its formal qualifications and certificates.

5. Finally, teaching often takes place in dysfunctional buildings not conducive to learning, in which it must be difficult to be at home and identify with one's own place of education. Moreover, children and young people are usually confronted with a completely predetermined learning environment. The social standardization of the rooms prevent them from being creatively adapted and adopted; the required conformity to a prescribed space structure produces a marked alienation from school buildings.

These critical views of school are expressed strongly in pupils' own judgments of schooling.

CONCLUDING REMARKS

Toward the end of the 1960s, analysis of school as an instrument of socialization brought about a changed view of pedagogical processes in the school. This perspective prompted intensive research activity and has led, over the years, to an extensive empirical body of knowledge beyond the scope of this chapter. Initially, the understandings of socialization outlined here, which exposed alongside the intended aspects of instructional processes the unplanned side effects, the hidden curriculum, of institutionalized learning, were judged and criticized as a sociological attack on pedagogy, but today the controversies have died down. Socialization theories have long become accepted as relevant for understanding school teaching and learning, and adopted for analyzing the development problems of children and young people. They have accordingly taken their place in the curricula of teacher training.

REFERENCES

Goffman, E. (1973), Interaktion, Spaß am Spiel, Rollendistanz, München.

Habermas, J. (1973), Stichworte zu einer Theorie der Sozialisation. Manuskriptdruck 1968, reprinted in: Kultur und Kritik, Frankfurt.

Holtappels, H. G. (1996), Schulische Sozialisation. In: Marotzki, W., Meyer, M., Wenzel, H. (Hrsg.): Erziehungswissenschaft für Gymnasiallehrer, Weinheim.

Mead, G. H. (1969), Geist, Identität und Gesellschaft, German ed., Frankfurt.

Parsons, T. (1968), Die Schulklasse als soziales System: Einige ihrer Funktionen in der amerikanischen Gesellschaft. In: Parsons, T. (1968), Sozialstruktur und Persönlichkeit. Frankfurt.

Tillmann, K.-J. (1993), Sozialisationstheorien, Reinbek.

Tillmann, K.-J. (1995), Schulische Sozialisationsforschung. In: Rolff, H.-G.: (Hrsg.): Zukunftsfelder der Schulforschung. Weinheim.

90 A Multicultural School
A Challenge for Postmodern Societies—The Case of Poland

Anna Lubecka

CONTENTS

Culture decides about everything.

David Landem, 2000, p. 516

INTRODUCTION

To say that postmodern societies are multicultural is a commonplace statement, which however does make the issue easier neither to understand nor to manage. Despite many efforts undertaken by various institutions on local, national, and international or even global levels, a boutique version of multiculturalism (Fish 1999) still prevails today and results in its being made equal to a fashionable policy or a catchy slogan. It is typically experienced in its most outer layer, and visual changes such as, e.g., the ubiquitous presence of McDonalds, Heinz, Microsoft, Ford, Nike, and other global brands; real and virtual human mobility; intensity as well as differentiation of interpersonal contacts and their subsequent socio-cultural aspects, flow and exchange of information across borders in real time due to news channels providing information 24 hours, interpersonal communication with an immediate effect via the Internet, e-mail, and social network sites, etc. Also everyday life banal activities, such as shopping and eating, have become a multicultural experience as they consist of buying products made in another country, trying imported foodstuffs, or shopping in a foreign owned store.

We have all become a *global village* according to McLuhan (1964) or a *McWorld* as posited by Barber (1995), but does it mean that our average understanding of the concepts has improved? The metaphor of an iceberg illustrates perfectly well the situation, which makes us realize that real changes are much deeper, affecting all important sectors of human life, such as education, social structures, management, politics, and the understanding of who we are on the international,

national, and individual level. They also challenge the quality of relationships in multicultural societies between the members of the mainstream culture and minorities as the latter come out of the socio-cultural margin to become active in the public sphere.

Thus culture-specific issues have become hot topics on all socio-political and economic agendas as they are often the underlying causes of most local and global unrests and tensions in the mutually diversified and connected societies. Consequently, they have also become a concern for educators and educational policy makers whose aim is to act in such a way as to make education meet new challenges and answer a growing need to prepare young people for global citizenry and an intercultural dialogue. It should be stressed that most schools have also become multicultural institutions themselves; they are often the only places that provide their learners and teachers alike with the first experience of dealing with strangers on a daily basis. Hence, the discussion is on the role of school in the postmodern reality and a vivid debate on how to prepare, first of all teachers, to enable them to assume the role of intercultural mediators.

In old democracies and also multicultural societies with a rather long tradition of cultural diversity, there have already been implemented various programs facilitating the contacts among cultures. Still, they are not very successful, which has been openly admitted by the German Chancellor Angela Merkel and the President of France Nicolas Sarcosy. The failure of their multicultural policy is evidenced by unsolved cultural problems and growing tensions among immigrants and the natives, even open outbursts of mutual misunderstanding and hatred in many culturally diverse French and German cities. Additionally, the most recent economic crisis has become another factor against admitting them to the European club of the still wealthy. Other countries also have problems with cultural minorities, e.g., Slovakia and Romania, which cannot successfully solve the Roma issue. The conflicts do not seem to be easier to deal with depending on the immigrants' culture, their length of stay in Europe, or their being holders of European passports. It should be also stressed that after the tragic date of the 11th of September 2001 in New York and other places in the United States followed by terrorist attacks in Madrid and London the chances for an authentic Muslim–non-Muslim dialogue have become very small.

The bleak picture becomes even bleaker because economic crises, ecological disasters, and political conflicts (cf. Arab Spring, the Caucasian region) result in a great number of illegal immigrants crossing the borders of European countries in search for safety, a decent life, and better living conditions. They bring with them their own world embodied in their customs, traditions, religions, languages, identities, world views, and understanding of reality and, more specifically, their own culture-specific values, which, as Ting-Toomey (1999) puts it, "influence their overall self-conception, and their self-conception, in turn, influences their behavior." Thus, when confronted with the European norms and their subsequent behaviors they look and feel strange and different. Their otherness is usually too difficult to be properly understood as it constitutes an intellectual and emotional test of tolerance asking for an effort to get out of the known, familiar, and friendly cognitive schemata. Many Europeans fail the test, and an ethnocentric perspective based on stereotypes and prejudice still belongs to their common attitude toward the less privileged strangers.

On the other hand, the cultural makeup is so strong that even long years spent in the host country cannot make the immigrant get rid of their ethnic stigma and "go native" even if they want it to happen. The comment about the multiculturalism of Great Britain by the British Conservative politician Norman Tebbit is very significant in this respect. In April 1990, he stated the following: "A large proportion of Britain's Asian population fail to pass the cricket test. Which side do they cheer for? It's an interesting test. Are you still harking back to where you came from or where you are?"

Since 1989, when Poland became again a democracy, and then since 2004, when it joined the European Community, multiculturalism has become its important socio-political and cultural issue. Due to the policy of the communist epoch, which did not promote an authentic multiculturalism, the Polish experience, both institutional and individual, is rather limited in the field, which means that the proper multicultural policy had to be created. This is also the case of educational policy, which has to take into consideration both the minorities living on the territory of Poland for centuries as well as the newcomers—refugees and economic and ecological immigrants. Consequently, multicultural issues in education address both learners and teachers, who themselves must appreciate cultural diversity and see it as an asset to be efficient and authentic in teaching its value. This, in turn, makes an emic and etic approach to a multicultural education an absolute must.

The present article focuses on the multiculturalism in Polish schools discussing its two facets. First, it deals with the legal frames accounting for the presence of multiculturalism in Polish schools, both in their structure and in their programs. Second, it argues that the competence to live in a transcultural society, which is a more advanced version of intercultural communication competence, is a chance to meet the aims of modern education but on the condition that it has become an inherent element of the education of teachers and their learners alike. Although both these groups should develop the new competence, the stress is put on the education of teachers who organize the teaching process and implement the teaching programs but, first of all, act as significant others to their learners. Their example as cultural bridges can significantly influence the learner's success in becoming a multiculturalist. The school in Niemce (the Lublin region) where there is a group of children from Chechnya illustrates the problems Polish schools encounter dealing with intercultural issues. The research has been conducted in 2010–2011 by Joanna Kotlarz within the program of the master's degree seminar I offered at the Institute of Public Affairs.

MULTICULTURALISM OR TRANSCULTURALISM: A NEW ENVIRONMENT FOR SCHOOLS

As already mentioned, cultural diversity has become an issue not only in old democracies and economically safe societies but also in new European Union members such as Poland. Central and Eastern Europe has become a target destination for many illegal economic immigrants, most often from Asia, who use it as a back door to Western Europe. New democracies have also become host countries for political dissidents, refugees, and asylum seekers. This is a new challenge for all kinds of their institutions dealing, directly or indirectly, with the issue but, first of all, for real people who, on a daily basis, work with strangers, e.g., social workers, teachers, local administration staff, etc. and must manage their own otherness as well as the otherness of the foreigners. In many cases, none of them has an experience of multiculturalism, knowledge, and understanding of the issue, and their encounters can turn into a traumatic experience either for both parties involved or for one of them. In most cases, the members of the host country are not encouraged by their superiors to open up to the unknown and to see it as an opportunity for personal growth and enrichment because the decision makers themselves often ignore multicultural issues or are as helpless when facing them as their subordinates.

There is also another important aspect of the postmodern multiculturalism conspicuous in the times of economic crises. Immigrants, but especially the economic ones, who used to be a cheap labor force performing low paid, nonprestigious, and unattractive jobs, are now treated as competitors on the job market endangering the economic safety of the natives. This may result in a perception of multiculturalism as merely a source of troubles, extra work, extra costs, etc., eventually leading to racism. The second, more optimistic attitude toward strangers tends to be lost and damaged by the fear of the unknown and a lack of competence in how to deal with it.

The challenge of contemporary multiculturalism has become so overwhelming because it is a very rich and complex concept embracing more categories of cultural diversity than 15 or even 10 years ago. Moreover, their number is continually growing; they have also become more fine grained, and additionally, their importance as identity markers have become bigger for each cultural minority.

Today, the categories of cultural diversity are constituted by both traditional and new concepts: ethnicity, together with race, religion, and nationality, often additionally combined with language function as its main and, by now conventional, criteria (Mamzer 2001). They also imply a strong correlation between a given ethnic culture and a geographic area where it has been developed; in most cases, it corresponds to the idea of a nation. For many years, national cultures have functioned as the key factors of cultural diversity. This approach to multiculturalism helps clarify the borders between cultures as well as the boundaries between the familiar and the unfamiliar world they stand for. Let me stress that the traditional approach to cultural diversity makes cultural strangeness communicated more explicitly and unambiguously, nearly in binary categories, which greatly facilitates interpersonal contacts.

However, the second half of the 20th century additionally multiplied the criteria of cultural diversity and freed them from their immediate geographical dependence. Age, gender, sexual orientation, physical and mental abilities and even diet (e.g., fast food vs. slow food eaters, partisans of genetically modified vs. ecological food), preferred leisure activities, watched soap operas, etc. constitute new identity criteria and create its new socio-cultural paradigms of a global value.

Counterbalanced with glocalization, contemporary multiculturalism demands a redefinition of social order and reality from one-sided to multiple sided. This, in turn, leads to a reconstruction of cultural identity, which becomes hybridized and multidimensional and also fragile and unidentified in the continual process of being created and recreated, constructed, deconstructed, and reconstructed. Continually challenged by the other, the stranger who is like its mirror, it undergoes many alterations on a daily basis, and it never attains the stage of being completed. Consequently, multiculturalism as an inherent constituent of postmodern societies, unavoidable and inevitable, is not only a political, social, and economic issue, but it concerns each and every individual, who, even against his or her will, is affected with it. Today, the stress is on interconnectedness and how to deal within its context with all kinds of differences, which have not only become visible but also have impacted upon definite individuals.

As Graddol (2007, p. 19) claims, we live in "the age of the multicultural speaker," but in their case, multiculturalism is not reduced to merely multilingualism. It is much more as it embraces a constant dialogue of messages, ideas, products, people, places, events, and the like, which, in turn, presupposes an ability to negotiate values, learn foreign languages, and suspend stereotypes and prejudices in order to live in the world that has developed around us. Living in a global society requires us to question the very essence of national values and beliefs and to seek to adapt to the requirements being tacitly and explicitly demanded from us in the international arena. The difficulty of the previous statement lies in its making us understand the impact of culture on each of us, which has been clearly stated by Edward Hall (1959, p. 183):

> Probably the most difficult point to make and make clearly is that not only is culture imposed upon man but it is man in a greatly expanded sense. Culture is the link between human beings and the means they have of interacting with other. The meaningful richness of human life is the result of the millions of possible combinations involved in a complex culture.

This is also why Ryszard Kapuściński (2006), a Polish journalist, writer, and traveler, claims that cultural strangeness belongs to the most difficult challenges of the 21st century. Meeting a stranger, interacting with him or her, creating any kind of rapport with the other is an overpowering and difficult experience, which involves intellectual, emotional, and even somatic capacities of the interactants. To become a multicultural person is an ambitious and difficult task. In 1992, Roberta Klitgaard (2003, p. 16), who studied the mutual relationship culture or multiculturalism and progress, claimed that despite a growing awareness of the interdependence between the two factors we still lack definite answers about what to do to explore the potential of multiculturalism and treat it as the most important asset of the global community. The same concern has been expressed by two supranational organizations: The European Union and UNECSO. Their answer to the issue was two initiatives aiming at promoting true multiculturalism. The first one was the initiative of the European Union which proclaimed the year 2008 as the *European Year of Intercultural Dialogue*; the second one was the idea of UNESCO to declare the year 2010 the *International Year for the Rapprochement of Cultures*.

As already stated, multiculturalism hides in itself a great potential for growth and richness of both individuals and whole societies as well as a tangible challenge of the unknown Report of the Advisory Committee on Cultural Diplomacy, U.S. Department of State (2005). Hence the relationships between the culturally different groups have always been shaped, on the one hand, by a strong drive toward openness, resulting in mutual cognition, and, on the other, by an equally powerful reticence, even hostility and aggression to annihilate the other culture. The above attitudes still visible today in intercultural relations are fueled by four types of intercultural policy promoted on the state level and impinging on the daily attitudes to multiculturalism. Studying them internationally, Polish scientist Paweł Boski (2009, pp. 56–60) has identified the following: the policy of exclusion, segregation, assimilation, or acculturation. Each of these creates a different balance between the cultures in touch from eliminating one culture, usually of the minority group from the mainstream society (exclusion) to a strong affirmation of each of them together with a full appreciation and an authentic recognition of their respective constitutive values (multiculturalism or transculturalism). Segregation, in turn, which consists of creating two different and separate social spaces for each cultural group legalizing the domination of the majority, constitutes one of the intermediary stages, only a step further from exclusion. It is followed by the policy of assimilation, whose idea is best expressed by the metaphor of a melting pot. It consists of replacing the culture of the minority group by the dominating culture to such an extent that, after some time, the first one gets totally dissolved in the second.

Today, especially, transculturalism becomes an important ideology as, following its concept coined by Welsch (1999), it goes far beyond a mere coexistence of cultures. It implies their active and creative interaction resulting in their hybridization with a final effect of synergy. For Welsch (1999), hybridization means such a mutual interdependence among cultures, which results in elements of one culture becoming internal constituents, either central or satellite, of the other. The process is continual and, on the one hand, leads to a leveling of culture-specific differences and homogenization of cultures but, on the other, to their glocalization. Consequently, each culture and language community stresses its individual and unique character. The tension between these two mutually exclusive trends accounts for a special dynamic of postmodernity and often also for conflicts and tensions. Repeating after Zygmunt Bauman, all of us have become global nomads and players. Thus, we live in the epoch of constant changes, in the *liquid* reality in which nothing is stable and given forever as what is marginal today becomes a center and a biding socio-cultural norm tomorrow (Bauman 1994).

Transculturalism constitutes a new environment, both internal and external, for schools. First, schools themselves have (or will) become transcultural because of the type of interactions among all the culturally diversified actors of the teaching-learning process. Second, an implementation of the principles of transculturalism into public life with its treatment as a social asset and an authentic value of postmodernity as well as a preparation of young people to a new kind of conscious multicultural citizenship is an obligation of all educational institutions, independent of their level.

SCHOOLS AS MULTICULTURAL INSTITUTIONS: THE CASE OF POLAND

In postmodern schools, there are at least three basic sources of multiculturalism: students, teachers, and school administration and programs. All the three elements constitute a microcosm reflection of the global changes.

A Cultural Profile of Learners in Polish Schools

As already stated, no contemporary society is monocultural. Thus learners will be also multicultural, and their multiculturalism will be created by both traditional and new categories of cultural diversity. (The new multiculturalism has not been researched in the context of schools yet, and thus only traditional multiculturalism will be discussed.) It is also true about Poland although it is stereotypically considered as a monocultural country. This picture derives from the fact that cultural minorities constitute not more than 3% of its whole population. Additionally, there is no detailed information, e.g., statistical data about refugees and immigrants in Poland, which neither shows the dynamics of the process nor facilitates its understanding by indicating its features. As for the ethnic and national groups living in Poland, they were silent and invisible under the communist regime, and it is only now that they have started the process of coming out and appearing in the public sphere as its important actors. Their presence is a result of at least three different factors. First, the rights of national minorities are guaranteed by bilateral agreements between Poland and their home countries as well as the Polish Constitution. Second, it is the global policy of multiculturalism and, more specifically, the policy of the European Union, which is bidding for Poland as its member and which promotes cultural diversity in Europe and in its individual countries. Third, these are minorities themselves whose members have been empowered by the EU multicultural policy and, feeling proud of their cultural heritage, insist on having the opportunity to cultivate it and to promote their own identity on a daily basis.

Traditional and old minorities constitute nine national groups in Poland (German, the most numerous; Belarussian; Ukrainian; Lithuanian; Russian; Slovakian; Jewish; Czech; and Armenian) and four ethnic communities (Lemkos, Roma, Tartars, and Karaims who count merely 43 members). There is also a distinct group of Kashubians whose language has been given the status of a regional language, which means that students can take their secondary school final examination in Kashubian. Following the data from the Polish National Census from 2002 (http://www.mswia.gov.pl/portal/pl/61/37/), national minorities comprise 234,202 people, and ethnic minorities embrace only 19,071 people. Kashubians constitute a small group, which amounts to 5062 people.

The new aspect of Polish multiculturalism is created by political and economic immigrants who have been seeking a refuge in Poland since 1989 when it again became a democracy and also since 2004 when, due to the access to the European Union, the Polish eastern border became its boundary. Most of the refugees come from the areas of political unrest in Russia, which is understandable, considering the geopolitical position of Poland. Moreover, having signed the Geneva Convention on the 20th of December 1991, Poland declared its help to all those whose lives are threatened in their home countries because of their race, religion, nationality, political convictions, or a membership in a social group.

The exact number of political immigrants is difficult to give for two reasons. First, Poland has created a larger definition of a refugee than the one proposed by the Geneva Convention as it also includes people with the tolerated sojourn who wait in one of the 14 centers for the foreigners for the decision allowing them to become refugees (Jasiakiewicz et al. 2006). Second, the centers keep statistical data only on those who have applied for the status of a refugee and those who have been granted such a status, which leaves out a big number of nonapplicants. In the period from the first of January 2010 to the 31st of December 2010, there were 6534 applications for the status of refugee in Poland with 4795 (73%) applications from Russians and more, exactly, Chechens; 1082 from Georgians (14%); and 107 from Armenians. In the same period of time, the status of the refugee was granted to a much smaller number of only 82 people, mainly from Russia (42), Belarus (19), Iran (5), and Afghanistan (4) (http://www.udsc.gov.pl/Zestawienia.roczne.233.html). There is also the third group of political immigrants who have regained their status in light of the changed definition of a foreigner from 2005, which uses the temporal criterion. To qualify as a minority, the group is required to live in the territory of Poland for at least 100 years. This is the case, for example, of the Greek minority who came to Poland in the 1950s (Kozień 2006).

Immigrants who have come to Poland for economic or educational reasons belong to the next category. Following the data from the Main Census Office (Główny Urząd Statystyczny) from 2008, immigrants come from 131 countries, but the data do not embrace a large group of those who work on the black labor market. The most numerous groups are from the Ukraine (26.6%), Belarus (9.2%), and China (4.2%). We should also mention the Vietnamese, who have settled mainly in Warsaw and Łódź. They live in communities, which helps them cultivate their own culture and religion, e.g., the

Vietnamese community in Warsaw in the Praga Quarter has built a kind of Vietnam Town even with a Buddhist temple.

Repatriates belong to the third distinct group although it is not very numerous. According to the data from 1990 to 2005, 5000 Poles, the descendants of Polish post-uprising political deportees from 1830, 1863, and from 1939, mainly from Kazakhstan and the Ukraine and generally from the former Soviet Union, decided to settle in Poland (Hut 2007). They feel Polish, and they cherish their Polish identity even more than the Poles residing in Poland. However, as they often speak Polish with a heavy Russian accent, they are taken for Russians, which results in their not always being warmly welcomed in the villages where they are to live. Additionally, considering the unemployment rate in Poland, they have problems with finding a good job, which often makes them disillusioned with Poland.

As far as the obligatory education is concerned, all children who live on the territory of Poland, be they Polish or foreign immigrants, refugees, or members of ethnic and national minorities, they must attend school until they are 18 (Dziennik 2004, no. 256, item 2572 and later changes, Constitution of the Republic of Poland, art. 70, reg. 1, Children Rights Convention, art. 22 and 28, the Regulation about Helping Foreigners on the Territory of the Republic of Poland, the Regulation of the Minister of National Education from October 4, 2001, about admitting children who are not Polish citizens to public kindergartens, schools, education centers, Dziennik 2001, no. 131, item 1458). The aptitudes of children of immigrants and refugees are assessed on the basis of either the documents from their schools or tests they are to take in Poland. Polish school curricula constitute the evaluation criteria. Education is for free up till the level of gymnasium, and the cost of a post-gymnasium education is 1200 € per year and 1500 € per year in higher education. The children of repatriates do not pay the fee; other children can pay a reduced fee or be exempt from it (Koryś 2006).

The children of immigrants and refugees who attend Polish schools bring with them all kinds of problems, which stem from their experience of belonging to one of these groups. Their status makes them differ from children of other minorities and holders of Polish citizenship for whom Poland is a home country and Polish is the language they speak fluently. They also know Polish culture and are its native users. Although their cultural identity is cleft, they can function very well in Polish reality.

The first important difference, which works against the kinds of immigrants and refugees, is their inability to speak Polish, which makes all their contacts very difficult and limited. If they come from the former Soviet Union, Russian, the language they use in formal contacts, becomes their additional stigma. Uncertainty and unpredictability of their situation as a reflection of the situation of their parents; the label of a stranger and foreigner; often prejudice toward them; trauma because of their previous experience; the feeling of loneliness, estrangement, and fear; the feeling of longing for their home and friends who stayed away; economic difficulties; limited contacts with the local population or the staff in the centers for foreigners, which delays their integration; climatic and dietary differences; and dependence on others even in small things are only the most often mentioned difficulties that impact their school experience in Poland (Czerniejewska 2008a,b). For many of them, Poland is the only experience of another culture for which they have been prepared neither intellectually nor emotionally, which accounts for more acute forms of culture shock. It should be also stressed that school and education belong to culture-bound concepts, and as such, children of foreigners may find it difficult to adapt to Polish schools, expectations of teachers toward them, teaching and learning styles, teaching materials, relationships with peers and teachers, etc. Even such components organizing work in the classroom as discipline, forms of address used by teachers to students and students to teachers, lesson scripts, turn-taking rules, etc. may differ from what they have been used to and thus affect their sense of safety, group belonging, acceptance, identification, and identity.

Following Polish author Jerzy Nikitorowicz (2009), all their problems that impinge upon their education and school experience can be assigned to three categories: (1) those that stem from the Polish legal system and the policy toward foreigners, immigrants, refugees, and minority members; (2) those that concern their identity concept and self-concept; and (3) those that are directly related to their living conditions in the host country, e.g., climate.

The challenge of multiculturalism depends on the type of school. It hardly exists in American and British schools in Warsaw and Krakow, which are attended by young Americans, British, or other foreigners whose parents have been delegated to work in Poland. Their Polish learners belong to the Polish elite, and both teachers and parents help them overcome all adaptation problems. The situation is also good in minority schools by children with Polish citizenship who, at the same time, are members of minority groups. In their case, teachers also tend to belong to respective minority groups and are native speakers of particular languages and native users of corresponding cultures because, according to *The Regulation about National and Ethnic Minorities and the Regional Language* from 2005 (Dziennik no. 17, item 141), members of ethnic and national minorities (Greeks have been recognized as a national minority) as well as Kashubians have the right to get education in their own languages and study their own history, culture, and literature. The point is that the number of such school tends to diminish because each year minority communities get older and fewer youngsters learn their minority language. Consequently, there are also fewer minority teachers. Finally, the number of minority leaders interested in promoting their language and culture diminishes conspicuously (Łodziński 2005). Considering the above, the costs of educating their own teachers and of running schools for minorities have become very high, and neither *gminas*, the local governments, which are responsible for their financing, nor the minorities themselves do dispose of proper funds. Thus, it is not a lack of proper legal regulations but mainly the economic situation of local authorities and the minority groups that accounts for minorities, e.g., Ukrainians and Germans complain that they do not learn Ukrainian and German as minority but foreign languages (Łodziński 2005).

Totally different problems are to be solved in schools with Polish teachers and Polish as the language of instruction when they are attended by the children of the refugees and economic immigrants.

Multiculturalism: A Challenge for Polish Teachers

Teachers as the main organizers of the teaching-learning process play a particularly important role in a multicultural school; in our case, a Polish school with the children of the refugees. Unfortunately, most often, they are to deal with many problems that have been caused by cultural diversity.

The first serious difficulty stems from language and culture differences, which from the very beginning hinders communication and often strengthens stereotypes resulting in a more acute form of culture shock of the foreign learners. At this point, it must be mentioned that that Poles are rather negative about the immigrants from the East—Ukrainians, Belarusians, Russians, Vietnamese, and Turks—which influences the adaptation process of the latter, so important at school (Attitude of Poles Towards Other Nations, Research Report, CBOS 2006). Consequently, there can be often observed an attitude of reserve and uncertainty toward the foreign learners as they are a challenge for teachers and school administration. Their cultural differences, which make them need more attention, teacher's work, help, sympathy, understanding, etc., often account for their common perception as troublemakers. As they do not speak proper Polish, they may have serious problems not only with integration with their Polish peers, but also with understanding the material that they study. Hence, they may be unjustly qualified as less intelligent, unable, and less hardworking, which creates another barrier in their process of overcoming culture shock. The main difficulty stems from the fact that most teachers, headmasters, and school administration staff are not prepared themselves to deal with multiculturalism and even more with its active form, transculturalism, and thus they do not see cultural diversity as an authentic value and a source of all kinds of benefits that result in developing human capital (Stańkowski 2004). In most cases, it is very difficult for them to change their mindset as well as their emotional attitude toward the strangers who are so close to them. Their own experience of multiculturalism, which may be none or very limited and based on stereotypes, is not enough to facilitate their role of cultural mediators in the classroom.

Let me stress once more that most of the teachers neither speak the language of their learners nor can they use English as a tool to relate to them. Moreover, they know very little or nothing about the foreign learners' country and even less about their educational system and the school culture that defines the mutual relationships, expectations, and perceptions of teachers and students. Teachers have to realize that participation in class activities depends not only on learners' willingness to become part of them, but on the cultural value embodied in power relationships. This was the case in the school in Niemce in which Chechen learners were active contributors to the classroom discussions only when they were straightforwardly addressed by their teacher while Polish kids did not need a direct request. Without this piece of knowledge, Chechens can be justly assessed as lazy, uninterested in the classroom activities, unintelligent, or disobedient.

The main difficulty for teachers stems from their inability to consider the above differences as crucial for the educational success of a learner, and they are rarely sensitive to their impact on the learning outcomes. A sense of exclusion and of a double estrangement—from the Polish culture in general and the school culture in particular—experienced by learners seems to be ignored also because the teachers themselves are faced with a culture shock they often can hardly manage on their own (Archer et al. 2002). The task is even more demanding because they tend to be left to themselves. Thus, they often experiment with their teaching methods and deal with all kinds of problems by trial and error.

This was the experience of the teachers from the school in Niemce who learned about their Chechen pupils only when they came to school. Thus, they were totally unprepared to work with them and stressed, which was not a good start for either side. As follows from the interview with the school headmaster, both the teachers and their learners were learning about each other, discovering basic things about themselves without any assistance from their superiors and local administration. It was a source of a very valuable experience, eventually extremely important and positive, but many mistakes might have been avoided, and the process could have been more effective and rewarding for both parties.

The above discussion leads to formulating the following conclusion: Polish teachers need formal preparation to develop intercultural skills allowing them to deal successfully with their multicultural students, e.g., children of the refugees and immigrants. Their good will is important, but it is not enough to account for an educational success.

Management of Multicultural Schools

As already mentioned, one of the areas in which multiculturalism is manifested in schools is their management. Religion and religious holidays, norms and beliefs as well as their subsequent behaviors constitute the most important factors that account for a visible manifestation of differences. Dietary restrictions due to religious norms may demand modifications in the canteen menu, e.g., pork-based dishes must be replaced with beef or poultry in the case of Muslim learners. The Ramadan rules of fasting are also very difficult to observe by Muslim children as their Polish schoolmates may neither understand them nor approve of fasting or else respect the tradition. The holiday itself may affect school discipline as Muslim learners may then stay at home. The same applies to the fact that Friday is a holy day for them while it is a regular working day in Poland. Finally, the dress code, e.g., the tradition of covering the head by girls older than 15 or a lack of approval for girls wearing trousers may create some controversies or misunderstanding or else result in questions or comments, sometimes nasty or mocking on the part of Polish learners. Even names can be a problem for the school administration, which was the case of Niemce because their Polish computer programs did not accept that Zhannat and Khalimat are girls.

A not less important area is that of the contacts with parents. They may be a source of difficulties and misunderstandings because, quite often, the refugees may have a culturally different idea about how their children should be educated. Once more, religion and religion-based tradition may play an important role in the process. Second, the mastery of the Polish language by adults is usually less successful than by children, and consequently communication problems cannot be ignored, either. Furthermore a more acute awareness of their situation in the center for the refugees may make them less confident, open, and trustful. All the above factors impinge on successful communication, which, in turn, determines the relationships between school, learner, and home (parents).

Finally, a proper organization of education for children of immigrants or refugees needs close cooperation between the local administration and the school, e.g., the *gmina* should inform the school headmaster about new non-Polish learners, which was not done by the local authorities in Niemce. Thus, both the school administration and the teachers were not given a chance to prepare themselves to work in a culturally new environment. It is also the *gmina* who often should provide the refugee children with school books because of their economic status and to support the school-inspired initiatives aiming at integrating the two communities.

Cultural Diversity in School Programs

The value of a multicultural education has been stressed in many documents of global organizations, which constitute a frame of reference for policy makers in the field of education, culture, sports, etc., e.g., the concept of the *European Year of Intercultural Dialogue* or the UNECSO strategy for the *International Year for the Rapprochement of Cultures*. The latter consists of four pillars:

1. Promoting reciprocal knowledge of cultural, ethnic, linguistic, and religious diversity
2. Building a framework for commonly shared values
3. Strengthening quality education and the building of intercultural competences (Bolded by AL)
4. Fostering dialogue for sustainable development (*UNESCO leaflet* 2010, p. 3)

As for Poland, the presence of cultural diversity in teaching programs directly corresponds to the directives from the Polish Ministry of National Education from December 23, 2008, which, in turn, has to respect the policy of the European Union in the field (Komorowski 2001). The European Union Convention on National Minorities signed in 1955 and ratified by Poland in 2000 clearly points at the role of education in democratic and culturally diversified, pluralistic societies, which consists of educating young people with respect for the ethnic, linguistic, cultural, and religious identity of individuals belonging to minorities and in creating for them proper conditions to promote it, preserve it, and develop it (Dziennik 2002, no. 22, item 209). The reform of the Polish educational system, which started on September 1, 1999, deals with the following aspects of multiculturalism. First, it focuses on creating an attitude of knowledge of and respect for Polish national heritage as self-respect and self-pride in one's own culture is the only way to teach respect for other cultures. Second, it recognizes tolerance for differences and openness as basic attitudes to be developed in students. They are absolutely necessary to bridge differences and enhance social coherence, which makes them constitute basic ingredients of the values proper to a civic society. Third, it stresses the importance of providing students with proper cultural knowledge, which is an indispensable prerequisite to understand and accept culturally pluralistic societies (Nikitorowicz 2009).

The aims of education should prepare the learner to meet global challenges. It should be thus the education that is based on the values built into intercultural dialogue, orientation to change, anticipation of new situations, and reaction to them by finding creative alternative solutions and active participation in global issues (Melosik 1989). Consequently, the teaching and learning process should enable the learner to do the following:

1. Identify universal elements and values of culture
2. See oneself and others as inherent elements of the global reality
3. Be an active and conscious member of an international society

4. Understand culture as a product of humanity
5. Develop an ethno-relative attitude, which will allow an understanding that members of different cultures see and value the world differently (Melosik 1996)

The above aims of formal education have been also clearly articulated in the Book of Orders from January 1, 2009 (Dziennik no. 4, January 15, 2009). According to the document, a school is to act against any manifestation of racism, xenophobia, and discrimination by constructing such teaching programs whose contents promote, among others, the following values: honesty, truthfulness, responsibility, stamina, cognitive curiosity, self-worth, personal culture, willingness to interact with other cultures, etc. All of them, if successfully developed by learners, constitute their symbolical capital to use for their personal growth and for the sustainable development of the global village in the future.

It is interesting to notice, that, on the one hand, an education should teach the modern concept of citizenry and strengthen the sense, both local and global, of a social membership, and, on the other, it is to free learners from social routines and culturally conventional ideas and acts by encouraging their creativity, feeding their entrepreneurial spirit, and inspiring their imagination, open-mindedness, curiosity, and search for innovative solutions. Although the above concept of education seems to be based on controversial ideas that should exclude each other, in practice, they are not only complementary but also mutually inspiring, resulting in synergy.

The biggest challenge for the programs of formal education is to provide learners with the proper tools to enable them to continue the process after having left the educational institutions when they themselves become responsible for who they have become, how well they can deal with everyday difficulties and contribute to the society and what kind of relationships they are able to build with others. Today both nonformal learning and informal learning have become alternative forms of education whose importance is not smaller than that of the formal one. However, their success depends on the capital learners get while being formally educated. Non-formal learning can be realized by NGOs, volunteers, and civic activities in the organization from the first sector; it can have various forms, and although its participants do not get any formal diplomas confirming the skills they have mastered, its role cannot be ignored in the process of preparing individuals to live in a postmodern society. It usually completes formal education and although its participants are volunteers, teachers tend to use it more and more often to meet their teaching aims, which are, in general terms, creating conscious global citizenry. Its strong points are engaging learners in the learning process by focusing on them and on their experience, defined needs, and individual predispositions as well as encouraging their creativity, initiative, enterprise, inventiveness, etc.

At the moment, two European programs, "Life long learning" and "Youth in action," deal, among other issues, with multiculturalism. The first program aims at encouraging some reflection on the value of cultural diversity and its impact on, for example, interpretations of history. The second one, by enabling contacts between young people of different ethnicity, religion, and social background and engaging them in common projects serve to prevent racism and xenophobia.

An even less recognized form of education is informal learning, which is a life-long process. It has a great impact on creating world views, understanding and evaluating reality, ways of relating to others, practicing civic virtues, being able to adjust to changing demands of the labor markets, etc. Literally everybody performs both the functions of teacher and learner in it; it is not organized and systematic, and in many cases, it is below the learner's awareness. This is the reason why informal education is often referred to as incidental. Based on a learner's direct experience, which activates their intellectual skills as well as their emotions, it treats education as a holistic process. Its positive outcomes are directly proportional to the knowledge, skills, and attitudes, which the learner has developed during the process of formal learning. The European Union promotes its idea, among others, by implementing the Lifelong Learning Programme, which, by means of the decision from July 8, 2003, has also become a part of the Polish document *The Strategy of Lifelong Learning till 2010* and also (*The Strategy for the Development of Education in Poland till 2020*).

Life-long learning is especially effective in preparing young people to live in culturally diversified societies. Thus, international exchange programs, e.g., Comenius, are used in all types of secondary schools to make learners get in touch with a European culture and directly experience its various manifestations in daily situations. Hopefully, the method of learning by experiencing both a cultural synergy and a culture shock will eventually lead to creating truly multicultural citizens.

COMPETENCE TO LIVE IN A TRANSCULTURAL SOCIETY: A BASIC SKILL OF POSTMODERN TEACHERS

Our discussion about various manifestations of cultural diversity in Polish schools makes us posit that many difficulties and tensions might be successfully overcome if teachers have mastered the competence to live in a transcultural society. The concept is new, but it heavily draws on the idea of intercultural communication competence, which is relatively old. Its definitions are many, and all of them deal with the mutual relationship between language and culture. Thus, following the simplest but, at the same time, the most comprehensive approach to it, we posit that intercultural communicative competence is an ability to understand and to be understood by interlocutors who belong to different cultures and who, in most cases, use a foreign language in the process, e.g., both of them use English, which has achieved the status of a modern *lingua franca*. Intercultural communication competence aims at working toward creating greater mutual understanding between groups of people and thus facilitating an intercultural dialogue at any level of meeting, from private encounters to official and formal conventions of the heads of states. The difficulty intercultural communicators are faced with is the necessity to translate not only linguistic items but also, much more difficult, cultural concepts of reality, of their interlocutor's identity, and of their own identity. These two aspects of intercultural communication competence have been stressed by Byram and his cooperators who posit that intercultural communication competence consists of an *ability to ensure a shared understanding by people of different social identities and their ability to interact with people as complex human beings with multiple identities and their own identity* (Byram et al. 2002). Alessandro Duranti (2001) focuses on the cultural aspect of meaning, rightly arguing that *linguistic forms, either because of their arbitrary nature (for Sapir) or because of their implicit worldview (for Whorf), are seen as constraints on the ways in which individual speakers as members of speech communities perceive reality or are able to represent it.* The same idea has been even more powerfully expressed by Claire Kramsch (2006), who claims that *language expresses cultural reality* [...] *language embodies cultural reality* [...] *language symbolizes cultural reality* [Italics in original]. Consequently, language is not only a basic tool of communication, but it plays a decisive role in mapping reality, interpreting it, creating identity and self-concept, shaping relationships with others as it accounts for their evaluation, and choice of verbal and nonverbal behaviors that correspond to it. Considering the above, Byram et al. (2002) are right to claim that intercultural communication competence is to *ensure a shared understanding by people of different social identities and their ability to interact with people as complex human beings with multiple identities and their own individual identity.*

Intercultural communication competence consists of three basic components: knowledge, attitudes, and competencies.

Proper intercultural knowledge serves to explain and understand existing differences. Attitudes account for creating tolerant and prejudice-free relationships with the other. Finally, competencies facilitate a recognition of the potential of the other. Consequently, intercultural knowledge, attitudes, and competencies allow interactants to build bridges over the fear of the unknown, the hatred of the different, and the uncertainty of the unfamiliar. Thus, when mastered, intercultural communication competence serves not only to give answers to questions about diversity but, even more important, by teaching sensitivity, encouraging curiosity and openness, and promoting an ethno-relative attitude, it teaches that way to ask the right questions to change the unknown into the known. Such a demand put on intercultural education corresponds to Fons Trompenaars' (1993), an unquestionable authority in the field of intercultural communication, understanding of the process of relating to the other against a multicultural setting. He claims that "there are no universal answers but there are universal questions and dilemmas we are to start with" in the process of domesticating otherness.

As rightly observed by Brislin and Yoshida (1994), theoretical knowledge alone is helpful, but it needs to be completed with proper skills and attitudes. Additionally, all of the components should be learned within two complementary perspectives: culture general and culture specific. The first group embraces such general knowledge, attitudes, and skills valid in any cultural context, which help to tolerate ambiguity, manage stress, establish realistic expectations about the other, and demonstrate flexibility and empathy. The second one is culture-specific as insufficient knowledge about a given culture may lead to the interactor being trapped in a stereotype, unable to see the uniqueness of particular individuals with whom they interact and using inappropriate communication skills in particular social contexts.

Skills, the second component of intercultural communication competence, have been differently classified by various researchers. Ruben (1989) has distinguished three basic distinct types according to their function in interpersonal contacts, which are the following:

1. Relational—Building and Maintaining Competence: Competence associated with the establishment and maintenance of positive relationships
2. Information—Transfer Competence: Competence associated with the transmission of information with minimum loss and distortion
3. Compliance—Gaining Competence: Competence associated with persuasion and securing an appropriate level of compliance and/or cooperation

In turn, Gudykunts and Kim (1992) is more specific, and his classification is more fine-grained as it contains seven categories of skills, all of which derive from cross-cultural awareness. He distinguishes (1) open-mindedness, (2) intercultural empathy, (3) accurate perception of differences and similarities, (4) nonjudgmentalness, (5) astute noncritical observation, (6) the ability to establish meaningful relationships, and (7) minimal ethnocentrism. They share many underlying values with the following classification by Ruben (1976), which also contains these seven types: (1) the ability to show respect, (2) the ability to show empathy, (3) the ability to take role behavior, (4) the ability to manage interaction, (5) the ability to tolerate ambiguity, (6) the ability to take appropriate interaction posture, and finally, (7) the ability to be oriented to knowledge. Their acquisition helps the cross-cultural communicator develop his or her personal autonomy and become more flexible and open.

Byram's (1997) understanding of skills and their subsequent classification into five *savoirs* is based on their progressive nature and a nontesting aspect, which was criticized by Sinicrope, Norris, and Watanabe (2007) as European and not valid for other cultures. Putting aside the controversy, we should stress the rational for Byram's work, which derives from the changing nature of societies and, consequently, the changing needs of communicators, which lead to *changes in certification. Existing certificates may be altered to emphasize different purposes and new certificates and curricula are developed to accommodate the ambitions of individuals and the predictions of governments* (Byram 1997). Byram highlights that his *savoirs*, apart from covering knowledge and learning, consists in applying them in practice by understanding, discovering, and relating information back to the learner's own experiences. His five *savoirs* are as follows (Byram 1997):

1. *Savoir être* (attitudes): curiosity and openness, readiness to suspend disbelief about other cultures and belief about one's own
2. *Savoirs* (knowledge): of social groups and their products and practices in one's own and in one's interlocutor's country and of the general processes of societal and individual interaction

3. *Savoir comprendre* (skills of interpreting and relating): ability to interpret a document or event from another culture, to explain it, and relate it to documents or events from one's own
4. *Savoir apprendre/faire* (skills of discovery and interaction): ability to acquire new knowledge of a culture and cultural practices and the ability to operate knowledge, attitudes, and skills under the constraints of real-time communication and interaction
5. *Savoir s'engager* (critical cultural awareness): an ability to evaluate, critically and on the basis of explicit criteria, perspectives, practices, and products in one's own and other cultures and countries

The short overview of the concept of intercultural communication competence entitles us to claim that, when developed, it serves both to understand better the stranger who is our interlocutor and ourselves as it encourages our *questioning and discovering, not simply accepting a transmitted account of a specific country and its dominant culture* (Byram 1997). As it aims at creating an awareness of our identity and of our own culture, it develops a critical analysis of ourselves and the attitude of self-awareness and mindfulness, which sensitizes us to the effect of our communication on others and its perception by others.

The learners who have mastered intercultural communication competence possess *the ability to see and manage the relationships between themselves and their own cultural beliefs, behaviors, and meanings as expressed in a foreign language and those of their interlocutors, expressed in the same language—or even a combination of languages* (van Ek 1986). When looked at more in depth, the results are not limited to communication sensu stricte but can be seen from the perspective of modernity that is global citizenry in civic societies. Its concept has been made more precise by James Appleby, the president of the American Association of State Colleges, for whom modern citizens *must be trained in global citizenship, and they must develop special competencies. They'll need strong interpersonal skills, broad cultural awareness, and advanced technical and communication proficiencies as well as lively intellectual curiosity to function effectively and live productively in a multicultural free-trade environment.*

Thus, intercultural communication competence, as a part of multicultural or even transcultural education, refers to an overall ability to live in a multicultural society, a society of the future. It should be stressed that although this kind of competence can be learned, its holistic nature impinging on every aspect of life makes it be treated rather as a lifestyle and life philosophy than a particular kind of competence.

CONCLUSION

The changing nature of modern multiculturalism, which has already become transculturalism due to the dynamic interactions of a variety of cultures, has been affecting the functioning of all public institutions. Considering the role of education in postmodern societies, schools, independently of their level, be it primary, middle, secondary, or high, are faced with the issue in a very specific way. As a source of formal education, which has a direct impact on life-long informal education, their main task and responsibility is to educate young people in the spirit of tolerance for cultural diversity and develop in them the necessary skills and competencies to make a multicultural dialogue their basic and authentic value they identify with and treat as their life philosophy and lifestyle. The school mission formulated in such a way is not easy to put into practice for many reasons, which stem from all the components of the teaching-learning process as well as the school management, which has to be adjusted to new educational objectives. The societies in which the tradition of multiculturalism and, consequently, such an education, which corresponds to the needs of their culturally diverse members, has been long and can rather successfully managed all the challenges of multicultural schools. In societies such as Poland, for example, which have discovered their own multiculturalism pretty recently and also due to political changes (the fall of communism in Central and eastern Europe), they have become a target destination for political and economic immigrants, and multicultural schools function as a new kind of educational institution. It means, in practice, that quite often they have to solve all the problems on their own, discovering the real meaning of transculturalism in the process.

Poland definitely belongs to the group of countries that is faced with all the problems discussed above, and its many schools, especially in small localities, are challenged on a daily basis with all kinds of intercultural issues. As argued in the article, all of them can be categorized into three distinct but mutually related groups on the basis of the main actors of the teaching-learning process: learners, teachers, programs, and school administration and management. All of them interact with each other as intercultural communicators, which means that their communication behaviors are filtered by a culture, which assigns to them new meanings and values. Hence, if not properly managed, culture accounts for many misunderstandings, which apply to both factual and relational levels of communication. Intercultural communication competence, which helps to overcome communication problems in intercultural encounters, seems to be the necessary skill to be mastered by all the actors in transcultural encounters. As it helps to properly relate to others and negotiate identity in the win-win process, it should be made an obligatory component in the education of all educators and school managers.

There are at least three reasons to encourage its inclusion in educational curricula for them. First, the school of the future is a multicultural institution with all kinds of cultural diversity embodied in its staff and learners, which has to be successfully managed to attain teaching objectives. Second, teachers act as role models and significant others whose system of values and their subsequent behaviors tend to be imitated by students. Third, a modern school objective is to educate young people not only by providing them with knowledge, but also by modeling their attitudes to prepare them to act as responsible citizens in the global world. Modern citizenry means thinking in terms of such values as public good; human capital; cultural diversity; social inclusion and solidarity; and the peaceful coexistence of strangers who use intercultural dialogue as a bridge over differences, ignorance, disrespect, and hatred. Although not directly manifested, all these values are built into intercultural communication competence as its sound fundamentals. Hence, we can venture an opinion that intercultural communication competence is not merely a communication strategy. In its deepest sense, it is much more, it is a new lifestyle and life philosophy, a system of values absolutely necessary to have by members of postmodern societies who themselves have become multicultural people even without knowing it. Briefly, it is a competence to live in a transcultural society we all belong to.

REFERENCES

Archer, L. et al. (2002). *Higher Education and Social Class: Issues of Inclusion and Exclusion.* London, Falmer Press.

Attitude of Poles Towards Other Nations, Research Report, Polish ed. *Stosunek Polaków do innych narodów. Komunikat z badań CBOS.* 2006. Warszawa.

Barber, B. R. (1995). *Jihad vs. McWold.* New York, Times Book.

Bauman, Z. (1994). *Liquid Modernity.* Cambridge, Blackwell Publishing Ltd & Polity Press.

Boski, P. (2009). *Multiculturalism and psychology in two cultures integration* Polish Ed. "Wielokulturowość i psychologia w dwukulturowej integracji," In: Nikitorowicz, J. Edukacja regionalna i międzykulturowa, PWN, Warszawa.

Brislin and Yoshida. (1994). *Intercultural Communication Training: An Introduction.* Thousand Oaks, CA, Sage Publications, Inc.

Byram, M. (1997). *Teaching and Assessing Intercultural Communicative Competence.* Clevedon, Multilingual Matters.

Byram, M., B. Gribkova and H. Starkey. (2002). *Developing the Intercultural Dimension in Language Teaching: A Practical Introduction for Teachers.* Strasbourg, The Council of Europe.

Czerniejewska, I. (2008a). *Educational strategies in multicultural environment,* Polish Ed. *Strategie edukacyjne w środowisku zróżnicowanym kulturowo. Kilka przypadków z praktyki.* Kultura współczesna 2: 56, pp. 180–203.

Czerniejewska, I. (2008b). *Education on immigrants*, Polish Ed. *Edukacja o uchodźcach*. In: Czerniejewska, I. and I. Main (eds.), *Uchodźcy: teoria i praktyka*, Poznań: Stowarzyszenie Jeden Świat, pp. 159–170.

Duranti, A. (ed.). (2001). *Linguistic Anthropology: A Reader*. Malden, MA, Blackwell Publishing.

Dziennik, U. (2001). *Journal of Law*, no. 131, item 1458, http://isap.sejm.gov.pl/DetailsServlet?id=WDU 20042562572, available on September 20, 2011.

Dziennik, U. (2002). No. 22, item 209.

Dziennik, U. (2004). *Journal of Law*, no. 246, item 2572, http://isap.sejm.gov.pl/DetailsServlet?id=WDU 20042562572, available on September 20, 2011.

Dziennik, U. (2005). No. 17, item 141.

Dziennik, U. (2009). No. 4, January 15, 2009.

Fish, S. (1999). *Boutique multiculturalism* [w:] S. Fish, *The trouble with the principle*. Harvard, Cambridge University Press.

Graddol, D. (2007). *English Next: Why Global English May Mean the End of "English as a Foreign Language."* London, British Council.

Gudykunst, W. B., and Y. Kim. (1992). *Communicating with Strangers: An Approach to Inter Cultural Communication*. New York, Random Press.

Hall, E. T. (1959). *The Silent Language*. New York, Random House.

Hut, P. (2007). *Report. State policy towards people of Polish origin abroad in 1989–2005*, Polish ed. *Raport. Polityka państwa polskiego wobec Polonii i Polaków za granicą w latach 1989–2005*, Warszawa: Kancelaria Prezesa Rady Ministrów, http://www.etnologia.pl/polska/teksty/mniejszosci-etniczne-w-polsce-kaszubi.php, available on September 30, 2011.

Jasiakiewicz, A., W. Klaus and B. Smoter. (2006). *Situation of temporary immigrants – recommendations for integration policy*, Polish Ed. *Sytuacja cudzoziemców posiadających zgodę na pobyt tolerowany – rekomendacje dla polityki integracyjnej*, Analizy i opinie, no. 60.

Kapuściński, R. (2006). *Meeting "the other" as a XXI century challenge*, Polish Ed. "Spotkanie z Innym jako wyzwanie XXI wieku," [w:] R. Kapuściński, 2006, *Ten inny*, Kraków, ss. 65–76.

Klitgaard, R. K. (2003). *Culture matters*. In: Harrisom, L. E. and S. P. Huntington (eds.), Polish ed. *Kultura ma znaczenie* [Poznań: Zysk – S-ka].

Komorowski, T. (2001). *Educational law in practice*, Polish Ed. *Prawo oświatowe w praktyce*, Warszawa: Oficyna Ekonomiczna.

Koryś, I. (2006). *Children of immigrants – new problem, new challenges*, Polish Ed. Dzieci imigrantów w Polsce – nowe zjawisko, nowe wyzwania. *Biuletyn Migracyjny* 7 (2).

Kozień, M. (ed.) (2006). *Multicultural word and migrations*, Polish ed. *Wielokulturowość a migracje. Materiały z konferencji "Współczesne procesy migracyjne a perspektywy wielokulturowości w Polsce,"* Ministerstwo Polityki Społecznej i Stowarzyszenie na Rzecz Integracji i Ochrony Cudzoziemców "Progenia," Warszawa.

Kramsch, C. (2006). *Language and Culture*. Oxford, Oxford University Press.

Landem, D. (2000). *Richness and poverty of nations*, Polish Ed. *Bogactwo i nędza narodów*, Warszawa: Wydawnictwo "Muza."

Łodziński, S. (2005). *Equality and difference*, Polish ed. *Równość i różnica. Mniejszości narodowe w porządku demokratycznym w Polsce po 1989 roku*, Warszawa: Wydawnictwo Naukowe Scholar.

Mamzer, H. (2001). *Multiculturalism – liberation from ethnocentrism?* Polish ed. "Wielokulturowość – czy wyzwolenie z więzów etnocentryzmu?" *Przegląd Bydgoski* XII: 33–43.

McLuhan, M. (1964). *Understanding Media*. London, Roitledge.

Melosik, Z. (1996). *Global education. Hopes and controversies*, Polish ed. "Edukacja globalna. Nadzieje i kontrowersje." In: Jaworska, T. and R. Leppert (eds.), *Wprowadzenie do pedagogiki: wybór tekstów*, Kraków.

Melosik, Z. (1989). *Education focused on the Word – educational ideal of the XXI century*, Polish ed. "Edukacja skierowana na świat – ideał wychowawczy XXI wieku." *Kwartalnik Pedagogiczny*, nr 3.

Mikołajczyk, B. (2004). *Immigrants – rights and responsibilities*, Polish ed. *Osoby ubiegające się o status uchodźcy. Ich prawa i standardy traktowania*, Katowice: Wydawnictwo Uniwersytetu Śląskiego.

Nikitorowicz, J. (2009). *Regional and multicultural education*, Polish ed. *Edukacja regionalne i międzykulturowa*, Warszawa.

Report of the Advisory Committee on Cultural Diplomacy, U.S. Department of State (2005). *Cultural Diplomacy is the Linchpin of Public Diplomacy*. September 2005, electr. doc., http://209.85.129.104/search?q=cache:2smHpUQeIHIJ:www.publicdiplomacywatch.com/091505Cultural-Diplomacy-Report.pdf+cultural+diplomacy&hl=pl&gl=pl&ct=clnk&cd=6, accessed on May 17, 2011.

Ruben, B. D. (1976). Assessing communication competency for intercultural adaptation. *Group and Organization Studies*, 1, pp. 334–354.
Ruben, B. D. (1989). The study of cross-cultural competence: Traditions and contemporary issues. *International Journal of Intercultural Relations*, 13, pp. 229–240.
Sinicrope, C., J. Norris and Y. Watanabe. (2007). *Understanding and Assessing Intercultural Competence: A Summary of Theory, Research and Practice* (Technical Report for the Foreign Language Program Evaluation Project). *Second Language Studies*, 26 (1), pp. 1–58.
Stańkowski, A. (2004). *Portrait of a teacher – pedagogical reflections*, Polish ed. "Portret nauczyciela wychowawcy – refleksje pedagoga." In: Etyka, kultura a multikulturowy dialog. Wyd. FHV UMB, Bańska Bystrzyca, pp. 24–33.
Ting-Toomey, S. (1999). *Communicating Across Cultures*. New York, Guilford.
Trompenaars, F. (1993). *Riding the Waves of Culture. Understanding Cultural Diversity in Business*. The Bath Press, The Economist Book. London, Nicholas Brealey Publishing.
UNESCO leaflet. 2010_leaflet_en. (2010, January 25). Retrieved 11 17, 2010, from UNESCO: http://www.unesco.org/culture/pdf/2010/2010_leaflet_en, available on September 20, 2011.
van Ek, J. A. (1986). "Objectives for foreign language learning," vol. 1: Scope. Starsbourg: Council of Europe. In: Byram, M. *Teaching and Assessing Intercultural Communicative Competence*. Clevedon: Multilingual Matters.
Welsch, W. (1999). "Transculturalism – The puzzling form of cultures today." In: Featherstone, M. and L. Scott (eds.), *Spaces of Culture: City, Nation, World*. London, Sage, pp. 194–213.
www.mswia.gov.pl/portal/pl/61/37/, available on September 20, 2011.
www.udsc.gov.pl/Zestawienia,roczne,233.html, available on September 20, 2011.

91 Computer and Multimedia Skills as Elements of the System of Modern Teacher's Competencies

Janusz Sasak and Iwona Ewa Waldzińska

CONTENTS

MODERN TEACHER'S COMPETENCIES

Social and economic changes shaping societies of developed countries in the 21st century are closely connected with common use of computer and multimedia technologies in communication processes. The economy that is created in these conditions is called a knowledge-based economy, and the society that builds it is called an information society. These conditions pose a new challenge to both the teacher and the school. One of the roles of a modern school is to prepare young people for life in an information society, which means that they need skills in gathering, processing, and using information with the help of computer techniques. Fast development of information systems and increasing efficiency of computer networks and hardware caused information to change into a multimedia format. This means that now information is available, for example, as a text, sound, graphics, animation, video recording, or a combination of elements saved in different formats.

New ways of accessing and transferring information change the traditional process of education into a new computer-based education. That is why modern buildings, equipped with computers and Internet access, are of considerable importance. Providing schools with computer and multimedia technologies is only one of the vital changes that must be implemented in teaching techniques. The teacher who knows how to use information and communication appropriately should remain the most important element of the education process. Only such a teacher is able to participate actively in the process of sharing knowledge. If a teacher lacks the necessary competence, he or she will only copy the Internet transfer and consequently will be defeated by attractive multimedia techniques. Thus, a modern teacher must have both a traditional set of competencies and the knowledge of how to use modern computer and multimedia technology.

Modern education, as any other field of human activity, changes according to the conditions in which it takes place. The directions of those changes follow educational ideas or changing social and technical conditions, which accompany education. In Poland, similarly to other countries, reforms that allow for the change of conditions have been implemented. The aim of those reforms is not only to change school as an institution. They are supposed to lead to the creation of a system supporting the realization of learners' educational intentions. As a result, a modern school should help students to have access to education according to their interests and capabilities and, at the same time, be adjusted to the needs of modern society.

Modern society demands that a teacher should have specific competencies. In the literature, the term "competence" is defined according to different aspects. The first of those aspects shows that the word "competence" is of Latin origin and is derived from the word *competentia*, which means responsibility or taking a certain position. The second aspect emphasizes the importance of the potential and capabilities of learners.

In the *Pedagogical Encyclopedia of the 21st century*, the entry "competence" is defined in the following way: "competence" – one of the most important pedagogical terms, which is difficult to define in a precise way (...). There are two ways of explaining its meaning. The first one defines "competence" as a person's potential, which allows him or her to adjust his or her actions to the conditions of his or her surroundings (...). The other way defines "competence" as person's potential, whose actions are liable to trigger creative modifications as a result of the interpretation of the context of those actions (*XXI Century Pedagogical Encyclopedia* 2008).

In recent years, the term competence has been referring to either an individual's capabilities or potential to act efficiently in a particular situation. The emphasis is put not only one the knowledge itself, but also on the ability to use it correctly. P. Perrenoud regards "competent teaching" as the one that encourages a person to use the knowledge they have in different situations that are complex and unpredictable. He defines "competence" in the following way: "*une capacité d'agir efficacement dans un type défini de situations, capacité qui s'appuie sur des connaissances, mais ne s'y réduit pas,*" which can be understood as "the ability to act efficiently in various situations; the ability based on but not limited to knowledge" [the text was translated from French by European Office *Eurydice*] (Perrenoud 2001, p. 7). Weinert, on the other hand, interprets "competence" as more or less a specialized system of abilities necessary or sufficient to achieve a particular aim (OECD 2001, p. 45).

The changes connected with countries joining the European Union and processes of globalization have led to the uninterrupted flow of people, goods, and capital, which has created a problem concerning the recognition of diplomas and certificates connected with different levels and systems of education. Therefore, the European Commission had decided to create a guideline of teaching competencies in the member countries of the European Union.

The first actions on the key competencies concerning European educational space were taken at a symposium organized in the Swiss town of Bern in 1996. The outcome of the meeting was a list of the common key competencies that European high-school graduates should have. The list of the Bern competencies includes the following (Sielatycki 2005, p. 15):

- Cooperating in groups
- Using modern means of information and communication
- Problem solving
- Using various sources of information
- Listening to and making use of other people's opinions
- Communicating in a few languages
- Combing through and ordering pieces of knowledge
- Accepting responsibility
- Organizing and assessing your own work
- Dealing with uncertainty and complexity

What needs to be emphasized is that those competencies are connected with the results of the findings published in 1994 in the report "Europe and the Global Information Society. Recommendations to the European Council," called the "Bangemann Report." The report shows the necessity to create an information society and describes the changes in ways of working, common work, and living together in the new society. Bangemann emphasizes the problem of creating new jobs using new technologies. The report presents the technological differences between Europe, the United States, and Japan. Moreover, it describes different ways of preparing the Europeans for life in an information society and new key competencies entwined with information technologies.

In the year 2002, the European Commission published a document entitled "Education and Training in Europe: Diverse Systems, Shared Goals for 2010," which lists the updated competencies of a teacher. The document is known as The Barcelonan Competence List. The European Commission presented the following understanding of the key competencies: a collection of knowledge, capabilities, and attitudes indispensable for the meaningful and active participation of an individual in society. A list of eight competencies was created:

1. Communicating in a native language
2. Communicating in foreign languages
3. Mathematics competencies and basic science-technical competencies
4. Information technology competencies
5. The ability to learn
6. Social and civil competencies
7. Initiative and enterprise
8. Cultural awareness and expression

As can be seen, the European Commission accepted that the development of educational focus on the future and to adjust to the needs of European society, creating a knowledge-based economy is a vital element of society's development. In 2005, a working group, appointed by the European Commission, produced a set of competencies required for a European teacher. The basic ones include the following (Key competences, Introduction on the level of compulsory education in Europe 2005, pp. 12–16):

1. Competencies connected with the process of learning and teaching
   - 1.1. The ability to work in a socially multicultural group
   - 1.2. The ability to create favorable conditions, which means that a teacher's role is to do the following:
     - Organize the learning process
     - Cause learners to become researchers
     - Design syllabi, educate himself or herself, improve his or her teaching methods, belong to different organizations
     - Organize cultural events in the region
   - 1.3. The ability to incorporate information-communicative technologies into students' everyday lives
   - 1.4. The ability to work in various teams
   - 1.5. The ability to participate in the creation of syllabi and organization of the process of education and assessment
   - 1.6. The ability to cooperate with people from the local environment and parents
   - 1.7. The ability to notice and tackle problems
   - 1.8. The ability to develop knowledge and skills
2. Competencies connected with shaping learners' attitudes
   - 2.1. The ability to shape learners' social and civil attitudes
   - 2.2. The ability to promote such a development of learners' competencies that will allow them to function successfully in an information society, which includes the following:
     - Motivation to learn not only what is compulsory
     - Critical information processing

- Creativity and innovation
- Problem solving
- Enterprise
- Cooperation with others
- Easy communication with others
- The ability to function in a visual culture

2.3. The ability to use the aforementioned competencies in the process of learning and teaching a certain subject

In the light of that set of competencies, the basic role of a teacher is to cause young people to learn for their whole lives and constantly gain new knowledge and abilities. Furthermore, learners need to become accustomed to using information resources on their own and know how to cooperate with others and solve problems.

What should be highlighted are the abilities connected with the use of computer and multimedia technologies, not only in the process of teaching, but also improving one's knowledge and communicating with learners and colleagues in order to share experience.

Because the standardization of the education process is a vital thing, the European Commission proposed common rules referring to the teaching profession. They describe a teacher's requirements and characteristics. A teaching job was defined as the following:

- Requiring higher education
- Connected with the idea of learning for life
- Mobile
- Based on partnership

In the modern world, the qualities of a teaching job require a person to use computer tools effectively. They not only make access to up-to-date information easier, but also enable its fast exchange. If computer tools are well chosen and appropriately used, they support partnership and increase the mobility of teachers. So a teacher should be able to make effective use of information, knowledge, and new technologies both in a traditional form and a multimedia one. A good knowledge of information and computer technologies will enable teachers to use it in the process of teaching and learning, for example, how to direct and support students in their attempts to gain and gather information by means of modern computer tools. The aim of using multimedia techniques in education processes is to improve communication, ways of conveying information and adjust them to the content and abilities of the receivers.

COMPUTER TECHNOLOGIES AS A MODERN TEACHER'S PEDAGOGICAL TOOLS

The development of hardware and information technology creates new possibilities for education processes in modern schools. These possibilities concern not only the information technology subject, but also the academic ones and vocational training. In case of the IT classes, the computer and software are the basic pedagogical tools. As far as the other subjects are concerned, the computer has a useful function, which enables the teacher to convey knowledge more effectively. In Polish schools, using the computer for teaching vocational subjects plays a vital role. The reason is the close connection between those subjects and the abilities of a person whose job requires them to use modern equipment, including computers. In the literature, there are two points of view on using information technology in education processes. According to the first one, using a computer does not always mean making lesson preparation easier. The problem might be the excess of information and means of communication, which a teacher must analyze and choose from. It should be used if other teaching methods are not sufficient or worse. The second idea is based on a statement that in some situations it is impossible to explain certain things to students without the use of the computer. Therefore, only if a teacher uses a computer can he or she achieve his or her educational aims.

It is important to bear in mind that if you teach the humanities or scientific subjects, the amount of information is so huge that it is impossible to obtain and process it without the help of computers. The introduction of the computer to teach those subjects makes it possible to improve the ways of presenting the material and assist the traditional teaching method. However, one needs to remember that using modern computer technologies cannot replace a human being, that is, a teacher in an educational process. A modern teacher can use computer technologies on the following levels:

- A basic level—the ability to use operational systems and office applications
- An advanced offline level—the ability to use the applications installed on a computer for research and simulation purposes, which improve teaching
- An advanced online level—a teacher aids the educational process by adding information and knowledge taken from the network sources or using e-learning tools
- A multimedia level—a teacher employs multimedia technologies to the fullest extent so as to prepare teaching materials by combining the Internet information and knowledge sources with the applications installed locally

Achieving the basic level allows a teacher to use the hardware and software to a limited extent. In such a case, a computer is mainly used for creating folders, the aim of which is to store the documents prepared by a teacher or downloaded from the Internet. An operational system allows creating the structure of catalogs, using basic tools such as a calculator, notepad, search engine, e-mail program, and a tool for searching for files on a hard disk. Additionally, basic level teachers use the simple functions of a word processor and spreadsheet. Those teachers do not pay attention to the problem of data protection. They do not know the tools for improving data security, coding, or making a backup. Hardly do they use user's profiles, which are connected with rights or passwords. Also, they lack the knowledge of alternative operational systems and software packages.

A teacher achieves the basic offline level the moment the computer begins to be used as a tool for making the process of acquiring knowledge easier. At this level, a teacher can make use of the advanced functions of a word processor and a spreadsheet as well as prepare presentations to a small extent. They use the software designed specially for teaching a particular subject. The computer often plays the role of a specialty measuring or simulation tool. Usually, the teacher does not use the Internet resources, and multimedia elements are not prepared by them, but come from ready-made applications.

As soon as the Internet resources are included in an educational process, a teacher reaches the advanced online level. Apart from the sources of online information, it is vital to include the following communication tools in an educational process: chat rooms, discussion forums, or videoconferences. The computer network is used as a tool for creation and integration of a group when solving problems and developing projects.

In order to achieve the multimedia level, a teacher has to acquire the ability to create and use the ideas conveyed by means of different communication channels. A teacher is required not only to gain knowledge on how to use a word processor, spreadsheet, and program to make presentations, but also the tools for creating graphics, animations, films, or sound processing. Thus, the materials are usually prepared by groups of specialists in multimedia transmission, methodologists, and those who have expertise in these matters.

Social changes consisting of the increasing amount of computer use in communications processes led to the creation of the idea of information technology or IT. It includes information management, computer science, telecommunications, and also some behavior patterns that are connected with the process of information and knowledge transmission. In this paper, information technology has been defined as a group of means, technologies, and tools, the aim of which is to use information freely (Bolter 1990; Friedrichs and Schaff 1987). Those means include computers, their input-output interfaces, and computer networks. Tools are the software whose aim is to enable information processing. The function of technology is to support the process of sending and receiving information.

The concept of information technology, which came into existence as a result of computer network development and communication software, has only recently become popular.

Combining the word "technology," which means a process or a way of doing something, with the word "information," which means an object having a particular way of recording explains the modern understanding of the form and ways of using information. Nowadays, information is accompanied by processes and activities during which both the information and the way of using it change. What is also different is how information is sought and gained. Information in the form of an encyclopedic entry, which is stored in one place only, is replaced by information stored in many places. Such information is shaped by a computer system on the basis of an inquiry made. Therefore, it is really crucial to know how to make inquires appropriately. This ability is of particular importance to teachers, who need to know not only how to gain knowledge and information from networks, but also necessary information from computer systems.

At present, information is entwined with various processes concerning its content and the extent to which it is used. As a result, one has to remember that the answer given by students or the solution suggested by them takes the form of a temporary state of those processes. This aspect of information started to exist together with information technology and possibilities of its use. Consequently, it is absolutely indispensable for a modern teacher to be capable of understanding and making use of the forms that include information and mechanisms and methods of utilizing it.

MULTIMEDIA IN THE TEACHING PROCESS

Multimedia techniques, known just as multimedia, are commonly used in teaching nowadays. Most often they are associated with a computer technique, that is, the casual use of micro- and macro-computers as tools supporting the process of sending information. Applying multimedia techniques enables the processing of information composed of sound, pictures, graphics, and text. A combination of a few media during the information transfer increases a person's ability to acquire information. Polish research has shown (Kiełtyka 2002) that if one only listens to information (in other words, if the information is conveyed only by means of sound), only 20% of what is conveyed is acquired. If, apart from hearing, a person additionally sees information, they remember 40% of what is conveyed. Finally, if one becomes emotionally engaged, the percentage of what they remember reaches 75%. Those results have been supported by other sociological research clearly showing that the use of multimedia enhances perception.

The major advantage of using multimedia in education is that it presents information in an attractive form, stimulating more than one sense. Another important feature of multimedia is the opportunity for a person who receives information to take part in creating the information transfer interactively.

In literature, there are two terms that are connected with multimedia: multimedia techniques, which are defined as technical means that influence human senses, and multimedia technologies, which operate information composed of sound, text, graphics, voice, and picture, both moving and stable. The function of multimedia techniques is to introduce combinations of elements mentioned above by means of very large files. In case of using multimedia in computer networks, it is important to define whether the transfer is made in the form of "save and send" or whether as a transfer in real time. In case of the former, the information does not have to be delivered immediately. Files are stored on the server; next the receiver downloads them on their computer where they are stored and later played. In such a situation, download speed plays a secondary role. Sending information in real time requires much higher capacity of tele-information connections. Those connections must allow a receiver to obtain the information in real time.

In order to explain the influence of multimedia on the process of knowledge acquisition, Kiełtyka (2002) uses the term "operational memory," which is also called the "brain board." Its essence consists of combined states of consciousness following each other simultaneously with restored fragments of knowledge so as to think in an abstract way, understand language, learn, and create new solutions. Operational memory allows retrieving symbolic information, such as a word acquired visually or parts of multimedia transfer acquired manually. Thanks to this operational memory, they are transformed into a concrete chain of motion or verbal reactions. The role of a modern teacher is to prepare multimedia transfer in a way that will guarantee the information that is being retrieved in the brain board will cause the reaction expected by a receiver.

Preparing appropriate multimedia transfer requires special abilities. In the literature, a teacher who specializes in preparing multimedia transfer is called a "media educator," whose main role is to create information and computer software that facilitate learning. Thus, a media educator is a person who transfers the information from traditional books into multimedia files in a way that will ensure the greatest possible value and the easiness of acquisition. This task requires a teacher to find the gaps in the knowledge conveyed in a traditional way and fill them with a combination of words, pictures, and sounds in order that a learner will remember as much as possible.

The job of a multimedia educator requires a person to know the basics of programming and how the current software tools available can be used. Technical knowledge must be supplemented with the pedagogical one concerning the age, the level of education, and the culture of the receiver. A person who uses multimedia in the educational process must be creative. This important feature allows them to seek and make use of new forms of knowledge transfer. Another thing is the capability of cooperation with specialists and technicians, which is indispensable in creating advanced multimedia materials. A multimedia educator should play the role of a coordinator of a team preparing multimedia transfer.

Nowadays in Poland, multimedia transfer only supplements the material of the lesson run in a traditional way. It is connected with the teacher training system, which does not prepare them fully for the role of a creator of multimedia transfer. Another problem is the insufficient knowledge of information technology by those who graduate from pedagogical schools, which makes it difficult to use multimedia transfer. Social changes that lead to the creation of an information society create the demand for the information in the multimedia form in all spheres of life. A growing need for multimedia information will cause curricula to change. Moreover, new methods, which let teachers acquire the ability to make multimedia transfer, will have to be devised.

DISTANCE LEARNING AS AN ELEMENT OF A TEACHER'S COMPETENCIES

Distance learning was used more than 300 years ago. At the beginning, it took the form of postal correspondence. The aim of correspondence courses was to improve knowledge connected with a job. In the materials sent to a learner, there were practical tasks to do and the instructions explaining how to do them. When a learner had done the tasks, he or she would send them back by post in order to get feedback. Once the tasks had been assessed, the learner would receive them again, along with new ones to do. One of the disadvantages of that system was the length of time necessary to send the materials between a teacher and a learner. Another problem was the lack of direct interaction between a teacher and a learner, which made it difficult to control the educational process and adjust the knowledge transfer to the learner's expectations. What also could be observed was that learners were less motivated or consistent.

The next step in the development of distance learning is connected with the invention of the radio. This invention enabled faster information transfer between a teacher and a learner and a new communication channel that made it possible to send sound messages. Two-channel information transfer allowed a learner to acquire knowledge more easily and caused him or her to learn much more. The information transfer by means of the radio required a learner to receive the information at a particular time. On the one hand, this motivated a learner to work systematically. On the other hand, a learner had to organize all other activities according to his learning process. The lessons that a learner missed were available only in a written form. The problem of access to sound transfer was solved when the forms of recording sound became more common. First, sound was recorded on phonograph records and later on tapes.

Once the television became more common, distance learning acquired an important tool, which made it possible to send pictures. The participants were able to observe the activities and, at the same time, listen to the comments. An additional channel caused teaching to become a multimedia process. The possibility of recording the lesson on VHS cassettes allowed the learners to watch it whenever they wanted, not only when it was being broadcast, and replay the video as many times as it was necessary.

Nowadays, distance learning is strictly connected with the use of computers and computer networks in an educational process. This method of teaching is understood as the one in which teachers and learners are not in the same place, and in order to convey information, they need to use modern telecommunications technologies that send sound, video picture, graphics, text, and other computer data.

The introduction of computer technology into distance learning enabled teaching in two modes: synchronous and asynchronous. The former requires the sender and the receiver to participate in the transfer at the same time. Thanks to computer technology, this transfer can be a fully multimedia one. The latter one allows replaying the transfer many times, which makes it easier to understand. Another significant advantage of modern technologies is direct contact between a teacher and a learner. This contact can be made by means of an audio channel, which transmits only sound, or a video channel, thanks to which both picture and sound are sent regardless of the distance between the teacher and the learner.

Distance learning requires a teacher to possess technical and pedagogical abilities of a higher level in comparison to the traditional or multimedia way of teaching. When preparing materials for self-study, a teacher uses the same computer tools as he or she does in the case of traditional ones. Basic skills are connected with the use of word processors, programs for creating presentations, and simple graphics software. The materials created in such a way are composed of text documents and presentations, including sources of information and graphics that illustrate the topic of discussion. Other types of materials used in distance learning are multimedia ones that include video sequences or sound transmission. In order to prepare such materials, one needs to know how to use both programs for multimedia transfer and devices, such as video cameras, sound recorders, cameras, and scanners. If a person lacks the ability to use software and multimedia equipment, he or she needs to cooperate with technicians who help produce multimedia materials. Multimedia transfer can be produced in an asynchronous mode. First, multimedia files are created, and then they become available to learners and can be used many times when learning. There also exists the opportunity to create a multimedia record of synchronous transmission and then make the files accessible. Such a way of recording allows replaying given parts of the lesson so as to understand them better.

The knowledge of and ability to use Internet tools and communication channels is another key element of the teaching profession. Basic communication tools are Web sites and e-mails. The typical feature of both channels is the high speed of information transfer.

In conventional understanding, e-mail is treated as a two-way communication channel whereas a website is one way. In practice, both a website and e-mail can be used as two-way channels; the difference is that a Web site is generally directed toward many users without differentiation, but e-mail is directed to a particular receiver. In distance learning, there are also supplementary communication channels such as chat rooms, discussion forums, chats over the Internet, and teleconferences. A modern teacher who participates in distance learning needs to use a few communication channels. What is mainly used in asynchronous communication is e-mail and discussion forums if a group has to solve a problem. Synchronous communication is achieved by means of chat rooms, which make it possible to exchange text messages or chats over the Internet when a teacher has a tutorial with a learner or conferences with which it is necessary to exchange sound and pictures among the members of a team.

Nowadays, distance learning is supported by specialized platforms. A teacher who uses such platforms must have all the abilities mentioned earlier and also use the software available on a particular platform. The role of a platform is to support the teaching process, knowledge acquisition, and recording of the learning process and communication management.

CONCLUSION

The common use of computers and creating an information society demand that modern teachers use computer and multimedia technologies in teaching. Minimal use of hardware and software involves the basic use of computer tools that allow creating simple text documents and presentations that illustrate the issues discussed.

Including elements of multimedia transfer in the didactic process increases the extent to which knowledge is acquired and also makes students focused on tasks for a much longer time span. Widespread use of the Internet, which is characterized by a specific way of presenting information, causes learners to expect that they will be provided with information in a similar way, more often by means of something they are familiar with thanks to the Internet. The consequence is that the traditional way of teaching must be enhanced by new forms of information transfer.

The increasing speed and availability of the Internet makes multimedia transmission possible in real time. This results in the opportunity to teach interactive lessons over the Internet. Both teacher and student can actively participate in the teaching process. Applying distance learning methods and platforms can be a supplement to traditional teaching. A definite advantage of this method is, first of all, the individualization of the way and pace of learning.

A modern teacher must be familiar with the possibilities of computer technologies and be able to implement them successfully. The role of these technologies is to support and record the teaching process and enable students to remember information easily. What is more, the aim of adopting computer technologies is to support communication processes and prepare learners to use computers and computer networks as tools for gaining knowledge and participating in the lives of local communities.

REFERENCES

Bolter, J. D. (1990), *Turing's Man*, Polish ed. Człowiek Turinga. PIW, Warszawa.

Education and Training in Europe: Diverse Systems, Shared Goals for 2010, (2002), Luxemburg, http://www.szok.rcez.pl/doc/ewewcdr2010.pdf, (accessed 15.12.2011).

Friedrichs, G., and Schaff, A., (eds.) (1987), *Microelectronics and Society*, Polish ed. Mikroelektronika i społeczeństwo. Na dobre czy na złe? Raport Klubu Rzymskiego, Książka i Wiedza, Warszawa.

Key competences, Introduction on the level of compulsory education in Europe, (2005), Euridice Network information on education in Europe, http://www.eurydice.org.pl/sites/eurydice.org.pl/files/kkomp_PL.pdf, (accessed 15.12.2011).

Kiełtyka, L., (2002), *Communication in Management, Techniques, to ols anf forms of information transfer,* Polish ed. Komunikacja w zarządzaniu, Techniki, Narzędzia i formy przekazu informacji, Warszawa, Placet.

OECD. (2001), *Defining and Selecting Key Competencies*. OECD, Paris.

Pedagogical Encyclopedia of XXI Century, Polish ed. Encyklopedia pedagogiczna XXI wieku, (2008), praca zbiorowa tom I i II, Żak Wydawnictwo Akademickie, Warszawa.

Perrenoud, P. (2001), *Developing Skills at School. Practical and Pedagogical Issues*, Polish ed. Construire des compétences dès l'école. Pratiques et enjeux pédagogiques. Paris, ESF éditeur.

Sielatycki, M. (2005), *Teachers Competences in European Union,* Polish ed. Kompetencje nauczyciela w Unii Europejskiej w: Trendy uczenie w XXI wieku, Internetowy magazyn CODN nr 3/2005.

Weinert, F. E. (Hrsg.) (2001), *The Evaluation of Competences at Schools*, Polish ed. Leistungsmessungen in Schulen. Weinheim, Beltz.

92 Teacher Competence in the Use of the Local Environment

Aleksander Noworol

CONTENTS

INTRODUCTION

Modern society is changing. Globalization, the increasing importance of information and communication technologies, and the growing role of the Internet in shaping social relationships require new attitudes and readiness to face unprecedented challenges. This presents some very particular challenges to the education system, which is gradually moving away from being primarily a carrier of knowledge and is becoming an instrument of preparation for active participation in the society of knowledge and complex, multilevel inter-organizational dependencies. The following text delineates the new challenges arising from schools playing a unique social role in the local environment and teacher competence resulting therefrom against the background of the requirements to be met by Polish schools.

EXPECTATIONS FOR SCHOOLS ARISING FROM PEDAGOGICAL SUPERVISION

The performance standards for Polish schools are reviewed in a formalized evaluation process, which—from a legal point of view—is conducted based on two major legislative acts:

- The Act of September 7, 1991, on the Education System (*Journal of Laws from 2004*, No. 256, item 2572, as amended, Referred to as *the Act* throughout this paper) and arising from its art. 35 paragraph 6.
- Ordinance of the Minister of National Education of October 7, 2009, on Pedagogical Supervision (*Journal of Laws from 2009*, No. 168, item 1324, Referred to as the *Ordinance* throughout this paper).

These acts introduce strict requirements (defined in the system of so-called external pedagogical supervision) for schools, their managerial staff, and teachers. The Act and the Ordinance specify the object, participants, and evaluation mode. The Ordinance is of particular importance to the presented issue. It stipulates, among other things, the detailed terms and procedures and the forms of exercising pedagogical supervision, the qualifications necessary for conducting pedagogical supervision, and qualifications of persons who may be charged with performing research and evaluation.

The Ordinance defines the principles and modes of pedagogical supervision. This supervision is exercised, taking into account the following:

- Free access to the requirements
- Cooperation between authorities exercising pedagogical supervision and authorities in charge of schools or institutions, head teachers, and teachers
- Creating conditions conducive to the development schools and educational institutions
- Obtaining information that ensures a full and objective assessment of the educational and social care activities as well as other statutory activities of a school or educational institution

Forms of pedagogical supervision include evaluation, monitoring, and support. Pedagogical supervision is implemented by performing tasks and activities referred to in art. 33 of the Act. It thus consists of the following:

- Assessing the state and conditions of activities of schools, institutions, and teachers in terms of education and social care
- Analyzing and evaluating the effects of teaching, educational, and social care–related activities as well as other statutory activities of schools and educational institutions
- Providing assistance to schools, institutions, and teachers in performing their educational and social care–related tasks
- Inspiring teachers to implement pedagogical, methodological, and organizational innovations

The requirements of the state for schools, subject to external pedagogical supervision, indicate strategic tasks and priorities selected in such a way that would help organize the activity of schools in order to address key civilizational issues. Those requirements should function as targets that educational institutions should strive to implement in order to enable students to succeed and also to reinforce the creation of modern society. The requirements relate to four areas of activity of schools and institutions:

- Area of effects—everything that is considered as a valuable and desirable result of the school's activity
- Area of processes—includes the processes leading to the emergence of the desired effects
- Area of the environment—covers the important aspects of cooperation with the local, regional, and global environment of the school
- Area of management—includes management activities affecting the three previous areas

External evaluation is carried out within the above-mentioned problem areas. This includes, at the same time, collecting and analyzing information in these areas, and determining the level of compliance with the requirements for each area as described below by the school or institution. Under the Ordinance, upon completion of the external evaluation, the results and conclusions of the evaluation are presented at the teaching staff meeting. The Ordinance also provides that within seven days of the teaching staff meeting during which the results and conclusions of the external evaluation process were presented the team or the person who conducted the evaluation draw up a report on the evaluation that should include the following:

- Evaluation of results
- Determination of the level of compliance with the requirements from A to E, according to the scheme presented above
- Conclusions of the evaluation

TABLE 92.1
Level of Compliance with the Requirements Contained in the Annex to the Ordinance in Relation to the Functioning of the School in the Local Environment

Requirements for the Functioning of the School or Institution in the Local Community	Level of Compliance with the Requirements
Environmental resources are used for mutual development	A–E
Information on further progress of the graduates is used	A–E
The value of education is promoted	A–E
Parents are partners of the school	A–E

Source: Author's own work based on the Ordinance.

The primary function of such a report is the cognitive function. It should provide its readers with information concerning the operation of educational institutions, in particular referring to the area of the functioning of the school in the local environment as indicated in the Ordinance. The report should also serve the development of schools and educational institutions because the information collected is to contribute to the decision-making processes for the improvement of the activities conducted by these institutions (and thus their development) and to the successful implementation of educational policy.

The report is concluded with the synthetic "conclusions of the evaluation" and a table showing the level of compliance with each of the requirements from A to E. Table 92.1 is a list of requirements for elementary schools, middle schools, high schools, art schools, institutions of learning, practical training centers, and training and professional development centers in relation to the area of the environment.

It is worthwhile to look at the requirements for schools and educational institutions in the area of their functioning in the local community in a broader context, which results from the creation of a modern education policy.

EDUCATION POLICY AND THE LOCAL ENVIRONMENT

In scholarly terms, the definition of educational policy was introduced by Dziewulak (1997), who described it as "making choices and taking decisions on matters of goal, direction, content and organization of education and upbringing, and the planning of educational activities" (p. 11). Education policy, like any public policy, is a collection of tools and principles to be adopted for the implementation of strategic objectives related to a specific area of socioeconomic activity. Educational policy, as an integral part of the social policy, defines the tools associated with the development of human capital and social capital. It encompasses the entirety of education issues, organization of the teaching process, and methods of teaching and education; it establishes rules for the education of children, adolescents, and adults. In terms set out by K. Przyszczypkowski, the main tasks of local authorities in the field of educational policy boil down to the following:

- Independent determination of the public kindergarten network and primary and secondary schools as well as the borders of their districts
- Supervision over the control of school obligation fulfillment by children and adolescents aged 3 to 18
- Cooperation with pedagogical supervision authorities to the extent specified by the Act on the education system and the implementation of the recommendations of the chief education officer

- Conducting human resources policy to appoint head teachers of primary schools, middle schools, high schools, kindergartens, and educational institutions
- Approval of annual organization sheet projects for primary schools, middle schools, high schools, kindergartens, and educational institutions
- Development of proposals for the maintenance of kindergartens, schools, and educational institutions and providing them with housing and material conditions for the implementation of their statutory tasks, including the dissolution of these units
- Performing works toward the implementation of the competence of the executive body of the local government in preparing the draft budget, control over its implementation, and compliance with budgetary discipline in the educational units (2010, pp. 43–44).

Taking into consideration the reflections and opinions of a number of authors (Łuczyński 2009; Noworól 2007; Przyszczypkowski 2010; Sielatycki 2005), the nature and elements of the local educational policy can be outlined. In a general sense these—sectoral policies—are systems for the implementation of strategic objectives. Policies should have an internal structure that allows directing the activities of the entities responsible for the implementation system. They should, therefore, against the background of the analysis of determinants for a specific strategic objective, present the principles and ways of solving the identified problems. Within such policies, alternative solutions are often analyzed, and consequence scenarios for the implementation of each option are tested. An essential element of policy is the recommendation of the selected variant of implementation of the strategic objective along with the description of the most favorable instruments of implementation (Noworól 2007, pp. 160–161). A summary of the education policy elements used by schools or educational institutions and the requirements set out in the external evaluation system is presented in Table 92.2.

Policies are implemented through a system of instruments. The key to the scheme is the four stages of the managing process of a territorial unit. There are, therefore, the following:

- Instruments of development of the objectives and tasks in the form of plans and programs, known as planning instruments
- Instruments for organizing the implementation of policy objectives, in which the key element is the creation of relevant institutions (organizational instruments)
- Instruments of current administration of the policy-implementing organizations
- Instruments for monitoring and control of the methods and techniques of the implementation of objective—rules—findings (Noworól 2007, pp. 94–95).

The educational policy of the local authorities consists of programming interventions in the four areas of education management: planning, organizing, implementation, and control.

The essence of the *planning aspect of the policy* is identifying local educational priorities and developing long-term action and financial decision plans in line with those priorities. The objectives of educational policy can specify various issues: From the general—ideological and philosophical to practical—guiding the operational measures. The synthetic goals of the local policies should not be in opposition to the "ideological" Preamble of the Act. Those goals are related—to a greater or lesser extent—to the social value that is the education of the young generation into thoroughly educated people, open to the understanding of the world, people who are physically, emotionally, and morally healthy. Educational programs should promote the values of respect for all human beings and their natural and cultural environment. They should encourage the understanding and promotion of the peace and harmony of life on Earth and prepare young people to exist in an open, multicultural, and "networked" society of the 21st century. Education policy must also take into account the economic and technological changes, requiring lifelong learning in a dynamically shifting social environment. J. Łuczyński suggests a system of educational objectives, including the following:

- Education—the transfer of knowledge
- Socialization—the process of the student's adaptation to the life in society
- Personal development of students
- Social change, defined as work for the reconstruction of social relations (2009, pp. 105–110).

One can argue that nowadays the main challenges and *objectives of education*, including education services, are the following:

- In the area of development through education and knowledge—imparting the necessity of creativity to students
- In the area of development through socialization—helping students develop the ability to "draw the line"—in terms of axiology and, especially, ethics—because such boundaries can be particularly hard to grasp in the network society
- In the area of social life—creating an awareness of the importance of active participation in civil society through one's own social activities, involvement in the affairs of the local community, and voluntary service.

The structure of educational objectives should be *correlated with other social policies*, including those aimed at building an innovative economy and fighting against exclusion. In the planning documents, it is advisable to also determine the *local educational standards*, covering systemically the desirable solutions within the framework of other policy tools as described below.

A very important element of the policy is *the organizational infrastructure of education*—the construction and modernization of the network of schools and educational institutions. There are a number of managerial issues related to that subject that require a separate reference.

First, the organizational system is implemented by a *network of public schools and institutions*. Its determination involves the standards of operation of the local education system basic to the effectiveness of the policy. Those standards, in turn, should be based on reasonable premises regarding the availability and conditions of education. Another organizational aspect is the determination of the model of operation for schools and educational institutions, which involves the exploration of the possibility of, for instance, expanding the scope of their activities (e.g., building an open school, changing the education profiles). Changes of this type are of particular importance in the conditions of demographic changes that force the resolution of many dilemmas related to the analysis of both the financial and social implications of closing down schools and educational institutions, including, inter alia, the restructuring of the local education network. *Here, an important element of the system can be the premise of implementing multipurpose architectural solutions, allowing for the use of school facilities in connection with other social functions* (such as community centers; public libraries and media libraries; and centers of social activity, sports and, recreation centers).

The organizational aspect of the local education policy remains primarily the responsibility of the authority in charge of the school or institution. In the synthetic perspective, those responsibilities include the following:

- Organization of the network of public schools and educational institutions
- Implementation of investment policies, which are to ensure adequate material base for the realization of educational objectives and acquisition of the best possible equipment for schools and educational institutions
- Creation of organizational interrelations (networks, agreements, contracts) with entities operating in the local environment, which could become partners of the district authorities in achieving the objectives of education policy.

The school, on the other hand, influences the degree of openness to the local environment and institutional relationships with civil society organizations (NGOs), which can operate under the

TABLE 92.2
Elements of Territorial Development and Selected Activities of the School

Processes Conditioning Territorial Development	Development of Students through Education (Dissemination of Knowledge)	Development of Students through Socialization and Social Activity	Managing the Education Process and the School or Institution
Increase in complexity, diversity, and flexibility of the subsystems	– Cooperation of the school with other entities (of three sectors), whose participation in the educational process will enhance the knowledge and skills of the students – Encouraging teachers and students to perform extracurricular intellectual activities – Promotion of the value of education and lifelong learning – Promotion of the knowledge of the complexity of the modern world: glocalization processes, multiculturalism, and environmental challenges	– School as an entity in the system of local partnership for development – Participation of teachers and students in the activities of other entities (apart from school) operating for the benefit of the environment – Supporting parents in raising children – Raising awareness of the importance of active participation in civil society through own social activities, involvement in the affairs of the local community, and voluntary service	– Diversification of ways of organizing and funding the operation of schools (organizational disaggregation of the education system) – Creation of the organizational conditions for lifelong learning – Creation of conditions (organizational, housing) for the performance by the school of a broader social role in the local environment: center of development and activity (culture, arts, sports, recreation)
Increase in the capacity for auto-regulation of subsystems	– Use of local environmental resources (natural, cultural, ethnic, social) in supplementing the core curriculum – Including local authorities (e.g., academic) in the education process – Knowledge of the problems of environmental management (local and global)	– Content-related cooperation between school and other local social policies organizations: community centers, sports centers, NGOs – Cooperation between the school and parents – Use of information concerning the careers of the graduates – Implementation of integration programs (battling exclusion)	– Determination of formal and financial principles of the school's cooperation with other local social policies organizations: community centers, sports centers, NGOs – Acquisition and implementation of parents' opinions – Monitoring of how collaboration with institutions and organizations working in the environment affects the educational development of students

Increase in innovation	– Use of local environmental resources (natural, cultural, ethnic, social) as inspiration for intellectual exploration for pupils (and teachers), which, in particular, relates to the identification of the endogenous drivers of local development – Promoting students' creativity	– Building various forms of relationships within the local community, based on the identified potential – Building local relationships in spite of cultural differences (if there are any) – Reducing the effects of digital exclusion	– Use of innovative concepts and management techniques for building environmental partnerships for schools – Creation of a platform for cooperation that allows the use of local innovation (e.g., in economy, cultural life, and tourism)
Intensification of contacts with the environment	– Opening of schools to collaboration with other entities and organizations spreading knowledge, not only from the local environment – Use of schools as a means of students' and teachers' participation in the global information society – School's involvement in promoting the knowledge of the local potential	– Shaping the attitudes of tolerance and openness to functioning in a multicultural and multi-religious society – Involvement of the school in existing and constructed network structures, including social networks – Partnership with nonlocal organizations and social policies organizations in the sphere of culture, education, and sport	– Providing financial and organizational conditions for promoting contact between students, teachers, and the environment – Creating opportunities for tourist use of local potentials discovered by the school
Adjustment of the dynamics of continuous and discrete changes, including the guarantee of security	– Ensuring adequate standards of knowledge acquired through external contacts, which involves, at the same time the following: – Being open to new, inspiring sources of knowledge – Controlling the standard and adequacy of the knowledge derived by the students through school education	– Developing students' ability to "draw the line" in terms of axiology and, especially, ethics, as such boundaries can be particularly hard to grasp in the network society – Informing students about the benefits and risks associated with participation in the network society – Assessment of students' behavior and undertaking educational activities aimed at eliminating risks and enhancing appropriate behavior	– Enabling parents to participate in the decision-making processes on matters concerning the school or institution, and to participate in the activities undertaken – Analyzing undertaken educational actions aimed at eliminating risks and enhancing appropriate behavior (evaluation of the effectiveness of actions, including students' initiatives)

Source: Author's own research.

"educational partnership" for development. The organizational task of the school is also creating a "systemic" platform for cooperation with the students' parents.

Aspects *of the current management of the area of local education policy* concern key education policy issues. It is a wide array of issues that go beyond the scope of this publication. Let us focus in this paper on the identification of those elements that are closely related to various aspects of the functioning of schools in the local environment. Among the *instruments for the development of students through education* (dissemination of knowledge), one needs to distinguish those that are connected with the need to provide a state-of-the-art range of educational content tailored to the needs of socioeconomic development. At the level of the tools employed, in relation to the local environment, meeting those requirements involves taking the following actions:

- Creating conditions conducive to students' gaining knowledge and skills, including, among other things, issues related to local problems and challenges on environmental, social, and economic issues; the educational offer should therefore correspond to the needs or interests of learners and also the labor market requirements.
- Creating conditions for the intellectual and cognitive activity of students, particularly in identifying local cultural and environmental values and the prospects of students' development associated with them.

The instruments for students' development through socialization and social activities also have an environmental dimension. Development through socialization requires taking such actions as the following:

- Creating conditions that induce students to abide by social norms, which relates to the operation in all communities in which the students are involved: classmates, peers from the neighborhood, and also participants in social networks, the participants of which the students become; activities covered by the instrument described involve the following:
  - Ensuring the safety of pupils or students at school (institution) and its surroundings
  - Informing students about the behavior that is expected of them
  - Informing students about the benefits and risks associated with the participation in the network society
  - Assessment of the behavior of students and taking action aimed at eliminating risks and enhancing appropriate behavior
  - Analyzing educational actions aimed at eliminating risks and enhancing appropriate behavior (evaluation of the effectiveness of initiatives involving students)
- Shaping the attitudes of students (promotion of tolerance and social integration of students of different environmental, cultural, and religious backgrounds) by undertaking educational activities adequate to the needs of students in a planned manner, taking into account their participation as well as the conclusions from the analysis of the effects of the educational work
- Cooperation between the school or educational institution and the social and economic environment for mutual development, which requires the following:
  - Pursuit of the broader concept of educational partnership, which consists of combining local current educational problems with long-term local development issues
  - Cooperation with institutions and organizations operating in the local environment
  - Taking into account the capabilities and needs of the local environment
  - Monitoring how the cooperation with institutions and organizations operating in the environment affects the educational development of learners

The instruments *of managing the education process and managing a school or educational institution* are related to managerial issues, enabling expedient and proper implementation of the basic instruments associated with the development of students. Among the instruments of management, it is necessary to distinguish those that are exclusively within the competence of the authority in charge of the school or institution and those whose implementation depends on the work of the head teachers and directors of schools and institutions.

In particular, managerial instruments relating to the local environment, which remain the responsibility of the authority in charge of the school or institution include the following measures:

- Providing a high level of professional, educational, and ethical management staff of schools and institutions as well as the teaching staff, which is dependent on the attractiveness of the location
- Providing the funding of education at a level allowing the implementation of other instruments of the education policy, which is associated with the level of affluence of the local community and its willingness to involve financially in educational processes
- Removing and preventing the emergence of disparities between the conditions of education in public schools within a territorial unit
- Creation of conditions (organizational, housing) for the performance by the school, in the wider social environment, of the role of the center for development and activity (in the areas of culture, arts, sports, and recreation)

The managerial instruments that remain the responsibility of the school or institution include primarily the cooperation with parents, treating them as partners of the school by doing the following:

- Obtaining and utilizing the parents' views on the operation of the school
- Supporting parents in the upbringing of children
- Enabling parents to co-decide on the matters important to the school or institution and to participate in the actions taken

As shown by the above review of the issues surrounding the interdependence between the school and the local environment, the issues of partnership between the education system and the management system for the operation and development of territorial units have many dimensions in common and concern the essential elements of governance at the local level. A synthetic interpretation of the conclusions of the evaluation reports on schools and educational institutions against the background of education policy instruments remaining within the competence of the authorities in charge, has been presented by the author in previous publications (Noworól 2011).

LOCAL ENVIRONMENT AND THE SCHOOL

Following from the above description, schools and educational institutions can and should take into account the issues related to the environment in which they operate virtually in all areas of their activity. The dissemination of knowledge and socialization of students, as well as the implementation and monitoring of educational policies, have an important component associated with the reality in which the school (institution) functions. This applies to virtually all aspects of a particular territorial unit, its potential, and also its deficits. So let us look at this environment in a more systematic manner or, rather, a systemic manner, in order to identify the problem areas and challenges that may be the subject of the creative intervention of a modern school.

Based on Ludwig von Bertalanffy's (1984) theory of open systems, it is proposed that territory—as the human environment—should be construed as a system of dependencies comprising the following four components (sub-systems):

- Spatial environment (animate and inanimate matter), which has the following aspects: geometric, technical, environmental-natural, and aesthetic, etc.
- Human capital—a human being, as a phenomenon, has the following aspects: health, psychological and social, artistic, religious, etc.
- Organization—the relationships between people and between people and the spatial environment, understood as their surroundings; those relationships have the following aspects: technological, informational, political, social, and economic, etc.; following J. Stoner, R. Freeman, and D. Gilbert an organization is, at the same time, construed as "two or more people cooperating within a specific framework of relations in order to achieve a specific purpose or set of purposes" (Stoner, Freeman, Gilbert 2001, p. 619)
- Territorial system management as a special component of the "organization" subsystem (Noworól 2007, pp. 18–26).

The territory is thus a phenomenon in which the above elements remain in a constant state of interdependence. Between these territorial subsystems, there is a tangle of interactions, taking various forms underpinned by the developing information technologies. The dynamic aspects of the territory are constantly developing it by changing its internal structure and its relationship with the environment. At the same time, the systemic changes of the qualitative character manifest themselves by doing the following:

- Entropy, as a process occurring in the system, consists of the destruction, during the operation of the system, of its elements and removing them to the environment, which results in a reduction in the viability and efficiency of the system
- Negentropy, meaning the creation of new configurations of elements from which the system has been built, which conditions its more efficient operation

A system's ability to reduce the entropy level and achieving the predominance of negentropic processes determines its continued existence and development in the environment. *Territorial development* will therefore be understood as such targeting of the change of the territorial system that, by stimulating driving forces, such as business, balancing the endo- and exogenous factors, and competitiveness, leads to negentropic processes, including the following:

- An increase in complexity, diversity, and flexibility of the subsystems, improving their capabilities and adaptability
- An increase in the capacity for the auto-regulation of the subsystems
- An increase in the innovation (creativity, proactivity) of the subsystems
- An adjustment of the dynamics of continuous and discrete changes in the subsystems to own needs and to the influence of the environment
- Intensification of contacts with the environment through communication, exchange of goods and values, and expansion (spatial, social, economic) with a monitored level of adaptation and aggressiveness (Noworól 2007, pp. 42–66)

One must perceive *the fundamental role of schools in the process of stimulating growth and development* in precisely such a context. Let us remain at the level of local relationships, analyzing the relationship between developmental processes and the elements of school activity that form the essence of environmental partnerships. It is schematically illustrated by Table 92.2.

The synthetic approach to the role of schools in the local environment presented above allows us to define teacher competence associated with that role, which shall be detailed in the following portion of this paper.

DESIRABLE TEACHER COMPETENCIES IN THE CONTEXT OF CHANGING LOCAL ENVIRONMENTS

The description of teacher competence in relation to the requirements of the local environment should be preceded by a reflection on the changes taking place in this environment.

The contemporary world is undergoing enormous and rapid changes. One can observe these changes in the economy, in the environment and the climate, and lastly in social and cultural behavior. Two fundamental processes, globalization and competitiveness, are having the biggest impact on territories and their educational organizations. The point is that globalization and the shift toward a knowledge-based economy have had diverse effects on territorial competitiveness (called glocalization). Competitiveness forces territories to contribute to the creation of economic activities and to attract people and capital in a competitive environment. According to M. Sudarskis (2010), the metropolitan challenges of today can be arranged as follows:

- Globalization as an economic challenge
- Sustainability as an environmental challenge
- Cohesion as political, social, and cultural challenges

So, priority fields for public intervention include such areas as control of urban development and urban regeneration, mobility and transport, resources for production, influencing the labor market, technology and financing, and finally, governance as an effective and democratic process.

These processes create an ever more challenging base for reference for the process of training and education of young people and adults. It leaves a significant mark on the requirements presented to the teachers today. For more than 10 years, it has been known that—as G. Miłkowska-Olejniczak puts it—the teacher in the 21st century, instead of providing information, "becomes a guide into the world of knowledge, the person who opens for the student the great gateway to the world of values, ideas, characters, thoughts, words, deeds, and scientific discoveries. 'Welcome!' he says, 'Enter and see all the treasures accumulated here over the centuries. Experience it and find something that you like!'" (Miłkowska-Olejniczak 1998, p. 96). It also involves the shift in the very notion of what teacher competence entails.

According to K. Stokking and M. Van der Schaaf (2006, pp. 15–16), referring to other authors, "traditionally the term teacher competence has often been perceived in terms of behaviour or in terms of individual psychological attributes, the recent concern for teacher competence has been expressed as a need for a more integrated concept denoting teachers' knowledge, skills, and attitudes in context while performing professional tasks." However, the situation has changed. The same authors underlie that "recently, scholars have been moving toward a more integrated approach, based on the notion that competence is a relational concept, bringing together the abilities of the individual and the professional tasks to be performed in particular situations" (Stokking and van der Schaaf 2006, p. 16). According to Gonczi (1996, p. 6), "... the integrated approach to competence-based education provides a conceptual base for the competency movement and a promising direction for educational reform for all levels of occupational education." So, an integrated approach recognizes the importance of teacher cognitions and of the context in which teaching takes place.

Let us examine, therefore, teacher competence in the context of the teacher's cognitions of developmental processes that should be the object of the process of the school's functioning in the local environment. Following M. Olczak (2009), we shall distinguish:

TABLE 92.3
Key Teacher Competencies in the Context of the School's Impact on the Local Environment

Type of Teacher Competence	Development of Students through Education (Dissemination of Knowledge)	Development of Students through Socialization and Social Activity
Abilities and communication skills - leadership, dialogue	– Cooperation of the school with other entities (of three sectors) – Use of schools as a means of students' and teachers' participation in the global information society – Use of local environmental resources (natural, cultural, ethnic, social) as inspiration for intellectual exploration for pupils (and teachers), which, in particular, relates to the identification of the endogenous drivers of local development	– School as an entity in the system of local partnership for development – Participation of teachers and students in the activities of other entities (apart from school) operating for the benefit of the environment – Building various forms of relationships within the local community, based on the identified potential – Building local relationships in spite of cultural differences (if there are any) – Shaping the attitudes of tolerance and openness to functioning in a multicultural and multi-religious society – Involvement of the school in existing and constructed network structures, including social networks – Partnership with nonlocal organizations and social policies organizations in the sphere of culture, education, and sport
Talent—potential of creative surpassing of the teachers' professional role	– Encouraging teachers and students to perform extracurricular intellectual activities – School's involvement in promoting the knowledge of the local potential – Ensuring appropriate standards of knowledge acquired through external contacts	– Content-related cooperation between school and other local social policies organizations: community centers, sports centers, NGOs – Reducing the effects of digital exclusion – Developing students' ability to "draw the line" in terms of axiology and, especially, ethics as such boundaries can be particularly hard to grasp in the network society
Operational styles—focus on cooperation	– Use of local environmental resources (natural, cultural, ethnic, social) in supplementing the core curriculum – Including local authorities (e.g., academic) in the education process – Opening of schools to collaboration with other entities and organizations spreading knowledge, not only from the local environment	– Supporting parents in raising children – Raising awareness of the importance of active participation in civil society – Implementation of integration programs (battling exclusion)
Other competencies	– Promotion of the value of education and lifelong learning – Promotion of the knowledge of the complexity of the modern world: glocalization processes, multiculturalism, and environmental challenges – Knowledge of the problems of environmental management (local and global) – Promoting students' creativity	– Use of information concerning the careers of the graduates – Informing students about the benefits and risks associated with participation in the network society – Assessment of students' behavior and undertaking educational activities aimed at eliminating risks and enhancing appropriate behavior

Source: Author's own research.

- *Knowledge-related competencies,* or being prepared to carry out specific tasks within the profession, specialization, position, or organization, connected with the "technical knowledge" that sets the goals to be achieved; provides information on methods, understood as a reliable and reproducible means of achieving goals; and finally, defines the measures and conditions that determine the chances of achieving the goals
- *Competencies related to skills and abilities,* referring to communication skills (including interpersonal skills, leadership skills, ability to enter into dialogue)
- *Talent-related competencies,* referring to the employee's potential, their capacity for development and creative surpassing of the professional role of a teacher, for instance, by solving problems of a practical, moral, and technical nature that are considered to be typical in the teacher's environment
- *Competencies associated with operational styles,* describing personality traits determining, e.g., one's social orientation, focus on cooperation, and ways of achieving goals
- *Competencies related to principles and values,* concerning the principles, values, and beliefs, allowing the determination of one's motives, which is associated with an education toward humanist values
- *Interest-related competencies,* indicating preferences for the tasks and the type of work and working environment
- *Physical competencies,* referring to the predispositions related to the physical requirements of the job, which entails the need for appropriate physical, sensory, and mental abilities (Olczak 2009)

The juxtaposition of the above considerations with the essence of the school's functioning within the local environment leads to the identification of the appropriate frames of reference. Key teacher competencies have been assigned actions that the school should take in order to function properly in the environment and promote local development. In the list, the competencies that apply to all areas of teacher activity (knowledge-related competencies, interest-related competencies, and physical competencies) have been integrated. It is the author's opinion that the remaining competencies can be defined as key to the issues discussed. Table 92.3 constitutes a proposal for school activities in the field of student development through education and socialization in relation to key environmental competencies.

CONCLUSION

The school plays a crucial role in the local environment. This role is not restricted merely to narrowly construed educational processes. It involves, as has been extensively discussed above, the activities that promote various aspects of local development. Key teacher competencies are, at the same time, related to a teacher's abilities and communication skills, especially in the areas of interpersonal skills, leadership, and the ability to enter into dialogue. Another prerequisite for the inclusion of schools in local partnership are the teacher's talents, relating to their capacity for creative surpassing of their professional role. It is also connected with the operational style focused on cooperation, which allows the opening up not only of a particular school, but also the environment's way of thinking, to the new challenges posed by the development of the network society.

REFERENCES

Dziewulak, D. (1997), *School Systems of the European Union*, Polish ed. *Systemy szkolne Unii Europejskiej*, Warszawa. Wydawnictwo Żak.

Gonczi, A. (1996), *Reconceptualising Competency-based Education and Training: With Particular Reference to Education for Occupations in Australia.* http://epress.lib.uts.edu.au/dspace/bitstream/handle/2100/303/01front.pdf?sequence=1.

Łuczyński, J. (2009), *The School and the Labour Market*, Polish ed. Szkoła a rynek pracy in: *Zarządzanie Publiczne, Zeszyty Naukowe Instytutu Spraw Publicznych UJ* (1(5)/2009).
Miłkowska-Olejniczak, G. (1998), *Teacher Education and the Educational Reform*, Polish ed. *Edukacja nauczycieli a reforma oświaty* Kultura i Edukacja 1998, no. 4.
Noworól, A. (2007), *Territorial Development Planning on the Regional and Local Scale*, Polish ed. *Planowanie rozwoju terytorialnego w skali regionalnej i lokalnej* Kraków: Wydawnictwo Uniwersytetu Jagiellońskiego.
Noworól, A. (2011), *How to Read a Report Following an External Evaluation of a School or Educational Institution*, Polish ed. *Jak czytać raport z ewaluacji zewnętrznej szkoły lub placówki oświatowej*, Warszawa, Ośrodek Rozwoju Edukacji.
Olczak, M. (2009), *Teacher Competence and Qualifications*, Polish ed. *Kwalifikacje i kompetencje nauczyciela* Edukacja i dialog, 4/2009.
Przyszczypkowski, K. (2010), *Local Governments' Education Policy*, Polish ed. Polityka oświatowa samorządów in: M. Korolewska, J. Osiecka-Chojnacka (Eds.), *Polityka oświatowa*. Warsaw: Biuro Analiz Sejmowych No. 2(22).
Sielatycki, M. (2005), *Teacher Competence in the European Union, Kompetencje nauczyciela w Unii Europejskiej* TRENDY - uczenie w XXI wieku. CODN online magazine, No. 3/2005. http://www.trendy.ore.edu.pl/struktura/czytelnia/artykuly/doc/kompetencje_nauczyciela_w_unii_europejskiej.pdf.
Stokking, K., van der Schaaf, M. (2006), *Assessing Teacher Competences in Teaching Students Research Skills*. http://igitur-archive.library.uu.nl/dissertations/2006-0103-200145/c2.pdf.
Stoner J., Freeman R., Gilbert D. (2001), *Management*, Polish ed. *Kierowanie*, Warsaw: Polskie Wydawnictwo Ekonomiczne.
Sudarskis, M. (2010), *Metropolis Now*, INTA International Symposium on Urbanism: *Cities in Metropolitan Regions, Gedafe*. http://www.ciudadesenregionesmetropolitanas.com/download/Sudarkis-metropolis.pdf.
von Bertalanffy, L. (1984), *General System Theory: Foundations Development Applications*, Polish ed. *Ogólna teoria systemów. Podstawy, Rozwój zastosowania*, PWN, Warszawa.

93 Frozen in Time, Left Out in the Cold

Teacher Training and Development and the Role of the Head Teacher—Considering Alternatives

Robin Precey

CONTENTS

INTRODUCTION

Most would agree that the core business of schools is learning. It is both curious and ironic, however, that we often invest too little in ensuring that our teachers and their leaders are fit for the purpose of making our business as successful as it could be. If we were in many other organizations—for example, managing hotels, producing oil, running an airline—we would be trying very hard to ensure that our workforce was well trained and that their knowledge and skills of the business were constantly kept up to date in order to continually improve levels of service (however, we chose to define these).

The way in which people prepare to be teachers varies greatly across countries. In some, such as Poland, it is University based and relatively academic. It is teacher education. In others such as England, it is school based and very much to do with learning the practicalities of teaching. It is teacher training. The continuing professional development of teachers also varies across the world. Some systems, such as in Canada and Australia generally, make it a priority and expectations and resources are relatively high. Elsewhere, it is lower down the educational pecking order and is very much up to the individual's motivation and resources. Furthermore, the professional development of school leaders is often *hit and miss*. England, for example, has had a National College for Teaching and Leadership of Schools and Children's Services for 12 years, and all head teachers have to graduate on a program (the *National Professional Qualification for Headship*) to be able to become a school leader. In Spain, preparation and continuing development for school leadership is more ad hoc. Obviously, these wide variations are a reflection of culture and context and the value placed upon teachers, their leaders, and on their development. This chapter argues that teacher and teacher leader development are essential and that there are ways for this to be carried out both efficiently and effectively. It examines the why, what, when, how, and where of such professional learning and also suggests ways to evaluate its effect.

WHY DO WE NEED TO DEVELOP OUR TEACHERS AND LEADERS?

> "Teaching…anyone can do it."
> "Those who cannot actually do anything become teachers. Those who cannot teach become teacher trainers."

There are a number of myths around teaching and teacher training that need to be shattered, and they quickly are for any non-teacher who has stood in front of a class of 30 adolescent students and tried to get them to learn. They soon realize that an effective teacher has very high-level skills. Teachers need some innate qualities that have to be constantly built upon to get and keep their expertise at a high level. Why is it important to do this? First, education makes a difference to people's life chances and is a significant key to our future. It is directly linked to economic success, the status of people in society, health, and well-being. As both Bill Clinton (United States) and Tony Blair (UK) said when they were their countries' leaders: "*Education, education, education.*" This mantra is repeated across the world from Rwanda to Chile, from Vietnam to New Zealand. If you want to get on, get educated.

Second, the education business is an ongoing process because our world is increasingly dynamic, interconnected, and complex. Political, economic, environmental, and technological changes are too great to list here, and they are continuing at a faster and faster pace. No longer can an individual's education be frozen in time unless he or she really does wish to be left out in the cold. Education is the key to not just surviving in the world. It is also the key to thriving in the world. We all need to know this.

Third, within our current systems of learning (and this may change) teachers are our main (and the most expensive) resource in the education process. If this resource is wasted, we are negligent. Research by, for example, Hattie (2009) based on the work of Viviane et al. (2009) shows the importance of effective continuous professional development in terms of improving students' learning. This needs to be ongoing, and trainee teachers' first experiences are critical in terms of their subsequent attitude and motivation to their chosen profession. We need to invest financial capital to enhance this important human capital.

Fourth, those responsible to schools need to lead and manage this valuable resource to have the maximum beneficial effect for our students. This is a moral imperative. Head teachers are the main source of leadership in their schools, and they contribute to student learning through a combination and accumulation of strategies and actions, such as encouraging the use of data and research and improving teaching policies and practices (Day et al. 2010)

Questions:

- Do our teachers realize and understand the significance of what they are doing?
- Is this understood, explained, and reinforced by those who lead and manage them?

WHAT IS OR SHOULD BE TEACHER TRAINING OR EDUCATION OR PROFESSIONAL DEVELOPMENT?

Language is important when considering the different approaches to developing teachers. "*Training*" is functional, task-focused, and, to some extent, routinized. At its most extreme, it reflects a Tayloresque view of the world proposing "*one best way*" to get the job done. To some extent, it is done to you and can be transactional. "*Education*" takes a broader view of teaching and learning. The word "*education*" is derived from the Latin word "*educare*" meaning to *bring or draw out* that which is *within*. Education thus means bringing out the values and talents within human beings and translating them into action. To be educated as a teacher means knowing and understanding the philosophical, psychological, sociological, and political aspects of the process as well as the pedagogy. It is complex, dynamic, non-routinized, creative, and relational, and there are many different ways to succeed, each of which has challenges and value. "*Development*" is ongoing and, in a profession such as teaching, it is a personal as well as an organizational responsibility.

Alexander puts a current context on the use of these concepts in different international systems. In England, "Teacher training requirements have shifted from a specification of the kinds of knowledge which it is assumed that teachers need, to instrumental but not very specific accounts of skill allied to policy-driven information which teachers are expected to know and which they are expected to comply. In contrast, European comparisons open up the well-developed fields of pedagogy and didactics, while transatlantic research has produced taxonomies which include both different domains of professional knowledge (of children, curriculum, pedagogy, pedagogical content, aims, contexts and so on) and professional knowledge of different kinds (public and personal, disciplinary and craft)" (2010, p. 430).

To be an effective teacher and leader of teachers and students one needs a bit of training, education, and continuous development. For example, a person who wants to become an effective teacher needs to be told how to do the basics, such as manage student behavior, yet also needs to understand theories of motivation and group dynamics and to continue to develop their own approach. Thus, teacher development involves an appropriate knowledge and understanding of theory and research, policy, and practice as well as the way in which these are interrelated (see Figure 93.1).

Perhaps a useful framework for thinking about professional development, including that of teachers and leaders is illustrated in Figure 93.2.

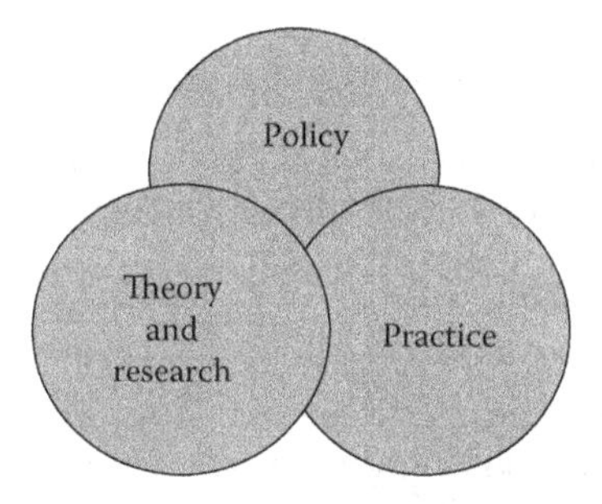

FIGURE 93.1 Areas of teacher knowledge and understanding.

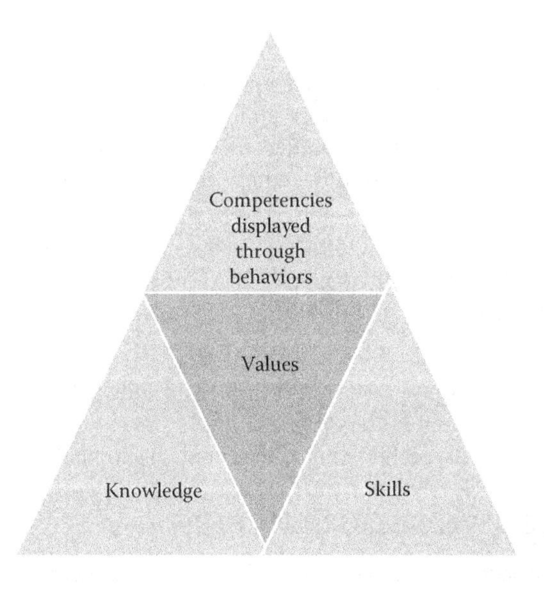

FIGURE 93.2 A framework for professional development.

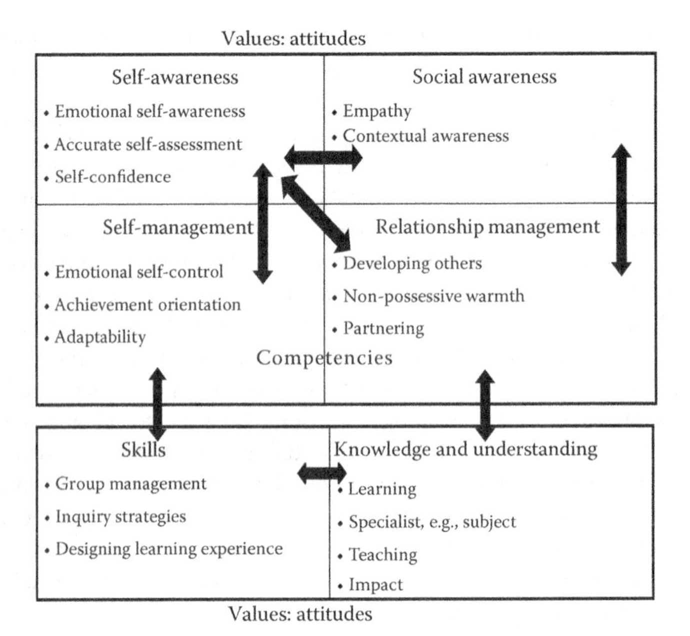

FIGURE 93.3 What do teachers need to know? (Based on the National College Competency Framework.)

These components (which, in combination, may be seen as *expertise*) are unpicked in Figure 93.3. In the case of teaching, the *knowledge* that is required is that of learning and teaching. This pedagogy may be seen as the correct use of instructive strategies or, in the case of critical pedagogy (Freire 1998), the instructor's own philosophical beliefs of instruction are harbored and governed by the pupil's background knowledge and experience, situation, and environment as well as learning goals set by the student and teacher. An effective teacher needs to know about the area in which they may be a specialist, e.g., the subject as well as knowing how to measure the "impact" of teaching on learning.

The *skills* required by a teacher are, in broad terms, those of group management, developing inquiry strategies, and designing learning experiences. The *competencies* (by which is meant habitual behaviors) needed by an effective teacher are those of emotional intelligence (Bar-On 1997; Goleman 2010; Boyatzis 1999). These are self-awareness (the starting point for emotional intelligence), self-management, social awareness, and relationship management. Each of these competency areas has specific competencies that teachers need to develop. Finally teachers need *attitude.* They need passion, self-motivation, resilience, stamina, and a desire to understand learning (metacognition) and to keep on keeping on learning, thus modeling life-long learning. These attitudes are in turn founded on *values*, particularly those of equity and social justice (Espinosa 2007).

Teaching is not a science in the sense that it is not completely rational, logical, and predictable. A highly effective teacher also manages himself or herself as an artist and is comfortable with complexity, diversity, unpredictability, creativity, and the unexpected. It is a profession dealing in the magic of human relationships.

Questions:

- How well do teachers feel prepared in terms of their knowledge, skills, and attitudes for teaching, and how can school leaders improve this?
- How is the development of teachers linked to school development plans?

WHEN SHOULD WE DEVELOP OUR TEACHERS?

The question of when we should develop our teachers may seem to be rhetorical because the answer may seem obvious—before they actually face their own class and ongoing while they are teaching.

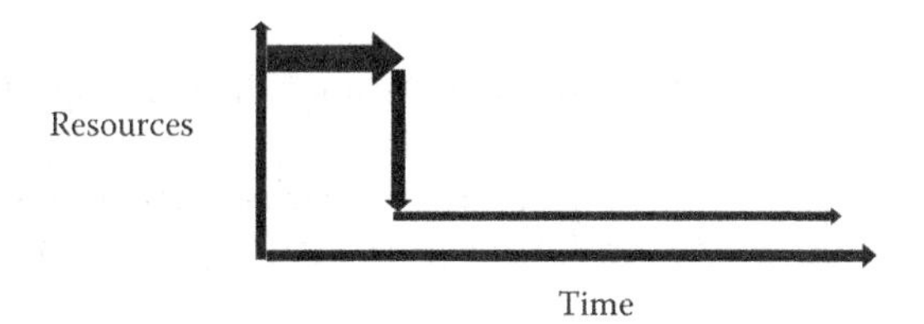

FIGURE 93.4 Patterns of professional development.

In reality, we often put most resources into the early stages of teacher training and far less with experienced teachers. Sometime those who need it most get less. The pattern is shown in Figure 93.4.

Moreover, the balance between training, education, and development may change with, rightly, the first dominating in the early stages (Alexander 2010). There can be great value for both teacher and learner in the use of skilled, experienced teachers, teaching trainees, and novice entrants to the profession. In England, these are called "*advanced skill teachers*" (ASTs) or, more recently, "*specialized leaders in education*" (SLE).

Questions:

- When do most of our efforts go into developing our teachers?
- Is professional development sustained over an individual teacher's career?
- Does its emphasis in terms of content and process change as a teacher's career develops?

HOW SHOULD WE DEVELOP OUR TEACHERS?

Education is the learning business. As leaders in education, we are the experts. We now know a great deal about learning and teaching, and we need to be more confident about articulating and proclaiming this. To what extent do we actually apply this to our practice of developing our teachers—our most valuable resource, by far, that we have to educate our young people? Learning and teaching start with motivation of the participants in the relationship. If learners and teachers are highly motivated (albeit by power, potential income, the desire to please others or whatever), they are more likely to develop traits, a self-image, or a social role that will enable them to develop the skills and knowledge to teach and learn successfully.

We also know that there are patterns to the chronology of learning. Kolb (1984, 1999), inspired by the work of Lewin (1942), outlines a cycle of learning. While learning and, thus, teaching are not neat and tidy in terms of timing, the model does help our understanding (Figure 93.5).

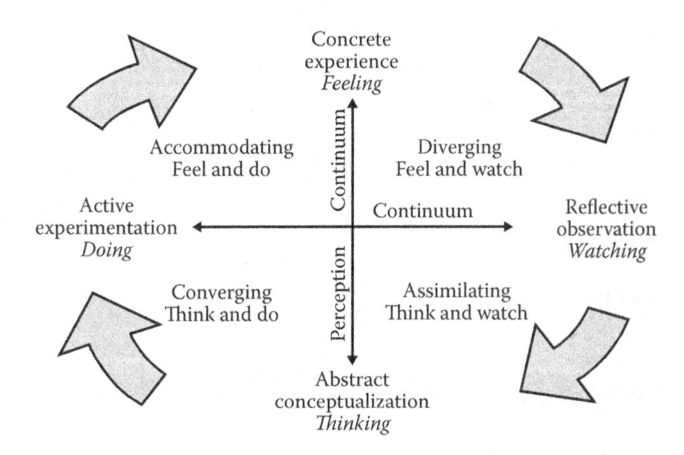

FIGURE 93.5 Kolb's learning cycle.

Glaser (1963) and Glaser and Strauss (1967) present a similar experiential learning cycle from the point of view of what the teacher needs to be doing. Again, although simplistic as presented below, we know that, in general, the process works (Figure 93.6).

A more refined approach (Precey 2008; Precey and Jackson 2009) based on learning theory looks at the components of transformational or deep learning (Figure 93.7).

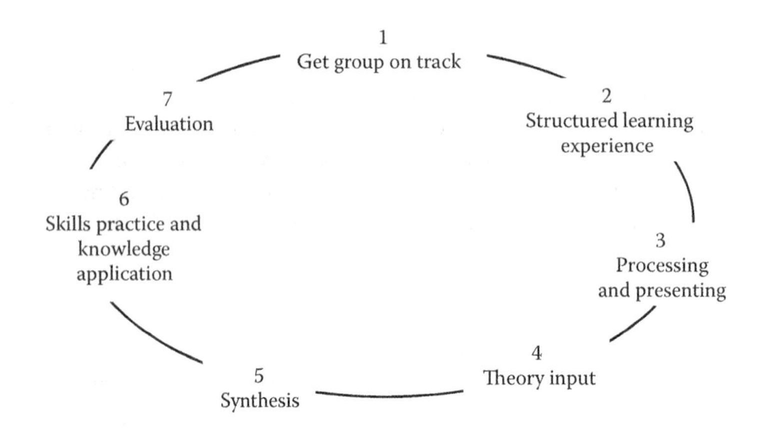

FIGURE 93.6 Experiential learning cycle (adapted from Glaser).

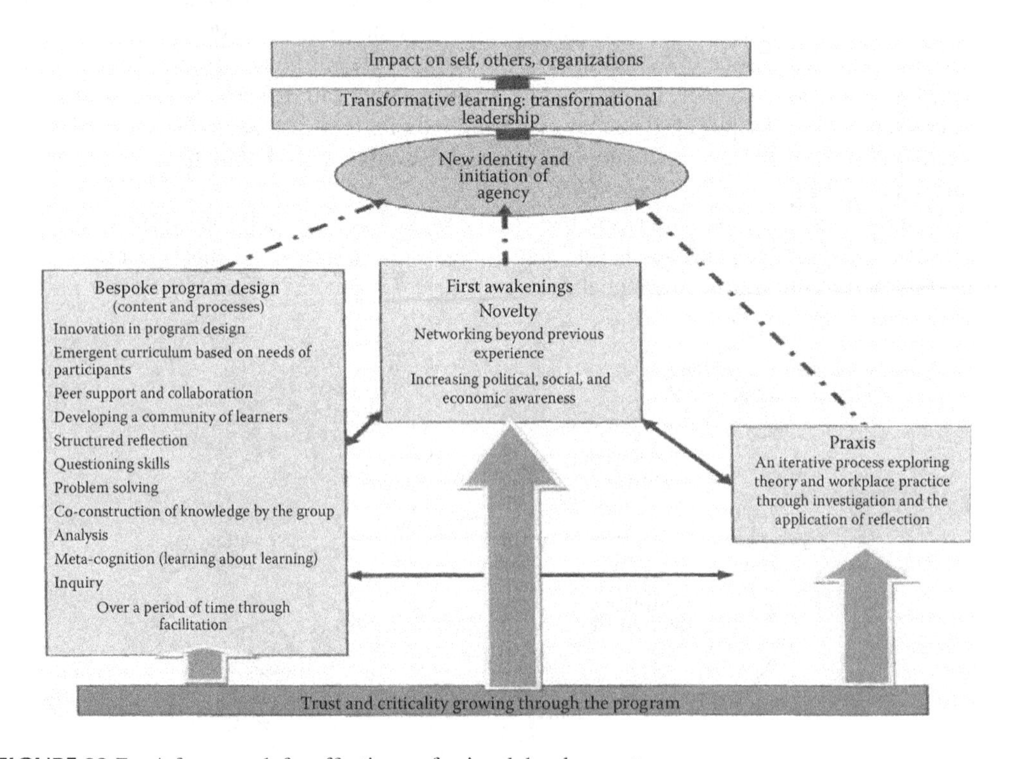

FIGURE 93.7 A framework for effective professional development.

Initially, the teacher needs to design, prepare, deliver, and evaluate a program *tailor made* to the group of learners. This program needs to contain as many as possible of the following elements:

- An emerging curriculum in that although there is a planned curriculum, it will be flexible enough both in terms of process and content to respond to real student needs.
- The focus continually on participant needs, not those of the teacher or even the organization.
- There are mechanisms developed for peer support, such as action learning sets, peer coaching.
- There is collaboration between students.
- A community of learners is cultivated by the teacher in order to move from dependent through independent to interdependent learning.
- Wherever possible, there is attention given to meta-cognition (learning about learning). In the learning business, we need to continually develop our knowledge, understanding, and skills of this, the core of our business.
- The teacher seeks to develop students questioning and inquiry to keep asking "Why?" and "So what?" in order to become critical (in a true sense) teachers.
- Practice in problem solving will enable teachers, whether trainees or experienced, to successfully face the growing challenges in education.
- There is co-construction of knowledge to create new meaning in recognition that learning is a social activity as well as the fact that previous paradigms may not be for this purpose in the future.
- Analysis is an important component of teacher development and is linked to inquiry and criticality.

Teacher education programs also need to be founded on *praxis,* seeking to use relevant theory and research to interrogate practice and vice versa. Creating spaces for active reflection can lead to new ways of seeing the world (*awakenings*). This, in turn, often enables the learning teacher to develop a new professional identity, enhancing his or her sense of wanting to make a real difference to his or her students (*agency*) most likely to have the greatest effect (*impact*). This framework has been tested on many leadership programs and in many countries including over six years with an Erasmus Leadership Intensive Program with participants from Poland, Turkey, Ireland, Norway, Spain, and England and on other programs in China, Pakistan, and England. In the light of this, it has been refined to provide a structure for the planning, preparation, facilitation, and evaluation of professional development.

Questions:

- How thorough and well planned are the ways in which our teachers learn both at the start of their careers and continuously throughout their careers?
- To what extent as school leaders do we use our knowledge about learning to develop our teachers?
- Do we keep that knowledge up to date? If so, how?

WHERE SHOULD OUR TEACHER DEVELOPMENT TAKE PLACE?

Some countries organize teacher development programs that are university based, particularly those involving initial teacher training (ITT). For example, for ITT in one university in Poland, students have 300 hours of academic subjects (psychology, pedagogy, and methodology of teaching), which is university based. They also have 150 hours of school practice, and usually they go to schools in September in large groups and mainly observe or/and teach. China has an emphasis on trainees observing experienced, skillful teachers just as a trainee chef just observes a master chef for a full year before they are allowed to cook. In contrast, in England, there is a growing shift toward school-based ITT and continuing professional development albeit in partnership with other agencies including universities. For example the *Graduate Teacher Program* (GTP) enables participants to gain their teaching qualification while working and training in a paid teacher role. The school pays the individual for this on-the-job training, which is one year full time. *Teach First* is a two-year teacher training and leadership development program, recruiting high-quality graduates to work in challenging, urban schools in five regions in England. The training salary comes from the host schools, and participants spend only six weeks out of school at Summer Institute training. Now, as a result of new legislation in England (2011), teacher training and professional development are following a teaching hospital model in the health service. An example of this is Old Ford Primary School in Bow in East London, which was recently (2011) designated as a teaching school and is now responsible for coordinating and planning professional developments as indicated below (see Figure 93.7).

Clearly, one can see benefits for both a university-based program (the provision of reflection spaces and times, criticality, challenge, deepening learning, connecting with theory and policy) as well as one located primarily in a school—the place of work (practical on-the-job learning). Learning theory would suggest a balanced combination of both is helpful. Who organizes this may be down to local politics, economics, and professional relationships.

Questions:

- Where does teacher development take place?
- Is the balance of locations right to maximize learning?
- If there is a school experience, how well is that led and managed?

HOW DO WE KNOW OUR TEACHER DEVELOPMENT IS WORKING?

The ultimate test of teacher professional development is the response to the question: "*Does it help to improve students' learning or not?*" In order to stand any chance of answering that difficult question (the Holy Grail of Education) satisfactorily, a number of factors need to be in place:

- What are we measuring the success or otherwise of teacher development against? National standards can be helpful here, but they must not reduce all that we measure to that which is easily measurable, e.g., simple test results, rather than considering the less simplistic concept of *expertise* (Alexander 2010, p. 410). Appraisal, target-setting, and performance management can, where skillfully led, be powerful to help teachers focus on improving student learning and capturing evidence that evaluates this.
- How do we measure it? Is there ongoing, formative monitoring, or do we just try to measure the effect at the end of the process in a summative way? Is it imposed, or is there an element of self-criticality for potentially deeper, more sustained development with peer observation and support.
- When are we measuring impact? If it is immediately after the teacher's learning experience, it may soon dissipate. Sustained long-term impact is more important, and measuring this requires an investment and commitment over time. This is often lacking in education systems, which may, for political reasons, be focused on short-term impact.

Questions:

- What systems do you have in place to check the quality of teaching in your school?
- How do you know what needs to be developed further?

LEADERSHIP DEVELOPMENT

If, as research suggests, school leaders are so important in relation to the quality of teaching and learning (Day et al. 2010), what can leaders do to encourage effective professional development of their staff, and how are leaders themselves developed? On the former, they can seek to influence external policy and to shape their own internal policies to develop best practice as outlined. In terms of their own development as the lead learner in a learning organization, it is important that they prepare for the realities of educational leadership today. School leaders need to understand and apply complexity theory to their lives. Running a school today is increasingly challenging as the dynamics of the political, social, and economic realities of schools change faster and faster. They need to develop so that they recognize critical points in their organization's development (points of *bifurcation*). Schools are complex systems that exist on the "edge of chaos." They do not enjoy equilibrium. As we move from one state to another, schools reorganize themselves at many levels and change and emerge organically as different (*self-organization* and *emergence*). School leaders need to learn any formal regularities may emerge in this turbulence (*attractor states* and *recursive symmetries*). They need to recognize and combat *lock-in* as schools can (often?) become fixed in their ways (inertia) and are less adaptable and able to change. Such development needs to assist school leaders to become comfortable living with ambiguity. Effective leaders have to learn to develop resilience and take care of themselves as well as others. The job of educational leaders is even more tough (intellectually, emotionally, and physically) and potentially even more rewarding than it has ever been.

In some countries, e.g., Poland, there is no national system of head teacher training. In England, there is a compulsory well-resourced program for aspiring head teachers and leadership at all levels from middle leaders to systems leaders through its National College.

Effective leadership development is underpinned by the same components as for teacher development (see Figures 93.1, 93.2, and 93.3). The professional development framework (Figure 93.7) has a similar structure for leaders as for teachers except that a school leader's knowledge needs to focus on leadership and management, and a classroom teacher needs to know about their subject(s) (Figures 93.8 and 93.9).

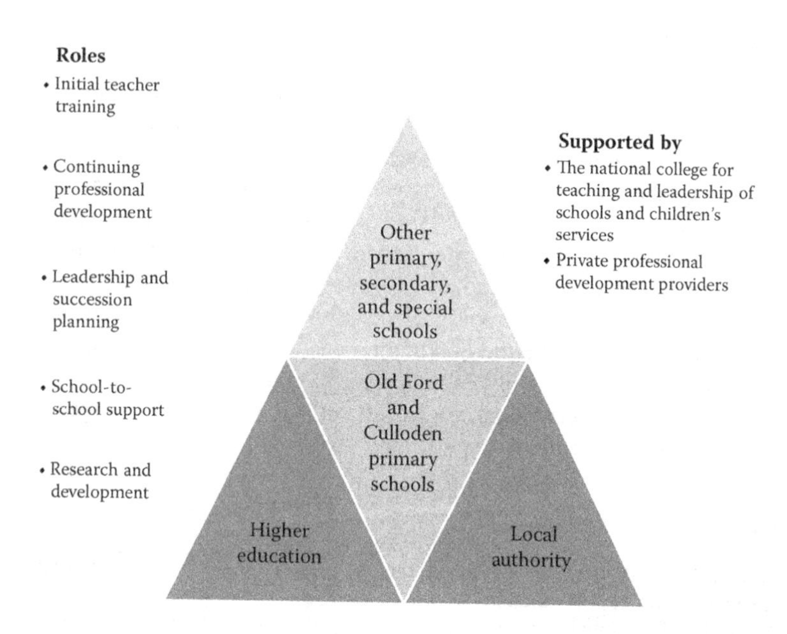

FIGURE 93.8 Old Ford Primary School, London: A new teaching school.

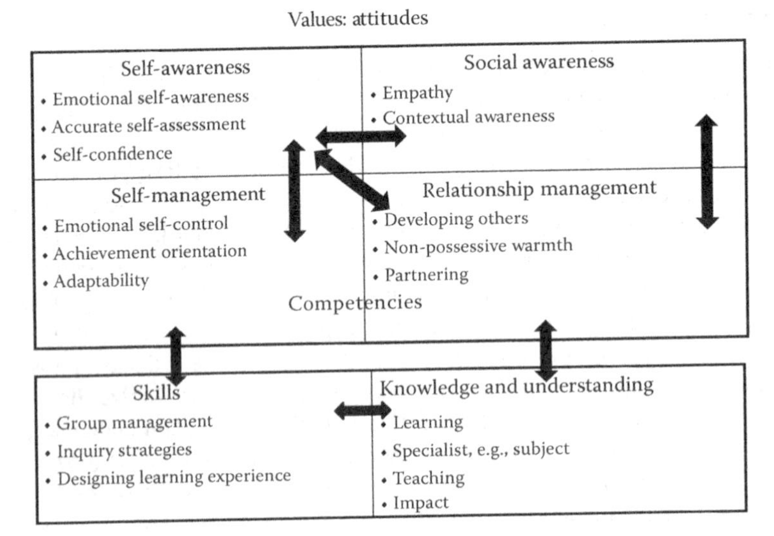

FIGURE 93.9 What do school leaders need to know? (Based on the National College Competency Framework.)

Questions:

- How well do you feel school leaders in your country are prepared and supported in terms of leading and managing the professional development of their teachers? What is working well? What could be even better?
- How well do you feel school leaders in your country are prepared and continue to be supported in relation to their own professional development?

CONCLUSION

This chapter has sought to explore and suggest why teacher and leader development are so important; what they should be like; when, where, and how they should take place; and, most importantly, the impact this might have. It has also raised issues to do with how our leaders learn to be leaders and how this learning should continue. Can anyone teach? Can anyone lead? No. Both require high levels of expertise—competencies, skills, and knowledge developed and honed over time. Education makes a difference to people's life chances and is a significant key to our future. This is an ongoing process in our dynamic, rapidly changing world. Teachers are our main (and most expensive) resource in the education process. Head teachers need to lead and manage this valuable resource and themselves to have the maximum beneficial effect for our students. Continuing professional development is not a luxury in education. It is an absolute necessity. To neglect this shortchanges our children. If, as educators, we are left out in the cold, we will indeed become frozen in time, and our students deserve better than this.

REFERENCES

Alexander, R. (ed) (2010): Children, their world and their education. Final report and recommendations of the Cambridge Primary Review. Routledge.

Bar-On, R. (1997): The Emotional Quotient Inventory (EQ-i): A test of emotional intelligence. Toronto, Canada, Multi-Health Systems Inc.

Boyatzis, R. E. (1999): Self-directed change and learning as a necessary meta-competency for success and effectiveness in the 21st century. In Sims, R. and Veres, J. G. (eds.). Keys to employee success in the coming decades. Westport, CN: Greenwood Publishing.
Day, C., Sammons, P., Hopkins, D., Harris, A., Leithwood, K., Gu, Q., and Brown, E. (2010): 10 Strong claims about successful school leadership (National College).
Espinosa, O. (2007): Solving the equity-equality conceptual dilemma: A new model for analysis of the educational process, Educational Research, vol 49, no 4, December 2007, pp. 343–363.
Freire, P. (1998): Politics and education UCLA Latin American studies; v. 83. Los Angeles, UCLA Latin American Center Publications.
Glaser, B. G. and Strauss, A. L. (1967): The discovery of grounded theory: Strategies for qualitative research. Chicago: Aldine Publishing Company.
Glaser, R. (1963): Instructional technology and the measurement of learning outcomes: Some questions. American Psychologist, vol 18, no 8, pp. 519–521.
Goleman, D. (2010): Leadership that gets results, Harvard Business Review, April 2010.
Hattie, J. (2009): Visible learning: A synthesis of over 800 meta-analyses in Education. Routledge.
Kolb, D. (1984): Experiential learning: Experience as the source of learning and development. New Jersey: Prentice-Hall.
Kolb, D. (1999): The Kolb Learning Style Inventory, Version 3. Boston: Hay Group.
Lewin, K. (1942): Field theory and learning in D. Cartwright (ed.) Field theory in social science: Selected theoretical papers, London; Social Science Paperbacks, 1951.
Precey, R. (2008): Transformational school leadership development: What works? Journal of Research in Professional Development, vol 1, pp. 4–20.
Precey, R. and Jackson, C. (2009): Transformational learning for transformational leadership. Professional Development Today, vol 12, pp. 46–51.
Viviane, R., Hohep, M., and Lloyd, C. (2009): School leadership and student outcomes: Identifying what works and why best evidence synthesis, (New Zealand Ministry of Education).

94 Creating Quality Teaching Professionals for Schools
Organic Professionalism

Trevor Davies

CONTENTS

INTRODUCTION

This chapter considers briefly some goals of education in the 21st century and reflects on whether the methods of training teachers in England are fit for this purpose. The issues associated with the development of quality teaching professionals are comparatively considered with some examples of where good and interesting practice might lie in the world.

GLOBAL CONTEXT FOR EDUCATIONAL CHALLENGE

The issues that challenge a more globalized environment are the following:

1. Failure in developing education systems to respond to the changing nature of society in developing and developed societies. See Tickly (2001), Marginson and Mollis (2002), and Crossley (2008).
2. The shifting identities of migrating peoples and populations. See Ugglee (2008).
3. The inequities of race, gender, disability, and marginalized groups. See Gundara (2003) and McKencie and Scheurich (2008).
4. The global inability in large parts of the world to create sustainable employment and social integration. See Castells (1998) and Falk (1995).
5. Intensification of technological and entrepreneurial enterprise on a multinational basis and its implications. See Castells (2001) and Barber (1995, 2007).

6. The reconfiguring of international global relationships in terms of power and the deliberative exercise of voice. See Landorf et al. (2008).
7. The complexity of the world for young people and the claims of multiple identity and citizenship. See Kress (2008).

It is against this background of issues that there has been an international response to provide new educational opportunities for future generations. It goes further than just the literacy and numeracy programs initiated by the United Nations as basic concerns of the millennium program. Rather there will be work to be completed to close the inequities in learning that will still be there even after basic literacy and numeracy in the developing countries is achieved. The creation of quality teaching professionals is central to developing new educational structures that are responsive, resilient, and forward looking.

TEACHER PROFESSIONALISM: THE BUILDING BLOCKS FOR CREATING DYNAMIC RELATIONSHIPS

Politicians in England have largely taken control of educational reform, and across the planet have been striving to promote advancement, largely linked to ultimately achieving economic success. However, as Davies (2006) found in a pan-European small-scale research study into the role of creativity in European education systems, it is a "holy grail" with little evidence for its existence in educational settings but with teachers desiring it. "Managerial" approaches to education have reduced the possibilities of teachers and learners building relationships in natural and empowering ways.

Essentially, what has been factored out of the professional equation is trust in the qualities a professional holds based upon their knowledge, skills, and capabilities (Frowe 2005). Decisions of any kind relating to public education can be held up for scrutiny and publically examined. Of course, it is right and proper that the uses of public money in education should be held up for scrutiny but using methods that are "fit for purpose" and recognizing the worth and value of professional knowledge and skill. The extent of the situations in schools in which teachers' decision making is trusted has been intentionally minimized through successive policy-making over the last 20 years. The culture of educational institutions, through the way they are governed, clearly convey a sense of deskilled and untrusted professionalism on the part of teachers.

Professionals in any area clearly need to be regulated and held accountable for their actions, but what is at the heart of good teaching? It is the trust that a learner has in the teacher's ability to prepare him or her for his or her future world—not the present one. Trust always involves vulnerability (Frowe 2005, p. 43), but if the basis for trust is built solidly into the way teachers are trained, opportunities for abrogation of trust are minimized and the discretionary powers given to a professional are amply rewarded as in other international high-performing systems, such as the Finnish one. Frowe (p. 52) also argues that innovation and experimentation is encouraged by the freedom granted by trust, leading to mutually beneficial opportunities that would be missed in the absence of it.

Judgments about the work of learners and teachers are made by many agencies, for example, educationalists, parents, government, industry and commerce, and other lobbies who interpret domains in ways that serve their interests and value systems. Each makes judgments based on what they perceive to be in the best interests of the learners in relation to their future roles as citizens and participants in complex communities and society as a whole. However, teachers are the mediators of learning in the settings of schools and classrooms. Teachers need to be in possession of the tools to plan meaningful classroom experiences to equip a "learning culture." They then need to be participative and reflective about the classroom activities in order to create the relationships that enable learners to be equipped with competencies that enable them to face uncertain futures, calculate and manage risks, and turn "problems into opportunities" in their lives.

Aubusson et al. (2008) recognize how important close nurturing is to generate and maintain professional learning communities (p. 134). Teacher (professional) learning is at the heart of this coupled with teacher research. Because of the complexity of schools and teacher communities, there are no simple methods to instigate or drive the process. Aubusson et al. (2008) describe how "action learning" that places an emphasis on inquiry-led action can help schools build the capacity to improve. Utilizing such approaches in a bolt-on way, however, can never achieve sustainable change. Within the context of an extensive literature of change management, such an approach can stimulate and contribute to sustainable change over time, resulting in large-scale shifts in values and attitudes.

BUILDING FOR THE FUTURE

What are the demands of new learning scenarios? It should, of course, be recognized that, in terms of longevity of life, children born today may easily achieve life-spans unknown to us today—certainly 100+. This is in the context of the planet not imploding due to some climate catastrophe, internecine warfare, or unhealthy living context. But it does sharply focus education into teaching that the nature of work and sustainable relationships is a long one.

This would mean the reconstructing of relationships is necessary because of the difficulties in preparing students not only for their formal education but for lifelong and informal education. Changes in society and global penetration gaining intensity in certain areas has led some researchers to see it as an age of uncertainty and insecurity. This provides an impetus for reform and a focus on the need to live together. Indeed Carneiro and Draxler (2008) feel that

> "Practical application of this principle in school systems is often assumed, neglected or limited to the early years of schooling before academic and practical learning for the economic sector begins" (p. 149).

Because of all these factors, it is evident that we need to look again at the operation of schooling and the relationships as found in schools. Recent research has stressed the need to look at the formal working time of students at school as well as the informal and contextual area of students' lives. Further, this must be seen in a context that ranges in dimension from the local to the global. The intensity of technological communication systems, 24-hour news, and different population and migratory groupings have led to a more multilayered society. There is also the preconception that there is a need to reform schools themselves in order to understand and act on these influences and change the very nature of the learning experience. At the heart of educational reform is redefining relationships and to define the framework with which that takes place—redefining professionalism for the educators. Without the educators, education will not happen.

CENTRALITY OF THE TEACHER

It is interesting to note across the planet that countries uniformly recognize the importance of the role of the teacher in reforming education, but different systems value them less or more in different ways, depending upon circumstances; they can also structure the profession less or more and define professionalism in different ways. It is important to note that every politician realizes that by controlling education you can control the future of your society. Perhaps a notion that doesn't crop up sufficiently in public discourse is that education is about preparing learners for the future, but no one knows what the future will bring, not even tomorrow! So who has the right to control education? Surely, it is the property of society, and in such times, a share between individual societies and global societies—a strong argument for including the global dimension in the education of every learner. So if the teacher is central, what can we say about a teacher's training? And should it be teacher training? Or teacher education? In this world in which we don't know what tomorrow brings, surely we want teachers to be effective decision makers with well-developed skills in critical thinking, creative thinking, and evaluative thinking who are centrally able to undertake and understand research methods and use evidence-based approaches to their work; this must be more about education than training. It also implies a continuous process of professional development—not a short spurt of initial teacher training (ITT) then a "sentence for life" in the classroom with the "point of exit" knowledge and skills used to create lesson plans forevermore. There is a need for continuous professional development that continues to challenge methodologies and processes and public local and global agendas as they evolve.

I would like to argue that teacher professionalism forms the building blocks for creating dynamic relationships of all kinds with stakeholders. Politicians have largely taken control of educational reform and across the planet have been striving to promote advancement, largely linked to achieving ultimate economic success. Aims are stated and orders are issued, but often without essential support mechanisms, achievements are hollow and often nonexistent.

REDEFINING PROFESSIONALISM

Evans (2008) argues that a hallmark of the modern era in redefining professionalism has been to set narrower, more procedurally based boundaries for teachers (that many would argue has de-professionalized teachers). Beck (2008) discusses in detail whether modern recent educational developments in England have been about re-professionalizing or de-professionalizing (pp. 119–143) and that an aim has been to create a compliant profession with the "discourse of standards" at the center with the intention to marginalize or even silence competing ideas about educational development and reform (p. 138). Frowe (2005, p. 49) believes that there has been an attempt to capture, in a series of propositions, something that cannot be exhaustively captured.

This restricts the number and levels of "teacher judgment" required. Evans (2008, p. 31) argues that educational reform in England sets out to change the professionalism of teachers and has succeeded in doing so, but it has only succeeded in a "functional way" and achieved limited success in raising standards and achievement within international settings. This should not be surprising as, in redefining professionalism, government demands that it controls the "fine detail" responses that constitute the day-to-day activities in classrooms. Teachers, therefore, constantly need to update themselves on what their expected response to these situations should be. It is this updating of their knowledge of "government-approved policy" that has comprised "accepted professional development." Usually undertaken by "approved agencies" to ensure consistency in the delivery of accepted policy. Hoyle and Wallace (2005, pp. 4–5), however, note that the expected success with government reform has not been realized and has led to widespread dissatisfaction. Typical of the procedural tools adopted to force school improvement has been school inspection. According to Ehren and Visscher (2008, p. 224), these have typically enabled schools to make small changes to their procedures but have been ineffective in facilitating the more complex changes usually required to "make a difference."

Evans (2008, p. 35) stresses that central to raising standards and increasing society growth capacity, must be professional development for teachers. This is echoed by Brighouse (2008), who summarizes the ways in which successive recent governments have spent most time and energy engineering ways to effect and control entry into the profession and have largely ignored the development needs of the existing teaching force.

Kennedy (2005) argues the case vigorously that teachers' professionalism has been forced into narrow channels of reactive responsibility while, at the same time, there are, in England, wider expectations that teachers should work with other stakeholders in society to "rectify society's ills." Agendas like those of "Every Child Matters" in England required teachers to act and engage with other professionals from other areas of public service until the act was repealed by the current coalition government. Teachers' professionalism does not prepare them or allow them to build the knowledge base of how to do it. In fact, schools have become more "isolationist and inward looking" since becoming so subject to the overloading and over-regulation from government. Kennedy argues that "civic professionalism" needs to be taken seriously, and civic knowledge and values need to be regarded as a core requirement for teacher education. Teachers and learners need to be part of their communities and share and collaborate over civic missions.

In examining new values of professionalism, Kogan (2005) feels that professionals have a duty to equip themselves with knowledge of all aspects of education and need to refresh this knowledge through professional training and contacts beyond their own school. Knowledge and participation in professional tasks substantiated a sense of individual identity for both new entrants and long-serving teachers, and this does not now happen in the English education system even though there are currently glimpses that the coalition government recognizes the need for teachers to have wider perspectives on education even though it is hampered by a lack of resources in the public sector to take any meaningful action.

A recent ambition expressed by government is that teaching should become a masters level profession reflecting a value placed on independence of thought and action. It is a premise for research in social science that it informs action that implies the need for those empowered to take action to understand the processes and mechanisms of research. For teachers, without an understanding of what constitutes evidence, judgment formation is randomized and beliefs and values have no substance (Sharpe 2005).

There needs to be a deepening of understanding as to what needs to be delivered as a curriculum to fashion and claim well-being for all in the 21st century. Many people are even more concerned about this since the beginning of the current economic crisis. The concept of well-being is seen as one that not only provides basic goods but rather those needed for a fuller life in line with Maslow's (1943) hierarchy of needs. Amartya Sen (1993) has developed this concept in a modern context and made the distinction between basic functioning of well-being and that required for a much fuller life. This work on well-being shows the impoverishment of allowing basic functioning to be a reliable guide to a life, which allows for options for choice.

This notion of well-being could influence a much broader definition of schooling than could be provided by the early school effectiveness research. This is to recognize the importance of the need to look more carefully at school reform not just in terms of the relationship between measurable outputs matched against pecuniary inputs but looked at in terms of promoting 'human' capacities. The context must therefore be set where society understands and values the development of a progressive, futures-oriented curriculum for learning, the hallmarks of which include local, national, and international opportunities. So how does this impact upon how we recruit, train, support, and maintain the professional growth of teachers?

TRAINING TO BE A TEACHER IN ENGLAND: AN OVERVIEW

Education is an expensive business, and while schools are seen as an essential part of this process, money spent on training teachers is, in certain societies such as England, currently seen as too expensive, particularly in times of economic hardship such as these. In the Coalition Government White Paper, the idea is promulgated in England that "higher education" education departments (HEIs) promote tired old theories of education generally and learning and the curriculum in particular and that the training of teachers should be more school-based. In the previous jurisdiction, when the Labor Party was in control, teacher trainer roles were defined in the following ways:

- Teacher training policy implementation was managed by a separate government agency—the Teacher Development Agency—which actively managed the processes that HEIs were used to implement teacher training, using funding and the Ofsted (inspection) process as key elements for control.
- A curriculum was set (comprising 36 "standards") that all HEIs were obliged to follow; all trainees were required to be assessed against and pass all of the standards before being awarded Qualified Teacher Status (QTS). These were in the general categories of professional attributes, professional knowledge, and professional skills.
- Every trainee was legally required to spend a set amount of time (approximately half of a training year) in school-based training.
- School mentors were to be trained by HEI staff.
- HEI tutors acted as quality assurance agents to hire and fire schools, mentors, and trainees.

The Teacher Development Agency (TDA) now operates with a greatly diminished role, and the current move is to shift the locus for training teachers onto schools, which will have influence over the way in which they work with HEIs rather than HEIs being able to select the schools they work with. Trainees are likely to be required to spend an even greater proportion of their time in schools.

STRUCTURES FOR TRAINING TEACHERS

It is useful to note that there are three main ways of training to teach in the UK: One is an undergraduate degree currently (often lasting four years but being slimmed to three as a result of fees introduction). Very few currently take this route.

A post-graduate certificate in education (PGCE), with which graduates in any subject are recruited and given a one-year course constructed around pedagogy, are HEI–based courses.

An employment-based route exists—the Graduate Teacher Program (GTP)—in which trainees are employed by the school but managed by HEIs and SCITS (school consortia). Other routes exist as well, but are currently minority providers. All routes require a mandated period of school-based experience.

There are also a range of initiatives to entice highly qualified people from other professions and industry into the profession: "Teach First" and "Teach Next."

Through the Ofsted inspection of HEI providers, they have the power to adjust the numbers HEIs are allowed to train depending upon the measured quality of the outputs.

There are, of course, perceived achievements that have been linked to the UK system, such as the development of clear and transparent "institutional structures" for education and clearly defined progression routes in professional pathways. The Blair government—by focusing on the ways teachers "make a difference" through high-profile press publicity—reversed major teacher shortages even in shortage subject areas such as sciences, mathematics, and modern languages and generally was thought, even temporarily, to raise the professional status of teachers. It must be mentioned, however, that financial inducements were used to recruit and retain teachers during this period. At current times, however, these funding streams are being withdrawn, and next year, students will be charged to train as teachers, bearing the cost themselves in popular areas, such as humanities and arts.

CURRENT AGENDAS

Promises have also been made by the secretary of state that more training with school-based training schools be established as the secretary of state seeks to marginalize HEIs (the first ones are already in existence even though some backtracking has occurred, requiring them to have links with HEIs). At the same time, the first free schools have been opened in which groups of parents, charities, faith groups, or any other interested stakeholders can open a school anywhere and be funded from the central government. Like independent schools in the UK, there will be no requirement to appoint teachers who have formal Qualified Teacher Status (QTS) qualifications. A new initiative regarding recruitment, however, is proposed that all teachers should have a second-class degree or better in their first (undergraduate) degree. Interestingly, this would have precluded some of our best-known personalities and educational gurus in the UK, such as Carol Vorderman, maths adviser to the government; Michael Morpurgo, a famous children's author; and many others. This, of course, is another example of neoliberal thinking about education, implying that a straightforward Taylorist approach can be adopted using quantitative metrics to every aspect of education. Additionally, politicians have the right to micromanage professional issues using economic and populist arguments to justify their actions, and immediately, governments change and the rules and protocols are rewritten, often to adopt opposite positions and values. There is a refusal to recognize that reference to complexity and chaos theories give greater clues as to how to manage educational development for the future.

CULTURE OF EDUCATION

Achieving the right culture in education is something we all care about; we know education matters to every child, every parent, every teacher, and every employer. But the world continuously changes, and schools and teachers, like everyone else, are required to take risks to survive. If tutors in universities and teachers in schools are narrowly prescribed as to what decisions they can make, of course, the results will reflect a narrowly defined set of competencies and certainly not those required for global learning.

In England, the extent of the situations in schools in which teachers' decision-making is trusted has been intentionally minimized through successive policy-making over the last 20 years. The culture of educational institutions, through the way they are governed, clearly conveys a sense of deskilled and untrusted professionalism on the part of teachers. It is the trust that a learner has in the teacher's ability to prepare him or her for his or her future world—not the present one. Some clear implications can be discerned from the UNICEF report card published in 2007 in which the UK performed uniformly badly across the six dimensions of child well-being defined by UNICEF using OECD statistics compared with the other 21 richest nations of the world. The categories were material well-being, health and safety, educational well-being, family and peer relationships, behaviors and risks, and subjective well-being. Poland, incidentally, faired well in the educational well-being and behaviors and risk categories. The countries that fared best were the Nordic countries, which performed consistently well across the categories.

BEST PRACTICE ON THE GLOBAL STAGE

I would now like to turn to what happened at an educational summit that took place in March 2010 in the United States: The International Summit on the Teaching Profession. It was convened by the U.S. Department of Education, and the participants were brought together to identify and elaborate on best practices from around the world for recruiting, preparing, and supporting teachers in ways that effectively enhance the teaching profession and, ultimately, elevate student performance. Host organizations included the U.S. Department of Education, the Organization for Economic Cooperation and Development (OECD), Education International (EI), National Education Association (NEA), American Federation of Teachers (AFT), Council of Chief State School Officers (CCSSO), Asia Society; and WNET (a primary station of, and program provider to, the public broadcasting service [PBS]. WNET.ORG, formerly known as the Educational Broadcasting Corporation).

The themes discussed included teacher recruitment and preparation; development, support, and retention of teachers; teacher evaluation and compensation; and teacher engagement in education reform.

The participants were education ministers, leaders of education unions and organizations, and teachers from countries with high-performing and rapidly improving education systems. The countries represented included Belgium, Brazil, Canada, China (Shanghai), Denmark, Estonia, Finland, Germany, Hong Kong, Japan, Netherlands, Norway, Poland, Singapore, Slovenia, the United Kingdom, the United States, and others.

The summit was informed by a comprehensive background report prepared by Andreas Schleicher, Head of the Indicators and Analysis Division of the OECD's directorate for education. Data was included from Programme for International Student Assessment (PISA), Teaching and Learning International Survey (TALIS), other OECD reports, and a meeting of OECD ministers in November 2010 to inform the summit.

As regards teacher recruitment into ITT, it is unsurprising that the most successful countries use intelligent incentive measures, including pay levels, raising status, real career prospects for teachers, and teachers having responsibility to lead educational reform. For instance, teacher education helps teachers to become innovators and researchers (China, Shanghai, is the core of the Shanghai reform agenda). Singapore is notable for a comprehensive approach to identifying and nurturing teaching talent—different career pathways—master teacher, curriculum or research specialist, leader.

Finland made teaching a sought-after occupation and raised entry standards, offering a high degree of responsibility for teachers as "action researchers."

For teacher development, support, careers, and employment conditions; it was identified that collaborative activity between teachers improves practice. As does mentoring and teachers having conditions of employment and career prospects matched to their aspirations. Greater career diversity is possible in Australia, England, Wales, Ireland, and Quebec.

TALIS reports noted that longer term professional development is far more successful than other forms of professional development activity such as courses, peer observation, and education conferences, but of course, they are far less likely to be supported because of cost implications. In the UK, in order to turn teaching into a masters level profession, the Labor Government set up a masters in teaching and learning that all newly qualified teachers (NQTs) and newly qualified heads of department were entitled to receive. After a year and a new government, money has been withdrawn from this initiative.

With teacher evaluation and compensation—an area that neoliberal reform associates with hard-edged inspection and league tables for performance of learners and schools it is suggested that

- Supportive appraisal and feedback (TALIS) are useful.
- Generally, appraisal is not used to identify good performance.
- Rewards link to performance (fairness test, multiple measures, and transparently applied).

In Sweden, pay is negotiated between the principal and the teacher (70% approval rating). In Denver, teachers get extra pay linked to professional improvement, good evaluation, and student progress.

Then, a contentious issue among many countries is teacher engagement in education reform. Fundamental changes trigger resistance. Without the active and willing engagement of teachers, reform fails (Hargreaves and Shirley 2009), and so there is evidence that the following are necessary:

- Effective consultation among stakeholders
- Willingness to compromise
- Involvement of teachers in the planning and implementation of reform
- Teaching professionals being in the lead

Examples of teachers in the lead cited include the following:

- Involving unions in reform in Australia
- Building trust in Finland
- The principle of consensus in Swedish decision-making
- Ontario—consultation with teachers by government. Implementation by teachers not bureaucrats (U.S. Department of Education 2011)

Hargreaves and Fullan (2009) in *Change Wars* suggest that fundamental reform takes 25 years to implement, but in many countries, like the UK, change takes place every five or fewer years. Of course, change does not stick.

So, in summary, the best advice seems to include the following: A central role for teachers who carry high levels of responsibility but are well rewarded. Transforming the profession from within so that it embeds on the basis of changing values—not just behavior. The smarter development of professionals, including integrated professional development into teacher's careers, school, and system changes. Professional learning that is directly linked to classroom development (pupil learning, action research). Then, finally, a constructive political process in which politicians and administrators share the goals of reform; collaborative models of reform are most effective.

REFERENCES

Asia Society (2011). Improving teacher education around the world: The international summit on the teaching profession. http://asiasociety.org/files/lwtw-teachersummitreport0611.pdf. Accessed on October 16, 2011.

Aubusson, P., Steele, F., Dinham, S. and Brady, L. (2008). Action learning in teacher learning community formation: Informative or transformative? *Teacher Development*. Vol. 11, No. 2, pp. 133–148. Taylor & Francis, Routledge.

Barber, B. (1995). *Jihad vs. McWorld*, Ballantine Books, New York.

Barber, B. R. (2007). *Consumed*, W. W. Norton and Company, New York, London.

Beck, J. (2008). Governmental professionalism: Re-professionalising or de-professionalising teachers in England. *British Journal of Educational Studies*. Vol. 56, No. 2, pp. 119–143. Blackwell Publishing Ltd., UK.

Brighouse, T. (2008). Putting professional development centre stage. *Oxford Review of Education*. Vol. 34, No. 3, pp. 313–323. Taylor & Francis, Routledge.

Carneiro, R. and Draxler, A. (2008). Education for 21C: Lessons and challenges. *European Journal of Education*. Vol. 43, No. 2.

Castells, M. (1998). *The Information Age Vol III: End of Millennium*, Blackwell, Oxford.

Castells, M. (2001). *The Internet Galaxy: Reflections on the Internet, Business and Society*, Oxford University Press, Oxford.

Crossley, M. (2008). Bridging cultures and tradition for educational and international development: Comparative research, dialogue and difference. *International Review of Education*. Vol. 54, No. 3–4, pp. 319–336.

Davies, T. C. (2006). Creative teaching and learning in Europe: Promoting a new paradigm. *The Curriculum Journal*. Vol. 17, No. 1, pp. 37–57. Taylor & Francis, Milton Keynes, UK.

Ehren, M. C. M. and Visscher, A. J. (2008). The relationship between school inspections, school characteristics and school improvement. *British Journal of Educational Studies*. Vol. 56, No. 2, pp. 205–227. Blackwell Publishing Ltd., UK.

Evans, L. (2008). Professionalism, professionality and the development of education professionals. *British Journal of Educational Studies*. Vol. 56, No. 1, pp. 20–38. Blackwell Publishing Ltd., UK.

Falk, B. (1995). *On Humane Governance*, Polity Press, Cambridge.

Frowe, I. (2005). Professional trust. *British Journal of Educational Studies*. Vol. 53, No. 1, pp. 34–53. Blackwell Publishing Ltd., UK.

Gundara, J. (2003). *Intercultural Education: World on the Brink?* Institute of Education, University of London, Hodder and Stoughton, London.

Hargreaves, A. and Fullan, M. (Eds.) (2009). *Change Wars*, Solution Tree, Bloomington.

Hargreaves, A. and Shirley, D. (2009). *The Fourth Way*, Corwin, London.

Hoyle, E. and Wallace, M. (2005). *Educational Leadership*, Sage, London.

Kennedy, K. (2005). Rethinking teachers' professional responsibilities: Towards a civic professionalism. *International Journal of Citizenship and Teacher Education*. Vol. 1, No. 1, July.

Kress, G. (2008). Meaning and learning in a world of instability and multiplicity. *Studies in Philosophy and Education*. Vol. 27, No. 4, pp. 253–266.

Landorf, H., Doscher, S. and Rocco, T. (2008). Education for sustainable human development: Towards a definition. *Theory and Research in Education*. Vol. 6, No. 2, pp. 221–236.

Marginson, S. and Mollis, M. (2002). The door opens and the tiger leaps: Theories and reflexivities of comparative education for a global millennium. *Comparative Education Review*. Vol. 45, No. 4, pp. 581–615.

Maslow, A. H. (1943). A Theory of Human Motivation, http://en.wikipedia.org/wiki/Maslow%27s_hierarchy_of_needs. Accessed on August 28, 2008.

McKencie, K. B. and Scheurich, J. J. (2008). Teacher resistance to improvement of schools with diverse students. *International Journal of Leadership in Education*. Vol. 2, No. 2, pp. 117–133.

OECD (2011). Building a High-Quality Teaching Profession, Lessons from around the World: Background Report for the International Summit on the Teaching Profession. http://asiasociety.org/files/lwtw-teachersummit.pdf. Accessed on April 16, 2011.

Sen, A. (1993). *Capability and Well-Being*, in *The Quality of Life*, ed. Nussbaum, M. and Sen, A. pp. 30–53, Oxford University Press, Oxford.

Sharpe, C. (2005). *Why Should Teachers Be Interested in Research?* Spark- *Secondary Practitioners Action Research* Knowsley. Vol. 1, No. 2, pp. 5–8.

Tickly, L. (2001). Globalisation and education in the postcolonial world: Towards a conceptual framework. *Comparative Education*. Vol. 37, pp. 151–171.

Ugglee, B. K. (2008). *Identity, Communication and Learning in an Age of Globalization, Studies in Philosophy and Education*. Vol. 27, No. 4.

U.S. Department of Education (2011). 2011 International Summit on the Teaching Profession. New York City, March 16–17, 2011. http://www2.ed.gov/about/inits/ed/internationaled/teaching-summit-2011.html. Accessed on October 14, 2011.

95 Lifelong Learning Policy for Europe as a Challenge for Teacher Training in Poland

Grażyna Prawelska-Skrzypek

CONTENTS

INTRODUCTION

In today's world, people's competencies and qualifications are perceived as capital, which enables one to face the challenges of changing technologies and globalization in the economic, social, and cultural dimensions. This belief causes the lifelong learning policy to be widely recognized as one of the most important instruments of development. The term Lifelong Learning (LLL) is defined as "all the learning activities undertaken throughout life, with the aim of improving knowledge, skills/competences and/or qualifications for personal, social and/or professional reasons" (Promotion and validation of non formal and informal learning: Examples in the fields of integration of people with disability and of intercultural mediation online document). According to the OECD definition, Lifelong Learning involves "individual and social development in all forms and contexts: in a formal sense, that is at schools, vocational education institutions, universities, adult education centres, as well as informal education systems—at home, at work and within the community" (OECD 2009). For a full understanding of the term, the definitions of formal, nonformal, and informal education need to be clarified.

1. Formal education or formal learning refers to any school-related forms of education and also qualification training required by law that is necessary for work in a profession, e.g., a university graduate from a nonteaching degree program must take post-graduate studies or professional courses conducted by authorized institutions in order to obtain a teaching qualification. In such a case, the courses taken will be, similarly to the university degree, regarded as an element of formal education. The completion of such training is confirmed by a formal document stating that a person has certain qualifications.
2. Nonformal education, or nonformal learning, refers to all institutional forms of learning organized outside of education and training programs leading to being awarded qualifications. It includes learning prior to school education and, for instance, postgraduate courses refreshing the knowledge in a particular field. A good example of this type of learning is a training course that an employer sends their employees to in order for them to master the operation of a new computer program implemented in the company or training undertaken

by teachers who want to gain or update their knowledge and skills to some extent, allowing them to maintain or increase their competitiveness on the labor market. Learners do not receive any specific qualifications or professional licenses as the result of such training programs, but they significantly expand their knowledge and skills, their understanding of some phenomena, and their attitude toward said phenomena. The sum of training courses completed may lead to the acquisition of new qualifications. There should also be a way to formally identify and recognize them.
3. Informal education (nonformal learning) is defined as not organized by institutions and carried out intentionally or unintentionally. In today's information society considerable information and knowledge resources remain as openly accessible to all, which allows those interested to use those resources in order to broaden their knowledge. Knowledge and, in particular, skills can also be developed through experience and professional practice. Good mastery of professional skills should enable a person, having obtained the necessary supplementary knowledge, to obtain formal confirmation of their qualifications. In the system of professional advancement of teachers, nonformal and informal learning is the necessary prerequisite for promotion.

The Lifelong Learning Policy includes formal education systems and nonformal learning and informal learning and is understood quite broadly, which is reflected in the fact that it is additionally referred to as Lifewide Learning. This follows from a desire to draw attention to the fact that an important aspect of learning throughout one's life is not only the time (lifelong) but also openness to the use of different learning opportunities in different places and forms. It should be noted that formal education, which has, until now, been de facto the only area of education policy, is—within the concept of Lifelong Learning—only one of the elements of the broader public education policy. This radically changes the role and importance of formal education and requires a profound reorientation of the system and, in particular, a clarification of the relationship between formal, nonformal, and informal education.

1. Lifelong learning has both a professional and nonprofessional dimension. Its potential effects include an increase in professional activity and capacity for employment, the development of adaptive potential among the employees of a company, raising the level of education, and reducing social exclusion areas (Drogosz-Zabłocka 2008, p. 16).
2. OECD emphasizes the fact that lifelong learning is not merely economic in its character, but realizes many noneconomic objectives, such as personal development, enlargement of one's knowledge, cultural, and social goals (Education Today. The OECD Perspective 2009, p. 60).

LIFELONG LEARNING POLICY FOR EUROPE

Striving to create a European area of lifelong learning has been derived from the records of the Lisbon Strategy (2000), in which the introduction of changes to education systems was promoted in order to create conditions for and support lifelong learning. This is to facilitate the free flow of learners and workers and the transfer and renewal of their qualifications, as well as the promotion of creativity and innovation and is to ultimately contribute to economic growth and increase in employment.

The Member States are in charge of the education and training policy. The EU's role is to support the improvement of national education systems through complementary instruments at the level of the entire EU as well as mutual learning and exchange of good practices. The Lifelong Learning Policy for Europe is determined by the seven basic principles, whose implementation has been recognized as a key strategic goal of European cooperation in education and training for the period up to 2020 (Mairesse 2010).

1. An appreciation for different forms of learning: formal, nonformal, and informal.
2. Directing attention to learning, not only of pupils, students, and participants of courses, but also of little children, working people, and those not professionally active—both temporarily unemployed and pensioners.
3. Cooperation as partners in the LLL processes between educational institutions, employers, learners, and the public administration as a result of qualitative changes in the understanding and organization of the learning process.
4. Valorization and recognition of qualifications, regardless of where, how, and when the learning took place. This principle is the cornerstone upon which the European Qualifications Framework was constructed.
5. Changing education and vocational career paths and facilitating access to them particularly for the purpose of retraining.
6. Focusing LLL policy on the learner, not the institutions or the educational system. The effectiveness of the policy should be measured by indicators pertaining to the achievements of learners.
7. Reorientation of the models of education and training funding toward the empowerment of learners.

In the process of building a European area of lifelong learning, the emphasis was initially on creating a LLL strategy, harmonization of formal education systems, the mobility of students, and the mutual recognition of the effects of formal education. Over time, more and more emphasis was put on the effects of learning outside the formal mode. At the same time, the European system for transferring the effects of education and training was being developed.

In 2008, a groundbreaking document was drawn up: the Recommendation of the European Parliament and of the Council on the establishment of the European Qualifications Framework (EQF) for lifelong learning on the 23rd of April 2008. Member States were encouraged to develop, by 2010, the National Qualifications Framework, whose levels would be correlated with the EQF. The implementation of this recommendation requires, inter alia, the description of qualifications through the effects of learning (both on the part of educational and training institutions and the expectations of employers). The European Qualifications Framework includes all existing qualifications in the perspective of lifelong learning and in all respects:

- From general education and vocational training to university education
- From elementary education to education of adults
- From formal to nonformal and informal learning

Qualifications are construed (in accordance with the act amending the Law on Higher Education Act of March 18, 2011, art. 1, point 18) as the effects of education, confirmed with a diploma, certificate, or other document issued by an authorized institution attesting to the fact of obtaining the intended learning effects. Learning effects are, according to this law, the knowledge, skills, and social competencies acquired through training by the learner (*Journal of Laws No. 84*, item 455).

Unsatisfactory progress of the implementation of LLL policy, including the EQF, caused a stronger articulation of the policy in the Europe 2020 Strategy. Among the critical factors for its implementation, those particularly emphasized include, inter alia, the need to develop adequate infrastructure for LLL and building up the capacities of educational and training institutions in this field. The importance of LLL in the Europe 2020 Strategy is stressed by the record related to the implementation of the "Strategic framework for European cooperation in education and training" (ET 2020). ET 2020 is to be implemented by actions taken at the EU level as well as at the level of countries, regions, and local communities with a strong inclusion of all interested parties. It is assumed those goals will be achieved by using the Open Method of Coordination, characterized by common setting of objectives, learning from each other, the development of European instruments and the support through the Lifelong Learning Program and the European Structural Funds. Among the seven flagship initiatives planned to be taken in the new programming period, two are related directly to the implementation of ET 2020: "Youth on the Move" and "Agenda for New Skills and Jobs." A good deal of importance is attached to the continuation of works on the European Qualifications Framework (EQF) and the National Qualifications Frameworks (NQFs): gearing learning processes toward learning effects (learning outcomes), visual learning paths, clearing the connections and possible transitions between education stages, types of education, and between education and work, and also toward improving and building systems for the recognition of qualifications obtained through the nonformal and informal mode and on improving the consulting system quality.

European LLL policy is part of the latest trend of public policies known as "evidence-based policy," which assumes that those implementing education policy, including those managing educational institutions, need to base their decisions more on a firm knowledge of the results achieved and on the evidence from reliable studies (Towards More Knowledge-Based Policy and Practice in Education and Training 2007). The introduction of a culture of evaluation has been proposed, in which a greater emphasis would be put in education on the outcomes of the learning process. Individualization of the learning process is promoted, and the necessity of directing it (both in the formal system and outside it) toward achieving specific learning outcomes is emphasized—in terms of knowledge, broadly defined attitude characteristics, skills, and the ability to use them in a changing social and economic environment. Learning outcomes are understood as a measure of the competencies acquired by graduates during the learning process and as a measure of their qualifications. These are a unique proof of the effectiveness of the learning process.

The core concept of lifelong learning is competence development. The importance of key competencies that every person should have and that should be developed at all stages of formal education is particularly emphasized. In many countries, the development of key competencies is central to the curriculum. Those competencies should be developed as part of every school subject.

WHAT ARE THE IMPLICATIONS OF LLL POLICY FOR THE TEACHER EDUCATION AND TRAINING SYSTEM?

Teachers are that professional group for whom the development of LLL system is a double challenge, and at the same time, it creates the opportunity for significant expansion and diversification of the potential labor market. The nonformal education sector is becoming increasingly important on the labor market alongside formal education. Today, we can distinguish the specific occupations that are forming there: coaching, mentoring, etc.; the profession of a governess is also entering its renaissance. It is to be expected that professions related to facilitation or supporting of cooperation between education and economic and social environment organizations will emerge.

Today's teachers are responsible for the development of a generation of learners who will remain lifelong learners. This new generation needs to be prepared through the implementation of an entirely different way of teaching and learning. Graduates should, first of all, have their key competencies developed in such a way as to enable them not only to find their way in society and on the labor market, but in such a way that those key competencies could become the basis for sustainable and continuous development and could counteract the exclusion from a full professional and social life. To cope with such a monumental task, teachers themselves must learn in a different way, i.e., by studying (rather than listening to lectures and obtaining marks) and continuously developing their skills and competencies following the completion of their formal education. In a few or a dozen years, they will function in a changed social environment. The Europe 2020 Strategy assumes a significant increase in society of the presence of university graduates (in 2020, 40% of Europeans aged 30–34 are expected to have a university degree), who will have the flexibility to adapt to the changing needs of the economic and social environment, including the requirements of the labor market. This necessitates the improvement of the quality and appropriateness of education and training programs—these are the tasks and expectations directed specifically at teachers and the systems of their training (Communication from the Commission to the European Parliament, the Council, the European Economic and Social Committee and Committee of the Regions, Supporting Growth and Jobs – an agenda for the modernization of Europe's higher education system 2011). The Strategy for Europe sets the priorities and shows how the EU can assist in the modernization of national education policies. These priorities for higher education are the following:

1. An increase in the number of graduates, attracting more of the population to higher education, and reducing the number of school dropouts.
2. Improving the quality and relevance of higher education so that education programs would reflect both individual needs and the requirements of the labor market and future careers and would also stimulate and reward excellence in education and research.
3. Providing more opportunities for students to acquire additional skills through study and training outside universities; encouraging cross-border cooperation to improve the results of higher education.
4. Training more researchers to prepare the ground for the economy of the future.
5. Strengthening the connections between education, research, and the economy to promote excellence and innovation.
6. Ensuring that funding is effective—the release of education management and investment in the quality of education in order to better meet the needs of the market.

The question of how to educate today's teacher and what qualifications he or she should possess remains open, and as shown by the Polish experience of recent years, opposing ideas tend to gain recognition every couple of years. The draft regulation on the education of teachers of 2006 placed a lot of emphasis on the general teacher training, general-educational subjects (such as psychology or pedagogy), but it was not accepted by the environment and did not enter the legislative process. It was primarily accused of inadequate preparation of teachers in terms of the subjects to be taught by them, which also resulted from the requirement to educate dual subject teachers.

The draft regulation of 2011 proposed a modular model of teacher education. The number of class hours dedicated to teacher-specific training was reduced in favor of a greater emphasis on the particular competencies. The scope of general pedagogical education was significantly reduced (psychology and pedagogy—only 90 h). Greater emphasis was placed on training in teaching (150 h), including the methodology, forms, and rules for working with children and young people and on the progress monitoring and evaluation process. Highlighted were the issues of facilitating the educational activities of children and young adults as well as the problems of discovering and developing their talents and abilities. As noted by Karwasz, great importance was attributed to the didactics of particular school subjects (Karwasz 2011, p. 56). The module covering training in didactics, under specific subject didactics, the importance of developing key competencies of children and adolescents at all stages of education was emphasized as well as the importance of the didactic competence of candidates for teachers. An important change is the departure from the hitherto existing dual-subject education in teacher training degree programs. Nowadays, one is free to choose whether or not to realize the module in preparation for teaching a second subject; it can be implemented alongside didactic training or after its completion. A similar flexibility is associated with the implementation of the module devoted to special education. Both of the above-mentioned modules can also be realized during a post-graduate program.

The draft regulation mentioned above is currently out of the ministerial and social consultation process, in which it was generally criticized for limiting the number of hours per the psychological-pedagogical preparation module; there were also pointed out some inconsistencies with the Ordinance of the Minister of Education of March 12, 2009, on the specific qualifications required of teachers. These reservations were included in the revised version of the Regulation, which should be released soon.

It appears that the difficulties in developing regulations for the training of teachers are due to the fact that it is necessary to reconcile the two directions of the evolution of education. It seems appropriate to agree with Karwasz on his statement that teachers must have the knowledge of their chosen subjects in order to direct the development of knowledge and skills of the learners. On the other hand, teachers should not do it ex cathedra but rather understand students' development process and so choose and modify the methodology in order to build on the potential already possessed by the student and use and develop his or her abilities in the most effective way but—above all—to develop the key competencies, especially social skills. The latter is often perceived by employers as the area of competence deficiency.

The best school systems are characterized by an individualized approach to each student, which most definitely does not result from the fact that one teacher is responsible for fewer students than in Poland. Statistics shows that Poland has a relatively low ratio of pupils per teacher (Society on the Way to Knowledge 2011, p. 181). The best school systems are also characterized by the emphasis on the development of key competencies. Key competencies were incorporated in Poland (September 1, 2009) to the core curriculum for general education and should be developed within each subject.

Competencies are defined and classified differently. According to the definition of the Ministry of Science and Higher Education adopted in the National Qualifications Framework, competencies are "everything that the person knows, understands and is able to do—the cumulative effects of learning." Competencies are revealed and verified in life, in action—at work, during social activity, or family life, etc. From the perspective of the education system, it is particularly important to support the development of competencies, particularly that portion of them that constitutes transferable competencies, i.e., those that are developed in one context (e.g., education) and used in other contexts (e.g., professional life or various forms of social life). It is what is known as general competencies or key (core) competencies. These include the following competencies, discussed in detail in the document referred to in this paper (Society on the Way to Knowledge 2011, p. 127):

1. Communication in the mother tongue, i.e., the ability to express and interpret thoughts, feelings, and facts
2. Communicating in a foreign language
3. Mathematical thinking, reasoning ability in terms of natural sciences and familiarity with technical issues
4. Ability to use ICTs and their responsible use on an everyday basis
5. Ability to learn
6. Social skills, cooperation with others, including civic and intercultural skills
7. Entrepreneurship
8. Cultural expression

In the education process, competencies are developed through the following:

1. *Different teaching techniques, educational and didactic forms* (lecture, seminar, teaching in small groups, workshops, classes, demonstrative exercise, problem-solving sessions, labs, internships, field exercises, etc.) (Tuning Educational Structures in Europe 2008, p. 85).
2. *Performing of specific activities by the learners*, under the subject syllabus (e.g., searching for materials, reading literature, summarizing read books, posing and solving problems relevant to the level of education and competence level, preparation and creation of written assignments and oral presentations, team work on problem-solving, formulating questions, etc.) (Tuning Educational Structures in Europe 2008, pp. 86–87). It seems that there is enormous, untapped potential for competence development here.
3. *Ways of assessing student achievement of learning outcomes.* Without diminishing the importance of summary assessment, formative assessment is particularly important,

especially when combined with motivational feedback—highlighting the results achieved and the pointing out of deficiencies and methods for improvement as well as with supporting the student throughout the improvement process. There are also a lot of forms here—from the tests of knowledge or skills, presentations, research reports, through written assignments, analyses, observation of practice, to professional portfolios (Tuning Educational Structures in Europe 2008, pp. 87–88).

4. *Career advice.* It is not just about the narrowly construed professional career, but also the responsibility for managing one's own development and life as a citizen, a member of society. From the perspective of a university teacher, I can say with full confidence that the infantilism of young people coming to college is often incapacitating. Each school subject should be an opportunity to shape key competencies. Teachers are burdened with a great responsibility for developing attitudes of readiness for lifelong learning and, generally speaking, for active learning with full awareness of one's responsibility for one's own future. Unfortunately, our social policies, not just educational policy, teach people to be passive or even drive them to passivity, which becomes a kind of addiction.

TEACHER QUALIFICATIONS FOR THE DEVELOPMENT OF KEY COMPETENCIES OF LEARNERS

The teacher training system and the methods of verification of their competencies leave much to be desired. General opinion, supported by studies contained in the Report on the State of Education in Poland of 2011, is that teachers are ill prepared for the challenges they are faced with in the modern world. Due to the very high requirements that are placed on teachers today, there also appeared a postulate (coinciding with the position of the European Network on Teacher Education Policies) for the separation of teaching studies as postgraduate studies (third stage), which would have similar prestige to doctoral studies. These topics are much discussed and written about.

In my opinion, the key issue that is still being neglected is the admission process for teaching degree programs and the recruitment into the teaching profession itself. Universities, by undertaking the training of teachers, i.e., creating conditions in which a learner could develop teaching skills, must very clearly define what kind of qualifications graduates will possess and conduct the education process in such a way that will ensure the development of those qualifications. Therefore, one should first determine the gap between the qualifications of a particular candidate and those of a projected graduate. During the studies, this gap is to be plugged, and this is the commitment taken on by the university. The key lies in determining the minimum conditions for entry—in other words, the eligibility of the candidates and the admission requirements in terms of knowledge, skills, and social competencies. A responsible university should not admit individuals who, at the recruitment stage, are found to possess an educational and skills gap the scale of which would render it impossible to master the competencies required of a graduate during the course of their studies. It means that a responsible university will not admit candidates without sufficiently developed key and specific competencies necessary to prepare for the teaching profession. This has not been the case so far. The basis for admission has been the results of the external high school leaving exam (matura), and universities have merely determined the point threshold for admission and the subjects to be taken into consideration for calculating the total number of points received by the candidate. The features and attitudes necessary for the successful performance of the teaching profession have gone completely unverified. Of course, I am not referring to matters concerning one's world view, but to the qualities that prove a true calling to the profession, an understanding and sharing of its mission. Universities offering degrees in teaching do not follow the professional careers of their graduates, which would allow for evaluation of the quality of teacher training. It is worth mentioning that in Poland there are no certificates or examinations for the admission into the teaching profession (Society on the Way to Knowledge 2011, p. 190).

Those systemic weaknesses should be mitigated because the new Law on Higher Education Act obliges higher education institutions to outcome-oriented teaching, monitoring graduates' professional careers, and close cooperation in the preparation and implementation of university programs with employers. They can also introduce additional requirements for admission, beyond the scope of the matura examination, as it no longer requires the permission of the Minister. They will also issue university, not state, diplomas of higher education, which increases the subjectivity and responsibility of universities.

Both in the opinion of teachers and head teachers, graduates embarking on the career of a teacher have large gaps in terms of the skills necessary to work in schools. According to the Report on the State of Education, teacher education provides too little practical training and does not show students how to use their theoretical knowledge in practice (Society on the Way to Knowledge 2011, p. 191). These more practical qualifications may be acquired in two ways, either in the process of formal education (through the system of student internships) or through nonformal or informal education. According to the data of the Education Information System, presented in the Report quoted, only a small percentage of teachers are involved in various forms of further education (in the academic year 2008–2009, just 12%). In such a context, the promotion procedures are not evaluated particularly favorably. Teachers themselves indicate that they develop their skills mainly through informal learning. According to the TALIS study, as many as 39% of Polish secondary school teachers said they would like to improve their professional competence but had not found adequate training programs that would suit their needs (Society on the Way to Knowledge 2011, p. 194).

Lifelong learning becomes particularly important in the context of the teaching profession. This view is generally embraced and expressed by teachers. It appears repeatedly in the documents of the European Network on Teacher Education Polices (ENTEP Conference in Luxemburg 2011). During a meeting in Luxembourg in March 2011, the directions in which teacher education is evolving in the European countries were presented. Particularly stressed was the need for placing more emphasis on the process of recruitment into the profession, the development of doctoral studies for teachers, and drawing attention the skills and competencies of teacher-training staff. In all those countries, the problem of teacher competence is a burning, critical issue.

Example: Austria. Toward the end of 2008, a group of experts was commissioned by two ministries, of science and of education, to develop a new system of teacher education, which would meet the demands of the new situation, related to the progress of the implementation of the Bologna Process and to the requirement of lifelong learning for teachers. The system that was designed was called "Innovative Teacher Education–the Future of the Teaching Profession" and was presented in March 2010. It provides a flexible development path. The key elements of the system are suitability for the profession and interest in the teaching profession. The following steps for the selection of candidates were proposed:

1. Creation of a profile of competencies, based on experience and the realities of the teaching profession.
2. Implementation of a systematic three-step program created in order to promote self-reflection and self-assessment.
3. Creation of an initial stage in teacher education at the university level, which enables the formation of transition scenarios, the selection of the program (modules, components), and deciding on the length of the studies in terms of time.

The concept leads to the modularization of the program and the three-stage model of studies. By adopting such a problem-oriented modularization, it was sought to increase the functionality of knowledge and skills (qualifications). The three-stage model consists of the preliminary stage, the stage of introduction into the profession, and the advanced stage—learning in the course of the job and continuous training. The architecture of the "Innovative Teacher Education" program (ITT) is designed so that it is a holistic model, which, at the same time, allows for diverse pathways of development; that model is important and effective for all teaching professions and has been adapted to the requirements of different areas of education. It includes the acquisition of key competencies common to all teaching professionals as well as specific competencies.

The preliminary stage leads to obtaining a bachelor's degree, which entitles graduates to pursue a career in teaching and/or fulfills the requirements of entry into the second stage, the phase of introduction into the profession. The idea is that the preliminary level constitutes the base for specialist knowledge, the didactics of specialist subjects and educational theories associated with different subjects and with a common focal area.

The second stage may lead to obtaining a master's degree. It qualifies graduates to take full teaching responsibility. In this phase, called the phase of introduction into the profession, teachers are introduced to the career path, they start out their professional career on the basis of a preliminary approval to teach as trainees, and during this internship period, they develop and expand their qualifications.

In the third stage, the emphasis is placed on lifelong learning. This stage also provides the chance for continuous improvement and the acquisition of skills and competences in the course of work, which may lead to the acquisition of new skills at a master's level. Strong emphasis is placed on the intensification of institutional cooperation in the regional clusters in order to better utilize teacher strengths and promote development.

This program (ITT) means the reorientation and a new orientation for the entire system of teacher education and is recognized as an important milestone in shaping educational policy. It is assumed that, after testing, the structures will be modified, and the cooperation of all interested parties remains the critical factor. Creating more flexible learning paths, looking at them through competencies and designing them with the perspective of lifelong learning in mind all create, in the opinion of the authors of the program, the possibility for the adaptation of the teaching profession to meet future challenges (ENTEP Conference in Luxemburg 2011).

The Austrian model presented here, due to its modularity and flexibility to choose the educational path, is reminiscent of the latest Polish proposal for teacher education. However, the two differ substantially in terms of the emphasis on lifelong learning and the possibility of choosing alternative career paths through the acquisition of skills by means of work experience that are later certified in the formal education system. Due to its characteristics, the Austrian model may be regarded as a mature adaptation of the principles of European LLL policy in relation to teacher education.

CONCLUSION

The key determinant of the development of modern societies is dynamic economic, social, and cultural change. It entails the need to develop adaptive competences that will enable individuals and communities to find fulfillment and satisfaction in different, changing conditions and social roles. It is a challenge for education systems supporting the development of individuals and societies and especially for teachers and the systems of teacher education. These systems are opening up to the idea of lifelong learning, and formal education, which, until recently, was de facto the only area of education policy, in the concept of lifelong learning becomes only one of the elements of a broader public education policy. This is radically changing the role and importance of formal education and requires a profound reorientation of the system and, particularly, a clarification of the relationship between formal, nonformal, and informal education.

The core concept of lifelong learning is the development of competencies. Particularly emphasized is the importance of key competencies that every person should possess and that should be developed at all stages of formal education. What are the key competencies every teacher should possess in order to support the learning process of a person who is capable of functioning in the environment that is characterized by permanent change? What should the education process of the modern or future teacher be? This paper aims to provoke a wider discussion on this subject because, in the opinion of the author, it is one of the most important problems, pivotal to the development of civilization. In European countries, the problem of teacher education is regarded as urgent. Many of those countries have introduced various types of changes: A stronger emphasis is placed on recruitment into the profession; doctoral studies for teacher are being developed; more attention is paid to the competencies of teacher training staff and to the need for lifelong learning; and modular education is being introduced along with flexible career paths that allow for the possibility of pursuing alternative career paths through the acquisition of skills by means of work experience that is, in turn, certified in the system of formal education. The example presented in the paper seems to be a good illustration of a mature adaptation of the principles of the European Policy on Lifelong Learning in relation to teacher education.

REFERENCES

Communication from the Commission to the European Parliament, the Council, the European Economic and Social Committee and Committee of the Regions, Supporting Growth and Jobs – an agenda for the modernization of Europe's higher education system, EC, Brussels, 20.09.2011, (SEC [2011] 1063 final).

Drogosz-Zabłocka, E., (2008), Lifelong Learning - a Model for the Future? Polish ed. Kształcenie ustawiczne – model na przyszłość? In: Górniak, J., Worek, B., (ed.), Lifelong Learning: A Prospect for Małopolska. Polish ed. Kształcenie przez całe życie: perspektywa Małopolski, WUP. Kraków, pp. 15–28.

Education Today. The OECD Perspective, OECD, Paris, 2009.

ENTEP Conference in Luxemburg, March 10–13, 2011. Conference materials.

Karwasz, G. (2011), Teacher Studies as an Additional Stage, Polish ed. Studia nauczycielskie jako dodatkowy etap. 2011. Forum Akademickie, pp. 56–57.

Lifelong Learning: The Contribution of Education Systems in the Member States of the European Union. 2002. Eurydice. FRSE. Warsaw.

Mairesse, P. (2010), Education and training are key to the Europe 2020 Strategy: An "Agenda for New Skills for Jobs" and "Youth on the Move." 8th FREREF Summer University. September 10, 2010, Brussels.

Promotion and validation of non formal and informal learning: Examples in the fields of integration of people with disability and of intercultural mediation. European Guide. Valid Info. The Leonardo da Vinci Project. Transfer of Innovation LLP-LdV-TOI-2008-BE2/01 (electronic document).

Society on the Way to Knowledge. Report on the State of Education 2010, 2011, Polish ed. Społeczeństwo w drodze do wiedzy. Raport o Stanie Edukacji 2010, 2011. Institute for Educational Research. Warszawa.

Towards More Knowledge-Based Policy and Practice in Education and Training. Commission Staff Working Document. SEC (2007) 1098. Brussels. 2007.

Tuning Educational Structures in Europe, Polish ed. Harmonizacja struktur kształcenia w Europie. 2008. Sokrates-Tempus. FRSE. Warszawa.

96 School Organizational Culture

Neglected Factor in Teacher Training and Continual Professional Development

Roman Dorczak

CONTENTS

INTRODUCTION

The concept of organizational culture became popular among those interested in education from the very beginning of its presence in the field of general management (Anderson 1982; Handy and Aitken 1986). Growing interest of both researchers and educational practitioners in the understanding of the school organizational culture concept and the capacity of its building in real schools to serve educational processes is no doubt clearly visible in the field, especially during the last decade (Walker 2010). It is because the theories of school organizational culture help to describe the complexity of schools as organizations with specific aims better than other concepts from the field of educational management (Bottery 1992; Bush 2011). School culture is also recognized as the main factor contributing to the development of learners through the support of a learning environment that helps to initiate and sustain educational processes in schools. It is then especially striking that the issue of school organizational culture is absent in discussions about teacher training and development as well as in existing programs of initial and continuing teacher training.

WHAT IS SCHOOL CULTURE AND WHY IS IT SO IMPORTANT FOR TEACHER TRAINING AND DEVELOPMENT?

There is no common definition of school organizational culture; even the concept of culture has been used synonymously with other concepts, such as "climate," "ethos," "organizational saga," "school atmosphere," or, even more specifically, "school socio-moral climate" (Power et al. 1989; Deal 1993).

Most theories see organizational culture as a very complex and multilevel phenomenon. Edgar Schein, in his theory, describes organizational culture as something that can be seen on three levels: basic assumptions, artifacts, and values. Basic organizational assumptions are the deeper and the invisible part of a culture, shaping unconsciously the ways people think and behave in an organization. In the educational organization's context, these are, for example, assumptions about the nature of individual development, the nature of learning, the nature of knowledge and truth, the assumption that students are equal or not, etc. Artifacts, such as school physical environment and material resources, walls in classrooms, student work displays, are partly visible manifestations of the assumptions that underlie them. The values that can be seen in statements of school documents, such as the mission and the vision of a school, but also in everyday actions and practices, such as the rules that we do something or do not, that we respect other people, that we talk or not about somebody's work, that we invite colleague teachers to our lessons or not, etc., are also strongly determined by the basic assumptions. In that respect, all aspects of school organizational culture are organically interrelated and cannot be separated from each other (Schein 1985). Similarly, most authors understand school culture broadly as a specific set of beliefs, attitudes, behaviors, and artifacts that characterizes the school and shapes its everyday life, being visible in the ways different groups in school treat each other, in the extent to which they feel included, and in the forms of collaboration and participation. For Sergiovanni, school culture includes values, symbols, beliefs, and shared meanings of all groups involved in the school life, such as parents, students, teachers, and others within the school community. The culture governs what is valuable for these groups and how their members think, feel, and act. The culture includes a school's customs and traditions, historical accounts, stated and unstated understandings, organizational habits, norms and expectations, common meanings, and shared assumptions that underlie organizational activities. The more commonly understood, accepted, and cohesive the organizational culture of a certain school is, the better for educational purposes (Sergiovanni 1995).

It must be stressed that an important aspect of school organizational culture is that it emerges or develops through the process of social construction of the system of meanings and becomes "the collectively constructed software of minds" (Hofstede 1991). School organizational culture evolves through interaction and is not possible without the communication between the people who constitute the school community. A social, interactive, and communicational dimension of school culture development seems to be the core of its specificity as schools should be organizations that can be called professional learning communities (Dufour et al. 2006). Another important aspect of school culture is that it is a result of a consensus about values and norms that is reached through professional discourse and should not be imposed by any individual.

This is because that brings the danger of being artificial for the culture that is not internalized and cannot be a sustainable source of professional action and professional development of teachers and others in schools (Richardson 2001).

Values and norms are the core of school organizational culture, but there are other dimensions of culture. An American educationalist connected with Kohlberg's group, Anderson (1982), describes four of them:

- **School environment**, understood as material conditions of school functioning, such as buildings, space, teaching resources, and other equipment, the physical environment of a school
- **Milieu of a school**, understood as accumulated characteristics of the teachers and students of a school, such as their social background, gender, age, life experience, etc.
- **Social system of a school**, described as organizational structure, the shape of decision making and communication processes
- **Normative system** that comprises the values, the norms, and the system of meaning that are characteristic for a certain school community

Those dimensions of school culture are interrelated, and they influence each other deciding about school reality and creating a certain learning environment.

The presented ways of defining school culture convince undoubtedly that the concept of school culture should be central for the teaching profession and should be reflected in teacher training as well as in teacher continuing professional development programs. School culture is the result of professional discussion and gives impulse for professional discussion that is crucial for the creation of learning conditions for learners in schools. If we want to prepare teachers who are able to take responsibility for educational processes, we must raise the issue of school culture in teacher training programs and give teachers the knowledge of how to build school culture and how important it is for educational processes within a school.

TYPES OF SCHOOL CULTURE

Of course, every school has an organizational culture; it can take different forms that are not always educationally positive. The school culture we are looking for is not the only one possible in educational reality. Numerous theories of organizational culture of schools gave descriptions of different cultures. The best known is the typology of Charles Handy, who describes four types of organizational cultures and names them using names of Greek gods: Zeus, Apollo, Athena, and Dionysus.

The Zeus culture is characterized by a centralized structure with the school leader controlling every aspect of the school life. All members of the school are controlled by the *Zeus*. Such culture is extremely individualistic or even egoistic (the *Zeus*'s interest valued only). Many school cultures take this form, having strong and charismatic head teachers who shape the school life according to their will and potential. The individual development of all as the main value of the school as an organization is, in such a strongly centrally controlled culture, in danger of becoming dominated by one person who suppresses others and controls their development.

The Apollo culture is built on rules and regulations that describe every single step of the members of the organization, including the school head. The roles and the duties of all are precisely described, and there is no space for autonomous activity. Such organizational cultures are called bureaucracies. They function perfectly when predictability can be assumed and in organizations that can work according to prepared algorithms, which is not the case of educational processes that have to have a creative, unpredictable, and individual nature. Unfortunately, most schools are such bureaucracies, especially within educational systems with centralized regulations and a strong position of the central or local educational authorities.

The Athenian culture builds on teamwork and the potential of all members of organization. It creates the conditions for professional communication and makes it the main tool of solving complex problems that schools deal with. It is, no doubt, the best possible (among those described by Handy and Aitken [1986]) organizational culture for educational institutions as it gives good conditions for professional cooperation and development and stimulates the individual development of all. Good schools develop toward such culture.

The Dionysian culture is called the culture of freedom, but it is unfortunately the culture of independent individuals that define their own methods of work on their own. If they work with each other, it happens only because it helps them achieve their individual goals. Dionysian culture gives freedom and requires no commitments or involvement in common activities. Because of that, it is attractive to independent and creative individuals, but it limits them to the levels of their individual potentials. Schools tend to be such cultures especially when individual success is valued more than cooperation, which is the case in many schools focused on student results as they count in school league tables, so popular in public discussions around education (Handy and Aitken 1986).

The described types of school organizational culture, except the Athenian one, are negative from the point of view of educational processes. The bureaucratic Apollo culture and the autocratic Zeus culture kill individuality, and the Dionysus culture gives individuality so much freedom that, at the same time, it creates the danger of the decomposition of the school life and the destruction of school culture that always has to be a socially agreed phenomenon. Deal and Peterson, in their work, also describe a negative form of school culture using the term of a toxic culture. The main characteristics of that type of negative school culture include such features as a negative attitude toward students seen as a source of problems, resistance to change and development, false self-confidence

TABLE 96.1
Results of Research on Polish Schools Organizational Culture

Type of Organizational Culture	Number of Subjects	Percentage
Zeus culture	402	13.33%
Apollo culture	980	32.50%
Athena culture	1026	34.03%
Dionysus culture	538	17.84%
Not determined	69	2.30%
Total	3015	100%

of teachers, concentration on problems and difficulties, habit of making complains the main issue in school conversations, lack of professional discussion and sharing of expertise, etc. (Deal and Peterson 1999).

In real schools, such "noneducational" cultures can be unfortunately quite frequently found. In one of the latest researches carried out during the period between January 2010 and September 2012 with participation of more than 3000 teachers from Polish schools of different types, almost two thirds of teachers described the school culture of their school as one of the three cultures that were named "noneducational," and only 34% assessed their schools as an Athenian culture of teamwork and cooperation. The detailed results of that research are shown in Table 96.1.

One of the main reasons for such a situation is probably the fact that teacher training and development courses are focused mainly on subject issues, the methodology of teaching, and the use of teaching resources. There is not enough room in such programs for the development of interpersonal skills and teamwork competencies and for real discussion about basic educational values, their understanding, and importance for learning and teaching processes. Such teacher training leads to the de-professionalization of the teaching profession, which is reduced to the knowledge of practical teaching techniques instead of being given a chance to develop a really professional understanding of a teacher's role and competencies, enabling them to take active part in educational discourse within a team of professionals. Lack of school organizational culture issue in the programs of initial teacher training and continual professional development is, in that light, a natural consequence. The problem is that not only do we need to introduce the issue of organizational culture to discussions and practice around the teaching profession, but we also have to define what kind of school culture we need from the point of view of specific educational aims that are centered around the value of autonomous individual development.

WHAT TYPE OF SCHOOL ORGANIZATIONAL CULTURE IS NEEDED?

As it was said before, schools have to build a very specific organizational culture, a professional culture that serves well the educational processes. The Athenian culture, as described by Handy and Aitken (1986) in his typology, seems to be a good starting point for describing such "educationally proper organizational culture," but the process of building school culture has to take into account something more (Dorczak 2011). School culture has to be built around some basic assumptions that are necessary if we want to constitute a real "educational organizational culture."

The first important element that has to be taken into consideration is the fact that autonomous personal individual development is (or should be) the core and fundamental value underlying the functioning of a school (Bottery 1990). The autonomy of individual development is especially important, and describing it within the teachers work context means that it cannot be reduced to or dominated by any other value, such as, for example, social good, needs of the state, economy, etc. Without putting that value in the heart of the school normative system and without understanding individual human development as an autonomous process being a value itself that cannot be reduced to any other value and organized according to the needs of others than the developing individual, we cannot imagine a real educational culture of school as an organization (Łuczyński 2011).

Second, it is necessary to underline that school culture has to be built on the value of interpersonal processes and teamwork. The nature of the educational process is interpersonal and relational; learning and development can only happen in interactions of different kind. The communication skills or competencies are probably the most important set of skills that support the development of such professional, interpersonal relations. Properly organized communication in teams of teachers gives them a chance to interpret, together with other teachers, the meaning of the basic educational values and through that, make school organizational culture more coherent.

Third, school culture needs a specific type of leadership. It requires the involvement of all members of the school community in leadership processes, which, in return, would give them the possibility to develop individually. Avery names such leadership an *organic leadership* and stresses that it is not easy to be built in traditional organizations in which formal division of power and responsibilities makes it difficult to involve everybody equally. At the same time, he stresses that the challenges of the contemporary world demand such organizational cultures to be successful (Avery 2004).

Fourth, school culture has to be built on the value of inclusion. For educational processes with autonomous development as their core value, it is important to create the possibility of expressing themselves and involve all members of the school community. Not only do they have to be encouraged to be involved, but it is necessary to build mechanisms supporting that involvement because only that can fully support the process of individual development of all members of the school community. The most important element helping to make inclusion in schools real is an inclusive educational leadership that engages all those who are a part of the school community on different levels of decision-making processes in school (Dorczak 2013).

HOW TEACHER TRAINING MUST BE TRANSFORMED TO SUPPORT EDUCATIONAL SCHOOL CULTURE?

The core feature of organizational school culture is that it develops in interaction between people in school as an organization. It seems that two things are important for the quality of such an interaction. First, this is the individual potential of teachers: knowledge, the understanding of educational concepts and values, communication skills, and teamwork competencies. Second, these are organizational conditions that emerge thanks to a certain style of leadership, such as planning, the division and organization of work, motivation, support, and resource management. Both are interrelated as it is not possible to have well-structured and organized processes in an organization without adequately prepared members of a team, and, at the same time, organizational conditions are important for the professional development of the members of the organization. Having this in mind lets us only concentrate on the first set of issues connected with teacher training and development.

Thinking about changes that are needed in programs of teacher training and continuing professional development to change the individual and group potential of teachers and help them as a profession in building real "educational school culture," we may see them in three main areas: the content of curricula of teacher training; the methods of the delivery of that content, the skills and competencies that are valued and developed (Sayer 1993). Changes in those areas should mainly include changes in the following:

- **Content** of the curricula of initial and continuing teacher training should be more focused on the contemporary problems of schools as organizations and the teaching profession. They need to be updated in all areas according to the knowledge present in the recent psychological and educational theory and research, but it is particularly important to include the knowledge about organizational culture and its importance for the educational process and individual development of the learner.

- **Methods** used in teacher training programs should stimulate the development of critical thinking and ability to understand, assess, and creatively transform educational theories and concepts. Passive reception of knowledge on educational theories should disappear from the practice of teacher training, and teachers have to become responsible for the creation of that knowledge through reflection on their own experience. It only can be achieved through a fundamental change of the teaching methodology used in teacher training programs at universities and in other places taking part in teacher training.
- **Skills** developed in teacher training courses so far are mostly skills of using methods of teaching and resources in teaching different subjects. That, of course, has to stay an important part of the training, but what should be more important on that list are two other sets of skills. First of all, the skills that are needed to become a good teacher are communication and, using a broader term, interpersonal skills. Second, these are skills of teamwork. Both are closely interrelated, and they build on each other; communication skills help in teamwork, and teamwork helps in the development of communication and interpersonal skills. The interpersonal nature of teaching and learning needs those skills as basic for being a good teacher. It is especially surprising that they are forgotten or, at least, not developed properly, especially as far as initial teacher training programs are concerned (Sayer 1995).

CONCLUSION

If the programs of teacher training can be transformed in those three areas, they may better serve teaching and learning, which are not processes of a simple transfer of knowledge but are highly creative and transformative (Freire 2001). Building such understanding of those processes from the beginning of preparation for the teaching profession, which starts during the initial teacher training at universities, makes it easier to continue such a way of professional development throughout the whole professional life of teachers even if they as individuals meet obstacles in their particular workplaces. Such a solid basis for individual and group professional growth may also help teachers to become more conscious and active members of their organizations, and it contributes to the establishment of organic leadership within schools (Mazurkiewicz 2011). Only that can make the educational system the real change agent and help to build a democratic knowledge-based and learning society of the 21st century.

REFERENCES

Anderson, C. S. (1982), The search of school climate: The review of a research, in: *Review of Educational Research,* vol. 52, pp. 368–420.

Avery, G. C. (2004), *Understanding leadership. Paradigms and Cases,* Sage Publications of London, Thousand Oaks and New Delhi.

Bottery, M. (1990), *The morality of the School. Theory and Practice of Values in Education.* Cassell, London.

Bottery, M. (1992), *The ethics of educational management,* Cassell, London.

Bush, T. (2011), *Theories of educational leadership and management,* Sage Publications Ltd, London.

Deal, T. E. (1993), The culture of schools, In: Sashkin M., Walberg H. J., (eds), *Educational Leadership and School Culture*, McCutchan Publishing, Berkeley, CA.

Deal, T. E., and Peterson, K. D. (1999), *Shaping school culture: The heart of leadership,* Jossey-Bass, San Francisco.

Dorczak, R. (2011), School organizational culture and inclusive educational leadership, in: *Contemporary Management Quarterly,* vol. 2, pp. 45–55.

Dorczak, R. (2013), Inclusion through the lens of school culture, in: MacRuairc, G., Ottesen, E., Precey, R., *Leadership for inclusive education. Values, vision, voices,* Sense Publishers, Rotterdam/Boston/Taipei, pp. 47–60.

Dufour, R., Eaker, R., Many, T. (eds.) (2006), *Learning by doing: A handbook for professional learning communities at work,* Solution Tree, Bloomington.

Freire, P. (2001), *Pedagogy of freedom, ethics, democracy and civic courage,* Rowman and Littlefield Publishers Inc., Boulder-Oxford-New York.

Handy, C., and Aitken, R. (1986), *Understanding schools as organizations,* Penguin Books, Harmondsworth.

Hofstede, G. (1991), *Cultures and organizations: Software of mind,* McGraw-Hill, London.

Łuczyński, J. (2011), *Educational management and student's education in schools,* Polish ed. *Zarządzanie w edukacji dla wychowania uczniów w szkole,* Wydawnictwo UJ, Kraków.

Mazurkiewicz, G. (2011), *Educational leadership.* Polish ed. Przywództwo edukacyjne, Kraków, Wydawnictwo UJ.

Power, C., Higgins, A., Kohlberg, L. (eds.) (1989), *Lawrence Kohlberg's approach to moral education,* Columbia University Press, New York.

Richardson, J. (2001), *Shared culture: A consensus of individual values. Results,* National Staff Development Council, Oxford, OH.

Sayer, J. (1993), *The training and development of teachers,* Interim Consultative Document, TEMPUS Project DSDE 1477, Oxford.

Sayer, J. (1995), *Developing schools for democracy in Europe, Oxford Studies in Comparative Education,* vol. 5. Triangle Books, Wallingford, UK.

Schein, E. (1985), *Organizational culture and leadership; A dynamic view,* Jossey-Bass, San Francisco.

Sergiovanni, T. J. (1995), *The principalship: A reflective practice perspective.* Allyn and Bacon, Needham Heights, MA.

Walker, A. (2010), *Building and leading learning cultures,* In: Bush, T., Bell, L., Middlewood, D., (eds.), *The principles of educational leadership and management,* Sage Publications Ltd, London.

97 A Thomistic-Social Profile of Karol Górski's Personalistic Pedagogy

Janina Kostkiewicz

CONTENTS

INTRODUCTION

The thought of personalistic education in Poland in 1918–1939 was characterized by an integral approach toward man and his existence while it developed social questions. Karol Górski (1903–1988) was a co-originator of this movement with his pedagogy influenced by Thomistic and social aspects. Even though Karol Górsky's personalistic pedagogy is affirmed by educationists, a presented attempt to clarify the classification is new. I am doing my best to clarify it with awareness that making one generally accepted classification of main pedagogy domains is a really difficult, even impossible, task (Śliwerski 2009). But without this attempt, as is written by Bogusław Śliwerski, it would be very difficult to capture the variety of pedagogy thoughts or the knowledge structure.

Prior to constituting his personalistic pedagogy, Karol Górski explored Thomism—first, at the Jagiellonian University with a philosophical and religious group called the "Revival" Academic Youth Association when, by listening to lectures, of Konstanty Michalski, he learned, as he says, to appreciate Thomism. Later, he developed his interest in Thomism by participating in "Revival" Social Weeks organized at the Catholic University of Lublin, where he took part in lectures of Jacek Woroniecki and Antoni Szymański. Regardless of his studies of Thomism at Polish universities, an encounter with J. Maritain in 1934 in Poznan, where they were both speakers during the International Congress of Thomistic Philosophy, significantly influenced his pedagogy (*Chronicle,* Ruch Katolicki [1934]). All these factors—from his first studies of Thomism to the role played by his mentors—contributed to the formation of his opinions as well as his Thomistic-social approach.

WHEN HUMAN NATURE REVEALS ITS SOCIAL ASPECTS

Górski distinguishes between the sensual and spiritual worlds. Of what is significant, the spiritual world comprises into a world of human freedom. This *world of freedom* proves to be not only an important criterion determining the rules of educational practice, but also a foundation *in terms of a social dimension of life and education.* Assuming that the spiritual world is characterized by intelligence and free will, freedom is perceived by Górski as a consequence (result) of spirituality:

Intelligence exists to get to know a being, yet "no being other than the Absolute can satisfy it. Hence, it is free in relation to everything that is not God, and so is will, which unreservedly desires or discards all that is not the Highest Good. This is the world of freedom" (Górski 1935). Thus, Górski outlines the scope of freedom although he has yet to define it. This area of freedom for the human spirit can be perceived as an instance enabling freedom to exist in the social dimension. In fact, it could be said that the freedom of the human spirit is the source of social freedom, which finds its substantiation and origin in the spiritual world.

Man belongs to the world of matter and the spiritual world. According to Górski, man, subject to the laws of the matter, has to live within an organization, within a society, in order to survive and develop culture. The good of the community (he does not use the term "the common good" yet) will always prevail over the good of man as an individual in the earthly order. The reasons are pragmatic: Man has to conform to the community and live for it as "part of it." Diagnosing his times, Górski regretfully notes that in the earthly order, the largest organization of the human community is the state: "Below it there are occupations, as inherent groups, and, finally, the lowest level is occupied by family" (Górski 1935). The contemporary role of family is insubstantial as it is placed at the very bottom of the social hierarchy. He also regrets to observe that the modern state (i.e., the time around 1935 in Europe) has no hesitation in utilizing material and spiritual forces of man to serve its own purposes. *The area to which the state is denied access, however much it would desire it, is the world of man's freedom.* After all, man is a person (His definition of the concept of a person: "A person is a being that can decide about itself. It is a spiritual being, endowed with intellect and free will, completed by self-awareness" [Górski 1935]). Therefore, Górski considers the acquisition of freedom as the aim of personal development. Consequently, freedom, and not dignity or any other spiritual dimension of man, should be the aim of educational development.

Why freedom? It seems that this rationalization is derived from its social aspects. The determinant manifests itself in two orders of the social function of man: (1) social order as such (with the state being placed behind social life) and (2) the order of the social development of personality (in order of importance: family, occupation, and then state) (Górski 1935).

Despite the omnipotent presence of the state in both orders, the innermost depth of a soul is beyond the reach of any violence or arsenal wielded by the state. No influence, whether material or spiritual, can violate "the temple of the human self, unless it willingly opens itself up" (Górski 1935). Only love has such power; it can impel man to undertake continuous efforts and direct them. Górski sees love as the path to effectively influence man as well as the method of developing social life and the feature of constructive social relationships. However, love alone is not enough to form both social orders, which will be discussed later.

An individual lives in society and, apart from family, needs a wider circle that is one's homeland. Karol Górski writes about the need for educating the new man, who looks to the future. Even if things that, at one time, used to be an important part of human culture have died, we cannot save that which is irretrievably lost (by which he means the "dead" national traditions, subdued throughout the Partitions). Therefore, he writes in the 1930s about the need for creating a new future and a new world. He thinks that Poland should have a clear and crystallized educational doctrine (Górski's term), which can only be constituted by personalistic education. He says: "For Poland to even exist amidst the two seas of evil [Soviet Communism and German National Socialism – JK] (...) the only, however difficult, warranty is a complete man" (Górski 1936). By complete, he means free, active, and aware of his identity and dignity (in forming his own educational system, Górski consciously based it on Thomism and its continuators, e.g., Gilson, Maritain, Gardeil. It should be noted that he writes about a conscious creation of an outline of educational doctrine called personalism. For Górski, it is not just the concept of education, it is also the concept of the world). The role of intellect also reveals the social dimension of Górski's idea. He assumes that human intellect has an unlimited capability of knowing the being, and will has an unlimited capability of desire. Intellect can be satisfied by the unlimited being and will by unlimited goodness; its only determinant is the thirst (desire) for God while it remains free in relation to other goods. He attributes two kinds of freedom to will: freedom of choice and freedom of autonomy. The former means the ability to incessantly choose between good and evil; the latter means freely doing what intellect indicates. Freedom of autonomy is not a stable attribute of man (as opposed to Kant's concept), but it can be acquired as stated by Socrates, Plato, Aristotle, and others (Górski 1936). This approach is associated with a conviction that the judgment of a single individual can prove fallible. What is infallible is the entire humankind throughout its history because its philosophy goes back to the laws instilled in the human soul by the Creator (Górski 1936). This turn toward the infallibility of judgments verified by the historical experience (knowledge) of society also makes one look for its justification in human nature. In personalism, the social character of man is derived from his nature, but at the same time, there is a conviction that man, due to his desire for perfection, searches for a pure community, connected solely by intellect and love. From a personalistic point of view, social life, its organization, technological development, and distribution of work, should all serve one purpose, namely that a human person could develop and strive for contemplation as the highest level of human existence (Górski 1936).

In both types of social orders, there is room for the methods of overcoming personal differences, designed for controlling experiences. If the result is the development of personality, it manifests itself in the release of intellect and will from the haze of emotions and moods (created in us by nerves, which seem to be a disputable statement on the part of Górski). This vortex of emotions and moods is perceived by Górski as "the hotbed of subjectivity." He calls this dimension a bleak well, which a person can never get rid of. However, one can build over it an "immovable edifice of the life of personality that is solely the life of intellect and will, where everything should be clear and transparent. (...) Purification of the life of intellect from vapours coming from beneath, freeing the will of the spell cast upon it with the lower part of our nature—this is what the work on personal development is. (...) No hesitations, no concessions, no compromises or pacts with the swamp" (Górski 1936). Such a radical approach to lower nature is rare among personalists. On one hand, it raises doubts concerning the personalistic nature of his idea; on the other hand, it allows one to feel its "coldness" (to be warmed by the category of "heart"). Assuming that intellect is necessary for personality development, drawing on his knowledge of mysticism, Górski draws attention to the disseminated opinions that allegedly mystics show disdain for knowledge, education, or intellectual development as superfluous in terms of self-development. He claims that mystics perfectly developed intellect, and any misunderstandings in this regard are the result of the inadequacy of the terms employed in philosophy and mysticism (Górski 1936). He summarizes it in the following manner: Who fails to cultivate intellect and finds no pleasure in searching for the truth will turn to vice because man cannot live without pleasure, he says, following Saint Thomas Aquinas. S. Kunowski writes about an overly emphasized role of intellect in Górski's thought, referring to it as an intellectual-Thomistic personalism (Kunowski 2000). This pedagogy is further characterized by a concern for people living without God, who despair when faced with death. Therefore, Górski believes that personalism without religion is not possible (Górski 1936).

AREAS OF SOCIAL EDUCATION AND THEIR PERSONALISTIC DIMENSION

Tradition—any references to it should be sagacious. Given that tradition has certain rigid forms that could compromise sagacity, Górski favors sagacity and places it over tradition. He perceives tradition as an external custom, which fails to enrich personality but rather protects it from risky solutions. It is, in a sense, external armor; consequently, it shields and burdens one at the same time; it gives protection and restricts freedom (Górski 1936; Wernerówna 1938). Górski distinguished three historical areas of the Polish educational tradition connected with the gentry, the bourgeoisie, and the peasantry—traditional education provided within each of them was based first and foremost on a set of virtues required for good and honest work in a chosen occupation. Górski's advice is to remain cautious about tradition. (Family is a special center of tradition. Górski believes it to be a powerful lever of personal development. It provides support and thus ensures the continuity of family.) Fighting it leads to the collapse of social life. Therefore, he advises to correct rather than eradicate tradition, and he believes that it is important to be able to grow out of the bonds of tradition so as to develop it by giving it new, more creative forms, thriving on freedom. They ensure the development of a person and the progress of the centers of tradition (Górski 1936).

Economic education. An integral perception of man was connected with the interest in his existence. This found its reflection in the development of social-economic education. Karol Górski is part of this movement, and in his attempt to define the essence of social education, he states that ultimately it comes down to instilling heroism and justice, applied above all to one's own life (Górski 1936). Both are very important even though many people will not be able to achieve them, and most will abandon both. However, according to Górski, if a handful will retain heroism and justice, the society will not disintegrate. He also indicates the common purpose and direction of social development: "The society should aim to transform itself into the unity of human beings who would give one another truth and goodness—like a family. This purpose cannot be achieved wholly, yet the mutual cordial relations among people can be achieved to some degree" (Górski 1936). The foundation of such an approach to social existence is formed by egoism, altruism, and love of your neighbor. Górski rejects the balancing of the whole issue by accepting the condition: a certain dose of [justified] egoism with a certain [degree] of altruism (Górski 1936). Altruism is not beneficial for the objectively perceived good. What distinguishes it from egoism is the direction of subjection with human activity being subjected to the good of a group, for which everything has to be sacrificed. As he says, thus understood altruism sanctifies the egoism of a group, be it a family, a clique, or a social group based on any given criterion. Altruism knows nothing about a person and his development; therefore, it is frequently adapted by collective movements. Thus Górski claims that perceived altruism does not favor the love of others (Górski 1936). This point of view is compatible with, and simultaneously completed by, the position of Kazimierz Kowalski expressed in "*O społeczeństwie, władzy i autorytecie* (Considerations on Society, Power and Authority)" (Kowalski 1932). Consequently, both egoism and altruism are questionable values of social life.

Górski associates the postulate of social love education with an issue of economic education. In 1936 (a time of intensive reconstruction of Poland after a century and a half of bondage) he warns against neglecting economic education, which could result in new calamities and disasters (Górski 1936). Economic education should start with "presenting the society clear criteria of ethical conduct in the economic area, which are generally missing, and the formation of an ideal of universal affluence, rather than wealth, as the aim of the state's economic policy" (Górski 1936). Górski asserts that the entire economic life should serve the needs of people rather than be used to boost production and increase capital; he further accentuates that the idea of serving the needs of people has different meanings in particular ideologies. Górski rejects the liberal and Marxist approaches; he also attempts to rebut popular ideas of the evangelical principle of poverty and the negative picture of a wealthy man (Górski 1936). The conclusion is rather obvious: There are no evangelical reasons why production, affluence, and wealth should be frowned upon. It is only necessary to connect it with a respect for the rigid principles of personalism. The personalistic approach is unjustifiably described by individualists as anti-productive; personalism emphasizes the importance of production; however, it also says that the producer needs to observe "the cult of poverty," understood as a means to ensuring the use of goods by all people. The excess should be earmarked for social, charity, and state purposes. Górski's personalism clearly specifies that wealth is not evil; in fact, once a person (observing the above mentioned rules) can rise above the wealth he possesses, keep internal freedom, known as "the freedom of autonomy" and keep the desire to achieve full humanity, he will see the economic work in terms of service (Górski 1936).

Family education. Górski sees an aspiration to set up a family as an innate feature of man. Happy families are those that use love as a medium of coexistence. This conviction is derived by Górski from Thomism; he believes that no other philosophy could provide a better basis for education. Thomistic sources give him a conviction that blood ties or common economic foundations, significant as they may be, are of secondary importance. Love, as a medium of family interactions, renders it superior in the hierarchy of human relationships in terms of education. He further asserts that in the order of education, family is preceded by the Church that knows the value of family and endeavors to strengthen it. In family education, bad inclinations of parents or wrong patterns can have a negative influence on the direction of activities with authoritarian upbringing seen as particularly wrong.

National education. In a social dimension of Górski's personalistic thought, national education is important insofar as it is related to the concept of a nation, nationality, the common good, fidelity, justice, and the observance of law. Polemics carried out in the interwar period between the supporters of state and national education had various contexts and was invariably a consequence of an "engaged" attitude toward the fate of the society. Considering this issue, Górski bases his reasoning on the assumption that the common good is explicitly an intentional being (intentionality is associated by Górski with the mode of existence. Each thing can exist in three modes: in the mind of the Creator, materially, and in the mind of man. The last mode of existence is called "intentional") as opposed to a real being (Górski 1936). In consequence, he offers the following argument in relation to national and state education: "The antinomy of nation and state arises on the level of naturalism. It disappears at the point where we reach personalism" (Górski 1936). Personalism can reconcile the love for one's country with a friendly attitude toward other nations and the obligations to the state. He further declares that the existence of a nation without the country is not a natural phenomenon and that because the idea of an order based on one nation came into being in the history of mankind, the state assumed the right to exist. Górski does not explicitly favor either the national or state movement; he tries to approach them integrally; nevertheless, his sympathies for state education come to the forefront more frequently. Placing love above justice, he is able to deal with an integral view of both. Also, the love of neighbor allows him to do so, given that—as he previously stated—it cannot surpass the love of oneself. However, there appears to be a gap that is significant in terms of social life; it is created by wrongs done to some nations and the domination, justified by various types of violence, of other nations. The submission of heart (emotions) and feelings to reason, advocated in philosophical foundations of Górski's concepts, seems to have found its practical use.

What is missing in Górski's personalism is the recognition of a person's dignity; it was already discussed at this time in the writings of Zygmunt Bielawski the first and later Jacek Woroniecki. Górski's reflections on state and nation are quite revealing in this respect: He acknowledges family and its rights, but is rather reserved when it comes to issues of nation (as the family of families) and national education. Rather than expressing national drives through actions within the state, he promotes the mysticism of suffering for the cause to which the action could refer. The mysticism of suffering, as presented by him, is a means of nonviolent resistance. It may have its value, but it is not typical of the Catholic personalists of the interwar period. Górski thus writes about the state built on the social personalistic thought: Here "the development of human personality will be the foundation. Such a state will be deeply homogeneous and cohesive if the external social homogeneity will correspond to another, invisible community: a mystic collective unit, living as if under the cover of a state organization. (...) The world of mysticism [and though it is true, as said later—JK] knows a powerful, terrifying and benevolent solidarity. Suffering borne by a pauper or a monk gives power to statesmen and army leaders" (Górski 1936). This proposal of mysticism is the proposal of the "act" that can be achieved only by exceptional individuals. Thus perceived social dimension of personalism fails to find its confirmation in the thoughts of other personalists. Górski also mentions that the order of the fallen country gives rise to custom and tradition, which, in turn, leads to the radicalism of national movements. "The concept of the welfare of the nation is essentially a certain order, where the past and the present extend a hand towards the future. Hence Jacek Woroniecki could see a custom as an essential feature of nationality" (Górski 1936). The concept of tradition and custom in the social sphere are further discussed in "Społeczne podstawy kultury (Social Basis of Culture)" (Górski 2006). Górski seems not to fully approve of this position as he says that for the nations without states, a custom is a substitute for a more perfect order. Even a custom constitutes a certain social area of life and social education. Górski's criticism of state education based on communism, racial nationalism, or liberalism, where the interest as part of the "idea" becomes the basis of social relationships within the state is quite explicit (Górski 1936). A social inclination is also reflected in Górski's commentary on the cultural crisis in the life of the state, caused by the partiality of individualism, which sees culture as a tool. The culture pervaded with individualism and based on the use leads to its own crisis (Górski 1936), and its impact is more visible in social life than in individual life.

A SOCIAL DIMENSION OF THE THOUGHT REFLECTED IN RELATION TO "THE SOCIAL QUESTION"

To reveal the social aspects of Górski's pedagogical thought, it is also necessary to refer to the way he perceives social issues. It was already stated that the philosophy of man assumed by Górski determines the perception of social questions from the perspective of love. It should be added that he also points out the need for the obligatory presence of justice and fairness (Górski 1929).

Górski's position regarding the question of solving social problems could be discussed by reference to the so-called "social question." According to Górski, the social question emerged when "liberalism of law deprived the working masses of protection" (Górski 1929). However, he believes that it is not the lack of state care and material poverty that constitute the essence of the social question. The essence lies in the "moral relations between the states" and is constituted by "the destruction of moral connections within different social layers. (...) Where there is no trust, love and kindness between employees and employers, the social question arises. (...) It is the lack of love and mutual aversion between different social layers" (Górski 1929). He says that its resolution is no longer conditional upon the lack of the intervention on the part of the state or the improved material status of workers. How, then, can the question be solved? asks Górski. The set of answers includes such terms as harm, the good of others, justice, concession, and Christian love. Unilaterality fails to provide a good solution as the attempts to advocate the love of the opponent could in fact deepen the social question. Therefore, the issue is by no means simple.

Using the principles of Christian virtues, Górski advocates that the resolution of the social question be based on the implementation of the three of them, namely *justice, love, and rightness,* located between the first two. Justice alone cannot resolve the social question as it needs to be supported by love, which gives rise to the concessions of one's own good to the good of others. The solution to the question would be even more ill suited if we would wait for love to prevail as the way to solve the problem. Love, not based on justice, even if possible, would only consolidate the evil. Consequently, Górski sees both of them, love and justice, as indispensable; however, the mode of their coexistence is not irrelevant. He recommends the following order: "justice is supplemented with love, in consequence the sacrifices are to be made not out of Christian love for the benefit of justice, but based on just laws for love, if those laws would be too harsh or rigid" (Górski 1929). However, even this arrangement of justice and love does not resolve the social question. What is needed here is rightness; its place is between justice and love. Its need is conditioned mainly by the changeability of social relationships. From the Christian thought, one should "take factual relations as the basis for the system of justice, hence the need for a transitional virtue. (...) It is rightness, giving to each according to the deserts, needs and requirements of the collective life" (Górski 1929). It gives all people what they rightly deserve, as Karol Wojtyła—John Paul II—used to say.

CONCLUSION

Karol Górski recommends his concept of resolving the social question to Catholic activists, and by so doing, he stands in opposition to a philanthropic approach, in which the only reason and direction of any action is charity. He also defies socialists opting for a detached concept of justice. Górski makes his thought on the resolution of the social question alive by including references to various occurrences of practical social life and to the creators of the social thought of the Catholic Church (e.g., Le Play, considered to be its creator) or to the *Rerum Novarum* encyclical, sanctioning the social importance of organizational work among the working classes.

Karol Górski's personalistic pedagogy is completed and closed by a basic issue: the one of a mutual relationship between personalism and religion. He poses a primary question here: "Can personalism, as an educational and philosophical doctrine, be visioned without religion" (Górski 1936). Letting alone the extensive explanations provided by our author, it should be stated that he believes that personalism is not possible without religion, including in its social dimension.

REFERENCES

Górski K. (1929). Sprawiedliwość i miłość w życiu społecznym, *Prąd*, v. 17, Lublin, 187–196 (Justice and Love in Social Life. *The trend*).

Górski K. (1935). *Rodzina a kultura współczesna,* General Institute of Catholic Action, Poznań (Family and Contemporary Culture).

Górski K. (1936). *Wychowanie personalistyczne*, General Institute of Catholic Action, Poznań (Personalistic Education).

Górski K. (2006). *Społeczne podstawy kultury,* (Edition and preface by W. Piasek), UMK Publishing House, Toruń (Social Basis of Culture).

Kowalski K. (1932). O społeczeństwie, władzy i autorytecie, *Ruch Katolicki*, II [1932], 102–117 (Considerations on Society, Power and Authority. *Catholic Movement*).

Kunowski S. (2000). *Problematyka współczesnych systemów wychowania*, Impuls Publishing House, Kraków (The Issues of Contemporary Educational Systems).

Katolicki R. (1934). *Sekcja: Kronika*, IV [1934], 351 (Section: Chronicle).

Śliwerski B. (2009). Klasyfikacje, typologie, mapy myśli pedagogicznej, in: Idem, *Współczesna myśl pedagogiczna. Znaczenia, klasyfikacje, badania*, Impuls Publishing House, Kraków (Classifications, Typologies and Maps of Pedagogical Thought. *Contemporary Pedagogical Thought. Interpretations, Classifications, Research*).

Wernerówna M. B. (1938). Postulat społecznego wychowania młodzieży, *Ateneum Kapłańskie*, 24 [1938], Volume 41, 370–381 (Demand for Social Education of the Youth. *Priests' Ateneum*).

Author Index

B

C

D

E

F

G

H

M

N

O

P

Q

R

S

T

U

V

W

X

Y

Z

Subject Index

Page numbers followed by f, t, b, and n indicate figures, tables, boxes, and notes, respectively.

C

E

F

G

I

L

M

O

Q

R

S

T

W

X

Y

Z